MATERIALS INTERFACES
Atomic-level structure and properties

Edited by
Dieter Wolf
Senior Physicist
Materials Science Division
Argonne National Laboratory

and

Sidney Yip
Professor
Department of Nuclear Engineering
Massachusetts Institute of Technology

CHAPMAN & HALL
London · Glasgow · New York · Tokyo · Melbourne · Madras

Published by Chapman & Hall, 2-6 Boundary Row, London SE1 8HN

Chapman & Hall, 2-6 Boundary Row, London SE1 8HN, UK

Blackie Academic & Professional, Wester Cleddens Road, Bishopbriggs, Glasgow G64 2NZ, UK

Chapman & Hall, 29 West 35th Street, New York NY10001, USA

Chapman & Hall Japan, Thomson Publishing Japan, Hirakawacho Nemoto Building, 6F, 1-7-11 Hirakawa-cho, Chiyoda-ku, Tokyo 102, Japan

Chapman & Hall Australia, Thomas Nelson Australia, 102 Dodds Street, South Melbourne, Victoria 3205, Australia

Chapman & Hall India, R. Seshadri, 32 Second Main Road, CIT East, Madras 600 035, India

First edition 1992

© 1992 The individual contributors

Typeset in Plantin 10/12 by Best-set Typesetter Ltd., Hong Kong

Printed in Great Britain at The University Press, Cambridge

ISBN 0 412 41270 5

A catalogue record for this book is available from the British Library

Library of Congress Cataloging-in-Publication data
Materials interfaces : atomic-level structure and properties / edited by D. Wolf and S. Yip – 1st ed.
    p.   cm.
  Includes indexes.
  ISBN 0–412–41270–5
  1. Surfaces (Physics) 2. Solids. I. Wolf, Dieter, 1942–   .
II. Yip, Sidney.
QC173.4.S94M39   1992
530.4'27–dc20   92–21645
                                                      CIP

Contents

Contributors   xi

Introduction   xiv

1   **Atomic-level geometry of crystalline interfaces**   1
    D. Wolf
    1.1   Introduction   1
    1.2   Basic terminology   3
    1.3   Microstructure   7
    1.4   Coherency, epitaxy and topotaxy   8
    1.5   Commensurability   13
    1.6   Degrees of freedom of crystalline interfaces   16
    1.7   Atomic-level geometry of planar stacking   29
    1.8   Grain boundaries   35
    1.9   Characterization of the atomic structure of interfaces   52

2   **Experimental investigation of internal interfaces in solids**   58
    D. N. Seidman
    2.1   Introduction   58
    2.2   Buried internal interfaces   59
    2.3   Chemistry of internal interfaces   76
    2.4   Outlook and future promise   80

PART ONE   **Bulk Interfaces**   85

3   **Correlation between the structure and energy of grain boundaries in metals**   87
    D. Wolf and K. L. Merkle
    3.1   Introduction   87
    3.2   Characterization of geometry and structure of GBs   88
    3.3   Investigation methods   94

3.4 Correlation between macroscopic degrees of freedom and energy   101
3.5 Correlation between microscopic degrees of freedom and energy   127
3.6 Correlation between the atomic structure and energy   135
3.7 Conclusions   146

4 Grain and interphase boundaries in ceramics and ceramic composites   151
M. G. Norton and C. B. Carter
4.1 Introduction   151
4.2 Illustrations of GBs in ceramic materials   159
4.3 Phase boundaries   174
4.4 Summary   187

5 Special properties of Σ grain boundaries   190
G. Palumbo and K. T. Aust
5.1 Introduction   190
5.2 CSL classification of GBs   190
5.3 Solute segregation   192
5.4 Energy of GBs   193
5.5 Kinetic properties   197
5.6 Mechanical and physical properties   201
5.7 Corrosion properties   203
5.8 Triple junctions   206
5.9 Summary   207

6 Grain boundary structure and migration   212
D. A. Smith
6.1 Introduction   212
6.2 Grain boundary structure   213
6.3 Steps and dislocations   215
6.4 Grain boundary migration   217
6.5 Observations by transmission electron microscopy   218
6.6 Grain structure of films   221
6.7 Discussion   223
6.8 Conclusion   225

7 Role of interfaces in melting and solid-state amorphization   228
S. R. Phillpot, S. Yip, P. R. Okamoto, and D. Wolf
7.1 Introduction   228
7.2 Simulation methods   229
7.3 Molecular-dynamics simulation of interface-induced   233
     thermodynamic melting
7.4 Mechanical melting   241
7.5 The $T-V$ phase diagram and its low-temperature extensions   244
7.6 Interpretation of solid-state amorphization experiments   247
7.7 Summary and outlook   251

Contents   vii

8   Wetting of surfaces and grain boundaries   255
D. R. Clarke and M. L. Gee
   8.1 Introduction   255
   8.2 Long-range forces   256
   8.3 Wetting of external surfaces   258
   8.4 Wetting of internal surfaces   262
   8.5 Effect of surface chemistry   270
   8.6 Closing remarks   271

PART TWO   Semi-bulk and Thin-film Interfaces   273

9   Structural, electronic, and magnetic properties of thin films and   275
     superlattices
A. Continenza, C. Li, and A. J. Freeman
   9.1 Introduction   275
   9.2 Computational method   276
   9.3 Transition metal surfaces and metal–metal interfaces   277
   9.4 Metal–ceramic interfaces   281
   9.5 Semiconductor interfaces and heterojunctions   287

10 Scanning tunneling microscopy of metals on semiconductors   299
R. J. Hamers
   10.1 Introduction   299
   10.2 Group III metals on silicon   299
   10.3 Group V metals on silicon   307
   10.4 Alkali metals on silicon   308
   10.5 Transition metals on silicon   309
   10.6 Metals on gallium arsenide   313
   10.7 Conclusions   314

11 Epitaxy of semiconductor thin films   316
J. L. Batstone
   11.1 Introduction   316
   11.2 Principles of epitaxy   317
   11.3 Misfit dislocations   326
   11.4 Phase stability in epitaxial systems   331
   11.5 Summary   332

12 Phase behavior of monolayers   336
S. G. J. Mochrie, D. Gibbs, and D. M. Zehner
   12.1 Introduction   336
   12.2 Scattering experiments   339
   12.3 Conclusion   352

13 Elastic and structural properties of superlattices   354
M. Grimsditch and I. K. Schuller
   13.1 Introduction   354

13.2 Background   354
13.3 Elastic properties   355
13.4 Structure   356
13.5 Discussion   360
13.6 Conclusions   361

14 Computer simulation of the elastic behavior of thin films and   364
      superlattices
      D. Wolf and J. A. Jaszczak
      14.1 Introduction   364
      14.2 Simulation concepts and techniques   368
      14.3 Thin films   372
      14.4 Composition-modulated superlattices   394
      14.5 Summary and conclusions   403

15 Interfaces within intercalation compounds   407
      M. S. Dresselhaus and G. Dresselhaus
      15.1 Introduction   407
      15.2 Background material   408
      15.3 Structure   412
      15.4 Relation between properties and interfaces   423
      15.5 Concluding remarks   427

16 Nanophase materials: structure–property correlations   431
      R. W. Siegel
      16.1 Introduction   431
      16.2 Synthesis   433
      16.3 Structure   439
      16.4 Properties   452
      16.5 Conclusions   457

PART THREE   Role of Interface Chemistry   461

17 Interfacial segregation, bonding, and reactions   463
      C. L. Briant
      17.1 Introduction   463
      17.2 The segregation process   464
      17.3 Grain boundary segregation   468
      17.4 Grain boundary reactions   475
      17.5 Applications   477
      17.6 Conclusions   478

18 Physics and chemistry of segregation at internal interfaces   481
      R. Kirchheim
      18.1 Introduction   481
      18.2 Distribution of segregation energies   482
      18.3 High concentrations with solute–solute interaction   484

Contents

18.4 On the correlation between segregation and solubility 485
18.5 Interstitial diffusion in GBs 486
18.6 Experimental results on H-segregation at GBs 487
18.7 Phase separation in nanocrystalline Pd-H 491
18.8 H-diffusion in nanocrystalline Pd 492
18.9 H-segregation at metal/oxide interfaces 493
18.10 Conclusion 495

19 **Atomic resolution study of solute-atom segregation at grain** 497
    boundaries: experiments and Monte Carlo simulations
    S. M. Foiles and D. N. Seidman
    19.1 Introduction 497
    19.2 Methodology 498
    19.3 Monte Carlo computer simulations 502
    19.4 Atom-probe observations of solute–atom segregation 509
    19.5 Conclusions 512

20 **Amorphization by interfacial reactions** 516
    W. L. Johnson
    20.1 Introduction 516
    20.2 Review of experimental results 517
    20.3 Phase formation and growth in diffusion couples – theoretical 531
          considerations
    20.4 Relationship to other melting phenomena 542

21 **Relationship between structural and electronic properties of** 550
    metal–semiconductor interfaces
    R. T. Tung
    21.1 Introduction 550
    21.2 Interface states 551
    21.3 Electron transport of Schottky barriers 554
    21.4 Evidence for SBH inhomogeneity: non-epitaxial MS interfaces 558
    21.5 Epitaxial MS interfaces 567
    21.6 Conclusions 586

22 **Electronic properties of semiconductor–semiconductor interfaces and** 592
    their control using interface chemistry
    D. W. Niles and G. Margaritondo
    22.1 Surface techniques in heterojunction physics 592
    22.2 Microscopic control of interface parameters 595
    22.3 Future directions 612

23 **Microscopic nature of metal–polymer interfaces** 616
    P. S. Ho, B. D. Silverman, and S.-L. Chiu
    23.1 Introduction 616
    23.2 Molecular structure and morphology of polyimides 618
    23.3 Surface and interface chemistry 620

23.4 Diffusion and interface formation   625
23.5 Thermal stress and interfacial fracture   631
23.6 Summary   635

PART FOUR   **Fracture Behavior**   639

24 **Tensile strength of interfaces**   641
    A. S. Argon and V. Gupta
    24.1 Introduction   641
    24.2 Cracks at interfaces   642
    24.3 Measurement of strength of interfaces   645
    24.4 Measurement of toughness of interfaces   650
    24.5 Measurement of strength and toughness of carbon fibers   651
    24.6 Conclusions   652

25 **Microstructure and fracture resistance of metal–ceramic interfaces**   654
    A. G. Evans and M. Rühle
    25.1 Introduction   654
    25.2 Atomistic structure   654
    25.3 Bonding models   655
    25.4 Defects at interfaces   656
    25.5 The work of adhesion   657
    25.6 Fracture resistance   659

26 **Role of interface dislocations and surface steps in the work of**   662
    adhesion
    D. Wolf and J. A. Jaszczak
    26.1 Introduction   662
    26.2 Interfacial decohesion: from interface dislocations to surface steps   664
    26.3 Computer simulations   671
    26.4 Core and elastic strain-field effects in free surfaces and GBs   679
    26.5 A broken-bond model for interfacial decohesion   685
    26.6 Summary and conclusions   689

27 **Microstructural and segregation effects in the fracture of polycrystals** 691
    D. J. Srolovitz, W. H. Yang, R. Najafabadi, H. Y. Wang, and R. LeSar
    27.1 Introduction   691
    27.2 Microstructural effects   692
    27.3 Segregation effects on GB fracture   698
    27.4 Final remarks   701

Materials index   703
Subject index   710

Contributors

Prof. Ali S. Argon
Department of Mechanical Engineering
Room 1-306
MIT
Cambridge, MA 02139, USA

Prof. K. T. Aust
Department of Metallurgy and
Materials Science
University of Toronto
Toronto, Ontario
Canada M55 1A4

Dr. Joanna L. Batstone
IBM T.J. Watson Research Center
PO Box 218
Yorktown Heights, NY 10598, USA

Dr. Clyde L. Briant
General Electric Research and Development
Center
PO Box 8, MB 269-K1
Schenectady, NY 12301, USA

Prof. C. Barry Carter
Department of Chemical Engineering and
Materials Science
University of Minnesota
205 Amundson Hall
421 Washington Ave S.E.
Minneapolis, MN 55455-0132, USA

Dr. Shih-Liang Chiu
IBM T.J. Watson Research Center
PO Box 218
Yorktown Heights, NY 10598, USA

Dr. David R. Clarke
Materials Department
University of California
Santa Barbara, CA 93106, USA

Dr. A. Continenza
Physics Department
Università L'Aquila
I-67010 Coppito, Italy

Dr. G. Dresselhaus
Francis Bitter National Magnetic Laboratory
MIT
Cambridge, MA 02139, USA

Prof. Mildred S. Dresselhaus
Department of Physics
Room 13-3005
MIT
Cambridge, MA 02139, USA

Prof. Anthony G. Evans
Materials Department
University of California
Santa Barbara, CA 93106, USA

Dr. Stephen M. Foiles
Division 8341
Sandia National Laboratories
PO Box 969
Livermore, CA 94550, USA

Prof. A. J. Freeman
Physics Department
Northwestern University
Evanston, IL 60208, USA

Prof. Michelle L. Gee
Department of Chemical Engineering
Princeton University
Princeton, NJ 08544, USA

Dr. D. Gibbs
Physics Department 510B
Brookhaven National Laboratory
Upton, NY 11973, USA

Dr. Marcos H. Grimsditch
Materials Science Division
Argonne National Laboratory
Argonne, IL 60439, USA

Prof. Vijay Gupta
Thayer School of Engineering
Dartmouth College
Hanover, NH 03755, USA

Prof. Robert J. Hamers
Department of Chemistry
University of Wisconsin
Madison, WI 53706, USA

Prof. Paul S. Ho
Center for Materials Science and Engineering
ETC 9. 104
University of Texas
Austin, TX 78712-1063, USA

Prof. John Jaszczak
Department of Physics
Michigan Technological University
Houghton, MI 49931, USA

Prof. William L. Johnson
Division of Applied Physics
California Institute of Technology
Pasadena, CA 91125, USA

Dr. Rainer Kirchheim
Max-Planck Institut für Metallforschung
Institut für Werkstoffwissenschaften
Seestr. 92
D 7000 Stuttgart
Germany

Dr. Richard A. LeSar
MS B262
Los Alamos National Laboratory
Los Alamos, NM 87545, USA

Dr. Chun Li
Physics Department
Northwestern University
Evanston, IL 60208, USA

Prof. Giorgio Margaritondo
Institut de Physique Appliquée
Ecole Polytechnique Federale de Lausanne
Ecublens
Switzerland

Dr. Karl L. Merkle
Materials Science Division
Argonne National Laboratory
Argonne, IL 60439, USA

Prof. Simon G. J. Mochrie
Department of Physics
MIT
Cambridge, MA 02139, USA

Dr. Reza Najafabadi
Department of Materials Science and Engineering
University of Michigan
Ann Arbor, MI 48109, USA

Dr. D. W. Niles
Synchrotron Radiation Center
University of Wisconsin – Madison
Stoughton, WI 53589, USA

Prof. M. Grant Norton
Department of Mechanical and Materials
Engineering
Washington State University
Pullman, WA 99164, USA

Dr. Paul R. Okamoto
Materials Science Division
Argonne National Laboratory
Argonne, IL 60439, USA

Dr. G. Palumbo
Ontario Hydro Research Division
800 Kipling Ave
Toronto, Canada M82 5S4

Dr. Simon R. Phillpot
Materials Science Division
Argonne National Laboratory
Argonne, IL 60439, USA

Prof. Manfred Rühle
Max-Planck Institut für Metallforschung
Institut für Werkstoffwissenschaften
Seestr. 92
D 7000 Stuttgart
Germany

Prof. Ivan K. Schuller
Department of Physics
University of California
La Jolla, CA 92093, USA

Prof. David N. Seidman
Materials Science and Engineering Department
and the Materials Research Center
Northwestern University
Evanston, IL 60208, USA

Dr. Richard W. Siegel
Materials Science Division
Argonne National Laboratory
Argonne, IL 60439, USA

Dr. B. D. Silverman
IBM T.J. Watson Research Center
PO Box 218
Yorktown Heights, NY 10598, USA

Dr. David A. Smith
IBM T.J. Watson Research Center
PO Box 218
Yorktown Heights, NY 10598, USA

Prof. David J. Srolovitz
Department of Materials Science and Engineering
University of Michigan
Ann Arbor, MI 48109, USA

Dr. Raymond T. Tung
A T & T Bell Laboratories
600 Mountain Ave.
Room 1T-102
Murray Hill, NJ 07974, USA

Dr. H. Y. Wang
Department of Materials Science and Engineering
University of Michigan
Ann Arbor, MI 48109, USA

Dr. Dieter Wolf
Materials Science Division
Argonne National Laboratory
Argonne, IL 60439, USA

Dr. Wuhua Yang
Department of Materials Science and Engineering
University of Michigan
Ann Arbor, MI 48109, USA

Prof. Sidney Yip
Department of Nuclear Engineering
Room 24-208
MIT
Cambridge, MA 02139, USA

Dr. D. M. Zehner
Solid State Division
Oak Ridge National Laboratory
Oak Ridge, TE 37831, USA

Introduction

Dieter Wolf and Sidney Yip

Some of the most important properties of materials in high-technology applications are strongly influenced or even controlled by the presence of solid interfaces. For example, interfaces are the critical element in fiber-reinforced structural ceramics with mechanical properties not imagined a decade or two ago. The entire electronics industry is based on the fascinating electrical properties of semiconductor interfaces, with ceramic–semiconductor, metal–semiconductor and metal–ceramic interfaces playing critical roles as well. Further examples are surface-modification techniques, designed to enhance the corrosion resistance of materials in hostile environments, or tailored for tribological or catalytic applications. In contrast to their enormous technological importance, our basic understanding of even the simplest interfaces, such as free surfaces and grain boundaries, is rudimentary at best. It is increasingly recognized, however, that truly significant technological advances can come from a better understanding and control of interfacial processes. To call attention to the considerable opportunities that exist in this lively area of materials research, in this book we are highlighting a number of promising recent developments primarily in the atomic-level understanding of solid interfaces.

It is widely appreciated that a highly interdisciplinary approach holds the greatest promise for providing novel insights into the fundamental physical, chemical, electronic, and mechanical processes at solid interfaces. As a whole, the interface community is as diverse as virtually any other in materials research, involving disciplines as disparate as solid-state, electronic-structure and thin-film physics, solid-state and surface chemistry, on the one hand, and fracture, physical metallurgy, and physical ceramics, on the other. Yet, it is rare that this diverse pool of expertise is ever brought together in a focused attack on interface-related phenomena and properties. It is our hope that this volume will give the reader a sense of the rich and fertile common ground which already exists (mostly, however, in a nucleation-like state), and to bring these very different communities closer together. To facilitate this highly necessary bridge building between these communities, in this Introduction and the first two chapters we will emphasize what we believe are the features in common to all the different types of interfaces.

The importance of interface materials is based primarily on their inherent inhomogeneity, i.e. the fact that physical properties at or near an interface can differ dramatically from those of the nearby bulk material. For example, the thermal expansion, electrical resistivity, or elastic response near an interface can be highly anisotropic in an otherwise isotropic material, and can differ by orders of magnitude from those of the adjacent bulk regions. Typically these gradients extend over only a few atomic layers; their investigation consequently requires experimental techniques capable of atomic-level resolution and detection coordinated with atomic- and/or electronic-level computer simulations. While for surfaces and thin films many suitable experimental techniques for the characterization and investigation of such gradient properties have become available in recent years, buried interfaces still pose a major challenge, because such a small fraction of atoms is actually affected by the presence of the interfaces (for details see Chapter 2).

A central theme in much of the work on solid interfaces concerns the interrelation between interfacial properties and the underlying structure, both geometrical and chemical. Since little control can be exerted over the **atomic-level** degrees of freedom (DOFs), any attempt to design interface materials with certain desired properties ('interface engineering') must be based on an understanding of the interrelation between the **macroscopic** structure (including the chemistry) of interfaces and their physical properties. However, in spite of various controlled bicrystal experiments performed to date, a systematic exploration of the misorientation phase space associated with the five macroscopic and three microscopic DOFs needed to characterize a single atomically flat interface (Chapter 1) is exceedingly difficult – if not impossible – for any type of interfacial system. The task appears to be even more difficult for epitaxial interfaces where virtually no control exists over some of the macroscopic DOFs, because, for a chosen substrate orientation, the epitaxial overlayer will attempt to choose a planar orientation which minimizes the free energy of the system. In computer simulations, by contrast, the macroscopic geometry of the interface can be controlled rather easily. An approach utilizing the complementary capabilities of computer simulation and experiment therefore seems to be particularly promising in the investigation of structure–property correlations. However, while a comparison between experiments and modelling results is absolutely essential, particularly in crucial test cases, we see the main strength of such simulations in their ability to provide unique atomic- and electronic-level insights

into structure–property correlations and into the underlying structure and gradient properties in the vicinity of the interfaces.

While a widespread recognition of the need to investigate structure–property correlations has evolved in recent years among the various parts of the interface community, at present there appears to be no commonly accepted terminology to characterize the macroscopic DOFs of even the simplest interfaces. For example, in much of the work on grain boundaries a terminology based on the so-called coincident-site lattice (CSL) is being used widely. However, such a terminology cannot be applied to incommensurate interfaces, for which a different terminology, based largely on the crystallographic planes joined at the interface, has evolved. Not surprisingly, although grain boundaries represent probably the best-studied type of interfacial system (other than free surfaces), relatively little of what has been learned on grain boundaries has cross-pollinated with work on other kinds of interfaces, such as epitaxial systems, strained-layer superlattices, and free surfaces. By bringing together in this book overlapping work on different types of interfaces, and by attempting to develop a common terminology to describe the geometry and underlying DOFs of all interfacial systems (Chapter 1), we hope to stimulate discussion on the features in common to all interfacially controlled materials phenomena.

In organizing this book we realize that any attempt to cover such a wide field is necessarily a highly selective endeavor. One can imagine many ways of classifying interfacial systems, for example, by distinguishing commensurate from incommensurate systems, coherent from incoherent interfaces, homophase (grain-boundary) materials from heterophase (dissimilar-material) interfaces, or internal interfaces from thin-film-type systems. The latter distinction between buried interfaces and systems composed of thin-film-type interfaces appears to be particularly meaningful for two reasons; first, because of the fundamentally different experimental techniques required for their investigation (Chapter 2) and second, because the lattice parameter(s), and hence the physical properties in the interfacial region, depend strongly on whether or not the interface is embedded between bulk material.

This classification is illustrated schematically in Fig. I.1 in which the interfacial region, assumed to be infinite in the x–y plane, is embedded in the z direction between two perfect, semi-infinite bulk crystals. The lower and upper halves of this bicrystal generally consist of different materials, A and B. By examining the different ways in which the interfacial region may or may not be sandwiched between bulk material, three types of interfacial systems can be distinguished. The **bulk interface** depicted in the figure, in which the interfacial region is on both sides surrounded by bulk material, is either a grain boundary (for A = B) or an interphase boundary (for A ≠ B). When one of

the bulk regions is removed, a **semi-bulk interface** is obtained. Finally, a **thin-film interface** is formed when both bulk regions are removed; since no bulk embedding is left, the material near the interface does not 'know' its bulk lattice parameter, and – in contrast to bulk and epitaxial interfaces – the lattice parameter in the interface is determined by the relaxation of surface or interfacial stresses.

In this scheme, a free surface or an epitaxial system is viewed as a semi-bulk interface, while a strained-layer superlattice, . . . |A|B|A|B| . . . , is viewed as a superlattice of thin films formed by periodically extending the thin-film sandwich, A|B, also in the z-direction. Stress-induced lattice-parameter changes are probably also important in nanophase materials due to their small grain size; we therefore regard them as essentially having the generic properties of thin-film materials. In a conventional polycrystal, by contrast, the lattice parameters at the interface are those of the surrounding bulk materials, thus lacking any of the stress-induced phenomena of the thin-film materials.

Following this distinction between three types of solid interfaces, the first half of the book, comprised of two parts, is devoted to a comparison of the structure and properties of these different interfacial systems. We will first focus on bulk interfaces, including grain boundaries (Part I), followed by semi-bulk and thin-film interfaces (Part II). By contrast with this focus on interfacial **systems** the second half, also comprised of two parts, is devoted to two kinds of interfacial **phenomena** which are important in all types of interfacial materials. We will first discuss interface chemistry (Part III), followed by a brief outlook on some fundamental aspects of interface fracture (Part IV).

The effects of interface chemistry, considered in Part III, are particularly striking in the electronic and mechanical properties of interface materials. In spite of the enormous importance of electronic and structural applications of interface materials, however, very little systematic work on the interrelation between such properties and the interface chemistry and structure has been performed. The experimental problems lie mainly in the extremely

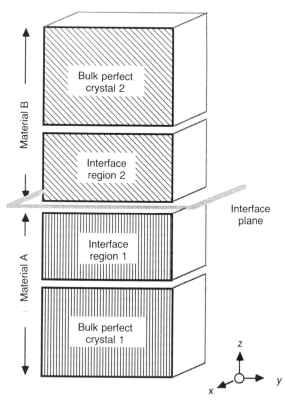

Fig. I.1 Distinction of three types of interfacial systems. Depending on whether the system is embedded in bulk material on both sides of the interface, on only one side or not at all, we distinguish 'bulk', 'epitaxial', and 'thin-film' interfaces. A and B are generally different materials.

high spatial resolution, coupled with high detection sensitivity, which are required to identify minute amounts of chemical species, charge states of atoms and point defects, and different phases which may be present in the narrow interface region. Since spatial resolution and detection sensitivity cannot normally be optimized simultaneously, the chemical characterization, particularly of buried interfaces, has evolved into a highly sophisticated experimental art. Theoretically, interface chemistry and its effect on properties represent a major challenge as well. The main problem here is that the nature of atomic and molecular bonding near an interface may not be the same as in a bulk environment. These interfacial electronic effects give rise to interesting electronic and magnetic phenomena and properties. Simulation methods based on an atomic-level description of interatomic forces must therefore be closely interconnected with insights gained at the electronic-structure level. Such a combination of electronic and atomic-level simulation methods holds considerable promise, although at this time only a few results are available for interfacial systems.

Mechanical properties, in particular interface fracture considered in Part IV, are the reason for many 'real-world' applications of interface materials, such as structural ceramics and composites. They also provide a rich area for the investigation of the effects of the local inhomogeneities near the interfaces, and their effect on the inter-relation between the structure and chemistry, on one hand, and the mechanical behavior, on the other. From much experimental work on grain-boundary fracture it seems that, with the exception of 'beneficial' segregants, the embrittlement potential of most impurities is governed by their propensity for segregation to the grain boundaries. Interface fracture and chemistry, the theme of the second half of the book, are therefore intimately connected, although relatively little knowledge has been accumulated on this complex interrelation.

The four parts of the book are preceded by two introductory chapters which, we hope, will be useful to all parts of the interface community (Chapters 1 and 2 mentioned above). In the first chapter an attempt is made to describe the geo-metry of crystalline interfaces at the atomic level from a common viewpoint, bringing together terminologies used almost exclusively in the grain-boundary area with the terminology usually applied to describe thin-film and epitaxial interfaces. It is our hope that one day a common terminology will emerge which is amenable to discussing all inter-facial systems and phenomena within a common geometrical framework. The second chapter offers an overview of the experimental techniques used to investigate the structure, chemistry, and physical properties of bulk, semi-bulk, and thin-film inter-faces in different types of materials. Such a review appears useful and desirable, particularly since the main body of the book focuses on interfacial systems, phenomena, and properties without drawing particular attention to the impressive (and to some perhaps bewildering) variety of experimental methods used in this area nor to the extreme difficulties in the investigation particularly of internal interfaces.

The inherent difficulty in the experimental investigation of buried interfaces is actually an advantage in the atomic- and electronic-level study of solid interfaces by means of computer-simulation techniques. Because the properties in the interfacial region are controlled by relatively few atoms, electronic and atomic-level computer simulations can provide a close-up view of the most critical part of the material. The limitations of such simulations are well known, pertaining to the approximate nature of what is known about the electronic structure and the interatomic interactions, the finite size of any simulation cell, its embedding in the surrounding material and the finite duration of any simulation, permitting only very fast dynamical phenomena to be studied (of duration typically less than 10^{-9} s). However, within this framework the local nature of atomic bonding, the positions and movements of atoms, local stresses, etc., can be investigated.

The unique and complementary features of such atomic- and electronic-level simulations offer a rich opportunity for a joint approach which combines atomic-level experimental techniques with computer-simulation methods. Although only in relatively few studies have these complementary capabilities been fully exploited, one of the factors motivating the particular choice of topics highlighted in this book is to illustrate the promises of such a joint approach.

We have benefited greatly from discussions with our colleagues at Argonne and MIT and have received many helpful suggestions from them as well as the contributors to this book. In particular, we would like to express our gratitude to

H. Wiedersich, M. Rühle, and A. J. Freeman for their encouragement and advice throughout this project. We would also like to acknowledge the long-standing support we have received from the US Department of Energy, BES Materials Sciences, under Contract W-31-109-Eng-38. One of us (SY) gratefully acknowledges the hospitality and support of the Materials Science Division at Argonne over many years.

CHAPTER ONE

Atomic-level geometry of crystalline interfaces

D. Wolf

Introduction · Basic terminology · Microstructure · Coherency, epitaxy and topotaxy · Commensurability · Degrees of freedom of crystalline interfaces · Atomic-level geometry of planar stacking · Grain boundaries · Characterization of the atomic structure of interfaces

1.1 INTRODUCTION

Any attempt to systematically investigate physical properties of solid interfaces, structure–property correlations in particular, should from the outset be based on a thorough understanding of their basic geometry. Moreover, a prerequisite for the atomic-level investigation of solid interfaces is a description of their basic structure and geometry not only in macroscopic terms but also at the atomic level.

The terms 'geometry' and 'structure' are often used interchangeably. However, when referring to the **geometry** of an interface, one usually thinks of the more **macroscopic** and purely crystallographic aspects of the structure, while by the **structure** one usually implies the **atomic** or **electronic** structure, including the chemical composition locally at the interface. The distinctions between, for example, coherent and incoherent interfaces, commensurate and incommensurate systems, homophase and heterophase interfaces, between 'special' and 'vicinal' surfaces, and between low-angle and high-angle (tilt or twist, symmetrical or asymmetrical, 'special' or general, twinned or non-twinned, coherent or incoherent) grain boundaries add to the terminology used to describe the geometry, structure, and chemistry of solid interfaces. This introductory chapter represents an attempt to clarify (or at least to collect) some of the termin-

ology used in this diverse area of materials research and to formulate a unified atomic-level geometrical description applicable to all crystalline interfaces, including internal (homophase or heterophase) interfaces and external surfaces (i.e. solid-vacuum interfaces). It is our hope that such an undertaking will facilitate and clarify the communication among the different groups of the materials research community, broadly known as the interface community.

By close analogy to the distinction between crystallography and physics, throughout this chapter a sharp distinction will be made between the 'geometry' and the 'structure' of an interface. The latter term will therefore be reserved for the atomic and/or electronic structure (including the local distribution of chemical species), which includes the fully relaxed positions of all the atoms in the system and contains all the detailed information on phenomena such as reconstruction, misfit localization or delocalization, elastic strains and interfacial dislocations, impurity segregation, interface reactions, etc. at the interface. In common to all these phenomena is their origin in the **physics** of the system (by contrast with its crystallography), as prescribed by the nature of the electronic and atomic bonding near the interface.

In our discussion of the 'geometry' of crystalline interfaces, the macroscopic crystallographic aspects will be distinguished from the atomic-level characterization of the basic interface geometry. The macroscopic geometry, to be discussed in section 1.6, includes all aspects of the geometry and crystallography determined by (i) the crystal structure(s) forming the interface and (ii) the degrees of freedom (DOFs) of the interfacial system, including the five macroscopic and three translational (or 'microscopic') degrees of freedom (sections 1.6.1 and 1.6.2). Taking the geometrical characterization further, down to the level of the atoms (albeit in their unrelaxed positions), naturally leads to the concept of the **atomic-level geometry** (section 1.7). Based on **unrelaxed** atom positions, i.e. strictly crystallographic concepts, this description provides information on the plane-by-plane arrangement of the atoms near the unrelaxed interface, most importantly on the size and shape of the planar unit cell (if the atomic structure is, indeed, periodic).

To illustrate these concepts by a simple example, we briefly consider a stacking fault (for details see sections 1.6.3 and 1.7.3). As illustrated in Figs. 1.1(a) and (b), the **macroscopic geometry** of a stacking fault is fully characterized by the two macroscopic DOFs associated with the fault $(x-y)$ plane (here characterized in terms of Miller indices, $\langle hkl \rangle$, associated with the interface-plane normal) and the two translational DOFs in the translation vector, $\boldsymbol{T} = (T_x, T_y)$, which characterizes the stacking discontinuity in the fault plane. Based on this information, it is apparent that the interface may be generated by first choosing a particular crystallographic plane in a perfect crystal (Fig. 1.1(a)) and subsequently translating one half of that crystal relative to the other half, parallel to the intended fault plane by the vector $\boldsymbol{T}$ (Fig. 1.1(b)). Generation of its **atomic-level geometry**, illustrated in Figs. 1.1(c) and (d), provides information on the unrelaxed atom arrangement in the defect, including information on (i) the size and shape of its planar unit cell (which are obviously identical to those of the perfect crystal in Fig. 1.1(c)), (ii) the spacing of atom planes, $d(hkl)$, parallel to the fault plane, and (iii) the number of lattice planes in the repeat stacking sequence, $P(hkl)$, in the direction of the fault normal. In some cases this strictly geometrical atomic-level information on the defect may be very useful in predicting some of its basic physical properties, but without a knowledge of the detailed (i.e. relaxed) atomic structure of the interface and its chemical composition, both of which are governed by the electronic-structure-based physics of the interactions between the atoms.

Because of the complexity of the basic geometry of grain boundaries (GBs), involving both macroscopic concepts (such as the distinction between tilt, twist and general boundaries, and between low- and high-angle, symmetrical and asymmetrical boundaries, to name only a few) and atomic-level concepts (based on the geometry of Bravais lattices), a consistent atomic-level characterization of their basic geometry is particularly desirable and useful. In section 1.8 we will make an attempt to clarify the GB terminology by defining the atomic-level geometry of GBs within the framework, applicable to all interfacial systems, of the macroscopic and atomic-level concepts developed earlier in sections 1.6 and 1.7, respectively. Such a characterization of GBs, although not commonly used in the GB community, naturally exposes their close geometrical relationship to other planar defects and interface systems. In particular, the similarity between GBs and free surfaces has not been fully appreciated in the past, a fact which might be one of the reasons why only relatively little is known about the ideal-cleavage energy (or work of adhesion) of even the simplest GBs. Efforts based on the recognition of the considerable similarities in the atomic-level geometries of GBs and free surfaces therefore appear particularly promising to better elucidate the basic physics of interfacial decohesion.

The chapter is organized as follows. Section 1.2 contains a collection of terms and concepts, some macroscopic and some atomistic, the clarification of which at the outset might be helpful. The concepts of macroscopic and atomic-level geometry are developed and illustrated in sections 1.6 and 1.7, respectively. The goal of these two sections is to formulate a unified geometrical framework applicable to all types of interface systems. Then, in section 1.8, the geometry of GBs is reviewed within this general framework; a good understanding of sections 1.6 and 1.7 is therefore

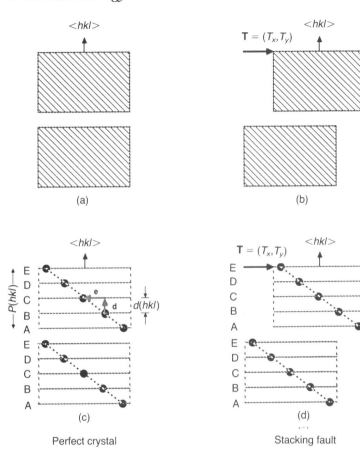

Fig. 1.1 'Macroscopic geometry' (top half) and 'atomic-level geometry' (bottom half) of a perfect crystal ((a) and (c)) and a stacking fault ((b) and (d)) on some crystal plane with normal $\langle hkl \rangle$. The concept of the stacking period, $P(hkl)$, the interplanar lattice spacing, $d(hkl)$, and the planar unit cell are discussed in detail in section 1.7.1.

required. It is our hope that this atomic-level view of the geometry of GBs, and the definition within this framework of much of the 'jargon' used to describe this geometry, will bring out the parallels that clearly exist with other types of crystalline interfaces. Finally, to emphasize and clarify the distinction between the **geometry** and **atomic structure** of solid interfaces, the chapter concludes with a brief review of the concepts and methods used to characterize the atomic structure of crystalline interfaces.

1.2 BASIC TERMINOLOGY

As discussed in the Editors' Introduction [1], three types of interfacial systems (labeled 'bulk', 'semi-bulk', and 'thin-film' interfaces) may be distinguished. This admittedly somewhat arbitrary classification is based on the fundamentally different effects of interfacial stresses and strains in the three types. Within the framework of this or any other classification, a variety of terms is used to describe the microstructure of polycrystalline materials as well as the macroscopic and atomic-level geometries and the atomic structure of individual interfaces. Although a common terminology to describe all structural aspects of crystalline interfaces has not evolved to date, in the following we will collect and define some of the more commonly used interface vocabulary.

1.2.1 Three basic types of interfaces

Depending on whether an interface is embedded in bulk material on both sides, on only one side, or not at all, we distinguish three basic types of interfaces, namely 'bulk', 'semi-bulk' and 'thin-film'

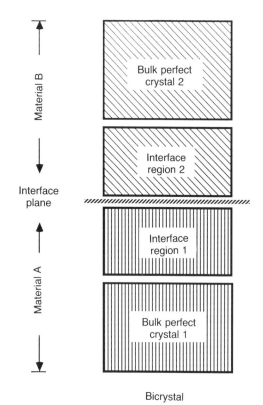

Bicrystal

Fig. 1.2 Distinction of three types of interfacial systems. Depending on whether the system is embedded in bulk material on both sides of the interface, on only one side, or not at all, we distinguish 'bulk' (or 'buried' or 'internal'), 'semi-bulk', and 'thin-film' interfaces. A and B are generally different materials [1].

interfaces (Fig 1.2). This (highly idealized and somewhat subjective) distinction is based on the observation that (i) lattice-parameter changes in the interfacial region, induced by interfacial stresses, may have a pronounced effect on the physical properties and chemical composition at or near an interface and (ii) these stress-induced effects are fundamentally different in these three types of systems.

1.2.2 Bulk interfaces

The first type, a bulk interface (sometimes also called a **buried** or **internal** interface; see Fig. 1.1) represents the greatest experimental challenge because (i) an extremely small fraction of the atoms

(typically about one in 10^{10}) actually experience the presence of the interface, and (ii) by contrast with a free surface, the disturbed atoms are sandwiched between (or buried in) bulk material. (For a review of the related experimental methods, see the following chapter in this volume [2].)

In the **bicrystal** shown in Fig. 1.2, the interface is conceptualized as being embedded between two well-oriented single crystals, and is hence characterized geometrically by five macroscopic and three translational ('microscopic') geometrical degrees of freedom (see section 1.6 below). Since crystallinity is only a necessary – but not a sufficient – condition for commensurability (section 1.4), a bicrystal may thus contain either a **commensurate** or an **incommensurate** interface in its center. If the two materials A and B forming the bicrystal are not the same (or at least represent different phases of the same material), the interface is usually referred to as an **interphase**, a **dissimilar-material** or a **bimaterial** interface, or as a **phase** or **heterophase** boundary, while the bicrystal contains a **grain boundary** (or **homophase** interface) or a **stacking fault** if A and B are identical materials and phases.

1.2.3 Semi-bulk interfaces

The second type, a semi-bulk interface (Fig. 1.3), is obtained by removing one of the two bulk semi-infinite crystals from Fig. 1.2. Containing both an external free surface and an internal interface, this type of interface may be viewed as consisting of a thin film of material B attached to a bulk substrate of material A. The term 'thin-film overlayer' therefore provides an alternate, equally descriptive characterization of this type of interface. A free surface is obviously included here as the case in which materials A and B are the same. According to Fig. 1.3, a 'bulk free surface' may be viewed conceptually as a homophase interface consisting of a strained (because of surface stresses) thin film which is attached to a bulk substrate.

We mention that the semi-bulk interface defined here is sometimes also referred to as an 'epitaxial' interface, or simply 'epitaxy'. As discussed further in section 1.4, this terminology arises from the manner in which a thin-film overlayer with a

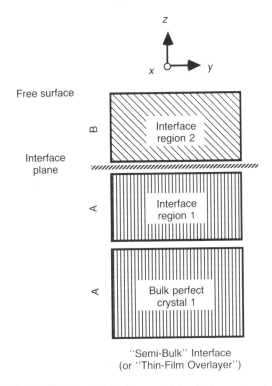

Fig. 1.3 'Semi-bulk' interface, containing both an external free surface and an internal interface. This type of interface may be viewed as consisting of a strained (because of surface stress) thin film of material B attached to a bulk (i.e. rigid) substrate of material A. This type of interface may conceptually be obtained from Fig. 1.2 by the removal of one of the two bulk regions.

mostly coherent epitaxial alignment relative to the substrate is usually produced. However, in many instances the interfaces ('epitaxy') obtained from 'epitaxial-growth' processes are coherent only up to some critical thickness, i.e. a 'perfect epitaxy' in the crystallographic sense [3] is not always obtained. To avoid confusion, in the above definition of a **semi-bulk** interface we mean simply an interface consisting of a thin film attached to a bulk, rigid substrate, thus avoiding any reference to the atomic-level quality of the thin-film overlayer. While the overlayer may stretch or contract to enable formation of a coherent interface (section 1.4), this definition includes incoherent interfaces and a possibly rotated (and even incommensurate) thin-film overlayer as well (section 1.5). Con-

sequently, while the term semi-bulk interface will be used to emphasize its distinction from a bulk or thin-film interface, the atomic-level geometry and structure of such interfaces will be characterized in terms of the concepts of coherency (and the related concepts of epitaxy and topotaxy) and commensurability, discussed in detail in sections 1.4 and 1.5.

1.2.4 Thin-film interfaces

Finally the third type, a thin-film interface (Fig. 1.4), is obtained from Fig. 1.3 by replacing the bulk substrate by a thin film itself (i.e. by removing both bulk regions from Fig. 1.1), thus creating a second free surface and, hence, an unsupported, free-standing thin film (or a **thin slab**).

The two free surfaces in Fig. 1.4(a) may be eliminated conceptually by periodically extending the geometry in Fig. 1.4(a) in the z-direction, thus creating the thin-film superlattice sketched in Fig. 1.4(b) in which – ideally – all interfaces are identical. Such a material is also known as a **composition-modulated** or a **dissimilar-material** superlattice. If, in spite of a lattice-parameter mismatch, coherent interfaces can be sustained in the superlattice, it is also called a **strained-layer** superlattice.

If materials A and B are identical, the system in Fig. 1.4(a) degenerates into a thin slab with a grain boundary or stacking fault (or no interface at all) in its center, while the system in Fig. 1.4(b) becomes a superlattice of homophase interfaces (or a **grain-boundary superlattice** [7]). Although such superlattices have not actually been investigated experimentally, by the elimination of interfacial chemistry as a factor, they represent ideal model systems for investigating, by means of computer simulations, the strictly structural aspects of the physical properties of superlattice materials, against which any effects due to interface chemistry can be probed.

1.2.5 Effects of interfacial stresses

As already mentioned, the above distinction between bulk, semi-bulk and thin-film interfaces is

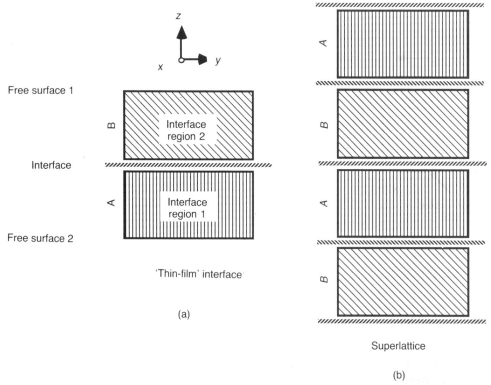

Fig. 1.4 'Thin-film' interface, generally consisting of two external and one internal surface (a). The geometry of this interface may be obtained from Fig. 1.3 by removing the remaining bulk region (Fig. 1.2). By periodically extending the geometry in (a) in the z direction, a 'thin-film superlattice' may be generated, thus replacing the external surfaces by internal interfaces. The latter may be coherent or incoherent, commensurate or incommensurate, homo- or heterophase interfaces. If all interfaces are coherent, the system is called a 'strained-layer' composition-modulated superlattice.

based on the observation that (i) lattice-parameter changes in the interfacial region, induced by interfacial stresses, may have a pronounced effect on the physical properties and chemical composition at or near an interface and (ii) these stress-induced effects are fundamentally different in these three types of systems.

It has been widely recognized in recent years that the surface-stress tensor, $\sigma_{\alpha\beta}$ (α, $\beta = x, y, z$), may play an important role not only in surface reconstruction [4] but also, for example, in the elastic response of thin films [5, 8, 9] and thin-film superlattices [6, 7, 9]. To illustrate the concept of interfacial stress, here we briefly consider a free surface. $\sigma_{\alpha\beta}$ is then defined as the variation of the specific surface energy, γ, as a function of the strain, $\varepsilon_{\alpha\beta}$, i.e. $\sigma_{\alpha\beta} = \delta\gamma/\delta\varepsilon_{\alpha\beta}$. In a fully relaxed

'bulk' free surface (i.e. one attached to a bulk substrate; Fig. 1.3), $\sigma_{\alpha\beta}$ is usually diagonal, with a vanishing component, σ_{zz}, in the direction of the surface normal (z direction). Its only non-zero elements, σ_{xx} and σ_{yy}, are usually tensile and of significant magnitude, favoring contraction in the (x–y) plane of the surface. However, in a bulk free surface this stress can only be relaxed by reconstruction; by contrast, a thin film may in addition contract, giving rise to a uniform reduction in the average lattice parameter(s) in the film plane, with a consequent Poisson expansion in the z direction (Fig. 1.5).

In the thin-film system sketched in Fig. 1.4(a), interfacial stresses arise not only from the two film surfaces but also from the interface between the films. By contrast, in the superlattice sketched in

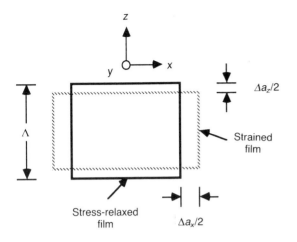

Fig. 1.5 Effect of the surface stresses, σ_{xx} and σ_{yy}, on the average lattice parameters of an unsupported thin film ('thin slab') parallel and perpendicular to the film plane. The in-plane contractions of the film, Δa_x and Δa_y (<0), are accompanied by a Poisson expansion, Δa_z (>0), in the direction of the film normal (z direction). Notice that σ_{zz} vanishes identically both in the bulk free surface and in the fully stress-relaxed film [8].

Fig. 1.3(b) all stresses originate from the presence of the interfaces between the two materials. Being entirely unconstrained by any bulk material, both the thin slab and the thin-film superlattice will adjust their lattice parameters to values governed by the relaxing surface and the interfacial stresses [7–9], hence creating favorable conditions for the formation of a coherent interface even if the corresponding bulk lattice parameters are mismatched.

To illustrate the role of interfacial stresses in a bulk interface (Fig. 1.2), we assume that both sides of the interface are indeed crystalline. Because the interface is embedded between bulk material, the lattice parameter(s) at or near the interface is (are) governed completely by that of the surrounding bulk material. Interfacial stresses associated with the atomic-level structural disorder at the interface can therefore not be relaxed, thus preventing any change in the average lattice parameter(s) near the interface. By contrast with a thin-film system, this constraint renders atomic-level reconstruction and/or a redistribution of chemical species (i.e. segregation) as the only possible stress-relaxation mechanisms.

The effects of interfacial stresses in the semi-

bulk system sketched in Fig. 1.3 differ qualitatively from those of the thin-film and the bulk interface systems. Because the substrate consists of bulk material, its lattice parameter is undisturbed by the presence of the thin overlayer. However, the thin overlayer may, in principle, be strained relative to its 'intrinsic' lattice parameter when not attached to a substrate (i.e. relative to the lattice parameter of the surface-stress relaxed thin slab in Fig. 1.5). This straining of the film as it is attached to the substrate usually enables the film to be in more or less perfect registry ('coherency' or 'commensurability'; sections 1.4 and 1.5 below) with the substrate. It therefore appears that the physical properties of this type of interface should lie somewhere between those of a bulk interface and those of a thin film, the latter usually being strained, however.

1.3 MICROSTRUCTURE

The above classification, leading to the distinction between three basic types of interfaces, is highly idealized for two reasons. First, a 'real' material may contain a multitude and variety of interfaces; and second, in many instances the interfaces are not atomically flat but contain steps, ledges, dislocations and/or voids, and may be roughened, amorphous, etc. Hence, before discussing the atomic-level concepts of 'coherency' and 'commensurability', we briefly clarify several terms commonly used to describe the microstructure of 'real' interface materials.

By contrast with the bicrystal sketched in Fig. 1.2, a **polycrystal** contains many interfaces separated by crystallites of various orientations. For a 'large' grain size (typically of the order of microns or larger), one can expect the structure, chemistry, and properties of individual interfaces in the polycrystal to be similar to those of a bulk interface. In addition, however, the poorly understood **triple junctions** (i.e. line defects along which three interfaces meet) play an important role as well.

If the grain size is very small, typically of atomic or nanometer dimensions, the polycrystal is referred to as a **nanocrystal** or a **nanophase** material. The interfaces may be either of a homophase or heterophase type; in the latter case the material is also called a **nanocomposite**.

The interfaces in these small-grained materials are not usually embedded between bulk material, thus permitting interfacial stresses to be relaxed. One would therefore expect their physical properties to be governed by both the presence of the interfaces and the deviation in their average lattice-parameter(s) from that of bulk material. Presumably their properties are therefore more similar to those of thin-film interface materials (section 1.2.4) than to those of bulk polycrystalline interface systems (section 1.2.2).

1.4 COHERENCY, EPITAXY AND TOPOTAXY

The epitaxial growth of one crystal on another is of considerable practical interest in the semiconductor industry which requires crystals free from dislocations and other defects. With the size of components in electronic devices rapidly approaching atomic-level dimensions, the need for atomic-level perfection of the crystals comprising such devices is ever increasing.

Central to the growth of nearly perfect epitaxial devices is the concept of a coherent (or dislocation-free) interface. Following Christian [10], a **coherent** interface between two crystals is defined as one for which corresponding atom planes and lines are continuous across the interface, i.e. one whose atomic structure is characterized by an atom-by-atom matching across the interface. Conversely, if there is no continuity of planes and lines across the interface, i.e. if a one-on-one atomic matching does not exist even locally, the interface is referred to as **incoherent**.

The terms epitaxial and coherent are sometimes used interchangeably to describe a planar defect with an atom-by-atom match across the interface. To illustrate the widespread confusion in the use of the term 'epitaxy', we here give two commonly used definitions. Webster's Dictionary, focusing on the growth process, defines epitaxy as 'the growth on a crystalline substrate of a crystalline substance that mimics the orientation of the substrate'. By

contrast, the International Union of Crystallography [3] focuses on strictly crystallographic factors, by defining epitaxy as 'the phenomenon of mutual orientation of two crystals of different species, with two-dimensional lattice control (mesh in common)'. While Webster's definition clearly includes the usual distinction between homo- and hetero-epitaxy and between a thin-film overlayer and a bulk interface, the second definition is limited to hetero-interfaces with bulk material on both sides; it is hence much more restrictive than Webster's definition. Also, the purely crystallographic definition does not seem to require the continuity of lines in Christian's definition of 'coherency'. The concept of epitaxy is therefore less restrictive than the concept of coherency, as evidenced for example by the distinction between 'perfect epitaxy' (i.e. presumably coherent) and 'rotated epitaxy' (such as Au on Cr) in which a one-on-one correspondence of atoms across the interface may not exist. To avoid confusion, we will use the term 'epitaxy' when referring to the growth process, while the atomic structure at the interface will be characterized in terms of the concepts of coherency and commensurability (section 1.5).

(We also mention that the International Union of Crystallography has recommended avoidance of the term 'epitaxial' in favor of the terms 'epitaxic' or 'epitactic', with preference given to the term 'epitaxic' [3]. Webster's Dictionary, by contrast, finds the term 'epitaxial' perfectly in order; because of its wide and common use, throughout this chapter we will therefore use the latter.)

Another concept sometimes used in this context is that of **topotaxy**, which is defined as 'the phenomenon of mutual orientation of two crystals of different species resulting from a solid-state transformation or chemical reaction' [3]. This is in contrast with epitaxy, where we imply a layer-by-layer growth.

In both the coherent and the incoherent semi-bulk interfaces illustrated in Figs. 1.6(a) and (b), respectively, the lattice parameter of the substrate (open circles) is that of the bulk material, a_A. In the epitaxy shown in Fig. 1.6(a), the thin overlayer, of thickness Λ, is strained relative to its bulk lattice parameter, a_B, to match that of the substrate. In the incoherent case sketched in Fig. 1.6(b), by

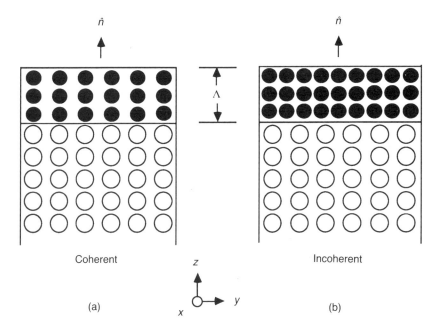

$\hat{n}$

Λ

Coherent

(a)

z

x — y

Incoherent

(b)

Fig. 1.6 (a) Coherent and (b) incoherent dissimilar-material interfaces, consisting of a thin-film overlayer attached to a bulk substrate. The coherent system represents an **epitaxy** in the crystallographic sense [3].

contrast, the lattice parameter of the overlayer is more similar to that of a fully relaxed, unstrained thin film.

To illustrate the concepts of epitaxy and topotaxy, experimental examples for both are shown in Figs. 1.7 (courtesy of K. L. Merkle) and 1.8 (courtesy of C. B. Carter). Figure 1.7 represents a high-resolution transmission-electron microscopy (TEM) image of a perfectly coherent epitaxial interface (dashed line) formed between a thin film of TiO_2 and an Al_2O_3 substrate [11]. Figure 1.8 shows a TEM image of spinel which has grown topotactically in an olivine matrix as a result of internal oxidation [12].

Because of the elastic energy involved, a coherent interface (i.e. 'epitaxy') can be formed only for a relatively small lattice-parameter mismatch,

$$f = (a_B - a_A)/a_A \qquad (1.1)$$

where A and B denote the substrate and thin-film overlayer, respectively. The 'critical mismatch', f_c, achievable for particular combinations of materials (involving, for example, the fcc and bcc lattices) is obviously not only a function of the interfacial geometry but also of the local thermo-elastic behavior at the interface which, in turn, is a function

[010] TiO_2

1 nm

[0001] Al_2O_3

Fig. 1.7 High-resolution transmission-electron micrograph of a perfectly coherent epitaxial interface (dashed line) formed between a thin film of TiO_2 and an Al_2O_3 substrate. Notice the small change in the 'tilt' angle across the interface, resulting from a virtually sudden change at the interface of the lattice parameter in the direction of the interface normal [11] (Courtesy of K. L. Merkle).

of the film thickness, Λ. Also, the energy difference between the coherent and incoherent structures sketched in Fig. 1.6 depends critically on the relative strength of the interaction between atoms across the interface, as well as the modulus for

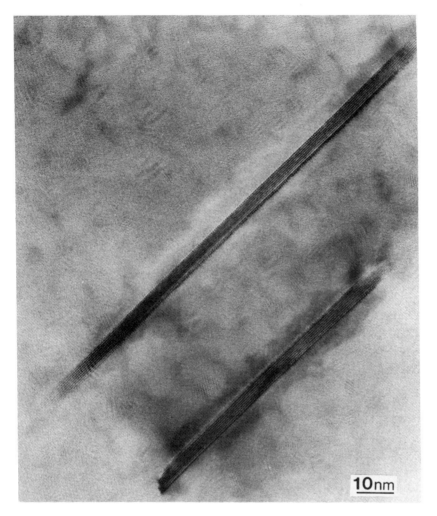

(a)

Fig. 1.8 (a) TEM image of spinel which has grown topotactically in an olivine matrix as a result of internal oxidation. The precipitate and the matrix are closely lattice-matched on one plane but the misfit is larger on the other – hence the elongated shape of the particles. (b) High-resolution image of one of the particles in (a). [12] (Courtesy of C. B. Carter).

shear locally at the interface [13, 14]. By contrast with strictly crystallographic geometrical concepts, the concept of coherency therefore also involves the elastic behavior of the material locally at the interface [14].

In practice, above a certain 'critical thickness', Λ_c, coherency cannot be sustained, and the interface becomes incoherent. The incoherent structure in Fig. 1.6(b) is therefore usually replaced by either of the **semi-coherent** types of structures sketched in Figs. 1.9 and 1.10 in which a strict one-to-one correspondence between atoms across the interface, as well as the continuity of lattice planes and

lines, exists only locally in various regions along the interface. By contrast with **extraneous** (i.e. regular lattice) dislocations, the so-called **inherent** (or **misfit**) dislocations [15] in Figs. 1.9 and 1.10 at a critical distance, Λ_c, from the interface are an integral part of the long-range interface structure and, hence, its geometry and crystallography.

Depending on whether or not the inherent dislocations are long-range ordered, we distinguish in Figs. 1.9 and 1.10 the two types of semi-coherent interfaces, formed by **commensurate** and **incommensurate** crystal lattices, respectively (section 1.5 below). In most practical cases, if the materials or

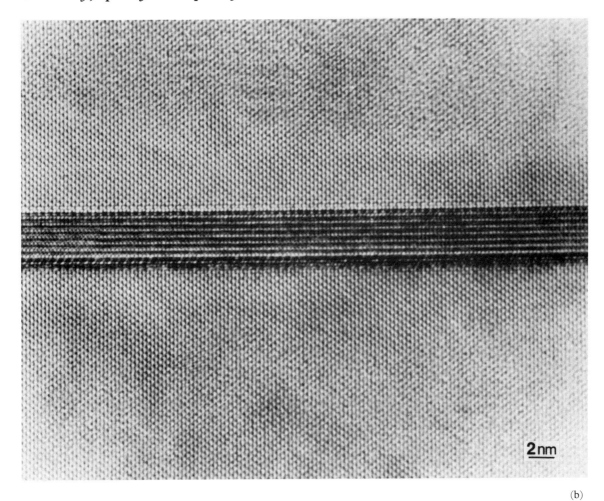

(b)

phases forming the interface are not the same, only the incommensurate structure in Fig. 1.10 exists, unless both the thin-film overlayer and the substrate can adjust their lattice parameters. While this is impossible for the bulk substrates sketched in Figs. 1.6, 1.9, and 1.10, in the thin-film super-lattices shown in Fig. 1.11 both a_A and a_B are adjustable due to the effect of interfacial stresses. The **strained-layer** (i.e. coherent) superlattice shown in Fig. 1.11(a) can therefore be expected to exhibit a wider range of stability against becoming incoherent; therefore, in Fig. 1.11(b) we assume a larger critical mismatch and critical thickness than in the corresponding **semi-bulk** epitaxial system in Fig. 1.6(a).

While eq. (1.1) provides a useful measure of the mismatch for a bulk-substrate epitaxial system, it is not very meaningful for thin-film systems (in which neither constituent 'knows' its bulk lattice parameter). The mismatch is then better characterized in terms of the average lattice parameter, $\bar{a} = (a_B + a_A)/2$, by defining [14]

$$\bar{f} = (a_B - a_A)/\bar{a} \qquad (1.2)$$

The definitions of the mismatch parameters in eqs. (1.1) and (1.2) for semi-bulk and thin-film interface systems, respectively, have proven useful in the determination of their critical mismatch and thickness, f_c and Λ_c, by means of continuum-elasticity theory [13, 14].

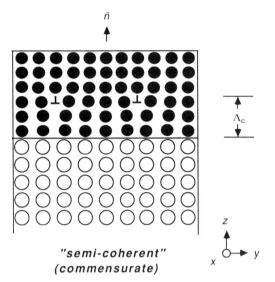

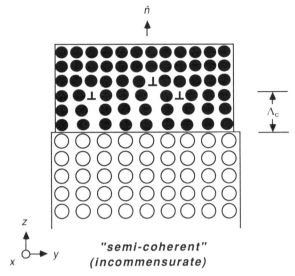

Fig. 1.9 'Semi-coherent' but commensurate (section 1.5) thin-film overlayer in which a strict one-to-one correspondence between atoms across the interface, as well as the continuity of lattice planes and lines, exists only locally in various regions along the interface.

Fig. 1.10 'Semi-coherent' but incommensurate (section 1.5) thin-film overlayer. A planar periodic unit cell cannot be defined for such a system, as indicated by the mismatch between the lateral dimensions on the two sides of the interface.

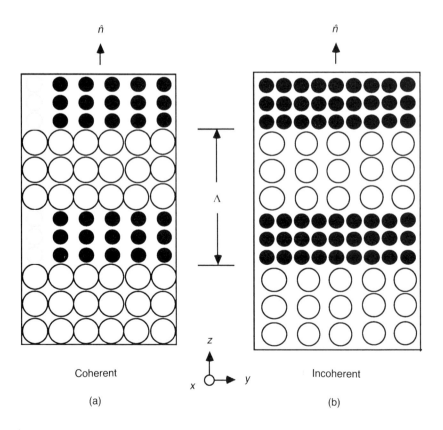

Fig. 1.11 (a) Coherent or 'strained-layer' and (b) incoherent thin-film superlattice. By contrast with the semi-bulk system in Fig. 1.6 (in which the substrate is a bulk material), in the thin-film system sketched here the lattice parameters parallel and perpendicular to the interfaces can adjust in response to interfacial stresses.

1.5 COMMENSURABILITY

In a **commensurate** interface the atomic positions on both sides of the interface are long-range ordered parallel to the plane of the defect, and a common planar unit cell exists which describes the periodic (i.e. crystalline) structure. Conversely, if the atomic structure is non-periodic, the interface is usually referred to as **incommensurate**. The concept of commensurability obviously requires crystalline long-range order in both sets of lattice planes forming the interface. An interface between a crystal and an amorphous material or a liquid can therefore never be commensurate. On the other hand, as evidenced by the existence of incommensurate GBs, the necessary requirement of crystalline order is not sufficient to guarantee the formation of a commensurate interface.

To formulate the criterion for commensurability between two lattice planes, we define two planar Bravais lattices by the Bravais vectors a_1, a_2 and b_1, b_2 (Fig. 1.12). The task then consists of finding the planar vectors, C_1 and C_2, which define the primitive unit cell of the common 2D superlattice – if it exists. As an example, Fig. 1.13 illustrates the formation of a planar superlattice in which four primitive unit cells of the first (a_1, a_2) plane are combined with nine of the other (b_1, b_2). At first sight one is tempted to require as the condition of commensurability that, in addition to the coordinate origin, there are infinitely many common points which satisfy the relation,

$$n_1a_1 + n_2a_2 = m_1b_1 + m_2b_2 \qquad (1.3)$$

with n_1, n_2, m_1, $m_2 = 0$, ± 1, ± 2, etc. (In the example of Fig. 1.13, $n_1 = n_2 = 2$ and $m_1 = m_2 = 3$.) Equation (1.3) would rule out lattice planes with irrational unit-cell dimensions from being commensurate. Yet, as the example in Fig. 1.14 shows, two square planar lattices with a $\sqrt{2}$ ratio of the unit-cell dimensions are commensurate. As illustrated in the right half of the figure, however, in spite of the fact that eq. (1.3) cannot be satisfied, the two lattices are actually commensurate (albeit incoherent) if one allows for a 45° rotation about the common plane normal. The condition (1.3) is hence too restrictive, as it does not permit for the rotation of the two planes relative to one another about their common normal. As discussed in section 1.6.1 below, such a 'twist' rotation by some angle θ is a macroscopic geometrical degree of freedom of the interface and hence should be included in the criterion for commensurability. Equation (1.3) then becomes

$$n_1a_{1x} + n_2a_{2x} = (m_1b_{1x} + m_2b_{2x})\sin\theta \\ - (m_1b_{1y} + m_2b_{2y})\cos\theta \quad (1.4a)$$

$$n_1a_{1y} + n_2a_{2y} = (m_1b_{1x} + m_2b_{2x})\cos\theta \\ + (m_1b_{1y} + m_2b_{2y})\sin\theta$$

$$(1.4b)$$

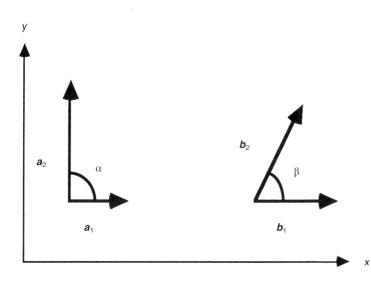

Fig. 1.12 Two sets of planar Bravais vectors a_1, a_2 and b_1, b_2.

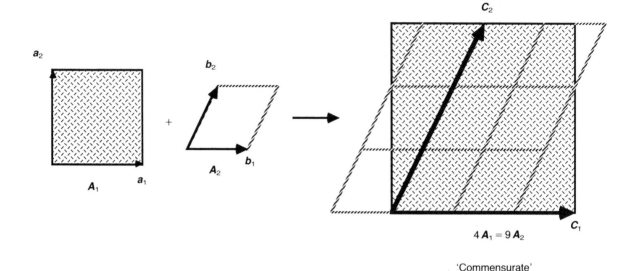

$4\,\boldsymbol{A}_1 = 9\,\boldsymbol{A}_2$

'Commensurate'

Fig. 1.13 Example illustrating the formation of a commensurate interface by combining four primitive unit cells of the first $(\boldsymbol{a}_1, \boldsymbol{a}_2)$ crystal with nine of the other $(\boldsymbol{b}_1, \boldsymbol{b}_2)$. $\boldsymbol{A}_1$ and $\boldsymbol{A}_2$ are the related planar unit-cell areas.

From eqs. (1.4a) and (b) it follows that

$$\tan\theta = \frac{\begin{array}{c}(n_1a_{1x} + n_2a_{2x})(m_1b_{1x} + m_2b_{2x}) \\ + (n_1a_{1y} + n_2a_{2y})(m_1b_{1y} + m_2b_{2y})\end{array}}{\begin{array}{c}(n_1a_{1y} + n_2a_{2y})(m_1b_{1x} + m_2b_{2x}) \\ - (n_1a_{1x} + n_2a_{2x})(m_1b_{1y} + m_2b_{2y})\end{array}} \quad (1.5)$$

For a given set of Bravais vectors $\boldsymbol{a}_1$, $\boldsymbol{a}_2$, and $\boldsymbol{b}_1$, $\boldsymbol{b}_2$, eq. (1.5) provides an infinite set of θ values associated with $n_1, n_2, m_1, m_2 = 0, \pm1, \pm2$, etc. This does not necessarily mean, however, that the two planes are commensurate since, for some allowed angle θ, eq. (1.5) merely describes a line of coincident points in common to the two planar Bravais lattices obtained when the two lattices are rotated relative to each-other. The first non-vanishing point on this line, given by the integers n_1^0, n_2^0, m_1^0 and m_2^0, defines one of the two primitive vectors, say $\boldsymbol{C}_1$, of the superlattice (expressed here in the unrotated x–y coordinate system in Fig. 1.14):

$$\boldsymbol{C}_1 = n_1^0\boldsymbol{a}_1 + n_2^0\boldsymbol{a}_2 \quad (1.6)$$

For the two planes to be commensurate, for the same angle θ a second vector,

$$\boldsymbol{C}_2 = n_1^1\boldsymbol{a}_1 + n_2^1\boldsymbol{a}_2 \quad (1.7)$$

which is not collinear with $\boldsymbol{C}_1$, must also exist. The condition of commensurability hence requires that, for the same value of θ, eq. (1.5) yields at least two vectors which satisfy the condition

$$|[\boldsymbol{C}_1 \times \boldsymbol{C}_2]| > 0 \quad (1.8)$$

As is well known, the vector product $[\boldsymbol{C}_1 \times \boldsymbol{C}_2]$ defines the area vector (parallel to the plane normal) of the plane of the common superlattice. Equation (1.8) hence expresses the requirement for the existence of a planar unit cell of the superlattice with a non-vanishing area.

We mention that $\boldsymbol{C}_1$ and $\boldsymbol{C}_2$ could have equally been expressed in the rotated x'–y' coordinate system in Fig. 1.14 as follows:

$$\boldsymbol{C}_1' = m_1^0\boldsymbol{b}_1' + m_2^0\boldsymbol{b}_2' \quad (1.9)$$

$$\boldsymbol{C}_2' = m_1^1\boldsymbol{b}_1' + m_2^1\boldsymbol{b}_2' \quad (1.10)$$

and the condition of commensurability becomes:

$$|[\boldsymbol{C}_1' \times \boldsymbol{C}_2']| > 0 \quad (1.11)$$

From the existence of a planar superlattice, it follows that the related primitive planar unit-cell areas, A_1 and A_2, of the underlying Bravais lattices are compatible with one another in that they are rational multiples of each other, i.e.

$$nA_1 = mA_2 \tag{1.12}$$

where n and m are positive integers. Expressing the areas in terms of the basis vectors, this condition becomes:

$$n(a_{1x}a_{2y} - a_{1y}a_{2x}) = m(b_{1x}b_{2y} - b_{1y}b_{2x}) \tag{1.13}$$

Although the compatibility of the primitive planar unit-cell areas is clearly necessary for two lattice planes to be commensurate, in order to form a superlattice, the two lattice planes also have to share infinitely many common (albeit not necessarily all) lattice points, i.e. satisfy eq. (1.5). (If they share all lattice points, the interface is 'coherent'; section 1.4.)

To illustrate the above expressions with a simple example, we consider the two square lattices of Fig. 1.14, with the Bravais vectors:

$$a_1 = (a, 0), a_2 = (0, a);$$
$$b_1 = (\sqrt{2}a, 0), b_2 = (0, \sqrt{2}a) \tag{1.14}$$

Inserting eq. (1.14) into eqs. (1.4a) and (b) gives

$$n_1 a = m_1 \sqrt{2}a \sin \theta - m_2 \sqrt{2}a \cos \theta \tag{1.15a}$$

$$n_2 a = m_1 \sqrt{2}a \cos \theta + m_2 \sqrt{2}a \sin \theta \tag{1.15b}$$

and, according to eq. (1.5), the angle θ is given by

$$\tan \theta = (n_1 m_1 + n_2 m_2)/(n_2 m_1 - n_1 m_2) \tag{1.16}$$

As already discussed, for $n_1 m_1 + n_2 m_2 = 0$ ($\tan \theta = 0$, i.e. $\theta = 0°$) or $n_2 m_1 - n_1 m_2 = 0$ ($\tan \theta = \infty$, i.e. $\theta = 90°$), a superlattice does not exist and, as readily verified, the condition (1.8) can therefore not be satisfied. The smallest integers for which eq. (1.16) yields a value of $\tan \theta$ that differs from zero and infinity and which satisfy the condition (1.8) are, for example, $n_1^0 = 1$, $n_2^0 = 1$, $m_1^0 = 1$ and $m_2^0 = 0$, for which eq. (1.16) yields $\tan \theta = 1$, or $\theta = 45°$. According to eq. (1.6) the first vector of the superlattice, C_1, expressed in the unrotated $(x-y)$ coordinate system in Fig. 1.14, is hence given by

$$C_1 = a_1 + a_2 \tag{1.17}$$

To determine C_2, for the same value of θ a second superlattice point has to be extracted from eq. (1.16) which is not collinear with C_1. One such point, also nearest to the origin, is obtained for example for $n_1^1 = 1$, $n_2^1 = -1$, $m_1^1 = 0$ and $m_2^1 = 1$; hence according to eq. (1.7),

$$C_2 = a_1 - a_2 \tag{1.18}$$

C_1 and C_2 could have been equally determined in the rotated coordinate system, i.e. in terms of the rotated vectors b_1' and b_2'. Inserting the values for m_1 and m_2 into eqs. (1.9) and (1.10), we find that (Fig. 1.14)

$$C_1' = b_1', C_2' = b_2' \tag{1.19}$$

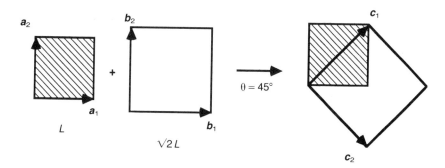

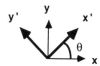

Fig. 1.14 Example illustrating that the formation of a commensurate interface may involve a rotation about the interface normal, in this case by an angle of 45°.

To summarize, two Bravais-lattice planes are commensurate if (a) under a general (twist) rotation, the two lattice planes share infinitely many common lattice points (eq. (1.4)), and (b) their primitive planar unit-cell areas are compatible (eq. (1.12)).

1.6 DEGREES OF FREEDOM OF CRYSTALLINE INTERFACES

The **macroscopic geometry** of interfaces to be discussed in this section includes all aspects of the geometry and crystallography determined by (i) the crystal structure(s) forming the interface and (ii) the degrees of freedom (DOFs) of the interfacial system, including the five macroscopic and the three translational (or 'microscopic') degrees of freedom.

The macroscopic geometrical description of crystalline interfaces has been an area of considerable activity during the past 30 years. Much of this work, particularly in the grain-boundary area, has focused on the formal description, in terms of linear algebra, of the misorientation relationship between the two crystal lattices forming the interface. The description of GB structures in terms of the coincident-site lattice (CSL), the displacement-shift-complete (DSC) lattice, and the 0-lattice are the main outcome of this work [16–19]. Within this framework the **macroscopic** DOFs are defined either within what we call the **CSL-misorientation scheme** or in terms of the **tilt-inclination scheme**; the underlying concepts will be discussed in detail in sections 1.8.1. and 1.8.2.

In common to these two methods for defining the macroscopic DOFs is their focus on how, hypothetically, a particular 2d interface structure can be generated by a single CSL rotation of two interpenetrating 3d crystal lattices. Considering the fact that solid interfaces are **planar** defects, this focus on a **three**-dimensional superlattice in common to the two crystals forming the interface appears somewhat surprising. Apart from the obvious limitation of such a description to **commensurate** interfaces, intuitively one would expect that the physically relevant geometrical features

of **crystalline** interfaces are related to (i) the crystallographic orientation of the interface **plane** and (ii) the size and/or shape of the **planar unit cell** (if the interface is commensurate).

Here we will therefore adopt a more widely applicable definition of the macroscopic degrees of freedom of solid interfaces [20], referred to as the **interface-plane scheme** [21], which is applicable to all types of crystalline (homophase and heterophase) interfaces, commensurate or incommensurate, and which enables a direct comparison of the geometry of all three basic types of interfaces from a common point of view. While the underlying terminology, to be developed in section 1.6.1, is rather commonly applied to semi-bulk and thin-film (coherent or incoherent) dissimilar-material interfaces, in the GB area it is not widely used; instead, the geometry of GBs is usually described in terms of the CSL-based terminology. As an example, the three simplest interface systems (from the point of view of their underlying number of DOFs) will be discussed in section 1.6.3. Later, in section 1.8, we will consider the conventional GB terminology within the framework of the interface-plane nomenclature. Apart from providing a basis for the geometrical description of all types of interfaces, the main advantages of the interface-plane terminology over the two CSL-based definitions of the macroscopic DOFs will then, hopefully, become apparent. The three principal advantages are the following:

1. The number of macroscopic DOFs of any particular type of GB is readily apparent, which is of considerable aid in structure–property investigations.
2. The geometrical resemblance between symmetrical and asymmetrical GBs is rather transparent, thus greatly facilitating the comparison of their physical properties.
3. The fact that – from a purely geometrical point of view – symmetrical and asymmetrical-**tilt** boundaries represent a special subset of symmetrical and asymmetrical-**twist** boundaries, respectively, is incorporated naturally into the interface-plane description of the macroscopic geometry.

1.6.1 Macroscopic DOFs ('interface-plane scheme')

As is well known, in addition to the crystal structure(s) and lattice parameter(s) eight geometrical parameters are needed to characterize the geometry of a single bicrystalline interface. These eight DOFs are usually subdivided into the five **macroscopic** and three **translational** or **microscopic** ones [18]. The latter, to be discussed in section 1.6.2, are usually represented by the components of a vector, T, associated with rigid-body translations parallel and perpendicular to the interface plane. By their very nature, the determination of the three components of T requires experimental methods capable of detecting 'microscopic' (i.e. atomic-level) translations. By contrast with the macroscopic DOFs and the atomic structure, these three translational DOFs are therefore often referred to as the **microscopic** DOFs of the GB (section 1.6.2 below).

A simple, unified method of defining the five **macroscopic** DOFs of an arbitrary (commensurate or incommensurate) bicrystalline interface is the following (Fig. 1.15): [20]

$$\{DOFs\} = \{\hat{n}_1, \hat{n}_2, \theta\} \quad \text{('interface-plane scheme')} \quad (1.20)$$

Here the unit vectors $\hat{n}_1$ and $\hat{n}_2$ represent the common interface-plane normal, $\hat{n}$, in the two halves (Fig. 1.15(b)), referred to the same principal coordinate system (Fig. 1.15(a)). For example, the (x, y, z) system in Fig. 1.15(a) might be aligned along the $\langle 100 \rangle$ principal cubic directions, relative to which the interface-plane normals in the two halves may be defined. In Fig. 1.15(b) the two normals are then aligned parallel to each other, as indicated by the planar structure in the figure, thus defining the (x_1, y_1, z_1) and (x_2, y_2, z_2) coordinate systems. (Naturally, for **crystalline** interfaces $\hat{n}_1$ and $\hat{n}_2$ have to be directions permitted by the particular crystal structure.) Since each unit vector contributes two DOFs, the interface **plane** therefore represents four DOFs. Having thus fixed the interface plane in the two semi-crystals, the only remaining DOF is the one associated with a so-called 'twist' rotation, by the angle θ, about the common interface-plane normal, because any other rotation would change $\hat{n}_1$ and $\hat{n}_2$ (Fig. 1.15(c)).

Because of the emphasis placed on the interface **plane** (by assigning to it four out of the five DOFs), we will refer to the definition in eq. (1.20) as the 'interface-plane scheme' for defining the macroscopic DOFs. As already mentioned, two other CSL-based definitions, referred to as the 'CSL-

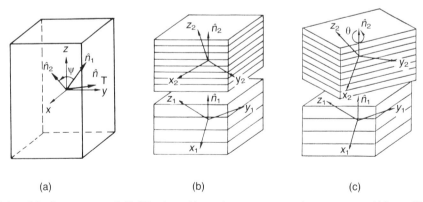

(a)   (b)   (c)

Fig. 1.15 A definition of the five **macroscopic** DOFs of an arbitrary (commensurate or incommensurate) bicrystalline interface [20, 21]. The unit vectors $\hat{n}_1$ and $\hat{n}_2$ represent the interface-plane normal, $\hat{n}$, in the two halves, however referred to the same principal coordinate system. (a) Illustrates the orientations of $\hat{n}_1$, $\hat{n}_2$, and $\hat{n}_T$ in a space-fixed (x, y, z) coordinate system (oriented, for example, along the principal cubic axes); $\hat{n}_T \equiv [\hat{n}_1 \times \hat{n}_2]/\sin \psi$. A tilt rotation of two identical (x, y, z) coordinate systems of (a) such that $\hat{n}_2 \parallel \hat{n}_1$ defines the (x_1, y_1, z_1) and (x_2, y_2, z_2) coordinate systems in (b). Finally, in (c) the twist component of the GB is introduced by rotating the top crystal in (b) about the GB-plane normal by the angle θ.

misorientation' and 'tilt-inclination' schemes, will be discussed in detail in sections 1.8.1 and 1.8.2 (eqs. (1.45) and (1.54), respectively). We mention that sometimes a sixth macroscopic DOF, representing the **position** of the interface plane in the direction of its normal, is added to the usual five defined here and in the two other schemes [18]; for a fixed (i.e. immobile) interface, however, this DOF is of no relevance.

In the case of GBs, a distinction is usually made between pure **tilt** and pure **twist** boundaries, with a **general** boundary having both tilt and twist components (and hence characterized by the full set of five DOFs; see section 1.8.2 below). This concept is rather useful as it provides information about the types of dislocations present in the GB structure, with the tilt and twist components defining, respectively, the edge- and screw-dislocation content. It can be adopted for other types of crystalline interfaces as well, including commensurate and incommensurate heterophase interfaces. As is common for GBs, we thus define a twist rotation as a rotation about an axis, described by the unit normal $\hat{n}_r$, which is parallel to the common interface-plane normal, $\hat{n}$ ($\hat{n}_r \parallel \hat{n}$). In a tilt rotation, by contrast, $\hat{n}_r$ and $\hat{n}$ are perpendicular to one another ($\hat{n}_r \perp \hat{n}$).

From these definitions, the tilt and twist components of a 'general' bicrystalline interface defined in eq. (1.20) are readily apparent. The angle θ, by definition, describes a twist rotation, since any other rotation would change the interface normal; the corresponding rotation matrix is denoted by $\boldsymbol{R}(\hat{n}_1, \theta)$. Provided we define $\theta = 0°$ as the angle for which the interface is of a pure tilt type (i.e. its structure contains only edge dislocations), the twist component of a general interface, defined by the rotation $\boldsymbol{R}(\hat{n}_1, \theta)$, is immediately obvious from the DOFs in eq. (1.20). The tilt component, characterized by the rotation matrix $\boldsymbol{R}(\hat{n}_T, \psi)$, is governed by the condition that $\hat{n}_T \perp \hat{n}_1, \hat{n}_2$ (Fig. 1.15(a)); hence [21]

$$\hat{n}_T = [\hat{n}_1 \times \hat{n}_2]/\sin \psi \qquad (1.21)$$

with

$$\sin \psi = |[\hat{n}_1 \times \hat{n}_2]| \qquad (1.22)$$

where $\hat{n}_T$ is a unit vector defining the orientation of the tilt axis, while ψ is the so-called tilt angle (Fig. 1.15(a)).

In much of the work on free surfaces, the tilt axis and tilt angle defined here are referred to as the **pole** axis and **pole** angle, respectively. This terminology is related to the **pole figure** in which the possible orientations of a unit vector are represented on the surface of a unit sphere (Fig. 1.16). This corresponds to defining the two DOFs of the unit vector in terms of spherical coordinates (section 1.6 below). The so-called **pole of a plane** thus represents, by its position on the unit sphere, the orientation of that plane. [22] A plane may also be represented by the great circle of the sphere which is perpendicular to the plane normal. For example, the great circles ABCD and KDMB in Fig. 1.16 represent, respectively, the planes with unit normals $\hat{n}_1$ and $\hat{n}_2$, whose poles are the points P_1 and P_2, respectively. The 'pole' axis and 'pole' angle, $\hat{n}_T$ and ψ, respectively, are the same as the 'tilt' axis and 'tilt' angle defined in eqs. (1.21) and (1.22).

Equations (1.21) and (1.22) illustrate that the tilt component of a **general interface** (i.e. one with all five DOFs, see Fig. 1.15(c)) is solely determined by the normals $\hat{n}_1$ and $\hat{n}_2$, i.e. by the interface **plane**. Irrespective of the value of the twist angle, θ, in eq. (1.20), an interface therefore has a tilt component whenever $\hat{n}_1$ and $\hat{n}_2$ represent different crystallographic directions or, if the interface is symmetrical (see below), different sets of crystallographically equivalent lattice planes. Because of the similarity of a general interface to a pure (i.e. symmetrical) twist boundary, a general interface will also be referred to as an **asymmetrical-twist** interface; however, because of the asymmetry in the interface plane, such an interface always has a tilt component (given by eqs. (1.21) and (1.22)), in contrast to the symmetrical-twist interface discussed below. In the special case for $\theta = 0°$ (and $\theta = 180°$; section 1.8.2), a pure **asymmetrical-tilt** interface with only 4 DOFs is obtained (see Fig. 1.15(b)). If the two lattice planes forming the GB are incommensurate (section 1.5), a unique definition of the origin of the twist rotation, i.e. of $\theta = 0°$, and of an asymmetrical-tilt configuration is not

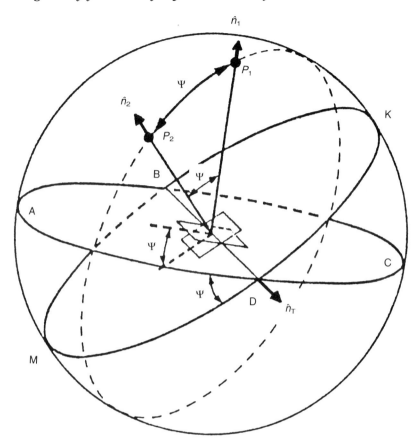

Fig. 1.16 Definition of the angle between two planes in terms of a pole figure [22]. The 'pole' axis and 'pole' angle, $\hat{n}_T$ and ψ, are identical to the 'tilt' axis and 'tilt' angle defined in eqs. (1.21) and (1.22). The great circles ABCD and KDMB represent, respectively, the planes with unit normals $\hat{n}_1$ and $\hat{n}_2$, whose poles are the points P_1 and P_2, respectively.

possible. Here, however, by considering only commensurate GBs, we avoid this issue altogether.

For example, a thin-film overlayer with a normal, say $\langle 100 \rangle$, which differs from that of the substrate, say with a $\langle 111 \rangle$ normal, may be assigned a tilt axis and angle expressing the rotation from the substrate normal to that of the overlayer. If the overlayer is perfectly aligned with the substrate (i.e. for $\theta = 0°$), the interface is of an **asymmetrical-tilt** type. If the overlayer is, in addition, rotated about the substrate normal (i.e. for $\theta \neq 0°$), the interface is a **general** or **asymmetrical-twist** interface (Fig. 1.15(c)).

Finally, in cubic crystals the normal $\hat{n}$ may be given in terms of the Miller indices, (hkl), according to

$$\hat{n} = (h^2 + k^2 + l^2)^{-\frac{1}{2}} \begin{pmatrix} h \\ k \\ l \end{pmatrix} \qquad (1.23)$$

and all relevant geometrical parameters may be expressed in terms of the h, k and l associated with a given plane. The definition in eq. (1.20) may then be rewritten as follows:

$$\{DOFs\} = \{(h_1, k_1, l_1), (h_2, k_2, l_2), \theta\}$$
$$(\text{'interface-plane scheme'}) \qquad (1.24)$$

where, in order to simplify the notation, the plane **normals** in eq. (1.20) have been replaced by the actual lattice **planes** forming the interface.

In a close parallel with the terminology commonly used for GBs (see section 1.8), one can classify the various types of interfaces described by eqs. (1.20) and (1.24). Thus, a **symmetrical** interface is defined as one for which $\hat{n}_1$ and $\hat{n}_2$ are linearly related, i.e. there exists a **linear** relationship,

$$\hat{n}_2 = L(\hat{n}_1) \qquad (1.25a)$$

in the sense that $\hat{n}_1$ and $\hat{n}_2$ represent crystallographically equivalent lattice planes, which reduces the number of DOFs from five to only three:

$$\{\text{DOFs}\} = \{\hat{n}_1, \hat{n}_2 = L(\hat{n}_1), \theta\}$$
(symmetrical interface)   (1.26a)

All other interfaces are hence **asymmetrical**.

As an example for eq. (1.25a) we consider the collinearity condition,

$$\hat{n}_2 = \pm\hat{n}_1 \qquad\qquad (1.25b)$$

Equation (1.26a) then becomes:

$$\{\text{DOFs}\} = \{\hat{n}_1, \pm\hat{n}_1, \theta\}$$
(symmetrical twist)   (1.26b)

An interface characterized by eq. (1.25b) obviously has no tilt component because $[\hat{n}_1 \times \hat{n}_2] = 0$. By analogy to the GB terminology, such an interface is therefore called a **symmetrical-twist** (or **pure twist**) interface. Analogously, as already mentioned, if the interface is asymmetrical (eq. (1.20)) we refer to it as an **asymmetrical-twist** interface.

For the $+$ sign in eq. (1.26b), the two sets of lattice planes forming the interface are obviously identical, while the $-$ sign describes a bicrystal in which one of its two halves has been turned upside down with respect to the other. To illustrate a simple relationship between the two sign choices, we consider a perfect crystal, which is obviously included in eq. (1.26b) as the special case in which

$$\{\text{DOFs}\} = \{\hat{n}_1, \hat{n}_1, \theta = 0°\} \qquad (1.27a)$$

In a crystal lattice with inversion symmetry, in which a 180° rotation about some crystallographically allowed rotation axis inverts any lattice vector, r (i.e. $r \rightarrow -r$; section 1.7.2), a perfect crystal is also obtained for

$$\{\text{DOFs}\} = \{\hat{n}_1, -\hat{n}_1, \theta = 180°\} \qquad (1.27b)$$

For any other value of θ, a **symmetrical-twist** interface is obtained. Since either of the two perfect-crystal configurations in eqs. (1.27a) and (1.27b) may be used as starting point for the twist rotation, the following relationship holds:

$$\begin{aligned}\{\text{DOFs}\} &= \{\hat{n}_1, \hat{n}_1, \theta\} \\ &= \{\hat{n}_1, -\hat{n}_1, 180° - \theta\}\end{aligned} \qquad (1.28)$$

We emphasize that this relation is only valid for a crystal lattice with inversion symmetry (section 1.7.2).

At the atomic level, the most characteristic geometrical feature of a symmetrical-**tilt** grain boundary (STGB) is the inverted stacking of the lattice planes on one side of the interface with respect to the other (section 1.8.3). In a crystal lattice with inversion symmetry, STGBs may therefore be viewed as special twist boundaries (for $\theta = 180°$) [20]. Similar to the perfect crystal considered in eqs. (1.27a) and (1.27b), the STGB configuration is thus situated at the endpoints of the twist-rotation range, and is characterized as follows:

$$\{\text{DOFs}\} = \{\hat{n}_1, \hat{n}_1, \theta = 180°\}$$
(symmetrical tilt)   (1.29a)

or

$$\{\text{DOFs}\} = \{\hat{n}_1, -\hat{n}_1, \theta = 0°\}$$
(symmetrical tilt)   (1.29b)

The fact that STGBs may be viewed as a subset of symmetrical-twist boundaries translates into a unique atomic-level geometry of STGBs, in comparison with all other GBs (sections 1.7.3 and 1.8.4).

This definition of STGBs as a subset of symmetrical-twist boundaries is in apparent conflict with the fact that, according to eqs. (1.21) and (1.22), these interfaces have no tilt component. Moreover, if we identify the type of a GB as tilt or twist by the edge or screw dislocations in its structure, then STGBs should be classified as tilt and not twist boundaries. This apparent discrepancy originates from the fact that the example of the symmetry relation (1.25a) given in eq. (1.25b) does not cover all sets of crystallographically equivalent lattice planes which satisfy eq. (1.25a) but not (1.25b). To formally assign a tilt component to a symmetrical interface, a non-collinear combination of $\hat{n}_1$ and $\hat{n}_2$ has to be found which characterizes the same set of crystallographically equivalent planes. In a cubic crystal such sets are readily identified. For example, a GB formed by the two crystallographically equivalent (albeit not identical) sets (h, k, l) and $(k, h, -l)$, would obviously be considered to be 'symmetrical', since on an atom-by-atom basis the two types of lattice planes are indistinguishable. Generally for cubic crystals, if we define $\hat{n}_1$ by the normal $[h_1, k_1, l_1]$, the general condition of symmetry [eq. (1.25a)] may be formulated as follows: [21]

$$[h_2, k_2, l_2] = \in \{\langle \pm h_1, \pm k_1, \pm l_1 \rangle\} \qquad (1.30)$$

where the angular brackets indicate that any permutation of the Miller indices, including their signs, is permitted. The set of normals, $\{\langle \pm h_1, \pm k_1, \pm l_1 \rangle\}$, obviously includes the collinear normals in eq. (1.25b) as the special cases in which $[h_2, k_2, l_2] = [h_1, k_1, l_1]$ or $[h_2, k_2, l_2] = [-h_1, -k_1, -l_1]$. Equation (1.30) thus represents the general condition of symmetry for cubic crystals.

1.6.2 Translational ('microscopic') DOFs

As already mentioned, the different types of interfaces discussed above still have three independent **translational** (or so-called microscopic) DOFs involving translations, $\boldsymbol{T} = (T_x, T_y, T_z)$, parallel (x, y) and perpendicular to the interface plane (z). From a thermodynamics point of view, the z component of $\boldsymbol{T}$ (parallel to the interface normal) is particularly important in that it accounts for any volume expansion at the interface. Such an excess free volume of the interface can be expected to (a) be closely related to its excess free energy and (b) give rise to stresses near the interface that are similar in nature to the well-known surface stress in free surfaces.

Similar to the definition of the excess free energy of the interface as the change in Gibbs free energy, G, with interface area, A, at constant temperature, T, pressure, p, and chemical composition (expressed in terms of the numbers of atoms, N_i, of each species), according to [23]

$$\gamma = (\partial G/\partial A)_{T, p, N_i} \qquad (1.31)$$

the so-called excess free volume per unit area of the interface is defined by [24]

$$\delta V = (\partial V/\partial A)_{T, p, N_i} \qquad (1.32)$$

By definition, δV is a volume expansion **per unit area** (and is conveniently given in units of the lattice parameter, a) and is to be distinguished from the overall three-dimensional thermodynamic volume, V.

According to eqs. (3.31) and (3.32), the Gibbsian excess free energy and free volume per unit area are to be determined while the temperature, pressure, and composition are held fixed. This constraint may pose particular conceptual problems in some computer simulations of these excess quantities. Following these definitions, simulations always require consideration of an appropriate interface-free reference system under the same conditions of T, p, and N_i as the interfacial system. For example, constant-volume simulations lead not only to numerically wrong values of γ for a given potential, but energies thus determined are conceptually not the true excess energy in the Gibbsian sense (which is the one usually determined experimentally).

In the case of GBs, the existence of translations parallel to the interface have been well established by means of high-resolution transmission-electron-microscopy (TEM) experiments, as well as computer simulations. These translations contribute to a lowering of the excess free energy of the system by avoiding energetically unfavorable translational states. They are also thought to play an important role during the process of GB migration [25, 26].

1.6.3 The three 'simplest' interface systems

Given that crystalline interfaces are **planar** defects, at least two macroscopic DOFs (namely the two associated with the interface normal) are required to characterize even the simplest interface. According to eq. (1.26b), such an interface is symmetrical and its twist angle must be fixed, or else be irrelevant altogether. There are in total three distinct types of **homophase** interfacial systems satisfying this condition, namely stacking faults, symmetrical-tilt grain boundaries (STGBs) and free surfaces (Fig. 1.17). Although not an internal interface, the free surface (a crystal–vacuum interface) is included here. Because of its importance as the final state in interfacial decohesion, a terminology and geometrical description in common to both internal interfaces and external

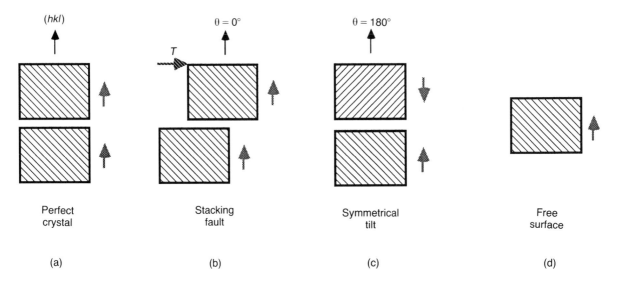

Fig. 1.17 From the point of view of the number of macroscopic DOFs involved, the stacking fault in (b), the symmetrical-tilt grain boundary (STGB: (c)) and the free surface in (d) represent the three simplest of all homophase planar defects (schematic). For comparison, a perfect crystal with the same crystallographic orientation is sketched in (a). The shaded arrows indicate the direction of planar stacking to be discussed in detail later. (An atomic-level view of these geometries is given in section 1.7.3; Fig. 1.30).

surfaces might facilitate a better understanding of interface fracture. From the point of view of the number of DOFs involved, these interfaces represent the simplest of all homophase interface systems. Among the **heterophase** systems, the coherent, incoherent and semi-coherent semi-bulk and thin-film (strained-layer) systems in Figs. 1.6, 1.10, and 1.11 and the topotaxy in Fig. 1.8 could also be included here. However, because of their geometrical similarity (particularly the coherent systems) with the three homophase systems described here, they will not be considered in this context.

First, a **stacking fault** may be generated on a given plane by any suitable in-plane translation, (T_x, T_y), of the perfect-crystal configuration in Fig. 1.17(a), thus destroying the perfect registry between the planes adjacent to the interface. Since no rotation is involved ($\theta = 0°$), in addition to these two translational DOFs, the stacking fault is characterized by only the two macroscopic DOFs associated with orientation of the fault plane. In principle, the translation may be accompanied by a volume expansion, hence requiring all three

microscopic DOFs to be specified for its full characterization.

Second, the **symmetrical-tilt** GB on the same plane, with its familiar 'twinned' inversion of the stacking sequence of lattice planes at the interface (Fig. 1.17(c)), is obtained by simply turning one of the two halves upside down. As discussed in more detail in sections 1.7.3 and 1.8.4 below, in crystal lattices with inversion symmetry, the STGB configuration on a given plane is obtained simply from a 180°-twist rotation of, say, the upper semicrystal. If the components of T are such that one plane is a mirror plane (i.e. shared by both semicrystals), this inverted configuration represents a **special twin**, to be distinguished from the **general-twin** configuration obtained for some arbitrary translation (and hence merely with an inverted planar stacking at the interface, but without a mirror plane; for details see section 1.8.3 and Fig. 1.30).

Third, the free surface on a given plane (Fig. 1.17(d)), also with only two macroscopic DOFs, differs from the stacking fault and the symmetrical-tilt configuration on the same plane by not having

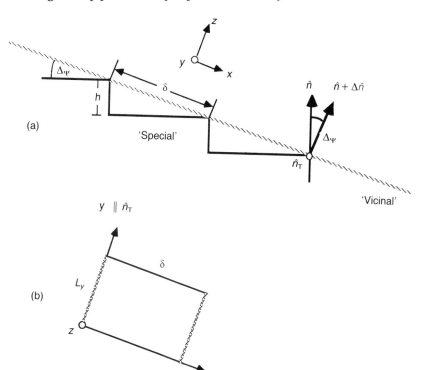

Fig. 1.18 Distinction between 'special' and 'vicinal' surfaces. While the special surface, with normal $\hat{n}$, represents a cusped minimum-energy orientation in the $\gamma(\hat{n})$ plot, the geometry, atomic structure, and physical properties of vicinal (i.e. nearby) surfaces are governed by those of the special surface and by the spacing between and the geometry of its steps.

any translational DOFs, as in the case of the perfect crystal. Formally, the free surface may be viewed as having been generated from the perfect crystal, the stacking fault, or the STGB as the limit in which $T_z \to \infty$, thus creating a symmetrical arrangement of two infinitely separated (i.e. non-interacting) surfaces. The magnitudes of T_x and T_y are then obviously irrelevant, as is the twist angle θ, and no microscopic DOFs are required to characterize the surface.

In spite of these geometrical similarities between the simplest three interface systems, which will be investigated further at the atomic level in section 1.4.3 below, to date little is known about any similarities in their physical properties [27].

1.6.4 'Vicinal' versus 'special' interfaces

While in the case of external surfaces a distinction between 'special' and 'vicinal' interface planes has been commonly made for almost a century [28], in the case of internal interfaces such a distinction has

only recently been suggested [29]. To illustrate these concepts for the case of free surfaces, we consider a 'special' low-index surface with normal $\hat{n}$ which is as nearly as possible atomically 'flat' (i.e. free of steps), and a second surface with a slightly different orientation, say, $\hat{n} + \Delta\hat{n}$ [27]. If $\Delta\hat{n}$ is small, the second surface will look just like the first one, except for the appearance of widely separated steps (Fig. 1.18). Because each step adds to the surface energy, $\gamma(\hat{n} + \Delta\hat{n}) - \gamma(\hat{n})$ will be positive for any orientation of $\Delta\hat{n}$ and, for small values of $\Delta\hat{n}$, will be asymptotically proportional to the density of steps. The surface with normal $\hat{n}$ hence represents a **cusped** minimum-energy orientation in a $\gamma(\hat{n})$ plot [27]. Such a surface is called a **special** surface; it is distinguished from the nearby **vicinal** surfaces whose geometry, atomic structure, and physical properties are governed by those of the special surface and by the vector difference, $\Delta\hat{n}$, in the phase space representing the two macroscopic DOFs associated with the surface normal, $\hat{n}$.

As illustrated in Fig. 1.18, the misorientation

between $\hat{n}$ and $\hat{n} + \Delta\hat{n}$ may be characterized by the rotation axis, $\hat{n}_T$, and rotation angle, $\Delta\psi$ (given by eqs. (1.21) and (1.22), respectively), and for a fixed rotation axis the surface energy may be expressed solely as a function of $\Delta\psi$. Assuming the steps to be far enough apart that their mutual interaction may be ignored, one can easily show that [27, 30].

$$\gamma(\Delta\psi) = \gamma_{\text{cusp}} \cos \Delta\psi + \Gamma^\infty \sin \Delta\psi/h \qquad (1.33)$$

where γ_{cusp} denotes the cusped energy of the special surface and Γ^∞ represents the energy per unit length of individual steps whose height is denoted by h (Fig. 1.18).

In the example shown in Fig. 1.19, the un-relaxed and relaxed zero-temperature energies of fcc surfaces perpendicular to a $\langle 110 \rangle$ pole axis are plotted against 2ψ which is defined as the angle of a particular surface with respect to the (110) plane. These energies were determined [31] by means of the simple Lennard-Jones (LJ) potential fitted to the lattice parameter and approximate melting point of Cu, although this potential function is usually thought to be more appropriate for noble-gas crystals. The fact that the unrelaxed and relaxed energies differ by only very little indicates that (i) the change in surface energy as a function of $\Delta\psi$ is governed by strictly crystallographic factors, namely the total length of the steps, and (ii) the interaction between steps, given by the relaxation energy, is, indeed, very small.

Figure 1.19 demonstrates the existence of three cusped 'special' orientations associated with the three principal planes in the fcc lattice, namely (111), (100), and (100). Being the densest planes in the fcc lattice, the smallest number of nearest-neighbor bonds is broken when one of these surfaces is created. This is thought to be the reason for the appearance of cusps for these particular surfaces, with the depths of the cusps decreasing rapidly with decreasing planar density, i.e. inter-planar lattice spacing, $d(hkl)$ (Table 1.1). A de-tailed analysis of the smooth variation of $\gamma(\Delta\psi)$ in the vicinity of the cusps [27] indicates that eq. (1.33) provides an excellent representation of the simulation data near all three cusps; such an analy-sis also provides values of the step energy per unit length for all three types of steps.

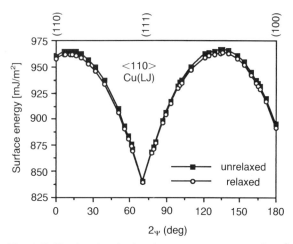

Fig. 1.19 Unrelaxed and relaxed zero-temperature energies of fcc surfaces (computed by means of a Lennard-Jones potential fitted for Cu), plotted against the tilt angle of each surface with respect to the (110) plane. All surfaces are perpendicular to the $\langle 110 \rangle$ pole axis [31].

Table 1.1 Interplanar spacing, $d(hkl)$ (in units of the lattice parameter a), and number of planes in the repeat stacking sequence, $P(hkl)$ (the so-called stacking period), for the most widely spaced planes in the fcc lattice. According to eq. (1.44), these planes also correspond to the ones with the highest planar density of atoms

No.	(hkl)	$(h^2 + k^2 + l^2)$	$P(hkl)$	$d(hkl)/a$
1	(111)	3	3	0.5774
2	(100)	1	2	0.5000
3	(110)	2	2	0.3535
4	(113)	11	11	0.3015
5	(331)	19	38	0.2294
6	(210)	5	10	0.2236
7	(112)	6	6	0.2041
8	(115)	27	27	0.1925

Plots similar to Fig. 1.19 have also been obtained for symmetrical and asymmetrical-tilt and twist GBs, suggesting that the distinction between 'special' and 'vicinal' interfaces is meaningful for GBs as well as for free surfaces. In particular, most (symmetrical or asymmetrical) **low-angle twist** boundaries may be viewed as vicinal to the cor-responding **tilt** boundaries (sections 1.8.2 and 1.8.3). A well known special case is that of sym-metrical low-angle twist boundaries which may be considered as vicinals of the perfect crystal. Similar to the steps in free surfaces, the dislocation structure of these GBs preserves areas of perfect crystal. (For further details, see [27] and [32] and sections 1.8.2 and 1.8.3.)

Finally, we mention that the distinction between special and vicinal interfaces is not possible on a strictly crystallographic basis alone because the identification of the special interfaces requires a knowledge of the relaxed energy of the system. Free surfaces represent somewhat of an exception in that geometrical concepts, based for example on the number of broken bonds per unit area, are very useful in predicting some of their basic properties [31].

1.6.5 Misorientation phase space

One of the main goals of structure-property investigations is the exploration of the so-called **misorientation phase space** represented by the five **macroscopic** DOFs of the interface. In principle, any property in this five-dimensional (5D) phase space, say, the interface energy γ, can therefore be represented as a 6D hypersurface, $\gamma(\hat{n}_1, \hat{n}_2, \theta)$. Needless to say, even a mere geometrical representation of such a hypersurface, perhaps via 3D cross-sections, is a non-trivial conceptual undertaking. Any symmetries which reduce the number of DOFs should therefore be fully exploited because they simplify the misorientation phase space considerably; a knowledge of the exact number of DOFs of the interfacial system under investigation is therefore imperative.

To illustrate possible ways of exploring misorientation phase space, we start with the simplest interface systems, characterized by only the two macroscopic DOFs associated with the crystallographic orientation of the interface normal, $\hat{n}$. The generally 6D structure-energy hypersurface, $\gamma(\hat{n}_1, \hat{n}_2, \theta)$, thus degenerates into a single 3D plot, $\gamma(\hat{n})$, which can be constructed as follows. For simplicity we limit ourselves to cubic crystals for which, when all symmetry operations are considered, all possible orientations of $\hat{n}$ fall into a triangle on the unit sphere of Fig. 1.16, with the principal cubic poles at its corners (see the shaded

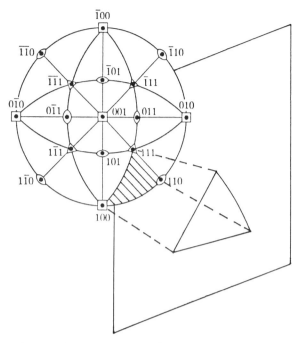

Fig. 1.20 Standard projection for cubic crystals of the unit sphere of Fig. 1.16 onto a plane, with the (001) pole in the center (so-called 'projection on (001)'). Due to the cubic symmetry, all possible orientations of a unit vector fall into a spherical triangle, such as the shaded one. (For details see Ref. [22]).

triangle in Fig. 1.20). Naturally, it is desirable to project this spherical triangle onto a plane (Fig. 1.20) by the so-called **stereographic projection** [22]. However, because of the highly non-linear nature of this projection, the angle scale of the **stereographic triangle** thus obtained is non-linear, which renders its use a somewhat complicated endeavor.

To simplify this projection while yet still capturing the essential features of the stereographic triangle, however, we express the orientation of $\hat{n}$ in terms of spherical coordinates, ϑ and φ, according to (eq. (1.23))

$$(h^2 + k^2 + l^2)^{-\frac{1}{2}} \begin{pmatrix} h \\ k \\ l \end{pmatrix} = \begin{pmatrix} \sin\vartheta\cos\varphi \\ \sin\vartheta\sin\varphi \\ \cos\vartheta \end{pmatrix} \quad (1.34)$$

For $h \geqslant k \geqslant l$, all orientations of $\hat{n} = (h^2 + k^2 + l^2)^{-\frac{1}{2}}(h, k, l)$ then fall into the **phase-space triangle**

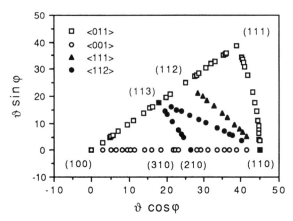

Fig. 1.21 Schematic plot, similar to the stereographic projection, of the spherical triangle of Fig. 1.20 onto a plane. ϑ and φ are the usual spherical coordinates which define the orientation of the interface normal according to eq. (1.34); ϑ is in degrees. The symbols indicate positions in the phase-space triangle at which the free-surface energy in Fig. 1.22(a) was determined [31].

sketched in Fig. 1.21, with the three principal cubic orientations at the corners [31]. This triangle differs slightly from the conventional stereographic triangle sketched in Fig. 1.20 in that its axes are scaled **linearly**. Given values of ϑ and φ, this simplification makes it a rather straightforward matter to locate its position within the triangle. As in the stereographic triangle, all orientations perpendicular to a $\langle 110 \rangle$ pole axis appear along the edge connecting (100), via (113), (112), and (111), with (110). Similarly, all orientations perpendicular to a $\langle 100 \rangle$ pole axis appear along the edge connecting (100) more directly with (110) via (310) and (210)], while plane orientations perpendicular to $\langle 111 \rangle$, $\langle 112 \rangle$, etc. cover the central regions of the triangle.

Using this phase-space triangle as the base plane, the surface energy, $\gamma(\hat{n}) \equiv \gamma(\vartheta, \varphi)$, obtained by means of the LJ potential is given by the 3D structure–energy plot shown in Fig. 1.22(a). By contrast with the 2D cross-section in Fig. 1.19 (along the edge connecting (100), via (113), (112), and (111), with (110)), this plot demonstrates the full extent in phase space of the three energy cusps at the corners of the triangle, with a remarkable absence of cusps in the central regions of phase space. Figure 1.22(b) represents a similar

plot for the number of broken nearest-neighbor bonds per unit surface area. The remarkable similarity of Figs. 1.22(a) and (b) demonstrates that the structure-energy correlation for free surfaces in fcc crystals is dominated by geometrical factors, with elastic effects associated with the interactions between the surface steps being relatively unimportant [27, 31]. We mention in passing that a similar plot is obtained for symmetrical-tilt GBs [33]. Combining the two plots permits one to determine the work of adhesion (i.e. the ideal cleavage-fracture energy)

$$E^{\mathrm{cl}}(\vartheta, \varphi) = 2\gamma(\vartheta, \varphi) - E^{\mathrm{STGB}}(\vartheta, \varphi) \qquad (1.35)$$

in the entire 2D phase space [27, 33].

If the twist angle, θ, is now added as a third DOF, every point in the 2D-phase-space triangle associated with the interface plane is unfolded into a infinite number of θ values which, in Fig. 1.21, are projected into a single point. Any structure–property correlation for symmetrical interfaces may hence be thought of as a 4D hypersurface.

To illustrate a method for constructing 3D cross-sections through this hypersurface, we briefly consider the case of symmetrical GBs [33]. As already mentioned, the 3D phase space for these boundaries contains infinitely many twist boundaries for every STGB (section 1.6.1). To gain some insight into what the corresponding 4D structure–energy hypersurface might look like and, in particular, to illustrate the distribution of the tilt and twist boundaries in this phase space, we consider cross-sections obtained as follows. In each 3D cross-section we limit ourselves to a well-defined subset of lattice planes, defined to be perpendicular to a particular pole (or tilt) axis, such as $\langle 110 \rangle$, $\langle 100 \rangle$, $\langle 111 \rangle$, $\langle 112 \rangle$, etc. By fixing the tilt axis, a single tilt angle, ψ, uniquely defines a given GB plane (see, for example, Fig. 1.21). Combined with the twist angle, θ, a 3D cross-section of the structure–energy phase space may thus be obtained for any given tilt axis.

One such cross-section, perpendicular to a $\langle 110 \rangle$ tilt axis, is shown in Fig. 1.23 for the same Cu(LJ) potential used above. On each plane, i.e. for each value of the tilt angle, the STGB configuration appears at the twist angle $\theta = 180°$ (while all other

Surface energy   Cu (LJ)

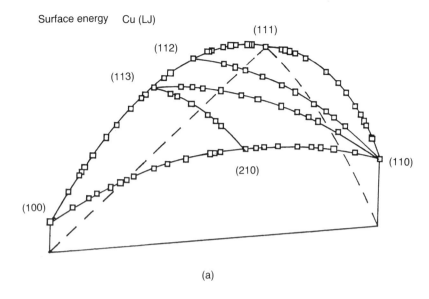

(a)

nn coordination   fcc

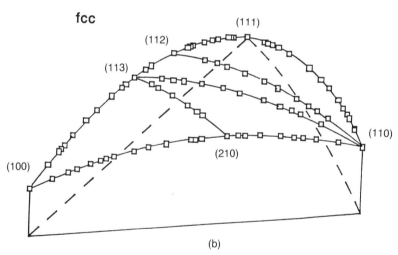

(b)

Fig. 1.22 (a) Three-dimensional structure–energy plot for free surfaces in fcc crystals (in arbitrary units, simulated by means of the Cu(LJ) potential), with the 2D-phase-space triangle of Fig. 1.21 as base plane [31]. (b) Nearest-neighbor miscoordination per unit area for the same surfaces.

GBs are pure twist boundaries), demonstrating the geometrical uniqueness of the STGBs in the phase space (sections 1.7.3 and 1.8.3). The deep energy cusps at the corresponding $\theta = 180°$ angles indicates that this unique geometry translates into a particularly low GB energy. The deep valley at the (111) plane demonstrates that the densest plane of the fcc lattice is a 'special' GB plane, with the slopes of the valley representing 'vicinal' GB-plane orientations. (Notice that the endpoint of the (111)

valley at $\theta = 180°$ represents the well-known (111) twin boundary.)

Finally, the GB and free surface energies in Figs. 1.22(a) and 1.23 can be combined to determine the variation of the ideal-cleavage energy,

$$E^{cl}(\vartheta, \varphi, \theta) = 2\gamma(\vartheta, \varphi) - E^{GB}(\vartheta, \varphi, \theta) \quad (1.36)$$

in the three-parameter phase space for all symmetrical GBs. A $\langle 1\bar{1}0 \rangle$ cross section through this 4D hypersurface, similar to Fig. 1.23, is shown in

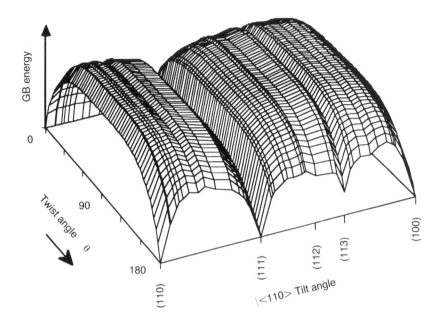

Fig. 1.23 Three-dimensional cross section of the 4D structure–energy plot for symmetrical GBs in fcc metals for the Cu(LJ) potential (in arbitrary units). In the particular cross section shown here, only GB planes perpendicular to a $\langle 110 \rangle$ tilt axis are considered (see also Fig. 1.21) [33].

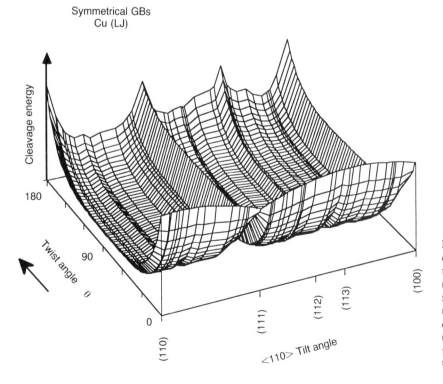

Fig. 1.24 Three-dimensional $\langle 110 \rangle$ cross section similar to Fig. 1.23 for the ideal cleavage-fracture energy (see eq. (1.36)) of symmetrical GBs in fcc metals for the Cu(LJ) potential (in arbitrary units). To show more clearly the bulk-cleavage energies, 2γ (at $\theta = 0°$), the direction of increasing twist angle was chosen opposite to that in Fig. 1.23 [33].

Fig. 1.24. Notice that, for reasons of clarity, the direction of increasing twist angle was chosen opposite to that in Fig. 1.23. We point out that, since for zero twist angle a perfect crystal is obtained (eq. (1.27a)), the cleavage energy at $\theta = 0°$, $E^{\text{cl}}(\vartheta, \varphi, 0°)$, represents the energy to cleave a perfect fcc crystal along a specific plane, i.e. twice the energy of the related free surface, $2\gamma(\vartheta, \varphi)$.

If the free-surface energy, γ, were isotropic (i.e. independent of ϑ and φ), Fig. 1.24 would represent merely an upside-down version of Fig. 1.23). However, as discussed earlier in Fig. 1.22(a), γ varies significantly in the two-parameter phase space, although its variation is not as pronounced as that of the GBs.

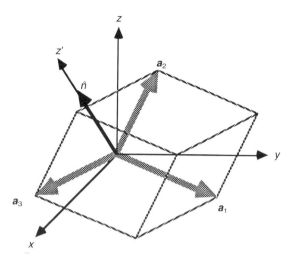

Fig. 1.25 Definition of a three-dimensional Bravais lattice by the Bravais vectors a_1, a_2, and a_3 in a Cartesian (x, y, z) coordinate system. z' defines some arbitrary direction in the lattice, with unit normal $\hat{n}$.

1.7 ATOMIC-LEVEL GEOMETRY OF PLANAR STACKING

Following the focus in the preceding section on the macroscopic geometrical description of crystalline interfaces, we are now ready to take the geometrical characterization one step further, down to the level of the atoms (albeit in their unrelaxed positions). This leads us naturally to the concept of the **atomic-level geometry** of solid interfaces, from which the concepts of coherency, commensurability, epitaxy and topotaxy discussed in section 1.2 follow logically. As an example, in section 1.7.3 the three simplest interface systems discussed earlier only at the macroscopic level (see section 1.6.3), will be revisited. Later on, in section 1.8, the atomic-level geometry of GBs will be discussed within the same framework.

1.7.1 Planar stacking in Bravais lattices

Since in both of its halves a crystalline interface contains stacks of well-defined lattice planes, in the following we briefly consider some basic geometrical definitions and properties associated with the planar stacking in a perfect Bravais lattice. Similar to Fig. 1.12, the underlying three-dimensional Bravais lattice is defined by the three Bravais vectors, a_1, a_2, and a_3, defined in Fig. 1.25 with

respect to a Cartesian (x, y, z) coordinate system. For simplicity we assume the basis attached to the Bravais lattice to contain only one atom such that the crystal lattice, identical to the Bravais lattice, is simply given by

$$r = la_1 + ma_2 + na_3 \qquad (1.37)$$

with $l, m, n = 0, \pm 1, \pm 2, \ldots$. The conventional geometrical description of the crystal lattice is then based on the primitive unit cell sketched in Fig. 1.25 which, together with its periodic images, defines the lattice completely.

In crystalline interface materials, a different choice of the primitive periodic Bravais unit cell, which emphasizes the plane-by-plane arrangement of the atoms and hence greatly facilitates the visualization of the atomic structure, is often advantageous. We start by defining an arbitrary (but rational) direction, z', defined in Fig. 1.25 by some unit vector $\hat{n}$, which represents the normal to a set of lattice planes which we will choose as the lattice planes of interest. The (x', y', z') coordinate system in Fig. 1.26(a) is then chosen such that the generally non-orthogonal x' and y' directions point along the edges c_1, c_2 of the primitive **planar** Bravais unit cell of the plane with normal $\hat{n}$, while the z' direction (orthogonal to the x'–y' plane) is defined in Fig. 1.25 to be parallel to the normal $\hat{n}$. In this new coordinate system, the atom positions are given by the new basis vectors c_1, c_2, and c_3, according to (Fig. 1.26)

$$r' = l'c_1 + m'c_2 + n'c_3 \qquad (1.38)$$

with $l', m', n' = 0, \pm 1, \pm 2, \ldots$ While the vectors c_1 and c_2 define the primitive **planar** unit cell (with plane normal $\hat{n}$), the out-of-plane vector c_3 enables one to proceed successively from one lattice plane to the next. It is important to recognize that the vector $c_3 = d + e$ does not generally point along the z' direction; this gives rise to a staggering of the planes. While its out-of-plane component, d (along z'), is determined by the interplanar spacing in the direction of $\hat{n}$, $d = |d|$, the relative translation of one plane relative to another is governed by e, the in-plane component of c_3. The 'natural' coordinate system associated with a given direction, $\hat{n}$, in the crystal is therefore the (x', y', z'') system shown in Fig. 1.26(a): Within the plane

the atom sites are given by the vectors c_1 and c_2 whereas the z''-direction, parallel to c_3, defines the 'direction of staggering' of one plane relative to a neighboring one. In this coordinate system all lattice planes are consequently identical, with no net in-plane (x', y') translation required as one proceeds from one plane to a neighboring one. The 3D unit cell of this lattice is sketched in Fig. 1.26(b).

To express the new, **plane-based** primitive Bravais vectors, c_1, c_2, and c_3, in terms of the primitive vectors, a_1, a_2, and a_3, of the **conventional** Bravais lattice and the normal $\hat{n}$ simply requires the projection of a_1, a_2, and a_3 onto the plane and the z' direction (parallel to $\hat{n}$). Since both the conventional (a_1, a_2, a_3) and the plane-based (c_1, c_2, c_3) Bravais unit cells are primitive (i.e. contain exactly one lattice site), with volume Ω, the two have identical volumes, i.e.

$$([a_1 \times a_2] \cdot a_3) = ([c_1 \times c_2] \cdot c_3) = \Omega \qquad (1.39)$$

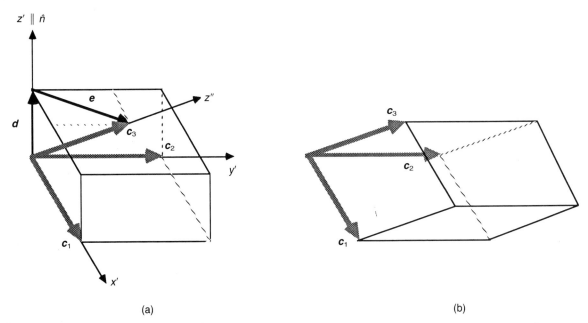

(a)

(b)

Fig. 1.26 Three-dimensional Bravais lattice of Fig. 1.25 projected into an (x', y', z') coordinate system chosen such that $z' \parallel \hat{n}$ while the (generally non-orthogonal) x' and y' directions point along the edges of the primitive **planar** Bravais unit cell of the plane with normal $\hat{n}$. The out-of-plane component of c_3, d (along z'), is determined by the interplanar spacing in the direction of $\hat{n}$. By contrast, the relative translation of one plane relative to another ('staggering of planes') is governed by the in-plane component of c_3, e. The primitive **volume** unit cell of the Bravais lattice, defined by c_1, c_2, and c_3, is sketched in (b).

A simple example of the two Bravais representations is shown in Fig. 1.27 for the (001) plane in the fcc lattice. Because both a_1 ($\parallel$ [110]) and a_2 ($\parallel$ [$\bar{1}$10]) lie in the (001) plane, in this case the vectors a_1, a_2, and a_3 of the well-known rhombohedral primitive fcc unit cell are identical to the vectors c_1, c_2, and c_3, with the vectors d and e along [001] and [010], respectively.

Given that the lattice planes are in general staggered, an obvious question concerns the number of lattice planes in the repeat stacking unit, sometimes referred to as the **stacking period**, P. The well-known . . . |ABC|ABC| . . . stacking of the (111) planes in the fcc lattice is an example of a three-plane stacking period; also, the two-plane . . . |AB|AB| . . . stacking period of the (001) and (011) planes is readily recognized. To determine P, one has to find the nearest lattice plane in the direction of $\hat{n}$ in the Cartesian coordinate system which is identical to the plane through the origin (i.e. entirely untranslated or translated by a multiple of the planar unit-cell dimensions); i.e. P is the smallest integer which satisfies the condition (eq. (1.38))

$$Pe = l'c_1 + m'c_2 \qquad (1.40)$$

with l' or $m' = 0$, ± 1, etc. and e, c_1, and c_2 as defined in Fig. 1.26(a).

To illustrate eq. (1.40) we consider a cubic crystal in which the normal $\hat{n}$ may be given in terms of the Miller indices (hkl) (eq. (1.24)). All relevant geometrical parameters, including P, may then be expressed in terms of the Miller indices. For example, in a Cartesian coordinate system the interplanar spacing, $|d| = d(hkl)$ (Fig. 1.26(a)), is given by the well-known expression

$$d(hkl) = \varepsilon a(h^2 + k^2 + l^2)^{-\frac{1}{2}},$$
$$(\varepsilon = 0.5 \text{ or } 1) \qquad (1.41)$$

where a is the cubic lattice parameter and where the value of $\varepsilon (=0.5$ or $1)$ depends on the particular combination of odd and even Miller indices. (For example, in the fcc lattice, $\varepsilon = 1$ if h, k, and l are all odd but 0.5 otherwise.) The period, $P \equiv P(hkl)$, is similarly given by

$$P(hkl) = \delta(h^2 + k^2 + l^2), \quad (\delta = 1 \text{ or } 2) \quad (1.42)$$

where the value of $\delta (=1$ or $2)$ also has to be determined in each case by inspection of the various odd and even combinations of Miller indices.

To illustrate these expressions, in Tables 1.1 and 1.2 the values of $d(hkl)$ and $P(hkl)$ are listed for the eight planes of the fcc and bcc lattices with the largest values of $d(hkl)$. Also, Fig. 1.28 shows schematically a **unit stack** of lattice planes, labeled . . . |AB . . IJ| . . ., in a direction with $P(hkl) = 10$ planes in the stacking period (such as the $\langle 210 \rangle$ direction of the fcc lattice; Table 1.1). The x' and z' axes, parallel to c_1 and d, respectively, are the same ones shown in Fig. 1.26(a),

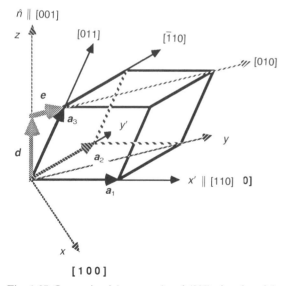

Fig. 1.27 Conventional (a_1, a_2, a_3) and (001)-plane-based (c_1, c_2, c_3) fcc Bravais lattices (see also Figs. 1.25 and 1.26). Since a_1 and a_2 lie in the (001) plane, $c_1 \equiv a_1$ and $c_2 \equiv a_2$ in this case.

Table 1.2 Same as Table 1.1, but for the bcc lattice

No.	(hkl)	$(h^2 + k^2 + l^2)$	$P(hkl)$	$d(hkl)/a$
1	(110)	2	2	0.7071
2	(100)	1	2	0.5000
3	(112)	6	6	0.4082
4	(310)	10	10	0.3162
5	(111)	3	3	0.2887
6	(321)	14	14	0.2673
7	(114)	18	18	0.2357
8	(210)	5	10	0.2236

and periodic border conditions are implied in all three dimensions. Figure 1.28 illustrates the constant in-plane translation, e, as one proceeds from one plane to another. Because of the periodicity parallel to the plane, the lattice site in plane F, which would normally fall outside of the unit cell as the vector e is added to the E plane, is reflected back into the unit cell.

The unit stack shown in Fig. 1.28 is obviously comprised of $P(hkl) = 10$ primitive plane-based Bravais unit cells c_1, c_2, and c_3. Since each plane contains exactly one atom (because we have chosen a basis containing only one atom), the planar unit-cell area, $A(hkl)$, and the interplanar spacing,

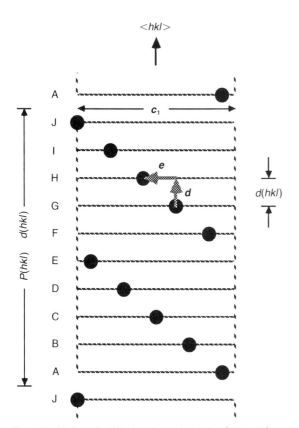

Fig. 1.28 Unit stack of lattice planes, labeled ... |AB ... IJ| ... , in a direction with $P(hkl) = 10$ planes in the stacking period (such as for the (210) plane in the fcc lattice; see Table 1.1). Illustrated is the constant in-plane translation, e, with the vector d allowing one to proceed from one plane to the nearest one (schematic).

$d(hkl)$, are related via the atomic volume, Ω, as follows:

$$A(hkl)d(hkl) = \Omega \qquad (1.43)$$

Based on this relationship, the most widely spaced planes correspond to the ones with the smallest planar unit-cell areas, i.e. the densest planes of the crystal lattice (also Tables 1.1 and 1.2).

As discussed in section 1.5.4, a necessary condition for two lattice planes to be commensurate is that their planar unit-cell areas are compatible, and hence satisfy eq. (1.12). Given eqs. (1.43) and (1.41), for the lattice planes in two different cubic crystal lattices (with lattice parameters a_1 and a_2, respectively) eq. (1.12) may be rewritten as follows:

$$\varepsilon_2^2 a_2^2 (h_1^2 + k_1^2 + l_1^2)/\varepsilon_1^2 a_1^2 (h_2^2 + k_2^2 + l_2^2) = m^2/n^2,$$
$$(\varepsilon_1, \varepsilon_2 = 0.5 \text{ or } 1) \qquad (1.44)$$

In the case of GBs (for which $a_1 = a_2$), the ratio of the sum of the squares of the Miller indices hence has to be a ratio of squares of integers. The lattice planes with the smallest unit cells which are then compatible with the (111), (001), and (011) planes are listed in Table 1.3. These are the only relatively low-index combinations of planes forming commensurate GB interfaces with one of the three densest planes on one side. All other combinations have to form aperiodic or quasiperiodic structures [32].

Finally, we mention that the validity of eq. (1.43) is not limited to cubic lattices, as readily seen from eq. (1.39). The unit-cell area is generally given by the vector $[c_1 \times c_2]$ which is parallel to $\hat{n}$ (Fig. 1.26(a)). According to eq. (1.39), the unit-cell volume is the dot product of the vector c_3 and the area vector $[c_1 \times c_2]$. Since the latter is parallel to $\hat{n}$ (Fig. 1.26(a)) and since the projection of c_3 onto $\hat{n}$ is identical to the interplanar spacing, $|d|$, eq. (1.43) is reproduced.

1.7.2 Stacking inversion by rotation

A basic property of all lattices with inversion symmetry, particularly all Bravais lattices, is that the stacking sequence in a particular (rational) direction, $\langle hkl \rangle$, may be inverted by a 180° rotation about the plane normal $\langle hkl \rangle$. This property

Table 1.3 Cubic lattice planes with the smallest unit cells which are commensurate with the (111), (001) and (011) planes, respectively (see eq. (1.44) for $a_1 = a_2$ and $\varepsilon_1 = \varepsilon_2$)

No.	$(h_2k_2l_2)$	$(h_1k_1l_1)$	m^2/n^2	$(h_2k_2l_2)$	$(h_1k_1l_1)$	m^2/n^2	$(h_2k_2l_2)$	$(h_1k_1l_1)$	m^2/n^2
1	(1 1 1)	(1 1 1)	1	(0 0 1)	(0 0 1)	1	(0 1 1)	(0 1 1)	1
2	(1 1 1)	(1 1 5)	9	(0 0 1)	(2 2 1)	9	(0 1 1)	(1 1 4)	9
3	(1 1 1)	(1 5 7)	25	(0 0 1)	(4 3 0)	25	(0 1 1)	(0 7 1)	25
4	(1 1 1)	(1 5 11)	49	(0 0 1)	(2 3 6)	49	(0 1 1)	(3 4 5)	25
5	(1 1 1)	(1 1 1 11)	81	(0 0 1)	(1 4 8)	81	(0 1 1)	(1 4 9)	49
6	(1 1 1)	(5 7 13)	81	(0 0 1)	(4 4 7)	81	(0 1 1)	(3 5 8)	49
7	(1 1 1)	(1 11 19)	121	(0 0 1)	(6 6 7)	121	(0 1 1)	(4 5 11)	81
8	(1 1 1)	(5 7 17)	121	(0 0 1)	(2 6 9)	121	(0 1 1)	(7 7 8)	81

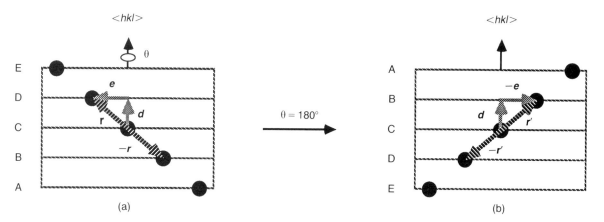

Fig. 1.29 Inversion of a (hypothetical) five-plane ideal-crystal stack of lattice planes, A–E, by a 180° twist rotation about the plane normal. (a) Before rotation; (b) after 180° rotation and application of inversion-symmetry operation (schematic).

is illustrated in Fig. 1.29 for the case of a (hypothetical) five-plane stacking period. The vector r in Fig. 1.29(a) represents the translation vector, $r = e + d$, with components (e_x, e_y, d) (Fig. 1.28), between atoms in neighboring planes of the stack. A 180° rotation about the plane normal, $\langle hkl \rangle$, transforms r into $r' = -e + d = (-e_x, -e_y, d)$. However, in a lattice with inversion symmetry a lattice point is found at $-r'$ if there is one at r'; the vector $-r' = e - d = (e_x, e_y, -d)$ therefore represents a lattice point as well.

So far we have not elaborated on the rotation center for the 180° rotation about the plane normal, $\langle hkl \rangle$, which inverts the stacking sequence. Obviously, for a direction with an **odd** period, any lattice point in the central plane of the unit stack (such as the C plane in Fig. 1.29(a)) represents a

possible rotation center. By contrast, for a direction with an **even** period, any half-way point between lattice sites in symmetrically related planes may serve as rotation center.

If the rotation axis for the unit-stack inversion has m-fold rotation symmetry, inversion is also achieved for the corresponding twist angles of $\theta = 180°/m \pm k\,360°/m$ $(k = 1, 2, \ldots, m - 1)$. Thus, although each individual (111) plane in a cubic crystal contains a six-fold rotation axis, the staggered three-plane $\langle 111 \rangle$ unit stack only has three-fold symmetry $(m = 3)$ because the center for this inversion rotation differs from that of the six-fold rotation symmetry. For similar reasons, $m = 2$ for a two-plane $\langle 100 \rangle$ unit stack, although individual (100) planes have four-fold rotation symmetry.

The property that in a crystal lattice with

inversion symmetry the stacking sequence may be inverted by a 180° rotation about the plane normal has important consequences for the atomic-level geometry of GBs. As illustrated further in section 1.8 below, it is responsible for the fact that in such crystal lattices (i) all symmetrical-tilt boundaries represent special 180° twist boundaries (section 1.6.1); (ii) asymmetrical-tilt boundaries (ATGBs) form a special subset of general (or asymmetrical-twist) boundaries; and (iii) for every asymmetrical combination of lattice planes, there are two distinct ATGB configurations (obtained for θ = 0° and 180°, respectively) which differ merely by the inversion of the stacking in one half relative to the other.

1.7.3 Example: stacking faults, free surfaces and symmetrical-tilt GBs in cubic crystals

In section 1.6 we described three interface systems which share the common property of having only two macroscopic DOFs, namely STGBs, stacking faults, and free surfaces. From the preceding discussion it is clear that they also share some important geometrical features. Most importantly, these three systems have identical planar unit-cell dimensions and areas (Figs. 1.30(b)–(d)), and

their planar unit cell projected onto the interface plane is identical to that of the perfect crystal on the same plane, sketched in Fig. 1.30(a). Also, a tilt axis and a tilt angle may be formally assigned to all three, although in the case of stacking faults this terminology is not commonly used, while in the case of the surfaces the **tilt** axis is usually referred to as the **pole** axis.

Because a stacking fault differs from the STGB on the same plane only by the inversion of the stacking sequence in the latter (Figs. 1.30(b) and (c)), but is otherwise so similar to the STGB configuration, one might expect their physical properties to be rather similar also. In fact, based on their close geometrical relationship, one would expect that STGBs have a lot more in common with stacking faults than with GBs; this is contrasted by the common practice of viewing STGBs as high-angle GBs. Hence, rather than referring to these simple planar defects as **grain boundaries**, it might be more illustrative to call them 'inverted' or 'tilted' stacking faults. This would merely require a broadening of the definition of a stacking fault to include not only translations parallel to the fault plane but also a possible inversion of the stacking sequence at the plane of the defect.

As a consequence of having the smallest planar

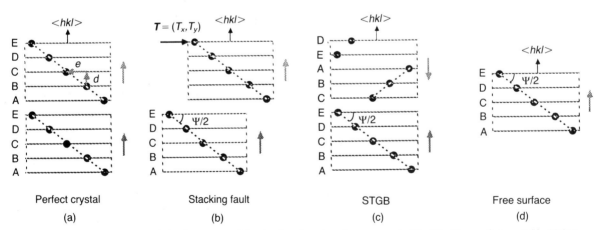

Fig. 1.30 Comparison of the atomic-level geometry of the simplest three interface systems in (b)–(d) with a perfect crystal in (a) (see also Fig. 1.1). Most importantly, the four systems sketched here have identical planar unit-cell dimensions and areas; this unit cell is the smallest possible for any atom arrangement involving this particular lattice plane, with a ⟨hkl⟩ normal. The shaded arrows indicate the direction of planar stacking. We note that Figs. 1.17 and 1.30 may be directly superimposed, illustrating the considerable similarity among these three systems and with the perfect crystal.

unit cell of any planar defect on a particular plane, the energies of stacking faults and STGBs are extremely sensitive functions of translations parallel to the interface, with a consequently large resistance towards shear parallel to the plane of the defect. This extreme sensitivity towards translations has been shown to be the cause for the appearance of energy cusps for symmetrical GBs whenever the STGB configuration on a particular plane is approached (i.e. as $\theta \to 180°$; Fig. 1.23) [32, 33].

It appears that the close geometrical similarity between STGBs and free surfaces (Figs. 1.30(c) and (d)) has not always been recognized in the past, which might be one of the reasons why only relatively little is known about the work of adhesion of even these simplest of all GBs (section 1.6.5 and Fig. 1.24). From a conceptual viewpoint it appears that when an STGB is cleaved, thus creating two identical free surfaces, the **dislocations** in the GB are transformed into **steps** in the surfaces (Fig. 1.18), while the planar unit-cell dimensions of the interface are left unchanged. It would therefore seem that a better understanding of ideal-cleavage fracture requires an investigation of the relationship between steps in surfaces, on the one hand, and edge dislocations in STGBs on the other. The realization of the close geometrical similarity between these two simple interface systems might be of some aid in elucidating this complex problem [27].

1.8 GRAIN BOUNDARIES

Grain boundaries (GBs) represent ideal model systems for the investigation of the strictly geometrical aspects of structure–property correlations for the following three reasons. First, the complexity due to the myriad of possible choices of material combinations forming the interface is avoided, thus enabling a focus on the distinct roles played by the GB geometry, on the one hand, and the atomic structure on the other. Second, because GBs are bulk interfaces, dimensional interface parameters (such as the modulation wavelength in strained-layer superlattices, or the thickness of epitaxial layers) do not enter into the problem.

Finally, the GB **energy** is thought to play a central role in various GB properties, such as impurity segregation, GB mobility and fracture, GB diffusion and cavitation, to name but a few. A better understanding of the correlation between the structure and the energy of GBs therefore promises to offer insights into more complex structure–property correlations as well. This correlation also represents a base line against which the effects of interfacial chemistry can be probed.

Although GBs probably represent the best-studied type of all interface systems, relatively little knowledge acquired on their physical behavior has filtered into other areas of interface research. It appears that one reason responsible for this unfortunate situation is related to the terminology used to describe their geometry, which differs fundamentally from that commonly used to describe, for example, epitaxial and thin-film interfaces. In this section we will review some of the GB 'jargon' within the framework of the concepts described in sections 1.6 and 1.7; a good understanding of these sections is therefore helpful. By defining these important defects in the context of the unified interface terminology, we hope to further elucidate the geometrical features in common to all interface systems and materials.

1.8.1 Coincident-site lattice description ('CSL-misorientation scheme')

As is well known, whether a GB is of a pure tilt, twist or mixed type, whether it is symmetrical or asymmetrical, or whether it is of a low- or high-angle type is fully determined by the choice of the five macroscopic DOFs of the interface. While, so far, we have reviewed one definition of these five geometrical variables (the interface-plane scheme in section 1.6.1), much of the terminology used to describe the geometry of GBs is based on the concept of the **coincident-site lattice** (**CSL**). By contrast with the interface-plane scheme in eq. (1.20), the CSL description of GBs focuses on the misorientation between the two grains, rather than on the plane of the defect. Also, because of the requirement that a superlattice must exist in common to the two halves of a bicrystal (at least prior to allowing for rigid-body translations associated with the three translational DOFs), the CSL terminology is limited to commensurate interfaces and is hence not usually applied to interfaces other than GBs, rendering a comparison of properties with those of other interfacial systems virtually impossible.

In this and the following section, we will discuss two CSL-based, and often intermixed, definitions of the five macroscopic DOFs (the **CSL-misorientation** and **tilt-inclination schemes**). We then explore the relationship between the three rather different definitions of the macroscopic DOFs of crystalline interfaces discussed in this chapter by investigating the mathematical connections between them.

Within the framework of the CSL description of GBs [16–19], three of the five macroscopic DOFs are identified with the CSL misorientation, and only the remaining two DOFs are assigned to the GB plane. (Although redundant, the inverse volume density of CSL sites, Σ, is usually added as a sixth parameter.) The misorientation between the crystal lattices associated with the two grains may be characterized by the rotation matrix $\mathbf{R}(\hat{n}_{CSL}, \phi_{CSL})$, with $\hat{n}_{CSL}$ and ϕ_{CSL} denoting the CSL rotation axis and angle, respectively (representing three DOFs). The five DOFs are then defined as follows:

$$\{DOFs\} = \{\hat{n}_{CSL}, \phi_{CSL}, \hat{n}_1\}$$
$$\text{('CSL-misorientation scheme')} \quad (1.45)$$

where, as in the interface-plane scheme in eq. (1.20), $\hat{n}_1$ represents the GB-plane normal in either of the two halves of the bicrystal (here chosen to be semicrystal 1). Both $\hat{n}_{CSL}$ and $\hat{n}_1$ are expressed in a space-fixed principal crystallographic coordinate system, such as the (x, y, z) system in Fig. 1.15(a). Because of its focus on the CSL misorientation between the two halves of the bicrystal, the definition in eq. (1.45) will be referred to as the 'CSL-misorientation scheme'. Given these five variables, the GB-plane normal in the second crystal, $\hat{n}_2$, is determined by [18, 21]

$$\hat{n}_2 = \mathbf{R}(\hat{n}_{CSL}, \phi_{CSL})\hat{n}_1 \quad (1.46a)$$

Similarly, the inverse density of CSL sites, Σ, is

governed completely by the three DOFs in the CSL rotation, i.e. $\Sigma = \Sigma(\mathbf{R}(\hat{n}_{CSL}, \phi_{CSL}))$.

We note that, prior to the CSL rotation, the two interpenetrating crystal lattices need not necessarily be identical. If they are identical, the CSL rotation results in the formation of a CSL for a **grain boundary**. Obviously, a superlattice in common to the two crystal lattices can not be formed unless the two are commensurate in all three dimensions (analogous to the concept of two-dimensional commensurability defined in section 1.5).

More explicitly, if one defines $\hat{n}_{CSL}$ by its direction cosines, say, in the (x, y, z) coordinate system in Fig. 1.15(a), according to $\hat{n}_{CSL} = (u, v, w)$ (with $u^2 + v^2 + w^2 = 1$), and if one uses the abbreviations $\cos \phi_{CSL} = c$, $\sin \phi_{CSL} = s$, the rotation matrix may be written as follows [18]:

$$\mathbf{R}(\hat{n}_{CSL}, \phi_{CSL}) = \mathbf{R}(u, v, w, c, s)$$

$$= (1 - c)\begin{pmatrix} u^2 & uv & uw \\ uv & v^2 & vw \\ uw & vw & w^2 \end{pmatrix}$$

$$+ \begin{pmatrix} c & -ws & vs \\ ws & c & -us \\ -vs & us & c \end{pmatrix} \quad (1.47)$$

A pure **tilt** boundary is defined by the condition that $\hat{n}_{CSL}$ be perpendicular to $\hat{n}_1$ in eq. (1.45), while a pure **twist** boundary is obtained whenever $\hat{n}_{CSL}$ is parallel to $\hat{n}_1$. As illustrated in Fig. 1.15, the total (CSL) misorientation between the two halves of the bicrystal may be viewed as consisting of a tilt rotation (Fig. 1.15(b)) followed by a twist rotation (Fig. 1.15(c)). Using the perfect crystal in Fig. 1.15(a) as starting point, the purpose of the tilt rotation is to align $\hat{n}_1$ and $\hat{n}_2$, thus defining the common GB-plane normal in Fig. 1.15(b). This rotation is followed by a twist rotation about this common normal, thus adding a twist component to the tilt component while forming a 'general' GB. The CSL rotation matrix, $\mathbf{R}(\hat{n}_{CSL}, \phi_{CSL})$, may therefore be decomposed into its tilt and twist components, $\mathbf{R}(\hat{n}_T, \psi)$ and $\mathbf{R}(\hat{n}_1, \theta)$, respectively (with the tilt and twist angles ψ and θ, respectively) according to [18]

$$\mathbf{R}(\hat{n}_{CSL}, \phi_{CSL}) = \mathbf{R}(\hat{n}_1, \theta)\,\mathbf{R}(\hat{n}_T, \psi) \quad (1.48)$$

Because rotations do not generally commute, interchanging the sequence of the tilt and twist rotations leads to a different final state, as is also evident from the matrix equation (1.48).

Finally, since the twist rotation does not alter the common GB normal, the CSL rotation in eq. (1.46a) may be replaced by its tilt component, and

$$\hat{n}_2 = \mathbf{R}(\hat{n}_T, \psi) \, \hat{n}_1 \qquad (1.46b)$$

Equation (1.46b) expresses the fact that the **plane** of a general GB is fully determined by its tilt component; conversely, the tilt component of a general GB is determined fully by the GB plane (section 1.6.1 and eqs. (1.21) and (1.22)).

A problem with the above CSL-based terminology is that within its framework the underlying number of macroscopic DOFs of a given GB is not always readily apparent. The three rotation matrices defined above involve a total of **nine** geometrical variables in $\hat{n}_{CSL}$, ϕ_{CSL}, $\hat{n}_T$, ψ, $\hat{n}_1$ and θ for an overall misorientation which is characterized by only the **three** DOFs in $\hat{n}_{CSL}$ and ϕ_{CSL}. Six relationships must therefore exist among these variables. While eq. (1.48) represents three of these, the remaining three may be obtained by expressing $\hat{n}_{CSL}$ and ϕ_{CSL} directly in terms of the twist and tilt axes and angles, according to [21]

$$\hat{n}_{CSL} = \beta^{-1}\{[(1 + \cos\theta)(1 - \cos\psi)]^{\frac{1}{2}} \, \hat{n}_T$$
$$+ [(1 - \cos\theta)(1 + \cos\psi)]^{\frac{1}{2}} \, \hat{n}_1$$
$$+ [(1 - \cos\theta)(1 + \cos\psi)]^{\frac{1}{2}} \, [\hat{n}_T \times \hat{n}_1]\} \qquad (1.49)$$

and

$$\cos\phi_{CSL} = (1 + \cos\theta)(1 + \cos\psi)/2 - 1 \qquad (1.50)$$

where

$$\beta = (3 - \cos\theta - \cos\psi - \cos\theta\cos\psi)^{\frac{1}{2}} \qquad (1.51)$$

While the (unit-) vector equation (1.49) connecting the three rotation axes represents two of the three additional relationships, the remaining one is eq. (1.50) which relates the three rotation angles to one another.

As is readily verified, eqs. (1.49)–(1.51) pass several trivial tests [21]:

(1) for $\psi \equiv 0$, $\hat{n}_{CSL} \equiv \hat{n}_1$ with $\phi_{CSL} \equiv \theta$ (pure twist);
(2) for $\theta \equiv 0$, $\hat{n}_{CSL} \equiv \hat{n}_T$ with $\phi_{CSL} \equiv \psi$ (pure tilt);
(3) for $\phi_{CSL} \equiv 0$ (no CSL misorientation), eq. (1.50) yields $\theta = \psi \equiv 0$; however, $\hat{n}_{CLS}$ becomes singular in this case because a rotation axis cannot be defined for this singularity of the CSL rotation.

Together with eqs. (1.21) and (1.22), the above expressions may be used to make the connection between the interface-plane and the CSL-misorientation schemes for the characterization of the same interface. For example, starting from the interface-plane definition of the five DOFs in eq. (1.20), the expressions (1.21) and (1.22) may be used to determine the tilt component, $(\hat{n}_T, \psi)$, of the interface. With its twist component, $(\hat{n}_1, \theta)$, apparent from the outset in eq. (1.20), eqs. (1.49) and (1.50) may then be used to determine the CSL misorientation, $(\hat{n}_{CSL}, \phi_{CSL})$. Conversely, starting from the CSL-based definition of the five DOFs in eq. (1.45), the expression (1.46a) may be used to determine $\hat{n}_2$. Given $\hat{n}_1$ and $\hat{n}_2$, the tilt component of the CSL boundary is simply given by eqs. (1.21) and (1.22); using this information in eq. (1.50), one may determine the twist angle θ needed in eq. (1.20).

The above CSL-based expressions describe rather elegantly, in terms of linear algebra, how a commensurate interface can be formed by a single rotation of two infinitely large, interpenetrating crystal lattices with respect to one another. From a practical viewpoint it is important to recognize that the CSL rotation axis and angle, as well as the GB-plane normal, are defined in the **unrotated** principal coordinate system associated with one of the two semicrystals (Fig. 1.15(a)). This makes it rather tedious at times to characterize a GB experimentally, starting from the already **rotated** positions of the atoms at hand. By contrast, the interface-plane scheme in eq. (1.20) focuses on the actual interface geometry at hand rather than on how this particular geometry may be thought of as having been generated by a single rotation of two interpenetrating crystal lattices with respect to one another.

Finally, we mention that the CSL-based terminology is further complicated by the concept of the boundary-plane **inclination**, α, which introduces a fourth rotation to the three already defined above (for details section 1.8.2). The number of CSL-based geometrical variables often used interchangeably to describe a single bicrystalline GB (with at most five DOFs) thus increases to a total of 13, including $\hat{n}_{CSL}$, ϕ_{CSL}, $\hat{n}_T$, ψ, $\hat{n}_1$, θ, $\hat{n}_2$, α, and Σ. This somewhat startling number emphasizes the importance in GB studies of knowing the correct number of independent degrees of freedom of the system, particularly when structure–property correlations are being investigated. As a practical matter, it might help in any such investigation to adhere strictly to any one of the three choices of DOFs given in eqs. (1.20), (1.45), and (1.54) below, from which all the others can, in principle, be derived, via expressions such as the ones given here and in section 1.6.1.

1.8.2 Asymmetrical GBs

In spite of indications for the preponderance of **asymmetrical** GBs in polycrystalline materials [34], our current understanding of structure–property interrelations is based largely on the investigation of symmetrical boundaries. With the development of high-resolution transmission-electron-microscopy methods (HREM) in 'edge-on' studies of tilt GBs, much atomic-level information concerning the orientation of the GB plane has become available during recent years. Much of this work shows that in both metals [35–37] and ceramic materials [38, 39] asymmetrical combinations of lattice planes and faceting occur rather commonly.

Two types of asymmetrical GBs are usually distinguished; these are known as **asymmetrical-tilt** boundaries (ATGBs) and **general** boundaries (Fig. 1.31). In this section we will attempt to describe their geometry within the framework of both the interface-plane scheme and the CSL-based terminology.

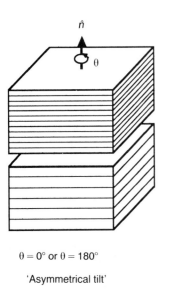

$\theta = 0°$ or $\theta = 180°$

'Asymmetrical tilt'

(a)

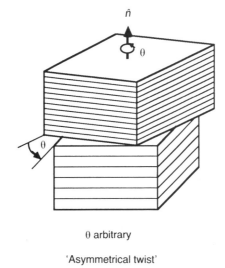

θ arbitrary

'Asymmetrical twist'

(b)

Fig. 1.31 Distinction between **asymmetrical-tilt** and **asymmetrical-twist** grain boundaries. While an **asymmetrical-tilt** boundary (ATGB) is obtained for $\theta = 0°$ or 180°, for some arbitrary twist angle, θ, an asymmetrical-twist or **general** GB, sketched in (b), is obtained. The alignment of the lattice planes is analogous to that in Figs. 1.15(b) and (c).

Geometry of general GBs

Applying the interface-plane based nomenclature in eq. (1.20) to asymmetrical GBs, one can readily distinguish the tilt and twist components of the interface. Its tilt component, and hence its edge-dislocation content, is completely given by the two sets of lattice planes forming the interface, i.e. by the GB **plane** (Fig. 1.31(a) and eqs. (1.21) and (1.22)). The twist angle θ introduces in addition a twist component, i.e. screw dislocations, into the boundary (Fig. 1.31(b)). The values of θ corresponding to the beginning and end of the twist-misorientation range (i.e. $\theta = 0°$ and $180°$; see Fig. 1.31(a)) are most appropriately chosen such that the GB has no twist component, i.e. that its structure contains edge dislocations only. The interface thus obtained for (eq. (1.20))

$$\{DOFs\} = \{\hat{n}_1, \hat{n}_2, \theta = 0° \text{ or } \theta = 180°\} \quad (1.52)$$

is therefore of a pure tilt type and is commonly referred to as an **asymmetrical-tilt boundary** (ATGB). Because of the absence of screw dislocations (which, similar to the case of low-angle twist boundaries, would increase the planar unit cell area of the GB), this interface has the smallest possible planar unit cell of any asymmetrical GB formed by a given set of lattice planes. Since the value of θ is fixed, an ATGB has only four DOFs,

by contrast with a **general** boundary which also has a twist component and, hence, five DOFs (Fig. 1.31(b)).

To relate this GB-plane-based picture of an asymmetrical boundary to the conventional CSL-based picture, Fig. 1.32 offers a different view of how the ATGB structure in Fig. 1.31(a) may be thought of as having been generated by a pure tilt rotation. The orientation of the tilt axis, $\hat{n}_T$, perpendicular to the GB-plane normal, $\hat{n}$, is illustrated in Fig. 1.32(a). Adding the tilt angle, ψ, to the two unit vectors, $\hat{n}_T$ and $\hat{n}$, according to Fig. 1.32(a) and eq. (1.45), it appears that five geometrical variables are necessary to specify an asymmetrical-tilt boundary, because now

$$\{DOFs\} = \{\hat{n}_T, \psi, \hat{n}\} \quad (1.53)$$

However, since $\hat{n}_T$ and ψ are fully governed by $\hat{n}_1$ and $\hat{n}_2$ (see eqs. (1.21) and (1.22)), ATGBs have only four DOFs; therefore they are a subset of asymmetrical-**twist** boundaries.

In the conventional 'edge-on' view (down the tilt axis) of an ATGB, the atoms in densest directions perpendicular to the tilt axis are usually connected (Fig. 1.32(c)). In Fig. 1.32(b) the same atoms are, instead, connected by lines parallel to the GB, making clear the connection with the ATGB configuration in Fig. 1.31(a). Finally, Fig. 1.32(d) provides a view onto the GB plane, showing the

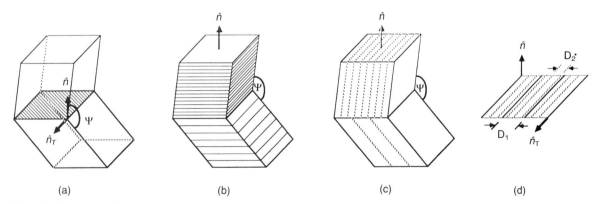

(a)   (b)   (c)   (d)

Fig. 1.32 Conventional CSL generation of an asymmetrical-tilt boundary by a pure tilt rotation (a). In the edge-on view in (c), the atoms in the densest directions perpendicular to the tilt axis are connected. To demonstrate the connection with the ATGB configuration in Fig. 1.31(a), in (b) the same atoms are, instead, connected by lines parallel to the GB-plane. Finally, (d) provides a view **onto** the GB plane, showing the two sets of edge dislocations, with spacings D_1 and D_2; in the edge-on view in (c), these dislocations correspond to the (dashed) lattice planes terminating at the GB.

two sets of edge dislocations, with spacings D_1 and D_2; in the edge-on view in Fig. 1.32(c), these dislocations correspond to the (dashed) lattice planes terminating at the GB.

It is important to recognize that for the same combination of lattice planes forming the asymmetrical GB, two ATGB configurations are obtained. To illustrate this little recognized geometrical feature of ATGBs, in Fig. 1.33 we have chosen two hypothetical orientations, $\hat{n}_1$ and $\hat{n}_2$, with $P(\hat{n}_1) = P_1 = 3$ and $P(\hat{n}_2) = P_2 = 11$ planes in the repeat stacking period defined in section 1.7.1. For the purpose of this illustration, we also assume that the planar unit cells are commensurate such that, say, nine planar unit cells of crystal 2 (3×3) match a single unit cell of crystal 1; the unit-cell areas are hence related by $A_2/A_1 = \frac{1}{9}$ (Fig. 1.33). As discussed in section 1.7.2, in crystal lattices

with inversion symmetry a $180°/m$ twist rotation (about the common GB-plane normal) leads to the inversion of the stacking sequence on one side of the interface with respect to the other. (Here m characterizes a possible rotation symmetry (for $m > 1$) in the planar unit cell of the GB; section 1.7.2.) In such a lattice the two ATGB configurations thus obtained for the twist angles of $\theta = 0°$ and $\theta = 180°/m$ differ merely by the inversion of the stacking sequence, while their unit-cell dimensions are identical (Fig. 1.33). Since only a relative rotation of the two halves is involved, starting from the ATGB in Fig. 1.33(a) the same ATGB configuration in (b) would have been obtained had the **lower** semicrystal been inverted (provided both crystal lattices have inversion symmetry). More generally, in a crystal lattice without inversion symmetry the two ATGB configurations

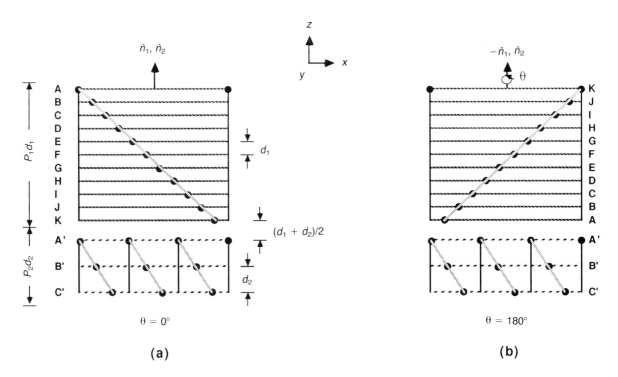

(a)   **(b)**

Fig. 1.33 Two **asymmetrical-tilt** configurations, with identical unit-cell dimensions but inverted stacking sequences in one with respect to the other, are obtained for $\theta = 0°$ and $180°$, respectively. In the case shown, the upper crystal was inverted (i.e. turned upside down) by the twist rotation. Since only a **relative** rotation of the two halves is involved, the same ATGB configuration in (b) would have been obtained had the lower semicrystal been inverted (provided the crystal lattices have inversion symmetry). $d_\alpha = d(\hat{n}_\alpha)$ and $P\alpha = P(\hat{n}_\alpha)$ ($\alpha = 1, 2$) denote, respectively, the interplanar lattice spacings and stacking periods in the two halves. To preserve the perfect-crystal density, the effective interplanar spacing at the interface is given by the arithmetic average, $d_{\text{eff}} = (d_1 + d_2)/2$.

in Fig. 1.33(a) and (b) would differ also by the inversion of stacking at the interface; however, to achieve that inversion, one semicrystal has to be turned upside down with respect to the other, an operation not possible by a pure twist rotation.

In a few high-symmetry cases, the two ATGB configurations generally possible for a particular combination of lattice planes may be identical. For example, the (100) and (110) planes in the fcc and bcc lattices exhibit only a two-plane stacking period, . . . |AB|AB| . . . . The inversion of these planes by a 90° twist rotation (because $m = 2$; section 1.7.2) about $\langle 100 \rangle$ or $\langle 110 \rangle$, respectively, can be undone simply by a rigid-body translation parallel to the plane, thus restoring the original unrotated perfect crystal (Fig. 1.39 below). Hence, in asymmetrical GBs in fcc or bcc metals with a (100) or (110) plane on one side of the interface, the two ATGB configurations are identical. (To better visualize the effect of a twist rotation by 180°, one may simply think of its net effect being the turning upside down of one of the semi-crystals in Fig. 1.33(a).) Generally, an asymmetrical GB must be formed by two sets of lattice planes, each with more than two planes in the stacking period, for the two ATGB configurations to be different.

As already mentioned, the introduction of screw dislocations, in addition to the edge dislocations already present in the ATGB, obviously increases the planar unit-cell area of the interface. The two ATGBs obtained for a given combination of planes therefore represent the asymmetrical GBs with the smallest planar unit-cell area of all the GBs that can be formed by these lattice planes. From a strictly geometrical point of view, asymmetrical-twist boundaries may therefore be viewed as 'vicinals' to the two 'special' ATGBs obtained for a given combination of planes (section 1.6.4). More-over, since the inversion of the stacking sequence preserves the planar unit-cell dimensions, the two ATGB configurations are unique geometrically in that they have identical planar unit-cell dimensions, with an area which is the smallest of all the GBs formed by the same combination of lattice planes. As illustrated in Chapter 3 of this volume [32], in both fcc and bcc metals this unique geometry translates into a particularly low energy of asym-

metrical-**tilt** boundaries, giving rise to energy cusps and, hence, indeed 'special' properties of these two configurations at the endpoints of the twist-misorientation range. Most asymmetrical-**twist** boundaries are therefore 'vicinal' boundaries, in the sense defined in section 1.6.4.

We finally mention that the interface-plane-based view of asymmetrical GBs brings out naturally that (i) the tilt component of a general boundary is solely responsible for the asymmetry in the GB plane, (ii) ATGBs are a special subset, with four DOFs, of general boundaries, (iii) there are gener-ally **two** ATGB configurations for every combi-nation of lattice planes, and (iv) there is considerable resemblance of a general (or asymmetrical-twist boundary) with a symmetrical-twist boundary, in that both may be generated by a rotation about the GB-plane normal by some angle θ (compare Fig. 1.31(b) with Fig. 1.35(b)), introducing screw dislocations into the GB. As illustrated in Chapter 3 of this volume [32], the incorporation of this geo-metrical similarity between general and symmetrical GBs in the interface-plane-based definition of the macroscopic DOFs is of considerable aid in the investigation of their properties. Finally, since the interface-plane-based description of GBs is not limited to commensurate interfaces, it facilitates a direct comparison of GBs with other types of inter-face systems.

The concept of the GB-plane inclination ('tilt-inclination scheme')

One CSL-based description of asymmetrical GBs uses the concept of the inclination of the GB plane, thus adding a fourth rotation to the CSL-, tilt- and twist rotations already discussed in section 1.8.1. The related choice of macroscopic DOFs (referred to as the 'tilt inclination scheme') has the advantage that it closely resembles an experimental situation in which the tilt misorientation between two grains is fixed, while the GB plane may choose whatever inclination may lead to a particularly low GB energy. Such a situation is encountered, for example, when a tilt bicrystal is grown from two preoriented seeds [37].

It may be useful to describe the concept of the

GB-plane inclination in a somewhat unconventional manner. We start with the observation that the tilt component, $(\hat{n}_T, \psi)$, of a general GB is fully determined by the GB-plane normals, $\hat{n}_1$ and $\hat{n}_2$ (eqs. (1.21) and (1.22)). Hence, starting from the **three** variables in $\hat{n}_T$ and ψ, in order to fix the GB **plane**, a single additional parameter, α, is required; α is called the inclination angle. Therefore, by replacing $\hat{n}_1$ and $\hat{n}_2$ in eq. (1.20) by the tilt misorientation and inclination angle, the five DOFs may be defined as follows:

$$\{DOFs\} = \{\hat{n}_T, \psi, \alpha, \theta\}$$
$$\text{('tilt-inclination scheme')} \qquad (1.54)$$

The geometrical meaning of α is illustrated in Fig. 1.34(a) [37] for the case of pure (symmetrical and asymmetrical) **tilt** boundaries in cubic crystals (eqs. (1.52) and (1.53)). For a given fixed tilt misorientation, $(\hat{n}_T, \psi)$, **symmetrical** boundary-plane configurations are obtained for the angles $\alpha = \psi/2$ (labeled STGB$_1$) and $\alpha = \psi/2 + 90°$ (labeled STGB$_2$), respectively. The asymmetrical combinations of GB planes may conveniently be charac-

terized by the inclination angle, α, with respect to the STGB$_1$ orientation (Fig. 1.34(a)). The two STGB configurations are then defined by $\alpha = 0°$ and $\alpha = 90°$, while for any other value of α an ATGB is obtained.

Figure 1.34(a) also illustrates that for every combination of $\hat{n}_T$ and ψ, at most two symmetrical, but infinitely many asymmetrical, GB-plane orientations are possible. (We should point out, however, that this is true only for a rational tilt axes in cubic crystals; for systems with less symmetry, a symmetrical configuration may not exist [37].)

A simple method for determining α is the following. Starting with the fixed tilt misorientation between two grains, the two STGBs are usually either well known (based on the value of Σ) or readily determined [19]; these define the angles $\alpha = 0°$ and $90°$. Given an ATGB in the same tilt bicrystal (i.e. with the same tilt axis and angle), $\cos \alpha$ is simply given by the dot product between corresponding STGB and ATGB plane normals in the same half of the bicrystal.

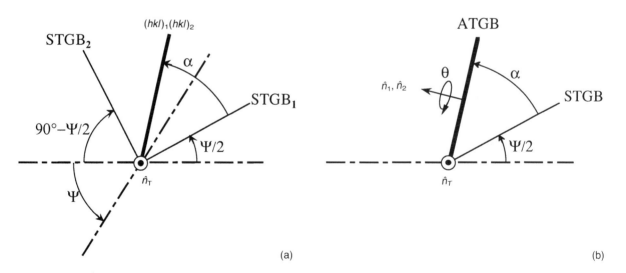

(a)   (b)

Fig. 1.34 (a) Definition of the two symmetrical-tilt configurations, STGB$_1$ and STGB$_2$, obtained for a given **tilt** misorientation, by the angle ψ, about the two-fold tilt axis with unit vector $\hat{n}_T$ (schematic) [37]. One of the mirror planes is indicated by dash-dotted lines for both crystal 1 and crystal 2. α is the inclination of the GB plane of some arbitrary **asymmetrical**-tilt configuration, combining $(h_1k_1l_1)$ and $(h_2k_2l_2)$ planes, measured with respect to STGB$_1$. The two **symmetrical** configurations correspond to inclinations $\alpha = \psi/2$ and $\alpha = \psi/2 + 90°$, respectively. (b) illustrates how a **twist** component may be introduced into the ATGB via a rotation, by the angle θ, about the common ATGB-plane normal, resulting in a **general** (or **asymmetrical-twist**) boundary, with both tilt and twist components. The meaning of θ is the same as in Figs. 1.15(c) and 1.31(b). (For further details, see Ref. [37]).

Within the tilt-inclination scheme, only one of the five DOFs (α) is actually assigned to the GB plane, and the determination of $\hat{n}_1$ and $\hat{n}_2$ represent a non-trivial undertaking. However, given the GB-plane normal on one side of the interface, $\hat{n}_1 = \hat{n}_1(\hat{n}_\mathrm{T}, \psi, \alpha)$, the GB-plane orientation in the other half is determined by (eq. (1.46b))

$$\hat{n}_2 = \mathbf{R}(\hat{n}_\mathrm{T}, \psi)\, \hat{n}_1(\hat{n}_\mathrm{T}, \psi, \alpha) \qquad (1.55)$$

To complete the definition of all five DOFs, a **twist** component ($\hat{n}_2$, θ), may be introduced into the ATGB in Fig. 1.34(a) via a rotation about the common GB-plane normal (Fig. 1.34(b)).

In the tilt-inclination scheme, a **symmetrical** interface is charactertized as follows:

$$\{\mathrm{DOFs}\} = \{\hat{n}_\mathrm{T}, \psi, \alpha = 0° \text{ or } \alpha = 90°, \theta\} \qquad (1.56)$$

At first sight, eq. (1.56) appears to require the four variables in $\hat{n}_\mathrm{T}$, ψ and θ for the characterization of an interface with only three DOFs. However, when the condition of symmetry, $\hat{n}_1 = \pm\hat{n}_2$ (eq. (1.25b)) is combined with eq. (1.55), a relationship between $\hat{n}_\mathrm{T}$ and ψ is obtained, according to $\hat{n}_1 = \pm\mathbf{R}(\hat{n}_\mathrm{T}, \psi)\,\hat{n}_1$, which permits ψ to be expressed as a function of $\hat{n}_\mathrm{T}$, $\psi = \psi(\hat{n}_\mathrm{T})$, thus reducing the number of independent variables in eq. (1.56) to only three.

As already mentioned, the choice of macroscopic DOFs in eqs. (1.54) and (1.56) has the advantage that it sometimes resembles an experimental situation rather closely [37]. On the other hand, particularly when the twist component of an asymmetrical GB is varied systematically, it is more advantageous to use the interface-plane definition in eq. (1.20) and the concept of the asymmetrical-twist boundary. Hence, while formally one choice is as good as another, any particular choice will depend on the particular experimental situation.

Finally, as an example we consider the two ATGB configurations that can be formed when bringing together two fcc grains with (557) and (771) faces, respectively [37]. Within the interface-plane terminology, the two ATGB configurations at $\theta = 0°$ and $\theta = 180°$ would be characterized by (eqs. (1.24) and (1.52))

$$\{\mathrm{DOFs}\} = \{(557), (771), \theta = 0° \text{ or } \theta = 180°\} \qquad (1.57)$$

and the fact that geometrically they differ merely by the inversion of the stacking sequence at the interface in one with respect to the other is obvious.

In the CSL-misorientation scheme, the two boundaries are characterized as follows (see also eqs. (1.45) and (1.53)) [37]:

$$\{\mathrm{DOFs}\} = \{\langle 1\bar{1}0\rangle, 38.94° \,(557)\} \quad (\Sigma = 9) \qquad (1.58a)$$

and

$$\{\mathrm{DOFs}\} = \{\langle 1\bar{1}0\rangle, 50.58° \,(557)\} \quad (\Sigma = 11) \qquad (1.58b)$$

where, for clarity, the related values of Σ are given in parentheses, indicating that the two ATGBs belong to different CSL systems. Given these definitions of their DOFs, the two GBs are readily identified as tilt boundaries (by contrast with general or pure twist boundaries, because $\hat{n}_\mathrm{CSL} \perp \hat{n}_1$). However, it is neither obvious that the two GBs are asymmetrical tilts nor that they are such close relatives of one another geometrically. Also, identifying the GB plane in the other half of the bicrystal as the (771) plane requires one to solve eq. (1.46) with the rotation matrix (1.47).

Finally, in the tilt-inclination scheme, the two ATGBs are characterized by (eq. 1.56)

$$\{\mathrm{DOFs}\} = \{\langle 1\bar{1}0\rangle, 38.94°, \alpha = 25.24°, \theta = 0°\} \\ (\Sigma = 9) \qquad (1.59a)$$

and

$$\{\mathrm{DOFs}\} = \{\langle 1\bar{1}0\rangle, 50.58°, \alpha = 70.53°, \theta = 0°\} \\ (\Sigma = 11) \qquad (1.59b)$$

Here the underlying values of α were determined from the knowledge of the corresponding symmetrical configurations ((114)(11$\bar{4}$) and (221)(22$\bar{1}$) in the $\Sigma = 9$ system, and (113)(11$\bar{3}$) and (332)(33$\bar{2}$) for $\Sigma = 11$), from which $\cos\alpha$ is simply given by the dot product of corresponding symmetrical and asymmetrical planes in the same half of the bicrystal.

Again, as in eqs. (1.58a) and (b), the two GBs are readily identified as pure tilt GBs (from the fact that $\theta = 0°$), and their asymmetry follows directly from the fact that $\alpha \neq 0°$ and $\alpha \neq 90°$. However, starting from these parameters, the identification of the actual GB plane from a determination of $\hat{n}_1$ and $\hat{n}_2$, is non-trivial. Also, as in their CSL-misorientation description, their close geometrical similarity is not very obvious, particularly that (i) they are formed by the same set of lattice planes, and (ii) they have identical planar unit-cell dimensions.

1.8.3 Symmetrical GBs

In section 1.6.1 we defined an interface as **symmetrical** if $\hat{n}_1$ and $\hat{n}_2$ are related **linearly**, i.e. if $\hat{n}_2 = L(\hat{n}_1)$, such as $\hat{n}_2 = \pm\hat{n}_1$ (eqs. (1.25a) and (b)), thus reducing the number of DOFs from five to only three (eq. (1.26b)). $\theta = 0°$ now corresponds to the perfect crystal while, for some arbitrary value of θ, a **symmetrical-twist** boundary (with three DOFs and no tilt component) is obtained (Fig. 1.35). This terminology emphasizes the similarity of Figs. 1.35(a) and (b) with the **asymmetrical-twist** boundary shown in Fig. 1.13(b).

As discussed in section 1.7.2, in crystals with inversion symmetry a twist rotation by $\theta = 180°$ (i.e. about the GB-plane normal, $\hat{n}_2 = \hat{n}_1$) inverts the stacking sequence on one side of the interface with respect to the other while preserving the perfect-crystal planar unit-cell dimensions (Figs. 1.29 and 1.36). Consequently, the symmetrical-**tilt**

boundary (STGB) thus obtained for $\theta = 180°$ (and $\hat{n}_2 = \hat{n}_1$) is fully determined by only the two DOFs associated with the GB plane, $\hat{n}_2 = \hat{n}_1$, and eqs. (1.29a) and (1.29b) are obtained, i.e.

$$\{\text{DOFs}\} = \{\hat{n}_1, \hat{n}_1, \theta = 180°\} \qquad (1.60a)$$

In crystals without inversion symmetry, a $180°$ twist rotation inverts only the Bravais planes but not the basis, resulting in an incomplete inversion of the crystal planes. More generally, the STGB may then be viewed as a perfect crystal ($\theta = 0°$) in which one half is turned upside down (i.e. $\hat{n}_2 = -\hat{n}_1$), according to

$$\{\text{DOFs}\} = \{\hat{n}_1, -\hat{n}_1, \theta = 0°\} \qquad (1.60b)$$

a characterization which does not require inversion symmetry.

As in the asymmetrical case, for any twist deviation from $\theta = 0°$ and $\theta = 180°$ in Figs. 1.36(a) and (b) the planar unit-cell area increases by introduction of screw dislocations. The crossed grid of screw dislocations is sketched schematically in Fig. 1.35(c). According to Frank's formula [10], the spacing, D, between these dislocations is given by

$$D = b/[2 \sin (\theta/2)] \approx b/\theta \qquad (1.61)$$

where b is the Burgers vector. The linearization in eq. (1.61) is valid only for **low-angle** twist boundaries; for larger twist angles this approximation is not possible, thus defining the regime of **high-angle** twist boundaries. Given eq. (1.61), the

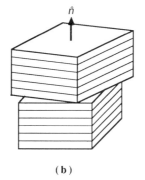

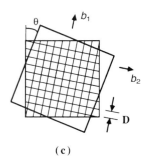

(a)   (b)   (c)

Fig. 1.35 (a) **Symmetrical-twist** boundary, obtained by rotation of two symmetrically aligned semicrystals ($\hat{n}_2 = \hat{n}_1$) about the common GB-plane normal, $\hat{n}$. The lattice planes shown in (b) illustrate the similarity with the **asymmetrical-twist** boundary, Fig. 1.31(b). (c) shows the crossed grid of screw dislocations characteristic for the structure of such an interface, with a spacing, D, given by eq. (1.61).

strain-field energy of the dislocation grid, with its well-known logarithmic cusp at $\theta = 0°$, was first derived by Read and Shockley [40] in terms of isotropic continuum-elasticity theory.

The increase in the planar unit-cell area of the GB as one rotates from the perfect-crystal or STGB configuration can be characterized by the inverse **planar** density of CSL sites, Γ. In contrast with Σ (which is the inverse **volume** density of CSL sites), the value of Γ indicates directly by how much the unit-cell area, $A(\theta)$, of a twist boundary exceeds that of the perfect-crystal or STGB configuration, $A(\theta = 0°) = A(\theta = 180°)$, on that plane because (eq. (1.28))

$$\begin{aligned}
A(\theta) &\equiv \Gamma(\theta)\, A(\theta = 0°) \\
&= \Gamma(180° - \theta)\, A(\theta = 180°) \\
&\equiv A(180° - \theta)
\end{aligned} \tag{1.62}$$

By definition,

$$\Gamma(\theta = 0°) = \Gamma(\theta = 180°) \equiv 1 \tag{1.63}$$

Equation (1.63) expresses the fact that the STGB and perfect-crystal configurations on a given plane are unique in that they share identical planar unit-cell dimensions, with an area which is the smallest possible for any planar defect on that plane (Fig. 1.36) [20]. In Chapter 3 of this volume [32] it will be shown that this unique geometry of the pure tilt configuration gives rise to a deep energy cusp at $\theta = 180°$ and hence to 'special' properties of STGBs. Most symmetrical-twist boundaries may therefore be viewed as 'vicinal' to either the STGB configuration on a given plane or to the perfect crystal; in the latter case, they are the well-known low-angle boundaries.

To relate the GB-plane-based picture of the STGB in Fig. 1.36 to the conventional CSL-based view, Fig. 1.37 shows the generation of the STGB structure in Fig. 1.36(b) by a pure tilt rotation. Figure 1.37(a) illustrates the orientation of the tilt axis, $\hat{n}_T$, perpendicular to the normal $\hat{n}$ of the GB; this normal is the same in the two semi-crystals. By contrast with the asymmetrical case in Fig. 1.32, the GB-plane normal and, therefore the tilt angle, ψ, are fixed by the condition of symmetry, leaving only two independent DOFs in the CSL characterization of these simple planar defects by the five variables in (eqs. (1.45) and (1.53))

$$\{DOFs\} = \{\hat{n}_T, \psi, \hat{n}\} \tag{1.64}$$

The fact that the CSL characterization of both STGBs and ATGBs is based equally on a set of five parameters (eqs. (1.63) and (1.64)) illustrates the difficulty in identifying a GB as either symmetrical or asymmetrical in this scheme.

In the conventional edge-on view of an STGB, the atom columns parallel to the tilt axis are usually connected (Fig. 1.37(c)). In Fig. 1.37(b) the same

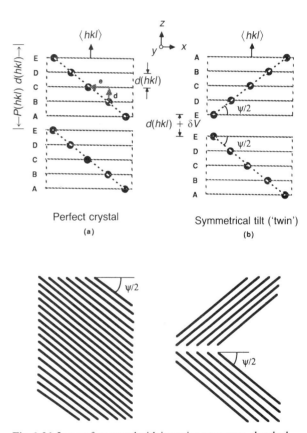

Fig. 1.36 In a perfect crystal with inversion symmetry, sketched schematically in (a) for a (hypothetical) five-plane stacking period, a rotation of one half about $\langle hkl \rangle$ generates the STGB configuration on the (hkl) plane sketched in (b). The latter is characterized by the familiar inversion of the stacking of the lattice planes at the interface ('twinning'), a feature in common to all STGBs. This inversion usually results in a volume expansion per unit GB area, δV. $d(hkl)$ and $P(hkl)$ are the interplanar spacing and repeat stacking period defined in section 1.7 (see eqs. (1.41) and (1.42)). The lower half represents an 'edge-on' view of the densest lattice directions, i.e. a view of atom columns parallel to the tilt axis. In this case, the tilt axis is parallel to the y direction; the tilt angle, ψ, is indicated.

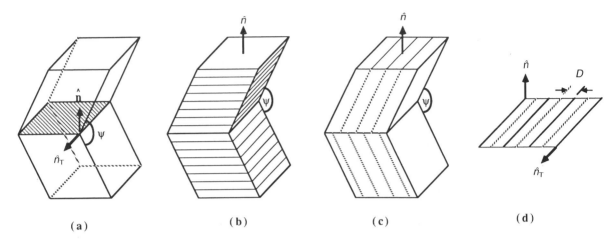

Fig. 1.37 Conventional CSL-based generation of the STGB structure in Fig. 1.36(b) by a pure tilt rotation. (a) Orientation of the tilt axis, $\hat{n}_T$, perpendicular to the normal, $\hat{n}$, of the GB; (b) lattice planes parallel to the GB illustrating the similarity with the **asymmetrical**-tilt boundary in Fig. 1.32(b). (c) shows the usual edge-on view down the tilt axis, in which the atom columns parallel to $\hat{n}_T$ are connected. (d) A view onto the GB plane shows two parallel sets of edge dislocations, with a spacing, D, given by eq. (1.61).

atoms are, instead, connected by lines parallel to the GB, making clear the connection with the STGB configuration in Fig. 1.36(b) and the ATGB in Fig. 1.32(b). Finally, the two sets of edge dislocations (one from each side of the interface) characteristic for this structure are sketched in the view onto the GB plane in Fig. 1.37(d). In contrast with the symmetrical-twist boundaries, the two sets of dislocations (identifiable, of course, only for small tilt angles) are parallel in this case. Interestingly, for strictly geometrical reasons in both cases the dislocation spacing is given by Frank's formula (1.61).

Finally, as an example we consider the characterization of the STGB configuration on the (332) plane of the fcc lattice, which belongs to the same $\Sigma = 11$ CSL system as one of the two ATGBs considered at the end of the preceding section. Within the interface-plane terminology, this boundary may be characterized in one of two ways. First, for $\hat{n}_2 = +\hat{n}_1$ the boundary would be characterized as a 180° twist boundary (which is permissible because the fcc lattice has inversion symmetry), according to (eqs. (1.24) and (1.60a))

$$\{\text{DOFs}\} = \{(332), (332), \theta = 180°\} \qquad (1.65a)$$

More generally, the boundary may be viewed as a

perfect crystal ($\theta = 0°$) in which one half is turned upside down (i.e. $\hat{n}_2 = -\hat{n}_1$), according to (eq. (1.60b))

$$\{\text{DOFs}\} = \{(332), (33\bar{2}), \theta = 0°\} \qquad (1.65b)$$

a characterization which does not utilize the inversion symmetry. In writing the last expression, we have taken into account the fact discussed in section 1.8.4 below that in a cubic crystal an arbitrary combination of Miller indices $(-h, -k, -l)$, can always be reduced, by a sequence of 90° rotations about $\langle 100 \rangle$ which transforms the crystal into itself, for example to a form $(h, k, -l)$, $(h, -k, l)$, or $(-h, k, l)$.

In the CSL-misorientation scheme, the same boundary would be characterized in terms of five parameters as follows (eqs. (1.45) and (1.53)) [41]:

$$\{\text{DOFs}\} = \{\langle 1\bar{1}0 \rangle, 50.58° (332)\} \quad (\Sigma = 11)$$
$$(1.66)$$

where the value of Σ is included in parentheses. The GB is readily identified as a **tilt** boundary (because $\hat{n}_{CSL} \perp \hat{n}_1$). However, a comparison with the asymmetrical boundary in eq. (1.58b) (which belongs to the same $\Sigma = 11$ CSL system) demonstrates the difficulties in identifying the GB as symmetrical or asymmetrical.

Finally, in the tilt-inclination scheme the boundary is characterized by (eq. (1.54))

$$\{\text{DOFs}\} = \{\langle 1\bar{1}0 \rangle, 50.58°, \alpha = 0°, \theta = 0°\}$$
$$(\Sigma = 11) \tag{1.67}$$

Again, as in the CSL scheme, the GB is readily identified as a pure **tilt** GB (from the fact that $\theta = 0°$), and its symmetry follows directly from the fact that $\alpha = 0°$. However, starting from these parameters, the identification of the crystallographic orientation of the GB plane is non-trivial.

We conclude this section by emphasizing an aspect of the conceptualization of symmetrical-tilt boundaries as special twist boundaries; this is of particular relevance to the computer simulation of GBs. In all GB simulations, in addition to the bicrystal at hand, a suitable undefected reference system containing the same number of atoms has to be considered in order to determine any excess quantity associated with the interface (such as the interface energy, thermal expansion, etc.). STGBs are conventionally generated by rotating two infinite, interpenetrating lattices with respect to one another, with the subsequent removal of corresponding atoms on both sides of the GB plane. This procedure gives rise to ambiguities regarding the undefected reference system with the same density and crystallographic orientation of the simulation cell as the bicrystal, rendering the determination of any excess quantities ambiguous.

1.8.4 Atomic-level geometry of symmetrical-tilt boundaries

From a strictly geometrical viewpoint, symmetrical-tilt boundaries are fascinating, yet little understood, objects. Since, like free surfaces and stacking faults, these simplest of all GBs have only two macroscopic DOFs (section 1.6.3), in this section we will discuss their atomic-level geometry in more detail.

As illustrated in Fig. 1.36(b), the atomic structure of an STBG is characterized by the familiar inversion of the stacking of the lattice planes at the interface ('twinning'), a feature in common to all STGBs. However, the directly inverted configuration in Fig. 1.36(b) is usually unstable because

two identical lattice planes are right on top of one other. We refer to this translational state, in which the GB-plane is a mirror plane but not an atom plane, as the **unstable** twin. Two energetically more favorable translational states are shown in Fig. 1.38. In the **special twin** in Fig. 1.38(b), the GB is both a mirror plane and an atom plane. In the fcc lattice, for example, there is only one special-twin configuration, namely the STGB on the (111) plane. With the ... |ABC| ... stacking of (111) planes (with a three-plane stacking sequence, $P(111) = 3$; section 1.7.1 and Table 1.1), the directly inverted, unstable-twin configuration would be characterized by ... |ABC|CBA| ..., while in the optimal translational state the C plane is shared by the two halves, according to ... |ABC|AB C BA|CBA| ....

On some arbitrary higher-index lattice plane, no physical reasons exist to favor the particular, high-symmetry special-twin configuration in Fig. 1.38(b). The STGB on such a plane usually exhibits some rather arbitrary rigid-body translation which is not generally a multiple of the in-plane stacking vector e in Fig. 1.28. We refer to such a translational state, sketched in Fig. 1.38(c), as a **general** twin, emphasizing the fact that its planar structure is still characterized by the inversion at the GB ('twin'), but that no special rigid-body translation exists.

Figure 1.39 illustrates that at least *three* planes are required in the repeat stacking sequence in the direction of $\hat{n}$, $P(\hat{n})$, for that plane to accommodate an STGB configuration. Consider, for example, the generation of an STGB on a set of lattice planes with only two planes in the repeat stacking sequences, such as the (100) and (110) planes in the fcc and bcc lattices. Starting with the perfect-crystal stacking ... |AB|AB| ... in Fig. 1.39(a), a 180° twist rotation yields the inverted configuration ... |AB|BA| ... (Fig. 1.39(b)). For reasons given above, this configuration of the STGB in which two planes are right on top of one another is usually unstable (Fig. 1.38(a)), giving rise to a rigid-body translation, $\boldsymbol{T} = (T_x, T_y)$, parallel to the interface. The obvious translation to minimize the mismatch across the interface is one in which B returns to A; such a translation, however, leads to

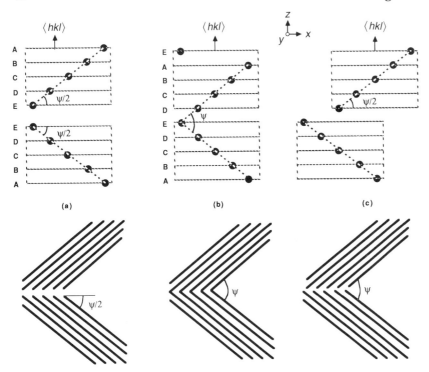

Fig. 1.38 Definition of three well-characterized translational states of an STGB. (a) Directly inverted (and usually unstable) configuration of Fig. 1.36(b) in which two identical (*hkl*) planes face each other across the interface. Notice that in this configuration the GB-plane is a twinning plane, but not an atom plane. (b) In the 'special' twin, the GB is both the twinning and an atom plane. (c) In most cases, an STGB exhibits some rather arbitrary rigid-body translation, referred to as a general twin, emphasizing the fact that its planar structure is still characterized by the inversion at the GB ('twin') but that no special rigid-body translation exists.

the re-establishment of the ideal-crystal stacking in the lower crystal, as illustrated in Fig. 1.39(c). Consequently, the 'STGB' on a set of lattice planes with two or only one plane in the repeat stacking sequence is identical to the ideal crystal.

With three planes in the repeat stacking sequence, the lowest-index plane in the fcc lattice which can actually accommodate an STGB is the (111) plane. Because of the three-fold symmetry axis of a unit stack of (111) planes discussed in section 1.7.2, a rotation by $60° \pm k*120°$ ($k = 0, 1, 2, \ldots$) about the $\langle 111 \rangle$ normal produces the well-known (111) twin boundary sketched in Fig. 1.40. (In the GB community, this geometrically most special of all STGBs in the fcc lattice is commonly known as the **coherent twin**; however, because in section 1.6.3 the concept of coherency was defined in a different sense, here we try to avoid this term.) All other STGBs in the fcc lattice involve at least a six-plane repeat stacking sequence (Table 1.1), and no other STGB has been observed to exist in the 'special-twin' configuration (in which the GB plane is a mirror plane).

For a more quantitative discussion of the geometry of STGBs, we now turn to cubic crystals in which all relevant geometrical parameters may be expressed explicitly in terms of the Miller indices, (*h*, *k*, *l*), associated with the GB plane [20].

Formally, according to eqs. (1.21) and (1.22), the interface characterized by eq. (1.60) has a vanishing tilt component because $[\hat{n}_1 \times \hat{n}_2] = 0$. As discussed in section 1.6.1, this apparent discrepancy originates from the fact that the symmetry relation, $\hat{n}_2 = \pm \hat{n}_1$ (eq. (1.25b)) does not cover all sets of crystallographically equivalent lattice planes (see eq. (1.30)). To formally assign a tilt component to a symmetrical interface, at least one non-collinear combination of $\hat{n}_1$ and $\hat{n}_2$ has to be found which characterizes the same set of crystallographically equivalent planes.

To illustrate the formal assignment of a tilt component to an STGB, we mention the well-known fact that in a cubic crystal an arbitrary combination of Miller indices ($\pm h$, $\pm k$, $\pm l$), can always be converted, by a sequence of 90° rotations about $\langle 100 \rangle$ which transforms the crystal into itself, for

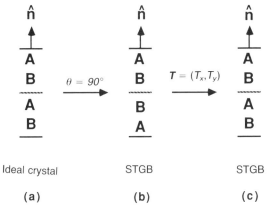

Fig. 1.39 The creation of an STGB on the (100) plane in the fcc or bcc lattice by a 90° twist rotation (for $m = 2$; section 1.7.2) illustrates that, upon suitable translation in (c), the STGB configuration in (b) becomes identical to the perfect crystal in (a).

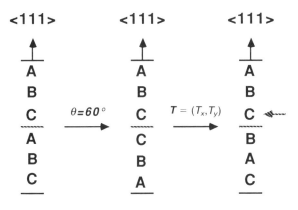

Fig. 1.40 Generation of the (111) twin boundary in the fcc lattice as a 60° twist boundary (with $m = 3$; section 1.7.2). In the interface-plane nomenclature, this STGB would thus be characterized by either (111) (111) 60° or (111) (11$\bar{1}$) 0° (eqs. (1.60a) and (b)). Its CSL-based characterization (eq. (1.64)) as a **twist** boundary would be ⟨111⟩ 60° (111); by contrast, as a **tilt** boundary it would be defined by ⟨110⟩ 70.53° (11$\bar{1}$) ($\Sigma = 3$).

example into a form $(h, k, \pm l)$, $(h, \pm k, l)$, or $(\pm h, k, l)$. Starting, for example, with $(-k, l, h)$, one can perform a 90° rotation about the y-axis (such that $y \to z, z \to -x, x \to z$) to obtain $(h, l, -k)$; next, a rotation about x by $-90°$ (such that $x \to x, z \to -y, y \to z$) finally yields the (h, k, l) configuration.

Let us now consider the case in which $\hat{n}_2 = -\hat{n}_1$, i.e. in Miller indices the interface is characterized by (eq. (1.60b))

$$\{\text{DOFs}\} = \{(h, k, l), (-h, -k, -l), \theta = 0°\} \tag{1.68}$$

Applying the cubic symmetry operations illustrated above, this expression may be rewritten, for example, as follows (Fig. 1.41):

$$\{\text{DOFs}\} = \{(h, k, l), (h, k, -l), \theta = 0°\} \tag{1.69}$$

where, in order to be specific, the z-direction was chosen as the symmetry direction. (Had we chosen, for example, the y-axis as the symmetry direction, $(h, k, -l)$ would be simply replaced by $(h, -k, l)$.) As illustrated in Fig. 1.41, the condition, $\hat{n}_2 = -\hat{n}_1$ may thus be equally written as $(\hat{n}_2)_z = -(\hat{n}_1)_z$.

To summarize, a non-collinear set of planes which is crystallographically equivalent to the collinear set for $\hat{n}_2 = -\hat{n}_1$ is given, for example, by the condition that $(\hat{n}_2)_z = -(\hat{n}_1)_z$. These expressions for $\hat{n}_1$ and $\hat{n}_2$ may be inserted into eqs. (1.21) and (1.22) to determine the tilt component of the interface, according to [21]

$$\hat{n}_{\text{T}} = (h^2 + k^2)^{-\frac{1}{2}} \begin{pmatrix} -k \\ h \\ 0 \end{pmatrix} \tag{1.70}$$

$$\sin \psi = 2 * l(h^2 + k^2)^{\frac{1}{2}}/(h^2 + k^2 + l^2) \tag{1.71}$$

i.e. the tilt axis and -angle are fully determined by the Miller indices associated with the STGB plane.

Given that an STGB has only *two* DOFs, but that $\hat{n}_{\text{T}}$ and Ψ represent three geometrical parameters, it is not surprising that the tilt axis and angle obtained from eqs. (1.70) and (1.71) are not unique, and other $(\hat{n}_{\text{T}}, \psi)$ combinations may be given for the same STGB on the (h, k, l) plane. The number of combinations is finite, however, because the number of crystallographically equivalent non-collinear combinations of Miller indices is finite. As an example, we consider the STGB on the (3, 4, 1) plane which, according to eqs. (1.70) and (1.71), may be considered as having been generated by a rotation about the ⟨−4, 3, 0⟩ axis by $\psi = 22.62°$. However, in deriving eqs. (1.70) and (1.71), we could have chosen, for

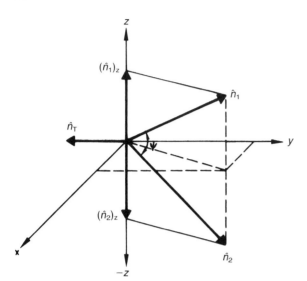

Fig. 1.41 Example for the formal assignment of a tilt axis and angle to a STGB. $\hat{n}_1$ and $\hat{n}_2$ characterize the GB-plane normals on the two sides of the interface in a cubic crystal in a fixed principal coordinate system (for example, with axes parallel to $\langle 100 \rangle$, as in Fig. 1.15(a)).

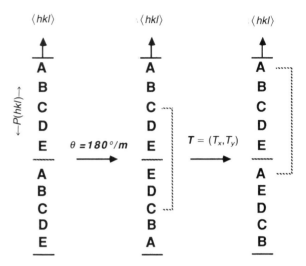

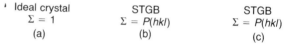

Fig. 1.42 Creation of an STGB on a plane with an odd (and hypothetical five-plane) stacking period, $P(hkl)$, starting with the perfect crystal in (a) As illustrated by the planes in (b) and (c) that are still in coincidence, even after the inversion of the stacking sequence, the value of Σ is given by $\Sigma = 2P(hkl)/2 = P(hkl)$. If the period were even, one would find, instead, that $\Sigma = P(hkl)/2$. [20].

example, the y direction in Fig. 1.41 as the symmetry direction, by writing $(\hat{n}_2)_y = -(\hat{n}_1)_y$; the tilt axis perpendicular to $\langle 3, 4, 1 \rangle$ then obtained would be the more familiar $\langle 1\bar{1}1 \rangle$ direction of this boundary, with a correspondingly different value of ψ, etc.

To express other geometrical properties of STGBs in terms of the Miller indices as well, we recall eqs. (1.41) and (1.42) which express the interplanar spacing, $d(hkl)$, and repeat stacking period, $P(hkl)$, of a perfect-crystal stack of planes in terms of (hkl). As seen from Fig. 1.42 for the case of an odd stacking period, the value of Σ of the STGB configuration is simply given by $P(hkl)$ since its creation involves simply the inversion of the planar stacking. While the two perfect-crystal stacks in Fig. 1.42(a) are in perfect coincidence with one another ($\Sigma = 1$), in the unstable STGB configuration in Fig. 1.42(b) the C planes remain in perfect registry while all other planes are out of coincidence, i.e. upon inversion of one of the two stacks, out of $2P(hkl)$ planes only two remain in perfect registry, and $\Sigma = 2P(hkl)/2 = P(hkl)$. As seen from Fig. 1.42(c), a rigid-body translation by

a multiple of the stacking vector, e (Fig. 1.28), does not change that result. A similar investigation of a stack with an even period shows that four out of $2P(hkl)$ planes then remain in perfect registry; hence $\Sigma = 2P(hkl)/4 = P(hkl)/2$ in that case, ensuring tht Σ is always odd. To summarize, given eq. (1.42) we may write

$$\begin{aligned} \Sigma = \beta P(hkl) &= \beta\,\delta(h^2 + k^2 + l^2), \\ &\qquad (\beta = 1 \text{ or } 0.5;\ \delta = 1 \text{ or } 2) \\ &= \beta'(h^2 + k^2 + l^2), \\ &\qquad (\beta' = 1 \text{ or } 0.5) \qquad (1.72) \end{aligned}$$

Given the possible values for β and δ, at first sight one would expect the three values 0.5, 1, and 2 for $\beta' = \beta\delta$. However, the last value is not permitted because for $\delta = 2$ the period $P(hkl)$ is even, with a consequent value of $\beta = 0.5$, resulting in $\beta' = 1$.

Sometimes it may also be desirable to relate Σ to $d(hkl)$. Combining eqs. (1.72) and (1.41) we readily obtain

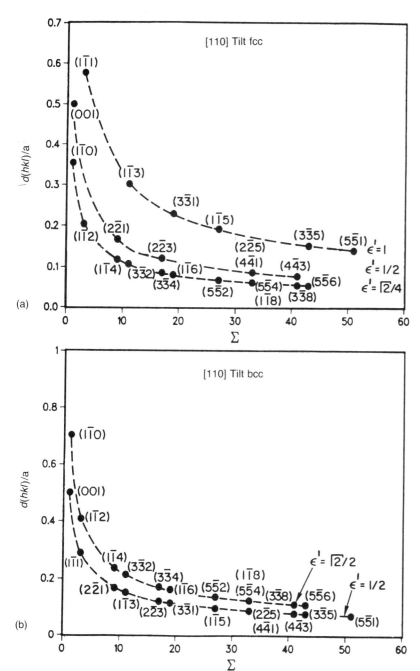

(a)

(b)

Fig. 1.43 Relationship according to eq. (1.74) between $\Sigma = \Sigma(hkl)$ and $d(hkl)$ for STGBs with a common $\langle 110 \rangle$ tilt axis in (a) the fcc and (b) the bcc lattice [20]. The figure demonstrates that a low value of Σ is necessary, but not sufficient, for a large value of $d(hkl)$. The value of $\varepsilon' = \varepsilon'(hkl)$ defined in eq. (1.74) also has to assume one of its largest values for $d(hkl)$ to be large.

$$\Sigma = \beta'\varepsilon^2 a^2 [d(hkl)]^{-2}, \quad (\beta', \varepsilon = 1 \text{ or } 0.5)$$
$$= \beta''a^2 [d(hkl)]^{-2},$$
$$(\beta'' = 1, 0.5, 0.25 \text{ or } 0.125) \qquad (1.73)$$

This relationship between $d(hkl)$ and the value of Σ of the STGB on that plane is illustrated in Figs. 1.43(a) and (b) for STGBs in the fcc and bcc lattices, respectively. Instead of plotting Σ as a function of $d(h, k, l)$, however, the figure shows the inverse relation (eq. (1.73))

$$d(h, k, l) = (\beta'')^{-\frac{1}{2}} \; a\Sigma^{-\frac{1}{2}} = \varepsilon' a\Sigma^{-\frac{1}{2}},$$
$$(\varepsilon' = 1, \sqrt{2}/2, 0.5, \text{ or } \sqrt{2}/4) \quad (1.74)$$

Finally, for a Bravais lattice with a basis of one atom (with atomic volume Ω), the planar unit-cell area of the STGB and perfect-crystal configurations on the (h, k, l) plane is given by (eqs. (1.41) and (1.43))

$$A(h, k, l) = \Omega/d(h, k, l)$$
$$= (\Omega/a)\varepsilon^{-1}(h^2 + k^2 + l^2)^{-\frac{1}{2}},$$
$$(\varepsilon = 0.5 \text{ or } 1) \quad (1.75)$$

For example, with $\Omega = a^3/4$ and $\Omega = a^3/2$ for the fcc and bcc lattices, respectively, the areal units Ω/a in eq. (1.75) are given by $a^2/4$ and $a^2/2$, respectively. Equation (1.75) is readily extended to twist angles other than $\theta = 0°$ and $180°$. Starting from the planar density of CSL sites for a given twist angle, $\Gamma(\theta)$, eq. (1.62) yields

$$A(\theta, h, k, l) = \Gamma(\theta) A(h, k, l)$$
$$= \Gamma(\theta)(\Omega/a)\varepsilon^{-1}(h^2 + k^2 + l^2)^{-\frac{1}{2}},$$
$$(\varepsilon = 0.5 \text{ or } 1) \quad (1.76)$$

We conclude by summarizing the following properties of symmetrical grain boundaries in cubic crystals.

1. Without loss of generality, *all* symmetrical boundaries in cubic Bravais lattices may be characterized as follows:

$$\text{DOFs} = \left\{ \hat{n}_1 = (h^2 + k^2 + l^2)^{-\frac{1}{2}} \begin{pmatrix} h \\ k \\ l \end{pmatrix}, \right.$$
$$\left. n_2 = (h^2 + k^2 + l^2)^{-\frac{1}{2}} \begin{pmatrix} h \\ k \\ \pm l \end{pmatrix}, \theta \right\} \quad (1.77)$$

Here the $+$ sign defines the set of symmetrical-**twist** boundaries in which the symmetrical-**tilt** boundaries are included for $\theta = 180°$. The $-$ sign is useful only if one wishes to formally assign a tilt component to the interfaces.

2. In the related three-parameter phase space, the STGBs are geometrically unique in that (a) they represent a special subset of twist boundaries, and (b) they represent the GBs with the smallest

planar unit-cell dimensions (identical to those of the perfect crystal). This unique geometry of the pure tilt configuration on a given plane gives rise to a deep energy cusp at $\theta = 180°$, and hence to 'special' properties of STBGs [33]. Most twist boundaries may therefore be viewed as 'vicinal' to either the STGB configuration on a given plane or to the perfect crystal.

3. Similar to free surfaces, the geometry of the tilt boundaries may be expressed entirely in terms of the two DOFs associated with the Miller indices of the GB plane. The three geometrical parameters in the tilt axis and angle are therefore not unique.

4. The value of Σ, the inverse volume density of CSL sites, is governed by the number of planes in the repeat stacking sequence, $P(hkl)$, which in turn is given by the Miller indices of the GB plane.

5. At least three planes are required in the repeat stacking sequence, $P(hkl)$, for a plane to accommodate a non-trivial STGB configuration. The 'STGB' on a set of lattice planes with $P(hkl) = 2$ or 1 is identical to the ideal crystal.

1.9 CHARACTERIZATION OF THE ATOMIC STRUCTURE OF INTERFACES

Throughout this chapter we have been concerned with the (macroscopic and atomic-level) **geometry** of solid interfaces, which we distinguish fundamentally from their **atomic structure**. The investigation of 'structure'–property correlations for interface materials usually involves both of these aspects of the interface structure, geometrical and physical, although the ultimate goal remains to correlate physical properties with the five macroscopic DOFs. To illustrate this distinction between the (crystallography-based) geometry and (physics-based) atomic structure of interfaces, in this final section we briefly review a few concepts that have evolved for the characterization of the atomic structure of solid interfaces.

The usefulness of any model for the atomic structure of solid interfaces should be assessed in terms of its ability to predict physical properties of the interface, in addition to providing a good description of its crystallography and/or atomic structure. The dislocation model of Read and Shockley [40], which predicts the structure, energy, mobility, and other properties of low-angle GBs in terms of a quantitative characterization of the atomic structure (via Frank's formula (1.61), is an excellent example for such a model.

The most pronounced structural feature of solid interfaces is the atomic-level disorder near the interface. This type of disorder is well characterized in terms of the radial distribution function, $G(r)$, or its Fourier transform, $S(k)$. As is well known, thermal disorder in an otherwise perfect crystal gives rise to two effects in $G(r)$: First, the δ-function-like zero-temperature peaks associated with the shells of nearest, second-nearest and more distant neighbors are broadened; second, because of the volume expansion, the peak centers are shifted towards larger distances.

Owing to the presence of planar defects, interface materials are structurally disordered even at zero temperature. However, because of its localization near the interface, this type of disorder is **inhomogeneous** – in contrast with thermal disorder [42]. To illustrate this inhomogeneity in the direction of the interface normal, it is useful to consider the radial distribution function for each of the atom planes near the interface (or, conversely, the planar structure factor). As seen from Fig. 1.44 for the case of a symmetrical (100) twist boundary in the fcc lattice, the amount of structural disorder decreases very rapidly from one (100) plane to another, indicating the existence of large gradients in many properties. Interestingly, the **structural** disorder at the interface gives rise to the same two effects even at zero temperature, as does **thermal** disorder in an otherwise perfect crystal, namely, a broadening of the peaks combined with a shift towards larger distances [42]. This shift, originating from the volume expansion near the interface, is seen from the shift of the arrows in Figs. 1.44(a)–(c), with the open arrows marking the average peak position in the bicrystal while the solid ones mark the corresponding perfect-crystal peaks.

The type of detailed atomic-level information contained in Fig. 1.44 is difficult to obtain experi-

mentally. Various models have therefore been proposed to describe the atomic structure at the interface without having to know the atom positions too precisely. In one such group of models, summarily known as **polyhedral-** or **structural-unit** models and primarily applied to GBs [41, 43–46], the atomic structure is described in terms of the stacking of polyhedra along the interface (Fig. 1.45; [47]). The requirements of space filling at the interface, and of the compatibility of the structural units with the adjoining grains, generally cannot be satisfied simultaneously, however, unless the polyhedra are elastically distorted. Hence, while the idea to describe the structure of GBs in terms of a few basic polyhedra is interesting, the systematic distortions of the polyhedra from one GB to another make it difficult to quantify the structure.

In a second group of models, known as **hard-sphere** models, the optimum translation parallel to the interface plane is assumed to be the one which minimizes the volume expansion at the GB [41, 48–50]. Although based on **unrelaxed** atomic structures, via the volume expansion these models provide at least a rough quantitative measure of the degree of structural disorder at the interface.

As discussed in detail elsewhere [29], both groups of models may be combined via the characterization of the atomic-level disorder in terms of the number of broken nearest, second-nearest and higher-order neighboring bonds. Such a quantified formulation of these models makes them now available for a systematic investigation of atomic-level structure–property correlations [29]. While such broken-bond models have been very successful for over 50 years in predicting physical properties of free surfaces from their relaxed or unrelaxed atomic structure [28], their application to solid interfaces is not yet very widespread.

In a broken-bond model the atomic structure is characterized quantitatively, for example, via the number of broken ν-th nearest-neighbor (nn) bonds per unit interface area, given by [29]

$$C(\nu) = \Sigma_n |K_{id}(\nu) - K_n(\nu)|/A \qquad (1.78)$$

where A is the planar unit-cell area. $K_n(\nu)$ denotes the number of ν-th nearest neighbors of atom n, while $K_{id}(\nu)$ is the related perfect-crystal value.

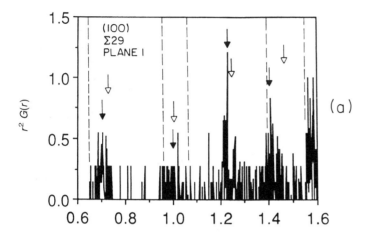

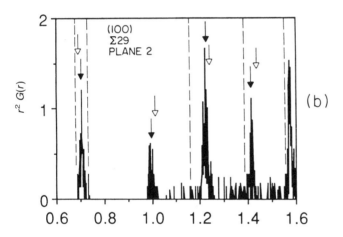

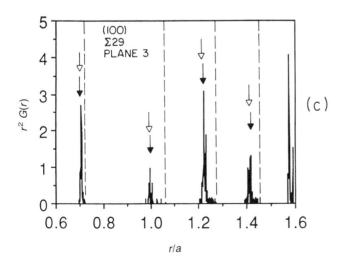

Fig. 1.44 Radial distribution function, $r^2G(r)$, for the three lattice planes nearest to the symmetrical (100) $\theta = 43.60°$ ($\Sigma = 29$) twist boundary in the fcc lattice. Solid arrows indicate the corresponding perfect-crystal peak positions; open arrows show the average value of r in a given atomic shell. The widths of these shells are indicated by dashed lines [51].

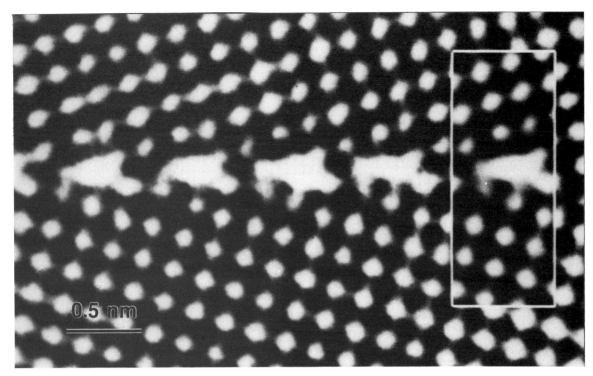

Fig. 1.45 HREM picture of the structural units in the $\langle 001 \rangle$ $\psi = 36.87°$ (310) ($\Sigma = 5$) STGB in NiO [47]. (In the interface-plane scheme, this GB would be referred to as the STGB on the (310) plane, (310) ($3\bar{1}0$) 0°; eq. (1.60)b.) The inset shows the image averaged over many structural units [47]. (Courtesy of K. L. Merkle).

Somewhat arbitrarily, the nearest neighbors are defined to include all those atoms situated between zero distance and the half-way point, $R_{12} = (R_1^{id} + R_2^{id})/2$, between the nearest and 2nd-nearest neighbors in the perfect crystal. $C(v = 1)$ is thus closely connected with the area under the nn peak in the radial distribution function. The 'miscoordination coefficient' $C(1)$ thus provides a convenient quantitative measure of how well the atoms are, on average, coordinated in the fully relaxed structure. $C(1)$ is usually given in units of a^2 [29].

This definition of broken nn bonds corresponds to the replacement of the overall plane-by-plane radial distribution functions near the GB, for example, in Figs. 1.44(a)–(c) by the integral under the nearest-neighbor peak in $G(r)$, with a subsequent summation over all the lattice planes near the interface. Because all the information contained in the detailed shape of the nn peak is thus lost, elastic strain-field effects associated with interface dislocations or surface steps are therefore

not included in $C(1)$, as is also the case for the polyhedral-unit and hard-sphere models. While in the case of the free surfaces this may not be a very severe limitation, in low-angle GBs the strain-fields associated with interface dislocations may dominate the behavior near the interface, thus severely limiting the use of these models. (For further details see [27] and [29].)

ACKNOWLEDGMENTS

I have benefited from many stimulating discussions with J. A. Jaszczak, K. L. Merkle, S. R. Phillpot, C. L. Wiley and S. Yip. I am particularly grateful to K. L. Merkle for making available copies of Figs. 1.7, 1.34 and 1.45, and to C. B. Carter for kindly providing Fig. 1.8. This work was supported by the US Department of Energy, BES Materials Sciences, under Contract No. W-31-109-Eng-38.

REFERENCES

1. See also D. Wolf and S. Yip, Interfaces. *MRS Bulletin*, **XV**(9) (Sept. 1990), 21.
2. D. N. Seidman, *Experimental Investigation of Solid Interfaces*, Chapter 2 in this volume.
3. W. Bailey, V. A. Frank-Kamenetskii, S. Goldsztaub, A. Kato, A. Pabst, H. Schulz, H. F. W. Taylor, M. Fleischer, and A. J. C. Wilson, *Acta Cryst.* **A33** (1977), 681.
4. See, for example, R. J. Needs, *Phys. Rev. Lett.*, **58** (1987) 53, and references therein.
5. See, for example, D. Wolf, *Surf. Sci.*, **225** (1990) 117, and references therein.
6. A. I. Jankowski and T. Tsakalakos, *J. Phys.*, **F15** (1985), 1279.
7. D. Wolf and J. F. Lutsko, *Phys. Rev. Lett.*, **60** (1988), 1170.
8. D. Wolf, *Appl. Phys. Lett.*, **58** (1991), 2081.
9. See also D. Wolf and J. Jaszczak, *Computer Simulation of the Elastic Behavior of Thin Films and Superlattices*, Chapter 14 in this volume, and references therein.
10. J. W. Christian, *Transformations in Metals and Alloys – Part I. Equilibrium and General Kinetic Theory*, Pergamon Press, Oxford (1975).
11. Y. Gao, K. L. Merkle, H. L. M. Chang, T. J. Zhang, and D. J. Lam, *Mat. Res. Soc. Symp. Proc.*, **221** (1991), 59, and *Phil. Mag. A*, (1992), in press.
12. C. B. Carter and H. Schmalzried, *Phil. Mag.* **A52** (1985), 207; C. B. Carter, E. G. Colgan, S. McKernan, Y. K. Simpson, S. R. Summerfelt, D. W. Susnitzky, and L. A. Tietz, *J. Phys. Colloque* **C5**(49) (1988), C5–239.
13. W. A. Jesser and J. H. van der Merwe, in *Dislocations in Solids*, Vol. 8 (ed F. R. N. Nabarro), North-Holland (1989), p. 421.
14. J. H. van der Merwe and W. A. Jesser, *J. Appl. Phys.*, **63** (1988) 1509; W. A. Jesser and J. H. van der Merwe, ibid., p. 1928.
15. G. B. Olson and Morris Cohen, *Acta Metall.*, **27** (1979) 1907.
16. D. G. Brandon, B. Ralph, S. Ranganathan, and M. S. Wald, *Acta Metall.*, **12** (1964), 813.
17. B. Chalmers and H. Gleiter, *Phil. Mag.*, **23** (1971), 1541.
18. See, for example, C. Goux, *Can. Metal. Quarterly*, **13** (1974), 9.
19. W. Bollmann, *Crystal Defects and Crystalline Interfaces*, Springer-Verlag, New York (1970).
20. D. Wolf, *J. Phys. Colloque*, **C4**(46) (1984), C4–197.
21. D. Wolf and J. F. Lutsko, *Z. Kristallographie*, **189** (1989), 239.
22. See, for example, B. D. Cullity, *Elements of X-Ray Diffraction*, Addison-Wesley, Reading, MA (1967).

23. See, for example, W. D. Kingery, H. K. Bowen, and D. R. Uhlmann, *Introduction to Ceramics*, 2nd edn, Wiley, New York (1976), p. 107.
24. See, for example, H. B. Aaron and G. F. Bolling, in *Grain Boundary Structure and Properties* (eds G. A. Chadwick and D. A. Smith), Academic Press, New York (1976), p. 107.
25. D. A. Smith, C. M. F. Rae and C. R. M. Grovenor, in *Grain-Boundary Structure and Kinetics*, ASM, Metals Park (1980), p. 337.
26. G. H. Bishop, R. J. Harrison, T. Kwok and S. Yip, *J. Appl. Phys.*, **53** (1982), 5596.
27. D. Wolf and J. Jaszczak, *Role of Interfacial Dislocations and Surface Steps* in the *Work of Adhesion*, Chapter 26 in this volume.
28. See, for example, C. Herring, in *Structure and Properties of Solid Surfaces* (eds R. Gomer and C. S. Smith), University of Chicago Press (1953), p. 4, and references therein.
29. D. Wolf, *J. Appl. Phys.*, **68** (1990), 3221.
30. See, for example, P. G. Shewman and W. M. Robertson, in *Structure and Properties of Solid Surfaces* (eds R. Gomer and C. S. Smith), University of Chicago Press (1953), p. 67.
31. D. Wolf, *Surf. Sci.*, **226** (1990), 389.
32. D. Wolf and K. L. Merkle, *Correlation between the Structure and Energy of Grain Boundaries in Metals*, Chapter 3 in this volume.
33. D. Wolf, *J. Mater. Res.*, **5** (1990), 1708.
34. See, for example, V. Randle, B. Ralph, and D. Dingley, *Acta Metall.*, **36** (1988), 2753; V. Randle and D. Dingley, *Scripta Metall.*, **23** (1989), 1565.
35. H. Ichinose and Y. Ishida, *J. Physique Colloque*, **C4**(46) (1985), C4–39.
36. K. L. Merkle, Proc. 46th Ann. Meeting of the Electron Microscopy Soc. of America, (1988), p. 588; *MRS Symp. Proc.*, **153**(83) (1989); *Colloque de Phys.*, **51** (1990), C1–251.
37. K. L. Merkle and D. Wolf, *Phil. Mag. A*, **65** (1992), 513.
38. K. L. Merkle and D. J. Smith, *Phys. Rev. Lett.*, **39** (1987), 2887, and *Ultramicroscopy*, **22** (1987), 57.
39. C. B. Carter, *Acta Metall.*, **36** (1988), 2753.
40. W. T. Read and W. Shockley, *Phys. Rev.*, **78** (1950), 275; see also W. T. Read and W. Shockley in *Imperfections in Nearly Perfect Crystals* (eds W. Shockley, J. H. Hollomon, R. Maurer and F. Seitz), Wiley, New York (1952), p. 352.
41. See, for example, H. J. Frost, M. F. Ashby and F. Spaepen, *A Catalogue of [100], [110], and [111] Symmetric Tilt Boundaries in Face-Centered Cubic Hard-Sphere Crystals*, Harvard University Press (1982).
42. S. R. Phillpot, D. Wolf and S. Yip, *MRS Bulletin*, **XV**(10) (Oct. 1990), 38.
43. M. Weins, B. Chalmers, H. Gleiter, and M. F. Ashby, *Scripta Metall.*, **3** (1969), 601.

44. M. Weins, H. Gleiter, and B. Chalmers, *Scripta Metall*, **4** (1970), 732, and *J. Appl. Phys.*, **42** (1971), 2639.

45. D. A. Smith, V. Vitek, and R. C. Pond, *Acta Metall.*, **25** (1977), 475 and ibid., **27** (1978), 235.

46. H. Gleiter, in *Atomistics of Fracture* (eds R. M. Latanision and J. R. Pickens), Plenum, New York (1983), p. 433.

47. K. L. Merkle, J. F. Reddy, C. L. Wiley and D. J. Smith, *J. Phys. Colloque*, **C5**(49) (1988), C5–251.

48. M. F. Ashby, F. Spaepen, and S. Williams, *Acta Metall.*, **26** (1978), 1647.

49. H. J. Frost, M. F. Ashby and F. Spaepen, *Scripta Metall.*, **14** (1980), 1051.

50. M. Koiwa, H. Seyazaki and T. Ogura, *Acta Metall.*, **32** (1984), 171.

51. D. Wolf and J. F. Lutsko, *Phys. Rev. Letters*, **60** (1988), 1170.

CHAPTER TWO

Experimental investigation of internal interfaces in solids

D. N. Seidman

Introduction · Buried internal interfaces · Chemistry of internal interfaces · Outlook and future promise

2.1 INTRODUCTION

If an internal interface is in thermodynamic equilibrium with the two crystals that adjoin the interface to make a bicrystal, then there is a local phase rule which describes the possible phase equilibria at this interface. This local phase was recently derived by Cahn [1], on the assumption that the five macroscopic degrees of freedom (DOFs) (see below, p. 59) of an internal interface are thermodynamic state variables, and also subject to the condition that the microscopic variables (see below, p. 66) of an interface are relaxed to their lowest energy values. It is also assumed that an interface behaves as an open thermodynamic system, i.e. it is free to exchange matter and energy with the adjoining grains. For a bicrystal the temperature and chemical potentials are constant throughout the system [2–4]. Therefore, for an internal interface the total number of state variables is six plus the number of chemical components; that is, the five macroscopic state variables, temperature, either one chemical potential for each of the components or the bulk composition and pressure. For example, for a binary single phase bulk alloy an eight dimensional hyperspace is required, and each point in this hyperspace represents a state of an internal interface. For the case of a single phase bulk alloy the local phase rule for an interface is given by:

$$F = C - P + 7 \qquad (2.1)$$

where F is the variance (number of DOFs), C is the number of chemical components, and P the number of nonequivalent phases at an internal interface. If there is more than one bulk phase present the local phase rule becomes:

$$F = C - (P + P_b) + 8 \qquad (2.2)$$

where P_b is the number of bulk phases. In eqs. (2.1) and (2.2) the five macroscopic geometric state variables are contained in the numbers 7 or 8. The latter implies that any consideration of phase equilibria at an interface requires a knowledge of these five geometric variables. For the case of an internal interface in a single component system it is possible for the interface to break up into a hill-and-valley type structure – i.e. to facet – and each facet can constitute a different phase.

Thus from an experimental point of view one must either fix these geometric variables or alternatively measure them. In addition, the microscopic variables must be determined, as one is often faced with the possibility that a metastable interface state is obtained rather than the true equilibrium structure or that the energies of different interface structures are degenerate. The microscopic variables can be described in terms of a rigid-body translation vector with components parallel and perpendicular to an interface plane. Next the atoms associated with an interface are relaxed in different directions which cannot be described by a simple rigid-body translations, thus one ultimately wants to know the vector displacement field of the individual atomic relaxations. Internal interfaces also contain a dislocation structure and this structure can play an important role in the phenomenon of solute-atom segregation at an interface. Therefore the three dimensional solute-atom distribution associated with an internal interface, on an atomic scale, should be measured to complete the characterization of an interface. In this chapter I systematically discuss the measurement of the basic physical parameters associated with an interface. First, in section 2.2 the structure of internal interfaces is discussed both in terms of macroscopic and microscopic DOFs; examples are given to illustrate the principles of each measurement. In section 2.3 the chemistry of internal interfaces is discussed with the emphasis being on atomic scale observations. In particular, it is shown how atom probe microscopy can be used to determine detailed atomic scale chemical information.

2.2 BURIED INTERNAL INTERFACES

2.2.1 Structure

The complete characterization of the structure of an internal interface in a solid requires obtaining many different types of information on different length scales on the same interface. First, it is necessary to determine the five macroscopic DOFs that fix the geometric relationship between the two crystals that reside to either side of an interfacial region. These five macroscopic DOFs determine the possible dislocation structures of an interface. Second, one therefore also wants to assess the dislocation structure of an interface, as many important physical properties, e.g. solute-atom

segregation, of a material are determined by its dislocation structure. For a general internal interface this should include the complete characterization of both the primary and secondary dislocations associated with it. Third, it is important to measure the three microscopic DOFs of an internal interface, as a bicrystal can decrease its Gibbs free energy because of local rigid-body translations both parallel and perpendicular to the plane of an interface. Fourth, one also wants to determine the final positions of **all** the atoms associated with an interface, as local atomic relaxations can result in a further reduction of the Gibbs free energy of a bicrystal; these atomic relaxations cannot be described by a single vector but rather each atom has a relaxation vector associated with it; and the complete specification of this displacement field requires that a complete set of vectors be specified for all the atoms in a bicrystal. To the best of my knowledge measurements of these quantities have not yet been performed completely on a single internal interface, although there are many measurements extant that deal with the characterization of the structure of internal interfaces to different degrees of completeness and on different length scales. From the above it is seen that the complete structural characterization of an internal interface represents a formidable experimental task.

Determination of the macroscopic DOFs of an internal interface

An internal interface is characterized by five macroscopic DOFs [5, 6]. The first two are specified by a unit vector c about which one block of crystalline material is rotated with respect to a second block of crystalline material (Fig. 2.1(a)). The third DOF is the rotation angle θ about c. Finally, the interface plane is characterized by an outward unit vector n that requires two additional DOFs, for a total of five macroscopic DOFs. (There is also a sixth macroscopic DOF that specifies the displacement of an interface plane parallel to itself. If this displacement is a true macroscopic displacement vector it does not result in any new information.) The unit vectors c and n need two direction cosines each to describe them, while θ is described by its

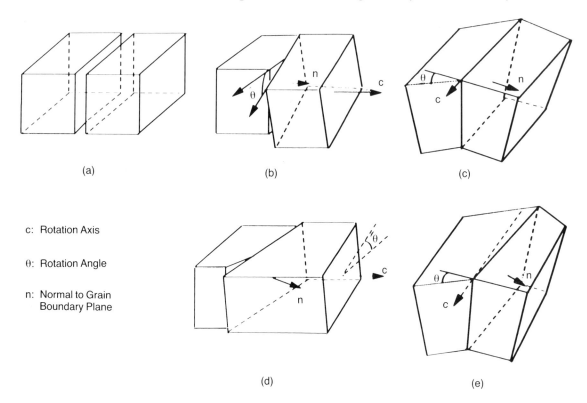

Fig. 2.1 The five macroscopic degrees of freedom of different types of interfaces. (a) two single crystal blocks of material at the same orientation. (b) A pure twist boundary (*c* is parallel to *n*). (c) A pure tilt boundary (*n* is perpendicular to *c*). (d) A predominantly twist boundary with a tilt component. (e) A predominantly tilt boundary with a twist component (J. G. Hu, Ph.D thesis, Northwestern University).

magnitude about *c*. The third direction cosine for *c* can be calculated from the equation:

$$c_1^2 + c_2^2 + c_3^2 = 1 \qquad (2.3)$$

A similar relation exists, of course, for the direction cosines of *n*. This is then the complete macroscopic description of an interface in terms of five macroscopic DOFs.

Figure 2.1 illustrates these five DOFs for a pure twist boundary (Fig. 2.1(b)); a pure tilt boundary (Fig. 2.1(c)); a twist boundary with a small tilt component (Fig. 2.1(d)); and a tilt boundary with a small twist component (Fig. 2.1(e)). A rotation axis/rotation angle pair (*c*/θ) defines the misorientation of one crystal with respect to a second crystal, or more concisely the misorientation of an internal interface in a bicrystal Alternatively, the

misorientation can be described by a value of Σ using the coincidence site lattice geometric description of interfaces; Σ is the ratio of the volume of the unit cell of the coincidence site lattice to the volume of the unit cell of the direct lattice [7]. In practice, however, a rotation matrix **R** is first determined experimentally and the *c*/θ pair is then calculated from **R**. The unit vector *n* is also determined experimentally, and methods for determining *c*/θ and *n* are presented below – that is, all five macroscopic DOFs.

The character of an internal interface is determined by the value of the dot product *c* · *n*. A pure twist boundary is obtained when *c* is parallel to *n* and therefore *c* · *n* = ±1 (Fig. 2.1(b)). Alternatively, a pure tilt boundary is obtained when *c* is perpendicular to *n* and therefore *c* · *n* = 0 (Fig.

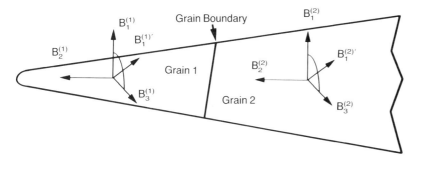

$$R = [B_1^{(2)}B_2^{(2)}B_3^{(2)}][B_1^{(1)}B_2^{(1)}B_3^{(1)}]^{-1}$$

Fig. 2.2 An internal interface in a field-ion microscope specimen. The vectors in each of the crystals are used to determine the experimental rotation matrix **R** from which the rotation axis (*c*) and rotation angle (θ) are readily calculated (J. G. Hu, Ph.D thesis, Northwestern University).

2.1(c)). A mixed boundary exists when $0 < |c \cdot n| < 1$; that is, when the angle between *c* and *n* is greater than 0° but less than 90°. Thus a mixed boundary has both tilt and twist components (Figs. 2.1(d) and (e)). The value of $|c \cdot n|$ determines whether a given internal interface is predominantly a twist or a tilt boundary.

For a cubic lattice, because of its high symmetry, there can be as many as 1152 equivalent *c*/θ descriptions of the same internal interface [8]. For example, a rotation of 60° about the [111] direction is equivalent to a rotation of 70.53° about the [110] direction. The so-called **disorientation** is the *c*/θ pair that yields the smallest value of θ whose *c* axis falls into the standard stereographic triangle: [100], [110], [111] [8]. This disorientation is the canonical form for the description of an interface in terms of a *c*/θ pair, although it is not always employed and therefore one must be cautious when reading the literature.

Let us now consider how one experimentally determines a *c*/θ pair for an internal interface. Transmission electron microscopy is the most widely used technique to characterize completely an interface. In our own research we have determined the crystallography of internal interfaces in field-ion microscope specimens, before the measurement of the chemical composition of the same internal interface via atom probe field-ion microscopy; the latter technique is described below (p. 76). The principles discussed here are general though the application is specific to the case of

an internal interface in a wire-shaped field-ion microscope specimen. Figure 2.2 exhibits a sketch of a wire-shaped field-ion microscope specimen containing a bicrystal separated by an interface; the various vectors used in the determination of an experimental rotation matrix **R** are also indicated. The vector $B_1^{(1)}$ is in the direction of an incident electron beam in the coordinate system of crystal 1 and vector $B_1^{(2)}$ is in the coordinate system of crystal 2 – the superscripts indicate crystal 1 or crystal 2. The $B_1^{(1)}$ or $B_1^{(2)}$ vectors are parallel to one another and are related by:

$$B_1^{(2)} = RB_1^{(1)} \tag{2.4}$$

The matrix **R** is a 3 × 3 matrix of direction cosines that accomplishes the transformation of a conformable vector or matrix from one frame of reference to a second one; here the transformation is between orthonormal coordinate system 1 in the two crystals.

Now let **G**1 be a 3 × 3 matrix whose three-column vectors are three orthonormal vectors in crystal 1 – **G**1 = $[B_1^{(1)}, B_2^{(1)}, B_3^{(1)}]$. Similarly, **G**2 is composed of three orthonormal vectors in crystal 2 – **G**2 = $[B_1^{(2)}, B_2^{(2)}, B_3^{(2)}]$ – that are parallel to those of **G**1. It follows from eq. (2.4) that:

$$G2 = RG1 \tag{2.5}$$

and therefore:

$$R = G2G1^{-1} \tag{2.6}$$

Because of the limited tilt range of a typical goniometer stage in a transmission electron microscope it is not possible to determine two beam directions that are 90° apart, e.g. $\boldsymbol{B}_1^{(i)}$ and $\boldsymbol{B}_3^{(i)}$ in Fig. 2.2, where $i = 1, 2$. In practice, therefore, two beam directions $\boldsymbol{B}_1^{(i)'}$ and $\boldsymbol{B}_1^{(i)}$ are determined, and $\boldsymbol{B}_2^{(i)}$ and $\boldsymbol{B}_3^{(i)}$ are calculated from the cross products $\boldsymbol{B}_2^{(i)} = \boldsymbol{B}_1^{(i)'} \times \boldsymbol{B}_1^{(i)}$ and $\boldsymbol{B}_3^{(i)} = \boldsymbol{B}_1^{(i)} \times \boldsymbol{B}_2^{(i)}$, as indicated in Fig. 2.2.

The full rotation matrix is given by:

$$\mathbf{R} = \begin{pmatrix} \mathbf{r}_{11} & \mathbf{r}_{12} & \mathbf{r}_{13} \\ \mathbf{r}_{21} & \mathbf{r}_{22} & \mathbf{r}_{23} \\ \mathbf{r}_{31} & \mathbf{r}_{32} & \mathbf{r}_{33} \end{pmatrix} \tag{2.7}$$

where the individual elements (r_{ik}) are given by:

$$\mathbf{r}_{ik} = \delta_{ik} \cos \theta + c_i c_k (1 - \cos \theta) + c_{ik} \sin \theta \tag{2.8}$$

$$c_{ik} = \begin{pmatrix} 0 & -c_3 & c_2 \\ c_3 & 0 & -c_1 \\ -c_2 & c_1 & \ddots \end{pmatrix} \tag{2.9}$$

where the c_i are the direction cosines of the unit vector c. Since experimentally one measures $\mathbf{R}$, as described below, we can calculate c and θ by inverting eq. (2.6); this yields:

$$\theta = \cos^{-1}[\tfrac{1}{2}(\mathbf{r}_{11} + \mathbf{r}_{22} + \mathbf{r}_{33} - 1)] \tag{2.10}$$

where $\mathbf{r}_{11} + \mathbf{r}_{22} + \mathbf{r}_{33}$ is the trace of $\mathbf{R}$. The c_i, except the case when $\theta = 180°$, are given by:

$$c_1 = \frac{\mathbf{r}_{32} - \mathbf{r}_{23}}{2 \sin \theta}$$

$$c_2 = \frac{\mathbf{r}_{13} - \mathbf{r}_{31}}{2 \sin \theta}$$

$$c_3 = \frac{\mathbf{r}_{21} - \mathbf{r}_{12}}{2 \sin \theta} \tag{2.11}$$

It is interesting that the rotation matrix $\mathbf{R}$ can be written as the sum of a diagonal matrix, a symmetrical matrix and an antisymmetrical matrix [9]:

$$\mathbf{R} = \cos \theta \begin{pmatrix} 1 & 0 & 0 \\ 0 & 1 & 0 \\ 0 & 0 & 1 \end{pmatrix}$$

$$+ (1 - \cos \theta) \begin{pmatrix} c_1 c_1 & c_1 c_2 & c_1 c_3 \\ c_2 c_1 & c_2 c_2 & c_2 c_3 \\ c_3 c_1 & c_3 c_2 & c_3 c_3 \end{pmatrix}$$

$$+ \sin \theta \begin{pmatrix} 0 & -c_3 & c_2 \\ c_3 & 0 & -c_1 \\ -c_2 & c_1 & 0 \end{pmatrix} \tag{2.12}$$

Figure 2.3(a) is a schematic drawing exhibiting three pairs of Kikuchi lines; each pair has a so-called excess line and a deficit line; the excess represents an enhancement over the background scattering, while the deficit line represents a diminution of the background scattering. The angular separation between an excess and a deficit line is twice the Bragg angle for the set of reflecting planes responsible for each pair of lines. The dashed line between a pair of Kikuchi lines is situated symmetrically with respect to the excess and deficit lines and is the intersection of the projection of the relevant reflecting plane with a phosphor screen. And the D_i spacing (D_1, D_2, D_3) is inversely proportional to the interplanar spacing of the reflection responsible for a given pair of Kikuchi lines. The angles ϕ_i in Fig. 2.3(a) are the angles between two nonparallel pairs of Kikuchi lines. Figure 2.3(b) is a three-dimensional drawing exhibiting the direction of the incident electron beam and three reflecting planes that cause the three pairs of Kikuchi lines shown in Fig. 2.3(b). The angles Ω_i in Fig. 2.3(b) are the angles between the direction of the incident beam and the vectors corresponding to the poles L, M, and N. Note that point B in Fig. 2.3 is the projection of B' and it is the direction of the incident electron beam – B in Fig. 2.4(b) and Fig. 2.5.

Three pairs of Kikuchi lines are needed to determine a beam direction. By performing careful tilting experiments it is possible to obtain a Kikuchi pattern with three pairs of lines present. In cases, however, where Kikuchi patterns have to be taken at certain positions – e.g. at the internal interface end-on position – three pairs of Kikuchi lines may not be present in a pattern. The use of computer-generated Kikuchi patterns circumvents this problem. Four Kikuchi patterns, two for each crystal, are required to determine the misorientation. In practice, however, three sets of Kikuchi patterns, whenever possible, are used to check for self consistency. The scatter between different sets of Kikuchi patterns is within ±0.5° and ±1.0° for θ and c respectively.

Thus the first step for determining the misorientation of a bicrystal is to determine the orientations of both crystals precisely. This is accomplished by

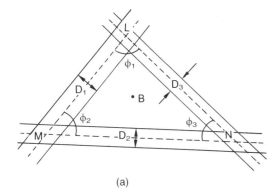

(a)

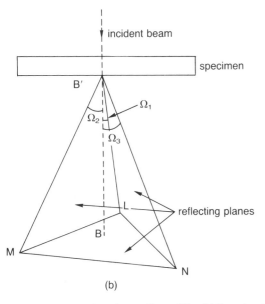

(b)

Fig. 2.3 (a) Three pairs of noncolinear Kikuchi lines for the three pole solution. The direction of the incident beam is indicated by the point B. And the angles ϕ_i are the angles between two nonparallel pairs of Kikuchi lines. The dashed line between a pair of Kikuchi lines is situated symmetrically with respect to the excess and deficit lines, and is the intersection of the projection of the relevant reflecting plane with the phosphor screen. (b) A three dimensional diagram illustrating how the three reflecting planes intersect the phosphor screen. The angles Ω_i are the angles between the direction of the incident electron beam and the poles L, M, N. The point B is the projection of B' and it is in the direction of the incident electron beam $-B$ in (a) (Courtesy of Ho Jang).

recording Kikuchi patterns of both crystals at the same tilt angle and then matching the experimentally obtained Kikuchi patterns with computer generated ones. Figure 2.3 exhibits an example of an experimental Kikuchi pattern compared with a computer-generated Kikuchi pattern; the specimen is a W-25 at.% Re bicrystal. The letter B shows the center of the Kikuchi pattern and represents the direction of the incident electron beam. The distances from B to three indexed poles are measured from a simulated pattern, and a beam direction is then calculated [10, 11].

Next an interface plane outward unit normal n is determined by tilting an interface to an end-on position, i.e. the specimen is tilted until the interface plane exhibits a minimum projected width on the phosphor screen of a transmission electron microscope. When this occurs the interface plane is parallel to the electron beam, and the internal interface normal is therefore perpendicular to both the beam direction B and a vector A in the internal interface plane, where A is determined from a point on the trace of an interface. This situation is illustrated in Fig. 2.5. The cross product $A \times B$ then yields the interface normal n; the unit vector n describes the inclination of an interface plane in terms of two direction cosines. For a symmetric interface the outward n's in each crystal are identical, while for an asymmetrical interface the outward n's are different. For wire field-ion microscope specimens it is always possible to tilt an internal interface to an end-on orientation, although it may involve rotating the specimen manually outside the transmission electron microscope using a pair of tweezers. Once an interface is in an end-on position an electron micrograph of the interface and Kikuchi patterns from both crystals are recorded without tilting the specimen. For a specimen in the form of a thin foil or a thin film the end-on method is not feasible, unless the angle between an interface plane and a specimen's surface is less than one half the range of the tilt angle of a transmission electron microscope goniometer stage. Otherwise the method is identical for wires, thin foils, or thin films.

Figures 2.6(a) and (b) exhibit transmission electron microscope micrographs of an interface

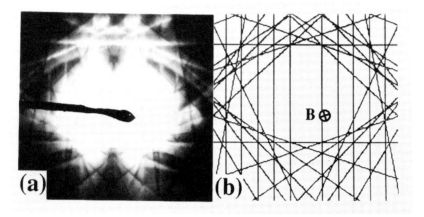

Fig. 2.4 The match of an experimental Kikuchi pattern with a computer generated Kikuchi pattern. (a) Kikuchi pattern of a W-25 at.% Re crystal obtained using an Hitachi H700 200 kV electron microscope. The center of the Kikuchi pattern was recorded using a double exposure. (b) A computer generated Kikuchi pattern (J. G. Hu, Ph.D thesis, Northwestern University).

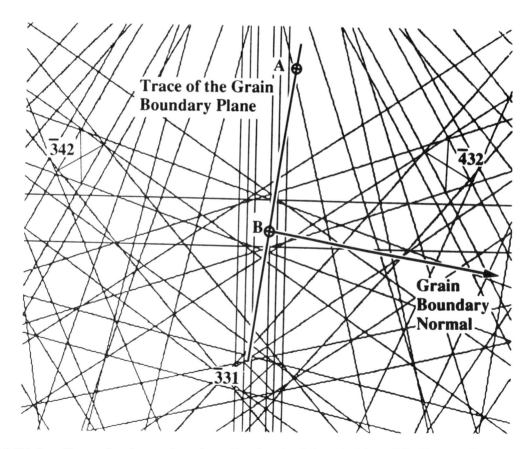

Fig. 2.5 This figure illustrates how the normal to an internal interface plane is determined from a Kikuchi pattern. The beam direction is determined by measuring the distances from **B** to three poles. The internal interface normal is given by the cross-product $A \times B$ (J. G. Hu, Ph.D thesis, Northwestern University).

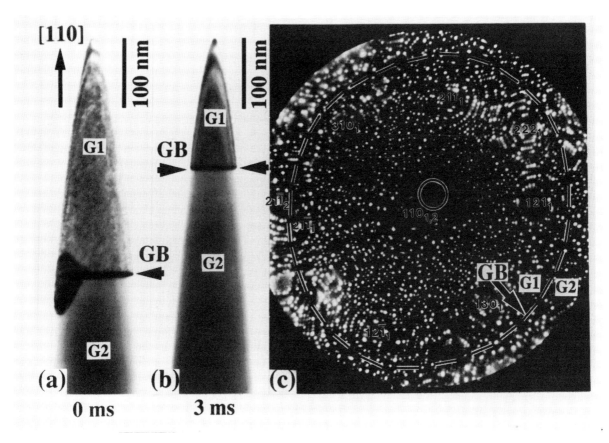

Fig. 2.6 (a) A transmission electron microscope electron micrography of an internal interface in a W-25 at.% Re field-ion microscope tip with the individual crystals denoted G1 and G2. (b) A transmission electron microscope micrograph of the same specimen as in (a) after 3 ms of electropolishing – note that the interface is now closer to the tip. (c) A field-ion microscope micrograph of the same specimen as in (a) and (b) at an imaging temperature of 30 K (J. G. Hu, Ph.D thesis, Northwestern University).

in a field-ion microscope specimen whose five DOFs were determined employing the procedure described above; the individual crystals are labeled G1 and G2. A field-ion microscope micrograph of the same specimen is shown in Fig. 2.6(c). The circular region on the periphery of the image labeled GB shows the intersection of the interface with the approximately hemispherical field-ion microscope tip. The projection of the probe hole of our atom probe field-ion microscope is the small circular region near the $110_{1,2}$ pole; below (p. 76). The presence of an interface in a field-ion microscope is detected by searching for broken symmetries, as a specimen without an internal interface

exhibits a perfectly symmetrical pattern. Some of the major poles in the two crystals are indexed, and the crystal to which they belong is denoted by a subscript 1 or 2. Note that the $110_{1,2}$ pole in the center of the image is common to both crystals G1 and G2. Figure 2.7 is a field-ion microscope micrograph of a W-25 at.% Re alloy that does not contain an interface and therefore exhibits the perfect symmetry of the $2\,mm$ crystallographic point group. Compare Figs. 2.6 and 2.7, and note that in both cases the 110 pole is in the center of each micrograph; also note that in Fig. 2.6 the symmetry is broken on the periphery of this micrograph due to the interface. The field-ion micro-

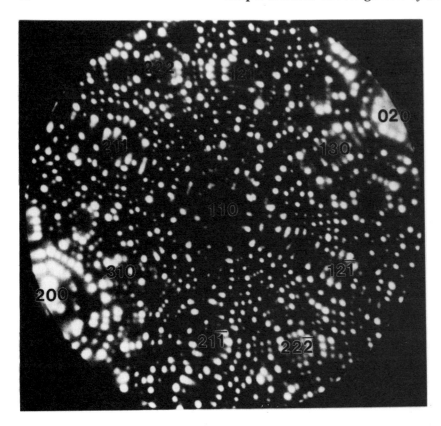

Fig. 2.7 A field-ion micrograph of a W-25 at.% Re alloy imaged at 30 K. The major poles are indexed. A two fold vertical rotation axis through the center of this pattern is apparent. The crystallographic point group of this 110 field-ion image is 2 *mm* (J. G. Hu, Ph.D thesis, Northwestern University).

scope images are used as a check on the transmission electron microscope determination of c/θ and n, but the most accurate measurement of these three quantities is obtained via transmission electron microscopy.

We conclude this section by stating that well developed procedures now exist for determining the five macroscopic degrees of an interface by transmission electron microscopy. And it is essential that these five DOFs be experimentally determined, as they constitute state variables of an interface and are therefore important in specifying the thermodynamic state of an interface – section 2.1.

Determination of the microscopic translational DOFs of an internal interface

In the early 1970s it was demonstrated – employing molecular statics computer simulations at 0 K – that the internal energy of a bicrystal is lowered if a relative rigid-body translation, without rotation, of two crystal lattices in a coincidence site lattice orientation is made parallel to an interface plane [12–15]. Thus, the coincidence site lattices are usually **not** continuous across the plane of an internal interface. This rigid-body translation creates a stacking fault in the coincidence site lattice, as shown in Fig. 2.8. In general, a rigid-body translation vector (t) consists of two vectorial components [16]. When the possibility of this t vector is considered the number of DOFs of an internal interface is increased from five to eight. The three DOFs associated with t are microscopic DOFs, as opposed to the five macroscopic DOFs discussed above (p. 59). The first component of t is denoted p, and corresponds to the situation when the density of the perfect crystal is preserved; in this case the rigid-body movement of one crystal with respect to the other may be parallel or perpendicular to the interface plane; the perpendicular component to a rational interface plane must be equal to an integral number of interplanar spacings.

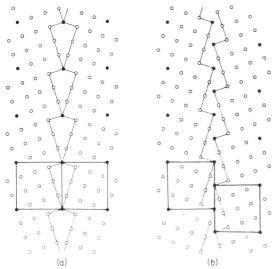

Fig. 2.8 (a) and (b) illustrate the effect of a rigid-body translation on the positions of coincidence sites at an internal interface. The coincidence sites are denoted by filled circles. (a) The coincidence-site lattice is continuous across the internal interface. (b) The two crystals have been rigidly translated with respect to one another parallel to the plane of the interface. Note that a stacking fault has been formed in the coincidence-site lattice at the crystal boundary (After J. W. Matthews and W. M. Stobbs).

The second component of t is denoted e and corresponds to a rigid-body translation perpendicular to the interface plane such that the density of the perfect crystal is not preserved, i.e. excess volume is created.

The crystallographic geometry of the first component p has been worked out in detail [9] based on the coincidence site lattice model. The geometric pattern produced by two interpenetrating lattices at a coincidence site lattice orientation is a periodic function of the so-called displacement shift complete vectors (d_{dsc}) of the coincidence site lattice, for the misorientation (Σ value) of concern. Thus identical geometric patterns are created by the displacements $p + d_{dsc}$, within the plane of the interface [17]. The reduced description of a translation t is denoted p', where p' is the vector with the smallest magnitude of all the equivalent descriptions $p + d_{dsc}$, i.e. the displacements that fall inside the Wigner–Seitz cell of the bicrystal lattice. Thus the rigid-body translation vector to be determined experimentally is written as:

$$t' = p' + e \qquad (2.13)$$

In principle both the rigid body translation p' and the internal interface excess volume component e can be determined experimentally using transmission electron microscopy.

There are many different transmission electron microscope techniques for determining the p' or e rigid-body translation vectors associated with an interface. They are as follows:

(a) displacement or α-fringe contrast;
(b) Moiré fringe contrast;
(c) high energy electron diffraction or high resolution transmission electron microscopy;
(d) convergent beam electron diffraction;
(e) Fresnel fringe contrast;
(f) interfacial dislocation structure.

In general one measures the vector t' and not the individual component vectors p' or e, although under certain conditions it is possible to measure both vectorial components. I now consider each of these techniques for determining t' and conclude with a general evaluation of the situation for determining microscopic variables of an internal interface.

Displacement or α-fringe contrast technique

Pond and Smith [18] were the first ones to show experimentally the existence of t', i.e. the presence of a microscopic DOF. They did this by observing stacking-fault like fringes at an annealing twin ($\Sigma = 3$) in high purity aluminum. The experimental basis of this technique is that for two crystals in a coincidence site lattice orientation there is an array of vectors (g_c) – in three dimensions – in the reciprocal lattice spaces that are common to the reciprocal lattice of the two crystals that are at an exact coincidence site lattice orientation. If a bicrystal is oriented such that only the reflections g_c are excited then the bicrystal should behave as though it is a single crystal, from a diffraction point-of-view, if the atomic planes are continuous across an interface, as shown in Fig. 2.8(a). If, however, a rigid-body translation is present, as illustrated in Fig. 2.8(b), then stacking-fault like scattering occurs at the boundary and fringes are present in a bright-field image of the interface. If only common reflections in the two crystals are excited then the fringes behave in similar manner to the fringes observed at stacking faults in single crystals. The periodicity of the fringes depends on the effective extinction distance for the common excited reflection. The intensity profile of the fringes depends on the deviation parameter, w, and the sign and magnitude of $\mathbf{g}_c \cdot \mathbf{t}'$. To measure the components of $\mathbf{t}'$ it is necessary to use three different and non-coplanar reflections, $\mathbf{g}_c$, and to compare microdensitometer traces of the experimental fringes with simulated intensity profiles. Figure 2.9 exhibits bright-field and dark-field electron microscope images, electron diffraction patterns, and experimental and simulated intensity profiles for an incoherent $\Sigma = 3/(1\bar{2}1)$ twin in an annealed aluminum specimen. Figures 2.9(a) and (b) are for the same $\mathbf{g}_c$ but different values of w; note the extremely strong differences in the intensity profiles between these two situations. Figure 2.9(c) is a dark field micrograph of the same image; note the strong asymmetry in the fringe contrast. (It is emphasized strongly that to find $\mathbf{t}'$ uniquely it is necessary to employ three independent $\mathbf{g}_c$.)

Pond [16] lists the experimental limitations of this technique to be:

1. The two crystals adjoining the interface must be very close to an exact coincidence site lattice orientation, otherwise the image contrast observed is affected by the presence of additional contrast effects from Moiré fringes, dislocations or so-called δ-fringes [19].
2. To have strong fringe intensity it is necessary to employ common reflections, $\mathbf{g}_c$, with low Miller indices; this is essential to match the experimental and simulated intensity profiles.
3. The excitation of noncommon beams must be avoided, as they can couple to the selected common beam and alter the image contrast [20]. The presence of, for example, thickness fringe contrast in addition to stacking fault contrast affects the determination of $\mathbf{g}_c \cdot \mathbf{t}'$ and therefore the determination of $\mathbf{t}'$ is affected [21].

Because of these experimental limitations the interfaces $\Sigma = 3$ and $\Sigma = 9$ are the ones for which Pond made a complete determination of $\mathbf{t}'$. Also, there are certain ambiguities that exist and these are discussed in detail by Pond [16]. However, it is possible to characterize these boundaries in reasonable detail. For example, Pond has shown that for the $(1\bar{2}1)$ incoherent $\Sigma = 3$ twin in Al that $\mathbf{g}_c \cdot \mathbf{t}' = -0.35 \pm 0.005$ for $\mathbf{g}_c = 111$. In addition, Pond showed that the component e has a magnitude of 0.02 nm.

Papon *et al.* [22] studied $\Sigma = 9/(1\bar{2}2)$ and $\Sigma = 11/(2\bar{3}3)$ symmetrical tilt boundaries in germanium – for both boundaries the tilt axis is [011] – using the α-fringe method. They found both boundaries to be dilated and to include a mirror glide symmetry; the latter precludes a rigid-body translation parallel to the interface. For the $\Sigma = 9$ symmetrical tilt boundary they determined $|e| = 0.010 \pm 0.004$ nm. For same $\Sigma = 9$ symmetrical tilt boundary they state that the dilatation is a few hundredths of a nanometer using $\mathbf{g} = [2\bar{2}4]$.

Moiré fringe contrast technique

Matthews and Stobbs [23] used Moiré fringe patterns, that form in electron microscope images when one crystal is placed on top of a different crystal, to find in an elegant manner $\mathbf{t}'$ [24]. The Moiré fringes are a result of the interference between the undeviated electron beam and one operating reflection in each of the two crystals involved. The particular coincidence, $\Sigma = 3$, interface they studied is the $\{112\}$ lateral twin boundary which is commonly found in $\{111\}$ epitaxial deposits of gold on MoS_2. To obtain these $\{112\}$ twins they evaporated gold on molybdenite (MoS_2) substrates at 200 or 250°C and studied the 'sandwich' of isolated three-dimensional gold islands on MoS_2; the incoherent $\{112\}$ twins form as a result of the coalescence of mobile three-dimensional gold islands. This sandwich corresponds to the case of two crystals differing only in lattice parameter, and therefore a series of Moiré fringes is obtained normal to an operating reflection vector $\mathbf{g}$ with a fringe spacing given by:

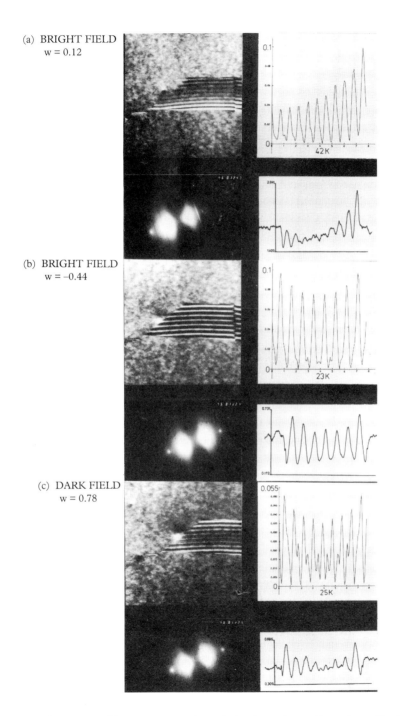

Fig. 2.9 Electron micrographs, diffraction patterns, theoretical and experimental intensity profiles for a $\Sigma = 3/(121)$ incoherent twin boundary in aluminum (Courtesy of R. C. Pond).

$$\Lambda = \frac{1}{1/d_1 - 1/d_2} \qquad (2.14)$$

where d_1 and d_2 are the relevant lattice spacings. The bright field images contain three sets of Moiré fringes that are a result of diffraction from the {220} gold planes and the {1120} MoS_2 planes.

First, they showed for the {112} lateral twin boundary in gold that the coincidence site lattice of this interface is not continuous across the boundary plane. This is consistent with early molecular statics computer simulations which showed that an internal interface can lower its energy, as a result of rigid-body shift parallel to the interface plane – see above (p. 66). Second, by measuring the offsets of the Moiré fringes at twin interfaces they extracted the component of t' parallel to the MoS_2 substrate. Matthews and Stobbs measured a displacement of 0.056 nm along the normal to the boundary plane. The sense of the displacement is such that the lattices of the two crystals moved towards one another. This can be considered to be the combination of the vector e plus the removal of one {224} plane at the interface.

Matthews and Stobbs obtained a bimodal distribution in the number of fringes displaced by an amount Λ, as a function of Λ, and they suggested that this corresponds to a real variation in the displacement of Moiré fringes. A possible explanation of this variation in Λ is that the value $\Sigma = 3$ is for an infinite crystal, whereas for finite crystals the density of coincidence sites in finite {112} boundaries may be equal to, larger, or smaller than the value $\Sigma = 3$ in infinite boundaries.

Matthews and Stobbs's method is extremely simple, as it does not require contrast calculations in order to extract t'. A question arises, however, as to whether or not the observed twin boundaries are fully relaxed. The answer to this question lies with whether or not the gold islands had enough mobility, after they coalesced, to reach a true minimum energy configuration. Another question concerns the effect of islands of unequal heights, that coalesce to form a boundary, on the spacings of the observed Moiré fringes. It would be interesting to study these same twins employing the α-fringe method (above, p. 67) to see if a similar result is obtained for t'.

High resolution electron microscopy

Phase contrast microscopy or high resolution microscopy has become an important technique for studying the atomic structure of internal interfaces. A crystalline specimen in the object plane of a magnetic lens has its corresponding diffraction pattern in the back focal plane of this lens. In so-called diffraction contrast microscopy an aperture is placed around the central undiffracted spot and a real space image is then observed in the image plane of the objective lens. The contrast in this so-called bright field image is a result of imperfections which locally change the periodicity of the perfect lattice. Alternatively, a dark field image may be formed by deflecting a diffracted beam in the back focal plane of the objective lens through an aperture in the same plane. Both bright and dark field images allow an experimentalist to readily observe individual dislocations and dislocations at internal interfaces.

Phase contrast or high resolution microscope images are formed by placing an aperture in the back focal plane of the objective lens which will accept the central undiffracted beam, as well as additional diffracted beams [25]. In this manner an image is formed which is a result of the interference of all these beams so that the phase of the diffracted beams contributes to the observed contrast, and not just their amplitude. If a specimen is not too thick – typically 20 to 40 nm – and certain electron optical conditions are met then the lattice images formed by phase contrast microscopy are interpretable [26–29]. The resolving power of modern high resolution microscopes is such that the individual atomic columns of all the diamond cubic semiconductors can now be imaged for the (100) and (111) zone axis.

High resolution electron microscopy has been employed to study rigid-body translations at twin boundaries in both germanium and silicon. Bourret and co-workers and Bacmann and co-workers [30–32] have performed careful studies of near-coincidence tilt boundaries over the past 15 years.

Their work is highly systematic with great attention given to obtaining the optimum experimental conditions under which the high resolution images are recorded. In addition, they also simulate the experimental images employing a multislice image simulation code [30, 31]. Their studies have become more and more refined, as the point-to-point resolution of commercial microscopes has decreased to 0.19 nm for a 200 kV instrument and better than 0.17 nm for a microscope operating at 300 kV and higher. In addition, in parallel with the experimental work they have performed molecular statics calculations to study the relative energies of different internal interface structures and also the electronic structure of interfaces [33].

In particular, a rigid-body translation can be determined from the positions of images of atoms in micrographs. The procedure is relatively simple and involves using a transparent grid which simulates the perfect lattice, at the same magnification as the micrograph containing an internal interface. A grid is then aligned with the atom positions in each crystal and an overlap of the two grids occurs due to the presence of an internal interface. The relative translation of the two crystals in the overlap zone is then determined to an accuracy of 0.020 nm. The components of the rigid-body translation have been measured for a $\Sigma = 3/\{211\}$ twin. The results of these investigations are summarized in Table 2.1; for the $\langle 211 \rangle$ direction the results are scattered because the $\langle 211 \rangle$ twins in germanium are found to contain defects.

Convergent beam electron diffraction

Convergent electron diffraction was discovered in 1939 by Kossel and Mollenstedt [34]. So-called convergent beam patterns are formed in an electron microscope by creating a highly convergent electron beam that is made to impinge on a specimen; the angle of convergence is essentially determined by the diameter of the second condenser aperture. The patterns formed in this manner are two-dimensional maps of the diffracted electron intensity, as a function of the angle between the incident beam and a crystallographic direction. A convergent beam pattern consists of a central disk

Table 2.1 Measured components of the rigid body translation vector for a $\Sigma = 3/\{211\}$ twin in silicon and germanium expressed in terms of the corresponding vector (after Bourret, Billard and Petit[a])

Reference	Material	$\langle 211 \rangle$	$\langle 111 \rangle$	$\langle 011 \rangle$
Fontaine and Smith[b]	Si	$+\frac{1}{18}$	$\frac{1}{9}$	0
Pond[c] and Vlachavas	Si	$+\frac{1}{100}$	$\frac{1}{8}$	0
Labidi and Rocher[d]	Si	$\frac{1}{18}$	$\frac{1}{8}$	0
Bacmann[e]	Ge	>0	?	0
Bourret, Billard and Petit[a]	Ge	$\frac{1}{35}$ to $\frac{1}{17}$	$\frac{1}{12}$ to $\frac{1}{10}$	0

[a] A. Bourret, L. Billard and M. Petit, *Inst. Phys. Conf. Ser.*, **76** (1985), 23.
[b] C. Fontaine and D. A. Smith, *Appl. Phys. Lett.*, **40** (1982) 153.
[c] Reference [39].
[d] M. Labidi and A. Rocher, *J. Phys. Appl. (Paris)* (1985).
[e] J.-J. Bacmann, *J. Phys. (Paris)*, **43** (1982), C6–93.

plus surrounding disks; the latter each correspond to a particular Bragg reflection. The detailed intensity variations within a disk carry a great deal of information [35]. In particular the symmetry of convergent beam patterns gives information about the 32 point groups and the 31 possible diffraction groups [36, 37].

Pond and Bollman [38] and Pond and Vlachavas [39] have developed the theory of bicrystallography for internal interfaces; the theory is for a planar interface between two crystals of the same or different form. For a given internal interface it is always possible to assign a point group, as well as a space group to the bicrystal. Convergent beam electron diffractions patterns have been used to search for rigid-body translations parallel to the plane of the internal interface. Figure 2.10 illustrates the reduction in overall symmetry expected from a rigid-body translation parallel to the plane of an interface, for a periodic interface that possesses an overall point group symmetry of 4 mm in the absence of a rigid-body translation. Figure 2.10(a) is a square with a point group symmetry of 4 mm and Fig. 2.10(b) is the exact superposition

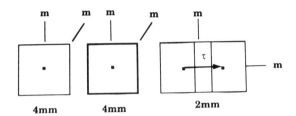

Fig. 2.10 An illustration of the effect of a rigid-body translation parallel to the plane of a periodic interface on symmetry of a bicrystal: (a) A square with the point group symmetry $4\,mm$; (b) exact superposition of an identical square without a translation yields the same point group symmetry $4\,mm$; (c) super-position of an identical square with a lateral translation reduces the composite symmetry to the point group $2\,mm$ (Courtesy of V. P. Dravid).

of a second square on the first square. Finally, in Fig. 2.10(c) the second square is superimposed on the first square but with a lateral translation of τ, note that the point group symmetry is reduced to $2\,mm$, as a result of τ. This symmetry lowering should be detectable in a bicrystal, via convergent beam electron diffraction patterns, if a rigid-body translation parallel to an interface is present.

As an example we consider the recent research of Dravid *et al.* [40] on a heterophase interface in a ceramic material – $NiO/ZrO_2(CaO)$. The crystallography of this interface is: [41]:

$[1\bar{1}0]NiO//[001]ZrO_2$
$[\bar{1}\bar{1}2]NiO//[010]ZrO_2$, and
$[111]NiO//[001]ZrO_2$

where the last relationship is for the interface normal. Also the $(111)NiO//(100)ZrO_2//$interface has two dimensional periodicity and corresponds to a cusp in the energy. Dravid *et al.* searched for a rigid-body translation at the NiO/ZrO_2 heterophase interface using plan-view convergent beam electron diffraction patterns. Figure 2.11 exhibits convergent beam electron diffraction patterns from a plan-view $(111)NiO//(001)ZrO_2$ bicrystal; Fig. 2.11(a) is a zero order Laue zone pattern which exhibits $2\,mm$ projection symmetry, and Fig. 2.11(b) is the whole pattern which exhibits m projection symmetry; the whole pattern –

Fig. 2.11(b) – yields the three-dimensional symmetry of the bicrystal. The explanation of the observed symmetries in Fig. 2.11 can be understood with the aid of Fig. 2.12. Figure 2.12(a) explains the effect of the intersection ($\cap$) of a point group with symmetry $4\,mm$ – $[100]ZrO_2$ – with a point group with symmetry $6\,mm$ – $[111]NiO$. Note that the resulting point group symmetry is lowered to $2\,mm$ and, therefore, only a vertical two fold rotation axis and two aligned mirrors survive the intersection process. The whole pattern symmetry of this composite, however, is given by the intersection of $3\,m$ (of $[111]NiO$) with $4\,mm$ (of $[111]ZrO_2$) to yield the point group m – see Fig. 2.12(b) – which is the only common symmetry element. Thus the convergent beam diffraction patterns of Figs. 2.11(a) and (b) correspond to the two situations illustrated in Fig. 2.12. From these patterns Dravid *et al.* concluded that there is no rigid-body translation parallel to the interface plane present at the $NiO–ZrO_2$ bicrystal. Finally they verified that only one mirror plane is present by tilting the bicrystal along the zero order Laue zone mirrors, and showing that the whole pattern symmetry contains only a single mirror plane. Dravid *et al.* [42] performed high resolution electron microscopy and electron diffraction analysis on the same interface to show that the rigid-body translation perpendicular – e in eq. 2.13 – to this interface is also zero.

Convergent beam diffraction patterns have been employed by Eaglesham *et al.* [43] to study epitaxial Al on GaAs and $NiSi_2$ on Si. They employed higher order Laue zone lines, as well the zero order Laue zone lines, and were able to show by this approach that the symmetry observed can be explained by arrays of misfit dislocations at the interface that relate symmetry equivalent sites of the Al–GaAs heterophase interface. In the case of Ni_2Si on Si they did not observe a rigid-body translation.

Schapink *et al.* [44] have applied convergent beam diffraction to study the bicrystallography of a (111) coherent twin in a gold crystal, and found three distinct areas with different symmetries.

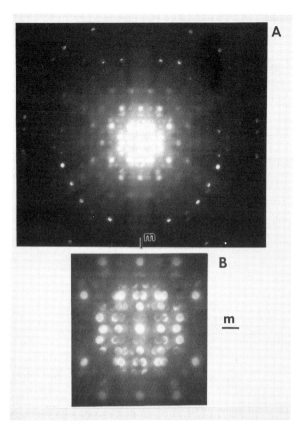

Fig. 2.11 Plan view convergent beam electron patterns from a (111)NiO//(100)ZrO$_2$ bicrystal. (a) Zero order Laue zone (ZOLZ) exhibiting $2\,mm$ projection symmetry. (b) The whole pattern (WP) – i.e. it includes higher order Laue zones (HOLZ) – exhibiting m symmetry (Z ≡ ZrO$_2$ and N ≡ NiO) (Courtesy of V. P. Dravid).

First, the highest symmetry observed was $6/m^3$ and they state that this most likely corresponds to a zero rigid-body translation. Second, the next lowest symmetry observed was $-6\,m^2$, and this implies a translation of unknown magnitude in the [111] direction. The lowest symmetry detected was mm^2 and this corresponds to a translation of $(\frac{1}{12})$ $\langle 112 \rangle$ parallel to the boundary, followed by a translation in the [111] direction. The exact physical origin of this mm^2 symmetry, however, is not clear.

Thus, the convergent beam electron diffraction technique appears to be an excellent complementary technique to the other techniques discussed for determining a rigid-body translation.

Fresnel fringe contrast technique

Boothroyd *et al.* [45] have developed and analyzed in great detail the intensity variations of Fresnel fringes that appear in dark-field images of boundaries with a displacement component e perpendicular to the plane of the boundary. They modeled a $\Sigma = 3/(111)$ coherent twin in copper using a supercell containing 42 atoms, projected along a $[1\bar{1}0]$ direction, and calculated the expected diffraction pattern and the expected image using a multislice computer code. The effects of specimen thickness, defocusing conditions, diffraction conditions, and the value of the magnitude of e, as a fraction (f) of the (111) interplanar spacing (d_{111}); f was varied between 0.01 to 0.1. The sensitivity of the dark field Fresnel contrast technique was experimentally tested employing images of $\Sigma = 3/(111)$ coherent twins in electron microscope specimens prepared from electrodeposited copper. The images obtained were recorded at the $(1\bar{1}0)$ normal with the 111 reflection strongly excited in the two crystals that are adjacent to the (111) twin. This method yields a value for the $\Sigma = 3/(111)$ coherent twin boundary displacement of $\approx 0.06 d_{111}$ or $\approx 0.012 \pm 0.002\,nm$. There are, however, systematic small differences between the results of the model calculations and the experimental data, as a function of both the defocus value and the thickness of the specimen, which they are unable to explain. In addition, the magnitude of e measured by this technique is a factor of ≈ 10 greater than the value measured for the same interface by the lattice α-fringe method. The authors conclude that there is some systematic error present in the Fresnel fringe method that they can not identify, or that there is a segregant present that affected the measured value of $|e|$. It is felt that the Fresnel fringe contrast technique is less reliable than the α-fringe, high resolution electron micro-

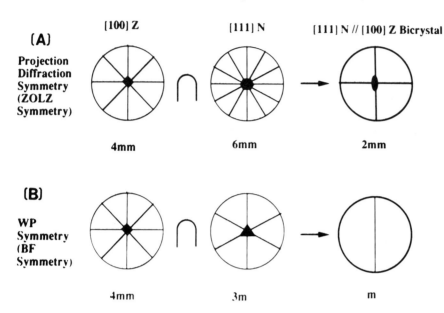

(A)

Projection Diffraction Symmetry (ZOLZ Symmetry)

[100] Z   [111] N   [111] N // [100] Z Bicrystal

4mm   6mm   2mm

(B)

WP Symmetry (BF Symmetry)

4mm   3m   m

Fig. 2.12 The schematic representation of the intersection symmetry of point groups due to superposition. (a) 4 *mm* projected diffraction symmetry of [100] ZrO_2 intersecting 6 *mm* projected diffracted symmetry of [111] NiO results in 2 *mm* composite symmetry. (b) 4 *mm* whole pattern – three dimensional – symmetry of [100] ZrO_2 intersecting 3 *m* whole pattern symmetry of [111] NiO. Note also that this results only in a single coincident mirror (Courtesy of V. P. Dravid).

scopy, or convergent beam diffraction techniques for determining rigid body translations. Perhaps further experimental research and simulations could improve this technique.

Interfacial dislocations

The presence of dislocations at internal interfaces may be used to infer a value of t'; more precisely the determination of the Burgers vectors of dislocations residing at an interface implies the presence or absence of a specific t'. This method is less direct than the ones discussed in the previous sections, as it depends on determining the Burgers vector of dislocations at an interface and then inferring t' from a putative dislocation model for an interface. The subject of determining Burgers vectors of dislocation at interfaces is not discussed here, as there are a number of detailed discussions of this subject in the literature [19, 24, 46].

First, we consider an interesting example which involves faceting of a planar interface. An internal interface can lower its energy by faceting, just as a solid/vacuum surface facets to lower its energy [1, 47]. Pond and Vitkek [48] have measured the difference in rigid body translations between two adjacent facets via a Burgers vector analysis. Two

immediately adjoining facets with displacements t'_1 and t'_2 must have a partial dislocation with a Burgers vector $b = t'_1 - t'_2$ that lies along the intersection of the two facets. Thus by determining the Burgers vector of this partial dislocation the difference in displacement between two facets is simultaneously obtained by measuring $\Delta t'(t'_1 - t'_2)$; the latter is physically the smallest magnitude difference between the translation states of the two adjacent facets.

Second, Sun and Balluffi [49] made a novel discovery with respect to t' during a study of secondary grain boundary (GB) dislocations in [001] twist boundaries in MgO. They studied, via weak-beam transmission electron microscopy, GB-dislocation structures accommodating twist deviations from exact coincidence site lattice orientations for the $\Sigma = 1, 5, 13, 17, 25, 29$ and 53 (001) boundaries. For all these boundaries, with the exception of $\Sigma = 13$, the dislocation network consisted of perfect screw dislocations; the Burgers vectors of these secondary GB dislocations were primitive vectors of the displacement shift complete lattice for the corresponding coincidence site lattice misorientation. In the case of the $\Sigma = 13$ ($\theta = 22.6°$) they detected square grids of partial GB dislocations. They came to this conclusion because

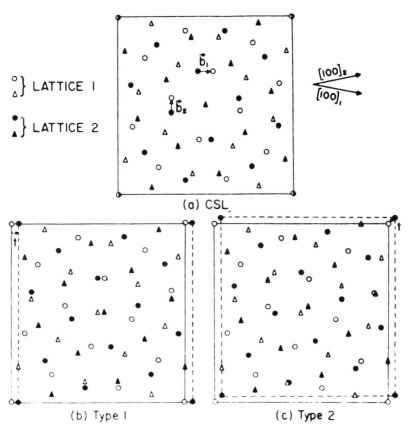

Fig. 2.13 The effect of rigid body translations on the structure of [001] $\Sigma = 13$ twist boundaries in an fcc structure. Lattice 1 is indicated by open circles and triangles and lattice 2 by solid circles and triangles. Two planes of each lattice above and below the interface plane are indicated, their positions are projected along the [001] direction. The vector t is a rigid body translation vector of lattice 2 with respect to lattice 1. (a) The two lattice are at the exact coincidence-site lattice orientation. The vectors b_1 and b_2 are primitive vectors of the displacement shift complete lattice. (b) Type-1 lattice produced by a translation of lattice 2 by $t = (\frac{1}{2})b_1$. (c) Type-2 structure produced by a translation of lattice 2 by $t = (\frac{1}{2})[b_1 + b_2]$ (Courtesy of R. W. Balluffi).

the GB dislocations observed were parallel to the $\langle 320 \rangle$ directions in the plane of the interface, which is 45° from the $\langle 510 \rangle$ – the expected direction for perfect screw GB dislocations. The explanation for these observations had been suggested by Bristowe and Crocker [50] and Smith [51], as Bristowe and Crocker had performed molecular statics calculations of the energies of $\Sigma = 5, 13, 17$ and 25 [001] twist boundaries in copper and nickel. Figure 2.13 exhibits Bristowe and Crocker's results for a $\Sigma = 13$ boundary; (a) the coincidence site lattice orientation; (b) Type-1 structure which is produced by a translation of lattice 2 by $t = (\frac{1}{2})b_1$; (c) Type-2 structure which is

produced by a translation of lattice 2 by $t = (\frac{1}{2})[b_1 + b_2]$. The vectors b_1 and b_2 in Fig. 2.13 are primitive vectors of the displacement shift complete lattice. Bristowe and Crocker found that these boundaries can exist with almost the same low energies. A square network of partial GB dislocations with Burgers vectors $(\frac{1}{2})[b_1 + b_2]$ and $(\frac{1}{2})[b_1 - b_2]$ can now be introduced into any of the three boundary structures exhibited in Fig. 2.13. These create different possible 'checkerboard' dislocation structures (Sun and Balluffi) that have a lower energy than an interface with perfect GB dislocations, because the elastic energy is lower with these partial dislocations. Thus, Sun and

Balluffi's electron microscope observations of partial GB dislocations in the $\Sigma = 13$ boundary are indicative of the presence of a rigid-body translation; although they were unable to distinguish which of the two possible t's was responsible for the observed dislocation networks.

Concluding remarks about the microscopic translational DOFs of an internal interface

From the discussion presented above of the many different transmission electron microscope methods that have been employed to determine the microscopic translational DOFs of an internal interface it is clear that no universal method is yet available. And furthermore it is extremely difficult to determine all the components of the rigid-body translational vector t' except in a few cases. This implies that if molecular statics simulations demonstrate that several geometrically possible structures are energetically degenerate, then there is no guarantee that experimental determinations of t' can always be used to find the lowest energy microscopic structure; the structures observed may be metastable, as opposed to the lowest energy structure.

2.3 CHEMISTRY OF INTERNAL INTERFACES

After completely characterizing the structure of an internal interface, as described in section 2.2, it is then necessary to determine the chemical identities of all atoms at an internal interface. This would ideally result in a complete three-dimensional spatial distribution of all atoms associated with an internal interface and their chemical identities. Technologically the means are almost available to arrive at this objective, although it has not yet been achieved in a single case. In addition to determining the chemical identity of each atom associated with an interface we would also like to measure the associated electronic structure of an interface. The electronic structure of individual internal interfaces has been calculated for a number of different cases using the fully linearized augmented plane wave methodology [52]. Unfortunately, the experimental research lags behind the theoretical research with respect to electronic structure of interfaces.

2.3.1 Atom-probe field-ion microscopy and its variants

An atom-probe field-ion microscope, or more simply an atom probe, consists of a field-ion microscope combined with a special time-of-flight mass spectrometer (Fig. 2.14). An atom probe permits the measurement, one at a time, of the chemical identity of any atom that appears in a field-ion microscope image [53]. Thus, it is possible to obtain an atomic resolution image of subnanometer scale microstructural features and to measure the mass-to-charge state (m/ne) ratio – where m is the mass of an atom, n its charge state, and e the charge on an electron – of single atoms associated with these features [54]. An atom probe with a straight time-of-flight tube has a mass resolution $(m/\Delta m)$ of ≈ 200 [55]; the resolution for this instrument is limited by the physical phenomenon of energy deficits that occurs when ions are removed using the technique of pulsed-field evaporation. In order to ameliorate the $m/\Delta m$ value atom probes have been constructed with either a toroidal (or Poschenrieder) lens [56, 57] – this lens yields an $m/\Delta m$ value of ≈ 2000 – or a reflectron lens [58–60]; the latter lens can yield an $m/\Delta m$ approaching 13 000, and therefore a reflectron lens appears to be the best solution, at present, for obtaining a large value of $m/\Delta m$ on a routine basis.

The operation of an atom probe can be understood with reference to Fig. 2.14 [61–65]. When a short (<25 ns) high voltage pulse (V_{pulse}) – with a subnanometer rise time – is superposed on a steady-state imaging voltage (V_{dc}), atoms on the surface of the specimen are field evaporated in the form of ions; the most common charge state of an ion is often two; however, a mass spectrum with several different charge states of the same element is often obtained – e.g. molybdenum field evaporates as Mo^{2+}, Mo^{3+} and Mo^{4+}. The ions that are projected into the probe hole – located at the center of an internal image intensification system that is based on a channel electron multiplier array plate – pass through the Poschenrieder energy compensator to a chevron channel plate detector. A chevron detector consists of two channel electron multiplier array plates in series, and it has a gain of $>10^6$. Each ion produces a single voltage pulse which stops a clock in a digital timer; the most common one in use is a LeCroy 4208 time-to-digital convertor; it has a resolution of 1 ns for a single event and 3 ns for multiple events and contains a total of eight timing chains. A computer is used in a multitasking mode to control the specimen voltage system and to analyze the time-of-flight data on line; in my laboratory we use a Macintosh IIfx computer for this purpose. The time-of-flights of ions and voltages (V_{dc} and V_{pulse}) applied to a specimen are stored in the computer after each evaporation pulse. The m/ne ratios of the ions are then calculated according to:

$$m/ne = 2e(V_{dc} + \alpha V_{pulse})(t - t_o)^2/d^2 \qquad (2.15)$$

where α is the so-called pulse factor, d the flight distance (>2.2 m), and ($t - t_o$) the actual time-of-flight of an ion. The quantity t is the observed time-of-flight and t_o is the total delay time for the system. In practice, a field-ion microscope specimen is pulsed at a frequency of approximately 10 to 180 Hz [66].

For semiconducting or ceramic materials, where the electric field penetration is significant, a laser is used to desorb ions; in the case of a metal the Debye screening length is <0.1 nm and hence the electric field is highly localized at the site of an atom on a surface, while for nonmetals the Debye screening length is significantly >0.1 nm and hence αV_{pulse} is ill-defined for an atom at a surface. In the case of pulsed-laser desorption a specimen is maintained at V_{dc} and a nitrogen laser beam is used to thermally excite an ion over a Schottky hump in

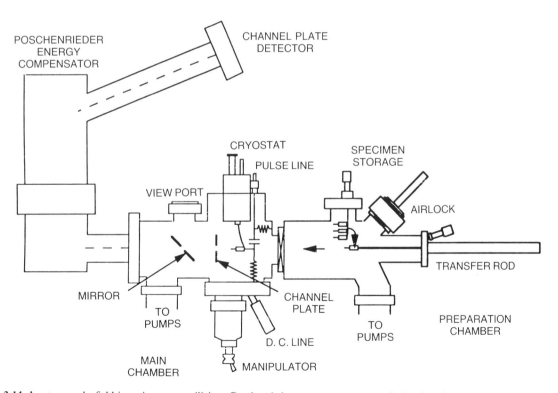

Fig. 2.14 An atom-probe field-ion microscope utilizing a Poschenrieder energy compensator. A sharply pointed tip points towards a channel plate; the field-ion image is viewed, with the aid of a mirror, on this channel plate. The Poschenrieder energy compensator is used to remove the energy dispersion in the pulsed-field evaporated ions and thereby improve the mass resolution of the atom probe.

an ionic potential energy curve that is created by V_{dc}; this results in the desorption of an ion [67].

It is also possible to produce an image of the atoms on the surface of a tip by using the field-desorbed ions themselves to create it; this instrument is called an imaging atom-probe [61–65, 68]. An imaging atom probe is achieved by employing a spherical chevron detector to image the individual field-desorbed ions with a flight distance of 0.10 to 0.20 m; the short flight distance is necessary to observe either the entire surface area of the tip or a significant fraction of it; the value of $m/\Delta m$ is ≈ 40 to 150 for this instrument and hence the materials to be studied via this technique must be carefully chosen. To form a field desorption image of an ion with a specific m/ne ratio the chevron detector is time-gated, so that only the specified m/ne ratio ion is detected. The spherical geometry guarantees that each ion will have the same flight distance, independent of its position on the surface of a specimen. An imaging atom-probe mass spectrometer can be used, for example, to determine elemental maps of the distribution of alloying elements in the vicinity of an interface.

A fairly new variant of the imaging atom-probe is called a position sensitive atom probe or a tomographic atom probe [69, 70]. The geometry of this instrument is similar to that of an imaging atom-probe, although a flat chevron, rather than a hemispherical detector, is employed. The tomographic atom-probe, however, utilizes a two-dimensional position sensitive detector (it sits immediately behind the flat chevron) in conjunction with a time-of-flight measurement to ultimately yield a three-dimensional map of the positions and chemical identities of all the atoms contained within some volume of a tip. However, it has not yet been demonstrated that it is possible to obtain a three-dimensional map with a resolution comparable to the interatomic spacing. It has already been applied to study the chemical concentration gradients in semiconducting superlattices, as well as precipitates.

The atom probe technique and its variants is of great real and potential utility in studying the compositions of buried internal interfaces – e.g. anti-phase boundaries, stacking faults, twins, sub-grain boundaries, GBs, and heterophase interfaces – in solids.

The principal **advantages** of the atom-probe technique are:

1. The interface to be analyzed is observed in direct lattice space with atomic resolution.
2. Crystallographic information about the interface can be determined.
3. The technique for measuring the chemical concentrations is a time-of-flight technique and, therefore, all the elements in the periodic table are detected almost simultaneously.
4. The m/ne ratio of an individual ion is measured one at a time.
5. The probability for detecting an ion of any mass is nearly 100%.
6. A mass spectrum of all the ion species is obtained within $\leq 30\,\mu s$ and, therefore, the pulse repetition frequency can be reasonably high.
7. The measurement of the concentrations of gaseous species in materials, e.g. hydrogen, helium, oxygen, is straightforward, as the $m/\Delta m$ value required is low for these light species in a typical matrix.
8. The correction of the chemical data obtained is minimal and in some cases there is no correction.
9. The spatial resolution, in depth, for chemical analysis is equal to the interatomic spacing of the planes along the crystallographic direction in which a specimen is analyzed – this can be less than 0.1 nm.
10. The pulsed field-evaporation technique is used to dissect an interface on an atom-by-atom basis and in so doing one determines the actual composition of an interface.
11. Conventional transmission electron microscopy can be used to determine the crystal misorientation and plane of the interface, as well as its dislocation structure, prior to performing an atom probe analysis on the same specimen.

The **disadvantages** of the atom probe technique and its variants are:

1. It is destructive, as the process of pulsed field-evaporation – which is used to uncover and analyze an interface – removes material from a specimen.

2. The maximum volume of material which can be analyzed is $\approx 10^{-22} \, \mathrm{m}^3$ per specimen.

3. The areal density of interface $(\mathrm{m}^2/\mathrm{m}^3)$ must be extremely high – $>10^9 \, \mathrm{m}^2/\mathrm{m}^3$ – if one is to find an interface in a specimen without *pre*locating it by transmission electron microscopy.

4. The process of combining the transmission electron and atom probe microscopy techniques to locate an interface and then backpolish a specimen so that a specific interface in a tip is tedious – it does, however, work.

5. Since the chemical composition is determined by measuring the m/ne of ions one at a time the counting statistics can be a problem for a species which is randomly distributed at a very low concentration $(<10^{-4} \, \mathrm{at.fr.})$.

6. If the material can not be prepared in the form of a wire then the field-ion microscope specimen must be 'hogged' out of a bulk specimen or prepared by the method of sharply-pointed shards.

7. The specimens can fail mechanically in the presence of the high electric fields – and their concomitant elastic stress fields – required to perform pulsed field-evaporation.

8. The field-ion microscope images of solid-solution alloys may be quite irregular and this may make it difficult to detect the presence of an internal interface.

Locating a buried internal interface for atom-probe spectroscopy

The general approach we are using at Northwestern University to study solute-atom segregation at internal interfaces is to draw wire specimens from alloys, then heat treat them in the bulk form to induce solute-atom segregation, and finally prepare a field-ion microscope specimen by electropolishing, electroetching or ion-beam milling, or a combination of these processes.

The specific approach used involves electrolytically backpolishing to an internal interface,

employing a versatile system for systematically preparing atom-probe specimens for the study of internal interfaces. This systematic approach incorporates ac electroetching and dc electropolishing in both automated and manual modes. The ac waveforms available are sine or triangular (0.002 Hz to 2.1 MHz) or square (10 Hz to 100 kHz) in either the one shot or continuous wave modes. Triggering and gating are accomplished manually or with a pulse generator. The dc electropolishing mode produces pulses with widths in the range $0.5 \, \mu s$ to 500 s. The power supply provides 0 to ± 48 Vac or Vdc at 1 A.

The atom probe specimen is examined utilizing a radically modified double-tilt stage for a Hitachi H-700H 200 kV transmission electron microscope; this stage is vibrationless at a magnification of $310\,000\times$ [73, 74]. It has a tilting range of $\pm 30°$ for the x-tilt and $\pm 27°$ for the y-tilt; this range is sufficient to allow us to analyze an internal interface in great detail. This double-tilt stage and transmission electron microscope are used to examine atom probe specimens during the course of a backpolishing treatment to place an internal interface in a tip. This approach has built on the pioneering work of Loberg and Nordén in Gothenburg, Sweden [75].

Chemical analysis of a buried internal interface by atom-probe spectroscopy

Figure 2.15 illustrates the principles of the chemical analysis of an internal interface (N.B., this figure is not to scale – it is highly schematic) [76–79]. An internal interface is shown intersecting a plane, i.e. a high-index pole; an atomic step is produced at the points where the interface intersects the surface (Fig. 2.15(a)). In Fig. 2.15(a) the internal interface is aligned so that the plane of an internal interface is centered symmetrically within the projection of the probe hole on the surface of a specimen, and also so that the axis of the cylinder of alloy chemically analyzed passes through the center of the probe hole (Fig. 2.15(b)). For this arrangement the mean composition of a cylinder of alloy is determined, as well as the mean composition of each plane that is perpendicular to the axis of this cylinder. In this case there is a simple geometric correction to the measured composition of the interface, due to the fact that the matrix contributes solute atoms as well as the internal interface. The magnitude of this correction depends on the projected area of the probe hole and the actual solute concentrations of the interface and the matrix. The diameter of the cylinder analyzed is typically 1 to 5 nm; smaller diameters can be used but the price is an increase, of course, in the time necessary to collect statistically significant data. This geometry is also useful for obtaining information about the spatial distribution of solute atoms within the plane of the interface, as a function of depth from the surface of the field-ion microscope specimen.

Figure 2.15(c) is an arrangement which is used to obtain information about the width (η) of the concentration profile normal to the interface. To perform this experiment an interface is placed as close as possible to the edge of the probe hole – typically λ is only a few tenths of a nm. The projection of the probe hole on the surface of the specimen is shown in Fig. 2.15(d).

Figure 2.16 shows how a chevron detector 'sees' the two situations. In both halves of this figure the concentration profile is projected onto the detector. The concentration profile obtained in both cases is a so-called integral profile (see Chapter 19). The quantity $\langle c_s \rangle$ is the mean concentration of solute in the matrix, $\langle c_s^{gb} \rangle^{\star}$ is the maximum concentration of solute at the interface, η is the half-width of a linear distribution measured at the point where the concentration at an internal interface becomes equal to $\langle c_s \rangle$, and D_a is a projected diameter of the probe hole. The arrangement in Fig. 2.16(a) corresponds to the situation described in Figs. 2.15(a) and (b), while Fig. 2.16(b) is for the situation in Figs. 2.15(c) and (d).

We have shown [77] that the value of $\langle C_s^{gb} \rangle^{*}$ is given by the approximate equation:

$$\langle C_s^{gb} \rangle^{*} \approx \langle C_s^{gb} \rangle_U \left(\frac{\pi D_a}{8\eta} \right) + \langle C_s \rangle \left(1 - \frac{\pi D_a}{8\eta} \right)$$

$$(2.16)$$

where $\langle C_s^{gb} \rangle_U$ is the uncorrected mean value of the interface concentration, i.e. the measured value. In

deriving this equation it was assumed that a linear approximation to the concentration profile normal to the interface gives a reasonable physical description of the situation; this assumption yields integrals with closed-form solutions and, therefore, a simple equation (eq. (2.16)). In reality the concentration profile normal to the interface has a ladder-like character, as the composition of each plane that is parallel to the interface decreases until the value in a given plane reaches the concentration value of the matrix $\langle C_s \rangle$. Equation (2.16) contains two unknown quantities, the values of $\langle C_s^{gb} \rangle^{*}$ and η, which can be determined by performing the two measurements illustrated in Fig. 2.15; it is assumed that the value of D_a can be determined from the field-ion microscope image. An analogous equation has been derived to determine the compositions of local concentration fluctuations – from integral profiles – within the plane of the interface [77].

It is also possible to determine a concentration profile of an interface if the geometry can be arranged such that the plane of the interface is perpendicular to the axis of the field-ion microscope specimen; for this case the researcher wants to work at as low a magnification as possible in order to maximize the area of interface which is analyzed, as this geometry implies that one can only record one measurement of the composition of the interface [71, 72, 80–83]. For this geometry it is not necessary to deconvolve the experimental data, and therefore it represents the most direct method for determining the chemical composition of a buried internal interface. There is no other technique that can yet achieve this result; see Chapter 19 for a further discussion of atom probe microscopy.

2.4 OUTLOOK AND FUTURE PROMISE

At present a wide range of sophisticated experimental techniques is available to investigate both the structure and the chemical compositions of internal interfaces in a wide range of materials. The full potential of these techniques has yet to be realized. The future emphasis should be placed on performing as many different experiments as

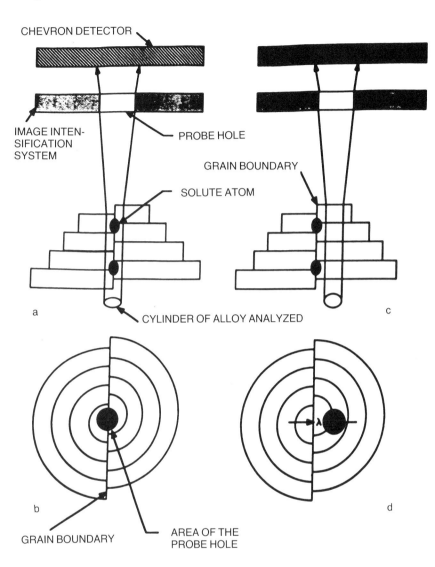

CHEVRON DETECTOR

IMAGE INTEN-
SIFICATION
SYSTEM

PROBE HOLE

GRAIN BOUNDARY

SOLUTE ATOM

a

CYLINDER OF ALLOY ANALYZED

c

b

GRAIN BOUNDARY

AREA OF THE
PROBE HOLE

d

Fig. 2.15 (a) and (b) Schematic diagrams illustrating the basic ideas involved in analyzing an internal interface by the atom-probe field-ion microscope technique. The internal interface is placed at the center of the projection of the probe hole on the surface of the field-ion microscope tip. And the axis of the atom probe field-ion microscope's flight tube passes through the center of the probe hole and the plane of the internal interface (c) and (d). The plane of the internal interface is placed at a distance λ away from the edge of the probe hole; λ is made as small as possible. The diameter of the cylinder of alloy analyzed is typically ≈ 1 to 3 nm. This drawing is *not* to scale. The chevron detector is ≈ 2200 mm from the surface of the tip and the image intensification system is at a distance of ≈ 40 mm.

possible on the same or similar specimens in order to understand in depth the behavior of model systems.

ACKNOWLEDGMENTS

This research is supported by the NSF (grant No. DMR-8819074 – Dr. B. MacDonald, grant officer). It made use of central facilities supported by the NSF through Northwestern's Materials Research Center, Grant No. DMR 85-8821571. I wish to thank J. G. Hu, Ho Jang and V. P. Dravid for stimulating discussions and figures, K. L. Merkle for stimulating discussions and D. Wolf and S. Yip for their kind and stimulating invitation to write this paper, and much forbearance.

REFERENCES

1. J. W. Cahn, *J. Phys.* (*Paris*), **43** (1982), C6–199.
2. J. W. Gibbs, *Collected Works*, Yale University Press, New Haven, CT Vol. 1, (1948), p. 223.

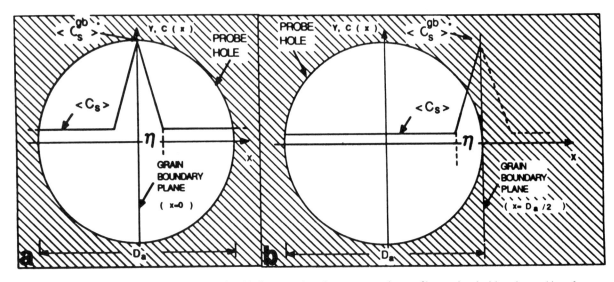

Fig. 2.16 A schematic diagram showing the relationship between the solute concentration profile associated with an internal interface and the probe hole. This is a 'bird's eye' view of what the Chevron detector 'sees'. The detector measures the average value of the concentration profile integrated over the area of the probe hole. (a) This is the situation corresponding to Figs. 2.15(a), (b). (b) This is the situation corresponding to Figs. 2.15(c), (d).

3. E. W. Hart, in *Nature and Behavior of Grain Boundaries* (ed Hsun Hu), Plenum, NY, (1972), p. 155.

4. J. W. Cahn, in *Interfacial Segregation* (eds W. C. Johnson and J. M. Blakely), American Society for Metals, Metals Park, OH (1979).

5. W. T. Read Jr. and W. Shockley, *Phys. Rev.*, **78** (1950), 275.

6. W. T. Read Jr., *Dislocations in Crystals*, McGraw-Hill, NY, (1953).

7. D. H. Warrington and P. Bufalini, *Scripta Metall.*, **5** (1971), 771.

8. H. Grimmer, *Acta Cryst. A*, **30** (1974), 685.

9. W. Bollmann, *Crystal Lattices, Interfaces, Matrices*, (Imprimerie des Bergues, Geneva), (1982).

10. C. J. Ball, *Phil. Mag. A*, **44** (1982), 1307.

11. F-R Chen and A. H. King, *J. Elect. Micros. Techn.*, **6** (1987), 55; F-R. Chen and A. H. King, *Phil. Mag. A*, **57** (1988), 431.

12. M. Weins, H. Gleiter and B. Chalmers, *Scripta Metall.*, **4** (1969), 235.

13. B. Chalmers and H. Gleiter, *Phil. Mag.*, **23** (1971), 1541.

14. G. Hasson, J.-Y. Boos, I. Herbeuval, M. Biscondi and C. Goux, *Surf. Sci.*, **31** (1972), 115.

15. M. J. Weins, *Surf. Sci.*, **31** (1972), 138.

16. R. C. Pond, *J. Micros.*, **116** (1979), 105; R. C. Pond in *Grain Boundary Structure and Kinetics*, American Society for Metals, Metals Park, OH (1980), pp. 13–42.

17. H. Grimmer, W. Bollmann and D. H. Warrington, *Acta Cryst. A*, **30** (1974), 197.

18. R. C. Pond and D. A. Smith, *Canad. Metall. Quart.*, **13** (1974), 39.

19. S. Amelinckx, *Surf. Sci.*, **31** (1972), 296.

20. A. P. Sutton and R. C. Pond, *Phys. Stat. Sol. (a)*, **45** (1978), 149.

21. P. H. Pumphrey, H. Gleiter, and P. J. Goodhew, *Phil. Mag. A*, **36** (1976), 1099.

22. A-M Papon, M. Petit, and J-J Bacmann, *Phil. Mag. A*, **49** (1984), 573.

23. J. W. Matthews and W. M. Stobbs, *Phil. Mag.*, **36** (1977), 373.

24. P. B. Hirsch, A. Howie, R. B. Nicholson, D. W. Pashley, and M. J. Whelan, *Electron Microscopy of Thin Crystals*, Butterworths, Washington, DC (1965), p. 169.

25. J. C. H. Spence, *Experimental High-Resolution Electron Microscopy*, 2nd. edn, Clarendon, Oxford (1989).

26. A. Bourret, *Coll. Phys. (Paris)*, **51** (1990), C1-1; A. Bourret and J. Desseaux, *Phil. Mag.*, **39** (1979), 405.

27. Y. Gao and K. L. Merkle, *J. Mater. Res.*, **5**, (1990), 1995; K. L. Merkle and D. J. Smith, *Ultramicr.*, **22** (1987), 57.

28. J. M. Gibson, *Ultramicr.*, **14** (1984), 1; J. M. Gibson, *MRS Bull.*, **16** (1991), 27.

29. D. Cherns, G. R. Anstis, J. L. Hutchison and J. C. H. Spence, *Phil. Mag. A*, **46** (1982), 849.

30. A. Bourret and J-J. Bacmann, *Surf. Sci.*, **162**, (1985), 495;

A. Bourret, J. L. Rouviere and J. M. Penisson, *Acta Crys. A*, **44** (1988), 838.

31. A. J. Skarnulis, *J. Microsc.*, **127** (1982), 39.

32. A. Bourret, J. L. Rouviere and J. M. Penisson, *Acta Crys. A*, **44** (1988), 838.

33. A. Mauger, J. C. Bourgoin, G. Allan, M. Lannoo, A. Bourret and L. Billard, *Phys. Rev. B*, **35** (1987), 1267.

34. W. Kossel and G. Mollenstedt, *Ann. Phys.*, **36** (1939), 113.

35. J. W. Steeds, in *Introduction to Analytical Electron Microscopy* (eds J. J. Hren, J. I. Goldstein and D. C. Joy), Plenum, NY (1979), pp. 387–422.

36. P. Goodman, *Acta Cryst. A*, **31** (1975), 804.

37. B. F. Buxton, J. A. Eades, J. W. Steeds and G. M. Rackham, *Phil. Trans. Roy. Soc. A*, **281** (1976), 171.

38. R. C. Pond and W. Bollmann, *Phil. Trans. Roy. Soc. A*, **292** (1979), 449.

39. R. C. Pond and Vlachavas, *Proc. Roy. Soc. A*, **386** (1983), 95.

40. V. P. Dravid, M. R. Notis, C. E. Lyman and A. Revcolevschi, *Mat. Res. Soc. Symp.*, **159** (1990), 95.

41. G. Dhalenne and Revcolevschi, *J. Cryst. Growth*, **69** (1984), 616.

42. V. P. Dravid, V. Ravikumar, G. Dhalenne and A. Revcolevshi, *Mat. Res. Soc. Symp. Proc.*, **238** (1992), 815; V. P. Dravid, C. E. Lyman, M. R. Notis and A. Revcolevshi, *Metall. Trans. A*, **21A** (1990), 2309.

43. D. J. Eaglesham, C. J. Kiely, D. Cherns and M. Missous, *Phil. Mag. A*, **60** (1989), 161.

44. F. W. Schapink, S. K. E. Forghany and B. F. Buxton, *Acta Cryst. A*, **39** (1983), 805; P. W. Schapink, F. and J. M. Mertens, *Scripta Metall.*, **15** (1981), 611.

45. C. B. Boothroyd, A. P. Crawley and W. M. Stobbs, *Phil. Mag. A*, **54** (1986), 663.

46. R. C. Pond, *J. Micros.*, **135** (1984), 213.

47. C. Herring, *Phys. Rev.*, **82** (1951), 87.

48. R. C. Pond and V. Vitek, *Proc. Roy. Soc. Lond. B*, **357** (1977), 453.

49. C. P. Sun and R. W. Balluffi, *Phil. Mag. A*, **46** (1982), 49.

50. P. D. Bristowe and A. G. Crocker, *Phil. Mag. A*, **38** (1978), 487.

51. D. A. Smith, *Scripta Metall.*, **14** (1980), 715.

52. A. J. Freeman, A. Continenza and C. Li, *MRS Bull.*, **15** (a) (1990), 27.

53. E. W. Müller, J. A. Panitz and S. B. McLane, *Rev. Sci. Instrum.*, **39** (1968), 83.

54. S. S. Brenner, *Surface Sci.*, **70** (1978), 427; S. S. Brenner and M. K. Miller, *J. Metals*, **35** (3) (1983), 54.

55. A. Wagner, T. M. Hall and D. N. Seidman, *Rev. Sci. Instrum.*, **46** (1975), 1032; T. M. Hall, A. Wagner and D. N. Seidman, *J. Phys. E: Sci. Instrum.*, **10** (1977), 884.

56. W. P. Poschenrieder, *Int. J. Mass Spectrom. Ion. Phys.*, **6** (1971), 413; **9** (1972), 357.

57. E. W. Müller and S. V. Krishnaswamy, *Rev. Sci. Instrum.*, **45** (1974), 1053.

58. B. A. Mamyrin, V. I. Karataev, D. V. Shmikk and V. A. Zagulin, *Sov. Phys. JETP*, **37** (1973), 45; B. A. Mamyrin and D. V. Shmikk, *Sov. Phys. JETP*, **49** (1979), 762.

59. M. Yang and J. P. Reilly, *Int. J. Mass Spectrom. Ion Proc.*, **75** (1987), 209.

60. W. Drachsel, L. V. Alvensleben and A. J. Melmed, *Colloq. Phys. (Paris)*, **50** (1989), C8–541.

61. R. Wagner, *Field-Ion Microscopy*, Springer, Berlin, (1982), Chapter 3.

62. D. N. Seidman, in *Encylopedia of Materials Science and Engineering* (ed. M. B. Bever), Pergamon, Oxford, (1986), pp. 1741–1744.

63. M. K. Miller and G. D. W. Smith, *Atom Probe Microanalysis*, Materials Research Society, Pittsburgh, PA, (1989), Chapters 1 and 4.

64. T. Sakurai, A. Sakai and H. W. Pickering, *Atom-Probe Field-Ion Microscopy and Its Applications*, Academic Press, New York, (1989), Chapter 3.

65. Tien T. Tsong, *Atom-Probe Field-Ion Microscopy*, Cambridge University Press, Cambridge, UK, (1990), Chapter 3.

66. U. Rolander and H.-O. Andrén, *Surf. Sci.*, **246** (1991), 390.

67. G. L. Kellogg and T. T. Tsong, *J. Appl. Phys.*, **51** (1980), 1184; G. K. Kellogg, *J. Chem. Phys.*, **74** (1991), 1479; G. L. Kellogg, *J. Appl. Phys.*, **52** (1981), 5320.

68. J. A. Panitz, *Rev. Sci. Instrum.*, **44** (1973), 1034; *J. Vac. Sci. Technol.*, **11** (1974), 206.

69. A. Cerezo, T. J. Godfrey and G. D. W. Smith, *Rev. Sci. Instrum.*, **59** (1988), 862.

70. A. Bostel, D. Blavette, A. Menand and J. M. Sarrau, *Colloq. Phys. (Paris)*, **50** (1989), C8–501.

71. J. G. Hu, Ph.D. Thesis, Northwestern University (1991).

72. B. W. Krakauer, Ph.D. Thesis, Northwestern University (1992).

73. B. W. Krakauer, J. G. Hu, S.-M. Kuo, R. L. Mallick, A. Seki, D. N. Seidman and J. P. Baker, *Rev. Sci. Instrum.*, **61** (1990), 3390.

74. B. W. Krakauer and D. N. Seidman, *Rev. Sci. Instrum.* (submitted).

75. B. Loberg and H. Hordén, *Ark Fys.*, **39** (1968), 383; L. Karlsson and H. Nordén, *Acta Metall.*, **36** (1988), 13; **36** (1988), 36.

76. R. Herschitz and D. N. Seidman, *Scripta Metall.*, **16** (1982), 849.

77. R. Herschitz and D. N. Seidman, *Acta Metall.*, **33** (1985), 1547; **33** (1985), 1565.

78. J. G. Hu, S. M. Kuo, A. Seki, B. W. Krakauer and D. N. Seidman, *Scripta Metall. Mater.*, **23** (1989), 2033; D. Udler, J. G. Hu, S.-M. Kuo, A. Seki, B. W. Krakauer and D. N.

Seidman, *Scripta Metall. Mater.*, **25** (1991), 841.

79. S.-M. Kuo, A. Seki, Y. Oh and D. N. Seidman, *Phys. Rev. Lett.*, **65** (1990), 199.

80. J. G. Hu and D. N. Seidman, *Phys. Rev. Lett.*, **65** (1990), 1615.

81. D. N. Seidman, J. G. Hu, S.-M. Kuo, B. W. Krakauer, Y. Oh and A. Seki, *Colloq. Phys. (Paris)*, **51** (1990), C1–47; D. N. Seidman, *Mater. Sci. Eng. A*, **137** (1991), 57.

82. J. G. Hu and D. N. Seidman, submitted for publication (1992).

83. H. Jang, D. N. Seidman and K. L. Merkle, *Scripta Metall. Mater.*, **26** (1992), 1493; submitted for publication (1992).

PART I:
Bulk Interfaces

CHAPTER THREE

Correlation between the structure and energy of grain boundaries in metals

D. Wolf and K. L. Merkle

Introduction · Characterization of geometry and structure of GBs · Investigation methods · Correlation between macroscopic DOFs and energy · Correlation between microscopic DOFs and energy · Correlation between the atomic structure and energy · Conclusions

3.1 INTRODUCTION

The investigation of structure–property correlations represents a rather complex endeavor, not only because interfacial systems are intrinsically inhomogeneous, with chemical composition and physical properties differing from the surrounding bulk material, but also since three different aspects of the geometrical structure are involved, namely the macroscopic, microscopic, and atomic structures. As is well known, in addition to the choice of the materials which form the interface, five **macroscopic** and three **microscopic** degrees of freedom (DOFs) are needed to characterize a single bicrystalline interface [1]. The importance of the **atomic** structure at the interface as well as the local interfacial chemistry, extrinsic (i.e. impurity segregation) or intrinsic (for example, via interfacial reactions or space-charge phenomena), greatly add to the complexity of the task.

Grain boundaries (GBs) in pure metals represent ideal model systems for the investigation of the strictly geometrical aspects of structure–property correlations for the following three reasons. First, the complexity due to the myriad of possible choices of materials combinations forming the

interface is avoided, thus enabling a focus on the different roles of the three distinct geometrical aspects of the structure. Second, because GBs are bulk interfaces, dimensional interface parameters (such as the modulation wavelength in strained-layer superlattices, or the thickness of epitaxial layers) do not enter into the problem. Finally, the GB **energy** is thought to play a central role in various GB properties, such as impurity segregation, GB mobility and fracture, GB diffusion and cavitation, to name a few. A better understanding of the correlation between the structure and **energy** of GBs therefore promises to offer insights into more complex structure–property correlations as well. Also, it represents a base line against which the effects of interfacial chemistry can be probed.

In spite of a variety of controlled bicrystal experiments performed to date, a systematic experimental exploration of the misorientation phase space associated with the five macroscopic and three microscopic DOFs of a single flat GB has not been performed. Because of the relative ease with which the macroscopic DOFs can be manipulated in the computer, an approach utilizing the complementary capabilities of atomic-level computer simulation and experiment seems to be particularly promising. However, while a comparison between experiments and modeling results in crucial test cases is absolutely essential, the main strength of such simulations lies in their ability to provide atomic-level insights into structure–property correlations. In the work described here we have combined HREM experiments with atomistic simulations in an attempt to demonstrate that a coordinated effort exploiting their complementary capabilities can provide considerable insight into all three aspects of the structure–energy correlation for GBs in metals.

In correspondence with the three types of structure–energy correlations to be considered below, three kinds of HREM studies combined with atomic-level computer simulations will be highlighted. First, concerning the correlation with the five macroscopic DOFs, the HREM identification of planar facets illustrates the dominating role of the densest planes in the crystal, in agreement with the simulation of GB energies (section 3.4). Second, among the three microscopic DOFs the volume expansion usually observed at metallic GBs is most directly related to the GB energy; this expansion can be determined via both simulation and HREM experiments (section 3.5). Finally, the atomic structures observed in HREM experiments typically consist of well-matched regions separated by highly localized areas of misfit. The origin of this tendency to maintain perfect-crystal coordination at the interface is elucidated by extensive simulations showing a direct correlation between the number of broken bonds per unit area of the GB and its energy (section 3.6).

Although relatively little controlled bicrystal work has been performed to date on bcc metals, much information has become available recently from atomistic simulations. These simulations have permitted some important conclusions drawn from extensive work on fcc metals to be tested, particularly those concerning the importance of the GB plane. Throughout this chapter we will therefore present the simulation results for fcc and bcc metals in parallel, in full awareness, however, that many of our modeling results have yet to be verified experimentally.

Finally, to gain some insight into the sensitivity of our simulation results to the choice of interatomic potential, and in particular into the importance of many-body effects, throughout this study both a many-body potential (either of the embedded-atom-method (EAM) [2] or Finnis–Sinclair (FS) type [3]) and a simple pair potential will be used for both the fcc and bcc metals. Such a comparison of conceptually different potentials will permit identification of phenomena which are entirely independent of the particular interatomic-force description chosen.

3.2 CHARACTERIZATION OF GEOMETRY AND STRUCTURE OF GBs

A systematic investigation of structure–property correlations for GBs should, from the outset, be based on a clear understanding of what is meant by the term 'structure'. As outlined in Chapter 1, the **atomic structure** is to be distinguished from the **macroscopic geometry** (characterized by the five macroscopic DOFs) and from the **microscopic structure** (characterized by the three 'microscopic' translational DOFs). In accordance with this distinction, to be discussed further in this section, we will discuss three fundamentally different aspects of the correlation between the geometry, structure, and energy of GBs.

3.2.1 Macroscopic DOF

As is well known, the type of the GB (i.e. whether a GB is of a pure tilt, twist or mixed (general) type, whether it is symmetrical or asymmetrical, or whether it is of a low- or high-angle type), is fully determined by the five macroscopic DOFs. (If the GB is mobile, a sixth macroscopic DOF is needed to characterize the position of the GB plane.) [1] Within the framework of the coincident-site-lattice (CSL) description of GBs [4–6] the five macroscopic DOFs of a general GB are chosen to consist of the CSL misorientation (three DOFs) and the GB-plane normal (two DOFs). The misorientation between the two grains may be characterized by the rotation matrix $\mathbf{R}(\hat{n}_r, \phi)$, with $\hat{n}_r$ and ϕ denoting the CSL rotation axis (a unit vector) and angle, respectively. The five DOFs are then defined as follows:

$$\{\text{DOFs}\} = \{\hat{n}_r, \phi, \hat{n}_1\} \quad (3.1)$$

where $\hat{n}_1$ represents the GB-plane normal expressed in the principal coordinate system associated with one of the two halves of the bicrystal (here chosen to be crystal 1). Given these five variables, the GB-plane orientation in the second crystal, $\hat{n}_2$, is given by

$$\hat{n}_2 = \mathbf{R}(\hat{n}_r, \phi)\hat{n}_1 \quad (3.2)$$

The inverse density of CSL sites, Σ, characterizing the volume unit cell of the superlattice, is similarly governed completely by the three DOFs in the CSL misorientation, i.e. $\Sigma = \Sigma(\hat{n}_r, \phi)$. As an example, in Fig. 3.1(a) the $\Sigma 5$ CSL in the fcc lattice is shown from which, for example, a symmetrical (310) (Fig. 3.1(b)) or an asymmetrical configuration combining the (100) and (430) planes (Fig. 3.1(c)) can be obtained.

The CSL terminology leads to a natural distinction between pure tilt (for $\hat{n}_r \perp \hat{n}_1$) and pure twist boundaries (for $\hat{n}_r \parallel \hat{n}_1$), with general GBs satisfying neither of these special conditions, and hence with both tilt and twist components. This distinction permits a natural description of the dislocation structure of the GB in terms of edges and screws, and hence essentially represents a crystallographic extension of the dislocation model of Read and Shockley [7]. Because of its focus on the misorientation, however, at least from a conceptual viewpoint the GB plane plays a relatively minor role only, as evidenced by the only two DOFs assigned to it (via the unit vector $\hat{n}_1$ in eq. (3.1)). In fact, within the CSL model the GB plane is usually assumed (or implied) to coincide with a densest (e.g. symmetrical) CSL plane, and hence not really treated as a truly independent variable (Fig. 3.1).

By contrast, much of the 'edge-on' HREM work on tilt GBs of recent years has offered strong evidence for the importance of the GB plane, as exemplified by (a) the existence of atomic-scale faceting in many cases and (b) the frequent preference for asymmetrical rather than symmetrical GB-plane configurations [8–12]. Much of this work shows that in both metals [8–10] and ceramic

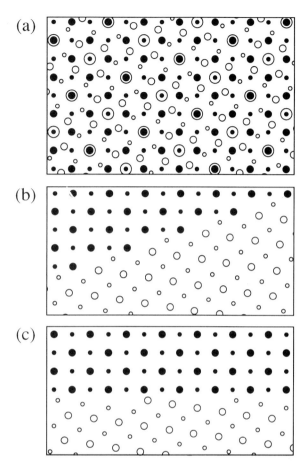

Fig. 3.1 Symmetrical and asymmetrical $\Sigma 5$ GBs in a rigid-lattice model. (a) $\Sigma 5$ CSL (ringed atom sites) formed by the two misoriented, interpenetrating lattices (marked by white and black atoms). Shown for the $\langle 001 \rangle$ projection are two (001) planes of the fcc lattice (distinguished by large and small symbols), misoriented by $\theta = 36.87°$ about $\langle 001 \rangle$. (b) A rigid model of the (310)(310) symmetric tilt grain boundary (STGB) is generated from (a) by removing black and white atoms on opposite sides of the (310) plane. (c) The asymmetric (100)(430) GB is generated by splitting (a) along (100). Both (b) and (c) illustrate that open spaces appear at the interface and while some atomic distances across the GB are larger than in the perfect crystal ('broken bonds'), other distances are smaller, illustrating that the rigid models are poor representations of real interfaces. The arrangement of atoms in the GB can be viewed as a packing of polyhedral units.

materials [11, 12] asymmetrical combinations of lattice planes and faceting occur rather commonly. Recent 'rotating-sphere-on-a-plate' experiments on fcc metals in which the GB-plane orientations were identified [13, 14] also show a substantial fraction of asymmetric GBs, a much larger number of which constitutes asymmetrical tilt rather than general GBs, however. These observations suggest that, by contrast with the implicit CSL assumption of symmetry, asymmetrical GBs may often have lower energies than symmetrical ones, and that pure asymmetrical tilt boundaries may be energetically favored over general GBs (with tilt and twist components).

Given the practical importance of asymmetrical GB-plane orientations, the fact that symmetrical and asymmetrical GBs are not readily distinguished must be considered a shortcoming (or at least an inconvenience) of the CSL terminology. On the other hand, the focus on the misorientation between the two grains is sometimes closely related to the experimental situation, particularly when preparing tilt bicrystals by fixing the misorientation ($\hat{n}_r$, ϕ) between the two grains while allowing the GB plane to choose an arbitrary orientation (section 3.4.6). The GB plane is then usually characterized by the inclination angle, α, with respect to the symmetrical configuration; in the case of pure tilt boundaries, symmetrical GB-plane orientations are then defined by $\alpha = 0°$ or $\alpha = 90°$ (assuming the tilt axis has two-fold rotational symmetry) while for all other values of α the tilt boundary is asymmetrical.

Another difficulty with the CSL characterization of GBs, more conceptual in nature, arises from the fact that the three translational DOFs associated with rigid-body translations (section 3.3.2) usually destroy the CSL sites in common to the two halves of a bicrystal – as evidenced, for example, by computer simulations and, more recently, high-resolution TEM observations. Because of their purely crystallographic nature, the main goal of CSL-based GB models is therefore less to predict physical properties rather then provide a dislocation-based description of their atomic structure [15]. Essentially the only macroscopic CSL variable which has been used in attempts to address **macroscopic** structure–property correlations is the inverse density of CSL sites, Σ, which has often been said to play some role in the properties of high-angle GBs.

Because of its focus on the misorientation between the two grains, rather than the plane of the defect, the CSL-based terminology is not usually applied to interfaces other than grain boundaries, thus rendering a comparison of properties of different interfacial systems virtually impossible. A more appropriate choice of the five macroscopic DOFs, applicable to both commensurate and incommensurate interfacial systems, focuses on the importance of the interface plane by assigning to it

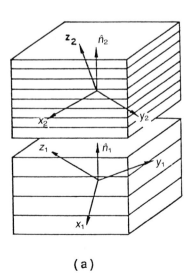

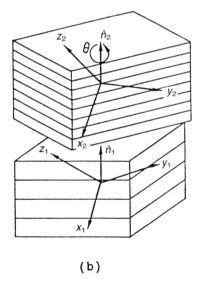

(a)   (b)

Fig. 3.2 Creation of an 'asymmetrical twist' (or general) GB by a twist rotation of an asymmetrical combination of lattice planes about the common GB-plane normal. Two asymmetrical tilt configurations (ATGBs), with identical unit-cell dimensions but inverted stacking sequences, are obtained for $\theta = 0°$ and $180°/m$ (sketched in (a)). (Here m characterizes a possible rotation symmetry (for $m > 1$) in the planar unit cell of the GB.) (b) For some arbitrary twist angle, θ, an asymmetrical twist GB is obtained.

four of the five DOFs as follows (Fig. 3.2) [16]:

$$\{DOFs\} = \{\hat{n}_1, \hat{n}_2, \theta\} \qquad (3.3)$$

Here the unit vectors $\hat{n}_1$ and $\hat{n}_2$ (i.e. four DOFs) represent the GB-plane normal, however rotated respectively into the principal coordinate systems of semi-crystals 1 and 2 forming the bicrystal. For example, $\hat{n}_1$ and $\hat{n}_2$ might be the normals of (100) and (115) planes, respectively. By definition, the angle θ (one DOF) describes a **twist** rotation (i.e. about the GB-plane normal; Fig. 3.2), because any other rotation would change $\hat{n}_1$ and $\hat{n}_2$ [16].

In cubic crystals it is often more convenient to use Miller indices (hkl), instead of the unit GB normal, $\hat{n}$. As is well known, the two are related by $\hat{n} = (h^2 + k^2 + l^2)^{-\frac{1}{2}}(hkl)$. Equation (3.3) may then be replaced by

$$\{DOFs\} = \{(hkl)_1, (hkl)_2, \theta\} \qquad (3.4)$$

The twist component, $(\hat{n}_1, \theta)$, of a general boundary described by eq. (3.3) is therefore governed by θ and by the GB-plane normal, while its tilt component, $(\hat{n}_T, \psi)$, governed by the condition that $\hat{n}_T \perp \hat{n}_1, \hat{n}_2$, is given by [17]

$$\hat{n}_T = [\hat{n}_1 \times \hat{n}_2]/\sin \psi \qquad (3.5)$$

$$\sin \psi = (|\hat{n}_1 \times \hat{n}_2|) \qquad (3.6)$$

where $\hat{n}_T$ is a unit vector defining the orientation of the tilt axis, while ψ denotes the tilt angle.

Equations (3.5) and (3.6) illustrate that the tilt component of a general boundary (i.e. the edge-dislocation content) is fully determined by the normals $\hat{n}_1$ and $\hat{n}_2$, i.e. by the GB plane (Fig. 3.2(a)). We therefore define $\theta = 0°$ as the angle for which a general boundary has no twist component, i.e. $\theta = 0°$ thus characterizes an asymmetrical tilt boundary (ATGB), with the smallest planar unit cell of all GBs formed by the same combination of planes. For a finite value of θ (Fig. 3.2(b)) screw dislocations are introduced in addition to the already present edge dislocations, with a consequent increase in the planar unit-cell area.

It is important to recognize that for any asymmetrical combination of lattice planes forming the GB, a second ATGB configuration is found at $\theta = 180°/m$. (Here m characterizes a possible rotation symmetry (for $m > 1$) in the planar unit cell of the GB.) The reason for the existence of **two** ATGB configurations is that in crystal lattices with inversion symmetry, a twist rotation by $\theta = 180°/m$ leads to an inversion of the stacking sequence of the lattice planes, say, in the top half of Fig. 3.2(a) with respect to the bottom half [16]. In the fcc and bcc lattices, there are two exceptions: Because of the $\ldots|AB|AB|\ldots$ stacking, the inversion of the (100) and (110) planes by a $180°/m$ twist rotation about $\langle 100 \rangle$ or $\langle 110 \rangle$ (with $m = 2$ and $m = 1$, respectively) can be undone simply by a rigid-body translation parallel to the plane, thus restoring the perfect-crystal configuration. Hence, in asymmetrical GBs with a (100) or (110) plane on one side of the interface, the two ATGBs are identical. To better visualize the effect of a twist rotation by $180°/m$, one may simply think of its net effect being the turning upside down of one of the two semi-crystals in Fig. 3.2(a). (For further details, see Chapter 1 in this volume.)

Since (i) the inversion of the stacking sequence preserves the planar unit-cell dimensions of the ATGB at $\theta = 0°$, and (ii) any twist deviation from $\theta = 0°$ and $\theta = 180°/m$ results in an increase of the planar unit cell of the GB, the two ATGB configurations are unique geometrically in that they have identical planar unit-cell dimensions with an area which is the smallest of all GBs formed by the same combination of lattice planes. As illustrated in section 3.4.5, this unique geometry translates into a particularly low energy, giving rise to 'special' properties of these two configurations at the end-points of the twist-misorientation range.

Symmetrical GBs are obviously characterized by the condition that $\hat{n}_2 = \pm\hat{n}_1$ in eq. (3.3), thus reducing the number of DOFs from five to only three (Fig. 3.2). Equation (3.3) then simplifies as follows:

$$\{DOFs\} = \{\hat{n}_1, \hat{n}_1, \theta\} \qquad (3.7)$$

Similarly, in cubic crystals eq. (3.4) now becomes

$$\{DOFs\} = \{(hkl)_1, (hkl)_1, \theta\} \qquad (3.8)$$

$\theta = 0°$ now corresponds to the perfect crystal while, for some arbitrary value of θ, a pure twist boundary (with three DOFs) is obtained.

As in the asymmetrical case, a twist rotation by $\theta = 180°/m$ inverts the stacking sequence on one side of the GB plane with respect to the other while preserving the perfect-crystal planar unit-cell dimensions [16]. Consequently, the symmetrical **tilt** boundary (STGB) thus obtained is fully determined by only the two DOFs associated with the GB-plane (because θ is fixed at $180°/m$). Its atomic structure is characterized by the familiar twinning of the lattice planes at the interface, a feature in common to all STGBs (see also Chapter 1 in this volume). In cubic crystal lattices, the related tilt axis, $\hat{n}_T$, and tilt angle, ψ, can be determined from the following expressions (compare eqs. (3.5) and (3.6)) [17]:

$$\hat{n}_T = (h^2 + k^2)^{-\frac{1}{2}}\begin{pmatrix} -k \\ h \\ 0 \end{pmatrix} \qquad (3.9)$$

$$\sin \psi = 21 \, (h^2 + k^2)^{\frac{1}{2}} \,/(h^2 + k^2 + l^2) \qquad (3.10)$$

As in the asymmetrical case, for any twist deviation from $\theta = 0°$ and $\theta = 180°/m$ the planar unit-cell area increases during the creation of a pure (i.e. symmetrical) twist boundary. The STGB and perfect-crystal configurations on a given plane are therefore unique in that they share identical planar unit-cell dimensions, with an area which is the smallest possible for any planar defect on that plane. (For further details, see Chapter 1 in this volume.) As illustrated below, this unique geometry of the pure tilt configuration gives rise to a deep energy cusp at $\theta = 180°/m$, and hence to 'special' properties of STGBs by comparison with twist boundaries. Also, since both STGBs and free surfaces are characterized by only the two DOFs associated with the interface plane, the properties of these two simplest types of all planar defects can be compared directly. However, because free surfaces lack the three translational DOFs of GBs, the related three-dimensional (3D) structure-energy plots show some interesting differences (section 3.4.4).

The choice of the five DOFs in eq. (3.3) has essentially three advantages over the CSL-based definition of the macroscopic DOFs. First, the resemblance of a general boundary (with five DOFs) with a pure twist boundary (with three DOFs) becomes apparent (Fig. 3.2(b)), in that (i) the beginning and endpoints of the twist rotation, at $\theta = 0°$ and $\theta = 180°/m$, represent pure tilt boundaries (Fig. 3.2(a)), with the smallest possible planar unit cells, and (ii) both may be viewed as having been generated by a twist rotation about the GB-plane normal. The term **asymmetrical twist** boundary therefore appears to reflect the geometry of a general boundary more succinctly [18]; its tilt and twist components are apparent in that any asymmetry in the GB plane automatically implies the existence of a tilt component (eqs. (3.5) and (3.6)). Second, incorporated naturally in the description is the fact that – from a purely geometrical point of view – **tilt** boundaries represent a special subset of symmetrical or asymmetrical **twist** boundaries, thus greatly simplifying the exploration of the five-parameter misorientation phase space associated with the macroscopic DOFs (section 3.4.4). Finally, the above choice of macroscopic DOFs is not limited to the description of commensurate systems and is, in fact, rather commonly applied to coherent or incoherent dissimilar-material interfaces, thus enabling a more direct comparison of the different basic types of interface systems defined in the Editors' Introduction.

3.2.2 Microscopic DOFs

As is well known, the three **microscopic** DOFs involve rigid-body translations, $\mathbf{T} = (T_x, T_y, T_z = \delta V)$, parallel and perpendicular to the GB plane. To focus attention on the importance of these three translational DOFs, hard-sphere models have been proposed in which the optimum translation parallel to the interface plane is assumed to be the one which minimizes the volume expansion at the GB (for example, Refs. [19–21]). Although based on **unrelaxed** atomic structures, these models have gained strong support from

(a) experiments probing the effect of hydrostatic pressure on the relative energies of different GBs [22];

(b) high-resolution TEM experiments which provide direct evidence for the existence of volume expansions and translations at GBs [11, 23–25];

(c) theoretical models based on electronic-structure theory [26, 27];

(d) recent computer simulations which show a practically linear relationship between the volume expansion at metallic GBs and their energy [28–31].

From a thermodynamic point of view, the z component of $\boldsymbol{T}$ (parallel to the GB normal) is particularly important in that it accounts for a volume expansion per unit area, δV, at the GB, with consequent stresses parallel to the GB which are similar in nature to the surface tension. In atomistic simulations it is important to allow for this expansion, because the definition of the GB energy as the change in Gibbs free energy, G, with GB area, A, at constant temperature, T, pressure, p, and number of particles, N (including composition), according to [32]

$$E^{\mathrm{GB}} = (\partial G/\partial A)_{T,p,N} \qquad (3.11)$$

requires constant-pressure simulations. Unless the related Helmholtz free energy is considered, constant-volume simulations, therefore, lead not only to numerically wrong values of E^{GB} for a given potential, but energies thus determined are conceptually not the true excess energy in the Gibbsian sense.

Consistent with the definition of E^{GB} in eq. (3.11) is the definition of the volume expansion per unit area, δV (to be distinguished from the overall volume, V) according to [33]

$$\delta V = (\partial V/\partial A)_{T,p,N} \qquad (3.12)$$

which indicates that the excess volume of a GB is to be determined, of course, under the same conditions as the excess energy.

In section 3.5, the correlation between the three microscopic DOFs and the GB energy will be investigated in some detail. Our main conclusion will be that, although translations parallel to the GB provide an important relaxation mechanism for reducing the GB energy, it is the volume expansion at the boundary which, ultimately, determines its energy.

3.2.3 Atomic structure

The usefulness of any structural GB model should be assessed in terms of its ability to predict physical properties of GBs, in addition to providing a good description of their crystallography and/or atomic structure. The dislocation model of Read and Shockley (RS) [7], predicting the energy, mobility and other properties of **low-angle** GBs in terms of a quantitative characterization of their atomic structure, is an excellent example of such a model. Models based on the CSL, O lattice and DSC lattice [4–6], including the so-called plane-matching model [34], represent strictly crystallographic extensions of the dislocation model.

Another group of models is concerned with the characterization of the **atomic** structure in terms of the stacking of polyhedral units along the GB plane (for example, Refs. [35–37]). The requirements of space filling at the interface, and of the compatibility of the structural units with the adjoining grains, can generally not be satisfied simultaneously, however, unless the polyhedra are elastically distorted by strains which differ from one GB to another. Within the framework of these models it is therefore difficult to quantify the structure (such as, for example, via a structure factor), and they consequently possess little or no predictive capability [38].

Because during the past 50 years broken-bond models have been very successful in predicting physical properties of free surfaces in metals from their relaxed or unrelaxed atomic structure [39], in this chapter we will characterize the atomic structure of GBs quantitatively via the number of broken nearest- (nn), second-nearest (2 nn) and higher-order neighbor bonds per unit area, given by [40]

$$C(\nu) = \sum_{n} |K_{id}(\nu) - K_n(\alpha)|/A,$$
$$(\nu = 1, 2, \dots) \qquad (3.13)$$

where A is the planar unit-cell area. $K_n(\nu)$ denotes the number of ν-th nearest neighbors of atom n, while $K_{id}(\nu)$ is the related perfect-crystal value. Somewhat arbitrarily, we define as the nearest neighbors all those atoms situated between zero distance and the half-way point, $R_{12} = (R_1^{id} + R_2^{id})/2$, between the nearest and second-nearest neighbors in the perfect crystal, at distances R_1^{id} and R_2^{id}, respectively. $C(v)$ is thus closely connected with the area under the vnn peak in the radial distribution function, $G(r)$. However, as discussed further in section 3.6, while $G(r)$ contains all the detailed information on the distribution of interatomic separations, the 'coordination coefficient' $C(v)$ defined in eq. (3.13) represents a much cruder – yet simpler and more convenient – quantitative measure for how well the atoms are, on average, miscoordinated in the fully relaxed structure. Throughout section 3.6, values of $C(v)$ will be given in units of a^{-2}, where a is the lattice parameter.

A detailed comparison of low- and high-angle GBs in section 3.6.2 demonstrates that only the dislocation cores, but not the surrounding long-range elastic strain fields, give rise to miscoordinated atoms at the interface [41]. The complementarity between the dislocation model of Read and Shockley [7], focusing on dislocation **strain fields**, and the broken-bond model [41], focusing on the dislocation **cores**, is thus elucidated, as are the reasons for the better success of broken-bond models in free surfaces than in GBs.

3.3 INVESTIGATION METHODS

While for surfaces and thin films many suitable experimental methods for the characterization and investigation of gradient properties near interfaces have become available in recent years, buried interfaces still pose a major challenge because such a small fraction of atoms is actually affected by the presence of the interfaces (Chapter 2 in this Volume). High-resolution transmission-electron microscopy (HREM) is one of only a handful of methods which permit a close-up view of at least some types of grain boundaries (namely tilt GBs). (For other methods, see Chapter 2 in this Volume.) This inherent difficulty in the experimental investigation of buried interfaces is actually an advantage in atomistic computer simulations of interfacial systems because the properties in the

interfacial region are controlled by relatively few atoms; computer simulations can hence provide a close-up view of the most critical part of the material. Also, because of the ease with which the macroscopic DOFs can be varied (and controlled) in the computer, such simulations offer probably the only hope of comprehensively exploring **macroscopic** structure–property correlations.

3.3.1 Computer-simulation techniques

A bulk interface, by its very nature, is composed of two coupled regions, the interfacial region and the surrounding bulk material. It is an inherent difficulty in atomistic modeling to formulate proper border conditions so that (i) the size of the simulation cell (containing the interface) can be kept to a minimum without artificially perturbing the interface by the action of the bulk embedding, and (ii) the physical response of the bulk surroundings to interfacial behavior (such as stresses, volume expansion, etc.) is treated realistically. In simulation studies of bulk interfaces, the embedding of the interface between bulk material is accomplished by surrounding the interfacial region, denoted as Region I in Fig. 3.3, by two semi-infinite bulk ideal crystals, denoted as Region II [42]. With the same two-dimensionally periodic border conditions (2-D PBCs) applied in the x–y plane to both Regions I and II, the system has no free surfaces. This border condition is appropriate for a commensurate bulk bicrystalline interface at zero temperature, which is characterized by a strictly periodic arrangement of atoms parallel to the GB.

An iterative energy-minimization algorithm ('lattice statics') is then used to compute the fully relaxed zero-temperature atomic structure and energy of the GB [43]. By computing the forces which the two halves of the bicrystal exert on each-other, translations parallel to the GB plane (T_x, T_y) are allowed while the atoms relax. Also, to enable the GB to expand or contract (related to T_z), the unit-cell dimension in the direction of the GB normal is allowed to increase or decrease in response to the internal stresses [43, 44]. By starting from a variety of initial rigid-body translational configurations, the GB energy may thus be minimized with respect to both the atomic positions and the three microscopic DOFs. Although not used in this chapter, the Region I–Region II embedding in Fig. 3.3 was recently extended to enable finite-temperature molecular dynamics simulations [45]. (For details, see Chapter 7 in this volume.)

The interatomic potentials used here come from two sources. To provide some insight into the role of many-body effects, throughout this study both a semi-empirical many-body potential and a conceptually simpler pair potential will be employed for both fcc and bcc metals. Many-body potentials, such as embedded-atom-method (EAM) [2] and

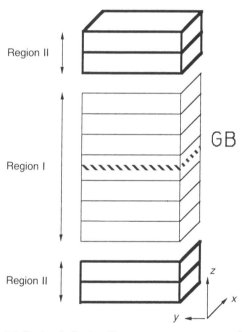

Fig. 3.3 Region I–Region II constant-pressure setup of the computational cell [42, 43]. The cell is periodic in the GB (x–y) plane, while in the z direction the interface is embedded between bulk perfect-crystal material. The atoms in Region I are relaxed explicitly; the two semi-infinite rigid perfect-crystal blocks forming Region II are permitted to move up or down in response to the pressure in Region I, thus enabling the GB to expand or contract [30]. By computing the forces which the two halves exert on each other, the blocks are allowed to slide parallel to the GB while the atoms relax. This border setup was recently extended to finite temperatures, enabling the molecular dynamics simulation of bulk interfaces [45].

Finnis–Sinclair (FS) potentials [3], have the advantage over pair potentials of incorporating, at least conceptually, the many-body nature of metallic bonding, while being relatively efficient computationally. In these potentials the strength of the interaction between atoms depends on the local volume or, in another interpretation, on the local electron density 'sensed' by every atom. Also, while any equilibrium pair potential at zero temperature automatically satisfies the Cauchy relation for the elastic constants, $C_{12} = C_{44}$, these many-body potentials permit all three elastic constants of a cubic metal to be determined (or used in the fitting).

Both types of many-body potentials employed here use the same mathematical expression as starting point, in which the total energy, U, of a system of N interacting metal atoms is subdivided into a bonding term (based on band-structure considerations) and a repulsive central-force term, according to [2, 3, 46]

$$U = - \sum_i F(\rho_i) + \sum_i \sum_{j>i} \phi(r_{ij}) \qquad (3.14)$$

Here $F(\rho_i)$ is the many-body energy gained when embedding atom i in the charge density,

$$\rho_i = \sum_j \rho_{ij}(r_{ij}) \qquad (3.15)$$

due to the surrounding atoms, while $\phi(r_{ij})$ is the central-force repulsive energy between atoms i and j, with $r_{ij} = |\mathbf{r}_i - \mathbf{r}_j|$. The EAM and FS potentials, fitted empirically to properties of Au [46] and Mo [3, 47], respectively, differ essentially only with respect to (i) the basic physical justification (or better perhaps: rationalization) of eqs. (3.14) and (3.15), (ii) the fact that FS offer analytical expressions for $F(\rho)$ and $\phi(r)$, in contrast to the numerical tables used in EAM potentials, and (iii) the procedure employed for fitting the basic expressions to empirical bulk-crystal data for a given material.

Ackland and Thetford [47] have recently modified some of the bcc potentials originally presented by Finnis and Sinclair [3] in order to enhance the repulsion for distances shorter than the nearest-neighbor distance. The potential for Mo used in our investigation is therefore the one parameterized by Ackland and Thetford although we refer to it as the Finnis–Sinclair potential, Mo(FS).

Formally eq. (3.14) contains a pair potential, $\phi(r)$, as the special case in which $F(\rho)$ vanishes identically. The simplest and best-known of all pair potentials is probably the Lennard–Jones (LJ) potential, with only two adjustable parameters, σ and ε, defining the length and energy scales, respectively. Although the LJ potential used here was fitted to the lattice parameter and melting point of Cu (with $\varepsilon = 0.167$ eV and $\sigma = 2.315$ Å; [45]), the relative energies of different GBs are the same for any fcc metal if all energies and distances are expressed in units of ε and σ, respectively. That the LJ potential actually stabilizes the fcc structure relative to the bcc structure is seen from Fig. 3.4(a) in which the perfect-crystal cohesive energy in the fcc and bcc lattices is plotted against the isotropic strain, $\Delta a/a$. As clearly seen from the figure, the fcc structure has a cohesive energy which is about 5% lower than that of the bcc structure described by exactly the same potential. A similar comparison for the Au(EAM) potential yields the same qualitative behavior.

One of the best-known pair potentials for bcc metals is probably Johnson's spline-fitted α-Fe potential [48]. Prior to choosing this and the Mo(FS) potential for a comparison of GBs in bcc metals, the question whether they do, indeed, stabilize the bcc structure over the fcc structure was investigated by determining, similar to Fig. 3.4(a) the cohesive energies at the related equilibrium lattice parameters for both potentials. It was found that the cohesive energy for bcc α-Fe of -1.536 eV (Table 3.1) is about 2% lower than for the fcc and hcp structures. (The same cohesive

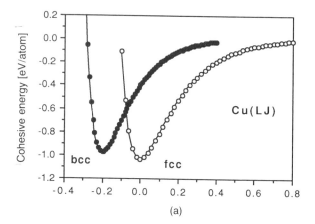

(a)

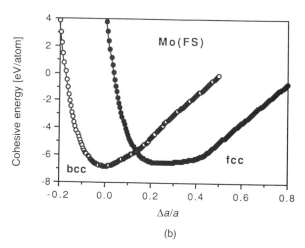

(b)

Fig. 3.4 Comparison of the relative stabilities of the fcc and bcc structures for the (a) Cu(LJ) and (b) Mo(FS) potentials. The corresponding perfect-crystal cohesive energies are plotted against the lattice-parameter change, $\Delta a/a$ for the two structures. Δa is the difference in the lattice parameter from its equilibrium value, a, at zero temperature. Qualitatively identical results were obtained for the Au(EAM) and Fe(Johnson) potentials.

Table 3.1 Perfect-crystal cohesive energy, E^{coh} (in eV/atom), zero-temperature equilibrium lattice parameter, a (in Å units), and cutoff radius, R_c (in units of a) for the four potentials used in this study. The number of neighboring atom shells included within the cutoff radius is also indicated as is the total number of atoms within these shells

Potential	Equilibrium structure	E^{coh} (eV)	a (Å)	R_c/a	Shells	No. of atoms
Cu(LJ)	fcc	-1.038	3.6160	1.49	4	66
Au(EAM)	fcc	-3.901	4.0810	1.32	3	40
α-Fe(Johnson)	bcc	-1.536	2.8600	1.20	2	14
Mo(FS)	bcc	-6.820	3.1472	1.31	2	14

energy was obtained for the latter two, owing to the short cutoff radius of Johnson's potential.) Similarly, as shown in Fig. 3.4(b), the cohesive energy for the Mo(FS) potential of -6.820 eV (Table 3.1) is about 4% lower than that for the fcc and hcp structures. For both potentials the related fcc equilibrium lattice parameter was found to be about 25% larger than for the related bcc structure. With an activation barrier between the two structures, the transition from bcc to fcc requires significantly more energy than suggested by the relatively small differences in the cohesive energies. Both potentials therefore yield bcc structures which are stable towards significant distortions before transforming into higher-energy metastable (at $T = 0$) fcc structures.

To avoid discontinuities in the energy and forces (and, in the case of the many-body potentials, the charge density), all potentials were shifted smoothly to zero at their corresponding cutoff radius (Table 3.1). The zero-temperature lattice parameters, a, determined by finding the zero-pressure configuration, and the related cohesive energies are also listed in Table 3.1. By comparison with potentials for fcc metals, the cutoff radii of both the bcc potentials between the second- and third-nearest neighbors must be considered extremely short, because in an ideal crystal each atom thus interacts directly with only 14 atoms (8 nearest- and six second-nearest neighbors; Table 3.1).

For later reference, in Fig. 3.5 the lattice-parameter dependences of (a) the fcc [18] and (b) bcc potentials [49] is compared. To emphasize the shape differences in these curves, the corresponding equilibrium (i.e. minimum) values of the cohesive energies were subtracted (Table 3.1). Later it will be seen that these shape differences are the cause for the quantitative differences obtained for the pair- and many-body potentials.

3.3.2 Experimental methods

Considerable progress has been made recently in several experimental techniques that are available for the characterization of interfaces (for an overview see the chapter by D. N. Seidman),

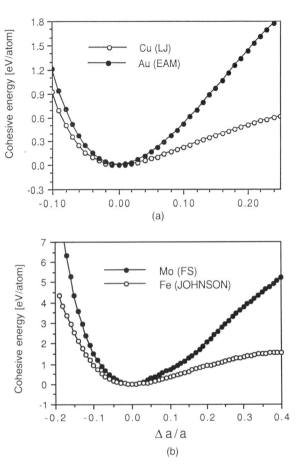

Fig. 3.5 Perfect-crystal cohesive-energy versus deviation, Δa, of the lattice-parameter from its equilibrium value for (a) the two fcc [18] and (b) the two bcc potentials [49] employed in this study. To better illustrate the shape differences between the two curves, the equilibrium values listed in Table 3.1 (i.e. the values at the related minima) were subtracted.

and particularly in those techniques that are capable of local characterizations of internal interfaces down to the atomic scale. The best-developed techniques include high-resolution electron microscopy (HREM), atom-probe field-ion microscopy (APFIM), and scanning-tunneling microscopy (STM). While the latter techniques are capable of directly providing atomic-scale information about external surfaces, recent applications have demonstrated their great potential for the study of internal interfaces [50–52]. In this chapter we shall focus on the application of HREM, since this technique is presently the most advanced for the investigation of the atomic structure of GBs and has recently led to considerable progress in our understanding of GBs. We should mention that X-ray diffraction techniques can also be used to derive atomic position information, albeit for very special geometries and for average structures of periodic GBs with small planar unit cells [53–55]. In view of the increased availability of very powerful synchrotron radiation sources considerable progress in the application of X-ray scattering techniques to GB studies is also expected.

The study of GBs requires the availability of suitable interface specimens. Since there are five macroscopic DOFs for a GB, a very selective control of the GB parameters is necessary for systematic investigations of GB structure and for comparisons to computer simulated boundaries. This presents a major experimental task, the different approaches to which will be briefly discussed after introducing electron microscopy (EM) characterization techniques of GBs.

EM Characterization techniques for GBs

Macroscopic characterization

The characterization of the five macroscopic DOFs of a GB requires, for example, the determination of the misorientation between the two crystals (three DOFs) and the determination of the GB plane (two DOFs). X-ray or electron scattering techniques can be used to determine the relative misorientation between the two crystals. The misorientation between even very small crystals ($<1\,\mu m$) can be observed in the scanning electron microscope (SEM) by backscattered Kikuchi patterns (to $\pm1°$) [56, 57] or in the transmission electron microscope (TEM) by transmitted Kikuchi patterns (to $\pm0.1°$) [58, 59]. The determination of the GB plane is generally not possible with macroscopic methods (such as X-ray or SEM), since the trace of the intersection of the GB with an external surface does not suffice to define the GB plane. Therefore, TEM is typically used, which allows the observation of the GB plane (and, for example, alignment of the boundary into the edge-on con-

figuration) and thus the determination of all five macroscopic DOFs of a GB.

Microscopic characterization

The three microscopic DOFs, which characterize the relative shift of the two crystals (rigid-body translation) are intimately connected with the atomic structure of the interface, as will be discussed in section 3.5. Nevertheless, in conventional TEM, relative translations between the two halves of a bicrystal can be determined with high accuracy by the so-called α-fringe method [60, 61]. This technique is however limited to the determination of shifts in relatively low Σ bicrystals, since it requires imaging of the GB by a common reflection in both crystals.

Other important aspects of the microscopic characterization of GBs by TEM is the observation of microscopic faceting, the identification of ledges and steps and the observation of structural periodicities by strain-contrast [62, 63], as well as by electron diffraction techniques [64, 65]. Secondary GB dislocations as well as interaction of matrix dislocations with GBs can be investigated by TEM. *In-situ* observations of the dynamic interaction of dislocations with GBs has led to a better understanding of deformation mechanisms [66–68].

Atomic-scale characterization

The transmission electron microscope can be used to form a lattice image from which in principle the atomic structure of an interface can be derived [69]. Interfaces suitable for HREM investigations are tilt GBs, whose tilt axes coincide with a low-index zone axis. Several advances in HREM techniques have been introduced in recent years, which make it now possible to obtain atomic-scale information about many interfaces of interest to materials science. The most important advance came with the introduction of modern HREM instruments capable of a point-to-point resolution better than $2\,\text{Å}$. The basic technique is concerned with the determination of the position of atoms. We should also mention that valuable extensions or modifications of the technique (using z-contrast scanning transmission electron microscopy, for example [70, 71]), can provide information on the chemical composition or electronic structure of interfaces [72, 73].

HREM techniques have been described in a number of publications [69, 74]. The standard HREM imaging method is axial illumination high-resolution electron microscopy, which is schematically illustrated in Fig. 3.6: A low-index zone axis of a thin specimen ($\leqslant 10$ nm) is aligned parallel to the electron beam. The diffracted beams form a diffraction pattern. A suitable objective aperture is selected and positioned concentric to the primary electron beam. The size of the aperture is generally chosen such that the cutoff in the transmitted spatial frequencies is well above the corresponding point resolution of the instrument.

Essential steps in forming an HREM image are: (i) the interaction of the incident electron beam with the sample and (ii) the magnification of the electron wave distribution at the specimen exit surface by the objective lens of the microscope. The image is formed through interference of the diffracted beams. Contrast reversals as a function of thickness and objective-lens defocus are typical. Therefore, an interpretation of the observed images must be based on detailed computer simulations of the processes in steps (i) and (ii).

In step (i) it is necessary to calculate the electron wave distribution at the specimen exit surface via n-beam dynamical diffraction theory. A number of computer codes are available based on the multislice formulation of Cowley and Moodie [75]. The contrast transfer by the objective lens is calculated by wave optics. Due to the spherical aberration of the objective lens, phase shifts are introduced between the various diffracted beams. The phase shift depends on the defocus ΔF and is a function of the scattering angle, i.e. the spatial frequency in the object. Near the optimum defocus $\Delta F_o = -(3C_s\lambda/2)^{\frac{1}{2}}$, where C_s is the spherical aberration coefficient of the objective lens and λ is the electron wavelength, a broad transfer band is obtained for which the phase of all diffracted beams has the same sign up to the maximum spatial frequency given by the point-to-point resolution. For thin specimens (<10 nm) directly interpretable images can be obtained at this defocus, provided the point-to-point resolution is better than the smallest spacing between atomic columns in the image. However, even for imaging near optimum defocus, image simulations are very desirable in order to obtain the sensitivity of GB structure images to changes in imaging conditions.

Figure 3.7 shows simulated images of a (221) symmetrical $\langle 1\bar{1}0 \rangle$ tilt boundary (with $\Sigma = 9$) in Au, based on the model structure derived by EAM calculations, at a thickness of 4.3 nm. In order to have periodic continuation of the calculational cell, two (identical) grain boundaries are introduced in the GB supercell (for clarity, three GB planar unit cells are shown horizontally in Fig. 3.7). Quite good correspondence between the loci of the atomic columns in the model structure and the centers of the bright contrast spots in Fig. 3.7 is found near optimum defocus.

Digital image processing is often used to help extract the atomic structure of a GB. Figure 3.8 shows a HREM micrograph of a (113)(113) STGB in Au. Since this GB is periodic, several GB units

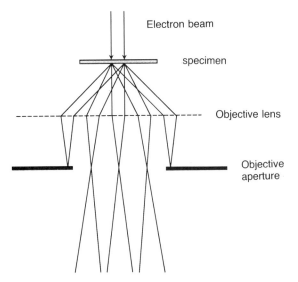

Fig. 3.6 Schematic ray diagram illustrating axial illumination high-resolution electron microscopy. A thin specimen ($\leqslant 10$ nm) is aligned with a low index zone axis parallel to the electron beam. Several of the diffracted beams are used to form the HREM image.

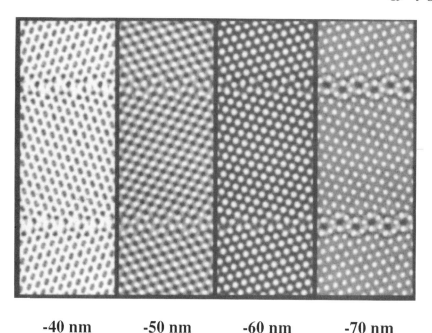

-40 nm　　**-50 nm**　　**-60 nm**　　**-70 nm**

Fig. 3.7 Image simulations are an important component of high-resolution electron microscopy studies. Atomic columns are imaged in an easily interpretable manner only for very special conditions. This focal series was calculated using a multislice algorithm. The strong dependence of the HREM image on the objective lens focus is readily apparent. This image simulation was based on a computer generated GB structure for the (221)(221) symmetric tilt GB in Au, a film thickness of 4.3 nm, and electron optical parameters of a 300 kV HREM. In the image near optimum defocus (-600 nm) the atomic columns appear as white dots.

along the GB can be averaged digitally. The image which is obtained by this procedure is shown in the inset to Fig. 3.8. The shot noise from the electrons and the photographic grains, as well as possible background due to amorphous surface layers is considerably reduced in the averaged image, which can now serve as basis for comparisons to model structures. Digital image processing techniques that employ Fourier filtering have to be approached with caution, since in these procedures essential image information may be lost and image artefacts could also be introduced.

In principle, the construction of an atomic model from HREM images proceeds as follows: An experimental through-focal series is matched to a series of simulations, based on iterative refinements of a proposed structure and on a knowledge of all of the relevant instrumental and specimen parameters. Moreover, since observation along a single low-index zone axis can at best give a two-dimensional projection of the structure, observation along an additional direction is necessary for obtaining a full three-dimensional model of the structure. In practice such a full set of data is almost never available. In very rare instances,

observations along two orthogonal directions have been made [76] and quite frequently important parameters, such as specimen thickness, are not accurately known. Nevertheless, even the semi-quantitative analysis of HREM observations has provided a wealth of important information concerning the atomic structure of interfaces. Moreover, when computer simulated GB structures are available, a direct comparison between observed images and images from simulations can be made. The accuracy to which the relative position of atomic columns can be determined under optimum conditions is presently believed to lie near 0.2 Å [76, 77], although greater precision may be achieved under certain circumstances [23].

Bicrystal manufacture

Two approaches have been taken to the manufacture of well-controlled interfaces. In one method bulk bicrystals or heterophase interfaces are produced by diffusion bonding of two single crystals under ultrahigh vacuum [78]. An alternative method, illustrated in Fig. 3.9, is suitable for the preparation of pure twist or tilt GBs in thin metallic

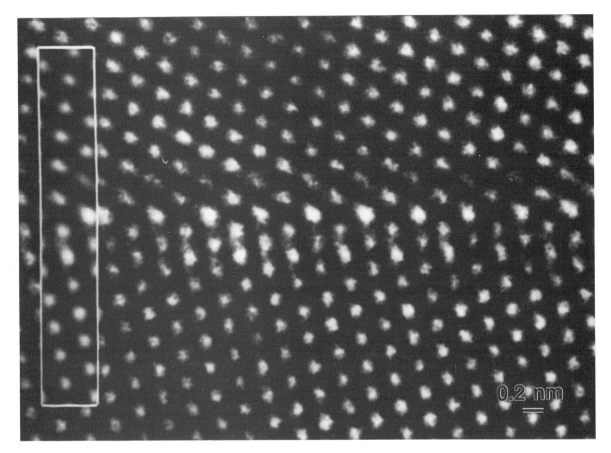

Fig. 3.8 Digital image processing can aid in the interpretation of GB structure images. Since the photographically recorded HREM images typically are associated with considerable background noise, which has its origin in electron shot noise, as well as possible amorphous surface layers, digital image processing is often applied. The translational symmetry along the GB in this image of a (113)(113) tilt GB in Au has been used to obtain a digitally averaged image of the GB structural unit cell, shown in the inset. It is readily seen that the GB phane does not coincide with a mirror symmetry plane.

films. In this case thin, epitaxially grown films are misoriented by a specified twist angle ψ and bonded by a suitable pressure/temperature treatment while still on the substrate. This so-called Schober–Balluffi technique produces a twist GB [79]. After removal from the substrate pure tilt GBs can be obtained upon further annealing due to boundary migration. The development of a columnar grain structure results in the production of transverse tilt GBs with the misorientation angle ψ [80]. Suitably thin sections for HREM can be prepared from such samples by ion-beam milling and a subsequent annealing treatment. Since the GBs in the resulting 'bicrystal' can assume all poss-

ible inclinations normal to the tilt axis, one can in this manner study the GB planes which are allowed for the predetermined misorientation between the crystals.

3.4 CORRELATION BETWEEN MACROSCOPIC DOFS AND ENERGY

Using the methods described above, and applying the terminology defined in section 3.2, we are now able to address the first of the three aspects of the structure-energy correlation, namely the variation of the GB energy in the five-parameter misorienta-

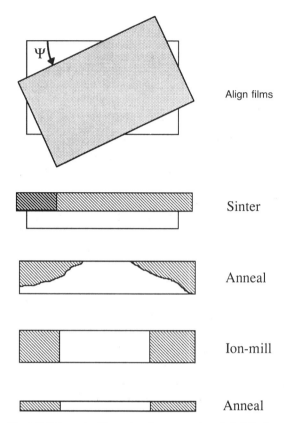

Align films

Sinter

Anneal

Ion-mill

Anneal

Fig. 3.9 Schematic, illustrating the preparation of tilt GBs from thin, epitaxially grown films. Thin Au films, for example, are grown on NaCl substrates. The films are aligned precisely to the desired misorientation angle ψ and sintered together (see plan and side views). After removal from the substrate, further annealing steps are used to convert the twist GB to tilt grain boundaries (see cross-section views). Ion-thinning is used to get suitably thin sections for HREM.

tion phase space defined in eqs. (3.1) and (3.3), respectively.

In reviewing the relative energies of GBs in the five-parameter misorientation phase space, we start with symmetrical tilt boundaries (STGBs) which, with only two DOFs, represent the simplest of all GBs. As discussed above in section 3.2.1, these two DOFs are the ones associated with the GB-plane normal, $\langle hkl \rangle$ (eqs. (3.8)–(3.10)). As in the case of free surfaces, the investigation of STGBs thus provides an opportunity to study the role of the interface plane in the energy and physical properties of these simple interfacial systems. Next,

the twist angle will be added as the third DOF (eq. (3.8)); this will enable us to compare directly the energies of symmetrical twist and tilt boundaries (sections 3.4.1–3.4.4). Finally, removing the constraint of symmetry leads to asymmetrical tilt and twist boundaries, with four and five DOFs, respectively (section 3.4.5). The section concludes with a review of HREM investigations of the importance of the GB plane in symmetrical and asymmetrical tilt boundaries (section 3.4.6).

3.4.1 Symmetrical GBs

Symmetrical tilt boundaries

The properties of symmetrical tilt boundaries are usually discussed in terms of the three CSL-based variables in the tilt axis, $\hat{n}_T$, and tilt angle, ψ (eqs. (3.9) and (3.10)). The specification of three parameters for a system with only two DOFs means that the system is overspecified, and a relationship must exist amongst them. The tilt axis and angle obtained from eqs. (3.9) and (3.10) are therefore not unique, and other ($\hat{n}_T$, ψ) combinations may be given for the same STGB on the (hkl) plane (see Chapter 1 in this volume). Because only two of the three variables in ($\hat{n}_T$, ψ) are independent DOFs, for a given choice of the tilt axis (by the condition that, for example, $\hat{n}_T \perp \hat{n}_1$), the tilt angle is thus fully determined. Hence, while eq. (3.9) appears to restrict $\hat{n}_T$ to directions with at least one vanishing Miller index, we point out that, in deriving eqs. (3.9) and (3.10) [17], one of the Miller indices in $\hat{n}_T$ may be chosen freely, and in this case l_T was chosen to vanish [17].

As an example, we consider the STGB on the (341) plane which, according to eqs. (3.9) and (3.10) may be considered as having been generated by a rotation about the $\langle \bar{4}30 \rangle$ axis by $\psi = 22.62°$. However, had we chosen, for example, $l_T = 1$ in formulating eqs. (3.9) and (3.10), the direction perpendicular to $\langle 341 \rangle$ obtained would be the $\langle 1\bar{1}1 \rangle$ direction, with the more familiar value of $\psi = 27.80°$. (For further details see Refs. [17], [18] and [82].)

For the moment, however, we will use the conventional, CSL-based terminology, and return to an analysis of STGBs based on their DOFs in section 3.4.4. Figures 3.10(a) and (b) show the variation of the energy, $E^{GB}(\psi)$, for STGBs perpendicular to a $\langle 110 \rangle$ tilt axis for both the fcc and bcc metals. In both cases, the many-body and pair potentials give a rather similar qualitative behavior. Since STGBs are fully characterized by the Miller indices of the GB plane (section 3.2.1 and Chapter 1 in this volume), the existence of deep energy

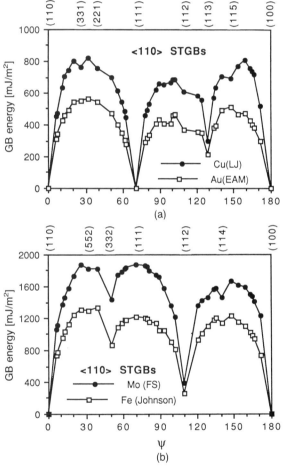

Fig. 3.10 Comparison of the energies (in mJ/m²) of symmetrical $\langle 011 \rangle$ tilt boundaries in the fcc lattice [81] (a) with those in the bcc lattice [82] (b) obtained by means of both potentials in each case. The STGB planes giving rise to energy cusps are indicated on the top of each figure. The existence of the primary cusps was verified experimentally by means of the thermal-grooving technique [83, 84]. The lines connecting the data points are merely a guide to the eye.

cusps indicates an important role played by the interface plane. Interestingly, however, while the fcc metals exhibit deep cusps at the (111) and (113) GB planes (with minor cusps at the (331), (221), (112), and (115) orientations), the bcc metals show primary cusps at the (112) and (332) planes (with minor cusps at the (552) and (114) orientations).

The existence of the primary fcc cusps in Fig. 3.10(a) has been verified experimentally via thermal-grooving experiments for Cu and Al by Hasson *et al.* [83], while Sedlacek *et al.*, in similar experiments on Mo, have confirmed the existence of a deep cusp at the (112) STGB orientation in Fig. 3.10(b) [84].

A comparison with the related interplanar lattice spacings, $d(hkl) \sim (h^2 + k^2 + l^2)^{-\frac{1}{2}}$, in Fig. 3.11(a) and (b) demonstrates a strong correlation between large values of $d(hkl)$ and the appearance of energy cusps [16]. We note, however, that the (100) and (110) planes are exceptions in both the fcc and bcc lattices because, with only two planes in the repeat stacking sequence (Table 3.2), the STGB configurations on these two planes are identical to the ideal crystal (see also Chapter 1 in this volume) [16]. The densest fcc planes which can sustain an

Table 3.2 Interplanar spacing, $d(hkl)$ (in units of the lattice parameter a), for the 11 most widely spaced planes in the fcc and bcc lattices. These planes also correspond to the ones with the highest planar density of atoms, i.e. the smallest planar repeat unit cells. Also listed is the number of planes in the repeat planar stacking unit, referred to as the 'period', $P(hkl)$ (see also Chapter 1 of this volume)

	fcc			bcc		
No.	*(hkl)*	*d(hkl)/a*	*P(hkl)*	*(hkl)*	*d(hkl)/a*	*P(hkl)*
1.	(111)	0.5774	3	(110)	0.7071	2
2.	(100)	0.5000	2	(100)	0.5000	2
3.	(110)	0.3535	2	(112)	0.4082	6
4.	(113)	0.3015	11	(310)	0.3162	10
5.	(331)	0.2294	38	(111)	0.2887	3
6.	(210)	0.2236	10	(321)	0.2673	14
7.	(112)	0.2041	6	(114)	0.2357	18
8.	(115)	0.1925	27	(210)	0.2236	10
9.	(513)	0.1690	35	(332)	0.2132	22
10.	(221)	0.1667	18	(510)	0.1961	26
11.	(310)	0.1581	10	(341)	0.1961	26

inverted stacking sequence upon translation, and hence accommodate an STGB configuration, are therefore the (111) and (113) planes, while the densest bcc planes with this property are the (112) and (310) planes (Table 3.2). (The latter plane is perpendicular to an $\langle 001 \rangle$ tilt axis, and also gives rise to an energy cusp [82].) A comparison with Table 3.2 shows that, with the exception of the (111) plane in the bcc lattice [82], both the major and minor cusps correspond to relatively large values of $d(hkl)$. One is therefore inclined to conclude that a large value of $d(hkl)$, indeed, favors a low GB energy [16, 28]. As discussed in the next section, however, this criterion applies only to the 'special' but not the 'vicinal' GB planes, i.e. to those giving rise to cusps but not those in the vicinity of the cusps; their energy and structure is governed by the angular deviation from the nearest 'special' (i.e. cusped) orientation [16].

Symmetrical twist boundaries

Adding now the twist angle, θ, as the third DOF, based on the above results one would expect that the symmetrical twist boundaries with the lowest energies are found on the densest planes of each crystal lattice. Figures 3.12(a) and (b) show this to be, indeed, the case for both fcc and bcc metals. (While these results were obtained for the pair potentials, qualitatively identical results are found for the many-body potentials [30, 49, 85, 86].) According to Fig. 3.12(a), in the fcc metals the (111) boundaries show by far the lowest energy, followed by the GBs on the (001) and (011) planes. By contrast, in the bcc lattice the lowest energies are observed for the (011) and (001) planes, with more or less equal energies on the less-dense planes.

By comparison with the $E^{GB}(\psi)$ (STGB) results in Fig. 3.10, the $E^{GB}(\theta)$ (twist-boundary) curves in Fig. 3.12 are relatively smooth. The latter show significant cusps only at the beginning and end-points of the twist-misorientation range (i.e. for $\theta = 0°$ and 180°) and for the (011) (011) $\theta = 70.53°$ twist boundary [85]. The latter is the so-called $\Sigma 3$ twist boundary, with a unit-cell area only three times larger than that of the perfect-crystal

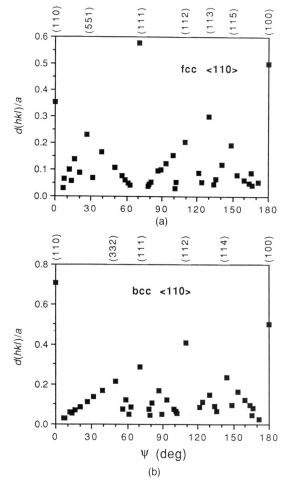

Fig. 3.11 Interplanar lattice spacing (in units of the lattice parameter) versus tilt rotation angle, ψ, about $\langle 110 \rangle$ for (a) the fcc and (b) the bcc lattice (Figs. 3.10(a) and (b)) [16, 81, 82].

configurations on the (011) plane at $\theta = 0°$ and 180°. All other twist boundaries have significantly larger planar repeat unit cells (with $\Sigma = 9, 11, 17, 19, 27$, etc. [85]). This suggests that, in the case of twist boundaries, a small planar unit-cell area favors a low GB energy as does a relatively dense GB plane.

The overall smoothness of the $E^{GB}(\theta)$ plot associated with the (001) twist boundaries in Fig. 3.12(a) was verified by Chan and Balluffi [87] in 'rotating-crystallite' experiments in Au, in which only a minor cusp, near $\theta = 36.87°$ ($\Sigma 5$), was detected. For smaller angles the rather smooth

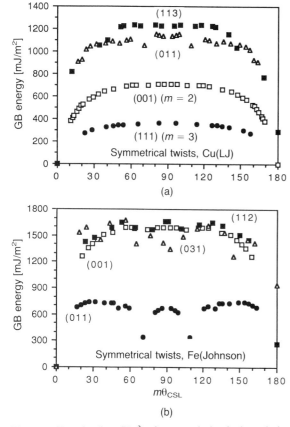

Fig. 3.12 Energies (in mJ/m^2) of symmetrical twist boundaries versus twist angle for the four densest planes (Table 3.1) of (a) Cu obtained by means of the LJ potential [30, 85] and (b) α-Fe [49, 86] simulated using Johnson's pair potential. (The factor m is related to the rotation symmetries of the (111) and (001) planes.) In both cases, the many-body results are qualitatively identical. Notice at the endpoints (i.e. for θ = 180°) the non-vanishing energies of the STGB configuration on the (113) plane of the fcc lattice (a), and on the (112) and (031) planes of the bcc lattice. These are the GBs with the smallest planar unit-cell areas.

decrease towards zero seen in Fig. 3.12(a) was confirmed while for larger angles the existence of a flat plateau region was deduced [87]. Also, the existence of the cusp associated with the (011)(011)θ = 70.53° (Σ3) twist boundary in fcc metals was verified by Maurer and Balluffi [13, 14] by means of 'rotating-ball' experiments [88, 89]. These experiments also demonstrated the existence of cusps associated with the STGBs on the densest planes of the fcc lattice.

From the computer investigation of symmetrical GBs we therefore conclude that the GBs with the lowest energies are those on the densest planes and with the smallest planar unit cells. This conclusion is in accord with simple intuition according to which one would expect that (a) the creation of a GB on a widely-spaced set of lattice planes (with a high planar density) creates the least amount of structural disorder across the interface [16], and (b) a small planar repeat unit cell leads to a better ordered and more coherent atomic structure, and hence a lower energy, than a large unit cell [16].

That for the STGBs in Fig. 3.10 these two criteria are actually identical is readily seen. As discussed in Chapter 1 of this volume (eq. (1.43)), $d(hkl)$ and the planar unit-cell area, $A(hkl)$, of a perfect-crystal stack of (hkl) planes are related via the atomic volume, Ω, by $A(hkl)\,d(hkl) = \Omega$ [16]. A large interplanar spacing therefore means a small planar unit cell and vice versa. Since the STGB- and perfect-crystal configurations on a given plane have identical planar unit-cell dimensions [16], the appearance of cusps in Figs. 3.10(a) and (b) is correlated with the smallest planar unit-cell areas (see also Figs. 3.11(a) and (b)), i.e. the densest lattice planes.

Within the framework of the CSL terminology it has often been said that it is the CSL unit-cell volume, $\Sigma\,\Omega$ (where Ω is the atomic volume), which must be small in order to achieve a low GB energy. As discussed in detail in Chapter 1 in this volume (Fig. 1.43), however, a small value of Σ does not necessarily translate into a small planar unit cell, although for STGBs the two are simply related by [16] (Chapter 1)

$$\varepsilon(hkl)\,A(hkl) = (\Omega/a)\,\Sigma^{\frac{1}{2}} \qquad (3.16)$$

where $\varepsilon(hkl)$ assumes one of the following four values (depending on the particular combinations of h, k and l): $\sqrt{\tfrac{2}{4}}$, $\tfrac{1}{2}$, $\sqrt{\tfrac{2}{2}}$, and 1. Equation (3.16) may alternatively be written in terms of $d(hkl)$ as follows [16] (Chapter 1, eq. (1.73)),

$$d(hkl)/a = \varepsilon(hkl)\,\Sigma^{-\frac{1}{2}} \qquad (3.17)$$

a relationship which is illustrated in Figs. 3.13(a) and (b) for the STGBs in Figs. 3.10(a) and (b), respectively. The conclusion drawn from this

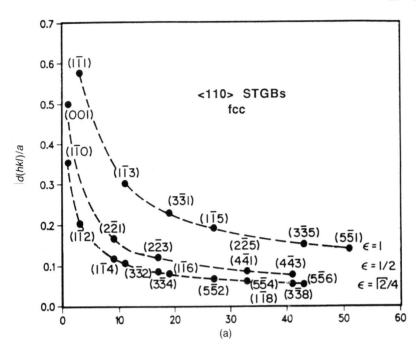

(a)

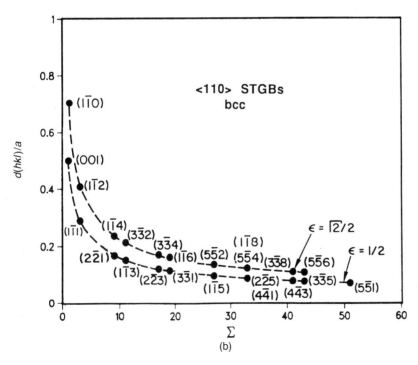

Σ

(b)

Fig. 3.13 Interplanar spacing versus value of Σ (eq. (3.17)) for the STGBs in Figs. 3.10(a) and (b) [16]. $d(hkl)$ is directly related to the **planar** unit-cell area of the GB (eq. (3.16)), whereas Σ governs the **volume** unit cell of the CSL. Although $\varepsilon(hkl)$ may, in principle, assume any of four different values, giving rise to four different curves (dashed lines), for the ⟨110⟩ misorientation axis considered here, not all four values are realized by the allowed combinations of Miller indices [16].

analysis is that a small **planar** unit cell of the GB does not necessarily correspond to a small **volume** unit cell of the CSL [90], unless the value of $\varepsilon(hkl)$ is large (eq. (3.16)). A small value of Σ is therefore a necessary but not sufficient requirement for a low STGB energy.

3.4.2 A Read–Shockley model for low- and high-angle twist boundaries

The dislocation model of Read and Shockley [7] represents one of the first successful attempts to incorporate analytically the variation of the GB energy as a function of the misorientation angle. In this model isotropic continuum elastic theory is used to subdivide the GB energy, $E^{GB}(\theta)$, of low-angle GBs into a contribution due to the dislocation cores, with energy per unit length, E_c, and one arising from their surrounding strain fields, with energy per unit length, E_s, according to [7]

$$E^{GB}(\theta) = \theta(E_c - E_s \ln \theta)/b, \quad (\theta \leqslant \theta_0) \quad (3.18)$$

where, traditionally, θ_0 is taken to be approximately 15°. In deriving eq. (3.18) from continuum-elastic theory, the dislocation-core energy was introduced on an ad-hoc basis [7], and the spacing between dislocations was assumed to be large enough, i.e. θ to be small enough, such that $\sin \theta \cong \theta$. Linear continuum elasticity theory may then be used to derive expressions for E_s; for twist boundaries, $E_s = Gb/2\pi$, with G and b denoting the shear modulus and the Burgers vector, respectively, while for tilt GBs, $E_s = Gb/[4\pi(1 - \nu)]$, where ν is Poisson's ratio. In practice, E_c and E_s may be viewed as adjustable parameters, to be determined from a fit to actual GB energies; their values, and hence the physics of low-angle GBs, are therefore governed by the core- and strain-field properties of isolated lattice dislocations.

Because of the overlap of the dislocation cores in **high-angle** GBs, continuum-elasticity theory cannot be used to describe the stresses and strain fields near a GB in terms of the overall (by contrast to the local) elastic behavior of the material. Various attempts have nevertheless been made to extend GB models based on lattice dislocations to high-angle boundaries, but no simple extension of

the functional form of eq. (3.18) for the entire range of θ values has evolved from these efforts (see, for example, Refs. [91–96]).

In a recent, strictly empirical attempt to extend the range of validity of eq. (3.18) to larger angles, it was demonstrated that for symmetrical twist, tilt and even asymmetrical GBs (discussed in section 3.4.5 below), the following simple extension of the Read–Shockley equation (3.18) is capable of describing computed GB energies and volume expansions extremely well over the entire range of θ values ($0 \leqslant \theta \leqslant 180°$) [97]:

$$E^{GB}(\theta) = \sin\theta \, [E_c^< - E_s^< \ln(\sin \theta)]/b,$$
$$(0 \leqslant \theta \leqslant 90°) \quad (3.19a)$$

$$E^{GB}(180° - \theta)$$
$$= E^{STGB} + \sin(180° - \theta)$$
$$\{E_c^> - E_s^> \ln[\sin(180° - \theta)]\}/b,$$
$$(90° \leqslant \theta \leqslant 180°) \quad (3.19b)$$

Here E^{STGB} is the energy of the symmetrical tilt configuration at $\theta = 180°/m$ which, because of its particularly small planar unit cell, always gives rise to a deep cusp (section 3.4.1; for the definition of m, section 3.2.1). The core and strain-field energies $E_c^<, E_s^<$ and $E_c^>, E_s^>$ associated with angles $\theta < 90°$ and $\theta > 90°$, respectively, are regarded as adjustable fitting parameters.

An interesting property of eqs. (3.19a) and (b) is that the derivatives, $dE^{GB}(\theta)/d\theta$, vanish for $\theta = 90°$, giving rise to a more or less wide plateau region in which the energy is independent of θ (Fig. 3.12). Moreover, equating $E^{GB}(90°)$ obtained for the two branches in eqs. (3.19a) and (b) (i.e. assuming that the GB energy is continuous at $\theta = 90°$) is equivalent with assuming that only a single dislocation-core energy, $E_c^< = E_c^> = E_c$, is required for the entire misorientation range ($0° \leqslant \theta \leqslant 180°$), thus reducing the number of fitting parameters from four to only three [97]. That this assumption is well satisfied was demonstrated not only for symmetrical twist boundaries (Fig. 3.12) but also for asymmetrical twist and tilt boundaries (although the physical implications of this fact are not clearly understood [97]). This enables, at least in principle, a description of the entire misorientation range, $0 \leqslant \theta \leqslant 180°$, by an analytical function with only three adjustable parameters ($E_s^<$, $E_s^>$, and E_c) which is continuous at $\theta = 90°$ in both its value and derivative. For any given (symmetrical or asymmetrical) GB plane, a simple three-parameter model based on eqs. (3.19) (and the physics of isolated lattice dislocations!), is thus capable of describing **analytically** the θ dependence of the GB energy. For obvious reasons, eqs. (3.19a) and (b) cannot be derived from continuum elasticity, although their basic validity can be rationalized to some degree, particularly the $\sin \theta$ variation of the core contribution [97, 98]; in practice, their only justification lies in their ability to provide a quantitative description of the GB energy (and volume expansion; section 3.5.2) for the entire range of misorientation angles and for different types of GBs.

How well eqs. (3.19a) and (b) actually represent simulated GB energies is illustrated in Fig. 3.14 for (a) symmetrical (111) twist boundaries in the fcc lattice (Fig. 3.12(a)) and (b) for symmetrical (001) twist boundaries in the bcc lattice (Fig. 3.12(b)) obtained by means of the two types of potentials in each case. The fitting parameters are listed in Table 3.3.

The fact that $E^{GB}(\theta)$ can be described by an analytical expression in the entire misorientation range ($0° \leqslant \theta \leqslant 180°$) strongly suggests that the underlying physics is the same for low- and high-angle boundaries [99]. Since the core energy seems to be practically independent of the spacing between the dislocations, even when they overlap completely (Li [99]), one is inclined to conclude that the details of the distribution of strain in the highly relaxed GB region is relatively unimportant from the GB-energy point of view. This conclusion is supported by the rotating-crystallite experiments

of Chan and Balluffi [87] for (100) twist boundaries in Au, in which secondary GB dislocations were observed near $\theta = 36.87°$ ($\Sigma 5$) and $\theta = 22.62°$ ($\Sigma 13$) orientations [79], with a remarkable absence, however, of any significant energy cusps at these orientations (Fig. 3.12(a)). As these 'special' orientations are approached (with particularly small planar unit-cell areas), a redistribution of strain, giving rise to the secondary GB dislocations [79], obviously takes place, with little effect on the energy, however. This raises the question as to whether such dislocations have much physical meaning or whether they are just microredistributed strain contrasts, with mostly crystallographic significance only.

As already mentioned (section 3.2.1), if the planar unit cell of a perfect-crystal stack of (hkl) planes contains an m-fold rotation axis (as is the case for the (100) and (111) planes in cubic crystals, with $m = 2$ and $m = 3$, respectively), the STGB cusp associated with pure twist misorientations appears at $\theta = 180°/m$ instead of $\theta = 180°$, as was assumed in writing eqs. (3.19a) and (b); the point of zero slope is scaled accordingly from 90° to $90°/m$. For eqs. 3.19 to be applicable in such cases, θ and $180° - \theta$ must be replaced by θ/m and $(180° - \theta)/m$, respectively. An alternative approach, followed in the original paper [97], is to interpret θ in eqs. (3.19) as the angle $\theta = m\theta_{CSL}$, where θ_{CSL} is the usual CSL misorientation angle (Fig. 3.12).

A shortcoming of eqs. (3.19a) and (b) is their inability to account for cusps in the $E^{GB}(\theta)$ curve (Figs. 3.12(a) and (b)). As discussed in the preceding section, such cusps generally appear whenever a GB with a particularly small planar unit cell, surrounded by very large ones, is approached. In

Table 3.3 Parameters E_s/b and E_c/b obtained from a least-squares fit of eqs. (3.19a and b) to the simulation data in Figs. 3.14(a) and (b) for symmetrical (111) twist boundaries in the fcc lattice [97] and for symmetrical (001) twist boundaries in the bcc lattice (in mJ/m²)

	E^{GB} ($\theta = 0°$)	E^{GB} ($\theta = 180°$)	E_s/b ($\theta < 90°$)	E_c/b	E_s/b ($\theta > 90°$)
(111) Twist Cu(LJ)	0	1	371	356	368
(111) Twist Au(EAM)	0	3	181	216	171
(001) Twist Fe(Johnson)	0	0	1835	1578	1835
(001) Twist Mo(FS)	0	0	2117	2199	2117

principle, a least-squares analysis based on eqs. (3.19a) and (b) would have to be performed for every cusp, requiring more simulation data in Figs. 3.12(a) and (b); hence, at present only the 'low-angle cusp' at $\theta = 0°$ and the 'STGB cusp' [100] at $\theta = 180°$ have been analyzed in terms of eqs. 3.19. (For further information, see Refs. [41], [97], and [98].)

3.4.3 'Special' versus 'vicinal' GBs

As discussed in section 3.4.1, the simulation results for STGBs and symmetrical tilt and twist boundaries (Figs. 3.10 and 12) suggest a strong correlation between the appearance of energy cusps and GB-plane orientations parallel to the densest lattice planes. One is therefore tempted to conclude that in general a larger interplanar spacing, indeed, means a lower GB energy. This suggestion is based on the physical reasoning that atoms are shoved more closely together when creating a twist boundary on a plane with a small rather than a large $d(hkl)$ value [16, 28, 30, 44]. As a consequence of the short-range repulsion between the atoms, the bicrystal then expands locally at the GB, and this expansion is larger for the GBs with the smaller $d(hkl)$ spacing, with consequently higher GB energy [16, 28, 30, 44]. In this section we discuss a simple model for high-angle **twist** boundaries which provides insight into the role of the GB plane in the GB energy.

A 'typical' high-angle twist boundary, in the sense of Read and Shockley [7], is characterized by a complete overlap of the cores of the screw dislocations introduced when rotating from $\theta = 0°$ [99, 100]. As illustrated in Figs. 3.12 and 3.14, the energy (and volume expansion; section 3.5.2) and probably other properties as well are then essentially independent of the twist angle, θ, and a simple model for **high-angle** GBs may be formulated which incorporates only the four macroscopic DOFs associated with the GB plane but in which θ is not a variable.

In the so-called random grain-boundary (RGB) model [44, 98, 101, 102] all interactions across the GB are assumed to be random, with the constraint, however, that – as in 'real' GBs – the atoms are

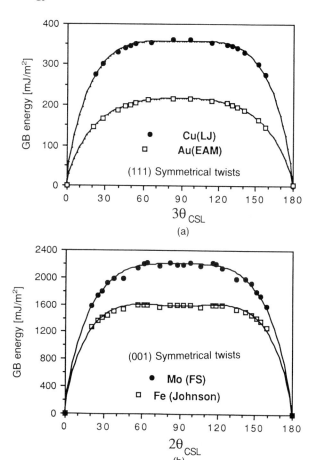

Fig. 3.14 Least-squares fits of the extended Read–Shockley eqs. 3.19(a) and (b) to the simulation data for (a) symmetrical (111) twist boundaries in the fcc lattice (Fig. 3.12(a)) and (b) symmetrical (001) twist boundaries in the bcc lattice (Fig. 3.12(b)) obtained by means of the two types of potentials in each case. The fitting parameters are listed in Table 3.3. Because of the 3-fold symmetry axis in the planar unit cells, in figure (a) the usual misorientation angle, θ_{CSL}, was replaced by $\theta = 3\theta_{CSL}$, thus scaling all angles to 180°. The (111) twin configuration, with an unrealistically low energy for both potentials (Table 3.3), is thus shifted from $\theta_{CSL} = 60°$ to $\theta = 180°$, as required by eqs. (3.19).

assumed to lie in well-defined lattice planes. As discussed in detail in Chapter 27 of this volume, this model provides an excellent description of (symmetrical and asymmetrical) high-angle twist boundaries with completely overlapping dislocation cores [99]. The GB energy is then independent of rigid-body translations parallel to the interface, with a consequently vanishing modulus for shear parallel to the interface [103]. The number of **microscopic** DOFs of the GB (section 3.2.2) is then reduced from three to only one, namely the volume expansion at the GB which is usually spread out over a number of lattice planes near the GB. In the RGB limit an asymmetrical high-angle twist GB is therefore characterized by only the **four macroscopic** DOFs associated with the GB plane and the **one microscopic** DOF associated with the volume expansion, hence enabling a systematic investigation of the role of the GB plane which is unencumbered by the twist angle, θ, and by rigid-body translations parallel to the GB. In the symmetrical case the number of macroscopic DOFs is thus reduced to only the **two** associated with the RGB plane, in close similarity to free surfaces and STGBs. This enables a systematic comparison of the role of $d(hkl)$ in the energies of these three simple interface systems.

In Figs. 3.15(a) and (b) the energies of RGBs, free surfaces and STGBs perpendicular to a $\langle 110 \rangle$ pole (or tilt) axis are plotted as a function of the tilt angle, ψ, for the many-body potentials. However, while all three systems show major cusps at the same interface-plane orientations, a more extensive investigation of the role of $d(hkl)$ in their energy [98, 102, 104] shows that in both the fcc and bcc lattices the simple interplanar-spacing criterion applies only to the four or five densest lattice planes (indicated on the tops of Figs. 3.15–3.17). For the smaller values of $d(hkl)$ the simulation data obtained for both the many-body and pair potentials (in Figs. 3.16 and 3.17, respectively) scatter widely, indicating that the GB energy is not exclusively governed by the interplanar spacing.

That the energy of high-angle twist GBs on some arbitrary higher-index GB plane **cannot** be governed by $d(hkl)$ becomes apparent if we consider, for example, GB planes in the vicinity of the (111) cusps in Figs. 3.15(a) and (b). The energy of GBs near this cusp are obviously governed by the energy (and, hence, by the interplanar spacing) of the nearby cusped (111) orientation. As the cusp is approached, the corresponding value of $d(hkl)$ approaches zero (Figs. 3.11(a) and (b)) while the energy remains finite (Figs. 3.15(a) and (b)). For example, a (98, 98, 99) plane is very close to the

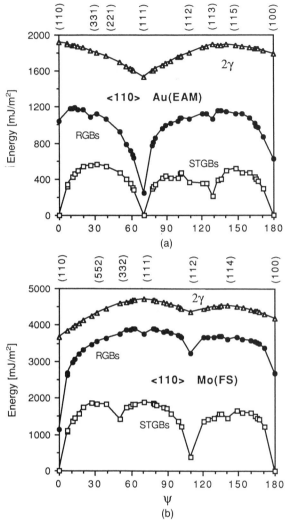

Fig. 3.15 Comparison of the energies (in mJ/m^2) obtained by means of the many-body potentials for symmetrical RGBs, free surfaces and the STGBs in Fig. 3.10 which can be generated by a rotation, by the angle ψ, about a $\langle 011 \rangle$ pole (or 'tilt') axis; (a) Au(EAM), (b) MO(FS) potential [86, 98]. The densest lattice planes (Table 3.2), giving rise to energy cusps, are indicated on the top of each figure. Notice that the energy of *two* free surfaces, 2γ, is plotted, corresponding to the RGB limit for $\delta V \to \infty$ [86, 98]. The pair-potential results are qualitatively identical.

(111) plane, with a practically vanishing value of $d(hkl)$ because the latter is proportional $(h^2 + k^2 + l^2)^{-\frac{1}{2}}$ (Chapter 1 in this volume). These very small $d(hkl)$ values, however, related to the same interface energy as the cusped (111) orientation, are

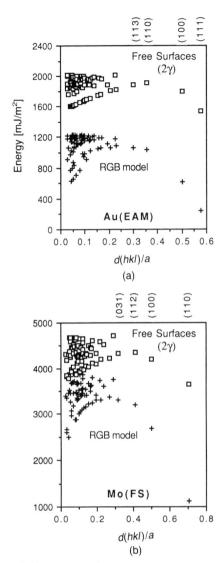

(a)

(b)

Fig. 3.16 RGB and related free-surface energies, 2γ (in units of mJ/m^2), as a function of the interplanar lattice spacing, $d(hkl)$, parallel to the interface (in units of the lattice parameter) for (a) the Au(EAM) and (b) Mo(FS) potential. In addition to the interfaces considered in Fig. 3.15 (with a $\langle 110\rangle$ pole axis), interfaces with a common $\langle 001\rangle$, $\langle 111\rangle$, or $\langle 112\rangle$ pole axis are included here [86, 98]. Similar plots are obtained for the STGBs in Figs. 3.10(a) and (b).

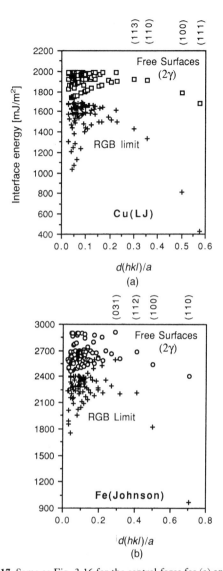

(a)

(b)

Fig. 3.17 Same as Fig. 3.16 for the central-force fcc (a) and bcc (b) potentials.

responsible for the data points in the far left of Figs. 3.16 and 3.17, and hence for the apparent scatter of the data.

Moreover, as is well known [35–37] the structure of the STGBs near a major cusp may be decomposed into the **polyhedral units** of the 'special' STGB causing the cusp. Similarly, the structure and energy of free surfaces in the vicinity of a cusped orientation (so-called 'vicinal surfaces') may be characterized by the number of and distance between the **steps** introduced into the nearby flat principal surface [105]. As one deviates further and further from the cusp, the distance between steps decreases, with a consequent increase in the interface energy (Figs. 3.15(a) and (b)). In the case of the STGBs, this increase as a function of $\Delta\psi$ is described by the Read–Shockley equations (3.18) and (3.19) (in which θ is replaced by $\Delta\psi$), while for free surfaces the logarithmic $\Delta\psi$ dependence is replaced by a much weaker $\sin\Delta\psi$ variation of the surface energy, γ [106] (see also Chapter 26).

It follows that in assessing the role of $d(hkl)$ in the GB and free-surface energies one should distinguish between the 'special' interfaces (in the case of free surfaces referred to as 'flat') giving rise to the cusps and 'vicinal' interfaces, whose properties are dominated by a nearby 'special' interface and the value of $\Delta\psi$. A detailed analysis of the data in Figs. 3.16 and eq. (3.17) presented elsewhere [98, 102, 104], in which all data points were identified in terms of the related cusps, reveals that the energy of the **special** surfaces and GBs is, indeed, governed by the magnitude of $d(hkl)$. By contrast, the value of $d(hkl)$ is irrelevant in the energies of the **vicinal** surfaces and GBs, because their energy is governed by the angular deviation, $\Delta\psi$ or $\Delta\theta$, from the cusped orientation. This distinction between vicinal and special GBs and free surfaces, and the different roles played by $d(hkl)$, thus provides a fundamental concept for the understanding of free surfaces and GBs, and probably heterophase interfaces as well.

We conclude this section with an analysis of how well the RGB model actually represents the simulated energies of 'true' high-angle twist GBs (in the Read–Shockley sense). As illustrated in Figs. 3.18(a) and (b), for both symmetrical and asymmetrical twist boundaries a reasonably good linear relation exists between the RGB energy and the corresponding plateau energy (in the symmetrical case extracted from the $E^{GB}(\theta)$ curves in Figs. 3.12(a) and (b), and for asymmetrical GBs discussed in section 3.4.5 below). The solid lines in Figs. 3.18(a) and (b), representing linear least-squares fits to all the data through the origin, indicate that the RGB energy exceeds the related plateau energy typically by 20–30%. (Similar values are obtained for the many-body potentials [86, 98].) The reason for this difference arises from the 'rumpling' of the atom positions within the lattice planes for the actual GBs, by contrast with the RGB model which assumes the lattice planes to

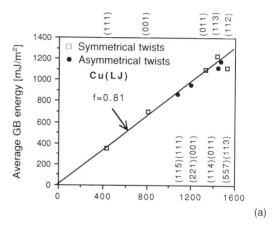

(a)

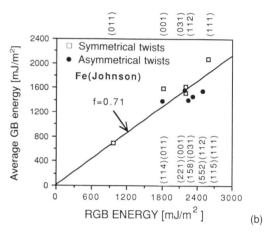

(b)

Fig. 3.18 Average GB energy in the plateau (i.e. high-angle) region versus energy obtained in the random-boundary (RGB) limit (in mJ/m^2) for symmetrical and asymmetrical twist boundaries involving the densest planes in (a) the fcc and (b) the bcc lattice. The straight lines represent least-squares fits through the origin; their slopes are indicated. The symmetrical plane orientations are indicated on the top; asymmetrical combinations of lattice planes are shown in the lower right of the figures. Similar plots are obtained for the many-body potentials [98, 102].

be completely flat [98]. This rumpling of the atom planes (or perhaps better called 'roughening'), represents a rather effective relaxation mechanism resulting in a lowering of the GB energy by comparison with the value obtained in the RGB limit.

3.4.4 Structure–energy phase space for all symmetrical GBs

As discussed above, STGBs and free surfaces represent the conceptually simplest of all crystalline interface systems because both are equally characterized by only the two macroscopic DOFs associated with the interface plane. Their macroscopic structure–energy correlation, which may thus be displayed in a single 3D plot, should therefore provide additional insight into the importance of the interface plane in their energy.

As is well known, in a cubic crystal all orientations, specified by the polar and azimuthal angles ϑ and φ, may be represented by the phase-space triangle sketched in Fig. 3.19 which is similar to the stereographic triangle. Figure 3.19 shows the distribution of the STGB and free-surface normals considered in our exploration, by means of computer simulation, of this two-parameter $\{\vartheta, \varphi\}$ phase space [81, 82, 98]. The figure also illustrates our motivation for including $\langle 111 \rangle$ and $\langle 112 \rangle$ tilt or pole axes, in addition to the more popular $\langle 110 \rangle$

and $\langle 100 \rangle$ orientations: the latter sample only the periphery of the triangle, whereas the $\langle 111 \rangle$ and $\langle 112 \rangle$ axes cover its center section.

The three-dimensional structure–energy correlation for STGBs, with the 2D phase-space triangle of Fig. 3.19 as the basis plane, is illustrated in Figs. 3.20(a) and (b) for the fcc metals (for the Cu(LJ) and Au(EAM) potentials, respectively [81]) and in Figs. 3.21(a) and (b) for the bcc metals (for the Fe(Johnson) and Mo(FS) potentials, respectively [82]). As clearly seen from these figures, in both crystal lattices the qualitative features are identical for both the central-force and the many-body potentials (compare also Figs. 3.16 and 3.17), with the deepest cusps (associated with the densest GB planes) along the edges of the triangle and relatively smooth central regions, with only minor cusps near the slightly less-dense STGB planes.

A comparison with similar simulations for free surfaces [40, 104] shows the disappearance of the minor cusps in Figs. 3.20 and 3.21 in the central regions of the triangle, leaving only those cusps associated with the two or three densest planes at the corners. Since free surfaces lack the

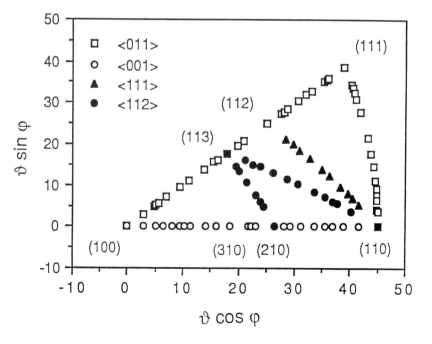

Fig. 3.19 Schematic plot, similar to the stereographic triangle, of the distribution of STGB planes simulated in order to explore the 2D phase space associated with the GB-plane normal. Orientations perpendicular to $\langle 001 \rangle$, $\langle 011 \rangle$, $\langle 111 \rangle$, and $\langle 112 \rangle$ tilt axes are included. ϑ and φ are the two DOFs associated with the polar and azimuthal angles characterizing the GB-plane normal. ϑ is in degrees [81, 82].

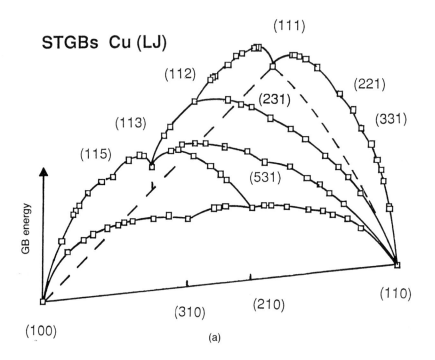

STGBs Cu (LJ)

(111)

(112)

(231)

(221)

(113)

(331)

(115)

(531)

GB energy

(100)

(310)

(210)

(110)

(a)

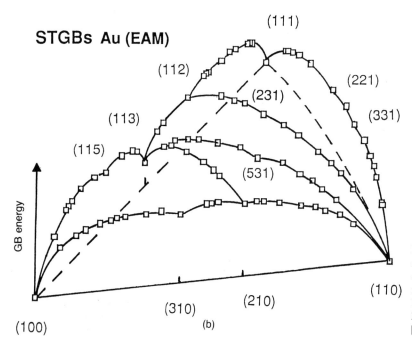

STGBs Au (EAM)

(111)

(112)

(231)

(221)

(113)

(331)

(115)

(531)

GB energy

(100)

(310)

(210)

(110)

(b)

Fig. 3.20 3D structure–energy plots illustrating the correlation between the energy and GB-plane normal for STGBs in fcc metals. The base plane is represented by the phase-space triangle defined in Fig. 3.19. (a) Cu(LJ) and (b) Au(EAM) potential [81].

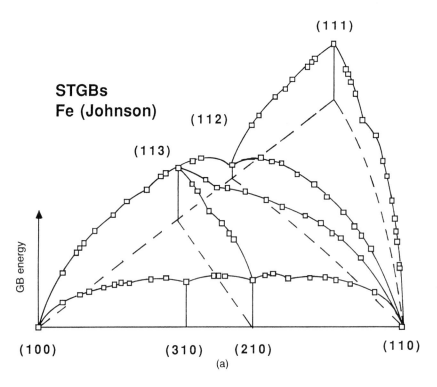

STGBs
Fe (Johnson)

(111)

(112)

(113)

GB energy

(100)   (310)   (210)   (110)

(a)

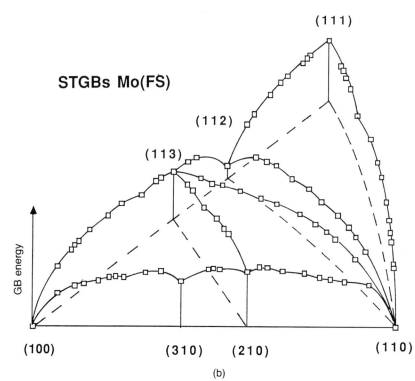

STGBs Mo(FS)

(111)

(112)

(113)

GB energy

(100)   (310)   (210)   (110)

(b)

Fig. 3.21 Same as Fig. 3.20 for bcc metals. (a) α-Fe pair potential and (b) Mo(FS) many-body potential [82].

three microscopic DOFs of GBs, this comparison illustrates the importance of translations at the interface in the minimization of the GB energy, particularly the energy of those boundaries with the smallest planar unit cells which consequently give rise to cusps. By subtracting the GB energies in Fig. 3.15 from those of free surfaces, γ, the ideal-cleavage energy associated with brittle decohesion at the GB, $2\gamma - E^{GB}$ (also known as the work of adhesion) is readily obtained. Without giving details here, it is clear from Fig. 3.15 and the corresponding (much smoother) plot for free surfaces that the GBs on the densest planes, as well as their vicinal orientations, also exhibit by far the highest ideal-cleavage energies. (For details, see Chapter 26 in this volume.)

If the twist angle is now added as the third DOF (eq. (3.1)), every point in the 2D phase space associated with the GB plane is unfolded into an infinite number of θ values which, in Fig. 3.19, are projected into a single point. The structure–energy plot for *all* symmetrical GBs may then be thought of as 4D hypersurface. To gain some understanding of what this plot might look like, one can consider three-dimensional cross sections obtained as follows [98]. By selecting a subset of planes perpendicular to a particular pole axis (for example, a $\langle 110 \rangle$ direction), a single tilt angle, ψ, uniquely defines a given symmetrical GB plane. If we now choose the twist angle, θ, as the second variable, a 3D cross section, $E^{GB}(\psi, \theta)$, of the 4D structure-energy hypersurface may thus be obtained which contains all symmetrical (tilt and twist) GBs with the same tilt axis.

For each of the fcc and bcc lattices two such cross-sections, perpendicular to $\langle 110 \rangle$ and $\langle 112 \rangle$, are shown in Figs. 3.22 (fcc) and 3.23 (bcc). (Other cross sections, perpendicular to $\langle 100 \rangle$ and $\langle 111 \rangle$, were considered elsewhere [98].) The energies of the STGBs, appearing for twist angles of 180°, are the same ones as those in Figs. 3.10(a) and (b), and the variation of the GB energy as a function of the twist angle θ is based on the extended Read–Shockley eqs. 3.19(a) and (b) [98]. Combined with the RGB model for the energies in the high-angle portions of each $E^{GB}(\theta)$ curve (Fig. 3.18 and Ref. [98]), a description of the entire 4D structure–

energy phase space for all symmetrical GBs is thus possible.

Figures 3.22 and 3.23 illustrate graphically that the STGBs are, indeed, a small – yet geometrically and energetically special – subset of twist boundaries (section 3.2.1 and Chapter 1 in this Volume). The cusped valleys in these figures demonstrate the importance of the densest lattice planes and of their vicinal orientations. (For further details, see Ref. [98].)

3.4.5 Asymmetrical GBs

A fundamental question concerns the relative energies of symmetrical and asymmetrical GBs. With the development of high-resolution transmission-electron-microscopy methods (HREM) in 'edge-on' studies of tilt GBs, much atomic-level information concerning the orientation of the GB plane has become available during recent years (section 3.4.6). Much of this work shows that asymmetrical combinations of lattice planes and faceting occur rather commonly. Moreover, recent 'rotating-sphere-on-a-plate' experiments [13, 14, 107] show a substantial fraction of asymmetric GBs. These observations suggest that asymmetrical GBs may often have lower energies than symmetrical ones, a question addressed by means of computer simulations in this section.

Figures 3.24(a) and (b) show typical variations of the GB energy, $E^{GB}(\theta)$, for two different symmetrical and asymmetrical combinations of lattice planes in the fcc lattice [18, 108] (here shown for the Cu(LJ) potential, although the Au(EAM) potential gives qualitatively identical results [18]). Analogous simulation results for bcc metals are shown in Figs. 3.25(a) and (b) [86].

We first compare the asymmetrical **twist** boundaries, obtained for arbitrary values of θ in these figures, with the asymmetrical **tilt** boundaries (ATGBs) at the endpoints of the misorientation range (i.e. obtained for $\theta = 0°$ and $180°/m$; see section 3.2.1). As in the symmetrical case (Figs. 3.12 and 14), the two ATGB configurations obtained for each combination of planes give rise to pronounced energy cusps, separated by more or less flat plateau regions in which the energy is

Symmetrical GBs
Cu(LJ)

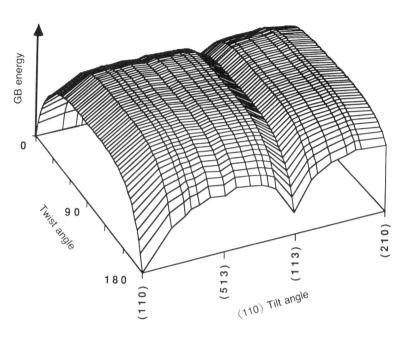

Symmetrical GBs
Cu(LJ)

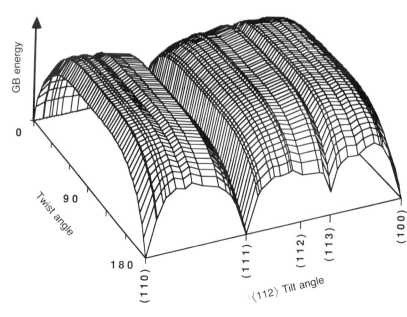

Fig. 3.22 3D cross section, for GB planes perpendicular to (a) ⟨110⟩ and (b) ⟨112⟩, of the 4D structure–energy hypersurface associated with **symmetrical** GBs in fcc metals. The STGBs are seen for the twist angles of 180°. The figure illustrates that in the 3D misorientation phase space associated with all symmetrical GBs, the symmetrical tilt boundaries represent an infinitely small subset of twist boundaries.

Symmetrical GBs
Fe (Johnson)

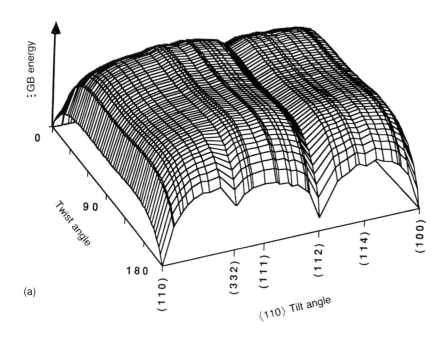

(a)

Symmetrical GBs
Fe (Johnson)

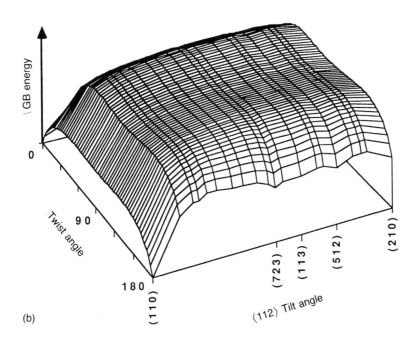

(b)

Fig. 3.23 Same as Fig. 3.22 but for symmetrical GBs in **bcc** metals.

practically independent of θ. As in the symmetrical case, the physical origin of these cusps lies in the especially small planar unit cells of the ATGBs (section 3.2.1), which enable a better local inter-locking of the atom positions across the GB than in the twist boundaries, thus permitting their energy to be minimized more effectively. This explains the preponderance of tilt boundaries (both symmetrical and asymmetrical) in recent 'rotating-sphere-on-a-plate' experiments in which the GB planes and types were identified, with only a small fraction representing twist (or general) GBs [14, 107].

Replacing one of the two sets of symmetrical planes, thus creating asymmetrical boundaries, has a profound effect on the GB energy in each of the two cases considered in Figs. 3.24 and 25. While the replacement of the densest (111) planes in the fcc lattice by less-dense (115) planes leads to a significant increase in the GB energy over the entire misorientation range (see Fig. 3.24(a)), the energy in the plateau region actually decreases when (557) planes are substituted for the fourth-densest (113) planes (Fig. 3.24(b)). Similarly, in the bcc lattice the replacement of the densest (110) planes by less-dense (114) planes leads to a significant increase in the GB energy over the entire misorientation range (Fig. 3.25(a)), while the energy in the plateau region actually decreases on average when (552) planes are substituted for the fourth-densest (112) planes (Fig. 3.25(b)). Interestingly, however, the (113) and (112) STGBs (in the fcc and bcc metals, respectively) have nevertheless substantially lower energies than the corresponding (557)(113) and (552)(112) ATGB configurations at θ = 180° and even at θ = 0° (Figs. 3.24(b) and 3.25(b)). This indicates an important role played not only by the densest lattice planes but also by the particular less-dense planes they are combined with to form asymmetrical boundaries. It also suggests that some asymmetrical GB-plane orientations may have a lower GB energy than related symmetrical ones, although the asymmetry may affect tilt and twist boundaries in different ways.

In the fcc lattice one should also notice, how-ever, that in spite of the sharp increase in energy upon substituting (115) planes for (111) planes, the energies of the (115)(111) boundaries are

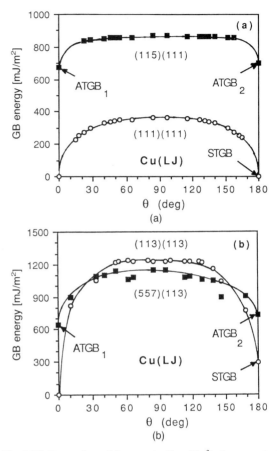

Fig. 3.24 Comparison of the energies (in mJ/m^2) of symmetrical and asymmetrical twist boundaries in Cu simulated by means of the LJ potential. (a) (111)(111) versus (115)(111); (b) (113)(113) versus (557)(113) [18, 107]. The ATGB and STGB configur-ations at the endpoints, with the smallest planar unit cells, are indicated by arrows. Because of the 3-fold symmetry axis in the planar unit cells, in (a) the usual misorientation angle, θ_{CSL}, was replaced by $\theta = 3\theta_{CSL}$, thus scaling all angles to 180°. The lines represent least-squares fits of the extended Read–Shockley eqs. 3.19(a) and (b) to the simulation data.

nevertheless significantly lower than those of the boundaries in Fig. 3.24(b) with a (113) plane on one or both sides of the interface. Similarly, in the bcc lattice the asymmetrical (114)(011) twist boundaries have a generally lower (or at most equal) energy than the GBs with a (112) plane on one or both sides of the GB.

As illustrated by the solid lines in Figs. 3.24(a) and (b), a quantitative analysis of the variation of

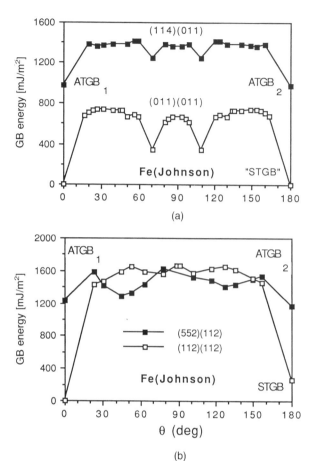

analysis of asymmetrical GBs in bcc metals has not been performed so far; the solid lines in Figs. 3.25(a) and (b) therefore represent merely guides to the eye.

3.4.6 HREM investigation of the role of the boundary plane

The fact that – due to their unique geometry – tilt boundaries are energetically preferred is a very fortunate circumstance because these are just the GBs whose atomic structure is readily accessible by HREM methods. With no twist components, the energy of these boundaries is governed completely by the four DOFs associated with the GB plane (eqs. (3.3) and (3.4)); tilt boundaries thus represent ideal model systems for an experimental investigation of the role of the GB plane. Such a study involves the preparation of a number of bicrystal samples with a range of different misorientations. Experimentally, the misorientation between the two halves of a given bicrystal is fixed, but unless geometrical constraints are imposed, the orientation of the GB plane usually can take on any energetically preferred orientation according to its two DOFs. However, since in tilt bicrystals the CSL misorientation axis must lie in the boundary plane, an additional constraint is imposed in this case; one parameter is therefore sufficient to describe the inclination α of the GB plane.

For thin-film tilt bicrystals manufactured by the process described above in section 3.2.2, the transverse orientation of the GBs (perpendicular to the film surfaces), and thus their tilt character, is maintained because this geometry minimizes the total GB area. Therefore, as the GB plane is free to choose any symmetrical or asymmetrical configuration which is energetically favored, the GBs observed in thermally equilibrated bicrystals should reveal the preferred (i.e. lowest-energy) boundary planes.

The edge-on views of tilt GBs in Fig. 3.26 display strong faceting, which is typical for well-annealed samples. Observation of planar facets shows (i) that GBs are indeed planar and (ii) that certain GB planes are preferred over others. While this conclusion may not be fully justified on the

Fig. 3.25 Comparison of the energies (in mJ/m²) of symmetrical and asymmetrical twist boundaries in α-Fe simulated by means of Johnson's pair potential. (a) (011)(011) versus (114)(011); (b) (112)(112) versus (552)(112). The ATGB and STGB configurations at the endpoints, with the smallest planar unit cells, are indicated. (We note that the 'STGB' configuration on the (011) plane is identical with the perfect crystal.) The lines are merely a guide to the eye [49, 86].

the GB energy as a function of θ may be based on the empirically extended Read–Shockley eqs. 3.19(a) and (b) discussed in section 3.4.2 (Ref. [97]). As in the symmetrical case (Fig. 3.14), the core and strain-field energies per unit length of the GB dislocations, E_c and E_s (in this case **screw** dislocations because E^{GB} varies as a function of the **twist** angle), are regarded as adjustable fitting parameters. Because of the much larger scatter of the simulation data, a similar Read–Shockley

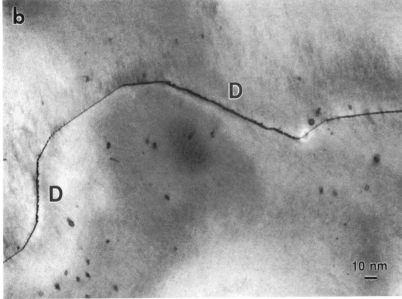

Fig. 3.26 Bright-field transmission electron microscopy images of $\langle 1\bar{1}0 \rangle$ tilt GBs in Au. Planar facets are typical for well-annealed specimens at all misorientations, including bicrystals whose misorientation allows the formation of GBs composed of two low-index planes, as in (a), $\psi = 55°$, or GBs at or near low Σ misorientations, as in (b), $\psi = 40°$. While such observations clearly suggest that low energy boundary configurations are planar and are comprised of a finite set, atomic-scale observations are necessary to establish that the interfaces are indeed planar and do not decompose into atomic scale facets of different orientations than their macroscopic shape suggests, and to reveal the atomic scale structure, which may take on different forms for the same macroscopic geometry. The facets labeled D in (b), for example have been shown to dissociate into triangular regions of Σ = three GBs, and the facets labeled S have been shown to exist in the form of two different atomic-scale structures (see below).

basis of conventional TEM observations, as shown in Figs. 3.26(a) and (b), it is clear, even from observations at low magnification, that certain boundary orientations are preferred and, therefore, lead to the formation of extended facets. However, HREM observations are necessary to establish whether or not the GBs are truly planar and which facets are present on an atomic scale.

In order to investigate the role of the GB plane in forming preferred GBs [109], a number of $\langle 1\bar{1}0 \rangle$ tilt bicrystal gold specimens were prepared by the thin-film method described in section 3.2.2. As

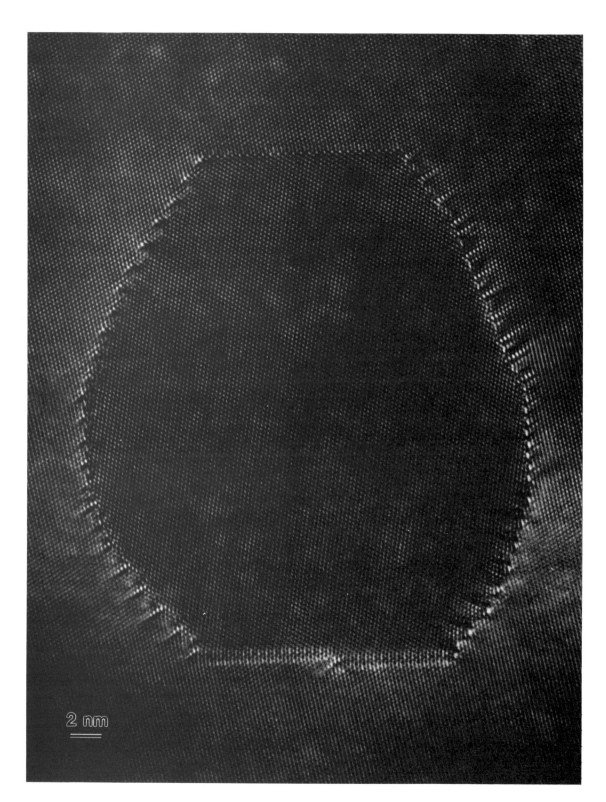

2 nm

noted above, analysis of the GB inclinations, the corresponding sets of GB planes, and the average length of different facets is expected to give information on the relative energies of the GBs in a given bicrystal. The examination of small island grains in a single-crystalline matrix is particularly well suited for this type of investigation. Figure 3.27 illustrates the formation of distinct facets in such a small grain, embedded in a larger crystal.

All of the bicrystals that were examined [109] exhibit a number of facets, corresponding to a finite set of inclination angles α. When viewed by conventional electron microscopy, at intermediate magnifications, extended, seemingly planar facets are frequently found (Fig. 3.26). However, in the HREM mode, atomic-level steps are quite common, and occasionally faceting on the atomic scale is also observed. A frequent geometrical preference in the formation of facets is the incorporation of one or two low-index, i.e. atomically rather dense-packed, planes into the GB. Examples of this are shown in Figs. 3.28 and 3.29. Figure 3.28(a) shows a $\langle 1\bar{1}0 \rangle$, $\psi = 55°$ tilt GB in Au. For a misorientation of 54.74° one set of (111) and (001) planes are parallel to each other in crystals 1 and 2, respectively. We note that this GB configuration appears to have a particularly low energy, since faceting along these planes does not only take place along the horizontal facets in Fig. 3.28(a), but also at the step between the two facets. Moreover, very long (111)(001) facets are commonly observed at this misorientation.

In Fig. 3.28(b) the misorientation angle is $\psi = 39°$ with the same $\langle 1\bar{1}0 \rangle$ tilt axis, as in (a); however, this misorientation does not allow two low-index planes to be parallel to each other. Nevertheless, the facets in Fig. 3.28(b) appear close to (111) and (110) planes in the top and bottom grains, respectively. The (111) and (110) planes could be parallel to each other at $\psi = 35.26°$. The accommodation of the orientational mismatch connected with the formation of (111)(110) facets

may be the cause for the considerable lattice strain which is evident in Fig. 3.28(b).

For some arbitrary misorientation two dense-packed planes can usually not be matched as a possible pair of GB planes. In such cases one often

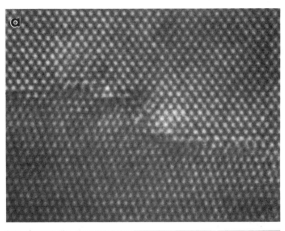

Fig. 3.28 Although extended facets have been observed that are planar on an atomic scale, a very common observation is the occurrence of steps or ledges, and faceting on an atomic scale. The faceting frequently takes place along planes that include a low-index configuration. In (a) faceting along (111) planes in the horizontal section as well as for the step is observed in a $\langle 1\bar{1}0 \rangle$, $\psi = 55°$ tilt bicrystal in Au (white atomic columns). Microscopic facets along (111) are also formed in (b) for a $\psi = 39°$ bicrystal. The facets are close to (110)(111) in this $\Sigma = 9$, $\langle 1\bar{1}0 \rangle$ tilt GB. Although this (incommensurate) interface deviates by a few degrees from $\Sigma = 9$, it appears that the GB has a tendency to incorporate such an interface formed from two atomically rather dense planes. Note strain contrasts, which may be associated with the dislocation or disclination character of the steps, that is needed to accommodate the 3.7° misorientation difference between $\Sigma = 9$ and the (110)(111) GB.

Fig. 3.27 $\langle 1\bar{1}0 \rangle$ tilt bicrystal viewed along the tilt axis. The small island grain of Au shows pronounced symmetric and asymmetric facets. Misorientation $\theta = 50°$, $\Sigma = 11$.

observes that GBs incorporate just one low-index plane which is paired with a high-index plane on the other side of the boundary. Such asymmetric boundaries are frequently found in ceramic oxides [12, 110, 111] and bcc metals [112]. From the work of Penisson on bcc metals Fig. 3.29 shows, as an example, an $\langle 001 \rangle$, $\psi = 32°$ tilt GB in Mo [112]. The dense-packed (110) plane in one crystal is opposed by a high-index plane in the other crystal. Inclusion of just one low-index plane into the GB is also common in fcc Au [109].

The observation of low-index planes (i.e. (111) and (001)) as part of many tilt boundaries for various misorientations suggests that the incorporation of dense-packed planes results in GBs of low energy. The special role of the densest-packed planes in the energy of twist, tilt and general boundaries has already been discussed above, based on the results from computer simulations (sections 3.4.1 and 3.4.5). Based on our observations for fcc metals, it appears that the two densest planes are, indeed, most likely to lead to low-energy interfaces when incorporated into a GB. Clearly, the (111)(111) combination (i.e. the coherent twin) gives the lowest interfacial energy. The next largest average interplanar spacing is given by the (111)(001) GB, and has been observed experimentally to give extended facets [10], attest-

ing to its low energy. Concerning the combination of a dense-packed plane with a general high-index plane, one may get some insight from the results of the random-boundary model (Figs. 3.16 and 3.17). Figures 3.16 and 3.17 suggest that, except for the two densest-packed planes in fcc and the densest plane in bcc, there is no significant energetic advantage to be expected when a low-index plane is combined with an arbitrary high-index one.

We now point our attention to the complete sets of symmetric and asymmetric tilt GBs that have been observed by HREM for specific bicrystals and try to examine their relation to GB energy. The faceting behavior and the structures observed by HREM in $\Sigma = 9$ and $\Sigma = 11$ tilt bicrystals have recently been compared to detailed calculations for these same GBs [109]. Within the low-Σ misorientations that were examined no particular GB inclination was clearly dominant over other facets, except for the **symmetrical** (111)(111) and (113)(113) tilt configurations in $\Sigma = 3$ and $\Sigma = 11$ bicrystals, respectively. The preferred formation of these facets involving the densest and fourth-densest fcc planes, respectively (and the two densest fcc planes on which STGBs can be accommodated), is due to the deep energy cusps that exist for these tilt misorientations (Fig. 3.10(a)).

A few examples for the coexistence of symmetric

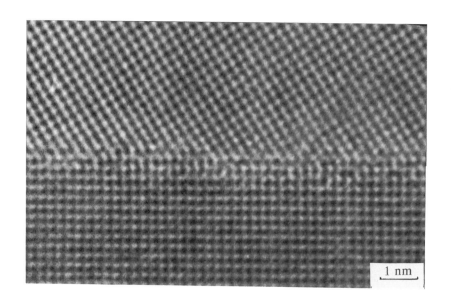

Fig. 3.29 Molybdenum $\langle 001 \rangle$, $\psi = 32°$ tilt GB observed by Penisson [112]. This asymmetric boundary includes the densest atomic plane in bcc Mo. The well developed (110) facet is opposed by a high-index plane.

1 nm

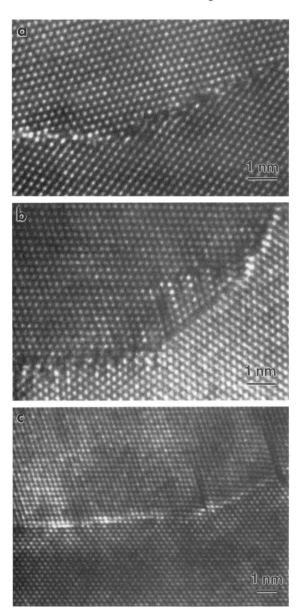

and asymmetric facets can be found in Fig. 3.30. The two possible symmetric facets for the $\Sigma = 9$ bicrystal ($\langle 1\bar{1}0 \rangle$, $\psi = 39°$) are shown in Figs. 3.30(a) and 3.30(b) together with several asymmetric facets. Extended asymmetric facets are observed in the bicrystal near $\Sigma = 41$ ($\langle 1\bar{1}0 \rangle$, $\psi = 55°$). Nevertheless, symmetric facets are also present in this bicrystal, as can be seen in Fig. 3.30(c). Coexistence of symmetric and asymmetric tilt GBs was found in all bicrystals and appears to be a general feature of cubic systems (including predominantly ionic metal oxides with rocksalt structure [110, 111]).

In earlier studies, correlations between the five macroscopic **CSL-based** DOFs (eq. (3.1)) and GB energy have often been sought. It is now well recognized, however, that no simple relationship between these macroscopic DOFs (nor the value of inverse **volume** density of CSL sites, Σ) and GB energy exists [113, 114]. In an investigation of GB faceting in Au, Goodhew *et al.* [115] attempted to relate their observed GBs to the inverse **planar** density of CSL sites, Γ. In this investigation it was assumed that low GB energy was associated with a high planar density of CSL sites (i.e. a low value of Γ). They found, however, that a number of facets did not obey the low-Γ criterion. This can be understood on the basis of our calculations of GB energies which clearly show that there is no direct correlation between Γ and GB energy for a given bicrystal. (For details see Refs. [109] and [111].)

Figure 3.31 shows GB energies (for the Au(EAM) and Cu(LJ) potentials; section 3.3.1) as a function of inclination angle, α, for the $\Sigma = 9$ bicrystal. Among the particular combinations of GB planes considered in the figure, only the (112)(1,1,22) GB has not been observed experi-

Fig. 3.30 Examples for the coexistence of symmetric and asymmetric grain boundaries. (a) HREM micrograph of $\langle 1\bar{1}0 \rangle$ tilt GB in Au, showing a short symmetric (221)(221) facet on the left side of the figure followed by an asymmetric (557)(771) facet ($\Sigma 9$, $\psi = 39°$). (b) For the same misorientation as in (a), the GB changes its inclination (starting at the left) from the (11, 11, 1)(111), to the only other symmetric GB for $\Sigma = 9$, the (114)(114) GB. Between these two boundaries a short (111)(115) facet has dissociated into a triangular arrangement of $\Sigma = 3$, (111)(111) and (112)(112) STGBs. Quite long facets of this configuration are present in Fig. 3.12(b) at D. (c) Extended

facets of (111)(001) ATGBs are found in $\langle 1\bar{1}0 \rangle$, $\psi = 55°$ ($\Sigma = 41$). A small part of one such facet is shown in (c), followed by the symmetric (338)(338) GB. The appearance of extended facets may be equated with low GB energies. The importance of asymmetric facets for polycrystalline materials is due to their observed abundance as well as their greater number of possible configurations. The coexistence of symmetric and asymmetric GBs indicates that any given bicrystal will have a number of GB inclinations, which must not differ substantially in their interfacial energies.

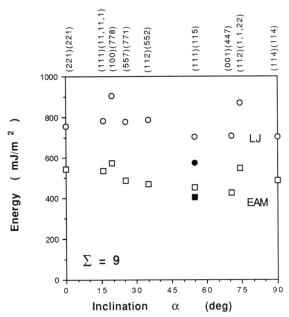

Fig. 3.31 Calculated GB energies as a function of inclination α for $\Sigma = 9$ bicrystals. The symmetric GBs are located at $\alpha = 0°$ and $\alpha = 90°$. Note that the GB energies lie within a relatively narrow band, and that many asymmetric GBs are lower in energy than their corresponding symmetric GBs. LJ and EAM potentials give qualitatively similar variations as a function of α. Except for (112) (1, 1, 22), all of the indicated GBs have been observed experimentally. These energies therefore represent local minima relative to vicinal GB inclinations. The filled symbols indicate the energies for the dissociated (111)(115) ATGB.

mentally. Several points should be noted about Fig. 3.31:

1. The results from LJ and EAM calculations are qualitatively very similar.
2. The GB energies lie in a relatively narrow band.
3. For those facets that have been observed experimentally, the data points in Fig. 3.31 must represent energy cusps.
4. Symmetric GBs, corresponding to the inclinations angles $\alpha = 0°$ and $\alpha = 90°$, are not necessarily preferred energetically.

The experimentally observed faceting behavior in $\Sigma = 9$ and $\Sigma = 11$ Au bicrystals [109] shows good agreement with the calculated energies. The fact that many, often asymmetric facets are present is thus in agreement with calculated GB energies.

The rather similar energies associated with different inclinations explains the multitude of facets that have been observed in this and other bicrystals. While the actual number of low-energy boundaries may vary with misorientation and may include quite low-energy symmetric GBs, such as (113)(113) for $\Sigma = 11$ bicrystals, it is clear that tilt GBs are not dominated by symmetric facets. In fact, because in contrast to the geometrically unlimited number of possible asymmetric configurations, at most two different symmetric GBs are possible for each misorientation [109], asymmetric GBs may, indeed, be expected to play a dominant role in polycrystalline samples. We should note that, since real bicrystals are produced at finite temperatures, entropic and kinetic effects may also play a role in determining the faceting of bicrystals. Also, because of the relative closeness between calculated 0 K energies for different inclinations, the approach to equilibrium grain shapes in the sense of a Wulff construction [116] should be extremely slow. Therefore even very small grains such as the one shown in Fig. 3.27 may not represent ideal equilibrium shapes. Nevertheless, the presence, absence, and length of facets should be correlated, in a qualitative way, to GB energy. In view of this, the agreement between observed facets and low-energy GBs as determined by 0 K calculations must be considered very good.

We shall now discuss a few special GB topologies that have been observed experimentally. The GB calculations described above refer to planar boundary configurations that are obtained by the relaxation of fully dense interface planes. The relaxed GBs typically conserve their planar configuration, and the structural disorder associated with a high-angle GB is generally localized within a few atomic planes of the interface. This is in agreement with most GB configurations that are observed by HREM. In a few instances, however, GBs have been found to take on a three-dimensional character, that can be due either to GB dissociation or to delocalization of the structural disorder associated with the GB. Figure 3.32(a) shows a $\Sigma = 9$ bicrystal for which a horizontal, planar facet would correspond to the (111)(115) GB. However, instead of forming this particular

asymmetric facet, the GB dissociates into triangular regions consisting of two **symmetric** (111)(111) GBs and one (112)(112) GB. This dissociation of the $\Sigma = 9$ into $\Sigma = 3$ tilt boundaries has been reported by several authors [117–119] and is quite remarkable since, according to GB simulations, the planar (111)(115) GB has one of the lowest energies for a planar interface in this bicrystal and is also the $\Sigma = 9$ GB with the smallest planar unit cell. Nevertheless, this dissociation is fully consistent with the simulations since, when the total energy of the dissociated boundary is added up and normalized to the planar, undissociated interface area, one finds that this energy is considerably smaller (by more than 10% for both the EAM and LJ potentials) than the energy of the asymmetric (111)(115) interface [109]. As a consequence, the dissociated (111)(115) ATGB becomes the facet with the lowest energy for $\Sigma = 9$ bicrystals (full symbols in Fig. 3.31). We note that the longest facets in Fig. 3.26 (labelled D) correspond to this configuration.

A different type of three-dimensional boundary topology has been found in a few instances, namely in $\langle 1\bar{1}0 \rangle$ tilt GBs in gold in which, on both sides of the GB, (111) planes are almost perpendicular to the interface and appear continuous across the GB, as seen in Fig. 3.32(b). In this case the structural disorder at the GB is distributed over a finite width. In contrast to a discrete dissociation into well-localized regions of misfit, the misfit appears to be distributed more or less continually over a finite width at the boundary. This type of grain-boundary relaxation is clearly not typical; since it appears to be associated with the generation of stacking disorder at (111) planes it may also be limited to materials with a low stacking fault energy.

3.5 CORRELATION BETWEEN MICROSCOPIC DOFS AND ENERGY

The second aspect of the structure–energy correlation involves the three **microscopic** DOFs. Rigid-body translations of the two halves of a bicrystal relative to each other, $\boldsymbol{T} = (T_x, T_y, T_z)$, can provide

an important relaxation mechanism for minimizing the GB energy. This has been realized already in the very early computer studies of GB relaxations [36]. Since translations parallel and perpendicular to the interface plane relate to quite different aspects of the structure and energy of GBs, it is useful and convenient to distinguish between these two components of the rigid-body translation.

3.5.1 Translations parallel to the boundary plane

As discussed in section 3.2.2, rigid-body translations parallel to the GB (x–y) plane of the two crystals in Fig. 3.3 represent an important part of the GB structure. When experimentally characterizing translations, a suitable reference frame has to be found relative to which rigid-body displacements can be measured. A further difficulty arises because, when GB mobility is considered, the actual location of the interface is intimately connected with the translational state of the GB. In fact, although redundant, the location of the GB plane has occasionally been introduced as an additional microscopic degree of freedom [1]. For commensurate boundaries, a convenient frame of reference is the CSL. In this case the rigid-body translation, which usually destroys the CSL points in common to the two crystal lattices, is measured relative to the **planar** CSL unit-cell, positioned in the GB plane.

The rigid-body translations parallel to the GB, which seek out the lowest-energy relaxed GB states, strongly affect the core structures of the GB. In symmetrical boundaries, for example, the GB plane could be a mirror plane. However, except for the coherent (111)(111) twin boundary illustrated in Fig. 3.33, mirror symmetry is almost never observed. For this boundary, the interplanar spacing of the (111) planes at the GB is practically unchanged, and the first nearest-neighbor distances are the same as in the ideal crystal. The reversal of the stacking sequence of the most densely-packed planes therefore results in an atomically well-matched interfacial region, without a rigid-body displacement that would destroy the mirror symmetry.

In contrast to this special situation for the densest

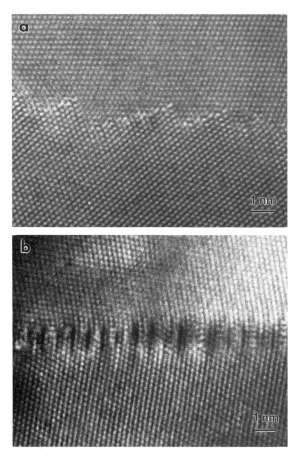

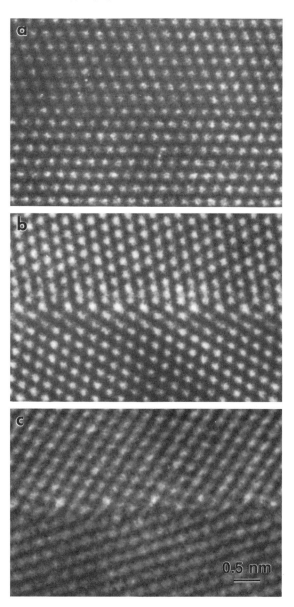

Fig. 3.32 Grain boundaries do not necessarily assume a strictly planar configuration, as suggested by the geometric models (Fig. 3.2). Two types of dissociations are shown here: (a) A sequence of triangular minigrains in place of a planar (111) (115) interface in this $\Sigma = 9$ Au bicrystal. The dissociation of the $\Sigma = 9$ boundary into two (111) (111) and one (112) $\Sigma = 3$ boundaries is energetically favored, in agreement with results from computer simulations (text and Fig. 3.31). (b) A different type of three-dimensional boundary topology is observed in this $\langle 1\bar{1}0 \rangle$, $\psi = 20°$ tilt GB in gold, where (111) planes on both sides of the GB are almost prependicular to the interface and appear continuous across the GB. In this $\Sigma = 33$ GB the structural disorder at the GB appears distributed over a finite width.

Fig. 3.33 Symmetric $\langle 1\bar{1}0 \rangle$ tilt GBs in Au. (a) The (111) (111) twin GB is the only tilt GB in fcc which displays mirror symmetry. Both the (113) (113) shown in (b) and the (221) (221) in (c) include rigid-body shifts parallel to the GB plane that destroy the mirror symmetry that is associated with the 'special' twin configuration in (a).

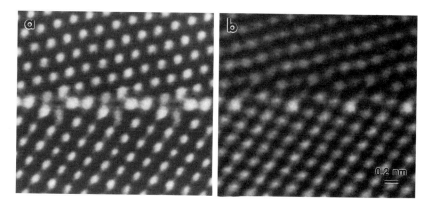

Fig. 3.34 Multiple translational states are frequently observed for the same macroscopic GB configurations. (a) displays a configuration with approximate mirror symmetry, which has only been observed in the form of a short facet. The configuration with glide-plane symmetry (b) has been observed in the form of extended facets, and is therefore expected to be the energetically favored one. These are in agreement with lattice-statics calculations which predict a significantly higher energy for the GB with mirror symmetry compared to the lowest energy configuration incorporating a rigid-body shift parallel to the GB plane.

planes, all higher-order symmetric GB planes (with larger Miller indices, and hence a much smaller interplanar lattice spacing; Table 3.2) necessarily bring some atoms in close proximity to each other across the GB when mirror symmetry is maintained, as for example, in Fig. 3.1(b). Consequently, when the GB is allowed to relax toward a minimum-energy configuration, typically a rigid-body translation is introduced that destroys the mirror symmetry, thus avoiding the strong short-range repulsion between atoms in very close contact. As examples of such boundaries, the symmetrical (113)(113) and (221)(221) tilt GBs are shown in Figs. 3.33(b) and (c), respectively.

For the (221)(221) STGB two different structures have been found. The structure of this boundary in Fig. 3.34(a) displays mirror symmetry, while the structure in Fig. 3.34(b) shows the typical zig-zag structure, which is also common in semiconductor GBs [120, 121]. Computer simulations of the (221)(221) STGB have found the mirror-symmetry configuration to be metastable in that its energy is higher than that of the translated configuration in Fig. 3.34(b). This finding is in agreement with HREM observations, which have shown extended facets of the latter configuration, but only a short segment of the former [108].

When computing the relaxed atomic structure of a GB, several metastable translational configurations, with more or less equal or fully degenerate energies are often found to coexist (section 3.4.1). Not surprisingly, when the relaxed energies associated with two translational states are reasonably close, multiple structures can be expected, as illustrated in Fig. 3.34. The most convincing evidence for the existence of energetically close, multiple translational states – and, hence, core structures – comes from the observation of GB facets that are separated by small steps, as first observed in NiO [11].

Figure 3.35 illustrates the existence of two different GB structures on the two sides of a small step for the (113)(113) ($\Sigma = 11$) STGB in Au. Since the step height is less than the interplanar spacing of the CSL for this plane, the two boundaries must have different rigid-body translations parallel to the GB plane. This is illustrated schematically in Fig. 3.35(b) which shows the corresponding CSL-model depictions of several translational states. When the GB is formed between different (113) planes within their periodic stacking sequence normal to the GB (the so-called stacking period, P(113); Table 3.2), the resultant GB structure may be viewed as resulting from a rigid-body translation parallel to the GB. Since there are $P(113) = 11$ (= Σ; Table 1.1 of Chapter 1 in this volume) possible **discrete**, nonequivalent translational states of the GB, the geometrical description of the possible CSL boundaries correspondingly will generally require $P(hkl)$ different rigid-body translations parallel to the GB. The actually observed rigid-body translation of the relaxed GB will, in general, be different from any of those. However, when two parallel facets exist that are separated by a small step (provided the step is not associated with an exactly compensating dislocation component parallel to the GB), then these two GB segments will have different rigid-body displacements (Fig. 3.35(b)), and consequently will also differ in their atomic structures. The physically important feature of these x–y translations is that, at least for certain boundaries, more than one low-energy translation exists, thus leading to multiple structures for the same macroscopic GB parameters.

It has been shown experimentally (section 3.6.3) that GBs can form on incommensurate planes. In this case, there is no strict translational periodicity in at least one direction within the GB plane. For such boundaries quasiperiodic structures, formed by means of misfit-dislocation-like defects, have been identified. Strictly speaking, however, one will thus have an infinite manifold of structures, since all these structures, although all similar to one another, are slightly different from each other when compared at the atomic scale.

To summarize, in spite of their importance as a relaxation mechanism for the GB energy, from a theoretical point of view even a very detailed knowledge of the rigid-body translations parallel to the GB offers little insight into the basic physical properties of the interface, not even its energy. Whether the GB has only one, or more than one, optimal translational state is of some interest, of course, as it may provide information on its response to shear stresses. Mostly, however, information on these translations, via a comparison with HREM observations, may be used to test the validity of the simulation procedure, in particular of the interatomic potentials employed.

3.5.2 Volume expansion

By contrast with the x–y translations, the z component of T should have a strong effect on the GB energy because $T_z = \delta V$ accounts for a possible volume expansion per unit area at the GB and volume is a thermodynamic variable. As pointed out over 30 years ago, the volume expansion usually found at metallic GBs should thus be related directly to the energy (and electrical resistivity) of the boundary [26].

It is important to recognize that the excess volume of the GB (measured relative to an interface-free perfect crystal) is spread out over a number of lattice planes near the GB. δV is therefore not simply related to the spacing of lattice planes on opposite sides of the GB. To determine δV, the distance between the far-away perfect-crystal material on both sides of the GB (between which the GB is sandwiched) must therefore be measured.

In computer simulations, by monitoring the displacement of the Region-II blocks relative to their positions in an interface-free perfect crystal (Fig. 3.3), the determination of δV is a relatively straightforward matter. Hence, in all our simulations the excess volume was determined in addition to the optimum x–y rigid-body translation(s).

As an example, Figs. 3.36(a) and (b) show the volume expansion per unit GB area determined for the symmetrical (111) and (001) twist boundaries in fcc and bcc metals, respectively. A comparison of these expansions as a function of the twist angle with the related GB energies in Figs. 3.12(a) and (b) shows remarkable similarities. Particularly remarkable is the equally steady variation as a function of the twist angle, with a leveling off into rather flat plateau regions for the larger angles. Similar to the quantitative analysis of the GB energies, an empirically extended Read–Shockley model, based on the expressions [31] (eqs. (3.19a) and (b))

$$\delta V(\theta) = \sin\theta\,[\delta V_c^< - \delta V_s^< \ln(\sin\theta)]/b,$$
$$(0 \le \theta \le 90°) \qquad (3.20a)$$

$$\delta V(180° - \theta)$$
$$= \delta V^{\mathrm{STGB}} + \sin(180° - \theta)$$

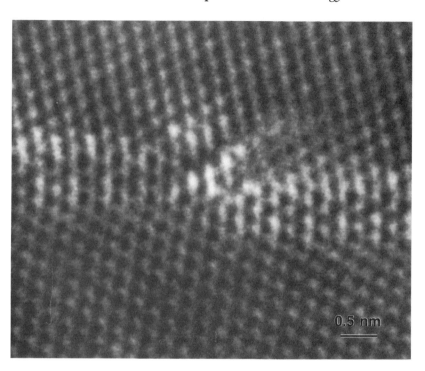

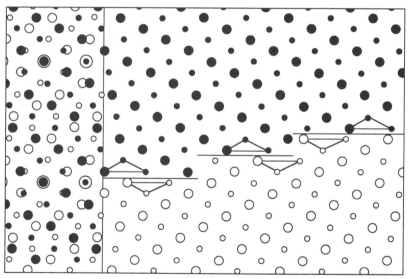

Fig. 3.35 The most convincing evidence for the formation of multiple translational states that are energetically degenerate or very close in energy comes from the observation of steps which are small in height compared to the planar stacking repeat period of the lattice perpendicular to the GB plane. Positioning of the interface at a different plane is equivalent to a rigid-body shift parallel to the GB, as schematically shown in (b) for a CSL boundary. The left hand side of (b) shows the CSL, while the corresponding GBs for different planes on the right have different rigid-body shifts parallel to the GB plane (compare relative positions of triangles). Note that for real crystals with a rigid-body translation, the additional shift due to the change in plane is the same as for the CSL model, and leads to a different core structure whenever the step is not a multiple of the interplanar spacing of the CSL.

$$\{\delta V_c^> - \delta V_s^> \ln[\sin(180° - \theta)]\}/b,$$
$$(90° \leqslant \theta \leqslant 180°). \qquad (3.20b)$$

account very well for the simulated volume expansions (see the solid lines in Figs. 3.36(a) and (b)). Analogous to eqs. (3.19a) and (b), δV^{STGB} is the

excess volume associated with the symmetrical tilt configuration at $\theta = 180°/m$ which, because of its particularly small planar unit cell, always gives rise to a deep cusp in both the energy and volume expansion. As in the case of the energy, the core and strain-field contributions $\delta V_c^<$, $\delta V_s^<$ and $\delta V_c^>$,

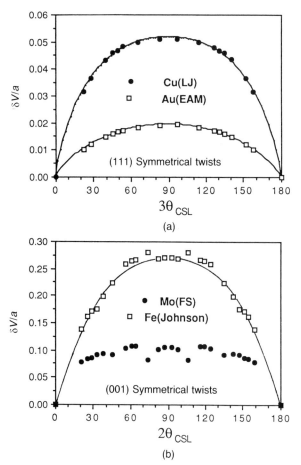

Fig. 3.36 Volume expansion per unit area, $\delta V/a$ (in units of the lattice parameter) for (a) symmetrical (111) twist boundaries in fcc metals and (b) (001) twist boundaries in bcc metals [31]. The solid lines are the result of a least-squares fit of eqs. 3.20(a) and (b) to the simulation data. A fit to the (001) data obtained for the Mo(FS) potential is not meaningful due to the pronounced cusps.

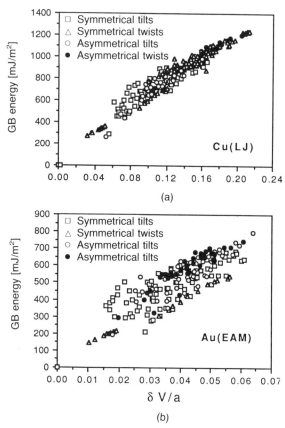

Fig. 3.37 Correlation between the GB energy and volume expansion per unit area, $\delta V/a$ (in units of the lattice parameter) for symmetrical and asymmetrical tilt and twist GBs in fcc metals. (a) Cu(LJ) and (b) Au(EAM) potential [31].

$\delta V_s^>$ associated with angles $\theta < 90°$ and $\theta > 90°$, respectively, are regarded as adjustable fitting parameters.

A least-squares fit of eqs. (3.20a) and (b) thus provides information on the density changes due to the strain fields [122] and cores [99] of **isolated lattice dislocations**. While in direct dislocation simulations this information is difficult to obtain due to the necessary embedding of the dislocation in a continuum-elastic medium [123], by considering the ordered arrays of the dislocations con-

stituting a GB this problem is avoided altogether because of the cancellation of all relevant divergences. Simulations of low- and high-angle GBs thus provide an excellent opportunity to investigate the basic physics of lattice dislocations.

Rather extensive recent computer simulations, using various pair- and many-body potentials for fcc [18, 30, 81, 98] and bcc metals [49, 82, 86, 102], have confirmed the existence of a more or less linear relationship between the volume expansion per unit area and the related GB energy [31]. As seen from Figs. 3.37 and 3.38, such a linear interrelation exists not only for symmetrical tilt and twist boundaries (with two and three macroscopic DOFs, respectively) but also for asymmetrical tilt

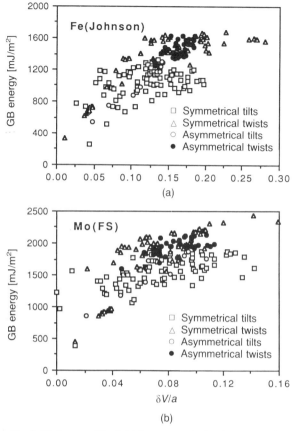

Fig. 3.38 Same as Fig. 3.37 for symmetrical and asymmetrical tilt and twist GBs in bcc metals. (a) α-Fe pair potential and (b) Mo(FS) many-body potential.

After selecting an interface plane, the bicrystal is generated by removing all black atoms from one side and all white atoms from the other side of the interface. We note that when this procedure is applied to asymmetric GBs, it does not necessarily maintain the density of the ideal lattice at the GB [24]. In general, one will get a higher than ideal-crystal density at the GB, which is unphysical as illustrated for the (111)(115) GB in Fig. 3.39(b). The bulk atomic density is formally maintained at the GB if the interplanar spacing at the unrelaxed interface, d_i, is given by $d_i = [d(hkl)_1 + d(hkl)_2]/2$, where $d(hkl)_1$ and $d(hkl)_2$ are the interplanar spacings of the $(hkl)_1$ and $(hkl)_2$ planes that form the GB. Similar to the determination of δV in computer simulations, this interface spacing thus provides a natural reference state for measuring the volume expansion at the fully relaxed GB. The necessary adjustments to obtain the correct density to zero-th order are illustrated for the hard-sphere (i.e. rigid) boundaries in Figs. 3.39(c) and (d). In Fig. 3.39(c) the normal rigid-body translation $\delta_0 = d(115) = d(111)/3$ is applied to restore the bulk density, while in Fig. 3.39(d) one (115) plane is removed. The latter can also be expressed in terms of a translation of the displacement-shift-complete (DSC) lattice; the latter represents the coarsest periodic lattice that contains all lattice points of the two crystal lattices forming the CSL [6].

Figure 3.39 illustrates another important aspect of measuring volume expansions. It is clear that the rigid-body displacement can not be uniquely defined just by measuring the relative positions of the rigid blocks of single crystal far away from the GB. In order to determine the physically meaningful quantity, namely the excess volume that is created at the GB, the core structure at the GB has to be known also. It is therefore necessary to ascertain how many atoms are at the GB core. For example, the removal of one atomic column from the core region would correspond to an expansion by the corresponding interplanar distance involved. In contrast to α-fringe techniques that can determine very accurately only the relative displacements for low-Σ boundaries [60, 61], HREM can thus provide the necessary atomic-level information on both the core struc-

and twist GBs (with four and five macroscopic DOFs, respectively).

Based on these simulation results, HREM observations of the volume expansion at the GB should provide a means for estimating its energy. The HREM measurement of such volume expansions involves determining the positions of atomic columns or planes (if the latter are relatively dense) in the unstrained region away from the boundary and comparison to the unrelaxed GB structure. As discussed in section 3.5.1 above, as the reference state we can choose the CSL structure. The standard procedure for generating a bicrystal from the two interpenetrating lattices of black and white atoms that define the CSL is shown in Fig. 3.1.

(a)

(b)

(c)

(d)

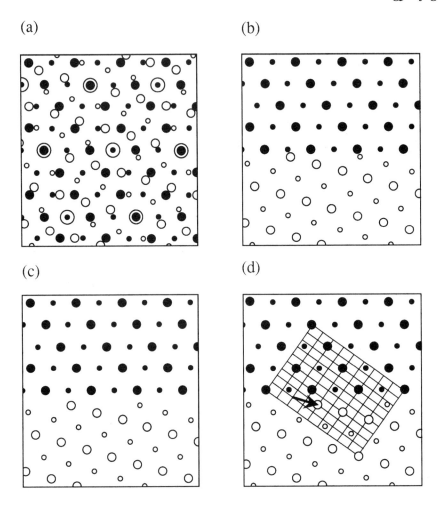

Fig. 3.39 Unrelaxed $\Sigma = 9$ grain boundary viewed along $\langle 1\bar{1}0 \rangle$ direction. A and B layers are distinguished by different size symbols. (a) CSL and interpenetrating lattices 1 and 2. (b) Black and white atoms removed on opposite sides of the (111)(115) GB plane. (c) Normal rigid-body translation to restore bulk density at GB. (d) one (115) plane is removed to get bulk atomic density at the GB. The body-centered orthorhombic DSC lattice and a DSC shift vector which is equivalent to the removal of an atomic plane are indicated.

ture and on the displacement normal to the GB.

While such measurements are in principle quite straightforward, the practical application encounters several difficulties. One problem is that the volume expansions in metals are typically only of the order of a few percent of the lattice parameter (Figs. 3.36–3.38). The routinely achieved precision in determining positions of atomic columns in HREM lies near 0.2 Å; greatly more precise methods have to be developed which are capable of determining rigid-body translations normal to the interface. Under certain circumstances, e.g. when a low-index plane is parallel to the interface, considerably higher accuracies (near 0.05 Å) have in fact been achieved [23, 124]. Work

is underway to extend the desired precision of better than 0.1 Å to more general boundaries [125].

In principle, the determination of the excess volume of a GB by HREM is only straightforward when the observed projection of atomic columns, indeed, represents fully occupied atomic columns. Any point defects, such as vacancies, present in the columns bounding the GB core have to be accounted for in the determination of the total excess volume. While in metals no large deviations from the ideal-column occupancy are expected, such effects have to be considered in the predominantly ionic metal-oxides [11, 126].

For non- or quasi-periodic interfaces, which typically have low-index planes parallel to each other, a determination of the volume expansion can

Fig. 3.40 A digitized HREM image of an asymmetric (111)(001) tilt GB is analyzed for its volume expansion by utilizing an intensity scan across several (100) and (111) planes away from the GB region.

be based on the observation of the position of planes. Figure 3.40 shows an example of a GB formed by two low-index planes, which permit such a determination of the volume expansion in this case. By extrapolation of the peak and valley positions in Fig. 3.40 to the GB plane, a quite accurate value for the volume expansion can be obtained. It should be noted, however, that the determination of the actual number of columns is not always straightforward in such a case, and particular attention also has to be given to the correct identification of the GB plane.

Few comparisons between measured and computed volume expansions at GBs in metals have been performed to date. While the measured volume expansions for Al agree well with EAM calculations [127], experimental values for Au are significantly larger (typically a factor of two) than the values obtained from EAM calculations [10, 125, 128]. The fact that in Au the LJ-potential predictions agree much better with the experimental values is particularly surprising since a local-volume dependence of the interatomic interactions is taken into account explicitly only in the EAM potential. The origin of this discrepancy is not known at the moment, although this could indicate that the EAM potential for Au may not be very well suited to describe the volume dependence of the GB energy. It is expected that systematic comparisons between observed and calculated rigid-body displacements for a range of different metallic systems and boundaries will become available in the near future. Detailed comparisons between the measured and computed rigid-body displacements for the $\Sigma = 3$, (112)(112) tilt boundary in Si and Ge have recently been reported [129].

3.6 CORRELATION BETWEEN THE ATOMIC STRUCTURE AND ENERGY

The third aspect of the structure–energy correlation involves the role of the local atomic structure at the GB. As discussed in section 3.6.1, experimentally a strong tendency for forming atomically well-matched regions at the interface, separated by highly localized regions which accommodate the atomic-level mismatch, is commonly observed. A quantitative description of the amount of atomic-level structural disorder at the GB, based on the radial distribution function and the number of broken bonds per unit GB area, is discussed in section 3.6.2, where the existence of a rather good correlation between computed energies and the number of broken bonds per unit GB area is demonstrated to exist for both GBs and external free surfaces. The underlying physical cause for these simple connections between energy and atomic structure of GBs lies in the localized nature of atomic relaxations. The latter appear to favor structures which maintain, as much as possible, the coordination of the ideal lattice even **locally** at the interface, even if it means destruction of the **long-range** structural periodicity. This tendency of the local structure to dictate the structural periodicity will be discussed in section 3.6.3.

3.6.1 HREM investigation of the role of atomic matching across the GB

It is well known that the structure of low-angle twist and tilt GBs consists of arrays of screw and edge dislocations, respectively, while general low-angle GBs exhibit dislocations of mixed type. In low-angle GBs with well-separated dislocation cores, it is clear that the two lattices must be connected via a smooth transition region in-between GB dislocations. Hence, while the interface may be commensurate or incommensurate, its atomic structure at the interface can be considered semi-coherent.

For high-angle GBs an important issue concerns the degree of coherency that is maintained between the lattices. The idea that large-angle GBs may consist of regions of good match, separated by regions of poor match, goes back to a suggestion by Mott [130]. Many GB models, including the CSL model, have in the past used the idea of atomic matching in one form or another. However, in some of the earliest computer simulations of GBs it was established that the geometric match between certain atomic sites in the GB, which exists for CSL orientations, is destroyed when the bicrystal undergoes a rigid-body translation [36, 131]. Although these translations associated with the three microscopic DOFs destroy the CSL, they do not alter the overall shape of the planar unit cell nor the overall periodicity. Based on a rigid model, high-angle CSL boundaries are therefore **incoherent** but commensurate (Chapter 1, section 1.4, in this volume). HREM observations however suggest that many GBs in metals as well as oxides are at least **semi-coherent** in that they show atomically well-matched regions for many high-angle GBs [132].

The semi-coherent atomic matching at GBs can be characterized in terms of the existence of a GB region in which the relaxed atomic structures smoothly connect the two crystal lattices which are misoriented with respect to one another. Such matching between the two lattices is best recognized experimentally by an apparent elastic continuation of low-index atomic planes across the GB. That atomically well-matched regions are

indicative of a low interfacial energy is suggested by the observation of extended facets in atomically well-matched interfaces for many tilt bicrystals in Au [109]. Figure 3.41 illustrates four different tilt GBs (two symmetrical and two asymmetrical) which are part of a small island grain in the $\Sigma = 11$ tilt bicrystal. A particularly well-matched structure is the (113)(113) STGB near the top of Fig. 3.41. Simulations using the Au(EAM) potential show that the energies of the remaining facets are larger, by more than a factor of two, than the energy of the (113)(113) STGB [109]. The latter shows good continuation of three sets of low-index lattice planes across the GB in addition to having a very small structural repeat distance. The two ATGBs have slightly lower energies than the (332)(332) STGB which runs perpendicular to the (113)(113) STGB. One notes that the asymmetric GBs in Fig. 3.41 still maintain some degree of coherency between lattice planes on opposite sides of the interface. By contrast, the (332)(332) STGB, with the highest energy of the four [109], has a compact core structure with no apparent continuation of dense-packed planes, and can therefore be considered incoherent.

Most GBs have considerably larger planar unit cells (or structural repeat periods) than the very small unit cells associated with the (113)(113) and (332)(332) STGBs in Fig. 3.41. This can be seen for the ATGBs in Fig. 3.41 and for the (443)(443) STGB in Fig. 3.42 (with $\psi \approx 55°$ and $\Sigma = 41$). Such boundaries are likely to display misfit localizations within their GB unit cells, which then permit the formation of well-matched regions. In the (443)(443) STGB in Fig. 3.42 we clearly see the misfit localization within the structural units of this boundary, as well as the continuation of low-index planes across a large fraction of the planar unit cell. In addition, the compressed image

---

Fig. 3.41 Faceted $\langle 1\bar{1}0 \rangle$ tilt microgram at bottom (tilt angle $\psi = 50°$, $\Sigma = 11$) shows four facets, ranging from the (113)(113) at the top of the figure, over (225)(441) and (557)(771) to the (332)(332) GB. Note that the (113)(113) is fully coherent, while both asymmetric GBs show extended strain contrasts in the top grain, with little evidence for elastic distortions in the bottom grain.

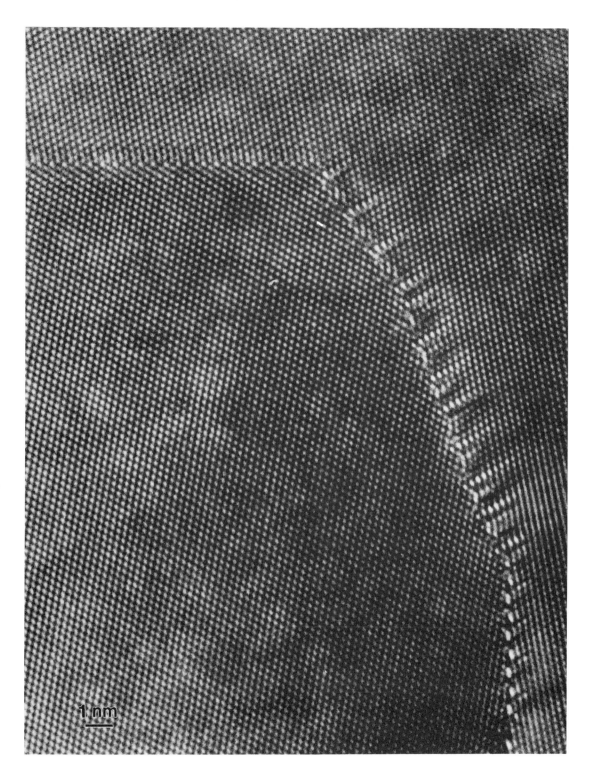

1 nm

Fig. 3.42 HREM image of the (443) symmetric tilt GB (tilt angle ψ = 55°, Σ = 41). Note the misfit localization and the strong tendency to maintain coherence between dense-packed planes. The compressed image in the bottom half clearly shows the correlations between atomic positions in the near-core region.

Fig. 3.43 Asymmetric (111) (001), ⟨1$\bar{1}$0⟩ tilt GB in Au. The GB shows misfit localization in the form of misfit-dislocation-like defects. The centers of misfit are arranged in a periodic fashion, but the GB is in fact quasiperiodic, since the atomic distances along the GB are incommensurate. The compressed image at the bottom of the figure indicates that the misfit localization is accompanied by slight elastic distortions in the surrounding lattice.

of this boundary at the bottom of Fig. 3.42 indicates a continuous arrangement of corrugated, dense-packed 'planes' on both sides of the core region of this boundary.

Another GB with a very well matched structure is shown in Fig. 3.43. This ATGB combines the (111) and (001) planes and exhibits exceptionally long facets (Fig. 3.30(c)), indicating a very low energy. Again, misfit localization is important for this boundary, as will be discussed in more detail in section 3.6.3. In all these observations there appears to be a tendency at the GB to retain as much as possible the atomic coordination of the ideal crystal. Also, these structures seem to have developed through relaxation of the relatively dense-packed lattice planes which are relatively nearby on both sides of the GB plane, i.e. the role of relaxa-

tions seems to be to establish structures that are 'vicinal' to certain low-index lattice planes.

To summarize, HREM observations of a number of fcc boundaries indicate that, whenever possible, atomically well-matched regions are formed. This tendency to maintain, as much as possible, a perfect-crystal atomic environment of the lattice locally at the interface, can be quantified utilizing HREM information on atomic positions at GBs. Such an investigation is presently underway. It is hoped that this work will enable establishment of a direct connection between the number of broken bonds per unit GB area (i.e. atomic miscoordination) and the GB energy. As discussed in the next section, such a broken-bond description of GBs holds considerable promise in structure–property investigations.

3.6.2 A broken-bond model for GBs and free surfaces

The role of the local coordination at the interface was also investigated by means of computer simulations. While for free surfaces a broken-bond description of the structure–energy correlation has been commonly used for over half a century [39], such an approach has been adopted for GBs only recently [41].

To introduce the relatively simple concept of broken bonds, we mention the well-known fact that a full characterization of the atomic structure of a GB would involve knowledge of the positions of all the atoms in the periodic unit cell of the defect (if the GB is, indeed, commensurate). A subset of this vast type of information, particularly appropriate for materials with a near-isotropic nature of interatomic bonding, is contained in the radial distribution function, $G(r)$, or its Fourier transform, $S(k)$. As is well known, the zero-temperature shape of $G(r)$ consists of a series of δ-function-like peaks associated with the shells of nearest-, second-nearest, and more distant neighbors. In an otherwise perfect crystal, thermal disorder is responsible for a broadening and shift towards larger distances of these peaks. Owing to the presence of planar defects, interface materials, by contrast, are structurally disordered even at zero temperature, and their zero-temperature radial distribution function shows the same two effects. However, because of its localization near the interface, this type of disorder is **inhomogeneous** – by contrast with thermal disorder.

To illustrate this inhomogeneity in the direction of the interface normal, it is useful to consider the radial distribution function (or the planar structure factor) for each of the atom planes near the interface. As seen from Fig. 3.44 for the case of a symmetrical (100) twist boundary in the fcc lattice, the amount of structural disorder decreases very rapidly from one (100) plane to another, indicating the existence of large gradients in many properties

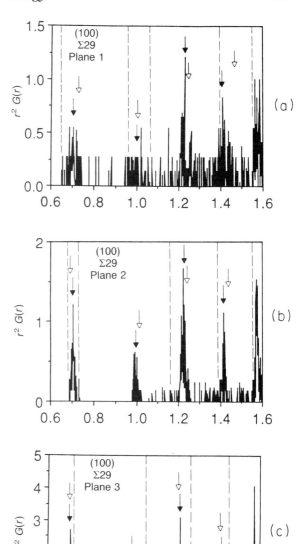

Fig. 3.44 Plane-by-plane zero-temperature radial distribution functions, $r^2 G(r)$, for the three lattice planes closest to the (001) (001) $\theta = 43.60°$ ($\Sigma 29$) symmetrical twist boundary simulated by means of an EAM potential for Cu [133]. The full arrows indicate the corresponding perfect-crystal δ-function peak positions; open arrows mark the average neighbor distance in each shell. The widths of these shells are indicated by dashed lines.

[133]. Comparing this effect for twist boundaries on the three densest planes of the fcc lattice, Fig. 3.45 illustrates that from one GB to another the detailed shape of $G(r)$ for the atom plane(s) immediately at the GB varies rather strongly [30, 85]. As discussed earlier, the energies of the particular boundaries considered in Fig. 3.45 differ significantly: according to Fig. 3.12, for the Cu(LJ) potential they range between $\approx 300 \, \mathrm{mJ/m^2}$ for the (111) plane, via $\approx 700 \, \mathrm{mJ/m^2}$ for the (100) plane to $\approx 1200 \, \mathrm{mJ/m^2}$ for the (110) plane. Since for a pair potential, $V(r)$, the GB energy is given simply by a plane-by-plane sum over integrals of the type $\int r^2 \, G(r) \, V(r) \, dr$, these energy differences are reflected by the related degree of broadening and disordering of the peaks in Figs. 3.45(a)–(c).

The type of detailed atomic-level information contained in Figs. 3.44 and 3.45 is difficult to obtain experimentally. Various models have therefore been proposed to describe the atomic structure at the interface without having to know the atom positions too precisely. Starting from the very detailed description of structural disorder in terms of the radial distribution function, a broken-bond model represents a particularly logical step towards simplification. By characterizing the atomic structure in terms of the number of broken v-th nearest-neighbor (αnn) bonds per unit area, $C(v)$ (see eq. (3.13)), essentially in a broken-bond model the detailed peak shapes in $G(r)$ are simply replaced by the areas under these peaks and subsequently integrated over the entire interface region and normalized to the unit-cell area of the interface. Because all the information contained in the detailed shapes of these peaks is thus lost, small long-range elastic strain-field effects associated, for example, with interface dislocations or surface steps are therefore not 'seen' in the coordination coefficient, $C(v)$. The consequent limitations of a broken-bond model will be a part of the discussion in this section.

Figure 3.46 demonstrates the correlation between the average number of broken nn bonds per unit GB area, $C(1)$, and the GB energy for the two fcc potentials [41]; for comparison, the corresponding free-surface results are also shown [40]. While, in principle, more distant neighbors

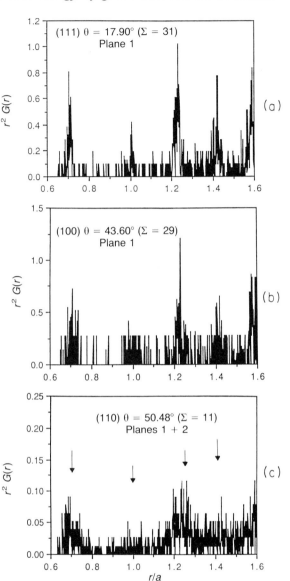

Fig. 3.45 Radial distribution functions, $r^2 G(r)$, for twist boundaries on the three densest planes of the fcc lattice. For the two densest planes, (a) and (b) [30], only atoms in the plane immediately at the GB were included in $G(r)$; by contrast, because of the smaller spacing of the (110) planes and the larger degree of atomic-level 'rumpling' in these planes, the *two* planes nearest to the GB were considered in the case of the (110) twist GB, (c) [85].

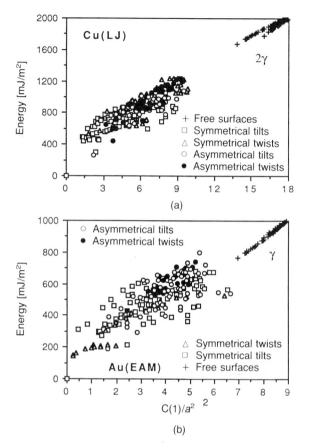

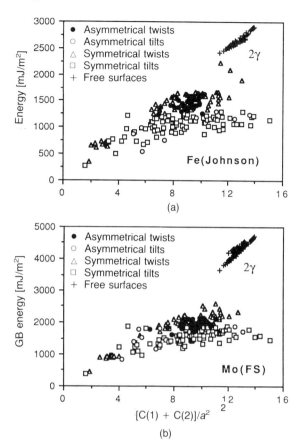

Fig. **3.46** GB energy (in mJ/m²) versus number of broken nearest-neighbor bonds per unit area, C(1) (in units of a⁻²; eq. (3.13)) for the two fcc potentials; (a) Cu(LJ), (b) Au(EAM). [41] For comparison, the related free-surface energies, γ, are also shown [40]. Notice that, to avoid an overlap of the data, in the LJ case 2γ is plotted.

Fig. **3.47** GB energy (in mJ/m²) versus sum of the number of broken nearest- and second- nearest-neighbor bonds per unit area, C(1) + C(2) (in units of a⁻²) for the two bcc potentials (a) a-Fe, (b) Mo(FS) [82]. For comparison, the related free-surface energies, 2γ, are also shown [104].

should also play a role, in fcc metals their contribution was found to be rather small (typically about 10–20% of the nn contribution for the LJ potential, and smaller yet for the EAM potential [40]). By contrast, in bcc metals the second-nearest neighbors were found to be practically as important as the nearest neighbors [104]; therefore in Fig. 3.47 the energies of GBs and free surfaces obtained for the two bcc potentials are plotted against C(1) + C(2), effectively treating the bcc lattice as having 14 nearest neighbors. That the volume expansion at the GB is governed equally by the number of broken bonds per unit area is

illustrated in Fig. 3.48 for the case of the two fcc potentials.

Although according to Figs. 3.46 and 3.47, for both free surfaces and GBs the interface energy and the degree of areal miscoordination are reasonably well correlated, the GB data scatter significantly. This scatter arises from an intrinsic limitation of a broken-bond characterization of the atomic structure of GBs, and is largely due to the fact that only the dislocation **cores**, but not their elastic strain fields, give rise to broken bonds. To investigate this limitation quantitatively, we recall that according to Read and Shockley [7] the structure of **low-angle**

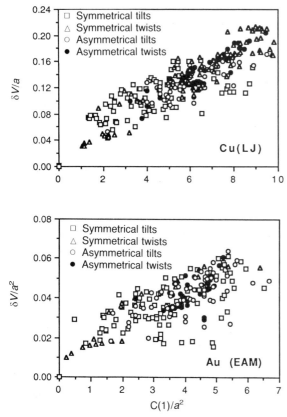

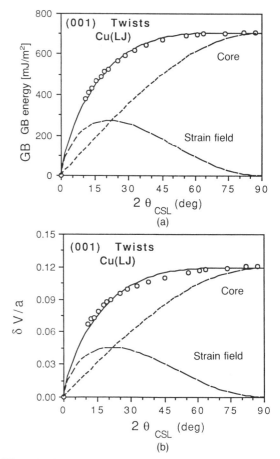

Fig. 3.48 Volume expansion per unit area, $\delta V/a$, versus $C(1)$ for the two fcc potentials [41].

Fig. 3.49 Core and strain-field contributions to (a) the energy and (b) the volume expansion per unit area for symmetrical (001) twist boundaries obtained for the LJ potential [41]. The two contributions were obtained from least-squares fits of eqs. (3.19) and (3.20), respectively, to the total GB energies (Fig. 3.12(a)) and free volumes (circles). θ_{CSL} is the usual coincident-site lattice misorientation angle; for reasons discussed above in connection with eqs. (3.19) and (3.20), the energies and volume expansions are plotted against $2\theta_{CSL}$ instead of θ_{CSL} in this case.

boundaries consists of ordered arrays of dislocation cores, separated by perfect-crystal like regions which are subject to the long-range strain fields surrounding the dislocation cores. The energy and excess volume of the GB may therefore be subdivided into a contribution due to the dislocation cores, $E_{core}(\theta)$, and one arising from the strain fields surrounding the dislocations, $E_{strain}(\theta)$, according to

$$E^{GB}(\theta) = E_{core}(\theta) + E_{strain}(\theta) \qquad (3.21a)$$

$$\delta V(\theta) = \delta V_{core}(\theta) + \delta V_{strain}(\theta) \qquad (3.21b)$$

Although continuum elasticity theory breaks down when the dislocation cores overlap (typically above $\theta \approx 15°$), the empirical extension of the RS model in eqs. (3.19) and (3.20) [31, 97] still permits a separation of core- and strain-field contributions

even in high-angle GBs. A 'typical' high-angle boundary (in the RS sense) thus consists entirely of overlapping dislocation cores, with consequently vanishing strain-field contributions in eqs. (3.19) and (3.20) [99].

This is illustrated for the LJ potential in Figs. 3.49 and 3.50 which show a breakdown of the energy and volume expansion as a function of the angle according to eqs. (3.19)–(3.21) for symmetrical (001) and (111) twist boundaries, respec-

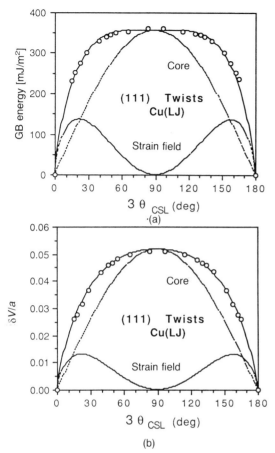

(a)

(b)

Fig. 3.50 Same as Fig. 3.49 for symmetrical (111) twist boundaries, plotted against $3\theta_{CSL}$ in this case [41].

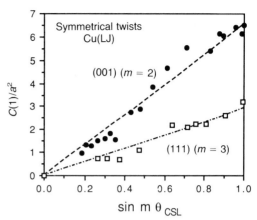

Fig. 3.51 Nearest-neighbor coordination coefficients (eq. (3.13)) versus the sines of the normalized misorientation angle for (001) and (111) twist boundaries obtained for the Cu(LJ) potential (Figs. 3.49 and 3.50). The dashed lines represent linear least-squares fits to the data through the origin [41].

tively [41]. These figures illustrate that while $E_{core}(\theta)$ and $\delta V_{core}(\theta)$ in eqs. (3.19)–(3.21) are proportional to the total length of dislocation core per unit area, i.e. to $\sin\theta$, the related strain-field contributions exhibit the familiar logarithmic $\log(\sin\theta)$ variation [7, 31, 97].

A broken-bond description of the Read–Shockley model starts with the observation that only the dislocation **cores**, but not their elastic strain fields, give rise to broken bonds. (The latter are typically no more than a few percent of the lattice parameter.) Hence, while the small elastic displacements of the atoms from their perfect-crystal sites should be visible in the peaks in $G(r)$, the coordination coefficient $C(\alpha)$ should be affected solely by the total length of the dislocation cores in

the unit cell of the GB which, according to the Frank formula [134] (eqs. (3.19) and (3.20)), is proportional to $\sin\theta$, i.e.

$$C(\nu) = C_{core}(\nu)/b\sin\theta, \quad (0 \leq \theta \leq 90°) \quad (3.22)$$

where C_{core} is the number of broken nn bonds per unit length of dislocation core. Hence, while the energy and volume expansion approach their respective cusps at $\theta = 0°$ with **infinite** slope (due to the domination of the logarithmic strain-field term in eqs. (3.19) and (3.20)), the slope in the coordination coefficient in eq. (3.22) remains finite (as does the slope of the core contribution to $E_{core}(\theta)$ and $\delta V_{core}(\theta)$ in eqs. (3.19)–(3.21); Figs. 3.49 and 3.50); its value of C_{core}/b provides information on the miscoordination per unit length of the dislocation cores in the GB.

The qualitatively different behavior of $C(1)$ in eq. (3.22) by comparison with the GB energy and volume expansion in Figs. 3.49 and 3.50 and eqs. (3.19) and (3.20) is illustrated in Fig. 3.51 for the boundaries of Figs. 3.49 and 3.50. The straight lines in this figure represent least-squares fits of eq. (3.22) to the simulation data through the origin. The larger scatter in the values of $C(1)$, by comparison with the scatter in the energies and free volumes in Figs. 3.49 and 3.50, is due to the much greater sensitivity of $C(1)$ to translations parallel to the GB plane, by contrast with E^{GB} and δV, observed for all our GB simulations (Ref. [81]).

The above analysis clearly establishes the limitation of a broken bond model to **high-angle** GBs, i.e. GBs whose structure is characterized by a virtually complete overlap of the dislocation cores. Considering that the simulation data in Figs. 3.46–3.48 include both high- *and* low-angle boundaries, thus including significantly varying core and strain-field contributions, their large scatter is not very surprising.

When comparing the structures of GBs and free surfaces, one is tempted to merely replace the dislocation cores by the surface ledges and the dislocation strain fields by those surrounding the steps. One is then led to conclude that the success of a broken-bond description for free surfaces [105] (Figs. 3.46 and 3.47) rests mainly on the fact that the strain fields surrounding the steps contribute rather little to the total surface energy, in contrast to dislocations. This conclusion is based on the well-known r^{-2} decrease of the step–step interaction energy as a function of the distance, r, between the steps [135], by contrast with the much slower logarithmic decay of the elastic strain-field energy associated with GB dislocations [136]. This leads to a domination of the elastic energy over the dislocation-core energy in low-angle GBs (Figs. 3.49 and 3.50). In surface steps or ledges, by contrast, for comparably large separations between these line defects the core energy dominates entirely over the practically vanishing elastic energy. From a broken-bond model point of view, the net result is that the free-surface energy can practically always be related to the miscoordination per unit area, while in GBs such a description is limited to the high-angle regime in which the dislocation-core overlap is complete. (For further details, see Ref. [136].)

The reasons for the failure of a broken-bond model for **low-angle** GBs (with substantial strain-field contributions to the energy and free volume) are directly related to the replacement of the overall radial distribution function of the GB by the integral under the nearest-neighbor peak; the actual atom displacements in the strain fields

surrounding the dislocations are too small to show up in the rather rough grid in terms of which a broken-bond model samples the actual radial distribution function. This observation suggests important limitations in the possibility of extracting quantitative information from HREM micrographs of interface structures. While such experiments can provide information on the local coordination of the atoms (section 3.6.1), it is exceedingly difficult to extract information on the elastic displacements near the dislocation cores in low-angle and vicinal GBs.

Finally, to relate the broken-bond model to other models for the atomic structure of GBs, we mention that the main criticism of many of these models has been that, unlike the Read–Shockley model, they do not permit quantification of the atomic structure, and therefore do not enable a quantitative investigation of structure–property correlations. Two such groups of models, in which elastic displacements of the atoms are systematically ignored anyway, seem particularly well suited for a quantification in terms of broken bonds. For example, in **polyhedral-** or **structural-unit** models [15, 35–38], the atom arrangement is described qualitatively in terms of the stacking of polyhedra along the interface. Characterizing each structural unit in terms of the underlying number of broken bonds per unit area would permit the structure to be characterized quantitatively, although such a description would not take into account the systematic elastic distortions of the structural units from one GB to another (resulting from the requirements of space filling at the interface, and of the compatibility of the structural units with the adjoining grains). Similarly, **hard-sphere** models (in which the optimum translation parallel to the interface plane is assumed to be the one which minimizes the volume expansion at the GB [19–21, 137]), could be quantified by characterizing the structure in terms of broken bonds, although, being based on **unrelaxed** atomic structures, again these models ignore elastic effects.

3.6.3 HREM investigation of structural periodicities in GBs

The importance of local atomic arrangements that maintain, as much as possible, the atomic coordination of the perfect lattice even near the GB also becomes apparent from the existence of well-coordinated but **incommensurate** or **quasiperiodic** structures. Perfectly periodic GBs (but extending over a finite distance in real crystals) are only possible for very special misorientations for which a CSL superlattice in common to the two grains exists. Aperiodic features may be introduced by deviations from the exact coincidence misorientation angle or by forming boundaries on irrational planes. Theoretically it has been shown that such irrational GBs can be considered quasiperiodic, in analogy with the structures of quasicrystalline materials [138–140].

Another type of quasiperiodicity is encountered when the atomic relaxations are such that smaller structural units are formed than those given by a usually rather large CSL period. For example, the $\langle 1\bar{1}0 \rangle$ tilt GB ($\psi = 55°$) in Fig. 3.43 is, on the one hand, very close to the $\Sigma = 41$ CSL misorientation (for which $\psi = 55.88°$) and, on the other, to an incommensurate interface in which the (111) and (001) planes are exactly parallel to each other ($\psi = 54.74°$). In one possible asymmetrical configuration of the $\Sigma = 41$ boundary, the (24, 24, 23) and (001) planes are combined, with a structural repeat period of 11.8 nm; the (24, 24, 23) (001) ($\Sigma = 41$) ATGB thus represents a commensurate CSL boundary close to the incommensurate (111)(001) ATGB. The compressed image of the latter in Fig. 3.43 illustrates that slight elastic distortions of the lattice are present in a periodic fashion on both sides of the GB. Regions of well-matched atomic columns alternate with centers of misfit. The period of the misfit localization is approximately 1.8 nm. These homophase defects are the exact analogue to the formation of misfit dislocations in heterophase boundaries. The misfit in the distances between atomic columns along the $\langle 110 \rangle$ and $\langle 121 \rangle$ directions for the (001) and (111) planes, respectively, is accommodated by misfit-dislocation-like defects. The misfit in this case is 14%, while the period, determined for a rigid hard-sphere model, is approximately 1.86 nm, in agreement with the observation.

This observation of one-dimensional quasi-periodicity in a tilt GB shows the ability of the lattice to seek out relaxation modes that do not necessarily maintain the commensurate symmetry of the bicrystal, but rather assume local atomic arrangements that maintain, as much as possible, the atomic coordination of the lattice, however, without being commensurate. When a GB consists of dense-packed planes, it most often forms an interface that is quasiperiodic in at least one direction. Moreover, the extended facets that have been observed for the (111)(001) interface indicate that this GB has quite low energy. Observation of this boundary geometry in other systems and the observation that (001) surfaces in Au reconstruct into the same geometry, supports this conclusion [12, 141].

While ATGBs may often be incommensurate, by virtue of combining two identical sets of lattice planes, STGBs are always commensurate; their structural repeat periods should therefore always be governed by the bicrystal geometry. Quasi-periodicities for tilt GBs would then only arise due to deviations from exact CSL orientations, and the quasiperiod would be long compared to the corresponding CSL unit. However, we have recently found a different kind of quasiperiodicity [10, 142], which we believe is due to thermodynamic fluctuations, possibly caused by entropic effects. Figure 3.52 shows a symmetric, high-angle GB that has an interesting three-dimensional nature given by the stacking disorder, which extends to a short distance from the GB for, on average, every 15 (111) planes crossing the GB. It can clearly be seen in Fig. 3.52 that the spacing between stacking faults is not strictly periodic, but varies by a few (111) planes. Since the GB is symmetric, and is within less than 1° of the $\Sigma = 41$ orientation, one would expect that the structural units of this GB would be identical in size to the distance between CSL points on the (338) plane. However, the structural repeat distance for the CSL boundary is 2.6 nm, compared to

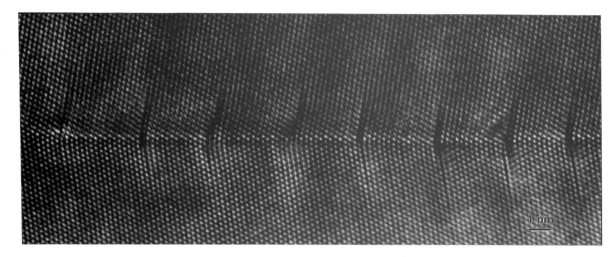

Fig. 3.52 Symmetric (338)(338) GB in Au (tilt axis $\langle 1\bar{1}0 \rangle$, tilt angle $\psi = 55°$, $\Sigma = 41$). The GB shows strong coherency between (111) planes crossing the interface. The misfit is accommodated by stacking faults which extend to both sides of the interface, where they are terminated by partial dislocations. The stacking faults are arranged in quasiperiodic fashion (for details, see the text).

the observed average distance between stacking faults of 3.6 nm. The structure of this GB is obviously *not* governed by the $\Sigma = 41$ geometry. Instead, this GB can be considered **vicinal** to the (113)(113) STGB (with $\psi = 50.5°$), with the additional 5° misorientation being accommodated by the stacking faults protruding from the interface. The reason why an incommensurate quasiperiodic structure is formed for this STGB, and not the structure commensurate with the $\Sigma = 41$ misorientation, lies probably in the quite low energies associated with the stacking disorder in Au. It appears that, during formation, possible residual stresses and entropic effects might be responsible for these deviations from periodicity.

In summary, although STGBs should, in principle, be commensurate interfaces, HREM has shown that, at least for certain misorientations and GB planes, the desire of the GB to maintain a perfect-crystal-like local coordination at the interface can lead to structures that are quasiperiodic and, hence, incommensurate. The formation of such quasiperiodic structures enables misfit localization along the interface, at the price, however, of giving up long-range periodicity.

3.7 CONCLUSIONS

In the above investigation we have attempted to illustrate how the complementary capabilities of HREM experiments and atomistic computer simulations can provide new insights into the correlation between the macroscopic, microscopic, and atomic structures of GBs and their energy.

An extensive exploration of the five-parameter misorientation phase space associated with the macroscopic DOFs is realistically possible only via simulation, while HREM experiments are limited to the tilt boundaries in this phase space. However, since the latter are fully governed by only the four DOFs associated with the GB plane, and because of their unique geometry, their HREM investigation provides valuable information on the role of the GB plane in their energy. In particular, the comparison of the relative lengths of symmetrical and asymmetrical facets provides semi-quantitative information on their relative energies, while the detailed crystallographic analysis of such facets suggests a dominating role of the densest planes in the crystal lattice.

HREM experiments also provide valuable information on the three microscopic DOFs associated with rigid-body translations at the GB. Extensive simulations for GBs in fcc and bcc metals, using both pair- and many-body potentials, demonstrate a practically linear relationship between the energy and volume expansion per unit GB area (i.e. translations perpendicular to the GB plane). Translations parallel to the GB, by contrast, play a relatively minor role in the relaxed GB energy, although their optimization represents an important relaxation mechanism. Combined with the simulations, HREM measurements of the volume expansion thus provide direct information on the GB energy, while a comparison of measured with computed rigid-body translations parallel to the boundary plane provide an important test for (i) the validity of the interatomic potentials and (ii) the relaxation procedures used in the simulations.

Finally, HREM experiments clearly demonstrate a tendency of the atomic structure to preserve a high degree of atomic-level coherency across the interface. As in the case of free surfaces the underlying causes, elucidated via simulation, are closely connected with the desire of the interface to minimize its number of broken bonds per unit area, i.e. its energy.

Throughout this chapter, qualitatively the same behavior was found for both pair- and many-body (EAM and FS) potentials. This qualitative agreement is encouraging as it indicates that in the investigation of structure–property correlations for metallic grain boundaries, the detailed nature of local atomic bonding at the interface may not be as important a factor as are, for example, the elastic strain-field and core effects associated with interface dislocations, effects which can be distinguished and separated for any interatomic-force description.

Throughout this chapter we have attempted to highlight the complementary type of information and insight which can be obtained from HREM experiments, on the one hand, and atomic-level simulations on the other. While a systematic experimental exploration of the misorientation phase space associated with the five macroscopic DOFs of a single GB is virtually impossible, because of the relative ease with which the macroscopic DOFs can be manipulated in computer simulations, an approach utilizing the complementary capabilities of simulation and experiment seems to be particularly promising for the investigation of structure–property correlations. HREM observations on the other hand, although limited to tilt boundaries, provide the opportunity to validate, at the atomic level, many of the predictions derived from the simulation of relatively simple, but well characterized model systems.

ACKNOWLEDGMENTS

We gratefully acknowledge discussions with S. R. Phillpot. We also would like to thank L. D. Marks for making available the H9000 in the Department of Materials Science and Engineering at Northwestern University. This work was supported by the US Department of Energy, BES Materials Sciences, under Contract No. W-31-109-Eng-38.

REFERENCES

1. See, for example, C. Goux, *Can. Metall. Quarterly*, **13** (1974), 9.
2. M. S. Daw and M. I. Baskes, *Phys. Rev. B*, **29** (1986), 6443.
3. M. W. Finnis and J. E. Sinclair, *Phil. Mag. A*, **50** (1984), 45.
4. D. G. Brandon, B. Ralph, S. Ranganathan and M. S. Wald, *Acta Metall.*, **12** (1964), 813.
5. B. Chalmers and H. Gleiter, *Phil. Mag.*, **23** (1971), 1541.
6. W. Bollmann, *Crystal Defects and Crystalline Interfaces*, Springer Verlag, New York, (1970).
7. W. T. Read and W. Shockley, *Phys. Rev.*, **78** (1950), 275.
8. H. Ichinose and Y. Ishida, *J. Physique Colloque C4*, **46** (1985), C4–39.
9. K. L. Merkle, *Proc. 46th Ann. Meeting of the Electron Microscopy Soc. of America*, (1988) p. 588; *MRS Symp. Proc.*, **153** (1989), 83.
10. K. L. Merkle, *Colloque de Phys.*, **51** (1990), C1–251.
11. K. L. Merkle and D. J. Smith, *Phys. Rev. Lett.*, **59** (1987), 2887, and *Ultramicroscopy*, **22** (1987), 57.
12. C. B. Carter, *Acta Metall.*, **36** (1988), 2753.
13. R. Maurer, *Acta Metall.*, **35** (1987), 2557.

14. R. W. Balluffi and R. Maurer, *Scripta Metall.*, **22** (1988), 709.

15. For a recent comprehensive review, see H. Gleiter, in *Atomistics of Fracture* (eds R. M. Latanision and J. R. Pickens), Plenum, New York (1983), p. 433.

16. D. Wolf, *J. de Physique*, **46** (1985), 197.

17. D. Wolf and J. F. Lutsko, *Z. Kristallographie*, **189** (1989), 239.

18. D. Wolf, *Acta Metall. Mater.*, **38** (1990), 791.

19. M. F. Ashby, F. Spaepen and S. Williams, *Acta Metall.*, **26** (1978), 1647.

20. H. J. Frost, M. F. Ashby and F. Spaepen, *Scripta Metall.*, **14** (1980), 1051.

21. M. Koiwa, H. Seyazaki and T. Ogura, *Acta Metall.*, **32** (1984) 171.

22. H. Meiser and H. Gleiter, *Scripta Metall.*, **14** (1980), 95.

23. G. J. Wood, W. M. Stobbs and D. J. Smith, *Phil. Mag. A*, **50** (1984), 375.

24. K. L. Merkle, *Scripta Metall.*, **23** (1989), 1487.

25. M. Cheikh, A. Hairie, F. Hairie, G. Nouet and E. Paumier, *Colloque de Phys.*, **51** (1990), C1–103.

26. A. Seeger and G. Schottky, *Acta Metall.*, **7** (1959), 495.

27. J. Ferrante and J. R. Smith, *Phys. Rev. B*, **31** (1985), 3427.

28. D. Wolf and S. R. Phillpot, *Mat. Sc. Eng. A*, **107** (1989), 3.

29. S. P. Chen, D. Srolovitz and A. F. Voter, *J. Mater. Res.*, **4** (1989), 62.

30. D. Wolf, *Acta Metall.*, **37** (1989), 1983.

31. D. Wolf, *Scripta Metall.*, **23** (1989), 1913.

32. See, for example, W. D. Kingery, H. K. Bowen and D. R. Uhlmann, *Introduction to Ceramics* (2nd edn), Wiley, New York (1976), p. 107.

33. H. B. Aaron and G. F. Bolling, in *Grain Boundary Structure and Properties* (eds G. A. Chadwick and D. A. Smith), Academic Press, New York (1976), p. 107.

34. P. H. Pumphrey, *Scripta Metall.*, **6** (1972), 107.

35. M. Weins, B. Chalmers, H. Gleiter and M. F. Ashby, *Scripta Metall.*, **3** (1969), 601.

36. M. Weins, H. Gleiter and B. Chalmers, *Scripta Metall.*, **4** (1970), 235, and *J. Appl. Phys.*, **42** (1971), 2639.

37. D. A. Smith, V. Vitek and R. C. Pond, *Acta Metall.*, **25**, 475 (1977), and ibid., **27** (1978), 235.

38. A. P. Sutton, *Phil. Mag. Lett.*, **59** (1989), 53.

39. See, for example, C. Herring, in *Structure and Properties of Solid Surfaces* (eds R. Gomer and C. S. Smith), University of Chicago Press (1953), p. 4.

40. D. Wolf, *Surf. Sci.*, **226** (1990), 389.

41. D. Wolf, *J. Appl. Phys.*, **68** (1990), 3221.

42. D. Wolf, *J. Am. Cer. Soc.*, **67** (1984), 1.

43. D. Wolf, *Physica*, **131B** (1985), 53.

44. D. Wolf, *Acta Metall.*, **32** (1984), 245.

45. J. F. Lutsko, D. Wolf, S. Yip, S. R. Phillpot and T. Nguyen, *Phys. Rev. B*, **38** (1988), 11572.

46. S. M. Foiles, M. I. Baskes and M. S. Daw, *Phys. Rev. B*, **33** (1986), 7983.

47. G. J. Ackland and R. Thetford, *Phil. Mag. A*, **56** (1987), 15.

48. R. A. Johnson, *Phys. Rev. A*, **134** (1964), 1329.

49. D. Wolf, *Phil. Mag. B*, **59** (1989), 667.

50. D. N. Seidman, J. G. Hu, S. Kuo, B. W. Krackauer, Y. Oh and A. Seki, *Colloque de Phys.*, **51** (1990), C1–47.

51. J. G. Hu and D. N. Seidman, *Phys. Rev. Lett.*, **65** (1990), 1615.

52. L. L. Kazmerski, in *Polycrystalline Semiconductors* (eds H. J. Möller, H. P. Strunk and J. H. Werner), *Springer Proceedings in Physics*, **35**, Springer-Verlag, Berlin (1989), 96.

53. M. R. Fitzsimmons and S. L. Sass, *Acta Metall.*, **36** (1988), 3103.

54. M. S. Taylor, I. Majid, P. D. Bristowe and R. W. Balluffi, *Phys. Rev. B*, **40** (1989), 2772.

55. I. Majid, P. D. Bristowe and R. W. Balluffi, *Phys. Rev. B*, **40** (1989), 2779.

56. J. A. Venables and R. Bin-Jaya, *Phil. Mag.*, **35** (1977), 1317.

57. D. J. Dingley, R. Mackenzie and K. Baba-Kishi, in *Microbeam Analysis 1989* (ed P. E. Russell), San Francisco Press (1989), p. 435.

58. R. Bonnet and F. Durand, *Phys. Stat. Sol. (a)*, **27** (1975), 543.

59. P. Heilmann, W. A. T. Clark and D. A. Rigney, *Ultramicroscopy*, **9** (1982), 365.

60. R. C. Pond and D. A. Smith, *Can. Metall. Quarterly*, **13** (1974), 39.

61. R. C. Pond, D. A. Smith and W. A. T. Clark, *J. Microsc.*, **102** (1974), 309.

62. F. Cosandey and C. L. Bauer, *Phil. Mag. A*, **44** (1981), 391.

63. E. P. Kvam and R. W. Balluffi, *Phil. Mag. A*, **56** (1987), 137.

64. C. B. Carter, A. Donald and S. L. Sass, *Phil. Mag. A*, **41** (1980), 467.

65. J. A. Eastman, M. D. Vaudin, K. L. Merkle and S.L. Sass, *Phil. Mag. A*, **59** (1989), 465.

66. W. A. T. Clark, C. E. Wise, Z. Shen and R. H. Wagoner, *Ultramicroscopy*, **30** (1989), 76.

67. T. C. Lee, I. M. Robertson and H. K. Birnbaum, *Ultramicroscopy*, **29** (1989), 212.

68. I. M. Robertson, T. C. Lee, P. Rozenak, G. M. Bond and H. K. Birnbaum, *Ultramicroscopy*, **30** (1989), 70.

69. J. C. H. Spence, *Experimental High-Resolution Electron*

Microscopy (2nd edn), Oxford University Press, New York, Oxford (1988).

70. S. J. Pennycook, D. E. Jesson and M. F. Chisholm, *Mat. Res. Soc. Symp. Proc.*, **183** (1990), 211.

71. S. J. Pennycook and D. E. Jesson, *Phys. Rev. Lett.*, **64** (1990), 938.

72. A. Ourmazd, D. W. Taylor, M. Bode and Y. Kim, *Science*, **246** (1989), 1571.

73. J. C. H. Spence, in *High-Resolution Transmission Electron Microscopy*, (eds P. Buseck, J. Cowley and L. Eyring), Oxford University Press, New York, Oxford (1988), p. 190.

74. *High-Resolution Electron Microscopy and Associated Techniques* (eds P. R. Buseck, J. M. Cowley and L. Eyring) Oxford University Press, New York, Oxford, (1988).

75. J. M. Cowley and A. F. Moodie, *Acta Cryst.*, **10** (1957), 609.

76. A. Bourret, *Colloque de Phys.*, **51** (1990), C1–1.

77. W. O. Saxton and D. J. Smith, *Ultramicroscopy*, **18** (1985), 39.

78. H. F. Fischmeister, W. Mader, B. Gibbesch and G. Elssner, *Mat. Res. Soc. Symp. Proc.*, **122** (1988), 529.

79. T. Schober and R. W. Balluffi, *Phil. Mag.*, **21** (1970), 109.

80. T. Y. Tan, J. C. M. Hwang, P. J. Goodhew and R. W. Balluffi, *Thin Solid Films*, **33** (1976), 1.

81. D. Wolf, *Acta Metall. Mater.*, **38** (1990), 781.

82. D. Wolf, *Phil. Mag. A*, **62** (1990), 447.

83. G. Hasson and C. Goux, *Scripta Metall.*, **5** (1971), 889.

84. L. Sedlacek, K. L. Merkle and N. L. Peterson, 1989 (private communication).

85. D. Wolf, *Acta Metall.*, **37** (1989), 2823.

86. D. Wolf, *J. Appl. Phys.*, **69** (1990), 185.

87. S.-W. Chan and R. W. Balluffi, *Acta Metall.*, **26** (1986), 113.

88. G. Herrmann, H. Gleiter, and G. Baro, *Acta Metall.*, **24** (1976), 353.

89. H. Sautter, H. Gleiter, and G. Baro, *Acta Metall.*, **25** (1977), 467.

90. D. A. Smith, *Scripta Metall.*, **8** (1974), 1197.

91. J. H. Van der Merwe, *Proc. R. Soc. London A*, **43** (1950), 616.

92. C. Rey and G. Saada, *Phil. Mag. A*, **33** (1977), 825.

93. R. Bonnet, *Phil. Mag. A*, **43** (1981), 1165.

94. A.-C. Shi, C. Rottman, and Yu He, *Phil. Mag. A*, **55** (1987), 499.

95. G.-J. Wang and V. Vitek, *Acta Metall.*, **34** (1986), 951.

96. V. Vitek, *Scripta Metall.*, **21** (1987), 711.

97. D. Wolf, *Scripta Metall.*, **23** (1989), 1713.

98. D. Wolf, *J. Mat. Res.*, **5** (1990), 1708.

99. J. C. Li, *J. Appl. Phys.*, **32** (1961), 525.

100. D. Wolf, *Scripta Metall.*, **23** (1989), 377.

101. A. Brokman and R. W. Balluffi, *Acta Metall.*, **29** (1981), 1703.

102. D. Wolf, *Phil. Mag. A*, **63** (1991), 1117.

103. M. D. Kluge, D. Wolf, J. F. Lutsko and S. R. Phillpot, *J. Appl. Phys.*, **67** (1990), 2370; D. Wolf and M. D. Kluge, *Scripta Metall.*, **24** (1990), 907.

104. D. Wolf, *Phil. Mag. A*, **63** (1991), 337.

105. See, for example, C. Herring, in *Structure and Properties of Solid Surfaces* (eds R. Gomer and C. S. Smith) University of Chicago Press, (1953), p. 4.

106. V. I. Marchenko and A. Y. Parshin, *Sov. Phys.-Jetp*, **52** (1980), 129.

107. W. Lojkowski, H. Gleiter and R. Maurer, *Acta Metall.*, **36** (1988), 69.

108. K. L. Merkle and D. Wolf, *MRS Bulletin*, Vol. XV (1990), p. 42.

109. K. L. Merkle and D. Wolf, *Phil. Mag. A*, **65** (1992), 513.

110. K. L. Merkle, J. F. Reddy, C. L. Wiley and D. J. Smith, in *Ceramic Microstructures '86* (eds Pask and Evans 1987), p. 241.

111. K. L. Merkle and D. J. Smith, *Ultramicroscopy*, **22** (1987), 57.

112. J. Penisson, *J. de Physique*, **49** (1988), C5–87.

113. A. P. Sutton and R. W. Balluffi, *Acta Metall.*, **35** (1987), 2177.

114. P. J. Goodhew, in *Grain Boundary Structure and Kinetics* (ed R. W. Balluffi 1980), p. 155.

115. P. J. Goodhew, T. Y. Tan and R. W. Balluffi, *Acta Metall.*, **26** (1978), 557.

116. C. Herring, *Phys. Rev.*, **82** (1951), 87.

117. L. M. Clarebrough and C. T. Forwood, *Phys. Stat. Sol. A*, **60** (1980), 51.

118. C. T. Forwood and L. M. Clarebrough, *Acta Metall.*, **32** (1984), 757.

119. W. Krakow and D. A. Smith, in *Grain Boundary Structure and Related Phenomena* (ed Y. Ishida), *Trans. Jap. Inst. Met.*, **27** (1986), 277.

120. J.-L. Rouviere and A. Bourret, *Colloque de Physique*, **51** (1990), C1–329.

121. J. L. Putaux and J. Thibault-Dessaux, *Colloque de Physique*, **51** (1990), C1-323.

122. A. Seeger and P. Haasen, *Phys. Stat. Sol.*, **3** (1958), 470.

123. See, for example, D. M. Esterling in *Computer Simulation in Materials Science* (eds R. J. Arsenault, J. R. Beeler and D. M. Esterling), ASM International, 1988, p. 149 ff, and R. J. Arsenault, ibid., p. 165 ff.

124. W. M. Stobbs, G. J. Wood, and D. J. Smith, *Ultramicroscopy*, **14** (1984), 145.

125. K. L. Merkle, *Ultramicroscopy*, **40** (1992), 281.

126. K. L. Merkle, J. F. Reddy, C. L. Wiley, and D. J. Smith, *Journ. de Physique*, **49** (1988), C5–251.

127. M. J. Mills and M. S. Daw, *Mat. Res. Symp. Proc.*, **183** (1990), 15, and M. J. Mills, private communication.

128. F. Cosandey, S.-W. Chan, and P. Stadelmann, *Colloque de Phys.*, **51** (1990), C1–109.

129. M. Cheikh, A. Hairie, F. Hairie, G. Nouet, and E. Paumier, *Colloque de Phys.*, **51** (1990), C1–103.

130. N. F. Mott, *Proc. Phys. Soc.*, **60** (1948), 391.

131. H. Gleiter, *Mater. Sci. Eng.*, **52** (1982), 91.

132. K. L. Merkle, *Mat. Res. Soc. Symp. Proc.*, **153** (1989), 83.

133. D. Wolf and J. F. Lutsko, *Phys. Rev. Lett.*, **60** (1988), 1170.

134. See, for example, W. T. Read and W. Shockley, in *Imperfections in Nearly Perfect Crystals* (eds W. Shockley, J. H. Hollomon, R. Maurer and F. Seitz), Wiley, New York (1952), p. 352.

135. V. I. Marchenko and A. Y. Parshin, *Sov. Phys.-Jetp*, **52** (1980), 129.

136. See, for example, D. Wolf and J. A. Jaszczak, *Role of Dislocations and Surface Steps in the Works of Adhesion*, Chapter 27 in this volume.

137. See, for example, H. J. Frost, M. F. Ashby and F. Spaepen, *A Catalogue of [100], [110], and [111] Symmetric Tilt Boundaries in Face-Centered Cubic Hard-Sphere Crystals*, Harvard University Press (1982).

138. N. Rivier and A. J. A. Lawrence, *Physica B*, **150** (1988), 190.

139. A. P. Sutton, *Acta Metall.*, **36** (1988), 1291.

140. A. P. Sutton, *Phase Transitions*, **16/17** (1989), 563.

141. T. Hasegawa, K. Kobayashi, N. Ikarashi, K. Takayanagi and K. Yagi, *Jap. J. Appl. Phys.*, **25** (1986), 366.

142. K. L. Merkle, in *Electron Microscopy 1990*, Proceedings of the XIIth Internatl. Congr. for Electron Microscopy, San Francisco Press (1990), Vol. 4, p. 332.

CHAPTER FOUR

Grain and interphase boundaries in ceramics and ceramic composites

M. G. Norton and C. B. Carter

Introduction · Illustrations of GBs in ceramic materials · Phase boundaries · Summary

4.1 INTRODUCTION

4.1.1 Atomic structure of ceramic materials

In this chapter some of the fundamental features of interfaces in ceramic materials will be discussed, with particular emphasis on atomic level structure of the boundaries. In order to discuss boundaries in ceramic materials it will be helpful to describe briefly bonding in ceramics since this is more complex than that in the more thoroughly studied metals.

The classic definition of a ceramic material is 'an inorganic non-metallic solid' [1]. In ceramics the bonding may be either ionic or covalent but is most often a mixture of the two. In crystals where the bonding is predominantly ionic, each positive ion is surrounded by several negative ions, and likewise each negative ion is surrounded by several positive ions. Stable structures are formed when an ion achieves the maximum number of nearest neighbors (coordination number) of opposite charge, so that densely packed structures are preferred. Compounds of metal ions with group VII anions are strongly ionic (e.g. NaCl, LiF). The electron distribution in ions is nearly spherical, and the interatomic bond, since it arises from coulombic forces, is nondirectional in nature. Surfaces and grain boundaries (GBs) in ionic crystals may carry an electric charge resulting from the presence of excess ions of one sign, this charge must then be compensated for by a space-charge region of the opposite sign adjacent to the boundary [2, 3]. For a pure material this charge arises if the energies to form anion and cation vacancies or interstitials at the boundary are different; the magnitude and sign of the boundary charge changes if there are aliovalent solutes present which alter the concentration of lattice defects in the crystal [1]. It should be noted that there appear to be no direct observations of this space-charge region.

Many metal halides and oxides are largely ionic in character (e.g. NaCl, Al_2O_3, ZrO_2, MgO – the I–VII, II–VI and III–VI compounds) although they do show some degree of covalency (~40% for Al_2O_3); as the electronegativity of the two elements becomes closer, the degree of ionic component in the bonding decreases. Covalent crystals are formed between atoms having a similar electronegativity but which are not close in electronic structure to the noble gas configuration, therefore many covalent crystals are made up of elements from group III, IV and V of the periodic table (the III–V and IV–IV compounds). The bonding in covalent materials is strongly directional, and ceramic materials with primarily covalent bonding (e.g. diamond, Si_3N_4) have high hardness, high melting points, and low atomic diffusivities.

4.1.2 What properties of ceramics are important?

The important uses of ceramics are primarily in areas where their mechanical strength at high temperature and their electrical properties can be utilized. The useful electrical properties of ceramics were, until 1986 [4], concerned with their high electrical resistivity, but with the discovery of high-temperature superconductivity in oxide ceramics the interest in electrical applications, and the definition of a ceramic material, have been somewhat extended. Examples and possible applications of ceramic materials have been summarized in Table 4.1.

The properties of a material are often controlled by the microstructure and in polycrystalline materials it is often the GBs which have the greatest effect on the properties. In this chapter, the correlation between the microstructure and properties of some materials which are technologically important will be discussed. The examples used will be the thermal conductivity of AlN substrates, the transport properties in epitactic thin films of the $YBa_2Cu_3O_{7-\delta}$ superconductor, and the improvement in toughness associated with the phase transformation occurring in ZrO_2-toughened Al_2O_3.

When GBs are fully understood, it will be possible to modify their properties in a controlled manner. Such **grain boundary engineering** is already being used for the improvement of high-temperature mechanical properties in Si_3N_4-based ceramics [5]. For example, it has been reported [6] that strengths as high as 965 MPa at 1200°C can be achieved with Si_3N_4 sintered with Y_2O_3 and Al_2O_3 additives. This improvement in the high-temperature strength is the result of crystallization of the GB phase. The development of such high-temperature materials is aimed at improving, simultaneously, the high-temperature strength and creep behavior while improving oxidation resistance [7]. In order to be able to do this it is essential that the GB is understood since the microstructure developed in these polycrystalline ceramics is a result of the sintering and forming processes.

4.1.3 Formation of interfaces

There are many ways to produce interfaces in ceramic materials. Examples are summarized in Table 4.2 and will now be discussed.

Table 4.1 Examples of properties and applications of ceramic materials

Property	Example	Applications
Mechanical strength	Si_3N_4	Automotive engine
Electrical		
1. Insulating	Al_2O_3, AlN	Spark plug bodies, electronic packaging
2. Superconducting	$YBa_2Cu_3O_{7-\delta}$	'SQUIDS'[a]
3. Mixed	ZnO	Varistors (surge protectors)

[a] Superconducting quantum interference devices

Table 4.2 Methods of formation of interfaces

Method	Atomic process	Example
Sintering		
1. With applied pressure	Atomic diffusion	Densification of Al_2O_3
2. Pressureless	Liquid-phase sintering	Densification of AlN with Y_2O_3
Reaction	Solid-state reaction	$Al_2O_3 \rightarrow$ spinel (involving chemical change)
	Phase transformation	ZrO_2 tetragonal $\rightarrow$ monoclinic (with no chemical change)
Heteroepitactic growth	Bond strength Lattice mismatch	$YBa_2Cu_3O_{7-\delta}$ on MgO

Sintering

Single crystals of many ceramic materials are prohibitively expensive for widespread use and, indeed, may not be suitable for a number of applications. In particular, for structural applications crack propagation through single-crystals can proceed quite rapidly because there is no feature present to deflect, or blunt, the crack tip. Most ceramics are therefore necessarily polycrystalline. Dense ceramics are usually obtained by the consolidation of fine powder compacts. Densification is achieved by heating to a high temperature: sintering either with an applied pressure (hotpressing) or without, as is the case for liquid-phase sintering. In either case, the mechanism for densification is the transport of matter from the bulk to the junction between adjoining particles. The transport processes involve diffusion by a number of different paths: across the surface, along phase or grain boundaries, and through the vapor [1]. Hot-pressing allows high densities and a fine grain microstructure to be achieved. The expense and difficulty of working at high temperatures has led to the widespread use of liquidphase sintering if this does not adversely affect the properties of the interfaces in the resulting material.

In liquid-phase sintering small, controlled amounts of other materials are added. The additives are usually oxides which become liquid during the sintering process and thus enhance densification by providing fast diffusion paths. For non-oxide ceramics densification is almost always achieved by use of a liquid-phase because the activation energy for self-diffusion is particularly large for these covalent materials [8]. Studies of a large number of systems indicate that densification will take place rapidly when the following conditions are satisfied:

1. An appreciable amount of liquid phase is formed.
2. The solubility of the solid in the liquid is significant.
3. The liquid wets the solid.

So it can be seen that the GB may not simply be an abrupt interface between two grains, but may include a thin layer of a second-phase material. For an introduction to the mechanisms of sintering ceramics the reader is referred to the textbook by Kingery *et al.* [1]. It is important to realize that there is a causal sequence between processing, the resultant microstructure and the properties which are then a function of the microstructure. The applications are then determined by the material properties [9, 10].

Solid-state reaction

Two classes of solid-state reaction can occur depending on whether or not a chemical change takes place at the interface. The formation of oxide spinels, e.g. NiO and Al_2O_3 reacting to form $NiAl_2O_4$, is an example where both the structure and the chemistry changes at the interface [11]. The reaction proceeds by the flux of charged particles across the phase boundaries and through the reaction product. The spinel–NiO interface can also form during precipitation as has been studied for the $NiO-NiFe_2O_4$ system, where the lattice misfit between the two oxygen sub-lattices is small, so that the interface may involve only a change in distribution of the cations [12]. The situation is illustrated in Fig. 4.1 which was produced by imaging using a spinel reflection which was not common to the NiO. The misfit strain between the oxygen sub-lattices may still be a factor in determining the particle morphology. In systems involving only a small lattice misfit, such as $NiO-NiFe_2O_4$, large coherent precipitates are frequently observed; for example, dendritic precipitates which may be $>1\,\mu m$ long.

The second example of a phase transformation involving a solid-state reaction is that of the martensitic transformation which occurs when tetragonal ZrO_2 transforms to the monoclinic form [13]. This transformation is diffusionless and requires only a shear of the parent structure to obtain the new phase. This type of transformation has recently been widely studied since it leads to twin formation in oxides with structures related to perovskite.

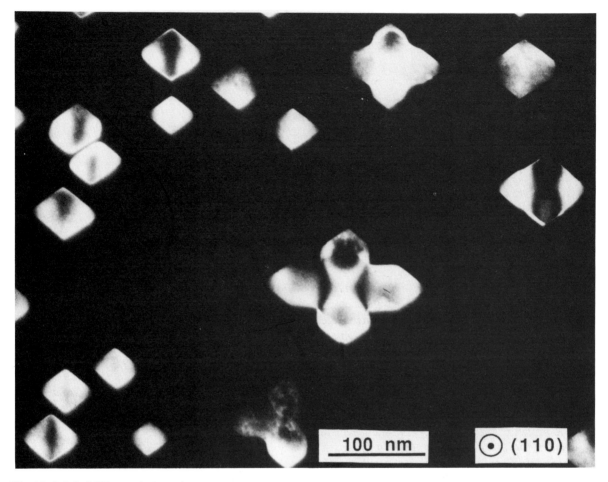

Fig. 4.1 A dark-field image of spinel particles embedded within a NiO matrix (courtesy S. R. Summerfelt).

Heteroepitactic growth of ceramic thin films

In recent years there has been considerable interest in preparing ceramic thin films for use in a number of applications including electronic, magnetic, and optical. In many such applications bulk single crystals of metal oxides are used, but often the use of thin crystalline films would have many advantages. Heteroepitactic growth of ceramic thin films, i.e. the growth of a thin film on a substrate which has either a different chemical composition or a different crystal structure, has now become an extremely active area of research as illustrated by the growth of superconducting thin films on a wide range of substrates [e.g. 14]. Because the mode of

growth of an epitactic film is important in determining defects found in the film, it is relevant to review briefly the prominent features of epitactic growth mechanisms. For a more detailed discussion on this subject the reader is referred to the book edited by Matthews [15].

The nucleation and initial growth stages of epitactic deposits are governed strongly by the bonding between deposit and substrate. It is only when the film atoms interact directly with the substrate that there will be a driving force for layer-by-layer growth (Fig. 4.2(a)) and even then only up to a limited film thickness for non-zero misfit. Therefore, for most film systems where there is a non-zero misfit, growth will occur by an island

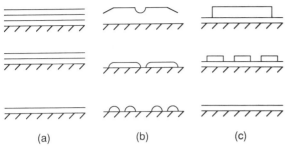

Fig. 4.2 Three forms of epitactic growth: (a) layer-by-layer growth; (b) formation and growth from discrete nuclei; (c) initial formation of a tightly bound layer followed by nucleation and growth (after M. J. Stowell [17]).

Table 4.3 Microscopy techniques for interface study in ceramics

Technique	Resolution	Depth of field
Visible-light microscopy (VLM)	250 nm	250 nm
Scanning electron microscopy (SEM)		
1. Secondary electron image mode (SEI)	5–15 nm	several mm
2. Backscattered electron image mode (BEI)	500–250 nm	
Transmission electron microscopy (TEM)		
1. Strong beam	10 nm (0.2 nm)	km
2. Weak beam dark-field (WBDF)	1 nm	
3. High-resolution electron microscopy (HREM)	0.16 nm (0.12 nm)	m

mechanism, where the islands form either on the bare substrate (Volmer–Weber mode, Fig. 4.2(b)) or on a few layers of uniform film (Stranski–Krastanow mode, Fig. 4.2(c)). Highly oriented films can still occur, however, for quite large misfits, e.g. films of $YBa_2Cu_3O_{7-\delta}$ on MgO, where the lattice mismatch is ~9% [16]. The three forms of epitactic growth are illustrated schematically in Figs. 4.2(a–c). In all the three cases, it is possible for the initial deposit to be strained elastically to match the substrate; this is referred to as pseudomorphic growth. The dominant condition for the occurrence of this mode of growth is that the strain necessary to ensure coherency at the substrate–deposit interface is small [17].

4.1.4 What techniques can be used to study interfaces in ceramics?

A number of techniques can be used to examine the structure of interfaces in ceramic materials. These have been listed in Table 4.3 along with their resolution and depth of field. Advantages and disadvantages of the techniques are discussed below.

1. Visible-light microscopy (VLM) is extremely valuable in studying interfaces in ceramics since many ceramics are transparent. Measurements are often made on GB grooves formed by thermal or chemical etching. Such observations are used to estimate GB energies. Many mechanisms using reflection, diffraction and inter-ference and polarization of light are available for enhancing contrast in the image. By use of interferometric–contrast methods it is possible to see steps as small as 5 nm and actual measurements have been made by multiple beam interference of steps only 0.3 nm high [18] so that, in principle, near-atomic scale detail can be studied by VLM. In comparison with the techniques that follow, VLM has the important advantage that the sample does not need to be coated to increase electrical conductivity.

2. Scanning electron microscopy (SEM) gives much greater depth resolution than is possible with visible-light and is particularly advantageous for examining the morphology of fine-grain sintered materials, characterizing features such as grain shape and grain size distribution and analyzing the segregation of second-phase materials. The technique is particularly straightforward when there is a large difference in the atomic number of the atoms contributing the primary and secondary phases. In the examples illustrated here, SEM has been used to examine the fracture surfaces and crack propagation in fiber-reinforced composite materials.

3. Transmission electron microscopy (TEM) is the essential tool for any study of interface structure. It enables examination of microstructure on a nanometer scale – a point-to-point resolution of better than 0.2 nm can be achieved. The morphology of interfaces and structure can be examined by imaging techniques and correlated with compositional detail obtained by use of energy-dispersive spectroscopy (EDS) or electron energy-loss spectroscopy (EELS). The major disadvantage of this technique is that the necessary sample preparation is almost always destructive since electron-transparent samples (~100 nm thick) must be prepared from the bulk material. This thinning process may, unfortunately, introduce artifacts. Preparation of composite materials where both phases are required to be thinned may also be arduous due to the different hardness values, and therefore thinning rates, of the two materials.

4.1.5 Grain boundaries in ceramics

A grain boundary is simply the interface between two grains of the same structure and composition. For ceramic materials the structures of GBs are more complex than those in pure metals since second phases are often present and may be of significant thickness, in which case two interfaces will actually be present. The term 'grain boundary' will therefore denote an interface between two crystalline grains; in ceramic materials three distinct types of GB can be identified.

1. Interfaces where the two grains are in intimate contact over their entire area; such interfaces correspond most closely to GBs which have been extensively studied in metals and semiconductors.
2. Interfaces where two grains are clearly separated by a layer of a second phase which may be crystalline or amorphous – which type may depend on the ambient temperatures.
3. Interfaces which are a combination of the first two kinds.

The purpose of this chapter is to review some of the fundamental features of interfaces in ceramic materials and to illustrate these features with experimental observations, obtained using transmission, scanning electron, or visible-light microscopy of interfaces from a wide range of different systems. A knowledge of the GBs in ceramics is increasingly necessary with the development of the new generation of ceramic materials and also in the development of **high-tech** composite materials. The particular GBs included in this discussion are

(a) low-angle GBs, where the structure of the interface can be related to the allowed lattice dislocations with both perfect and partial Burger's vectors;
(b) twin-boundaries where either the whole of the crystal is affected by the presence of the interface or where, to a first approximation, only one sublattice is affected;
(c) high-angle GBs.

The structure of defects in ceramic materials can be more complex that those of similar defects in pure metals because of the presence of two, or more, ionic species on two sub-lattices. For example, one of the most striking structural phenomena that results from the presence of two types of ions is the phenomenon of dissociation by climb of both lattice and grain boundary dislocations. The phenomenon has been recorded for such materials as spinel, alumina, and garnet (for reviews see [19–21]). The partial dislocations created in these oxides by the dissociation process are 'perfect' dislocations of the oxygen sub-lattice; the stacking fault thus formed is therefore only a fault on the cation sub-lattice. Similar processes can occur in GBs in oxides.

It has also been reported [22, 23] that the plane adopted by the interface is a very important factor in any discussion of GBs. Indeed, it has often been observed that GBs in ceramic materials show a particularly strong tendency to facet. The facet plane is almost invariably parallel to a low-index plane in one, or both, grains. Reasons for this phenomenon have been discussed by Wolf [22]; the basic argument is that low-index planes are the most widely separated planes and thus involve less energy when a 'stacking error' is present in the crystal. A particularly striking illustration is provided by the $\Sigma = 99$ and the $\Sigma = 41$ boundaries in spinel where, by chance, the interface can facet parallel to several (different) low-index planes in both grains simultaneously, as illustrated in Figs. 4.3(a–c). A similar situation exists for many phase boundaries.

The macroscopic interface plane and the planes adopted on an atomistic level may well be different as can be appreciated from the example shown in Fig. 4.4, where the interface plane is actually composed of small-scale atomic steps [24]. This type of interface structure is directly analogous to the formation of vicinal surfaces on ceramic materials [25], i.e. it is a vicinal internal interface.

Any vicinal surface can be considered in terms of terraces of low-index with low-surface energy separated by small steps. These steps, which are analogous to facets in a GB, will certainly influence the defect properties such as segregation and they may provide vacancy sources and sinks that determine the space charge in ionic materials [26].

4.1.6 Grain boundary terminology

Early theories were based on the concept that adjacent crystals were joined by an amorphous layer [e.g. 27]. An alternative description viewed the GB as a transition lattice with a regular arrangement of atoms [28]. Although the latter

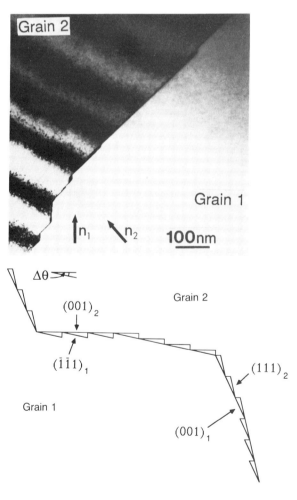

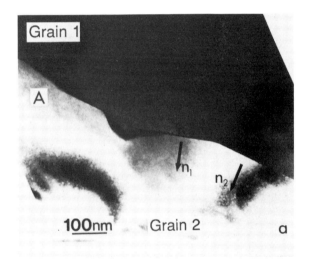

Fig. 4.3 (a) $\Sigma = 99$ grain boundary, strong-beam, bright-field image showing the faceted nature of the interface. n_2 is $[1\bar{1}\bar{1}]$ in grain 1; n_1 is the common $[1\bar{1}0]$ direction. (b) The $\Sigma = 41$ grain boundary, strong-beam, bright-field image showing the faceted nature of the interface. n_1 is $[\bar{1}\bar{1}\bar{1}]$ in grain 1. n_2 is $[\bar{1}\bar{1}1]$ in grain 2. (c) A schematic diagram to illustrate how the small misorientation from the exactly parallel relationship may be accommodated. The angle $\Delta\theta$ would be $0.58°$ and $1.4°$ for $\Sigma = 99$ and $\Sigma = 41$, respectively.

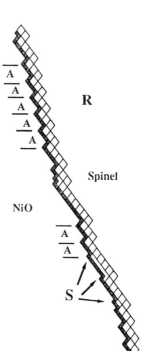

Fig. 4.4 (a) HREM image showing the interface between NiO and spinel (b) corresponding schematic showing how the interface consists of small scale atomic steps (courtesy S. R. Summerfelt).

description is found to describe GBs in metals, both types of interfaces are found to exist in ceramic materials. With the development of dislocation theory, the GB was subsequently described in terms of an array of dislocation lines [29], so that, for example, a tilt boundary is described in terms of a periodic array of edge dislocations. Many geometrical models have been proposed to describe GBs and the terminology associated with these models is tabulated in Table 4.4.

The coincidence-site-lattice (CSL) model was discussed by Kronberg and Wilson [30]. The CSL can be obtained by imagining that the two misoriented crystal lattices, A and B, interpenetrate one another so that certain lattice points of each

crystal coincide to give a three-dimensional lattice. The fraction of lattice sites in either grain which are common to A and B is defined as $1/\Sigma$. A GB can then be constructed by passing a plane through the interpenetrating crystals and removing all of the A points on one side of the boundary plane and all the B points on the other side. The two-dimensional CSL in the boundary plane describes the actual periodicity of the GB structure even when the atoms move from their perfect crystal sites to take on a lower energy configuration.

It has also been pointed out that there are additional locations beside the CSL points where the two crystal lattices are in good coincidence [31]. These locations can be used to define another

Table 4.4 Grain boundary terminology and definitions

Terminology	Definition
CSL	The lattice of coincident individual lattice sites
Σ	The inverse of the fraction of individual-lattice sites which are common to the two adjoining grains
O-Lattice	The lattice of 'origins'
DSCL	The displacement-shift-complete lattice

Table 4.5 Types of GB

Boundary	Definition
Tilt	The rotation axis is contained within the GB plane
Twist	The rotation axis is normal to the GB plane
Mixed	When neither of the above special conditions occurs
Twin	When the adjoining grains are mirror related across the GB. This definition includes not only the classic $\Sigma = 3$, (111) but also such interfaces as the $\Sigma = 5$, (120) GB

lattice which is a sub-lattice of the CSL and is referred to as the O-lattice. A third lattice, the displacement-shift-complete lattice (DSCL), is so named because it is a lattice of vectors which describe possible relative translations which will not change the structure (or pattern) of the boundary. For a more detailed discussion of both the O-lattice theory and the DSCL model the reader is referred to the article by Smith and Pond [32]. What use do these terminologies have in our understanding of GBs in ceramic materials? The Σ notation can predict whether the formation of low-index symmetric boundaries are possible; it also provides a convenient shorthand notation – it does not predict relative energies. The DSCL construction can be used to derive the allowed Burger's vectors of GB dislocations. The O-lattice analysis gives a method for analyzing the geometry of a GB and deducing the DSCL vectors, but gives no indication of atomistic structure of the boundary.

Grain boundaries are often said to fit into one of four categories; namely, twist, tilt, mixed, and twin boundaries. The definition of these terms is summarized in Table 4.5. There has, however, been much confusion over the importance of this classification [21]. The difficulty can be illustrated by the faceting twin boundary (found in cubic metals, semiconductors, and insulators) shown in Fig. 4.5. Is this boundary a twist or tilt boundary? It can be seen from the diagram that both descriptions are correct. More complex twist boundaries occur and these can be easily generated by mirroring the structure across a plane other than the {111}/{112} plane (or the (0001) plane in hcp

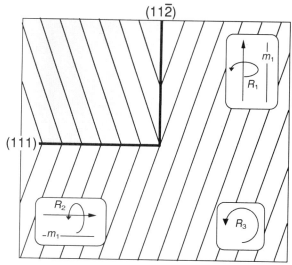

Fig. 4.5 The rotation axes (R_1, R_2, and R_3) and mirror planes (m_1, m_2) which can be used to produce the 'coherent' {111} and 'incoherent' (or lateral) {112} twin boundaries in fcc materials.

materials). Note that none of this discussion has considered the actual atomic structure of the GBs.

4.2 ILLUSTRATIONS OF GBS IN CERAMIC MATERIALS

4.2.1 Low-angle

There have been many studies of low-angle GBs in materials such as alumina [33], NiO, and spinel [34, 35]; such studies not only provide new understanding of GBs themselves but can also give new insight into the structure and behavior of dis-

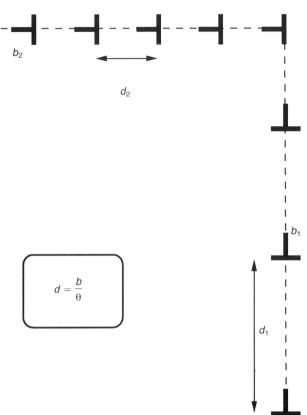

Fig. 4.6 The change in dislocation spacing with a change in boundary plane for a given pure-tilt boundary. The equation relating d, θ, and b is the same for both facet planes.

locations in these materials [22]. Low-angle GBs are composed of arrays of dislocations, twist boundaries are usually envisioned as arrays of screw dislocations, while tilt boundaries contain at least one array of edge dislocations. Such structures assume that the grains are in intimate contact over their entire area. If the dislocations forming the GB are in equilibrium they must be regularly spaced. However, the Burger's vectors of the dislocations present in the low-angle GB, and hence their spacing in the interface, will be different for the different planes. This effect is shown in Fig. 4.6.

To appreciate the complexities of GBs in ceramic materials, consider the same $(1\bar{1}0)$ tilt boundary in spinel and NiO (ignoring any difference in lattice parameter and assuming the same misorientation, θ). The magnitude of the Burger's vector of the dislocations in spinel is twice that of the dislocation

in NiO and hence, according to the formula in Fig. 4.6, the dislocations must be twice as far apart in spinel as they are in NiO. However, the $\frac{1}{2}[110]$ lattice dislocations in spinel can dissociate into two partial dislocations with Burger's vectors $\frac{1}{4}[110]$ which are separated by a stacking fault on the (110) plane. The resulting configuration is that shown in Fig. 4.7. The lattice dislocations in the spinel GB are then said to be climb dissociated since their Burger's vectors do not lie in the plane of the stacking fault. Other clear illustrations are the dissociation of an $\langle 1\bar{1}00 \rangle$ dislocation in Al_2O_3 into three $\frac{1}{3}\langle 1\bar{1}00 \rangle$ partial dislocations and the dissociation of the dislocations in the low-angle (001) twist boundary in spinel shown in Fig. 4.8. In this example, the situation is actually even more complex since there are two structurally different (001) planes for the stacking fault, and these two planar defects have a different energy per unit area.

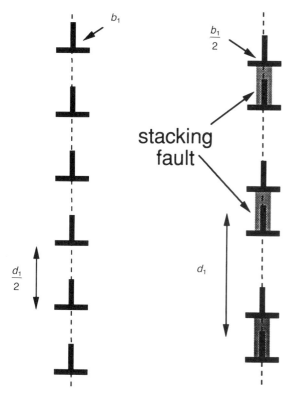

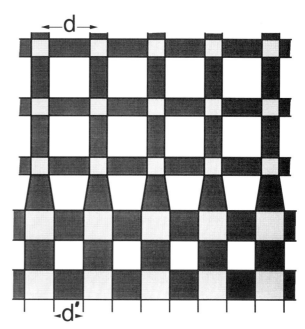

Fig. 4.8 A network of orthogonal screw dislocations producing a small-angle twist boundary in spinel. d is the spacing of the perfect, lattice dislocations, d′ is the periodicity of the screw partial dislocations.

Fig. 4.7 The spacing of dislocations in NiO(a) and spinel (b) for pure-tilt boundaries with the same misorientation.

Faceting, due to elastic interactions of dislocations in low-angle GBs, is widely observed [36, 37].

4.2.2 High-angle

Structured high-angle GBs have been analyzed in several, though not many, ceramic materials such as MgO, Al_2O_3, and spinel. The study of GBs in MgO by Sun and Balluffi [38] was made on specially prepared bicrystals and revealed periodic arrays of secondary dislocations. The high-angle GBs in Al_2O_3 were found in polycrystalline alumina samples and the presence of secondary dislocations was a rare occurrence although line defects were observed in certain special cases. It was noted, however, that a characteristic feature of such interfaces was that they tended to facet parallel to the basal plane, or other densely packed planes, in one grain or the other. A high-angle boundary

formed by hot-pressing two alumina monocrystals is illustrated in Fig. 4.9. This GB is close to the $\Sigma = 13$ coincidence relation with the interface lying parallel to the basal plane in both grains. Note how in the lower figure the interface is faceted, at atomic scale, parallel to the common (0001) plane.

4.2.3 Twin boundaries

Twin boundaries have been found in many ceramic materials including MgO [39, 40], spinel [41, 42], Al_2O_3 [43, 44], Fe_2O_3 [45], and the rare earth oxides [46]. The observation of so many twin boundaries is interesting because their presence in a crystal can introduce new polyhedral sites which may be occupied by cations which could not fit so well into the perfect crystal lattice. This ability of such interfaces to accommodate impurity ions has been developed [47] into the model of chemical

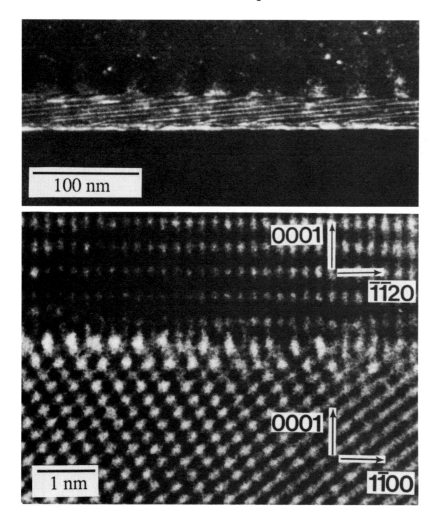

Fig. 4.9 (upper) WBDF image from $\Sigma = 13$ twist boundary recorded with $g = (\bar{1}\bar{1}20)$ reflection from upper crystal, (lower) HREM image from $\Sigma = 13$ twist boundary showing crystallographic continuity across interface (courtesy D. W. Susnitzky).

twinning, wherein apparently different crystal structures can be related to one another by the periodic repetition of a pair of twin interfaces. It was subsequently [43] demonstrated that this concept could then be used to understand the actual mechanism of a phase transformation in a ceramic system.

The first-order twin boundary in spinel ($\Sigma = 3$) has been particularly interesting since, crystallographically, the misorientation of the adjoining grain is then the same as occurs for twin boundaries in fcc metals and Si. In spinel the structure has been found to be very different from that found in metals and Si–Ge as illustrated in Fig. 4.10. The interface structure consists of an array of so-called 'white-spot defects' which actually correspond to triangular pyramids of less dense material running in $\langle 110 \rangle$ directions. The size of the pyramids is such as to translate the interface normal to its original plane by three $\{111\}$ spinel planes. This defect is thus essentially dislocation free and causes no long-range shearing of the sample. In fact the 'coherent twin interface' in this material is also interesting since its structure is not unique, i.e. unlike the coherent twin interface in Al, Si etc., it can exist with several different translation states.

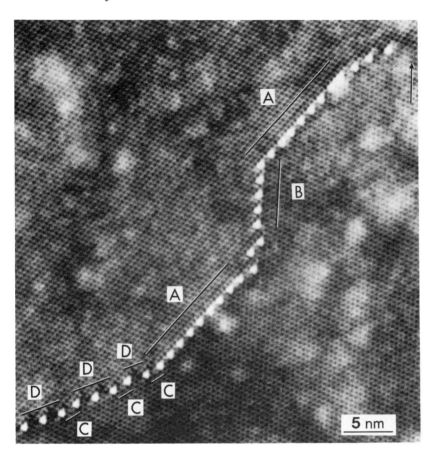

Fig. 4.10 A segment of a lateral twin boundary in spinel showing that the interface structure is actually a series of 'white-spot' defects which can be arranged in different arrays to give the impression of micro-faceting.

The basal twin boundary

The basal twin boundary in Al_2O_3 bears an interesting resemblance to the stacking fault in that material: there is no detectable translation and the oxygen sublattice is not (to a very good approximation) disturbed. However, there are three configurations for the interface when it lies parallel to the common (0001) plane: a mirror configuration and two glide configurations. A schematic diagram for the perfect Al_2O_3 structure and for the basal twin boundary lying parallel to the (0001) and ($1\bar{1}00$) plane is shown in Fig. 4.11 and Fig. 4.12, then shows this same interface faceting from the (0001) plane to the $\{11\bar{2}0\}$ plane; for simplicity, only one pair of oxygen planes parallel to $\{1\bar{1}00\}$ is shown. The faceting of the grain boundary can be

attributed directly to the influence of the cations. The step, shown in Fig. 4.12, is one unit cell high, and is thus dislocation-free. Steps on such interfaces may then show many interesting features. For example, different distributions of cations at a unit-cell-high step are possible. Furthermore, if the step is not a unit-cell-high step, then although the orientations of the 'matrix' and 'twin' grains will be the same on either side of the step, the interface structure will change from 'mirror' to 'glide'. The faceting of surfaces of ceramic materials is very closely related to the subject of GBs since surface faceting can directly influence which GBs form during sintering. Thus roughening transitions may occur for GBs but it will be extremely difficult to observe the phenomenon in these high-temperature materials.

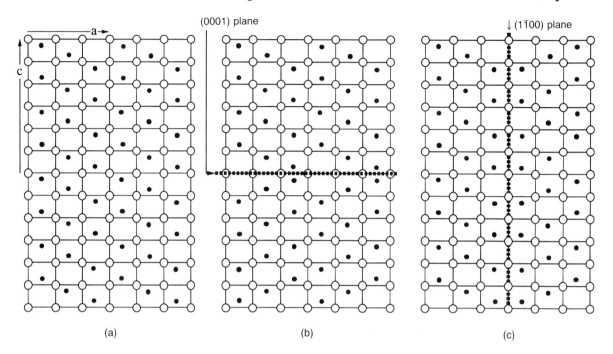

Fig. 4.11 Schematic diagrams of the basal twin boundary in alumina. (a) The perfect material; (b) parallel to the (0001) plane; (c) parallel to the $\{1\bar{1}00\}$ plane.

The observations made on the basal twin boundary in Al_2O_3 are of course special in the sense that these interfaces had sufficient time to reach an equilibrium, or low-energy configuration. An example of a faceted basal twin boundary in Al_2O_3 is shown in Fig. 4.13. Facets associated with a GB can facilitate boundary migration. Faceting allows the boundary to keep segments parallel to low-energy planes. In the case of the perfect basal twin boundary, migration can take place by the movement of the Al ions alone. When the Al ions move the facet plane moves. The twin interface effectively moves from one basal plane to another in this case. This process must continue along the vertical facet to allow the step to advance – i.e. a kink must move along the step.

Other twin boundaries in Al_2O_3

Another special interface in alumina is the $\{11\bar{2}3\}$ twin boundary. Fig. 4.14 shows an image of such an interface which has formed in a sintered poly-crystalline alumina sample. The interface has been referred to as a $\{11\bar{2}3\}$ twin interface because it can be created by mirroring the alumina structure across the $\{11\bar{2}3\}$ twin plane as illustrated in the schematic in Fig. 4.14(b). This description of the boundary is of course consistent with the definition of a twin boundary given in Table 4.5.

Twin boundaries involving a small displacive transformation

A class of twin boundaries which are being increasingly encountered in many ceramic materials can be illustrated by examples in materials with the 'perovskite-type' structures such as $BaTiO_3$, $LaAlO_3$, and the superconducting oxide, $YBa_2Cu_3O_{7-\delta}$. The common feature of these materials is that they undergo a symmetry-lowering phase change on cooling to room temperature. Such transformations between the different polymorphs require no breaking of bonds; if the displacements are small there may also be no change

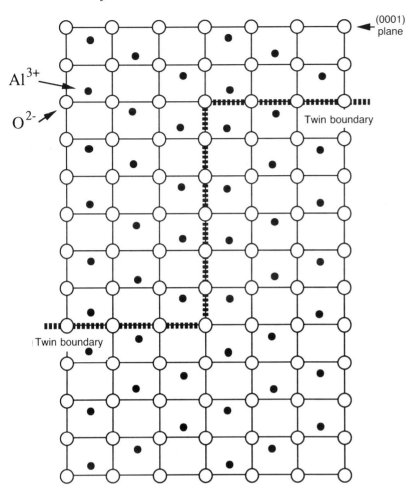

Fig. 4.12 A schematic of a basal twin boundary faceting from the (0001) plane onto the (1̄100) plane and back to the (0001) plane. The step is one unit cell high.

in the number of nearest neighbors. The transformation is diffusionless and takes place by a mechanical shear of the parent structure. Because no bond breaking is required this transformation can occur very rapidly.

BaTiO$_3$ is isostructural with the mineral perovskite (CaTiO$_3$) and is therefore referred to as a 'perovskite.' BaTiO$_3$ has a cubic unit cell above its Curie point (~130°C), as illustrated in Fig. 4.15. Below the Curie point the structure is slightly distorted to the tetragonal form giving a dipole moment along the c-direction (and hence the ferroelectric properties). Other transformations occur at temperatures close to 0°C and −80°C; the transformations are illustrated in Fig. 4.16. The approximate displacements of the ions associated with the cubic-tetragonal distortion in BaTiO$_3$ are shown schematically in Fig. 4.17. Rare-earth aluminates having a perovskite-like structure, such as LaAlO$_3$, undergo a similar phase transition [48] as illustrated in Fig. 4.18. The transition temperature is close to 435°C and the structure goes from the cubic (high-temperature) form to the rhombohedral (low-temperature) form.

A more recently discovered, though much more publicized, example of twinning through a displacive phase transformation is that found in the YBa$_2$Cu$_3$O$_{7-\delta}$ superconductor. The semiconducting, tetragonal phase of the YBa$_2$Cu$_3$O$_{7-\delta}$ is stable at high temperatures. However, on cooling through

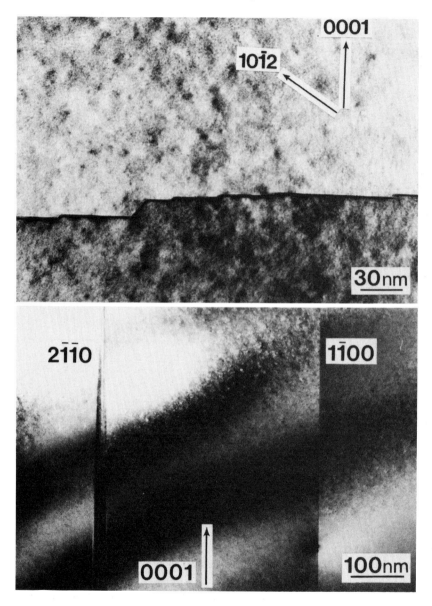

Fig. 4.13 Faceted basal twin boundaries in Al_2O_3.

the phase transition temperature the structure transforms to the superconducting orthorhombic phase. This phase transition is accompanied by the formation of twins on the {110} planes. Because of the crystallography of this process, the twin boundaries can form on two symmetry-related planes, i.e. (110) and (1$\bar{1}$0) – these variants can form in a single grain (Fig. 4.19).

The formation of microtwins (thin twin-related regions) in the orthorhombic phase can lead to a local reduction of the stresses induced by the change in lattice parameters during the tetragonal to orthorhombic phase transition as has been shown by *in-situ* electron microscopy [49]. These microtwins form both in bulk superconducting materials and in epitactic thin films. The orthorhombic–

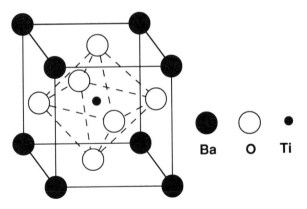

500nm

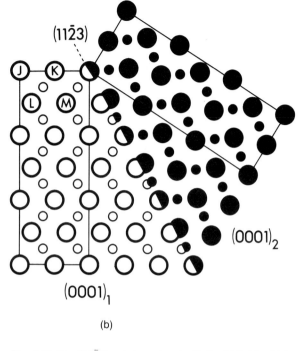

(b)

Fig. 4.14 The {11$\bar{2}$3} twin boundary in Al$_2$O$_3$. (a) Bright-field image and (b) schematic of the structure.

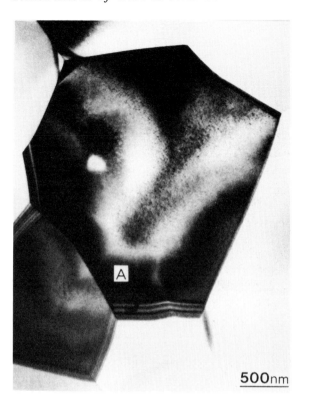

Fig. 4.15 The unit cell of BaTiO$_3$ (after A. J. Moulson and J. M. Herbert, *Electroceramics*, Chapman and Hall, 1990).

tetragonal transformation is dependent upon both temperature (T_p) and oxygen partial pressure (pO$_2$) [50, 51] in the ambient – when the oxygen partial pressure is decreased, T_p is also decreased (at pO$_2$ = 1 atm, T_p = 700°C, at pO$_2$ = 0.2 atm T_p =

670°C and at pO$_2$ = 0.02 atm T_p = 620°C). The value of δ in this partial pressure range is ~0.5. When pO$_2$ decreases T_p decreases while δ increases, as is shown graphically in Fig. 4.20.

4.2.4 Effects of GBs on properties

In this section the effects of GBs on physical properties will be discussed. Three particular examples have been chosen which represent technologically important areas where considerable research activity is in progress: the microstructure of polycrystalline aluminum nitride ceramic substrates, the effect of GB misorientation on the transport properties of heteroepitactic YBa$_2$Cu$_3$O$_{7-\delta}$ thin films, and the relationship between microstructure and mechanical properties of structural ceramics.

The final part of this section describes a model which has been used to study the effect of particular properties as a function of GB misorientation.

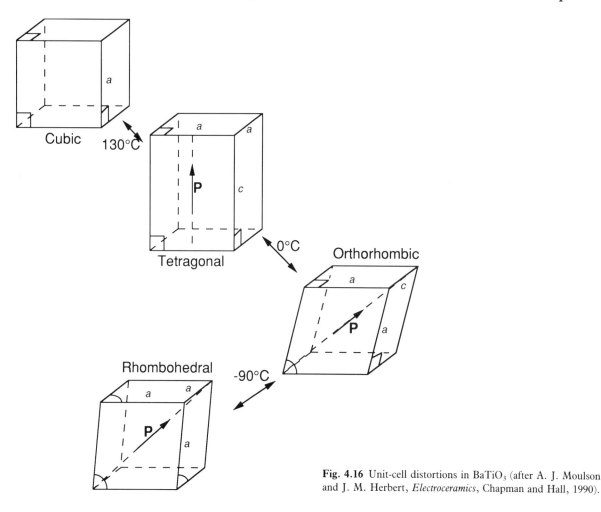

Fig. 4.16 Unit-cell distortions in BaTiO$_3$ (after A. J. Moulson and J. M. Herbert, *Electroceramics*, Chapman and Hall, 1990).

Microstructure of aluminum nitride ceramic substrates

A number of parameters have been identified for material structures possessing a high thermal conductivity; these include low atomic mass, strong interatomic bonding, and simple crystal structure. Among these, aluminum nitride is currently under investigation as a substrate material for use in microcircuit applications. The theoretical thermal conductivity of AlN is 320 W/m K [52], but the maximum value reported for a polycrystalline material is 260 W/m K [53]. The majority of commercially available substrates actually have much lower thermal conductivities.

In a non-metallic solid heat transfer occurs by phonon transport and there are a number of mechanisms which operate to limit the mean free path of these phonons, hence reducing thermal conductivity; these include scattering by defects and GBs in the specimen. The AlN substrates possessing the highest conductivity contain only small amounts of second-phase material, low concentrations of oxygen impurities, 'clean' grain boundaries, and low defect concentrations [54].

The morphology of an AlN substrate having a thermal conductivity in the range 70–90 W/m K is shown in Fig. 4.21. The image is a backscattered electron image: the bright areas prevalent at the grain boundaries correspond to a second-phase

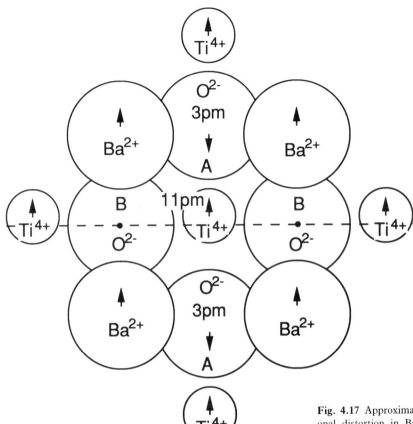

Fig. 4.17 Approximate ion displacements in the cubic-tetragonal distortion in BaTiO$_3$ (after A. J. Moulson and J. M. Herbert, *Electroceramics*, Chapman and Hall, 1990).

material containing elements having a high atomic number (electron yield increases with increasing atomic number). These areas have been determined, by X-ray diffraction, to be yttrium aluminum garnet (3Y$_2$O$_3$ · 5Al$_2$O$_3$, YAG) which formed as a result of reactions between the Y$_2$O$_3$ additives used in the sintering process and the oxide layer on the AlN powders. A secondary electron image of an AlN substrate having twice this thermal conductivity is shown in Fig. 4.22. Both substrates consist of densely packed polyhedral grains; however, in the latter sample the only additional phase detected was an oxynitride phase (Al$_{10}$N$_8$O$_3$) which has a structure based on that of AlN; the hexagonal unit cell has a *c*-axis of 5.310 nm [55]. Many of the

GBs in this AlN appear to be 'clean', as illustrated in Fig. 4.23. Defects are often observed within individual AlN grains. Many of the defects have been identified as antiphase boundaries (APBs) [56]; an example of such a defect can be seen in Fig. 4.23. These defects appear to be associated with oxygen incorporation [57]. An example of such a defect is the so called 'D' defect which is comprised of two planar defects consisting a basal and curved fault-segment which are connected. A high-resolution image of part of a 'D' defect recorded at the {11$\bar{2}$0} pole is shown in Fig. 4.25. It has been proposed that the basal fault-segment consists of a layer of alumina and that this is formed during powder processing [58]. These defects will

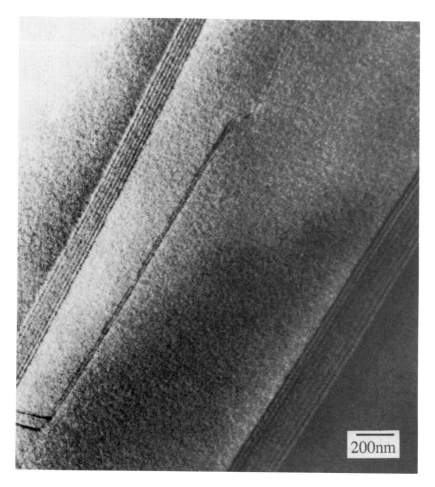

Fig. 4.18 Bright-field image
showing twin-boundaries in LaAlO₃.

presumably have a detrimental effect on thermal conductivity although not as large as that produced by second-phase material at the GBs. Also identified in some AlN substrate materials are GBs which correspond to special GBs. The boundary shown in Fig. 4.25 is viewed along a common [11$\bar{2}$0] direction. The (0002) and {10$\bar{1}$1} planes have only a 0.5% misfit and are at ~59° to each other. The faceting of the GB as it curves away from this orientation indicates that it is indeed a low-energy configuration.

Critical currents in YBa₂Cu₃O₇₋δ

Low critical current densities ($\mathcal{J}_c$) of polycrystalline YBa₂Cu₃O₇₋δ, have been attributed to a combination of critical-current anisotropy and weak coupling across GBs [59, 60]. The relationship between the superconducting transport properties and the structural properties of tilt boundaries was examined by measuring the critical current density as a function of misorientation angle [61]. The data are summarized in Fig. 4.26, for small misorientation angles; only a small decrease in critical current density ($\mathcal{J}_c$) is produced, however, for large angles a large attenuation of the critical current occurs. Therefore, high-angle GBs behave as weak-links with a suppressed $\mathcal{J}_c$. A superconducting weak-link is a region in which the order parameter is strongly depressed but which can nevertheless support a supercurrent [62]. Weak-links in a superconductor can be either

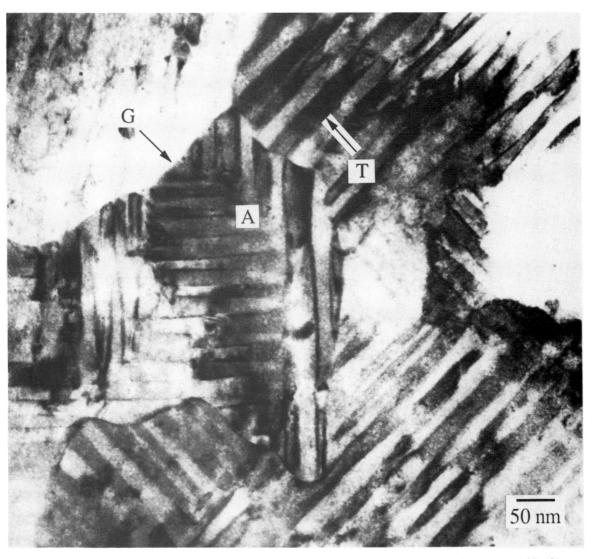

Fig. 4.19 Bright-field image of a YBa$_2$Cu$_3$O$_{7-\delta}$ thin-film formed on (001)-oriented MgO. The presence of twins (T) and GBs (G) are indicated. Area A shows where the two twin variants have formed.

inherent or can be produced externally. For example, in heteroepitactic films of YBa$_2$Cu$_3$O$_{7-\delta}$, grown on MgO where the lattice mismatch is large (~9%), high-angle tilt GB$_s$ are numerous in the microstructure [63]. These boundaries result due to the formation of rotationally misaligned islands during the early stage of film growth [16]. A number of different tilt boundaries corresponding to rotations about [001] have been identified and these are summarized in Table 4.6.

The GB structure in these films has not been established although the properties of such boundaries on the transport properties has been determined. However, the boundaries can be considered as regions of structural disorder leading to suppressing of the order parameter [62]. Further research is required to fully establish the relationship between GB structure and critical current density. For very thin insulating barriers (~10^{-9} m), the supercurrent is due to tunneling [64] – such superconductor – insulator – superconductor (SIS) junctions are referred to as tunnel junctions. These junctions are the basis of important devices and are an example where the properties of a single GB may, in principle, be used to create a device.

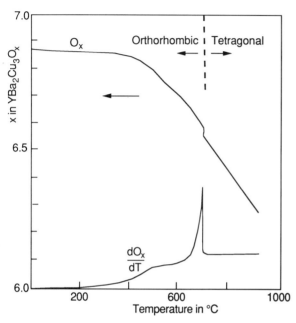

Fig. 4.20 Graph showing variation in oxygen content in $YBa_2Cu_3O_7$ on cooling from the sintering temperature in 1 atm. of O_2 ambient (after C. J. Jon and J. Washburn [51]).

Mechanical strength of polycrystals

Once a crack has been initiated in brittle materials such as glass and many ceramic single-crystals the stress conditions will usually favor its propagation – there is no energy-absorbing process comparable with the plastic deformation which occurs in ductile materials. The crack continues to propagate and causes complete failure in a uniform stress field. In polycrystalline materials, obstacles such as GBs can hinder crack propagation and prevent fracture of a body. It has been generally found that the strength of a polycrystal will increase as the grain size decreases because the effect of the GBs will be more pronounced in a fine grain microstructure [65]. The relationship between grain size (d) and yield point (σ yield) has been described by the Hall–Petch equation

$$\sigma_{\text{yield}} = \sigma_0 + \frac{K}{\sqrt{d}}$$

where σ and K are constants.

A typical microstructure of dense sintered material is shown in Fig. 4.27. The microstructure is characterized by an average grain size of $3-5\,\mu m$ and <2 vol. % porosity; the pores have an average diameter of $<2\,\mu m$ and are located primarily at grain intersections. In order to obtain a high den-

3μm

Fig. 4.21 Backscattered-electron image of a low thermal conductivity AlN ceramic showing yttrium-rich second phase regions at many GBs and triple points.

sity microstructure, the processing conditions must be carefully controlled. In Fig. 4.27, the relatively equiaxed grains of the predominantly 6H polytype α-SiC are associated with the low fault density within the grains [66]. The grain size has been kept small by an excess of carbon which is located at the triple junctions and thus inhibits grain growth.

The nature of the GB is also an important factor in controlling the mechanical properties of sintered ceramics. The liquid phases which are often used to enhance the densification of polycrystalline materials [e.g. 67] may remain at internal interfaces in some materials. For example, GBs in Si_3N_4 may contain a very thin (1–5 nm) amorphous film [68, 69]. The presence of this layer at the GB will limit the high temperature mechanical strength of the ceramics. Much effort has thus been expended on modification of this GB phase to improve the high temperature properties of silicon nitride and related nitrogen ceramics [5].

It must be appreciated, however, when predicting or assessing failure, that failure occurs at the most severe, not the average, flaw in the body. The most severe flaws in a ceramic compact are usually pores (or pore agglomerates), abnormally large grains (which give a much greater local grain size), and inclusions (particularly if they are large or if the interface with the matrix is weak).

Relating GB properties to macroscopic properties

An approach to relating GB properties to macroscopic properties is illustrated by the modeling of properties of polycrystalline superconductors [70, 71]. The polycrystalline material is first approximated as a statistical array of GBs of differing

Fig. 4.22 Secondary-electron image showing densely packed polyhedral grains in a high thermal conductivity AlN ceramic.

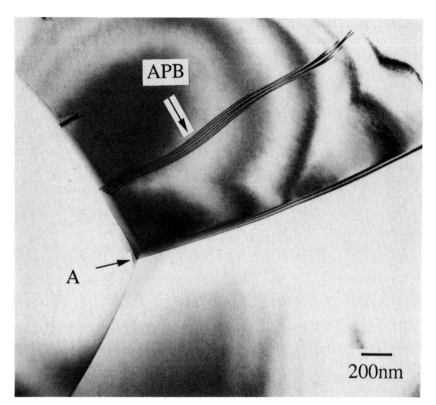

Fig. 4.23 Bright-field image showing 'clean' triple point (A) in a high thermal conductivity AlN ceramic. An anti-phase boundary is visible in one of the grains.

Table 4.6 Summary of GBs observed in YBa$_2$Cu$_3$O$_{7-\delta}$ thin films

Type	Rotation axis	Rotation (measured)	Σ cubic/angle [80]
Twin	(110), ($\bar{1}$10) twin plane	1°	
Low-angle	[001]	5, 5.5°	Δ (Σ17-Σ13)
Low-angle	[001]	6.5	Δ (Σ29-Σ5)
Low-angle	[001]	9.0°	Δ (Σ5-Σ17)
High-angle	[001]	20°	
High-angle	[001]	29°	Σ17/28.07
High-angle	[001]	45°	Σ29/43.6

misorientation angle. Properties are assigned to the GBs according to the crystallographic misorientation across the individual boundaries (resistance low for θ < 10°, otherwise large). The overall effect of the GBs is then determined. In the simplest approach, the boundaries are distinguished simply as either low-angle or high-angle boundaries; the grains are connected which are joined by low-angle boundaries as illustrated in Fig. 4.28. The new microstructure can be thought of as consisting of a number of entities much larger than the grain size of the material. Using this approach, the problem of describing a microstructure becomes a percolation problem and enables use of established data on percolation theory and scaling properties of percolation clusters. Similar approaches have been used to describe electrical properties of polycrystals [72] and the penetration of glass along GBs [73].

4.3 PHASE BOUNDARIES

The important characteristics of heterophase boundaries can best be illustrated by a number of examples. This type of interface will occur when a solid-state reaction takes place or when two different polytypes (or polytypoids) of a material can

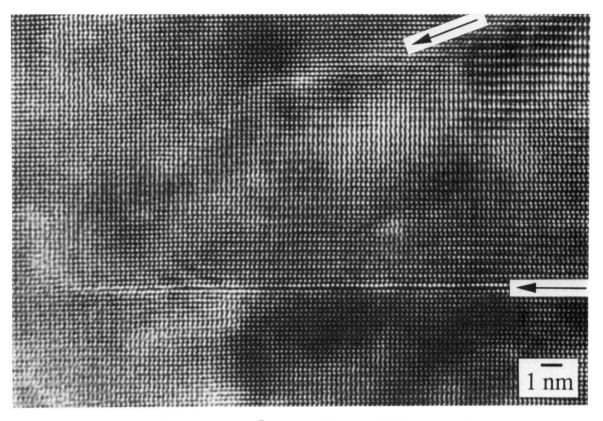

Fig. 4.24 HREM image of a 'D' defect recorded at the [11$\bar{2}$0] pole. The horizontal and inclined arrows indicate the basal and curved fault-segments respectively (courtesy S. McKernan).

coexist. Until recently, these two types of phase boundary were the only ones which were widely encountered in ceramic materials. Recently, new processing techniques have given phase boundaries a new importance through the development of ceramic composites and other polyphase ceramics. Most atomic features which are important for GBs are also found to be important for phase boundaries (misorientation, interface plane, 'cleanliness,' etc.).

The heterophase boundary occurring between a thin film and a substrate is directly analogous to the flat GBs used in many systematic studies. In many applications, it is desirable to have a small lattice mismatch between the film and substrate. Two examples will be used to illustrate such heterophase boundaries; the growth of $YBa_2Cu_3O_{7-\delta}$ thin films on MgO and the growth of hematite on sapphire.

A further example will consider where meta-stable tetragonal ZrO_2 particles can be constrained in an Al_2O_3 matrix – i.e. where the interface is a closed surface in three dimensions. Stresses produced ahead of a crack tip can release the matrix constraint of the tetragonal particles which then transform to monoclinic symmetry. This mechanism is used to improve fracture toughness of Al_2O_3 ceramics.

A final illustration of phase boundaries, which will continue the topic of improving mechanical strength, will be the boundary between a matrix and a fiber in a composite material, where the nature of the interphase bonding is important in determining the fracture mechanism of a composite under applied stress.

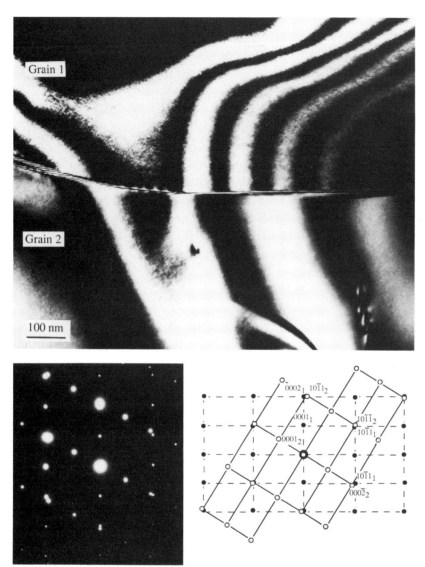

Fig. 4.25 (a) Bright-field image of faceted GB in AlN; (b) diffraction pattern and schematic of boundary in A (courtesy S. McKernan).

4.3.1 Al$_2$O$_3$/spinel

The interface plane is important for phase boundaries just as it was for GBs. Fig. 4.29 shows a HREM image from a special phase boundary in the spinel/Al$_2$O$_3$ system [74]. The interface plane lies parallel to the $(11\bar{2}3)$ and $(\bar{1}11)$ in alumina and spinel, respectively. The important features to notice are that the interface is atomically flat with no facets, and that one set of the {111} planes in

spinel runs smoothly into the (0001) planes in alumina although the plane is rotated through ∼9°. No misfit dislocations are present in this projection; instead, the lattice misfit is accommodated by the 9° rotation.

4.3.2 Polytypes in Mg-SiAlON

The sialons are a series of compounds based on cation and anion substitutions in Si$_3$N$_4$, and have

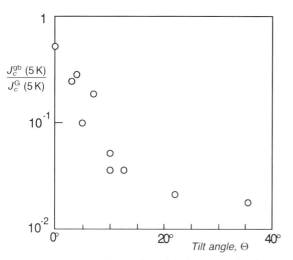

Fig. 4.26 Variation of the GB critical current density to the average value of the critical current density in the two grains versus the misorientation angle in the basal plane (after D. Dimos *et al.* [61]).

another example of crystal–crystal phase boundaries when different polytypes intergrow as shown in Fig. 4.30. These polytypes can be described as a regular insertion of a cubic-stacked layer into the basic hexagonally stacked wurtzite structure. Investigation of the microstructure of 12H (Ramsdell notation) Mg-SiAlON formed by hot-pressing the constituent oxides and nitrides showed that the majority of grains had the 12H structure; several grains having a 21R polytype structure were also observed. In the region to the left in Fig. 4.30, a regular periodicity of 1.89 nm has been imaged which was confirmed to be the well-ordered 21R polytype region by use of electron diffraction. The pattern of the right identified the 12H polytype. In the image, strain contrast can be seen at several points along the boundary, implying that there are misfit dislocations present to accommodate this mismatch. These intergrowths are probably undesirable for mechanical properties, since the strained interface could act as a possible source for crack initiation. However, these intergrowths may be difficult to avoid in hot-pressed material since the polytypes exist over such a narrow range of homogeneity with respect to changes in the metal-to-non-metal ratio [75].

been investigated with the view to their use in high-temperature structural applications [75, 76]. X-ray analysis has revealed the presence of a class of compositionally dependent polytypes based on the wurtzite structure. Such materials thus illustrate

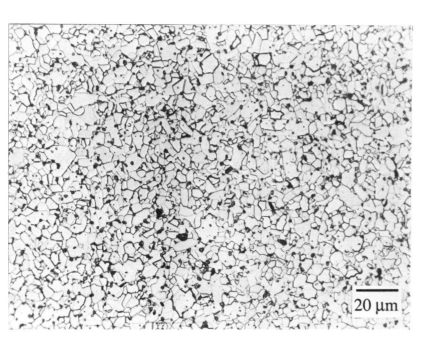

Fig. 4.27 Microstructure of α-SiC sintered with 0.35% B and 2.8% C (courtesy C. Greskovich and S. Prochazka).

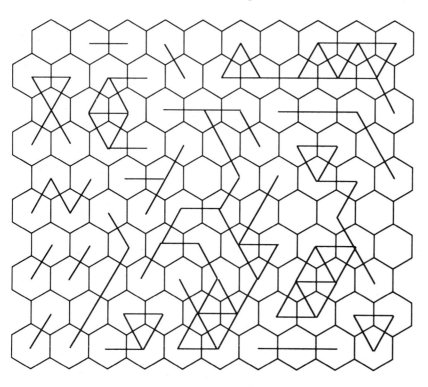

Lines only cross low-angle grain boundaries

Fig. 4.28 An example of clusters of grains connected by low-angle GBs in a simulated two-dimensional microstructure (after D. R. Clarke [70].

4.3.3 Heteroepitactic growth of ceramic thin films

A new approach has been developed to investigate the early stages of heteroepitactic growth of ceramic thin films using TEM [77, 78]. Electron-transparent thin-foils, which have been carefully cleaned and annealed to produce a well-defined surface morphology, act as the substrate for the subsequent deposition process. This technique can be used to study both the growth topology, e.g. thin $YBa_2Cu_3O_{7-\delta}$ films on MgO and $SrTiO_3$, and the introduction of misfit dislocations, e.g. in Fe_2O_3/Al_2O_3 heterojunctions.

Fig. 4.31 is a bright-field image recorded close to the [001] zone axis showing a thin (~ 12 nm) $YBa_2Cu_3O_{7-\delta}$ film deposited onto a (001)-$SrTiO_3$ thin-foil substrate. The film appears to be continuous and smooth. No Moiré fringes or dislocations are apparent which implies that the

film is pseudomorphic at this stage of growth. A schematic illustration of a coherent interface with slight misorientation is shown in Fig. 4.32. The strains that are associated with a coherent interface raise the total energy of the system, and for a sufficiently large misfit or at a sufficient film thickness it becomes energetically favorable to introduce misfit dislocations at the interface, thereby losing coherency. The incorporation of misfit dislocations into the interface is shown schematically in Fig. 4.33.

The presence of misfit dislocations can be observed by weak-beam imaging as illustrated in Fig. 4.34 for the heterojunctions between hematite (Fe_2O_3) and sapphire formed by chemical vapor deposition [80]. The lattice misfit in this system is 5.5% and could, in principle, be accommodated by one of two different hexagonal dislocation networks: $b = \frac{1}{3}\langle \bar{1}210 \rangle$ or $b = \frac{1}{3}\langle \bar{1}010 \rangle$ (partial dislocations in the perfect crystal). The image

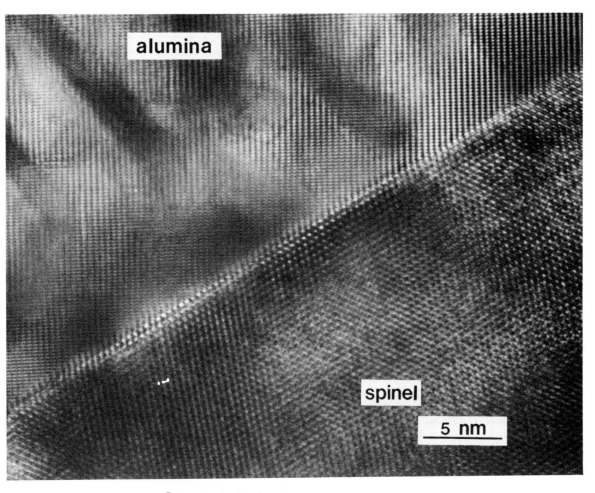

Fig. 4.29 A HREM of the $\{111\}/\{11\bar{2}3\}$ spinel–alumina phase boundary.

reveals that both types of dislocation can be present in the interface: one widely spaced, the other narrowly-spaced. Models of the misfit dislocations are shown in Fig. 4.35(a, b).

During epitactic growth, the substrate imposes constraints on the deposit and this can lead to the introduction of defects. An example of this is shown by the growth of $YBa_2Cu_3O_{7-\delta}$ thin films on MgO by pulsed-laser ablation [80, 81]. It has been found that the earliest observed stage of growth consists of the formation of discrete nuclei; these nuclei are $\sim 30\,nm$ in diameter. It has also been observed that steps on the substrate surface act as preferential sites for nucleation [80]. Further

deposition results in island growth and subsequent coalescence to form an interconnected network structure as shown in Fig. 4.36.

Fig. 4.36 is a bright-field image of a thin $YBa_2Cu_3O_{7-\delta}$ film grown on a (001)-MgO thin-foil substrate. Moiré fringes are visible in the images of the islands and it can be seen that a number of the islands are rotationally misaligned. The presence of Moiré fringes in the image indicates that the film–substrate interface cannot be coherent and that misfit dislocations must be present at the interface; alternatively, a thin amorphous region may be present at the interface; however, this latter possibility is discounted as such interfaces have

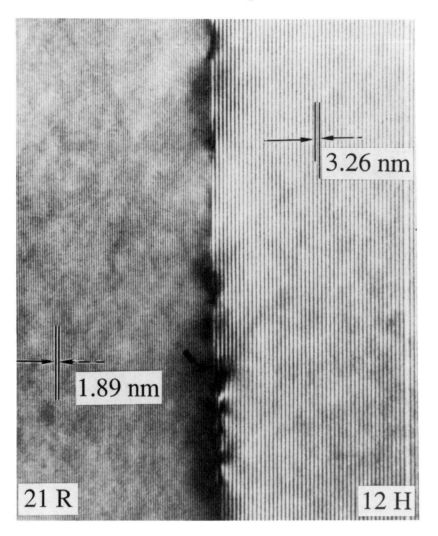

Fig. 4.30 An intergrowth between the 21R polytype (left) and the 12H polytype (right) in which the polytype spacing has been directly imaged (courtesy D. R. Clarke).

been determined to be crystalline using cross-sectional TEM.

Moiré fringes can also be used to identify microstructural defects in epitactic deposits [82]. In particular the presence of dislocations generated during film growth will be revealed in the Moiré fringe pattern. If a dislocation is present in one of the layers it will be visible as a terminating fringe. An example of a dislocation revealed by the Moiré pattern is shown in area A in Fig. 4.36. Dislocations may also be formed by the coalescence of rotationally misaligned islands; an area where three such islands have coalesced is illustrated by B in Fig.

4.36. These threading dislocations are actually very short segments of tilt grain boundaries intersecting the phase boundary. The resulting mosaic structure has been observed in $YBa_2Cu_3O_{7-\delta}$ films grown on substrates where there is a large lattice mismatch. In some areas of the interconnected film, holes remain as the film thickens. Incipient dislocations are often observed to be associated with such holes; for example, area C, in Fig. 4.37. The number of Moiré fringes terminating at such a hole is directly related to the Burger's vector(s) of the dislocations(s) which would form if the hole were to be eliminated; the formation of such dis-

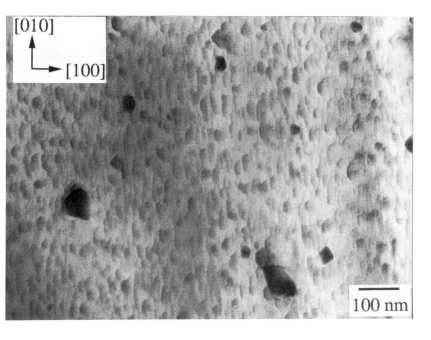

Fig. 4.31 Bright-field image showing a thin $YBa_2Cu_3O_{7-\delta}$ film on (001)-$SrTiO_3$.

locations may be energetically unfavorable. A similar area where terminating fringes are present is area D in Fig. 4.36; in this example the fringes around the hole are seen to curve as they are followed away from the center of the hole. If further film growth occurred it may be difficult to fill this hole, thus leaving a tubular void in the film [83].

4.3.4 Toughening of ceramics through incorporation of ZrO_2

The enhancement of toughness in ceramics can be obtained by several mechanisms which act so as to hinder crack propagation. It was first realized by Garvie *et al.* [84] that zirconia could potentially be used to increase the strength and toughness of ceramics by utilizing the tetragonal (t) to monoclinic (m) phase transformation of metastable tetragonal particles. Toughening occurs through the development of a zone of transformed material ('process zone') around or ahead of the advancing crack tip [85, 86]. The predominant mechanisms of toughening ceramics include transformation toughening [87, 88], microcrack toughening [89, 90], and crack-deflection toughening [91].

Crack-deflection can occur in two-phase ceramics. The two mechanisms inherent to partially stabilized ZrO_2 ceramics and ceramic matrices containing ZrO_2 particles are microcrack toughening and transformation toughening. Microcrack toughening can be affected by incorporating ZrO_2 particles in a ceramic matrix. On cooling through the t–m transformation temperature, the volume expansion (3–5%) which occurs in the ZrO_2 particles, results in crack formation at the particle–matrix interface. Hence, a crack which propagates into the particle will be deviated and become separated thereby leading to an increase in the measured fracture resistance. It is, however, possible to retain the ZrO_2 particle in the metastable tetragonal form within the matrix material. If a crack is made to extend under stress, large tensile stresses are generated around the crack, especially ahead of the crack tip [87]. These stresses release the matrix constraint of the t-ZrO_2 particles and, if sufficiently large, could result in a net tensile stress on the particle, which under the new conditions will transform to monoclinic symmetry. The volume expansion and shear strain (~1–7%) developed in the particle causes the martensitic reaction, with a resultant compressive strain being

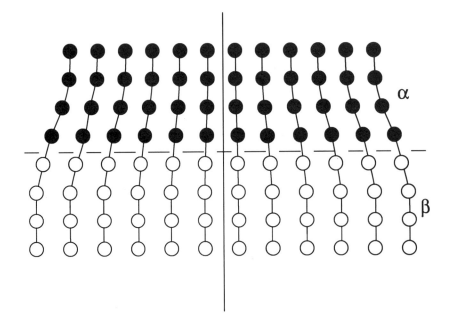

Fig. 4.32 A coherent interface with slight mismatch leads to coherency strains in the adjoining lattices.

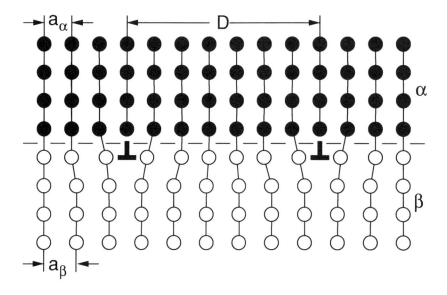

Fig. 4.33 A semicoherent interface. The misfit parallel to the interface is accommodated by a series of edge dislocations.

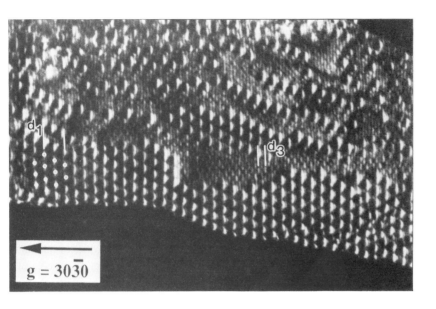

Fig. 4.34 Weak-beam image of a (0001) hematite (alumina interface recorded under the g–$4g$ condition where $g = (30\bar{3}0)$. Two types of interface misfit dislocation networks are visible. d1 and d3 correspond to the dimensions indicated in Fig. 4.35(a) and (b), respectively (courtesy L. A. Tietz).

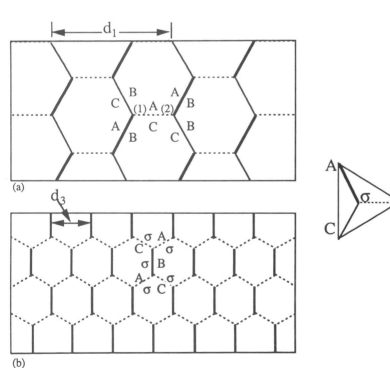

Fig. 4.35 (a) and (b) Models of misfit dislocation networks shown in Fig. 4.34. (a) $b = \frac{1}{3}\langle 1\bar{2}10\rangle$ edge dislocations, d_1 (calculated) = 15.8 nm. (b) = $\frac{1}{3}$ edge dislocations, d_3 (calculated) = 5.1 nm. AC, AB, BC = $\frac{1}{3}\langle\bar{1}2\bar{1}0\rangle$; A$\sigma$, B$\sigma$, C$\sigma$ = $\frac{1}{3}\langle 10\bar{1}0\rangle$.

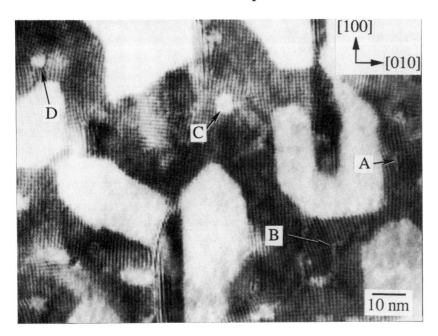

Fig. 4.36 Bright-field image of a YBa$_2$Cu$_3$O$_{7-\delta}$ thin film on MgO. Moiré fringes are visible in the image of the film areas. Various different defects in the film can be identified by discontinuities in the Moiré fringes and are indicated by arrows in the figure. The description of these defects is found in the text.

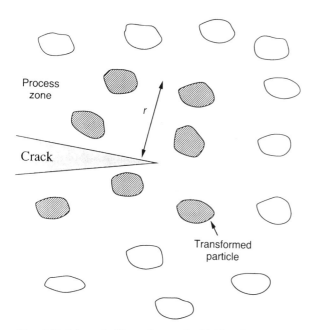

Fig. 4.37 Schematic illustrating crack shielding by a transformation zone.

generated in the matrix. Since this occurs in the vicinity of the crack, extra work would be required to move the crack through the ceramic, accounting for the increase in toughness and hence strength. This process is shown schematically in Fig. 4.37.

If the zirconia particles are smaller than a critical size, they will not transform, but if the size is larger than another critical size they will transform spontaneously to the monoclinic phase [92]. Zirconia-toughened alumina can contain amounts of ZrO$_2$ up to 30% dispersed in a fine-grained alumina matrix (Al$_2$O$_3$). An example of an intragranular tetragonal particle in 10 wt-% ZrO$_2$ toughened Al$_2$O$_3$ is shown in Fig. 4.38 [93].

4.3.5 Fiber-reinforced composites

Improvements in the mechanical properties of ceramics can be achieved by fiber reinforcement. It has been demonstrated that the fracture energy of a glass is increased by about three orders of mag-

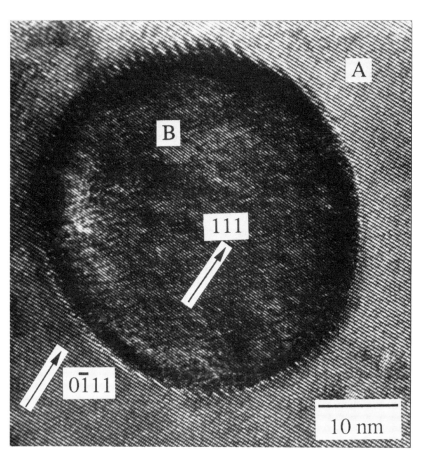

Fig. 4.38 HREM image showing an intragranular t-ZrO_2 particle (B) in an Al_2O_3 matrix (A) (courtesy T. E. Mitchell).

nitude by carbon-fiber reinforcement [94]. This improvement results from the controlled fracture behavior, compared with the catastrophic failure of unreinforced glass.

A number of important requirements must be met when fibers are incorporated into ceramic matrices [95]. These requirements can be summarized as follows:

1. A suitable source of fiber at an economically attractive price.
2. Fibers that generally are strong and stiff compared with the matrix phase.
3. An appropriate fabrication route which does not lead to degradation of the properties of the matrix or to damage of the fibers.
4. Chemical compatibility between the fibers and matrix both during fabrication and in service.
5. Physical compatibility between the fiber and matrix, in terms of, for example, relative coefficients of thermal expansion.
6. An interface between the fiber and matrix that induces a fibrous type of fracture.

Fibers presently in use include glass, carbon, boron, silicon carbide, silicon nitride, and alumina; the matrix material may be glass, carbon, carbides, nitrides, silicides, or oxides. The general features of the materials can be illustrated by two examples: (i) SiC fiber in alumina; (ii) SiC fiber in calcium aluminosilicate glass.

Alumina–SiC composites have high fracture toughness and strength up to elevated temperatures and consequently have great potential for use in structural components [96]. Their increased fracture toughness is the result of crack interaction with the fiber in the matrix. Crack propagation is impeded not only by such mechanisms as crack pinning and crack deflection but also crack bridging and fiber pull-out [96].

If a crack propagating through the matrix interacts with one of the fibers, two situations can arise which depend on the interfacial strength [97]. If the bonding between the fiber and matrix is sufficiently strong, the crack may propagate across the interface and continue its propagation into the fiber. Such a situation leads to a catastrophic failure of the composite. The toughness of the composite then depends on that of the tougher material. If the interfacial bonding is weak the crack will not propagate through the fiber, but will be diverted at the fiber–crack interface.

Examination of a typical fracture surface of an alumina–SiC fiber composite with high toughness is shown in Fig. 4.39(a) and illustrates how the fibers interact with a propagating crack to increase the fracture toughness. As shown in Fig. 4.39(a), the resulting fracture surface is very rough and fibers can be observed. When the fracture surface of an alumina–SiC-fiber composite with low toughness is examined, no similar interaction between the fiber and the propagating crack is observed, as illustrated in Fig. 4.39(b). The surface is relatively smooth with no fibers visible.

From a composite fabrication point of view, the glass matrix composites, as compared to the other ceramic candidates, probably offer the greatest commercial potential regarding ease of densification, low cost and also achievement of high performance [98, 99]. An example of crack propagation through a calcium aluminosilicate glass–SiC fiber composite is shown in Fig. 4.40. Note that the crack advancing through the glass matrix is deflected around the fibers and, due to the weak bonding between fiber and matrix, does not propagate through the fiber.

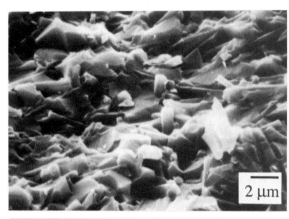

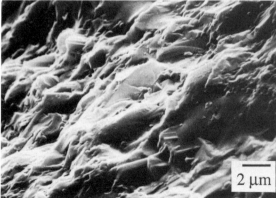

Fig. 4.39 Typical fracture surface of alumina – 20 vol% SiC fiber composite having (a) high-fracture toughness and (b) low fracture toughness (courtesy T. N. Tiegs).

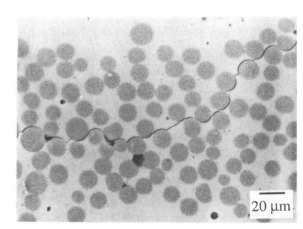

Fig. 4.40 Crack propagation through a calcium aluminosilicate glass–SiC fiber composite (courtesy S. McKernan and P. Kotula).

4.4 SUMMARY

Grain boundaries in ceramic materials may be either those where the two grains are in intimate contact over their entire area or where the two grains are separated by a layer of either an amorphous material or a crystalline phase having a different structure or chemical composition. This latter type may for example be a liquid-phase sintered material or a two-phase composite. The type of interfaces produced in a ceramic material depend largely upon the processing parameters used to form a dense microstructure. The interfaces in ceramics are often much more complex than those observed in metals due, in part, to the different bonding characteristics and also due to the presence of two or more, ionic species on two sub-lattices.

The boundary plane is an important consideration in any study of interfaces and the 'macroscopic' boundary may indeed, on an atomic scale, be highly faceted as has been illustrated by the NiO–spinel interface. Such faceting may also occur in an interface between a crystalline grain and a glassy GB phase. The idea of GB plane is particularly relevant in discussions of twin boundaries, which are important in a wide range of technologically important materials, where the adjoining regions are mirror related.

The atomic-level structure and composition of the GBs in a polycrystalline material are important in influencing a number of physical properties of the material, but as has been realized not all tilt GBs produce the same modulating effect on the macroscopic property and this realization has been utilized in simulations and modeling of the effects of GBs on a number of transport properties.

As the number of possible applications of ceramic materials and composites increases, the understanding of the GB will continue to be an important consideration in materials development; an example is in the area of GB engineering: the boundary must be understood before it can be engineered.

ACKNOWLEDGMENTS

The authors would like to thank colleagues who kindly contributed some of the images used in this paper. Support from NSF (DMR-8901218 and the Materials Science Center), DoE (DE-FG02-89ER45381), ONR (N000 14-89J-1692) and Corning Inc. is acknowledged.

REFERENCES

1. W. D. Kingery, H. K. Bowen and D. R. Uhlmann, *Introduction to Ceramics* (2nd edn), John Wiley and Sons, New York (1970).
2. J. Frenkel, *Kinetic Theory of Liquids*, Oxford University Press, Fair Lawn NJ (1946).
3. K. Lehovec, *J. Chem. Phys.*, **21** (1953), 1123.
4. J. G. Bednorz and K. A. Müller, *Z. Phys. B.*, *Condensed Matter*, **64** (1986), 189.
5. R. N. Katz and G. E. Gazza, in *Processing of Crystalline Ceramics*, *Materials Science Research II* (eds H. Palmour III, R. F. Davis and T. M. Hare) (1978), p. 547.
6. A. Tsuge and K. Nishida, *Amer. Ceram. Soc. Bull.*, **57** (1978), 424.
7. R. N. Katz, in *Progress in Nitrogen Ceramics* (ed F. L. Riley) Martinus Nijhoff Netherlands, **65** (1983), 3.
8. C. Greskovich and J. H. Rosolowski, *J. Amer. Ceram. Soc.*, **59** (1976), 336.
9. R. J. Brook, in *Surfaces and Interfaces of Ceramic Materials*, (eds L-C. Dufour, C. Monty and G. Petot-Ervas), NATO ASI Series E 173, Kluwer Academic Publishers, Dordrecht (1989).
10. W. D. Kingery, *J. Am. Ceram. Soc.*, **57** (1974), 1.
11. H. Schmalzried, Solid State Reactions, *Monographs in Modern Chemistry 12*, Verlag Chemie Weinheim (1981).
12. S. R. Summerfelt and C. B. Carter, *Ultramicroscopy*, **30** (1989), 150.
13. A. H. Heuer and M. Rühle, *Acta Metall.*, **33** (1985), 2101.
14. J. Geerk, G. Linker and O. Meyer, *Mat. Sci. Rep.*, **4** (1989), 193.
15. J. W. Matthews (ed), *Epitaxial Growth, Parts A and B*, Academic Press, New York (1975).
16. M. G. Norton, L. A. Tietz, S. R. Summerfelt and C. B. Carter, *Appl. Phys. Lett.*, **55** (1989), 2348.
17. M. J. Stowell in [15], p. 439.
18. D. K. Bowen and C. R. Hall, *Microscopy of Materials*, John Wiley and Sons, New York (1975).
19. T. E. Mitchell, W. T. Donlon, K. P. D. Lagerlof and A. H. Heuer, *Mat. Sci. Res.*, **18** (1984), 125.

20. A. H. Heuer and J. Castaing, *J. Adv. Ceram.*, **10** (1984), 238.

21. C. B. Carter, *Proc. Int. Colloq. on Dislocations* (eds P. Veyssière, L. P. Kubin and J. Castaing) CNRS Editions (1984), p. 227.

22. D. Wolf, *J. de Phys.*, **46** (1985), 197.

23. C. B. Carter, *Acta Metall.*, **36** (1988), 2753.

24. S. R. Summerfelt, Ph.D thesis, Cornell University (1990).

25. P. W. Tasker and D. M. Duffy, *Surface Sci.*, **137** (1984), 91.

26. R. B. Poeppel and J. M. Blakely, *Surface Sci.*, **15** (1969), 507.

27. R. King and B. Chalmers, *Progress in Metal Physics* Butterworths Scientific Publishers, London (1949), p. 127.

28. F. Hargreaves and R. J. Hills, *J. Inst. Met.*, **41** (1929), 257.

29. J. M. Burgers, *Proc. Phys. Soc.*, **52** (1940), 23.

30. M. L. Kronberg and F. H. Wilson, *Trans AIME*, **185** (1949), 501.

31. W. Bollman, *Crystal Defects and Crystalline Interfaces*, Springer-Verlag, New York (1970).

32. D. A. Smith and R. C. Pond, *Inter. Met. Rev.*, **205** (1976), 61.

33. D. J. Barber and N. J. Tighe, *Phil. Mag.*, **14** (1966), 531.

34. M. H. Lewis, *Phil. Mag.*, **14** (1966), 1003.

35. P. Veyssière, J. Rabier, H. Graém and J. Grihlé, *Phil. Mag.*, **33** (1976), 143.

36. H. Schmidt, M. Rühle and N. L. Peterson, *Mat. Sci. Res.*, **14** (1981), 177.

37. J. Eastman, F. Schmückle, M. D. Vaudin and S. L. Sass, *Adv. Ceram.*, **10** (1984), 324.

38. C. P. Sun and R. W. Balluffi, *Phil. Mag. A*, **46** (1982), 63.

39. J. M. Cowley, R. L. Segall, R. St. C. Smart, and P. S. Turner, *Phil. Mag. A*, **39** (1979), 163.

40. S. B. Newcomb, D. J. Smith and W. M. Stobbs, *J. Microscopy*, **130** (1983), 137.

41. T. M. Shaw and C. B. Carter, *Scripta Metall*, **16** (1982), 1431.

42. C. B. Carter, Z. Elgat and T. M. Shaw, *Phil. Mag. A*, **55** (1986), 1.

43. K. J. Morrissey and C. B. Carter, *J. Amer. Ceram. Soc.*, **67** (1984), 292.

44. S. C. Hansen and D. S. Phillips, *Phil. Mag. A*, **47** (1983), 209.

45. L. A. Bursill and R. L. Withers, *Phil. Mag. A*, **40** (1979), 213.

46. B. Yangui, C. Boulesteix, A. Bourret, G. Nihoul and G. Schiffmacher, *Phil. Mag. A*, **45** (1982), 443.

47. B. G. Hyde, S. Andersson, M. Bakker, C. M. Plug and M. O'Keeffe, *Prog. Solid State Chem.*, **12** (1974), 273.

48. S. Geller and V. B. Bala, *Acta Cryst*, **9** (1956), 1019.

49. G. Van Tenderloo, H. W. Zandbergen, T. Okabe, and S. Amelinckx, *Solid State Comm.*, **63** (1987), 603.

50. J. D. Jorgensen, M. A. Geno, D. G. Hinks, L. Sederholm, K. J. Volin, R. L. Hitterman, J. D. Grace, I. K. Schuller, C. U. Segre, K. Zhang and M. S. Kleefisch, *Phys. Rev. B*, **36** (1987), 3608.

51. C. J. Jon and J. Washburn, *J. Mater. Res.*, **4** (1989), 795.

52. M. P. Borom, G. A. Slack and J. W. Szymaszek, *Amer. Ceram. Soc. Bull.*, **51** (1972), 852.

53. N. Kuramoto, H. Taniguchi and I. Aso, *IEEE Trans. Comp. Hybrid and Manuf. Tech.*, **CHMT-9** (1986), 386.

54. M. G. Norton, *Hybrid Circuits*, **20** (1989), 18.

55. S. F. Bartram and G. A. Slack, *Acta. Cryst.*, **835** (1979), 2281.

56. S. McKernan and C. B. Carter, *Mat. Res. Soc. Symp. Proc.*, **167** (1989), in press.

57. A. D. Westwood and M. R. Notis, *Adv. Ceram.*, **26** (1989), 171.

58. S. McKernan, M. G. Norton and C. B. Carter, *Mat. Res. Soc. Symp. Proc.*, **183** (1990), in press.

59. P. Chaudhari, F. Le Goues, and A. Segumuller, *Science*, **238** (1987), 342.

60. R. A. Campo, J. E. Evetts, B. A. Glowachi, S. B. Newcomb, R. E. Somekh and W. M. Stobbs, *Nature*, **329** (1987), 229.

61. D. Dimos, P. Chaudhari, J. Mannhart and F. K. Le Goues, *Phys. Rev. Lett.*, **61** (1988), 219.

62. D. Dimos and D. R. Clarke, in *Surfaces and Interfaces of Ceramic Materials* (eds L-C. Dufour, C. Monty and G. Petot-Ervas), NATO ASI Series E 173, Kluwer Academic Publishers, Dordrecht (1989).

63. M. G. Norton, L. A. Tietz, C. B. Carter, B. H. Moeckly, S. E. Russek and R. A. Buhrman, *Mat. Res. Soc. Symp. Proc.*, **169** (1989), in press.

64. B. D. Josephson, *Phys. Lett.*, **1** (1962), 252.

65. A. G. Evans and R. W. Davidge, *Phil. Mag.*, **20** (1969), 373.

66. C. Greskovich and S. Prochazka, in *Ceramic Microstructures 86: Role of Interfaces, Materials Science Research 11* (eds J. A. Pask and A. G. Evans), Plenum, New York (1987).

67. J. S. Moya, W. M. Kriven and J. A. Pask, *Mater. Sci. Res.*, **14** (1981), 317.

68. D. R. Clarke, *Ultramicroscopy*, **4** (1979), 33.

69. D. R. Clarke, *J. Am. Ceram. Soc.*, **70** (1987), 15.

70. D. R. Clarke, *Amer. Ceram. Soc. Bull.*, **69** (1990), 681.

71. C. S. Nichols, D. R. Clarke and D. A. Smith, *AIME Proceedings on Computer Simulation of Grain Growth* (in press).

72. D. S. McLachlan, M. Blaszkiewicz and R. E. Newnham, *J. Am. Ceram. Soc.*, **73**(8) (1990), 2187.

73. T. M. Shaw (personal communication).

74. Y. Kouh Simpson and C. B. Carter, *Mat. Res. Soc. Symp.*

Proc., **94** (1987), 45.

75. D. R. Clarke and T. M. Shaw, in *Processing of Crystalline Ceramics, Materials Science Research 11* (eds H. Palmour III, R. F. Davis and T. M. Hare) (1978), p. 589.
76. K. H. Jack, *J. Mater. Sci.*, **11** (1976), 1135.
77. C. B. Carter, S. R. Summerfelt, L. A. Tietz, M. G. Norton and D. W. Susnitzky, *Inst. Phys. Conf. Cer.*, **98** (1989), 415.
78. M. G. Norton, S. R. Summerfelt, C. B. Carter, *Appl. Phys. Lett.*, **56** (1990), 2246.
79. L. A. Tietz, S. R. Summerfelt, C. B. Carter, *Mat. Res. Soc. Symp. Proc.*, **159** (1989), in press.
80. M. G. Norton, C. B. Carter, B. H. Moeckly, S. E. Russek, and R. A. Buhrman, *Proc. 2nd Conf. on Science and Technology of Thin Film Superconductors* (eds R. Nouffi and R. McConnell) (1990).
81. M. G. Norton and C. B. Carter, *Proc. XII Int. Cong. on Electron Microsc.* (1990), p. 18.
82. D. W. Pashley, in *Thin Films*, American Society for Metals, Metals Park, Ohio (1964).
83. F. R. N. Nabarro, *Theory of Crystal Dislocations*, Dover Publications, New York (1967), p. 303.
84. R. C. Garvie, R. H. J. Hannink and R. T. Pascoe, Nature, **258** (1975), 703.
85. A. G. Evans, D. B. Marshall and H. H. Burlingame, in *Science and Technology of Zirconia Advances in Ceramics*, Vol. 3, American Ceramic Society, Ohio (1981), p. 202.
86. A. G. Evans, in *Science and Technology of Zirconia Advances in Ceramics*, Vol. 12, American Ceramic Society, Ohio (1984).
87. A. G. Evans and R. M. Cannon, *Acta. Metall.*, **34** (1986),

761.

88. N. Claussen and M. Rühle, in *Science and Technology of Zirconia Advances in Ceramics*, Vol. 3, American Ceramic Society, Ohio (1981), p. 137.
89. N. Claussen, *J. Am. Ceram. Soc.*, **59** (1976), 49.
90. A. G. Evans and K. T. Faber, *J. Am. Ceram. Soc.*, **67** (1984), 255.
91. K. T. Faber and A. G. Evans, *Acta. Metall.*, **3** (1983), 565.
92. F. F. Lange and D. J. Green, in *Science and Technology of Zirconia, Advances in Ceramics*, Vol. 3, American Ceramic Society, Ohio (1981), p. 217.
93. S. P. Kraus, MS thesis, Case Western Reserve University (1986).
94. D. H. Bowen, D. C. Phillips, R. A. J. Sambell and A. Briggs, in *Proc. Int. Conf. on Mechanical Behaviour of Materials*, Society of Materials Science, Japan (1972), p. 123.
95. R. W. Davidge, *Mechanical Behavior of Ceramics*, Cambridge University Press, Cambridge (1979).
96. T. N. Tiegs, P. F. Becher and L. A. Harris, *Ceramic Microstructures '86: Role of Interfaces, Mat. Sci. Res.*, **21**, Plenum, New York (1987), p. 911.
97. P. Pirouz, G. Moscher and J. Chyung, in *Surfaces and Interfaces in Ceramic Materials*, NATO ASI Series E, Kluwer Academic Publishers, Dordrecht (1989), p. 173.
98. K. M. Prowo, in *Tailoring Multiphase and Composite Ceramics, Mat. Sci. Res.*, **20** (1986), p. 529.
99. R. E. Tiegles, G. L. Messing, C. G. Pontano and R. E. Newnham (eds), *Tailoring Multiphase and Composite Ceramics, Mat. Sci. Res.*, **20** (1986).

CHAPTER FIVE

Special properties of Σ grain boundaries

G. Palumbo and K. T. Aust

Introduction · CSL classification of GBs · Solute segregation · Energy of GBs · Kinetic properties · Mechanical and physical properties · Corrosion properties · Triple junctions · Summary

5.1 INTRODUCTION

The various properties of Σ grain boundaries (GBs) are discussed in relation to their interface structure (CSL–DSC). The effects of deviation from exact low Σ CSL relationships are presented in terms of both theoretical and experimental considerations. The influence of boundary plane on interface properties is assessed on the basis of experimental data derived from both controlled bicrystal and conventional polycrystal studies.

The specific GB properties considered are: solute segregation, energy, kinetic (i.e. diffusion, mobility), mechanical, physical, and chemical. Recent experimental observations are also presented on the properties of triple line defects produced by the interaction of Σ and non-Σ GBs.

5.2 CSL CLASSIFICATION OF GBs

Classification of GBs on the basis of the coincidence site lattice model is strictly based upon the relative misorientation of adjoining crystals. The coincidence site lattice corresponds to the specific rotation of one lattice relative to the other yielding a three-dimensional atomic pattern in which a certain fraction of lattice points coincide. The reciprocal

of this fraction is defined as the value Σ. However, complete geometric characterization also involves the determination of boundary plane indices (referenced to both adjoining lattices). The interrelationship between the Σ-criterion and boundary plane indices was first demonstrated by Ranganathan [1], who showed that for specific twinning operations (i.e. 180° rotations) about rational indices [*hkl*], a three-dimensional coincidence site lattice, having

$$\Sigma = h^2 + k^2 + l^2 \qquad (5.1)$$

and a boundary (i.e. twinning) plane {*hkl*} is generated. All low Σ (i.e. Σ ≤ 29) CSLs can be generated by such twinning operations, with Σ51c being the lowest Σ non-twinned CSL [2].

Although all GBs can be represented by exact CSL relationships, Σ may achieve very high values of questionable physical significance. As described by the displacement shift complete (DSC) lattice [3], small angular deviations (Δθ) from low Σ CSLs result in a relative translation of the two lattices at the boundary which conserves the periodicity of the CSL and yields a network of intrinsic GB dislocations (IGBD). Thus the occurrence of IGBDs is an indication that regions of 'good fit' exist at the boundary, and that such an interface may display 'special' properties. Both theoretical [3] and experimental [4] investigations have demonstrated that the existence of discrete IGBDs is primarily dependent on the structural parameters Σ and $\Delta\theta$ as described by the CSL–DSC lattices. Figure 5.1 shows the maximum angular deviations ($\Delta\theta$) from specific CSLs (i.e. Σ) at which IGBDs (i.e. primary dislocations for $\Sigma = 1$, secondary GB dislocations (SGBD) for $\Sigma \geq 3$) have been observed by transmission electron microscopy. Also shown in Fig. 5.1 are the various criteria proposed for the angular deviation limit of the CSL. As shown in this figure, resolvable IGBDs are identified with interfaces lying well within the structural limits imposed by both the commonly applied criterion of Brandon [5] (i.e. $\Delta\theta \leq 15°\Sigma^{-\frac{1}{2}}$), and the recently proposed criterion of Deschamps *et al.* [6] (i.e. $\Delta\theta \leq 15°\Sigma^{-\frac{2}{3}}$); however, the maximum deviation angles tend to extend beyond the more restrictive criterion proposed by Ishida and McLean [7] (i.e. $\Delta\theta \leq 15°\Sigma^{-1}$).

The various criteria for the maximum allowable angular deviation from exact CSLs are largely derived from the low angle approximation of the Read–Shockley relation [8],

$$\theta \approx b/d \qquad (5.2)$$

where θ is the misorientation angle, b is the magnitude of the Burger's vector of grain boundary dislocations, and d is the dislocation spacing. Deschamps *et al.* [6], in developing their criterion, considered a dislocation spacing (d) proportional to a mean edge (p) of the CSL lattice [9] where

$$d \approx p \propto \Sigma^{\frac{1}{3}} \qquad (5.3)$$

and Burger's vector magnitudes (b) varying on average as the mean edge of the DSC lattice [9], such that

$$b \propto \Sigma^{-\frac{1}{3}} \qquad (5.4)$$

However, as demonstrated by Grimmer *et al.* [9], two of the three Burger's vectors described by the DSC lattice vary as $\Sigma^{-\frac{1}{2}}$; the third is independent of Σ and varies only as the interplanar spacing along the rotation axis. Thus the only variation of b with Σ expected would be of the type, $\Sigma^{-\frac{1}{2}}$. By considering $d \propto \Sigma^{\frac{1}{3}}$, $b \propto \Sigma^{-\frac{1}{2}}$, and an upper angular deviation limit of 15° for low angle boundaries (i.e. $\Sigma 1$), the following geometric criterion is obtained [10]:

$$\Delta\theta \leq 15°\Sigma^{-\frac{5}{6}} \qquad (5.5)$$

This geometric criterion is shown to be most consistent with the experimental observations of discrete IGBDs summarized in Fig. 5.1.

In evaluating the applicability of the CSL/DSC model to interfacial properties, it is also important to consider an upper limit for the Σ criterion beyond which special properties would not be expected. As shown in Fig. 5.1, discrete IGBDs have been identified with CSLs having very high Σ values (i.e. >49), indicating that some structural order can exist at these interfaces. However, it should be noted that, on the basis of experimental studies concerning intergranular corrosion and fracture in polycrystalline materials, Watanabe [11] has suggested that special properties would not be expected of interfaces having $\Sigma > 29$.

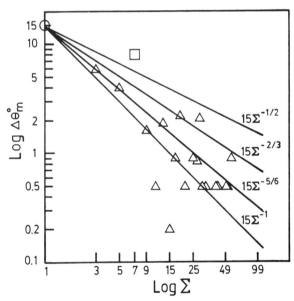

Fig. 5.1 Maximum deviation angle ($\Delta\theta$) from any CSL (Σ) at which discrete intrinsic grain boundary dislocations (IGBD) have been identified by TEM [10]; triangles, [12]; squares, [13]; circles, [14].

5.3 SOLUTE SEGREGATION

Although many experimental observations indicate an enhanced concentration of solute at GBs, only a few direct experiments exist that have shown differences in solute concentration between CSL-related boundaries and non-CSL-related boundaries. On a GB model having a periodic structure, one might expect corresponding periodicity in the distribution of solute segregation even within the same grain boundary plane. There is increasing evidence, based on Auger electron spectroscopy (AES) and computer modeling, for GB structure-dependent segregation [15–17]. The degree of segregation is largely dependent upon the degree of GB coherency or on the free volume associated with the boundary.

The cleavage method of revealing GB planes for study by AES may tend to select only GBs having a particular low cohesion, i.e. maximum segregation. Auger data were obtained for Fe-6.14 at .% Si bicrystal containing a 37° $\langle 100 \rangle$ boundary ($\Sigma = 5$, $\{013\}$ plane) [18]. A scatter in segregation of Si, P, and N was observed comparable to that usually found for various facets of polycrystalline materials and attributed to segregation anisotropy. It was shown that scatter of measured data across a single GB is due to the damage of the GB surface caused by the fracture process. Thus, the average segregation value cannot be attributed to true GB segregation, which must be obtained only from the undamaged regions of a GB fracture surface.

Hofmann [19] has suggested that a higher segregation enthalpy (ΔH) is expected for random boundaries than for special boundaries with special orientation relationships of the two adjacent lattices. This was experimentally shown in an AES study for indium segregation to GBs in Ni-1-at.% In [20]. Randomly oriented boundaries in polycrystalline samples gave a value of $\Delta H = 55 \, \text{kJ mol}^{-1}$, whereas $\Delta H = 39 \, \text{kJ mol}^{-1}$ was determined for symmetrical $\langle 110 \rangle$ tilt boundaries near a $\Sigma = 19$ CSL orientation relationship. However, in another paper [21] it was assumed that the lower segregation for the $\langle 110 \rangle$ tilt boundaries was due to the symmetrical atomic arrangement and not due to a CSL orientation relationship. It is clear that further such

studies are required to distinguish segregation properties among different large-angle Σ and non-Σ GBs.

An early field ion microscope study of segregation of oxygen to GBs in iridium appears to indicate differences in segregation between Σ and non-Σ boundaries [22]. With a matrix or volume concentration of about 500 ppm of oxygen, a level of six times this amount is found at a GB that deviates from a high density coincidence site relationship. However, no oxygen segregation is evident at a $\Sigma = 11$ coincidence boundary. Howell *et al.* [23, 24], using the field ion microscope, found that Cr atoms in tungsten GBs were distributed at random in the cores of several boundaries at concentration levels of about 6% and 12%. The misorientations of these boundaries were approximately 20° and 44° about $\langle 110 \rangle$ respectively. The former boundary, with less solute segregation, appears to be within 2° of a $\Sigma = 33$ CSL misorientation, while the latter boundary is close to a $\Sigma = 57$ relationship. These results are consistent with an increase in the degree of segregation as the boundary becomes more general.

Hu *et al.* [25] have presented results for grain boundary segregation of Re in a Mo-5.4 at % Re alloy, which was studied using classical transmission electron microscopy (CTEM) and atom probe field ion microscopy (APFIM). The specimen was recrystallized at 1773 K and annealed at 1273 K for 35 h to induce solute segregation to the GBs. The specific GB studied was within 0.4° of a $\Sigma = 9$ CSL orientation, mainly a tilt boundary with a twist component. The APFIM measurements showed that the minimum Re concentration at the boundary was 1.75 × greater than its concentration in the single-phase matrix. This result is consistent with the authors Monte Carlo simulations of solute segregation to small Σ [001] twist and $\Sigma = 5$ tilt boundaries in a Pt-1 at .% Au alloy, with solute enhancement factors in the range of 1.2 to 3.5. In these studies, the solute segregation appeared to be largely to the cores of GB dislocations. Brenner and Ming-Jian [26], using the APFIM, studied segregation in eight boundaries of Ni_3Al. B and C segregation was found at seven general, high-angle boundaries, but no enrichment of either B or C was found at a $\Sigma = 11$ GB. It may be interesting to extend these studies to higher Σ GBs in the same solvent–solute system.

Bouchet *et al.* [27, 28] have studied the poly-crystalline Ni–S system using an intergranular microetching method to reveal S segregation at GBs. It was concluded that the S segregation depends essentially on the interplanar spacing of the atomic planes parallel to the GB plane of low Σ boundaries ($\Sigma \leq 19$), i.e. the planar atomic density. The higher the interplanar spacing of the GB plane the less the amount of segregated impurities in the boundary. This conclusion is largely based on, first, studies of the $\Sigma = 3$ twin boundary where it is well known that its properties strongly depend upon the boundary plane orientation. The twin boundary is a first order twin relationship and should be distinguished from high-angle GBs. Second, the Σ GBs, with Σ from 5 to 19, are distinguished from the non-Σ GBs using the Brandon criterion. However, as was discussed in section 5.2, the angular deviation limit for Σ defined by $\Delta\theta = 15° \Sigma^{-\frac{5}{6}}$ is physically more realistic than that of the Brandon criterion, $\Delta\theta = 15° \Sigma^{-\frac{1}{2}}$, based on TEM observations of intrinsic GB dislocations, and the geometric characteristics of the CSL model of interface structure.

Swiatnicki *et al.* [29] studied the segregation of S at a $\Sigma 11$ (311) GB in a nickel bicrystal containing <10 ppm S. The presence of S at the boundary was confirmed by Auger analysis. An intergranular microetching technique was used to determine the segregation tendency as a function of GB plane orientation. The symmetrical part of the $\Sigma 11$ (311) GB with $\Delta\theta = 0.5°$ showed no evidence of segregation. The asymmetrical part of the $\Sigma 11$ boundary with $\Delta\theta \sim 2°$ from $\Sigma 11$ showed segregation dependent upon the planar atomic density. It should be noted that sub-boundaries were present in the bicrystal studied, which would produce not only changes in GB curvature but also changes in the $\Delta\theta$ value. In terms of the $\Delta\theta = 15° \Sigma^{-\frac{5}{6}}$ criterion, the $\Sigma 11$, $\Delta\theta \sim 2°$ boundary would correspond to an upper limit of allowable deviation from a $\Sigma 11$ boundary. This work clearly shows the importance of deviation from an exact Σ

GB on solute segregation. Further studies, with this in mind, are needed.

Solute segregation is not only dependent upon GB structure, but there is increasing evidence that the boundary structure itself may be changed by the solute concentrated there. Sickafus and Sass [30] have used analytical electron microscopy and Rutherford back scattering spectroscopy to show that segregation of Au to small angle (001) twist boundaries in Fe-rich Fe–Au alloys causes a change in the dislocation structure from that present in pure Fe. Lin and Sass [31] also demonstrated that the dislocation structure of small angle and large angle (near $\Sigma = 5$) [001] twist boundaries in Fe–S alloys can be changed by implantation of S in the vicinity of the interface. Bouchet *et al.* [28] have concluded that S segregation decreases the planar atomic density of $\Sigma = 3$ boundary, thereby indirectly supporting the previous observations of Sass *et al.* Additional studies by Lin *et al.* [32] using electron and ion irradiation on the small and large angle [100] twist boundaries in Fe–S alloys indicate that the observed change of structure is due primarily to the presence of S. However, non-equilibrium effects due to the transport of solute by point defect fluxes may affect the kinetics of the structural transformation.

A stronger segregation of solute atoms at non-Σ boundaries than at low-Σ boundaries is observed, indirectly, in studies of the effect of solute on different properties of GBs such as their energy, mobility, and corrosion. These latter observations are discussed in subsequent sections of this chapter.

5.4 ENERGY OF GBs

Wolf and Merkle [33] have discussed the correlation between the structure and energy of GBs using atomic level computer simulations combined with high-resolution electron microscopy experiments. In this section, several direct experimental studies are presented to determine if energy differences exist between Σ and non-Σ GBs, and the experimental conditions under which such differences may be observed.

Pumphrey [34], in a detailed review, found that there is considerable experimental evidence indicating that some of the 'special' high angle GBs generated by geometrical models (such as Σ and low periodicity of boundary atoms) have lower energies than those of 'random' or general boundaries. However, the models did not appear to correlate well with the extent to which the energies are lowered. In another review [35], it was indicated that a one-to-one correspondence between energy and Σ models of boundary structure is only expected where the energy is due to displacement of the atoms at the boundary from the positions they would occupy in a perfect lattice. For example, the energy of GBs with high Σ values ($\Sigma = 33$ and 43) depends on the electron structure, whereas in low Σ boundaries ($\Sigma = 11$) the energy is controlled by the geometry of the atomic arrangement in the boundary [36]. Experimental observations also indicate that solute segregation can result in two types of low energy orientation relationship: solute sensitive boundaries and solute insensitive boundaries [37]. The latter type of boundaries corresponds to those with the apparent electron insensitive structure in FCC materials. Both of these studies [36, 37] were conducted using a sintering technique involving the rotation of small single crystals in contact with other crystals, whereby the positions of low energy GB states are revealed but not the extent of energy decrease.

It appears that the energy of large angle GBs is, therefore, in some cases sensitive to the electron bond structure and to solute segregation. In addition, other factors such as electrostatic energy in ionic materials [38] and saturation of bonds in covalent solids [39] may control GB energy. Thus, it is not surprising that Sutton and Balluffi [39] find little support for the general usefulness of purely geometrical models in determining GB energy.

The latter authors [39] considered some of the measured GB energy (γ) versus misorientation (θ) data for symmetrical tilt and twist GBs in Al, Cu, and NiO. In some cases, clearly defined energy cusps were experimentally revealed for $\Sigma = 5$, 9, 11, 13a and 17a (in addition to the $\Sigma 3$ twin boundary). However, not all the expected low Σ ($\leqslant 25$) energy cusps were found. It should be noted

that the Al used in these studies contained from 6 to 20 ppm of impurities, and the Cu contained about 200 ppm of impurities. Also, many of the Al specimens were equilibrated at temperatures as low as $0.55\,Tm$ (Tm is absolute melting temperature). As was previously concluded [35, 37], solute segregation at GBs may drastically reduce the number of low energy GBs. Boundary entropy effects in copper also appear to influence the number and width of cusps in the (γ) versus (θ) curve [40].

In a discussion of the rotating crystallite experiments of Maurer [41] using Cu single crystal balls on (100) Cu substrates, it was noted [39] that of the 54 boundaries with $\Sigma \leqslant 33$, only 31 low energy states were detected. However, every low Σ relationship from $\Sigma 5$ to $\Sigma 19$ was found in these experiments [41] with the exception of $\Sigma 15$. In this connection, it should be noted that McLean [42] determined the GB energies of individual boundaries in Cu at $1030°C$ using the analysis of shapes of GB grooves, where the effects of surface energy anisotropy were fully taken into account. It was found that boundaries associated with low Σ orientation relationships, $\Sigma 3$ (incoherent twin) and $\Sigma 11$, have about 34% and 19% respectively, lower energies than average or general high angle boundaries; and that this difference of GB energy is less than 10% for boundaries in copper with $\Sigma > 11$. Solute segregation to the boundaries would further minimize these energy differences and thus make the detection of higher Σ boundaries more difficult.

Recently, Miura *et al.* [43] have determined the boundary energy of symmetric [011] tilt, and [011] and [001] twist boundaries in Cu using a new method. This involved a determination of boundary energies from the lenticular shapes of boundary SiO_2 particles observed by TEM in internally oxidized bicrystals of Cu-0.06% Si. After internal oxidation of the Si, the bicrystals were degassed at $1273\,K$ for $24\,h$ in vacuum. Energy cusps in the [011] tilt series were observed at $\Sigma 3\{111\}\{111\}$, $\Sigma 11\{311\}\{311\}$, and possibly at $\Sigma 9\{112\}\{112\}$. In the γ versus θ curve for the [011] twist boundaries, energy cusps again appear for the $\Sigma 3$ and $\Sigma 11$ boundaries and at $\Sigma 27$, $\Sigma 33$, and possibly $\Sigma 57$. In the γ (θ) curve for the [001] twist boundaries, Fig. 5.2, the existence of energy cusps is evident for $\Sigma 5$,

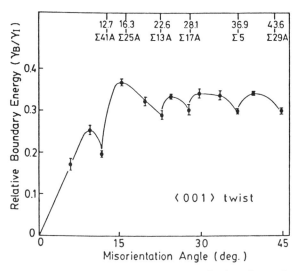

Fig. 5.2 Relative boundary energy versus misorientation angle for [001] twist boundaries in Cu [43].

$\Sigma 13$, $\Sigma 17$, $\Sigma 29$, and $\Sigma 41$. It was suggested [43] that the discovery of the new cusps, especially for the higher Σ boundaries ($\Sigma = 27, 29, 33, 41$ and 57), is due to better sensitivity of the experimental method. It was also found that without the degassing treatment, the boundary energies were higher; this was attributed primarily to oxygen segregation in the Cu GBs [43]. The higher energies may, however, be due to the presence of solute impurities in the form of oxides instead of as solute atoms.

Rotating crystallite experiments by Maurer and Gleiter [44] for Ni balls on (100) Ni substrates revealed only a $\Sigma 3$ low energy state. Since the latter result was quite different from that of Cu [41], this appeared to rule out purely crystallographic criteria to predict the orientation relationship corresponding to boundaries of low energy. However, the Ni used in this study [44] contained about 100 ppm of impurities. The present authors [45], using 99.999% Ni containing 0.3, 3, and 10 ppm S, find that the frequency of low Σ boundaries ($\Sigma 5$ to $\Sigma 25$) in recrystallized material is strongly dependent on the bulk S concentration. For example, a significant influence of twinning in generating low Σ boundaries (other than $\Sigma 3^n$, i.e. $\Sigma 9$ and $\Sigma 27$) was observed, with the effect being enhanced with decreasing S concentration. These solute effects

were rationalized in terms of selective impurity (S) segregation, whereby increasing impurity content results in diminished energy differences between CSL and non-CSL related interfaces.

Additional rotating crystallite studies in metals revealed the $\Sigma 5$ GB for Au crystallites around [100] on (100) Au substrates [46]. Similar studies using vapor-grown bcc Fe bicrystals with $\langle 100 \rangle$ tilt boundaries indicated the presence of $\Sigma = 5, 13, 17, 25$ and several near-coincidence boundaries possessing relatively low index planes parallel to the boundary [47]. It was suggested that the detection of these energy cusps (especially with large Σ values) may be due to the relatively low temperature of formation, ~0.5 Tm [47]. Thus, the entropy effects which can minimize energy differences between Σ and non-Σ boundaries are decreased [40].

By using rotating single crystal tablets of superconducting $YBa_2Cu_3O_7$, to form a population of c-axis (i.e. $\langle 001 \rangle$) twist boundaries, Smith et al. [48] were able to identify specific energy cusps (i.e. population maxima) within ~1° of $\Sigma 5$, $\Sigma 13a$, $\Sigma 17a$, $\Sigma 29a$, and $\Sigma 41a$, CSL orientation relationships.

The inclination of a high angle GB, for a given orientation relationship between two crystals, may exert an effect on the energy of the boundary [35]. Energy cusps are obtained in some cases for inclinations which permit the best atomic fit in the boundary. This is especially evident for a coherent twin boundary and for some GBs near other low Σ relations. Surprisingly, only a very small energy cusp [49], or none at all [42], is found for the variation of incoherent twin boundary energy with boundary inclination. Sutton and Balluffi [39] find that many GB faceting results appear to violate geometric criteria for low interfacial energy. Donald and Brown [50] conclude that GB faceting in an embrittled Cu-0.01% Bi alloy cannot be explained by the geometrical requirements of the coincident site lattice. They advanced a model in which the faceting is most likely to occur when there is a large size difference between the matrix and impurity atoms. Results have also demonstrated the existence of a reversible GB faceting phase transition driven by the addition, or removal, of Bi from a dilute solution in Cu [51]. Again, the role of solute atoms should be considered in any interpretation of GB energy observations based on GB faceting.

In many systems studied, a strong preference for Σ9, Σ27, and Σ81 is observed. The preference for these interfaces can be largely attributed to geometric interaction, not energy considerations [45]. These interfaces can be generated through the sole interaction of strongly preferred Σ3 boundaries, i.e. $Σ3^n$. These $Σ3^n$ boundaries are referred to as twin-related variants by Randle and Brown [52]. Figure 5.3 shows a section of Ni containing 0.3 ppm S where the interaction of non-coherent segments of a partial twin band in the same crystal leads to the formation of a small Σ9 segment, i.e. $Σ3 × Σ3 = Σ9$. Repeated interactions of this type can readily account for the high fraction of Σ3 related boundaries observed in many studies. Such geometric contributions to the presence of Σ GBs in polycrystalline systems can be considered to be independent of solute content [45].

GB dissociation experiments may provide information on relative GB energies if the total energy of the new interfaces is lower than the energy of the orginal boundary. As was shown by Forwood and Clarebrough [53], Σ9 boundaries are only observed to dissociate to form lower Σ boundaries, i.e. Σ3. However, unlike the Σ9 case, the dissociation of both Σ27a and Σ81d boundaries always involved

the formation of new intermediate grains which were bounded by higher valued Σ boundaries. In order to understand why there is a forward sense of dissociation, it is necessary to consider not only the change in interfacial energy, but also the kinetics of the dissociation process, i.e. the ease with which atomic movements can occur during boundary dissociation. All of these Σ interfaces, Σ3, Σ9, Σ27a, and Σ81d, are of the type of $Σ3^n$, whose introduction into the material is largely dependent upon geometric interaction considerations, not energetic. Similarly, the dissociation of Σ11 and Σ99 boundaries into Σ3 and Σ33a boundaries [54], which are multiples of 3 and 11, may not be reliable for comparison of energy with models.

It is evident that a comparison of experimentally measured energies of high angle GBs with Σ or geometric boundary models should be conducted using a very pure material at a very high temperature, to minimize solute segregation effects. One of the first experiments conducted on a pure material which indicated the importance of the Σ characterization of interface energy was the study of annealing twins in zone refined metals: Pb [55], Al [56], and Cu [57]. It was observed that where an annealing twin is formed during GB motion, a large angle 'random' or non-Σ boundary is replaced at the growth front by a large angle Σ boundary. Also, in some cases for Pb [55] and Cu [57], twinning occurred a second time during boundary migration, thereby replacing a higher Σ boundary with a lower Σ boundary. In terms of the energetic considerations of Fullman and Fisher [58] and energy measurements of Σ3 interfaces, the twinning results indicate that these GB energy differences may be as high as 10% or more, depending upon the Σ value of the interface. The observed Σ GBs resulting from twin formation are: Σ = 3 incoherent, 5, 7, 9, 11, 19a, 19b, 25a, 41c, in zone refined Pb; also Σ33a and Σ37a in zone refined Cu, and Σ3 incoherent, Σ7, Σ9, and Σ11 in zone refined Al. The fact that only very low Σ GBs are obtained by twinning in the pure Al is consistent with its higher twin boundary energy, as compared with that in Pb and Cu, and with the theory of Fullman and Fisher [58].

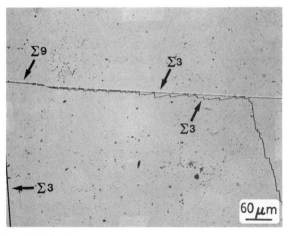

Fig. 5.3 Optical micrograph showing the interaction of two partial twin bands in 99.999% Ni yielding a small Σ9 segment (i.e. Σ3 + Σ3 = Σ9) [45].

Additional experiments [59–62] were conducted on zone refined Pb where the energies of several Σ boundaries were determined in terms of the energy of a non Σ-GB, in the temperature range of 322–326.5°C. As shown in Table 5.1, the relative boundary energy of Σ interfaces decreases with decreasing size of the periodic unit, in terms of the dimensions of the unit lattice cell, or increasing planar coincidence site density. The results given in Table 5.1 would be quite different if a sufficient concentration of strongly segregating solute is present in the solvent. For example, the relative energy value of 0.68 for the incoherent twin boundary increases to 0.88 with the addition of about 0.015 at.%Cu to the lead [59], due to a greater solute segregation at the non-Σ or reference boundary than at the low Σ boundary. If solute segregation effects are not taken into account, a 1:1 correlation between energy and Σ models would not be observed in the data of Table 5.1.

It appears, therefore, that many of the Σ GBs in pure metals with $\Sigma \leqslant 33$ are characterized by lower energies. This is in general agreement with the basic assumption in the CSL/DSC model, namely, an interface may be of relatively low energy when the two adjoining lattices are at a misorientation corresponding to a relatively dense CSL, i.e. a low value of Σ.

5.5 KINETIC PROPERTIES

5.5.1 Diffusion

Numerous experiments involving diffusional mass transport along boundaries during diffusional

creep, sintering, electromigration, and thermo-migration demonstrate clearly that GB diffusion must occur by a defect or vacancy mechanism which allows a net diffusional flux of atoms to occur [63]. It appears that the rate of point defect diffusion in the boundary may be reduced for low Σ grain boundaries. In addition, there is some experimental evidence which indicates that certain low Σ boundaries ($\Sigma 3$ to $\Sigma 15$) are practically inoperative as vacancy sources or sinks at low vacancy chemical potentials, even though more general GBs are operative [34]. The GB sources or sinks needed to support divergences in the atom or vacancy fluxes are GB dislocations [64].

Haynes and Smoluchowski [65] examined GB diffusion in $\langle 110 \rangle$ bicrystals of bcc Fe containing 3% Si. They found that a broad minimum occurred around 50° in the plot of penetration against θ about a $\langle 110 \rangle$ axis, corresponding to a $\Sigma 11$ relationship. The diffusion of Bi in $\langle 100 \rangle$ tilt boundaries of Cu has revealed diffusion cusps within 3° of $\Sigma 13$, $\Sigma 17$, and $\Sigma 25$ [66]. GB diffusion of Cr in Nb bicrystals with $\langle 100 \rangle$ symmetric tilt boundaries was studied at 1273°C using electron-probe microanalysis [67]. Increase in misorientation of the bicrystal enhances the boundary diffusion, except near the CSL boundary where the diffusion rate is a minimum.

Herbeuval *et al.* [68] measured relative diffusion rates along the tilt axis of [110] symmetric tilt boundaries in Al. Minimum diffusion rates were indicated for the $\Sigma 3$ and $\Sigma 11$ boundaries. Vacancy migration energies in the core of the $\Sigma 3$ coherent twin boundary are similar to that in the crystal lattice [69]. The $\Sigma 11$ boundary also showed a slower diffusion rate than a $\Sigma 27$ boundary [68]. The existence of cusps in the diffusion curves is revealed again for self diffusion in [110] tilt boundaries in Al bicrystals (<0.002% impurities) [70]. The GB diffusivity was calculated from the concentration profile determined from the sectioning method by using Whipple's solution. However, since the tilt axis of the Al bicrystals [70] deviated between 5° and 11° from the exact [110] orientation, the observed cusps can not be correlated with low Σ orientation relationships.

Boundary diffusion study of ^{65}Zn at 100 and 150°C in coarse-grained Al (99.999%) revealed

Table 5.1 Relative energies for boundaries in zone refined lead, equilibrated at 322–326°C

Misorientation	Size of periodic unit (Xa^{-2})	Relative energy $(\gamma\Sigma/\gamma\ non\text{-}\Sigma)$	Reference
$\theta = 38° \langle 111 \rangle, \Sigma 7$	3.0	0.85	[60]
$\theta = 36.9° \langle 100 \rangle, \Sigma 5$	1.1	0.71	[61]
$\theta = 70.5° \langle 110 \rangle, \Sigma 3$ (incoherent)	0.87	0.68	[59]
$\theta = 60° \langle 111 \rangle, \Sigma 3$ (coherent)	0.43	0.05	[62]

slower diffusion in the Σ GBs, such as Σ7 and Σ11 and in other boundaries with Σ < 19 [71]. These results suggest that only special ordered boundary structures can be significant barriers to diffusion and that boundary dislocations serve as the high diffusivity path. Additional studies of Zn diffusion in ⟨100⟩ and ⟨110⟩ twist boundaries in Al also indicate that Σ boundaries have minimum values of boundary diffusivity [72].

In general, the GB diffusion studies indicate the existence of the kind of directionality that the CSL–DSC boundary model would require. This is particularly evident in the observed anisotropy of boundary diffusion in tilt boundaries [70, 73, 74]. The structural unit/grain boundary dislocation model [75, 76] provides a basis for understanding diffusion anisotropy over a wide range of misorientation. The latter model is essentially equivalent to the more limited 'dislocation pipe' diffusion model originally proposed by Turnbull and Hoffman [77]. The important factor in controlling GB diffusion rates is the 'bad' material making up the cores of GBs and structural units, rather than the cores of intrinsic GB dislocations [63]. The structural unit model, i.e. small characteristic groups of atoms in ordered arrays, is an extension of the model originally proposed by Bishop and Chalmers [78].

The presence of solute or impurity atoms in the GB may considerably modify the boundary self diffusion coefficient. For example, Gupta and Rosenberg [79] observed that about 1 at. % of Ta in Au increased the activation energy of GB solvent diffusion by about 0.3 e.v. They also noted a crossover in the temperature-dependence data for GB diffusion at about 320°C, below which the diffusion rate in the Au–Ta becomes substantially smaller than in pure Au. This cross-over may explain some conflicting observations with regard to the effect of solute on GB diffusion based on measurements over a narrow temperature range (e.g. ref. [80]). Solute (Mg or Cu) in Al thin films [81] and Be in Cu thin films [82] decrease the rate of GB self diffusion. Additional studies also indicate that impurity segregation to the GB causes the GB self diffusion rate to decrease [83–85]. Solute in the GB may segregate to the open channel cores and reduce

their effectiveness by increasing the diffusion activation energy [79]. It is also suggested that the observed diffusion effects may be due to changes in GB structure caused by the solute atoms [86].

For a 16° ⟨100⟩ tilt GB in Ag (near Σ25a), values of activation energy for GB self diffusion are reported to be 19.7 kcal /g at. in 99.98% Ag [77] and only 11.8 kcal /g at. in 99.9999% Ag [87]. The segregation of impurity atoms may fill the regions of high diffusivity in the boundary, this effect being larger at lower temperatures due to a greater amount of impurity segregation at the lower temperature, thereby increasing the apparent activation energy. On this basis, the diffusion behavior of a low Σ boundary might then be expected to be less influenced by solute or impurity atoms than a high or non-Σ boundary. However, little or no data appear to exist on this point.

5.5.2 Mobility

The early lead bicrystal studies of Aust and Rutter [88, 89] demonstrated that certain large angle GBs, having orientation relationships at or close to those of the CSL migrate, more rapidly than other GBs in the presence of solute atoms and at lower temperatures. These favored boundaries, with higher mobilities and lower activation energies, correspond to interfaces with Σ = 5, 7, 13, and 17. Further study of a Σ5 GB in zone refined Al [90] showed that its velocity decreased, and its activation energy increased, with increasing deviation (up to 5°) from the ideal coincidence misorientation. Such differences in migration behaviour among different large angle boundaries are largely controlled by the selective interaction of residual impurity or solute atoms in the pure metal with the different grain boundary structures [89]. With little or no solute, or too much solute, this difference in migration behavior between Σ and non-Σ boundaries is not observed. In fact, in a pure zone refined metal at a temperature near the melting point of the metal, a non-Σ boundary may move faster than a Σ boundary [89].

This solute dependence was clearly evident in the work of Fridman *et al.* [91] where the activation energy of boundary motion versus θ ⟨100⟩ tilt

(Fig. 5.4) show minima at $\Sigma = 5$, 13, and 17 in 99.9992% Al (as was also observed in zone refined Pb [89]), but no minima are observed in still higher purity Al (99.99995%). Work by Molodov *et al.* [92] on the effect of pressure on the migration of ⟨001⟩ tilt GBs in tin bicrystals containing 10 ppm impurities indicated that the $\Sigma5$, $\Sigma13$, and $\Sigma17$ boundaries have minima in the activation energy and in the activation volume of the migration process.

As was discussed previously [35], when two adjacent crystals are exactly at the ideal coincidence relationship there is little, if any, elastic strain outside the core region of the boundary. Solute atoms would, therefore, be adsorbed only in the core region, diffusion of solute atoms is relatively easy, and the solute atoms are able to migrate with the boundary without retarding it seriously. However, departure from the ideal Σ orientation relationship gives rise to a strain field extending outward from the boundary into the crystals. This strain field interacts with solutes in a region where the lattice structure is only slightly disturbed. The much slower diffusion of solute atoms in this region leads to a substantial drag on the moving boundary [93, 94]. This effect would arise gradually with progressive departure from ideal coincidence as is observed in the mobility studies.

Viswanathan and Bauer [95, 96], in a study of GB migration in bicrystals of 99.999% Cu, also observed higher mobilities for a $\Sigma5$ ($\theta = 37 \pm 1°$ [001]) than those for non-Σ boundaries. However, the activation energy for migration of the $\Sigma5$ boundary (42 kcal/mol) was larger than those for the non-Σ boundaries (26–30 kcal/mol). Ferran *et al.* [57] reported a value of 20 kcal/mol for non-Σ boundaries in zone refined Cu. The studies on zone refined Cu [57] and zone refined Pb [88, 89] were conducted in the temperature range 0.75 Tm to 0.99 Tm as compared to only 0.47 Tm to 0.63 Tm for the 99.999% Cu bicrystals [95, 96]. It was suggested [96] that a smaller activation energy might be observed for a Σ boundary in Cu at higher temperatures (>0.6 Tm). In this regard, as seen in Fig. 5.5, the temperature dependence of a $\Sigma5$ GB in zone refined Al containing 40 ppm Cu is 14.9 kcal/mol at higher temperatures (0.83 Tm– 0.99 Tm) and 30.5 kcal/mol at lower temperatures

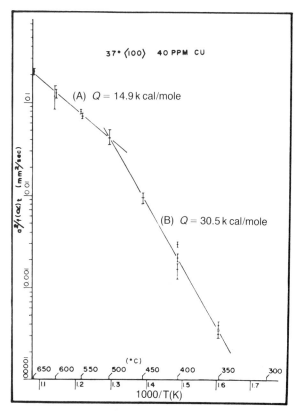

Fig. 5.5 Temperature dependence of boundary motion in 37° ⟨100⟩ bicrystals of Al containing 40 at. ppm Cu [90].

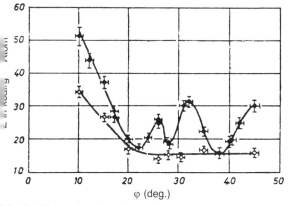

Fig. 5.4 Orientation dependence of the activation energy for the migration of ⟨100⟩ tilt boundaries in 99.9992% Al (filled circles) and 99.99995% Al (open circles) [91].

(0.61 Tm–0.83 Tm) [90]. Such a transition in the temperature dependence is not found for a Σ5 boundary in zone refined Al with no Cu addition [90]. Similar studies of non-Σ boundaries in lead showed an abrupt change in the activation energy at about 0.8 Tm [97]. The observed transition in boundary migration with temperature, e.g. Fig. 5.5, was attributed to a solute-induced transformation in the boundary structure [90]. As was discussed in section 5.3, such a boundary transformation is now confirmed directly in several studies [30–32].

The boundary orientation or inclination also has a strong influence on GB migration. These anisotropic migration effects appear to be more pronounced at low than at high temperatures and are generally observed for GBs at or near low Σ coincidence relationship. The obvious case is the Σ3 twin boundary where the coherent segment has very low or zero mobility, whereas the non-coherent segment has high mobility [98]. For GBs within 2° of Σ5, Σ7 and Σ13 in Al [99, 100], the pure tilt boundary segments have a higher migration rate than do either the mixed or pure twist segments. In terms of the coincidence–dislocation model for the boundary structure, the twist boundary with screw dislocations has little or no free volume, whereas the tilt boundary with edge dislocations has free volume. Therefore, the easy diffusive motion of solutes and, hence, smaller solute drag effects should apply only to the tilt case.

Studies of migration of large angle GBs in ice bicrystals [101] and 99.998% Al bicrystals [102] revealed that low Σ boundaries (e.g. Σ = 3, 9, and 11) had a strong tendency to assume complex faceted shapes which are characterized by lower mobilities than the non-faceted segments. These anisotropic migration rates may be interpreted in terms of a difference in the density of steps: the non-faceted boundary with many steps moves faster than the faceted boundary with fewer steps [103]. Dislocation and step mechanisms of boundary migration have been shown to be consistent with many observations of GB mobility [104]. These observations indicate that high coincidence density (or low Σ value) is not a

sufficient condition for high macroscopic grain boundary mobility [105] and low activation energy, as was also shown for the temperature and solute effects discussed previously.

A solute may not only exert a drag on a moving GB, but at higher concentrations may also provide a driving force for the migration of GBs [106]. GB motion can be induced by changing the chemical composition as a result of GB diffusion [107], i.e. so-called 'DIGM'. Since low Σ boundaries are generally believed to exhibit low diffusivity [108] (and as discussed in section 5.5.1), one would expect such boundaries to show lower migration rates in DIGM than do general or high Σ boundaries. Chen and King [109] have studied DIGM as a function of misorientation for symmetrical [001] tilt GBs in the Cu–Zn system. As expected, relatively low migration rates are observed for misorientations close to low Σ boundaries namely Σ5, Σ17, and Σ25. However, presumably because of their slower migration, high solute (Zn) penetration depths and high Zn concentrations deposited by the moving boundary were observed at these Σ boundaries as compared to the non-Σ boundaries [109]. The small migration rates of the low Σ boundaries were rationalized in terms of the dislocation climb model for DIGM [110, 111], where the ratio of the step height to the Burger's vector length for the responsible dislocations is minimized for low Σ boundaries and maximized for high Σ and non-Σ boundaries [112].

Additional work was conducted on DIGM in asymmetrical tilt [001] and [110] boundaries in Cu–Zn [113]. The Σ5 and Σ11 asymmetrical boundaries have a large deposited solute (Zn) concentration as it did for the symmetrical case, but showed a large migration distance and a small penetration depth, both of which are contrary to the result on symmetrical interfaces [109]. The Σ5 and Σ11 asymmetrical interfaces showed an incubation period followed by rapid migration. It was suggested [113] that in these interfaces the initial GB structure is immobile, but the increase in solute concentration in the GB transforms the boundary structure to one that has higher mobility and/or a higher diffusivity.

5.6 MECHANICAL AND PHYSICAL PROPERTIES

5.6.1 Grain boundary sliding

GB sliding, which involves the relative translation of two adjacent grains by a shear parallel to the boundary, is important in deformation of metals at temperatures above ~0.4 Tm. As concluded by Gleiter and Chalmers [80], very low sliding rates appear to be characteristic of low Σ GBs. For example, Biscondi and Goux [114, 115] observed that the sliding rate of a Σ5 boundary ($\theta = 37°$ [100]) was about 10% that of the sliding rate of a 40° (100) boundary under the same conditions of load and temperature. Additional studies of GB sliding in bicrystals of fcc metals and alloys have also revealed that GBs of exact or slightly off coincidence orientations do not slide easily [116–118].

Watanabe *et al.* [119, 120] investigated the misorientation dependence of GB sliding in zinc (99.999%) bicrystals and found a minimum in the sliding rate for a Σ17 near-coincidence boundary ($\theta = 49.5°$, $\langle 11\bar{2}0 \rangle$ tilt) and for a Σ9 near coincidence boundary (within 2° of 56.6° $\langle 10\bar{1}0 \rangle$ tilt). It was also found that increasing deviation in misorientation from the Σ9 near-coincidence boundary in zinc results in a greater sliding rate [121]. This latter study also indicated an abrupt change in the temperature dependence of sliding on $\langle 10\bar{1}0 \rangle$ tilt boundaries in Zn, ranging from 0.7 to 0.9 Tm depending on the GB misorientation. The highest transition temperature is found for the boundary within 2° of the 56.6° $\langle 10\bar{1}0 \rangle$ Σ9 boundary. The observed transition temperature is attributed to a GB structural transformation.

These results were interpreted in terms of a model for GB sliding based on the climb and glide of crystal lattice dislocations dissociated in the GB [119, 120]. As was seen in section 5.5.1, GB diffusion, which is important for the climb of dislocations in the boundary, is more difficult in the Σ GBs. Also, the dissociation of crystal lattice dislocations is more difficult in low Σ boundaries [122, 123].

Kokawa *et al.* [124] systematically analyzed the possible dislocation reactions in coincidence boundaries with Σ ⩽ 19 in fcc and bcc crystals, in terms of energetics according to the CSL model. The elastic energy reduction during dissociation of lattice dislocations suggests that they are more easily absorbed into CSL boundaries with large Σ values. TEM observations also showed a large difference in the absorption rate between CSL and random boundaries in specimens under creep conditions [124]. Lim and Raj [125] examined slip traces in iso-axial symmetrical tilt bicrystals of Ni, deformed in simple compression at 537 K to 2% total strain. They found that the low Σ boundaries show a higher degree of slip continuity of screw dislocations than high Σ boundaries. This structural dependence of continuity of screw bands was interpreted in terms of reactions between lattice and GB dislocations. Since the primitive DSC vectors are smaller in high Σ boundaries [126], these boundaries are more efficient traps for lattice dislocations. However, transfer of screw dislocations across a Σ51 symmetrical tilt boundary ($\theta = 16°$ [011]) has been observed in Ge crept at 1173 K at low stress [127]. It was also found in this latter work that deviations from the Σ51 coincidence structure are accommodated by secondary dislocations. Recently, GB sliding has been noted in a Σ9 boundary, but not a Σ3 boundary, in bicrystals of Cu-9 at.%Al, at about 0.8 Tm [128].

5.6.2 Grain boundary fracture and cavitation

Low Σ GBs are not only resistant to high temperature GB sliding, but also to high temperature fracture, while high angle random boundaries slide and fracture easily [129]. The stronger embrittlement of random GBs is believed to be due to an increasing number of segregation sites with an increasing number of secondary dislocations in the interface [59]. For example, Roy *et al.* [130] found the least embrittled GBs in Cu + 0.1% Bi alloy to be of lowest energies (i.e. presumably, some low Σ boundaries). Also, the further the boundary deviates from a low energy position, the more embrittled it is. This was explained in terms of a greater degree of segregation (Bi) at or near the secondary or misfit dislocations characterizing the off-low energy boundaries. It was also noted that intergranular fracture preferentially occurred at non-Σ boundaries in β-brass polycrystals during studies of liquid Ga-induced intergranular fracture [131]. However, CSL and low angle boundaries were difficult to break. *In-situ* SEM observations revealed that the change in fracture mode depends on the type of GB which the propagating crack meets in β-brass polycrystals. Coincidence boundaries can be strong obstacles to the propagation of cracks so that the intergranular fracture will stop at these Σ boundaries; the crack then enters the grain interior, changing the fracture mode. These results appeared to be independent of the orientation of the boundary plane in the polycrystalline β-brass alloy [131].

Lim and Raj [132] have shown that lower Σ boundaries in nickel polycrystals with smaller deviation angles from exact CSL misorientations are more resistant than other boundaries to intergranular cavitation in low cycle fatigue at 573 K. Don and Majundar [133] observed similar effects concerning intergranular cavitation of type 304 stainless steel during creep tests at 866 K under an applied stress of 172 MPa. Based on a review of many such experimental observations on structure-dependent intergranular fracture in metallic and ceramic materials in various environments, Watanabe [134] recently concluded that low angle ($\Sigma = 1$) and low-Σ coincidence boundaries are resistant to fracture, while high angle random boundaries are preferential sites for intergranular fracture.

5.6.3 Grain boundary hardening

Low Σ interfaces are observed to give less boundary hardening than non-Σ boundaries in zone refined lead containing different solutes, e.g. up to 50 atom ppm Sn for $\Sigma3$ and $\Sigma7$ boundaries, and up to 1.5 atom ppm Au for $\Sigma3$, $\Sigma9$ and $\Sigma25a$ boundaries [135]. Hardening cusps were also observed at Σ GBs in $\langle 100 \rangle$ bicrystals of Nb [136, 137]. The GB hardening, as determined from microhardness

testing at room temperature, was found to increase with misorientation of the Nb bicrystals, except at the CSL boundaries where the hardening is reduced to a minimum. The observed hardening in the Nb was attributed to impurity segregation. From microhardness measurements and Auger electron spectroscopy, GB hardening in αFe–Sn alloy was shown to be due to the Sn segregation to GBs [138].

A non-equilibrium GB segregation model, based on vacancy–solute interactions, was proposed to explain the excess boundary hardening observed in the presence of solute [139]. The smaller boundary hardening found at Σ GBs is attributed to their smaller interaction with solute atoms and their smaller efficiency as vacancy sinks. This model of non-equilibrium segregation has received considerable direct experimental support in recent years (e.g. [21, 140–142]).

5.6.4 Electrical properties

Lormand [143] has calculated GB resistivity using a pseudopotential method and concluded that the resistivity should vary with misorientation (θ) and show minima for CSL boundaries. Nakamichi [144] measured the electrical resistivity at 4.2 K of individual boundaries in zone refined Al. The specific grain boundary resistivity (ρ_{gb}) was found to be: $\rho_{gb} = A \sin(\theta/2) + B \sin(\phi/2)$, where θ is the tilt angle and ϕ is the twist angle. For large angle general boundaries with misorientation $>26°$, $A = 0.69 \times 10^{-15} \Omega m^2$ and $B = 1.3 \times 10^{-15} \Omega m^2$. For CSL boundaries with $\Sigma > 3$ and <31, $A = 0.57 \times 10^{-15} \Omega m^2$ and $B = 0.37 \times 10^{-15} \Omega m^2$. These results are illustrated in Fig. 5.6, showing the difference in electrical resistivity between Σ and non-Σ GBs. The boundary resistivity is attributed to GB dislocations, arising largely from the scattering of electrons by the dislocation core region. However, in Al containing 0.005 at.%Ag, the relationship found for pure Al no longer holds and the specific GB resistivity is much larger than in the pure Al. The increased resistivity is attributed to Ag segregation in the GB which appears to mask the intrinsic boundary resistivity due to structure [144].

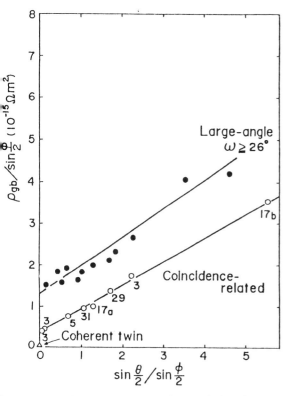

Fig. 5.6 Plot of $\rho_{gb}/\sin(\phi/2)$ versus $\sin(\phi/2)\sin(\phi/2)$ for zone refined Al. ρ_{gb} is the GB electrical resistivity and θ and ϕ are the GB tilt and twist angles, respectively. Numbers beside the data points are Σ values [144].

Low energy, low Σ GBs in semiconductors, Si, Ge, appear to exhibit no electrical activity unless they are contaminated with impurities [145, 146]. Poullain *et al.* [147] studied the structural and electrical properties of GBs in Si bicrystals. They found that the $\Sigma 9$ and $\Sigma 13$ boundaries provide no potential barriers, whereas the $\Sigma 27$ and general boundaries are always electrically active. Studies of semiconducting ceramics [148, 149] reveal that general boundaries in MnZn ferrites give rise to large segregation of Ca^{2+} coupled with a depletion of Fe^{2+}. These general boundaries display a high resistivity, about 10^5 times higher than within the grains. However, low Σ boundaries do not exhibit a strong segregation and have a low resistivity. Recent studies [150, 151] have also reported on the possible benefits of Σ GBs in superconducting materials.

5.7 CORROSION PROPERTIES

5.7.1 Bicrystal studies

Correlations between GB structure and susceptibility to IGC have been largely derived from corrosion groove depth measurements involving symmetrical bicrystals having controlled misorientation, fixed boundary plane indices (i.e. resulting from the inherent symmetry), and boundary planes perpendicular to the free surface. With materials of relatively high purity, susceptibility to IGC has generally been correlated with interfacial energy (section 5.4).

Arora and Metzger [152] investigated the corrosion behavior of $\langle 100 \rangle$ tilt and twist bicrystals of 99.999% Al in 16% HCl. Intergranular groove depths were measured with reference to a $\{100\}$ surface, and were found to increase with increasing crystal misorientation up to ~15°. With further increases in misorientation, the penetration depth was found to remain constant except for tilt misorientations of ~23° and ~37° which displayed groove depths of approximately one-half of other large angle boundaries. These misorientations correspond to CSLs of $\Sigma 13$ and $\Sigma 5$ respectively. A particularly low penetration depth was also noted for a 16° twist boundary ($\Sigma 25a$). A 54° twist boundary corresponding to ~$\Sigma 5$ $\{100\}$ was not observed to have a lower corrosion rate than the other general boundaries. As discussed by Aust and Iwao [153], the absence of 'special' behavior at this interface may have been the result of the relatively large deviation from the exact CSL (i.e. 3–4°).

Penetration cusps were also observed by Qian and Chou [154] with bicrystals of 99.85% Nb corroded in 70% HNO_3 + 30% HF. Among the large angle boundaries investigated, minima in penetration depth were observed at 23° $\langle 100 \rangle$ ($\Sigma 13$), 28° $\langle 100 \rangle$ ($\Sigma 17$), and 29° $\langle 110 \rangle$ ($\Sigma 19$).

Boos and Goux [155] investigated the corrosion behavior of 99.998% Al tilt bicrystals in pressurized water (150°C). Minima in intergranular penetration for large angle GBs were not observed with $\langle 100 \rangle$ bicrystals; however, cusps were noted with $\Sigma 3$ and $\Sigma 11$ orientation relationships in the

⟨110⟩ bicrystal series. The observed minima in penetration at these interfaces were attributed to the high planar packing densities associated with their respective symmetry planes (i.e. {111} and {311}), resulting in less hydrogen accumulation at the boundary, and lower corrosion rates.

Froment and Vignaud [156, 157] investigated ⟨100⟩ and ⟨110⟩ tilt bicrystals of 99.99+% Al in 10% HCl. In the ⟨110⟩ series, low rates of corrosion were observed with Σ3, Σ9, and Σ11 interfaces. With the ⟨100⟩ series, a penetration cusp was only observed with a $\theta = 37°$ Σ5{310} boundary. In this same series, an apparent maximum penetration was observed with a $\theta = 53°$ Σ5{210} interface.

Beaunier *et al.* [158, 159] determined that certain low Σ CSL relationships in ⟨100⟩ tilt bicrystals of 17%Cr–13%Ni austenitic stainless steel displayed high susceptibility to IGC at transpassive potentials in 2N H_2SO_4. 'Maxima' of corrosion attack (as assessed by groove widths) were observed at misorientations of 37° and 63° corresponding to Σ5{310} and Σ13{320} interfaces respectively. In this same series, a minimum was observed at a misorientation of 53° (Σ5{210}). These observations are contrary to those reported by Froment and Vignaud [156, 157] for ⟨100⟩ boundaries in Al.

Yamashita *et al.* [160] investigated the intergranular corrosion behavior of ⟨110⟩ bicrystals of Cu-9at.%Al in a dislocation etchant (2 ml Br_2 + 60 ml HCl + 20 ml CH_3COOH + 100 ml H_2O). Distinct penetration cusps were evident at orientation relationships corresponding to Σ9{221}, Σ3{111}, and Σ11{311}. Cusps were not observed at Σ17b{322} and Σ9{411}. The observed corrosion behavior was attributed to energy differences between these interfaces.

In these bicrystal studies, the observation of 'special' GB properties has been exclusive to interfaces at or near low Σ CSL relationships; however, a direct correlation between corrosion susceptibility and Σ value cannot be established. Table 5.2 summarizes the results of these corrosion studies with symmetrical tilt bicrystals in terms of the presence or absence of observed corrosion rate minima (i.e. cusps) at the specific CSL interfaces

Table 5.2 Summary of results from symmetrical tilt bicrystal corrosion studies

Boundary plane	$\rho\,(a^{-2})$	Σ	Corrosion minima	No corrosion minima
{111}	2.31	3	b, c, f	–
{311}	1.21	11	b, c, f	–
{320}	1.11	13	e	–
{410}	0.97	17	e	–
{331}	0.92	19	e	b, c
{420}	0.89	5	d	c
{442}	0.67	9	c, f	b
{620}	0.63	5	a, c	b, d
{640}	0.55	13	a	b, d
{644}	0.49	17	–	f
{820}	0.49	17	–	a, b, c, d
{822}	0.47	9	–	f

a: Al [152].
b: Al [155].
c: Al [156, 157].
d: Fe–17%Cr–13%Ni [158, 159].
e: Nb [154].
f: Cu–9%Al [160].

examined. The data are tabulated in descending order of planar packing density (ρ) for the associated boundary (i.e. twinning) planes. Three distinct regions of behavior can be identified on the basis of boundary plane character. Interfaces having boundary planes with $\rho \geqslant 0.97$ display penetration cusps in all investigations. With $\rho \leqslant 0.49$, all investigations demonstrate the absence of cusps. At intermediate values of ρ, variations in behavior are noted between studies. This dependence on boundary plane character is consistent with the 'interplanar Spacing/Planar Packing Density' model advanced by Wolf [161] and Paidar [162], which has been previously shown [161] to account for energy variations in symmetrical tilt bicrystals.

5.7.2 Polycrystal studies

The most direct correlation between GB energy and susceptibility to intergranular corrosion was obtained by Erb *et al.* [163] in sintering studies on 99.999% Cu. Thousands of randomly oriented single crystal Cu spheres (~50 μm diameter) were sintered on to the {110} surface of a single crystal plate of Cu, and exposed to a 16% HCl + 1 g/l FeCl₃ solution, so that preferential interfacial corrosion would result in selective sphere detachment from the plate. Subsequent X-ray diffraction analysis (i.e. ⟨111⟩ pole figure) displayed intensity peaks at certain orientations (i.e. remaining spheres), demonstrating that specific orientation relationships provide a high resistance to intergranular corrosion. The diffraction patterns obtained with corroded samples were found to correspond to those generated by long sintering times, where the spheres rotate into low energy orientation relationships, corresponding in this study to Σ3 and Σ11 CSLs. The observed diffraction peaks with corroded specimens were observed to be much narrower than those associated with long sintering times, covering a maximum angular range of ≤1°. These observations indicate that small angular deviations from the exact (low energy) CSL relationships can significantly alter IGC susceptibility. The authors [163] further investigated this phenomenon by applying a plastic strain of ~0.2% to the plate. Under these conditions, all the spheres became detached during the corrosion treatment. The observed corrosion behavior was attributed to closely spaced dislocations providing continuous channels for corrosion at the sphere–plate interface; thus emphasizing the importance of intrinsic GB dislocations (e.g. SGBDs on deviation from exact CSL relationships), and extrinsic dislocations (i.e. lattice dislocations entering the boundary during plastic deformation of the plate), in compromising the corrosion resistance of the GBs. Environmental/electrochemical influences were also noted in this investigation, as the application of potentiostatic polarization in various de-aerated solutions did not allow for any sphere detachment [163].

The applicability of the CSL–DSC descriptions to corrosion properties in conventional polycrystalline materials has also been investigated. Visitserngtrakul *et al.* [164] assessed the structural characteristics of 75 GBs in stabilized 310 stainless steel, and evaluated corrosion susceptibility on the basis of intergranular groove depths following light

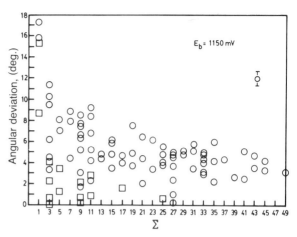

Fig. 5.7 Structure-dependence (Σ, Δθ) of intergranular corrosion morphology for 99.999% Ni containing 0.3 ppm wt. S following potentiostatic etching in 2N H₂SO₄ at a potential $E = 1150$ mV, SCE to a constant anodic charge transfer $Q = 5$ C/cm² [10]. □, Immunity; ○, pitting corrosion.

etching and a successive corrosion treatment (20 min.) in boiling 5N HNO₃ containing 3 g/l potassium dichromate. Among the high angle boundaries, the Σ3, Σ11, and Σ13b CSL interfaces displayed the highest resistance to intergranular corrosion.

The influence of deviation angle on the corrosion morphology associated with CSL-related GBs in polycrystalline 99.999% Ni, following potentiostatic and coulostatic etching at passive potentials in 2N H₂SO₄, is shown in Fig. 5.7 [10]. As shown in this figure, GBs close to low Σ CSL relationships tend to display selective immunity to corrosion (i.e. resistance to passive film breakdown). It was determined [10, 165] that the field of immunity (defined by Δθ and Σ) could be expanded with decreasing applied electrochemical potential; however, a limiting structural field defined as not extending beyond Σ25, and having an upper angular deviation limit defined by a relation of the type $\Delta\theta \leqslant 15°\Sigma^{-\frac{5}{6}}$ (Fig. 5.8), was experimentally determined. The characteristic potentials for structure-dependent intergranular corrosion behavior were also determined [10, 165] to be strongly dependent on the concentration of trace amounts of S (i.e. 0.3 to 50 ppm). These

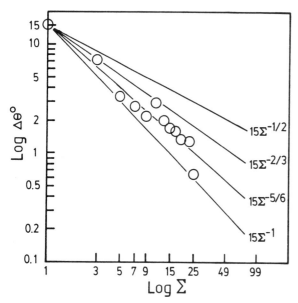

Fig. 5.8 Maximum deviation angles ($\Delta\theta$) from any CSL (Σ) in 99.999% Ni displaying selective immunity to intergranular corrosion in 2N H_2SO_4 [10].

results were attributed to a dislocation mechanism whereby resistance to intergranular passive film breakdown is dependent on the specific distribution (CSL–DSC) and local chemistry (solute segregation) of intrinsic grain boundary dislocations (IGBD).

5.8 TRIPLE JUNCTIONS

The presence of triple junctions (i.e. intersection line of three (or more [166]) GBs) in polycrystalline materials has been shown to influence several material properties, including: ductility [167, 168]; GB anelasticity [169]; diffusion induced GB migration (DIGM) and recrystallization (DIR) [170]; GB sliding [171]; intergranular corrosion [172]; and internal friction [173]. A thermal etching study [174] conducted with 99.999% Cu has demonstrated that triple lines can have a substantial line energy. Recent studies [168, 175] have further demonstrated that triple junctions can comprise a significant fraction of the bulk volume

of nanocrystalline materials, and thus may be responsible for their unique properties.

Several complementary models for the structure of triple junctions have been advanced, each being an extension of existing interfacial models. Bollmann [176–178] first demonstrated the defect character of triple junctions through an extension of the O-lattice model for interface structure [3]. Bollmann demonstrated [176–178] that under conditions whereby nodal balancing of adjoining intrinsic GB dislocation (IGBD) arrays cannot occur due to crystallographic constraints (i.e. Bollmann's U-lines), the triple line can be regarded as a distinct crystalline defect – a **disclination**. When a nodal balance can be achieved (Bollmann's I-lines), the Burgers vectors of adjoining IGBD arrays balance to zero, and the triple junction can be considered an imaginary line. Bollmann's analysis completes the representation of polycrystalline materials as a balanced network of dislocations (i.e. lattice dislocations, GB dislocations, triple line disclinations) [178].

Experimental support for Bollmann's structural treatment has been obtained. Triple junctions having the crystallographic characteristics of Bollmann's I- and U-lines have been experimentally identified in Ni [172], and Cu-8.5%Al [179]. The present authors [180, 181] determined that preferential corrosion at triple junctions (i.e. relative to adjoining GBs and crystal surfaces) in 99.999% Ni occurs exclusively at triple junctions having disclination character (i.e. Bollmann's U-lines) (Fig. 5.9). A computer simulation study [182] of intrinsic stress distributions at triple junctions has shown that uncompensated stresses exist at triple lines having disclination character (i.e. Bollmann's U-lines). Clarebrough and Forwood [183] experimentally identified (by TEM) an IGBD balance at a $\Sigma3$-$\Sigma9$-$\Sigma27$b 'I-type' triple junction in Cu-3%Si. In this latter study [183], the IGBDs were resolvable secondary grain boundary dislocations (SGBD) arising from CSL–DSC relaxations.

The significance of CSL–DSC relaxations to triple junctions has been discussed by Bollmann [178] and further emphasized in the 'special' triple junction [184] and coincident axial direction [185] approaches to triple line structure. Doni and Bleris [184] have evaluated the geometric characteristics of triple junctions composed of low Σ CSL boundaries, and have shown that these 'special' triple junctions can be structurally characterized through the symmetry of the CSL (i.e. CSL–DSC relaxations). The present authors [185] have shown that a one-dimensional periodicity along the triple junction, arising through low index planar continuity, can be achieved when the sequence of crystallographic rotations describing the junction can be represented about a low index axis. This model is analogous to the coincident axial direction [186] and plane matching [187] models of interface structure, and provides a one-dimensional limit to the three-dimensional CSL–DSC (e.g. $\Sigma \Rightarrow \infty$) and O-lattice descriptions. The significance of the CSL–DSC (and associated models) to the structure of GB intersections is further demonstrated by the characteristics of 'quadruple' and 'quintuple' junctions. Such multiple junctions, which are expected to be thermodynamically unstable, have been experimentally identified in Ni [45], Cu [166], and 304 stainless steel [166] to consist entirely of Σ boundaries; particularly boundaries of the type $\Sigma 3^n$.

5.9 SUMMARY

The applicability of the Σ-criterion to GB properties has been demonstrated in numerous studies, and, as outlined here, is primarily due to the importance of Σ (and $\Delta\theta$) in defining interfacial dislocation (IGBD) structures through the relaxations associated with the CSL–DSC. The boundary plane (i.e. inclination) has also been shown to be of significance; however, its influence on interface properties has been generally observed to be greatest when the GB is already in a low Σ CSL orientation relationship (e.g. coherent and incoherent segments of $\Sigma 3$ twin boundary). Although direct correlations between the magnitude of Σ and interface properties are not generally observed, 'special' properties (e.g. solute segregation, energy, diffusion, mobility, sliding, fracture, cavitation, hardening, resistivity, and

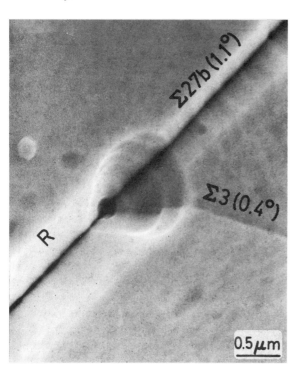

(b)

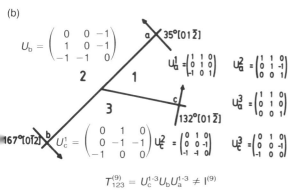

$$T_{123}^{(9)} = U_c^{1\text{-}3} U_b U_a^{1\text{-}3} \neq I^{(9)}$$

Fig. 5.9 (a) Scanning electron micrograph of preferential triple junction corrosion (following potentiostatic etching in 2N H_2SO_4) at a $\Sigma 3 - \Sigma 27b - R(\Sigma > 49)$ intersection in 99.999% Ni. (b) Corresponding orientation relationships (determined by electron channeling) and calculated U-matrices [176–178] which yield the most widely spaced (i.e. energetically favored) IGBD arrays for the adjoining interfaces. Of the nine possible values of the characteristic tensor (T_{123}) of the triple junction (i.e. composed of the anti-clockwise sequence of possible U-matrices), none is determined to be the identity matrix; indicating that nodal balancing of adjoining IGBD arrays is not possible (Bollmann's U-Line).

corrosion) are almost exclusively associated with interfaces at, or deviating slightly from, low-Σ CSL relationships. Theoretical and experimental considerations discussed in this work indicate that special properties of Σ GBs are generally observed when $\Sigma \leq 29$, and the angular deviation from Σ is given by $\Delta\theta \leq 15°\Sigma^{-\frac{5}{6}}$. The importance of CSL–DSC relaxations (and associated Σ-criterion) to the structural characterization of triple (and multiple) junctions in polycrystalline materials has been shown. Consideration of triple junctions as distinct structural defects, described by the interaction of adjoining intrinsic GB dislocation arrays, will possibly lead to a better understanding of the bulk properties of conventional polycrystalline materials and the unique properties of nanocrystalline systems.

ACKNOWLEDGMENT

Financial support from the Natural Sciences and Engineering Research Council of Canada is gratefully acknowledged.

REFERENCES

1. S. Ranganathan, *Acta Crystall.*, **21** (1966), 197.
2. H. Mykura, *Grain Boundary Structure and Kinetics* (ed R. W. Balluffi), ASM Metals Park OH (1979), p. 445.
3. W. Bollmann, *Crystal Defects and Crystalline Interfaces*, Springer-Verlag, Berlin (1970).
4. T. Schober and R. W. Balluffi, *Phil. Mag.*, **20** (1970), 511.
5. D. G. Brandon, *Acta Metall.*, **14** (1966), 1479.
6. M. Deschamps, F. Baribier and A. Marrouche, *Acta Metall.*, **35** (1987), 101.
7. Y. Ishida and M. McLean, *Phil. Mag.*, **27** (1973), 1125.
8. W. T. Read and W. Shockley, *Phys. Rev.*, **78** (1950), 275.
9. H. Grimmer, W. Bollmann and H. Warrington, *Acta Crystall. A*, **30** (1974), 197.
10. G. Palumbo and K. T. Aust, *Acta Metall.*, **38** (1990), 2343.
11. T. Watanabe, *J. Phys.*, **46** (1985) C4–555.
12. L. S. Shvindlerman and B. Straumal, *Acta Metall.*, **33** (1985), 735.
13. G. L. Bleris, J. G. Antonopoulos, T. H. Karakostas and P. Delavignette, *Phys. Stat. Sol.*, **67** (1981), 249.
14. K. L. Merkle, J. F. Reddy and C. L. Wiley, *J. Phys.*, **C4** (1985), 95.
15. B. D. Powell and D. P. Woodruff, *Phil. Mag.*, **34** (1976), 1679.
16. T. Ogura and C. J. McMahon, Jr., *Acta Metall.*, **26** (1978), 1317.
17. T. Watanabe, T. Murakami and S. Karashima, *Scripta Metall.*, **12** (1978), 361.
18. P. Lejcek and S. Hofmann, *J. Mat. Sci. Letters*, **7** (1988), 646.
19. S. Hofmann, *Vacuum*, **40** (1990), 9.
20. T. Muschik, S. Hofmann, W. Gust and B. Predl, *Appl. Surf. Sci.*, **37** (1989), 439.
21. T. Muschik, W. Gust, S. Hofmann and B. Predl, *Acta Metall.*, **37** (1989), 2917.
22. M. A. Fortes and B. Ralph, *Acta Metall.*, **15** (1967), 707.
23. P. R. Howell, D. E. Fleet, T. F. Page and B. Ralph, *Proc. 3rd Int. Conf. Strength of Metals and Alloys*, Vol. 1, Cambridge (1973), p. 149.
24. P. R. Howell, D. E. Fleet, A. Hildon and B. Ralph, *J. Micros.*, **107** (1976), 155.
25. J. G. Hu, S. M. Kuo, A. Seki, B. W. Krakauer and D. N. Seidman, *Scripta Metall.*, **23** (1989), 2033.
26. S. S. Brenner and Hua Ming-Jian, *Scripta Metall.*, **24** (1990), 667.
27. D. Bouchet and L. Priester, *Scripta Metall.*, **21** (1987), 475.
28. D. Bouchet, B. Aufray and L. Priester, *J. Phys.*, **49** (1988), C5–417.
29. W. Swiatnicki, S. Lartigue, M. Biscondi and D. Bouchet, *J. Phys.*, **51** (1990), C-341.
30. K. E. Sickafus and S. L. Sass, *Acta Metall.* **35** (1987), 69: also *Scripta Metall.*, **18** (1984), 165.
31. C-H. Lin and S. L. Sass, *Scripta Metall.*, **22** (1988), 1569.
32. C-H. Lin, S. L. Sass, C. W. Allen and L. E. Rehn, *Acta Metall.*, **38** (1990), 619.
33. D. Wolf and K. L. Merkle, in *Atomic-Level Properties of Interface Materials*, Ch. 1 (eds D. Wolf and S. Yip) Chapman and Hall, London, (1991).
34. P. H. Pumphrey, in *Grain Boundary Structure and Properties* (eds G. A. Chadwick and D. A. Smith), Academic Press, NY (1976), p. 139.
35. K. T. Aust, in *Chalmers Anniversary Volume* (eds J. W. Christian, P. Haasen and T. B. Massalski), *Prog. Mat. Sci.*, Pergamon Press (1981), p. 27.
36. G. Herrmann, H. Gleiter and G. Baro, *Acta Metall.*, **24** (1976), 353.
37. H. Sautter, H. Gleiter and G. Baro, *Acta Metall.*, **25** (1977), 467.
38. P. Chaudhari and H. Charbnau, *Surf. Sci.*, **31** (1972), 104.

39. A. P. Sutton and R. W. Balluffi, *Acta Metall.*, **35** (1987), 2177.

40. U. Erb and H. Gleiter, *Scripta Metall.*, **13** (1979), 61.

41. R. Maurer, *Acta Metall.*, **35** (1987), 2557.

42. M. McLean, *J. Mater. Sci.*, **8** (1973), 571.

43. H. Miura, M. Kato and T. Mori, *Colloque de Phys.*, *C1*, **51** (1990), 263.

44. R. Maurer and H. Gleiter, *Scripta Metall.*, **19** (1985), 1009.

45. G. Palumbo and K. T. Aust, in *Recrystallization 90* (ed T. Chandra), TMS (1990), p. 101.

46. S. W. Chan and R. W. Balluffi, *Acta Metall.*, **33** (1985), 1113.

47. Y. Ishida and T. Yamamoto, *Trans. Japan Inst. Met.*, **18** (1977), 221.

48. D. A. Smith, M. F. Chisholm and J. Clabes, *Appl. Phys. Lett.*, **53** (1988), 2344.

49. H. Gleiter, *Acta Metall.*, **18** (1970), 23.

50. A. M. Donald and L. M. Brown, *Acta Metall.*, **27** (1979), 59.

51. T. G. Ference and R. W. Balluffi, *Scripta Metall.*, **22** (1988), 1929.

52. V. Randle and A. Brown, *Phil. Mag.*, **59** (1989), 1075.

53. C. T. Forwood and L. M. Clarebrough, *Metals Forum*, **8** (1985), 132.

54. P. J. Goodhew, *Metal Sci.*, **13** (1979), 108.

55. K. T. Aust and J. W. Rutter, *Trans. Met. Soc. AIME*, **218** (1960), 1023; also **224** (1962), 111.

56. K. T. Aust, *Trans. Met. Soc. AIME*, **221** (1961), 758.

57. G. Ferran, G. Cizeron and K. T. Aust, *Mem. Sci. Rev. Met.*, **54** (1967), 1067.

58. R. L. Fullman and J. C. Fisher, *J. Appl. Phys.*, **22** (1951), 1350.

59. H. Gleiter, *Acta Metall.*, **18** (1970), 117.

60. J. W. Rutter and K. T. Aust, referred to by K. T. Aust in *Surfaces and Interfaces, I Chemical and Physical Characteristics* (eds J. J. Burke, N. L. Reid and V. Weiss), Syracuse University Press (1967), p. 435.

61. G. Dimou and K. T. Aust, *Acta Metall.*, **22** (1974), 27.

62. G. F. Bolling and W. C. Winegard, *J. Inst. Met.*, **86** (1958), 492.

63. R. W. Balluffi, *Met. Trans.*, **13A** (1982), 2069.

64. R. W. Balluffi, in *Grain Boundary Structure and Kinetics*, ASM, Metals Park, OH (1980), p. 297.

65. C. W. Haynes and R. Smoluchowski, *Acta Metall.*, **3** (1955), 130.

66. S. Yukawa and M. J. Sinnott, *Trans. AIME*, **215** (1959), 338.

67. X. R. Qian and Y. T. Chou, *Phil. Mag.*, **52** (1985), 213.

68. I. Herbeuval, M. Biscondi and C. Goux, *Mem. Sci. Rev. Met.*, **70** (1973), 39.

69. B. A. Faridi and A. G. Crocker, *Phil. Mag. A*, **41** (1980), 137.

70. W. Lange and M. Jurich, quoted in Ref. [80].

71. Y. Ishida, K. Inou, T. Yamamoto and M. Mori, *Met. Science*, **10** (1976), 424.

72. A. L. Petelin and L. S. Shvindlerman, *Acad. Nauk. Metallofiz.*, **2** (1980), 83.

73. S. R. L. Couling and R. Smoluchowski, *J. Appl. Phys.*, **25** (1954), 1538.

74. R. E. Hoffman, *Acta Metall.*, **4** (1956), 97.

75. A. P. Sutton and V. Vitek, *Phil. Trans. Roy. Soc. Lond. A*, **309** (1983), 1, 37, 55.

76. R. W. Balluffi and P. D. Bristowe, *Surf. Sci.*, **144** (1984), 28.

77. D. Turnbull and R. E. Hoffmann, *Acta Metall.*, **2** (1954), 419.

78. G. H. Bishop and B. Chalmers, *Scripta Metall.*, **2** (1968), 133.

79. D. Gupta and R. Rosenberg, *Thin Solid Films*, **25** (1975), 171.

80. H. Gleiter and B. Chalmers, *Prog. Mater. Sci.*, **16** (1972), 77.

81. F. M. d'Heurle, A. Gangulee, C. F. Aliotta and V. A. Ranieri, *J. Electronic Mat.*, **4** (1975), 497.

82. F. M. d'Heurle and A. Gangulee, *Thin Solid Films*, **25** (1975), 531.

83. B. Aubray, P. Gas., J. Bernardini and F. Cabane-Brouty, *Scripta Metall.*, **14** (1980), 1279.

84. R. A. Padgett and C. L. White, *Scripta Metall.*, **18** (1984), 459.

85. H. Hansel, L. Stratmann, H. Keller and H. J. Grabke, *Acta Metall.*, **33** (1985), 659.

86. J. Juve-Duc, D. Treheux and P. Guiraldenq, *Scripta Metall.*, **12** (1978), 1107.

87. J. T. Robinson and N. L. Peterson, *Surf. Sci.*, **31** (1972), 586.

88. K. T. Aust and J. W. Rutter, *Trans. TMS-AIME*, **215** (1959), 119 and 820.

89. J. W. Rutter and K. T. Aust, *Acta Metall.*, **13** (1965), 181.

90. D. W. Demianczuk and K. T. Aust, *Acta Metall.* **23** (1975), 1149.

91. E. M. Fridman, C. V. Kopezky and L. S. Schvindlerman, *Z. Metallkd.*, **66** (1975), 533.

92. D. A. Molodov, B. B. Straumal and L. S. Schvindlerman, *Scripta Metall.*, **18** (1984), 207.

93. K. Lücke and K. Detert, *Acta Metall.*, **5** (1957), 628.

94. J. W. Cahn, *Acta Metall.*, **10** (1962), 789.

95. R. Viswanathan and C. L. Bauer, *Acta Metall.*, **21** (1973), 1099.

96. R. Viswanathan and C. L. Bauer, *Trans. TMS-AIME*, **5** (1974), 1691.

97. K. T. Aust, *Can. Met. Quart.*, **8** (1969), 173.

98. C. J. Simpson, K. T. Aust and W. C. Winegard, *Scripta Metall.*, **3** (1969), 171.

99. S. Kohara, M. N. Parthasarathi and P. A. Beck, *Trans. TMS-AIME*, **212** (1958), 875.

100. R. B. Rath and H. Hu, *Trans. TMS-AIME*, **236** (1966), 1193.

101. T. Hondoh and A. Higashi, *Phil. Mag.*, **39** (1979), 137.

102. M. S. Masteller and C. L. Bauer, *Acta Metall.*, **27** (1979), 483.

103. H. Gleiter, *Acta Metall.*, **17** (1969), 565.

104. D. A. Smith and C. M. F. Rae, *Met. Sci. J.*, **13** (1979), 101.

105. C. L. Bauer, *Defect and Diffusion Forum*, **66–69** (1989), 749.

106. M. Hillert, *Met. Sci. J.*, **13** (1979), 118.

107. M. Hillert and G. R. Purdy, *Acta Metall.*, **26** (1978), 333.

108. N. Peterson, in *Grain Boundary Structure and Kinetics* (ed R. W. Balluffi), ASM Metals Park, OH (1980), p. 209.

109. F. S. Chen and A. H. King, *Acta Metall.*, **36** (1988), 2827.

110. D. A. Smith and A. H. King, *Phil. Mag., A*, **44** (1981), 333.

111. R. W. Balluffi and J. W. Cahn, *Acta Metall.*, **29** (1981), 493.

112. A. H. King, *Acta Metall.*, **30** (1982), 419.

113. A. H. King and G. Dixit, *Colloque de Physique, C1*, **51** (1990), C-545.

114. M. Biscondi and C. Goux, *C. R. Acad. Sci.*, **258** (1964), 2806.

115. M. Biscondi and C. Goux, *Mem. Scient. Rev. Metall.*, **65** (1968), 167.

116. P. Lagarde and M. Biscondi, *Can. Met. Quart.*, **13** (1974), 245.

117. P. Lagarde and M. Biscondi, *Mem. Scient. Rev. Metall.*, **71** (1974), 121.

118. B. Michaut, A. Silvent and G. Sainfort, *Mem Scient. Rev. Metall.*, **71** (1974), 527.

119. T. Watanabe, N. Kuriyama and S. Karashima, *Proc. 4th Int. Conf. Strength of Metals and Alloys, Nancy*, (1976), p. 383.

120. T. Watanabe, M. Yamada, S. Shima and S. Karashima, *Phil. Mag. A*, **40** (1979), 667.

121. T. Watanabe, S. Kimura and S. Karashima, *Phil. Mag. A*, **49** (1984), 845.

122. R. C. Pond and D. A. Smith, *Phil. Mag. A*, **36** (1977), 353.

123. H. Kokawa, T. Watanabe and S. Karashima, *Phil. Mag. A*, **44** (1981), 1239.

124. H. Kokawa, T. Watanabe and S. Karashima, *J. Mater. Sci.*, **18** (1983), 1183.

125. L. C. Lim and R. Raj, *Acta Metall.*, **33** (1985), 1577.

126. H. Grimmer, *Scripta Metall.*, **8** (1974), 1221.

127. W. Skrotzki, H. Wendt, C. B. Carter and D. L. Kohlstedt, *Acta Metall.*, **36** (1988), 983.

128. S. Miura, Y. Okada, S. Onaka and S. Hashimoto, *Colloq. de Physique, C1*, **51** (1990), C-587.

129. T. Watanabe, *Met. Trans. A*, **14** (1983), 531.

130. A. Roy, U. Erb and H. Gleiter, *Acta Metall.*, **30** (1982), 1847.

131. T. Watanabe, M. Tanaka and S. Karashima, *Symp. Embrittlement of Liquid and Solid Metals* (ed H. Kandar), AIME, Warrendale, (1984), p. 183.

132. L. C. Lim and R. Raj, *Acta Metall.*, **32** (1984), 1183.

133. J. Don and S. Majundar, *Acta Metall.*, **34** (1986), 961.

134. T. Watanabe, *Mat. Sci. Forum*, **46** (1989), 25.

135. J. H. Westbrook and K. T. Aust, *Acta Metall.*, **11** (1963), 1151.

136. Y. T. Chou, B. C. Cai, A. D. Romig and L. S. Lin, *Phil. Mag.*, **47** (1983), 363.

137. Z. Q. Zhou and Y. T. Chou, *J. Less Common Met.*, **114** (1985), 323.

138. T. Watanabe, S. Kitamura and S. Karashima, *Acta Metall.*, **28** (1980), 455.

139. K. T. Aust, R. E. Hanneman, P. Nissen and J. H. Westbrook, *Acta Metall.*, **16** (1968), 291.

140. L. Karlsson, H. Norden and H. Odeluis, *Acta Metall.*, **36** (1988), 1, 13, 25 and 35.

141. F. Ferhat, D. Roptin and G. Saindrenan, *Scripta Metall.*, **22** (1988), 223.

142. X. L. He, Y. Y. Chu and J. J. Jonas, *Acta Metall.*, **37** (1989), 2905.

143. G. Lormand, *Scripta Metall.*, **13** (1979), 27.

144. I. Nakamichi, *J. Sci. Hiroshima Univ. A*, **54** (1990), 49.

145. X. J. Wu, N. Szkielko and P. Haasen, *J. Phys.*, **43** (1982), C1.

146. W. Szkielko and G. Petermann, *Poly-Microcrystalline and Amorphous Semiconductors*, Les editions de Physique (1984), 379.

147. G. Poullain, A. Bary, B. Mercey, P. Lay, J. L. Chermant and G. Nouet, *Trans. Japan Inst. Met.*, **27** (1986), 1069.

148. J. Y. Laval and M. H. Pinet, *Trans. Japan Inst. Met.*, **27** (1986), 1021.

149. M. H. Berger and J. Y. Laval, *Colloq. de Physique C1*, **51** (1990), C-965.

150. J. Y. Laval, C. Delamarre, M. H. Berger and C. Cabanal, *Colloq. de Physique C1*, **51** (1990), C-991.

151. Y. Takahashi, N. Tomita, M. Mori and Y. Ishida, *Colloq. de Physique C1*, **51** (1990), C-1049.

152. O. P. Arora and M. Metzger, *Trans. TMS-AIME*, **236** (1966), 1205.

153. K. T. Aust and O. Iwao, *Localized Corrosion*, NACE-3 (1974), 62.

154. X. R. Qian and Y. T. Chou, *Phil. Mag.*, **45** (1982), 1377.

155. J. Y. Boos and C. Goux, *Localized Corrosion* NACE-3 (1974), 556.

156. M. Froment and C. Vignaud, *C. R. Acad. Sc. Paris, C*, **272** (1971), 165.

157. M. Froment and C. Vignaud, *C. R. Acad. Sc. Paris, C*, **275** (1972), 25.

158. L. Beaunier, M. Froment and C. Vignaud, *J. Electroanal. Chem.*, **119** (1981), 125.

159. L. Beaunier, M. Froment and C. Vignaud, *Phys. Chem. Solid State: Appl. to Metals and their Compounds* (ed. P. Lacombe), B. V. Amsterdam (1984), p. 335.

160. M. Yamashita, T. Mimaki, S. Hashimoto and S. Miura, *Scripta Metall.*, **22** (1988), 1087.

161. D. Wolf, *Mat. Res. Soc. Symp. Proc.*, **40** (1985), 341.

162. V. Paidar, *Phys. Stat. Sol.*, **92** (1985), 115.

163. U. Erb., H. Gleiter and G. Schwitzgebel, *Acta Metall.*, **30** (1982), 1377.

164. S. Visitserngtrakul, S. Hashimoto, S. Miura and M. Okubo, *Mem. Faculty of Eng. Kyoto Univ.*, **52** (1990), 68.

165. G. Palumbo and K. T. Aust, *J. de Phys.*, **49** (1988), C5-569.

166. G. D. Sukhomlin, CH. V. Kopetskiy and A. V. Andreyeva, *Phys. Met. Metalloved.*, **62** (1986), 349.

167. V. B. Rabukhin, *Phys. Met. Metalloved.*, **61** (1986), 996.

168. G. Palumbo, U. Erb and K. T. Aust, *Scripta Metall.*, **24** (1990), 2347.

169. V. B. Rabukhin, *Phys. Met. Metalloved.*, **55** (1983), 178.

170. S. A. Hackney, *Scripta Metall.*, **22** (1988), 1255.

171. S. Miura, S. Hashimoto and T. K. Fujii, *J. de Phys., C*, **5–49** (1988), 599.

172. G. Palumbo and K. T. Aust, *Scripta Metall.*, **22** (1988), 847.

173. G. M. Ashmarin, M. Yu. Golubev, N. I. Naumova and A. V. Shalimova, *Phys. Met. Metalloved.*, **67** (1989), 110.

174. P. Fortier, G. Palumbo, G. D. Bruce, W. A. Miller and K. T. Aust, *Scripta Metall.*, **25** (1990), 177.

175. G. Palumbo, S. J. Thorpe and K. T. Aust, *Scripta Metall.*, **24** (1990), 1347.

176. W. Bollmann, *Phil. Mag. A.*, **49** (1984), 73.

177. W. Bollmann, *Phil. Mag. A.*, **57** (1988), 637.

178. W. Bollmann, *Mat. Sci. Eng. A*, **113** (1989), 129.

179. W. Bollmann and H. Guo, *Scripta Metall.*, **24** (1990), 709.

180. G. Palumbo and K. T. Aust, *Advanced Structural Materials* (ed D. Wilkinson), Pergamon, Oxford (1989), p. 227.

181. G. Palumbo and K. T. Aust, *Mat. Sci. Eng. A*, **113** (1989), 139.

182. W. Bollmann, *Mat. Sci. Eng. A*, **113** (1989), 129.

183. L. M. Clarebrough and C. T. Forwood, *Phil. Mag. A*, **55** (1987), 217.

184. E. G. Doni and G. L. Bleris, *Phys. Stat. Sol. (a)*, **110** (1988), 393.

185. G. Palumbo and K. T. Aust, *Scripta Metall.*, **24** (1990), 1771.

186. V. Randle and B. Ralph, *J. Mater. Sci.*, **23** (1988), 934.

187. P. H. Pumphrey, *Scripta Metall.*, **6** (1972), 107.

CHAPTER SIX

Grain boundary structure and migration

D. A. Smith

Introduction · Grain boundary structure · Steps and dislocations · Grain boundary migration · Observations by transmission electron microscopy · Grain structure of films · Discussion · Conclusion

6.1 INTRODUCTION

Grain growth and grain boundary (GB) migration are major factors in the evolution of the microstructure of materials during processing of materials. This statement applies not only to purely solid state processes such as dynamic recrystallization or diffusion induced GB migration but also to solidification, growth from the vapor, and electrochemical deposition. For a great many applications of materials the grain structure is the single most important microstructural variable. Consequently GB migration has been the subject of many experimental investigations. The field has been reviewed a number of times. The present chapter is substantially limited to developments since 1980. The state of understanding and the experimental data prior to that time are described in [1] and references cited therein.

Perhaps paradoxically the study of GB migration has given as much insight to GB structure as vice-versa! For example, one of the earliest pieces of evidence that all high angle GBs were not identical came from the investigation of GB migration in the context of the formation of recrystallization textures [2]. Deformation ordinarily results in a preferred orientation; this is elegantly shown together with the corresponding microstructure for the case

of copper single crystals in [3]. Subsequent primary or secondary recrystallization also produces a grain population which has a preferred orientation which need not be the same as that of the precursor structure. It is found that the formation of some textures could be understood in terms of a dominant contribution from the motion of particular, or special, GBs. Multiple twinning is found to make a particularly significant contribution to the final orientation distribution of the grains [4, 5]. The consequence of this process is that the product texture can be related to its predecessor by a rotation operation; this lattice rotation is effected by the motion of the special GBs. For the case of multiple twinning the recrystallization texture is related to the deformation texture by the rotation corresponding to $\Sigma = 3^n$ where n is the number of twinning operations. However, the initial correlation [2] concerned the primary and secondary recrystallization textures of copper which are related by variants of a $38°$ rotation about [111]; this rotation generates the $\Sigma = 7$ coincidence site lattice. This result stimulated research into the crystallography of GBs and the nature and behavior of defects in GBs. Nevertheless, there is still no accepted atomistically explicit model of GB migration. GB migration is usually modeled as a stochastic process and analyzed in terms of rate theory [6, 7]. Such models account satisfactorily for the qualitative features of the kinetics and the retarding effects of solute on grain growth but do not address the structure sensitive aspects of GB migration. To some extent this would be overly ambitious anyway, because there is also no generally applicable and accepted description of GB structure. The purpose of this chapter is to relate what is known about GB structure to the body of knowledge concerning GB migration with some illustrations from the behavior of thin films where the structure is strongly affected by GB migration.

6.2 GRAIN BOUNDARY STRUCTURE

Grain boundary structure is treated in detail elsewhere in this book; what follows is a summary of the salient points in the context of mechanisms of GB migration.

There is ample evidence for the view developed by Read and Shockley [8] that GBs can be described as arrays of dislocations providing that the dislocation separation is large relative to the dimensions of the cores of the dislocations [9]. Frank's formula [10] permits the Burger's vector content of such low angle GBs to be deduced from knowledge of the axis and angle of rotation and the interface plane. It must be emphasized that the connection between Burger's vector content and the configuration of dislocations is not always well known even for the deceptively simple case of the low angle boundary. It has been found that the Burger's vector content may be partitioned as lattice vectors, usually the shortest possible or as familiar partial lattice translations of the Frank and Shockley types [9, 11]. The limits to the validity of the low angle boundary model depend somewhat on definition. One upper limit on rotation angle is that angle above which the dislocation cores are no longer separated by any atoms in a crystal-like coordination shell. Alternatively, that angle below which the dislocation separation is such that the dislocations can respond individually to an external stress might be selected as an operational definition. Irrespective of the problem of definition, there is a large portion of 'misorientation' space in which

Frank's formula has only a formal meaning. This regime is commonly assigned to high angle GBs. However, this definition contains no information other than that the boundary is not comprised of individual lattice or partial dislocations separated by 'good' crystal. Extensive investigations by transmission electron microscopy (TEM) show that dislocations are common features of high angle GBs (Fig. 6.1), although the dislocations observed are frequently insufficient to accommodate the total grain misorientation [12–14]. These observations can be explained by postulating that there exist subsidiary minima in the energy-misorientation function similar to but shallower than that associated with the zero misorientation, i.e. single crystal configuration. If two grains are constrained to a misorientation which deviates from this local minimum the observation of dislocation-like features is interpreted as implying that high angle boundaries close in misorientation to local energy minima are structurally similar to low angle boundaries [15]. Explicitly this means that the structure of such high angle boundaries consists of perfect GB dislocations which bound patches of 'good' GB. The term 'good' has the same connotations as when it is used in the definition of the Burger's circuit for a crystal lattice dislocation. All the local minima in GB energy (for cubic materials) correspond to misorientations where a coincidence site lattice (CSL) exists. To each CSL there is a corresponding DSC lattice, the translation vectors of which are the possible Burger's vectors for perfect GB dislocations. Such perfect GB dislocations have been characterized systematically in a wide range of $\langle 100 \rangle$ twist boundaries for gold and magnesium oxide [12, 16]. Similar observations have been made in more general boundaries and in specimens prepared from conventionally processed bulk materials [13, 14]. There is thus good reason to propose the generality of the CSL-dislocation model for a subset of the boundaries possible irrespective of the (cubic) material involved. For other crystal systems some development of the concept of the misfit dislocation in generalized epitaxial interfaces may be appropriate. However, in the present context the key result is the verification of the crystallographically-derived result

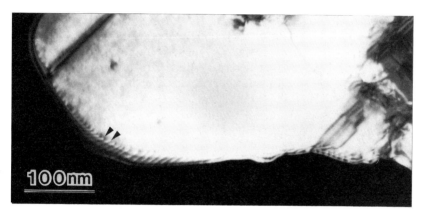

Fig. 6.1 A dark field transmission
electron micrograph showing some
but not all of the geometrically
necessary dislocations in a migrating
boundary in gold. The rotation axis is
near [001] and the rotation angle is
measured as 31° from selected area
diffraction patterns. The image is
dominated by those dislocations with
large Burger's vectors even though
the remaining dislocations should not
be invisible according to the GB
criterion.

that appropriately oriented DSC dislocations have
steps at their cores [15]. The DSC dislocation
belongs to the set of transformation dislocations
and shares crystallographic features with the transformation dislocation shown in Fig. 6.2 which
represents a Shockley dislocation in the hexagonal–
cubic interface for a material such as cobalt. Thus
when DSC dislocations move, GB migration also
occurs. It is established that GB dislocations can
move and multiply and that they retain compact
cores at elevated temperatures [17, 18]. A familiar
example of the association between GB dislocations
and steps is the $\frac{1}{6}\langle 112 \rangle$ twinning dislocation in fcc
materials. Such a dislocation moves by glide in
$\{111\}$ planes and the twin plane advances by one
$\{111\}$ planar spacing with the passage of each
dislocation. Suppose that the interface plane is
$\{1mn\}$, then the dislocation cannot glide but a
combined glide and climb motion is possible and
will result in migration of the twin boundary by
the component of the step height in the direction
normal to $\{1mn\}$. Although this crystallographic
result has not been verified for the general case, it
has been shown to hold for $\{111\}$ twins in gold [11]
and gallium arsenide and the related case of a
cubic–hexagonal interphase boundary [19]. Figure
6.3 is a high resolution electron micrograph which
illustrates the dislocation and step character of a
$\frac{1}{6}\langle 112 \rangle$ twinning dislocation in GaAs. Such dual
nature is a property of DSC dislocations in general
according to theory and has also been observed for
a $\frac{1}{17}\langle 410 \rangle$ dislocation in a $\Sigma = 17$ tilt boundary in
gold [20]. In this case the Burger's vector of the

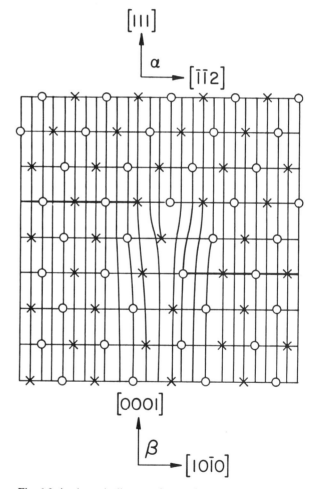

Fig. 6.2 A schematic diagram of a transformation dislocation in
the interface between an fcc and an hcp material. The different
symbols identify the two layers of the (110) and (1012) stacking
sequences.

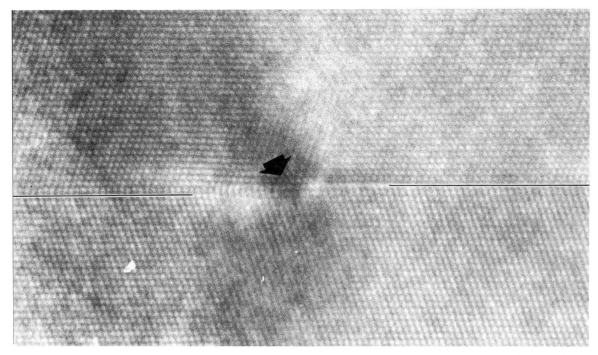

Fig. 6.3 A twinning dislocation with associated step in a {111} 1st order twin in GaAs. The lines indicate the change of level resulting from the twinning dislocation which is arrowed.

dislocation and the height of the step at the dislocation core have a well established relationship. In the more general cases, although the core step height is constrained by the boundary crystallography, there are multiple values allowed for the step height.

6.3 STEPS AND DISLOCATIONS

The nature and interrelationship of steps and dislocations are illustrated in Fig. 6.4 for the case of an epitaxial interphase boundary. The transformation of a strain free surface **step** into an edge **dislocation** in the [21] α-phase is shown schematically. Figure 6.4(a) shows a step on the surface of the α-phase and Fig. 6.4(b) its subsequent propagation into the α-phase followed by the motion of the dislocation across the α–β-phase boundary by a reaction:

$$b_\alpha \rightarrow b_\beta + \Delta b \qquad (6.1)$$

Figure 6.4(d) shows the motion of the dislocation

in the β-phase followed by the formation of a slip step on the surface of the β-phase, (Fig. 6.4(e)). The α–β interface becomes stepped as a result of these processes (Figs. 6.4(d) and (e)), and the step has an associated elastic field which gives it dislocation character (Fig. 6.4(e)). An interface structure identical with that in Fig. 6.4(e) results when the slabs of α- and β-phases shown in Fig. 6.4(f) are joined and relaxed. The terms dislocation and step may thus be used to convey information about the **elastic** and **topographic** properties of a defect, respectively. Linear features, in interphase boundaries particularly, may exhibit dual character. In practice interfacial line defects show a spectrum of characteristics from a purely topographic feature such as a surface step to a purely topological feature such as a screw dislocation in a twist boundary. The image matching and extinction techniques [22] which are used to determine Burger's vectors by electron microscopy are most readily applicable to the problem of identifying a Burger's vector which belongs to known set, usually with low Miller

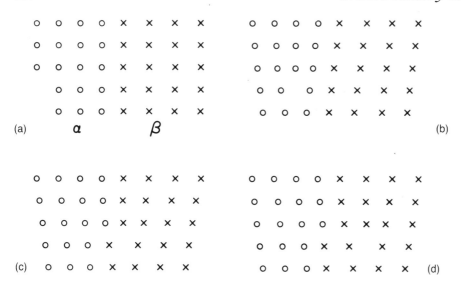

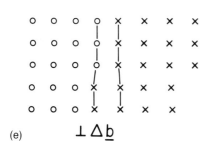

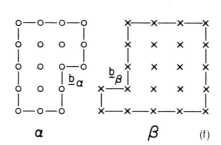

Fig. 6.4 shows the transformation of a strain free surface **step** into an edge **dislocation** in the α-phase, subsequent propagation of the dislocation across the α–β phase boundary and the formation of a slip step on the surface of the β-phase. The α–β interface becomes stepped as a result of these processes and the step has an associated elastic field which gives it dislocation character.

indices; in addition it is the **direction** not the **magnitude** of the vector which is determined. For grain and interphase boundaries the *a priori* limitations on direction and magnitude of the Burger's vector are not well defined. Consequently it is no surprise that there are very few complete determinations of the Burger's vectors of interfacial dislocations. Usually it is impossible to set up two extinction conditions; Burger's vector determination then depends on other arguments such as correlation with the interface crystallography or dislocation interactions involving an identifiable lattice dislocation [14]. The option of constructing a Burger's circuit for a high resolution image of a dislocation is of limited value because (a) there is no information in any particular image concerning displacements in the beam direction and (b) lattice

images can only be obtained under the restrictive conditions that the beam direction has low Miller indices in both phases and is the zone axis of the interface plane or planes.

The distinction of the topographic and topological features of a defect by observation in the electron microscope remains a difficult task. The effect of an abrupt change in the thickness of the diffracting crystal is to cause a discrete offset of the thickness fringes [23]. Although the geometry of this effect is straightforward it must be remembered that interfacial dislocations also give rise to perturbations to thickness fringes by changing the value of the deviation parameters locally. The step character of grain boundary dislocations can sometimes be elucidated directly from lattice fringe images [24].

6.4 GRAIN BOUNDARY MIGRATION

Movement of GBs is a thermally activated process which involves the transfer of an atom or possibly a group of atoms from one grain to a neighbor across a GB. The kinetics of GB migration can be analyzed in terms of absolute rate theory [25]. On a one dimensional argument the velocity, V, of the interface is given by the expression:

$$V = rA\nu \left(\frac{\Delta\mu}{kT}\right) \exp\left(-\frac{\Delta g}{kT}\right) \qquad (6.2)$$

where r is the distance moved per activated event, ν is the Debye frequency, A is a geometric factor, $\Delta\mu$ the chemical potential change driving the process, Δg the free energy of activation and kT has its usual meaning; r, A, ν, and Δg are all in principle structure dependent. The above expression is in fact valid for the movement of any interface providing the kinetics of attachment are rate limiting. The rate theory approach can readily be extended to a two-dimensional situation in which it is supposed that migration occurs by the attachment of individual atoms or groups of atoms to interfacial defects. Then an equation similar to eq. (6.2) describes the thermally activated motion of an individual defect and the motion of the interface through the collective motion of a density ρ of defects each moving at a velocity v and having a step height h results in an interface velocity:

$$V = h\rho v \qquad (6.3)$$

This approach is not satisfactorily explicit about the nature of the sites to which atoms are attached during the migration process. The closely related theory of diffusional phase transformations is framed in terms of attachment of atoms to steps and ledges [26]. However, it is becoming clear that steps and dislocations are not separate entities but two aspects of the character of a rather general kind of defect in partially coherent interfaces as discussed in the previous section.

A considerable body of experiments provides a reference against which a detailed theory of GB migration can be assessed. Even though the observations represent the outcome of extensive series of experiments it must be noted that in reality only a very small subset of materials and the possible boundaries have been studied. The key observations are enumerated below.

1. In lead, the $\Sigma = 5$, 13 and 17 tilt [001] boundaries migrate with activation energies which are local minima [27].
2. The mobility of random boundaries is sensitive to impurities whilst that of special (CSL) boundaries is not [28]. These observations refer mainly to the lead–tin system. It is interesting to observe that in silicon, n-type dopants **enhance** GB mobility [29].
3. The mobility of $\langle 111 \rangle$ boundaries exhibits a maximum at a misorientation angle of 40 **degrees** [30]. This observation refers to aluminum.
4. The mobility of the $\Sigma = 7$ tilt boundary is greater than that of the $\Sigma = 7$ twist boundary [31]. This observation refers to copper and indicates the important orientation of the GB plane in addition to the crystal misorientation.
5. The mobility of low angle boundaries is less than that of high angle boundaries [30]. This is a rather general observation but the cited reference concerns aluminum.
6. A vacancy flux or supersaturation enhances GB mobility [32, 33]. The vacancy supersaturations were the result of prior neutron irradiation and simultaneous ion bombardment respectively.
7. A twin boundary migrates at a low velocity during recrystallization and at a high velocity in response to a shear stress [30, 34]. In both instances there is a large effect of anisotropic mobility which results in a characteristic lenticular or prismatic morphology respectively as a result of the slow motion of the 'coherent' interfaces which in fcc materials are of the $\{111\}$ form.

Some insight to the mechanisms underlying the phenomenology of GB migration detailed above may be gained by interpreting the rate eqs. (6.2) and (6.3) in terms of GB dislocation processes. The special boundaries which migrate with low activation energies are some of those for which glide motion of GB dislocations provides a plausible mechanism for GB migration. Thermal activation is likely to be necessary to overcome a Peierls barrier to glide (note that the extended core structure of a DSC dislocation with a step at its core implies a substantial Peierls barrier) and to allow climb around steps in a largely planar boundary. The sensitivity of GB mobility to solutes is an expected, but not unique, corollary of a dislocation mechanism for GB migration. The origins of the differences in sensitivity of special and random boundaries are not immediately obvious but could lie in the dominance of glide in the former case in contrast to climb in the latter. The high mobility of the boundaries created by a rotation of 40° about $\langle 111 \rangle$ remains obscure but the higher mobility of tilt as against twist boundaries may be correlated with the possibility of migration by dislocation glide only in the former case. An intriguing possibility is that at this misorientation dislocations lose their individual identities and the boundary becomes a high mobility slab! The motion of most low angle boundaries requires some transport, by lattice diffusion, between the constituent dislocations. This does not occur for migration of high angle GBs. The role of excess point defects in facilitating GB motion may be interpreted in terms of the well known mediation of lattice self diffusion by vacancies, in the context of low angle GBs, and an analogous effect, but on GB diffusion, in high angle boundaries.

6.5 OBSERVATIONS BY TRANSMISSION ELECTRON MICROSCOPY

The experimental data already summarized necessarily lack an explicit and validated microstructural interpretation. The behavior of steps is the basis of the propagation of interphase boundaries in diffusion controlled phase transformations, whilst for deformation twinning dislocation mechanisms have been proposed. Developments in the theory and observation of GB structure provided compelling grounds for suspecting that GB dislocations would provide the mechanistic link between the known crystallographic sensitivity of the GB migration process and the boundary structure. Indeed, some of the discussion thus far has been framed in terms of a dislocation mechanism

for GB migration. However, this putative connexion, as will now be discussed, seems not to provide a model which is entirely consistent with the observations.

The fundamental challenge to the experimentalist in the context of the study of interface migration is to deduce an atomistic mechanism from observations which do not have true single atom sensitivity and, in addition, to justify the extension of this mechanism to systems which, at least by present techniques, are inaccessible to observation.

It is well known that a symmetrical, low angle tilt boundary comprised of a single set of dislocations can migrate in a plane perpendicular to itself by glide of its constituent dislocations [35]. Movement of more general boundaries requires diffusion. It is possible to cause low angle boundaries to move *in situ* by heating. These boundaries can move in response to a variety of sources of chemical potential which it must however be noted are difficult to quantify satisfactorily for a particular experiment. In a thin foil heated in a transmission electron microscope, image forces, temperature gradients and surface grooving in addition to the usual driving forces need to be considered. Providing the objective is to elucidate a mechanism rather than to make a study of kinetics, however, the magnitude of the driving force is a secondary consideration. Qualitatively, it is clear from *in situ* studies that motion of a low angle GB occurs by the consecutive motion of the constituent dislocations; this is illustrated in Fig. 6.5, which is a series of micrographs showing the motion of a low angle GB at 600°C in gold. It can be seen that the boundary does not simply translate but changes its conformation. This process involves the movement of individual lattice dislocations which are separated by good crystal. The pure translational motion of a GB leaves both the Burger's vector content and the Burger's vector distribution invariant. Consequently there is zero net flux to the boundary. However, the individual dislocations move by a mixture of glide and climb so that local diffusional fluxes may flow between dislocations in the boundary; note that this is lattice diffusion in the case of a low angle boundary. This process has been

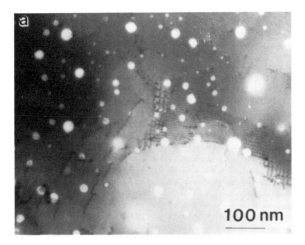

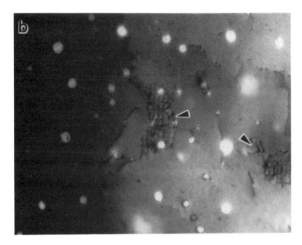

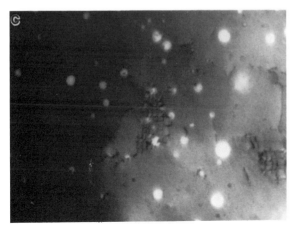

observed and interpreted similarly by Lanxner and Bauer [36]. Some long range diffusion seems necessary when a boundary changes from tilt to twist character; this is to accommodate the excess volume associated with the dilatational component of the displacement fields of edge dislocations.

The extension of the description developed above to high angle boundaries appears trivial in theory but in practice is confronted by numerous difficulties. The case of the $\Sigma = 3$ twin is considered first. The paradox of this example is that deformation twinning is known as a very rapid process and yet the {111} annealing twin, which of course has the same lattice rotation and consequently the same underlying structure, is a boundary of exceptionally low mobility. The resolution of this paradox begins with further discussion of Fig. 6.3. It is clear that the hard sphere model of the {111} twin boundary structure is substantially correct. The atomic environment at this interface can be deduced to be virtually unchanged relative to that of the crystal interior at the first nearest neighbor level. This defect free interface is sessile for the same fundamental reason that close packed surfaces are slow growing. In order to grow it is necessary to overcome the nucleation barrier associated with the formation of a step. Similarly, the motion of a twin requires the formation of defects which are not part of the equilibrium structure; Shockley dislocations, of which one is arrowed in Fig. 6.3, have a high mobility and thus, once they are formed, a twin can propagate rapidly. In the context of plastic deformation the applied stress, especially in the presence of a stress raiser, can readily overcome the nucleation barrier. However, defect nucleation during thermally activated motion in response to the modest driving forces found in recrystallization must rely on heterogeneous effects such as thread-

Fig. 6.5 A series of transmission electron micrographs showing the motion of low angle boundaries by the loosely coupled movement of individual lattice dislocations; the observations were made *in situ* at 600° in a gold specimen. The twist boundary configuration seen in (a) rotates into a partial tilt configuration in (b) and (c).

ing dislocations [37] and, consequently, if the supply is limited annealing twins adopt a characteristic lamellar habit bounded by low mobility {111} surfaces. This is characteristic of the (multiple) twins which are formed during conventional recrystallization and diffusion induced recrystallization. An example of the latter behavior for the case of copper diffusing into gold is illustrated in Fig. 6.6; the material in the field of view is a single phase dilute solid solution of copper in gold. It is in fact commonplace for the defects which are necessary to mediate the motion of an interface not to be part of the (local) equilibrium structure of the interface [38]. The first order twin is the most familiar

example and one of the clearest since the {111} interface may be regarded as dislocation free in its local minimum energy state; however, GB migration or the growth of a second phase involves the formation and movement of defects which may be dislocations and which are not part of this structure. The overall geometrically necessary Burger's vector is necessarily perpendicular to the rotation axis in GBs but there is no such restriction on the Burger's vectors of the dislocations which might be involved in migration. Even for the case where the necessary dislocations are part of the locally equilibrated structure the dislocation network resists the independent motion of individual

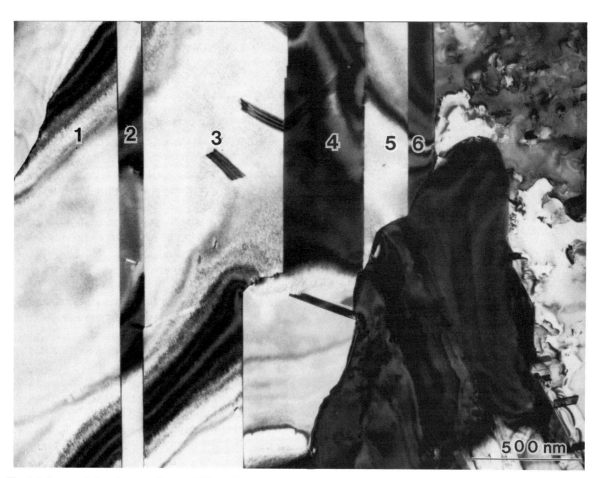

Fig. 6.6 A transmission electron micrograph illustrating the morphology of the multiple twins formed during diffusion induced recrystallization in gold. The direction of motion of the recrystallization interface is from right to left and the numerals designate successive twinning operations.

constituent dislocations or the generation of further dislocations so that there is a threshold which must be exceeded before a dislocation mediated process can occur in a GB [39].

From a crystallographic point of view the description of twin boundary migration seems satisfactory. In principle eq. (6.3) is applicable to the description of the growth kinetics of deformation and annealing twins but thus far has not been used successfully. Indeed, the known values of V imply extremely large, possibly implausibly large, values of v. A different problem arises in the case of annealing twins and other GBs; it is that there has never been an observation of the number of defects expected in order to balance eq. (6.3). There is strong evidence from the observation of dislocations in GBs at elevated temperatures including within a degree of the melting point for aluminum [40] and relief left behind a moving boundary [18] that GB dislocations do retain their character in the context of dynamic processes at elevated temperatures. Babcock and Balluffi [41, 42] attempted to make a direct correlation between the density and velocity of moving defects in well characterized GBs in gold but found two distinct behaviors. The first involved an observation of migration by a dislocation mechanism reminiscent of twinning, i.e. consistent with eq. (6.3) and the foregoing discussion. The several observations of coupled GB sliding and migration in the context of creep deformation are likely also to be dislocation mediated [43, 44]. However, there also appears to be another migration mechanism which is not related to the motion of observed line defects in the GB. Related computer calculations are equally ambiguous; Monte Carlo calculations in tilt boundaries are suggestive of the dislocation model [45] whilst similar calculations for twist boundaries are more compatible with a thermally activated cooperative rotation process leading to GB migration [46]. At present it remains unclear when and if the motion of GB dislocations is a mechanism or merely a byproduct of GB migration.

It is clear how solute drag and the enhancement of GB mobility by fluxes of point defects could both be the GB dislocation analogs of well known behavior of crystal dislocations [1].

6.6 GRAIN STRUCTURE OF FILMS

The discussion so far has concentrated on the behavior of individual boundaries. However, the main practical relevance of the study of GB migration concerns the processing of polycrystalline materials with the intention of engineering desirable properties through control of the microstructure. The classical example of this approach is the minimization of hysteresis losses in Fe-3%Si transformer laminations through texture control. In modern technology the control of the properties of materials in thin film form is critical. The challenge is to develop materials which simultaneously optimize resistivity, creep resistance, and electro-migration lifetime. The dominant microstructural variable is the grain structure which can be controlled either through choice of the substrate temperature or by post-deposition annealing. In both situations the key process is GB migration. The microstructure of a deposited film represents the outcome of a number of processes: adsorption, surface diffusion, nucleation clustering, coarsening, and grain growth, with the latter dominating the final microstructure. In addition, the thin film context offers a number of significant insights to the GB migration process together with an opportunity to compare modeling with experiment.

The general trends relating microstructure and substrate temperature were recognized for evaporated films by Movchan and Demchishin [47] and embodied in their zone model; an updated version is shown as Fig. 6.7 [48]. The diagram indicates the nature of the grain structure of an evaporated film as a function of the substrate temperature expressed in terms of the homologous temperature of the deposit. The diagram describes the behavior of all pure metals but must be modified for concentrated alloys and non-metals. There are essentially four different microstructures observed and each is associated with a particular range, or zone of homologous temperatures. These zones are designated I, T, II and III in ascending order of homologous temperature. Thornton [49] made a similar correlation for sputtered films and included the pressure of the sputter gas as a second independent variable.

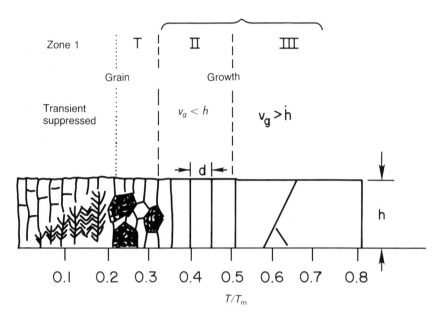

Fig. 6.7 A diagram indicating the nature of the grain structure of an evaporated film as a function of the substrate temperature expressed in terms of the homologous temperature of the deposit. There are essentially four different microstructures observed and each is associated with a particular range, or zone of homologous temperatures. These zones are designated I, T, II, and III in ascending order of homologous temperature.

Zone I,   random columns of subgrains, $d \sim 20$ nm;

zone I,   dendrites;

zone T,   bimodal;

zone 2 $h/d > 1$;

zone 3 $h/d < 1$

The microstructural characteristics of the four zones are now described. It is generally observed that pure metallic films deposited from the vapor onto substrates maintained at or below $0.2\,T_m$ are polycrystalline and consist of randomly oriented grains which are typically 5–20 nm in diameter. Normal GB migration is too slow to account for the zone I grain size and too fast to be consistent with the zone III grain size. The zone I kinetics may be enhanced by excess point defects or perhaps, more properly, be attributed to a crystallization rather than a grain growth process. The zone III kinetics are likely to be retarded by GB grooving. The bimodal grain structure characteristic of zone T is generally undesirable because of its lack of homogeneity but is important because it provided

very direct evidence of the importance of grain growth *during* film deposition [50]. The overall temperature dependence of microstructure for the case of substrates which do not act as templates can be described as follows. Deposited films are polycrystalline and consist of grains which have an aspect ratio h/d which decreases with increasing substrate temperature. The aspect ratio of the grains depends on the ratio of the deposition rate and the rate of grain growth. In this sense zones II and III are not physically distinct. The perfection of the internal structure of these grains increases as the substrate temperature increases. In zone I the dendritic incompletely densified structure is associated with the incorporation of impurities and the decrease in surface diffusion. The two key processes involved are thus the nucleation of successive layers by surface diffusion and grain growth during the deposition process. The role of surface energy in biasing the growth of certain grains to give the growth textures which are characteristic of zones II and III is illustrated in Fig. 6.8. Figures 6.8(a), (b) and (c) are bright field transmission electron micrographs of gold deposited at 294 K and then heated to 873 K, followed by gold deposited at 473 K. The substrate is amorphous silicon nitride and the film is capped with a thin SiO$_2$ layer to minimize pinning of GBs by surface grooving during post deposition anneals [51]. The insets in each figure are the corresponding selected area diffraction patterns. Figures 6.8(d), (e) and (f) are histograms representing the distribution of grain sizes corresponding to the structures shown in Figs. 6.8(a), (b) and (c), respectively, of the large grains. The grain size distribution is clearly bimodal. The anomalously intense 220 diffraction rings and the convergent beam diffraction experiments are evidence for a {111} fiber texture. Note the characteristic bimodal grain size distribution. The {111} fiber texture becomes more pronounced as a result of post-deposition heating. Evidently surface energy anisotropy affects the grain structure both during deposition and in subsequent annealing.

6.7 DISCUSSION

The preceding exposition of the phenomenology of GB migration and the inconclusive outcome of attempts to understand its atomistic basis point to the innate complexity of the problem together with some limitations of the theoretical and experimental techniques which have so far been applied. The database is limited to a small number of materials and to a small proportion of the possible interfaces. The difficulty of establishing a general theory of the structure for GBs is a consequence of the intrinsically large number of possible configurations for a defect with five macroscopic degrees of freedom, the diversity of crystal structures exhibited by polycrystalline materials, the subtleties of interatomic bonding which are likely to cause

variations of structure even amongst materials with the same structure, and the limitations of electron microscopy which nevertheless seems to offer the best prospect of elucidating the mechanism of migration. Conventional high resolution electron microscopy can only be applied to determining the structure of interfaces which have a simple stacking (such as ABAB.. or ABCABC) in the beam direction and are viewed edge on. This effectively limits the technique to tilt boundaries. Structure images have been obtained by *in situ* high resolution microscopy of moving GBs [52]. Computer modelling, although initially applied to the same subset of GBs [53, 54], has the potential to explore the static and dynamic behavior of a broader range of grain boundaries [55–57]. The cases where the electron microscope data are consistent with computer calculations [58, 59] lend credibility to both techniques, and give a certain confidence in extension of the use of atomistic modeling to those boundaries which are inaccessible to electron microscopy.

The modeling of the evolution of microstructure, rather than the atomic configuration of particular defects has emerged as a powerful technique which permits the exploration of such questions as the relationship between the initial and final grain size [60, 61] distribution, the necessity for grains with sizes or mobility which diverge from some average distribution for secondary grain growth to occur [62, 63] and the relative effectiveness of different morphologies of second phases in the pinning of GBs [64]. This sub-field is one where there is a clear potential to compare quantitatively the results of simulations and *in situ* electron microscopy experiments.

The early and systematic investigations of Aust and Rutter [27, 28] established unequivocally that solute atoms exert a profound effect on GB migration which furthermore varied from boundary to boundary. However, very little has been done experimentally to establish how solutes interact with GBs. The potential implications are substantial since there are indications that solute segregation changes the kinetics of GB processes, usually slowing them down, and can change the structure of the GB core [65]. In addition, it has been shown how coherency stresses resulting from alloying by

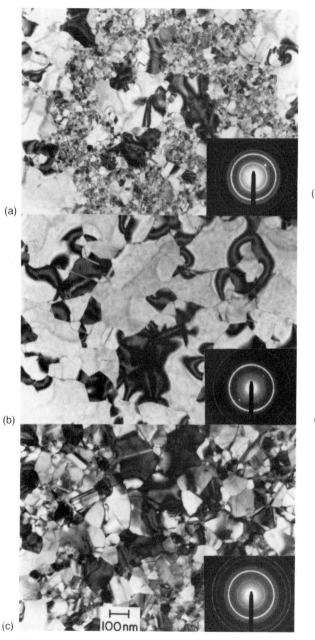

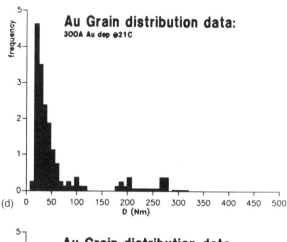

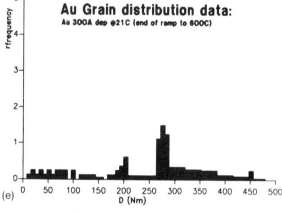

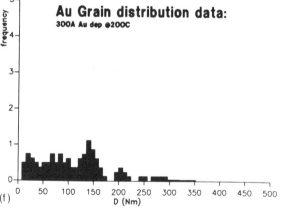

Fig. 6.8 (a), (b) and (c) are bright field transmission electron micrographs of gold deposited at 294 K and then heated to 873 K, followed by gold deposited at 473 K. The substrate is amorphous silicon nitride. The insets in each figure are the corresponding selected area diffraction patterns. Figs. 6.8 (d), (e) and (f) are histograms representing the distribution of grain sizes corresponding to the structures shown in Figs. 6.8 (a), (b) and (c) respectively of the large grains. The grain size distribution is clearly bimodal. The anomalously intense 220 diffraction rings are evidence for a {111} fiber texture.

GB diffusion can stimulate GB migration [66, 67].

It is evident that the task of correlating the structure and migration behavior of individual interfaces is fraught with difficulty; the difficulty is generic but some limited success has been achieved in other contexts. Perhaps the most satisfying examples are the misorientation dependencies of the anisotropy of GB diffusion [68, 69] and the critical current density in the ceramic high temperature superconductors [70–72]. However, for polycrystalline materials the problem is to combine the divergent behavior of many boundaries and their interactions to account for the properties of a polycrystalline body. The classical and remarkably successful approach is to connect the properties of the polycrystal to an average GB property and a single length scale, the average grain size as has been done for creep rate [73, 74], yield stress [75] and electrical resistivity [76]. This approach cannot possibly include variations in properties from boundary to boundary; such variations are significant in the context of texture formation, for example. Further developments in understanding hinge on progress in experimental characterization and novel theoretical treatments. In the context of migration the excess volume at GBs may be the single most significant microstructural variable and it is accessible for a wide class of interfaces through the method of Fresnel imaging [77]. On the interpretive side, the concepts of percolation hold promise as a means to include some differentiation of boundary properties in a model for the behavior of a polycrystalline ensemble [78–80].

6.8 CONCLUSION

Grain boundary migration is an **interface controlled process** which in special cases can occur by shear but in general requires diffusion-like thermally activated processes. Movement of GB dislocations provides a plausible mechanism for migration of low angle boundaries and tilt related boundaries together with an insight to the dependence of the migration process on GB crystallography, solutes and excess point defects. However, this mechanism does not account for all the observations. For many, probably most, boundaries the migration process seems to involve thermally activated transport across the core and the kinetics integrated over time are likely to depend more on the local excess volume than on the atomistic details of the structure.

Further elucidation of the proposed role for GB dislocations and other mechanisms requires the development of techniques for characterizing arbitrary boundaries and the refinement of methods for *in situ* observation of dynamic processes at high resolution including computer modeling.

Grain boundary migration is the major process affecting the microstructure of thin films. However, the bias of driving force for the grain growth process resulting from anisotropy of the surface energy rather than the crystallographic variation of grain boundary properties accounts for the textures observed in as-deposited or post-deposition annealed films.

REFERENCES

1. D. A. Smith, C. M. F. Rae and C. R. M. Grovenor, *Grain Boundary Structure and Kinetics* (ed R. W. Balluffi), ASM, Metals Park, Ohio (1980), p. 337.
2. M. L. Kronberg and F. H. Wilson, *Trans. AIME*, **185** (1949), 50.
3. P-J. Wilbrandt and P. Haasen, *Z. Met.*, **71** (1980), 273.
4. P-J. Wilbrandt and P. Haasen, *Z. Met.*, **71** (1980), 385.
5. C. R. M. Grovenor, D. A. Smith and M. J. Goringe, *Thin Solid Films*, **74** (1980), 257; ibid. 269.
6. I-Wei Chen, *Acta Met.*, **36** (1987), 1723.
7. K. Lucke and K. Detert, *Acta Metall.*, **5** (1957), 628.
8. W. T. Read and W. Shockley, *Phys. Rev.*, **78** (1950), 275.
9. S. Amelinckx and W. Dekeyser, *Sol. Stat. Phys.*, **8** (1959), 325.
10. W. T. Read, *Dislocations in Crystals*, McGraw-Hill, New York (1953).
11. W. Krakow and D. A. Smith, *Ultramicros.*, **22** (1987), 47.
12. R. W. Balluffi, Y. Komem and T. Schober, *Surf. Sci.*, **32** (1972), 68.
13. W. A. T. Clark and D. A. Smith, *Phil. Mag.*, **38** (1978), 367.
14. W. Bollmann, B. Michaut and G. Sainfort, *Phys. Stat. Sol(a)*, **13** (1972), 637.
15. W. Bollmann, *Crystal Defects and Crystalline Interfaces*, Springer, Berlin (1970).

16. C. P. Sun and R. W. Balluffi, *Phil. Mag. A*, **46** (1982), 49.
17. D. J. Dingley and R. C. Pond, *Acta Metall.*, **28** (1979), 667.
18. C. M. F. Rae, *Phil. Mag. A*, **44** (1981), 1395.
19. J. M. Howe, H. I. Aaronson and R. Gronsky, *Acta Metall.*, **33** (1985), 649.
20. W. Krakow and D. A. Smith, *Mat. Res. Soc. Symp. Proc.*, **57** (1987), 357.
21. D. A. Smith, *Mat. Res. Soc. Symp. Proc.*, **31** (1984), 223.
22. P. Humble and C. T. Forwood, *Phil. Mag.*, **31** (1975), 1011.
23. P. B. Hirsch, A. Howie, R. B. Nicholson, D. W. Pashley and M. J. Whelan, *Electron Microscopy of Thin Crystals*, Butterworth, London (1965).
24. P. J. Goodhew and D. A. Smith, *Phil. Mag. A*, **46** (1982), 161.
25. D. Turnbull, *Trans. Met. Soc. AIME*, **191** (1951), 661.
26. H. I. Aaronson, C. Laird and K. R. Kinsman, ch. 8 in *Phase Transformations*, ASM Metals Park, OH (1970).
27. J. W. Rutter and K. T. Aust, *Acta Metall.*, **13** (1965), 181.
28. K. T. Aust and J. W. Rutter, *Ultra High Purity Metals*, ASM Metals Park, OH (1962), p. 115.
29. Y. Wada and S. Nishimatsu, *J. Electrochem. Soc.*, **125** (1978), 1499.
30. B. Liebmann, K. Lucke and G. Masing, *Z. Met.*, **47** (1956), 57.
31. G. Gottstein, H. C. Murmann, G. Renner, C. J. Simpson and K. Lucke, *Int. Conf. Textures, Aachen* (1978), p. 521.
32. W. In der Schmitten, P. Haasen and F. Haessner, *Z. Met.*, **51** (1960), 101.
33. Atwater, H. A., Thompson, C. V. and Smith, H. I. *J. Appl. Phys.*, **64** (1988), 2337.
34. W. P. Mason, H. J. McScimin and W. Shockley, *Phys. Rev.*, **73** (1948), 1213.
35. J. Washburn and E. R. Parker, *Trans. AIME*, **194** (1952), 1076.
36. M. Lanxner and C. L. Bauer, *Suppl. Trans. Jap. Inst. Met.*, **27** (1986), 411.
37. J. W. Christian, *Theory of Transformations in Metals and Alloys*, Oxford, Pergamon Press (1965).
38. D. A. Smith, in *Dislocation Modelling of Physical Systems* (eds M. F. Ashby, R. Bullough, C. S. Hartley and J. P. Hirth), Pergamon, Oxford (1981), p. 569.
39. A. H. King and D. A. Smith, *Metal Sci.*, **14** (1980), 57.
40. T. E. Hsieh and R. W. Balluffi, *Acta Metall.*, **37** (1989), 1637.
41. S. E. Babcock and R. W. Balluffi, *Acta Metall.*, **37** (1989), 2357.
42. S. E. Babcock and R. W. Balluffi, *Acta Metall.*, **37** (1989), 2367.
43. S. Hashimoto and B. Baudelet, *Scripta Metall.*, **23** (1989), 1855.
44. H. Ando, J. Sugita, S. Onaka and S. Miura, *J. Mat. Sci. Letts.*, **9** (1990), 314.
45. G. H. Bishop, R. J. Harrison, T. Kwok and S. Yip, *Grain Boundary Structure and Kinetics*, (ed R. W. Balluffi), ASM, Metals Park, OH (1980), p. 373.
46. P. D. Bristowe and R. J. Jhan, *Scripta Metall. and Mater.*, **24** (1990), 1313.
47. B. A. Movchan and A. V. Demchishin, *Fiz. Met. Metalloved.*, **28** (1969), 83.
48. D. A. Smith, *Proceedings of EMAG-MICRO 89* (ed P. J. Goodhew), Bristol, Institute of Physics.
49. J. A. Thornton, *J. Vac. Sci. and Tech.*, **12** (1975), 830–835.
50. C. R. M. Grovenor, H. T. G. Hentzell and D. A. Smith, *Acta Metallurgica*, **32** (1984), 773–781.
51. W. W. Mullins, *Acta Metallurgica*, **6** (1958), 414–427.
52. H. Ichinose and Y. Ishida, *Phil. Mag. A*, **60** (1989), 555.
53. D. A. Smith, V. Vitek and R. C. Pond, *Acta Metall.*, **25** (1977), 475.
54. A. P. Sutton and V. Vitek, *Phil. Trans. Roy. Soc., Lond.*, *A*, **309** (1983), 1.
55. I. Majid, P. D. Bristowe and R. W. Balluffi, *Phys. Rev. B*, **40** (1989), 2779.
56. D. Wolf, *Scripta Metall.*, **23** (1989), 377.
57. D. Wolf, *J. Mater. Res.*, **5** (1990), 1708.
58. R. C. Pond and V. Vitek, *Proc. Roy. Soc. Lond. A*, **357** (1977), 453.
59. W. Krakow, J. T. Wetzel and D. A. Smith, *Phil. Mag. A*, **53** (1986), 739; H. J. Frost, C. V. Thompson, C. L. Howe and J. Whang, *Scripta Metall.*, **22** (1988), 65.
60. D. J. Srolovitz, G. S. Grest and M. P. Anderson, *Acta Metall.*, **33** (1986), 2233.
61. G. S. Grest, M. P. Anderson, D. J. Srolovitz and A. D. Rollett, *Scripta Metall. and Mater.* **24** (1990), 661.
62. H. J. Frost, C. V. Thompson and D. T. Walton, *Acta Metall. and Mater.*, **38** (1990), 1455.
63. A. D. Rollett, D. J. Srolovitz and M. P. Anderson, *Acta Metall.*, **37** (1989), 1227.
64. G. N. Hassold, E. A. Holm and D. J. Srolovitz, *Scripta Metall.*, **24** (1990), 101.
65. K. Sickafus and S. L. Sass, *Acta Metall.*, **35** (1987), 69.
66. D. N. Yoon, *Ann. Rev. Mater. Sci.*, **19** (1989), 43.
67. A. H. King, *Int. Met. Rev.*, **32** (1987), 173.
68. S. Hofmann and P. Lejcek, *J. Phys.*, **51** (1990), C1–179.
69. D. A. Smith and W. Krakow, *Mat. Res. Soc. Symposium*, **82** (1986), 349.
70. D. Dimos, P. Chaudhari and J. Mannhart, *Phys. Rev. B*, **41** (1990), 4038.
71. M. F. Chisholm and D. A. Smith, *Phil. Mag. A*, **59** (1989), 181.
72. Y. Gao, K. L. Merkle, G. Bai, H. L. M. Chang and D. J. Lam, *Physica C*, in press.

73. R. L. Coble, *J. Appl. Phys.*, **34** (1963), 1679.

74. F. R. N. Nabarro, *Conf. Strength Solids*, *Phys. Soc. (London)*, 1948, p. 75.

75. N. J. Petch, *J. Iron Steel Inst. London*, **174** (1953), 25.

76. K. Fuchs, *Proc. Camb. Phil. Soc.*, **34** (1938), 100.

77. W. M. Stobbs and F. M. Ross, *Phil. Mag. A*, **63** (1991), 1; ibid. 37.

78. C. S. Nichols, D. R. Clarke and D. A. Smith, in *Simulation and Theory of Evolving Microstructures and Textures* (eds M. P. Anderson and A. Rollett), TMS, Warrendale, PA (1990).

79. C. S. Nichols, R. F. Cook, D. R. Clarke and D. A. Smith, *Acta Metall.*, in press.

80. C. S. Nichols, R. F. Cook, D. R. Clarke and D. A. Smith, *Acta Metall.*, in press.

81. C. S. Nichols and D. R. Clarke, *Acta Metall.*, in press.

CHAPTER SEVEN

Role of interfaces in melting and solid-state amorphization

S. R. Phillpot, S. Yip, P. R. Okamoto, and D. Wolf

Introduction · Simulation methods · Molecular–dynamics simulation of interface-induced thermodynamic melting · Mechanical melting · The *T-V* phase diagram and its low-temperature extensions · Interpretation of solid-state amorphization experiments · Summary and outlook

7.1 INTRODUCTION

Among structural phase transitions which evolve from an initially crystalline state, melting is the most common and probably the most extensively studied. Another transformation that produces a disordered final state is solid-state amorphization. While the connection between the two final states, a liquid on the one hand and a glass on the other, has long been studied in the context of vitrification, the analogies between the processes of solid-state amorphization and melting themselves are only now being recognized [1–5]. It is the purpose of this chapter to compare the underlying thermodynamic and kinetic features of these two phenomena and to clarify the role played by internal and external interfaces in their atomic-level mechanisms. By focusing on insights derived from recent molecular-dynamics simulations on the mechanisms of melting and applying them to interpret a variety of amorphization data, we are led quite naturally to a broadened view of melting. In this view crystal disordering can take place in two different forms; while melting at high temperature is the manifestation of the conventional crystal-to-

liquid (C → L) transition, the crystal-to-amorphous (C → A) transition of solid-state amorphization may be interpreted as melting at low temperatures.

Despite its obvious fundamental significance, the mechanism by which disorder is initiated during melting is not yet well understood. A variety of models have been developed for the C → L transition which emphasize the role of intrinsic lattice defects [6]; however, they do not address the experimentally known fact that melting generally occurs at extrinsic defects such as free surfaces, grain boundaries (GBs), and voids [7]. From the standpoint of actually observing the mechanism of the C → L transition, molecular-dynamics (MD) simulation [8, 9] offers the most direct means because in addition to having precise control of the interface and the manner in which the sample is heated one can also easily distinguish the onset of disorder. Recent MD simulations have shown that the C → L transition can occur by two distinct mechanisms [10–12]. As we shall see below, in a system with an extrinsic defect, melting is a **heterogeneous** process in which disorder is first initiated at an interface and then propagates through the system [10, 11]. This process is referred to as thermodynamic melting. By contrast, in a system with no extrinsic defects, melting is a **homogeneous** process in which the crystal lattice becomes mechanically unstable to shear deformation [12]. This process is referred to as mechanical melting. While heterogeneous melting takes place at the thermodynamic melting point, T_m, which can be independently determined by equating the free energies of the crystal and liquid, homogeneous melting generally occurs at a higher temperature, T_s, which can be determined from an analysis of the elastic constants of the crystal [13]. (Additionally, entropy arguments may be used to predict crystalline instabilities at temperatures above the thermodynamic melting temperature [4, 14].)

Because conventional melting is usually induced by heating, the attendant high atomic mobility associated with the C $\rightarrow$ L transition causes the heterogeneous process of nucleation and growth of disorder at an interface to be predominant over the homogeneous process. Indeed, this is what is generally observed [7]. If, however, thermodynamic melting at the crystal surface can be suppressed by some means, then the solid can be substantially superheated, as has been observed in the case of crystalline spheres of silver coated with gold [15] and as we will demonstrate in the simulations of mechanical melting.

In the case of the C $\rightarrow$ A transition observed only at low temperatures [1], the heterogeneous process no longer dominates because of the considerably reduced atomic mobility, and one then has the possibility of observing the transformation by the homogeneous mechanism. It turns out that, in comparison with C $\rightarrow$ L transitions, the interpretation of solid-state amorphization data is less clear-cut because issues such as the nature of the driving force and the characterization of the amorphous state now have to be considered. As discussed in this chapter, despite this difficulty, it is possible to incorporate both the C $\rightarrow$ L and C $\rightarrow$ A transitions into a single description by focusing on the role of volume expansion in the disordering process and in the two mechanisms of melting [12].

In terms of an ordinary thermodynamic phase diagram, conventional melting has a clear and simple representation. When the equation of state is projected onto the temperature–volume plane, the melting curve is seen to terminate at the triple point temperature, T_t. Conceptually one may think of an extension of the melting curve below T_t as suggesting the existence of a metastable crystalline phase in which sublimation, a thermally activated process, is kinetically suppressed. Following this point of view one can thereby construct an effective phase diagram which also describes solid-state amorphization in the sense of a combined representation of thermodynamics and kinetics.

The chapter is organized as follows. We begin in section 7.2 with a review of the key features of molecular-dynamics-simulation methodology relevant to the study of interfacial disorder. The simulation results of interface-induced thermodynamic melting in model systems of silicon and copper are discussed in section 7.3, where we also address the issues of 'premelting' at GBs and free surfaces. Simulation results on the mechanical melting in the absence of extrinsic defects are presented in section 7.4, followed in the next section by a comparison of the characteristics of the two melting mechanisms. In section 7.6 we consider an extended phase diagram which incorporates the kinetic effects of crystal disordering and we discuss the parallels between solid-state amorphization and melting. We close with a brief summary and outlook.

7.2 SIMULATION METHODS

Molecular dynamics, the realization of the classical-dynamics description of a model system through the Newtonian equations of motion, is the ideal simulation tool for the study of both the thermodynamics and the kinetics of systems at elevated temperatures (i.e. above the Debye temperature). There are three basic ingredients to an MD simulation [8, 9]. First, an interatomic potential must be specified which, in contrast to an electronic-level description of the atom–atom interactions, is assumed to depend only on the atomic positions $\{r_N\}$ of the N particles in the system. Second, the geometry of the N-particle system must be defined. Finally, a border condition must be invoked, which describes how the N-particle system interacts with its environment. This border condition may, for example, consist of a periodic extension of the simulation cell containing the N-particle system in three, two, or one dimensions, the particular choice being determined by the geometry of the system of interest, and the phenomena being investigated. As discussed below, a new border condition has recently been developed for the study of isolated GBs, i.e. GBs embedded between bulk perfect-crystal material.

7.2.1 Interatomic potentials

Among the many interatomic potential functions which have been developed for atomistic simulations we are particularly concerned with two types: those which describe simple metals such as copper and those which describe covalent semiconductors such as silicon.

There is a long history for the description of metals by central-force potentials, in which only two-body interactions are incorporated. The simplest and best known of all pair potentials is probably the Lennard–Jones (LJ) potential, with only two adjustable parameters, σ and ε, defining the length and energy scales, respectively. Central-force potentials have certain intrinsic limitations, however. For example, all equilibrium central-force potentials automatically satisfy the Cauchy relation $C_{12} = C_{44}$, thus yielding only two independent elastic constants rather than three for a perfect cubic crystal. This shortcoming of central-force potentials has been a major motivation in the development of a broader theoretical framework for the description of interatomic interactions in metals.

The two types of many-body potentials, known as embedded-atom-method (EAM) [16] and Finnis–Sinclair (FS) potentials [17], have the advantage over pair potentials of incorporating, at least conceptually, the many-body nature of metallic bonding, while being relatively efficient computationally. In these potentials the strength of the interaction between atoms depends on the local volume or, in another interpretation, on the local electron density 'sensed' by every atom. Also, these potentials do not automatically satisfy the Cauchy relation for the elastic constants, thereby allowing all three elastic constants of a cubic metal to be determined (or used in the fitting).

Both the EAM and FS potentials use the same mathematical expression as the starting point. The total energy, U, of a system of N interacting metal atoms is subdivided into a bonding term (based on band-structure considerations) and a repulsive central-force term, according to [16, 17]

$$U = -\Sigma_i F(\rho_i) + \Sigma_i \Sigma_{j>i} \phi(r_{ij}) \qquad (7.1)$$

Here $F(\rho_i)$ is the many-body energy gained when embedding atom i in the charge density,

$$\rho_i = \Sigma_j \rho_{ij}(r_{ij}) \qquad (7.2)$$

due to the surrounding atoms, while $\phi(r_{ij})$ is the central-force repulsive energy between atoms i and j, with $r_{ij} = |\mathbf{r}_i - \mathbf{r}_j|$. The EAM and FS potentials differ essentially only with respect to (i) the basic physical justification, or rationalization, of eqs. (7.1) and (7.2), (ii) the fact that FS offer analytical expressions for $F(\rho)$ and $\phi(r)$, in contrast to the numerical tables used in EAM potentials, and (iii) the procedure employed for fitting the basic expressions to empirical bulk-crystal data for a given material. Formally eq. (7.1) contains a pair potential, $\phi(r)$, as the special case in which $F(\rho)$ vanishes identically.

Despite the many-body effects characteristic of metals incorporated in the EAM potentials, in most cases the qualitative features of the structure, elastic, and thermodynamic properties of bulk interfaces, thin films and superlattices of fcc materials have been found to be independent of whether the systems were described by a Lennard–Jones pair potential or an EAM potential [18]. In the following discussion of the high-temperature properties of metallic GBs, we will examine results obtained using both types of potential. The EAM potential used here is parameterized to Cu in a Cu/Ni alloy, while the LJ potential, also parameterized to Cu, is characterized by $\sigma = 2.315 \text{ Å}$ and $\varepsilon = 0.167$ eV. These LJ parameters correspond to $a_0 = 3.616 \text{ Å}$ and a melting temperature close to 1200 K. (In this chapter we denote the zero-temperature lattice parameter a_0, while the lattice parameter at the simulation temperature is denoted a.)

For covalent systems it is necessary to consider the strong directional bonding by including interactions amongst three or more atoms. Stillinger and Weber (SW) [19] have developed a potential containing both two- and three-body components. The two-body part increases strongly at small atomic separations, has a minimum at the nearest neighbor distance, and goes smoothly to zero at a distance d, which is a little less than the zero-temperature ideal-crystal second-nearest-neighbor distance. The three-body part has the form [19]:

$$\Phi_3 \sim \sum_{ijk} \exp(\gamma/(r_{ij} - d)) \exp(\gamma/(r_{ik} - d))$$

$$(\cos \theta_{jik} + \tfrac{1}{3})^2, \qquad (7.3)$$

where r_{ij} and r_{ik} are the lengths of the vectors joining particles i and j, and i and k, respectively, and γ is a constant. θ_{jik} is the angle between these vectors. Φ_3 is zero for ideal tetrahedral angles (cos $\theta_t = -\tfrac{1}{3}$) and positive otherwise.

The six disposable parameters of the SW potential were chosen [19] to make the diamond lattice the lowest-energy crystal structure, to give the correct zero-temperature lattice parameter of $a_0 = 5.43\,\text{Å}$ and to yield both a reasonable value for the silicon melting temperature and reasonable agreement with the liquid radial distribution function observed experimentally. The SW potential has been successfully used to describe silicon in a wide variety of environments, ranging from small clusters [20] and GBs [21], to the amorphous phase [22, 23] and solid–liquid interfaces [24, 25].

7.2.2 Interface geometry

Two different high-angle twist boundaries were chosen for the studies of GB-induced melting in Si and in Cu. For the silicon simulations the GB was the (110) $\theta = 50.48°$ ($\Sigma 11$) high-angle twist boundary. This boundary is obtained by rotating two perfect semicrystals with (110) faces by $\theta = 50.48°$ with respect to each other about the interface-plane normal. The area of the rectangular planar unit cell is $\Sigma = 11$ times that of the corresponding primitive planar unit cell ($\Sigma = 1$) on the (110) plane; each

lattice plane thus contains 22 atoms. For the Cu simulations, the GB chosen was the (001) $\theta = 43.60°$ ($\Sigma 29$) twist boundary, with 29 atoms in the planar unit cell. In each case the computational cell was chosen to contain 32 atomic planes, for a total of 704 atoms for the Si simulation cell and 928 atoms for the Cu simulation cell. In both cases, the GB initially lay at the center of the cell between planes 16 and 17.

The primary considerations underlying the choice of GB geometries were (a) their relatively large planar unit cell (thus representing what we consider 'typical' high-angle GBs), (b) the large spacing of lattice planes parallel to the GB plane ($d(110) = 0.354a$, $d(001) = 0.5a$), allowing behavior of atoms within an individual plane to be isolated from that of the others, and (c) the relative insensitivity of their zero-temperature energies towards translations parallel to the GB plane.

7.2.3 Border conditions for bicrystal simulations

Since GBs are extended objects in two dimensions it is natural to construct a simulation cell which is periodic in the plane of the GB (the x–y-plane). In the z-direction the border condition should simulate the bulk media on either side of the interface. If a periodic border is also imposed in this direction (i.e. imposing 3-D periodic borders on the whole system), there will be two interfaces in the simulation cell. This is unsatisfactory since the two interfaces interact and, at sufficiently high temperatures, GB migration can lead to mutual annihilation [26].

For the simulations discussed below, a new border condition based on a Region I–Region II approach, shown schematically in Fig. 7.1, was recently developed [27]. Under these border conditions the system is 2-D periodic in the x–y-plane and there is only one GB in the simulation cell. Atoms in Region I move according to Newtonian equations of motion, while Region II consists of two rigid blocks of atoms. The extent of these blocks in the z-direction is determined by the range (i.e. cutoff radius) of the potential. Each rigid Region II block is allowed to move independently, thereby allowing all three translational DOFs to be

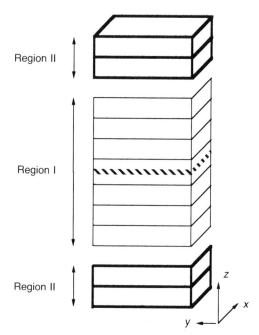

Fig. 7.1 Schematic of the simulation cell for bicrystal studies. Region I contains 32 atomic planes. Initially the GB lies between planes 16 and 17.

dynamically optimized throughout the simulation [27]. Such a strategy with movable Region-II blocks is commonly used in zero-temperature (lattice-statics) simulations of GBs [28]. The movement of the Region-II blocks in the z-direction is governed by the **pressure** exerted on each block by the atoms in Region I through a modified Parrinello–Rahman scheme [29]; the sliding of each block in the x–y-plane is controlled by the **force** exerted on the block across the border of the two regions. While expansions and contractions of the simulation cell are allowed in the z-direction, when simulating an isolated interface embedded between two semi-infinite blocks of atoms, areal expansions (in the x–y-plane) are forbidden. Thus the x–y dimensions of our simulation cell are fixed by the lattice parameter of the bulk region, determined independently from a constant-pressure simulation of a bulk ideal crystal (with 3-D periodic borders) at the desired temperature.

The so-called 'fixed-border' condition used by previous workers [30] was also based on a Region I–Region II scheme. However, unlike the new

border condition described above, the center of mass of the simulation cell was fixed, thereby constraining the two halves of the bicrystal to translate in equal and opposite directions in the interface plane. This constraint on the sliding of the GB in both the fixed-border and 3-D periodic border schemes is particularly significant since GB migration has long been known to be coupled to sliding [31].

7.2.4 Characterization of planar disorder and melting

A commonly used measure for the characterization of melting is the atomic mean-squared displacement (MSD). While in the solid the MSD is essentially time independent, in the liquid it increases approximately linearly with time, with a proportionality constant that is a direct measure of the liquid diffusion constant. However, as we shall see in studying the structural integrity of grain boundaries at temperatures below the melting temperature, a reliance on the MSD alone can lead one to draw incorrect conclusions. Additional methods of analyzing the melting process are therefore essential.

As a measure of the breakdown of crystalline order upon melting, we define the square of the magnitude of the static structure factor, $S(\boldsymbol{k})$, which for brevity we denote simply by $S^2(\boldsymbol{k})$:

$$S^2(\boldsymbol{k}) \equiv |S(\boldsymbol{k})|^2 = \left[\frac{1}{N} \sum_{i=1}^{N} \cos(\boldsymbol{k} \cdot \boldsymbol{r}_i) \right]^2$$

$$+ \left[\frac{1}{N} \sum_{i=1}^{N} \sin(\boldsymbol{k} \cdot \boldsymbol{r}_i) \right]^2 \quad (7.4)$$

where $\boldsymbol{r}_i$ is the position of atom i. For the **overall** $S^2(\boldsymbol{k})$, all the atoms in the simulation cell are included in the sums in eq. (7.4). However, because a GB is a planar defect, it is essential to also monitor spatial variations of the structure factor along the normal to the interface plane. To obtain such information, we divide Region I into slices along the z-direction. Here each slice is chosen to contain a single atomic plane in the crystal. Thus we define the **planar** structure factor, $S_p^2(\boldsymbol{k})$, for which only atoms in a given lattice plane are considered. For an ideal crystal at zero temperature, $S_p^2(\boldsymbol{k})$ then equals unity for any wave vector, $\boldsymbol{k}$, which is a reciprocal lattice vector in the plane p. By contrast, in the liquid state (without long-range order in plane p), $S_p^2(\boldsymbol{k})$ fluctuates near zero. As the two halves of a bicrystal are rotated with respect to each other about the GB-plane normal, two different wave vectors, $\boldsymbol{k}_1$ and $\boldsymbol{k}_2$, are required, each corresponding to a principal direction in the related half. For a well-defined crystalline lattice plane, say in semicrystal 1, $S_p^2(\boldsymbol{k}_1)$ then fluctuates near a finite value (≈ 1) appropriate for that temperature, whereas $S_p^2(\boldsymbol{k}_2) \approx 0$. (Reduction of the static structure factor due to thermal motion is familiar in crystal diffraction studies, the effect of which is generally taken into account through the Debye–Waller factor.) In the GB region one also expects somewhat lower values for $S_p^2(\boldsymbol{k}_1)$ due to the local disorder. By monitoring $S_p^2(\boldsymbol{k}_1)$ and $S_p^2(\boldsymbol{k}_2)$ every slice may be characterized as (a) belonging to semicrystal 1 (for $S_p^2(\boldsymbol{k}_1)$ finite, $S_p^2(\boldsymbol{k}_2) \approx 0$), (b) belonging to semicrystal 2 (for $S_p^2(\boldsymbol{k}_1) \approx 0$, $S_p^2(\boldsymbol{k}_2)$ finite) or (c) disordered or liquid (for $S_p^2(\boldsymbol{k}_1) \approx 0$, $S_p^2(\boldsymbol{k}_2) \approx 0$). The vectors $\boldsymbol{k}_1$ and $\boldsymbol{k}_2$ were chosen to represent reciprocal lattice vectors along the $\langle 110 \rangle$ direction in each half of the Si bicrystal and along the $\langle 100 \rangle$ direction in the two halves of the Cu bicrystal.

As a further measure of the overall disorder in the system, we introduce the quantity N_{def}, which represents the number of defected atoms in the system. An atom is considered defected if its number of nearest neighbors differs from the ideal-crystal coordination number. The nearest-neighbor shell is defined to end half way between the ideal-crystal first- and second-nearest-neighbor distances ($0.577a$ in Si and $0.854a$ in Cu). In this definition the value of N_{def} is zero for an ideal crystal, even at elevated temperatures. For a liquid, on the other hand, at almost all instants in time the value of N_{def} is equal to the number of atoms in the system.

7.3 MOLECULAR-DYNAMICS SIMULATION OF INTERFACE-INDUCED THERMODYNAMIC MELTING

In any simulation study of melting, knowledge of the thermodynamic melting point, T_m, of the model system is of primary importance. Since the model is specified by a particular interatomic potential function, it is essential to determine T_m for that potential, which may or may not turn out to agree with the experimental value. In principle, T_m is defined by the equality of the free energies of the solid and liquid phases. Figure 7.2 shows that for Cu described by an EAM potential, the crossing of the free-energy curves yields a value of $T_m = 1171 \pm 30$ K, which is somewhat different from the experimental value for elemental Cu. The melting temperature of the LJ potential used here is also a little below 1200 K. For Si described by the Stillinger–Weber potential, Broughton and Li obtained a value of $T_m = 1691 \pm 20$ K [32], which is close to the experimental value of 1683 K to which it was fitted [19].

One can imagine that an interface might have its own distinct melting point which, by virtue of the local structural disorder, might be expected to be lower than that of the bulk. This would then imply that a premelting transition would exist where the interface would become liquid at a temperature

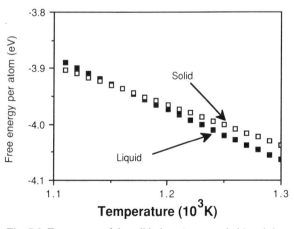

Fig. 7.2 Free energy of the solid phase (open symbols) and the liquid phase (closed symbols) of Cu (described by an EAM potential) as a function of temperature near the melting point.

distinctly below the bulk melting point [33]. The possibility of premelting at a GB, with an attendant loss of local shear resistance, clearly would have very significant implications on the mechanical properties. Understandably, attempts have been made to look for premelting effects in GBs experimentally [33, 34]; as will be discussed below, simulation can contribute to the clarification of this basic issue.

The possibility of GB premelting and the well-known high mobility of GBs in metals are two factors which can complicate the study of GB induced melting. By contrast to metals, GBs in undoped Si are known to have very low mobilities [35]. Thus to investigate the disordering associated with melting, however without the complicating effects of any possible GB migration, we have first considered the melting of a bicrystal of silicon [10] (section 7.3.1). We subsequently discuss similar studies for Cu in section 7.3.2. Since we find that the kinetics of melting are qualitatively identical in the two cases, the conclusions that we draw may be expected to have rather general validity.

7.3.1 Thermodynamic melting of silicon

Our silicon bicrystal simulations were performed at seven different temperatures: 1650, 1750, 1800, 1900, 2000, 2100, and 2200 K. At the beginning of each simulation run, atoms were given random velocities corresponding to a temperature of 1200 K. To reach the desired simulation temperature the system was then heated by 100 K every 200 time steps. (In all the simulations on silicon the time step was 1.15×10^{-15} s, for which, in the microcanonical ensemble, energy was found to be conserved to six significant figures for simulations of several thousand time steps.) As we are primarily interested in phase transitions which involve a latent heat, a thermostat was applied by rescaling the particle velocities.

Figure 7.3 shows the plane-by-plane profile of the instantaneous planar structure factor (defined in section 7.2.4) after 9000 time steps at 1650 K (41 K below T_m for this potential). The sharp variation over a width of only 2–3 (110) lattice planes of $S_P^2(\mathbf{k}_1)$ between almost zero and unity,

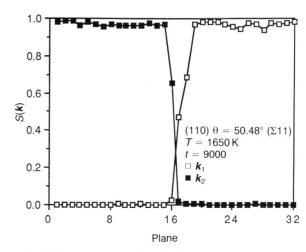

Fig. 7.3 Instantaneous values of $S_P^2(\mathbf{k}_1)$ and $S_P^2(\mathbf{k}_2)$ for the 32 atomic planes parallel to the Si (110) $\theta = 50.48°$ ($\Sigma 11$) GB after 9000 time steps at 1650 K.

coupled with a corresponding increase in $S_P^2(\mathbf{k}_2)$, demonstrates the existence of crystalline order in the entire bicrystal. A comparison of these results with those after only the first 2000 time steps shows neither any increase in the width of the GB region nor any change in the location of the GB plane. The latter indicates that the GB is, indeed, immobile at this temperature during the entire simulation. The near-unity values of $S_P^2(\mathbf{k})$ away from the GB, even at this elevated temperature, reflect the relative stiffness of this covalent material. Corresponding to the stability of the bicrystal, reflected by these structure factors, the plane-by-plane potential energy and mean-squared displacement were also found to be time independent as was the total number of defected atoms in the system.

At all the higher temperatures the system became unstable because melting was initiated at the GB. Figures 7.4(a) and (b) show two instantaneous atomic configurations in a bicrystal of silicon at 2200 K [10]. With the atoms shaded according to their average nearest-neighbor coordination, C, it is clear that the interface region is highly disordered and that, as the system evolves, the disordering spreads into the bulk regions. On the left, the two related planar structure factors are shown; they too indicate that the (110) planes are disordered at the interface but well ordered in the

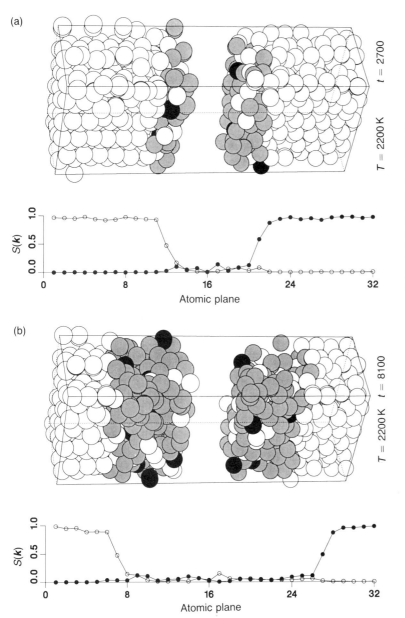

Fig. 7.4 Heterogeneous 'thermodynamic melting' of a silicon bicrystal containing in its center the (110) $\theta = 50.48°$ ($\Sigma 11$) twist boundary. (The two semicrystals are pulled apart to facilitate visualization of the structural disorder.) After the bicrystal was heated from 1600 K ($T < T_m$) to 2200 K ($T > T_m$) over a period of 600 time steps (1000 steps corresponds to 1.15 psec of real time), the simulation time was set to $t = 0$. The shading of atoms indicates their nearest-neighbor coordination, C, with white, gray and black circles denoting, respectively, $C = 4$, $C \geqslant 5$ and $C \leqslant 3$.

(a) After 2700 time steps, a number of planes on either side of the GB plane has melted. The near-zero values of the structure factors $S(k_1)$ and $S(k_2)$ show that long-range order has now broken down in approximately seven (110) planes closest to the GB. By contrast, at $t = 0$ (not shown) only a few atoms at the GB had coordination greater than four, and the structure-factor profiles showed a well-defined GB region consisting of about four (110) planes (b) After 8100 time steps, over half of the system has melted; long-range order has been lost in the 20 central planes of the system.

bulk regions. However, knowing the behavior of $S_p^2(k)$ alone is not sufficient to tell us whether the spreading disordered region is a disordered solid or liquid. One can look to the atomic mobility, as measured by the MSD, to make this distinction. Figure 7.5 shows that the MSD of two planes of atoms, taken from the center of the disordered region, attains a linear behavior that is characteristic of diffusive motion. Based on these data, we determine the diffusion constant to be $\approx 7 \times 10^{-9}\,\mathrm{m^2\,s^{-1}}$. This is comparable to the value obtained from a direct simulation of the liquid at this temperature [32]. Further verification of the liquid nature of the disordered region comes

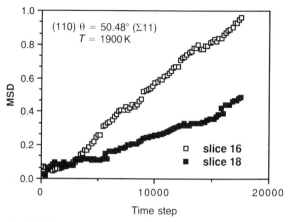

Fig. 7.5 Mean-squared displacement for planes 16 and 18 of the silicon bicrystal as a function of time (in number of time steps) at 1900 K.

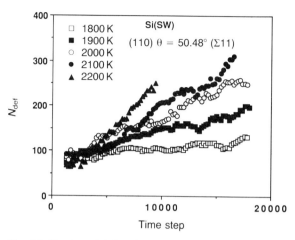

Fig. 7.6 Number of defected atoms in the silicon bicrystal as a function of time at five temperatures above T_m.

from the related volume contraction of the system, which was found to agree well with that extracted from an independent simulation of liquid silicon [10].

Figure 7.6 shows the number of defected atoms, N_{def}, in the bicrystal as a function of time at five temperatures above T_m. At any given temperature N_{def} grows approximately linearly with time. Knowing the number of atoms that were in each of the melted atomic planes (22) and the interplanar spacing ($d = 0.354a$), the associated propagation velocities of the solid-liquid interfaces into the crystalline regions, v, can be extracted. In Fig. 7.7 these velocities are plotted as a function of temperature; they are in good agreement with experimental values [36]. Extrapolation to zero velocity should yield an estimate of the coexistence temperature, T_m, at which the crystal and liquid are in thermodynamic equilibrium, and at which a solid–liquid interface will not propagate. The temperature so obtained from Fig. 7.7 is 1710 ± 30 K, which is remarkably close to the value of $T_m = 1691 \pm 20$ K obtained from the free-energy analysis. From these simulations we conclude that the propagation of the solid–liquid interface seen in all but the 1650 K simulation is due to thermodynamic melting of the bicrystal nucleated at the GB.

Simulations on the (110) free surface of Si show that thermodynamic melting is also nucleated at the

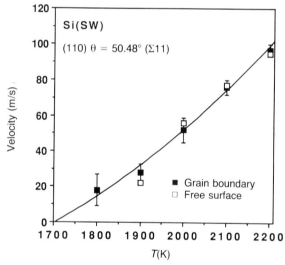

Fig. 7.7 Propagation velocity of the silicon solid–liquid interface as a function of temperature after nucleation from a GB (solid squares with error bars) and from a free surface (open squares). The curve, representing a quadratic fit to the data points, extrapolates to zero velocity at $T = 1710 \pm 30$ K.

surface at the thermodynamic melting temperature [10]. As is also shown in Fig. 7.7, the propagation velocity at a given temperature for solid–liquid interfaces nucleated from the (110) free surface is the same as that for solid–liquid interfaces nucleated at the GB on the (110) plane. Interestingly though, as illustrated in Fig. 7.8, the

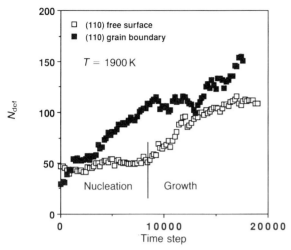

Fig. 7.8 Number of defected atoms in the silicon bicrystal (open squares) and in the free-surface terminated thin slab (solid squares) as a function of time at $T = 1900\,$K.

nucleation time for the melting process is considerably shorter at the GB than at the free surface. This is consistent with the greater degree of structural disorder at the GB.

7.3.2 Grain-boundary migration and interface-induced melting in Cu

Experimentally, GBs in metals are known to be highly mobile [37]. Grain boundaries have also been seen to be mobile in simulation studies using the bicrystal border condition discussed in section 7.2.3 [11, 27, 38]. Figure 7.9 shows the instantaneous plane-by-plane structure factor for a (001) $\theta = 43.60°$ ($\Sigma 29$) GB, described by an EAM potential for Cu, after 1000 and 16 400 time steps at 1100 K, which is 71 K below T_m for this potential [11]. The profiles are very similar, except that the position of the GB (determined from the cross-over in the structure factors) has shifted, i.e. the grain boundary has migrated. The evolution of the structure factor of one of the planes next to the GB during its migration is shown in Fig. 7.10(a) for a simulation using the LJ potential [27]. One can see that up to time step 4000 (~8 picoseconds of real time) this plane is clearly part of the upper half of the bicrystal. As time progresses, however, the

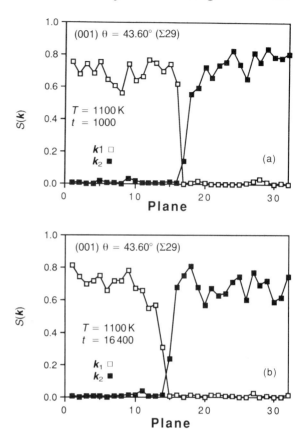

Fig. 7.9 Instantaneous values of $S_P^2(k_1)$ and $S_P^2(k_2)$ for the 32 planes parallel to the (001) $\theta = 50.48°$ ($\Sigma 29$) GB in Cu (described by an EAM potential) after (a) 1000 and (b) 16 400 time steps at 1100 K.

symmetry of the plane changes to that of the lower half of the bicrystal, indicating the migration of the GB. The migration process seems to involve two steps. In the first step, a plane adjacent to the GB becomes disordered and may remain so for some time. In the second step, it recrystallizes into the opposite-symmetry state. As is shown in Fig. 7.10(b), during migration the instantaneous MSD for atoms in the migrating plane increases almost linearly in time. After the migration is completed the MSD again becomes constant in time. This linear increase in the MSD with time should not be interpreted as showing that the intermediate state during migration is liquid, however, since substantial atomic rearrangements must take place as the plane changes symmetry. Indeed, recent simu-

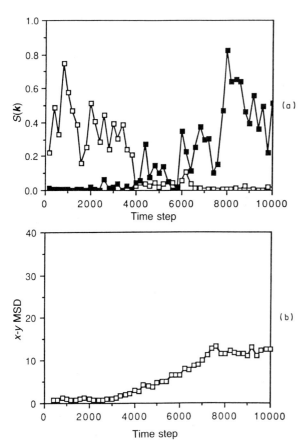

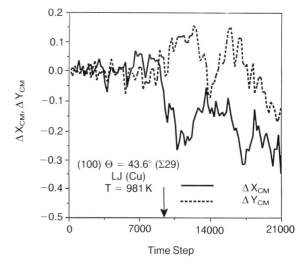

Fig. 7.11 The relative displacement of the x and y components of the center of mass of the migrating plane of a copper bicrystal (described by an LJ potential) relative to a plane initially in the same half of the bicrystal. The arrow indicates the approximate migration time.

Fig. 7.10 (a) Time evolution at 1100 K of the instantaneous $S(k_1)$ and $S(k_2)$ for the 16th plane in the (001) $\theta = 43.60°$ ($\Sigma 29$) GB in Cu, as described by a Lennard–Jones potential. The two wave vectors corresponding to $\langle 100 \rangle$ directions in the lower and upper halves of the bicrystal are given by open and solid symbols respectively. (b) The instantaneous x–y mean-squared displacement (in units of $10^{-2} a_0^2$) as a function of time for the same plane.

lations of the mechanism of GB migration have shown that a considerable amount of short-ranged atomic order, in the form of domain-like structures, is present in the migrating plane throughout the migration process [38].

These latter simulations have also explicitly demonstrated the strong coupling between sliding and migration [38]. Figure 7.11 shows the location of the x and y components of the center of mass of the migrating plane with respect to a plane initially in the same half of the bicrystal. The two planes move in unison until the migration occurs

(at approximately 7000 time steps), after which the migrating plane detaches from one half of the bicrystal and attaches to the other half, with the result that the motion of their centers of mass are no longer correlated.

Despite the complicating effects of GB migration and sliding, the kinetics of GB-induced melting on the (001) plane of Cu were found to be qualitatively identical to those seen on the (110) plane of Si [11]. We briefly illustrate this with results from simulations using the EAM potential. Figures 7.12(a) and (b) show the plane-by-plane structure factor after 5000 and 10 000 steps respectively at 1300 K (129 K above T_m for this potential). A region of disorder forms at the GB and spreads rapidly away from it. (Compare with the structure factors for melting in Si in Fig. 7.4 and that of Cu below the melting temperature in Fig. 7.10.) Figure 7.13 shows the instantaneous MSD, averaged over the approximately 116 atoms in planes 15–18 as a function of time over the last 10 000 time steps. The MSD increases approximately linearly in time, and from its slope a diffusion constant of $\approx 4 \times 10^{-9} \, \mathrm{m^2 \, s^{-1}}$ was calculated, which is of the correct order of magnitude for a liquid metal.

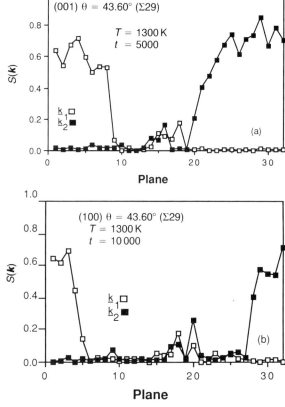

Fig. 7.12 Instantaneous values of $S_P^2(k_1)$ and $S_P^2(k_2)$ for the 32 planes parallel to the (001) $\theta = 43.60°$ ($\Sigma 29$) GB in Cu (described by an EAM potential) after (a) 5000 and (b) 10 000 time steps at 1300 K.

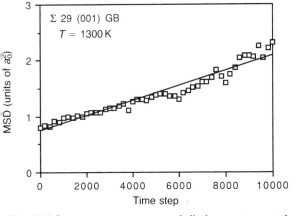

Fig. 7.13 Instantaneous mean-squared displacement averaged over four slices in the disordered region as a function of time over the last 10 000 time steps of the simulation.

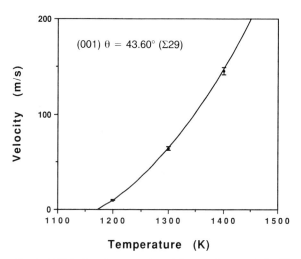

Fig. 7.14 Velocity of propagation of the two Cu crystal–liquid interfaces nucleated at the GB as function of temperature above the thermodynamic melting temperature T_m. The solid line shows a quadratic fit to the data points.

As in the case of silicon, the number of defected atoms in the system was found to increase approximately linearly with time, thereby allowing a propagation velocity for the solid–liquid interfaces to be calculated. Figure 7.14 shows this propagation velocity as a function of temperature. A quadratic extrapolation to zero temperature yields a temperature of 1179 ± 20 K – in striking agreement with the thermodynamic melting point ($T_m = 1171 \pm 30$ K). Again, the obvious conclusion is that thermodynamic melting has been nucleated at the GB. Thermodynamic melting was also found to be induced at both (001) free surfaces and at atomic voids in Cu. In each case, the kinetics were very similar to those of melting nucleated at the GB [11].

7.3.3 On 'premelting' at GBs and free surfaces

The question of whether a liquid-like layer can form at the GB at temperatures distinctly below the melting point, T_m, is of long-standing interest [33]. A number of MD simulations have been performed to address this issue. While all of the studies indicated that significant structural disordering can take place at $T \leq T_m$, they have drawn conflicting conclusions concerning the existence of GB 'premelting'. Pontikis has reviewed the various results giving rise to this controversy, and pointed out several aspects of the simulation procedure which could account for the apparent discrepancy [39].

In none of our simulations of the high-temperature stability of GBs have we seen any evidence for a premelting transition or a stable disordered phase below the melting temperature. This finding is in agreement with the results of Ciccotti *et al.* [40], who studied the (310) symmetrical tilt boundary using the Lennard–Jones potential and a 3-D border condition. It is also consistent with recent TEM studies of aluminum bicrystals which showed no boundary melting up to $0.999\,T_m$ [34]. Further support for the conclusion that there is no credible simulation evidence for GB premelting is provided by another study of the (310) symmetrical tilt boundary, in this case using an EAM potential for aluminum and 2-D periodic borders, with free surfaces in the third direction [41].

In the context of the previous MD studies which have led to conflicting results, we make three observations on the basis of the work presented here, which are pertinent to the resolution of the controversy. First the thermodynamic melting temperature, T_m, should not be taken to be the temperature at which a single, defect-free crystal melts [27]. As illustrated in section 7.4, in the absence of a surface or internal interface to nucleate the melting process, a single crystal can be substantially superheated, up to about 20–30% above T_m. Second, neither the 3-D periodic border condition [40, 42] nor the 2-D periodic fixed border condition [30] are suitable for simulating dynamical effects in GBs because they tend to suppress the sliding associated with GB migration. The effect that this has can be seen explicitly in Fig. 7.15, which shows the evolution of a plane next to a (001) $\theta = 43.60°$ (Σ29) GB under 3-D periodic border conditions. Although the temperature and lattice parameters for this system were identical to those of the system shown in Fig. 7.10, the evolution of the two systems is entirely different. In particular, in Fig. 7.15(a) both structure factors are near zero throughout the simulations and the x–y MSD in Fig. 7.15(b) grows approximately linearly with time. These two signatures are typical of the liquid

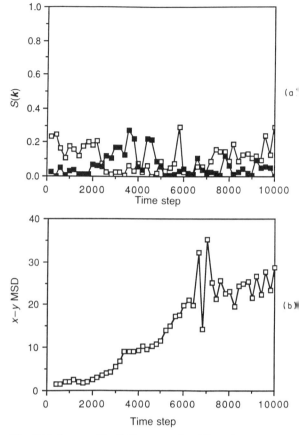

Fig. 7.15 Same as Fig. 7.10, except that 3-D periodic borders were applied to the computational cell.

state and they could lead to an interpretation in terms of the occurrence of premelting. This would be incorrect, however, since both the disorder in the structure factors and the increasing MSD are the direct results of the elimination of the two microscopic degrees of freedom associated with the sliding of the two halves of the bicrystal with respect to each other. Lastly, special care is needed in interpreting the data which often involved quantities averaged over periods of time long compared with the time typically associated with migration.

In contrast to GBs, there is considerable experimental evidence for stable structural disordering at free surfaces at temperatures below T_m. Surface melting has been directly observed by scattering protons from an atomically clean Pb (110) surface [43], the process occurring at approximately $0.75 T_m$. The transition begins with partial disordering of the surface region and progresses to a completely disordered film whose thickness increases rapidly as the temperature approaches T_m. Surface disordering and melting in various model systems have been studied by molecular dynamics [44–46]. For example, melting of the (111) and (100) surfaces of silicon induced by laser-pulse heating has been simulated [45]. Surface disordering below the thermodynamic melting temperature on the (110) and (111) surfaces of aluminum (using potential functions derived using effective-medium approximations similar to the EAM) and on the (110) surface of Ni (using an EAM potential) have also been investigated [46].

If this surface disordering is to be viewed as a true thermodynamic phase transition, to describe it as surface 'premelting' appears to be a misnomer. Figure 7.16 shows the thermodynamic phase diagram of a monatomic solid in the volume–temperature plane. The usual phase boundaries delineating the single-phase regions of crystal (C), liquid (L), and vapor (Vap) are indicated. The condition for thermodynamic melting is expressed by the melting curve, $T_m(V)$, which terminates at the triple-point temperature, T_t. This is the lowest temperature, according to equilibrium thermodynamics, at which the crystal, at a volume V_t^C, can coexist with the liquid (and also with the vapor). The freezing curve, $T_f(V)$ lies more or less parallel to the melting curve, and also terminates at T_t, where the liquid has a volume V_t^L. In the context of surface disordering, it should be noted that in most materials the thermodynamic melting temperature is only a little higher than the triple point temperature (often by less than 1 K). Thus almost any experiment performed at $T < T_m$ is also performed at $T < T_t$. Consequently, if the crystalline surface is thermodynamically unstable it must be unstable with respect to the **vapor phase** (by contrast with the liquid phase) and the phase transition should be more accurately dubbed surface 'presublimation'.

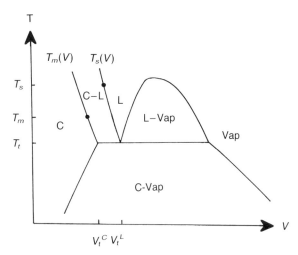

Fig. 7.16 Schematic temperature–volume phase diagram of a monatomic substance showing the single-phase regions of crystal (C), liquid (L), and vapor (Vap), and the various two-phase regions. On the horizontal triple line, at temperature T_t, the crystal (at volume V_t^C) and the liquid (at volume V_t^L) coexist with the vapor. The points on the thermodynamic melting line, $T_m(V)$, and the freezing curve, $T_f(V)$, indicate conditions of ambient pressure. As discussed in section 7.5, to a good approximation, the freezing curve and the mechanical-stability line, $T_s(V)$, coincide.

7.4 MECHANICAL MELTING

In the absence of crystal defects such as GBs, free surfaces, voids or dislocations, a mechanism of melting different from that of the thermodynamic melting described previously above would have to exist. In this section, by simulating the melting of a perfect defect-free crystal of Cu, we show that melting takes place **homogeneously** on an extremely rapid time scale set by the (inverse of the) characteristic phonon frequency and that it is driven by an intrinsic elastic instability in the system [12].

7.4.1 The Born instability

Thermodynamic melting is easily suppressed in computer simulations by the elimination of extended defects, i.e. by the application of 3-D periodic borders to a perfect crystal [10, 11]. Experimentally, by contrast, due to the presence of dislocations, superheating is extremely difficult

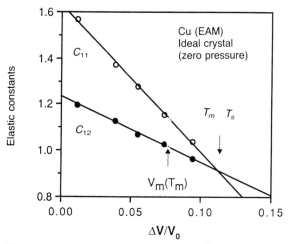

Fig. 7.17 Elastic constants C_{11} and C_{12} (in 10^{12} dyn/cm^2) versus temperature for a perfect crystal of Cu containing no extended defects, which can therefore, in principle, be superheated to the maximum superheating limit, $T_s \sim 1432 \pm 12$ K. The solid lines are straight-line fits to the data points.

melting, this type of transition is not governed by the free energies of two competing phases; in fact, the free energy of the final state does not enter into the instability criterion at all.

Figure 7.17 illustrates a method for determining T_s from the elastic constants of a superheated perfect crystal of Cu. From a least-squares fit to the decrease of the elastic constants C_{11} and C_{12} with increasing temperature, a value of $T_s = 1432 \pm 12$ K is obtained, which is about 260 K higher than the thermodynamic melting point, $T_m = 1171 \pm 30$ K, obtained for the same interatomic potential. The elastic constants in Fig. 7.12 were obtained from constant-pressure molecular dynamics simulations [27] with evaluation of both the Born and fluctuation contributions [50].

7.4.2 Molecular-dynamics simulation of mechanical melting of Cu

In practice it is very difficult, even in simulations, to reach the maximum superheating temperature, T_s, because of statistical fluctuations in the volume and temperature of the sample. However, by gradually stepping up the simulation temperature, we were able to superheat a perfect crystal of Cu to within about 80 K of T_s; beyond this temperature the crystal could not be stabilized [12]. After a step increase of the simulation temperature above T_s to 1700 K, only about 500 MD time steps (or about 10–20 lattice vibration periods) are required to destroy the long-range order within the (001) planes in the crystal. Figure 7.18 shows a plane-by-plane profile of the planar static structure factor for a perfect crystal consisting of 32 (001) planes, each containing 29 atoms. This unit-cell geometry is the same as that used in our GB-induced melting simulations, however without the GB. The plane-by-plane instantaneous profile after 0, 200, and 1000 time steps shows that planar order is lost simultaneously in all parts of the crystal. This suggests that the liquid phase is formed **homogeneously**, as one would expect from a phonon or elastic instability.

In the above discussion the role of the lattice instability in establishing a maximum superheating temperature at zero external pressure (Fig. 7.17)

[15, 47–49] even in the most favorable cases. Over 50 years ago Born [13] pointed out the existence of an absolute limit to superheating for any crystalline structure. By considering the volume dependence of the normal modes of a crystal lattice he demonstrated the existence of a phonon instability at a certain critical volume expansion, V_s. By couching the discussion in terms of the elastic response of a crystal lattice under isotropic tension, Born's phonon instability can be shown to correspond to an elastic instability in the shear constant, C'. In a cubic crystal

$$C' = (C_{11} - C_{12})/2 \qquad (7.5)$$

is the minimal shear constant, i.e. C_{44} expressed in a coordinate system with the z-axis parallel to $\langle 001 \rangle$ and the x–y-axes parallel to the $\langle 110 \rangle$ and $\langle 1\bar{1}0 \rangle$ directions. In this coordinate system, C_{44} assumes its smallest value for any cubic crystal. Born's criterion, when applied to a superheated crystal lattice, establishes the existence of a maximum volume, V_s, coupled with a maximum superheating temperature, T_s, above which the crystal becomes mechanically unstable and therefore has to undergo a phase transformation (to the liquid state or to some other crystalline or non-crystalline solid structure). By contrast with thermodynamic

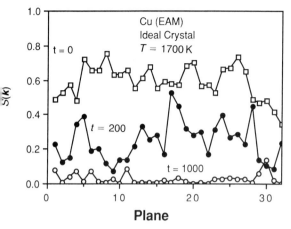

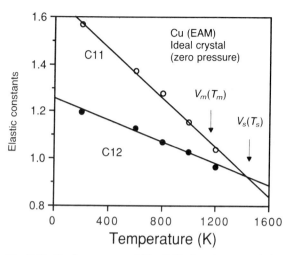

Fig. 7.18 Slice-by-slice spatial distribution of the square of the planar static structure factor defined in eq. (7.4) after 0, 200, and 1000 time steps following the sudden increase of the simulation temperature from 1300 to 1700 K. Apart from the scatter of the data points, the figure shows the crystalline order to disappear homogeneously and rapidly for this case of mechanical melting.

Fig. 7.19 Elastic constants of Fig. 7.17 plotted versus relative volume change, $\Delta V/V_0$ (with $V_0 = V(T = 0)$). $V_s = V(T = T_s)$ is the maximum volume up to which, at $T = T_s$, crystalline order can be sustained.

has been emphasized. Because of the thermal expansion accompanying any temperature change, however, effects of both temperature and volume change are present in the results in Fig. 7.17. To see the variation with volume alone, these results are plotted in Fig. 7.19 from which the critical stability volume, $V_s(T_s)$, can be extracted [12]. At this combination of volume and temperature, and under the condition of zero external pressure, the system melts homogeneously.

There is supporting evidence from our simulation results on a perfect 3-D periodic Si crystal pointing to the homogeneous nature of mechanical melting. Examination of $S_p^2(\mathbf{k})$ shows that the system melts homogeneously at between 2400 K and 2600 K, a value which is consistent with previous results [19, 32] and is considerably higher than the thermodynamic melting temperature of 1691 K [51].

7.4.3 Comparison of the characteristics of thermodynamic and mechanical melting

The above simulations illustrate that every crystal, in principle, has two melting points, T_m and T_s. Conceptually the two transitions have distinct physical origins: whereas thermodynamic melting is governed by the free energies of the liquid and the solid phases, mechanical melting is based on a phonon instability. Since the volume expansion required for mechanical melting is always larger than that associated with thermodynamic melting, the free energy always favors thermodynamic over mechanical melting, i.e. $T_s > T_m$. However, as illustrated above, the former requires atomic mobility and therefore may be kinetically hindered by slow atomic diffusion in the liquid phase. If a crystal is melted under atmospheric conditions, the thermodynamic state variables usually will be such that high atom mobility in the liquid enables the nucleation and growth of the liquid phase at extended defects. However, if a crystal is disordered at lower temperature (for example, by uniformly expanding the crystal), the limited atom mobility may hinder the thermodynamic phase transition because atomic mobilities decrease **exponentially** with decreasing temperature but increase only approximately **linearly** with increasing volume. The crystal may therefore not be able to disorder at the volume specified by equilibrium thermodynamics until a larger volume is reached. The largest possible volume is the instability volume, V_s', associated with the ultimate stability limit, where the crystal structure breaks down without change in volume.

To conclude, we summarize the four main distinguishing characteristics of thermodynamic and mechanical melting.

1. Whereas thermodynamic melting, characterized by (T_m, V_m), is based on the free energies of both the crystalline and liquid sates, mechanical melting is triggered by a phonon instability in the crystal lattice at a critical volume, V_s, associated with T_s.
2. Thermodynamic melting requires the existence of extended defects at which the liquid phase can nucleate. By contrast, mechanical melting can occur with and without the presence of such defects, although one would expect that the combination (T_s, V_s) of thermodynamic state variables needed to trigger the phonon instability depends somewhat on whether or not the system contains extended defects and on the nature of these defects.
3. The growth of the liquid phase into the crystal (by propagation of solid–liquid interfaces) requires thermally-activated diffusion in the liquid. Mechanical melting, by contrast, is caused by a phonon instability, and therefore happens typically within a few dozen lattice vibration periods, i.e. typically within about 10^{-12} s, and without requiring thermally-activated atom mobility.
4. As a consequence of (c), thermodynamic melting involves relatively slow kinetic processes in contrast to mechanical melting. Also, thermodynamic melting is a heterogeneous process, involving nucleation and growth of the liquid phase, whereas mechanical melting takes place homogeneously, without the need for the presence of lattice defects.

7.5 THE $T-V$ PHASE DIAGRAM AND ITS LOW-TEMPERATURE EXTENSIONS

Our discussions of thermodynamic and mechanical melting can be summarized by referring to the typical phase diagram of a monatomic system in the $T-V$ plane, shown in Fig. 7.16. This phase-diagram representation is useful not only for expressing the relation between the thermodynamic variables at the two C → L transitions, but also for discussing the underlying thermodynamic basis of the connection between C → L and C → A transitions.

In Fig. 7.16 the usual phase boundaries delineating the single-phase regions of crystal (C), liquid (L), and vapor (Vap) require no comment with perhaps one exception. The condition for thermodynamic melting is expressed by the melting curve, $T_m(V)$, which terminates at the triple-point temperature, T_t. This is the lowest temperature, according to equilibrium thermodynamics, at which the crystal, at a volume V_t^C, can coexist with the liquid (and also with the vapor). The freezing curve, $T_f(V)$, also terminates at T_t, where the liquid has a volume V_t^L. As we will suggest below, the freezing curve and the mechanical stability curve, $T_s(V)$, can be considered to be effectively the same.

While we know of no fundamental argument showing whether the stability and freezing curves are the same, we believe they should lie rather close to each other. Indeed, there is experimental evidence pointing to similar values for V_s and V_f [52]. We recall from section 7.4.1 that the onset of mechanical instability is measured by the vanishing of the shear modulus C'. Figure 7.20 shows that for a number of metals C' is observed to decrease with volume increase during heating from absolute zero to melting at constant pressure [52]. A linear extrapolation of the data beyond the indicated melting point, T_m, gives the critical volume $V_s(T_s)$ at which C' vanishes. In each case the result is seen to be very close to the volume of the corresponding melt, shown by an open symbol, the latter being obtained as $V_f(T_m) = V_m(T_m) + \Delta V(T_m)$, where $\Delta V(T_m)$ is the latent volume of fusion. Since, by definition, V_f lies on the freezing curve, $T_f(V_f)$, and V_s lies on the instability curve for mechanical melting, the comparison in Fig. 7.20 suggests that these two curves are quite close to each other. Because in this work we are not particularly concerned with the phenomenon of freezing, we will not distinguish between these two curves.

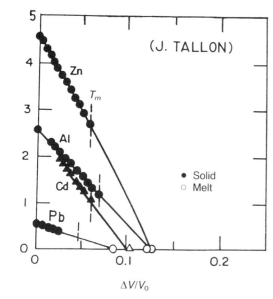

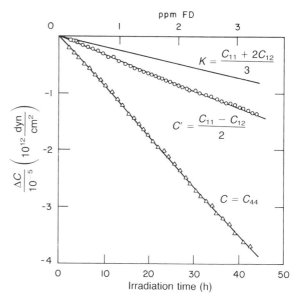

Fig. 7.20 Volume dependence of the shear modulus associated with the heating to melting of various metals at constant (zero) pressure. Dashed lines and open circles denote, respectively, the volume expansion of the crystal and of the melt at the thermodynamic melting point. All measurements are relative to the specific volume of the crystal at $T = 0$ (from Tallon, Ref. [52]).

Fig. 7.21 Change in the elastic constants of a copper single crystal during thermal-neutron irradiation at $4\,K$ (after Rehn *et al.*, Ref. [53]).

In view of the linear behavior of the shear modulus in Fig. 7.20, one may represent the volume dependence by [52]

$$C'(T) = C'(0)[1 - \gamma_G \Delta V(T)/V(0)] \qquad (7.6)$$

where $\Delta V(T)/V(0) = 1/\gamma_G$ is the critical volume expansion, with γ_G being the effective Grüneisen parameter which has been tabulated for a number of materials [52]. For copper $\gamma_G = 9.47$, which gives an isobaric critical volume expansion of 10.6%. For comparison, we may consider the elastic constants of a copper single crystal irradiated at $4\,K$ by thermal neutrons. Figure 7.21 shows the shear modulus decreasing essentially linearly with irradiation time and concentration of Frenkel defects [53, 54]. By extrapolating the results for C' to zero, one obtains a defect concentration of $C_d = 7.1 \times 10^{-2}$ which can be converted to a volume expansion through the expression [55] $\Delta V/V_0 = C_d(V_{rel}/\Omega)$, where V_{rel}/Ω is the relaxation volume of a Frenkel defect and Ω is the atomic volume. For copper $V_{rel}/\Omega = 1.5$ [55], therefore, the isothermal critical

volume expansion is $\Delta V/V_0 = 10.7\%$. Since the estimates involve linear extrapolation of the experimental data, the numerical values obtained for the isobaric and isothermal critical volume expansions should be regarded only as lower bounds. Nevertheless, they show a certain self-consistency and suggest that the Born stability criterion is at most weakly dependent on temperature.

7.5.1 Low-temperature extension of the mechanical-melting curve

Although the equilibrium phase diagram in Fig. 7.16 indicates that there is no thermodynamic C → L transition below the triple-point temperature, one can define for $T < T_t$ a critical volume $V_s(T)$ at which the crystal, upon uniform volume expansion, becomes mechanically unstable. This means that the instability curve for mechanical melting in Fig. 7.16 can be extended to arbitrarily low temperatures. To demonstrate that the Born stability criterion remains well defined, we have calculated C' at zero temperature using lattice dynamics [12]. The results, given as open circles in Fig. 7.22, show that C' indeed vanishes at a volume

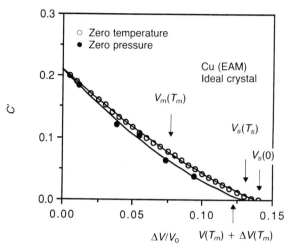

Fig. 7.22 Decrease of the minimal shear constant (in 10^{12} dyn/cm^2) defined in eq. (7.5) as function of a relative volume increase, $\Delta V/V_0$, introduced either by heating towards mechanical melting (full circles; MD results) or by isotropic hydrostatic-stress induced volume expansion at $T = 0$ (open circles; lattice-dynamics results).

$V_s(0)$. This then is the volume obtained when the instability curve for mechanical melting is extended to zero temperature. Figure 7.22 also shows corresponding results (filled circles) which are calculated using molecular dynamics by heating at zero pressure [12]. One sees that the volume dependence of C' for the two expansion processes is very nearly the same. Moreover, the critical volume expansions calculated by simulation are in reasonable agreement with those estimated from the experimental data in Figs. 7.20 and 7.21. Taking all these results together we infer that neither the precise temperature dependence of the mechanical-stability curve nor the specific process by which the volume expansion is produced need to be taken into account in considering the qualitative significance of this proposed extension.

A direct implication of the low-temperature extension of the mechanical-melting curve is that a crystal, under volume expansion, can be maintained in its crystalline phase so long as its volume is less than $V_s(T)$. Since equilibrium thermodynamics requires the system to be in the two-phase region upon crossing the sublimation curve

(Fig. 7.16), the extension is clearly only valid on a time scale short compared to the kinetics of sublimation. The extended curve, $V'_s(T)$, therefore implies the existence of a metastable crystalline phase in which the thermally activated process of sublimation is kinetically suppressed.

7.5.2 Low-temperature extension of the thermodynamic melting curve

When sublimation is kinetically suppressed, a similar extension of the thermodynamic melting curve $T_m(V)$ should also be considered. Figure 7.23 shows schematically the two extended curves $T'_m(V)$ and $T'_s(V)$ [12]. Notice that neither the triple line separating the C–L region from the C–Vap region nor the sublimation curve appear in this diagram. While the region lying to the left of $T'_m(V)$ now becomes the effective single-phase region for the crystalline state, it is important to keep in mind that the region between the original sublimation curve and the extended part of the thermodynamic melting curve represents a metastable overexpanded solid. Similarly, the extended two-phase region below the triple-point temperature represents the region where the metastable over-expanded solid can co-exist with the metastable supercooled liquid.

The effective phase diagram proposed in Fig. 7.23 has immediate implications for solid-state amorphization. In this picture, C $\rightarrow$ A transformation is viewed as an isothermal melting process driven by volume expansion at $T < T_m$, and the analogy with C $\rightarrow$ L transition is that volume expansion driven by external forces at constant temperature plays the same role as thermal expansion during heating to melting at constant pressure. When the expansion is large enough to cross the curve $T'_m(V)$, the crystal becomes unstable against topological disordering, and heterogeneous amorphization may take place. However, if in addition to sublimation the heterogeneous mode of amorphization is also kinetically suppressed, then the expansion may continue up to the curve $T'_s(V)$, at which point amorphization occurs by a homogeneous process triggered by the Born instability.

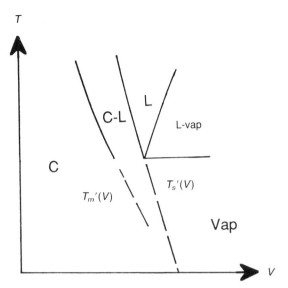

Fig. 7.23 Generalized T–V phase diagram showing the extensions of both the thermodynamic-melting and the mechanical-stability (or freezing) curves below the triple-point temperature. Along the extension of the former, an expanded crystal becomes unstable against the disordered phase by heterogeneous amorphization; by contrast, along the extension of the mechanical-melting curve, homogeneous amorphization can occur.

7.6 INTERPRETATION OF SOLID-STATE AMORPHIZATION EXPERIMENTS

The two traditional methods of producing amorphous metallic alloys are rapid solidification of a melt and quenching from the vapor state. However, during the past decade a variety of new methods based on solid-state processes have been discovered [1]. These include:

(a) radiation-driven processes such as ion beam mixing, ion implantation and the amorphization of intermetallic compounds by electron and ion beams [54];
(b) crystal-to-glass transformations driven by chemical interdiffusion reactions [1];
(c) amorphization of intermetallic compounds during reaction with gaseous hydrogen [56, 57];
(d) mechanically-driven processes such as ball-milling of bimetallic mixtures of metal powders and mechanical deformation of intermetallic compounds [1, 58, 59];

(e) pressure-induced amorphization processes [60, 61].

In this section we discuss the experimental evidence from solid-state amorphization, supplemented by computer simulation results, which demonstrate the parallels between C → L and C → A transitions.

We have previously noted that in a conventional melting experiment, atomic diffusion at elevated temperatures is the factor which makes it virtually impossible to observe the homogeneous process of mechanical melting. Because solid-state-amorphization experiments are usually carried out at temperatures where point-defect mobility is limited, both heterogeneous and homogeneous C → A transitions should be observable. Indeed irradiation and hydrogenation experiments show that this is the case. In irradiation studies, the amorphous phase has been found to nucleate at defect sites when the temperature is close to the threshold temperature for amorphization, whereas at lower temperatures homogeneous transformation in the irradiated volume was observed [62, 63]. In hydrogenation studies, distinctly heterogeneous or homogeneous processes were observed to occur in regions that are respectively poor or rich in hydrogen [56]. These observations constitute evidence that the two mechanisms of crystal disordering which operate in the C → L transition, discussed in sections 7.3 and 7.4, occur also in C → A transformations.

The diversity of external driving forces capable of inducing amorphization notwithstanding, a common feature which can be identified in all these processes is the volume expansion accompanying the onset of structural disorder [54, 56]. The essential role of volume expansion is underscored by the results of hydrogen absorption experiments [56] where little or no chemical disordering, believed to be the primary cause of radiation-induced amorphization [62–64], takes place. Moreover, the existence of a critical volume expansion underlying both melting and amorphization is clearly seen in the elastic behavior of the crystal; the observed variation of the shear modulus C' with volume expansion in heating measurements is found to be very similar to the decrease of C' with lattice expansion in solid-state amorphization.

7.6.1 Heterogeneous and homogeneous amorphization

Experimental evidence showing the existence of heterogeneous and homogeneous modes of solid-state amorphization can be found from two types of experiments, irradiation and hydrogenation. For clarity each kind of experiment will be discussed separately. In electron- and ion-irradiation studies a threshold temperature T_a has been found above which the system could not be amorphized, presumably because of temperature-dependent radiation-enhanced diffusion processes. *In situ* high-voltage electron-microscopy studies on the ordered intermetallic compounds of NiTi show $T_a \sim 0.2\,T_m$ [62–65]. Below the threshold, the onset of amorphization occurs in different ways. When a compound is irradiated in the narrow temperature interval of about $15-20$ K centered on or just below T_a, the amorphous phase nucleates heterogeneously at the extended defects, such as antiphase boundaries, dislocations, and free surfaces [63, 65]. The interfaces between the amorphous and crystalline regions are well defined, and the transformation proceeds by subsequent migration of the interface into the crystalline matrix. In contrast, below this temperature interval, or $T \lesssim 0.16\,T_m$, the entire irradiated volume undergoes a rapid homogeneous transformation to the amorphous state, despite the presence of dislocations and GBs [63, 65].

Corresponding to the two modes of C → A transitions, the total dose required for amorphization was found to vary markedly, with the heterogeneous process requiring greater doses by two orders of magnitude or more [62–64]. These observations are consistent with our previously stated expectation that the heterogeneous mode can be kinetically suppressed through a lowering of the temperature. Furthermore, they indicate that, in the case of NiTi, $T_a \sim 0.16\,T_m$ represents a low-temperature limit to the extension of the thermodynamic melting-curve in Fig. 7.23, whereas no such limit exists for the mechanical melting curve.

In ion-bombardment experiments, polycrystalline silicon films have been amorphized when irradiated by 1.5 MeV Xe ions [66]. These observations indicate that in polycrystalline Si the amorphous phase is nucleated heterogeneously at GBs and triple junctions prior to homogeneous nucleation in the bulk crystal. Specifically, for an ion flux of 3.5×10^{12} cm^{-2} s^{-1} the maximum temperature at which amorphization can take place is 498 K. At a temperature slightly below this, amorphization was found to be nucleated preferentially at high-angle GBs and triple junctions. Preferential nucleation of amorphous Si at low-angle (i.e. $<20°$) $\langle 110 \rangle$ tilt boundaries was also observed, but at temperatures lower than that required for heterogeneous amorphization at high-angle GBs. By contrast, coherent twin boundaries appear to be stable against amorphization. Variations in the widths of the amorphous layers observed at different grains suggest that the kinetics of amorphization is strongly dependent on GB structure. In the case of single crystals the amorphous phase is initiated at the crystal surface, and only where steep damage gradients exist (i.e. at the end of the incident-ion range) can the amorphous phase be nucleated homogeneously.

There is also evidence that hydrogen-induced amorphization can proceed by both heterogeneous and homogeneous nucleation modes [56]. Complementary neutron scattering, X-ray, and TEM studies on hydrogenated Zr_3Al showed that the compound undergoes phase separation into a hydrogen-poor and a hydrogen-rich phase. Very little lattice expansion occurs in the hydrogen-poor phase, indicating that only a small fraction of the absorbed hydrogen enters grain-interior regions. By contrast, the hydrogen-rich phase undergoes a large homogeneous expansion relative to hydrogen-free Zr_3Al. TEM observations showed that heterogeneous nucleation of the amorphous Zr_3Al at GBs and triple-junctions occurred only in the hydrogen-poor phase, whereas the amorphous phase nucleated homogeneously throughout the grain-interior regions of the hydrogen-rich phase. The interfaces between the crystalline and amorphous phases in the hydrogen-poor phase are sharp and well defined as in the earlier observations of heterogeneous amorphization in hydrogenated Zr_3Rh [57]. Further evidence of homogeneous amorphization during absorption of hydrogen has been recently obtained from studies of the intermetallic compound Fe_2Er [67].

7.6.2 Role of volume expansion

In studies of radiation-induced amorphization of intermetallic compounds it is generally acknowledged that chemical disordering is a necessary condition for the C → A transformation [62–64]. However, recent irradiation studies [5, 54] as well as hydrogen-absorption experiments [56] suggest that the driving force for amorphization may be related in a more fundamental way to defect-induced volume expansion, rather than chemical disorder *per se*. This is a view which is supported by recent ion-implantation experiments [68] and studies of amorphous-phase formation in mechanically deformed intermetallic compounds [59], as well as by a recent molecular dynamics study of the effects of chemical disorder on the stability of an ordered intermetallic compound [69]. Indeed, lattice expansion is likely to be the initial manifestation of the internal strains associated with extrinsic defect structures, such as Frenkel pairs, anti-site defects, dislocations, and impurities. With the exception of pressure-induced amorphization processes, where a density increase is expected, one can anticipate lattice expansion to be a fundamental response to any driving force in C → A transformations.

In the present context, the results of hydrogenation experiments on Zr_3Al, which have been mentioned above, allow the most clear-cut interpretation regarding the role of volume expansion because no significant chemical disordering is produced during hydrogen charging [56]. The critical volume expansions measured in these experiments are found to be comparable to those which occurred during irradiation of intermetallic compounds. This is the primary evidence showing that volume expansion is the most generic characteristic of C → A transformations.

Although the presence of chemical disorder observed in irradiation measurements, as well as

in amorphization by mechanical deformation [58], precludes interpreting these experiments in terms of the effects of only volume expansion, it is nevertheless relevant to note that chemical disordering is always accompanied by volume expansion. As shown in Fig. 7.24, at the onset of amorphization during ion irradiation, chemical disordering is accompanied by a volume expansion of about 2.5% [5]. This magnitude is similar to that observed during hydrogenation just preceding the onset of homogeneous amorphization of the hydrogen-rich hydride phase [56]. Another study which demonstrates the relation between chemical disordering and volume expansion is a recent MD simulation, carried out using an empirical tight-binding potential fitted to the properties of $NiZr_2$ [69]. In this work, randomly chosen Ni and Zr atoms were

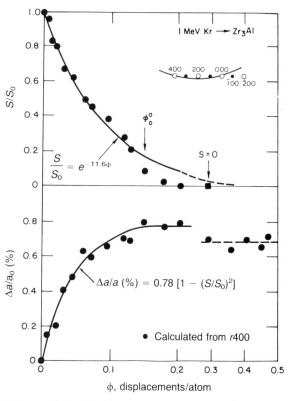

Fig. 7.24 Plots showing changes in the degree of long-range order and lattice dilatation of Zr_3Al as a function of dose during room-temperature bombardment with Kr ions (after Okamoto *et al.*, Ref. [5]).

instantaneously interchanged in an equilibrated configuration of the system at $T = 300\,\text{K}$, and the level of chemical disorder was measured in terms of the Bragg–Williams long-range order parameter S. It was found that amorphization occurred for initial values of S less than 0.6. The calculated volume expansion of 2% obtained at $S = 0.2$ is rather close to the critical volume expansion and to the experimental value of S observed at the onset of amorphization in the intermetallic compound, Zr_3Al [5, 56].

7.6.3 Elastic behavior during amorphization

To further explore the parallels between C → L and C → A transitions, we next consider the behavior of the shear modulus during amorphization as determined by recent Brillouin-scattering and electron-diffraction studies [5]. For the ion-irradiated polycrystalline samples of Zr_3Al already discussed, Fig. 7.25 shows that the shear modulus decreases by about 50% at the onset of amorphization as indicated by the dashed vertical line. In this case it is also known from *in situ* HVEM observations that amorphization occurs homogeneously throughout the irradiated volume despite the presence of dislocations and GBs. Comparing Fig. 7.25 to Fig. 7.20, one sees a rather striking similarity in the variation of C′ with volume expansion during amorphization at constant temperature and during heating to melting at constant pressure. This again demonstrates the parallel between melting and amorphization and the fundamental role of volume expansion. Additional information on the elastic behavior during irradiation is provided by the observation of a comparable decrease in the polycrystalline shear modulus in the B2 compound, FeTi, and the A-15 compound, Nb_3Ir, when amorphized by ion irradiation at low temperatures, whereas such a decrease did not occur in the B2 compounds NiAl and FeAl which remain crystalline under identical irradiation conditions [59, 70]. It appears therefore that a large pre-transition elastic softening similar to that induced by thermal expansion is a characteristic feature of radiation-induced amorphization.

In noting the similarities between Figs. 7.20 and

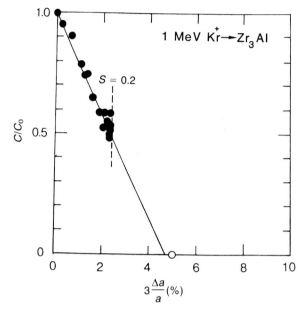

Fig. 7.25 Volume dependence of the average shear modulus during radiation-induced disordering and amorphization of polycrystalline Zr_3Al. The dashed vertical line denotes the maximum volume expansion due to disordering. The open circle on the dilatation axis denotes the volume expansion of amorphous Zr_3Al relative to the fully ordered polycrystalline compound (from Ref. [5]).

7.25, one sees that if the shear modulus is extrapolated linearly to zero one obtains a specific volume equal to that of the melt at its freezing temperature or, in the case of solid-state amorphization, the specific volume of the amorphous state. Both the onset of melting and amorphization seem to occur when the lattice is expanded to a specific volume where the modulus is reduced to about 50% of its initial value. Despite these similarities one should not infer that the physical processes involved are the same because of the existence of two modes of crystal disordering; the radiation-induced amorphization in Fig. 7.25 occurs homogeneously (and is therefore due to a mechanical instability), whereas Fig. 7.20 shows heating to thermodynamic melting which is known to take place heterogeneously.

If the amorphization in Fig. 7.25 is indeed caused by a mechanical instability, it would appear that C′ should vanish at the transition according to

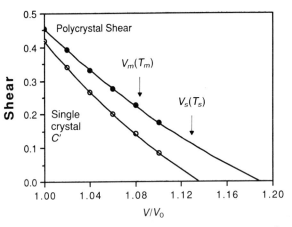

Fig. 7.26 Comparison of the minimal zero-temperature static shear constant, C', defined in eq. (7.5), for a single crystal (see Fig. 7.19, open circles) and for a polycrystal (full circles) determined from the single-crystal results in Fig. 7.19 using an expression given by Kroner [71].

the Born criterion and as illustrated by the MD results in Fig. 7.22. However, there is a reason why C' in fact should not vanish. One should first notice that the elastic constants in Fig. 7.25 were measured in a polycrystalline material whereas the results given in Figs. 7.20 and 7.22 refer to single crystals. We recall that the equality of C' and $C44$ in eq. (7.5) holds only in a coordinate system where one of its axes is aligned parallel to $\langle 001 \rangle$, and the other two parallel to $\langle 110 \rangle$ and $\langle 1\bar{1}0 \rangle$. In such a coordinate system, according to the Born criterion [13], $C' = C44$ vanishes when the critical stability volume is reached by expansion of a single crystal. For a polycrystal the observed shear constant is a complicated, non-linear average over grains having all possible orientations. Following Kroner [71], the average shear constant of a polycrystal was evaluated in terms of the elastic-constant tensor for a single crystal [12] (Figs. 7.17 and 7.19). As may be expected, the shear constant of a polycrystal does not vanish at the critical volume expansion (Fig. 7.26), although in each crystallite C' vanishes in the special coordinate system defined above.

Besides the distinction between single-crystal and polycrystal elastic constants, there is a conceptual issue in the interpretation of the measured shear modulus of a glass. Since the glass is a system

characterized by very long relaxation times, it appears that any measurement over a finite-time interval will result in the high-frequency modulus, $C'(\infty)$. (In liquids, $C'(\infty)$ is well known to be finite in spite of a vanishing static shear constant, $C'(0)$.) By contrast, the calculated moduli in Figs. 7.17, 7.19 and 7.22 are static shear moduli, as are the elastic constants appearing in the Born criterion. A direct quantitative comparison of the experimental and calculated values of C' in Figs. 7.25 and 7.26, respectively, is therefore not meaningful.

7.7 SUMMARY AND OUTLOOK

Conceptually the most direct way to study the role of interfaces in melting is to prepare two crystal samples identical in every way except that only one contains an interface and to heat them uniformly to elevated temperatures. To observe the spatial characteristics of the onset of melting, one must follow the real-time evolution of the local microstructural order in the two samples. While this apparently straightforward procedure is not feasible in the laboratory in several respects, we have shown in this chapter that it can be realized through molecular dynamics simulation. What one finds is that the interfacial region, if present, invariably melts first and the melt then propagates through the bulk. By monitoring the propagation velocity and extrapolating to a temperature where the propagation velocity vanishes, one obtains the melting temperature T_m where the crystal and liquid free energies can be shown to be equal. In the absence of an interface the crystal melts by a distinctly different mechanism at a temperature $T_s > T_m$; the structural disordering occurs uniformly and quite abruptly in the system at the critical temperature and volume determined by the vanishing of the shear modulus. These results constitute compelling evidence, arguably the most clear-cut to date, that thermodynamic melting is a heterogeneous process involving nucleation and growth at the interface, and that a second mechanism, mechanical melting, is a homogeneous process associated with the onset of a shear instability.

In order to observe mechanical melting one must

have a means of suppressing thermodynamic melting. In simulation this is achieved by eliminating the nucleation site, the interface. In an actual experiment, one can keep the system at low temperatures while inducing structural disordering by various other means. This, in effect, is what takes place in solid-state amorphization. Thus amorphization experiments involving electron and ion irradiations and hydrogen absorption provide a considerable body of data which demonstrate the existence of heterogeneous and homogeneous modes of crystal disordering. The larger implication is that one may regard both the $C \rightarrow L$ and the $C \rightarrow A$ transitions as generalized melting in the sense of destruction of crystalline order through volume expansion, with temperature (and the corresponding level of atomic mobility) as the distinguishing variable relating the two.

In this work we have formulated an extended phase diagram to describe both melting and solid-state amorphization by focusing on the independent thermodynamic state variables of temperature and volume. The essence of the phase diagram lies in the extension of the two melting curves to below the triple-point temperature, on the basis that sublimation can be kinetically suppressed. An obvious advantage of the formulation is its physical simplicity, since both state variables are directly accessible to measurement and calculation. The analogy between melting and amorphization also can be discussed in terms of the Gibbs free energy or the entropy, quantities which may be viewed as dependent functions of T and V [4, 14]. While one could, in principle, deconvolute the functional dependence and obtain formally equivalent descriptions, free-energy and entropy variations are not generally available from either measurement or calculation, and their physical content is relatively more difficult to visualize. Nevertheless, it would be very desirable to examine the implications of our proposed phase diagram in the context of the free energy.

It is important to emphasize that our conclusions regarding the effects of lattice expansion on amorphization apply only to materials which expand on melting. This would exclude materials which densify on melting, such as silicon or ice. From our $T-V$ diagram point of view such materials, usually covalently bonded, are expected to amorphize when forced to undergo an isothermal compression rather than an expansion, as confirmed by recent experiments of Clarke *et al.* [60] who amorphized silicon by application of external pressure. This phenomenon has also been demonstrated in the work of Mishima *et al.* [61] which shows that the 1 h phase of ice undergoes a $C \rightarrow A$ transformation when hydrostatically compressed to 10 kbar at 77 K. In this context, it may be relevant that silicon can be amorphized by heavy ions but not by electrons, even when irradiated at temperatures as low as 10 K [72]. This well-known difference in the damage response of silicon has never been fully explained. However, from the viewpoint of expansion effects, the explanation may well be that silicon undergoes no measurable volume change under electron irradiation [73], whereas under argon-ion bombardment it contracts [74]. By contrast, most metals expand on melting *and* upon the introduction of Frenkel pairs [56].

A final issue which deserves comment here is the question of the glass-transition temperature, T_g. Within the framework of our proposed phase diagram, T_g does not appear as a natural concept. For several other basic reasons we find it difficult to assign fundamental meaning, in the thermodynamic sense, to the concept of the glass transition. First, since the process involves non-equilibrium kinetics, the glass transition is not a concept based on equilibrium thermodynamics. Secondly, different definitions of T_g are commonly being used, depending on the system properties of interest and on details of how the glass was quenched from the liquid. Finally, the nature of the glass transition itself is not understood, and is very much a subject of current study.

In concluding, we acknowledge that our view of amorphization is unconventional and admittedly overly simplistic. When the role of chemical disorder, the free energy and entropy, and the different ways of producing volume expansion to induce glass formation are considered more explicitly, modifications of our basic picture may be needed. Nevertheless, our hope is to provoke renewed discussion of the underlying thermodynamic basis of solid-state amorphization, and to stimulate additional work in both simulation and experiment in this fascinating area.

ACKNOWLEDGMENT

This work was supported by the US Department of Energy, BES Materials Sciences, under Contract W-31-109-Eng-38.

REFERENCES

1. For recent reviews, see W. L. Johnson, *Prog. Mater. Sci.*, **30**, 81 (1986); *Solid-State Amorphizing Transformations*, (eds R. B. Schwartz and W. L. Johnson), Elsevier Sequoia, Netherlands (1988); *J. Less-Common Met.*, **140** (1988).
2. R. W. Cahn and W. L. Johnson, *J. Mater. Res.*, **1** (1986), 724.
3. P. Richet, *Nature*, **331** (1988), 56.
4. H. J. Fecht and W. L. Johnson, *Nature*, **334** (1988), 50.
5. P. R. Okamoto, L. E. Rehn, J. Pearson, R. Bhadra, and M. Grimsditch, *J. Less-Common Met.*, **140** (1988), 231.
6. A. R. Ubbelohde, *Molten State of Matter: Melting and Crystal Structure*, Wiley, Chichester (1978).
7. R. W. Cahn, *Nature*, **323** (1986), 668.
8. M. P. Allen and D. J. Tildesley, *Computer Simulation of Liquids*, Oxford Science Publications, Clarendon Press, Oxford (1987).
9. F. F. Abraham, *Adv. Phys.*, **35** (1986), 1.
10. S. R. Phillpot, J. F. Lutsko, D. Wolf, and S. Yip, *Phys. Rev.*, **40** (1989), 2831.
11. J. F. Lutsko, D. Wolf, S. R. Phillpot and S. Yip, *Phys. Rev. B*, **40** (1989), 2841.
12. D. Wolf, P. R. Okomoto, S. Yip, J. F. Lutsko and M. Kluge, *J. Mater. Res.*, **5** (1990), 286.
13. M. Born and K. Huang, *Dynamical Theory of Crystal Lattices*, Clarendon Press, Oxford (1954).
14. J. L. Tallon, *Nature*, **342** (1989), 658.
15. J. Daeges, H. Gleiter, and J. H. Perepezko, *Phys. Lett. A*, **119** (1986), 79.
16. M. S. Daw and M. I. Baskes, *Phys. Rev. Lett.*, **50** (1983), 1285; *Phys. Rev. B*, **29** (1984), 6443.
17. M. W. Finnis and J. E. Sinclair, *Phil. Mag. A*, **50** (1984), 45.
18. D. Wolf and K. L. Merkle, in this volume; see also D. Wolf, J. F. Lutsko and M. Kluge, in *Atomistic Simulation of Materials* (eds V. Vitek and D. J. Srolovitz), Plenum Press, New York (1989), p. 245.
19. F. H. Stillinger and T. A. Weber *Phys. Rev. B*, **31** (1985), 5262.
20. B. P. Feuston, R. K. Kalia and P. Vashishta, *Phys. Rev. B*, **35** (1987), 6222.
21. S. R. Phillpot and D. Wolf, *MRS Symp. Proc.*, **122** (1988), 129; *Phil. Mag. A*, **60** (1989), 545.
22. W. D. Luedtke and U. Landman, *Phys. Rev. B*, **37** (1988), 4656.
23. M. D. Kluge, J. R. Ray and A. Rahman, *Phys. Rev. B*, **36** (1987), 4234.
24. U. Landman, W. D. Leudtke, R. Barnett, C. L. Cleveland, M. W. Ribarsky, E. Arnold, S. Ramesh, H. Baumgart, A. Martinez and B. Kahn, *Phys. Rev. Lett.*, **56** (1986), 155; U. Landman, W. D. Luedtke, M. W. Ribarsky, R. N. Barnett, and C. L. Cleveland, *Phys. Rev. B*, **37** (1988), 4637; W. D. Luedtke, U. Landman, M. S. Ribarsky, R. N. Barnett and C. L. Cleveland, *Phys. Rev. B*, **37** (1988), 4647.
25. M. D. Kluge and J. R. Ray, *Phys. Rev. B*, **39** (1989), 1738.
26. G. H. Bishop, R. J. Harrison, T. Kwok, and S. Yip, *J. Appl. Phys.*, **53** (1982), 5596.
27. J. F. Lutsko, D. Wolf, S. Yip, S. R. Phillpot and T. Nguyen, *Phys. Rev. B*, **38** (1988), 11752.
28. D. Wolf, *J. Am. Cer. Soc.*, **67** (1984), 1, and in *Computer Simulations in Materials Science* (eds R. J. Arsenault, J. R. Beeler and D. M. Esterling), ASM, Metals Park, OH (1987), p. 111.
29. M. Parrinello and A. Rahman, *J. Appl. Phys.*, **52** (1981), 7182.
30. P. S. Ho, T. Kwok, T. Nguyen, C. Nitta, and S. Yip, *Scripta Metall.*, **19** (1985), 993; T. Nguyen, P. S. Ho, T. Kwok, C. Nitta, and S. Yip, *Phys. Rev. Lett.*, **57** (1986), 1919.
31. M. J. Ashby, *Surf. Sci.*, **31** (1972), 498.
32. J. Q. Broughton and X. P. Li, *Phys. Rev. B*, **35** (1987), 9120.
33. H. Gleiter and B. Chalmers, *High-Angle Grain Boundaries*, Pergamon Press, Oxford (1972), p. 113.
34. T. E. Hsieh and R. W. Balluffi, *Acta Metall.*, **37** (1989), 1637.
35. D. A. Smith and T. Y. Tan in *Grain Boundaries in Semiconductors* (eds H. J. Leamy, G. E. Pike and C. H. Seeger), North-Holland (1982).
36. J. Y. Tsao, M. J. Aziz, M. O. Thompson and P. S. Peercy, *Phys. Rev. Lett.*, **56** (1986), 2712.
37. For a review of grain-boundary migration, see the chapter by D. A. Smith in this volume.
38. J. M. Rickman, S. R. Phillpot and D. Wolf, *MRS Symposium Proceedings*, **193** (1990), 325; J. M. Rickman, S. R. Phillpot, D. Wolf, D. Woodraska and S. Yip, *J. Mater. Res.*, **6** (1991), 2291.

39. V. Pontikis, *J. de Phys.*, **49** (1988), C5–327.

40. G. Ciccotti, M. Guillope, and V. Pontikis, *Phys. Rev. B*, **27** (1983), 5576; M. Guillope, G. Ciccotti, and V. Pontikis, *Surf. Sci.*, **144** (1984), 67.

41. T. Nguyen, Ph.D thesis, MIT, 1988.

42. P. Deymier, A. Taiwo and G. Kalonji, *Acta Metall.*, **35** (1987), 2719; J. Q. Broughton and G. H. Gilmer, *Phys. Rev. Lett.*, **56** (1986), 2692.

43. J. W. M. Frenken, P. M. Maree, and J. F. van der Veen, *Phys. Rev. B*, **34** (1986), 7506.

44. R. M. J. Cotterill, *Phil. Mag.*, **32** (1975), 1283; F. F. Abraham, *Phys. Rev. B*, **23** (1981), 6145; J. Q. Broughton and G. Gilmer, *J. Chem. Phys.*, **79** (1983), 5105, 5119; *Acta Metall.*, **31** (1983) 845; W. Schommers and P. von Banckenhagen, *Vacuum*, **33** (1983), 733; *Surf. Sci.*, **162** (1985), 144; V. Pontikis and V. Rosato, *Surf. Sci.*, **162** (1985), 150; V. Rosato, G. Ciccotti and V. Pontikis, *Phys. Rev. B*, **33** (1986), 1860.

45. F. F. Abraham and J. Q. Broughton, *Phys. Rev. Lett.*, **56** (1986), 734.

46. P. Stoltze, J. K. Norskov, and U. Landman, *Phys. Rev. Lett.*, **61** (1988), 440; E. T. Chen, R. N. Barnett and U. Landman, *Phys. Rev. B*, **41** (1990), 439.

47. R. L. Cormia, J. D. Mackenzie, and D. Turnbull, *J. Appl. Phys.*, **34** (1963), 2239.

48. J. B. Boyce and M. Stutzmann, *Phys. Rev. Lett.*, **54** (1985), 562.

49. C. J. Rossouw and S. E. Donnelly, *Phys. Rev. Lett.*, **55** (1985), 2960.

50. J. Ray and A. Rahman, *J. Chem. Phys.*, **80** (1984), 4423, and *Phys. Rev. B*, **32** (1985), 733.

51. S. R. Phillpot, S. Yip and D. Wolf, *Computers in Physics*, **3** (1989), 20.

52. J. L. Tallon, *Phil. Mag.*, **39** (1979), 151; J. L. Tallon and W. H. Robinson, *Phil. Mag.*, **36** (1977), 741; J. L. Tallon, *J. Phys. Chem. Solids*, **41** (1984), 837.

53. L. E. Rehn, J. Holder, A. V. Granato, R. R. Coltman and F. W. Young, Jr., *Phys. Rev. B*, **10** (1974), 349.

54. P. R. Okamoto and M. Meshii, in *Science of Advanced Materials* (eds H. Wiedersich and M. Meshii), ASM International, Metals Park, OH (1990), p. 33.

55. P. Ehrhart, K. H. Robrock and H. Schober, in *Physics of Radiation Effects in Crystals* (eds R. A. Johnson and A. N.

Orlov), Elsevier, North-Holland, Amsterdam (1986), p. 7.

56. W. J. Meng, P. R. Okamoto, L. J. Thompson, B. J. Kestel, and L. E. Rehn, *Appl. Phys. Lett.*, **53** (1988), 1820; W. J. Meng, P. R. Okamoto, and L. E. Rehn, in *Science of Advanced Materials* (eds H. Wiedersich and M. Meshii), ASM International, Metals Park, OH (1990).

57. X. L. Yeh, K. Samwer and W. L. Johnson, *Appl. Phys. Lett.*, **42** (1983), 242.

58. R. B. Schwartz and R. R. Petrich, *J. Less-Common. Met.*, **140** (1988), 189.

59. J. Koike, D. M. Parkins and M. Nastasi, *Phil. Mag. Lett.*, **62** (1990), 257.

60. D. R. Clarke, M. C. Kroll, P. D. Kirchner, R. F. Cook and B. J. Hockey, *Phys. Rev. Lett.*, **60** (1988), 2156.

61. O. Mishima, L. D. Calvert and E. Whaley, *Nature*, **310** (1984), 3935.

62. D. E. Luzzi, H. Mori, H. Fujita and M. Meshii, *Scripta Metall.*, **19** (1985), 897.

63. H. Fujita, H. Mori and M. Fujita, *Proc. 7th Int. Conf. on High Voltage Electron Microscopy* (eds R. M. Fisher, R. Gronsky and K. H. Westmacott), University of California Press, Berkeley (1984), p. 233.

64. H. Mori, H. Fujita, M. Tendo and M. Fujita, *Scripta Metall.*, **18** (1985), 783.

65. H. Mori and H. Fujita, *Proc. Yamada Conf. VII on Dislocations in Solids* (ed H. Suzuki), University of Tokyo Press, Tokyo (1985), p. 563.

66. H. A. Atwater and W. L. Brown, *Appl. Phys. Lett.*, **56** (1990), 1.

67. H. Fecht, Z. Fu and W. L. Johnson, *Phys. Rev. Lett.*, **64** (1990), 1753.

68. A. Seidel, G. Linker and O. Meyer, *J. Less-Common Met.*, **145** (1988), 189.

69. C. Massobrio, V. Pontikis and G. Martin, *Phys. Rev. Lett.*, **62** (1989) 1142.

70. J. Koike, P. R. Okamoto, L. E. Rehn, R. Bhadra, M. Grimsditch and M. Meshii, *MRS Symp. Proc.*, **157** (1990), 777.

71. E. Kroner, *J. Eng. Mech. Div.*, **106** (1980), 890.

72. D. N. Seidman, R. S. Averback, P. R. Okamoto and A. C. Bailey, *MRS Symp. Proc.*, **51** (1987), 349.

73. V. Vook, *Phys. Rev.*, **125** (1962), 855.

74. S. Andersson, *Phys. Letters A*, **33** (1970), 455.

CHAPTER EIGHT

Wetting of surfaces and grain boundaries

D. R. Clarke and M. L. Gee

Introduction · Long-range forces · Wetting of external surfaces · Wetting of internal surfaces · Effect of surface chemistry · Closing remarks

8.1 INTRODUCTION

In the last decade or so the investigation of various aspects of wetting has seen an enormous growth driven by both a need to understand in greater detail the processes involved in epitaxial growth of semiconductors and the opportunity to exploit new-found experimental and computational capabilities. The results of this endcavor arc illustrated by a number of the contributions to this volume. For instance, the various growth modes in epitaxy are described by Batstone, the rich variety of phase transitions exhibited in monolayers adsorbed on graphite substrates, investigated by X-ray scattering, is described by Mochrie and colleagues, and the equally fascinating early stages of growth as revealed by scanning tunneling microscopy are detailed by Hamers.

In this contribution we describe the wetting behavior of free surfaces, under ambient atmospheres, and of grain boundaries (GBs). Although it can be expected to be an even richer subject than those mentioned above, it is, at the present time, a much less developed science. By way of contrast to the other chapters, where the emphasis is on short-range interatomic interactions, and because of our own interests, the approach we have chosen to take here is to emphasize the role of long-range inter-

atomic forces in wetting. Although these forces are usually much weaker they can nevertheless have profound effects on the wetting of surfaces by liquids, and it is believed that they are also important in the wetting of GBs in a number of ceramic materials.

Wetting of free surfaces has a number of rather obvious applications, ranging from the spreading of paints and inks to the coating of metals. However, the wetting of internal surfaces, such as GBs, is no less important. For instance, the densification of many ceramic materials from their powders proceeds by mass transport through a high temperature, liquid phase. If it remains wetting the resulting GBs, it can cause degradation in the elevated temperature mechanical (and electrical) properties. In addition, a number of electrical ceramics derive their peculiar electrical properties from a wetting phase. For instance, barrier layer capacitors have a continuous, insulating, second phase along their GBs from which they derive unusually high capacitances. A second example is the zinc oxide varistor ('variable resistance') material. In this class of material a high temperature wetting phase retracts from the GBs, localizing at the three-grain junctions, yet leaves an electrically active layer of bismuth (and other ions) on the boundaries to provide a Schottky barrier to electrical transport at the boundaries.

We start by describing the principal long-range forces now known to exist at surfaces between liquids and solids. Then follows a rather brief summary of some of the principal features of the continuum wetting of free surfaces. In section 8.4 we attempt to describe some of the observations that have been made of the wetting of GBs. Since little work has been done in this area under controlled conditions, this section essentially represents a phenomenology of wetting of GBs. Finally, we make mention of how alterations in surface chemistry can dramatically alter the wetting of those surfaces.

8.2 LONG-RANGE FORCES

A variety of long-range forces is now known to exist [1] and can affect the wetting behavior of materials under appropriate conditions. In this section, the principal ones that occur in the majority of solid–liquid interactions are summarized. In section 8.3, the effect of these long-range forces on wetting is described.

8.2.1 Van der Waals forces

The van der Waals forces are truly ubiquitous and have their origin in the electromagnetic interaction between molecules in a material. Electronic fluctuations result in momentary charge delocalizations and so, at any instant, a molecule possesses a temporary dipole. These dipoles create transient electric fields in their vicinity which, in turn, induce charge delocalizations in the surrounding medium. This occurs continuously and so within a material each point is simultaneously a generator and recipient of electromagnetic fields. The correlation between the random electric fields and the induced electric fields leads to an overall reduction of the total electromagnetic energy of the material. Consequently, there is an attractive force between two materials, even when separated by a third material. These induced dipole-induced dipole interactions are generally termed dispersion or London interactions. The same type of correlations

also occur between two permanent dipoles and between permanent and induced dipoles. Collectively, these three forces are now known as van der Waals forces and follow a $1/r^6$ dependence with distance for microscopic distances. Dispersion forces are, perhaps, the most important of the three since they are always present and generally dominate, except in the case of small, highly polar molecules.

The relevance of van der Waals forces between microscopic particles to the interactions between macroscopic bodies was first realized by Hamaker, but his theory included only pairwise interactions between molecules in the two macroscopic bodies. Many-body dipole interaction effects were first successfully accounted for by Lifshitz. The Lifshitz theory ignores atomic structure and treats materials as continuous media which best absorb and radiate energy at frequencies which correspond to their natural frequencies [2, 3]. Therefore, van der Waals forces can be related to the spectral absorption characteristics of the materials under consideration. According to the Lifshitz theory, the interaction energy per unit area $E_{132}(h)$ between half spaces with dielectric responses $\varepsilon_1(w)$ and $\varepsilon_2(w)$ separated by a thickness, h, of material having dielectric response $\varepsilon_3(w)$ is, in a non-retarded approximation, given by

$$E_{132}(h) = \frac{-A_{132}}{12\pi h^2} \tag{8.1}$$

where A_{132} is a Hamaker constant defined in terms of the dielectric responses at specific frequencies by

$$A_{132} = \frac{-3kT}{2} \sum_{n=0}^{\infty}{}' \int_0^{\infty} dx\, x \ln\left[1 - \Delta_{13}\Delta_{23}e^{-x}\right]$$

where

$$\Delta_{kj} = \frac{\varepsilon_k(i\xi_n) - \varepsilon_j(i\xi_n)}{\varepsilon_k(i\xi_n) + \varepsilon_j(i\xi_n)}$$

and the frequencies are

$$\xi_n = n\left[\frac{2\pi kT}{\hbar}\right]$$

The prime on the summation indicates that the $n = 0$ term should be divided by 2. The van der Waals force is given by the derivative of the interaction energy,

$$\frac{-\mathrm{d}E_{132}(h)}{\mathrm{d}h} = \Pi_{\mathrm{vdw}} \qquad (8.2)$$

where $\Pi_{\mathrm{vdw}}(h)$ is the van der Waals component to the disjoining pressure (section 8.3.2). The methodology for computing the van der Waals forces from the above parameters is described in detail by Hough and White [4].

8.2.2 Electrostatic double layer forces

When a material comes into contact with water or some other liquid of high dielectric constant, the chemical groups at the surface of the material can ionize (e.g. carboxylic acid groups dissociate according to $-COOH \Leftrightarrow -COO^- + H^+$), leaving the surface with an overall net charge. The surface charge, σ, is balanced by an equal but oppositely charged atmosphere of counterions concentrated in a diffuse double layer. The potential in the double layer, $\psi(x)$, is given by the solution to the Poisson–Boltzmann equation, and decays away from the surface such that at $x = \infty$, $\psi(\infty) = 0$ and $\frac{\mathrm{d}\psi}{\mathrm{d}x} = 0$.

In the Gouy–Chapman theory, the effective thickness of the double layer is $1/\kappa$, the Debye length, where

$$\kappa = \left[\frac{8\,\pi\varepsilon^2 n_0}{\varepsilon kT}\right]^{\frac{1}{2}}$$

and ε is the dielectric constant of the liquid and n_0 is number density of ions in the bulk solution [1].

Now, if two similarly charged surfaces approach each other such that their respective double layers overlap, there exists a repulsion between the surfaces. The simplest case for analyzing overlapping double layers is that of two flat plates which are at the same surface potential, ψ_s, and are separated by a distance h such that at the midplane the potential is ψ_d and its gradient is zero. As a consequence, the ion concentration between the two plates is different to that of the adjacent bulk medium and there is an osmotic pressure force acting on the plates, in addition to the force from

the electric field. The electric field $\mathrm{d}\psi/\mathrm{d}x$ is symmetrical about the midplane, which implies that the total force across the plates is given by the net osmotic pressure at this plane, since at equilibrium the total force must be a constant everywhere in the region between the surfaces. From these conditions, it follows the total force, P, is given by the equation

$$P = 2kTn_o\left[\cosh\left(\frac{e\psi_d}{kT}\right) - 1\right] \qquad (8.3)$$

where

$$\psi_d \approx \frac{4kT}{e}\cdot\tanh\left(\frac{e\Psi_s}{4kT}\right)\cdot e^{-\frac{\kappa h}{2}} \qquad (8.4)$$

This pressure due to the overlap of the diffuse double layers corresponds to the electrostatic component Π_{edl} of the total disjoining pressure (section 8.3.2) of a thin film. For an asymmetric system such as an adsorbed film in equilibrium with its vapor, the film/vapor interface is taken to be equivalent to the midplane of the above situation and so the above equations remain applicable.

An example of the combined van der Waals and electrostatic double layer forces is shown in Fig. 8.1. It compares experimental results for the alumina/water system with the force distance curves predicted using an attractive van der Waals and repulsive double layer forces.

8.2.3 Solvation forces

The intermolecular surface forces discussed thus far, i.e. van der Waals forces and double layer forces, are both described in terms of continuum theories. In both cases the thin film is regarded as a structureless continuum having properties of the bulk liquid such as dielectric constant and density. However, thin films are sometimes of thicknesses comparable to molecular dimensions and so the discreteness of the molecules in the film cannot be ignored.

Certain bulk liquids exhibit a small degree of molecular correlation. Taking a single reference molecule, the density profile, $\rho(r)$, of the surrounding liquid extending radially away from the

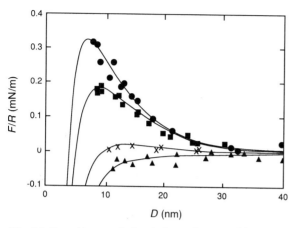

Fig. 8.1 Force between the basal planes of two sapphire crystal platelets immersed in 10^{-3} M NaCl solution plotted as a function of separation. The continuous lines are the theoretical DLVO forces, i.e. the sum of a van der Waals attraction and an electrical double-layer repulsion. By changing the pH, the total ionic strength and hence the electrical double-layer repulsion is systematically varied. pH 6.7, triangles; pH 9, pluses and crosses; pH 10, squares; pH 11, circles.

reference molecule is a damped oscillatory function of distance, r, and approaches the density of the bulk liquid at large distances. $\rho(r)$ is called the radial distribution function. The distance between successive maxima and minima corresponds to one molecular diameter. The magnitude and range of the oscillations of $\rho(r)$ generally depend on the shape of the liquid molecules. Small, rigid, symmetric molecules exhibit more order than large, flexible, asymmetric molecules that have a greater number of configurational degrees of freedom (DOFs).

It is not surprising that this type of ordering of liquid molecules also exists in thin films. For a thin liquid film adjacent to a molecularly smooth solid surface, the interactions between the surface and the liquid molecules and the geometric constraint the surface imposes on the liquid give rise to density oscillations, ρ_s, extending away from the surface [5]. However, density oscillations are not expected to occur at a vapor–liquid interface so for a solid–film–vapor system ρ_s decays to the bulk liquid density.

When a thin film is constrained between two smooth solid surfaces, the effects of liquid ordering

can have a dramatic effect on the interaction energies. At large surface separations ($\geqslant 10$ liquid molecular diameters) ρ_s approaches that for a film on an isolated surface. However, at small surface separations, there is overlap of the structured regions from each wall so the intervening film no longer has the same density as the bulk liquid. As the surface separation changes the liquid molecules reorder so as to be accommodated between the two surfaces. The variation of this ordering with surface separation gives rise to what is termed a solvation force (Fig. 8.2). The solvation force varies between attraction and repulsion as a function of surface separation depending on whether or not the gap between the surfaces can accommodate an integral number of molecular layers of the liquid film [6]. Disruption of the liquid ordering results in a repulsive solvation force. It should be pointed out that the van der Waals and solvation forces are not additive. Rather, the solvation force **is** the van der Waals force for very thin films where the molecular structure cannot be ignored.

8.3 WETTING OF EXTERNAL SURFACES

The subject of wetting of external surfaces exposed to a vapor species is an enormous one but is, as pointed out by de Gennes in a seminal review [7], in many respects poorly understood. Exploring the intricacies is beyond the scope of this short contribution, instead the interested reader is referred to de Gennes' review. Rather, the purpose of this section is to introduce some of the principal concepts involved that relate to the role of long-range forces in wetting so that comparisons can be made with the wetting behavior of internal surfaces described in the following section.

8.3.1 Macroscopic equilibrium

When a small liquid droplet is placed on a flat, non-deformable surface, two distinct regimes of behavior may be observed (Fig. 8.3). The liquid can spread out to completely wet the surface or form an equilibrium shape droplet with a finite

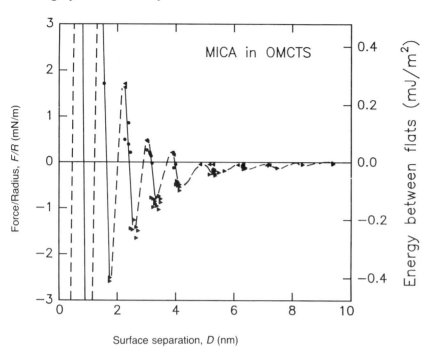

Fig. 8.2 An example of the structural force arising from the constrained packing of molecules (octamethylcyclotetrasiloxane in this case). The data show an alternating series of maxima and minima as the surface separation is varied with the force oscillating between attraction and repulsion. The period of oscillation is approximately equal to the size of the molecule. The continuum van der Waals force, if that alone were operating, would be given by the dotted line. Reproduced courtesy of R. G. Horn.

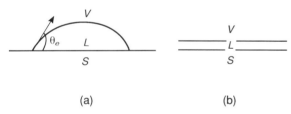

Fig. 8.3 Schematic of the shape of a small droplet on a surface and the macroscopically defined contact angle. (a) partial wetting, $\theta_e > 0$, and (b) complete wetting, $\theta_c = 0$.

contact angle, θ_e. The former is commonly referred to as 'complete wetting' and the latter as 'partial wetting.' (The transition from one to the other is described in section 8.3.4.) For the case of partial wetting, the contact line corresponds to where all three phases, the solid, liquid, and equilibrium vapor, are in contact. Little is known about the actual shape of the liquid at the contact line but it is possible, as originally shown by Young, to relate the macroscopic contact angle, θ_e, to the bulk values of the interfacial free energies, γ_{ij}, without any knowledge of the immediate 'core' region around the contact line. By invoking a virtual work

argument to an incremental displacement of the contact line along the surface of the solid, the equilibrium contact angle is given by

$$\gamma_{SV} - \gamma_{SL} - \gamma_{LV} \cos \theta_e = 0 \qquad (8.5)$$

the familiar Young equation. Where the contact angle, θ_e, is zero, the droplet completely wets the surface.

Although, as mentioned above, a detailed knowledge of the core region around the contact line is not necessary to derive the conditions for mechanical equilibrium of a partial wetting droplet, the shape is dependent on the nature and range of the long-range surface forces and should provide a sensitive observational test of these forces. This has yet to be done and also the shapes themselves have to be computed in detail for different forces but clearly promises to be a fertile research activity. An indication of the shapes under the influence of different types of forces has been presented by de Gennes and is reproduced schematically in Fig. 8.4.

On placing the liquid droplet on the surface, or when the vapor pressure is changed, the droplet

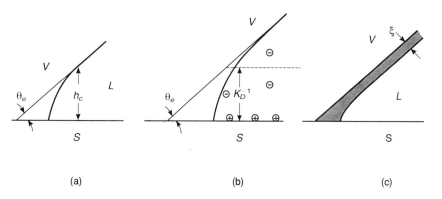

Fig. 8.4 Examples of the microscopic profile of the liquid at a contact line (schematic). (a) Effect of an attractive van der Waals force. For $\theta_e \ll 1$, the shape is hyperbolic and the height is of the order of $A/\gamma\theta_e$ where A is Hamaker constant. (b) Effect of an electrical double-layer potential. (c) Near a liquid/vapor triple point, the interface becomes diffuse. Redrawn after de Gennes.

will spread in order to attain the equilibrium contact angle. The driving force for spreading can be expressed in terms of the bulk interfacial free energies by the spreading coefficient, S:

$$S = \gamma_{SV} - \gamma_{SL} - \gamma_{LV} \qquad (8.6)$$

This coefficient was originally introduced near the turn of the century as a measure of the tendency for a droplet to spread over a surface.

8.3.2 Wetting by adsorption and the equilibrium thickness of films

In many physical systems, wetting occurs by the adsorption of atoms or molecules from the vapor phase onto a (chemically) non-reactive substrate to form a condensed monolayer or multilayer of thickness, h. The film thickness is given by the experimentally measured adsorption isotherm in terms of the vapor pressure, P, relative to the saturated vapor pressure, P_{sat}. Following the original suggestion of Polanyi [8], this is related to a potential function $\Delta E(h)$ associated with the interaction of the adsorbed molecules with the surface. This can be expressed formally as

$$\frac{kT}{\Omega}\ln\frac{P_{\text{sat}}}{P} = \left(\frac{\partial \Delta E}{\partial h}\right)_T = \Pi(h) \qquad (8.7)$$

and is equal to the total disjoining pressure $\Pi(h)$ [9]. The disjoining pressure, $\Pi(h)$, represents the sum of all the surface forces, including those described in the previous section and can also be expressed in terms of the change in chemical

potential of the molecules as they absorb to the surface

$$\Pi(h) = -(\mu_f - \mu_b)/\Omega \qquad (8.8)$$

where μ_b is their chemical potential in the bulk liquid. From this definition, it can be seen that a positive disjoining pressure favors film thickening since this corresponds to a decrease in chemical potential on adsorption, whereas a negative disjoining pressure causes the film to thin.

The equilibrium thickness of the adsorbed film is then obtained by minimizing the total energy of the film, including that due to the surface forces and to any gravitational interactions. As an example, the wetting of a number of different alkanes on quartz (Fig. 8.5) leads to film thicknesses which are in agreement with those predicted from eq. (8.2) where the disjoining pressure is solely due to dispersion forces given by the Lifshitz theory [2].

8.3.3 Spreading equilibrium

The final shape that a spreading, but not completely wetting, droplet adopts on a surface is dependent on not only the bulk interfacial energies through the spreading coefficient but also the surface force contributions $\Pi(h)$, the curvature at the contact line and any gravitational and hydrostatic effects, $G(h)$. The total free energy may be expressed, in one-dimension, x, by the functional

$$f = f_o - \int_{+a}^{b} dx\left[S - \frac{\gamma_{LV}}{2}\left(\frac{dh}{dx}\right)^2 - \Pi(h) - G(h)\right]$$

$$(8.9)$$

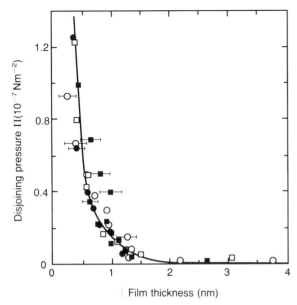

Fig. 8.5 Comparison between the Lifshitz prediction and experiment for the adsorption of n-alkanes on quartz, ● pentane, □ hexane, ■ heptane, ○ octane.

where the integral is evaluated over the extent of the droplet and $h(x)$ is the local thickness of the film. Minimization with respect to the film thickness, $h(x)$, leads to a condition for equilibrium of

$$-\gamma_{LV}\frac{d^2h}{dx^2} + \frac{d\Pi}{dh} + \frac{dG}{dh} = 0 \qquad (8.10)$$

which on integration gives

$$\frac{\gamma_{LV}}{2}\left(\frac{dh}{dx}\right)^2 = \Pi(h) + G(h) - S \qquad (8.11)$$

Thus, knowing the surface force contribution, the spreading coefficient, and the gravitational contribution, the shape of the film is completely defined. As an example, the shape of the contact line for a partially wetting liquid when only a dispersion force is acting can be simply obtained as follows. For a non-zero contact angle $S = -\frac{1}{2}\gamma_{LV}\theta_e^2$. Assuming that gravitational and hydrostatic effects can be ignored, eq. (8.11) reduces to

$$\left(\frac{dh}{dx}\right)^2 - \theta_e^2 = 2\Pi_{vdw}(h)/\gamma_{LV} \qquad (8.12)$$

Writing the dispersion force in its non-retarded form as

$$\Pi_{vdw}(h) = \frac{A}{12\pi h^2} \qquad (8.13)$$

the shape of the film has a hyperbolic form

$$h^2 = (x\theta_c)^2 + \frac{A}{6\pi\gamma_{LV}\theta_e^2} \qquad (8.14)$$

as sketched in Fig. 8.4(a). Similar calculations can be carried out for other surface forces but, as mentioned earlier, there have not been any thorough investigations of the contact shape to compare with.

The analysis above is a continuum one, yet the thicknesses predicted in the presence of short-range surface forces can be of the order of molecular dimensions. In fact, in recent work on the spreading of polydimethylsiloxane polymer droplets on surfaces, the advancing profile is seen to consist of layers each corresponding in height to the size of the polymer molecule [10].

8.3.4 The wetting transition

In many systems a partially wetting phase can become wetting, or vice versa, when a thermodynamic variable such as temperature, pressure or composition is altered. The change from non-wetting to wetting behavior occurs under well defined conditions corresponding to attaining thermodynamic co-existence of the phases involved. This transition occurs close to, but not generally at, a critical point, the temperature and pressure at which all three phases can exist simultaneously, and has been termed critical point wetting by Cahn, who first analyzed the transition in terms of critical behavior [11].

The principal features of the wetting transition and its relation to critical point mixing are illustrated by the experiments performed by Moldover and Cahn [12] using a mixture of cyclohexane and methanol. They are described here since they can be reproduced simply and vividly demonstrate the wetting transition. The cyclohexane–methanol mixture has a bulk critical point at a temperature,

T_c, of about 46°C, and additions of water are known to raise the critical temperature. In the absence of water, and at all temperatures up to 46°C, Moldover and Cahn found that the methanol-rich liquid completely wet the cyclohexane-rich liquid, separating it from both the walls of a glass cuvette and the saturated vapor. The formation of a thin film of the methanol-rich phase on the surface was despite the fact that it is the denser of the two liquids. As water was added to the mixture, shifting the composition away from the critical point, and the temperature lowered from the critical temperature, a transition from complete wetting to partial wetting occurred with a macroscopic contact angle forming between the methanol-rich liquid, the cyclohexane-rich liquid, and the vapor at the surface of the liquids. The transition is reversible by simply increasing the temperature. Similar examples of the wetting transition have now been studied in a variety of systems including alcohol–fluorocarbon and nitromethane–carbon disulphide mixtures and methane and ethylene on graphite.

For a number of years there was considerable discussion in the literature as to whether the wetting transition was a first or second order transition, i.e. whether the transition was accompanied by an abrupt, discontinuous increase in thickness of the wetting phase on the surface or not. However, it is now recognized as a result of a number of careful, predominantly continuum, analyses [13] that long range surface forces can alter the transition from being second to first order. Furthermore, long range forces that cause partial wetting (as distinct from promoting complete wetting) can suppress the wetting transition altogether until the critical point is reached [14]. Experimental evidence consistent with the occurrence of the wetting transition being due to the interplay between short range and long range forces has been obtained and is described in section 8.4.

8.4 WETTING OF INTERNAL SURFACES

The wetting of internal interfaces is, as might be expected from the considerable greater difficulties involved in experimentation, far less investigated than that of external surfaces described above. As a result, the phenomenology of internal surface wetting is relatively unexplored and much of the microstructural evidence for wetting comes from studies not specifically directed towards understanding the physics or the systematics of wetting, *per se*. The underlying physics of internal surfaces wetting is, of course, likely to be similar to that outlined earlier, but in wetting internal surfaces in solids at least two additional factors are involved. One is the strain energy accompanying the spreading of a wetting phase along the interface. The other is that the kinetics associated with spreading may well be quite different since accommodation will be by diffusional processes involving the solid phase, including such processes as solution-reprecipitation. These will be discussed later.

In the majority of cases of interest to the materials community the wetting is of grain boundaries by a chemically compatible liquid formed at temperatures above the solidus temperature. In contrast to many of the systems from which the physics of external surface wetting have been elucidated, the materials are usually multicomponent ones. Furthermore, the GBs within a polycrystalline material can exhibit a variety of DOFs and consequently have a spectrum of interfacial energies rather than a single one.

8.4.1 Microstructural distribution

Before describing some of the variety of wetting behavior of GBs revealed by SEM and TEM, we start with the simplest case, the morphology of wetting GBs in an isotropic solid by a melt as first discussed by C. S. Smith [15] in the context of the equilibrium microstructure of metals. By drawing an analogy with the wetting of fluids, Smith argued that GBs in crystalline solids behave in much the same way as interfaces in fluids; namely they are continuous, have a definite free energy per unit area and are extendable. He further argued that, this being so, the shapes of grains in a polycrystal are determined by the force equilibrium between the surface tensions of intersecting GBs (the equivalent of Young's equation). On the basis of these assumptions, he cataloged the shape of a second phase, such as a solidified melt phase, formed at the intersection of isotropic grains as a function of dihedral angle, the included angle subtended by the contact lines, of the melt. Thus, he was able to show that for dihedral angles of 60° or smaller, the second phase would, at equilibrium, be stable along the three-grain edges. And when the dihedral angle was zero complete wetting of all the GBs would occur, separating each grain by a layer of the wetting phase. Smith presented optical micrographs of cross sections of a number of metals and alloys illustrating the variety of wetting behaviors he cataloged. The range of shapes of the wetting phase as a function of both volume fraction and dihedral angle has more recently been plotted by Wray (Fig. 8.6) following Smith's methodology [16]. In today's nomenclature, all but the zero dihedral angle case (and 60° for the three-grain intersection line) would be termed partial wetting. In the complete wetting case, with isotropic grains, grain boundary wetting layers would meet at 120° as do the Plateau borders in a soap froth. Since metals and ceramics are rarely truly isotropic, the angle between wet grain boundaries only approximate to 120°. One case in which the wetting

boundaries do indeed appear to intersect at 120° is reported to be the wetting of solid helium 4 fcc by liquid helium [17]. The effect of crystallographic anisotropy introducing anisotropic GB energies was subsequent to the work of Smith analyzed by Herring [18]. He showed that mechanical equilibrium would be satisfied at the contact point provided

$$\gamma_{SS}(\theta) = 2\gamma_{SL}(\phi) \cos \psi + 2\frac{\partial \gamma_{SL}}{\partial \phi} \sin \psi \quad (8.15)$$

where $\gamma_{SS}(\theta)$ is the angular dependence of the grain boundary energy, ϕ is the orientation of the solid–liquid interface and ψ is the dihedral angle (the included angle of contact).

8.4.2 Complete wetting

By balancing the macroscopic surface tensions, all the two-grain junctions in a polycrystalline solid will be wet provided the inequality

$$2\gamma_{SL} < \gamma_{SS}$$

is satisfied. In such a case, the total volume fraction of liquid will be spread uniformly over all the GBs. This appears to occur, in addition to the He–He example mentioned above, in metal systems, such as partially melted Ag–Cu alloys, when rapidly quenched from above the solidus, at least when examined in the optical microscope. It is also known to occur in certain refractories.

In the last decade or so, complete wetting of GBs has been reported in a variety of ceramic materials primarily as a result of high resolution transmission electron microscopy investigations undertaken as part of characterization studies [19]. An example of the wetting of grain boundaries in a hot-pressed silicon nitride by a siliceous intergranular film is reproduced in Fig. 8.7. From micrographs such as this, it has been noted that (a) the majority of the volume fraction of the wetting phase resides at the three- and four-grain junctions, and (b) the thickness of the intergranular phase is approximately constant, varying from material to material but not substantially from one boundary to the next in a material. In only one instance to date has the former observation been systematically studied as

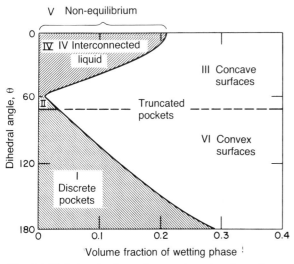

Fig. 8.6 Variations in shape of a second wetting phase, β, as a function of its dihedral angle and volume fraction. Redrawn after Wray.

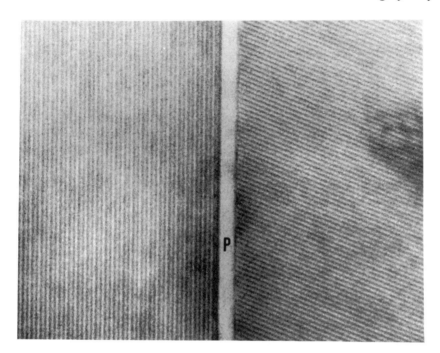

Fig. 8.7 High resolution transmission electron micrograph of a hot-pressed aluminum substituted silicon nitride (SiAlON) alloy illustrating the presence of a continuous, wetting intergranular phase, P, along a GB. The lattice spacing of the vertical fringes is 1.1 nm.

the volume fraction of the wetting phase has been altered. Greil and Weiss [20] investigated a Si—Al—O—N ceramic, where changes in composition correspond linearly to the volume fraction of the liquid phase. They found, using transmission electron microscopy, that as the liquid phase volume was increased, the volume of second phase at the three-grain junctions also increased, whereas the thickness along the GBs remained approximately constant.

In the majority of ceramics examined, the thickness of the wetting phase is microscopic and often as small as 0.8–1.5 nm. For siliceous wetting phases, this latter thickness is almost of molecular dimensions. (The size of the SiO_4 tetrahedra in silica is ∼0.3 nm.) This length scale has suggested that the origin of such an equilibrium thickness is a result of the interplay of long-range interfacial forces acting between the grains and across the intergranular phase at temperatures at which it is liquid [21]. According to this argument, the force acting between grains during liquid phase sintering, when the liquid phase is formed, is given by the disjoining pressure augmented by the

capillary pressure, P_{cap}, together with any externally applied stress P_{app}, namely

$$\Pi = P_{cap} + P_{app} + \Pi_{vdw} + \Pi_{edl} + \Pi_{st} \quad (8.16)$$

Π_{vdw} is the van der Waals pressure, Π_{edl} is the osmotic pressure due to the presence of any electrical double layer and Π_{st} is the structural disjoining pressure.

The structural disjoining pressure referred to here has been attributed to the repulsion between misoriented grains arising from the ordering correlation of SiO_4 tetrahedra imposed by the first layer of tetrahedra, on each grain, adopting an epitaxial orientation [21]. Calculations of the equilibrium thickness of a siliceous wetting phase in a variety of ceramics from the balance between van der Waals forces and the structural disjoining pressure are similar to those observed experimentally. However, it remains to be established whether these thin films are truly wetting or prewetting films, whose thickness is determined by combined long and short range forces, or whether they are adsorption films.

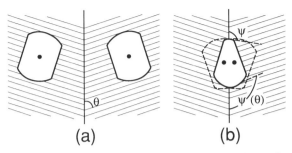

Fig. 8.8 Graphical construction for determining the shape of a partial wetting phase at a GB having an angular misorientation, θ. The details of the construction are described in the text. Redrawn after Rottman.

8.4.3 Partial wetting and GB spreading

The shape of a partial wetting phase, away from any core region, along a grain boundary can be found using a rather simple graphical construction provided the equilibrium shape of the liquid in the form of a trapped inclusion is known. The construction is due to Johnson and has been described in detail by Rottman [22]. The first step, shown in Fig. 8.8, is to draw in the shape of the liquid inclusion, appropriately oriented, in both grains with the center of each inclusion equidistant from the boundary and on a line perpendicular to the boundary plane. In the second step, the inclusion shapes are brought together along the line and pass one another until their centers are a distance γ_{SS} apart. The final step is to erase the portion of each inclusion shape on the same side of the boundary that its center lies. The remnant lines indicate the shape of the partial wetting phase, with the slopes of the lines determining the dihedral angle, ψ. The construction is a simple one and points to the need of comprehensive observations of the shapes of liquid inclusions in grains. Unfortunately few have been made in materials where the wetting of grain boundaries is of interest.

The spreading along grain boundaries of a liquid phase located at triple grain junctions as the temperature is raised or equivalently the retraction of a wetting phase to the triple junctions on cooling is analogous to the wetting transition against a surface studied by Cahn and referred to in section 8.3.4. Because it requires a redistribution not only of the liquid phase but also of the solid phase, the kinetics and the transition temperature itself may well be different than against a free surface. An example of a transmission electron micrograph of a partially retracted liquid phase is reproduced in Fig. 8.9. The micrograph is of a GB region in a $ZnO–Bi_2O_3$ oxide material cooled rapidly from a temperature at which the ZnO grains are completely wet by a bismuth oxide liquid phase. Because the volume of material is conserved, it is evident that retraction to a partially wetting state has involved transport of the bismuth oxide phase and diffusion of zinc oxide to replace the bismuth oxide that had previously coated the boundary. One of the intriguing aspects of the retractions of the bismuth oxide phase from being a wetting phase is that on retracting the liquid leaves either a thin bismuth-rich film or a bismuth segregation along the GB. Whether either represents a non-equilibrium effect or a pre-wetting transition remains to be established.

8.4.4 Macroscopic versus microscopic contact angle

In materials containing thin wetting films, it is frequently observed that the contact angle at the three-grain junction measured at macroscopic dimensions, e.g. in the optical or scanning electron microscope, is greater than zero. Such a measurement would suggest that a wetting film will not be present. However, examination at higher magnification reveals that there is no definite contact angle but rather a continuously varying one as shown schematically in Fig. 8.9. Similar observations have been reported in the emulsions literature.

A continuously varying contact angle can be a consequence of long-range, distance dependent interfacial forces. This is seen from the analysis first presented by Feijter and his colleagues [23]. They considered the force balance acting at an arbitrary cross-section (Fig. 8.10) of the film whose local thickness is dependent on the excess free energy $\Delta E(h)$ due to long-range interfacial forces.

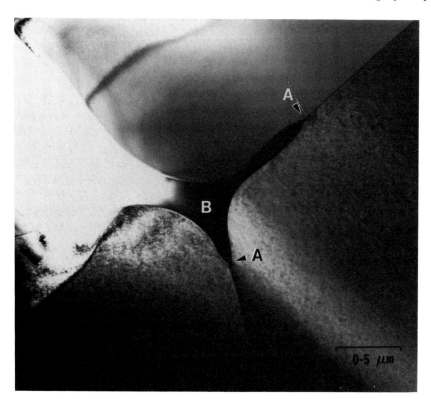

Fig. 8.9 Transmission electron micrograph of a partially wetting bismuth-rich phase, B, in a commercial ZnO varistor material. Ahead of the contact points, A, the GBs are enriched in bismuth.

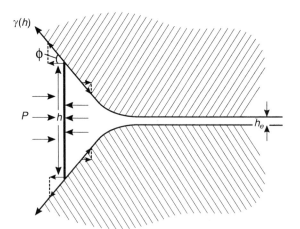

Fig. 8.10 Horizontal forces acting on an arbitrary cross-section of an intergranular liquid whose thickness is determined by surface forces. The macroscopic dihedral angle is $2\theta_e$ and the local one is $2\phi(h)$. Redrawn after de Feijter and Vrij.

Their analysis established the local slope of the solid–liquid interface in terms of the surface force contribution evaluated at both that point, h, and at its equilibrium thickness, h_e, well away from the triple point:

$$\cos\phi(h) = \frac{2\gamma_{SL} + \Delta E(h_e) + (h - h_e)P}{2\gamma_{SL} + \Delta E(h)} \quad (8.17)$$

P, as earlier, denotes the hydrostatic pressure in the liquid. By extrapolating the profile of the solid–liquid interface, far from the triple point, $h \to \infty$, to a contact point at $h = h_e$ the macroscopic dihedral angle can be defined. This has the value

$$\cos\theta = \cos\phi(h_e) = \frac{2\gamma_{SL} + \Delta E(h_e)}{2\gamma_{SL}} \quad (8.18)$$

Thus, as in the partial wetting of a free surface, the

local deviation in contact angle in the vicinity of the contact line provides information about the surface force contribution to the interfacial energy.

8.4.5 Grain boundary penetration

A distinction is made in this chapter between GB spreading and GB penetration. The former, as discussed above, is the change in wetting of GBs under conditions of constant total volume of material, and is the microstructural redistribution of the wetting phase. In contrast, the latter relates to the wetting of pre-existing GBs by an externally applied wetting phase. For instance, one method for the manufacture of GB layer strontium titanate capacitors is to first prepare dense, polycrystalline strontium titanate, then paint a layer of the dopant oxides on the surface and, finally, heat the material so that the dopant oxides melt and penetrate down the grain boundaries. A second example, of more historical interest now, is the penetration down GBs of metals by a second, liquid metal causing embrittlement, for instance gallium attack of aluminum. (There remain FAA regulations against the air transportation of such embrittling metals.)

Grain boundary penetration of a dense, polycrystalline material by a wetting phase must lead to a volume increase of the material. The manner in which this is accommodated will affect the kinetics of wetting and conceivably also alter the thermodynamics of wetting through the work that must be expended in changing the volume. Two extreme cases can be considered. In one, the penetrating liquid dissolves away the solid along the boundaries to the extent necessary to form the requisite final thickness of the liquid phase at the boundaries. The dissolved material is then transported to the outer surface where it can plate out on the surface grains. The volume of the material is increased but the grain centers have not moved relative to one another. The other extreme is that the penetrating liquid wedges apart the grains so that their centers move apart. During the penetration, when the penetration has not proceeded throughout the material, the expansion will be constrained by the unpenetrated, unexpanded material and as a result elastic strains will be generated that will act

to retard the rate of penetration. An analogy may be drawn to non-equilibrium crack growth in a material in an environment that lowers its surface energy: The material can lower its free energy by cracking in the presence of the environment but the rate at which the crack can propagate is determined by the instantaneous balance between the rate of release of strain energy (caused by the presence of the crack) and the rate at which surface energy is gained. Thus, the crack propagation rate can be altered by stresses that in turn increase or decrease the strain energy release rate. In principle the crack can be stopped or even reversed by an appropriate stress. Similarly, it can be expected that stress can aid the rate of penetration along GBs, as appears to be the case in liquid metal embrittlement, or slow the penetration. No analysis of this possibility yet exists but is desirable to help define the important parameters and regimes of behavior. Perhaps the analysis closest to pertaining is that of creep cavity growth along GBs at high temperatures [24].

In the majority of cases reported in the literature, penetration occurs under manifestly non-equilibrium conditions where there is a chemical driving force aiding wetting plus the possibility of residual or applied stresses so that the net spreading force is much different to that given by S. Even in the more controlled experiments the wetting liquid is not in chemical equilibrium with the solid phase. Two studies of penetration in which the wetting phase is in equilibrium with the crystalline phase have been undertaken [25, 26]. In both, the liquid phase, a calcium aluminum silicate liquid, was first saturated by heating to the penetration temperature in the presence of excess, single crystal aluminum oxide. The work of Flaitz and Pask [25] clearly showed that the saturated liquid penetrated the grain boundaries in a high-purity ('Lucalox') aluminum oxide. In more recent work, on the same system, Shaw was able to measure the volumetric expansions accompanying the penetration and showed that the penetration was linear with time rather than parabolic, suggesting that diffusion was not rate-limiting. A feature of both sets of experiments was that the polycrystalline aluminum oxide was fully immersed in the liquid phase thereby avoiding any capillary pressure due to incomplete wetting of the solid forming menisci at the surface. This work, and the findings to date, suggest that the $Al_2O_3-CaO \cdot Al_2O_3 \cdot SiO_2$ system may well be an excellent, albeit non-cubic, model system for wetting studies of GBs.

Finally, mention may be made of the experiments of Rabkin *et al.* [27], who have investigated the wetting of $43° \langle 100 \rangle$ symmetrical tilt GBs in bicrystals of Fe–6 a/o Si by zinc and tin. Although the liquid phase, molten zinc or tin, was not saturated with respect to either iron or silicon, they observed penetration with zinc but not with tin. Intriguingly, they found by X-ray microanalysis penetration contact point observed in the microscope. They attributed this to a phenomenon of prewetting of the boundaries and found that there was a pressure dependence. Confirmation of their results, preferably by a higher resolution technique, is necessary.

8.4.6 Grain boundary anisotropy and faceting

It has long been known from experiments, such as on GB grooving, and from dislocation modeling that the energy of a GB is dependent on its crystallographic nature. This key notion has been quantified and related to interatomic potential over the past 30 years as discussed in a companion contribution to this volume (Wolf). A consequence of an anisotropic GB energy is that wetting may be dependent on the GB misorientation. Such an example is reproduced in Fig. 8.11 of an aluminum oxide ceramic. The wetting phase, a strontium barium aluminum silicate, appearing light in the micrograph, completely wets one of the GBs but only partially wets another, forming 'lenses' along the boundary. The third boundary is wet but has clearly faceted, probably during wetting.

One way of rationalizing which GBs will be wet by second phase is to compare the angular misorientation of the GB interfacial energy with that of the orientation dependence of the solid–liquid interfacial energy. This cannot be done in any quantitative manner at the present time and must await interatomic potential based computations of all the energies involved. Nevertheless, the comparison can be shown schematically as in Fig. 8.12.

The GB surface tension as a function of misorientation has a dependence as indicated with a sharply rising portion (given by $\Delta E/\Delta E_o = \theta(A - \ln \theta)$ according to the Read–Shockley dislocation model [28]) and a series of cusps corresponding to 'special' GBs, ones having low symmetry, such as twin-related boundaries, and large proportions of atoms from either lattice in common. The orientation dependent solid–liquid interfacial tension is less clearly defined, not being the subject of much investigation. It can, however, be expected to vary with orientation and some indication of its functional form can be gained from examination of the shape of liquid inclusions trapped within grains. It cannot be reasonably expected to go to zero however, whereas the GB interfacial does. Comparing the two interfacial tensions, certain GB orientations can lower their interfacial tensions by being wet whereas others cannot. As a result, an anisotropy of wetting behavior arises. Whilst specific predictions as to which boundaries will be wet cannot at present be made, it is likely that the low angle and certain 'special' GBs will not be since they have the lowest interfacial tensions. This is consistent with the relatively few observations that have been made to date, but much more needs to be done.

The foregoing was couched in terms of the bulk solid–liquid interfacial tension, γ_{SL}. In practice, for thin-film wetting, the angular dependence of the long-range interfacial force contribution may dominate. Unfortunately, there is almost no data as to its magnitude. The one exception appears to be the work of McGuiggan and Israelachvili [29], who measured, using the Surface Force Apparatus, the work required to separate, in a liquid, two mica sheets misoriented by a twist angle, θ_o. An example of their results, for separation in a 1 mM KCl solution, is reproduced in Fig. 8.13. The work of separation is seen to be strongly dependent on the twist angle, having a peak at zero misorientation, and as illustrated in the figure strongly influenced by the equilibrium separation of the mica sheets. In the 1 mM KCl solution, there are a number of equilibrium separations, corresponding to minima in the long-range interfacial energy function, due to discrete layers of water molecules.

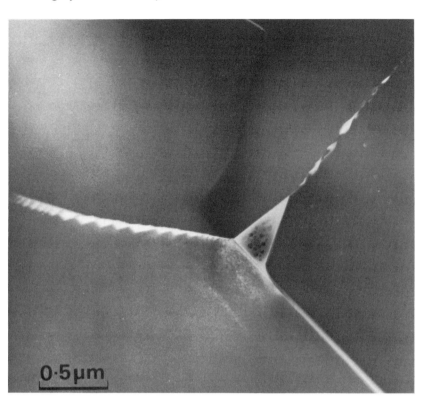

Fig. 8.11 A three-grain junction region in an aluminum oxide ceramic illustrating a different wetting behavior along each of the three GBs.

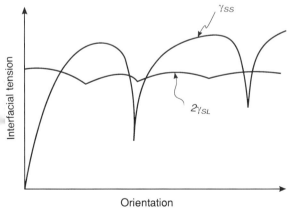

Fig. 8.12 Comparison of the expected GB interfacial tension, γ_{ss}, as a function of angular misorientation and the orientation dependence of the grain surface–liquid interfacial tension, $2\gamma_{SL}$, as a function of the surface orientation.

These findings of McGuigan and Israelachvili are notable because the lattice orientation effect is present even while the interaction energy has an oscillatory form and because, in the mica/KCl system at least, the effects of misorientation are detectable over separation distances of several molecular distances. Also, somewhat intriguingly, the misorientation dependence of the interaction energy fits, functionally, the Read–Shockley description of the energy of a grain boundary modeled as a set of low-angle GB dislocations. This, despite the lack of any recognizable dislocation type of defect in the ordered KCl solution to accommodate the misorientation of the mica lattices.

A less direct but nevertheless pertinent experiment illustrating the effect of GB misorientation

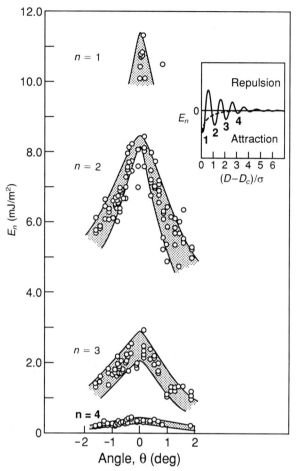

Fig. 8.13 Measured adhesion energies versus twist angle between two mica sheets immersed in a 1 mM KCl solution. The measurements were made from distances corresponding to successive energy minima in the oscillatory force–distance curve shown in the insert. The minima occur at distances corresponding to the diameter of the water molecule ($\sigma = 0.25$ nm). Figure courtesy of McGuiggan and Israelachvili.

8.5 EFFECT OF SURFACE CHEMISTRY

Common experience suggests that the wetting behavior of surfaces can be radically altered by chemical modification of the surfaces. Despite this, relatively few well-defined studies have been carried out either on the wetting transition or on the adsorption behavior. Since rather dramatic effects can be produced on free surfaces, it can reasonably be expected that changes in wetting of internal surfaces might also be produced by chemical adsorption to interfaces.

The wetting of water on quartz provides a striking example of the role of surface chemistry [31]. The surface of quartz, exposed to moist air, is typically terminated by hydroxyl (OH) groups. In such a condition, termed fully hydroxylated, the contact angle of water is zero. By careful heating, the hydroxyl groups can be removed, yielding a progressively more hydrophobic surface until, when completely dehydroxylated, the contact angle becomes ~45°. Recent work has shown that water films adsorb on partially heat-dehydroxylated quartz in a step-wise manner (Fig. 8.14a) as a function of the relative vapor pressure and degree of hydrophobicity, with the size of the individual steps corresponding to the size of a water molecule. In the dehydroxylated state, the absence of the hydroxyl groups eliminates the possibility of any surface ionization effects and thus any electrical double layer, hydration forces or hydrogen bonding forces affecting the wetting behavior. In marked contrast, when the quartz surface is hydroxylated, the equilibrium thickness of the film is much greater and the surface is completely wet, even when exposed to the unsaturated vapor (Fig. 8.14b).

Altering the nature of the solid surface can also affect the wetting transition as well as the contact angle. Abeysuriya *et al.* [32] studied the wetting transition of a liquid mixture of 40% nitromethane and 60% carbon disulphide in a borosilicate tube whose surface was treated to change the degree of hydroxylation by absorbing non-polar methyl groups from a silylating agent. The fully hydroxylated surface is completely wet, below the transition temperature, by the nitromethane-rich liquid

has been performed by Kagot and Albright [30]. They measured the fracture resistance of a series of pure aluminum, symmetric $\langle 110 \rangle$ tilt bicrystals embrittled with a liquid Hg-3% Ga alloy as a function of their crystallographic misorientation. They found that the fracture resistance was inversely proportional to the GB energy, being a maximum for low-angle and 'special' boundaries and a minimum for high-angle GBs.

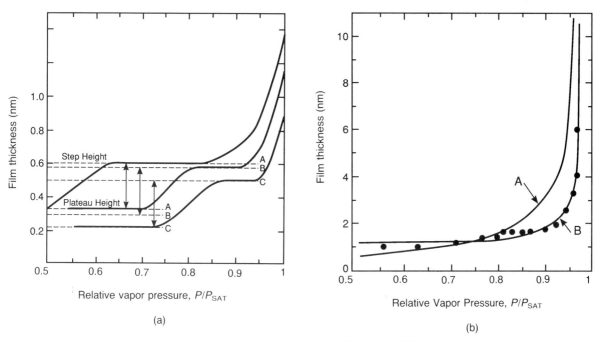

Fig. 8.14 (a) Water film thickness on quartz surfaces having varying degrees of hydrophobicities as a function of relative vapor pressure. The water absorbs in layers with the step height corresponding to the size of a water molecule. Curves A, B, and C correspond to hydrophobicities that give contact angles of 22°, 30°, and 43° (b) Water film thickness on a fully hydroxylated quartz surface as a function of relative vapor pressure. The contact angle is zero. The solid line A corresponds to a predicted isotherm based on BET behavior and B is a line through the data. Thickness measurements by ellipsometry.

and the mixture exhibits a wetting transition at 63.47°C. As the degree of hydroxylation is decreased by silylation, the wetting phase changes to the carbon disulphide-rich liquid and the transition is suppressed until 1 K of the critical mixing temperature.

8.6 CLOSING REMARKS

The relationship between the wetting behavior of free surfaces by liquids, the macroscopic interfacial energies, and long-range surface forces has reached a certain stage of maturity despite the absence of detailed computations using inter-atomic force laws or indeed atomic level models for the interfacial structures. By contrast, the wetting behavior of grain boundaries by a liquid phase is poorly understood even in continuum descriptions.

Furthermore, it is not yet apparent that the full systematics of GB wetting have been revealed by experiment. One of our motivations in writing this chapter has been to present the paucity of experimental observations of GB wetting in the hope that it might spur others to also investigate this area of wetting.

It should be clear that an atomic level understanding of the wetting of both external and internal surfaces is still at a rather rudimentary level. Yet, as briefly discussed in the previous section, wetting is not determined solely by long-range forces but can be significantly affected by short range interactions too. In fact one of the attractions of the subject, as well as one of the more significant complications, is that, except for certain specialized situations, the wetting behavior of surfaces appears to be dictated by a balance between long-range and short-range interactions.

REFERENCES

1. J. N. Israelachvili, *Intermolecular and Surface Forces*, Academic Press (1985).
2. I. E. Dzyaloshinskii, E. M. Lifshitz and L. P. Petaeviskii, *Adv. Phys.*, **10** (1961), 165.
3. J. Mahanty and B. W. Ninham, *Dispersion Forces*, Academic Press (1976).
4. D. B. Hough and L. R. White, *Adv. Colloid. Interface Sci.*, **14** (1980), 3.
5. F. F. Abraham, *J. Chem. Phys.*, **68** (1978), 3713.
6. R. G. Horn and J. N. Israelachvili, *J. Chem. Phys.*, **75** (1981), 1400.
7. P. G. de Gennes, *Rev. Mod. Physics*, **57** (1985), 827.
8. See, for instance, A. W. Adamson, *Physical Chemistry of Surfaces*, Wiley (1982), p. 543.
9. B. V. Derjaguin, *Theory of Stability of Colloids and Thin Films*, Consultants Bureau, New York (1982).
10. F. Heslot, A. M. Cazabat, P. Levinson and N. Fraysse, *Phy. Rev. Lett.*, **65** (1990), 599.
11. J. W. Cahn, *J. Chem. Phys.*, **66** (1977), 3667.
12. M. R. Moldover and J. W. Cahn, *Science*, **207** (1980), 1073.
13. See, for instance, G. F. Teletzke, L. E. Scriven and H. T. Davis, *J. Chem. Phys.*, **77** (1982), 5794.
14. P. G. de Gennes, *C. R. Acad. Sci. II*, **297** (1983), 9.
15. C. S. Smith, *Trans. A.I.M.E.*, **175** (1948), 15.
16. P. J. Wray, *Acta Metall.*, **24** (1976), 125.
17. J. P. Franck, J. Gleeson, K. E. Jornelsen, J. R. Manuel and K. A. McGreer, *J. Low Temp. Phys.*, **58** (1985), 153.
18. C. Herring, *The Physics of Powder Metallurgy* (ed W. E. Kingston), McGraw-Hill (1951).
19. D. R. Clarke, *Annual Review of Mater. Sci.*, **17** (1987), 57.
20. P. Greil and J. Weiss, *J. Mater. Sci.*, **17** (1982), 1571.
21. D. R. Clarke, *J. Am. Ceram. Soc.*, **70** (1987), 15.
22. C. Rottman, *Scripta Metall.*, **19** (1985), 43.
23. J. A. Feijter and A. Virj, *J. Electro. Anal. Chem.*, **37** (1972), 9.
24. T. J. Chuang, K. I. Kagawa, J. R. Rice and L. Sills, *Acta Metall.*, **27** (1979), 265.
25. P. L. Flaitz and J. Pask, *J. Am. Ceram. Soc.*, **70** (1987), 449.
26. T. M. Shaw, *J. Am. Ceram. Soc.*, **74** (1991), 2495.
27. E. L. Rabkin, L. S. Shvindlerman and B. B. Straumal, *Coll. de Physique*, **4** (1990), 599.
28. See, for instance, J. P. Hirth and J. Lothe, *Theory of Crystal Dislocations*, McGraw-Hill (1968).
29. P. M. McGuiggan and J. N. Israelachvili, *J. Mater. Res.*, **5** (1990), 2232.
30. J. A. Kargot and D. L. Albright, *Met. Trans. A*, **8** (1977), 27.
31. M. L. Gee, T. W. Healy and L. R. White, *J. Colloid. Interface Sci.*, **140** (1990), 450.
32. K. Abeysuriya, X-I. Wu and C. Franck, *Phys. Rev. B*, **35** (1987), 6771.

PART II:
Semi-bulk and Thin-film Interfaces

Structural, electronic, and magnetic properties of thin films and superlattices

A. Continenza, Chun Li, and A. J. Freeman

Introduction · Computational method · Transition metal surfaces and metal–metal interfaces
Metal–ceramic interfaces · Semiconductor interfaces and heterojunctions

9.1 INTRODUCTION

The promise of various high technological applications has sparked deep interest in solid state heterophase interfaces. Interfaces either control or strongly affect many chemical, electronic, and mechanical properties of materials, which are, of course, important technologically. The science and the technology of interfaces and nanostructures in general has been recently growing at a remarkable pace, boosted by the constantly increasing demand for 'better' materials suitable for device applications. This has caused dramatic advances in both condensed matter theory and experimental physics.

Experimentalists are now able to obtain high purity new materials, using highly sophisticated synthesis techniques, such as molecular beam epitaxy (MBE) and organo-metallic-vapor–phase deposition (OMVPD), and by means of refined and advanced techniques, such as low energy electron diffraction (LEED), X-ray photoemission (XPS), surface extended X-ray adsorption fine structure (SEXAFS), photo-luminescence (PL), to name just a few, are also able to obtain a complete and careful characterization of materials which do not exist in nature.

On the other hand, theorists have been able to follow the experimentalists' pace by implementing and developing new formalisms and theoretical concepts to be applied to real materials (rather than to restricted elemental models) and thereby giving precise answers and interpretations to the experimental data. Very often, they have even managed to lead the way by proposing new structures, simulating new materials for study and providing predictions and data, on the atomic scale, that are often impossible or hard to obtain experimentally.

The issues that are of particular relevant interest for these peculiar systems (such as interfaces, overlayers, and nanostructures) are many and span a wide variety of different fundamental physics aspects: structural properties, magnetic properties, electronic and transport properties (including potential barriers). On the other hand, the ultimate goal underlying the study is to achieve the capability of modifying these basic properties and eventually to 'tune' them at will in order to design the *ad hoc* material for each different device application. Of course, the basic requirement for that is the complete knowledge and understanding of each single property and how it can be affected and modified by changing several different external factors, such as growth condition, thickness of overlayers, strain, choice of substrate, interface reactivity, etc.; all this remains a formidable task.

The kinds of answers that a computational approach can give are various and comprise quantities which are experimentally accessible, such as density of states, and energy level transitions which are accessible to angle-resolved photoemission spectroscopy, as well as some that are not, such as, for example, in-layer bond-length relaxations. More specifically, such studies can give precise information regarding the interface morphology (bond length, atomic distances, and surface relaxation), the electronic properties (such as band gap, energy transitions, carrier effective mass as well as magnetic moments, hyperfine fields and magnetic anisotropy, potential line up and Fermi level pinning), and the stability properties (stability of nanostructures against disproportionation in the constituents, stability against chemical reaction or diffusion at the interface). These kinds of information are of remarkable relevance since they have been shown to be closely related to the final electrical and magnetic behavior of the interface. Moreover, understanding the electronic properties of a given system is at the basis of the full comprehension of its electric, thermal as well as mechanical properties; clearly, band structure calculations give us a powerful tool to explore in detail these properties.

The advantage of a computational approach consists in the ability to predict basic properties and to study 'clean' systems in which interface intermixing or reaction is absent (and that are not so easily accessible experimentally) in order to single out each different contribution. In recent years, highly precise LDF (local density functional theory) and LSDF (local spin density functional theory) methods have been used to obtain structural, electronic, and magnetic properties of many different materials, demonstrating how these calculations can give a clear understanding of the experimental results and lead to predictions on novel systems not yet made in the laboratory. In particular, the magnetization, including its anisotropic behavior, and charge and spin densities obtained from these spin-polarized calculations can

be compared directly with experiments, including conversion electron Mössbauer spectroscopy and other measurements of hyperfine fields. For excitations – like spin-polarized photoemission and inverse photoemissions – spin polarized photocurrents can be calculated from the LSDF results.

In this chapter, we discuss a few selected examples aimed at illustrating which kinds of information are reasonable to expect from a computational approach and their reliability when compared with experiments. We will first present a brief description of the computational approach used and, because of space limitations, we will then describe some of the systems we have been recently studying and which exemplify some of the interface and overlayer properties.

9.2 COMPUTATIONAL METHOD

One of the major problems that any band structure calculation has to face, within the Born–Oppenheimer approximation, is how to solve the electronic Hamiltonian that describes the system taking also into account the many-body electron–electron interactions. This makes the problem particularly complicated and hard to solve since one is forced to deal with many-body wavefunctions that are dependent on the coordinates of all the N electrons in the system. In this regard, density functional theory (DFT) (and more specifically the local density approximation or LDA) represents a powerful tool which simplifies the original many-body problem and reduces it to the solution of a one-electron Hamiltonian. This is achieved by making use of the Hoenberg–Kohn theorem [1] which states that the ground state energy of the many-body system with charge density $\rho(r)$ and spin density $n(r)$ can be expressed as a unique functional of the ground state charge density. This leads to Schrödinger-like one-particle equations (Kohn–Sham equations) containing an effective potential energy operator which is determined by the self-consistent charge distribution; thus, the one-particle equations have to be solved iteratively.

One of the most accurate and powerful schemes

to solve the LSDF one-particle equations is our all-electron full-potential linearized-augmented-plane-wave [2] (FLAPW) method which can be applied to bulk and superlattice as well as to thin film systems. The basic idea in this variational method is the partition of real space into three different regions, namely spheres around the nuclei, vacuum regions on both sides of the slab in the case of film calculations, and the remaining interstitial region.

In the FLAPW method no shape approximations are made to the charge density and the potential. Both the charge density and the effective one-electron potential are represented by the same analytical expansions, i.e. a Fourier representation in the interstitial region, an expansion in spherical harmonics inside the spheres, and a two-dimensional Fourier series in each vacuum plane parallel to the surface. The generality of the potential requires a method to solve Poisson's equation for a density and potential without shape approximations; this is achieved by a new scheme recently implemented [3].

Thus, the FLAPW method allows a fully self-consistent solution of the LSDF single-particle equations and yields charge densities and spin densities close to the LSDF limit. Besides the total charge density, the key quantity in density-functional theory is the total energy corresponding to the ground state charge density. We have also implemented a scheme to calculate accurate and stable all-electron density functional total energies [2] and have applied it within the FLAPW method. The capability of total energy calculations for various geometrical arrangements provides us with a powerful theoretical tool to study the energetics and, at least in principle, the dynamics of bulk solids and surfaces.

9.3 TRANSITION METAL SURFACES AND METAL–METAL INTERFACES

The magnetism of the three-dimensional (3D) ferromagnetic transition metals at surfaces and interfaces are of special interest because of their possible use in high-technology information storage devices. Because of the breakdown of the three-dimensional crystal symmetry, the electronic and magnetic properties are expected to be significantly different from that seen in the bulk solids. The interface features strongly control the magnetism of the material at least a few angstroms deep into the bulk. This interface influence on magnetism depends on the atomic structure and the nature of the materials on both sides of the interface.

9.3.1 Surface and interface magnetism

Unusual properties of surface magnetism (viewed as an interface between solid and vacuum) have been obtained from theoretical studies of various ferromagnetic transition metal surfaces; these include surface magnetic moments enhanced over the bulk values, narrowed 3D bandwidths for the states in the surface layer and largely reduced surface hyperfine magnetic fields (except for bcc Cr, for which the hyperfine field is enhanced). Listed in Table 9.1 are the calculated magnetic moments of 3D-transition metal surfaces using the FLAPW method[4–14]. As a general phenomenon, enhanced magnetic moments at the surface layer (with respect to bulk moments) are predicted. The magnetic moments calculated are comparable with experimental findings. Table 9.2 contains the FLAPW results for magnetic overlayers on Cu, Ag, and Au substrates[15–19]. Giant magnetic moments for monolayer Fe overlayers on Cu, Ag,

Table 9.1 Calculated magnetic moments of 3D-transition metal bulk and surfaces

			M_{surf} μ_B	M_{bulk} μ_B	*Ref.*
Ni	fcc	(100)	0.68	0.56	[4]
		(110)	0.63		[5]
Cr	bcc	(100)	2.49 (FM)	0.89 (AFM)	[6]
				0.59 (Exp.)	
Fe	bcc	(100)	2.98	2.15	[7]
		(110)	2.65		[8]
Fe	fcc	(100)	2.85	1.99	[9]
Co	fcc	(100)	1.86	1.65	[10]
Co	hcp	(0001)	1.76	1.65	[11]
Co	bcc	(100)	1.95	1.76	[12]
	bcc	(110)	1.84		[13]

Table 9.2 Calculated magnetic moments of 3D-transition metal thin overlayers on noble metals

Thin films	Moment (μ_B)	Ref.	Thin films	Moment (μ_B)	Ref.
1V/Au(001)	1.75	[14]	AFM 2Fe/Cu(001)	2.38 (S)	[15]
1V/Ag(001)	1.98	[14]		−2.22 (S − 1)	[15]
2V/Ag(001)	1.15 (S)	[14]	1Fe/Ag(001)	2.96	[14]
	<0.05 (S − 1)	[14]	2Fe/Ag(001)	2.94 (S)	[14]
Cr monolayer	4.12	[14]		2.63 (S − 1)	[14]
1Cr/Au(001)	3.70	[14]	1Fe/Au(001)	2.97	[16]
2Cr/Au(001)	2.90 (S)	[14]	1Fe/W(110)	2.18	[17]
	−2.30 (S − 1)	[14]	1Co/Cu(001)	1.79	[18]
Fe monolayer	3.20	[14]	1Ni/Ag(001)	0.57	[19]
1Fe/Cu(001)	2.85	[14]	Cr/Fe/Au(001)	3.10 (Cr)	[14]
FM 2Fe/Cu(001)	2.85 (S)	[15]		−1.96 (Fe)	[14]
	2.60 (S − 1)	[15]	Fe/Cr/Ag(001)	2.30 (Fe)	[14]
				−2.40 (Cr)	[14]

and Au(001) substrates are predicted. The effect of the magnetic properties of the substrate on the magnetic films is also carefully studied.

To illustrate the electronic and magnetic properties of transition metal surfaces and interfaces, we present briefly the FLAPW computational studies of the fcc Co(001) surface and of a Co overlayer on a Cu(001) substrate. The charge and spin densities in the (100) plane of a nine-layer fcc Co(001) slab is shown in Figs. 9.1(a) and (b) respectively. Features similar to those of the other ferromagnetic transition metal surfaces are seen in that the effect of the surface on the charge density is mostly localized to the surface layer. The other layers, including even the sub-surface layer, appear to be very much bulk-like. This is also confirmed by examining the detailed charge population (not shown) within each muffin-tin (MT) sphere of the atom; the charge distribution of the non-surface layer atoms is found to be very much the same.

Unlike the charge density distribution, which has a larger spherical-like component for each atom, the spin density shows strong anisotropic features around the atoms. Similar features have also been found in ferromagnetic metal surfaces of Fe, Ni, and Co. The charge and spin densities in the (100) plane of a Co overlayer on a fcc Cu(001) substrate are shown in Figs. 9.2(a) and (b), respectively. The charge density at the overlayer Co atoms shows few anisotropic features in the contour shapes. The interface Cu atom charge density

contours are slightly different compared with those of the other Cu atoms. The sub-interface (S − 2) and center layer (C) Cu atoms have primarily the same charge distribution, and are thus quite bulk-like. The spin density contours for the overlayer magnetic atoms shows stronger anisotropic features than do the charge density contours. The interface Cu atom also has a region of positive spin density distribution around the center of the atom. The spin density on the non-interface Cu atoms is relatively much smaller. These features have also been seen in other interfaces between ferromagnetic and noble metals.

The hyperfine magnetic field – an important property of magnetic transition metals which can be directly measured by Mössbauer effect and NMR experiments – provides information related to the magnetism which is sensitive to the environment of the atom. The contact part of the hyperfine field (listed in Table 9.3) are those of bcc Co(001), fcc Co(001), and hcp Co(001) surfaces, and the Co overlayer on Cu(001) substrate) is determined by the spin density at the nucleus [20]: this is mostly enhanced by the magnetism of the valence *d*-electrons and so is strongly dependent on the total magnetic moment. The contribution to the hyperfine field by core electrons is found to be rather precisely proportional to the magnetic moment, whereas the total hyperfine field varies from ~−96 kGauss for the overlayer Co atom in Co−Cu(001) to −314 kGauss for the bulk-like

(a) Charge density

(b) Spin density

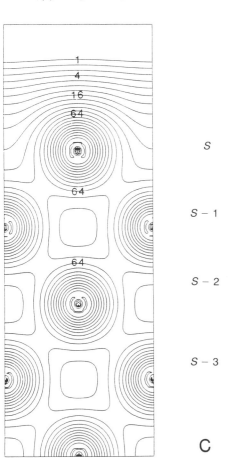

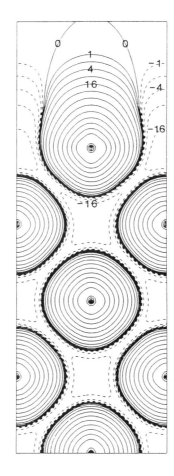

S

S – 1

S – 2

S – 3

Fig. 9.1 (a) Charge and (b) spin density contour plots for a nine-layer fcc-Co(001) slab in the (100) plane. For (a), indicated numbers are in units of $10^{-3}\,e/(a.u.)^3$; subsequent contour lines differ by a factor of $\sqrt{2}$. For (b), indicated numbers are in units of $10^{-4}\,e/(a.u.)^3$; subsequent contour lines differ by a factor of 2.

(center layer) atoms in fcc Co. The almost identical constant value ($-144 \pm 2\%\,\text{kGauss}/\mu_B$) of the ratio of the core hyperfine field and the magnetic moment is evidence that the core polarization coupling constant is not sensitive to the atomic environment.

9.3.2 Magnetic anisotropy of Fe thin films

The magnetic anisotropy properties of ferromagnetic thin-films (Fe, Co, Ni, etc.) on various substrates have been studied via surface magneto-optic Kerr effect (SMOKE) [21], ferromagnetic resonance (FMR), spin-polarized photoemission, etc. In ultra-thin Fe(001) films on Ag(001) (less than 2.5 monolayers), the magnetization is found to lie along the surface normal [22–24]. Both in-plane [25] and perpendicular anisotropy [26] are observed for monolayer range Fe(001) on Au(001). Liu and Bader *et al.* [26] reported a universal behavior of perpendicular spin orientation below a critical thickness of six monolayers or less in fcc Fe(100) on Cu(100), fcc Fe(111) on Ru(0001), bct Fe(100) on Pd(100), and bcc Fe(100) on Au(100).

The origin of magnetic anisotropy of 3D ferromagnetic materials was proposed by van Vleck [27] more than 50 years ago to be the spin-orbit interaction. Still today, the theoretical understanding of the magnetic anisotropy in realistic systems remains a great challenge, because it is necessary to know in precise detail both the electronic structure and the total energy (the latter to $\pm 10^{-5}\,\text{eV}$).

(a) Charge density   (b) Spin density

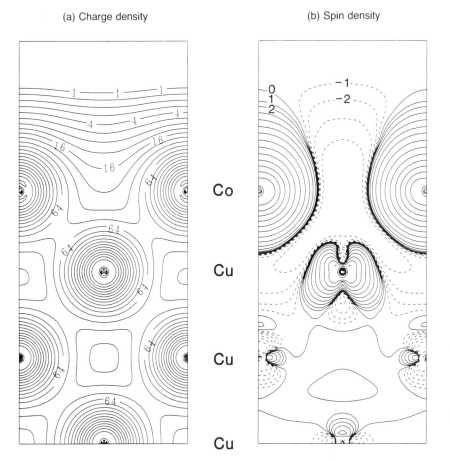

Fig. 9.2 (a) Charge and (b) spin density contour plots for Co overlayer on a Cu(001) substrate in the (100) plane. For (a), indicated numbers are in units of $10^{-3}\,e/$ (a.u.)^3; subsequent contour lines differ by a factor of $\sqrt{2}$. For (b), indicated numbers are in units of $10^{-4}\,e/$(a.u.)^3; subsequent contour lines differ by a factor of 2.

To study the magnetic anisotropy of metal surfaces and thin-films, we developed a so-called 'second variation' method based on our highly-precise total energy FLAPW approach, i.e. to solve the relativistic Dirac equation of the electronic system with a charge density obtained from a previous semi-relativistic self-consistent calculation [28]. As the first test of this method for calculating magnetic anisotropy, the free standing Fe monolayer with a square lattice structure and with lattice constants matching either fcc Ag(001) $(a_{Ag} = 4.086\,\text{Å})$ or fcc Au(001) $(a_{Au} = 4.078\,\text{Å})$ was studied. The charge density of monolayer Fe is determined with the semi-relativistic FLAPW method. The magnetic anisotropy energy is estimated by the second variation procedure. In both systems, a small in-plane preference of the spin orientation of order $\sim3 \times 10^{-5}\,\text{eV/atom}$ (or

an equivalent anisotropy magnetic field of ~1.7 kGauss along the in-plane direction) is obtained from the calculations [28]. This result is in contrast with the result of Gay and Richter [29] who obtained a perpendicular spin orientation in an unsupported Fe monolayer. We also calculated the anisotropy between the two in-plane directions, namely $(\theta = \pi/2, \phi = 0)$ and $(\theta = \pi/2, \phi = \pi/4)$. This anisotropy energy is found to be about two orders of magnitude smaller than the perpendicular/in-plane anisotropy, i.e. $\sim1\%$ of the anisotropy energy between the $(\theta = 0, \phi = 0)$ and $(\theta = \pi/2, \phi = 0)$ directions.

Similar calculations were then carried out for Fe monolayers on Au, Ag, and Pd(001) substrates. For these theoretical studies, the Fe/metal systems were modeled by a single-slab geometry with one layer of p(1 × 1) Fe coupled with one layer of Ag

Table 9.3 The contact hyperfine field (in kGauss) and the ratio of core electron contribution and effective magnetic moments of bcc Co(001), fcc Co(001), and hcp Co(001) surfaces and Co overlayer on a Cu(001) substrate.

	Atom	Core	Conduction	Total	Core/M	Moment (μ_B)
bcc Co(001) surface (9-layer slab model)	S	−285	+104	−181	−147	1.94
	S − 1	−255	−12.1	−268	−144	1.78
	S − 2	−252	+13.6	−238	−144	1.76
	S − 3	−253	+6.3	−247	−144	1.76
	C	−257	+18.3	−238	−146	1.76
fcc Co(001) surface (9-layer slab model)	S	−266	+55.5	−211	−144	1.86
	S − 1	−236	−75.1	−311	−144	1.64
	S − 2	−237	−70.5	−307	−144	1.65
	S − 3	−236	−69.1	−305	−144	1.64
	C	−237	−69.5	−306	−144	1.65
hcp Co(0001) surface (7-layer slab model)	S	−257	−31.5	−287	−146	1.76
	S − 1	−244	−84.7	−297	−144	1.68
	S − 2	−237	−76.1	−314	−145	1.65
	C	−239	−75.5	−314	−145	1.64
Co–Cu(001) interface		−257	161	−96	−144	1.79

(or Au or Pd) atoms, and with the Fe atoms located at the four-fold hollow site of the Ag (or Au or Pd) square lattice. The experimental Ag, Au, and Pd lattice constants were assumed. The computational results can be summarized as: (a) in all three systems, i.e. 1Fe–1Au, 1Fe–1Ag, and 1Fe–1Pd, the easy direction of the spin-orientation is perpendicular to the surface of the film, in contrast with the results for the free standing Fe(001) monolayer; (b) the values of the magnetic anisotropy energy for these systems are considerably larger than the results for the free standing Fe monolayer, i.e. 0.5 meV/atom for 1Fe–1Au, 0.1 meV/atom for 1Fe–1Ag, and 0.4 meV/atom for 1Fe–1Pd. In this case, the result for 1Fe–1Ag is in agreement with the preliminary result reported by Gay and Richter [29] for an Fe–Ag(001) slab with five Ag layers sandwiched by a monolayer of Fe on either side.

9.4 METAL–CERAMIC INTERFACES

One of the most important heterophase interfaces is that composed of metal and ceramic materials. They are essential in metallurgy, ultra-thin micro-electronics, metal–ceramic joining, metal oxidation process, metal coating, catalysis, etc. Furthermore, novel materials with unique properties may be developed from multi-layer ceramic–metal structures. Often, the bonding and adhesion between oxide and metal is critical to the performance of an electronic component. In each case, the interface geometry and chemistry play an important, and possibly even dominant, role in determining the electrical and mechanical properties of the system.

The study of metal–ceramic interfaces poses some formidable challenges both experimentally and theoretically. The problems of special importance include:

1. Interface formation and atomic structure.
2. The nature and role of interdiffusion (and in general of the chemistry) at the interface.
3. The nature and strength of bonding.
4. The work of adhesion.
5. Interface failure, e.g. fracture resistance, mechanics of crack propagation, and the role of different thermal expansion coefficients in the metal and ceramic which subject the system to residual stress.

Fortunately, a number of sophisticated synthesis and experimental techniques, including HREM, CTEM, SIMS, AEM, APFIM and STM, have

(a)   (b)

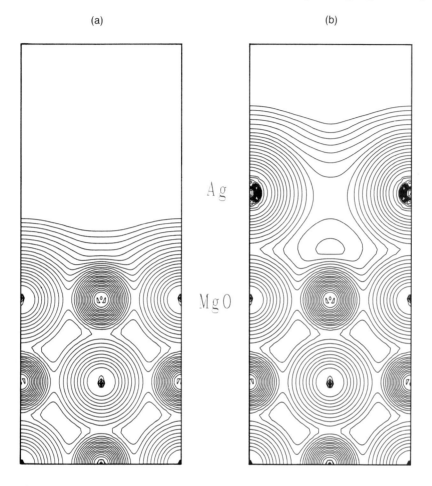

Ag

MgO

Fig. 9.3 Charge density distribution for (a) the clean 5 layer MgO(001) slab and (b) the Ag/MgO(001) (Ag above O site) slab (plotted for the upper half of a single slab unit cell in the (001) plane). Indicated numbers are in units of 10^{-3} e/(a.u.)^3; subsequent contour lines differ by a factor of $\sqrt{2}$.

been developed which allow high resolution characterization and property measurements to be made on carefully synthesized (and controlled) systems, reviewed recently by Balluffi *et al.* [30].

9.4.1 Ag–MgO(001)

A FLAPW computational study of the metal–ceramic interface problem was initiated by means of an investigation of metallic Ag on an MgO substrate. Now, as is well known, pure MgO has the NaCl crystal structure, one of the most convenient for model calculations. The (001) surface is by far the most stable surface of most alkaline earth oxides, which is almost an unrelaxed termination of the bulk lattice [31]. Ag has simple metal features and is rather predictable in its behavior when

interfaced with other materials. Also, the short range of the surface-interface of the noble metal (less than one atomic layer) [32] provided a sound basis for emphasizing the real metal–ceramic interface by including only a small number of atomic layers of Ag in the model calculations. The smallness of the lattice mismatch between the MgO(001) surface and fcc Ag (4%) is a necessary condition for a simple structure low energy interface between these two materials.

The metal/ceramic interface between Ag and MgO is represented by a monolayer of Ag adsorbed on the surfaces of the MgO(001) slab. Three possible adsorption sites of Ag atoms on the MgO(001) surface, i.e. above the O atom, the Mg atom or the hollow site, were considered in the calculations. The total energy versus interlayer

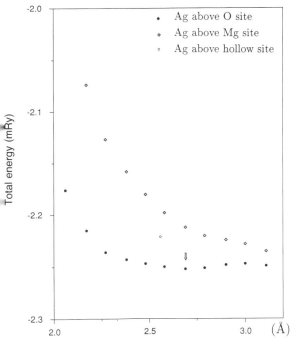

Fig. 9.4 Total energy versus interlayer spacing for the three possible positions for Ag overlayer atoms on the MgO(001) surface: Ag above O site, A above Mg site, and Ag above the hollow site.

Table 9.4 Charge population of the calculated systems: clean MgO(001) and Ag – MgO

	MgO 5 layer clean surface					
	Mg			O		
Layer	3s	2p	Total	2s	2p	Total
Center	0.077	5.850	6.067	1.634	3.851	5.489
S − 1	0.078	5.850	6.069	1.634	3.851	5.490
Surface	0.075	5.859	6.057	1.635	3.802	5.440
	Ag above the O site					
	Mg			O		
Center	0.076	5.848	6.065	1.635	3.847	5.486
I − 1	0.077	5.848	6.066	1.636	3.846	5.486
Interface	0.073	5.851	6.047	1.642	3.782	5.429
	Ag in Ag/MgO					
	5s	5p	4d	Total		
	0.237	0.055	8.657	8.953		
	Free standing Ag monolayer					
	0.240	0.040	8.652	8.936		

spacing of these three possible positions of a Ag overlayer on MgO(001) surface is plotted in Fig. 9.4. They show that the site above O for Ag is preferred over the others, with an optimized interlayer spacing of 5.1 a.u. (2.70 Å). (In the following discussions, the properties of the Ag–MgO(001) system are for the case of the Ag above O site obtained with this optimized interlayer spacing.)

Since the electronic charge density is the fundamental feature of density functional theory utilized in this computational study, it is given as a contour plot in the (001) plane of the Ag–MgO system in Fig. 9.3(b). Its charge population is listed in Table 9.4. Compared with the clean MgO 5-layer slab, the electron populations at the Mg and O at the center and the sub-interface (I − 1) layers remain primarily unchanged. At the interface layer, both Mg and O atoms show a net decrease of total charge, of the order of 0.01 electrons, compared with the surface layer of the

clean MgO five-layer slab. This net decrease is caused by the loss of Mg 3p and O 2p electrons. Compared with the charge population of the free Ag monolayer, the charge population of the Ag atom in the Ag–MgO system is only slightly larger (by ~0.02 electrons), indicating no significant charge transfer. By comparing the total energies of the calculated systems, i.e. Ag–MgO, the MgO(001) clean surface and the unsupported Ag monolayer, we obtain the binding energy of Ag atoms on a clean MgO surface, as −0.022 Ry/atom (0.64 J/m^2). Since no significant charge transfer between the Ag and the MgO(001) structure is found in the calculated electronic charge density distributions, the binding energy between the overlayer Ag atom and the MgO(001) surface seems to be due to the Coulomb dipole charge attraction.

9.4.2 Fe–MgO(001)

Our basic understanding of the simplest metal–ceramic interface falls far short of their obvious importance. As an illustration of the first steps taken to study the metal–ceramic interfaces from first principles, we cite some recent work on Fe on MgO(001). Some remarkable results have been obtained which show that MgO is an extremely non-interactive substrate for metal overlayers and, hence, that it might be an ideal two-dimensional substrate for studying magnetization, magnetic anisotropy, phase transitions, and critical behavior. Therefore, we present computational predictions on the electronic and magnetic properties of the Fe–MgO(001) interface system, in the hopes of stimulating future research interest both theoretically and experimentally.

To determine the interface structure between Fe metal and MgO(001), we studied two different positions for monolayer Fe on MgO(001), i.e. above the O site and above the Mg site. The total energy study determined that the position of Fe atoms above the O atom site is the preferred position, similar to the results of the previous Ag–MgO(001) calculations. The interlayer spacing between monolayer Fe and the MgO(001) substrate, 2.30 Å, was also determined by our total energy calculations. (In the following, the properties of the Fe–MgO(001) system correspond to the case of the Fe above O site with this optimized interlayer spacing.) The same interface structure between Fe and MgO, i.e. the interface Fe atoms sitting at 2.30 Å above the O sites, was assumed in our model calculations for two layers of Fe, i.e. 2Fe–MgO(001).

The charge density contour plot of ferromagnetic 1Fe–MgO(001) is shown in Fig. 9.5(a). As expected, the charge density of the MgO substrate remains essentially unchanged in comparison with clean MgO. The charge density contours for the O and Mg atoms have spherical shape, and a small interface effect is seen at the interface O atoms. The Fe atom shows a surface-like charge distribution, i.e. with the charge smoothly extended into the vacuum region. The charge population of the

Charge density

Spin density

(a)

(b)

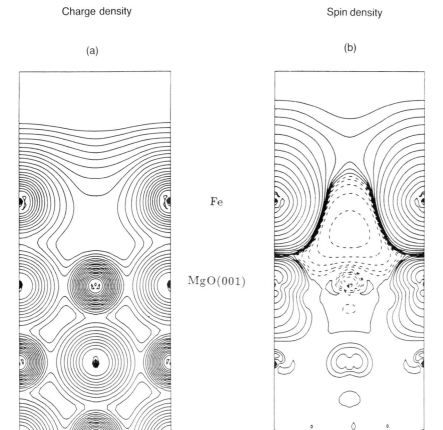

Fe

MgO(001)

Fig. 9.5 (a) Charge and (b) spin density distributions for Fe–MgO(001) (Fe above O site) plotted for the upper half of a single slab unit cell in (001) plane. For (a), indicated numbers are in units of 10^{-3} e/(a.u.)^3; subsequent contour lines differ by a factor of $\sqrt{2}$. For (b), indicated numbers are in units of 10^{-4} e/(a.u.)^3; subsequent contour lines differ by a factor of 2.

atoms is listed in Table 9.5. The overlayer Fe atom has a total charge of 6.31, which is equal to that of the free standing monolayer Fe, in spite of the MgO(001) substrate. Also interestingly, the MgO electron population in 1Fe–MgO(001) differs from that of the clean MgO(001) surface by only ~0.02 electron at the interface O atom, while the population of the other atoms remains almost unchanged. Thus, any strong direct reaction between Fe and MgO can be ruled out.

Shown in Fig. 9.5(b) is the spin density contour plot of ferromagnetic 1Fe–MgO(001) in the (100) plane. Large positive spin density is seen at the Fe

Table 9.5 Layer and *l*-projected muffin-tin charge populations and magnetic moments of O, Mg and Fe: clean MgO(001), 1Fe–MgO, 2Fe–MgO. (The M.T. radii used are 2.1 a.u., 1.7 a.u., and 2.2 a.u. for O, Mg, and Fe, respectively)

Atom	Layer		s	p	d	Total	$M(\mu_B)$
			1Fe–MgO(001)				
Fe	Overlayer	↑	0.15	0.07	4.48	6.31	3.07
		↓	0.12	0.05	1.45		
O	Interface		1.84	4.60	0.02	6.47	0.03
	$I-1$		1.84	4.68	0.02	6.54	0.00
	Center		1.84	4.68	0.02	6.54	0.00
Mg	Interface		0.08	5.89	0.04	6.01	0.00
	$I-1$		0.08	5.91	0.05	6.04	0.00
	Center		0.08	5.90	0.05	6.03	0.00
			2Fe–MgO(001)				
Fe	Surface	↑	0.16	0.08	4.44	6.40	2.96
		↓	0.15	0.08	1.49		
Fe	$S-1$	↑	0.15	0.09	4.40	6.43	2.85
		↓	0.14	0.10	1.55		
O	Interface		1.84	4.61	0.02	6.45	0.03
	$I-1$		1.84	4.68	0.02	6.53	0.00
	Center		1.84	4.68	0.02	6.53	0.00
Mg	Interface		0.08	5.89	0.04	6.02	0.00
	$I-1$		0.08	5.91	0.05	6.04	0.00
	Center		0.08	5.90	0.05	6.04	0.00
			clean 5-layer MgO(001)				
O	Surface		1.83	4.59	0.02	6.44	–
	$S-1$		1.84	4.67	0.02	6.53	–
	Center		1.84	4.68	0.02	6.53	–
Mg	Surface		0.08	5.90	0.04	6.02	–
	$S-1$		0.08	5.91	0.05	6.04	–
	Center		0.08	5.91	0.05	6.04	–
			Free standing Fe(001)				
Fe		↑	0.16	0.04	4.50	6.31	3.10
		↓	0.13	0.03	1.45		

overlayer, showing a *d*-like shape. Some negative spin density appears in the interstitial region between the overlayer Fe atoms (i.e. the region above the interface Mg atom). At the interface O atom, the positive net spin density has a *p*-like shape, indicating the slight influence of the magnetic Fe overlayer on the *p* electronic states of the interface O. The other atoms, i.e. all Mg atoms and the non-interface O atoms in the MgO(001) substrate, show almost no net spin density. The calculated magnetic moment, $3.07\,\mu_B$, is very close to that of a free standing Fe monolayer ($3.10\,\mu_B$), and hence is larger than that of the bcc Fe free surface ($2.98\,\mu_B$) [7] or of Fe monolayers on noble metal (Au and Ag(001)) surfaces [16, 33].

The charge density contour plot of 2Fe–MgO(001) in the (100) plane is shown in Fig. 9.6(a). The shape of the charge density in the MgO substrate in 2Fe–MgO is close to that seen in both 1Fe–MgO and clean MgO, a result of the lack of interaction between Fe metal and the MgO substrate. The charge density of the surface Fe layer shows surface-like features typical of transition metal surfaces, while that of the interface Fe ($S-1$) atom shows some influence from the substrate. The layer-projected charge and spin populations of 2Fe–MgO(001) are listed in Table 9.5. The surface Fe atom has a total muffin-tin (M.T.) charge of 6.40 electrons or about ~0.1 electrons more than that of Fe in 1Fe–MgO. The interface Fe ($S-1$) atom has a total M.T. charge of 6.43 electrons, a value very close to that of the Fe surface atom. Again, the charge population of the MgO substrate atoms is almost unchanged from that of clean MgO or 1Fe–MgO.

The spin density contour plot in the (100) plane of the 2Fe–MgO(001) slab is shown in Fig. 9.6(b). Large positive spin densities are seen at the two Fe atoms, with a strongly anisotropic shape. Positive spin density appears also at the interface O atom, similar to the 1Fe–MgO(001) results. Negative spin density is seen in the interstitial regions between the Fe atoms. The total magnetic moment of the surface Fe in 2Fe–MgO is $2.96\,\mu_B$, close in value to that of the free bcc Fe(001) surface atom, $2.98\,\mu_B$ [7]. Very interestingly, the interface Fe atom has a magnetic moment $2.85\,\mu_B$, which

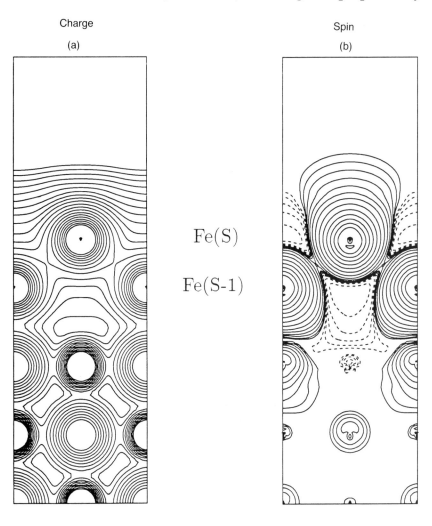

Charge

(a)

Spin

(b)

Fe(S)

Fe(S-1)

Fig. 9.6 (a) Charge and (b) spin density of the conduction electrons in 2Fe–MgO(001). For (a), indicated numbers are in units of 10^{-3} e/(a.u.)^3; subsequent contour lines differ by a factor of $\sqrt{2}$. For (b), indicated numbers are in units of 10^{-4} e/(a.u.)^3; subsequent contour lines differ by a factor of 2.

is smaller than that of the surface Fe by only $\sim 0.1\,\mu_B$. Compared with the calculated magnetic moment of the bulk-like bcc Fe atom, $2.25\,\mu_B$, both the surface and interface Fe atoms have largely enhanced moments (by 32% and 27%, respectively).

The Fermi contact terms of the magnetic hyperfine field were calculated for the Fe nuclei from the spin-polarized results. Listed in Table 9.6 are the hyperfine field results for 1Fe–MgO(001), 2Fe–MgO(001), and free standing Fe monolayer. Also listed in Table 9.6 are the results by Ohnishi *et al.* [7] for Fe atoms in a seven-layer bcc Fe(001) slab. Large negative core electron contributions to the hyperfine field are seen in all the systems listed.

As expected, the valence electron (VE) contribution is much more sensitive to the atomic environment.

In 1Fe–MgO(001), the VE contribution gives a large positive hyperfine field ($+396\,\text{kGauss}$), resulting in a small negative net value of the total hyperfine field of only $\sim -42\,\text{kGauss}$. This result is remarkably close to that of a free standing Fe monolayer, but in sharp contrast with results for Fe overlayers on Ag(001) or Au(001) substrates [16, 33], in which the total hyperfine field is ~ -150 and $\sim -210\,\text{kGauss}$ for Fe–Ag and Fe–Au, respectively.

In 2Fe–MgO(001), the surface (S) and the (S − 1) Fe atoms have rather different hyperfine

Table 9.6 The Fermi-contact term of the hyperfine magnetic field at Fe nuclei of the calculated systems: 1Fe–MgO and 2Fe–MgO

	Total	Valence (kGauss)	Core	Moment (μ_B)
Fe in 1Fe–MgO(001)	−41.9	+396.0	−437.9	3.07
Fe(S) in 2Fe–MgO	−284.2	+135.9	−420.1	2.96
Fe($S - 1$) in 2Fe–MgO	−367.9	+36.0	−403.1	2.85
Free standing Fe(001) monolayer	−50.8	+393.0	−443.8	3.10
Fe in 1Fe–Au(001)[16]	−213	+200	−413	2.97
Surface Fe in bcc Fe(001)[7]	−252	+143	−398	2.98
bulk-like bcc Fe[7]	−366	−75	−291	2.25

fields. For the Fe(S), the total hyperfine magnetic field is −284 kGauss, a combination of the VE (+136 kGauss) and core (−420 kGauss) contributions. This result is somewhat similar to that seen for the bcc Fe(001) surface. The Fe ($S - 1$) atom has a large value of the total hyperfine field, −368 kGauss, the result of a much smaller VE contribution (+36 kGauss). Once again, the hyperfine field is seen not to be proportional to the magnetic moment. Thus, these calculational values are essential for any correlations with experiments.

9.5 SEMICONDUCTOR INTERFACES AND HETEROJUNCTIONS

9.5.1 Background perspective

Among the various semiconductor properties that have been the object of careful and thorough investigation, the properties of semiconductor junctions (metal–semiconductor and semiconductor–semiconductor heterojunctions) are still not fully understood because of the complexity and the variety of the problems related to them. The interest in these particular issues is obviously related to the rectifying properties that metal–semiconductor junctions and heterojunctions in general show and to their importance for device applications; the aim of the great majority of the studies is therefore focused on the effort to single out the basic factors that can affect the potential barrier height at the interface.

Before we address these issues from a practical point of view and discuss some of the systems we have been recently studying, let us first quickly give a brief survey of the theoretical work and some experimental evidence that is available for metal–semiconductor and semiconductor–semiconductor junctions. The two interfaces just mentioned are basically different, from an electronic point of view, but show many similarities and, for this reason, have been often studied together using very similar models.

The main feature of a metal–semiconductor junction is that usually the Fermi level of the metal is 'pinned' (i.e. fixed) in a narrow energy range and its energy position is then independent of the doping of the semiconductor. When the two materials are put in contact, some charge will flow from the metal into the semiconductor; as a result, a small interfacial dipole is generated together with a bending of the bands which extends for several atomic layers inside the semiconductor. As a result of this charge rearrangement, metal electrons flowing toward the conduction band in the semiconductor have to overcome a potential barrier (the so-called Schottky barrier) determined by the difference in energy between the bottom of the semiconductor conduction band, for an *n*-type semiconductor, and the common Fermi level.

A very similar phenomenon occurs at the interface between two semiconductors; the main difference is due to the presence of a gap on both sides of the interface. The relevant quantity in this case is the relative position of the band gaps which determines the potential steps – valence band offset and conduction band offset – that the different carriers – electrons and holes – have to overcome in order to permit conductivity across the interface.

In both cases, Schottky barriers and band offsets, the relative alignment of the potential in the two different materials in contact is determined by some intrinsic bulk properties, i.e. the position of the Fermi level or of the conduction band with respect to the vacuum level, and by the charge readjustment at the interface (interface dipole). The way these two different contributions affect the final line-up and determine the barrier height is not a trivial matter and is the central object of most of the studies.

Many theories have been proposed after the early studies on the Schottky barrier formation and on its characteristics. Among these, we mention three fundamental models which roughly group most of the theoretical work done [34]. Schottky [35] initially proposed that the height of the barrier is linearly related to the difference between the metal work function Φ_m and the semiconductor electron affinity χ_s (the analogous quantities for the heterojunction case would be the electron affinities of the two semiconductors χ_s^1, χ_s^2). However, it was soon shown that in silicon, for example, the barrier height was completely independent of the metal work function.

Later, Bardeen [36] pointed out the importance of the formation of intrinsic localized states at the semiconductor surface with very high density of states that would easily accommodate the charge flowing from the metal towards the semiconductor without further displacing the position of the Fermi level. Bardeen's hypothesis was later revised by Heine [37] and Tersoff [38] who observed that the surface states, due to the clean surface, do not survive after the interface formation: in their place, metal induced gap states (MIGS) originate in the semiconductor band gap from the tails of the metal wave functions which tunnel into the semiconductor; the barrier is then determined by the alignment of the dielectric midgap energy (DME), i.e. the last filled MIGS level which assures charge neutrality, and the Fermi level of the metal. The final model we will mention points out the possi-

bility of local chemical bonding between metal and semiconductor (Phillips [39], Brillson [40]), making the Schottky barrier height very sensitive to the heat of formation of intermixed compounds and to the reactivity of the interface.

A comparable amount of theoretical effort has been devoted to the study of the semiconductor heterojunction band offset. Again, we can divide the models into two basic groups on the basis of the different reference levels used. The first class makes use essentially of the vacuum level as common reference energy level, in analogy with the earlier Schottky model, to which the conduction band position in the two semiconductors can be related. In this category we can include the original electron-affinity rule [41], the later Harrison model, which used atomic energy levels as reference for the valence and conduction band edges, and the more recent model solid approach by Van de Walle and Martin [42] in which the reference level is provided by the electrostatic average potential of the superposition of neutral atomic overlapping charge densities.

The second class can be defined as consisting of all the models that are more concerned with the effect of charge readjustment at the interface and that focus on the role played by the dipole that forms at the interface. In this category we can include the model which introduces the so-called charge neutrality point [38, 43] (strictly related to the previously defined DME); according to its definition, the states within the band gap change from 'conduction-like' into 'valence-like' in such a way that filling states above or taking out electrons below this level will result in a charge imbalance and therefore in a dipole. The relative alignment of these levels in the two different semiconductors gives directly the band offset. A similar approach that explicitly takes into account the effects of the interface dipole is the one recently proposed by Lambrecht and Segall [44] which combines first principles calculations for the bulk semiconductors with a study of the interface bonds in terms of a bond-orbital model.

The model we have used is a hybrid of those previously mentioned in the sense that it takes into account bulk properties as well as interface properties, and is conceptually very close to the main idea underlying the XPS experimental procedure to determine the offset in solids. First, a complete self-consistent calculation is performed for the particular interface under study; then, from this calculation, the energy difference between two core levels of different atoms is evaluated. This value is used as reference energy level with respect to which the valence band maximum (or the Fermi level) of the bulk compounds is aligned. This procedure, to be discussed in more detail later on, has been used successfully to predict and explain the valence band offset in different structures [45, 46] and seems to take into account both the bulk effect and the interface dipole contribution to the band line up. However, as we will see, there are still many factors that need to be accommodated for in the model and that can greatly affect the final band line up. These problems are mainly related to how close to the real interface is the model junction that we calculate.

One of the main reasons which makes the study of these systems particularly complex is that many different effects, such as the interfacial morphology and the stability of the interface with respect to chemical reaction and diffusion, are present in the real experimental situation. These effects cannot be easily simulated in a calculation. For this reason the study of a sharp, clean, well-matched junction is highly desirable in order to isolate and to look for the essential phenomena which determine the properties of the interface. Unfortunately, this is not always the case for real materials and we have examined cases in which the strain, the magnetism or the chemistry – more properly, the polarity – of the interface can dramatically affect the transport properties.

As a last remark, let us point out the geometrical assumptions we make in order to approach the interface problem. Due to the chemical nature of the compounds we are interested in, the study of a free surface is neither feasible nor realistic: as is well known, cleaved semiconductor surfaces are not stable and they always undergo a 'reconstruction' process in which all the unsaturated bonds are somehow 'readjusted' by oxidation, bond dimerization or other physical and/or chemical reactions.

Furthermore, wide technological applications are based on semiconductor superlattices (SL) and quantum wells – artificial structures in which two or more different semiconductors are grown one on top of the other in a regular sequence. We will therefore concentrate on this peculiar geometry, in which interface characteristics can still be extracted from an essentially bulk calculation.

9.5.2 Metal–semiconductor interfaces

We now briefly discuss a few selected examples of the systems we have recently studied, focusing, in particular, on the Fe–ZnSe [47, 48] and Fe–GaAs(001) systems that have been recently grown by G. Prinz and collaborators [23, 49–53] via MBE.

Due to its magnetic (easy switching) [23, 49–53] and optical properties (ZnSe has a wide band gap and therefore it is a good high frequency emitter), the Fe–ZnSe system appears to be very useful for lasers and other magneto-optical applications. There is, therefore, technological interest in exploring such properties and in studying the possibility of incorporating magnetic films into semiconductor structures in order to realize planar microelectronic configurations with the desired magneto-optical properties. Furthermore, the Fe–ZnSe interface presents particularly interesting features, such as small mismatch of their lattice constants and high stability with respect to chemical reaction, [23] that makes it highly desirable for a theoretical calculation and for a consequent meaningful comparison with experiment.

The focus of the studies were the electronic and magnetic properties of several superlattices of Fe on ZnSe with different layer thickness and Fe–GaAs. In particular, superlattices of Fe_n–$(ZnSe)_m(001)$ for $n = 1, 3$ and $m = 1, 2$ and the monolayer Fe–GaAs were investigated using structures that are basically derived from the zincblende structure, i.e. replacing one cation plane with one or more Fe planes in the [001] direction. The atomic arrangement of the Fe atoms is such that four Fe unit cells are needed to fit on a (001) zincblende plane; the mismatch in this configuration is only 1.1% and 1.4% for ZnSe and GaAs, respectively.

The interface characteristics of both the Fe–Se and Fe–As interfaces were determined from a total energy calculation. The difference between the two is remarkable and confirms the experimental results. While the interface between Fe and Se is well established, clean, and non-reactive, the Fe–As interface is highly reactive. We found that the equilibrium distance between Fe and Se is approximately the average of the Fe–Fe and Zn–Se bond length and agrees with the experimentally measured tetragonal distortion induced by Fe alloying in the $Zn_{1-x}SeFe_x$ dilute magnetic semiconductor [23].

In the case of Fe–GaAs we found that the Fe–As interface bond-length deviates quite significantly from that in bulk GaAs and Fe, leading to a much shorter bond-length. We found $d_{Fe-As} = 2.30$ Å to be compared with $d_{Fe-Se} = 2.46$ Å. Such different behavior correlates well with experiment: while Fe forms a well defined surface on ZnSe, it interdiffuses with the GaAs substrate forming intermixed compounds.

The magnetism of the Fe in the 1×1 Fe–ZnSe superlattice showed a remarkable enhancement of its magnetic moment ($\mu = 2.6 \mu_B$) with respect to its bulk value [47]. Due to the different site coordination of Fe in the different atomic layer this enhancement decreases as the number of Fe layers is increased: for three Fe layers the Fe magnetic moment decreases to its bulk value $\mu = 2.2 \mu_B$ [48]. Furthermore, due to the high directionality of the Se bonds, we observed a different magnetic behavior in the in-plane [110] direction with respect to the [1$\bar{1}$0]. This anisotropy is shown to decrease as the Fe thickness is increased, in agreement with the experiment. The calculated magnetic moments for the Fe sites in the thicker superlattices, see Table 9.7, are strongly reduced with respect to the results calculated for the 1×1 superlattice not only for the central layer, but also for the interface layers with the exception of the Feb2 site; however, they are still larger when compared with the magnetic moment of bulk Fe. The trend of a reduced total magnetization as a result of

an increased number of Fe layers has been also experimentally observed [53].

The enhanced magnetization observed on the Feb2 site is to be attributed to the different site coordination between Feb1 and Feb2 [48]. Unlike that of Feb1 in the 3×2 and Feb in the 3×1 superlattice, Feb2 has a smaller number of next nearest-neighbors but, most importantly, the closest neighbor along the (001) direction is Seb which is four atomic layers away. As a result, the electrons with negative polarization are mainly located in the interstitial region and far away from the muffin-tin sphere, thus enhancing the total magnetic polarization on that site. It needs to be emphasized that this effect is actually enhanced by the superlattice geometry we considered and in particular by the infinite repetition of the unit cell. In a thicker superlattice, this effect would be partially suppressed because of the presence of inner Fe layers.

A comparison of the results obtained for the 1×1, 3×1, and the 3×2 superlattices clearly shows that the magnetic moments of the larger

Table 9.7 Hyperfine field in (kGauss) and magnetic moment μ in (μ_B) for the Fe-sites in the superlattices considered

| | μ (μ_B) | *Hyperfine field (kG)* | | | |
		Core	C.E.	H_c(tot)	H_c/μ(kG/μ_B)
1×1					
FeI1	2.76	−386	+218	−169	61.23
FeI2	2.67	−371	+119	−252	94.38
3×1					
Feb	2.25	−318	−70	−389	172.89
FeI1	2.17	−307	+14	−293	135.02
FeI2	2.22	−312	−17	−329	148.20
3×2					
Feb1	2.25	−311	−98	−409	181.78
Feb2	2.66	−370	+28	−342	128.57
FeI1	2.33	−321	+9	−312	133.91
FeI2	2.40	−334	+1	−333	138.75
Fe – slab [7]					
surface	2.98	−398	+143	−252	84.56
center	2.25	−291	−75	−366	162.67
Fe-bct					
Fe	2.13	−299	−48	−347	162.91

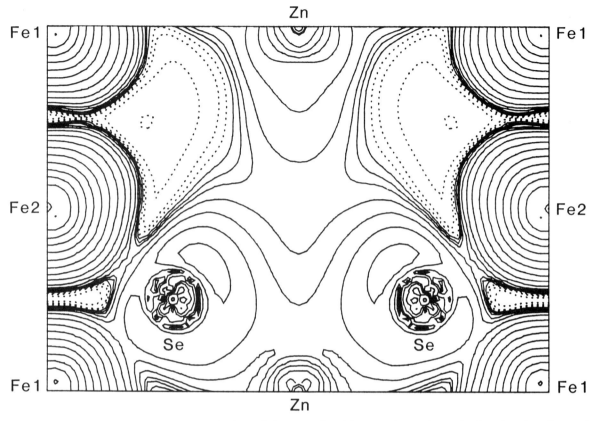

Fig. 9.7 Spin density distribution in the (110) plane of the Fe–ZnSe superlattice. Solid (dashed) lines indicate contours of positive (negative) spin density in units of $10^{-4} e/(a.u.)^3$. Successive contours shown differ by a factor of 2.

superlattices are reduced even for the Fe interface layers as the Fe thickness is increased. If the direct interaction between adjacent layers of Fe were the major contribution to the suppression of the surface-like magnetic enhancement of Fe at the interface, then still considerably increased magnetic moments for these layers, compared to the bulk layer, should result. This does not seem to happen since the magnetic moments in the two layers (bulk and interface) are very close (only a 2% difference). It appears that this is probably due to the fact that the Fe interface sites have an increased coordination number compared to the smaller superlattice. As a consequence, their behavior is more bulk-like than surface-like.

Furthermore, interaction among non-adjacent Fe layers through the semiconductor interlayer might contribute in suppressing the Fe magnetic properties. This might be confirmed by (i) the noticeable delocalization of the charge with negative spin polarization into the semiconductor region for the 3 × 1 SL, and (ii) by the fact that the interface Fe-sites have a higher magnetic moment in the 3 × 2 than in the 3 × 1 SL. The spin density distribution within the different unit cells is plotted in Figs. 9.7, 9.8, and 9.9. From these plots it is clear that in the 3 × 2 case the magnetism is much more strongly confined to the Fe region than in the 3 × 1 case. This results in a complete suppression of any magnetic interaction between non-adjacent Fe layers and in an enhancement of about 7% ($0.17 \mu_B$) in the magnetic moment at the interface. Actually, in addition to the highly anisotropic spin density distribution, the 3 × 1 superlattice has a

Fig. 9.8 Spin density distribution in (110) plane of $Fe_3-(ZnSe)_2$. Solid (dashed) lines indicate contours of positive (negative) spin density in units of $10^{-4}e/(a.u.)^3$. Successive contours shown differ by a factor of 3.

negative spin density which spreads from the interface layer over the entire crystal and is not localized only in the zincblende hollow sites as it is observed in the 1×1 case (Fig. 9.7).

The results for the Fermi-contact term of the hyperfine field (separated into core and conduction electron (CE) contributions) are shown in Table 9.7 and compared with the calculated results obtained for a seven-layer Fe slab [7] and bulk bct-Fe. As expected [54], the core contribution scales with the magnetic moment and the CE contribution is strongly reduced in going from the 1×1 to the 3×2, indicating that the outermost Fe s-like charge is affected by the indirect polarization with the d-states. The slightly positive value of the CE contribution for FeI1 is a result of its more favorable bonding position compared with FeI2, whereas the positive CE contribution found for Feb2 is the result of a strong positive polarization caused by the enhanced magnetic moment. The hyperfine field values (as well as the magnetic moment) calculated for Feb1 and Feb are very close to those calculated for the central layer of the Fe slab and for bct-Fe, showing that the bulk behavior is well recovered in both cases.

The electric properties of the interface and in particular the Schottky barrier height have also been determined. A study of the planar average charge density in the thicker (3×2) superlattice as a function of the (001) coordinate revealed that about 0.4 electrons are transferred at the interface from the Fe layers into the ZnSe side, that the bulk

Fig. 9.9 Spin density distribution in the (110) plane of $Fe_3-(ZnSe)_2$. Solid (dashed) lines indicate contours of positive (negative) spin density in units of $10^{-4} e/(a.u.)^3$. Successive contours shown differ by a factor of 3.

Table 9.8 Schottky barrier height (Φ_B), Pauling electronegativity (χ), Ref. [57] and work function (W), Ref. [58] for different metal–ZnSe junctions

Metal	χ	$W\,(eV)$	$\Phi_B\,(eV)$	Ref.
Au	2.4	5.15	1.29–1.36	[59, 61]
Pt	2.2	5.7	1.4	[59]
Cu	1.9	4.7	1.1	[59]
Co	1.8	4.63	1.72	[61]
Fe	1.8	4.5–4.3	1.09	(present work)

calculation and using the corresponding binding energy of these same levels in the reference bulk systems (Fe-bct and ZnSe). We found that the potential barrier at the interface is $\Phi_B = 1.09 \pm 0.09\,eV$. Since, to the best of our knowledge the Schottky barrier height for such a junction has not yet been measured, we compare our result with some measurements and data [57–61] obtained for metal–ZnSe interfaces, such as Au [60], Co [61], Cu [59], and Pt [59] junctions on ZnSe, shown in Table 9.8.

The Schottky model predictions are also shown in Table 9.8 for comparison; since the substrate is the same, one would expect a higher SBH for the metal with higher work function. From Table 9.8, we see that this general trend is actually quite well reproduced by the ZnSe surface, if we exclude the Co and Pt cases where chemical interactions between metal and substrate cannot be neglected. However, even for the Au and Fe cases, the bare Schottky model does not hold exactly (we recall the ZnSe electron affinity: $\chi = 4.06\,eV$).

Recently Xu *et al.* [60] found, for the low reactive Au–ZnSe interface, strong evidence that the Schottky model is obeyed; in particular they observed that as the coverage is increased, the work function changes in parallel with the band bending, thus supporting the idea that the barrier is essentially determined by the metal work function and the semiconductor affinity. On the other hand, the SBH they measured showed a constant deviation from predictions of the Schottky model; they attribute such deviations to a dipole formation at the clean ZnSe surface which is then reduced by the metal electronegativity.

charge densities are well recovered in both sides of the interface in the inner layers, and that a small induced electric dipole moment is observed on the Zn site in the layer next to the interface.

By considering the core levels as reference levels we were able to determine the Schottky barrier height (SBH), in analogy with the well known procedure used by XPS experiments and other theoretical calculations in the evaluation of the valence band offset in semiconductor heterojunctions [55, 56, 45, 46]. In this scheme, the discontinuity (Φ_B) between the Fermi level in the Fe side and the conduction band in the semiconductor side of the interface is evaluated by calculating the difference between an Fe and a Se core level, as obtained from the superlattice

In our case, due to the lower electronegativity of Fe with respect to Au, we expect Fe to behave differently from Au. In particular, since we found that charge flows from Fe into the more ionic ZnSe substrate, we expect a dipole contribution that tends to increase, rather than decrease, the metal work function. Within this sort of qualitative consideration, our calculation is in reasonable agreement with the experiment and is consistent with the picture that describes the SBH as dependent on two different contributions combined: (i) the intrinsic properties of the two compounds (i.e. metal work function, electron affinity, and/or metal electronegativity and semiconductor ionicity) and (ii) the characteristics of the interface (dipole formation, interface states). Unfortunately, the determination of such contributions is still not trivial overall if effects due to the non-perfect abruptness of the interface have to be taken into account.

Let us make a final remark: although LDA has been shown to successfully reproduce the band-offset at semiconductor heterojunctions, we cannot rule out a possible failure in the present case where the different dielectric and screening properties of the two materials at the interface (metal and semiconductor) can give rise to a non-exact cancellation of non-local effects which are usually neglected within LDA. This would imply that, in order to obtain more realistic results, non-local effects have to be considered in the superlattice calculation. This task has been only partially addressed [62] so far, and should be more carefully explored.

9.5.3 Semiconductor–semiconductor interfaces

Technological interest in semiconductors is focused on device applications while scientific interest is mainly devoted to understanding some characteristic properties, such as potential barriers, carrier properties, and band gaps, and how these can be modified by changing different external factors, such as epitaxial growth, strain effects, junctions, and doping. A complete knowledge and understanding of these complex issues is, in fact,

the basic requirement necessary in order to achieve the ability to 'tune' basic properties 'at will' and to design *ad hoc* materials for different device applications.

Let us consider a few examples. Among the most important achievements of computational physics in this field, we mention the prediction that the valence band offset is independent of the interface orientation for homopolar and unstrained interfaces (GaAs–AlAs being the 'classical' and most studied example). Unfortunately this does not hold for heteropolar and strained interfaces, where the interface morphology and internal strain plays a crucial role in determining how the charge redistributes at the interface and the consequent shape of the potential across the junction. This is among the most important issues still under theoretical investigation and for which the experiments have not been able to give precise answers. It must be noticed, in fact, that due to the particular nature of these phenomena, which are so much dependent on the detailed interface morphology on the atomic monolayer scale, the experiments are often not capable of resolving all the details and to single out each individual problem.

On the other hand, these kinds of information are partially obtainable through accurate computational studies. We have performed extensive studies of several different semiconductor heterojunctions in order to explore some of the crucial issues involved in the potential line-up problem; our studies on GaAs–AlAs [45], InAs–InP [67, 68] and recently α-Sn–CdTe together with several other theoretical studies performed by other groups (see Refs. [46, 42, 44] just to name a few of them) have shown that such calculations can give remarkable insights into interface formation, stability, and potential line-up.

In order to better illustrate how this work fits into the more general framework, we discuss, here, an example of how external factors (in this case the strain) can affect the transport and electronic properties of a particular heterojunction and how our calculations relate and compare with experiments. We will briefly present some recent results on heterostructures based on In compounds, since these have recently been attracting increasing interest because of their optical and transport properties [63–65].

The InAs–InP system in particular has been the focus of many experimental works; the mismatch between the two semiconductors is not particularly high (only 3.2%) and, moreover, it can be adjusted 'at will' when using InP substrates, by adding small amounts of Ga or P to the InAs matrix. The growth of $InAs_{1-x}Ga_x$–InP by metal–organic–chemical–vapor deposition (MOCVD) and by molecular beam epitaxy (MBE) has been reported by many groups [63–65, 55], as has the growth of InAs–InP quantum wells [64].

In our study, we focused on the InAs–InP (001) systems taking into account the effect of strain. To this end, we first studied the properties of the pure binary compounds and then performed total energy calculations in order to determine all the unknown lattice parameters of the 1×1 free-standing mode (FSM) structure [66, 67], i.e. the in-plane lattice constant, and the In–P and In–As distances. In order to compare with the available experiments and to study the effect of strain on the electronic properties of this structure, we also studied InAs as if grown on an InP substrate; in this case, the in-plane lattice constant and also the In–P bond-length are considered fixed and equal to their InP equilibrium values, whereas the In–As bond-length is determined by a total energy calculation.

We found that in both cases the ultra-thin 1×1 superlattice is unstable (the heat of formation is ~20 meV/4-atoms and 30 meV/4-atoms, respectively) and that the major contribution to this instability is due to the excess of elastic energy associated with the deformation of the two bulk semiconductor compounds. In the case in which no constraints are assumed on the in-plane lattice constant, the ordered phase of In_2AsP is seen to follow Vegard's law (within 0.02%), whereas the equilibrium anion–cation distances in the superlattice follow the bond-length conservation rule already observed in several other semiconductor alloy systems. We calculated that the charge transfer at the interface is rather small (less than 0.04 electrons are transferred from the InAs side into the more stable InP bonds) and the corresponding energy term is negative, i.e. it con-

stitutes a stabilizing energy term in the formation of the superlattice.

The study of the effect of strain on the electronic properties of the 1×1 structures revealed that the hole effective masses and the transport properties are very sensitive to the uniaxial distortion [67], making this system a good candidate for tailoring the band discontinuity and the carriers' properties.

We found, in fact, that the different choice of the growth conditions can result in a remarkable difference in the carrier mass and therefore in the transport properties of the superlattice. In particular, in the case of no-substrate, the top of the valence band shows a heavy mass for the in-plane hole and a light longitudinal hole mass. In the case of an InP substrate, the situation is reversed, leading to a configuration in which the hole mobility is enhanced for two-dimensional transport. In this latter case, in-plane transport is favored over longitudinal transport by a factor of 3 in the effective mass. This effect is also reflected in the charge density distribution at the top of the valence band. From a study of the strain effects in the thicker superlattices we were able to show that the hole becomes more and more confined within the InAs region as the in-plane strain is increased and that the band gap of the structure increases as the strain is relieved.

Total energy calculations performed on thicker superlattices aimed at exploring the effect of bond length relaxation on the layers away from the interface showed that the interface contribution to the potential line-up is not significantly affected by the atomic positions in the layers underneath the interface layer [68], provided that the interface atomic positions are left unchanged, and that the bulk characteristics are very well recovered in the layers just below the interface.

By looking at the charge profile of the planar average charge density and comparing it with the same profile in the pure bulk compounds, we were able to establish that the charge readjustment at the interface is rather small, that it does not significantly contribute to the potential line-up nor is it sensibly affected by the strain conditions.

The study of the potential discontinuity at the interface was performed, in analogy with the

X-ray spectroscopy experiments, by looking at the relative alignment of the core levels in both sides of the interface and comparing them with their respective positions in the bulk compounds. In the structures considered, there is one In atom at the interface (In(i)) and two independent 'bulk' In sites on the two sides of the interface: In(bAs) and In(bP) respectively. The evaluation of the band offset is performed by considering the relative alignment of the In(bAs) and In(bP) core levels (Δb) and the binding energy differences (ΔE_b) of the same core levels (relative to the valence band top) as calculated in the pure binary compounds. The total valence band offset is therefore:

$$\Delta E_v = \Delta b + \Delta E_b \qquad (9.1)$$

We found that the relative alignment of different core levels in the two sides of the interface was almost the same, resulting in a very small contribution to the overall potential line-up (Δb); this term is one order of magnitude smaller than the potential step due to the so-called 'band' term, i.e. the relative alignment of the two valence band tops in the pure constituents ΔE_b. This is shown in Table 9.9 where we report our results regarding the valence band offset in these structures and the values corrected for the spin–orbit interaction, which lifts the top of the valence band in the InAs site by 0.08 eV, i.e. one third of the difference of the spin–orbit corrections in InAs and InP [66]. The strain is seen to mainly affect the energy position of the valence band maximum in the two constituents, i.e. the term ΔE_b, rather than the interface dipole potential, i.e. the term Δb.

These results are very interesting since they show that the effect of the charge and potential readjustment at the interface is very small and that they do not change significantly as the strain conditions and the atomic positions are varied. On the other hand, the most important contribution to the band offset comes from the binding energy difference of these levels in the bulk materials which are much more sensitive to the strain conditions.

Finally, the present results show that InAs–InP junctions are particularly suitable for the design of junctions with tailored valence band discontinuities, since such a discontinuity can be easily estimated from a knowledge of quantities mainly related to the pure bulk compounds, the interface contribution to the band line-up being negligible.

The calculated valence band offset, obtained by considering the core levels as reference energy levels, is in very good agreement with the experimental results available to date for the strained InAs–InP interface [68] and it perfectly matches the XPS result by Waldrop *et al.* [55] if we consider as binary reference structures for the evaluation of the ΔE_b term the same equilibrium structures considered in the experiments: in this last case we obtain $\Delta E_v = 0.31 \pm 0.03$ eV.

ACKNOWLEDGMENTS

This work was supported by the Office of Naval Research (Grant No. N00014-89-J-1290), the National Science Foundation (Grant No. DMR88-

Table 9.9 Core energy differences and corresponding valence band offset, ΔE_v, (in eV) for the $n = 2, 3$ superlattices considered. ΔE_v^{so} denotes the spin-orbit corrected ΔE_v values

Core state		2×2				3×3			
		Δb	ΔE_b	ΔE_v	ΔE_v^{so}	Δb	ΔE_b	ΔE_v	ΔE_v^{so}
FSM	1s	0.02	0.21	0.23	0.31	0.01	0.21	0.22	0.30
	2s	0.01	0.21	0.22	0.30	0.02	0.21	0.23	0.31
	3s	0.02	0.21	0.23	0.31	0.02	0.21	0.23	0.31
InP-sub.	1s	0.03	0.39	0.42	0.50	0.03	0.39	0.42	0.50
	2s	0.02	0.40	0.42	0.50	0.03	0.40	0.43	0.50
	3s	0.02	0.40	0.42	0.50	0.03	0.40	0.43	0.50

21571 through the Northwestern University Materials Research Center) and the Department of Energy (Grant No. 88-ER45372).

REFERENCES

1. W. Kohn and L. J Sham, *Phys. Rev. A*, **140** (1965), 1133; L. J. Sham and W. Kohn, *Phys. Rev.*, **145** (1966), 561.
2. H. J. F. Jansen and A. J. Freeman, *Phys. Rev. B*, **30** (1984), 561.
3. M. Weinert, *J. Math. Phys.*, **22** (1981), 2433.
4. E. Wimmer, A. J. Freeman and H. Krakauer, *Phys. Rev. B*, **30** (1984), 3113.
5. A. J. Freeman, C. L. Fu and T. Oguchi, *Mat. Res. Soc. Symp. Proc.*, **63** (1985), 1.
6. C. L. Fu and A. J. Freeman, *Phys. Rev. B*, **33** (1986), 1755.
7. S. Ohnishi, A. J. Freeman and M. Weinert, *Phys. Rev. B*, **28** (1983), 6741.
8. C. L. Fu and A. J. Freeman, *J. Mag. Magn. Mats.*, **69** (1987), L1.
9. C. L. Fu and A. J. Freeman, *Phys. Rev. B*, **35** (1987), 925.
10. C. Li, A. J. Freeman and C. L. Fu, *J. Magn. Magn. Mats.*, **75** (1988), 53.
11. C. Li, A. J. Freeman and C. L. Fu, *J. Magn. Magn. Mats.* (in press) (1990).
12. J. I. Lee, C. L. Fu and A. J. Freeman, *J. Mag. Magn. Mats.*, **62** (1986), 93.
13. J. I. Lee, C. L. Fu and A. J. Freeman, *J. Magn. Magn. Mats.*, to be published.
14. C. L. Fu, A. J. Freeman and T. Oguchi, *Phys. Rev. Lett.*, **54** (1985), 2700; A. J. Freeman, C. L. Fu and T. Oguchi, *Mat. Res. Soc. Symp. Proc.*, **63** (1985), 1.
15. C. L. Fu and A. J. Freeman, *Phys. Rev. B*, **33** (1986), 1611.
16. C. Li, A. J. Freeman and C. L. Fu, *J. Magn. Magn. Mats.*, **75** (1988), 201.
17. S. C. Hong, A. J. Freeman and C. L. Fu, *J. de Physique, Coll.*, **C8** (1988), 1683; *Phys. Rev. B*, **38** (1988), 12156.
18. C. Li, A. J. Freeman and C. L. Fu, *J. Magn. Magn. Mats.*, **83** (1990), 51.
19. S. C. Hong, A. J. Freeman and C. L. Fu, *Phys. Rev. B*, **39** (1989), 5719.
20. C. L. Fu, A. J. Freeman and E. Wimmer, *Hyperfine Interactions*, **33** (1987), 53.
21. S. D. Bader, E. R. Moog and P. Grunberg, *J. Magn. Magn. Matls.*, **53** (1986), L295.
22. B. Heinrich, K. B. Urquhart, A. S. Arrott, J. F. Cochran, K. Myrtle and S. T. Purcell, *Phys. Rev. Lett.*, **59** (1987), 1756.
23. B. T. Jonker, K. H. Walker, E. Kisker, G. A. Prinz, and C. Carbone, *Phys. Rev. Lett.*, **57** (1986), 142; B. J. Jonker, J. J. Krebs, G. A. Prinz and S. B. Qadri, *J. Cryst. Growth*, **81** (1987), 524.
24. N. C. Koon *et al.*, *Phys. Rev. Lett.*, **59** (1987), 2463.
25. W. Dúrr *et al.*, *Phys. Rev. Lett.*, **62** (1989), 206.
26. C. Liu and S. D. Bader, submitted to *J. of Vac. Sci. and Tech. A*.
27. J. H. Van Vleck, *Phys. Rev.*, **52** (1937), 1178.
28. C. Li, A. J. Freeman, H. J. F. Jansen and C. L. Fu, *Phys. Rev. B* (in press).
29. J. G. Gay and R. Richter, *Phys. Rev. Lett.*, **56** (1986), 2728; *J. Appl. Phys.*, **61** (1987), 3362.
30. R. W. Balluffi, M. Rühle and A. P. Sutton, *Mater. Sci. Engin.*, **89** (1987), 1.
31. V. E. Henrich, *Reports on Progress in Physics*, **48** (1985), 1481.
32. H. Erschbaumer and A. J. Freeman (to be published).
33. A. J. Freeman, C. L. Fu, M. Weinert and S. Ohnishi, *Hyperfine Interactions*, **33** (1987), 53–68.
34. For a complete review, see: M. Schluter, *Thin Solid Film*, **93** (1982), 3 and L. J. Brillson, *Surf. Sci. Rep.*, **2** (1982), 123.
35. W. Schottky, *Z. Physik.*, **113** (1939), 367.
36. J. Bardeen, *Phys. Rev.*, **71** (1947), 717.
37. V. Heine, *Phys. Rev. A*, **138** (1965), 1689.
38. J. Tersoff, *Phys. Rev. Lett.*, **52** (1984), 465.
39. J. C. Phillips, *J. Vac. Sci. Technol.*, **11** (1974), 946.
40. L. J. Brillson, *Phys. Rev. Lett.*, **40** (1978), 260.
41. R. L. Anderson, *Solid State Electron.*, **5** (1962), 341.
42. C. G. Van de Walle and R. Martin, *Phys. Rev. B*, **37** (1988), 4801 and references therein.
43. J. Tersoff and W. A. Harrison, *J. Vac. Sci. Technol. B*, **5** (1987), 1221.
44. W. R. L. Lambrecht and B. Segall, *Phys. Rev. B*, **41** (1990), 2832.
45. S. Massidda, B. I. Min and A. J. Freeman, *Phys. Rev. B*, **35** (1987), 9871.
46. S. H. Wei and A. Zunger, *Phys. Rev. Lett.*, **59** (1987), 144.
47. A. Continenza, S. Massidda and A. J. Freeman, *J. Magn. Magn. Mat.*, **78** (1989), 195.
48. A. Continenza, S. Massidda and A. J. Freeman, *Phys. Rev. B*, **42** (1990), 2904.
49. G. A. Prinz, B. T. Jonker, J. J. Krebs, J. M. Ferrari and F. Kovanic. *Appl. Phys. Lett.*, **48** (1986), 1756.
50. G. A. Prinz and J. Krebs, *Appl. Phys. Lett.*, **39** (1981), 397.
51. G. A. Prinz, *Phys. Rev. Lett.*, **54** (1985), 1051.
52. G. A. Prinz, B. T. Jonker and J. J. Krebs, *Bull. of Am. Phys. Soc.*, **33** (1988), 562.
53. J. J. Krebs, B. T. Jonker and G. A. Prinz, *J. Appl. Phys.*, **61** (1987), 374.
54. A. J. Freeman and R. E. Watson, in *Magnetism*, Vol. IIA

(eds G. T. Rado and H. Suhl) Academic Press, New York (1966).

55. J. R. Waldrop, R. W. Grant and E. A. Kraut, *Appl. Phys. Lett.*, **54** (1989), 1878.

56. C. G. Van de Walle and R. Martin, *Phys. Rev. B*, **35** (1987), 9871.

57. L. Pauling, in *The Chemical Bond*, Cornell University Press, New York (1967).

58. A. R. Miedema, F. R. de Boer and P. F. de Chatel, *J. Phys.*, **F3** (1973), 1558.

59. C. A. Mead, *Solid State Electr.*, **9** (1966), 1023; C. A. Mead, *Phys. Lett.*, **18** (1966), 218.

60. F. Xu, M. Vos, J. H. Weaver and H. Cheng, *Phys. Rev. B*, **38** (1988), 13418.

61. S. G. Anderson, F. Xu, M. Vos, J. H. Weaver and H. Cheng, *Phys. Rev. B*, **39** (1989), 5079.

62. G. P. Das, P. Blöchl, O. K. Andersen, N. E. Christensen and O. Gunnarsson, *Phys. Rev. Lett.*, **63** (1989), 1168.

63. R. P. Schneider, Jr. and B. W. Wessels, *Appl. Phys. Lett.*, **54** (1989), 1142.

64. K. Huang and B. W. Wessels, *J. Appl. Phys.*, **64** (1988), 6770.

65. D. V. Lang, M. B. Panish, F. Capasso, J. Allam, R. A. Hamm, A. M. Sergent and W. T. Tsang, *J. Vac. Sci. Technol. B*, **5** (1987), 1215.

66. S. Massidda, A. Continenza, A. J. Freeman, T. M. De Pascale, F. Meloni and M. Serra, *Phys. Rev. B*, **41** (1990), 12079.

67. A. Continenza, S. Massidda and A. J. Freeman, *Phys. Rev. B*, **41** (1990), 12013.

68. A. Continenza, S. Massidda and A. J. Freeman, *Phys. Rev. B*, **42** (1990), 3469.

CHAPTER TEN

Scanning tunneling microscopy of metals on semiconductors

R. J. Hamers

Introduction · Group III metals on silicon · Group V metals on silicon · Alkali metals on silicon · Transition metals on silicon · Metals on gallium arsenide · Conclusions

10.1 INTRODUCTION

While many tools have been used to study the chemical and physical properties of metal–semiconductor interfaces, scanning tunneling microscopy (STM) has recently enjoyed an enormous growth due to its ability to probe not only the geometric structure but also the electronic structure of metal–semiconductor systems. Several review articles have been published describing STM and its applications [1, 2, 3, 4], and the reader is referred to these publications for a more general discussion of STM.

Much of the effort in the scanning tunneling microscopy area has focused on its application to the initial stages of metallization at semiconductor surfaces. Since STM tends to probe only the outermost few layers of atoms, such studies have tended to concentrate on the atomic structures of very thin metal overlayers on semiconductors. In most cases, the studies have concentrated on the structures adopted at metal coverages of one monolayer or less. Yet these studies demonstrate that metal overlayers on semiconductors exhibit a wide variety of interesting phenomena. This chapter will review some of these recent results. Since silicon has been

the most popular substrate and group III metals have been the most popular metals for STM investigations, this system is described before moving on to less well-studied systems.

10.2 GROUP III METALS ON SILICON

10.2.1 Al, Ga, and In on Si(111)

The atomic structure of ordered overlayers and thin films of group III semiconductors on Si(111) has been extensively studied. Like many other metals, the Group III metals Al, Ga, and In form ordered structures with $(\sqrt{3} \times \sqrt{3})$ symmetry at $\frac{1}{3}$ monolayer metal coverage. For the group III metals, this overlayer is generally believed to arise from simple adatom structures, in which the metal atom adsorbs on top of a bulk-like Si(111) lattice. This can occur in two different geometries. In the 'T$_4$' geometry the adsorbed metal atom sits above a three-fold hollow site of the outermost layer directly above a four-fold coordinated atom of the second atomic layer, thereby achieving partial four-fold coordination. In the 'H$_3$' geometry, the metal atom again sits above a three-fold hollow site, but without a second-layer atom underneath. The major distinction between these two different geometries is the extra coordination provided in the T_4 geometry between the metal atom and the second-layer Si atom; this can be observed in Fig. 10.1, which shows top view and side views of adatoms in the T_4 configuration.

Theoretical calculations by Northrup [5] have addressed the energetics of this bonding, finding that the T_4 geometry is considerably more stable than the H_3 geometry. The additional stability arises in large part from bonding interactions between the adatom and the atom directly beneath it. This conclusion is supported by a number of scanning tunneling microscopy experiments.

Although the ideal $\sqrt{3}$ structure arises at $\frac{1}{3}$

monolayer coverage, an ordered LEED pattern is observed at even lower coverage. STM images [6, 7, 8] show that when starting with a clean Si(111)–(7 × 7) surface, very low coverages of group III metals induce disorder up to coverages of approximately $\frac{1}{6}$ monolayer. At coverages above $\frac{1}{6}$ monolayer, the adsorbed Group III metal atoms begin to form ordered islands with $(\sqrt{3} \times \sqrt{3})$ symmetry ($\frac{1}{3}$ monolayer local metal coverage); since this requires $\frac{1}{3}$ monolayer of adatoms, the 'vacant' T_4 sites are occupied by silicon adatoms instead of metal atoms. Thus, the surface layer becomes essentially a random alloy of metal and silicon atoms, all adsorbed in T_4 sites atop the bulk-like Si(111) lattice. Due to the difference in electronic structure between aluminum and

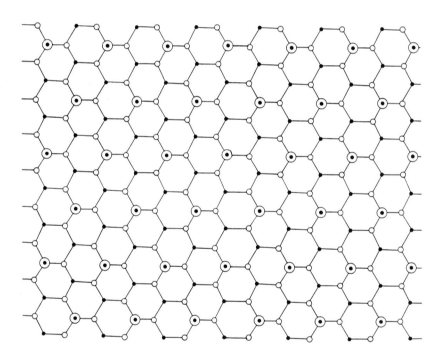

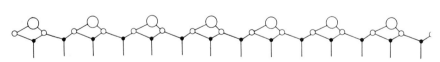

Fig. 10.1 Structural model for metal atoms adsorbed in the T_4 configuration on a Si(111) surface, forming a $(\sqrt{3} \times \sqrt{3})$ overlayer. Large, open circles represent metal adatoms; smaller open and closed circles represent silicon atoms in the outermost double-layer of the Si(111) surface.

silicon adatoms, the STM observes an apparent height difference of nearly 1 Å between the silicon and aluminum adatoms. Figure 10.2 shows STM images of identical regions of a Si(111) surface with $\frac{1}{6}$ monolayer of Al adatoms. The surface appears as a random alloy of aluminum and silicon adatoms, whose identity can be distinguished by apparent height differences and from changes in the appearance of the image with differing sample bias voltage. At positive sample bias (Fig. 10.2(a)) electrons tunnel from the tip into an empty 3p state on the Al atoms, making them appear as

protrusions while the silicon atoms appear lower; at negative sample bias (Fig. 10.2(b)), electrons tunnel from the partially filled 3p state of the silicon atoms into the tip, making the **silicon** atoms appear as protrusions. As the surface coverage is increased above $\frac{1}{3}$ monolayer, the number of aluminum atoms in the surface layer decreases, and the surface approaches a 'perfect' $\frac{1}{3}$ monolayer $\sqrt{3}$ structure. Similar behavior has also been reported for indium on Si(111) by Nogami *et al.* [9, 10].

At higher coverage, the group III metals exhibit a wider variety of overlayer structures. At greater

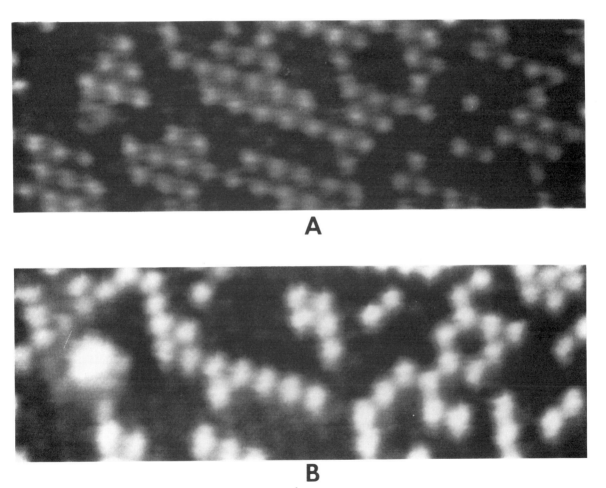

A

B

Fig. 10.2 STM image obtained from deposition of approximately $\frac{1}{6}$ monolayer Al on Si(111) surface. The surface consists of aluminum and silicon adatoms adsorbed atop the Si(111), but with the chemical identity of the adatom varying in a random manner. At positive sample bias (a), the STM preferentially images the aluminum adatoms; at negative sample bias (b), the STM preferentially images the silicon adatoms.

than $\frac{1}{3}$ monolayer coverage, aluminum forms a new ordered structure with $(\sqrt{7} \times \sqrt{7})R19.2°$ symmetry and a local coverage of $\frac{3}{7}$ (0.43) monolayer. Since the $\sqrt{3}$ structure has a local coverage of $\frac{1}{3}$ monolayer, at all coverages between 0.33 and 0.43 monolayer the surface is locally ordered into domains having $\sqrt{3}$ and $\sqrt{7}$ symmetry; the relative areas of the domains vary in order to accommodate global and local variations in the surface coverage. The structure of this overlayer was determined by Hamers using voltage-dependent STM imaging, tunneling spectroscopy, and lattice superposition analysis techniques [7]. The nature of this overlayer can be understood by remembering first that the $\sqrt{3}$ overlayer at $\frac{1}{3}$ monolayer coverage satisfies all Si dangling bonds and has all Al atoms threefold coordinated. Therefore, the $\sqrt{3}$ is expected to be a very stable structure. As more aluminum atoms are deposited beyond $\frac{1}{3}$ monolayer, the aluminum atoms must adopt less-favorable bonding configurations. First, one must determine the exact aluminum coverage. Since the area of the $\sqrt{7}$ unit cell is seven times that of the primitive (1×1) unit cell and the coverage must be greater than $\frac{1}{3}$ monolayer, we conclude that the local coverage must be exactly $\frac{3}{7}$ monolayer Al and there must be three Al atoms per $\sqrt{7}$ unit cell. This is also consistent with tunneling spectroscopy measurements which show that the structure has a large bandgap; electron-counting rules indicate that $\frac{3}{7}$ monolayer Al should be semiconducting, while $\frac{2}{7}$ and $\frac{4}{7}$ monolayer should be metallic.

The STM images of this overlayer show strong changes depending on the **sign** of the applied bias, as shown in Fig. 10.3. When a positive bias is applied to the sample (so that electrons tunnel from unoccupied electronic states), we see three strong protrusions per $\sqrt{7}$ unit cell. At negative sample bias, the STM images show only a single strong protrusion per unit cell. In order to understand how the STM images relate to the atomic structure of the overlayer, we remember that in the case of the $\sqrt{3}$ overlayer, the Al adatoms were characterized by a single **unoccupied** p_z-like orbital extending into the vacuum. As a result of this electronic structure, STM images at **positive** sample bias (sampling the unoccupied

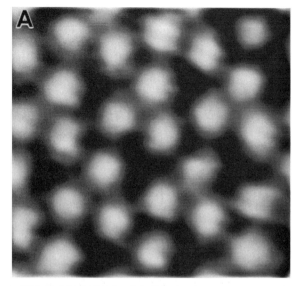

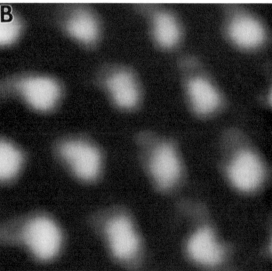

Fig. 10.3 STM image of the Si(111)-$(\sqrt{7} \times \sqrt{7})R19.2°$ Al overlayer at positive sample bias (a) and negative sample bias (b). Note the striking change in appearance resulting from a strong spatial separation of occupied and unoccupied electronic states.

states) revealed the positions of the Al adatoms, while images at negative sample bias sampled the occupied states, which in the case of $\sqrt{3}$–Al/Si arose from the Al–Si backbonds. In the case of the $\sqrt{7}$ overlayer, an analogous interpretation leads to the conclusion that the three protrusions observed

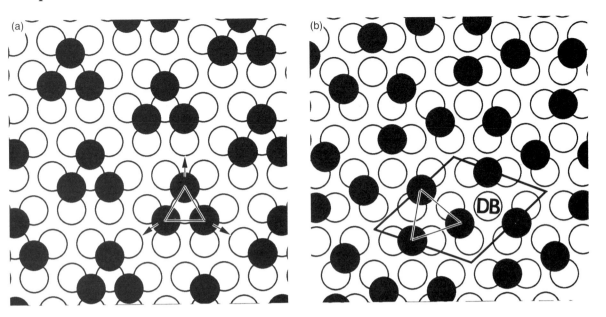

Fig. 10.4 Structural models for Si(111)-($\sqrt{7} \times \sqrt{7}$)R19.2° Al overlayer. In the model of Hansson (a), aluminum adatoms adsorb to form trimers, but leave the central Si atom coordinated to six atoms (three Al and three Si atoms). By radially displacing the Al adatoms outward from the center of each trimer onto two-fold sites, the arrangement of adatoms shown in (b) is produced. This model again predicts trimers of Al adatoms, but the trimers are larger and rotated with respect to the underlying (111) surface, in agreement with the STM images; also predicted is the occurrence of a single Si 'dangling bond' in each unit cell, which is labeled 'DB'.

at positive sample bias give the positions of the Al adatoms, while the single protrusion imaged at negative sample bias arises from some filled dangling bond state.

By careful measurements at domain boundaries between regions of $\sqrt{3}$ and $\sqrt{7}$ symmetries, it is possible to directly establish the atomic structure of the $\sqrt{7}$ overlayer. An earlier model for the $\sqrt{7}$ overlayer proposed by Hansson [11], shown in Fig. 10.4(a), had triangular clusters of Al adatoms on three-fold sites. However, this model predicts clusters oriented with their bases parallel to [1$\bar{1}$0]; this is in disagreement with the STM images, which clearly show that the triangular clusters are rotated. Moreover, the model shown in Fig. 10.4(a) also has six-fold coordination for the Si atom situated in the center of each Al trimer, while Si generally has a strong preference for four-fold coordination. A slight modification of this model, however, gives a model which is in good agreement with the experimental results. The modification is

based on the need to reduce the six-fold coordination of the central silicon atom through a radial outward displacement of the Al atoms in each cluster, as illustrated in Fig 10.4(a). The result, as shown in Fig. 10.4(b), is a model in which Al adatoms again form clusters, but with the clusters slightly larger and rotated with respect to [1$\bar{1}$0], in agreement with the experiment. Additionally, there is predicted to be one 'unique', Si atom in each unit cell (marked DB in Fig. 10.4(b)), which is expected to show the 'dangling bond' character evident in the STM images at negative sample bias. Thus, the STM images at positive sample bias reveal the positions of the Al adatoms in their triangular cluster, and the images at negative bias reveal the position of the single 'dangling bond' from the first-layer Si atom. This interpretation is further confirmed through theoretical calculations by Nelson and Batra [12]. Their calculations predicted this structure to have a very low total energy, and local density of states calculations

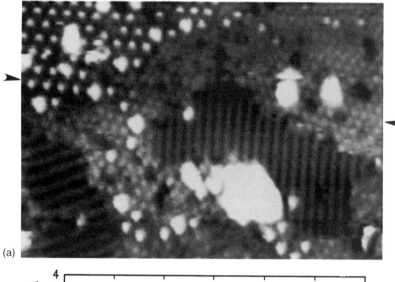

(a)

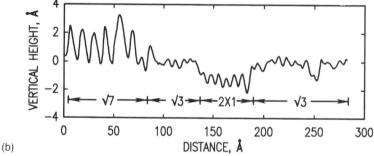

(b)

Fig. 10.5 STM image of aluminum overlayers on Si(111) surface exhibiting three co-existing superlattices. At the upper left (a), a $(\sqrt{7} \times \sqrt{7})R19.2°$ overlayer is observed; at the upper right (a), a $(\sqrt{3} \times \sqrt{3})$ overlayer is observed. Near the center and at lower left (a), a (2×1) reconstruction is observed, which is believed to arise from small patches of clean Si(111) as a result of desorption of Al during a short, hot anneal. The lower panel (b) shows the corrugations observed along the line indicated by the arrows in the upper figure.

predict empty and filled surface states concentrated in the same locations as the protrusions observed in the STM.

One surprising aspect of Al overlayers on Si(111) is that it is possible (in fact, rather common) for several different reconstructions to co-exist in relatively close proximity. The $(\sqrt{3} \times \sqrt{3})$ and $(\sqrt{7} \times \sqrt{7})R19.2°$ overlayers coexist as well-ordered islands over a substantial range of aluminum coverages. Heating this surface can also produce small regions with a (2×1) reconstruction, which is believed to arise from a patch of **bare** silicon where the activation barrier to formation of the thermodynamically-stable (7×7) reconstruction is too high. Figure 10.5 shows a typical larger-scale STM image where these $(\sqrt{7} \times \sqrt{7})R19.2°$, $(\sqrt{3} \times \sqrt{3})$, and (2×1) reconstructions coexist.

At coverages above one-half monolayer, the sur-

face structure becomes much less well-ordered. Near one monolayer Al coverage, STM images show an irregular (7×7) reconstruction shown in Fig. 10.5. Here, large-scale STM images (as in Fig. 10.6(a)) reveal irregular, triangular-shaped regions; a two-dimensional Fourier transform of this image, however, shows a well-defined average periodicity of approximately 7.0 ± 0.2 lattice constants as shown in Fig. 10.6(b). A closer look at the STM images shows that at positive sample bias (sampling empty surface states), as in Fig. 10.6(c), the individual triangular subunits appear slightly 'puckered', but are very smooth. At negative sample bias (probing filled surface states), (Fig. 10.6(d)) the surface appears remarkably flat, except for very small depressions at the corners of the triangular regions. In all cases, the '7×7' reconstruction shows a relatively high degree of disorder.

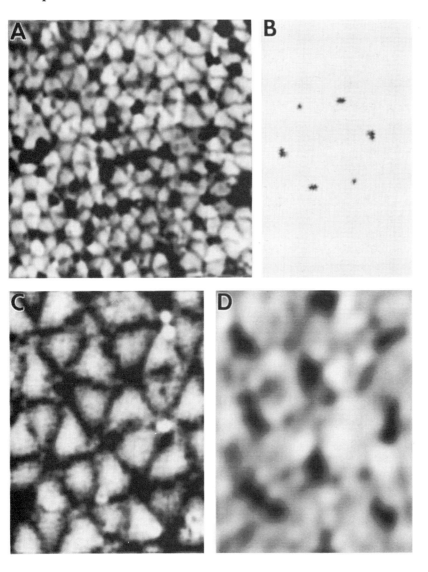

Fig. 10.6 The '7 × 7' reconstruction of aluminum on Si(111) near one monolayer Al coverage. Large-scale images (a) reveal an irregular packing of triangular-shaped regions. The average periodicity of the triangular regions is 7.0 ± 0.2 lattice constants, as shown by the two-dimensional Fourier transform of this image, shown in (b). (c) shows the puckered appearance of the individual triangular-shaped subunits at positive sample bias; at negative sample bias, (d), the surface appears perfectly flat except for very weak depressions at the corners of the unit cells.

Since the '7 × 7' overlayer arises near one monolayer Al coverage, the most likely locations for the aluminum atoms are either substituting for all Si atoms in the outermost atomic layer (as proposed by Lander [13]) or as adatoms directly above the Si atoms of the outmost double layer [14]. The major difference between these geometries is the coordination of the aluminum atoms; the former model involves bonding of each Al atom to three Si atoms, while the latter model involves only one bond between each Al atom and the underlying Si surface. At **bulk** Al–Si interfaces, both these configurations have been detected in transmission electron microscopy studies [15]. At the much lower Al coverages used in the STM experiments, tunneling spectroscopy data favor the model in which Al atoms substitute into the outermost double-layer. The origin of the relatively disordered (7 × 7) periodicity is likely a strain effect. Since Al is normally sp^2 hybridized, the lowest-energy bonding configuration likely has the Al almost coplanar with the second-layer Si atoms, distorting the Si lattice from its normal tetrahedral geometry. The presence of occasional vacancies or surface dislocations to relieve this strain are believed to be responsible for the dark troughs delineating the edges of each (7 × 7) unit cell.

Indium likewise shows a number of interesting reconstructions; Nogami et al. [16, 17] and Park et al. [18] observed structures having ($\sqrt{31} \times \sqrt{31}$), ($\sqrt{7} \times \sqrt{3}$), and (4 × 1) symmetries using STM. At the very lowest coverages, Nogami et al. showed that indium atoms simply replace individual Si adatoms of the (7 × 7) reconstruction of Si(111). At these low coverages, the indium shows no tendency to cluster, and the indium atoms do not show any preferential adsorption between faulted and unfaulted halves of the unit cell. At slightly higher coverage, gallium begins forming a $\sqrt{3}$-structure which is identical to the $\sqrt{3}$-structure discussed above for Al. However, in the case of indium it was possible to observe co-existing regions of metal-induced $\sqrt{3}$-symmetry coexisting with the (7 × 7) reconstruction characteristic of pure Si. Using the (7 × 7) reconstruction at its well-established Dimer-adatom-stacking fault (DAS) model [19] as a template, it was possible to definitely establish that the indium adatoms indeed reside on the T_4 sites and not on the H_3 sites, in agreement with total energy calculations by Northrup [5]. At higher coverages, indium behaves much like aluminum, forming a ($\sqrt{3} \times \sqrt{3}$) overlayer which initially consists of a nearly random alloy of gallium and silicon adatoms atop the Si(111) lattice; as the coverage increases, the ratio of In to Si adatoms increases until near $\frac{1}{3}$ monolayer virtually all surface atoms are In adatoms.

The ($\sqrt{31} \times \sqrt{31}$) structure [18] occurs with a local coverage of slightly greater than 0.52 monolayer. This structure has a diamond-shaped unit cell; when tunneling into the unoccupied electronic states of the sample, the internal structure of the ($\sqrt{31} \times \sqrt{31}$) structure shows an arrangement of 16 protrusions with an apparent (1 × 1) spacing. Assuming that the each protrusion arising from an indium atom gives coverage of $\frac{16}{31}$, or approximately 0.52, Park et al. [18] noted that this coverage was lower than the coverage determined by other means; this suggests that there may be one or two more indium atoms per unit cell, possibly located in sub-surface positions. Due to

strongly voltage-dependent changes in the images, separating the electronic from geometry contributions in these larger reconstructions is difficult, and it was not possible to determine the actual arrangement of atoms within the unit cell.

At coverages between 0.7 and 1.1 monolayer, indium on Si(111) shows a (4 × 1) reconstruction. The principal features of the (4 × 1) reconstruction are ridges running in the [$\bar{1}$10] direction separated by deep furrows. The separation between the furrows is 4× the (111) lattice constant and defines the long axis of the (4 × 1) reconstruction. The ridges have a secondary depression near their center, along with structure along the rows with a periodicity of 1 lattice constant (3.84 Å). Previous models for the (4 × 1) reconstruction had proposed that it was formed from pairs of (2 × 1) rows with minor differences, such as small atomic displacements, between alternate (2 × 1) rows, modulating the (2 × 1) symmetry and leading to a larger (4 × 1) unit cell. The symmetry of the (4 × 1) unit cell observed by Nogami, however, rules out these prior models and instead suggests a model with strong (4 × 1) character which is symmetric across the midpoint of the long axis of the unit cell. As the bias voltage between sample and tip was changed, Nogami noted that the appearance of the STM image changed also, indicating a spatial separation of filled and empty electronic states. Due to these voltage-dependent changes, determining the atomic positions from the STM image was not possible, and so no atomic model for the (4 × 1) reconstruction was proposed.

10.2.2 Gallium on Si(111)

At low coverages, gallium atoms replace the Si adatoms within the Si(111)–(7 × 7) reconstruction; at $\frac{1}{3}$ monolayer, it forms a ($\sqrt{3} \times \sqrt{3}$) reconstruction like Al and In. At coverages near one monolayer, gallium on Si(111) shows an **incommensurate** overlayer structure with a periodicity of approximately (6.3 × 6.3). Initial studies of this reconstruction were performed by Park et al. [18]. They observed irregular triangular-shaped unit cells; the internal structure of each unit cell showed an apparent (1 × 1) periodicity, and the edge of the triangles varied from four to seven lattice constants. Although a two-dimensional fast Fourier transform of their STM images showed an average periodicity of 5 ± 0.5 lattice constants, it is likely that this is the '6.3 × 6.3' incommensurate structure. A more detailed study of this structure was performed by Chen *et al.* [20]. They concluded that the periodicity is essentially a stress effect; they proposed that gallium and silicon form two nearly coplanar interpenetrating lattices, forming an inert graphite-like structure. The overall periodicity of this structure (6.3 lattice constants, or approximately 24 Å) was attributed to misfit dislocations arising from the different lattice constant of the Ga–Si overlayer structure and the bulk Si lattice. Similar misfit dislocations were also found for Al overlayers on Si(111), which near 1 monolayer forms a relatively disordered structure with an average period of close to seven lattice constants. Both the (7 × 7) Al and (6.3 × 6.3) Ga structures show ragged boundaries defining comparatively featureless triangular-shaped regions with an **average** spacing of 6.3 (for Ga) or 7.0 (for Al) lattice constants.

10.2.3 Gallium on Si(001)

Metal adsorption on the Si(001) surface is somewhat more complicated than on the Si(111) surface. Baski *et al.* [21] and Nogami *et al.* [16, 17] found that gallium adsorption on Si(001) resulted in the formation of Ga–Ga dimers. Below 0.3 monolayer coverage, these dimers arranged into a (2 × 3) local symmetry, while at 0.5 monolayer they arranged into a (2 × 2) symmetry. The STM results suggest that both the Ga adsorption leaves the Si–Si dimers intact, and that the (2 × 2) and (2 × 3) symmetries arise from ordered arrangements of Ga–Ga dimers on the dimerized Si(001) surface. The bond axis of the Ga–Ga dimers is rotated 90 degrees with respect to that of the underlying Si–Si dimers, consistent with the tetrahedral coordination of the Si(001) surface and similar in many respects to the 90 degree rotation observed in epitaxial growth of silicon on Si(001). This Ga-dimer model was proposed earlier by Bourguignon *et al.* [22]. At the lowest coverages, a (2 × 3) structure is formed; at

coverages approaching 0.5 monolayer, a mixture of local (2 × 3) and (2 × 2) arrangements is observed; at slightly above 0.5 monolayer, the surface consists almost entirely of Ga dimers in a (2 × 2) arrangement. At higher coverages, the STM results show a (n × 8) phase where n usually equals four or five. The appearance of this new phase suggests a pronounced change in the Si–Ga bonding at coverages higher than 0.5 monolayer, coinciding with the saturation of all Si dangling bonds. Unfortunately, the STM results were not sufficiently resolved to determine the detailed nature of this phase.

10.3 GROUP V METALS ON SILICON

The adsorption of the group V metals As and Sb on silicon and germanium has received much attention due to the valence electronic structure and possible application to heteroepitaxy of II–V semiconductors on group IV semiconductors. In the bulk-truncated (unreconstructed) Si(001) surface, each silicon atom is two-fold coordinated; the dimerization of the Si(001) surface is then best described as the formation of a double bond. This 'double-bond' has two associated occupied bonding states. The most stable of these states has σ-character and lies approximately 9 eV below E_F. The second state, however, has π-character and lies only about 0.3 eV below E_F. Due to the close proximity of this state to the Fermi energy, the Si(001) surface is chemically reactive and can still be said to have a 'dangling bond'. Arsenic and antimony, however, have one more electron than silicon. Therefore, instead of forming a strong σ- and a weak π-bond, arsenic and antimony are expected to form **only** a σ-bond; the weak π-bond of the Si–Si dimer is now replaced with two much more stable Si–As (or Si–Sb) bonds. The removal of electronic states from near the Fermi energy is expected to make the As- and Sb-terminated Si(001) surface chemically unreactive. Likewise, on the Si(111) surface, one would expect that substitution of three-fold coordinated As or Sb atoms for the three-fold coordinated silicon atoms exposed at this surface should also produce energetically stable structures, with the half-filled 'dangling bonds' of the exposed silicon atoms replaced by fully-occupied, low-lying 'lone-pair' orbitals on the As and Sb atoms.

Copel *et al.* [23] and Becker *et al.* [24] studied arsenic on Si(111) and observed regions of approximately 50–100 Å diameter having well-defined local (1 × 1) order. On longer distance scales, however, the surfaces had a high degree of disorder. Both studies showed that the **local** order was extremely good, with no evidence for stacking faults or other remnant features of the starting Si(111)–(7 × 7) surface. Becker *et al.* [24] performed a detailed analysis of both the geometric and electronic structure of this surface, showing that the STM images were consistent with arsenic substitution for the Si in the outermost half-double layer of the (111) lattice. The tunneling spectra showed a surface-state gap of 1.9–2.3 eV with the edges of the surface state bands at 0.6 eV above E_F. Both Copel *et al.* [23] and Becker *et al.* [24] observed the highest corrugation with a sample bias near 0.6 V, corresponding to tunneling into the edge of the empty surface-state band.

On the Si(001) surface, Becker *et al.* [25] found that arsenic adsorption preserved the (2 × 1) symmetry but produced a surface with much greater perfection that the starting Si(001) surface. Using miscut wafers which produce a nearly single-domain (2 × 1) reconstruction, Becker *et al.* [25] showed that at low adsorption temperatures the arsenic adsorbs on top of the Si(001) dimers, forming a new terrace with dimers rotated 90° from the original substrate. At higher temperatures, the arsenic dimers were able to displace the outermost Si layer, producing a (1 × 2) reconstruction with domains commensurate with the original reconstructed Si lattice. Similar behavior has been observed by Rich *et al.* [26] for antimony adsorption on Si(001).

10.4 ALKALI METALS ON SILICON

The study of alkali metals on silicon surfaces is complicated by the fact that surface diffusion is significant on the time scale of the STM measurement itself. Adsorption of the alkali metals is, in many respects, expected to be similar to adsorption of hydrogen, since in both cases the interaction with the surface occurs primarily through *s*-orbitals.

At coverages of only a few hundredths of a monolayer, Hasegawa *et al.* [27] observed that lithium atoms on the Si(111) surface form small clusters. Surprisingly, the STM images showed these trimers diffuse as an intact unit across the surface, suggesting strong bonding interactions between the lithium atoms within the cluster. The Li clusters were preferentially located in the faulted half of the (7 × 7) unit cell.

At higher coverages, six-atom and nine-atom clusters were also observed. These clusters also exhibited an 8:1 preference for adsorbing in the faulted half of the unit cell versus the unfaulted half. From detailed analysis of the STM images, it was concluded that the lithium atoms desorbed the original silicon adatoms, and that the lithium atoms were adsorbed in the sites corresponding to the 'rest atom' sites of the (7 × 7) superstructure.

The adsorption of alkali metals on Si(001) surfaces shows some surprising features. At very low metal coverages, the STM images show that lithium and potassium ions favor adsorption on top of **one** of the two Si atoms comprising a Si–Si dimer. This asymmetric bonding therefore breaks the mirror-plane symmetry of the Si–Si dimer and causes the dimers to tilt, or 'buckle'. Similar buckling has been observed on the clean surface near steps or defects which create an asymmetric potential well for the dimers as a function of tilt angle. The buckling induced by alkali metal adsorption propagates for approximately 50 Å away from the adsorption site.

At higher coverage, one might expect that due to the large amount of charge exchange between metal and semiconductor and the concurrent formation of a large surface dipole, individual alkali metals would repel one another. However, the STM images clearly show clustering of lithium and potassium atoms even at very low coverage on Si(001); this suggests that the amount of charge transfer with the bulk might be significantly smaller than anticipated.

Perhaps the most surprising result of STM studies of alkali metal adsorption on Si(001) is the asymmetric growth observed. A number of papers had previously proposed that at full monolayer coverage alkali metal adsorption occurred through the creation of one-dimensional 'chains' of atoms running parallel to the dimer rows of the Si(001) substrate. However, the STM images show that at low coverage lithium and potassium form clusters oriented **perpendicular** to the dimer rows of the substrate. This growth direction is the same as that observed in STM studies of Ga, Si, and Ag growth on Si(001).

10.5 TRANSITION METALS ON SILICON

The behavior of transition metals on silicon surfaces is generally complicated. Many of these have large bulk solubilities, and the equilibrium between metal at the surface, in the near-surface region, and in the bulk can be very complicated and strongly temperature-dependent. Due to its low chemical reactivity with silicon, silver has been one of the most widely-studied metallic adsorbates, although other metals have been studied as well.

10.5.1 Silver on Si(111)

Tosch and Neddermeyer [28, 29] found that after room-temperature adsorption of silver on Si(111)–(7 × 7) at room temperature, the silver atoms formed triangular-shaped islands preferentially located in the faulted half of the (7 × 7) unit cell. The edges of these triangular structures consist of oblong protrusions, with a triangular depression in the center. The inner adatoms of the (7 × 7) unit cell halves are incorporated into the triangular structure, but the outer adatoms (i.e. those adjacent to the corner holes) are not incorporated and seem to be unaffected. Tosch and Neddermeyer proposed that the triangular structures arose from three silver adatoms, each of which is two-fold coordinated by bonding to one of the central adatoms and one of the 'rest atoms'. Thus, each silver adatom is essentially two-fold coordinated (as in many bulk silver compounds), and the number of dangling bonds of the Si(111)–(7 × 7) surface is reduced. The absence of partially-completed

triangles (i.e. structures resulting from adsorption of only one or two Ag atoms in a (7 × 7) half-unit-cell) indicates that this triangular structure represents the critical nucleus for Ag cluster formation. As the silver coverage was increased further, the size of the clusters increased until they filled one-half the (7 × 7) unit cell, forming a triangular, two-dimensional metal island. These larger islands presumably consist of close-packed silver atoms, which would indicate that each cluster contains on the order of 45 silver atoms.

When annealed, silver-covered Si(111) surfaces form a ($\sqrt{3} \times \sqrt{3}$) reconstruction; the nature of this reconstruction has been highly controversial. Unlike the simple $\sqrt{3}$ structure formed by group III metals, the Ag-induced $\sqrt{3}$ structure has a honeycomb shape, with two protrusions in each unit cell. The difficulty in directly determining the atomic structure of metal–semiconductor structures from STM images alone is demonstrated by the fact that simultaneous STM studies by van Loenen *et al.* [30] and by Wilson and Chiang [31, 32] of the ($\sqrt{3} \times \sqrt{3}$)Ag–Si(111) reconstruction made similar experimental observations, yet came to opposite conclusions. Based on electron-counting and the observation of a semiconducting electronic structure, van Loenen *et al.* [30] concluded that the Ag atoms formed an embedded honeycomb and that the surface atoms were Si adatoms, while Wilson and Chiang [31, 32] concluded that the observed protrusions were the Ag atoms themselves. A later study [32] showed that the protrusions were located at sites corresponding to three-fold sites of the Si(111) lattice, but many questions remain regarding the nature of the $\sqrt{3}$Ag–Si(111) surface.

At coverages of several monolayers, the STM results confirmed the Stranski–Krastanov growth mode, and revealed large, flat islands with edges coinciding with the principal lattice directions of the Si(111) surface.

10.5.2 Silver on Si(001)

The growth of Ag on Si(001) has been studied independently by several groups. Unlike many other more strongly-bonded metals, Ag is only rather weakly bonded to Si(001). As a result, the tunneling images are susceptible to instabilities caused by diffusion of Ag atoms on the surface. The growth is strongly anisotropic. At very low coverage, silver forms rows of dimers which are oriented perpendicular to the underlying dimer rows. At higher coverage, the studies by Brodde *et al.* [33] and by Hashizume *et al.* [34] disagree; Hashizume *et al.* [34] observed no pronounced changes in the surface structure, while Brodde *et al.* [33] found that the Ag atoms formed structures aligned **parallel** to the dimer rows of the underlying substrate. One possible explanation for this rotations is that nickel contamination, which causes dimer vacancies to align in a (2×8) configuration, may introduce a new periodicity in the underlying structure.

10.5.3 Nickel on Si(111)

At low coverage, nickel on Si(111) results in the formation of a $(\sqrt{19} \times \sqrt{19})R23.4°$ structure

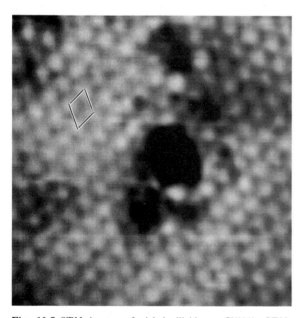

Fig. 10.7 STM images of nickel silicide on Si(111). STM images show these nickel silicide surfaces to be well ordered and remarkably flat over micron-sized areas. Occasional defects arise in the films, most of which appear to be vacancies of five or six surface atoms, as shown here.

which was studied by Wilson and Chiang [50]. They found that the $\sqrt{19}$ structure occurs with just one nickel atom in each unit cell. The proposed a model for the reconstruction in which the single Ni atom in each cell is in a sub-surface position, bonded to six silicon adatoms. In general, this surface tends to be relatively disordered, most likely due to the fact that the Ni atom is not just a simple adatom structure and activation barriers for diffusion are then very high.

At higher coverage, nickel on Si(111) forms nickel silicides. Markert and Hamers (unpublished) studied nickel silicide formation on Si(111). Nickel silicide is formed in large, nearly perfect platelets. Many of these are well-ordered and flat over distances of several thousand ångstroms in diameter. The atomic-scale perfection of these platelets is very high, as shown by the STM image in Fig. 10.7 [35]. The most common defect observed is the apparent vacancy of a surface atom. From the STM results alone, it is impossible to determine whether the surface atoms are nickel or silicon.

10.5.4 Nickel on Si(001)

On the Si(001) surface, nickel does not adsorb on **top** of the surface; instead, it occupies sub-surface positions. The equilibrium between nickel dissolved in the bulk and nickel in the sub-surface region is complicated and strongly temperature-dependent. Even trace quantities of nickel have relatively strong effects on the ordering of the Si(001)–(2×1) surface. The usual methods for preparing the Si(001)–(2×1) surface result in a defect density on the order of 5% of the surface. Most of these surface defects appear to be vacancies of one of more dimers from the outermost surface layer. On a 'clean' surface, these defects are usually randomly distributed on the surface. Small quantities of nickel, however, cause these dimer vacancies to align, forming an impurity-stabilized (2×8) structure [36].

10.5.5 Gold on Si(111)

Baski *et al.* [37] have studied the adsorption of gold on Si(111)–(7×7), which produces a (5×1)

overlayer at coverages of 0.1–0.4 monolayer Au and a mixture of (5×1) and $(\sqrt{3} \times \sqrt{3})$ reconstructions for coverages between 0.5 and 0.8 monolayer Au coverage. Baski *et al.* [37] focused their efforts on the (5×1) structure; high-resolution STM images show that the (5×1) unit cell has a rather complicated internal structure, in disagreement with simple models where strings of Au atoms with the Si unit-cell spacing sit in a regular array along $\langle 1\bar{1}0 \rangle$ directions. Due to the complicated internal structure and strong changes in the images depending on the sample-tip bias, it was not possible to determine the atomic configuration of the (5×1) superstructure.

10.5.6 Copper on Si(111)

Copper on Si(111) is an interesting system due to the formation of an incommensurate '5×5' structure. Wilson *et al.* [38] and Demuth *et al.* [39] performed simultaneous, independent studies of this system using STM. Both studies showed that the overlayer consists of small (5×5) sub-units which packed together in a nearly repetitive manner over lateral distances of five to seven unit cells (approximately 100–175 Å). After this distance, there is a phase boundary which disrupts the long-range order. The individual (5×5) subunits show a very complicated structure which is not completely periodic even between adjacent unit cells. There are several characteristic features of the (5×5) unit cells. Both studies identified several characteristic features of the (5×5) structure:

(a) deep depressions arranged in a hexagonal (5×5) array, one per unit cell;
(b) two kinds of inequivalent, nearly-flat triangular regions with a weak height modulation corresponding to a (1×1) periodicity;
(c) Striped regions.

Demuth *et al.* [39] performed measurements at various bias voltages, showing that there were strong variations in the local electronic structure, and that these variations also showed a weak (5×5) periodicity when measured over short length scales. Due to the complicated internal structure, however, neither of the STM studies were able to determine the detailed atomic structure of this overlayer.

10.5.7 Palladium on Si(111)

Kohler *et al.* [40] studied the nucleation of Pd on the Si(111)–(7×7). This study shows a remarkable spatial selectivity in the initial nucleation. At a coverage of 0.25 monolayer, the STM images reveal that palladium nucleates into small, three-dimensional islands containing about 13 atoms each. At a coverage of 0.25 monolayer, almost every (7×7) unit cell of the clean substrate containes **exactly one** palladium silicide island. Perhaps even more remarkable is the fact that 95% of these small islands are found on the **faulted** half of the Si(111)–(7×7) unit cell! This surprising spatial selectivity in the adsorption process suggests that impinging palladium atoms are relatively mobile and are able to move at least 20–25 Å across the surface in order to find the most preferable binding site. This strong preference for the faulted half of the (7×7) unit cell is rather surprising, since the arrangement of atoms in the outermost atomic layer is the same in the faulted and unfaulted halves and only differs in the lower bilayer. However, tunneling spectroscopy measurements [41] show that the density of states near the Fermi energy is much higher in the faulted than unfaulted half of the (7×7) unit cell. Therefore, it is likely that the strong preferential nucleation of Pd in the faulted half is driven by the increased density of states there, rather than by the geometric arrangement of atoms.

While at low temperatures the palladium atoms nucleate as small islands centered in the faulted halves of (7×7) unit cells, at higher temperatures the atoms diffuse and form ordered silicide overlayers with $(\sqrt{3} \times \sqrt{3})$ and $(3\sqrt{3} \times 3\sqrt{3})$ periodicities. This can be observed in Fig. 10.8, which shows STM images of palladium atoms deposited on Si(111)–(7×7) at room temperature (Fig. 10.8(a)) and the same surface after annealing and the formation of a $(3\sqrt{3} \times 3\sqrt{3})$ overlayer (Fig. 10.8(b)) [42]. Some islands exhibited a $(\sqrt{3} \times \sqrt{3})$ periodicity, which can be observed in Fig. 10.9. In this image, the $(\sqrt{3} \times \sqrt{3})$ period-

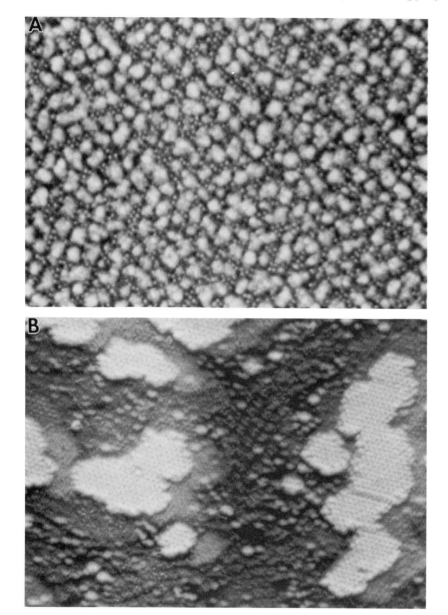

Fig. 10.8 Palladium atoms on Si(111)-(7 × 7) substrate. After room temperature deposition (a), the palladium atoms nucleate into small islands centered on the faulted halves of the (7 × 7) unit cells. After annealing (b), the atoms diffuse and form large islands exhibiting ($\sqrt{3}$ × $\sqrt{3}$) and/or (3$\sqrt{3}$ × 3$\sqrt{3}$) symmetry.

icity of the palladium silicide overlayer can be accurately measured because the (7 × 7) periodicity of the substrate can also be observed.

In the case of room-temperature deposition, by carefully calibrating the palladium coverage using Rutherford backscattering and measuring the height and area of the palladium silicide islands, it was also possible to directly observe several dif-

ferent growth regimes. For islands with an area smaller than 100 Å^2, the island volume varies with island area as $V \propto A^n$, where $n = 1.6 \pm 0.1$. This can be compared with a value of n of 1.5 expected for uniform three-dimensional growth of the islands. Islands larger than 100 Å^2 in area display a linear relation between island area and volume; in this regime, palladium silicide formation occurs by

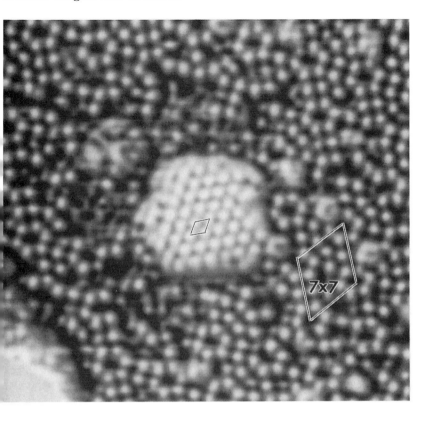

Fig. 10.9 Small islands of palladium silicide nucleated on Si(111)–(7 × 7) substrate.

lateral growth so that the ratio of volume to area is constant. Finally, islands larger than $300\,\text{Å}^2$ area also show a linear relation between volume and area due to coalescence of the islands.

10.6 METALS ON GALLIUM ARSENIDE

Whereas most of the STM work studying metals on silicon has emphasized the structural aspects of the metal–semiconductor systems, most of the work on GaAs has concentrated on the electronic properties of metal clusters and the metal–semiconductor interface. GaAs is a better choice for such studies because the surface reconstruction results in empty and filled states lying **outside** the bulk bandgap, and because GaAs has a larger, direct gap.

Martensson and Feenstra [43] studied that adsorption of antimony on GaAs(110). The geometry of the surface antimony atoms is similar to the geometry of the silicon atoms in the pi-bonded chain model of the Si(111)–(2 × 1) surface, except that the 'dangling bonds' are half-filled on the Si surface but are completely filled on the Sb/GaAs surface. There are two inequivalent Sb atoms per unit cell; they observed both atoms at low bias voltages, but at high bias voltages on the Sb atoms bonded to the underlying Ga atoms were observed. Ludeke *et al.* [44] studied adsorption of bismuth on GaAs(110). At coverages between 0.5 and 1 monolayer, they found that the Bi atoms formed zig-zag chains which were located both between and on top of the GaAs chains. Near 1 monolayer coverage the bismuth atoms formed dislocations spaced approximately 25 Å apart, which gave rise to unoccupied acceptor states within the bulk bandgap, and that these acceptor states pinned the Fermi level. Feenstra [45] also studied gold adsorbed on GaAs(110) and found that the behavior of Au was similar to that of Sb and Bi; all three generate empty states which act as acceptors and filled states which act as donors.

First *et al.* [46] studied cesium on GaAs(100) and found the Cs atoms exhibited a remarkable ordering, forming long one-dimensional chains. High resolution images of these chains show that they are composed of a repeating unit of two inequivalent Cs atoms; this inequivalence arises from the underlying GaAs lattice and is not a spontaneous symmetry-breaking. At higher coverage, they observed chains consisting of three inequivalent Cs atoms. The formation of such close-packed ordered structures indicates that charge transfer between Cs and GaAs is smaller than originally predicted, since large amounts of charge transfer would produce significant repulsive interactions between the Cs atoms. It should be noted that a similar effect was observed for alkali metals on silicon surfaces, where clustering of alkali metals was observed even at low coverages.

Several studies have been conducted on the morphology and electronic properties of iron clusters on GaAs(100) [47, 48, 49]. First *et al.* [46] performed STM and tunneling spectroscopy measurements on small iron clusters on GaAs(110). They found that iron clusters containing approximately 13 atoms were non-metallic, while clusters with more than 35 atoms showed indications of metallic character and a well-defined Fermi energy. Stroscio *et al.* [49] measured how the tunneling I–V curves on the GaAs changed as a function of distance away from individual Fe clusters. On the clusters, a 'metallic' I–V curve was observed, indicating a high density of states at the Fermi level. Away from the Fe cluster, the tunneling conductance in the gap region decayed exponentially with increasing lateral separation. From detailed analysis of these tunneling I–V curves and from measurements of the local tunneling barrier height, it was shown that the spatial dependence of the tunneling conductivity arises from wavefunction-matching between the semiconductor and the metal. In the metal, the density of states is continuous and the wavefunctions are propagating free-electron states; in the semiconductor bandgap, however, there are no allowed states, and so the wavefunctions must be imaginary

and decay exponentially with distance into the semiconductor. Thus, the tunneling I–V curves are 'metallic' directly on the Fe clusters, but also show metallic behavior on the GaAs surface near the clusters, since the high density of states near E_F of the clusters decays away with a characteristic length determined by the complex band structure of the semiconductor in the bandgap region. Measurements of the effective decay length of the states involved in tunneling also showed a discontinuity at the valence and conduction band edges, since here the complex (decaying) semiconductor states match the freely-propagating states of the metallic clusters.

10.7 CONCLUSIONS

Scanning tunneling microscopy studies clearly demonstrate the wide variety of behavior exhibited in the adsorption and nucleation of metals on semiconductors. While in many cases it is difficult to develop a detailed structural model based on the STM results alone, development of a detailed structural model from the interpretation of the results is difficult due to the strongly localized nature of semiconductor bonding; STM alone provides much new information regarding the internal symmetry and local electronic structure of metals on semiconductors. Further developments in the field continue to reveal new insight into the nature of metal–semiconductor bonding.

REFERENCES

1. G. Binnig and H. Rohrer, *Rev. Mod. Phys.*, **59** (1987), 615.
2. R. J. Hamers, *Annual Review of Physical Chemistry 1989*, **40** (1989), 531.
3. Y. Kuk and P. J. Silverman, *Rev. Sci. Inst.*, **60** (1989), 165.
4. P. K. Hansma and J. Tersoff, *J. Appl. Phys.*, **61** (1987), R1.
5. J. E. Northrup, *Phys. Rev. Lett.*, **57** (1986), 154.
6. R. J. Hamers, *J. Vac. Sci. Tech. B*, **6** (1988), 1462.
7. R. J. Hamers, *Phys. Rev. B*, **40** (1989), 1657.
8. R. J. Hamers and J. E. Demuth, *Phys. Rev. Lett.*, **60** (1988), 2527.

9. J. Nogami, S. Park, and C. F. Quate, *Phys. Rev. B*, **36** (1987), 6221.

10. J. Nogami and S. I. Park, *J. Vac. Sci. Tech. B*, **6** (1988), 1479.

11. G. V. Hansson, R. Z. Bachrach, R. S. Bauer, and P. Chiaradia, *Phys. Rev. Lett.*, **46** (1981), 1033.

12. J. S. Nelson and I. P. Batra, *Proceedings of the NATO Advanced Research Workshop on Metallization and Metal–Semiconductor Interfaces* (Plenum, New York, in press).

13. J. J. Lander and J. Morrison, *Surf. Sci.*, **2** (1964), 553.

14. H. I. Zhang and M. Schluter, *Phys. Rev. B*, **18** (1978), 1923.

15. R. K. LeGoues, W. Krakow, and P. S. Ho, *Phil. Mag. A*, **53** (1986), 833.

16. J. Nogami, S. I. Park, and C. F. Quate, *Phys. Rev. B*, **36** (1987), 6221.

17. J. Nogami, Sang-il Park, and C. F. Quate, *J. Vac. Sci. Tech. B*, **6** (1988), 1479.

18. Sang-il Park, J. Nogami, and C. F. Quate, *J. Microscopy*, **152** (1988), 727.

19. K. Takayanagi, Y. Tanishiro, M. Takahashi, and S. Takahashi, *J. Vac. Sci. Tech. A*, **3** (1985), 1502.

20. D. M. Chen, J. A. Golovchenko, P. Bedrossian, and K. Mortensen, *Phys. Rev. Lett.*, **61** (1988), 2867.

21. A. A. Baski, J. Nogami, and C. F. Quate, *J. Vac. Sci. Tech. A*, **8** (1990), 245.

22. B. Bourguignon, R. V. Smilgys, and S. R. Leone, *Surf. Sci.* **204** (1988), 473.

23. M. Copel, R. M. Tromp, and U. K. Kohler, *Phys. Rev. B*, **37** (1988), 10756.

24. R. S. Becker, T. Klitsner, and J. S. Vickers, *J. Microscopy*, **152** (1988), 157.

25. R. S. Becker, B. S. Swartzentruber, J. S. Vickers, M. S. Hybertson, and S. G. Louie, *Phys. Rev. Lett.*, **60** (1988), 116.

26. D. H. Rich, F. M. Leibsle, A. Samsavar, E. S. Hirschhorn, T. Miller, and T. Chiang, *Phys. Rev. B*, **39** (1989), 12758.

27. Y. Hasegawa, I. Kamiya, T. Hashizume, T. Sakurai, T. Tochihara, M. Kubota, and Y. Murata, *J. Vac. Sci. Tech. A*, **8** (1990), 238.

28. St. Tosch and H. Neddermeyer, *Phys. Rev. Lett.*, **61** (1988), 349.

29. St. Tosch and H. Neddermeyer, *J. Microscopy*, **152** (1988), 415.

30. E. J. Van Loenen, J. E. Demuth, R. M. Tromp, and R. J. Hamers, *Phys. Rev. Lett.*, **58** (1987), 373.

31. R. J. Wilson and S. Chiang, *Phys. Rev. Lett.*, **58** (1987), 369.

32. R. J. Wilson and S. Chiang, *Phys. Rev. Lett.*, **59** (1987), 2329.

33. A. Brodde, D. Badt, St. Tosch, and H. Neddermeyer, *J. Vac. Sci. Tech. A*, **8** (1990), 251.

34. T. Hashizume, R. J. Hamers, J. E. Demuth, and K. Markert, *J. Vac. Sci. Tech. A*, **8** (1990), 249.

35. K. Markert and R. J. Hamers, unpublished.

36. H. Niehus, U. K. Kohler, M. Copel, and J. E. Demuth, *J. Microscopy*, **152** (1988), 735.

37. A. A. Baski, J. Nogami, and C. F. Quate, *Phys, Rev. B*, **41** (1990), 10247.

38. R. J. Wilson, S. Chiang, and F. Salvan, *Phys. Rev. B*, **38** (1988), 12696.

39. J. E. Demuth, U. K. Koehler, R. J. Hamers, and P. Kaplan, *Phys. Rev. Lett.*, **62** (1989), 641.

40. U. K. Kohler, J. E. Demuth, and R. J. Hamers, *Phys. Rev. Lett.*, **60** (1988), 2499.

41. R. J. Hamers, R. M. Tromp, and J. E. Demuth, *Phys. Rev. Lett.*, **56** (1986), 1972.

42. U. K. Köhler, R. J. Hamers, and J. E. Demuth, unpublished work.

43. P. Martensson and R. M. Feenstra, *J. Microscopy*, **152** (1988), 761.

44. R. Ludeke, A. Taleb-Ibrahimi, R. M. Feenstra, and A. B. McLean, *J. Vac. Sci. Tech. B*, **7** (1989), 936.

45. R. M. Feenstra, *Phys. Rev. Lett.*, **63** (1989), 1412.

46. P. N. First, J. A. Stroscio, R. A. Dragoset, D. T. Pierce, and R. J. Celotta, *Phys. Rev. Lett.*, **63** (1989), 1416.

47. P. N. First, J. A. Stroscio, R. A. Dragoset, D. T. Pierce, and R. J. Celotta, *Phys. Rev. Lett.*, **63** (1989), 1416.

48. R. A. Dragoset, P. N. First, J. A. Stroscio, D. T. Pierce, and R. J. Celotta, *Material. Res. Soc. Symp. Proc.*, **151** (1989), 193.

49. J. A. Stroscio, P. N. First, R. A. Dragoset, L. J. Whitman, D. T. Pierce, and R. J. Celotta, *J. Vac. Sci. Tech. A*, **8** (1990), 284.

50. R. J. Wilson and S. Chiang, *Phys. Rev. Lett.* **58**, 2575 (1987).

CHAPTER ELEVEN

Epitaxy of semiconductor thin films

J. L. Batstone

Introduction · Principles of epitaxy · Misfit dislocations · Phase stability in epitaxial systems · Summary

11.1 INTRODUCTION

Epitaxy refers to the growth of one crystal on another with a fixed orientation relationship between the two crystals. The word 'epitaxy' has origins in Greek with επι, or epi, meaning 'on' and ταξιζ, or taxis, meaning 'arrangement' [1]. Epitaxial growth of a thin film requires that a definite and unique epitaxial relationship exists resulting in a single crystal film on a substrate. Crystallographic variants can occur, leading to domain structures. Homoepitaxy refers to growth of a crystal on itself, for example GaAs–GaAs. Although this does not seem distinct from conventional 'bulk' crystal growth, it provides a convenient method of obtaining abrupt doping concentration profiles in epitaxial device structures. Heteroepitaxy implies the growth of thin films of different crystal structures or composition on crystalline substrates, for example bcc/fcc and Ge on Si. Factors involved in epitaxy include the relative surface free energies of the substrate and deposit, the interface interaction and the effects of strain induced in the thin film due to differences in lattice parameter between the two crystals.

Historically, interest in epitaxial film growth was dominated by studies of metal films on alkali halide crystals [2] although the first recorded case of oriented growth of one crystal type on another was reported by Frankenheim in 1836 who successfully grew sodium nitrate on calcite from solution [3]. The advent of X-ray diffraction and subsequently electron diffraction led to rapid advances in understanding the principles of epitaxy due to the ability to determine orientation relationships from diffraction data rather than external crystal morphologies. The impact of diffraction experiments and the development of the science of epitaxy has previously been reviewed by Pashley [4].

In recent years, the driving force for our increasing understanding of factors controlling epitaxy has come from the semiconductor industry, where novel device structures have been fabricated by epitaxial techniques. Development of new growth techniques such as molecular beam epitaxy (MBE) and organometallic vapour phase epitaxy (OMVPE) as well as the hybrid chemical beam epitaxy (also known as metal–organic molecular beam epitaxy) have resulted in control of epitaxy at the atomic level enabling growth of 'quantum' structures such as quantum wells, wires, and dots where the dimensions of such structures are sufficiently small that quantum mechanical effects give rise to unusual device properties [5, 6]. Development of *in-situ* monitoring techniques have also advanced such that diffusive processes such as atom migration on surfaces can be studied as well as the appearance of surface reconstructions [7]. Enhancement of conventional epitaxial techniques has also appeared with the addition of laser and ion beam irradiation of the growing surface which, in conjunction with patterning, leads to selective area epitaxy [7]. Growth of heterostructures involving combinations of metals, semiconductors and insulators have been demonstrated as an approach to three-dimensional device integration [8, 9]. Such structures allow the study of epitaxy between dissimilar materials with varying degrees of ionic, metallic, and covalent bonding where interfacial structure can control electrical properties in a device [10].

This review will illustrate important features of epitaxial growth, taking examples from semiconductor thin film systems. Three main areas of thin film growth are considered: the initial stages of epitaxy, misfit dislocations, and phase stability in alloy systems. Initially, the importance of 'fit', or alignment of lattice planes, resulting in a coherent, dislocation-free interface between thin film and deposit will be considered. The mechanisms of thin film growth will then be examined, taking into account the effects of surface and interface free energies, surface reconstructions, and lattice strain. Two-dimensional and three-dimensional thin film growth will be described. The relative stability of different growth modes for films with finite strain energy is discussed. Growth of a film with finite strain due to the difference in lattice parameter between film and substrate requires the introduction of misfit-dislocations at a finite, **critical** thickness to relieve the strain. Models for misfit-dislocation introduction are summarized and discrepancies between experimental observations and theoretical predictions are discussed in terms of deviations from thermodynamic equilibrium and kinetic factors affecting defect nucleation and motion. Reduction of defect densities in the active region of device structures built in semiconductor thin films is extremely important and has led to extensive research in efforts of understanding defect nucleation and multiplication mechanisms, as well as methods to reduce the total dislocation density. Attempts to minimize threading dis-

location densities have followed two different approaches, namely the generation of strained layer superlattices and reduction of the interface area by growth on contact pads or mesas. The effects of crystal symmetry on the allowed deposit/substrate orientations and interfacial defect-types are also outlined. The final section describes thermodynamic phase stability in alloy systems and modifications to thin film 'phase diagrams' which occur due to the effects of interface structure and energy.

11.2 PRINCIPLES OF EPITAXY

Growth of a deposit on a substrate requires an understanding of both nucleation and growth processes. The initial stages of epitaxy can be thought of as belonging to the field of 'surface science' where the degree of 'wetting' and the orientation of 0–5 monolayers (ML) of deposit can be monitored by surface-sensitive techniques such as low energy electron diffraction (LEED), auger electron spectroscopy (AES), and scanning tunneling microscopy (STM). These techniques are ideally suited to nucleation studies and post-nucleation growth. As the deposit thickness increases, factors such as fit, interface coherence, and strain become increasingly important. Detection of interface structure, thin film orientations, and the nature of crystallographic defects within the deposit require three-dimensional diffraction techniques such as reflection high energy electron diffraction (RHEED) and transmission electron diffraction. Ideally, studies of epitaxy are performed *in situ* and improvements in ultrahigh vacuum design for electron microscopes have resulted in new insight into nucleation phenomena, phase formation, and island mobility and coalescence [11–14].

11.2.1 Coherence

The suggestion that oriented growth requires parallelism of lattice planes of closely similar spacings came from the work of Royer (1928). From X-ray diffraction data he noted that epitaxy occurs only if the misfit is no greater than about 15% where misfit ε was defined as $\varepsilon = (b - a)/a$, where a and b are the in-plane lattice parameters of the substrate and deposit respectively [15]. The energy of the interface formed between the two crystals is dependent on the misfit between the lattices. Calculations of the dependence of interfacial energy with atomic fit have been dominated by the work of Frank and van der Merwe (FM) [16], and this work has been extensively reviewed by Matthews [17]. Frank and van der Merwe introduced some fundamental concepts to the field of epitaxy based on the simple case of a misfitting monolayer on a rigid substrate. The theoretical description showed that for small values of ε, a thin deposit will strain elastically so as to reach coherence with the substrate to reduce the total energy of the interface between them. This occurs up to a critical value of ε, ε_c, above which the interface is divided into regions of 'good' registry separated from one another by regions of 'bad' registry. These regions of bad registry (or fit) consist of a periodic array of crystal dislocations which are described as misfit or interfacial dislocations. The idea of a critical strain leads to the concept of a critical thickness; ε_c is a function of film thickness, thus an initially coherent epitaxial film will become structurally unstable with increasing elastic energy as the thickness increases. Strain relief occurs at a critical thickness, h_c, where interfacial dislocations are introduced to accommodate the misfit. For a film of finite thickness, h, where $h > h_c$, partial relief of the misfit occurs by plastic deformation, generating dislocations. The remaining strain energy is accommodated elastically.

The details of atomic structure at an interface depend on a number of factors such as bonding and chemistry. Theoretical predictions and analyses of epitaxy also include the full crystal symmetries of the system which can place constraints on allowed orientations as well as defect type and methods of defect introduction, in addition to the possibility of atomic displacements or reconstructions at an interface. Symmetry considerations of interface structure developed from the crystallography of bicrystals and internal grain boundaries (GBs) where low-index orientations are controlled by a sufficient density of coincident sites. The applica-

bility of these ideas to epitaxy has been reviewed by Pond [18].

It seems appropriate at this stage to address the question of notation required to classify interface structures in epitaxial systems. For a system with zero or small misfit where all atom positions at the interface are in exact register, the interface is described as **coherent** or **commensurate**. Thin films which are strained elastically to match the substrate lattice parameter are often described as **pseudomorphic**. As coherence breaks down due to increasing strain energy, the interface accommodates the mismatch with a periodic array of dislocations and the interface can be described as **semi-coherent** or **discommensurate**. An alternative situation can arise where two crystals with different lattice spacings meet at an interface with no apparent coincidence or relaxation; both crystals retain their bulk lattice parameters and atomic registry becomes a linear function of position. Such an interface may be termed **incoherent** or **incommensurate** and implies weak interfacial interaction, perhaps for reasons of chemistry, although this does not seem consistent with the ability to form an unique epitaxially oriented film. Examples of all three types of interface have been found experimentally. Coherent films of $CoSi_2$–Si(111) become semi-coherent at $h_c \approx 30$ Å [19]. The alkaline-earth fluorides CaF_2, SrF_2, and BaF_2 are observed to grow epitaxially on cubic semiconductors and the interface structure has been shown to be dependent on the magnitude of the misfit [20]. For example, films of BaF_2 on InP(001) with a 5% misfit were found to be semi-coherent whereas films of BaF_2 on Ge(111) were found to be incoherent. The misfit for BaF_2–Ge is 10% and the BaF_2 was found to have retained its bulk lattice parameter to within 5 Å from the interface. Such observations are rare and are thought to be related to the presence of steps or some other periodic surface-topographical feature. The phenomenon has been described as **graphoepitaxy** where alignment of islands with topographic features occurs. This has been demonstrated for the case of Au epitaxy on NaCl [21] and is thought to play a role in some aspects of Al epitaxy on GaAs. Al–GaAs is an interesting example of a metallic fcc film on a cubic zinc blende semiconductor where the lattice misfit due to alignment of (100) planes in both cubic crystals is large, $\varepsilon \approx 28\%$. However, the interfacial misfit can be minimized by a 45° rotation to allow alignment of the (200) metal and (220) semiconductor planes enabling coincidence of lattice spacings to be reached. The misfit is reduced to 1.4% and a low energy interface is formed [22, 23].

The importance of substrate orientation on the quality of epitaxial films has also been demonstrated for films of Mo and Nb grown on GaAs (001) and (111) substrates [24]. Both metals are bcc with similar lattice parameters; however, Mo grows on GaAs with a (111) orientation on both (001) and (111) substrates, whereas Nb grows with a (001) orientation on GaAs(001) and no single orientation on GaAs(111).

11.2.2 Initial stages of epitaxy

The initial stage of epitaxy is the formation of 'clusters' of deposit atoms or molecules on the substrate surface which can grow and coalesce to form a continuous film. A comprehensive review of nucleation and growth mechanisms has been given by Venables and Price [25], who consider the kinetics of nucleation in thin films. Epitaxy is sometimes regarded as a post-nucleation phenomenon where the rearrangement of 'stable' clusters can occur via rotation and migration [24]. One important consideration is whether crystal growth is two-dimensional (2D) or three-dimensional (3D).

There are three modes of growth [26] which have been identified in heteroepitaxial systems; 2D layer-by-layer growth which was described by Frank and van der Merwe (FM) [16], 3D island growth identified by Volmer and Weber [27] (VW), and a combination of these mechanisms which was identified by Stranski and Krastanow (SK) [28] where layer-by-layer growth occurs initially followed by islanding. The transition from 2D to 3D growth can be described using the analogy of a liquid droplet on a surface. If the droplet **wets** the surface, a continuous 2D film will form, whereas **non-wetting** results in 3D drops. More rigorously, if the thermodynamics of the film and substrate are considered, the equilibrium growth mode is determined by the relative free energies of the substrate surface (σ_s), the deposit surface (σ_d) and the substrate-deposit interface (σ_i), providing that the strain energy in the system is neglected (this will, of course, influence the interface energy). Island growth is driven by a high interface energy and low substrate surface energy such that it is more favorable for deposit atoms to stick together rather than wet the substrate. Figure 11.1 schematically illustrates island growth for a finite contact angle, θ, where

$$\sigma_s = \sigma_d + \sigma_i \cos \theta \qquad (11.1)$$

For $\pi > \theta > 0$, island growth occurs. If $\theta = 0$, the film wets the substrate and growth is layer-by-layer, consistent with FM growth. SK growth generally occurs when there is wetting of the substrate for $\theta \gtrsim 0$, and large strain energy in the film. The early stages of SK growth are not well understood. For Ge on Si(100) with a 4% misfit, growth of approximately 3 ML of Ge can be achieved layer-by-layer before islands of Ge start to form. As the film thickness increases, dislocations which can relax the strain are observed underneath the

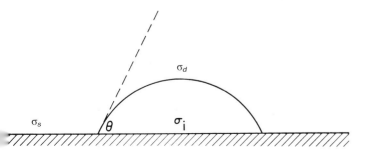

Fig. 11.1 Schematic diagram of a thin epitaxial island with surface free energy σ_d, with finite contact angle θ growing on a substrate with surface free energy σ_s and an interface energy σ_i.

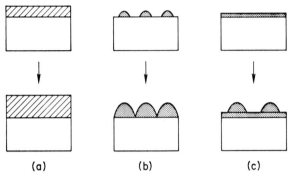

Fig. 11.2 Schematic diagram depicting three modes of epitaxial growth. (a) Two-dimensional layer-by-layer growth (FM), (b) Three-dimensional island growth (VW) and (c) layer-by-layer growth followed by islanding (SK).

islands. These three modes of growth are illustrated schematically in Fig. 11.2.

Determination of the growth mode can be inferred during growth from *in-situ* diffraction methods such as RHEED where measurement of RHEED oscillations can be used to distinguish between 2D and 3D growth, and between step or terrace nucleation in real time. Incorporation of RHEED into III–V MBE chambers has enabled direct measurement of surface diffusion rates for Group III species [29, 30]. *Ex-situ* characterization after growth can be performed by TEM both in plan view and cross-sectional orientations to detect islanding and by measurement of contact angles at the deposit/substrate interface. Figure 11.3 shows an island of epitaxial $CoSi_2$ on Si(111) grown by Co deposition on clean Si in UHV [19]. The $CoSi_2$ grows in a type B orientation where the silicide is rotated 180° with respect to the substrate normal. A type A interface would have the same orientation as the substrate. Preferential nucleation has occurred at surface steps where the step height of $3d_{111}$ avoids the need for dislocation introduction at the interface in this coherently strained island. The contact angle between $CoSi_2$ and Si is measured to be 50° consistent with island growth. Contact angles of close to 90° have been observed for hemispherical cap-like islands of GaAs on Si(001) [31].

These elementary ideas of wetting are of significance for the growth of multiple layers in a

heteroepitaxial structure, e.g. a superlattice. For two crystals A and B, a heterostructure could consist of alternate layers A/B/A/B on a substrate A. However, by necessity, one crystal will have a lower surface energy. Thus if B grows on the substrate A via FM or SK mechanisms, then A will grow on B via VW island growth. This phenomenon has been observed in both metal, semiconductor and semiconductor/semiconductor systems. A good example is the growth of Ge on Si, where Ge grows on Si(001) in a SK mode [32] and Si grows on Ge(001) in a VW mode [33]. As a result, growth of Si–Ge–Si quantum-well structures results in island-formation during growth of the Si capping layer [34, 35]. The traditional approach to avoid islanding of the surface capping layer has been to restrict the growth kinetics by decreasing the growth temperature or increasing the growth rate. A metastable 2D film can thus be grown away from thermodynamic equilibrium. However, Copel *et al.* [36] have recently described a novel method of island formation suppression by the use of a surface segregating surfactant to manipulate or control the surface free energies of both substrate and deposit. This elegant work shows that the 2D to 3D growth transformation for Ge on Si(001) can be inhibited by passivating the Si(001) surface with 1 ML of a third element, in this case As, prior to Ge growth. Segregation of the surfactant to the growing surface occurs resulting in relatively small amounts of As incorporation in the film. *In-situ* measurements using medium energy ion scattering revealed that for Ge growth on Si, approximately 3 ML of Ge grows in a 2D manner before islands are observed, whereas in the presence of the As surfactant films as thick as 15 ML were achieved. The technique also works for growth of Si on Ge on Si(001), enabling one to bury a thin film of Ge beneath single crystal Si. This example demonstrates that the energetics of growth can be altered by modification of surface energies. The As layer is thought to create a stable termination on the clean reconstructed Si surface by contributing a valence electron to the dangling bonds expected on the (2×1) surface.

Island growth of $CoSi_2$ on Si can also be suppressed by modification of the surface struc-

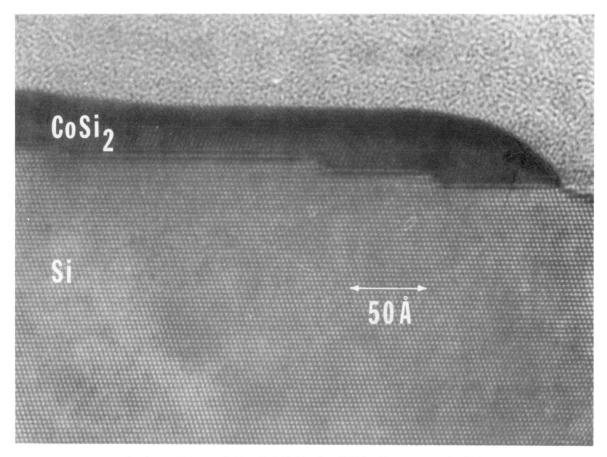

CoSi$_2$

Si

50 Å

Fig. 11.3 Cross-sectional $\langle 110 \rangle$ HREM image of a Type B CoSi$_2$ island on Si(111) with a contact angle of 50°.

ture which is thought to affect the surface energy of CoSi$_2$. *In-situ* UHV TEM experiments of CoSi$_2$ growth on Si showed that continuous films of silicide were formed initially at low growth temperatures, typically <500°C, but that the film developed pinholes as the temperature increased to $\geqslant$550°C [37]. Growth of a Si cap on CoSi$_2$ to form a buried metal layer requires temperatures typically of 600°C; thus the silicide layer **must** be stable to these higher temperatures. This was achieved by Tung and co-workers [38, 39] who determined that the surface structure of the silicide was very sensitive to stoichiometry. The low-temperature surface of CoSi$_2$ is Co-terminated and has been termed CoSi$_2$–C [38], whereas the high-temperature surface is terminated with a bilayer of Si and has

been termed CoSi$_2$–S. Formation of CoSi$_2$–S requires additional Si which can be provided either by opening up pinholes to expose the Si substrate which is shown in Fig. 11.4 or by deposition of 2 ML of Si to the CoSi$_2$ surface to suppress pinhole formation [39]. Modification of the surface structure then allows growth of Si–CoSi$_2$–Si double heterostructures where the orientation of the Si surface layer can be controlled [40]. Growth of a Si cap on CoSi$_2$–S surfaces results in the Si–CoSi$_2$–Si structure shown in Fig. 11.5(a) where the surface Si has the same orientation as the CoSi$_2$. If Si growth occurs on the CoSi$_2$–C surface, the Si layer has the same orientation as the substrate and forms a type B interface, as shown in Fig. 11.5(b). Manipulation of CoSi$_2$ surface structures allows

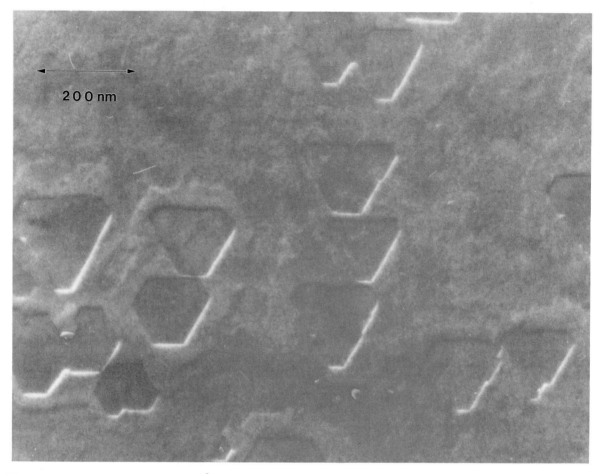

Fig. 11.4 Plan view weak beam image of a 12 Å thick coherent film of $CoSi_2$ on Si(111). Triangular pinholes open up to satisfy $CoSi_2$ surface stoichiometry [19].

superlattices to be grown and one example of this is shown in Fig. 11.6 where each Si–$CoSi_2$ interface is type B [40]. Up to seven period superlattices have been demonstrated [42]. The structure and electronic properties of these silicide films are reviewed in detail by Tung in Chapter 21 of this book.

Progress has also been made towards the realization of III–V based metal/semiconductor heterostructures. MBE growth of GaAs–$Sc_{1-x}Er_xAs$–GaAs structures has been demonstrated although a high density of microtwins and stacking faults are observed in the surface GaAs [43, 44].

Surface structure and reconstructions have important ramifications in epitaxial growth and their effects have been realized since the arrival of the STM to the field of surface science. Applications of the STM to studies of epitaxy are reviewed by Hamers in Chapter 10 of this book. Examples of some effects of surface structure on epitaxial orientations and growth modes are reviewed in brief here. The initial stages of epitaxy on clean surfaces will be affected by the presence of a surface reconstruction. Gossmann [45] has reviewed the effects of surface reconstruction on both Si(001) and Si(111) surfaces. Deposition of atoms upon a

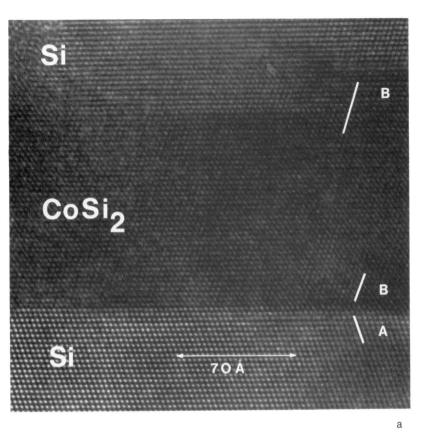

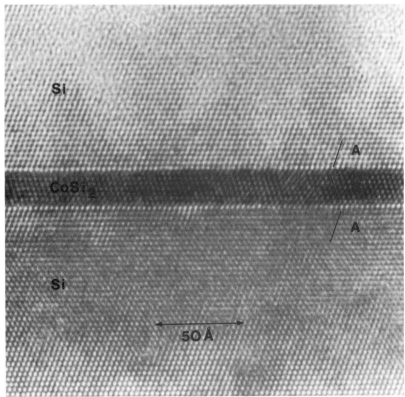

Fig. 11.5 Cross-sectional HREM images of $Si–CoSi_2–Si$ heterostructures where the interface orientation of the $CoSi_2$ and Si layers varies as (a) A/B/B and (b) A/B/A. Modification of the $CoSi_2$ surface by addition of Co or Si to form the $CoSi_2$–C or $CoSi_2$–S surfaces allows the surface Si orientation to be controlled [40].

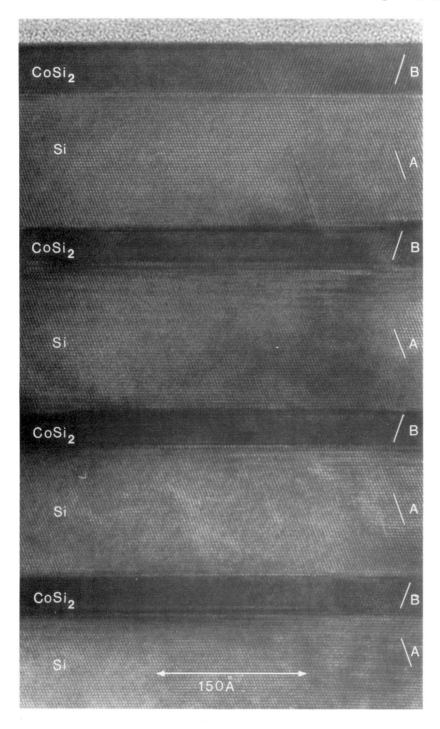

Fig. 11.6 Cross-sectional HREM image of a Si/CoSi$_2$ superlattice where islanding of CoSi$_2$ on Si and Si on CoSi$_2$ has been suppressed by modification of the surface structures [41].

reconstruction requires reordering of the substrate surface for perfect growth. Si(001) (2 × 1) and Si(111) (7 × 7) surfaces were found to behave very differently after room temperature deposition of Si, Ge, and Sn; the (7 × 7) did not order, whereas the (2 × 1) ordered, which means that the atoms originally displaced due to the reconstruction must have been brought back to their bulk positions. This ordering can be thought of as the formation of an ideal **template** as a necessary condition for perfect growth. Templates will be shown to be extremely important in the growth of metal silicides on Si [46]. Reordering of the Si (2 × 1) requires bond-breaking to disrupt the dimer chains whereas reordering of the Si (7 × 7) would require substantial atomic motion. The Si (7 × 7) structure consists of deep 'holes' at the corners of the unit cell, stacking faults and adatoms [47] and is a relatively stable surface [48]. Ordering in SiGe alloy films on Si(001) has also been shown to depend on a surface reconstruction during growth [49].

The surface reconstruction is a crucial part of the initial stages of **interface formation** in epitaxial thin films which can be important in determining the epitaxial orientation of the deposit. Headrick *et al.* [50] have recently demonstrated that the orientation of homoepitaxial Si films on Si(111) can be altered in the presence of boron. Homoepitaxy is usually considered to be indistinguishable from conventional crystal growth in that the orientation and composition of the deposit is **exactly** the same as the substrate and the interface (in a clean system) cannot be detected. In reality, homoepitaxy is a convenient way of controlling doping profiles in a device structure. In this example, the crystallographic orientation of the homoepitaxial layer can be controlled by a layer of boron. Deposition of $\frac{1}{3}$ML boron on Si(111) leads to a $(\sqrt{3} \times \sqrt{3})$ reconstruction [50]. Low temperature (400°C) growth of Si on the Si(111):B($\sqrt{3} \times \sqrt{3}$) surface results in an epitaxial film which is rotated by 180° with respect to the surface substrate normal. The surface reconstruction is partially preserved beneath the epilayer rather than reordered as observed for Si (2 × 1) surfaces. Si films grown in the absence of boron preserve the substrate orienta-

tion with no rotation. This phenomenon differs from the surfactant results of Copel *et al.* [36] in that boron remains buried at the original surface rather than segregating to the growing surface. In both cases, epitaxy has been altered by the addition of a foreign species which affects either the surface or interface energies of the growing system.

Nucleation of metals on Si is also affected by the surface reconstruction. This has been beautifully demonstrated by STM for palladium silicide on Si(111) [51]. Room temperature deposition of Pd on Si(111) (7 × 7) results in nucleation of the silicide which is spatially localized and occurs only in the faulted half of the (7 × 7) unit cell.

Surface reconstructions have been shown to influence both nucleation and orientation of an epitaxial film and are likely to influence the final interface structure. Buried interface reconstructions have only recently been observed which have conceptual similarity to surface reconstructions [52–54]. A well-ordered (1 × 2) dimer reconstruction has been observed at the buried interface in MBE grown $CoSi_2$–Si(001) [52] and $NiSi_2$–Si(001) [53] films. The reconstruction was identified by a combination of transmission electron diffraction and imaging and cross-sectional high resolution imaging. A buried interface reconstruction had previously been observed in bicrystals of Ge containing a $\Sigma 3\{112\}$ symmetrical tilt boundary [54].

11.2.3 Stability of epitaxial thin films

Epitaxial layers are often not in thermodynamic equilibrium but exist in a metastable state where the kinetic factors of nucleation and growth can determine epitaxial relationships. The structure and stability of these thin films can be examined using sophisticated molecular dynamics computer simulations [7]. Of particular interest is the stability of a 2D film relative to 3D clusters or islands. An extensive review of computer simulations and applications is given by Wolf and Jaszczak in Chapter 14 of this volume.

The transition from 2D to 3D growth for Ge on Si(001) has been studied by a number of groups where the transition mechanism has been of interest. SK growth is known to occur in strained systems where island formation occurs after a few monolayers. The exposed island edges facilitate the introduction of misfit dislocations underneath the islands to relieve strain. (Mechanisms of misfit introduction will be discussed in the next section.) Grabow and Gilmer [55] have shown that a uniform film is never the lowest energy state for a system with finite misfit. Nevertheless 2D growth can persist in a metastable state even for substantial misfits as a result of a large nucleation barrier to the formation of islands. In addition, coherent films remain in metastable equilibrium far beyond a critical thickness due to a large free energy barrier inhibiting the introduction of dislocations. Molecular dynamics simulations have shown that, for a system with finite misfit, a uniform film whose thickness exceeds a few ML is unstable with respect to cluster formation resulting in either SK or VW growth [55]. The initial stages of SK growth can also be coherent due to elastic deformation of both the island and substrate [56]. This has been demonstrated for Ge on Si(001) where islands of Ge are initially dislocation-free. Elastic deformation stabilizes coherent Ge growth by allowing the island and substrate to deform closer to their bulk lattice parameters, thus partially accommodating the misfit. The transition to dislocated SK growth is related to a critical island coverage; substrate deformation was observed to occur over dimensions $\approx 2\times$ the island diameter [56]. For small island spacings, coherent SK growth was not observed and the islands were relaxed by misfit dislocations rather than substrate strain. STM has shown that for Ge on Si(001), a metastable 3D cluster phase is found with a well-defined shape and structure which precedes the transition to SK growth [57].

11.3 MISFIT DISLOCATIONS

Predictions of critical thicknesses often result in an underestimate of the attainable dislocation-free film thickness, particularly for non-equilibrium growth. Although good estimates for h_c for $In_xGa_{1-x}As$–GaAs have been obtained [58] for a wide range of x, agreement is poor for $Si_{1-x}Ge_x$–Si where values for

h_c can exceed theoretical predictions by up to a factor of 4 [59, 60]. Metastable coherent films will occur for $h > h_c$ where an activation barrier for nucleation of dislocations exists. This is extremely important for device applications and has led to increased effort in our understanding of misfit accommodation, dislocation sources and multiplication mechanisms.

11.3.1 Prediction of a critical thickness

For strained layer epitaxy, the critical thickness h_c at which it becomes energetically favorable to introduce dislocations was first recognized by van der Merwe [61]. Predictions of h_c have been made by a number of workers [16, 17] and Jesser and van der Merwe have recently summarized the different approaches [62]. A brief review of current theories will be presented here.

The FM theory of growth and defect introduction relies on energy minimization by balancing the elastic strain energy in the film against the interface energy. The model considers initially a 1D dislocation in a misfitting monolayer which is then extended to consider a 2D network of interfacial dislocations. Assuming a non-interacting grid of edge dislocations with Burgers vectors in the interface plane, the FM formulation calculates the equilibrium separation of misfit dislocations required to relax the misfit strain. Since the strain energy is thickness dependent, the elastic misfit strain in the film will be minimized by introduction of dislocations at the critical thickness, h_c. The model has several limitations when applied to experimental thin film growth and defect nucleation. Experimental values for h_c deviate from predictions when screw and 60° dislocations occur at the interface, when growth is clearly not 2D and for crystals which are not necessarily elastically isotropic. Perhaps the most important failing of this approach is the omission of dislocation nucleation energetics.

A second approach taken by Matthews and Blakeslee (MB) [58] considers the forces acting on an existing dislocation (presumably threading from the substrate or previous layer) due to elastic stresses in superlattices of $GaAs$–$GaAs_{0.5}P_{0.5}$. The elastic force required to bend the dislocation into the interface plane is balanced by the line tension of the dislocation. If the line tension is less than the elastic strain, the dislocation will bow and bend into the interface generating misfit dislocations. If the two forces are balanced, an expression for h_c can be found:

$$h_c = \frac{b(1 - \nu \cos^2 \alpha)}{8\pi\varepsilon(1 + \nu)\cos\lambda}\left(\ln\frac{h_c}{b} + 1\right) \qquad (11.2)$$

where ν is Poisson's ratio, b is the dislocation Burger's vector, λ is the angle between the slip direction of the dislocation and the slip plane normal projected into the interface, α is the angle between the dislocation line and its Burgers vector. The same result is obtained using the FM formulation. The MB model is frequently used to predict h_c for a variety of semiconductor thin films and superlattices.

Both the FM and MB theories are equilibrium theories which take no account of dislocation introduction mechanisms or energy barriers to nucleation and propagation such as the Peierls barrier. People and Bean [60] (PB) modified the MB model to take into account dislocation nucleation at Si–SiGe interfaces and obtained an expression for h_c,

$$h_c \sim \left(\frac{1 - \nu}{1 + \nu}\right)\left(\frac{b^2}{16\pi\alpha\varepsilon^2\sqrt{2}}\right)\ln\left(\frac{h_c}{b}\right) \qquad (11.3)$$

which gave a qualitatively better fit to experimentally determined values of h_c. The PB theory assumes that the film is initially dislocation-free and that misfit dislocations are generated when the areal strain energy of the film of thickness h exceeds the energy of an isolated dislocation (screw, edge, loop etc.). The mechanisms and kinetics of dislocation nucleation are ignored. Marée et al. [63] included the effects of dissociated 60° dislocations as well as compressive and tensile lattice mismatch. Since films can be grown which are metastable with $h > h_c$, the time-dependence of the relaxation mechanism must be taken into account. Strain relaxation occurs via dislocation glide and multiplication where the glide velocity is dependent on the excess stress acting on the dislocation [64]. Further

agreement between theory and experiment was thus demonstrated by Dodson and Tsao [65] who developed a full kinematic theory of strain relaxation based on a phenomenological model of dislocation dynamics and plastic flow [66]. Chidambarrao et al. [67] have included the effects of Peierls barriers and the orientation of threading dislocations to obtain an expression for h_c which attempts to account for material-dependent parameters such as the highest processing temperature the film has undergone (often the growth temperature), which will affect the energetics of dislocation introduction and strain relaxation.

The above discussion shows the evolution of simple energy and force balance ideas to develop sophisticated models for misfit relief. Nevertheless, discrepancies are observed in semiconductors between measured and theoretical values of h_c [58–60]. The agreement between theory and experiment is much better for metals. This has recently been explained in terms of frictional effects. Fox and Jesser have modified these critical-thickness models to incorporate both equilibrium and kinetic factors of misfit accommodation [68]. The discrepancies for metals and semiconductors are linked to the magnitude of frictional forces; frictional barriers are low in fcc metals and h_c is expected to be close to the values predicted from the above theories, whereas high frictional forces in semiconductors result in deviations in h_c. To resolve this discrepancy, Fox and Jesser consider experimental conditions which could determine whether thin films (semiconductor or metal) are in the equilibrium or kinetic regimes of misfit accommodation and have obtained an expression for the growth rate r, below which the epitaxial film remains in equilibrium during growth;

$$r \leqslant \frac{2Dvh^2 n_a b \cos\lambda}{qC[\ln(h/b) - 1]} \qquad (11.4)$$

for $h > h_c$, where n_a is the number of inclined dislocations which are mobile, q is an integer equal to the number of possible trace directions of the slip plane on the interfacial plane (for (001) epitaxy with {111} slip planes, $q = 2$), v is the dislocation velocity and

$$C \equiv \frac{G_o G_s b^2 (1 - \nu \cos^2 \psi)}{2\pi (G_o + G_s)(1 - \nu)}$$

where G_o and G_s are the shear moduli of the deposit and substrate respectively.

The following section gives examples where deviations from h_c are observed in semiconductors.

11.3.2 Exceeding a critical thickness

The presence of dislocations in the active areas of semiconductor device structures is inherently undesirable. Tailoring the band gap and lattice parameter in a number of lattice-mismatched alloy systems [69], e.g. $Si_{1-x}Ge_x/Si$ and $In_xGa_{1-x}As/GaAs$, leads to the development of novel optoelectronic devices, providing that dislocations can be eliminated. In this section, dislocation sources and misfit-relieving defects in $Si_{1-x}Ge_x/Si$ and $In_xGa_{1-x}As/GaAs$ will be examined.

Misfit-relieving defect nucleation

Sources for misfit dislocation formation which have previously been identified include the glide of threading dislocations, multiplication of misfit dislocations, and the formation of surface dislocation half-loops which can glide to the interface [17, 58, 70]. The critical thickness transition will be affected by both nucleation and propagation events. Nucleation processes can be sub-divided into homogeneous and heterogeneous events. Homogeneous nucleation of half-loops at the growth surface requires that the loop exceed a critical radius, r_c, lest the loop 'shrink' under its line tension rather than grow and glide to the interface as a result of the misfit. The activation energy associated with this process increases with increasing r_c and with decreasing film stress. For cubic semiconductors, surface half-loops nucleate on $\{111\}$ planes and E_{act} will be determined by the energy to create the loop and the strain and surface energy released by the loop. Growth of the loop occurs by glide of threading screw or $60°$ segments under the influence of strain and this has been observed directly by Hull *et al.* during *in-situ* TEM annealing experiments on GeSi–Si(001) films [71].

Calculations for E_{act} for semicircular and semi-hexagonal loops have been performed which suggest that homogeneous nucleation of loop will occur for epilayer strains $\approx 6\%$ [72, 73]. Activation energies for heterogeneous nucleation processes such as dislocation multiplication are expected to be lower.

Dislocation multiplication can occur via the Hagen–Strunk mechanism [70]. Two misfit dislocations with the same Burgers vectors on different glide planes crossing at $90°$ experience an annihilation reaction at the point of intersection resulting in the formation of two angular dislocations. The apex of one dislocation at the intersection will glide to the surface due to repulsion from the intersection and attraction to the surface by a surface image force, thus creating two new separate dislocation segments. The new dislocations can subsequently glide to form more misfit dislocations which can then undergo the same multiplication process. This mechanism has been observed in epitaxial films of Ge–GaAs [70], InGaAs–GaAs [72], and GeSi–Si [73].

For low values of misfit-strain in films which grow in a 2D mode, additional low E_{act} mechanisms must be operative to account for the dislocation densities observed. If growth is 3D, dislocations can enter the deposit at the edges of growing islands [74, 75]. Two new misfit-relieving defect sources have recently been identified in the Ge–Si system. For growth of Ge on Si(001) where growth is constrained to be 2D with the aid of a surfactant [36], misfit dislocations are suppressed and a novel mode of strain relief occurs via the formation of new defects. Using a combination of high resolution electron microscopy (HREM) and multislice image simulations [76], LeGoues *et al.* [77] have shown that these defects are wedge-shaped, consisting of two Σ9 boundaries and a twin, with the Σ9 boundaries forming a V with its tip at the SiGe interface. An example of this V-shaped defect is shown in Fig. 11.7. The defect formation mechanism is thought to be similar to a martensitic transformation where local atomic rearrangement occurs in the absence of diffusive events. A defect separation of 85 Å would be required to relieve the 4% misfit strain completely.

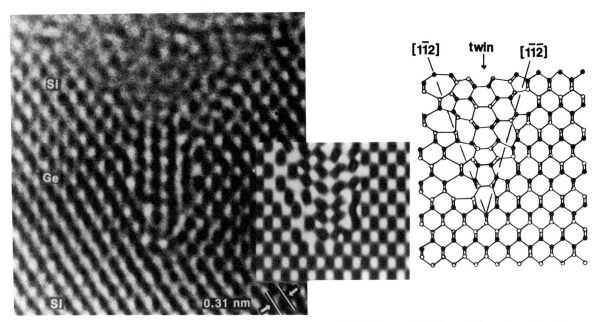

Fig. 11.7 Cross-sectional view of V-shaped defect in a 12 ML film of Ge buried in single crystal Si, shown with atomic model and corresponding image simulation (courtesy of F. K. LeGoues [77]).

In $Si_{1-x}Ge$–$Si(001)$ films with low strain, $\lesssim 2\%$, a diamond-shaped stacking fault has been shown to act as a source for emission of glissile dislocations. Eaglesham *et al.* [73] identified the Burgers vector for the diamond defect to be $a/6\langle 114\rangle$. The origin of the defect has not been established. The stacking fault acts as a dislocation source emitting dislocations by a dissociation process, resulting in the formation of $a/2\langle 110\rangle$ dislocations and $a/6\langle 112\rangle$ partial dislocations. The glissile $a/2\langle 110\rangle$ segments bow out due to misfit stress to form half-loops attached to $a/6\langle 112\rangle$ partials. Closure of the propagating loop regenerates the original diamond defect in a manner analogous to a Frank–Read source with a series of co-planar glissile half-loops being generated.

'Dislocation-free' epitaxial growth in lattice-mismatched systems

The previous descriptions of misfit dislocations assume that dislocation introduction is necessary at h_c for films grown on large substrates with effec- tively infinite lateral dimensions. Dislocations occur at the substrate/deposit interface when the strain energy/unit area of film is greater than the areal energy density associated with an isolated dislocation [60]. However, the strain energy in an epitaxial film can be reduced by restricting the strained zone to a narrow layer next to the interface. Luryi and Suhir [78] used this idea to propose a new approach to strained layer growth, namely, heteroepitaxial growth on small 'seed' pads of lateral dimension, l, with a uniform crystal orientation over the entire wafer. This leads to the concept of critical height/critical width arguments for dislocation introduction. For a given misfit, the critical film thickness is strongly dependent on l. For a sufficiently small seed pad size, l_{min}, it was predicted that the total strain energy/unit area of the film would never exceed the critical strain for dislocation generation. In other words, if $l \lesssim l_{min}$, the elastic stress in the film is accommodated without dislocations. This can be understood more clearly by reference to the work of Vincent and Jesser and van der Merwe [75, 79] who demonstrated a saw-tooth behaviour of average strain in a growing island as dislocations are incorporated. If the pad dimension shrinks to less than the equilibrium separation required for misfit dislocations, the presence of dislocations can result in a net increase in the strain energy. For example, growth of Ge on Si results in a 4% misfit which requires $a/2\langle 110 \rangle$ type dislocations to be spaced at ≈ 100 Å to relieve the strain. If the island/pad size is <100 Å, dislocation introduction will result in compressional rather than tensile strain [31, 78].

Luryi and Suhir found that a reduction factor $\phi(l/h)$ was needed in the expression for h_c, where

$$\phi^2\left(\frac{l}{h}\right) = \left(1 - \text{sech}\,\frac{\xi l}{h\phi^2(l/h)}\right)^2 (1 - e^{-\pi h/l})\frac{l}{\pi h}$$
$$(11.5)$$

where h is the height and l the length of a strained epitaxial island (or contact pad), and ξ is given by $\sqrt{3}(1-v)/\sqrt{2}(1+v)$. The reduction factor relates the critical dimensions, $h_c(l)$ in the strained 3D island/pad to the planar critical thickness, h_c, which is given by

$$h_c(l) = h_c[\phi(l/h)\varepsilon]$$
$$(11.6)$$

where ε is the lattice mismatch; the finite island size modifies the effective strain in the system. For large pad dimensions, $l > h$, and $\phi \rightarrow 1$, thus $h_c^l \equiv h_c$. As l decreases, h_c^l deviates from the predicted values of h_c. The theory predicts that films of arbitrary thickness can be grown epitaxially on a lattice-mismatched substrate where the films are dislocation-free provided the pad-size is sufficiently small. These ideas have been applied to growth of Ge on porous Si, GaAs on Si, and $In_xGa_{1-x}As$ on GaAs [31, 78, 80, 81]. The characteristic seed-pad size for Ge on Si was predicted to be ≈ 100 Å which is currently not attainable by lithographic techniques. Porous Si has been suggested as a viable 'patterned' substrate [78, 80].

A reduction in contact area at the interface also results in a reduction of the number of threading dislocations which can propagate. This has been clearly demonstrated for growth of InGaAs on both square and round mesas of GaAs [82]. The strain distribution in the patterned growth resulted in suppression of homogeneous half-loop nucleation.

For small growth areas, the total number of operative dislocation sources is also reduced. Each source can only generate a small, finite length of misfit dislocation (defined by the mesa dimensions) resulting in a reduction in dislocation interactions and multiplication mechanisms.

An additional method of producing 'dislocation-free' epitaxial films is the generation of a **strained layer superlattice** (SLS), which, although a popular technique, has had only moderate success in 'blocking' dislocations from the active area of a device structure [83, 84]. Alternating layers of crystals with different lattice parameters are grown on either high dislocation density substrates or thin films to act as a filter. The SLS generates an oscillating strain gradient which is intended to bend the threading dislocations into the interfacial plane forming segments of misfit dislocation which are forced to the edge of the wafer or substrate. If the average lattice parameter of the SLS does not match that of the substrate, a superlattice critical thickness also exists [85]. If the elastic constants are similar and Vegard's law applies for alloy materials, for a superlattice of, for example, $GaAs_{1-x}P_x$–GaAs or $Si_{1-x}Ge_x$–Si, the critical thickness is that of a single layer of the same thickness formed by uniform interdiffusion of the superlattice to form an average composition. In high dislocation density material, threading dislocations which have been bent into the interface plane need to travel large distances (inches) to reach the edge of the wafer, thus the likelihood for dislocation interaction and multiplication is high and additional threading segments can be created.

11.3.3 Symmetry considerations

Epitaxial growth of deposits with crystal structures and symmetries which are different from the substrate can lead to the introduction of interfacial line-defects which are dictated by symmetry arguments rather than required for misfit-relief. Classification of interfacial defects in both epitaxial interfaces and bicrystals has been reviewed by Pond [18] who gives the full mathematical description of the relationships between coincident symmetry and formation of dislocations, disclinations, and dispirations. The type of defect formed is defined by the differences in symmetry between the crystals and whether or not translational or rotational symmetry is broken at the interface.

Defects which occur due to symmetry constraints have been observed in epitaxial heterostructures where the defects separate regions of interface which are degenerate in structure and energy. Detection of $a/4\langle 111 \rangle$ dislocations at the $NiSi_2$–$Si(001)$ interface is one example of this [86]. The crystal space groups for this system are $Fm\bar{3}m$ and $Fd\bar{3}m$ for the cubic fluorite structure $NiSi_2$ and diamond-cubic Si respectively. The differences in space group lead to broken symmetry of the fourfold axis parallel to [001]. As a result, $a/4\langle 111 \rangle$ dislocations are introduced in association with steps $a/4[001]$ in height, which separate energetically degenerate regions of interface. This result is also applicable to films of $CoSi_2$–$Si(001)$ and CaF_2–$Si(001)$ where the same crystallographic relationships exist [87, 88].

Antiphase domain boundaries or inversion domains [18] in films of GaAs–Si(001) or GaAs–Ge(001) are another example of symmetry-related defects which arise due to broken four-fold symmetry of the Si(001) surface [89].

11.4 PHASE STABILITY IN EPITAXIAL SYSTEMS

For thin epitaxial films, the effects of the interface structure and energy can permit the growth of phases which usually would not be thermodynamically stable at the growth temperature. Epitaxial constraints can allow growth of very metastable structures if large kinetic barriers to relaxation to the stable phase exist and the 'phase diagrams' must be modified to incorporate the effects of interfacial stability. This has been demonstrated for CsCl/NaCl [90] and α-Sn/InSb [91]. The effect is most pertinent in the growth of alloy structures such as GaAsSb/GaAs, AlGaAs/GaAs, and SiGe/Si where spinodal decomposition might be expected to occur. However, the tendency for the alloy to phase separate can be sup-

pressed and ordered alloys are observed in both III–V and IV alloys [92–95]. Surface growth kinetics play a dominant role during epitaxy of $Si_{1-x}Ge_x(x = 0.5)$–$Si(001)$ alloys where atomic scale stresses in the reconstructed (001) growth surface result in double-layer segregation of Ge and Si [49].

Strain stabilization also affects phase formation sequences in metal silicide/Si thin films. For Ni deposition on Si, the expected reaction sequence according to the equilibrium phase diagram is orthorhombic δ-Ni_2Si $\Rightarrow$ NiSi $\Rightarrow$ $NiSi_2$ [96]. However, for thin films, epitaxial constraints dominate because of the large contribution of both surface and interface effects to the total free energy of the system. For nickel silicide films on Si(111) the phase formation sequence observed for metal deposition on clean Si(111) 7 × 7 surfaces is intimately related to the thickness of deposited Ni [97]. For room temperature depositions, Ni coverages $\lesssim 10$ Å thick result in the formation of type B $NiSi_2$. For increasing coverages up to ~ 25 Å, $NiSi_2$ is covered with unreacted fcc Ni. After annealing at ~ 350–$400°C$, the phase θ-Ni_2Si forms as an intermediate metal-rich phase prior to type A $NiSi_2$ formation at higher temperatures. This phase was identified as θ-Ni_2Si and not δ-Ni_2Si. The bulk phase diagram predicts that this phase is stable only above 850°C although it was previously detected in rapidly quenched films at room temperature [98]. Low temperature stabilization of this phase occurs due to a good lattice match to Si(111), minimizing strain energy. θ-Ni_2Si is hexagonal and grows epitaxially on the pseudo-hexagonal Si(111) surface. This phase is not observed for Ni deposition on Si(001) [99] due to a lack of reasonable interfacial atomic match between hexagonal θ-Ni_2Si and Si(001) with fourfold symmetry. Epitaxial phase stabilization can be understood by consideration of the free energies of the competing phases. If the free energies of two phases (at a finite, fixed temperature) are expressed as $U^1 t$ and $U^2 t$, per unit area with t the thickness of the film, and the surface and interface energies for the two phases written as σ_s^1, σ_i^1, σ_s^2 and σ_i^2, then the relative stabilities of the two phases can be examined. If $U^1 < U^2$ but $\sigma_i^1 > \sigma_i^2$, then phase 1

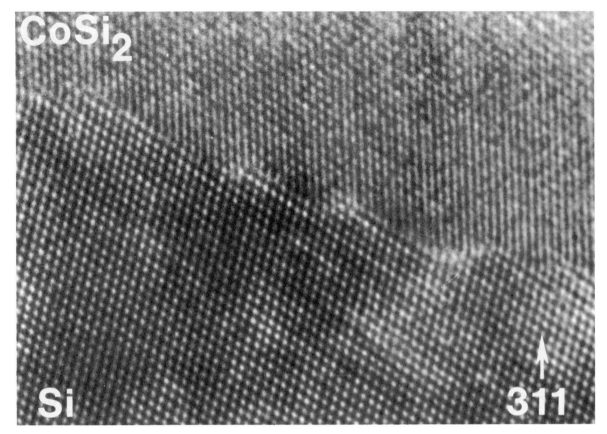

Fig. 11.8 Cross-sectional HREM image of the interface between $CoSi_2$ and Si for a substrate orientation (311); faceting occurs onto the lower energy (111) plane [103].

can be in thermodynamic equilibrium for thicknesses less than a critical thickness t_c [97],

$$t_c = \frac{\sigma_s^1 - \sigma_i^2}{U^2 - U^1} \qquad (11.7)$$

where t_c is observed to be several monolayers for both Ni and Co silicides on Si. For Co deposition on Si, qualitatively similar phenomena are observed, with type B $CoSi_2$ nucleating for thicknesses $\leqslant 10$ Å with the metal-rich phases occurring for thicknesses 10–25 Å of Co [100]. However, the phase sequence of cobalt silicide formation ($t > 10$ Å) was found to be δ-$Co_2Si \Rightarrow CoSi \Rightarrow CoSi_2$ with CoSi stabilized by the epitaxial relationship $CoSi(111)/Si(111)$ and $CoSi[112]/Si[110]$ (Ref. [101]). Growth of $CoSi_2$ on Si(111), Si(211), and

Si(311) (Ref. [102, 103]) also demonstrates the strong influence of interface energy and symmetry constraints on epitaxy. For films on Si(211) and Si(311), faceting at the interface occurs so that the $CoSi_2$/Si(111) local interface structure is preserved. An example of this faceting is shown in Fig. 11.8.

11.5 SUMMARY

The basic principles of epitaxy have been outlined with references to examples found in semiconductor thin films. The field of epitaxy clearly extends to metal/metal, metal/ceramic, metal/insulator and superconductor materials as well. The concepts introduced here are generic and applicable to other thin film systems. Recent advances in our understanding of epitaxial processes have been driven by the desire to fabricate novel device structures, often requiring 3D integration. As devices become smaller, control of epitaxy at the atomic level is crucial. Future progress will surely come from *in-situ* diagnostics during growth, both in the gas phase and on growing surfaces. The initial stages of epitaxy, misfit dislocation nucleation, novel defects, phase stability or ordering in alloy systems as well as crystal symmetry will continue to dominate experimental efforts in the field of epitaxy.

REFERENCES

1. J. D. H. Donnay, *J. Mater. Res.*, **1** (1986).
2. H. Lassen, *Phys. Z.*, **35** (1935), 172.
3. M. L. Frankenheim, *Ann. Phys.*, **37** (1836), 516.
4. D. W. Pashley, in *Epitaxial Growth Part A* (ed J. W. Matthews), Academic Press (1975), pp. 1–27.
5. M. Tsuchiya, J. M. Gaines, R. H. Yan, R. J. Simes, P. O. Holz, L. A. Coldren and P. M. Petroff, *Phys. Rev. Lett.*, **63** (1989), 466.
6. T. Fukui and S. Ando, *Electron. Lett.*, **25** (1989) (1986), 410.
7. E. G. Bauer, B. W. Dodson, D. J. Ehrlich, L. C. Feldman, C. P. Flynn, M. W. Geis, J. P. Harbison, R. J. Matyi, P. S. Peercy, P. M. Petroff, J. M. Phillips, G. B. Stringfellow and A. Zangwill, *J. Mater. Res.*, **5** (1990), 852.
8. J. M. Phillips and W. Augustyniak, *Appl. Phys. Lett.*, **48** (1986), 463.
9. R. W. Fathauer, B. D. Hunt, L. J. Schowalter, M. Okamoto and S. Hashimoto, *Appl. Phys. Lett.*, **49** (1986), 64.
10. J. L. Batstone, J. M. Phillips and E. C. Hunke, *Phys. Rev. Lett.*, **60** (1988), 1394.
11. R. J. Wilson and P. M. Petroff, *Rev. Sci. Instr.*, **54** (1983), 1534.
12. M. L. McDonald, J. M. Gibson and F. C. Unterwald, *Rev. Sci. Instr.*, **60** (1989), 700.
13. K. Takayanagi, T. Tanishiro, K. Kobayashi, K. Akiyama and K. Yagi, *J. Appl. Phys.*, **26** (1987), L957.
14. G. G. Hembree, P. A. Crozier, J. S. Drucker, M. Krishnamurthy, J. A. Venables and J. M. Cowley, *Ultramicroscopy*, **31** (1989), 111.
15. L. Royer, *Bull. Soc. Fr. Mineral. Cryst.*, **51** (1982), 7.
16. F. C. Frank and J. H. van der Merwe, *Proc. Roy. Soc. A*, **198** (1949), 205; **198** (1949), 216; **200** (1949), 125.
17. J. W. Matthews, in *Epitaxial Growth*, Part B (ed J. W. Matthews), Academic Press (1975), pp. 560–610.
18. R. C. Pond, in *Dislocations in Solids* (ed F. R. N. Nabarro), Vol. 8, North-Holland (1989), pp. 1–66.
19. J. L. Batstone, J. M. Phillips and J. M. Gibson, *Mat. Res. Soc. Symp. Proc.*, **91** (1987), 445.
20. J. M. Gibson and J. M. Phillips, *Appl. Phys. Lett.*, **43** (1983), 828.
21. D. A. Smith and M. M. J. Treacy, *Appl. Surf. Sci.*, **11/12** (1982), 131.
22. P. M. Petroff, L. C. Feldman, A. Y. Cho and R. S. Williams, *Appl. Phys. Lett.*, **35** (1981), 7317.
23. J. L. Batstone, J. M. Gibson, R. T. Tung, A. F. J. Levi and C. A. Outten, *Mat. Res. Soc. Symp. Proc.*, **82** (1987), 335.
24. M. Eizenberg, A. Sebmüller, M. Heiblum and D. A. Smith, *J. Appl. Phys.*, **62** (1987), 466.
25. J. A. Venables and G. L. Price, in *Epitaxial Growth* (ed J. W. Matthews), Part B, Academic Press (1975), pp. 382–430.
26. E. Bauer, *Zeitschrift für Kristallographie*, **110** (1958), 372.
27. M. Volmer and A. Weber, *Z. Phys. Chem.*, **119** (1926), 277.
28. N. Stranski and von L. Krastanow, *Akad. Wiss. Lit. Mainz Math.-Natur.*, Kl. IIb **146** (1939), 797.
29. J. M. Cowley, in *Reflection High Energy Diffraction and Reflection Electron Imaging of Surfaces*, (eds P. K. Larsen and P. J. Dobson), Plenum Press, NY, (1988), p. 261.
30. J. H. Neave, P. J. Dobson, B. A. Joyce and J. Zhang, *Appl. Phys. Lett.*, **47** (1985), 100.
31. R. Hull and A. Fischer-Colbrie, *Appl. Phys. Lett.*, **50** (1987), 851.
32. T. Narusawa and W. M. Gibson, *Phys. Rev. Lett.*, **47** (1981), 1459.
33. P. M. J. Marée, K. Nakagawa, F. M. Mulders and J. F. van der Veen, *Surf. Sci.*, **191** (1987), 305.
34. S. S. Iyer, J. C. Tsang, M. W. Copel, P. R. Pukite and R. M. Tromp, *Appl. Phys. Lett.*, **54** (1989), 219.
35. E. Kasper and H. Jorke, in *Chemistry and Physics of Solid Surfaces*, (ed R. Vanselow and R. F. Howe), Springer Verlag, Berlin (1988), p. 557.
36. M. Copel, M. C. Reuter, E. Kaxiras and R. M. Tromp, *Phys. Rev. Lett.*, **63** (1989), 632; M. Copel, M. C. Reuter, M. Horn von Hoegen and R. M. Tromp, *Phys. Rev. B*, **42** (1990), 11682.
37. J. M. Gibson, J. L. Batstone and R. T. Tung, *Appl. Phys. Lett.*, **51** (1987), 45.
38. F. Hellman and R. T. Tung. *Phys. Rev. B*, **37** (1988), 10786.

39. R. T. Tung and J. L. Batstone, *Appl. Phys. Lett.*, **52** (1988), 648.

40. R. T. Tung and J. L. Batstone, *Appl. Phys. Lett.*, **52** (1988), 1611.

41. J. L. Batstone, R. T. Tung and A. F. J. Levi, unpublished.

42. J. Henz, M. Ospelt and H. Von Känel, *Surf. Sci.*, **211/212** (1989), 716.

43. C. J. Palmstrom, S. Mounier, T. G. Finstad and P. F. Micelli, *Appl. Phys. Lett.*, **56** (1990), 382.

44. J. Zhu, C. B. Carter, C. J. Palmstrom and S. Mounier, *Appl. Phys. Lett.*, **56** (1990), 1323.

45. H.-J. Gossmann, *Mat. Res. Soc. Symp. Proc.*, **92** (1987), 53.

46. Chapter 21 by R. T. Tung, this book.

47. K. Takyanagi, Y. Tanishiro, S. Takahashi and M. Takahashi, *Surf. Sci.*, **164** (1985), 367.

48. J. M. Gibson, M. L. McDonald and F. C. Unterwald, *Phys. Rev. Lett.*, **55** (1985), 1765.

49. F. K. LeGoues, V. P. Keasn, S. S. Iyer, J. Tersoff and R. Tromp, *Phys. Rev. Lett.*, **64** (1990), 2038.

50. R. L. Headrick, B. E. Wier, J. Bevk, B. S. Freer, D. J. Eaglesham and L. C. Feldman, *Phys. Rev. Lett.*, **65** (1990), 1128.

51. U. K. Köhler, J. E. Demuth and R. J. Hamers, *Phys. Rev. Lett.*, **60** (1988), 2499.

52. D. Loretto, J. M. Gibson and S. M. Yalisove, *Phys. Rev. Lett.*, **63** (1989), 298.

53. J. L. Batstone, J. M. Gibson, R. T. Tung and D. Loretto, *Proceedings of the 44th Annual Meeting of the Electron Microscopy Society of America* (1989), p. 460.

54. A. J. Bourret and J. J. Bacmann, *Surf. Sci.*, **162** (1985), 495.

55. M. H. Grabow and G. H. Gilmer, *Mat. Res. Soc. Proc.*, **94** (1987), 15.

56. D. J. Eaglesham and M. Cerullo, *Phys. Rev. Lett.*, **64** (1990), 1943.

57. Y.-W. Mo, D. E. Savage, B. S. Swartzentruber and M. G. Lagally, *Phys. Rev. Lett.*, **65** (1990), 1020.

58. J. W. Matthews and A. E. Blakeslee, *J. Cryst. Growth*, **27** (1974), 118.

59. E. Kasper, H. J. Herzog and H. Kibbel, *Appl. Phys.*, **8** (1975), 199.

60. R. People and J. C. Bean, *Appl. Phys. Lett.*, **47** (1985), 322; **49** (1989), 229.

61. J. H. van der Merwe, *J. Appl. Phys.*, **34** (1963), 123.

62. W. A. Jesser and J. H. van der Merwe, in *Dislocations in Solids*, (ed F. R. N. Nabarro) Vol. 8, Elsevier (1989), pp. 421–460 and references therein.

63. P. M. Marée, J. C. Barbour, J. F. van der Veen, K. L. Kavanagh, C. W. T. Bulle-Lieuwma and M. P. A. Viegers, *J. Appl. Phys.*, **62** (1987), 4730.

64. J. W. Matthews, S. Mader and T. B. Light, *J. Appl. Phys.*, **41** (1970), 3800.

65. B. W. Dodson and J. Y. Tsao, *Appl. Phys. Lett.*, **51** (1987), 1325; **52** (1988), 852.

66. H. Alexander and P. Haasen, in *Solid State Physics*, Vol. 22, Academic Press, New York (1968).

67. D. Chidambarrao, G. R. Srinivasan, B. Cunningham and C. S. Murthy, *Appl. Phys. Lett.*, **57** (1990), 1001.

68. B. A. Fox and W. A. Jesser, *J. Appl. Phys.*, **68** (1990), 2801.

69. R. People, *IEEE J. Quantum Electron.*, **22** (1986), 1696.

70. W. Hagen and H. Strunk, *Appl. Phys.*, **17** (1978), 85.

71. R. Hull, J. C. Bean, D. J. Eaglesham, J. M. Bonar and C. Buescher, *Thin Solid Films*, **183** (1989), 117.

72. E. A. Fitzgerald, G. P. Watson, R. E. Proano, D. G. Ast, P. D. Kirchner, G. D. Petit and J. M. Woodall, *J. Appl. Phys.*, **65** (1989), 2220.

73. D. J. Eaglesham, D. M. Maher, E. P. Kvam, J. C. Bean and C. J. Humphreys, *Phys. Rev. Lett.*, **62** (1989), 187.

74. M. J. Stowell, in *Epitaxial Growth Part B* (ed J. W. Matthews), Academic Press (1975), pp. 437–492.

75. R. Vincent, *Phil. Mag.*, **19** (1969), 1127.

76. P. A. Stadelmann, *Ultramicroscopy*, **21** (1987), 131.

77. F. K. LeGoues, M. Copel and R. Tromp, *Phys. Rev. Lett.*, **63** (1989), 1826; *Phys. Rev. B*, **42** (1990), 11690.

78. S. Luryi and E. Suhir, *Appl. Phys. Lett.*, **49** (1986), 140.

79. W. A. Jesser and J. H. van der Merwe, *Phil. Mag.*, **24** (1971), 295.

80. Y. H. Xie and J. C. Bean, *J. Vac. Sci. Technol. B*, **8** (1990), 227.

81. S. Guha, A. Madhukar, K. Kaviani and R. Kapre, *J. Vac. Sci. Technol. B*, **8** (1990), 149.

82. E. A. Fitzgerald, P. D. Kirchner, R. Proano, G. D. Petit, J. M. Woodall and D. G. Ast, *Appl. Phys. Lett.*, **52** (1988), 1496.

83. G. C. Osbourn, *J. Appl. Phys.*, **53** (1982), 1586.

84. R. M. Biefeld, *J. Cryst. Growth*, **56** (1982), 382.

85. R. Hull, J. C. Bean, F. Cerdeira, A. T. Fiory and J. M. Gibson, *Appl. Phys. Lett.*, **48** (1986), 56.

86. R. C. Pond and D. Cherns, *Surf. Sci.*, **152** (1985), 1197.

87. S. M. Yalisove, R. T. Tung and J. L. Batstone, *Mater. Res. Soc. Symp. Proc.*, **116** (1988), 439.

88. J. L. Batstone and J. M. Phillips, *Mat. Res. Soc. Symp. Proc.*, **139** (1989), 351.

89. H. J. Kroemer, *J. Cryst. Growth*, **81** (1985), 193.

90. L. G. Schulz, *Acta Cryst.*, **4** (1951), 487.

91. R. F. C. Farrow, D. S. Robertson, G. M. Williams, A. G. Cullis, G. R. Jones, I. M. Young and P. N. J. Dennis, *J. Cryst. Growth*, **54** (1981), 507.

92. D. M. Wood and A. Zunger, *Phys. Rev. Lett.*, **61** (1988), 1501.

93. T. S. Kuan, T. E. Kuech, W. I. Wang and E. L. Wilkie, *Phys. Rev. Lett.*, **54** (1985), 201.

94. A. Ourmazd and J. C. Bean, *Phys. Rev. Lett.*, **55** (1985), 765.

95. F. K. LeGoues, V. P. Kesan and S. S. Iyer, *Phys. Rev. Lett.*, **64** (1990), 40.

96. H. Foll, P. S. Ho and K. N. Tu, *Phil. Mag.*, **45** (1982), 31.

97. J. M. Gibson and J. L. Batstone, *Surf. Sci.*, **208** (1989), 317.

98. K. Toman, *Acta Crystallog.*, 5 (1952), 329.

99. J. L. Batstone, J. M. Gibson, R. T. Tung and A. F. J. Levi, *Appl. Phys. Lett.*, **52** (1987) 828.

100. J. M. Phillips, J. L. Batstone, J. C. Hensel, M. Cerullo and F. C. Unterwald, *J. Mater. Res.*, **4** (1989), 144.

101. J. L. Batstone, A. C. Daykin, J. M. Phillips and J. C. Hensel, *Inst. Phys. Conf. Ser.*, **100** (1989), 641.

102. A. C. Daykin, C. J. Kiely, R. C. Pond and J. L. Batstone, *Proceedings of the 12th International Congress on Electron Microscopy*, Vol. 4 (1990), pp. 574.

103. J. M. Phillips, J. L. Batstone, J. C. Hensel, I. Yu and M. Cerullo, *J. Mater. Res.*, **5** (1990), 1032.

CHAPTER TWELVE

Phase behavior of monolayers

S. G. J. Mochrie, D. Gibbs, and D. M. Zehner

Introduction · Scattering Experiments · Conclusion

12.1 INTRODUCTION

It has long been realized that the thermal and structural properties of two-dimensional systems may differ dramatically from those of their three-dimensional counterparts. This is a direct result of the increased importance of fluctuations in one and two dimensions [1]. Experiments, however, are necessarily performed in the three-dimensional world. Nevertheless, it is possible to approach two-dimensional behavior in a monolayer of atoms or molecules confined to a surface or at the interface between two different media. It is even possible to produce a freely-suspended film of liquid crystal material that is only two molecular layers thick. In this chapter, we describe what has been learned empirically about the structure and phase behavior of several different monolayer systems. To this end, many experimental techniques have proven valuable. Here, we primarily focus on the results of recent X-ray scattering studies. Synchrotron X-ray scattering techniques are uniquely well suited to the investigation of spatial correlations from ångstrom to micron length scales, and, therefore, to the characterization of phase transformations and phase behavior. We briefly summarize what kinds of information can be obtained from mea-surements of the X-ray scattering cross-section in section 12.2.

During the last 15 years, synchrotron based X-ray scattering experiments have been performed on a wide variety of systems including rare gases adsorbed on graphite substrates, simple adsorbates on metallic substrates, intercalates, and Langmuir films. In this chapter, we have chosen to describe a few well-understood systems in detail. Where appropriate, we emphasize where theoretical results match the experimental results and where they do not. The first example we discuss concerns a monolayer of Pb adsorbed onto a Ge(111) sub-strate [2–4]. Pb on Ge(111) exhibits two distinct, ordered monolayer phases. In each case, the adsorbed atoms are located at sites commensurate with the Ge(111) substrate, that is, the reciprocal lattice vectors characterizing the overlayer structure are rational fractions of lateral substrate reciprocal lattice vectors. Specifically, both ordered monolayer phases of Pb on Ge(111) form hexagonal, commensurate $(\sqrt{3} \times \sqrt{3})$ structures, with a lattice constant $\sqrt{3}$- times that of the substrate. In one case there are four atoms per unit cell and in the other there is one. For such a structure, there are three equivalent domains or sub-lattices, which may be transformed onto one another by translation through a substrate lattice vector. For Pb on Ge(111), we describe the monolayer phase diagram in section 12.2.1.

The structure of an interfacial monolayer need not be commensurate with the substrate. If the interactions among the constituent particles of the monolayer are dominant over their interactions with the substrate, the monolayer may assume a structure with a lattice constant, which is incommensurate with the substrate. In this case, the reciprocal lattice vectors of the overlayer are unrelated to those of the substrate. It is possible, however, for the monolayer to transform from a commensurate to an incommensurate structure as some external parameter, such as the temperature or the chemical potential, is varied [5]. The concept of the discommensuration is central to the present understanding of commensurate–incommensurate transitions [6–8]. In two dimensions, a discommensuration is a linear defect at which one commensurate sub-lattice of the monolayer changes to the next, and thereby increases or decreases the average monolayer density. A commensurate–incommensurate transition involves the creation of discommensurations, which may themselves form a regular lattice. The deviation of the wave vector from a commensurate value, defined as the incommensurability, is proportional to the density of discommensurations. The arrangement of discommensurations may exhibit one-dimensional, hexagonal, or some more complicated symmetry, depending on the particular system in question. The most thoroughly understood case, it turns out, is that of non-intersecting linear discommensurations [9–11]. The corresponding structure, which is incommensurate in only one direction, is often called a stripe-domain structure. A two-dimensional example of this class of material is provided by bromine layers, which have been intercalated into a graphite host [12–14]. Because the bromine layers are widely separated from each other, their lateral interlayer interactions are weak, and each behaves as an independent, two-dimensional monolayer. In section 12.2.2, we describe the uniaxial, commensurate–incommensurate transition of bromine intercalated graphite.

The most studied commensurate–incommensurate transition is that of krypton adsorbed on the basal plane of graphite (Gr(001)) [15–19]. At low coverages, the Kr layer shows a commensurate $(\sqrt{3} \times \sqrt{3})$ structure. With increasing coverage, this is transformed into an incommensurate structure with a lattice constant near that of bulk Kr. The hexagonal symmetry is preserved through the commensurate–incommensurate transition. In section 12.2.3, we show that a particularly striking feature of this transformation is that the weakly incommensurate overlayer does not exhibit long-range translational order at seemingly low temperatures, whereas at larger incommensurabilities the overlayer does show long-range order. Weakly-incommensurate Kr on Gr is believed to be an example of a discommensuration-fluid phase, for which the microscopic atomic arrangement is well ordered but the arrangement of discommensurations is disordered [20–21].

In three dimensions, melting from a solid phase, in which translational correlations extend to macroscopic length scales, to a liquid, with only short-range translational order, is an abrupt transition. In contrast, in two dimensions, it is possible for melting to be continuous with the extent of translational order decreasing smoothly with temperatures increasing above the melting point [22–24]. An important prediction of the theory of two-dimensional melting [23] is that the continuous melting of a hexagonal monolayer occurs in two steps. The first transformation is to a phase with short-range translational order but with long-range orientational order. More precisely, while the average separation between distant molecules is not well-defined, the **direction** to a nearest-neighbor is well defined throughout the monolayer. This is called a hexatic phase. For increasing temperature there follows a transformation into a liquid phase with short-range translational and short-range orientational order. Incommensurate hexagonal monolayers provide realizations of two-dimensional solids for which the predictions of theories of two-dimensional melting ([23] and [24]) may be appropriate. In section 12.2.4, we describe measurements of the melting transition of an incommensurate Xe monolayer on

Gr(001) [25, 26]. In this case, melting is experimentally observed to be continuous. There are, however, also examples of discontinuous two-dimensional melting transitions, notably in freely-suspended liquid crystal films [28]. In the studies of adsorbed Xe, the question of the existence of a true hexatic phase is complicated by the six-fold symmetry of the substrate, which also imposes long-range orientational order on the overlayers [26, 27]. No such ambiguity exists for freely-suspended liquid crystal films [28].

Even though there is no translational registry between an incommensurate overlayer and a substrate, there may exist a definite orientational relationship between the two lattices. One might anticipate that the lowest energy configuration would occur in the case that a high symmetry direction of the overlayer is aligned with a high symmetry direction of the substrate. However, the possibility that the high symmetry directions of an incommensurate overlayer may be rotated away from the high symmetry directions of the substrate was originally predicted by Novaco and McTague [29], and subsequently observed for a monolayer of Ar on Gr(001) by low energy electron diffraction (LEED) [30]. In fact, just as it is possible for an overlayer to transform from a commensurate to an incommensurate structure, an incommensurate structure may further transform from an aligned to a rotated structure. The aligned-to-rotated transitions of the rare gases on graphite [19, 30, 31, 32] and of alkali metals on transition metal substrates [33, 34] are described in section 12.2.5.

In all of the above examples, the substrate is rigid when compared to the overlayer. For very many monolayer systems, this condition is not fulfilled. A specific example is provided by the reconstruction of the Au(111) surface. The ideal (111) surface of an elemental face-centered cubic material is hexagonally close-packed. This appears to be the lowest-energy configuration of the (111) surface of most fcc metals. In contrast, at 300 K the Au(111) surface layer is denser than the bulk (111)-planes. In particular, the uniaxial reconstruction is comprised of a series of linear discommensurations, which separate surface regions with the correct fcc ABC-stacking sequence from faulted regions with

an ABA-stacking sequence [33, 34]. However, there also exists an equilibrium density of kinks between rotationally equivalent domains of the uniaxial reconstruction [35, 36, 37]. (The (111)-surface of a cubic material has a three-fold rotational symmetry.) The kinks are themselves ordered and produce a structure in which two of the three possible rotationally equivalent domains of the uniaxial reconstruction alternate periodically across the surface. We note in section 12.2.6 that the origin of the domain formation is the accompanying reduction of elastic stress induced in the bulk crystal [38]. Other candidates for displaying this phenomenon include the (2 × 1) reconstructed Si(001) surface [39] and, indeed, any adsorbed monolayer which forms a structure that breaks a rotational symmetry of the surface and induces an elastic deformation in the substrate. In this instance, we see that interaction with the substrate can profoundly effect the structure of a monolayer. At still higher temperatures, the Au(111) surface exhibits a discommensuration-fluid phase reminiscent of that of Kr on Gr(001), although the disordering transformation into the discommensuration-fluid phase of the Au(111) surface is first-order [35] and there are only two commensurate sublattices.

A second example where elastic deformations of the substrate are important is provided by the reconstruction of the (001) surfaces of Au(001) and Pt(001). Despite the atomic planes of square symmetry lying immediately beneath, these surfaces exhibit a room temperature reconstruction in which the surface atoms form a close-packed, hexagonal monolayer. For both the Au(001) [42] and the Pt(001) [43, 44] surfaces, the hexagonal overlayer is incommensurate, with a smaller nearest-neighbor separation than in the crystal interior. A key feature of this structure is the existence of a surface corrugation which propagates, exponentially-damped, several layers into the bulk [42]. In section 12.2.7, we describe the aligned-to-rotated transitions of these overlayers. There are significant differences with the rotational behavior shown by the systems described in section 12.2.5. Another surprising aspect of the behavior shown by both the Au(001) [42] and the

Pt(001) [44] surfaces is decreasing translational order with decreasing temperature. It is possible that this behavior occurs because of the resultant reduction of elastic stress in the bulk, as in the case of domain formation on the Au(111) surface [37–41].

Finally, we note that three-dimensional behavior may intrude in a number of additional ways. For example, liquid and solid surfaces at high temperatures may not be ideally flat. Thermally-excited capillary modes may produce roughness [45–47]. Nevertheless, Langmuir films of a monolayer of organic molecules suspended at the (rough) air–water interface [48] are well ordered and may undergo structural phase transformations from one ordered structure to another [49].

12.2 SCATTERING EXPERIMENTS

Consider a scattering process in which an X-ray with initial wave vector K_i is scattered by a sample and exits with a wave vector K_f. The wave vector transfer to the sample is then $Q = K_f - K_i$. The differential cross-section, $d\sigma/d\Omega$, for scattering into an element of solid angle, $d\Omega$, is [48]

$$\frac{d\sigma}{d\Omega} = r_0^2 \langle n(Q)n(-Q) \rangle = r_0^2 S(Q) \qquad (12.1)$$

Where $n(Q) = \Sigma_j e^{iQ \cdot R_j}$, with R_j the position of the jth electron and r_0 the Thomson radius of the electron. Equation (12.1) defines the scattering function, $S(Q)$. The quantity $n(Q)$ is the Fourier transform of the sample number density, $n(r) = \Sigma_j \delta(r - R_j)$ [49]. The X-ray scattering cross-section (eq. (12.1)) may therefore be seen to be the product of two factors, the first of which (r_0^2) depends on the interaction of the electromagnetic field with the electronic charge, and the second of which ($S(Q)$) depends exclusively on the structure and properties of the sample. The significance of the scattering function is most readily appreciated by expressing it in terms of the number density:

$$S(Q) = V \int d^3r e^{iQ \cdot r} \langle n(r)n(0) \rangle \qquad (12.2)$$

where V is the illuminated sample volume. The density–density correlation function, $\langle n(r)n(0) \rangle$, gives the thermal average value of the number density at position r, assuming that the number density at the origin is $n(0)$ [49]. It is clear, therefore, that measurements of the X-ray scattering cross-section versus wave vector transfer directly probe translational correlations within the sample. As a concrete example, consider a solid in first three, then two, and finally one dimension. For a crystalline phase, the density–density correlation function may be approximated as

$$\langle n(r)n(0) \rangle$$
$$= n_0^2 + \sum_{G \neq 0} |A_G|^2 e^{iG \cdot r} \langle e^{-iG \cdot (u(r)-u(0))} \rangle \qquad (12.3)$$

where n_0 is the average number density, A_G is the amplitude of the density wave with wave vector G, and $u(r)$ is its local phase. It may further be shown that the function $C_G(r) = \langle e^{iG \cdot (u(r)-u(0))} \rangle$ depends qualitatively on dimension [52]:

$$C_G(r) = \text{const} \qquad 3D$$
$$C_G(r) = (r/a)^{-\eta} \qquad 2D$$
$$C_G(r) = e^{-r/\xi} \qquad 1D \qquad (12.4)$$

where $\eta = k_B T G^2/2\pi\kappa$ and $\xi = \kappa/G^2 k_B T$. T is the temperature, k_B is Boltzmann's constant, κ is an elastic constant, and a the nearest-neighbor separation. In three dimensions, the fact that correlation function is a constant indicates that there is long-range order. In one dimension, the correlation function decays exponentially with distance, indicating that translational correlations are short-ranged. In two dimensions the correlation function decays algebraically with distance. This is called quasi-long range order. The corresponding scattering functions are

$$S(Q) \approx N(2\pi)^3 \sum_G |A_G|^2 \delta(Q - G) \quad 3D$$
$$S(Q) \approx N \sum_G |A_G|^2 |Q - G|^{-2+\eta} \quad 2D$$
$$S(Q) \approx \sum_G \frac{N|A_G|^2}{1 + \xi^2(Q - G)^2} \quad 1D \qquad (12.5)$$

In contrast to the three-dimensional case, where the scattering from a solid is composed of a sum of δ-function peaks, in two dimensions the scattering function is composed of a sum of power-law singularities ($\eta < 2$) or cusps ($\eta > 2$). The scattering function in one dimension has a Lorentzian form, characteristic of short-range translational order. Fluid phases in two and three dimensions are also expected to exhibit Lorentzian scattering functions. Incommensurate two-dimensional solids are expected to exhibit a power-law scattering function. However, two-dimensional commensurate structures are expected to show δ-function peaks.

While we have pursued this discussion explicitly for X-ray scattering, it carries over to other probes, such as electrons or helium atoms, provided two important criteria are fulfilled. Firstly, the interaction between the sample and the probe must be sufficiently weak that the Born approximation is accurate. If the scattering is too intense, there will be multiple scattering. Secondly, the scattering must be quasi-elastic, so that the energy transfer to or from the sample is small compared to the energy of the probe. Then, one measures an instantaneous correlation function to obtain a 'snapshot' of the sample at a given instant of time. Because X-rays interact weakly with matter and because the energy of the X-rays typically employed for scattering experiments ($E = 10\,\text{keV}$) is much larger than the energies of typical thermally-excited modes ($k_B T = 0.025\,\text{eV}$), both conditions are satisfied in X-ray scattering experiments. In contrast, the Born approximation is usually inapplicable to low energy electron scattering, while the energy of the atoms employed for scattering experiments is comparable with thermal energies, so that quasi-elastic approximation is not automatically valid. Finally, it is important to note that the reciprocal space resolution which may be achieved in X-ray scattering experiments is typically an order of magnitude finer than possible in LEED and atom scattering experiments.

12.2.1 Pb on Ge(111)

Figure 12.1 shows the phase diagram of Pb on Ge(111), as deduced from the X-ray scattering study of Ref. [4]. There are three monolayer phases: two distinct, ($\sqrt{3} \times \sqrt{3}$) commensurate

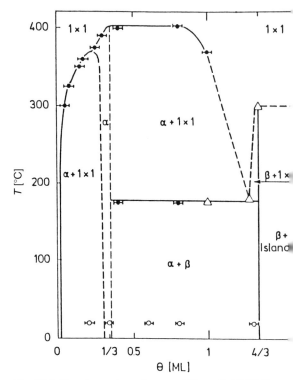

Fig. 12.1 Monolayer phase diagram of Pb on Ge(111) [4].

structures and a disordered (1×1) fluid phase. The low-coverage ($\sqrt{3} \times \sqrt{3}$) structure is called the α-phase, the high-coverage structure is called the β-phase. All of the phase transformations between the various phases appear to be first-order. Crystallographic analysis of the intensities of Bragg reflections from these two structures has shown that there is one Pb atom per ($\sqrt{3} \times \sqrt{3}$)-unit cell in the α-phase and four Pb atoms per unit cell in the β-phase. The two structures yield Bragg peaks at identical commensurate positions in reciprocal space. However, the intensity of a given Bragg peak is different for the two structures. Therefore it is possible to deduce the surface phase diagram by measuring the variations of a given Bragg reflection with temperature and Pb coverage (Fig. 12.1). Below a coverage of $\frac{1}{3}$, there is coexistence between islands of the α-phase and a disordered (1×1)-phase. In the coverage range between and $\frac{4}{3}$, the diagram of Fig. 12.1 is remarkably similar to the phase diagram of a three-dimensional

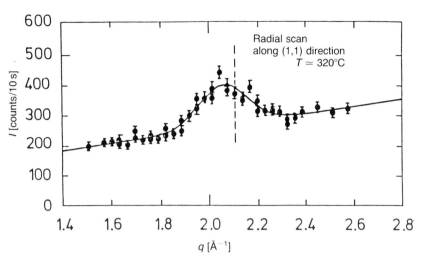

Fig. 12.2 Scattering function of a liquid monolayer of Pb on Ge(111) [4].

eutectic, with the α–β-liquid line at 175°C and the eutectic coverage at 1.25 ± 0.1 monolayers.

The character of the disordered phase at a coverage of 4/3 has been investigated by reflection high-energy electron diffraction (RHEED) measurements [3] and by diffuse X-ray scattering measurements [4]. Figure 12.2 shows the X-ray diffraction profile of a radial scan in reciprocal space. Similar scans at different azimuthal angles and different values of the surface normal momentum transfer reveal that the scattering function in this phase is described by a diffuse cylinder. The dashed line in Fig. 12.2 indicates the commensurate position. Evidently, the scattering is centered at an incommensurate wave vector. Furthermore, the incommensurate peak position is close to that expected for a two-dimensional liquid with the same nearest neighbor distance as in three-dimensional liquid Pb [4]. Thus, it seems as though the disordered phase at high coverages may best be described as a two-dimensional liquid.

12.2.2 Bromine intercalated graphite

There are many excellent reviews of the theory of the two-dimensional, uniaxial commensurate–incommensurate transition [5, 9–11]. Here, we give a brief summary. For an overlayer subject to the periodic potential of a substrate, there is a competition between the natural periodicity of the overlayer and that of the substrate. For small misfit between these two periodicities and a strong overlayer–substrate coupling, the overlayer conforms to the substrate periodicity. However, at a critical value of the natural misfit, a commensurate–incommensurate transition occurs. In the uniaxial case this involves the creation of regularly-spaced, linear discommensurations, separating locally commensurate regions. Thus, the overlayer remains commensurate in one direction and is characterized by a single incommensurate wave vector in the other direction. The incommensurate wave vector is proportional to the discommensuration density. In the absence of thermal fluctuations, the incommensurability of the overlayer near the commensurate–incommensurate transition is predicted to increase logarithmically with increasing natural misfit; far from the commensurate–incommensurate transition, the overlayer assumes its natural periodicity [6–8]. At finite temperatures, for widely separated discommensurations, the location of a discommensuration fluctuates, and is limited only by collisions with its neighbors. (It is assumed that discommensurations cannot cross each other.) As a result, the effective interaction between discommensurations is quite different at finite temperatures than at zero temperature. This in turn leads to a predicted one-half power-law

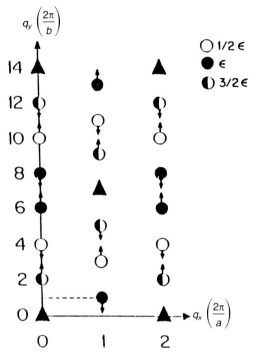

Fig. 12.3 HK-plane of reciprocal space for bromine-intercalated-graphite [12].

dependence of the incommensurability on the natural misfit [9–11].

Intercalation compounds are formed by the periodic insertion of layers of a reactive species (e.g. bromine) into a layered host material (e.g. graphite). In the case of dilute acceptor compounds, such as bromine-intercalated-graphite, there may be no lateral interlayer correlations and the material may be viewed as a stack of independent monolayers. Below $T_c = 342$ K, bromine layers intercalated into a graphite host exhibit a commensurate structure with a centered ($\sqrt{3} \times 7$) rectangular unit cell [12]. Accordingly, the primary order parameter is characterized by a commensurate wave vector equal to one-seventh of a graphite reciprocal lattice vector. There are seven possible commensurate sublattices. Figure 12.3 shows a quadrant of the HK-plane of reciprocal space for bromine-intercalated-graphite. The solid triangles correspond to lateral periodicities of the graphite host. The open, solid, and half-filled circles correspond to the

periodicities of the bromine layers. Figure 12.4 shows the results of diffraction scans parallel to the K-direction (seven-fold) direction through the (1, 5) and the (0, 6) peaks at three different temperatures. From these scans (and corresponding scans at intermediate temperatures), it emerges that for temperatures increasing above 342 K the profiles may be understood as two peaks – one more intense, one less intense – with peak positions displaced from the commensurate position. The pattern of the displacements of the stronger peak is illustrated in Fig. 12.3, where open circles correspond to peaks shifted by an amount $\varepsilon/2$, solid circles to peaks shifted by ε, and half-filled circles to peaks shifted by $3\varepsilon/2$. In Ref. [12], it is shown that this pattern of peak displacements is uniquely explained by a periodic array of uniaxial discommensurations which are separated by a distance $L = 4\pi/7\varepsilon$, and at each of which the bromine lattice shifts by $\frac{2}{7}$ of a commensurate lattice constant along the seven-fold direction. Figure 12.5 shows the continuous variation of the incommensurability (peak shift) of several reflections as the temperature is varied. The solid lines in Fig. 12.5 show the results of a least-mean-squares comparison of the data to the power-law form: $\varepsilon = \varepsilon_0 ((T - T_c)/T_c)_\beta$. The best-fit value of the exponent is $\beta = 0.50 \pm 0.02$, in excellent agreement with the theoretical prediction [9–11].

A second prediction for the behavior at non-zero temperatures is that the exponent, η, which is expected to characterize the algebraic decay of correlations in a uniaxial, weakly-p incommensurate structure (eq. (12.4)), is independent of temperature. Specifically, in the case that there are commensurate sublattices, the value predicted is $\eta = 2/p^2$ for the primary order parameter. For the *n*th order parameter, $\eta_n = 2n^2/p^2$. The displacement of a given peak from a commensurate position determines the appropriate n. For peaks that are displaced by $\varepsilon/2$, n equals 1; for peaks that are displaced by ε, n equals 2; etc. Fine resolution synchrotron X-ray scans of several bromine peaks at 56.6°C, below the transition, and 75.1°C, above the transition, are shown in Fig. 12.6. Evident is a clear difference between the profiles in the two phases. The further a peak is displaced from the

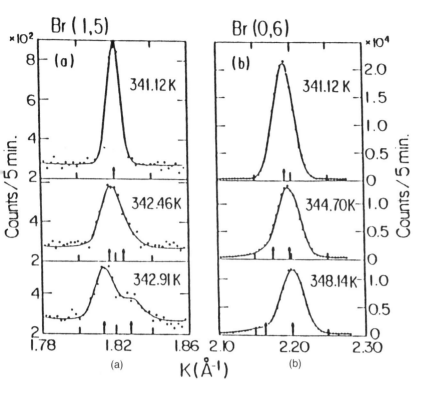

Fig. 12.4 Diffraction profiles through the $(1, 5)$ and $(0, 6)$ peaks of Br-intercalated graphite [12].

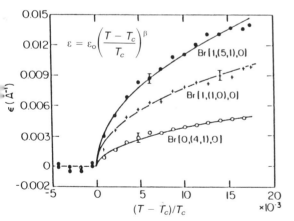

Fig. 12.5 Incommensurability versus temperature for Br-intercalated-Gr [12].

commensurate position, the greater is the ratio of the intensity in the tails to that at the peak. Increasing η increases the ratio. Indeed, the solid lines through the data of Fig. 12.7 correspond to a power-law lineshape, with each peak assigned the expected value of η. Thus, for the more intense peak at $(0, 6)$, which has an incommensurability of ε, η is 8/49, while for its less intense partner, η is 50/49. The excellent agreement between the predicted profiles and the experimental data is further confirmation that bromine-intercalated-graphite may be understood within the context of the theories of Refs. [9–11].

12.2.3 Kr on Gr(001)

Monolayers of Ar [53–55], N_2 [56], Kr [15–19], and Xe [25–27] adsorbed on the graphite basal plane constitute some of the earliest systems studied with the goal of investigating two-dimensional structure and phase transformations. In these systems, the interatomic interactions are simple and well understood (composed of a short-range repulsion and a long-range van der Waals attraction). Therefore, calculations and computer simulations

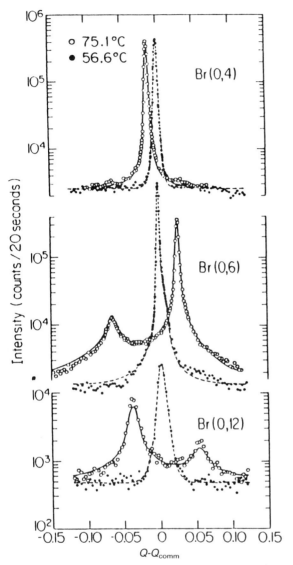

Fig. 12.6 Commensurate and incommensurate lineshapes of Br-intercalated-Gr [13].

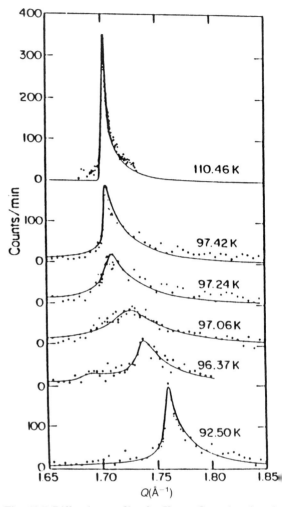

Fig. 12.7 Diffraction profiles for Kr on Gr undergoing th commensurate–incommensurate transition [18].

may be carried out using realistic potentials. It is also possible to obtain Gr(001) substrates which are flat and uniform over regions up to $0.5\,\mu m$ in size. As a result of these features, an extensive literature now exists concerning these systems.

In this section we describe the commensurate–incommensurate transition of krypton on Gr(001)

[15–19]. Below a critical Kr chemical potential, th adsorbed Kr monolayer shows a triangular $(\sqrt{3} \times \sqrt{3})$ structure. At higher Kr chemical potential, i forms an incommensurate triangular lattice with nearest-neighbor distance close to that of bulk Kr The first structural study of the commensurate-incommensurate transition in this system was per formed by Chinn and Fain [16] using LEED. The showed that the hexagonal symmetry is maintaine through the transition, and, furthermore, tha to the limits of their resolution, the incommensur ability evolves continuously. It was proposed tha the structure of the weakly incommensurate phase consists of a honeycomb network of discommensurations, with the size of the honeycomb increasing as the commensurate phase is approached. In fact, in the temperature range between 50 K and 95 K, the incommensurability (ε) is well described by the power-law: $\varepsilon = \varepsilon_0\{[\mu - \mu_c(T)]/\mu_c(T)\}_\beta$, with $\beta = 0.33 \pm 0.03$, and where μ is the Kr chemical potential and $\mu_c(T)$ is its critical value at the temperature T [16–18]. At higher temperatures $\beta = 0.44 \pm 0.06$ [19] seemed more appropriate [19]. At lower temperatures (40 K), the transition was shown to be first-order, with a 1.8% jump in the Kr lattice constant, when the Kr monolayer was compressed by coadsorption of deuterium [57].

New insight into the character of the transition followed synchrotron X-ray scattering experiments [18]. A series of profiles from Ref. [18] is shown in Fig. 12.7. At 110.46 K the profile corresponds to a well-ordered commensurate structure. On cooling below 97.24 K the commensurate diffraction profile disappears, and an incommensurate profile appears. The most striking feature of the weakly incommensurate phase is that the diffraction profiles are much broader than in the commensurate phase, indicative of short-range translational correlations. On further cooling, the profiles initially become broader as the incommensurability increases (97.06 K), but then they become narrower with further increase in the incommensurability (96.37 and 92.50 K). Detailed analysis of these data shows that the ratio of the incommensurability to the peak width approaches unity as the commensurate phase is approached. Within the context of a discommensuration description of the weakly incommensurate phase, this implies that the ratio of the mean surface periodicity to the correlation length is $4\pi/\sqrt{3} = 7.25$.

There are several theories which attempt to account for the observed behavior [21, 58–60]. The most transparent is that of Coppersmith *et al.* [21], who consider the properties of a honeycomb discommensuration lattice. There are two principal results of their analysis: firstly, for sufficiently widely-spaced discommensurations, the discommensuration lattice is unstable with respect to the formation of dislocations (see section 12.2.4) and,

therefore, is a fluid at all temperatures. This prediction is qualitatively in agreement with the experimental results of Ref. [18]. Secondly, they predict that the transition from the commensurate phase is first-order, although the jump in lattice constant may be small. At the present there is not any experimental evidence for such a jump. An essential feature of the picture presented in Ref. [21] is 'breathing' of the honeycomb array. It turns out that an individual hexagon may change its size ('breathe' in or out) without changing the total length of its constituent discommensurations or the number of discommensuration crossings. Direct evidence for such discommensuration motion is provided by molecular dynamics simulations [61]. Finally, it is worth nothing that in Ref. [60] it is suggested that the ratio of the incommensurability to the peak width should approach a constant for a certain class of continuous phase transformations.

12.2.4 Xe on graphite

The most familiar phase transformation in three dimensions is melting, which is a discontinuous transformation. No completely convincing theoretical description of three-dimensional melting yet exists. On the other hand, there is an elegant theory for two-dimensional melting, which predicts that it is a continuous transition [23–25]. In this scenario, the important excitations are dislocations. (In two dimensions dislocations are point defects.) For temperatures below the melting point, any dislocations are bound together in pairs. However, above the melting transition, there is a thermal equilibrium density of unbound dislocations and, as a result, translational correlations are short-ranged. The extent of positional order is determined by the mean separation between unbound dislocations. While the presence of unbound dislocations limits the extent of translational order, it does not limit orientational order. Thus, a second remarkable prediction of Ref. [23] for hexagonal structures is that two-dimensional melting may occur in two steps: from a crystal to an orientationally ordered fluid, and then to an isotropic liquid. The orientationally ordered fluid is called hexatic.

There are a number of studies of hexagonal, incommensurate adsorbed monolayers. The most detailed experiments concern Xe on Gr, and we describe those experiments in this section [25–27]. There are two important questions concerning the melting of Xe on Gr: (i) Is there a true hexatic phase? and (ii) Does the translational correlation length increase within the fluid phase as the transition is approached from above? In the following, we describe two classes of measurements: (1) measurements of the powder diffraction pattern of Xe on Gr(001), from which no orientational information is directly obtained; and (2) more recent measurements on single crystal Gr(001) substrates.

Figure 12.8 shows several powder diffraction profiles through the primary (1, 0) hexagonal peak of a saturated monolayer of Xe on Gr(001) for temperatures increasing from 135.0 K to 160.0 K [25]. As the temperature is increased from 135 K to 151.3 K, the intensity of the scattering in the tail of the profile increases relative to that at the peak. Above 151.3 K, the peak width increases rapidly, but smoothly. The apparently continuous evolution of the peak width is consistent with the development of longer- and longer-range translational correlations within the fluid phase ($T > 151.3$ K). Detailed analysis reveals that, for temperatures greater than 151.3 K, the diffraction profiles can be described by a single Lorentzian lineshape, as expected for a fluid. Furthermore, it is not possible to describe the profiles for temperatures greater than 151.3 K as the sum of a broad Lorentzian, corresponding to a poorly correlated fluid, together with a narrow profile, corresponding to a solid. Thus, any simple two-phase coexistence model for the transformation is ruled out. In this regard, it is worth noting that the peak position also evolves smoothly and continuously through the transition. At a first-order transition, one would expect a jump in the lattice constant. Below 151.3 K, in contrast to the situation above 151.3 K, the diffraction profile may be described by a power-law appropriate for an incommensurate, two-dimensional solid (eq. (12.5)). Between 135 K and 150 K, the exponent η is roughly constant with the value 0.25. Beyond 150 K, however, η increases noticeably so at the melting temperature $\eta = 0.32$. In this regard, it is noteworthy that the theories of two-

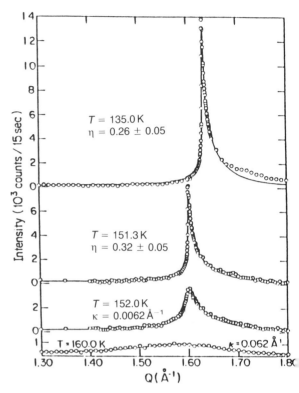

Fig. 12.8 Diffraction profile for Xe on Gr undergoing a melting transition [25].

dimensional melting ([23] and [24]) predict that the value of η at the dislocation unbinding transition is between 0.25 and 0.33, with the exact value depending on microscopic details.

Figure 12.9 shows the fitted inverse correlation length (κ) versus the peak amplitude (A) of the (Lorentzian) scattering function in the fluid phase. Evident from the figure is a power-law relationship between these two quantities. A further prediction of Ref. [23] is that $A = A_0 \kappa^{\eta-1}$. The best-fit to this functional form yields $\eta = 0.28$. The (approximate) consistency between the values of η, determined independently above and below the transition, seems to be particularly convincing confirmation that the transformation is continuous. In addition, the actual value of η is consistent with that predicted in Refs. [23] and [24]. Molecular-dynamics simulations have also been performed for xenon on graphite, producing remarkable, quantitative

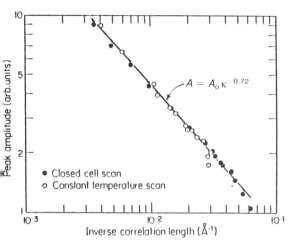

Fig. 12.9 Peak intensity versus inverse correlation length for the fluid phase of Xe on Gr, near melting [25].

agreement between the experimental liquid scattering function and that deduced from the simulations [62]. However, in the molecular dynamics simulations, it is possible to observe atomic diffusion. On this basis, coexistence of liquid (rapid diffusion) and solid (slow diffusion) regions is observed, and the character of the transition is identified as first-order.

More recent X-ray experiments performed on single crystal substrates also find a continuous evolution of the fluid phase scattering function [26, 27]. In addition, the use of a single crystal permits orientational information to be obtained. In this regard, it is found that the scattering function exhibits a six-fold rotational symmetry, in contrast to the cylindrically symmetric function expected for an isotropic fluid. It turns out that the azimuthal widths of the peaks are approximately six times broader than their width in the radial direction. As noted above, the expected hexatic phase exhibits (quasi-) long-range orientational order. Thus, the observed fluid scattering function is consistent with a hexatic. However, because the Gr(001) substrate has a six-fold rotational symmetry, even an intrinsically isotropic phase would display a six-fold symmetric scattering function [4]. Therefore, in the absence of further information, no definite conclusion can be drawn.

It is found, however, that a fluid phase of Xe on Ag(111) [63], which is believed to be a much smoother substrate than Gr(001), also shows the same ratio of azimuthal to radial widths as on the Gr(001) substrate. As a result, it was suggested that the observed ratio was a property of the Xe fluid alone. The existence of two-dimensional hexatic phases has been unambiguously established in liquid crystal films [28, 64].

12.2.5 Novaco–McTague rotational epitaxy

A basic question in the growth of one crystalline material on another concerns the orientational relationship between the two lattices: How are the high symmetry directions of the two crystals aligned with respect to each other? The analogue of this question for monolayers was discussed first in Ref. [29], where it was predicted that monolayers of the rare gases adsorbed on Gr(001) would be 'rotated', in the sense that the overlayer reciprocal lattice vectors would not be aligned parallel to the reciprocal lattice vectors of the substrate. A rotated phase of Ar on Gr was observed shortly thereafter [30]. The results of Ref. [29] may be understood as follows. The elastic susceptibility (χ) of a solid to a periodic potential with wave vector $\boldsymbol{G}$ is given by

$$\chi(\boldsymbol{G}) = \sum_s \frac{(\boldsymbol{G} \cdot \boldsymbol{e}_s(\boldsymbol{G}))^2}{M\omega_s^2(\boldsymbol{G})} \tag{12.6}$$

where $\omega_S(\boldsymbol{G})$ is the dispersion relation of the sth phonon branch and $\boldsymbol{e}_s(\boldsymbol{G})$ is its eigenvector. M is the mass of an overlayer atom. For a two-dimensional solid, there is one longitudinal branch and one transverse branch. When the overlayer and substrate reciprocal lattice vectors are aligned, the factor of $(\boldsymbol{G} \cdot \boldsymbol{e}_s(\boldsymbol{G}))^2$ ensures that the susceptibility is determined by the frequency of longitudinal phonons. On the other hand, a rotation of the overlayer with respect to the substrate allows transverse phonons to contribute. Since transverse phonon frequencies are usually smaller than those of longitudinal phonons with the same wave vector, the susceptibility increases as the rotation angle increases from zero. However, increasing the rotation angle also increases the wave vector of

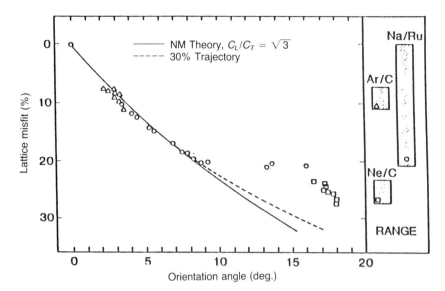

Fig. 12.10 Rotation angle versus lattice misfit for Na on Ru(001) [34].

the contributing phonons, thus increasing their frequency. The susceptibility is therefore at a maximum – the energy correspondingly at a minimum – at a finite rotation angle, which is determined by the compromise between these two competing effects. In fact, for modest incommensurabilities, eq. (12.6) suggests that the rotation angle should depend approximately linearly on the incommensurability, with the proportionality constant dependent on the ratio of the frequency of the transverse mode to that of the longitudinal mode.

It turns out that sodium adsorbed on the hexagonal Ru(001) surface at 80 K is an ideal system in which to test the predictions of Ref. [29]. A LEED study of this system is described in Ref. [34]. In LEED experiments, as well as in X-ray scattering experiments, the appearance of rotated domains on the surface is signaled by a rotation of the corresponding overlayer diffraction pattern. The sodium atoms, which form a hexagonal lattice, are strongly bound to the substrate and are partially ionized. Therefore, there is a strong, electrostatic repulsion between the atoms of the sodium monolayer. As a result, it is possible to vary the overlayer lattice constant over a wide range, simply by varying the sodium coverage. In addition, for materials in which the interatomic interactions are electrostatic, the ratio of longitudinal to transverse phonon

frequencies is expected to be $\sqrt{3}$. Fig. 12.10 shows the rotation angle versus lattice misfit from the commensurate $(\sqrt{3} \times \sqrt{3})$ structure for Na on Ru(001) over a wide range of lattice misfits from 8% to nearly 30%. Also shown is the prediction of Ref. [29]. Evidently, the agreement is excellent over a wide range, lending strong support to the basic assertion of Ref. [29].

The theory of Ref. [29] is not expected to be quantitatively correct near the commensurate–incommensurate transition. More elaborate theories, however, incorporate the non-linear response of the overlayer in the case that it is nearly commensurate with the substrate. In Ref. [65], a hexagonally incommensurate overlayer on an hexagonal substrate is examined with the resultant prediction that the overlayer remains aligned for incommensurabilities smaller than a critical value. For further increase in the incommensurability, the rotation angle may increase continuously, or it may jump to a non-zero value, depending on the overlayer dispersion relations. For large incommensurabilities, the rotation angle approaches the linear response result of Ref. [29] (as determined by the maximum of eq. (12.5)). The predictions of Ref. [65] have been borne out in the context of weakly incommensurate Kr on Gr(001) [19].

12.2.6 Reconstructed Au(111)

Perhaps the most unusual reconstruction of a metal surface is that of the close-packed (111) face of gold. This surface had been studied extensively at 300 K by many techniques [35, 36]. As a result, it was believed that its reconstruction might be described as a sequence of uniaxial discommensurations, separating surface regions with correct ABC-stacking from regions with faulted ABA-stacking (($p \times \sqrt{3}$) reconstruction, with $p = 23$). Recent X-ray [37] and scanning tunneling microscopy [38, 39] measurements reveal, however, that the structure, in fact, involves an equilibrium density of kinks separating domains whose discommensuration orientations differ by 120°. The kinks are themselves ordered and produce a structure in which two of the three possible rotationally equivalent domains of the ($p \times \sqrt{3}$) reconstruction alternate periodically across the surface. This was called the chevron phase in Ref. [37] and the herringbone phase in Ref. [38].

The existence of an equilibrium density of kinks immediately suggests that competing interactions compromise to produce a structure characterized by the new, incommensurate wave vector. Recently, Alerhand *et al.* [40] have proposed that high symmetry surfaces, which reconstruct into one of several, rotationally equivalent structures, are unstable to the formation of domains. This is because of the accompanying reduction of the elastic stress induced in the bulk crystal. In the context of the Au(111) surface, the elastic energy is smaller for smaller kink separations, but this must be balanced against the energetic cost of disrupting the favorable ($p \times \sqrt{3}$) structure in the neighborhood of each kink. The competition between these two contributions determines the kink wave vector. It is amusing to note that, since there are three rotationally equivalent domains of the chevron reconstruction, it may also be unstable to domain formation.

The chevron structure is stable between 300 and 865 K [37]. There is at higher temperatures a first-order phase transformation to a partially disordered structure, which is characterized by the scattering function shown in Fig. 12.11. From the radial width of these peaks a correlation length of the order of $\xi = 34a$ (~ 100 Å) is inferred. Remarkably, the correlation length remains unchanged between 950 and 1250 K and is only a factor of 1.4 larger than the surface periodicity. Analysis of the X-ray reflectivity shows that the ratio of regions with ABA-stacking to regions with ABC-stacking is preserved for all temperatures between 300 and 1250 K. This suggests that while atomic positions are well defined on a microscopic scale, the arrangement of kinks and discommensurations is very disordered for temperatures greater than 865 K. This structure, then, seems to provide a realization of a discommensuration fluid phase [21]. However, the calculations of Ref. [21] are carried out for widely separated discommensurations and the possibility of 120° kinks is explicitly excluded. Therefore, their application to the Au(111) surface is far from straightforward. Isolated dislocations, where two discommensurations meet and end are evident in the STM measurements of Ref. [39].

12.2.7 Au(001) and Pt(001)

The reconstructions of the clean Au(001) [42] and Pt(001) [43, 44] surfaces are particularly striking. In both cases, the topmost atomic layer forms a close-packed, hexagonal lattice (the overlayer) on top of the bulk planes of square symmetry (the substrate). The overlayer and the substrate are incommensurate; there is, however, a definite orientational relationship between the two lattices. This orientational epitaxy has been characterized as a function of temperature. The phase behavior of both surfaces is remarkably similar. At the highest temperatures, near the bulk melting temperature (T_m), each surface is disordered. On cooling to $\sim 0.9\,T_m$, there is a first-order phase transformation into the hexagonal phase. In addition to the peaks due to the bulk and to the hexagonal overlayer, the X-ray diffraction pattern includes peaks which can be indexed by a wave vector (δ, with $\delta = 0.206a^*$ for Au(001) and $\delta = 0.202a^*$ for Pt(001)) exactly equal to the difference between a particular substrate reciprocal lattice vector and a particular overlayer reciprocal lattice vector. This suggests that each lattice modulates by the other. In fact,

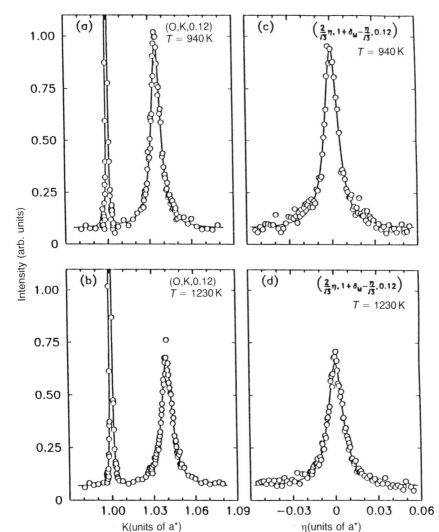

Fig. 12.11 Scattering function for the discommensuration fluid phase of the reconstructed Au(111) surface [37].

there is a surface corrugation of wave vector δ, which propagates several layers into the bulk. Immediately below the transition temperature, the two lattices are aligned in the sense that the two wave vectors that generate δ are parallel. (The two lattices have a high symmetry direction in common.) With decreasing temperature, there is a range of temperature in which the lattices remain aligned. However, at $\sim 0.8\,T_m$, there is an onset of rotation. In the case of the Au(001) surface,

the rotational transition is discontinuous. Rotated domains appear at a fixed rotation angle of $0.8°$. On further cooling, the population of rotated domains grows at the expense of unrotated population, but there is always coexistence between rotated and the unrotated domains. Figure 12.12 shows a series of angular scans through the principal hexagonal peak of the hexagonal overlayer obtained for temperatures decreasing from 1000 K to 300 K. In contrast, although it also occurs at $T_c \cong 0.8\,T_{m}$,

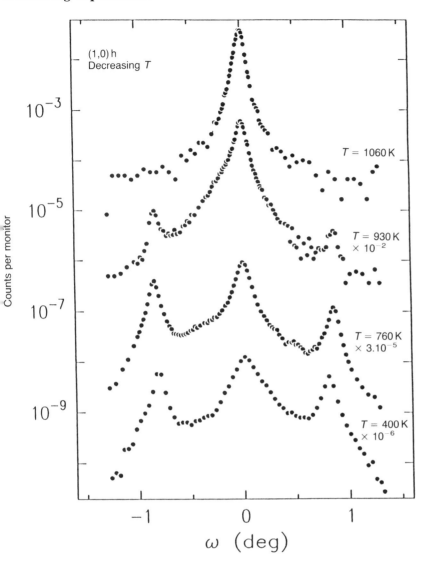

Fig. 12.12 Angular scans through the principal hexagonal peak of the hexagonally-reconstructed Au(001) surface.

the rotational transition of the Pt(001) surface is continuous. Indeed, close to the transition, the rotation angle (θ) obeys a one-half power-law:

$$\theta = \theta_o ((T_c - T)/T_c)^{\frac{1}{2}}$$

In this regard, it is important to note that overlayer lattice constants are only weakly temperature dependent between room temperature and the disordering transition. For Pt(001), there is no unrotated component below T_c. At lower temperatures, the rotation angle saturates at 0.75°. However, careful examination of the diffraction pattern reveals that in this regime there are peaks displaced from the substrate high symmetry direction by two distinct angles, indicating the coexistence of domains with two possible rotation angles. The second rotation angle for Pt(001) is 0.7°. Domains with this orientation appear discontinuously at ~0.7 T_m.

For rare gases adsorbed in graphite and for

alkali metals adsorbed on transition metals, the adsorbate–adsorbate interactions and the substrate–adsorbate interactions are quite well understood. In contrast, the predicted rotation angle for the Au(001) and Pt(001) overlayers is $\sim 7°$, which is approximately ten times the observed room temperature value. While the magnitude of the rotation angle remains unexplained, as does the existence of more than one stable rotation angle, it is possible to understand the temperature dependence of the rotation angle of the Pt(001) surface with a simple mean-field theory. Therefore, the transitions of Pt(001) and Au(001) seem to provide the first clear examples of temperature-driven rotational transitions.

12.3 CONCLUSION

A wide variety of interfacial systems exhibit phase transformations from one structure to another. Early studies focused on simple, model systems, such as the rare gases on graphite. As a result, a great deal is known about their behavior. More recently, it has proven possible to carry out detailed studies of the phase behavior of surface reconstructions and chemisorbed monolayers. Certain aspects of these systems show much in common with the rare gases. However, the structures involved are often much more elaborate and much remains to be learned. This will stimulate many future studies of the phase behavior of monolayers.

REFERENCES

1. See, for example *Ordering in Two Dimensions* (ed S. K. Sinha), North-Holland, New York (1980); and *Ordering in Strongly Fluctuating Condensed Matter Systems* (ed T. Riste), Plenum, New York (1980).
2. J. J. Metois, and G. LeLay, *Surf. Sci.* **133** (1983), 422.
3. T. Ichikawa, *Sol. Stat. Comm.*, **46** (1983), 827.
4. F. Grey, Thesis (Riso National Laboratory, Denmark, 1988) and references therein; F. Grey, R. Feidenhaus'l, J. Skov Petersen, M. Nielson and R. L. Johnson (unpublished).
5. For a comprehensive review, see P. Bak. *Rept. Prog. Phys.*, **45** (1982), 587.
6. F. C. Frank and J. H. van der Merwe, *Proc. R. Soc. London, Ser. A*, (1949), 198, 205.
7. W. L. McMillan, *Phys. Rev. B*, **12** (1975), 1187.
8. P. Bak and V. J. Emery, *Phys. Rev. Lett*, **36** (1976), 978.
9. V. L. Pokrovsky and A. L. Talapov, *Phys. Rev. Lett*, **42** (1979), 65.
10. V. L. Pokrovsky and A. L. Talapov, *Sov. Phys. JETP*, **51** (1980), 134.
11. M. E. Fisher and D. S. Fisher, *Phys. Rev. B*, **25** (1982), 3192.
12. A. R. Kortan, A. Erbil, R. J. Birgeneau, and M. S. Dresselhaus, *Phys. Rev. Lett.*, **49** (1982), 1427; A. Erbil, A. R. Kortan, R. J. Birgeneau, and M. S. Dresselhaus, *Phys. Rev. B*, **28** (1983), 6329.
13. S. G. J. Mochrie, A. R. Kortan, R. J. Birgeneau and P. M. Horn, *Phys. Rev. Lett.* **53** (1984), 985; S. G. J. Mochrie, A. R. Kortan, R. J. Birgeneau and P. M. Horn, *Z. Phys B*, **62** (1985), 79.
14. S. G. J. Mochrie, A. R. Kortan, R. J. Birgeneau and P. M. Horn, *Phys. Rev. Lett.* **58** (1987), 690.
15. A. Thomy and X. Duval, *J. Chem. Phys.* **67** (1970), 1101.
16. M. D. Chinn and S. C. Fain, Jr., *Phys. Rev. Lett.* **39** (1977), 146.
17. P. W. Stephens, P. Heiney, R. J. Birgeneau and P. M. Horn, *Phys. Rev. Lett.* **43** (1979), 47.
18. D. E. Moncton, P. W. Stephens, R. J. Birgeneau, P. M. Horn and G. S. Brown, *Phys. Rev. Lett.* **46** (1981) 1533; P. W. Stephens, P. A. Heiney, R. J. Birgeneau, P. M. Horn, D. E. Moncton and G. S. Brown, *Phys. Rev. B*, **29** (1984), 3512.
19. K. L. D'Amico, D. E. Moncton, E. D. Specht, R. J. Birgeneau, S. E. Nagles, P. M. Horn, *Phys. Rev. Lett*, **53** (1984), 2250; E. D. Specht, A. Mak, C. Peters, M. Sutton, R. J. Birgeneau, K. L. D'Amico, D. E. Moncton, S. E. Nagles and P. M. Horn, *Z. Phys. B*, **69** (1987), 347.
20. J. Villain and P. Bak, *J. Physique* **62** (1981), 657.
21. S. N. Coppersmith, D. S. Fisher, B. I. Halperin, P. A. Lee and W. F. Brinkman, *Phys. Rev. Lett*, **46** (1981), 549; S. N. Coppersmith, D. S. Fisher, B. I. Halperin, P. A. Lee and W. F. Brinkman, *Phys. Rev. B*, **25** (1982), 349.
22. J. M. Kosterlitz and D. J. Thouless, *J. Phys. C* 5 (1972), L124; **6** (1973), 1181; **1** (1974), 1046.
23. B. I. Halperin and D. R. Nelson, *Phys. Rev. Lett.*, **41** (1978), 121; **41** (1978), 519(E); D. R. Nelson and B. I. Halperin, *Phys. Rev. B*, **18** (1978) 2318; **19** (1979), 2457.
24. A. P. Young, *Phys. Rev. B*, **19** (1979), 1855.
25. P. A. Heiney, R. J. Birgeneau, G. S. Brown, P. M. Horn, D. E. Moncton and P. W. Stephens, *Phys. Rev. Lett.*, **48** (1982), 104; P. A. Heiney, R. J. Birgeneau, G. S. Brown,

P. M. Horn, D. E. Moncton and P. W. Stephens, *Phys. Rev. B*, **28** (1983), 6416.

26. T. F. Rosenbaum, S. E. Nagler, P. M. Horn and R. Clarke, *Phys. Rev. Lett.*, **50** (1983), 1791; S. E. Nagler, P. M. Horn, T. F. Rosenbaum, R. J. Birgeneau, M. Sutton, S. G. J. Mochrie, D. E. Moncton and R. Clarke, *Phys. Rev. B*, **32** (1985), 7373.

27. E. D. Specht, R. J. Birgeneau, K. L. D'Amico, D. E. Moncton, S. E. Nagler, and P. M. Horn, *J. Physique Lett.*, **46** (1985), L-561.

28. R. Pindak, D. E. Moncton, S. C. Davey and J. W. Goodby, *Phys. Rev. Lett.*, **46** (1981), 1135.

29. A. D. Novaco and J. P. McTague, *Phys. Rev. Lett.*, **38** (1977), 1286; J. P. McTague and A. D. Novaco, *Phys. Rev. B*, **19** (1979), 5299.

30. C. G. Shaw, S. C. Fain, Jr., and M. D. Chinn, *Phys. Rev. Lett.*, **41** (1978), 955.

31. H. Hong, C. J. Peters, A. Mak, R. J. Birgeneau, P. M. Horn and H. Suematsu, *Phys. Rev. B*, **36** (1987), 7311.

32. K. L. D'Amico, J. Bohr, D. E. Moncton and D. Gibbs, *Phys. Rev. B*, **41** (1990), 4368.

33. T. Aruga, H. Tochihara and Y. Murata, *Phys. Rev. Lett.*, **52** (1984), 1794.

34. D. L. Doering and S. Semancik, *Phys. Rev. Lett.*, **53** (1984), 66.

35. U. Harten, A. M. Lahee, J. P. Toennies and Ch. Wöll, *Phys. Rev. Lett.*, **54** (1985), 2619.

36. Ch. Wöll, S. Chiang, R. J. Wilson and P. H. Lippel, *Phys. Rev. B*, **29** (1989), 7988.

37. K. G. Huang, D. Gibbs, D. M. Zehner, A. R. Sandy and S. G. J. Mochrie, unpublished; K. G. Huang *et al.*, *Bulletin of the American Physical Society*, **35** (1990), 252. A. R. Sandy, S. G. J. Mochrie, D. M. Zehner, K. G. Huang and D. Gibbs (unpublished).

38. D. D. Chambliss and R. J. Wilson (unpublished).

39. J. V. Barth, H. Brune, G. Ertl and R. J. Behm (unpublished).

40. O. L. Alerhand, D. Vanderbilt, R. D. Meade and J. D. Joannopoulos, *Phys. Rev. Lett.*, **61** (1988), 1979. See also V. I. Marchenko, *Zh. Eskp. Teor, Fiz.*, **81** (1981), 1141; *Sov. Phys. JETP*, **54** (1981), 605.

41. F. K. Men, W. E. Packard and M. B. Webb, *Phys. Rev. Lett*, **61** (1988), 2469.

42. S. G. J. Mochrie, D. M. Zehner, B. M. Ocko and D. Gibbs, *Phys. Rev. Lett.*, **64** (1990), 2925; D. Gibbs, B. M. Ocko, D. M. Zehner and S. G. J. Mochrie, *Phys. Rev. B* in press; B. M. Ocko, D. Gibbs, K. G. Huang, D. M. Zehner and S. G. J. Mochrie (unpublished).

43. P. Heilmann, K. Heinz and K. Muller, *Surf. Sci.*, **83** (1979), 487.

44. D. Gibbs, G. Grubel, D. M. Zehner, D. L. Abernathy, S. G. J. Mochrie (unpublished); D. L. Abernathy, S. G. J. Mochrie, D. M. Zehner, G. Grubel, D. Gibbs (unpublished).

45. A. Braslau, P. S. Pershan, G. Swislow, B. M. Ocko and J. Als-Nielsen, *Phys. Rev. A*, **38** (1988), 2457.

46. D. Schwartz, M. L. Schlossman, E. H. Kawamoto, G. J. Kellogg, P. S. Pershan and B. M. Ocko, *Phys. Rev. A*, **47** (1990), 5687.

47. M. Sanyal, S. K. Sinha, K. G. Huang and B. M. Ocko (unpublished).

48. See, for example, J. Als-Nielsen and K. Kjaer, in *Phase Transitions in Soft Condensed Matter*, (ed T. Riste and D. Shernington), Plenum, NY (1989).

49. L. D. Landau and E. M. Lifshitz *Statistical Physics, Part 1*, Pergamon, Oxford (1980). p. 437.

50. J. D. Jackson, *Classical Electrodynamics* Wiley, New York (1974).

51. S. W. Lovesy, *Theory of Neutron Scattering from Condensed Matter*, Oxford, New York (1922).

52. See the articles by J. D. Axe and A. P. Young in Ref. [1].

53. H. Taub, L. Passell, J. K. Kjems, K. Carneiro, J. P. McTague, and J. G. Dash, *Phys. Rev. Lett.*, **34** (1975), 654; H. Taub, K. Carneiro, J. K. Kjems, L. Passell and J. P. McTague, *Phys. Rev. B*, **16**, 4551.

54. J. P. McTague, J. Als-Nielsen, J. Bohr, and M. Nielsen, *Phys. Rev. B*, **25** (1982), 7765.

55. M. Nielsen, J. Als-Nielsen, J. Bohr, J. P. McTague, D. E. Moncton, and P. W. Stephens, *Phys. Rev. B*, **35** (1987), 1419.

56. J. K. Kjems, L. Passell, H. Taub, and J. G. Dash, *Phys. Rev. Lett.*, **32** (1974), 724; J. K. Kjems, L. Passell, H. Taub, J. G. Dash and A. D. Novaco, *Phys. Rev. B*, **13** (1976), 1446.

57. M. Nielsen, J. Als-Nielsen, J. Bohr and J. P. McTague, *Phys. Rev. Lett.*, **47** (1981), 582.

58. P. Bak, D. Mukamel, J. Villain, and K. Wentowska, *Phys. Rev. B*, **19** (1979), 1610.

59. M. Kardar and A. N. Berker, *Phys. Rev. Lett.*, **48** (1982), 1552.

60. D. Huse and M. E. Fisher, *Phys. Rev. Lett.*, **49** (1982), 793; D. Huse and M. E. Fisher, *Phys. Rev. B*, **29** (1984), 239.

61. F. F. Abraham, S. W. Koch, and W. E. Rudge, *Phys. Rev. Lett.*, **49** (1982), 1830.

62. F. F. Abraham, *Phys. Rev. B*, **29** (1984), 2906.

63. N. Greises, G. A. Held, R. L. Frahm, R. L. Greene, P. M. Horn and R. M. Suter, *Phys. Rev. Lett.*, **59** (1987), 1706.

64. S. B. Dierker, R. Pindak and R. B. Meyer, *Phys. Rev. Lett.*, **56** (1986), 1819.

65. H. Shiba, *J. Phys. Soc. Jpn.*, **48** (1980), 211.

CHAPTER THIRTEEN

Elastic and structural properties of superlattices

M. Grimsditch and I. K. Schuller

Introduction · Background · Elastic properties · Structure · Discussion · Conclusions

13.1 INTRODUCTION

This chapter is devoted to the study of elastic and structural properties of materials which are composed of many layers of two or more different materials: these materials are broadly referred to as superlattices [1, 2]. Many superlattices have potential applications and hence have received considerable attention [3, 4]. Superlattices are interesting since the cumulative effects of many interfaces enables measurements to be performed, whereas the effects produced by a single interface are often not detectable with existing probes. From a basic science viewpoint these materials provide interesting systems in which to study the effect of interfaces on fundamental properties.

Within the general definition of superlattices given above one can define sub-classes depending on the nature of the interfaces. In a strict sense the word superlattice should only be used for a system in which the atomic planes on either side of an interface are coherently stacked. Lattice matching in the plane of the film may be present as in GaAs/AlAs or completely absent as in the mismatched Nb(bcc)/Cu(fcc). An intermediate type can be fabricated from materials that alloy and hence, although they are in registry, have broad or diffuse interfaces; these are often referred to as com-

positionally modulated alloys and an example is Cu/Ni. However, the term superlattices has been more broadly used for any layered material with a periodicity larger than the interatomic spacing as is the case for the amorphous Si superlattices.

Here we will discuss the structural and elastic properties of systems that are layered at an atomic scale. This discussion includes a review of the results expected if the interfaces played a negligible role, and the behavior which is found experimentally. The latter requires a review of the different techniques which can be used to study the elastic properties of thin films and also of the experimental probes which can be used to characterize their microscopic structure. The final section is devoted to a summary of the various theoretical models which have been proposed to explain the experimental results.

13.2 BACKGROUND

Structural characteristics of materials grown on a substrate have long been a subject of intense activity [5]. In the case of metals it is found that in many cases they grow along a normal to the most densely packed atomic plane [111] for fcc, [110] for bcc and [001] for hcp. It is also known that apart

Table 13.1 Experimental techniques available to study elastic properties of thin films

Technique	Constant measured	Remove substrate	Drawbacks
Bulge tester [9]	Biaxial modulus	Yes	Warping, clamping
Brillouin scattering [10]	Surface wave (C_{44})	No	Require excellent surface
Picosecond reflectivity [11]	C_{33}	No	Smooth film–substrate interface
Vibrating reed [12]	Young's and flexural	Yes	Warping
Surface acoustic waves [13]	Surface wave (C_{44})	No	Piezoelectric substrate
			Large contribution from substrate
Vibrating membrane [14]	Biaxial modulus	Yes	Large samples
Nanoindenter [15]	?	No	Plastic deformations
Continuous ultrasonic waves [16]	Young's, biaxial, flexural and C_{66}	Yes	Tensions applied to film

from the well defined crystallographic orientation of the surface normal, there are well defined relations between in-plane crystallographic orientations of the substrate and film. These relationships depend on crystal structure, lattice constant, mismatch of atomic radii, etc. [5].

Most of the information given in the preceding paragraph has been obtained from measurements on 'thick' films. The actual microscopic arrangement of the atoms at an interface is of course not obtained from such studies. A discussion of the microscopic structure close to an interface is, in part, the object of this treatise. In this chapter we will review the elastic behavior of systems with many 'interfaces'. In many of these systems we will find that the elastic properties are not those predicted on the basis of continuum elasticity theory: from this it can be concluded that certain physical properties of matter close to an interface, which may also include the structure, are different from those in the bulk.

The effective elastic constants of a superlattice are defined as the long wavelength limit of the stress–strain relation. For a layered material (in which the interfaces are assumed to have zero thickness) they have been investigated theoretically by a number of authors [6–8]. Two important conclusions can be drawn from all these theoretical studies: the effective elastic constants depend only on the fraction of each material (not on the repeat distance) and their numerical values are averages of the elastic constants of the constituents.

Experimentally the determination of elastic properties of materials which can only be prepared in the form of thin films poses very severe difficulties. Conventional ultrasonic techniques normally used to measure elastic constants are no longer effective for samples whose thicknesses are less than the ultrasonic wavelength ($\geqslant 20\,\mu m$). Because of this, special techniques have been developed to study thin films, but in many cases these techniques are hard to implement and have led to controversial results. The most commonly used techniques to study thin films are: bulge tester [9], Brillouin scattering [10], picosecond-reflectance [11], vibrating reed [12], surface acoustic waves [13], vibrating membrane [14], nano-indenter [15], and continuous ultrasonic waves [16]. Many of these techniques are highly sophisticated and cannot be described in adequate detail here. Table 13.1 lists the above mentioned techniques; in the table we have indicated if the method requires removal of the film from the substrate, any special characteristics needed for the samples and also if there are any other specific difficulties encountered in analyzing the experimental results.

13.3 ELASTIC PROPERTIES

Table 13.2 summarizes the results which have been obtained on the elastic properties of superlattices to date. This table contains the system investigated,

Table 13.2 Superlattices in which elastic constants have been determined

System	Elastic constant	Anomaly	Correlation structure	Reference
Cu/Ni	Y_B	Yes	–	[17]
	Y, F	Yes	–	[18]
	Y_B, Y, F, C_{66}	No	–	[19]
	C_{44}	No	–	[20]
Cu/Pd	Y_B	Yes	–	[9]
	Y_B, Y, F, C_{66}	No	–	[16]
	C_{44}	Yes	–	[21]
Mo/Ni	C_{44}	Yes	Yes	[22]
	C_{33}	Yes	Yes	[11]
Pt/Ni	C_{33}	Yes	Yes	[11]
Ti/Ni	C_{33}	Yes	Yes	[11]
Cu/Nb	C_{44}	Yes	Yes	[23, 24]
	C_{44}, C_{33}	Yes		[25]
	Y_B	?	Yes	[26]
	C_{12}	?	?	[27]
NbN/AlN	C_{44}	No	Yes	[28]
GaAs/AlAs	C_{33}, C_{44}	No	Yes	[29]
Nb/Si	C_{44}	Yes	Yes	[30]
Au/Cr	C_{44}	Yes	Yes	[31]
Ag/Pd	Y_B	Yes	–	[32]
	C_{11}, C_{44}	Yes	–	[60]
Au/Ni	Y_B	Yes	–	[9]
Cu/Au	Y_B	No	–	[32]
Cu/Al	?	Yes	–	[33]
V/Ni	C_{44}	Yes	Yes	[13]
Fe/Pd	C_{44}, C_{11}?	Yes	–	[34]
Co/Ag	C_{44}	Yes	Yes	[35, 36]
Mo/Ta	C_{44}	Yes	Yes	[37]
	C_{33}	No	No	[37]
Co/Cu	C_{44}	No	–	[38]
Fe/Cu	C_{44}	Yes	Yes	[39]
ZrN/AlN	C_{33}	Yes	–	[40]

elastic constant measured, presence or absence of an anomaly, and if concomitant changes were observed in structural data. In this table 'anomaly' indicates an elastic property which depends on the modulation wavelength of the superlattice and which is therefore not predicted by the continuum theories of Refs. [6–8]. In spite of some discrepancies between the results of different experiments on the same system (e.g. Cu/Ni) a careful analysis of all the data in Table 13.2 shows that the elastic properties of many superlattices are indeed 'anomalous'. Figure 13.1(a,b) shows the results for a shear [22] and a compressional [11] elastic

constant of Mo/Ni superlattices which were prepared by different groups and investigated with different techniques. Figure 13.2 shows results for identical V/Ni superlattices [13] which were investigated by two different measurement techniques; there is a clear cut agreement between measurements performed in different laboratories under well controlled experimental conditions. The most common feature which has been observed in superlattices is a softening of a shear modulus as seen in Figs. 13.1 and 13.2. One notable exception to this rule is Au/Cr [31] which shows a hardening of a shear constant as shown in Fig. 13.3. Another typical behavior is shown in Fig. 13.4 for Cu/Ni. Although many systems studied using the bulge tester have been claimed to show an anomaly, this type of measurement has been questioned regarding the data analysis [41]. It is interesting to note that the list of materials showing anomalous behavior in Table 13.2 show no correlation with the crystal structure of the constituents.

The existence of anomalous behavior in a particular superlattice system is irrefutable evidence that the continuum theory is inadequate. Since the two basic assumptions of the theory are: (i) bonding between the layers, and (ii) each layer retains its **bulk** properties, and since there is no experimental evidence for lack of bonding between most metals, it is reasonable to conclude that the individual layers no longer maintain their bulk elastic properties. Note however that it is impossible to conclude from the elastic behavior alone whether the changes are confined to the interface region or whether they extend substantially into the bulk of each layer.

13.4 STRUCTURE

In order to establish if the changes in elastic properties are due to structural effects it is imperative to obtain detailed and accurate structural information. This information can in principle be obtained using electron microscopy, low energy electron diffraction (LEED), high energy electron diffraction (HEED), extended X-ray absorption fine structure (EXAFS), and conventional X-ray

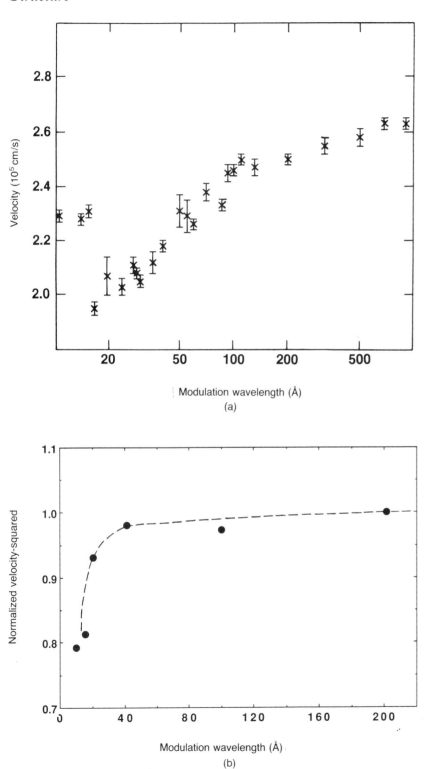

Fig. 13.1 Elastic properties of Mo/Ni superlattices fabricated and measured by two independent research groups. (a) shows the surface wave velocity [22]; (b) shows the longitudinal velocity perpendicular to the layers [11].

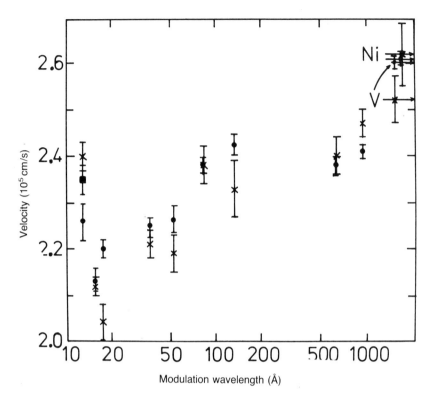

Fig. 13.2 Surface acoustic wave velocity measured on the same V/Ni films using two different techniques: Brillouin scattering and surface acoustic wave (SAW) techniques [13].

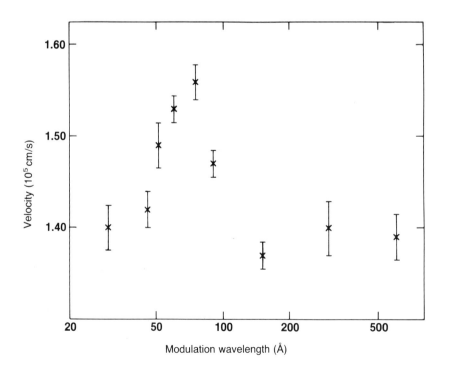

Fig. 13.3 Surface wave velocity in Au/Cr superlattices showing an unusual increase in the velocity as the modulation wavelength is reduced [31].

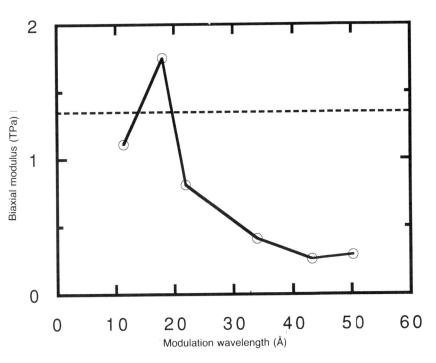

Fig. 13.4 Biaxial modulus of Cu/Ni films measured with a bulge tester [17].

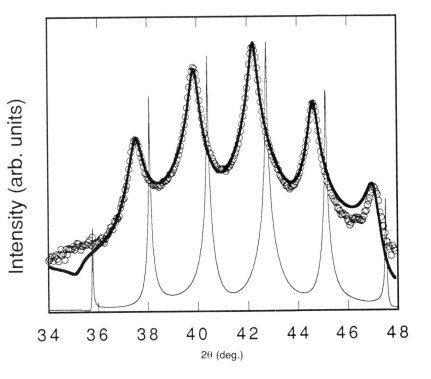

Fig. 13.5 Experimental (dots) and calculated (full line) X-ray spectra of a 40 Å Mo/Ni superlattice. The thin line is a calculation using bulk lattice constants for both Mo and Ni; the thick line is a fit with a model containing seven fit parameters.

scattering. Although all these techniques have been applied to superlattices, most of them yield only qualitative information. Almost all quantitative information has been obtained from X-ray scattering experiments but even these investigations usually only provide data pertaining to the structure perpendicular to the layers. In spite of the fact that X-ray spectra yield quantitative results related to the structure of a given superlattice, extracting physical parameters from the data is non-trivial since it requires extensive modeling. Figure 13.5 shows an X-ray spectrum obtained from a Mo/Ni superlattice (dots), the full line is a fit obtained using a nonlinear optimization technique on a model with seven adjustable parameters [42]. Although it is clear that excellent agreement can be obtained between model and experiment, it can always be argued that the particular parameters determined in the fit are not reliable. A determination of the reliability of X-ray diffraction fits using nonlinear optimization techniques requires the study of a large number of systems and comparison with other structural studies. These type of studies are presently underway and indicate that a full crystallographic characterization of superlattices may indeed be within reach soon. The single parameter to which the above arguments do not apply is the average lattice constant perpendicular to the layering [31]; this can be obtained from the experimental X-ray data without resorting to a specific model.

Figure 13.6 shows the average lattice constant [24] of Nb/Cu superlattices as a function of modulation wavelength, and also the measured shear elastic constant [23]. The strong correlation between the two measurements indicates that the two effects are intimately related. A molecular dynamics study [43] has indeed shown that, given the expansion of the lattice, the elastic softening observed in Mo/Ni can be predicted quantitatively. Table 13.2 lists all the systems in which a correlation has been found between structural and elastic properties. We are not aware of a single system for which both detailed X-rays and elastic properties have been measured, that do not exhibit such a correlation.

From the preceding paragraph it appears likely that both the elastic anomalies and the changes in lattice constant have the same origin. The simple picture claiming that defects introduced during the fabrication are responsible for the elastic anomalies, appears to be in contradiction with the results obtained on ion irradiated Ag/Co superlattices [35, 36]. We recall that in most materials [44–47] damage introduced during ion irradiation leads to a softening of the elastic constants; in Ag/Co superlattices, on the other hand, the softening observed as a function of Λ disappears when the samples are irradiated, i.e. the elastic constant hardens under irradiation. This fact, together with the scaling of the effect as the ratio of the constituents is changed [22], are two of the more intricate aspects of this strange anomalous elastic behavior of superlattices which must be addressed by theories attempting to explain the effect.

13.5 DISCUSSION

There have been a number of different explanations [48–57] proposed to account for the anomalous elastic behavior as well as some calculations which rule out certain approaches [58, 59]. The most severe difficulty encountered when attempting to identify which (if any) of the proposed mechanisms is correct, is the lack of specific predictions for a given superlattice by **any** of the models. Without exception, all explanations use the experimentally measured elastic anomaly to infer the presence of the driving mechanism (e.g. folding of electronic bands, coherency strains, grain boundaries (GBs), and surface tensions) depending on the model.

It is unfair however to place all the proposed explanations at the same level of sophistication. The GB explanation [52–55] is perhaps the most complete at the present time and is presented in detail in the next chapter. It has been used to calculate quantitative changes in the C_{ij} for specific GBs and has been successful in accounting for large decreases in C_{44} and small increases in the biaxial modulus (Y_B): it has not yet explained the increases in C_{44} such as those found in Au/Cr [31] and

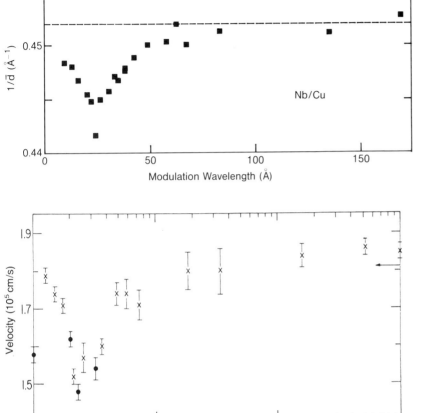

Fig. 13.6 The average lattice constant of Nb/Cu superlattices as a function of modulation wavelength is shown in the upper portion. The lower portion shows the surface wave velocity measured in the same films.

Nb/Si [30]. The electron transfer model [50] predicted that no anomaly should be found in superlattices in which at least one constituent was a non-metal: the results on ZrN/AlN [40] contradict this expectation. Since the prediction of other models have been less quantitative, there is no evident discrepancy between them and experiment.

13.6 CONCLUSIONS

In spite of the fact that individual reports of elastic anomalies in superlattices may be questionable, there is sufficient experimental evidence to show clearly that elastic properties of superlattices are anomalous, and also that elastic anomalies are invariably associated with changes in lattice constants. At present, information on the structural aspects is incomplete and we believe that it is in this area that the most concentrated effort is required. It appears that little is to be gained from studies where only the presence or absence of an anomaly is investigated; investigations must include a comprehensive structural characterization.

A number of explanations have been proposed to account for the elastic anomalies. Due to a combination of incomplete experimental results and the lack of predictive power of the models, it is not possible at present to identify uniquely the correct explanation.

This work has been supported by the Office of Naval Research Grant No. N00014-88-K-0210 and the US Department of Energy, BES-Materials Sciences, under contract No. W-31-109-ENG-38.

REFERENCES

1. For an early book see for instance *Synthetically Modulated Structures*, (eds L. L. Chang and B. C. Giessen), Academic Press, New York (1985).
2. See for instance *Interfaces, Superlattices and Thin Films* (eds J. D. Dow and I. K. Schuller), Materials Research Society, Vol. 77, Pittsburgh (1987).
3. For a recent book see for instance *Physics, Fabrication and Applications of Multilayered Structures* (eds C. Weisbuch and P. Dhez), Plenum Publishing Co, NY (1988).
4. See for instance *X-Ray Multilayered Optics* (ed G. F. Marshall), Optical Engineering SPIE, Vol. 25, Bellingham (1986).
5. See for instance J. A. Venables, G. D. T. Spiller and M. Hanbuchen *Rep. Prog. Phys.*, **7** (1984), 399; J. H. van der Merwe and M. W. H. Braun, *Appl., Surf. Sci.*, **22/23** (1985), 545; L. A. Bruce and J. Jaeger, *Phil. Mag. A*, **38** (1978), 223 and references therein.
6. S. M. Rytov, *Akust. Zh.*, **2** (1956), 71 [*Sov. Phys.-Acoust.*, **2** (1956), 68].
7. J. Sapriel, B. Djafari-Rouhani and L. Dobrzynski, *Surf. Sci.*, **126** (1983), 197.
8. M. Grimsditch, *Phys. Rev. B*, **31** (1985), 6818 and M. Grimsditch and F. Nizzoli, *Phys. Rev. B*, **33** (1986), 5891.
9. W. M. C. Yang, T. Tsakalakos and J. E. Hilliard, *J. Appl. Phys.*, **48** (1977), 876.
10. J. Sandercock, in *Topics in Applied Physics*, Vol. 51, *Light Scattering in Solids III* (eds M. Cardona and G. Guntherodt), Springer, NY (1982), p. 173.
11. B. Clemens and G. Eesley, *Phys. Rev. Lett.*, **61** (1988), 2356.
12. M. Barmatz, L. R. Testardi and F. J. DiSalvo, *Phys. Rev. B*, **12** (1975), 4367 and B. S. Berry and W. C. Pritchet, *IBM J. Res. Develop.*, **19** (1975), 334.
13. R. Danner, R. P. Huebener, C. S. L. Chun, M. Grimsditch and Ivan K. Schuller, *Phys. Rev. B*, **33** (1986), 3696.
14. A. Fartash, I. K. Schuller and M. Grimsditch, *Appl. Phys. Lett.*, **55** (1990), 2614.
15. R. J. Bower, *Appl. Phys. Lett.*, **23** (1973), 99 and C. Y. Ting and B. L. Crowder, *J. Electrochem. Soc.*, **129** (1982), 2590.

16. A. Moreau, J. B. Ketterson and B. C. Davis, *J. Appl. Phys.*, **68** (1990), 1622.
17. T. Tsakalakos and J. E. Hilliard, *J. Appl. Phys.*, **54** (1982), 734.
18. L. R. Testardi, R. M. Willens, J. T. Krause, D. B. McWhan and S. Nakahara, *J. Appl. Phys.*, **52** (1981), 510.
19. A. Moreau, J. B. Ketterson and J. Mattson, *Appl. Phys. Lett.*, **56** (1990), 1959.
20. J. Mattson, R. Bhadra, J. B. Ketterson, M. B. Brodsky and M. Grimsditch, *J. Appl. Phys.*, **67** (1990), 2873.
21. J. R. Dutcher, S. Lee, J. Kim, G. Stegeman and C. M. Falco, *Proc. Mater. Res. Soc.*, **160** (1990), 179; *J. Mater. Sci. Eng. A*, **127** (in press); and *J. Mater. Sci. Eng. B*, **6** (1990), 199.
22. R. Khan, C. S. L. Chun, G. P. Felcher, M. Grimsditch, A. Kueny, C. M. Falco and Ivan K. Schuller, *Phys. Rev. B*, **27** (1983), 7186.
23. A. Kueny, M. Grimsditch, K. Miyano, I. Banerjee, C. M. Falco and I. K. Schuller, *Phys. Rev. Lett.*, **48** (1982), 166.
24. I. K. Schuller and M. Grimsditch, *J. Vac. Sci. Technol. B*, **4** (1986), 1444.
25. J. A. Bell, W. R. Bennett, R. Zanoni, G. I. Stegeman, C. M. Falco and C. T. Seaton, *Sol. St. Comm.*, **64** (1987), 1339.
26. A. Fartash, E. E. Fullerton, I. K. Schuller, S. E. Bobin, J. W. Wagner, R. C. Cammarata, S. Kumar and M. Grimsditch, *Phys. Rev. B*, **44** (1991), 13760.
27. J. A. Bell, R. J. Zanoni, C. T. Seaton, G. I. Stegeman and C. M. Falco, *Appl. Phys. Lett.*, **51** (1987), 652.
28. R. Bhadra, M. Grimsditch, J. Murduck and I. K. Schuller, *Appl. Phys. Lett.*, **54** (1989), 1409.
29. M. Grimsditch, R. Bhadra, I. K. Schuller, F. Chambers and G. Devane, *Phys. Rev. B*, **42** (1990), 2923.
30. E. Fullerton, I. K. Schuller and M. Grimsditch (to be published).
31. P. Bisanti, M. B. Brodsky, G. P. Felcher, M. Grimsditch and L. R. Sill, *Phys. Rev. B*, **35** (1987), 7813.
32. G. E. Henein and J. E. Hilliard, *J. Appl. Phys.*, **54** (1983), 728.
33. V. S. Kopan and A. V. Lysenko, *Fiz. Metal. Metalloved.*, **29** (1970), 183.
34. P. Baumgart, B. Hillebrands, R. Mock and G. Guntherodt, *Phys. Rev. B*, **34** (1986), 9004.
35. S. M. Hues, R. Bhadra, M. Grimsditch, E. Fullerton and Ivan K. Schuller, *Phys. Rev. B*, **39** (1989), 12966.
36. E. E. Fullerton, I. K. Schuller, R. Bhadra, M. Grimsditch and S. M. Hues, *J. Mater. Sci. Eng. A*, **126** (1990), 19.
37. J. A. Bell, W. R. Bennett, R. Zanoni, G. I. Stegeman, C. M. Falco and F. Nizzoli, *Phys. Rev. B*, **35** (1987), 4127.
38. J. R. Dutcher, S.-M. Lee, C. O. England, G. I. Stegeman, and C. M. Falco, *J. Mater. Sci. A*, **126** (1990), 13.

39. E. E. Fullerton, I. K. Schuller, F. T. Parker III, K. A. Svinarich, G. L. Eesley, R. Bhadra and M. Grimsditch (to be published).

40. W. J. Meng, G. L. Eesley and K. A. Svinarich, *Phys. Rev. B*, **42** (1990), 4881.

41. H. Itozaki, Thesis, Northwestern University, 1982 (unpublished).

42. E. E. Fullerton, I. K. Schuller, H. Vanderstraeten, and Y. Bruynseraede, *Phys. Rev. B*, **42** (1992).

43. I. K. Schuller and A. Rahman, *Phys. Rev. Lett.*, **50** (1983), 1377.

44. M. Grimsditch, K. E. Gray, R. Bhadra, R. T. Kampwirth and L. Rehn, *Phys. Rev. B*, **35** (1987), 833.

45. R. Bhadra, J. Pearson, P. Okamoto, L. Rehn and M. Grimsditch, *Phys. Rev. B*, **38** (1988), 12656.

46. L. Rehn, P. Okamoto, J. Pearson, R. Bhadra and M. Grimsditch, *Phys. Rev. Lett.*, **59** (1987), 2987.

47. R. P. Sharma, R. Bhadra, L. E. Rehn, P. M. Baldo and M. Grimsditch, *J. Appl. Phys.*, **66** (1989), 152.

48. T. B. Wu, *J. Appl. Phys.*, **53** (1982), 5265.

49. W. E. Pickett, *J. Phys. F.*, **12** (1982), 2195.

50. A. F. Jankowski and T. Tsakalakos, *J. Phys. F.*, **15** (1985), 1279.

51. A. F. Jankowski, *J. Phys. F.*, **18** (1988), 413.

52. D. Wolf and J. F. Lutsko, *Phys. Rev. Lett.*, **60** (1988), 1170.

53. D. Wolf and J. F. Lutsko, *J. Appl. Phys.*, **66** (1989), 1961.

54. D. Wolf and J. F. Lutsko, *J. Mater. Res.*, **4** (1989), 1427.

55. D. Wolf, *Surf. Sci.*, **225** (1990), 117.

56. M. L. Hueberman and M. Grimsditch, *Phys. Rev. Lett.*, **66** (1989), 1403.

57. R. C. Cammarata and K. Sieradzki, *Phys. Rev. Lett.*, **62** (1989), 2005.

58. A. Banerjea and J. R. Smith, *Phys. Rev. B*, **35** (1987), 5413.

59. B. W. Dodson, *Phys. Rev. B*, **37** (1988), 727.

60. J. R. Dutcher, S. Lee, J. Kim, G. Stegeman and C. M. Falco, *Phys. Rev. Lett.*, **65** (1990), 1231.

CHAPTER FOURTEEN

Computer simulation of the elastic behavior of thin films and superlattices

D. Wolf and J. A. Jaszczak

Introduction · Simulation concepts and techniques · Thin films · Composition-modulated superlattices · Summary and conclusions

14.1 INTRODUCTION

The importance of interface materials is based largely on their inherent inhomogeneity, i.e. the fact that the chemical composition and physical properties at or near an interface can differ dramatically from those of the nearby bulk material. For example, the propagation of a crack along an interface – rather than through the surrounding bulk material – indicates a different mechanical strength near the interface. Also, the elastic response and thermal behavior near an interface can be highly anisotropic in an otherwise isotropic material, and can differ by orders of magnitude from those of the adjacent bulk regions. Typically these gradients extend over only a few atomic distances.

Because the properties in the interfacial region are controlled by relatively few atoms, the inherent difficulty in the experimental investigation of buried interfaces is actually an advantage in the atomic-level study of solid interfaces by means of computer-simulation techniques. While the limitations of such simulations are well known (pertaining mostly to the finite size of the simulation cell, its proper embedding in a suitable

environment, and to the atomic-level description of the electronic-structure-based interactions between the atoms), in this chapter we hope to demonstrate the unique insights which they can provide on the elastic behavior of both buried and thin-film interfaces.

The elastic response of materials is known to be particularly strongly affected by small changes in the volume, or in the underlying interatomic distances. For example, the thermal expansion of most materials, typically only a few percent between absolute zero and melting, causes an elastic softening typically by about 50%. As illustrated below, the changes in interatomic distances due to the presence of interfaces may be considerably larger by comparison, suggesting that the elastic response near an interface may differ dramatically from that of the nearby bulk perfect-crystal material even at zero temperature.

The supermodulus effect in composition modulated strained-layer superlattices (see the preceding chapter by Grimsditch and Schuller) offers a glimpse of the different elastic moduli of interface materials. The discovery of this effect [1] and the intensive investigation in recent years of anomalies in the elastic response of multilayer film

[2, 3] has raised hopes that one day it may be possible to develop synthetic interface materials for engineering applications with elastic and mechanical properties not otherwise achievable in bulk materials. While in some instances an elastic softening of multilayer films has been reported [2–4], in other instances a stiffening has been observed [1, 5–9]. In spite of much controversy about the very existence and physical nature of the supermodulus effect, it represents one of the motivating factors for the fundamental investigation of interface elasticity (or, generally, the elastic behavior of inhomogeneous systems), as does the desire to gain insight into the local nature of crack-tip processes in grain boundaries (GBs) and composite interfaces [10–13].

The most common form of structural disorder in a crystal lattice is due to the thermal movements of the atoms or molecules, usually giving rise to thermal expansion. As is well known, this **homogeneous** type of disorder, and the consequent volume increase, originates in the anharmonicity of the interactions between the atoms. Owing to the presence of planar defects, interface materials are structurally disordered even at zero temperature; however, because of its localization near the interface, this type of disorder is **inhomogeneous**. The effect on the volume is nevertheless similar, usually giving rise to a volume expansion locally at the interface. In a way, volume expansion (or the underlying stresses) may thus be viewed as a measure of the amount of structural disorder in the system, both homogeneous and inhomogeneous. Another measure is the interface energy which, in the case of GBs in metals, has been shown to be directly related to the local volume expansion at these interfaces [14–16] (see also Chapter 3 by Wolf and Merkle in this volume).

A direct consequence of the local structural disorder at an interface is the existence of interfacial stresses localized near the plane of the interface. (The out-of-plane component of this stress is responsible for the volume expansion just mentioned.) The stress in a bulk-free surface is an example of such an interfacial stress. However, in bulk interface systems (such as GBs or bulk free surfaces; see the Editors' Introduction) these interfacial stresses can only be relaxed by atomic reconstruction at the interface because of its attachment to bulk perfect-crystal material. By contrast, a thin-film interface system (such as a multilayer material or a free-standing thin film) may in addition contract, giving rise to a uniform reduction in the average lattice parameter(s) in the interface plane, with a consequent Poisson expansion in the direction of the interface normal.

The usual interface-stress induced decrease in the average atomic volume has been suggested to be the cause of a strengthening of at least some elastic moduli of thin films and superlattices [17–21], as one would expect for a homogeneous crystalline solid. However, as illustrated in this chapter, because of the complex, highly non-linear interplay between the localized atomic-level structural disorder, the stresses it gives rise to, and the consequent anisotropic lattice-parameter changes parallel and perpendicular to the interface, physical intuition gained from the investigation of homogeneous systems cannot usually be transferred to interface materials.

For example, in dissimilar-material superlattices, electronic causes due to the different electronic properties of the components have been invoked to explain the observed elastic anomalies and changes in average lattice parameters [22, 23]. In one [22], a finite-size effect giving rise to a folding back of the Brillouin zone is assumed to be responsible. In the other [23], the different electronic properties of the constituents (such as band structure, Fermi energy, work function, etc.) are assumed to produce strains in the direction of the interface normal (z-direction) which are supposedly distributed homogeneously throughout the bulk of the multilayer film. Both models thus assume the anomalies to be due to electronic effects. However, recent experimental evidence [4, 24] strongly suggests that the lattice-parameter changes in the z-direction are localized at the interfaces and that the elastic behavior is governed by the presence of the interfaces [4]. Moreover, elastic anomalies have also been observed in nanocrystalline metals [25] in which arguments based on electronic differences between the constituents of the superlattice cannot be invoked.

The recent observation [25] of elastic moduli of nanocrystalline Mg and Pd which are nearly the same as those of bulk polycrystals must be considered a surprise in view of the lower density of these materials, with a large fraction of atoms situated in or near GBs.

It may be helpful from the outset to clarify the terminology used throughout. Based on the distinction between the physical origin of interface stresses and their effect on the structure and elastic behavior, we distinguish two fundamentally different aspects of the structure of interface materials, namely the **atomic structure** and the stress-induced changes in the **average lattice parameters** parallel and perpendicular to the interface plane.

As is well known, at the atomic level structural disorder is best characterized by the radial distribution function, $G(r)$. For example, the homogeneous thermal disorder in an otherwise perfect crystal gives rise to two effects in $G(r)$: First, the δ-function like zero-temperature peaks associated with the shells of nearest-, second-nearest and more distant neighbors are broadened; second, because of the volume expansion the peak centers are shifted towards larger distances. As illustrated in Fig. 14.1, the inhomogeneous structural disorder at a solid interface (in this case a GB) gives rise to the same two effects even at zero temperature. Figures 14.1(a) and (b) show the computed zero-temperature radial distribution functions, $r^2 G(r)$, for atoms in the planes nearest and next-nearest to a high-angle twist boundary on the (001) plane of an fcc metal [14]. While in the plane closest to the GB the perfect-crystal δ-function peak structure (indicated by the arrows in Fig. 14.1(a)) has been replaced by a broad distribution of interatomic distances, in the second-closest plane the ideal-crystal peaks have largely been recovered (see Fig. 14.1(b)), illustrating the highly localized nature of the atomic-level structural disorder at the interface. Figure 14.1(a) also illustrates the cause for the expansion locally at the GB: The distances to the left of the arrows represent atoms shoved more closely together than in the perfect crystal; because of the anharmonicity in the interatomic interactions, these atoms repel each other par-

ticularly strongly, resulting in a local expansion at the interface.

By contrast with the highly inhomogeneous atomic-level structural disorder, the stress-induced changes in lattice parameters are homogeneous in nature, albeit anisotropic. One can therefore anticipate rather different effects on the elastic behavior due to the atomic-level structural disorder, on one hand, and the consequent lattice-parameter changes on the other, a basic theme we will address throughout this entire chapter. Such an investigation of simple model interface systems thus provides an opportunity to study the interplay between the inhomogeneous effects of the local disorder at the interfaces and the homogeneous effects due to the resulting volume change.

In an attempt to elucidate the physical origin(s) of the elastic anomalies observed in interface materials, in the present chapter we ignore electronic-structure effects altogether and, instead, focus exclusively on the intrinsic role of the **structural disorder** due to the presence of interfaces. Following a review in section 14.2 of the simulation concepts and techniques applied throughout this chapter, the effects due to stress-induced lattice-parameter changes in thin-film materials are considered for the 'simple' case of free-standing, unsupported thin films (section 14.3). Arranging the thin films periodically, separated however by GBs, leads to a superlattice of thin films (sometimes also referred to as a GB superlattice [26]), the structure and elastic behavior of which are also briefly discussed in section 14.3. Finally, dissimilar-material superlattices are investigated in section 14.4, with particular emphasis on the role of atomic-level coherency at the interfaces.

Since there are as yet only preliminary simulation studies of the thermo-elastically coupled effects of structural and thermal disorder on the elastic behavior of interfacial materials [27], the emphasis in this chapter will be on the zero-temperature elastic behavior of interface materials. One can anticipate that the effects of increasing temperature will significantly affect not only the elastic but also the thermodynamic properties near the interface. As is well known, this thermo-elastic

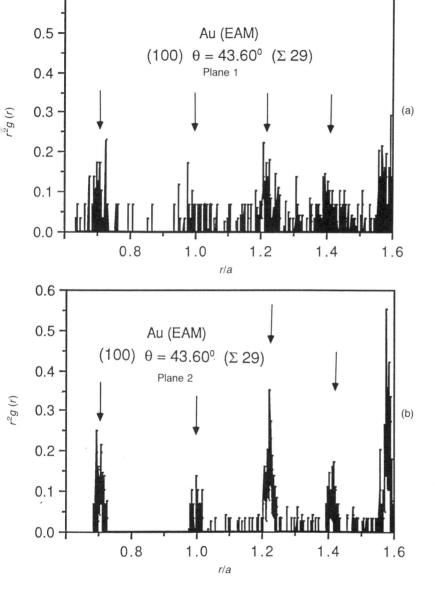

Fig. 14.1 Radial distribution function, $r^2G(r)$, for the two planes nearest to a (001) $\theta = 43.60$ (so-called $\Sigma 29$) twist GB as described by an embedded-atom-method potential for Au discussed in section 14.3.1. The arrows indicate the corresponding perfect-crystal δ-function like peak positions at zero temperature. Whereas atoms in the plane nearest to the interface (a) are very strongly affected by the the presence of the interface, the atoms in the second-nearest plane (b) have an environment much closer to that of an ideal crystal.

oupling is described by the Grüneisen relation according to which the thermal expansion of a **homogeneous** material is inversely proportional to its bulk modulus. In an **inhomogeneous** system one would similarly expect the **local** thermal-expansion tensor to be governed by the **local** elastic-constant tensor, with a consequently very different, highly anisotropic thermal expansion near the interface.

Throughout this study we will consider extremely simple model systems whose properties may not be those of real interface materials. In particular, while in our simulations ideal defect-free interfaces will be considered, point defects and, in particular, defect clusters due, for example, to impurity segregation or the presence of voids, might be important in real materials. We nevertheless hope to demonstrate that an investigation of relatively simple model interface systems provides an opportunity to study the interplay between the inhomogeneous effects of the local structural disorder at the interfaces and the homogeneous effects due to the resulting changes in the lattice parameters, i.e. in the average atomic volume. We also hope to demonstrate that the capability of computer simulations to probe a system in ways not possible experimentally can provide unique insights into the underlying atomic-level causes. In the investigation of interface elasticity, the distinct effects of structural disorder and accompanying volume changes can thus be separated.

14.2 SIMULATION CONCEPTS AND TECHNIQUES

Although much molecular-dynamics simulation work has been performed on the finite-temperature elastic behavior of homogeneous crystals [28], to date a similar investigation of inhomogeneous systems is not available. Throughout this chapter we will therefore limit ourselves essentially to discussing the zero-temperature elastic behavior of the interfacial materials. As described in detail below, the elastic constants will be determined by first relaxing the structure of the system under zero external stress and subsequently evaluating the elastic-constant and elastic-compliance tensors, from which all elastic moduli of interest can be extracted [29]. For a given interatomic potential (section 14.2.1), the constant-stress relaxation algorithm provides the lattice parameters a_x and a_y parallel to the interface (x–y) plane, and a_z in the direction of the interface normal (defining the z-direction).

14.2.1 Interatomic potentials

The interatomic potentials used here come from two sources. To provide some insight into the role

of many-body effects, throughout this study both semi-empirical many-body potential and a conceptually simpler pair potential will be employed. Many-body potentials, known as embedded-atom method (EAM) [30] or Finnis–Sinclair potential [31], have the advantage over pair potentials of incorporating, at least conceptually, the many-body nature of metallic bonding, while being relatively efficient computationally. In these potentials the strength of the interaction between atoms depends on the local volume or, in another interpretation, on the local electron density 'sensed' by every atom [31]. An important difference between pair- and many-body potentials lies in the fact that while an equilibrium pair potential automatically satisfies the Cauchy relation for the elastic constants, $C_{12} = C_{44}$, the many-body potentials permit all three elastic constants of a cubic metal to be determined (or used in the fitting).

Both the EAM and Finnis–Sinclair types of many-body potentials use the same mathematical expression as starting point, in which the total energy, U, of a system of N interacting metal atoms is subdivided into a bonding term (based on band structure considerations) and a repulsive central force term, according to [30, 31]

$$U = -\sum_i F(\rho_i) + \sum_i \sum_{j>i} \phi(r_{ij}) \qquad (14.1)$$

Here $F(\rho_i)$ is the many-body energy gained when embedding atom i in the charge density,

$$\rho_i = \sum_j \rho_{ij}(r_{ij}) \qquad (14.2)$$

due to the surrounding atoms, while $\phi(r_{ij})$ is the central-force repulsive energy between atoms i and j, with $r_{ij} = |r_i - r_j|$.

The EAM potential used below was fitted empirically to five properties of Au [32]. Formally eq (14.1) contains a pair potential, $\phi(r)$, as the special case in which $F(\rho)$ vanishes identically. The simplest and best-known of all pair potentials is probably the Lennard–Jones (LJ) potential, with only two adjustable parameters, σ and ε, defining the length and energy scales, respectively. Although the LJ potential was fitted to the lattice parameter and melting point of Cu (with $\varepsilon =$

0.167 eV and σ = 2.315 Å [33]), it may be viewed as a generic potential for an fcc material if all energies and distances are expressed in units of ε and σ, respectively.

To avoid discontinuities in the energy and forces (and, in the case of the many-body potentials, the charge density), both potentials were shifted smoothly to zero at the cutoff radius (R_c/a = 1.32 1.49) for the EAM (LJ) potential). The zero-temperature perfect-crystal lattice parameters, a, for these potentials were determined to be 4.0828 Å (EAM) and 3.6160 Å (LJ). In the principal cubic coordinate system (with x, y, z ∥ $\langle 100 \rangle$), the corresponding Young's moduli, Y_o, were found to be 0.346 × 10^{12} (EAM) and 1.08 × 10^{12} (LJ) dyn/cm², whereas the related shear moduli, G_o, are 0.440 × 10^{12} (EAM) and 1.01 × 10^{12} (LJ) dyn/cm². Finally, the Poisson ratios for these potentials are 0.465 (EAM) and 0.360 (LJ), and the biaxial moduli (for isotropic stretching of the x–y plane) are 0.647 × 10^{12} (EAM) and 1.681 × 10^{12} (LJ) dyn/cm².

Although the Cu(LJ) and Au(EAM) potentials are parameterized to describe specific fcc metals, it is not the aim of this study to compare the elastic properties of these particular interface materials. Rather, it is intended that by employing two very different potentials, generic features of the elastic behavior of model interface systems can be separated from effects depending on the interatomic potential. Such a comparison will also enable us to assess whether, and to what degree, the explicit local volume dependence of the EAM potential, and the fact that it does not satisfy the Cauchy relation, affect the local elastic behavior near a GB.

For later reference, in Fig. 14.2 the lattice-parameter dependences of the perfect-crystal cohesive energy for the two potentials are compared. To emphasize the shape differences in these curves, the corresponding equilibrium (i.e. minimum) values of the cohesive energies were subtracted. Later it will be seen that these shape differences are the cause for some of the differences observed for the pair- and many-body potentials. In particular, it will be seen that because of its much stronger volume dependence, the EAM potential gives rise to much larger interfacial

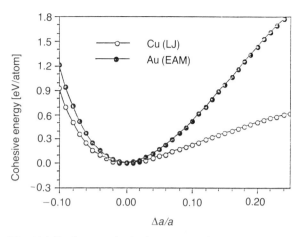

Fig. 14.2 Perfect-crystal cohesive energy against deviation, Δa, of the lattice parameter from its equilibrium value for the two fcc potentials employed in this study. To better illustrate the shape differences between the two curves, the cohesive energies at the equilibrium lattice parameter [−1.038 eV (LJ) and −3.901 eV (EAM)] were subtracted.

stresses; the consequently much larger lattice-parameter changes in thin-film systems (in which these stresses can relax) will be shown to be the major cause for any differences obtained for the two potentials.

14.2.2 Atomic structure and the effects of interfacial stresses

An iterative energy-minimization algorithm ('lattice statics') is used to compute the fully relaxed zero-temperature atomic structure and energy of the interfacial system [14, 34]. While the atoms relax, it is important to simultaneously relax the three microscopic degrees of freedom (DOFs) of the interface associated with translations, $T = (T_x, T_y, T_z)$, parallel and perpendicular to the interface [35]. By computing the forces across the interface(s), translations parallel to the interface plane (T_x, T_y) are thus permitted while the atoms relax. Also, to enable the interface(s) to expand or contract (related to T_z), the unit-cell size in the direction of the interface normal is allowed to increase or decrease in response to the internal stress [14]. By starting from a variety of initial rigid-body translational configurations, the interface energy may thus be minimized with respect to both the atomic positions and the three microscopic DOFs.

An inherent difficulty in the atomistic modelling of inhomogeneous systems is concerned with the formulation of proper border conditions so that the physical response of the surroundings to interfacial behavior (such as stresses, volume expansion, etc.) is treated realistically while keeping the size of the simulation cell (containing the interface(s)) to a minimum. In simulation studies of bulk interfaces, the embedding of the interface between bulk material is accomplished by surrounding the interfacial region by two semi-infinite bulk ideal-crystal blocks [34, 36]. Because of this bulk embedding, tensile stresses usually present in both bulk interfaces and free surfaces cannot relax, and the lattice parameter in the x–y interface plane is determined by that of the surrounding bulk material. When the bulk material is removed these tensile stresses can actually relax [26], resulting usually in a contraction in the GB plane. In the case of the free surface, the driving force for the contraction in the film plane is the familiar surface tension [21, 26], i.e. the stress parallel to the film plane. (In dissimilar-material superlattices, the interfacial stress is usually tensile in one constituent and compressive in the other.)

The fully relaxed 'structure' of thin films and superlattices therefore includes both the atomic-level DOFs and the changes in the average lattice parameters parallel and perpendicular to the interface. As illustrated in detail in this chapter, because of the combined – and distinct – effects of the **atomic structure** and the stress-induced **lattice-parameter changes**, the elastic behavior of thin-film materials is considerably more complex than that of bulk interfaces.

14.2.3 Simulation of elastic properties

A non-trivial conceptual problem in the atomistic calculation of elastic constants for interfacial systems arises from the fact that, due to the local variation of the elastic constants, an inhomogeneous system cannot be strained homogeneously, i.e. following the application of a homogeneous external strain or stress to the system, internal atomic

relaxations generally occur. This relaxation effect absent when deforming a simple **homogeneous** system (such as a perfect monatomic cubic crystal) greatly complicates the evaluation of the elastic constants of interfacial systems.

Another conceptual problem concerns the proper definition of **local** elastic constants for inhomogeneous systems. While in this chapter we discuss only the **average** (or 'global') elastic response of interface materials, we mention that the necessary modifications for the calculation of local elastic constants were discussed in detail by Kluge et al. [36]. By their very nature, the atomistic simulation techniques used here to study the **global** properties of interfacial materials only probe system averages of the elastic behavior. Two methods for calculating global elastic constants at $T = 0$ referred to as the lattice-dynamics (LD) and finite-strain (FS) methods, will be discussed below. Following the complete relaxation of the system, these methods permit the direct evaluation of the corresponding 6×6 average elastic-constant and elastic-compliance tensors of the system from which all elastic moduli of interest can be determined [29].

Lattice-dynamics versus finite-strain methods

We consider a system of N interacting atoms situated within a unit cell in the shape of a parallelepiped centered at the origin, whose edges are described by the vectors a, b, and c. The size and shape of the system may be conveniently described by a second-rank tensor H, whose columns are the vectors a, b, and c [37]. The atoms interact via potential-energy function $U(q_{ij})$ which, although not necessarily a pair potential, is assumed to depend only on the interparticle separations, q_{ij} and the appropriate border conditions are imposed on the system. In what follows, it will be convenient to introduce so-called reduced coordinates s_i, for each atom i, which are related to the real space atomic coordinates, q_i, by [37]

$$q_{i\alpha} = H_{\alpha\beta}s_{i\beta} \qquad (14.$$

where α and β are Cartesian indices, and a summation over β is implied. The application of homogeneous strain to the system corresponds to change in H from an initial value, H_0, to a final value of $H = H_0 + \delta H$. Following Ray and Rahman [28], the global strain and stress tensors are then given by

$$e_{\alpha\beta} = [(H_0^{-1})_{\gamma\alpha} H_{\psi\gamma} H_{\psi\rho} (H_0^{-1})_{\rho\beta} - \delta_{\alpha\beta}]/2 \tag{14.4}$$

$$t_{\alpha\beta} = 1/\Omega \sum_{i<j} (\partial U/\partial q_{ij\alpha}) q_{ij\beta} \tag{14.5}$$

where $\Omega = \det(H)$ is the volume of the system, and $q_{ij} = q_i - q_j$ is the vector joining atoms i and j. These definitions are in accord with those commonly used in continuum-elasticity theory.

The elastic-constant tensor, $C_{\alpha\beta\gamma\phi}$, is then defined as the derivative of global stress with respect to global strain,

$$C_{\alpha\beta\gamma\phi} = -(\partial t_{\alpha\beta}/\partial e_{\gamma\phi})|_{(F_i=0)} \tag{14.6}$$

where F_i is the total force acting on each atom i. The derivative in eq. (14.6) is to be carried out at vanishing constant force F_i. Physically this means that an infinitesimal, homogeneous strain is applied to the initially fully relaxed system; then, following the application of the strain, the atoms are to be allowed to again relax to positions of zero force.

Lutsko [38] has shown that the operator $(\partial/\partial e_{\alpha\beta})$ in eq. (14.6) may be broken down into the following two contributions:

$$(\partial/\partial e_{\alpha\beta})|_{(F_i=0)}$$
$$= (\partial/\partial e_{\alpha\beta})|_{(s_i)}$$
$$- (H^{-1})_{\nu\mu}(D^{-1})_{i\mu j\gamma} L^{\gamma}_{j\alpha\beta}(\partial/\partial s_{i\nu})|_{(e)} \tag{14.7}$$

where D is the force-constant matrix of the system, $D_{i\alpha j\beta} = \partial^2 U/\partial q_{i\alpha} \, \partial q_{j\beta}$, and $L^{\gamma}_{j\alpha\beta} = \partial \tau_{\alpha\beta}/\partial q_{j\gamma}$. This yields a more explicit expression for the global elastic-constant tensor [38]

$$C_{\alpha\beta\gamma\phi} = \sum_{i<j,k<l} \chi(i,j,k,l) - L^{\mu}_{i\alpha\beta}(D^{-1})_{i\mu j\nu} L^{\nu}_{j\gamma\phi}$$
$$+ (\delta_{\beta\phi}t_{\alpha\gamma} + \delta_{\beta\gamma}t_{\alpha\phi} + \delta_{\alpha\gamma}t_{\beta\phi} + \delta_{\alpha\phi}t_{\beta\gamma})/2 \tag{14.8}$$

where

$$\Omega \, \chi(i,j,k,l) = [(\partial^2 U/\partial q_{ij} \, \partial q_{kl})$$
$$- (\partial U/\partial q_{ij})\delta_{ijkl}/q_{ij}]$$
$$\times [s_{ij\alpha} \, s_{ij\beta} \, s_{kl\gamma} \, s_{kl\phi}/(q_{ij} \, q_{kl})] \tag{14.9}$$

Equation (14.9) is the well-known Born term [39], which describes the part of the elastic-constant tensor due to homogeneous deformation, while the second term in eq. (14.8) arises from the inhomogeneity of the system. The latter, known as the 'relaxation term' [38], is due to the atomic relaxations under strain in an inhomogeneous system, and is the zero-temperature limit of the stress–stress fluctuation term [40] in the calculation of elastic constants at non-zero temperature. Obviously important only for inhomogeneous systems (or for an inhomogeneous deformation of a homogeneous system), the relaxation term vanishes in the zero-termperature limit for an ideal crystal having only one atom per unit cell since, for such systems, all deformations are homogeneous. We note that eq. (14.8) is evaluated with $H = H_0$, i.e. at zero global strain.

The evaluation of the $3N \times 3N$ force-constant matrix for a system of N atoms, needed for the evaluation of the relaxation-term elastic-constant tensor by means of the above lattice-dynamics based method, is limited to relatively small systems, containing typically only several hundred atoms. Although this does not represent a problem in most systems considered here, in significantly larger systems a less complex method has to be used to evaluate either the average or the local elastic constants [36, 41]. In the so-called finite-strain (FS) method, instead of evaluating second derivatives associated with the interatomic potential, a small external strain is applied to the initially relaxed system, and the atoms are subsequently allowed to relax, subject to the applied global strain. The stresses and strains are then calculated, and the elastic-constant tensor is determined from the corresponding stress–strain curve [36, 41].

The FS method obviously corresponds more closely to an experimental situation. From a computational point of view, its main advantage over the LD method is that only the **forces** on the atoms are needed, instead of the **force constants**; because of the consequently smaller computer-storage requirement, this method can therefore be applied to larger systems. It suffers the disadvantage, however, that the full stress–strain curve has to be evaluated separately for each element of the elastic-constant tensor. Also, it cannot be used reliably when an elastic constant is rather small, because the linear portion of the related stress–strain curve may not be wide enough to extract a reliable slope. Kluge *et al.* [36] have recently presented a detailed comparison of the lattice-dynamics and finite-strain methods.

14.3 THIN FILMS

Unsupported thin films may be viewed as the basic building blocks of composition-modulated super-lattices, epitaxial systems, and even as a basic com-ponent of nanocrystalline materials. A thorough understanding of their physical properties there-fore offers hope for new insights into fundamental properties of much more complex interfacial systems.

Another motivation for studying thin films lies in the fact that they represent the simplest of all interface systems, thus providing ideal model systems to investigate the role of the surface-induced inhomogeneities. While within the film plane the material is homogeneous, the interplanar separations are distributed inhomogeneously. The atomic structure and elastic behavior of un-supported thin films are therefore of considerable theoretical interest not only in their own right as inhomogeneous – yet simple – interface systems, but also in the context of the role they play as basic building blocks of more complex interface materials. Single-crystal films, typically less than 10 nm thick, are not easily investigated exper-imentally. The use of atomic-level computer simulations hence provides an opportunity to systematically investigate the basic physical proper-ties of these basic interfacial building blocks. In this section we summarize some recent efforts at elucidating their basic structure and elastic behavior.

It has been widely recognized in recent years that the surface-stress tensor, $\sigma_{\alpha\beta}$ ($\alpha,\beta = x, y, z$), may play an important role not only in surface re-construction [42–44] but also in the elastic re-sponse of thin films and thin-film superlattices [26, 45–48]. $\sigma_{\alpha\beta}$ is defined as the variation of the specific surface energy as a function of the strain,

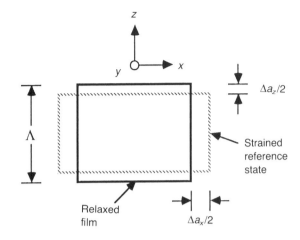

Fig. 14.3 Schematic depiction of the Poisson effect in thin films. The surface-stress-induced in-plane ($x–y$) contraction Δa_x and Δa_y (<0), are accompanied by a Poisson expansion Δa_z (>0), in the direction of the film normal (z-direction). These lattice-parameter changes are referred to a hypothetical strained reference state (dashed lines) in which only the atom positions but not the surface stresses, σ_{xx} and σ_{yy}, are relaxed (i.e. with perfect-crystal lattice parameter, a, in the film plane). Notice that, because the atom positions are fully relaxed, σ vanishes identically in both systems. For the (isotropic) (001) and (111) films, $\Delta a_x = \Delta a_y$.

$\varepsilon_{\alpha\beta}$, i.e. $\sigma_{\alpha\beta} = \delta\gamma/\delta\varepsilon_{\alpha\beta}$ [49]. In a fully relaxed surface, $\sigma_{\alpha\beta}$ is usually diagonal with a vanishing component, σ_{zz}, in the direction of the surface normal (z-direction). In many cases its only non-zero elements, σ_{xx} and σ_{yy}, are tensile (indicated by negative values), and of significant magnitude favoring contraction in the ($x–y$) plane of the surface.

While in a bulk free surface this stress can only be relaxed by reconstruction, an unsupported thin film may in addition contract, giving rise to a uni-form reduction in the average lattice parameter(s) in the film plane, and a consequent Poisson expan-sion in the z-direction (Fig. 14.3). The resulting decrease in the average atomic volume has been suggested to give rise to a strengthening of at least some elastic moduli [47, 48, 50–52], as one would expect for a homogeneous solid. As demonstrated in this section, however, in spite of the fact that the effect of the surface stress on the film's average

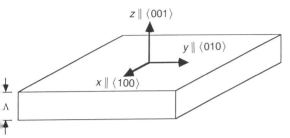

Fig. 14.4 Geometry of a thin slab of thickness Λ with $\langle 001 \rangle$ surface normal.

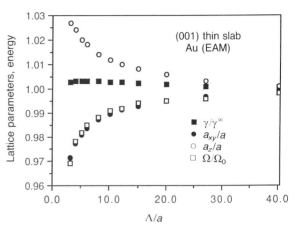

Fig. 14.5 Changes in the average lattice parameters and surface energy as a function of Λ for a thin (001) slab described by the Au(EAM) potential. a_{xy} is the lattice parameter in the plane of the film and a_z represents the average lattice parameter in the direction of the film normal. Both are normalized to the bulk perfect-crystal lattice parameter a. The volume change, Ω/Ω_0, is proportional to $a_{xy}{}^2 a_z$. γ is the energy of each of the two slab surfaces, normalized to the energy of a bulk free surface, γ^∞, listed in Table 14.1.

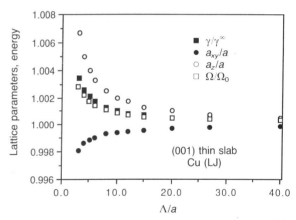

Fig. 14.6 Same as Fig. 14.5, however, for the Cu(LJ) potential (for details see Fig. 14.5).

lattice parameters (and hence the increase in their average density) can be accounted for by linear elasticity theory, due to their inhomogeneity in the z-direction unsupported thin films generally do not exhibit strengthened elastic moduli.

To illustrate the relationships between the atomic structure of the film surfaces, the average lattice parameters, and the elastic behavior, we first compare in some detail the behavior of films with $\langle 001 \rangle$ surface normal as predicted by both a pair- and a many-body potential [45]. This in-depth discussion of (001) films is followed by a comparison with similar simulations for (111) and (011) films [53].

14.3.1 Structure of (001) films

Using the two conceptually rather different (Cu(LJ) and Au(EAM)) interatomic potentials for fcc metals discussed in section 14.3.1, the computational procedure described in section 14.2.2 was employed to systematically investigate the atomic structure and surface-stress-induced changes in the average lattice parameter of unsupported thin (001) films as a function of the film thickness, Λ (Fig. 14.4). Λ is usually expressed in units of the bulk equilibrium lattice parameter, a. For example, $\Lambda/a = 3$ characterizes an unrelaxed thin slab containing six (001) planes, given the perfect-crystal interplanar lattice spacing of $d(001) = 0.5a$.

Some basic results obtained for the two potentials are compared in Figs. 14.5 and 14.6, according to which both yield a decrease in the lattice

parameter, $a_x \equiv a_y = a_{xy}$, parallel to the film plane, with decreasing Λ. The effect is much larger, however, for the EAM potential: For the thinnest film considered (containing four (001) planes, below which the film becomes mechanically unstable), a_{xy} decreases by about 3% for the EAM potential, in contrast to the much smaller 0.2% contraction for the LJ potential. Also, both potentials yield an **outward** relaxation, Δz, of the outermost lattice planes, as illustrated by the increase with decreasing Λ in the average lattice parameter in the z-direction, a_z. The combination of contraction in the film plane and expansion in the direction of the surface normal (Fig. 14.3) results in a change in the average atomic volume, Ω, which is proportional to $a_{xy}^2 a_z$. Figures 14.5 and 14.6 include plots of Ω, normalized to the perfect-crystal volume, $\Omega_o \sim a^3$, as function of Λ. Whereas the LJ potential yields a small increase in Ω (up to about 0.3%), the EAM potential yields a pronounced decrease (up to about 3%) with decreasing Λ. Finally, Figs. 14.5 and 14.6 also show the change in the energy, γ, of each of the two free surfaces of the slab normalized to the energy, $\gamma^\infty = \gamma(\Lambda \rightarrow \infty)$, of the related bulk free surface (Table 14.1).

As already mentioned, when the surrounding bulk material is removed the usually tensile surface stress can relax, resulting in a contraction in the film plane. The driving force for this contraction is thus the familiar surface tension, i.e. the stress component parallel to the film plane. To better understand the behavior displayed by Figs. 14.5 and 14.6, we have calculated the surface stresses for the fully relaxed bulk free surfaces (i.e. for the $\Lambda \rightarrow \infty$ limit) associated with the two potentials. According to Table 14.1, the EAM potential shows a surface stress almost ten times as large as the LJ potential, giving rise to the much larger contraction parallel to the film plane discussed above.

This dramatic difference in the magnitudes of the surface stresses obtained for the two potentials is closely related to the shapes of the cohesive-energy curves as a function of the lattice parameter shown in Fig. 14.2. Since stress, in general, is related to the derivatives of such curves [54, 55], it is apparent that the EAM potential yields generally much larger stresses. Figure 14.2 also illustrates that the cost in energy associated with a change in lattice parameter (or atomic volume) is much greater for the EAM potential than for the LJ potential. In response to the contraction parallel to the film plane, the LJ potential can thus sustain a small increase in the average atomic volume (by

Table 14.1 Comparison of some properties of the fully relaxed bulk (001) free surfaces obtained by means of the two potentials [45]. γ^∞ is the surface energy and $\sigma_{xx}^\infty = \sigma_{yy}^\infty$ (<0) the tensile surface stress (both in units of mJ/m^2 or, equivalently, erg/cm^2) [54]. Δz_1 is the relaxation displacement, relative to an unrelaxed free surface, of the outermost plane; positive (negative) sign means outward (inward) relaxation. $\Delta(1)$ is the change in interplanar spacing at the surface (see eq. (14.10)). Both Δz_1 and $\Delta(1)$ are in units of the perfect-crystal lattice parameter, a [45, 54]

	γ^∞	$\sigma_{xx}^\infty = \sigma_{yy}^\infty$	$\Delta z_1/a$	$\Delta(1)/a$
Cu (LJ)	982	−184	+0.0063	+0.0069
Au (EAM)	897	−1550	−0.0289	−0.0319

outward relaxation of the outermost (001) planes, whereas the EAM potential yields a decrease in the volume to avoid a more costly increase in energy.

As is well known, EAM potentials usually yield an **inward** relaxation at a bulk free surface [56], in contrast to pair potentials [57]. According to Figs. 14.5 and 14.6, however, the surface planes of a thin (001) film relax **outward** for both potentials. This outward relaxation is primarily a consequence of the Poisson effect (Fig. 14.3): The contraction in the $x-y$ plane is coupled with an expansion in the z-direction. (Poisson's ratio governing this effect will be discussed in the next section.) Another contribution arises from the interaction between the two free surfaces of the thin film. The effective force between these surfaces is given by the derivatives of the energy curves in Figs. 14.5 and 14.6. With the exception of the smallest Λ values for the EAM potential, the film surfaces slightly repel one another, which also contributes toward an outward relaxation. To separate the two effects, the thin films were relaxed while suppressing the contraction in the film plane. According to Fig. 14.7, the outermost planes now relax **inward** for the EAM potential, and the surface energy is practically independent of Λ, with the exception of the thinnest film, indicating no repulsion between the surfaces. The comparison between Figs. 14.7 and 14.5 demonstrates that the contraction in the plane of the film is, indeed, the main cause for the outward relaxation of the free surfaces. A similar comparison was performed for the LJ potential

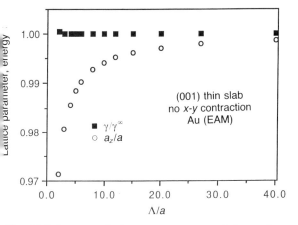

Fig. 14.7 Average z lattice parameter, a_z, and free-surface energies, γ, for the thin slabs of Fig. 14.5 in which the x–y contraction was suppressed (i.e. $a_{xy} \equiv a$). The volume change, Ω/Ω_0, is now proportional to a_z.

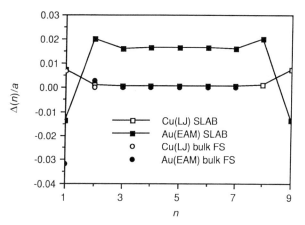

Fig. 14.8 Change in interplanar spacing, $\Delta(n)$, between planes n and $n - 1$ (see eq. (14.10)) for a slab consisting of 10 (001) planes. Two cases are considered for each potential: Thin slab with fully relaxed lateral contraction in the x–y plane, and a free surface attached to bulk material (so-called 'bulk free surface').

with the result that the surface atoms relax **outward** regardless of whether or not we allow for a change in sample lattice parameters in the film plane, a result not very surprising considering the very small magnitude of the lateral contraction obtained for this potential (see Fig. 14.6).

The inhomogeneous distribution of interplanar lattice spacings in the direction of the surface normal is illustrated in Figs. 14.8 and 14.9 for slabs containing ten and six (001) planes, respectively. Shown is the deviation

$$\Delta(n) = z(n) - z(n - 1) - d(001) \qquad (14.10)$$

of the interplanar spacing from that of the perfect-crystal (001) planes, $d(001)$. Here $z(n)$ denotes the z coordinate of the atoms in plane n. For comparison, the values for $\Delta(n)$ for a bulk free surface are also included in Fig. 14.8. For both the thin film and the bulk free surface, $\Delta(n)$ varies significantly only within two planes of the free surface. For a bulk free surface $\Delta(n)$ then vanishes, whereas for a thin film, because of the lateral contraction in the x–y plane, an expanded interplanar spacing is obtained for the bulk-like regions between the surfaces. Again, for reasons associated with the very different magnitudes of the lateral contractions obtained for the two potentials, this increase in interplanar spacing in the bulk-like

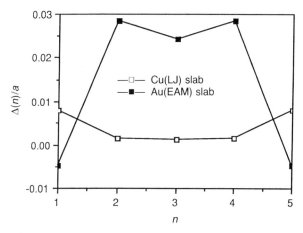

Fig. 14.9 $\Delta(n)$ according to eq. (14.10) for a thin slab consisting of only six (001) planes for the LJ and EAM potentials.

regions is substantially larger for the EAM potential. A comparison of Figs. 14.8 and 14.9 shows that this expansion becomes larger with decreasing film thickness, as one would expect from the increasing lateral contraction as Λ decreases. Figure 14.9 illustrates that even for a film consisting of only six (001) planes, the LJ potential, with its very small change in lateral dimensions, still permits the distinction between a bulk-like region

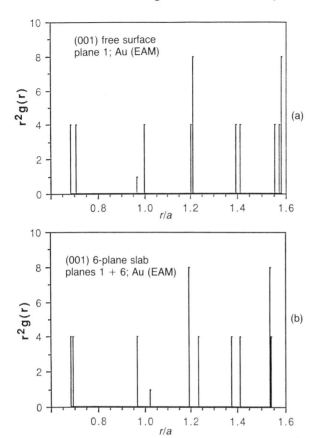

Fig. 14.10 Radial distribution functions, $r^2g(r)$, for the outermost (001) plane of a bulk free surface (top) and for the slab of Fig. 14.9 containing six (001) planes (bottom).

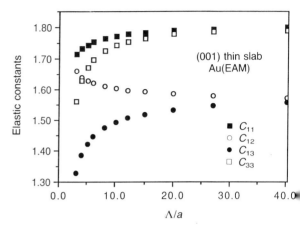

Fig. 14.11 Four of the six independent elastic constants (in 10^{12} dyn/cm^2) of a thin (001) film as a function of film thickness, Λ, for the EAM potential.

and a surface region, thus permitting the formulation of a simple two-state model (section 14.3.5 below).

The elastic response of the system is not determined by interplanar spacings, however, but by the actual separations between the atoms. In Fig. 14.10 the radial distribution functions, $r^2g(r)$, for the outermost planes of a bulk free surface and the six-plane film of Fig. 14.9 are compared. According to Fig. 14.10, the δ-function perfect-crystal peaks (of heights 12, 6, 24 etc., respectively, at the nearest-, 2nd and 3rd nearest-neighbor separations of $(\sqrt{2}/2)a$, a, $(\sqrt{5}/2)a$), are split for both the free surface and the thin film. However, whereas in the (effectively strained) **bulk** free surface those distances associated with interactions between atoms

within the same plane are the same as in the perfect crystal, in the (stress-relaxed) thin film no such perfect-crystal separations remain. Notice, for example, that for the outermost planes of the thin film both lines of the eight-member nearest-neighbor peak have been shifted towards shorter distances, an effect arising from the contraction in the x–y plane. As illustrated below, this shift towards shorter distances gives rise to a strengthened elastic response of the thin film, by contrast with the bulk free surface.

14.3.2 Elastic behavior of (001) films

When actually evaluating the relaxation-term elastic-constant contribution for a thin slab using the lattice-dynamics (LD) method (see above p. 370), the contributions due to the surface atoms of the slab must be eliminated from the force constant matrix prior to application of Lutsko's algorithm [38]. In the finite-strain (FS) method this would correspond to fixing the positions of the surface atoms following the imposition of an external strain to the thin film. Without this constraint the atoms could, in certain cases, relax back to their original sites in the unstrained slab, with a consequently vanishing elastic constant for such a strain. The elastic-constant contribution of these atoms, therefore, consists of the Born term only.

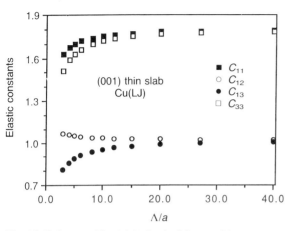

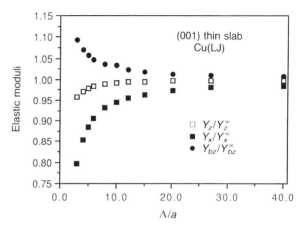

Fig. 14.12 Same as Fig. 14.11 for the LJ potential.

Fig. 14.14 Same moduli as in Fig. 14.13, however for the LJ potential. The related lattice-parameter changes are shown in Fig. 14.6.

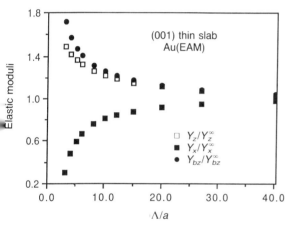

Fig. 14.13 Variation of Young's moduli, Y_z and $Y_x = Y_y$, and the biaxial modulus for isotropic stretching in the $x–y$ plane, Y_{bz}, as a function of Λ for the Au(EAM) potential. The related moduli in the $\Lambda \to \infty$ limit are listed in Table 14.2. The underlying lattice-parameter changes of these films are shown in Fig. 14.5.

The elastic-constant tensor of an (001) film contains six independent elements. Using the LD method described above, this tensor was evaluated for the EAM and LJ potentials; Figs. 14.11 and 14.12 show the four elastic constants in the upper left quadrant thus obtained. The two potentials yield the same qualitative behavior for all four elastic constants: Whereas C_{11}, C_{13}, and C_{33} decrease with decreasing Λ, C_{12} increases slightly. In addition to the shear constants, $C_{44} \equiv C_{55} \equiv G_{yz} \equiv$

G_{xz} and $C_{66} \equiv G_{xy} \equiv G_{yx}$, these are the only independent elastic constants of the film.

Using the elastic constants in Figs. 14.11 and 14.12, the moduli of the films were determined. The Young's moduli and the biaxial modulus thus obtained for the two potentials are shown in Figs. 14.13 and 14.14. While the Young's modulus in the film plane, $Y_x \equiv Y_y$, is found to decrease for both potentials with decreasing Λ, Y_{bz} increases. Notice, however, that the effects are much larger for the EAM potential. Inserting the values of the Poisson ratios discussed below (Figs. 14.15 and 14.16) into the relation

$$Y_{bz} \equiv \frac{Y_x}{(1 - \nu_{xy})} \equiv \frac{Y_y}{(1 - \nu_{xy})} \tag{14.11}$$

it is obvious that the increasing difference with decreasing Λ between Y_{bz} and Y_x exhibited in Figs. 14.13 and 14.14 is due to the rapid increase in the Poisson ratio, ν_{xy}. The two potentials yield opposite variations for Y_z, however, which increases for the EAM but decreases for the LJ potential; reasons for this qualitative difference will be discussed in the next section.

The shear moduli and Poisson ratios obtained for the two potentials are summarized and compared in Figs. 14.15 and 14.16. Both potentials yield the same qualitative behavior in all cases, although the

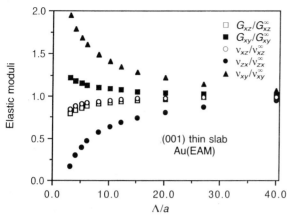

Fig. 14.15 Variation of the shear moduli, $G_{xz} \equiv G_{yz}$ and G_{xy}, and of the Poisson ratios v_{xz}, v_{zx} and v_{xy} (EAM potential). $v_{\alpha\beta}$ denotes the ratio for stress in the β direction with consequent lattice-parameter change in the α direction (section 14.2). The related lattice-parameter changes of these films are shown in Fig. 14.5.

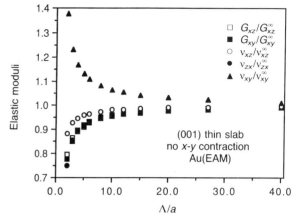

Fig. 14.17 Shear moduli and Poisson ratios of Fig. 14.15 in which, however, the $x–y$ contraction was suppressed in the relaxation (Au(EAM)) potential). These films are therefore under tensile stress parallel to the film plane.

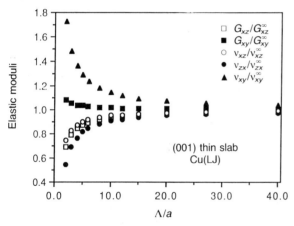

Fig. 14.16 Same as Fig. 14.15, however for the LJ potential. The related lattice-parameter changes are shown in Fig. 14.6.

effects are, again, more pronounced for the EAM potential. The modulus for shear parallel to the film plane, $G_{xz} \equiv G_{yz}$ ($\equiv C_{44} \equiv C_{55}$), decreases with decreasing Λ while the modulus for shear parallel to the edges of the film, G_{xy} increases moderately. Particularly pronounced are (a) the decrease in $v_{zx} \equiv v_{zy}$, the Poisson ratio associated with a compressive stress in the film plane and consequent expansion in the direction of the film

normal, and (b) the increase in $v_{xy} \equiv v_{yx}$, the Poisson ratio associated with stress and strain in the film $(x–y)$ plane. Interestingly, $v_{xz} \equiv v_{yz}$, the Poisson ratio governing the contraction in the $x–y$ plane caused by a stress in the z-direction, does not vary as strongly as a function of Λ.

The softening of the modulus for shear parallel to the film plane is readily understood. The on-average expanded interplanar spacing in the z-direction (Figs. 14.8 and 14.9) makes it slightly easier to force lattice planes partially on top of each other following the application of a shear stress, with consequent decrease in G_{xz}. For the same reason, the application of a tensile stress, say, in the x-direction causes a greater Poisson response in the y-direction, with consequently enhanced Poisson ratio v_{xy} in Figs. 14.15 and 14.16. The expanded lattice spacing in the z-direction can also account for the reduced Poisson response in the x (or z) direction following the application of a stress in z (or x), with consequent decrease in v_{xz} and v_{zx}, respectively.

This interpretation is confirmed by the results in Fig. 14.17: When the $x–y$ contraction is suppressed (as, for example, in the bulk free surface), the spacing of the (001) planes cannot increase as it does when the $x–y$ contraction is allowed for (Fig. 14.8). All anomalies in Fig. 14.15 therefore become

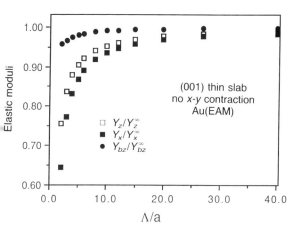

Fig. 14.18 Young's and biaxial moduli of Fig. 14.13 in which, however, no *x–y* contraction was permitted during relaxation (Au(EAM) potential).

smaller; however, the fact that they do not vanish completely (Fig. 14.17) indicates an intrinsic effect of structural disorder, due to the very presence of the free surfaces, on all moduli.

14.3.3 Relationship among structural disorder, lattice-parameter changes, and the elastic behavior of (001) films

As illustrated above, particularly for the EAM potential the contraction in the film plane has a strong effect on the relaxation of the outermost lattice planes of the slab and on the spacing of (001) planes in its bulk-like region. A comparison of Figs. 14.18 and 14.13 shows that the *x–y* contraction is the major cause for the enhancement of Y_z and Y_{bz} in the thin slab. Without such a contraction Y_z and Y_{bz} actually **soften**, indicating that in a bulk free surface these moduli are not enhanced – in contrast to the thin film. A similar calculation for the LJ potential yields qualitatively identical results.

A comparison of Figs. 14.17 and 14.15 illustrates that the anomalies in the shear moduli and Poisson ratios are also much smaller when the contraction in the film plane, accompanied by an increase in (001) interplanar lattice spacings, is suppressed. A similar comparison for the LJ potential shows the same qualitative results, in spite of the fact that the

outermost (001) planes of the film relax outwards even when the *x–y* contraction is suppressed (section 14.3.1). This comparison indicates that the role of the anisotropic **lattice-parameter changes** of the thin film is to **increase** the anomalies in all moduli, with the exception of Y_z and Y_{bz} which show a very different behavior, as discussed below.

The radial distribution functions in Figs. 14.10(a) and (b) give an indication for the cause for the strengthening of $Y_z(\Lambda)$ and $Y_{bz}(\Lambda)$ in the fully stress-relaxed thin film but not in the strained film (or bulk free surface; Fig. 14.3). In contrast to the outermost (001) plane of the free surface (Fig. 14.10(a)), the peaks in $r^2g(r)$ associated with the thin film (Fig. 14.10(b)) are not only split, but significant density is also shifted towards shorter distances. Bringing atoms closer together strengthens their local elastic response, in contrast to increasing their separation. On balance, more distances have been shortened in the thin slab than in the bulk free surface, because of the lateral contraction. Apparently, the net effect of a complex averaging process is a strengthening of those elastic moduli which depend on the overall radial distribution of interatomic distances, in contrast to the softening of those moduli which are governed by the interplanar spacing of (001) planes. The effects leading to a strengthening of $Y_z(\Lambda)$ and $Y_{bz}(\Lambda)$ are obviously not as pronounced for the LJ potential with its much smaller contraction in the *x–y* plane. This was verified by an inspection of the radial distribution functions, analogous to Figs. 14.10(a) and (b). Obviously, according to Fig. 14.14, the small *x–y* contraction is sufficient to slightly strengthen $Y_{bz}(\Lambda)$ but not $Y_z(\Lambda)$. If one were to apply a compressive stress in the *x–y* plane, however (as is the case in some epitaxially bonded thin films), based on the above analysis one would expect a strengthening of $Y_z(\Lambda)$ even for the LJ potential.

14.3.4 Surface-stress-induced structure and elastic behavior: comparison of (001), (011), and (111) films

The above analysis has enabled us to separate two types of phenomena which strongly influence the elastic response of thin films with ⟨001⟩ surface normals. First, the presence of the free surfaces of the thin slab gives rise to **structural disorder**, as evidenced by the broadened peaks in the radial distribution functions in Figs. 14.10(a) and (b). In a bulk free surface this disorder gives rise to a generally weakened elastic response (Figs. 14.17 and 14.18). Second, the effect of the **lattice-parameter changes**, due to the surface stress, is two-fold, namely (a) they enhance the structural disorder evidenced in the radial distribution function (compare Figs. 14.10(a) and (b)), and (b) they cause an increase in the spacing of (001) planes in the bulk-like region of the slab, with a consequent change in the average atomic volume of the film (Figs. 14.5, 14.6 and 14.8).

By comparing the structure and elastic behavior of the (001) films with those bordered by (111) and (011) surfaces, in this section we will investigate in more detail the effect of the surface stress on the average lattice parameters and the consequent changes in the elastic behavior. Since all stress-induced phenomena are much more pronounced for the EAM potential (Fig. 14.2), we will limit ourselves to reviewing the results obtained for the many-body potential, although a qualitatively identical – albeit less dramatic – behavior is obtained for the pair potential.

Prior to comparing the elastic response of these films, we investigate in some detail the surface-stress-induced lattice-parameter changes parallel and perpendicular to the film plane. In Fig. 14.19 the surface-stress-induced in-plane contractions, Δa_x and Δa_y, are compared for the three films (Fig. 14.5). For symmetry reasons, for the (001) and (111) films $\Delta a_x = \Delta a_y$. As expected, all three slabs contract in the x–y plane with decreasing film thickness, Λ. In zero-th order, for at least the larger values of Λ, the underlying stresses, $\sigma_{\alpha\beta}(\Lambda)$, and compliances, $S_{\alpha\beta}(\Lambda)$, may be approximated by the **bulk-surface** stresses, $\sigma_{\alpha\beta}^{\infty}$, and compliances, $S_{\alpha\beta}^{\infty}$ (obtained in the $\Lambda \to \infty$ limit). According to linear elasticity theory, in zero-th order the relative contractions (i.e. strains), $\Delta a_x/a$ and $\Delta a_y/a$, are then given by [53]

$$\frac{\Delta a_x}{a} = [\sigma_{xx}(\Lambda)\, S_{11}(\Lambda) + \sigma_{yy}(\Lambda)\, S_{12}(\Lambda)]/\Lambda$$

$$\approx [\sigma_{xx}^{\infty}\, S_{11}^{\infty} + \sigma_{yy}^{\infty}\, S_{12}^{\infty}]/\Lambda \qquad (14.12)$$

$$\frac{\Delta a_y}{a} = [\sigma_{xx}(\Lambda)\, S_{12}(\Lambda) + \sigma_{yy}(\Lambda)\, S_{22}(\Lambda)]/\Lambda$$

$$\approx [\sigma_{xx}^{\infty}\, S_{12}^{\infty} + \sigma_{yy}^{\infty}\, S_{22}^{\infty}]/\Lambda \qquad (14.13)$$

where we have taken into account that the overall stress tensor, Σ (i.e. the force per unit area), is related to the surface-stress tensor, σ, by $\Sigma = (1/\Lambda)\,\sigma$. Because of symmetry, for the (111) and (001) planes $\sigma_{xx} = \sigma_{yy}$, and therefore $C_{11} = C_{22}$ and $S_{11} = S_{22}$; eqs. (14.12) and (14.13) then become identical, and

$$\frac{\Delta a_x}{a} = \frac{\Delta a_y}{a} \approx \sigma_{xx}^{\infty}[S_{11}^{\infty} + S_{12}^{\infty}]/\Lambda = \frac{1}{\Lambda}\left(\frac{\sigma_{xx}^{\infty}}{Y_b^{\infty}}\right)$$

$$(14.14)$$

where Y_b^{∞} is the biaxial modulus in the $\Lambda \to \infty$ limit. A tensile surface stress (i.e. $\sigma_{xx}^{\infty} < 0$) therefore gives rise to an in-plane contraction of the film (Δa_x, $\Delta a_y < 0$).

As seen from Table 14.2, in spite of a smaller absolute value of its surface stress, $\sigma_{xx}^{\infty} = \sigma_{yy}^{\infty}$, the approximately three times smaller value of Y_b^{∞} for the (001) film is responsible for its much larger contraction by comparison with the (111) film. For the (011) film, because of the x–y asymmetry, an isotropic biaxial modulus cannot be defined; its changes in lattice-parameters are therefore

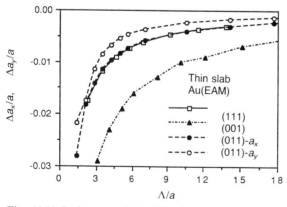

Fig. 14.19 Surface-stress-induced in-plane contractions, Δa_x and Δa_y (<0), of unsupported thin films of Au, referred to the ideal-crystal lattice parameter, a. For the (001) and (111) films, $\Delta a_x \equiv \Delta a_y$ (Fig. 14.15).

Table 14.2 Tensile stresses σ_{xx}^{∞} and σ_{yy}^{∞} (<0) of fully relaxed bulk free surfaces, with energies γ^{∞}, obtained by means of the Au(EAM) potential, in units of 10^{-3} J/m^2 [54]. δz is the relaxation displacement of the outermost surface plane relative to the unrelaxed bulk free surface (in units of the bulk lattice parameter, a); the negative sign indicates inward relaxation obtained for this potential [54, 56]. Also listed are the bulk ideal-crystal Young's and biaxial moduli, Y_x^{∞}, Y_y^{∞}, Y_z^{∞}, and Y_b^{∞}, and the Poisson ratios, v_{xy}^{∞} and v_{zx}^{∞}, obtained in the limit for $\Lambda \to \infty$ (in units of 10^{12} dyn/cm^2)

Film	x	y	z	σ_{xx}^{∞}	σ_{yy}^{∞}	Y_b^{∞}	v_{xy}^{∞}	v_{zx}^{∞}	Y_x^{∞}	Y_y^{∞}	Y_z^{∞}	γ^{∞}	$\delta z/a$
(111)	$[1\bar{1}0]$	$[11\bar{2}]$	$[111]$	-1739	-1739	1.947	0.617	0.232	0.745	0.745	1.212	767	-0.0217
(001)	$[100]$	$[010]$	$[001]$	-1550	-1550	0.647	0.465	0.465	0.346	0.346	0.346	897	-0.0289
(011)	$[100]$	$[0\bar{1}1]$	$[011]$	-1516	-862	–	–	–	0.346	0.745	0.745	957	-0.0480

governed by the full elastic-constant tensor. We note, however, that the smaller reduction in a_y than in a_x scales approximately with the smaller absolute value of σ_{yy}^{∞} by comparison with σ_{xx}^{∞} computed for this film (Table 14.2).

As a consequence of the Poisson effect, the in-plane contraction has a pronounced effect on the film structure in the z-direction (Fig. 14.3). Considering that σ_{zz} vanishes identically for any value of Λ, analogous to eqs. (14.12) and (14.13) the Poisson strain, $\Delta a_z/a$, is given by (again in zero-th approximation)

$$\frac{\Delta a_z}{a} = [\sigma_{xx}(\Lambda)\,S_{13}(\Lambda) + \sigma_{yy}(\Lambda)\,S_{23}(\Lambda)]/\Lambda$$

$$\approx [\sigma_{xx}^{\infty}S_{13}^{\infty} + \sigma_{yy}^{\infty}S_{23}^{\infty}]/\Lambda \quad (14.15)$$

For the (111) and (001) planes, with $S_{13} = S_{23}$, eq. (14.15) can again be simplified. Using eq. (14.14) to express σ_{xx}^{∞} in terms of $\Delta a_x/a$, and defining the Poisson ratios $v_{xy} = -S_{12}/S_{11}$ and $v_{zx} = -S_{13}/S_{11}$, eq. (14.15) becomes

$$\frac{\Delta a_z}{a} = -2\left(\frac{\Delta a_x}{a}\right)\frac{v_{zx}^{\infty}}{1 - v_{xy}^{\infty}} \quad (14.16)$$

As is to be expected for the Poisson effect, the signs of Δa_z and Δa_x are opposite, i.e. for $\Delta a_x < 0$ an **outward** relaxation of the film surfaces is expected (i.e. $\Delta a_z > 0$).

In writing eqs. (14.15) and (14.16), it is implicitly assumed that Δa_z is measured relative to a strained reference state (Fig. 14.3), consisting of a thin film bordered by two fully relaxed surfaces, but in which the x–y contractions (giving rise to non-vanishing values of $\Delta a_z > 0$) are not permitted. Δa_z therefore represents the deviation of the average lattice parameter in the z-direction, a_z, from that in the strained reference film, a_{zo}, i.e. $\Delta a_z = a_z - a_{zo}$. However, because of the **inward** relaxation in the **bulk** free surfaces (Table 14.2), a_{zo} is less than a. (By contrast, in the x–y plane the reference film has the bulk ideal-crystal lattice parameter, a.) The outward relaxation predicted by eq. (14.16) is therefore measured **relative to the surfaces in the strained reference film**. The determination of Δa_z hence requires first a simulation of the strained reference state, relative to which the surface displacements in the stress-relaxed film can be measured (Fig. 14.3).

Not surprisingly, when only the atom positions but not the average lattice parameters are allowed to relax, the film surfaces relax **inward**, by virtually the same amount as in the bulk free surface (Table 14.2). The strained reference films sketched in Fig. 14.3 hence consist essentially of two bulk free surfaces, separated by perfect-crystal regions, as also evidenced by practically Λ-independent surface energies and stresses, down to relatively small values of Λ ($\geqslant 4a$).

According to Fig. 14.20, the values of $\Delta a_z/a$ determined in the manner described above indeed increase proportionally to the in-plane contraction, except for the largest values of $\Delta a_x/a$. The two solid lines in the figure, with slopes $-2v_{zx}^{\infty}/(1 - v_{xy}^{\infty})$ $= -1.121$ and -1.738, respectively, represent the predictions of eq. (14.16) for the (111) and (001) planes based on the bulk stresses and moduli listed in Table 14.2. Their excellent representation of the simulation data leads us to our first conclusion: With the exception of the largest contractions (i.e.

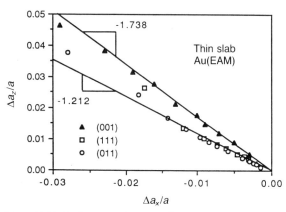

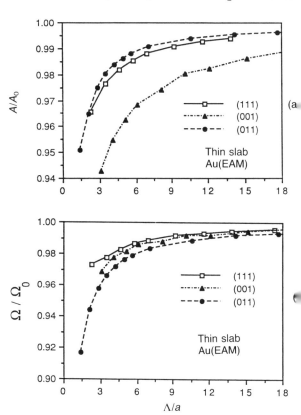

Fig. 14.20 Surface-stress-induced Poisson expansion, Δa_z (>0), in the direction of the surface normal. The straight lines are the curves predicted from eq. (14.15) with the bulk moduli in Table 14.2 for the (001) and (111) films (Au(EAM) potential).

Fig. 14.21 Normalized decrease in (a) the planar unit-cell area, $A = a_x a_y$, and (b) the average atomic volume, $\Omega = a_x a_y a_z/4$ (for the fcc lattice), against film thickness, Λ. The decrease in Ω for the (011) film indicates an up to 8% increase in the density of the material. The values of A and Ω are normalized to the related perfect-crystal values, $A_0 = a^2$ and $\Omega_0 = a^3/4$, respectively (Fig. 14.5).

the smallest values of Λ, typically $\Lambda \geqslant 4a$), the zero-th order linear-elastic equations (14.12)–(14.16) permit a prediction of the average lattice parameters based entirely on the knowledge of the **bulk**-surface stresses and the **perfect-crystal** moduli.

As a consequence of the lattice-parameter changes in Figs. 14.19 and 14.20, the planar unit-cell area, A, and the **average** atomic volume, defined by $\Omega = a_x a_y a_z/4$ (for the fcc lattice), also decrease with decreasing Λ (Fig. 14.21; we note that the (001) data is the same as in Fig. 14.5). The decrease in Ω indicates a substantially higher density of the thin films, up to about 8% for the (011) slab, than the perfect crystal. One would therefore expect at least some degree of elastic strengthening, particularly in the film plane where the material is homogeneous. However, as evidenced by the related Young's moduli shown in Fig. 14.22, while some moduli are, indeed, strengthened, others are weakened. For example, in spite of the largest in-plane contraction observed for the (001) film (Fig. 14.19), the related modulus Y_x ($=Y_y$) is weakened significantly (Fig. 14.22(a)) while, surprisingly, Y_z simultaneously strengthens (Fig. 14.22(b)). By contrast, the behavior of the (111) film is more like that of a homogeneous material: the in-plane contraction is accompanied

by a strengthening in Y_x, while the z expansion gives rise to a softening of Y_z. Interestingly, in the (011) film Y_x is nearly independent of Λ while both Y_y and Y_z soften substantially (Fig. 14.22), despite the smallest reduction in A observed for this film (Fig. 14.21(a)).

In exploring the origin of this anomalous elastic behavior, it is important to recognize that – even without allowing for any stress-induced changes in the average lattice parameters – the presence of the structurally disordered film surfaces alone alters the average elastic response of the film; this response is then further modified by the stress-induced lattice-parameter changes. As in the case

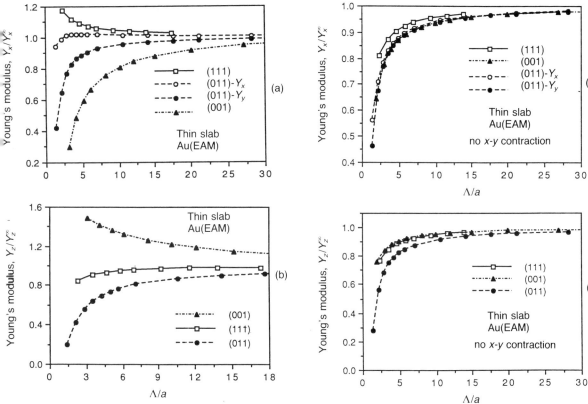

Fig. 14.22 Young's moduli, Y_x, Y_y and Y_z of fully-relaxed single-crystal films of Au (EAM), normalized to the related bulk values (Table 14.2). For the (001) and (111) films $Y_x \equiv Y_y$ (Fig. 14.13).

Fig. 14.23 Normalized Young's moduli of the strained reference slabs defined in Fig. 14.3 in which only the atom positions but not the surface stresses, σ_{xx} and σ_{yy}, are relaxed.

of Δa_z discussed above, two contributions to the net elastic behavior therefore have to be distinguished. These can be separated by first determining the elastic behavior of the strained reference system (Fig. 14.3), against which the additional effects due to the in-plane contractions can then be probed.

When the in-plane contractions are suppressed during the simulation, all three Young's moduli decrease steadily, as illustrated in Fig. 14.23. If now the surface stresses are also permitted to relax, the distinct effect of the in-plane contractions alone is given simply by the differences ΔY_x, ΔY_y, and ΔY_z between corresponding moduli in Figs. 14.22 and 14.23, shown in Fig. 14.24. According to Fig. 14.24(a), however, the

effect of the in-plane contractions alone leads to a **softening** of $Y_x = Y_y$ for the (001) film but a **stiffening** for the (111) film, with the (011) film falling in between. By contrast, Y_z is strengthened for both the (001) and (111) films, in spite of the Poisson expansion, Δa_z (>0) (Fig. 14.24(b)). Our second major conclusion is therefore that no direct relation exists between the surface-stress-induced lattice-parameter changes and either the overall elastic moduli (Fig. 14.22) or their surface-stress-induced portion (see Fig. 14.24). This lack of a direct correlation is particularly evident from the plots of ΔY_x, ΔY_y, and ΔY_z as a function of Δa_x, Δa_y, and Δa_z (Figs. 14.25(a) and (b)).

The rather different bulk-free-surface energies in Table 14.2 indicate that already without any

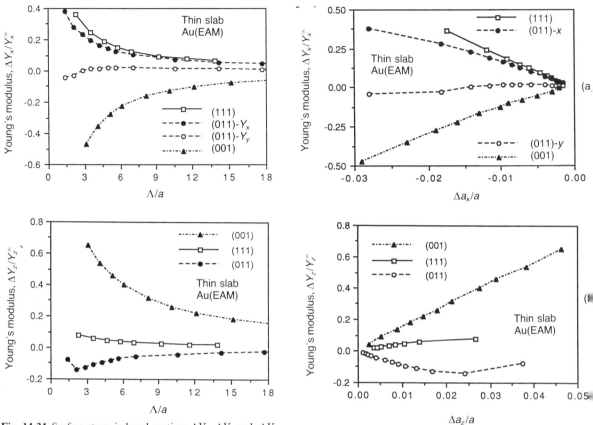

Fig. 14.24 Surface-stress-induced portions ΔY_x, ΔY_y and, ΔY_z of the Young's moduli in Fig. 14.22, normalized to the corresponding bulk moduli (Table 14.2).

Fig. 14.25 (a) ΔY_x and ΔY_y of Fig. 14.24(a) as a function of the in-plane strain $\Delta a_x/a$; (b) ΔY_z of Fig. 14.24(b) plotted against the Poisson strain, $\Delta a_z/a$ of Fig. 14.20.

changes in the average lattice parameters, the three films exhibit very different amounts of inhomogeneous structural disorder. It is not surprising then that upon contraction this disorder is modified in rather different ways, giving rise to a highly unpredictable elastic behavior. For example, the pronounced reduction in Y_x observed for the (001) film in spite of the by far largest in-plane contraction is clearly connected with the very large Poisson expansion in the z-direction (which, interestingly, **strengthens** Y_z). Moreover, a comparison of Figs. 14.22(a) and 14.23(a) demonstrates that when the in-plane contractions are prohibited, Y_x recovers substantially, while simultaneously the large enhancement in Y_z

changes into a significant softening. This example demonstrates a rather complex interplay between the in-plane contraction and consequent Poisson expansion, on the one hand, and between the related in-plane and out-of-plane elastic responses on the other.

This coupling between the in-plane contractions and the consequent yielding of the material in the z-direction leads to a continuous modification of the detailed atomic structure of the film surfaces as Λ decreases, i.e. to a continuous evolution of the nature of the inhomogeneity near the film surfaces. A focus on the elastic **constants** (by contrast with the **moduli**) might offer some hope to deconvolute the effects parallel (x, y) and perpendicular (z)

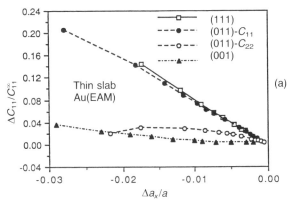

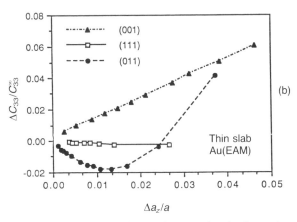

Fig. 14.26 Plot similar to Fig. 14.25, except that the changes in the elastic constants are plotted instead of those in the moduli. (a) Normalized difference, ΔC_{11} and ΔC_{22}, between the in-plane elastic constants, C_{11} and C_{22}, of the fully relaxed films and the strained reference system in Fig. 14.3 against the in-plane strain $\Delta a_x/a$ and $\Delta a_y/a$, respectively; (b) surface-stress-induced change in the out-of-plane elastic constant, ΔC_{33}, plotted against the Poisson strain, $\Delta a_z/a$. Normalization factors are given in Table 14.3.

to the surface. The latter are related to inverse compliances, and hence represent combinations of various (diagonal and off-diagonal, in-plane and out-of-plane) elastic constants; with the exception of the shear moduli it is then difficult, or impossible, to distinguish the in-plane (stress-induced) effects from those in the direction of the surface normal.

Analogous to Fig. 14.25, Fig. 14.26 shows the surface-stress-induced changes in the elastic

constants, (a) ΔC_{11} and ΔC_{22} plotted against the in-plane strain $\Delta a_x/a$, and (b) ΔC_{33}, plotted against the Poisson strain, $\Delta a_z/a$. As already mentioned, in a homogeneous system a reduction in the atomic volume or the planar unit-cell area (Figs. 14.21(a) and (b)) should give rise to an elastic strengthening. According to Fig. 14.26(a), that is, indeed, the case for the in-plane elastic constants. Upon contraction in the film plane the atoms are forced more closely together, with a consequent yielding in the z-direction (the manifestation of the Poisson effect). The magnitudes of the enhancements in C_{11} and C_{22} are therefore coupled with the response of the system in the z-direction in that a smaller expansion, $\Delta a_z = a_z - a_{zo}$ (with a_z shown in Fig. 14.20) is expected to cause more strengthening of the in-plane elastic constants. Figure 14.26(a) illustrates that the much larger enhancements in C_{11} and C_{22} for the (111) and (011) films are, indeed, accompanied by much smaller z-expansions of these films compared to the (001) film. A quantitative description of this coupling would involve the off-diagonal elements of the elastic-constant tensor, C_{13} and C_{23}, and the related Poisson ratios.

Finally, the effect of the expansions on C_{33} is shown in Fig. 14.26(b) where, in analogy to Fig. 14.24(b), ΔC_{33} is plotted as a function of the corresponding values of Δa_z. Surprisingly, the out-of-plane response still defies conventional intuition gained from the study of homogeneous systems: in spite of a large expansion, the (001) film actually strengthens, while the (111) film is virtually unaffected by this expansion. The (011) film, by contrast, displays a highly non-linear yet continuous behavior, with a softening for the smaller expansions followed by a strengthening for the larger ones.

In summary, although the surface-stress-induced lattice-parameter changes can be predicted – at least semi-quantitatively – via linear elasticity theory (as one would expect for a homogeneous system), the relationship between these changes and the elastic behavior is considerably more complex. The above analysis illustrates how computer simulations, in which the lattice-parameter changes can be selectively prevented, permit deconvolution of the distinct effects in the elastic behavior of thin films due to (i) the anisotropic changes in the average lattice parameters and (ii) the structurally disordered surfaces. While the surface-stress induced contraction and consequent elastic strengthening in the film plane are a characteristic of a **homogeneous** system, the behavior in the direction of the film normal shows typical characteristics of an **inhomogeneous** system in which a net expansion may sometimes give rise to an elastic **strengthening**, depending on the complex, highly non-linear interplay between structural disorder and consequent volume change.

14.3.5 On the applicability of a simple two-state model

The analysis in section 14.3.1 of the atomic structure and lattice-parameter changes of the (001) film has suggested the distinction of surface regions from bulk-like regions. At first sight, one might expect that this distinction should allow the formulation of a simple two-state (mean-field) model in which distinct physical properties are assigned to the interface region which differ from those of the bulk-like regions. As proposed by Clemens and Eesley [4] for the case of strained-layer superlattices, in such a model $1/\Lambda$ is a natural variable. $1/\Lambda$ is also a natural variable when determining the effective elastic constants of a superlattice of any symmetry [58, 59].

To illustrate the origin of the simple $1/\Lambda$ behavior, the thin slab is subdivided into two interfacial regions, each of width d (containing the free surfaces), and by a bulk region of width $\Lambda\text{-}2d$ (Fig. 14.27). Assuming the interface region to be characterized by an effective z lattice parameter, a_z^i, then this simple two-state model predicts that the **average** z lattice parameter of the slab, a_z, is given by the volume-weighted contributions from the two regions, according to [4]

$$a_z = \left(\frac{2d}{\Lambda}\right) a_z^i + \left(\frac{\Lambda - 2d}{\Lambda}\right) a_z^b \qquad (14.17)$$

$$\frac{a_z}{a_z^b} = 1 + \left(\frac{2d}{\Lambda}\right)\left(\frac{a_z^i}{a_z^b} - 1\right) \qquad (14.18)$$

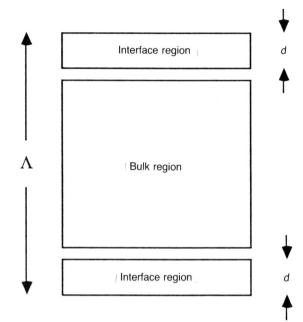

Fig. 14.27 Simple two-state (mean-field) model of a thin film in which the two interface regions of thickness d (each containing a free surface) are distinguished from the bulk-like center region of the slab (schematic).

where a_z^b is the bulk lattice parameter in the z-direction. (For example, for (001) planes $a_z^b = 0.5a$.) A similar equation should also hold for the x–y lattice parameters and for the interfacial energy.

However, as illustrated in Fig. 14.28 for the (001) film, neither the geometrical parameters nor the surface energy of the thin slab show good linearity in $1/\Lambda$ in the $\Lambda \to \infty$ ($1/\Lambda \to 0$) limit, although in some cases the deviations from linearity are not as pronounced as in others. While the two-state model is rather appealing for its simplicity, it contains an untenable assumption: it assumes that the properties of the bulk regions are independent of Λ. This cannot be the case because of the Λ-dependent Poisson expansion in the z-direction (Figs. 14.8, 14.9, 14.15, and 16), which necessarily gives rise to a change in the properties of the bulk region as a function of Λ. One might, however, expect the model to apply when the x–y contraction is suppressed, because the lattice par-

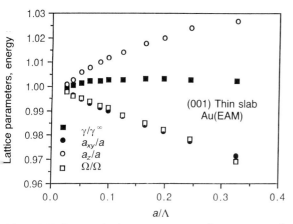

Fig. 14.28 Average lattice parameters, surface energy and average atomic volume of the (001) film of Figs. 14.4 and 14.5 plotted against $1/\Lambda$ for the many-body potential.

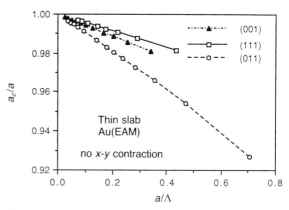

Fig. 14.29 Average z lattice parameters against $1/\Lambda$ for the (001), (011) and (111) films in which, however, no $x-y$ contraction was permitted (Au(EAM) potential) (Fig. 14.7).

ameter in the bulk region is then practically independent of Λ and equal to its perfect-crystal value (Figs. 14.8 and 14.9). Indeed, as demonstrated in Fig. 14.29 for the case of a_z, all three films then show a reasonably good $1/\Lambda$ dependence, which is also observed for the remaining geometrical parameters and the energy in Fig. 14.28.

The above comparison demonstrates that a linear $1/\Lambda$ dependence can only be expected for quantities which are relatively insensitive functions of the Poisson expansion. For this reason, the LJ potential generally shows much smaller deviations from a $1/\Lambda$ variation in many quantities than the many-body potential because the much smaller surface stresses give rise to a much less pronounced Poisson effect (section 14.3.1). Another possible cause of deviations from a simple $1/\Lambda$ behavior lies in the continuous modification of the free-surface atomic structure with decreasing Λ, even when the $x-y$ contraction is suppressed (see also the radial distribution functions in Figs. 14.10(a) and (b)). For the same reason, the surface relaxations of a thin film *cannot* be simply considered as a superposition of those associated with two bulk surfaces. The suggestion of Chen *et al.* [60] that such a simple interference model may account for the atomic structure of a thin film is therefore, at best, valid in the case in which the $x-y$ contraction of the film is suppressed.

Let us now consider the elastic properties. Denoting the elastic-constant tensors in the interfacial and bulk regions (Fig. 14.27) by $C^i_{\alpha\beta}$ and $C^b_{\alpha\beta}$, respectively, the general expressions for the *average* elastic constants of such a system [58, 59] may be applied to determine the net average elastic constants within the framework of the two-state model. Because the derivations are tedious, but straightforward, we will only quote the results of the application of these formulae to the two-state model [46]. While, similar to eq. (14.17), for C_{66} one finds the volume-weighted average,

$$C_{66} = C^b_{66} + \left(\frac{2d}{\Lambda}\right)(C^i_{66} - C^b_{66}) \qquad (14.19)$$

for C_{33}, C_{44}, and C_{55}, the *inverse* elastic constants must be averaged, according to [58, 59]

$$1/C_{kk} = 1/C^b_{kk} + \left(\frac{2d}{\Lambda}\right)(1/C^i_{kk} - 1/C^b_{kk}),$$
$$(k = 3, 4, 5) \qquad (14.20)$$

For C_{13}, by contrast, it is found that

$$C_{13}/C_{33} = C^b_{13}/C^b_{33} + \left(\frac{2d}{\Lambda}\right)(C^i_{13}/C^i_{33} - C^b_{13}/C^b_{33}) \qquad (14.21)$$

The expressions for C_{11} and C_{12} are slightly more complicated [46, 58]:

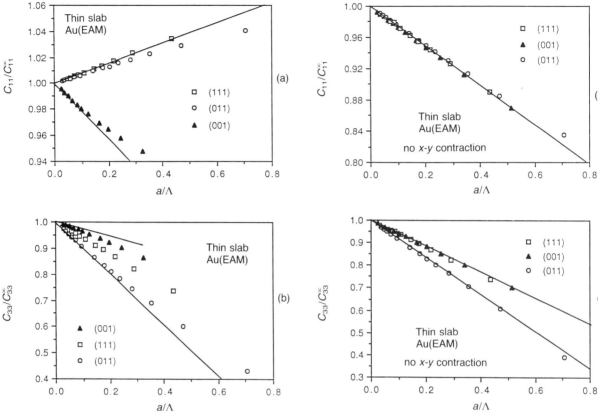

Fig. 14.30 Normalized values of (a) C_{11} and (b) C_{33} versus $1/\Lambda$ for the (001), (011) and (111) slabs. The straight lines represent linear fits to the large-Λ (small $1/\Lambda$) data.

Fig. 14.31 Same as Fig. 14.30, however, for films in which the x–y contraction was suppressed during the relaxation of the atoms. The straight lines represent linear fits to the small $1/\Lambda$ data.

$$A_{lk} = A_{lk}^b + \left(\frac{2d}{\Lambda}\right)(A_{lk}^i - A_{lk}^b), \quad (k = 1, 2)$$

$$(14.22)$$

where

$$A_{lk}^b = C_{lk}^b - \left[\frac{(C_{13}^b)^2}{C_{33}^b}\right] \qquad (14.23)$$

with similar expressions for A_{lk}^i.

The above expressions can, in principle, be used to determine the effective width and elastic constants of the surface regions of the film, provided a linear $1/\Lambda$ variation is, indeed, observed for the appropriate quantity. As illustrated in Figs. 14.30(a) and (b) for the case of C_{11} and C_{33}, however, systematic deviations from a linear

behavior are more the rule than the exception indicating that, as for the film structure, the two-state model fails because the properties of both the bulk-like and surface regions of the slab depend systematically on Λ. Again, however, when the x–y contractions are suppressed, much better linearity in $1/\Lambda$ is obtained, as shown in Fig. 14.31.

In conclusion, neither the geometrical parameters nor the surface energy of the thin films show good linearity in $1/\Lambda$ in the $\Lambda \to \infty$ ($1/\Lambda \to 0$) limit when the x–y contractions, with consequent Poisson expansion in the z-direction, is considered, although in some cases the deviations from straight lines are not as pronounced as in others. The surface-stress-induced x–y contraction causes an expansion of the bulk-like region of the film, the properties of which are therefore a function of Λ. However, when the $x–y$ contraction of the film is suppressed, the linearity in the $1/\Lambda$ behavior of the geometrical parameters and elastic constants and moduli is, indeed, better.

14.3.6 Grain-boundary superlattices (GBSLs)

By replacing the two free surfaces of a thin film by GBs, in this section we will illustrate how the qualitative features of the so-called supermodulus effect can be understood by considering as model systems superlattices of thin films separated by GBs ('grain-boundary superlattices', GBSLs [26]). In actual superlattice materials, interfacial chemistry and reactions and/or complex electronic-bonding effects considerably complicate a systematic theoretical investigation of this effect. So as to eliminate the effects of interfacial chemistry, and yet still capture the essential interfacial phenomena of inhomogeneous structural disorder coupled with anisotropic lattice-parameter changes, as a first step we investigate the elastic behavior of superlattices of GBs.

As illustrated in Fig. 14.32, these idealized and somewhat hypothetical layered materials consist of a periodic arrangement, $\ldots|A|A'|A|A'|\ldots$, of thin slabs A and A' of equal thickness, $\Lambda/2$. In contrast to a composition-modulated superlattice, however, A and A' consist of the same material, and are merely rotated with respect to each other about the interface normal by an angle θ (between $A|A'$) and $-\theta$ (between $A'|A$). We first consider GBSLs composed of (001) twist GBs in fcc metals as compared to the (001) thin films, followed by a comparison with GBSLs on the (111) and (011) planes. We hope to demonstrate that, similar to the thin films discussed so far in this section, the observed elastic anomalies of these simple model systems are governed by the competition between the effects of structural disorder at the interfaces and the accompanying lattice-parameter (i.e. volume) changes of the system. The insights gained from the study of these GBSLs leads us naturally to predictions for dissimilar-material

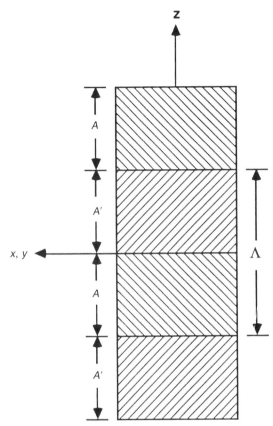

Fig. 14.32 Periodic arrangement of thin slabs, A and A', to form a 'grain-boundary superlattice' (GBSL). A and A' are thin slabs of the same material, each of thickness $\Lambda/2$, rotated about the GB-plane normal ($\|z$) to form a periodic array of twist GBs in the $x–y$ plane. The overall arrangement of atoms is thus periodic in all three dimensions.

superlattices, which will be investigated in section 14.4.

(001) Grain boundary superlattices

In superlattice materials in which detailed X-ray studies exist, the elastic anomalies were found to be accompanied by lattice-parameter changes [2–5]. In general, an expansion in the z-direction (parallel to the interface-plane normal) is observed that is accompanied by lattice-parameter changes in the interface ($x–y$) plane. Whereas the expansion in the z-direction can explain [17] the observed

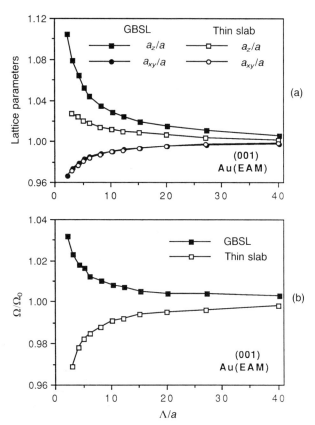

Fig. 14.33 Average lattice parameters perpendicular (a_z) and parallel ($a_x = a_y \equiv a_{xy}$) to the interface planes, and the average atomic volume (Ω/Ω_0) versus modulation wavelength, Λ (in units of the bulk lattice parameter, a) for (001) GBSLs as compared to (001) thin slabs (Au(EAM) potential). Ω/Ω_0 is the atomic volume of the perfect crystal.

Table 14.3 Bulk ideal-crystal elastic constants in units of 10^{12} dyn/cm^2 (Au(EAM) potential) for various orientations. These are also the values for the thin films in the $\Lambda \to \infty$ limit

Film	x	y	z	C_{11}^{∞}	C_{22}^{∞}	C_{33}^{∞}
(111)	$[1\bar{1}0]$	$[11\bar{2}]$	$[111]$	2.129	2.129	2.236
(001)	$[100]$	$[010]$	$[001]$	1.807	1.807	1.807
(011)	$[100]$	$[0\bar{1}1]$	$[011]$	1.807	2.129	2.129

volumes of fully relaxed (001) GBSLs and (001) thin slabs are compared in Fig. 14.33 as a function of Λ (the modulation wavelength for GBSLs and the thickness of the slabs). Both the GBSLs and the thin slabs contract in the x–y plane with decreasing Λ, due to their similar tensile interfacial stresses (Tables 14.2 and 14.4). Both also show an overall expansion along z with decreasing Λ due to a Poisson response to the x–y contraction. However, whereas the thin slabs show an overall volume decrease (Fig. 14.5), GBSLs show an overall volume **expansion** due to the large initial expansion already present at each individual GB [14, 15]. From these volume changes alone, one might naïvely expect that the thin films are elastically stiffer than the GBSLs. As evidenced by radial distribution functions, however, the GBSLs exhibit much more atomic-level structural disorder than the thin slabs [45, 46] (see also Fig. 14.1).

The elastic behavior of the (001) GBSLs is summarized in Fig. 14.34(a), which shows representative elastic moduli (for other moduli and constants see Ref. [46]) as a function of Λ by comparison with those of the (001) thin slabs. According to Fig. 14.34, a significant increase in the Young's modulus in the z-direction, Y_z, is coupled with a dramatic decrease in the shear modulus ($G_{xz} \equiv C_{55}$) for both the GBSLs and the thin slabs. Contrary to what one might expect based on the behavior of homogeneous systems, Y_z increases faster for the GBSLs than for thin slabs as Λ decreases, despite the larger volume expansion in the GBSLs. Due to the loss of the perfect-crystal stacking of the atomic positions across the GB (e.g. Fig. 14.1) and the consequently small shear resistance of GBs [36, 61], in the GBSLs G_{xz} is dramatically reduced compared to the thin slab

softening of the elastic constants C_{44} (for shear parallel to the interface plane) [3] and C_{33} (for strain parallel to z) [4], the **strengthening** reported in the Young's and the biaxial **moduli** [1, 8] and also the shear constant C_{55} [9] appears to be in conflict with conventional intuition since it is well known that in bulk crystals a lattice expansion is usually accompanied by a **softened** elastic response.

To investigate the origin of these elastic anomalies, we first investigate the atomic structure and anisotropic lattice-parameter changes of GBSLs consisting of (001) $\theta = 36.87°$ (so-called Σ5) twist GBs. The average lattice parameters and atomic

Table 14.4 Tensile stresses σ_{xx}^{∞} and σ_{yy}^{∞} (<0) of fully relaxed bulk GBs, with energies E_{GB}^{∞}, obtained by means of the Au(EAM) potential, in units of 10^{-3} J/m^2 [54]. δV is the volume expansion (in units of the bulk lattice parameter, a). Also listed are the Young's and biaxial moduli, Y_x^{∞}, Y_y^{∞}, Y_z^{∞}, and Y_b^{∞}, and the Poisson ratios, v_{xy}^{∞} and v_{zx}^{∞}, obtained in the $\Lambda \to \infty$ limit (in units of 10^{12} dyn/cm^2). The x and y directions are parallel to the directions listed in Table 14.2, for the corresponding thin slabs, of one of the crystal halves of each GBSL (the other crystal half of each GBSL is rotated by the twist angle θ with respect to the first)

GB or GBSL	θ	σ_{xx}^{∞}	σ_{yy}^{∞}	Y_b^{∞}	v_{xy}^{∞}	v_{zx}^{∞}	Y_x^{∞}	Y_y^{∞}	Y_z^{∞}	E_{GB}^{∞}	$\delta V/a$
111)	21.79°	−702	−702	1.947	0.617	0.232	0.745	0.745	1.212	212	0.018
001)	36.87°	−1494	−1494	0.647	0.109	0.775	0.576	0.576	0.346	524	0.056
011)	50.48°	850	−676	–	–	–	1.004	0.779	0.948	685	0.042

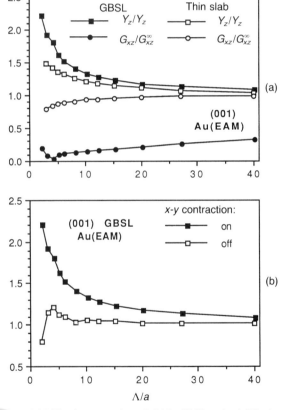

Fig. 14.34 Elastic properties of (001) GBSLs. (top) Elastic moduli, normalized to the $\Lambda \to \infty$ limiting values (Tables 14.2–14.5), for (001) GBSLs and (001) thin slabs as a function of the modulation wavelength (or film thickness), Λ, as described by the Au(EAM) potential. (bottom) Normalized Young's modulus, Y_z, a function of Λ for the (001) GBSLs with and without allowing for the stress-induced changes in the x–y lattice parameters.

[46, 62]. This comparison of the GBSLs with the thin slabs indicates that the amount of structural disorder in the systems is strongly correlated with the magnitude of the elastic anomalies [26].

The distinct effects due to atomic-level interfacial structural disorder and anisotropic lattice-parameter changes can be separated by comparing the elastic response of fully stress-relaxed GBSLs with identical GBSLs constrained to have the bulk x–y lattice parameters. Shown in Fig. 14.34(b) are the Young's moduli, Y_z, in each case [46]. When suppressing the x–y Poisson contractions, Y_z shows only a modest strengthening with decreasing Λ by comparison with the GBSLs in which the x–y contractions are permitted. This comparison illustrates that the stress-induced lattice-parameter changes serve to enhance the elastic anomalies which are already present due to the interfacial structural disorder alone. Thus, while the generic elastic anomalies of the interface system are caused by the structural disorder, the effect of the lattice parameter changes is to enhance these anomalies further. In addition, the maximum in Y_z observed for the case in which no x–y contraction is permitted is no longer present when the x–y contraction takes place. This effect, originating from the interaction between neighboring interfaces in the superlattice, was discussed in detail in Ref. [62].

According to the above results, the net elastic response of an interfacial system appears to be the result of two competing influences [26]: Whereas atoms near an interface may be pushed rather closely together, by up to 10% of the perfect crystal positions [46], causing a **strengthening** of the elastic constants, others are moved further apart (due to the usual expansion in the z-direction), giving rise to a **weakened** elastic response. Because of the anharmonicity of all interatomic interactions, contributions from atoms pushed more closely together are weighted more heavily than those from atoms moved further apart, with the net result that some elastic moduli may actually be strengthened in spite of the increase in the average distance between the atoms [26]. The structural disorder at the interfaces therefore provides causes for both a strengthening and a softening of the elastic response.

Comparison of (001), (111), and (011) superlattices

While the above discussion illustrates the very different roles played by the **structural disorder** and the consequent anisotropic **lattice-parameter changes** in the GBSLs, here we attempt to separate the two phenomena even further. If the structural disorder at the interfaces is, indeed, the main cause for the anomalous elastic behavior of the system, one would expect a strong dependence of the magnitude of these anomalies on the detailed atomic structure of (and chemistry at) the interfaces. More disorder, as measured, for example, via the interface energy, would be expected to enhance the elastic anomalies while less disorder would be expected to decrease them, provided the effects of disorder dominate the accompanying volume expansions.

To test this prediction, we compare the behavior of GBSLs on the (111) and (001) planes, with twist angles of 21.79° (so called Σ7 GB) and 50.48° (the so called Σ11 GB), respectively, with the above results for the (001) GBSLs. Our choice of (111) and (011) twist boundaries is motivated by the fact that the GB energy is known to depend very strongly on the interface plane [63], with the denser (i.e. more widely spaced) (111) planes giving rise to a much lower energy (typically by 50%) [14, 34] than the (001) plane. By contrast, twist boundaries on the less dense (011) planes exhibit a much higher energy than the GBs on the (001) plane (Table 14.4). Based on their lower

energies, and thus a lesser degree of interfacial structural disorder, one would expect the elastic anomalies of the (111) GBSLs to be significantly smaller than those of the (001) GBSLs [26, 46, 64]. Conversely, the higher energies of GBs on (011) indicate that the (011) GBSLs are more disordered than the (001) GBSLs, and should therefore exhibit even greater elastic anomalies. However, since lattice-parameter (and volume) changes are also important, it is difficult to predict exactly what to expect.

Based on the interfacial stresses listed in Table 14.4, the GBSLs on the (isotropic) (001) and (111) planes show monotonic contractions in the x–y plane with decreasing Λ, while the GBSLs on the (anisotropic) (011) plane expand in one direction and contract in the other. As a consequence of these anisotropic lattice-parameters changes, the average planar unit-cell area, A, decreases in all three cases while the average atomic volume, Ω, increases with decreasing Λ (Fig. 14.35). Based on the behavior of homogeneous systems, one might expect at least some strengthening of the in-plane elastic behavior, and a softening of the out-of-plane elastic behavior. However, as seen from Fig. 14.36, only the (111) GBSLs behave approximately in this homogeneous manner in that the in-plane Young's modulus, Y_x, indeed strengthens; however, the out-of-plane Young's modulus, Y_z, slightly strengthens with decreasing Λ, despite the z expansion, and in contrast to (111) thin slabs. By contrast, although the (001) GBSL show the largest increases in Ω and a_z, Y_z nevertheless strengthens the most. Also, all Young' moduli are substantially **softened** for the (011) GBSLs, despite their having the highest GB energy of the three superlattice types. It appears that the large volume expansion of the (011) GBSLs accompanied by a relatively small in-plane contraction, is sufficient to overcome any strengthening effects due to the structural disorder. Based on this comparison, it seems very difficult to predict the magnitude, or even the sign, of the elastic anomalies, even based on a knowledge of the GB energies and the stress-induced lattice parameter changes.

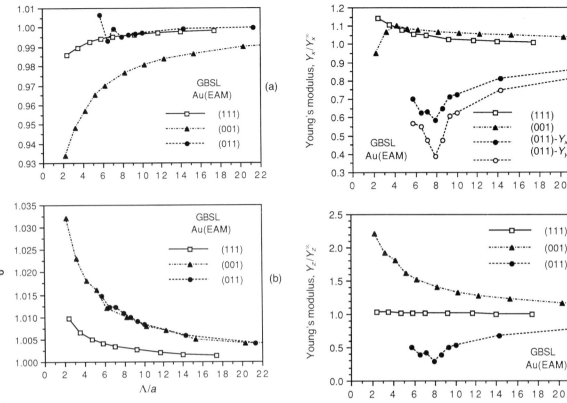

Fig. 14.35 Normalized (top) planar unit-cell area, A, and (bottom) average atomic volume, Ω, as a function of Λ, for GBSLs on (001), (111) and (011) planes using the Au(EAM) potential. Compare Fig. 14.21 for the corresponding thin slab results.

Fig. 14.36 Young's moduli, Y_x, Y_y and Y_z of fully-relaxed Au(EAM) GBSLs, normalized to the related bulk values (Table 14.4). For the (001) and (111) films $Y_x = Y_y$. Compare Fig. 14.22 for the corresponding thin slab results.

14.3.7 Conclusions

We conclude with a comparison of the two potentials employed throughout this investigation. As mentioned at the outset, it is not clear whether even the most sophisticated interatomic potentials represent a good description of those physical properties of a material to which they were not fitted. All we can hope to accomplish is an interpretation of the simulation results in terms of the input potentials.

In spite of the rather different structural relaxations observed for the two interatomic potentials used here, the generic elastic behavior predicted is remarkably similar. The only qualitative differences observed were shown to arise from the almost ten times larger stresses and consequent contractions in the x–y plane obtained for the EAM potential. The main reason for this qualitatively rather similar behavior seems to lie in the fact that the elastic behavior of thin films and thin-film superlattices is governed by the very presence of the interfaces as structural defects (giving rise to broken bonds and interfacial stresses), whereas the actual relaxation displacements of the atoms are of minor consequence. This qualitative agreement between the two potentials is particularly noteworthy if we recall that (i) the LJ potential automatically satisfies the Cauchy relation, $C_{12} = C_{44}$, and (ii) no elastic

Table 14.5 Elastic constants and shear moduli of GBSLs using the Au(EAM) potential (in units of 10^{12} dyn/cm^2) corresponding to the GBSLs in the $\Lambda \to \infty$ limit. The x and y directions are parallel to the directions listed in Table 14.2 (for the corresponding thin slabs) of one of the crystal halves of each GBSL (the other crystal half of each GBSL is rotated by the twist angle θ with respect to the first)

GBSL	θ	C_{11}^{∞}	C_{22}^{∞}	C_{33}^{∞}	G_{yz}^{∞}	G_{xz}^{∞}
(111)	21.79°	2.027	2.027	2.236	0.156	0.156
(001)	36.87°	1.955	1.955	1.807	0.440	0.440
(011)	50.48°	2.150	2.019	2.129	0.205	0.149

constants were used in the fitting of the two LJ parameters.

Utilizing the unique capability of computer simulations to probe the response of a system in ways often not possible experimentally, the above investigation of simple model interface materials has elucidated the distinct roles of the atomic-level structural disorder at the interfaces, and the consequent anisotropic changes in lattice parameters, in their elastic behavior. More interfacial structural disorder clearly yields larger elastic anomalies, provided the consequent volume expansion (and the accompanying elastic softening) does not dominate.

Based on these insights gained from the study of thin films and GBSLs, one would expect that the interfacial system containing the least amount of structural disorder, namely a composition-modulated superlattice with **coherent** interfaces, should exhibit the smallest elastic anomalies. Then, when structural disorder is reintroduced by removing the constraint of coherency at the interface (i.e. by introducing interface dislocations), the elastic anomalies should increase towards their magnitudes in the GBSLs. These predictions will be tested in the next section.

14.4 COMPOSITION-MODULATED SUPERLATTICES

Several attempts at explaining the elastic anomalies of superlattices, such as those focusing on elec-

tronic effects [22, 23] and lattice-parameter changes [18, 20, 21, 23], assume the anomalies to be due largely to bulk effects; however, the GBSL and thin-film results discussed in the preceding two sections show that the structural disorder due to the presence of the interfaces is of central importance to the elastic response, and that the observed elastic anomalies are largely an interface effect [26, 46, 62, 64]. Since the GBSLs are, to a degree, only model systems and the thin films are only basic building blocks, it is desirable to investigate composition-modulated superlattices whose 'structures' can range from coherent to incoherent. In particular, the incoherent super lattices can have significant, inhomogeneous atomic-level disorder and are therefore expected to show similar or even greater elastic anomalies than the GBSLs.

Dissimilar-material superlattices have been studied by means of computer simulations only rather recently. Coherent (strained-layer) fcc superlattices, modeled by means of EAM potentials [30], have been simulated by Dodson [65] at three small modulation wavelengths with the composition modulated along [001], and also by Jones and Mintmire [66] with the composition modulated along [111]. Due to the one-to-one correspondence of atoms directly across the interfaces, these coherent superlattices (COHSLs) have relatively little structural disorder and, hence showed only small elastic anomalies. Similar results have been obtained for COHSLs modeled by Lennard–Jones potentials [67, 68].

By contrast to COHSLs, incoherent super lattices (INCSLs) are unstrained as $\Lambda \to \infty$ and do not have the same number of atoms in the planes directly across the interfaces (Fig. 14.37). Since they are clearly more structurally disordered, the INCSLs should show significant elastic anomalies, based on the predictions of the section 14.3. Sasajima *et al.* [69] and Imafuku *et al.* [70] have investigated the elastic properties of INCSLs with composition modulation along [111] by using empirical Morse potentials. However, the implications of their results are difficult to assess since (i) the simulations do not appear to have been done at constant pressure along [111], (ii) they find no effect of changing

[001]   [001]

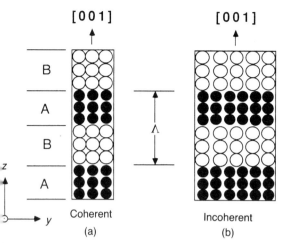

z

y

Coherent   Incoherent
(a)   (b)

Fig. 14.37 Schematic illustration of coherent and incoherent superlattice geometries. In our simulations, material B (open circles) is chosen to have a 20% larger lattice parameter (not shown to scale) than material A (solid circles). In the coherent superlattice there is, by definition, a one-to-one correspondence between atoms directly across the interfaces. The modulation wavelength is Λ, and the modulation direction is [001].

the relative lattice-parameter mismatch of the modulating mterials, and (iii) they do not present any physical interpretation of the results. Elastic moduli have recently been determined for Pd–Cu multilayers by Gilmore and Provenzano at several small modulation wavelengths using EAM potentials [71]. Since their simulations do not, *a priori*, impose a coherent or incoherent structure, they can probe the coherent–incoherent transition (see below); however, their elastic calculations only contain the Born-term contributions, and do not take into account the additional contributions from atomic relaxations.

In this section, we systematically study the role of atomic-level disorder on the elastic behavior of superlattices by investigating three types of composition-modulated superlattices with varying amounts of interfacial disorder [72]. After investigating their structures, we discuss their elastic behavior and the role of coherency. A comparison is made with GBSLs and thin films, and finally, the role of the lattice-parameter mismatch on the elastic properties of the superlattices is discussed.

14.4.1 Structure

The lattice-parameter mismatch $f = (a_B - a_A)/a_A$ between modulating materials A and B, modeled by LJ potentials, is chosen to be 20%. Although smaller lattice-parameter mismatches are typically of greater experimental interest, f is chosen to be large in order to provide a high degree of structural disorder at the incoherent interfaces. In addition, COHSLs and INCSLs with a smaller mismatch ($f = 10\%$) have also been studied [68], and will be discussed below.

Since we are primarily concerned with understanding the generic elastic behavior of the superlattices, rather than reproducing the elastic constants of a particular system, material B was chosen to have the same cohesive energy (i.e. $\varepsilon_A = \varepsilon_B$) but a 20% larger lattice parameter ($\sigma_B = 1.2\sigma_A$). For simplicity, the energy and length scales of the A–B interaction potential were taken in the usual fashion [73] as $\varepsilon_{AB} = (\varepsilon_A \varepsilon_B)^{1/2} = \varepsilon_A$, $\sigma_{AB} = (\frac{1}{2})(\sigma_A + \sigma_B)$, and with $a_A = 3.6160\,\text{Å}$ the average lattice parameter, $\bar{a} = (a_A + a_B)/2$, becomes $3.978\,\text{Å}$. As discussed in section 14.3.1, numerous studies have found LJ potentials to yield qualitatively the same physical behavior as the many-body EAM potentials for isolated interfaces [14, 16, 36], in GBSLs [26, 46, 62, 64], and in thin films [45]. Furthermore, the elastic properties of composition-modulated COHSLs are qualitatively the same using either LJ [67] or EAM [65, 66] potentials. These studies suggest that LJ potentials are adequate for investigating the qualitative elastic behavior of composition-modulated superlattices.

Since GBSLs modulated along [001] show larger elastic anomalies than those modulated along [111] (see above, p. 392), we choose [001] as the modulation direction for the superlattices of this study. We consider both COHSLs and INCSLs, each having all crystallographic directions of the modulating materials aligned in parallel (Fig. 14.37). In addition, a third type of superlattice will be considered in which the modulating materials are rotated with respect to each other about the interface normal by a rotation angle such as to generate overall commensurate structures but with maximal structural disorder at the interfaces.

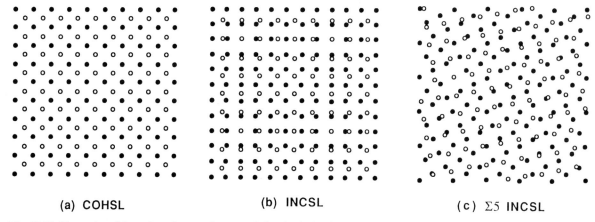

(a) COHSL   **(b) INCSL**   **(c) Σ5 INCSL**

Fig. 14.38 Illustration of the various degrees of structural disorder in the three superlattice types considered in this study, each having eight planes of each material per modulation wavelength. Shown are the relaxed atom positions, at zero temperature as viewed along [001], of one plane of material A (solid circles) and one plane of material B (open circles), bordering an interface of (a) the coherent superlattice (COHSL), (b) the incoherent superlattice (INCSL), and (c) and Σ5 incoherent superlattice (Σ5 INCSL).

These structures can be thought of as being derived from the INCSLs by introducing twist angles of $\pm 16.26°$ between alternating A and B layers, and will be designated Σ5 INCSLs (where here Σ5 simply denotes that the planar unit cell is five times larger than that of the INCSLs).

In all cases, materials A and B are taken to be separated by sharp (001) interfaces (i.e. each (001) plane was composed of only type A or only type B atoms), and an equal number of planes of each material is assumed. While the COHSLs contain only one atom per plane in the simulation cell, the INCSLs (Σ5 INCSLs) consist of $6 \times 6 = 36$ $(5 \times 6 \times 6 = 180)$ atoms per layer of material A for every $5 \times 5 = 25$ $(5 \times 5 \times 5 = 125)$ atoms per layer of material B. In contrast to the COHSLs in which the one-to-one correspondence of atoms across the interfaces produces an (unphysical and hence hypothetical) strained-layer superlattice even as $\Lambda \to \infty$, the INCSLs and Σ5 INCSLs are unstrained as $\Lambda \to \infty$.

The various degrees of disorder at the interfaces in the three superlattice types are illustrated in Fig. 14.38, where the the relaxed zero-temperature atom positions in the two planes adjacent to an

interface are shown for systems with eight planes of each material per modulation wavelength. As is evident from the figure, the COHSLs are perfectly ordered at the interfaces. The INCSLs are related to the COHSLs by the introduction of edge misfit dislocations [74], which relieve the long-range strains. These dislocations are visible as the square network of disordered atoms surrounding regions of relatively good fit. By introducing relative rotations about [001] between the modulating materials in the INCSLs, screw dislocations are introduced, thus producing highly disordered interfaces in the Σ5 INCSLs. Containing both edge and screw dislocations, the Σ5 INCSLs are the most disordered systems in this study and are hence expected to exhibit the greatest elastic anomalies.

The inhomogeneous distribution of interplanar spacings across the three superlattice types is shown in Fig. 14.39 for superlattices consisting of 16 (001) planes per Λ. In all three cases the interface regions are typically only two atomic layers thick. Since in the COHSL, material A is under tension in the $x-y$ plane while material B is under compression, the Poisson effect leads to an inter-

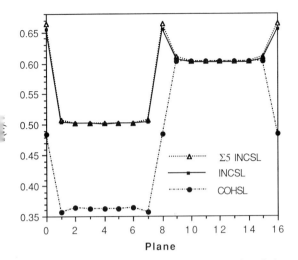

Fig. 14.39 Interplanar spacings along [001], in units of the lattice parameter of material A (a^A), for the three superlattice types, each with eight (001) planes of each material per modulation wavelength. The spacings at plane sixteen are periodic images of the spacings at plane zero.

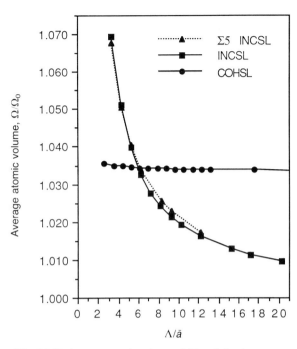

Fig. 14.40 Average atomic volume, Ω/Ω_0, of the three superlattice types as a function of the modulation wavelength Λ. The average atomic volumes of the unstrained systems in the $\Lambda \to \infty$ limit at Ω_0.

planar expansion in material A but compression in material B. By contrast, since in the INCSLs and in the $\Sigma 5$ INCSLs materials A and B are nearly stress free in the x–y plane, these superlattices have nearly the ideal-crystal interplanar spacings away from the interfaces. Whereas these superlattices show a pronounced expansion at the interfaces due to the interfacial disorder [16, 68, 72], the COHSLs contract. The volume expansions thus correlate well with the degree of interfacial disorder in each superlattice type. Similar to the GBSLs, the expansion at the interfaces in the INCSLs and $\Sigma 5$ INCSLs results in a Poisson contraction in the x–y plane, and due to the imposed uniform planar-unit-cell size, this x–y contraction leads to a small expansion of both materials in the z-direction away from the interfaces as well.

The overall volume and lattice-parameter changes are summarized in Figs. 14.40 to 14.42. Since the fraction of atoms experiencing the presence of the interfaces increases with decreasing Λ, the net effect is a smooth overall increase in the average atomic volume for the INCSLs and $\Sigma 5$ INCSLs (Fig. 14.40). Since there is already a significant amount of disorder in the INCSLs in the form of misfit dislocations, the introduction of further disorder by formation of the $\Sigma 5$ INCSLs leads to an overall volume increase which is only slightly greater than that in the INCSLs. On this scale, in the COHSLs the average atomic volume is practically independent of Λ. The x–y lattice-parameter changes are summarized in Fig. 14.41, according to which the disordered systems contract with decreasing Λ while the COHSLs expand. However, since these contractions are relatively small, the related changes in the average atomic volumes are dominated by the much larger changes in the average z lattice parameter.

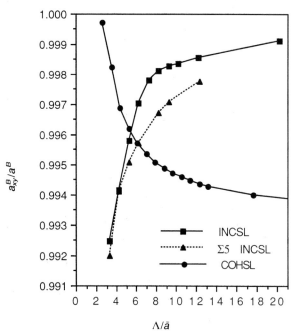

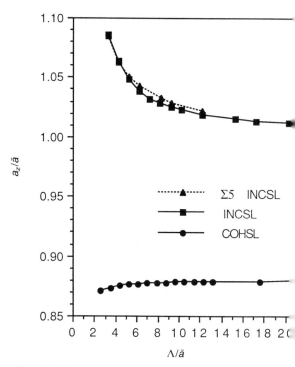

Fig. 14.41 Normalized average lattice parameter $a_x^B/a^B = a_y^B/a^B \equiv a_{xy}^B/a^B$, for material B in each superlattice type of Fig. 14.38, as a function of the modulation wavelength Λ. For the INCSLs and $\Sigma 5$ INCSLs, $a_{xy}^A/a^A = a_{xy}^B/a^B$ and for the COHSLs, $a_{xy}^A/a^A = 1.2 a_{xy}^B/a^B$.

Fig. 14.42 Lattice parameter $a_z/\bar{a}$, averaged over the whol superlattice, as a function of the modulation wavelength Λ fo the three superlattice types. The average lattice parameter $\bar{a}$ i given in Table 14.7.

14.4.2 Elastic properties

The dramatic effects on the elastic behavior of varying the degree of interfacial disorder are illustrated in Figs. 14.43(a–d) in which the three different superlattice types of Fig. 14.38 are compared. In general, the anomalous strengthening of certain Young's moduli and the softening of certain shear moduli correlate clearly with the degree of structural disorder.

In particular, according to Fig. 14.43(a) the two INCSL types show a large strengthening of Y_z (up to 32% for the $\Sigma 5$ INCSLs and 26% for the INCSLs), over their $\Lambda \to \infty$ value (Table 14.6) with decreasing Λ. As expected, the even more disordered $\Sigma 5$ INCSLs show the greatest strengthening of all three superlattice types. The COHSLs, by contrast, exhibit an overall softening for all values of Λ; as discussed earlier, this is due

to their being strained, even as $\Lambda \to \infty$ (Refs. [67] [68]). Overall, Fig. 14.43(a) clearly shows tha increased structural disorder at the interfaces ca lead to increased elastic strengthening of Y_z Following the same trends as those in Fig. 14.43(a for Y_z, 14.43(b) shows the elastic constant C_3 as a function of Λ for each superlattice type. Fo comparison, the biaxial modulus, Y_{bz}, is shown i Fig. 14.43(c). Here the $\Sigma 5$ INCSL shows a 13% strengthening, compared to the 9% strengthenin for the INCSL. By contrast, Y_{bz} for the COHSLs i again softened for all values of Λ.

The related shear moduli, $G_{xz} (=G_{yz})$, are show in Fig. 14.43(d). While for the COHSLs G_{xz} i relatively constant (and in fact shows a ver slight strengthening due to its contraction in th z-direction with decreasing Λ), for both INCSI types G_{xz} is very small at all Λ values investigate due to their high degree of interfacial disorder

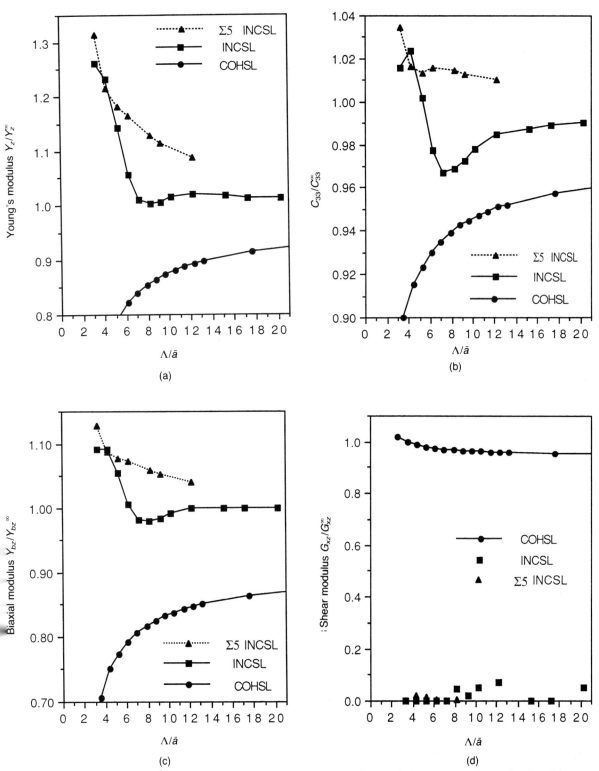

Fig. 14.43 (a) Young's modulus Y_z, (b) elastic constant C_{33}, (c) biaxial modulus Y_{bz} and (d) shear modulus G_{xz} as a function of the modulation wavelength Λ for the three superlattice types of Fig. 14.38. The normalization constants, which are the $\Lambda \rightarrow \infty$ values for the unstrained systems (i.e. the INCSLs), are given in Table 14.6.

Table 14.6 Normalization factors for the coherent superlattices (COHSLs), incoherent superlattices (INCSLs), and the INCSL derivative from introducing relative rotations between composition layers ($\Sigma 5$ INCSLs) in the dissimilar-material systems modeled by LJ potentials. The IFSL is an interface-free superlattice (see text) corresponding to the $\Sigma 5$ INCSLs. Also shown are the normalization factors for the monatomic (001) grain-boundary superlattices (GBSLs), and thin slabs. All of the factors are the $\Lambda \rightarrow \infty$ limiting values for the unstrained systems, and $\bar{a}$ is the average lattice parameter of the materials in a system

	20% lattice mismatch: COHSL, INCSL $\Sigma 5$ INCSL, IFSL	10% lattice mismatch: COHSL, INCSL, IFSL	No mismatch: (001) GBSL, slab, (001) $\theta = 36.87°$ GB in SLAB
Y_z^∞ (10^{12} dyn/cm^2)	0.793	0.925	1.076
C_{33}^∞ (10^{12} dyn/cm^2)	1.294	1.540	1.808
Y_{bz}^∞ (10^{12} dyn/cm^2)	1.295	1.462	1.681
G_{xz}^∞ (10^{12} dyn/cm^2)	0.727	0.866	1.016
$\bar{a}$ (Å)	3.9776	3.7968	3.6160

Similar to the GBSLs, the physical origin of the dramatic shear softening in the INCSLs and $\Sigma 5$ INCSLs can be attributed to the near-zero shear resistance of the disordered, volume-expanded interfaces [46, 68].

14.4.3 Comparison with strained crystals, thin films, and GBSLs

As in the case of the GBSLs [64], the effects exclusively due to the presence of the interfaces in the superlattices may be identified by comparing the elastic response of a given superlattice with that of corresponding interface-free perfect-crystal reference systems (IFSL) containing an identical number of atoms and with an identical simulation-cell shape and volume [72]. Figures 14.44 and 14.45 show such a comparison of the Young's and shear moduli for the case of the $\Sigma 5$ INCSLs, which showed the most anomalous elastic behavior of the three composition-modulated superlattices. According to Fig. 14.44, the dramatic strengthening of Y_z in the $\Sigma 5$ INCSLs is not nearly accounted for by the homogeneously strained IFSLs. Similarly, Fig. 14.45 shows that the IFSLs also cannot account for the extreme softening of G_{xz} in the $\Sigma 5$ INCSLs. This comparison clearly demonstrates, as a similar comparison did for GBSLs [64], that the enhancement of Y_z and the softening of G_{xz} are intimately connected with

the presence of structurally disordered interfaces in the system.

Another means of delineating the role of structural disorder on elastic properties is to compare the composition-modulated superlattices with previously studied homophase systems [26, 45, 46, 62, 64] that contain varying degrees of structural disorder. Such a comparison is also made in Figs. 14.44 and 14.45 between (a) the $\Sigma 5$ INCSLs, (b) GBSLs consisting of (001) $\theta = 36.87°$ ($\Sigma 5$) GBs, (c) a thin slab bounded by (001) free surfaces and (d) the same thin slab with a single (001) $\theta = 36.87°$ ($\Sigma 5$) GB in its middle. Both the Young's modulus Y_z (Fig. 14.44) and the shear modulus, G_{xz} (Fig. 14.45) display a nearly perfect correlation between the degree of interfacial disorder and anomalous elastic behavior. For example, the thin slabs, which represent the most structurally ordered systems, show a softening of both moduli as Λ decreases. Next, incorporation of a GB into the thin slabs introduces more structural disorder and results in up to a 17% **strengthening** of Y_z and a further softening of G_{xz} as Λ decreases. The (001) GBSLs, with yet even more structural disorder show an even greater strengthening of Y_z and softening of G_{xz} as compared to the thin slabs. Finally, the $\Sigma 5$ INCSLs, which are the most disordered of the systems, show the largest strengthening of Y_z (up to 32%) and the greatest softening of G_{xz}.

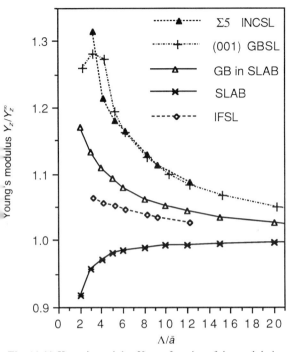

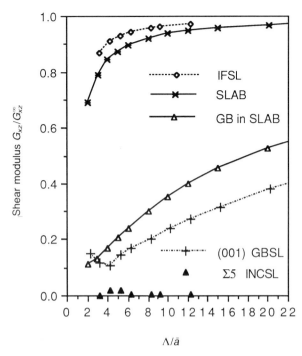

Fig. 14.44 Young's modulus Y_z as a function of the modulation wavelength Λ for the $\Sigma5$ INCSLs as compared to (001) GBSLs, thin slabs (SLABS) with and without an (001) $\theta = 36.87°$ ($\Sigma5$) GB, and an interface-free composition-modulated superlattice (IFSL) with the lattice-parameter changes of the $\Sigma5$ INCSLs (using LJ potentials). The normalization factors Y_z^∞ and $\bar{a}$ are given in Table 14.6.

Fig. 14.45 Shear modulus G_{xz} as a function of the modulation wavelength Λ for the interface-free composition-modulated superlattice (IFSL) with the lattice-parameter changes of the $\Sigma5$ INCSLs, the (001) GBSLs, and thin slabs with and without an (001) $\theta = 36.87°$ ($\Sigma5$) GB (using LJ potentials). The normalization factors are given in Table 14.6.

The Young's moduli (Fig. 14.44) of the GBSLs and the $\Sigma5$ INCSLs differ most at small Λ, where the former show a maximum and a subsequent decrease in Y_z with decreasing Λ, while the latter do not. The origin of the maximum in the GBSLs was shown to be due to the attractive nature of the interaction between the GBs [62], as evidenced by the GB energy which decreases at small Λ (Fig. 14.46). By contrast, as evidenced by the rising interface energy with decreasing Λ, the interfaces in the INCSLs and $\Sigma5$ INCSLs actually **repel** each other, and therefore do not show maxima in Y_z (see Fig. 14.43(a)). This qualitatively different behavior is paralleled by the shear moduli, which recover at small Λ for the GBSLs, but do not seem to recover for the INCSLs (Refs. [62] and [68]).

In summary, these results further suggest that the elastic strength of layered systems can be controlled by the nature of, and interactions between, the interfaces.

14.4.4 Role of the lattice-parameter mismatch

Based on the interpretation of the above results, one would expect that increasing the lattice-parameter mismatch should increase the elastic anomalies of INCSLs due to the increasing disorder in misfit dislocations. To test this expectation, in Fig. 14.47(a) the Young's modulus, Y_z, for the INCSLs with $f = 20\%$ is compared with that obtained for $f = 10\%$ [68]. Indeed, the $f = 20\%$ INCSLs are found to be stiffer in Y_z than the $f = 10\%$ INCSLs for most of the range in Λ, despite the 10% greater volume expansion of the

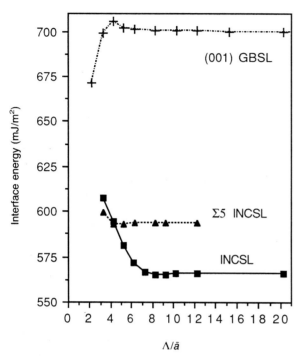

Fig. 14.46 Interface energies of the INCSLs and the (001) GBSL as a function of the modulation wavelength Λ. The average lattice parameters $\bar{a}$ are given in Table 14.6.

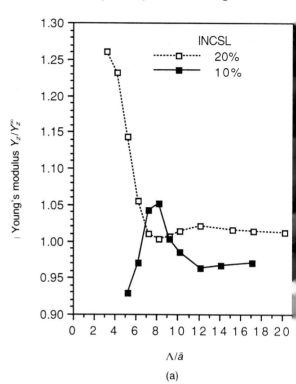

(a)

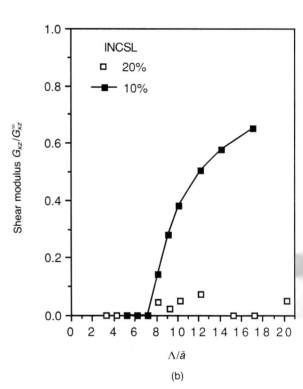

(b)

$f = 20\%$ INCSLs [79]. Furthermore, although G_{xz} is very small for both the $f = 10\%$ and 20% INCSLs (Fig. 14.47(b)), it shows a more rapid recovery in magnitude for $\Lambda > 8\bar{a}$ in the $f = 10\%$ superlattices [75]. Thus, as expected, the INCSLs with the larger lattice-parameter-mismatch show the greater elastic anomalies.

The $f = 10\%$ INCSLs show a maximum in Y_z near $\Lambda = 8\bar{a}$, and then softening at smaller Λ, while the $f = 20\%$ INCSLs show a continued strengthening at the smallest Λ (Fig. 14.47(a)). In contrast to the GBSLs, whose extrema (Figs. 14.44 and 14.45) in moduli is attributed to the attraction between interfaces [70], in both the $f = 10\%$ and

Fig. 14.47 (a) Young's modulus, Y_z, and (b) shear modulus, G_{xz}, as a function of the modulation wavelength Λ for INCSLs with a 10% and a 20% lattice-parameter mismatch between modulating materials. The normalization values in the $\Lambda \to \infty$ limit are given in Table 14.7.

20% INCSLs the interfaces repel each other (Fig. 14.46 and Ref. [68]). It appears that the maximum in Y_z in the $f = 10\%$ INCSLs is simply due to the continually increasing volume that, as Λ decreases, overcomes the strengthening due to the disorder. On the other hand, despite an even larger volume expansion than the $f = 10\%$ INCSLs, the $f = 20\%$ INCSLs do not show a maximum but the greatest strengthening in Y_z below $\Lambda = 7\bar{a}$. Apparently, the disorder in the $f = 20\%$ INCSLs is sufficient to always cause a strengthening, despite the increasing volume with decreasing Λ.

14.4.5 Conclusions

While the COHSLs are not stable at large Λ and the INCSLs are not stable at small Λ [72, 74], both structures can be studied at least metastably in computer simulations for a widely overlapping range of Λ values. The elastic properties of both types of superlattices can therefore be probed in experimentally unfeasible regimes, thus permitting the elucidation of the underlying behavior.

The comparison of the hetero- and homophase types of superlattices has demonstrated that the introduction of inhomogeneous structural disorder into the system, via edge and screw dislocations at the interfaces, can lead to significant elastic anomalies (i.e. behavior not seen in homogeneous systems) whose magnitudes correlate well with the degree of disorder. Most pronounced are the strengthening of the Young's and biaxial moduli and the dramatic softening of the moduli for shear parallel to the interfaces, as a function of decreasing modulation wavelength Λ. However, the price to be paid for the introduction of disorder into the systems is an overall volume expansion of the superlattices which, in some cases, can dominate the disorder effects and lead to a softening of the elastic response. That the disorder is introduced inhomogeneously in our layered systems seems to be of importance since, in general, there is an expansion along the modulation direction and a consequent Poisson contraction normal to it. The latter has been shown (p. 391) to further enhance the elastic strengthening already present due to the

interfaces [64], provided the expansion in the z-direction is not too large.

14.5 SUMMARY AND CONCLUSIONS

We have shown by means of computer simulations that any understanding of the elastic behavior of interface materials requires distinction between primary effects, arising from **inhomogeneous structural disorder** associated with the very presence of the interfaces, and secondary effects, arising from the interface-stress-induced **homogeneous**, albeit anisotropic, **lattice-parameter changes** of the system. By selectively suppressing the latter, it was possible to demonstrate that the basic anomalies associated with 'supermodulus behavior' (namely the enhancement in the Young's and biaxial moduli, the softening of the shear modulus, and the appearance of extremes in these moduli) arise from the structural disorder. By contrast, the lattice-parameter changes were shown to merely modify the intrinsic anomalies due to the very presence of the structurally disordered interfaces.

The effects due to the **in-plane** lattice-parameter changes conform with conventional intuition, namely that a contraction produces strengthened elastic behavior and vice versa. By contrast, the elastic response arising from the **out-of-plane** lattice-parameter changes as well as the structural disorder at the interfaces defy intuition: In spite of an overall expansion in the average lattice parameter in the direction of the interface normal, certain moduli may actually strengthen. It therefore appears that the highly nonlinear behavior known as the 'supermodulus effect' arises from the atomic-level structural disorder at the interfaces, including the interaction between different interfaces. This intricate connection between structural disorder and anomalous elastic behavior suggests that, for example, in nanocrystalline materials the connection between the average lattice parameter or density and the elastic behavior is considerably more complex than our intuition gained from homogeneous systems would suggest.

Throughout this study we have considered

extremely simple model systems whose properties may not be those of a real interface materials (in which the interfaces may, for example, not be structurally and chemically sharp). Also, while we have considered ideal defect-free interfaces mostly at zero temperature, entropic effects as well as point defects and defect clusters due, for example, to impurity segregation or the presence of voids, might be important in real materials. We nevertheless hope to have demonstrated that an investigation of relatively simple model interface systems provides an opportunity to study the interplay between the inhomogeneous effects of the local structural disorder at the interfaces and the homogeneous – yet anisotropic – effects due to the resulting changes in the average lattice parameters.

Our investigation of the distinct roles due to the interfacial structural disorder and the accompanying anisotropic lattice-parameter changes has relied largely on the unique capability of computer simulation to explore relatively simple, well-characterized model systems in ways not usually possible experimentally. This approach permits the essential physics of the model system to be exposed and provides unique insights into the underlying atomic-level causes. Four such studies have enabled elucidation of the supermodulus effect.

1. The elastic constants of superlattices were compared with those of a homogeneous perfect-crystal systems with identical unit-cell dimensions but no interfaces, thus permitting identification of the effects exclusively due to the interfaces.
2. The extremes in certain moduli were shown to arise from the mutual attraction between the interfaces in the superlattice, as evidenced by a rapidly decreasing interface energy for small modulation wavelengths.
3. By selectively suppressing the contraction in the interface planes, a manipulation not possible experimentally but straightforward in simulation, it was found that these changes in sample dimensions always lead to increased disordering of the system and, hence, larger elastic anomalies.

4. Although a coherent superlattice is not stable at large values of Λ while an incoherent superlattice is often not stable at small Λ, in our simulations both structures were investigated over a widely overlapping range of metastability; the elastic properties of both types of superlattices could thus be probed in experimentally inaccessible regimes.

We finally mention that throughout this chapter, qualitatively the same behavior was found for both pair- and many-body potentials used to describe metallic systems. This qualitative agreement is particularly noteworthy in view of the fact that pair potentials automatically satisfy the Cauchy relation, $C_{12} = C_{44}$, while many-body potentials yield three distinct elastic constants of a perfect cubic crystal. This qualitative agreement is encouraging as it indicates that in the investigation of structure–property correlations for metallic interface materials, the detailed nature of local atomic bonding at the interface may not be as important a factor as are, for example, the elastic strain-field and core effects associated with interface dislocations, effects which are present for any interatomic-force description.

ACKNOWLEDGMENTS

We have benefited from many discussions with S. R. Phillpot and J. M. Rickman. DW was supported by the US Department of Energy, BES Materials Sciences, under Contract No. W-31-109-Eng-38; JAJ received support from the Office of Naval Research, under Contract No. N00014-88-F-0019.

REFERENCES

1. W. M. C. Yang, T. Tsakalakos, and J. E. Hilliard, *J. Appl. Phys.*, **48** (1977), 876.
2. See, for example, I. K. Schuller, in *The Institute of Electrical and Electronics Engineers Ultrasonics Symposium, 1985* (ed B. R. McAvoy), IEEE, New York (1985), p. 1093; I. K. Schuller, A. Fartash and M. Grimsditch, *MRS Bulletin*, **XV**, (10) (1990) 33.
3. See, for example, M. Grimsditch, in *Light Scattering in Solids V* (ed M. Cardona and G. Guntherodt), Springer, Heidelberg (1990).

4. B. M. Clemens and G. L. Eesley, *Phys. Rev. Lett.*, **61** (1988), 2356.

5. D. Baral, J. B. Ketterson and J. E. Hilliard, *J. Appl. Phys.*, **57** (1985), 1076.

6. G. E. Henein and J. E. Hilliard, *J. Appl. Phys.*, **54** (1983), 728.

7. T. Tsakalakos and J. E. Hilliard, *J. Appl. Phys.*, **54** (1985), 734.

8. U. Helmersson, S. Todorova, S. A. Barnett, and J.-E. Sundgren, *J. Appl. Phys.*, **62** (1987), 491.

9. J. R. Dutcher, S. Lee, J. Kim, G. I. Stegeman, and C. M. Falco, *Phys. Rev. Lett.*, **65** (1990), 1231.

10. F. Erdogan, *ASME J. Appl. Mech.* **52** (1985), 823.

11. See, for example, *Interfacial Phenomena in Composites: Processing, Characterization and Mechanical Properties* (eds S. Suresh and A. Needleman), *Mater. Sci. Eng. A*, **107** (1989).

12. See, for example, *Metal–Ceramic Interfaces* (eds M. Ruhle, A. G. Evans, M. F. Ashby and J. P. Hirth), *Acta-Scripta Met. Proc.*, **4** (1990).

13. See, for example, *Science of Composite Interfaces* (eds R. G. Brandt and S. Fishman), *Mater. Sci. Eng. B*, **126** (1990).

14. D. Wolf, *Acta Metall.* **37** (1989), 1983.

15. D. Wolf, *Scripta Metall.* **23** (1989), 1913.

16. D. Wolf, J. F. Lutsko and M. Kluge, in *Atomistic Simulation of Materials* (eds V. Vitek and D. J. Srolovitz), Plenum Press, New York (1989), p. 245.

17. I. K. Schuller and A. Rahman, *Phys. Rev. Lett.*, **50** (1983), 1377.

18. A. I. Jankowski and T. Tsakalakos, *J. Phys. F.*, **15** (1985), 1279.

19. A. Banerjea and J. R. Smith, *Phys. Rev. B*, **35** (1987), 5413.

20. A. I. Jankowski, *J. Phys. F.*, **18** (1988), 413.

21. R. C. Cammarata and K. Sieradzki, *Phys. Rev. Lett.*, **62** (1989), 2005.

22. T. B. Wu, *J. Appl. Phys.*, **53** (1982), 5265.

23. M. L. Hubermann and M. Grimsditch, *Phys. Rev. Lett.*, **62** (1989), 1403.

24. W. R. Bennett, J. A. Leavitt, and C. M. Falco, *Phys. Rev. B*, **35** (1987), 4199.

25. D. Korn, A. Morsch, R. Birringer, W. Arnold and H. Gleiter, *J. de Physique Colloque C5*, **40** (1988), C5–769.

26. D. Wolf and J. F. Lutsko, *Phys. Rev. Lett.*, **60** (1988), 1170.

27. J. A. Jaszczak and D. Wolf, *Mater. Res. Soc. Symp. Proc.*, **229** (1990), 85.

28. J. Ray and A. Rahman, *J. Chem. Phys.*, **80** (1984), 4423, and *Phys. Rev. B*, **32** (1985), 733.

29. See, for example, A. G. Guy, *Elements of Physical Metallurgy* Addison-Wesley (1960) and D. C. Wallace, *Thermodynamics of Crystals* Wiley, New York (1972).

30. M. S. Daw and M. I. Baskes, *Phys. Rev. Lett.*, **50** (1983),

31. M. W. Finnis and J. E. Sinclair, *Phil. Mag. A*, **50** (1984), 45.

32. M. S. Daw and M. I. Baskes, *Phys. Rev. B*, **29** (1984), 6443.

33. J. F. Lutsko, D. Wolf, S. Yip, S. R. Phillpot and T. Nguyen, *Phys. Rev. B*, **38** (1988), 11572.

34. D. Wolf, *J. Am. Ceram. Soc.*, **67** (1984), 1, and *Physica B*, **131** (1985), 53.

35. See, for example, C. Goux, *Can. Metall. Quarterly*, **13** (1974), 9.

36. M. Kluge, D. Wolf, J. F. Lutsko and S. R. Phillpot, *J. Appl. Phys.*, **67** (1990), 2370; D. Wolf and M. Kluge, *Scripta Metall.*, **24** (1990), 907.

37. M. Parrinello and A. Rahman, *J. Appl. Phys.*, **52** (1981), 7182.

38. J. F. Lutsko, *J. Appl. Phys.*, **65** (1989), 2991.

39. M. Born and K. Huang, *Dynamical Theory of Crystal Lattices* Clarendon Press, Oxford (1954).

40. J. Ray, *Computer Physics Reports*, **8** (1988), 111.

41. J. B. Adams, W. G. Wolfer and S. M. Foiles, *Phys. Rev. B*, **40** (1989), 9479.

42. R. J. Needs, *Phys. Rev. Lett.*, **58** (1987), 53.

43. D. Vanderbilt, *Phys. Rev. Lett.*, **59** (1987), 1456.

44. B. W. Dodson, *Phys. Rev. Lett.*, **60** (1988), 2288.

45. D. Wolf, *Surf. Sci.*, **225** (1990), 117 and *Acta-Scr. Metall. Conf. Proc. Series*, **4** (1990), 52.

46. D. Wolf and J. F. Lutsko, *J. Mater. Res.*, **4** (1989), 1427.

47. R. C. Cammarata and K. Sieradzki, *Phys. Rev. Lett.*, **62** (1989), 2005; F. H. Streitz, K. Sieradzki and R. C. Cammarata, *Phys. Rev. B*, **41** (1990), 12285.

48. F. H. Streitz, K. Sieradzki and R. C. Cammarata, *Phys. Rev. B.*, **41** (1990), 12285.

49. See, for example, J. W. Cahn, *Acta Metall.*, **28** (1990), 1333.

50. A. I. Jankowski and T. Tsakalakos, *J. Phys. F*, **15** (1985), 1279.

51. J. R. Smith and A. Banerjea, *Phys. Rev. Lett.*, **59** (1987), 2451.

52. A. I. Jankowski, *J. Phys. F*, **18** (1988), 413.

53. D. Wolf, *Appl. Phys. Lett*, **58** (1991), 2081.

54. D. Wolf, *Surf. Sci.*, **226** (1990), 389, and *Phil. Mag. A*, **63** (1991), 337.

55. G. J. Ackland and M. W. Finnis, *Phil. Mag. A.*, **59** (1986), 301.

56. S. M. Foiles, M. I. Baskes, and M. S. Daw, *Phys. Rev. B.*, **33** (1986), 7983.

57. R. Benedek, *J. Phys. F.*, **8** (1978), 1119.

58. M. H. Grimsditch, *Phys. Rev. B.*, **31** (1985), 6818.

59. M. Grimsditch and F. Nizzoli, *Phys. Rev. B.*, **33** (1986), 5891.

60. S. P. Chen, A. F. Voter, and R. C. Albers, *Phys. Rev. B.*, **39** (1989), 1395.

1285, and *Phys. Rev. B*, **29** (1984), 6443.

61. D. Wolf and M. Kluge, *Scripta Metall.*, **24** (1990), 907.
62. D. Wolf and J. F. Lutsko, *J. Appl. Phys.*, **66** (1989), 1961.
63. D. Wolf, *J. Phys. Colloque C4*, **46** (1985), C4–197; D. Wolf and S. R. Phillpot, *Mater. Sci. Eng. A.*, **107** (1989), 3.
64. D. Wolf, Mater. *Sci. Eng. A.*, **126** (1990), 1.
65. B. W. Dodson, *Phys. Rev. B.*, **37** (1988), 727.
66. R. S. Jones and J. W. Mintmire, private communication; see also *Bull. Am. Phys. Soc.*, **35**, (1990), 780; J. W. Mintmire, *Mater. Sci. Eng. A*, **126** (1990), 29.
67. S. R. Phillpot and D. Wolf, *Scripta Metall. Mater.*, **24** (1990), 1109.
68. J. A. Jaszczak, S. R. Phillpot, and D. Wolf, *J. Appl. Phys.*, **68** (1990), 4573.
69. Y. Sasajima, M. Imafuku, R. Yamamoto, and M. Doyama, *J. Phys. F*, **14** (1984), L167.
70. M. Imafuku, Y. Sasajima, R. Yamamoto, and M. Doyama, *J. Phys. F*, **16** (1986), 823.
71. C. M. Gilmore and V. Provenzano, *Phys. Rev. B*, **42** (1990), 6899.
72. J. A. Jaszczak and D. Wolf, *J. Mater. Res.*, **6** (1991), 1207.
73. I. R. McDonald, *Mol. Phys.*, **23** (1972), 41.
74. W. A. Jesser and J. H. van der Merwe, in *Dislocations in Solids* (ed F. R. N. Nabarro), Elsevier, Amsterdam (1989), p. 421.

CHAPTER FIFTEEN

Interfaces within intercalation compounds

M. S. Dresselhaus and G. Dresselhaus

Introduction · Background material · Structure · Relation between properties and interfaces · Concluding remarks

15.1 INTRODUCTION

Intercalation compounds in general are formed by the insertion of atomic or molecular layers of a guest chemical species, called the intercalate, between layers of a host material. Thus in the ideal intercalation compounds the intercalate is separated by planar interfaces from the host material. Since intercalation compounds can be prepared such that alternate atomic layers correspond to host and intercalate layers, interfaces are ubiquitous in intercalation compounds.

To allow for the insertion of entire layers of guest species, the host material is itself required to be a layered material, with very strong in-plane bonding and very weak interplanar bonding. There are many examples of host materials which support intercalation, including graphite [1, 2, 3, 4, 5] (which has been studied most extensively), transition metal dichalcogenides [6, 7], some silicates and metal chlorides [6], some clays [8, 9, 10], and some polymers and gels [11].

The interfaces in intercalation compounds are unique in several ways. Firstly, the interfaces tend to be abrupt with the intercalate on one side of the interface retaining the basic properties of a monolayer of the bulk parent material while the host layer on the other side of the interface retains the

basic properties of the host material. Although the interface of an intercalation compound at room temperature is usually a solid–solid interface, melting can occur in the intercalate layer while the host layer remains rigid, giving rise to an internal solid–liquid interface [12]. Phase transitions from solid–solid to solid–liquid interfaces have been studied extensively [12]. In addition, at the surface of an intercalation compound is a solid–gas interface which behaves differently from the solid–solid and solid–liquid interfaces mentioned above.

With regard to the solid–solid interfaces, they can be of three types depending on the choice of host material and intercalate species: commensurate, incommensurate, and discommensurate; these types of interfaces are described and discussed in section 15.3. Since the adjacent interfaces are close enough spatially to interact with one another, structural relations between neighboring interfaces can be established.

Intercalation of a host material gives rise to the modification of its properties, resulting in materials that are more conducting, or more anisotropic, or magnetic, or superconducting, to give some examples. The nature of the interface significantly affects many of the properties of the resulting intercalation compound. Correspondingly, intercalation also results in the modification of the properties of the intercalate and because of the bonding characteristics between the intercalate and the host material at the interface, the chemical reactivity of the intercalate can be changed drastically. For example, $FeCl_3$ is a strongly hygroscopic material, but when intercalated into graphite the resulting $FeCl_3$–GIC is stable in the presence of water vapor.

In this chapter we start by providing some background material on intercalation compounds (section 15.2), followed by a discussion of the structural properties of intercalation compounds as they relate to the interface properties (section 15.3). The connection between the interface structures and the various properties of the intercalation compounds will then be presented (section 15.4).

15.2 BACKGROUND MATERIAL

As stated above, there are many host materials which support intercalation, and for each host material there are specific chemical species that can be intercalated, thereby giving rise to more than 1000 different intercalation compounds. A variety of chemical and structural considerations determine the facility with which a particular intercalate enters a given host material [13]. Although nature has synthesized intercalation compounds in the form of clays and minerals over geological time periods, and scientists have synthesized intercalation compounds for over 150 years [14], it is only recently that synthesis methods have been perfected to the point that one can prepare intercalation compounds with well-defined and reproducible interfaces [2, 3, 15].

The host materials which can be intercalated are conveniently classified according to their rigidity. For the class I compounds (Fig. 15.1) (e.g. graphite and BN), the host material has a basic structural unit that is only a single monolayer thick (i.e. 3.35 Å for graphite), thereby giving rise to a relatively thin basic structural unit for the class I compounds. This thin basic structural unit in turn gives rise to a small spatial separation between sequential interfaces, and in the case of the graphite

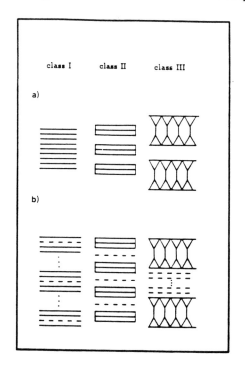

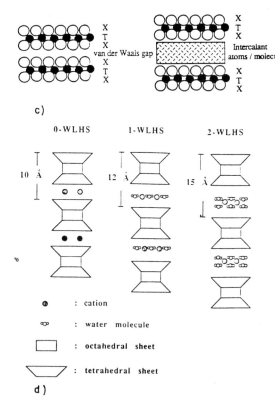

host material results in a long range structural order and a high degree of correlation between the interfaces. For graphite intercalation compounds (GICs), the nearest neighbor separation between interfaces can either occur across the intercalate layer (as for the alkali metal intercalates) or across the graphite host layer (as for metal chloride intercalates).

The basic structural unit of the host material for the transition metal dichalcogenide intercalation compounds consists of a trilayer (XMX), where M denotes the transition metal layer flanked on either side by a chalcogen layer denoted by X (Fig. 15.1(c)). This trilayer structural unit falls into the class II category, and is intermediate between class I and class III with regard to layer rigidity. Other examples of host materials in the class II category are the group III–VI (III: Ga, In; VI: S, Se) layered compound semiconductors, where the metal (M) and chalcogenide (X) layers are arranged in the sequence XMMX to form a four-layer slab with strong covalent bonding between each of the four layers and weak van der Waals bonding between these slabs [16]. In the case of class II compounds, the nearest neighbor interfaces tend to be across the intercalate layers, as is also the case for the class III materials. Examples of the rigid class III materials are the silicate clays, where the basic structural unit is typically ~10 Å thick, or thicker when the waters of hydration are included, as shown in Fig. 15.1(d) [9].

Of the various host materials, it is only graphite which forms well ordered superlattices, and this

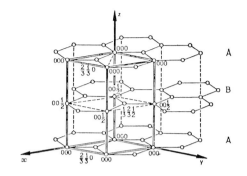

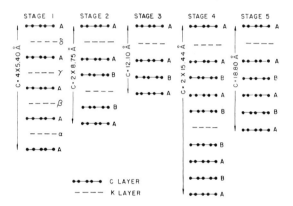

Fig. 15.2 Schematic diagram illustrating the staging phenomenon in graphite intercalation compounds for stages $1 \leq n \leq 4$. In this figure the potassium intercalate layers are indicated by dashed lines and the graphite layers by solid lines connecting open circles, and indicating schematically a projection of the carbon atom positions. The ...ABAB... graphite layer stacking (see graphite honeycomb structure) for stages $n \geq 2$ is maintained between intercalate layers, although a rhombohedral stacking arrangement (ABC) appears across intercalate layers. The stacking ordering is well-confirmed by X-ray diffraction (00l) patterns. For each stage, the distance I_c between adjacent intercalate layers is indicated. For first stage C_8K, the unit cell includes intercalate layers with stacking indices α, β, γ, δ [17].

Fig. 15.1 Schematic classification of (a) pristine layered host materials and (b) the intercalated forms of (a). In (b) the dashed (solid) lines represent guest (host) layers. Three classes of intercalation compounds are shown [8], class I having the lowest layer rigidity and class III having the greatest rigidity. (c) Schematic representation of transition metal dichalcogenide TX$_2$ sandwiches and the interlayer van der Waals gap before and after addition of an intercalation complex [7] between layers of TX$_2$, where T is the transition metal and X is the chalcogenide (i.e. S, Se, or Te). (d) Schematic representation of the vermiculite clay structure for 0-, 1-, and 2-water layer hydration states (WLHS). Typical values for the basal spacing are indicated.

relates to the staging phenomenon, by which a constant number of layers of a host material are sandwiched between sequential intercalate layers, as shown in Fig. 15.2. The staging phenomenon thus gives rise to a periodic array of intercalate sheets (a 1D superlattice normal to the layer planes) that can independently undergo 2D order–disorder and structural phase transitions within the layer planes. Staging allows control of the separation between sequential intercalate layers, thereby providing a means for control of the separation between the interfaces and of the magnitude of the interplanar interactions. This control is exploited in studies of a variety of 2D phenomena such as quasi-2D magnetism (section 15.4.3). The two graphite layers between which the intercalate layer is sandwiched are called the graphite **bounding** layers while the graphite layers which have graphite layers on either side are called graphite **interior** layers. Thus the interfaces in GICs occur between the graphite bounding layers and the intercalate layer.

Although graphite intercalation compounds (GICs) exhibit the highest degree of staging periodicity, some transition metal dichalcogenide intercalation compounds also exhibit evidence for staging, although this occurs only for low stage (i.e. $n = 1, 2$) compounds, and these compounds characteristically have low staging fidelity. Clay intercalation compounds (CICs) do not generally show evidence for staging, except for stage 1 compounds. On the other hand, clay host materials permit the intercalation of multiple sequential intercalate layers, in contrast to the graphite and transition metal dichalcogenide host materials which only permit intercalation of a single intercalate unit (either a monolayer or a trilayer unit as discussed below). Because of their high structural order, and the large variety of interfaces exhibited by the graphite intercalation compounds, this chapter will focus on the GICs.

Even for a single host material, the distance separating two host layers between which the intercalate is sandwiched, d_s, and the corresponding interfacial separation, can vary over wide ranges. One example of a GIC where d_s deviates only slightly from that of pristine graphite ($\bar{c}_0 = 3.35\,\text{Å}$) is the case of the Li donor intercalate where $d_s = 3.70\,\text{Å}$ and the interfacial separation is a mere $0.35\,\text{Å}$, while d_s can be greater than $3\bar{c}_0$ for trilayered intercalates such as transition metal chlorides (Table 15.1) [2, 3]. For the transition metal chloride GICs the interfacial separation

Table 15.1 Intercalate sandwich thickness d_s for several graphite intercalation compounds (GICs) showing that the repeat distance I_c for a low stage GIC can be varied over a wide range by proper choice of intercalate

Intercalate	$d_s\,(\text{Å})$
Li	3.706
Eu	4.87
K	5.35
Rb	5.65
Cs	5.94
Br_2	7.04
HNO_3	7.84
AsF_5	8.15
SbF_5	8.46
$NiCl_2$	9.30
$FeCl_3$	9.37
$SbCl_5$	9.42
$CoCl_2$	9.50
$FeCl_2$	9.51
$AlCl_3$	9.54
KHg	10.22
$ReCl_4$	11.78

across the intercalate layer will exceed that across the graphene layers for the stage 1 compounds. (A graphene layer is an isolated 2D layer of graphite that is not coupled to other layers in the graphite structure.)

The separation between specific intercalate layers can be further increased by the synthesis of bi-intercalation compounds [18, 19, 20], which are, for example, prepared by the intercalation of species M_2 into a stage 2 M_1–GIC, thereby forming a sequence $\ldots M_2 C M_1 C M_2 \ldots$. For such bi-intercalation compounds, the nearest neighbor interfaces are dissimilar. In a bi-intercalation metal chloride compound, the separation between sequential M_2 layers is close to $20\,\text{Å}$ and contains four interfaces. Such compounds are of particular interest for the synthesis of quasi 2D-magnetic systems [21] and in this connection the bi-intercalation compounds $CoCl_2$–$FeCl_3$–GIC and $CoCl_2$–$AlCl_3$–GIC have been synthesized and studied magnetically [22, 23].

With regard to the intercalation process itself, sufficient chemical activity is required to: (1) transfer charge from the intercalate to the adjacent host layers in order to provide the electrostatic attraction between the intercalate and the host layers that is necessary for bonding and (2) supply the elastic energy necessary to separate adjacent layers of the host material to accommodate the intercalate. The insertion of isolated intercalate molecules introduces large local strains into the host material. These strains are greatly reduced by the insertion of a second similar intercalate molecule nearby. Thus both elastic forces and normal chemical bonding forces in solids give rise to an attractive in-plane intercalate–intercalate interaction, so that once nucleation of an intercalate layer is initiated, the subsequent growth of the intercalate layer is rapid, exploiting the enhanced in-plane diffusion coefficients of layered materials.

For a given host material, some species can easily be intercalated and other species cannot. A strong electrostatic interaction and a large amount of charge transfer favors intercalation, while the separation of two sequential layers of the host materials to accept the intercalate layer requires the expenditure of energy. Thus a negative free energy $(G < 0)$ for the intercalation reaction favors intercalation, while a positive free energy $(G > 0)$ for the lattice strain suppresses intercalation. Thus for graphite or other host materials that permit intercalation and staging, the synthesis of a compound of a given stage depends on the sign $(G < 0)$ and magnitude of G. Thus for the Br_2–GICs, the stage 1 compound is energetically unfavorable while the stability of higher stage compounds increases with increasing stage index [13].

When an intercalation compound is energetically stable, it can often be prepared by a variety of techniques, such as vapor transport, liquid phase growth, growth from solution, or by electrochemical techniques [13, 24]. Of these various possibilities, electrochemical techniques offer perhaps the greatest number of possibilities for preparing intercalation compounds [25, 26, 27], while vapor transport often results in materials with the greatest order and staging fidelity [13].

In recent years, a great deal of attention has been given to the preparation of ternary compounds [28, 29, 30]. There are basically two different categories of ternary compounds: intercalates introduced as alloys, and intercalates introduced sequentially. An example of an intercalate alloy system is $K_x Rb_{1-x}$, which, to first order, intercalates into graphite as a substitutional random alloy that is homogeneous on each intercalate layer [28], so that there are no interfaces between the alkali metals. The interfaces only occur between the alkali alloy metal layer and the graphite bounding layer.

The more widely studied classes of ternary compounds are the sequentially intercalated compounds [28] which allow the intercalation of species M_2 which could not itself be intercalated into the host material. In such cases, the prior intercalation of species M_1, often an alkali metal, permits subsequent intercalation of species M_2. Examples of such ternary compounds, formed with M_1 as an alkali metal species, include as the M_2 species: tetrahydrofuran (THF), an organic species; NH_3, an inorganic species; and Hg and Tl, which form superconducting GICs. In the sequentially intercalated ternary GICs, interfaces form between the layers of the intercalate species, such as between the K and Hg layers in the KHg–GIC ternary compound shown in Fig. 15.3.

For the hydrogen ternary compounds, both chemisorbed and physisorbed varieties are possible. The structure of the hydrogen chemisorbed compound is similar to that for the KHg–GICs (Fig. 15.3) [30], while the physisorbed variety does not form the trilayer structure in the intercalate layer.

While energetics determine whether or not an intercalation compound is formed, kinetic considerations are important for determining the time necessary to form a given compound. To account for relatively fast staging transitions that are observed, Daumas and Hérold introduced a model (Fig. 15.4), which provides a simple explanation for a staging transition from stage n to stage $(n - 1)$ [31, 32, 33]. The Daumas and Hérold model thus introduces interfaces between regions of different stages for the same intercalation compound (section 15.3.3). *In-situ* X-ray $(00l)$ diffraction studies of the staging transition provide detailed information on the staging kinetics [34] and the motion of the Daumas and Hérold interfaces during the staging transition.

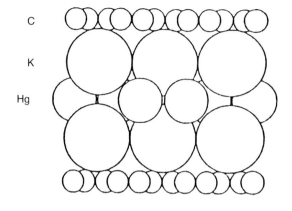

(a)

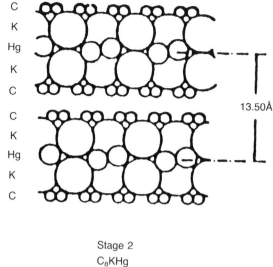

Stage 2
C_8KHg

(c)

Fig. 15.3 (a) Structure of a triple layer in C_4KHg-GIC [31]. (b) The layer stacking order for stage 1 (C_4KHg) showing the AA stacking of the graphene layers, and (c) the layer stacking order of the stage 2 (C_8KHg) showing the AB stacking of the adjacent graphene layers. The stage 2 compound still retains the AA stacking of the graphene bounding layers between which the intercalate is sandwiched. The ideal chemical formulae for the two compounds are also given.

C
K
Hg
K
C
K
Hg
K
C

10.15Å

Stage 1
C_4KHg

(b)

15.3 STRUCTURE

15.3.1 Overview

The structural arrangement within the intercalation compound governs the type of interfaces that are formed. In this section we review the structural properties of the graphite intercalation compounds with specific reference to the different kinds of interfaces that can be found [35, 36]. In section 15.3.2, we discuss the unusual properties of the interfaces between the different constituents of the intercalation compound and the effect of the interfaces on the stacking of the graphite bounding layers. In section 15.3.3 we discuss staging further

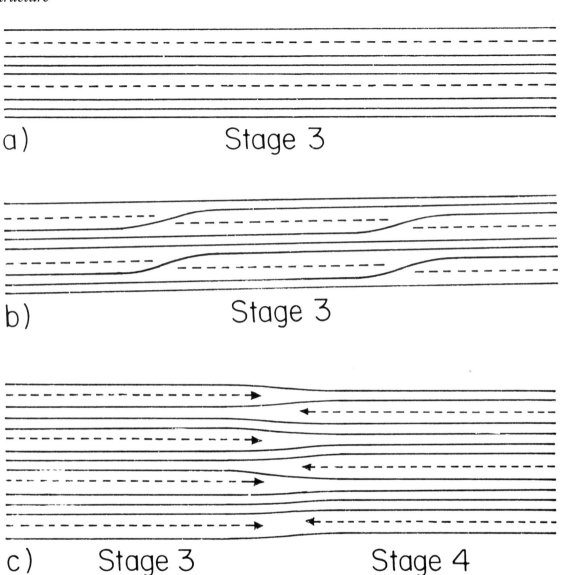

Fig. 15.4 (a) Single domain model for a stage 3 compound with every third layer occupied by the intercalate species. (b) Daumas–Hérold domain model for a stage 3 compound with intercalate species contained between all graphite planes. (c) Interchange of fourth and third stage regions as might occur during a stage transformation [32].

and the interfaces arising from staging at the Daumas–Hérold domain boundary.

Graphite intercalation compounds not only show long range staging fidelity with regard to their c-axis ordering (1D superlattice), but GICs also show various in-plane orderings, which relate directly to the interface between the intercalate and the graphite bounding layer. Two major classes of in-plane structures are the commensurate and the incommensurate arrangements of the

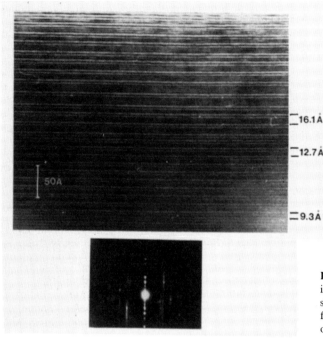

Fig. 15.5 High resolution TEM picture of the *c*-axis lattice images of a vapor grown graphite fiber intercalated with $CuCl_2$ showing regions with stage infidelity (marked by arrows). The figure below is an optical diffractogram taken from the negative of the TEM figure [37].

intercalate layers with respect to the graphite host layers, discussed in sections 15.3.4 and 15.3.5, respectively. Section 15.3.6 discusses discommensurate interfaces and the special case of the stripe domain phase. The long range site correlation between interfaces arising from the long range site ordering of the intercalate layers is described in section 15.3.7, while section 15.3.8 mentions the disproportionation phenomenon and possible phase boundaries implied by disproportionation. Internal solid–liquid interfaces and external surfaces are discussed in sections 15.3.9 and 15.3.10.

15.3.2 Interfaces between constituents

What is remarkable about the *c*-axis periodicity in GICs is the sharpness of the interfaces and the relative absence of intermixing between the intercalate and graphite layers (Fig. 15.5). One strong piece of evidence for these sharp interface properties comes from $(00l)$ X-ray diffraction measurements which show that the *c*-axis repeat distance I_c between sequential intercalate layers follows the simple relation $I_c = d_s + (n - 1)\bar{c}_0$

where $\bar{c}_0$ is the interlayer separation in pristine graphite. The validity of this simple relation down to stage 1 compounds indicates that, to first order, GICs can be considered as a linear superposition of graphite and intercalate layers with sharp interfaces and a negligible amount of intermixing. The sharp interfaces are confirmed by high resolution TEM lattice images of GICs, as shown for example in Fig. 15.5. Although sharp interfaces are exhibited by most GICs, one notable exception is fluorine-intercalated graphite where for fluorine concentrations greater than that corresponding to a stage 2 compound, some semi-ionic or covalent sp^3 bonding begins to form, and the ratio of the sp^3 bonding to the ionic sp^2 bonding characteristic of graphite and its intercalation compounds increases with increasing fluorine concentration [38]. The presence of sp^3 bonding results in a corrugation of the graphite bounding layer, giving rise to a waviness in the lattice fringes of fluorinated GICs.

The sharp interface between the intercalate layer and the adjacent graphite bounding layers represents a potential translational glide plane between the two graphite bounding layers. Prior to intercalation the two graphite bounding layers were in AB registry in the graphite structure (Fig. 15.2) [39], but after intercalation these two graphite layers usually assume an AA registry. This AA registry generally occurs for GICs having intercalate layers with mirror symmetry in the z-direction relative to the two graphite bounding planes, as for example in the alkali metal GICs [36]. Those GICs that do not have such a mirror symmetry plane tend to preserve the AB stacking of the graphite bounding layers [2]. The HNO_3–GICs are an example of a GIC which lacks mirror symmetry and has AB stacking of the two adjacent graphite bounding layers [2].

The scanning tunneling microscope (STM) allows investigation of the registration between the graphene and intercalate layers at the surface on an atomic scale. STM studies of single crystal graphite show a 3-fold pattern (Fig. 15.6(a)) [41], that differentiates between the a and b sites of the graphite structure (Fig. 15.2). However, after intercalation with an acceptor such as the first stage metal chloride $CuCl_2$–GIC, each carbon atom in the layer plane becomes equivalent because of the incommensurate relation between the graphene and intercalate layers [40] (section 15.3.5). As a result, the 6-fold pattern for the graphene layer (Fig. 15.6(b)) is observed. By changing the polarity of the GIC sample with respect to the tungsten tip from $+0.01\,V$ to $-0.01\,V$, the intercalate layer (rather than the graphene layer) is imaged (Fig. 15.6(c)), showing a centered distorted orthorhombic in-plane unit cell with lattice constants $a = 3.4 \pm 0.2\,\text{Å}$, $b = 3.9 \pm 0.2\,\text{Å}$ and an angle of $\gamma = 121°$ between them [40], consistent within experimental error with the X-ray result of $a = b = 3.81\,\text{Å}$ and $\gamma = 128°$ [42].

15.3.3 Interfaces introduced by staging

The structural ordering in the GICs is dominated by the staging phenomena, providing the basis for the c-axis periodicity (Fig. 15.2), and the interfaces associated with the superlattice. Many other host materials exhibit *some* short range staging correlations. For a few hosts, long range or quasi-long range staging is observed. For the case of graphite intercalation compounds (GICs), high resolution

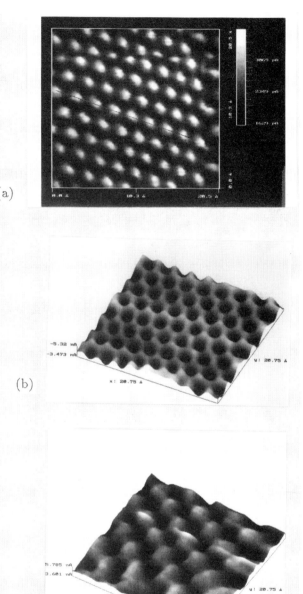

(a)

(b)

(c)

Fig. 15.6 STM images of dimensions $20.75\,\text{Å} \times 20.75\,\text{Å}$ [40] showing (a) a 3-fold symmetry pattern characteristic of pristine graphite where the a and b carbon atomic sites differ, (b) a 6-fold symmetry pattern seen for the graphene layer in stage 1 $CuCl_2$-GIC at a voltage of $+0.01\,V$ with respect to the tungsten tip where the adjacent incommensurate intercalate layer removes the distinction between carbon sites in the graphene layer, and (c) the STM image of the intercalate layer obtained by a voltage reversal from $+0.01\,V$ to $-0.01\,V$ [40].

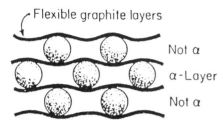

Fig. 15.7 Flexible graphite layer model for intercalate interlayer correlation. If the intercalate atom is on a site in the α-layer (possibly incommensurate), then the intercalate atom most probably will not lie on α-sites on adjacent intercalate layers [56]. Bringing the intercalate species closer together tends to flatten the graphene sheets, thereby providing an additional in-plane intercalate–intercalate attractive force.

TEM (transmission electron microscope) lattice fringe images show that a high degree of staging perfection (over a scale of ~100 Å) can be achieved [43] under favorable circumstances.

The classical model for staging (Fig. 15.2) cannot readily explain the rearrangement of the intercalate necessary to execute a staging transformation from stage n to stage $n - 1$. To address this issue, Daumas and Hérold introduced the concept that a GIC consists of well staged domains separated by an interface (Fig. 15.4) consisting of deformable layers of the host material that are arranged, so that between any two adjacent host layers, 2D islands of intercalate can be found [32]. Thus, a staging transformation from stage n to stage $n - 1$ corresponds to the sliding of the intercalate islands from one Daumas–Hérold domain to another, as shown in Fig. 15.4(c). To minimize the bending of the graphene layers (Fig. 15.7), the intercalate species tend to cluster in islands as shown in the figure. Calculations [44] show that the staggered islands bind to each other, thereby stabilizing the Daumas–Hérold interfaces (Fig. 15.4). Calculations [45] further show that for some intercalate species (e.g. K and Br$_2$) the binding energies of the Daumas–Hérold interfaces are small, so that the intercalate front moves easily, explaining why these species readily intercalate. Other species such as Na and Ca have high binding energies [45] and are difficult to intercalate. Thus we see that the intercalation process is closely connected to the behavior of the Daumas–Hérold

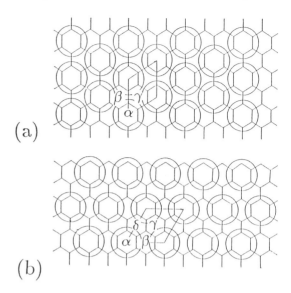

Fig. 15.8 Schematic diagram of an in-plane unit cell for the commensurate (a) $p(\sqrt{3} \times \sqrt{3})R30°$ and (b) $p(2 \times 2)R0$ structures showing the carbon atoms at the corners of the hexagons and the intercalate species as the large open circles. In this diagram the intercalate and graphene layers are projected onto a single plane. For (a) there are three equivalent sites labeled α, β, γ and for (b) there are four equivalent intercalate sites labeled α, β, γ, δ. On a given intercalate layer only one of these sites is occupied. For C$_8$K and C$_8$Rb the intercalate is arranged on sequential α, β, γ, δ sites [36].

interfaces. There are a number of microscopic techniques that have given detailed information on these interfaces as well as bulk properties which can be explained by the presence of such interfaces [33, 36].

15.3.4 Commensurate registration

In the commensurate structures (Fig. 15.8), the intercalate atoms are in registry with the graphite host, forming intercalate unit cells with areas that are simple multiples of the areas of the graphite unit cell (2D superlattices). Some examples of commensurate structures with regard to the graphite host material are: $(\sqrt{3} \times \sqrt{3})R30°$, $(2 \times 2)R0°$, $(\sqrt{7} \times \sqrt{7})R19.11°$, $(\sqrt{3} \times 2)R(30°, 0°)$ where the notation lists the lengths and angles of the intercalate basis vectors relative to those for the

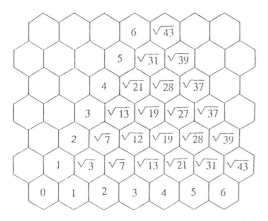

Fig. 15.9 A schematic sketch of a graphite honeycomb layer. Carbon atoms are located at the intersections of line segments. The numbers represent the commensurate distances, i.e. the distances from the centers of hexagons marked with the numbers to the center of the hexagon marked with 0, in units of the graphite in-plane lattice constant, $a_0 = 2.46$ Å [36].

graphite lattice, and angular redundancies are not repeated. Commensurate structures are found in a wide variety of model GIC systems, including: stage 1 binary donor compounds, such as C_6Li, C_6Eu, C_8K; coplanar molecular acceptor compounds, such as $C_{7n}Br_2$ for stages $n \geq 2$; triple layer sandwich acceptors, such as the $SbCl_5$–GICs; and triple layer sandwich donor ternary systems, such as KHg–GICs and KH–GICs [2, 3, 36]. Figure 15.9 shows many possible intercalate placements for commensurate interfaces, and many of these have been observed in GICs [36].

The formation of commensurate in-plane structures has a close analogy to the growth of strained layer superlattices by the molecular beam epitaxy (MBE) method or by metal–organo chemical vapor deposition (MOCVD) [46]. Because the intercalate layer in an intercalation compound is so thin (a single mono-layer in the case of an alkali metal donor GIC or a single trilayer sandwich in the case of an acceptor metal chloride), commensurate in-plane structures can be formed without the need for misfit dislocations to relieve local strains. The very high bulk modulus of the graphene layer makes that layer almost incompressible in the basal plane, so that the stretching (or contraction) of the

strained layer superlattice takes place almost entirely in the intercalate layer with almost no lattice expansion or contraction in the graphene layers. The small changes in the in-plane lattice constant in the graphene plane resulting from intercalation can be seen either in high resolution X-ray diffraction measurements [47, 48], or in the shift in the central frequency of the Raman-active in-plane E_{2g_2} graphitic mode [49], as discussed further in section 15.4.

The lowest energy state for an intercalate atom is obtained by placing it over (or below) the center of a graphite hexagon, thereby achieving the greatest packing density [2, 13]. A close-packed in-plane configuration for the intercalate maximizes its chemical bonding and minimizes the strain energy introduced into the adjacent graphite layers from insertion of the intercalate. When both registry with the graphite hexagons and close packing of the intercalate layers can be simultaneously satisfied with only a minor expansion or contraction of the intercalate layer, the commensurate registration of intercalate layers with the graphite bounding layers is favored. Otherwise, the close-packed configuration of the intercalate layer implied by its own chemical binding considerations prevails and an incommensurate interface is formed, as discussed in section 15.3.5.

15.3.5 Incommensurate interfaces

For incommensurate GICs (Fig. 15.10), the in-plane intercalate crystal structure is nearly the same as that of the parent material and the in-plane intercalate lattice constants are usually close to those prior to intercalation. The consequences of these similarities are far-reaching, allowing us to relate many of the properties of the constituents prior to intercalation to the properties expected for the intercalation compound (section 15.4). Common examples of intercalates forming incommensurate GICs are transition metal chlorides and bromides [51, 52].

Intercalate registration (the tendency for an intercalate atom to be located in a position of minimum energy with respect to the adjacent graphene layers) affects the structural ordering of GICs in

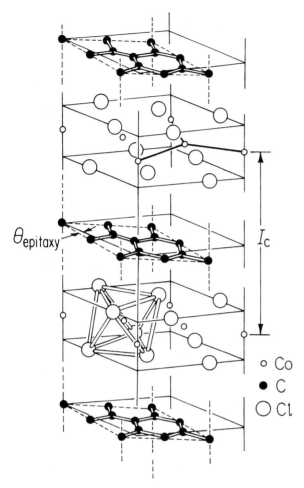

Fig. 15.10 Schematic crystal structure of stage 1 metal-chloride–GICs showing incommensurate registry of the intercalate layer with respect to the adjacent graphene layer. The rotational epitaxy angle is indicated by $\theta_{epitaxy}$. For stage 1 $CoCl_2$–GICs, $\theta_{epitaxy} \sim 0°$, the intercalate layers are stacked in an ... $\alpha\beta\gamma\alpha\beta\gamma$ ... arrangement (in contrast to the schematic diagram showing ... $\alpha\alpha$ ... stacking), and the graphene layers are stacked ABC [50].

many ways besides the formation of commensurate GICs, as discussed above. For the case of incommensurate GICs, local registration of the intercalate with respect to the graphite layers leads to **orientational** alignment of the intercalate, as evidenced by the spot patterns observed by ($hk0$) diffraction spectra for many incommensurate GICs [53, 54]. For example, the stage 1 $CoCl_2$–GIC exhibits alignment between the (110) axes of the

intercalate and the graphite bounding layers while for the stage 2 compound the angle between these (110) axes is 30°. Orientational alignment effects are very common for incommensurate GICs. In addition, there can be considerable interlayer correlation across incommensurate interfaces (section 15.3.7).

For many incommensurate intercalates (e.g. transition metal chlorides), the filling factor is significantly less than unity (e.g. ~80%). Recent high resolution TEM studies [55] show that, for the transition metal chloride intercalates, voids form in the intercalate layer, giving rise to another type of interface. The reason why the vacancies tend to cluster together is again related to the high bulk modulus of the graphene planes, resulting in an additional in-plane attractive force that enhances the binding of the nearest neighbor intercalate species. In viewing Fig. 15.7, we note that by clustering the intercalate the graphene layers are flattened, which is energetically favored by the high modulus of the graphene layers. For filling factor ~0.8, long range continuity of the intercalate within each layer can be maintained, despite the clustering of the vacancies to form voids [50].

15.3.6 Discommensurate interfaces

In contrast to the commensurate and incommensurate registrations between the intercalate and the surrounding graphite bounding layers, some GICs exhibit a discommensurate registration whereby local commensurate behavior is found for n_c unit cells with a distorted slip region of length after which the pattern is repeated. In this way long range incommensurate behavior is realized at the interface, while at the same time reaping the benefits of the local commensurate registration for minimizing the free energy.

Discommensurate interfaces and discommensuration domains are for example observed in high stage ($n \geqslant 2$) alkali metal GICs (Fig. 15.11) [61]. For the high stage ($n \geqslant 2$) alkali metals, detailed X-ray diffraction studies show that a commensurate ($\sqrt{7} \times \sqrt{7})R19.11°$ registration is inferred for the central atom and its nearest neighbors, but as we move further out into the discommensuration

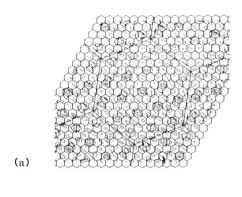

(a)

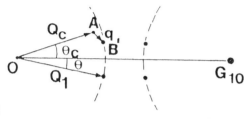

(b)

Fig. 15.11 Discommensurate domains for high stage ($n = 2$) alkali metal GICs. (a) Calculation of the alkali metal density distribution for $C_{12}Cs$ in a second stage compound with nominal stoichiometry $C_{24}Cs$ [57] showing commensurate registration for the central Cs atom and its six nearest neighbors, and incommensurate registration for Cs atoms near the interfaces of the discommensurate domains. (b) Important wave vectors in the discommensurate domain model [58, 59], including Q_c, the wave vector of the registered $(\sqrt{7} \times \sqrt{7})R19.11°$ structure, Q_1, the observed wave vector, q_1, a modulation satellite, G_{10}, the graphite (100) reciprocal lattice vector [60]. Here θ is the observed rotation angle and $\theta_c = 19.11°$.

domain, the alkali metal density becomes more smeared out [58, 59, 62], as noted in Fig. 15.11(a). By relaxing the commensurate registration within the domain, the measured in-plane alkali metal density of about $C_{12}M$ is achieved, rather than the $C_{14}M$ implied by the commensurate registration. The resulting modulated diffraction pattern is determined by the primary wave vector Q_1 observed in the diffraction pattern and the modulation wave vector q_1 which relates to the commensurate wave vector Q_c of the $(\sqrt{7} \times \sqrt{7})R19.11°$ structure. From Fig. 15.11(b) it is seen that $q_1^2 = Q_c^2 + Q_1^2 - 2Q_1Q_c \cos(\theta_c - \theta)$ where the angles are defined in the figure. The length L_d of the dis-

commensuration domain in Fig. 15.11(a) is given by $L_d = |G_{10}|/(\sqrt{7}|q_1|)$ where G_{10} is the (10) in-plane reciprocal lattice vector of the graphene layer. For the stage 2 Cs–GIC, the length $L_d = 11.04a_0$ and the rotation angle θ in Fig. 15.11(b) is $\theta = 8.93°$, so that each discommensuration domain in Fig. 15.11(a) contains between 18 and 19 Cs atoms [57, 60]. Although the interfaces between the discommensuration domains in Fig. 15.11(a) are indicated by sharp mathematical boundaries, these interfaces can physically be considered as the regions where the discommensuration takes place.

A particularly interesting example of a discommensurate in-plane interface, and qualitatively different from the one described above for the high stage alkali metal GICs, has been observed in a stage 4 Br_2–GIC with nominal stoichiometry $C_{28}Br_2$ [12]. At room temperature this compound exhibits a commensurate $(7 \times \sqrt{3})R(0°, 30°)$ in-plane structure with a rectangular unit cell containing four Br_2 molecules (Fig. 15.12(a)) [63]. The driving mechanism for a phase transition in this system is the differential thermal expansion between the graphite basal plane (which has essentially zero thermal expansion) and that for the intercalate bromine layer (which has a thermal expansion coefficient that is typical of molecular solids). For the Br_2–GIC system, the coupling between the intercalate and the graphite bounding layers is sufficiently large so that the structure remains commensurate over a large temperature range. At a temperature $T_c = 69°C$, a phase transition occurs, whereby above T_c the graphite and bromine layers retain long range registry along the $\sqrt{3}$-direction, but form a discommensurate interface along the 7-fold direction. This type of 1D discommensurate structure (remaining commensurate along one basis vector but becoming discommensurate along the other basis vector) is called a stripe domain phase (Fig. 15.12(b)). As the temperature increases above T_c, the average bromine lattice constant along the 7-fold direction increases, though the graphite and bromine remain locally commensurate over a domain containing n_c unit cells (Fig. 15.12(b)) [63]. These domains are arranged in a periodic array with lattice constant $(n_cb + \tau)$ where τ is the phase slip across the

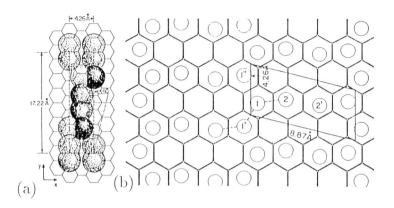

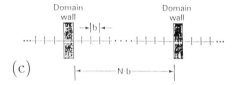

Fig. 15.12 (a) Commensurate structure of the bromine layer for C_7Br_2 showing the centered ($\sqrt{3} \times 7$) unit cell description, where the structure can be viewed as a packing of Br_2 molecules [63]. In (b) an oblique primitive cell is chosen (its area is half that of the ($\sqrt{3} \times 7$) unit cell) [36]. (c) Schematic diagram of the stripe domain phase formed along the large $7a_0$ unit cell vector. The intercalate lattice remains commensurate with the graphite bounding layer except in the domain wall where a phase slip occurs every n_c unit cells [63].

domain wall, and $\tau = 2b/7$ in which $b = 7a_0$ and a_0 is the graphite in-plane lattice constant. Both the temperature dependence of the incommensurability $\varepsilon = \varepsilon_0[1 - t]^{\frac{1}{2}}$ where $t = (T - T_c)/T_c$ and the lineshape of the observed X-ray reflections give strong evidence for a modulated stripe domain phase in stage 4 Br_2–GICs. At higher temperatures ($T_m = 102°C$), the bromine layers undergo a 2D melting transition, thereby forming an internal solid–liquid interface (see section 15.3.9).

15.3.7 Interlayer correlation between planar interfaces

Interlayer correlation between the planar interfaces resulting from the 3D stacking order of the intercalate layers with respect to one another is also observed in some cases, as, for example, in stage 1 alkali metal–GICs [2, 13] and in stage 1 metal chlorides [50]. Because the unit cells for GICs tend to be much larger than that for unintercalated graphite, there are several equivalent sites for the intercalate atoms in a commensurate GIC. For example, the $(2 \times 2)R0°$ in-plane lattice for first stage C_8K and C_8Rb (Fig. 15.8), shows four equivalent intercalate sites ($\alpha, \beta, \gamma, \delta$), but once one of these sites is occupied by an intercalate atom (e.g. an α site), then the intercalate atoms on the same layer (within a single domain) will also tend to occupy the same type of site (an α site) thereby giving the maximum packing density for the intercalate [2]. Introducing an intercalate atom on site introduces a local strain which is illustrated by the bending of the interface, as shown in Fig. 15.7. As a result of this local strain, the intercalate atom of the adjacent intercalate layers will be less likely be located on an α site than on a non-α site. The interlayer exclusion argument implies that when there is sufficient intercalate–intercalate interlayer correlation mediated by the intervening graphene layer (or layers), interlayer stacking order will result, as for example the interlayer stacking order (α, β, γ, δ) observed for the first stage compounds C_8K and C_8Rb and mentioned above. Interlayer stacking effects have more recently been also observed in commensurate ternary compounds [28, 30] and in incommensurate transition metal chloride–GICs [50]. It is interesting that although the intercalate and graphite bounding layers may be incommensurate, the orientational locking effect discussed in section 15.3.6 is sufficient in a stage 1 compound to give rise to interlayer stacking order for both the intercalate (α, β, γ) and the graphene (A, B, C) layers, as is observed in stage 1 $CoCl_2$–GICs [40].

Even when there is no long range interplanar stacking order (as for example in high stage compounds), there is often short range interplanar correlation based on the interplanar intercalate exclusion argument illustrated in Fig. 15.7. Typically this correlation extends from one intercalate layer to its nearest neighbors, thereby affecting three pairs of interfaces. Because the interplanar intercalate–intercalate interaction energy is small, thermal excitation at room temperature is often sufficient to destroy interplanar intercalate correlation, thereby producing quasi-2D intercalate layers, uncorrelated in the c-direction. For example, in the fourth stage $C_{28}Br_2$ intercalation compound [63, 64], the interplanar stacking correlation is lost above ~325 K, though commensurate in-plane ordering persists up to ~360 K.

15.3.8 Internal intercalate phase boundaries and disproportionation

Internal intercalate boundaries can arise from various sources, such as internal boundaries between the intercalate constituents in ternary compounds (e.g. C_4KHg) or in compounds showing disproportionation (e.g. $SbCl_5$–GIC). Many ternary compounds show a trilayer intercalate sandwich rather than a single intercalate layer (Fig. 15.3). In ternary GICs, such as KH–GICs prior to the hydrogen introduction, there is only a single intercalate layer, and even after very small additions of hydrogen only one intercalate layer is found within each unit cell. The introduction of more hydrogen (along with additional potassium) soon results in the formation of a three-layer sandwich, with the potassium layers charged positively with respect to the surrounding hydrogen and graphite layers, which become negatively charged. Thus a charged interface is introduced between the constituents of the intercalate sandwich in a ternary GIC (Fig. 15.3 and section 15.4.1). In this case the electrostatic energy gained by forming the charged interfaces with the intercalate trilayer exceeds the elastic energy associated with the c-axis expansion to form the trilayer intercalate sandwich.

Heavy (more dense) and light (less dense) walls also arise at the interface between regions of different site occupation in a commensurate in-plane superlattice (e.g. the boundary between an α domain and a β domain in the same intercalate layer of a ($\sqrt{3} \times \sqrt{3}$)$R30°$ superlattice). These interface phenomena have been widely studied for adsorbed rare gas phases on graphite [12, 65].

For many GICs, the chemical species within the intercalate layer is the same as the intercalate prior to intercalation, except for the charge transferred between the intercalate and the graphite layers through the intercalation process. For other systems, such as the $SbCl_5$–GICs, the intercalation process results in a chemical reaction called disproportionation [66], giving rise to different species in the intercalate layer relative to the parent intercalate species, as for example the species $SbCl_3$, $SbCl_6^-$, $SbCl_5$ and $SbCl_4^-$, in contrast to the single $SbCl_5$ compound prior to intercalation [66]. Interfaces between these intercalate constituents may give rise to another type of interface in GICs.

15.3.9 Solid–liquid interfaces

Since graphite does not itself undergo structural phase transitions up to very high temperatures (~4000°C), the intercalate layers in GICs exhibit a variety of structural phase transitions relative to the rigid layers of the host lattice, the most interesting general phenomena being the special type of two-dimensional (2D) melting that occurs in the intercalate layer. What is meant by a 2D liquid in

this context is that the rapid diffusion of the intercalate species is confined to the two dimensions of
the intercalate layer. However, because of the coupling between the intercalate and the surrounding
graphene layers, the intercalate atoms and molecules
have a higher probability to be found in registry
with the adjacent graphene layers. Such effects
have been observed in Rutherford backscattering–
channeling studies [67] of high stage ($n > 2$) alkali
metal GICs.

Extensive studies of 2D melting in GICs have
been carried out by many workers using X-ray and
neutron diffraction studies and this large literature
has recently been reviewed by Moss and Moret
[36], with the majority of the work having been
done on the high stage ($n \geq 2$) alkali metal GICs,
with a smaller literature on the acceptor molecular
intercalates. In 2D, the Fourier transform of the
lattice vibrations of an infinite vibrating sheet
exhibits a power law decay in the vicinity of the
melting point, arising from the $r^{-\eta}$ decay of the real
space pair correlations. The width of the diffraction
peaks also exhibit an η-dependent decrease with
increasing wave vector [63].

In general, experimental studies of 2D melting
have focused on the observation of a hexatic phase
[68, 69], but in none of the GIC studies has such a
phase been observed, presumably because of the
modulation of the liquid motion by the periodic
potential of the bounding graphite layers. The
characteristic diffraction patterns exhibited by the
2D liquid phase indicate a mean interatomic intercalate separation that is incommensurate with the
bounding graphite layers, but showing an enhanced
probability for local commensurate behavior
through the modulation potential [70].

Though most of the effort thus far has gone into
studies on the high stage alkali metals, a fair
amount of effort has also gone into studies on
molecular acceptors where the modulation potential is relatively weaker because of the strong intra-
molecular potentials of the intercalate molecules.
In some cases (such as stage 2 HNO_3–GIC and
stage 4 Br_2–GIC), anisotropic melting of the intercalate associated with the modulation potential of
the adjacent graphene layers has been observed
[12, 71]. However, for some molecular acceptors,

no evidence for a modulation by the graphene layer
potential has been observed, as for example for
the $GaCl_3$–GICs [72]. These results show that
modulation effects in the intercalate melting can
vary widely from one acceptor GIC to another.

15.3.10 External interfaces

The external surfaces of graphite display a high
degree of anisotropy from a chemical, mechanical
and electronic point of view. The c-face of graphite
having only weak π-bonds exposed at the surface, is
chemically inert, and only supports physisorption.
In contrast, the a-face has dangling σ-bonds which
make this face chemically active. We further know
that intercalation proceeds by propagation of the
intercalation front along the layer planes [33]. On
the basis of these arguments, we infer that the
external surfaces of graphite intercalation compounds likewise behave anisotropically from a
chemical standpoint.

From a mechanical standpoint, the external
surfaces are also anisotropic. Since graphite and
many of its intercalation compounds cleave easily,
highly planar c-faces can easily be prepared, and
these have been heavily utilized in the development
of the scanning tunneling microscope [73, 74]. In
contrast, it has been difficult to prepare reliable
a-faces.

Electronically, the external surfaces of GICs
differ from internal surfaces. Because of the competition between the electrostatic interlayer attraction (acting between the intercalate and the
graphite bounding layers (section 15.4.2)) and
the elastic interlayer repulsion (associated with the
lattice expansion to accommodate the intercalate
layer), donor intercalate species are found in higher
concentrations near the surface than in the bulk of
the GIC, and the opposite is true for acceptor
compounds [75]. Thus, for surface sensitive experiments, the surface of an alkali metal–GIC
usually behaves like a stage 1 intercalation compound. Recently, the cleavage of GICs at low
temperature (e.g. $\sim 77\,K$) has demonstrated the
possibility of preparing surfaces having intercalate
concentrations typical of the bulk for long enough

times (minutes) to permit quantitative properties measurements to be carried out [76].

15.4 RELATION BETWEEN PROPERTIES AND INTERFACES

Whereas in section 15.3 the structures of the various interfaces occurring in the GICs were discussed, section 15.4 focuses on structure–property relations in GICs that bear on interfaces. In section 15.4.1, we discuss charge transfer across the intercalate–graphite bounding layer interface, and in section 15.4.2 the effect of this charge transfer and of the abruptness of the interface on the electronic, lattice, and transport properties of GICs. In section 15.4.3 and section 15.4.4 we discuss the interface structures in connection with the magnetic and superconducting properties of magnetic and superconducting GICs, respectively.

15.4.1 Charge transfer across interfaces

Charge transfer across the intercalate–graphite bounding layer interface is the driving mechanism for intercalation in most intercalation compounds and is responsible for many of their observed properties. This charge transfer across an interface is most easily illustrated in the alkali metal GICs (Fig. 15.13). Let us suppose that each alkali metal atom contains one nearly free electron. If this electron is donated to the graphite layers, the alkali metal layers become positively charged, thereby attracting the negative charges in the graphite layers. Thus the majority of the electrons transferred from the intercalate to the graphite host will be located in the graphite bounding layers adjacent to the intercalate layers, as shown in Fig. 15.13. Thus we conclude that the interface in such an intercalation compound is a dipole layer and is electrically active. As a consequence of this high polarization, the intensities of the infrared and Raman spectra of graphite intercalation compounds are enhanced relative to the corresponding spectra from the constituent species.

Since the carrier concentration in graphite is very low ($\sim 2 \times 10^{-4}$ carriers/C atom at room tempera-

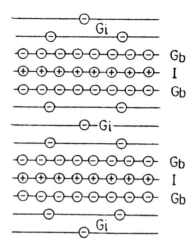

Fig. 15.13 Schematic representation of charge transfer in graphite intercalation compounds, showing the intercalate layer (I) for a donor compound transferring electrons to the graphite bounding (G_b) layer, with a small amount of charge also transferred to the graphite interior (G_i) layers.

ture and $\sim 6 \times 10^{-5}$ carriers/C atom at low temperature), the charge transferred to the graphite layers through intercalation causes a major increase in the mobile carrier concentration in the graphite layers [2, 3]. An estimate of about three orders of magnitude increase in carrier concentration follows from assuming a charge transfer of one carrier per intercalate unit to the graphite layers and assuming an intercalate concentration one order of magnitude smaller than that for graphite. Since the mobility of pristine graphite is very high (13 000 cm^2/V s at 300 K), intercalation compounds with very high in-plane electrical conductivity can be synthesized. On the other hand, the conductivity in the intercalate layers is expected to be low because of the low mobility of ions generally, and the low density and low mobility of electrons (or holes) in the intercalate layers. Thus electrical transport in the basal planes is mainly by graphitic π-band electrons and holes. Intercalation with donors raises the Fermi level and conduction is predominantly by electrons, while intercalation with acceptors lowers the Fermi level, giving rise to conduction by holes [2].

The charge transfer phenomenon in GICs is analogous to modulation doping in the semiconductor superlattices (Fig. 15.14). For these superlattices, the doping impurities are introduced in the wide bandgap region and the carriers are transferred to the highly ordered narrow gap region, which contains an extremely low defect density. In this way, exceptionally high carrier mobilities well in excess of $10^6 \, cm^2/V \, sec$ can be achieved in the GaAs layers [46, 77, 78]. The interface in the case of the semiconductor superlattices tends to extend over one or two unit cells whereas in the GICs the interface is abrupt on an atomic scale.

Dangling bonds on internal intercalate surfaces may give rise to charge accumulation. In this connection, charge transfer in ternary GICs has interesting implications for interfaces. For example, the addition of Hg or H to the first stage compound C_8K reduces the electron concentration in the graphite layers by a back donation process, while the Hg (or H) layers become negatively charged. This can be more easily understood by referring to Fig. 15.3 where the trilayer structure of the intercalate sandwich for the ternary compound KHg–GIC is shown. Thus the Hg or H layers behave like an acceptor intercalate, though residing in an otherwise donor compound.

15.4.2 Sharp interfaces implied by electronic and lattice properties

The weakness of the interaction between the intercalate and the graphite bounding layers plays a dominant role in determining the electronic and lattice properties of GICs and, by the same token, study of the electronic and lattice properties of GICs can be used to provide information on the intercalate–graphene layer interactions and on the sharpness of the interfaces between these layers. For example, a qualitative picture of the electronic structure of GICs can be obtained by the superposition of Hamiltonians for the parent graphene and intercalate layers, taking into account the new periodicity to the lattice and the associated zone folding effects [79]. Such models imply sharp interfaces and relatively weak intercalate–graphite

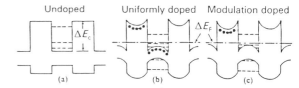

Fig. 15.14 Modulation doping causes charge transfer without introduction of charged impurity scatterers. A semiconductor heterojunction superlattice with various methods of doping: (a) undoped, (b) uniformly doped, and (c) modulation doped. The large black dots denote donor states and the dashed lines denote electron bound quantum well states. Doping causes the Fermi level to rise into the conduction band. Thus charge transfer during intercalation is analogous to modulation doping in semiconductor superlattices [77].

interactions. Although the intercalate–intercalate interplanar interactions are weak for high stage ($n \geqslant 2$) GICs, these interactions in the case of the stage 1 compounds are strong enough to produce long range interlayer stacking ordering [50] (section 15.3.7). This is not surprising since the intercalate–intercalate interlayer separations are often equal to what they were in the parent compound, despite the presence of an intervening graphene layer.

Likewise, the lattice properties of the GICs are strongly correlated with what they are in the parent materials prior to intercalation. Lattice dynamics calculations based on the lattice modes of the parent graphite and intercalate materials have been successful in accounting for the observed Raman and IR spectra as well as the velocity of sound and elastic properties [15, 80]. Specifically, the lattice properties of the GICs as measured by either the Raman or IR spectra indicate that the observed graphitic modes are only slightly shifted from pristine graphite, except for the stage 1 alkali metal GICs and the fluorine GICs. Furthermore, the Raman and IR spectra for GICs show little change in lineshape or linewidth resulting from intercalation (except for some stage 1 compounds) [49]. The absence of line broadening in the Raman allowed lines and the absence of disorder-induced Raman lines indicate that no great amount of disorder is introduced by intercalation. In a few cases Raman lines associated with the intercalate layer (e.g. Br_2–GICs) have been observed [81], and the shift in the Raman frequencies can be used to determine the amount of bromine–carbon coupling at the interface between the intercalate and graphite bounding layers. The observation of zone-folded modes in the alkali metal GICs [49] indicates a well ordered abrupt interface.

On the other hand, the transport properties are changed dramatically by intercalation as a result of the large intercalation-induced changes in charge carrier density. However, the more modest intercalation-induced decrease of the in-plane carrier mobility indicates that the interface scattering for most GICs is quite small. The case of fluorine intercalated graphite is however an exception, where the disorder at the interface is sufficiently large so that the conductivity falls far below that of pristine graphite for high fluorine concentrations [82, 83]. The larger amount of disorder at the fluorine–graphene layer interface results from the formation of some covalent C–F bonds rather than the ionic C–F bonds found in the low stage compounds [84, 85, 86, 87].

15.4.3 Magnetic interfaces

When magnetic species are intercalated into a host material such as graphite, magnetic intercalation compounds exhibiting magnetically ordered phases and magnetic phase transitions can be synthesized [21]. For such compounds, the structural interlayer interfaces are abrupt on an atomic scale (see section 15.3.4), though their magnetic analogs may not be. From the standpoint of studying low dimensional magnetism, the availability of sharp structural interfaces in magnetic GICs allows for quantitative studies of the interaction between isolated magnetic monolayers. In this section we discuss this distinction between the magnetic and structural interlayer interfaces, followed by a discussion of the effect of commensurate and incommensurate interface registration on the magnetic properties of magnetic GICs.

To illustrate the distinction between the structural and magnetic interfaces in GICs, we consider two different situations, first for the donor magnetic compound, stage 1 C_6Eu, where the magnetic interface becomes highly extended relative to the structural interface, and second for the high stage ($n \geqslant 2$) acceptor transition metal chloride GICs, where the magnetic interfaces are more localized.

For C_6Eu, the interlayer magnetic coupling is through the conduction electrons of the graphene layers, which follows the RKKY mechanism, and gives rise to a broad magnetic interface [21, 88]. On the other hand, the modified indirect exchange interaction in the $CoCl_2$–GICs involves firstly a super-exchange coupling of the magnetic species on the Co layers through the localized bonding orbitals of one chlorine layer, an indirect exchange through the π-bonds of the various graphene layers, followed by a similar indirect exchange interaction between the graphite bounding layer and the adjacent chlorine layer, and finally a super-exchange interaction with the adjacent magnetic layer [21]. Thus the magnetic interaction in the acceptor stage 1 compounds is relatively weak, and rapidly becomes weaker (it falls off exponentially) as more graphite layers are added by increasing the stage index [21]. Beyond stage $n \sim 3$, the only magnetic interaction of importance is a very weak dipolar interaction. For both donor and acceptor stage 1 compounds, magnetic long range ordering in the c-direction has been observed, in accordance with the coupling mechanisms outlined above. However, for the higher stage acceptor compounds ($n \geqslant 2$), only short range magnetic correlation (between one magnetic layer and its nearest neighbors) is achieved [21].

The in-plane ordering of the intercalate layer determines the relation between the magnetic behavior of the GIC relative to that of the magnetic species prior to intercalation [21]. For example, the incommensurate transition metal chloride GICs retain the in-plane structure and lattice constants of a layer of the pristine magnetic material prior to intercalation [50], so that the in-plane magnetic exchange interactions are also retained. Thus, the intercalation process allows isolation of single layers of the magnetic transition metal chloride compounds and the separation of these layers by controllable distances through insertion of non-magnetic spacer layers through the staging mechanism. In this way we can control the magnetic interlayer exchange interaction $\mathcal{J}'$ by several orders

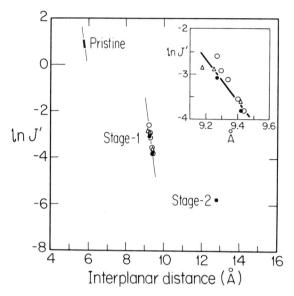

Fig. 15.15 The interlayer antiferromagnetic exchange interaction $\ln J'$ as a function of the c-axis spacing I_c for pristine $CoCl_2$, [89] stage-1, and stage-2 $CoCl_2$–GICs [90]. The fits are extrapolated beyond the experimental points with thin solid lines. The inset shows in detail the $\ln J'$ data points derived from resistivity data [90] ($\triangle$), the magnitude of $1/\chi'$ at 4.2 K ($\bigcirc$), and χ' (H) measurements ($\bullet$) for the stage-1 compound [90].

of magnitude (Fig. 15.15). The use of external pressure allows fine tuning of J' for each stage compound. The reason why the data points for pristine $CoCl_2$, stage 1 $CoCl_2$–GIC, and stage 2 $CoCl_2$–GIC in Fig. 15.15 do not join onto a smooth curve is due to the different magnetic interaction mechanisms operative for each compound: the super-exchange for pristine $CoCl_2$, the modified indirect exchange for the stage 1 and stage 2 compounds (discussed above), and the increased importance of the dipolar interaction for the stage 2 compound [90].

In the case of the donor compound C_6Eu, the intercalate species orders in a $(\sqrt{3} \times \sqrt{3})R30°$ structure commensurate with the graphite bounding layers, thereby giving rise to a magnetic Eu layer with a different structure and different lattice constants than the crystalline Eu metal. From the standpoint of magnetism, C_6Eu is an interesting material insofar as new magnetic phases are observed (i.e. the 4-ring spin phase) [91, 92]; these magnetic phases are not observed in metallic Eu,

nor have they been observed in any other magnetic material previously investigated.

15.4.4 Superconducting interfaces

The study of superconductivity in intercalation compounds offers the possibility of separating the intercalate layers (or the host layers) by large enough distances so that their interlayer interactions become negligible and two-dimensional (2D) superconductivity can be observed and studied. In general, larger interlayer separations are needed to observe 2D superconductivity compared with 2D magnetism, because of the much larger superconducting coherence distance relative to its magnetic counterpart. The possibility of 2D superconductivity was experimentally demonstrated by Gamble *et al.* [93], who intercalated octadecylamine into TaS_2 to achieve a 57 Å separation between the superconducting TaS_2 layers [93]. A more systematic study of this phenomenon was later made by Prober *et al.* [94] using the organic intercalates, collidine, pyridine, and aniline. The identification of 2D superconducting behavior was achieved through studies of the functional form of the temperature dependence of the upper critical field $H_{c2}(T)$ near T_c for various angles of the magnetic field with respect to the crystalline c-axis, as shown in Fig. 15.16. This early work on the intercalated transition metal dichalcogenides stimulated a number of detailed and definitive studies of 2D superconductivity using multilayer samples, especially samples of alternating superconducting and semiconducting constituents (e.g. Nb/Ge [95]) as well as alternating superconducting and normal metal constituents [96].

Despite the success and the importance of the multilayers for studying 2D superconductivity, the superconducting intercalation compounds nevertheless play a unique role in the overall framework of these low dimensional superconductivity studies [46]. It is only with GICs that single layers of a superconducting intercalate material with atomically abrupt interfaces can be prepared and studied.

Some of the unexpected phenomena observed in the superconducting GICs are the following. An increase in T_c is observed in going from a stage 1 to

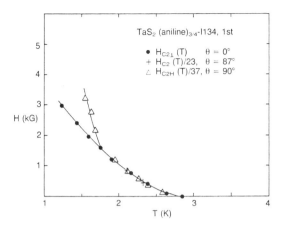

$$TaS_2 \text{ (aniline)}_{3/4}\text{-I134, 1st}$$

- • $H_{C2\perp}$ (T)   θ = 0°
- + H_{C2} (T)/23,   θ = 87°
- ∠ H_{C2H} (T)/37, θ = 90°

Fig. 15.16 Upper critical fields H_{c2} near T_c for various angles θ with respect to the *c*-axis for $TaS_2(\text{aniline})_{\frac{3}{4}}$. The upward rise of $H_{c2\parallel}$ at low temperatures has been identified with a Josephson tunneling mechanism coupling two-dimensional sheets. To emphasize the functional form for $H_{c2}(T)$ the θ = 87° and 90° data are scaled to fit near T_c [94].

a stage 2 KHg–GIC, despite the large decrease in the density of states at the Fermi level, E_F [97]. More generally, intercalation seems to enhance superconductivity, despite the smaller volume fraction occupied by the intercalate in the GICs [97]. Because of the high anisotropy of the structure and properties of superconducting GICs, it is possible to observe both type II and type I superconductivity in the same sample by varying the direction of the magnetic field relative to the crystalline *c*-axis (e.g. for $H \parallel$ *c*-axis, type I behavior is found, while for $H \perp$ *c*-axis, type II behavior is found in KHg–GICs) [98]. Perhaps the most striking of the unexpected phenomena is the observation of superconductivity in GICs having no superconducting constituents [99], such as the stage 1 alkali metal-GIC C_8K [100, 101]. These unusual observations led to the development of a theory for superconductivity in GICs, since the standard BCS model [102] with only occupied graphite π-bands cannot explain the observed transition temperatures because the π-band electron density of states is too small. On the other hand, a model with only heavy mass intercalate-related bands cannot explain the large observed

anisotropy in the superconducting properties of GICs. Al-Jishi [103] therefore proposed a model for superconductivity in GICs in which both graphite π-band electron states and heavy mass intercalate bands are occupied, and both types of electrons form Cooper pairs. This model accounts for most of the current observations in GICs. Another theory by Shimizu and Kamimura [104] for superconductivity in C_8K is based on the tight binding band structure of Ohno *et al.* [105, 106]. The same basic conclusion however is reached [104], namely, that it is necessary to have at least two types of occupied bands (intercalate bands and graphitic II bands) or strong hybridization between them to produce superconductivity in this class of compounds.

15.5 CONCLUDING REMARKS

This review has examined in detail the interfaces in intercalation compounds. These interfaces relate strongly to interfaces in many types of layered materials as well as to adsorbed species on graphite.

In addition to the structural interfaces, magnetic and superconducting interfaces can also be found in intercalation compounds and can be examined with regard to the connection between their structural and electronic (or magnetic) interface.

Particularly interesting in the intercalation compounds is the abruptness of the GIC interface and the many types of ordering that are possible at the interface.

Note added in proof: Since the writing of this chapter, two new classes of intercalation compounds have been discovered: high T_c cuprate superconductors and fullerenes. In each case, novel interface phenomena, not covered in this chapter, are being studied.

ACKNOWLEDGMENTS

The authors would like to express their gratitude to numerous collaborators and to various members of their research group for valuable discussions. The preparation of this manuscript was supported by the National Science Foundation Grant No. 88-19896DMR.

REFERENCES

1. M. S. Dresselhaus, *Physics Today*, **37** (1984), 60.
2. M. S. Dresselhaus and G. Dresselhaus, *Advances in Physics*, **30** (1981), 139.
3. S. A. Solin, *Adv. Chem. Phys.*, **49** (1982), 455.
4. H. Kamimura, *Physics Today*, **40** (1987), 64.
5. in H. Zabel and S. A. Solin, editors, *Graphite Intercalation Compounds I: Structure and Dynamics*, Springer Series in Materials Science. Springer-Verlag, Berlin, Vol. 14, (1990).
6. F. R. Gamble and T. H. Geballe, *Treatise on Solid State Chemistry*, Plenum Press, New York, Vol. 3, (1976), p. 89.
7. W. Y. Liang, Intercalation in layered materials, in *Intercalation in Layered Materials* (ed M. S. Dresselhaus), Vol. 148 of *NATO ASI Series B: Physics*, Plenum Press, New York, NY (1987), p. 31.
8. S. A. Solin, Intercalation in layered materials, in M. S. Dresselhaus, editor, *Intercalation in Layered Materials*, Vol. 148 of *NATO ASI Series B: Physics*, Plenum Press, New York, (1987), p. 145.
9. N. Wada, M. Suzuki, D. R. Hines, K. Koga, and H. Nishihara, *J. Mater. Res.*, **2**, (1987), 864.
10. A. C. D. Newman, in *Chemistry of Clays and Clay Minerals*, (ed A. C. D. Newman), John Wiley & Sons, New York, (1987).
11. P. Aldebert, N. Baffier, J. J. Legendre, and J. Livage, *Revue de Chimie Minérale*, **19** (1982), 485.
12. S. Mochrie, (this volume).
13. A. Hérold, Crystallo-chemistry of carbon intercalated compounds, in F. Lévy, editor, *Physics and Chemistry of Materials with Layered Structures*, Dordrecht Reidel, New York, Vol. 6, (1979), p. 323.
14. P. Schaffäutl, *J. Prakt. Chem.*, **21** (1841), 155.
15. H. Zabel, in *Graphite Intercalation Compounds I: Structure and Dynamics*, (eds H. Zabel and S. A. Solin), Springer Series in Materials Science. Springer-Verlag, Berlin, Vol. 14, (1990).
16. C. Julien, E. Hatzikraniotis, K. Paraskevopoulos, A. Chevy, and M. Balkanski, *Solid State Ionics*, **18–19** (1986), 859.
17. W. Rüdorff and R. Zeller, *Z. Anorg. allg. Chem.*, **279** (1955), 182.
18. P. Lagrange, A. Métrot, and A. Hérold, C.R. *Acad. Sci. Paris*, **278** (1974), 701.
19. B. R. York, S. K. Hark, and S. A. Solin, *Phys. Rev. Lett.*, **50** (1983), 1470.
20. A. Hérold, G. Furdin, D. Guérard, L. Hachim, M. Lelaurain, N.-E. Nadi, and R. Vangelisti, *Synth. Met.*,

12 (1985), 11.
21. G. Dresselhaus, J. T. Nicholls, and M. S. Dresselhaus, *Magnetic Intercalation Compounds of Graphite*, Springer Series in Materials Science. Springer-Verlag, Berlin, Vol. 18, Chapter 7, p. 247 (1992).
22. M. Suzuki, I. Oguro, and Y. Jinzaki, *Extended Abstracts of the 1984 MRS Symposium on Graphite Intercalation Compounds*, Materials Research Society, Pittsburgh, PA (1984), p. 91.
23. J. T. Nicholls and G. Dresselhaus, *J. Phys. Cond. Matter*, **2**, 8391 (1990).
24. P. Lagrange and R. Setton, in *Graphite Intercalation Compounds I: Structure and Dynamics*, (eds H. Zabel and S. A. Solin), Springer Series in Materials Science. Springer-Verlag, Berlin, Vol. 14, (1990).
25. R. Brec, in *Intercalation in Layered Materials*, (ed M. S. Dresselhaus), Plenum Press, New York, (1987), p. 93.
26. J. O. Besenhard, H. Moewald, and J. J. Nickl, *Synth. Met.*, **3** (1981), 187.
27. J. O. Besenhard, E. Wudy, H. Moewald, J. J. Nickl, W. Biberacher, and W. Foag, *Synth. Met.*, **7** (1983), 185.
28. S. A. Solin and H. Zabel, *Adv. Phys.*, **37** (1988), 87.
29. S. A. Solin, Phonon properties of graphite intercalation compounds, in *Intercalation in Layered Materials*, (ed M. S. Dresselhaus), Vol. 148 of *NATO ASI Series B: Physics*, Plenum Press, New York, (1987), p. 313.
30. R. Setton, in *Graphite Intercalation Compounds I: Structure and Dynamics*, (eds H. Zabel and S. A. Solin), Springer Series in Materials Science. Springer-Verlag, Berlin, Vol. 14, (1990).
31. P. Lagrange, M. El Makrini, and A. Bendriss, *Synth. Met.*, **7** (1983), 33.
32. N. Daumas and A. Hérold, *Compt. Rend. Acad. Sci. (Paris)*, C, **268** (1969), 373.
33. G. Kirczenow, in *Graphite Intercalation Compounds I: Structure and Dynamics*, (eds H. Zabel and S. A. Solin), Springer Series in Materials Science. Springer-Verlag, Berlin, Vol. 14, (1990).
34. H. Suematsu, K. Ohmatsu, T. Sakakibara, K. Suda, and N. Metoki, *Synth. Met.*, **23** (1988), 7.
35. R. Moret, in *Intercalation in Layered Materials*, (ed M. S. Dresselhaus), Plenum Press, New York, (1987), p. 185.
36. S. C. Moss and R. Moret, in *Graphite Intercalation Compounds I: Structure and Dynamics*, (eds H. Zabel and S. A. Solin), Springer Series in Materials Science. Springer-Verlag, Berlin, Vol. 14, (1990).
37. L. Salamanca-Riba, Structural Studies of Graphite Intercalation Compounds and Ion Implanted Graphite, Ph.D thesis, Massachusetts Institute of Technology, July 1985, Department of Physics.
38. N. Watanabe, T. Nakajima, and H. Touhara, Graphite

fluorides, in *Studies in Inorganic Chemistry*, Vol. 8, Elsevier, Amsterdam (1988).

39. R. W. G. Wyckoff, *Crystal Structures*, Vol. 1, Interscience: New York, (1964).

40. C. H. Olk, J. P. Heremans, M. S. Dresselhaus, J. S. Speck, and J. T. Nicholls, *Phys. Rev. B*, **42** 7524 (1991).

41. I. P. Batra, N. Garcia, H. Rohrer, H. Salemink, E. Stoll, and S. Ciraci, *Surf. Sci.*, **181** (1987), 126.

42. James Stephen Speck, Structural correlations in graphite and its layer compounds, Ph.D thesis, Massachusetts Institute of Technology, August 1989, Department of Materials Science and Engineering.

43. G. Timp and M. S. Dresselhaus, *J. Phys. C: Solid State*, **17** (1984), 2641.

44. G. Kirczenow, *Phys. Rev. Lett.*, **49** (1982), 1853.

45. S. E. Ulloa and G. Kirczenow, *Phys. Rev. B*, **33** (1986), 1360.

46. M. S. Dresselhaus, Superlattices and intercalation compounds, in *Intercalation in Layered Materials*, (ed M. S. Dresselhaus), Plenum Press, New York, (1987), p. 1.

47. P. A. Heiney, M. E. Huster, V. B. Cajipe, and J. E. Fischer, *Extended Abstracts of the Symposium on Graphite Intercalation Compounds*, at the Materials Research Society Meeting, Boston, Pittsburgh, PA, (1984), p. 131.

48. F. Rousseaux, R. Moret, D. Guérard, P. Lagrange, and M. Lelaurain, *J. Phys. Lett.*, **45** (1984), L1111.

49. M. S. Dresselhaus and G. Dresselhaus, *Light Scattering in Solids III*, (eds M. Cardona and G. Güntherodt), Springer-Verlag, Berlin, Topics in Applied Physics, Vol. 51, (1982), p. 3.

50. J. S. Speck, J. T. Nicholls, B. J. Wuensch, J. M. Delgado, M. S. Dresselhaus, and H. Miyazaki, *Phil. Mag.*, **64**, 181 (1991).

51. F. L. Vogel and A. Hérold, editors, *Proceedings of the Conference on Intercalation Compounds of Graphite*, Vol. 31, Elsevier Sequoia, Lausanne (1977); La Napoule, France 1977.

52. A. Hérold and D. Guérard, (eds), *Synth. Met.*, **8** (1983).

53. M. S. Dresselhaus and J. S. Speck, Microscopic studies of intercalated graphite, in *Intercalation in Layered Materials*, (ed M. S. Dresselhaus), NATO Summer School, Plenum Press, New York, (1987), p. 213.

54. S. Flandrois, J. M. Masson, J. C. Rouillon, J. Gaultier, and C. Hauw, *Synth. Met.*, **3** (1981), 1.

55. J. S. Speck and M. S. Dresselhaus, *Synth. Met.*, **34** (1989), 211.

56. M. S. Dresselhaus, N. Kambe, A. N. Berker, and G. Dresselhaus, *Synth. Met.*, **2** (1980), 121–131, Proceedings of the 2nd International Conference on Intercalation Compounds of Graphite, 1980.

57. M. Suzuki, *Phys. Rev. B*, **33** (1986), 1386.

58. Y. Yamada and I. Naiki, *J. Phys. Soc. Jpn.*, **51** (1982), 2174.

59. M. Suzuki and H. Suematsu, *J. Phys. Soc. Jpn.*, **52** (1983), 2761.

60. F. Rousseaux, R. Moret, D. Guerard, P. Lagrange, and M. Lelaurain, *Ann. de Phys. Colloque 2*, **11** (1986), 85.

61. H. Zabel, S. E. Hardcastle, D. A. Neumann, M. Suzuki, and A. Magerl, *Phys. Rev. Lett.*, **57** (1986), 2041.

62. R. Clarke, J. N. Gray, H. Homma, and M. J. Winokur, *Phys. Rev. Lett.*, **47** (1981), 1407.

63. A. Erbil, A. R. Kortan, R. J. Birgeneau, and M. S. Dresselhaus, *Phys. Rev. B*, **28** (1983), 6329.

64. S. G. J. Mochrie, A. R. Kortan, R. J. Birgeneau, and P. M. Horn, *Phys. Rev. Lett.*, **53** (1984), 985.

65. P. A. Heiney, R. J. Birgeneau, G. S. Brown, P. M. Horn, D. E. Moncton, and P. W. Stephans, *Phys. Rev. Lett.*, **48** (1982), 104.

66. P. Boolchand, W. J. Bresser, D. McDaniel, K. Sisson, V. Yeh, and P. C. Eklund, *Solid State Commun.*, **40** (1981), 1049.

67. G. Braunstein, B. S. Elman, J. Steinbeck, M. S. Dresselhaus, T. Venkatesan, and B. Wilkens, Analysis of structural properties of graphite intercalation compounds using the Rutherford backscattering-channeling technique, in *Extended Abstracts at the Symposium on Intercalated Graphite at the Materials Research Society Meeting, Boston* (1984), p. 168.

68. K. J. Strandberg, *Rev. Mod. Phys.*, **60** (1988), 161.

69. C. A. Murray and D. H. Van Winkle, *Phys. Rev. Lett.*, **58** (1987), 1200.

70. G. Reiter and S. C. Moss, *Phys. Rev. B*, **33** (1986), 7209.

71. F. Aberkane, F. Rousseaux, E. J. Samuelsen, and R. Moret, *Ann. de Phys. Colloque 2*, **11** (1986), 95.

72. N. E. Nadi, M. Lelaurin, A. Hérold, and F. Rousseaux, *Synth. Met.*, **23** (1988), 75.

73. S. Park and C. F. Quate, *Appl. Phys. Lett.*, **48** (1986), 2.

74. J. M. Soler, A. M. Baro, N. Garcia, and H. Rohrer, *Phys. Rev. Lett.*, **57** (1986), 444.

75. M. Laguës, X. Hao, and M. S. Dresselhaus, *Phys. Rev. B*, **38** (1988), 967.

76. R. Schlögl, in *International Colloquium on Layered Compounds* (eds D. Guérard and P. Lagrange), (1988), p. 237, Pont-a-Mousson Conference.

77. R. Dingle, H. L. Störmer, A. C. Gossard, and W. Wiegmann, *Appl. Phys. Lett.*, **33** (1978), 665.

78. J. H. English, A. C. Gossard, H. L. Störmer, and K. W. Baldwin, *Appl. Phys. Lett.*, **50** (1987), 1826.

79. J. Blinowski and C. Rigaux, *J. Phys. (Paris)*, **45** (1984), 545.

80. R. Al-Jishi and G. Dresselhaus, *Phys. Rev. B*, **26** (1982), 4523.

81. P. C. Eklund, N. Kambe, G. Dresselhaus, and M. S. Dresselhaus, *Phys. Rev. B*, **18** (1978), 7069.

82. I. Ohana, I. Palchan, Y. Yacoby, D. Davidov, and H. Selig, *Phys. Rev. B*, **38** (1988), 12627.

83. D. Vaknin, I. Palchan, D. Davidov, H. Selig, and D. Moses, *Synth. Met.*, **16** (1986), 349.

84. T. M. Mallouk and N. Bartlett, *J. Chem. Soc., Chem. Comm.*, **103** (1983).

85. T. M. Mallouk, B. L. Hawkins, M. P. Conrad, K. Zilm, G. E. Maciel, and N. Bartlett, *Phil. Trans. Roy. Soc. Lond. A*, **314** (1985), 179.

86. J. Kouvetakis, R. B. Kaner, M. L. Sattler, and N. Bartlett, *J. Chem. Soc., Chem. Commun.*, (1986), p. 1758.

87. J. Kouvetakis, R. B. Kaner, M. L. Sattler, and N. Bartlett, *Mat. Sci. Bull.*, **22** (1987), 399.

88. H. Akera and H. Kamimura, *Solid State Commun.*, **48** (1983), 467.

89. S. N. Lukin, P. V. Vodolazskii, and S. M. Ryabchenko, *Fiz. Nizk. Temp.* (USSR), **3** (1977), 1465, [*Sov. J. Low Temp. Phys.*, **3** (1977), 705].

90. J. T. Nicholls, C. Murayama, H. Takahashi, N. Mōri, T. Tamegai, Y. Iye, and G. Dresselhaus, *J. Appl. Phys.*, **67** (1990), 5746.

91. M. Date, T. Sakakibara, K. Sugiyama, and H. Suematsu, in *High Field Magnetism*, (ed M. Date), North-Holland, (1983), p. 41.

92. S. T. Chen, M. S. Dresselhaus, G. Dresselhaus, H. Suematsu, H. Minemoto, K. Ohmatsu, and Y. Yosida, *Phys. Rev. B*, **34** (1986), 423.

93. F. R. Gamble, F. J. DiSalvo, R. A. Klemm, and T. H. Geballe, *Science*, **168** (1970), 568.

94. D. E. Prober, R. E. Schwall, and M. R. Beasley, *Phys. Rev. B*, **21** (1980), 2717.

95. S. Ruggiero, T. W. Barbee, Jr., and M. R. Beasley, *Phys. Rev. B*, **26** (1982), 4894.

96. M. B. Salamon, S. Sinha, J. J. Rhyne, J. E. Cunningham, R. W. Erwin, J. Borchers, and C. P. Flynn, *Phys. Rev. Lett.*, **56** (1986), 259.

97. Y. Iye and S. Tanuma, Graphite intercalation compounds, in *Summary report of the Special Distinguished Research Project*, Tanuma and H. Kamimura, supported by the Ministry of Education, Japan, (eds S. Tanuma and H. Kamimura eds), (1984), 256.

98. A. Chaiken, M. S. Dresselhaus, T. P. Orlando, G. Dresselhaus, P. M. Tedrow, D. A. Neumann, and W. A. Kamitakahara, *Phys. Rev. B*, **41** (1990), 71.

99. N. B. Hannay, T. H. Geballe, B. T. Matthias, K. Andres, P. Schmidt, and D. MacNair, *Phys. Rev. Lett.*, **14** (1965), 255.

100. Y. Koike and S. Tanuma, *J. Phys. Chem. Solids*, **41** (1980), 1111.

101. Y. Koike and S. Tanuma, *J. Phys. Soc. Jpn.*, **50** (1981), 1964.

102. J. Bardeen, L. N. Cooper, and J. R. Schrieffer, *Phys. Rev.*, **108** (1957), 1175.

103. R. Al-Jishi, *Phys. Rev. B*, **28** (1983), 112.

104. A. Shimizu and H. Kamimura, *Synth. Met.*, **5** (1983), 301.

105. T. Ohno, K. Nakao, and H. Kamimura, *J. Phys. Soc. Jpn.*, **47** (1979), 1125.

106. T. Ohno and H. Kamimura, *J. Phys. Soc. Jpn.*, **52** (1983), 223.

CHAPTER SIXTEEN

Nanophase materials: structure–property correlations

Richard W. Siegel

Introduction · Synthesis · Structure · Properties · Conclusions

16.1 INTRODUCTION

Grain boundaries (GBs) and other interfaces are well known to affect the properties of materials with conventional grain sizes in a variety of important ways. These effects are quite often deleterious, as in the embrittlement of ceramic or intermetallic compounds. However, in other cases they can be enormously useful, as in the example of the GB-controlled current–voltage characteristics of oxide varistors. The effects of interfaces on the properties of materials can be expected to be amplified as the sizes of the grains or phases comprising a material are reduced. This has become especially true in recent years as materials with grain or phase structures on a nanometer scale have been synthesized by means of a number of methods.

Increasing interest has focused on a variety of synthetic nanostructured materials during the past several years with the anticipation that their properties will be different from, and often superior to, those of conventional materials that have phase or grain structures on coarser size scales [1]. This interest has been stimulated not only by the considerable recent effort and success in synthesizing a variety of one-dimensionally modulated, multilayered materials with nanometer scale modulations, but also by the recently developed possibilities for synthesizing three-dimensionally analogous bulk materials via the assembly of clusters of atoms [2].

The synthesis of nanometer size atom clusters of metals and ceramics by means of atom-cluster condensation in a gas, followed by the *in situ* consolidation of the clusters under high-vacuum conditions, has resulted in the past few years in a new class of ultrafine-grained, interface materials. These nanophase materials, with average grain sizes that presently range from 100 nm down to about 5 nm, exhibit properties that are often rather different and improved relative to those of conventional coarse-grained materials. Cluster-assembled nanophase materials, with their interesting structure and properties, will be reviewed here. Their novel properties appear to result from both the scale of their grain or phase structures and the concomitant presence of exceedingly large numbers of interfaces.

Nanophase materials synthesis by means of cluster assembly under controlled conditions should enable the design of a variety of materials heretofore unavailable with improved or unique properties. In addition, nanophase materials, with their uniquely large number of GBs and interfaces,

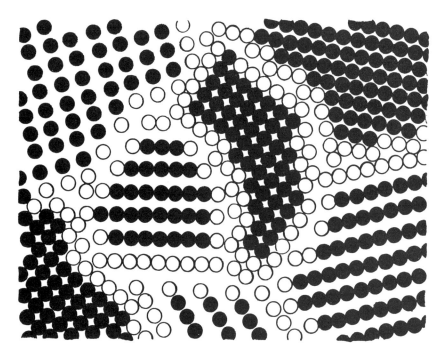

Fig. 16.1 Schematic representation of a nanophase material. The black circles represent atoms in regular lattice positions within the grains; the white circles, atoms that may be expected to relax in the GBs. No chemical differences between the atoms are implied. After [7].

should allow for the first time the study of the average properties of such interfaces. It is thus quite likely that the combination of new capabilities to synthesize, characterize, and engineer the properties of materials based upon the assembly of atom clusters will have a significant impact on materials science and engineering in the coming years [1, 2].

16.1.1 Background

The creation of materials by the consolidation of gas-condensed atom clusters is as old as our universe. It is thought to be the manner in which primordial condensed matter formed from our solar nebula during the cooling period following the 'big bang', as evidenced by the structure of the earliest meteorites [3–5]. The modern synthesis of ultrafine-grained materials by the *in situ* consolidation of nanometer size gas-condensed ultrafine particles or atom clusters was, however, first suggested by Gleiter [6]. By consolidating such clusters, materials with a large fraction of their atoms in GBs could be formed, as shown sche-

matically in Fig. 16.1. The surface atoms of each grain would relax into GB configurations or form intergrain porosity. The degree to which the percentage of atoms associated with GBs varies with grain size in the nanophase regime (<100 nm) is shown in Fig. 16.2, based on a simple GB model without porosity and not distinguishing between GBs and their junctions.

The considerable body of earlier research into the production of ultrafine particles by means of the gas-condensation method, as well as the previously assembled knowledge on powder metallurgy and ceramics, provided a solid basis upon which Gleiter's suggestion [6] could grow to fruition. The earlier research on the gas-condensation method and on the resulting atomic clusters [9–11], to a large extent, defined the various parameters that control the sizes of the clusters formed in the conventional gas-condensation method (primarily type of gas, gas pressure, and evaporation rate) that are used to synthesize nanophase materials. The application of this method in recent years [7, 12–18] to the synthesis of a variety of nanophase metals and ceramics has built upon this base.

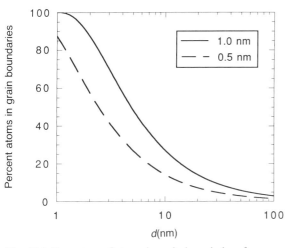

Fig. 16.2 Percentage of atoms in grain boundaries of a nanophase material as a function of grain diameter, assuming that the average grain boundary thickness ranges from 0.5 to 1.0 nm (*ca.* 2 to 4 atomic planes wide). From [8].

16.1.2 Advantages of cluster assembly

The gas-condensation method for the synthesis of nanophase materials appears to have great flexibility and control for engineering new forms of bulk nanostructured materials. Some unique advantages of the assembly of nanophase materials from gas-condensed clusters, and the nanophase processing method based on this, are as follows:

1. The ultrafine sizes of the atom clusters and their surface cleanliness allow conventional restrictions of phase equilibria and kinetics to be overcome during material synthesis and processing by the combination of short diffusion distances, high driving forces, and uncontaminated surfaces and interfaces.
2. The large fraction of atoms residing in the GBs and interfaces of these materials allow for interface atomic arrangements to constitute significant volume fractions of material, and thus novel materials properties may result.
3. The reduced size scale and large surface-to-volume ratios of the individual nanophase grains can be predetermined and can alter and enhance a variety of physical and chemical properties.
4. A wide range of materials can be produced

in this manner including metals and alloys, intermetallic compounds, oxide and nonoxide ceramics, and semiconductors. It is also apparent that nanophases can be formed with crystalline, quasicrystalline, or amorphous structures.

5. The possibilities for reacting, coating, and mixing *in situ* various types, sizes, and morphologies of clusters create a significant potential for the synthesis of a variety of new multicomponent composites with nanometer-sized microstructures and engineered properties. Most of the research carried out to date, nevertheless, has concentrated on single-phase metals and ceramics.

The following section describes the method of material synthesis and processing via gas-condensation that leads to ultrafine-grained polycrystalline metals and ceramics with controlled mean grain sizes below 100 nm and a number of the attributes just cited.

16.2 SYNTHESIS

16.2.1 Basic principles

In any of the methods for the synthesis of nanostructured materials, control of the size or sizes of the phases being assembled is paramount. Beyond this, chemical control of the phases and cleanliness of the interfaces between phases need to be carefully addressed. In bulk nanophase materials, with their large volume fraction of GBs, these issues take on considerable importance.

The predominant feature of nanophase materials, as depicted in Figs. 16.1 and 16.2 and shown in the electron micrograph of nanophase Pd in Fig. 16.3 [19], is their ultrafine grain size and, hence, the large fraction of their atoms that reside in GBs or interfaces. For example, as indicated in Fig. 16.2, a nanophase material with a 5 nm average grain size will have from about 27–49% of its atoms associated with GBs, assuming a very simple GB picture and an average GB thickness of about 0.5–1.0 nm (*ca.* 2–4 nearest-neighbor distances). This

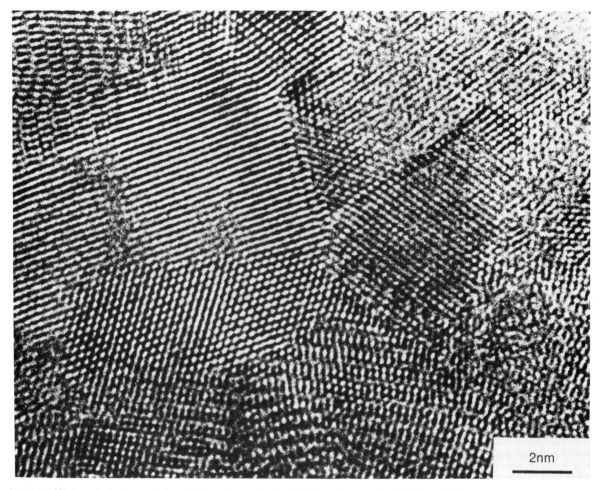

Fig. 16.3 High resolution transmission electron micrograph of a typical area in nanophase Pd. From [19].

percentage falls to about 14–27% for a 10 nm grain size, but is as low as 1–3% for a 100 nm grain size. The interface volume fraction is, of course, essentially negligible for conventional grain sizes of 1 μm and above. Therefore, the properties of nanophase materials can be expected to be strongly influenced by their grain sizes and the nature (atomic and electronic structure) of their internal boundaries, simply because of the very large number density of these boundaries. As such, the control over the various processes involved in the synthesis of nanophase materials takes on an even greater role

than usual. A high degree of control is available with the gas-condensation method.

16.2.2 Gas-condensation method

Research carried out on the gas-condensation method and on the resulting atom clusters during the 1960s and 1970s [9–11] essentially defined the various parameters that control the sizes of the clusters formed in the conventional gas-condensation method that are used to synthesize nanophase materials. The pioneering work of Uyeda

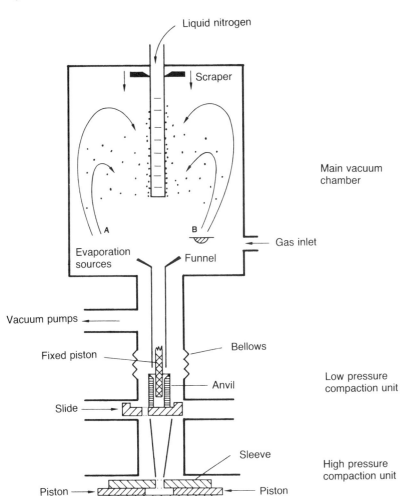

Main vacuum
chamber

Low pressure
compaction unit

High pressure
compaction unit

Fig. 16.4 Schematic drawing of a gas-condensation chamber for the synthesis of nanophase materials. Precursor material evaporated from sources A and/or B condenses in the gas and is transported via convection to the liquid-N_2 filled cold finger. The clusters are then scraped from the cold finger, collected via the funnel, and consolidated first in the low-pressure compaction device and then in the high-pressure compaction device, all in vacuum. From [20].

and coworkers [9] demonstrated that a wide range of metallic ultrafine particles (between 10 and 100 nm in diameter) could be condensed in a low pressure Ar atmosphere and that their sizes could be controlled by varying the gas pressure in the range of about 1–30 torr (0.13–4 kPa). Subsequent work by Thölén [11] extended these investigations to additional metals and to a study of the nucleation and coalescence of the clusters. The most detailed study to date of the conventional gas-condensation process for forming ultrafine metal particles or clusters via condensation in inert gases (He, Ar, or

Xe) was carried out by Granqvist and Buhrman [10]. It was these early studies that elucidated the essential parameters (gas type and pressure, precursor evaporation rate) that control the formation of atom clusters via condensation in naturally convecting gases that have enabled the synthesis of cluster-assembled nanophase materials.

A typical apparatus for the synthesis of nanophase materials via the *in situ* consolidation of gas-condensed clusters is shown schematically in Fig. 16.4 [20]. It is comprised of an ultrahigh-vacuum (UHV) system fitted with two resistively-heated evaporation sources, a cluster collection device (liquid nitrogen-filled cold finger) and scraper assembly, and *in situ* compaction devices for consolidating the powders produced and collected in the chamber. Before making the powders, the UHV system is first evacuated by means of a turbo-molecular pump to below 10^{-5} Pa and then back-filled with a controlled high-purity gas atmosphere at pressures of about a few hundred Pa. For producing metal or other elemental powders this is usually an inert gas, such as He, but it can alternatively be a reactive gas or a gas mixture if, for example, clusters of a compound are desired. A number of nanophase compounds, ceramics and others, have been synthesized via gas condensation. They have been formed from compound precursors directly or by subsequent reaction of metal clusters with gases. A more extensive review of the synthesis and processing of nanophase materials has recently appeared elsewhere [18].

During evaporation of the starting precursor material (or materials) from which the nanophase material will be synthesized, atoms condense in the supersaturated region close to the Joule-heated source. The clusters are continuously transported via entrainment in the naturally convecting gas to the liquid nitrogen-filled cold finger, where they are collected via thermophoresis. The type and pressure of the gas and the precursor evaporation rate, all of which are readily controlled, largely determine the resulting particle-size distributions in such an apparatus [10].

The smallest cluster sizes for a given metal are obtained for a low precursor evaporation rate and condensation in a low pressure of a light inert gas, such as He. These conditions lead to a lower super-saturation of precursor atoms in the gas, slower energy removal from the evaporated atoms (via lighter gas atoms at lower pressure), and more rapid convective gas flow owing also to the lower gas pressure. The more rapid gas flow leads to speedier removal of the condensed clusters from the supersaturated region in which, if they remained, they could grow further by atom accretion. For the smallest cluster sizes and narrow size distributions to be maintained, cluster–cluster coalescence also needs to be minimized.

The clusters that are collected on the surface of the cold finger form very open fractal structures as seen by transmission electron microscopy. A view of such a fractal collection of nanophase TiO_2 taken from the cold finger is shown in Fig. 16.5(a) [21]. The clusters are held there weakly and can be easily removed from this collection surface by means of a Teflon scraper. Upon removal, the clusters are funneled into a set of compaction devices (Fig. 16.4) capable of consolidation pressures up to about 1–2 GPa, in which the nanophase compacts are formed at room temperature, or at elevated temperatures if needed. An example of the grain morphology that results from the consolidation of powders such as those shown in Fig. 16.5(a) is presented in Fig. 16.5(b) [21]. The grain size distribution for the as-consolidated TiO_2 is shown in Fig. 16.6 [21]. It is quite narrow and has the log-normal shape typical of clusters formed via gas-condensation [10]. This shape is rather typical for the grain size distribution in any of the nanophase materials thus far produced by the gas-condensation method.

The pellets formed in the conventional research apparatus depicted in Fig. 16.4 are typically about 9 mm in diameter and 0.1–0.5 mm thick. The sizes of these research samples have been more a matter of convenience and history than any real limitation of the gas-condensation method itself. The scraping and consolidation are performed under UHV conditions after removal of the inert or reactive gases from the chamber, in order to maximize the cleanliness of the particle surfaces and the interfaces that are subsequently formed and to minimize the possibilities of trapping remnants of these gases in the nanophase compact.

Since the as-collected clusters are generally aggregated in rather open fractal arrays [9, 11], their consolidation at pressures of 1–2 GPa is easily accomplished. The difficulties in consolidating the hard equiaxed agglomerates of fine powders resulting from conventional wet chemistry synthesis routes are mostly avoided. The sample densities resulting from cluster consolidation at room temperature have ranged up to about 97% of theoretical for nanophase metals and up to about 75–85% of theoretical for nanophase oxide ceramics. This

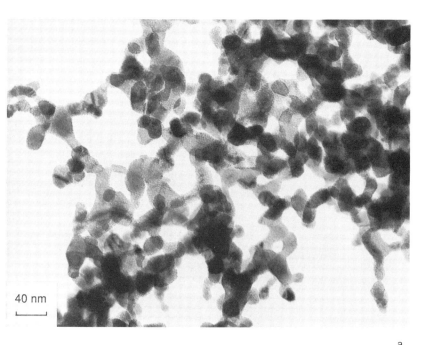

40 nm

a

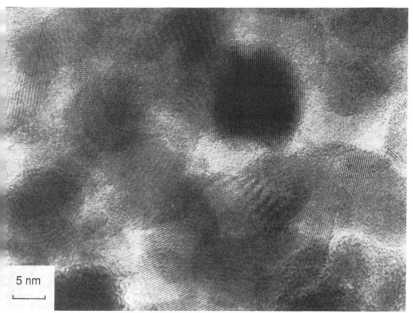

5 nm

b

Fig. 16.5 Transmission electron micrographs of (a) as-collected and oxidized TiO$_2$ clusters synthesized in the apparatus shown in Fig. 16.4 and (b) nanophase TiO$_2$ (rutile) after *in situ* consolidation at room temperature and 1.4 GPa pressure in the apparatus shown in Fig. 16.4, followed by sintering in air for 0.5 h at 500°C. After [21].

green-state porosity will be considered further below, but it probably represents (at least in part) a manifestation of powder agglomeration leading to void-like flaws. Fortunately, these appear to be capable of being removed by means of cluster consolidation at elevated temperatures and pressures without significant attendant grain growth.

16.2.3 Other methods

Using the conventional gas-condensation method described above, which utilizes natural convective gas flow, the average cluster diameters produced presently range down to about 5 nm, yielding nanophase materials with such minimum average grain sizes and the type of narrow size distribution shown in Fig. 16.6. However, transport of the condensed atom clusters via natural gas convection can be improved upon by using instead the motion of a forced gas. The use of a forced gas allows the gas flow rate to be independent of gas pressure (in contrast to natural convection), giving greater freedom of control of the important cluster condensation parameters.

Forced gas flow is already used in more sophisticated cluster synthesis methods by cluster chemists and physicists to produce low yields of even smaller atomic clusters with very narrow, and even monosized, size distributions [2]. It can be expected that new cluster generation systems based on similar principles [22, 23, 24] will be available in the future for the controlled synthesis of larger amounts of material than are normally produced (less than a few hundred milligrams) in the type of apparatus shown in Fig. 16.4.

Most of the atom clusters assembled into nanophase materials to date have been generated from Joule-heated evaporation sources. However, such sources have limitations that can be avoided by a wide variety of other available sources. The primary limitations are source-precursor incompatability, temperature range, uniformity and control, and dissimilar evaporation rates for different constituents in an alloy or compound precursor. Each of these limitations can be avoided by a host of alternative sources that have been developed over the years of ultrafine particle

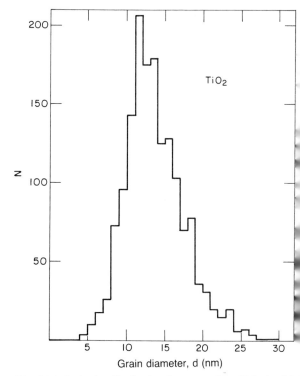

Fig. 16.6 Grain size distribution for a nanophase TiO_2 (rutile) sample compacted to 1.4 GPa at room temperature determined by dark-field TEM. From [21].

research, but which are just beginning to enter the field of nanophase materials synthesis. Among these are sputtering [25–29], electron beam heating [30–33], laser ablation [34], and plasma methods [35].

It should be clear that this wide variety of evaporation methods will allow for greatly increased flexibility in the use of refractory or reactive precursors for clusters, and will be especially useful as one moves toward synthesizing technological quantities of more complex multicomponent or composite nanophase materials in the future.

Before turning to the structure and properties of nanophase materials, it might be useful to mention that a number of other methods for producing nanostructured materials exist. Among the physical methods, the gas-condensation method for the synthesis of nanophase materials appears to have the greatest flexibility and control for engineering new forms of bulk nanostructured materials. However,

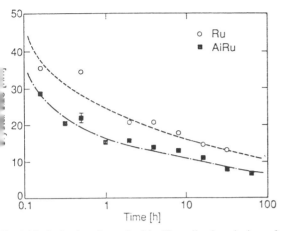

Fig. 16.7 Grain size, determined by X-ray line broadening, of Ru and AlRu produced via mechanical attrition as a function of ball-milling time. From [37].

ther physical methods, such as spark erosion [36] and mechanical attrition [37], can complement this approach. The first of these methods is an alternative to gas-condensation in that it can yield individual clusters which can be subsequently assembled via consolidation. The second method, however, produces its nanostructures by means of what is essentially a mechanical decomposition of coarser-grained structures, so that the individual grains are never isolated clusters and much synthesis and processing flexibility is lost. Nevertheless, grain sizes down into the nanometer regime can be accessed, at least in relatively hard materials, as shown in Fig. 16.7 [37]. The various chemical and biological methods for the synthesis of nanophase materials, such as sol–gel synthesis, spray conversion, and biomimetic methods, will not be covered here as they are essentially outside the scope of this chapter. Some recent references to these methods, however, can be found in References [1], [2], [38–41].

16.3 STRUCTURE

16.3.1 Grains and porosity

The structures of nanophase materials have been investigated by a number of direct and indirect

methods including transmission electron microscopy (TEM), X-ray and neutron scattering, and Mössbauer, Raman, and positron annihilation spectroscopy (PAS). TEM has shown that the grains in nanophase compacts are rather equiaxed, as shown in Figs. 16.3 and 16.5, and that they retain the narrow log-normal size distributions (Fig. 16.6) typical of the clusters formed in the gas-condensation method [10].

In addition to their ultrafine grain sizes, all of the nanophase materials consolidated at room temperature to date have invariably possessed a degree of porosity ranging from about 25% to less than 5%, with the larger values for ceramics and the smaller ones for metals. Clear evidence of this porosity has been obtained by PAS [21, 42, 43] and precise densitometry and porosimetry [44, 45] measurements. These measurements have shown that the porosity in as-consolidated nanophase metals and ceramics is primarily in the less than 100 nm size regime (although some larger porous flaws are observed), but that it is to a great extent interconnected and intersects with the specimen surfaces.

Porosimetry data [45] for as-consolidated nanophase TiO_2, and after sintering at elevated temperatures, are shown in Fig. 16.8. The porosity measurements were made using the BET (Brunauer–Emmett–Teller) N_2 adsorption method [46]. An example of the PAS evidence is shown in Figs. 16.9 and 16.10, where positron lifetime data [21] from nanophase TiO_2 are presented as a function of sintering anneals, which eventually remove the porosity and hence the positron trapping at voids that leads to the long positron lifetimes observed (τ_2). This will be discussed further in section 16.4.1. However, consolidation at elevated temperatures can remove this porosity without sacrificing the ultrafine grain sizes in these materials.

16.3.2 Grain boundaries

Since a large fraction of their atoms reside in the GBs of nanophase materials, the interface structures can play a significant role in determining the properties of these materials. A number of earlier

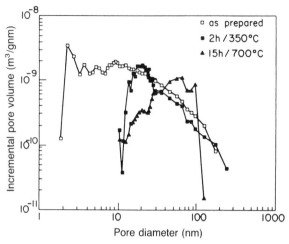

Fig. 16.8 Pore size distributions in nanophase TiO_2 in the as-prepared state (compaction at 150°C and 2 GPa for 2 h) and after sintering in air at atmospheric pressure at 350°C and 700°C. Measurements were made using the BET N_2 adsorption method. From [45].

X-ray absorption fine structure (EXAFS) [49, 50] have been interpreted in terms of GB atomic structures that may be random, rather than possessing either the short-range or long-range order normally found in the GBs of conventional coarser-grained polycrystalline materials. This randomness has been variously associated [17] with either the local structure of individual boundaries (as seen by a local probe such as EXAFS or Mössbauer spectroscopy) or the structural coordination among boundaries (as might be seen by X-ray diffraction). A rather confusing picture has emerged that needs further clarification, particularly with respect to the porosity present in the as-consolidated samples and the atomic relaxations that pertain to free surfaces and to conventional high-angle GBs.

X-ray diffraction

The coherent scattering interference function obtained from X-ray diffraction data [47] from nanocrystalline Fe with an average grain size of about 6 nm (measured by TEM) and a density of about 83% that of conventional α-Fe are shown in Fig. 16.11. The diffraction data indicate consider-

investigations on nanocrystalline metals by Gleiter and coworkers [15], including X-ray diffraction [47], Mössbauer spectroscopy [48], positron lifetime studies [42, 43], and more recently extended

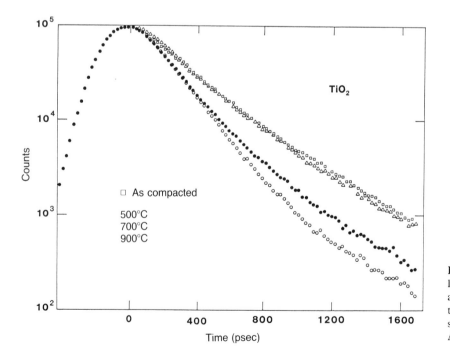

Fig. 16.9 Positron annihilation lifetime spectra for nanophase TiO_2 as a function of sintering temperature. The successive sintering anneals were for 0.5 h in air. After [21].

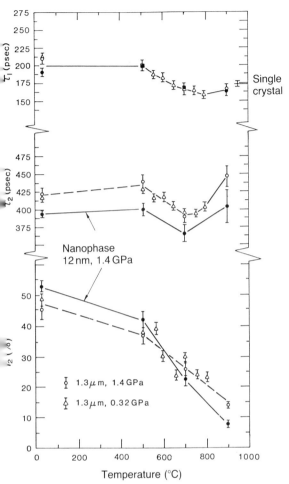

Fig. 16.10 Results of two-component (τ_1, τ_2) lifetime fits to positron annihilation data, similar to those shown in Fig. 16.9, from three TiO_2 samples as a function of sintering temperature. A 12 nm grain size nanophase sample (filled circles) compacted at 1.4 GPa is compared to 1.3 μm grain size samples compacted at 1.4 GPa (open circles) and 0.32 GPa (triangles) from commercial powder. The PAS data were taken at room temperature; no sintering aids were used. From [21].

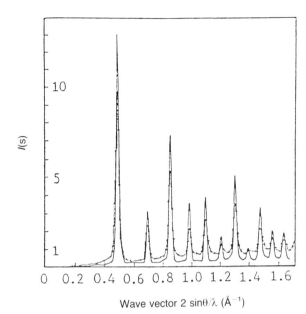

Fig. 16.11 Comparison of measured (--+--) and computed (——) X-ray interference functions for nanocrystalline Fe. The computational model system assumes a GB structure consisting of four atomic layers in which the atoms on the inner two (the boundary core) are randomly displaced by 0.15 nnd and those in the outer two by 0.07 nnd, where nnd is the equilibrium nearest neighbor distance in Fe. From [47].

ble diffuse scattering between the Bragg peaks in comparison with the computed interference function shown for a model which represents a modest degree of random atomic relaxations within the interfaces. However, direct comparison was not made with experimental scattering data from coarse-grained Fe. Zhu *et al.* [47] concluded from a number of such model comparisons that their data were best fit by a four-atom layer thick GB model in which the atoms of the inner two layers (planes) are randomly displaced by 0.5 nnd and those of the outer two by 0.25 nnd, where nnd is the nearest neighbor distance in bulk bcc α-Fe. Such a relaxed structure was shown to yield a radial distribution function that resembled a structure without short-range order, which it has been suggested could relate either to correlations among the large number of GB structures in the material or local structures of individual nanophase boundaries. This diffraction behavior does not appear to be ubiquitous for nanocrystalline metals, however, since very recent X-ray diffraction measurements [51] on approximately 8 nm grain size and about 80% dense Pd samples have shown no increase in diffuse scattering over that measured for a coarse-grained Pd control sample, as shown in Fig. 16.12.

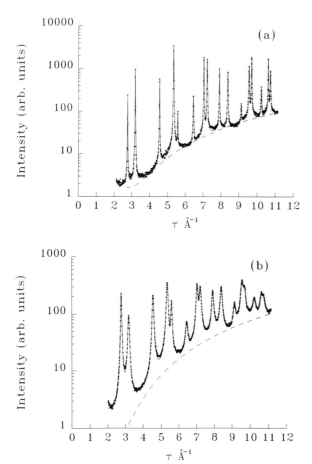

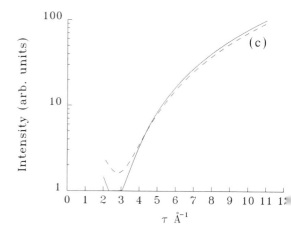

Fig. 16.12 Corrected X-ray intensity data from (a) coarse grained and (b) nanocrystalline Pd plotted on a logarithmic scale versus scattering vector magnitude, $\tau = 4\pi \sin \theta / \lambda$. Dashed lines indicate the intensity in each case unaccounted for by the Lorentzian-shaped Bragg reflections. Quadratic polynomial representations of the background intensities from (dashed line) coarse-grained and (solid line) nanocrystalline Pd are shown in (c). From [51].

Mössbauer spectroscopy

Further support for a 'gas-like' model [12, 15] of nanophase boundaries has been drawn from Mössbauer spectroscopy results on nanocrystalline Fe samples [48], as shown in Fig. 16.13. These samples, similar in grain size to those used in the X-ray scattering study by Zhu *et al.* [47], had a density of about 75% that of coarse grained poly-crystalline Fe. The measured spectrum was separated into two subspectra shown in Fig. 16.13: one (1), the well known bcc crystalline Fe spectrum; the other (2), a significantly broadened and shifted subspectrum, attributed to the sample's GBs and found to disappear with grain growth at elevated temperatures. It was concluded that this latter

interfacial component was compatible with the idea of a new type of solid state structure in nanophase GBs (characterized by a wider spectrum of inter-atomic spacings than in glasses or crystals of the chemically identical material), as deduced from the X-ray results [47].

Similar conclusions have been drawn from recent Mössbauer spectroscopy measurements [52] on nanophase samples (10 nm average grain size) of the ionic compound FeF_2. As in the case of the nanocrystalline Fe [48], a subspectrum with a broadened hyperfine field distribution was attributed to the sample's GBs and taken to indicate a non-lattice arrangement of atoms in these inter-faces. It was concluded that there is little difference in the local arrangement of atoms in the GBs of

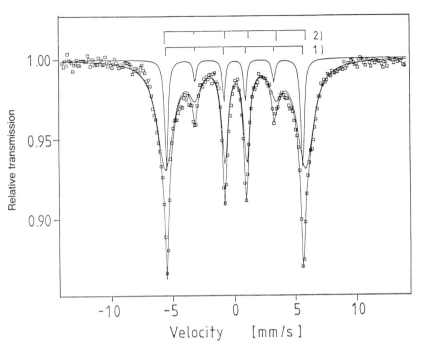

Fig. 16.13 Mössbauer spectrum of a nanocrystalline Fe sample at 77 K. Two subspectra (1: sharp lines; 2: broad lines) were used to fit the experimental data (squares). From [48].

nanocrystalline ionic and metallic systems. Positron lifetime behavior [42, 43] in nanocrystalline Fe with 6 nm grain size and densities ranging from about 51–80%, which was found to be intermediate between that for uncompacted (but agglomerated) 6 nm powder and that for either a glassy Fe alloy ($Fe_{85}B_{15}$) or polycrystalline bulk Fe, has also been taken as support for the hypothesis of an interfacial structure in nanocrystalline metals with a wide distribution of interatomic distances.

EXAFS

EXAFS measurements [49, 50] on mechanically powdered ($<5 \mu m$) nanocrystalline samples of Pd and Cu with average grain sizes in the range 10–24 nm have also been performed. Some of the results on Cu are shown in Fig. 16.14; the Pd results are very similar. The amplitudes of the weighted EXAFS oscillations and their Fourier transform were observed to be reduced in the nanocrystalline material relative to those from a coarse grained polycrystalline control sample, and it was found that this intensity reduction increased

with increasing coordination shell distance. This intensity reduction was attributed to the nanophase GBs and thought to be primarily due to a wide distribution of interatomic bond lengths in these internal interfaces.

However, very recent EXAFS measurements [53] have been carried out on unconsolidated and consolidated nanocrystalline Pd, as well as a Pd foil reference sample, which yield similar results to the earlier work [49, 50], but show that the EXAFS amplitude reduction is even greater in the unconsolidated powder than in the consolidated nanocrystalline sample. Since the consolidated sample, even with its approximately 80% density, has more grain boundary area and less free surface area than the unconsolidated (albeit agglomerated) powder, it seems clear that the attribution of such amplitude reductions to the GBs alone is an oversimplification. Indeed, the frequent lack of recognition of significant surface and porosity (and probably adsorbed impurity, as well) contributions to many of the surface-sensitive structural measurements on nanophase materials that have been used to deduce information about their GB structures has

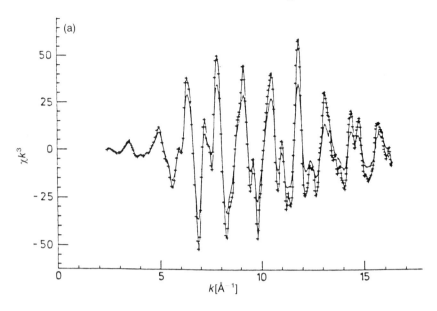

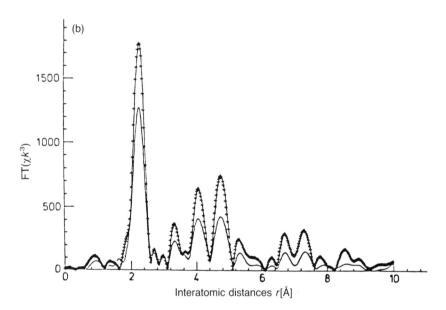

Fig. 16.14 The weighted EXAFS χk^3 (a) and its Fourier transform FTχk^3 (b) with phase shift not included of nanocrystalline Cu with a 10 nm grain size (———) and of coarse-grained polycrystalline Cu (++++) are compared. From [49].

undoubtedly exacerbated the problem of elucidating their nature.

Raman spectroscopy

Recent investigations of nanophase TiO_2 by Raman spectroscopy [54–56] and of nanophase Pd by

atomic resolution TEM [19, 57, 58] indicate that the GB structures in these materials are rather similar to those in coarser grained, conventional materials. These studies indicate that the nanophase GBs contain short-range ordered structural units representative of the bulk material and distortions that are localized to about ±0.2 nm on

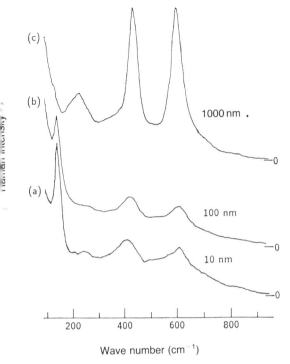

Fig. 16.15 Raman spectra of (a) an as-consolidated nanophase TiO2 sample with average grain diameter *ca.* 10 nm, (b) an as-consolidated nanophase TiO$_2$ sample with average grain diameter *ca.* 100 nm, and (c) the sample shown in (b) after annealing in air for 0.5 h at 900°C. The numbers shown for each spectrum are the approximate mean grain sizes in the respective samples. From [54].

ither side of the GB plane. These conclusions are consistent with the results from complementary small-angle neutron scattering measurements (SANS) [59, 60] (section 16.4.1), and with the expectations for the structures of conventional GBs from condensed matter theory [61–63].

Raman scattering [54] from nanophase TiO$_2$ as function of grain size (and hence GB volume fraction; Fig. 16.2) showed that the Raman spectra of as-consolidated samples were independent of grain size in their main features. Some results from this study are shown in Fig. 16.15. The Raman bands at about 600 and 418 cm^{-1} from the predominant rutile structure in this material were found to be significantly broadened in the as-consolidated (about 75% dense) nanophase samples

relative to single-crystal or coarse-grained (air-annealed) rutile, but the degree of broadening (thought to be due to oxygen deficiency) was found to be independent of grain size, as was the scattered intensity in the broad background. The broadening, shifting, and intensity reduction of Raman bands in TiO$_2$ are known to result from a number of causes, including stress and deviations from crystallinity [64, 65], but it was clearly shown in subsequent studies [55, 56] that the observed band effects from both the rutile and anatase phases of the nanophase samples, shown in Fig. 16.16, resulted from oxygen stoichiometry deviations. It was concluded that no 'gas-like' structural components exist in the GBs of nanophase TiO$_2$.

SANS

The structure of nanophase TiO$_2$ in its as-consolidated state, and as a function of sintering in air, has also been followed by SANS [59]. SANS can yield information regarding the nature of the intergrain nanophase boundaries, particularly the GB thickness and their average density vis-à-vis the grain density [59, 60, 66]. It can also yield valuable information about the presence of voids (or porosity), and also void removal during the sintering process (section 16.4.1). The results of a maximum entropy analysis of SANS data (Fig. 16.17) on TiO$_2$ are shown in Fig. 16.18, where the distributions of scattering centers for various times during sintering are presented. At least the first peak in these distributions appears to result from voids, whose number diminishes with sintering. The presence of such voids, which has been clearly demonstrated by PAS [21, 42, 43] (section 16.3.1), can reduce the apparent average density of GBs deduced from SANS data [59, 60], and may lead to erroneous conclusions regarding their structure. The low apparent densities in nanophase TiO$_2$ and Pd GBs (below 70% of bulk density) deduced from SANS measurements [59, 60, 66] may thus be partly a result of sample porosity which has not been taken fully into account. A clear separation of these effects needs to be accomplished before reliable comparisons between experiment and theory can be made.

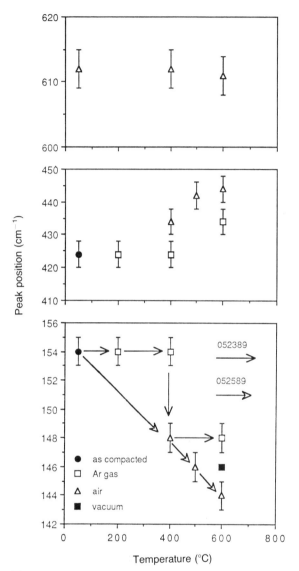

Fig. 16.16 Peak position of the three predominant Raman bands in nanophase TiO_2 as a function of annealing temperature in the various atmospheres denoted in the legend. The two sets of arrows indicate the annealing sequences (1 h at each temperature) for the two initially identical samples indicated. From [55].

junctions in the interpretation of data from other methods. Typical GBs in nanophase palladium are shown in Fig. 16.3; a higher magnification view of one such boundary is shown in Fig. 16.19. This HREM study [19, 57, 58], which included both experimental observations and complementary image simulations, indicated no manifestations of GB structures with random displacements of the type or extent suggested by earlier X-ray studies on nanophase Fe, Pd, and Cu [47, 49, 50]. For example, contrast features at the GB shown in Fig. 16.19 that may be associated with disorder do not appear wider than 0.4 nm on this micrograph, indicating that any structural disorder which may be present essentially extends no further than the planes immediately adjacent to the boundary plane. Such lattice relaxation features are of course typical of structures found in coarse-grained metals. HREM investigations of GBs in nanophase Cu [67] and Fe alloys [68] produced by surface wear and high-energy ball milling, respectively, appear to support this view, while HREM of gas-condensed nanocrystalline Pd [69] was less conclusive regarding this issue.

The HREM image simulations [19, 57, 58] examples of which are shown in Fig. 16.20, indicate that random atomic displacements of average magnitude greater than about 12% of a nearest neighbor distance, if present, could be readily observable by HREM for the assumed contrast conditions. The localized nature of the experimentally observed contrast changes found argues against the possibility that the interface atomic structure in nanophase materials can be fundamentally different from that observed in coarser-grained poly crystals. Such a fundamental difference could only be caused if atom positions were determined by their interactions with more than one boundary.

Since the displacements observed by HREM appear to fall off rapidly for distances much smaller than even the small grain diameters of the nanophase materials thus far investigated, the atomic relaxations must be dominated by the influence of only the closest boundaries, as they are in conventional polycrystals. This implies that the action of thinning the HREM foil, and hence removing the grains above and below the volume under

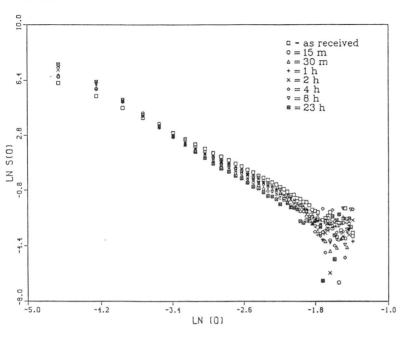

Fig. 16.17 Absolute SANS intensity S(q) as a function of scattering vector magnitude q = 4πsinθ/λ for as-consolidated nanophase TiO_2 and after sintering at 500°C in air for the times indicated. From [59].

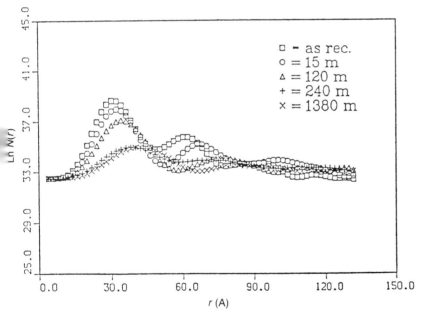

Fig. 16.18 Semi-logarithmic plot of the size distribution N(r) of scatterers obtained via a maximum entropy analysis of the SANS data in Fig. 16.17 from nanophase TiO_2 compacts sintered at 550°C in air for the indicated times. From [60].

bservation, does not in itself significantly affect he structures observed [69]. Recent calculations 70] of the possible effects of HREM specimen urfaces on GB structure further support the re-

liability of such structural observations. Indeed, as shown in Figs. 16.3 and 16.19, the nanophase GBs appear to be rather low energy configurations exhibiting flat facets interspersed with steps. Such

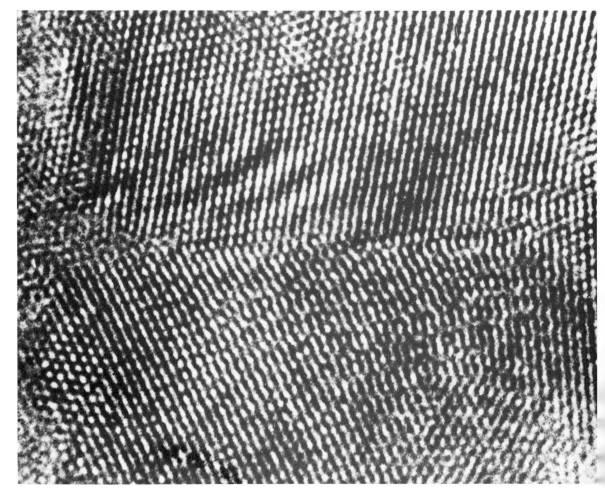

Fig. 16.19 High resolution transmission electron micrograph of a GB in nanophase Pd from an area as shown in Fig. 16.3. The magnification is indicated by the lattice fringe spacings of 0.225 nm for (111) planes. From [19, 57].

structures could only arise if sufficient local atomic motion occurred during the cluster consolidation process to allow the system to reach at least a local energy minimum.

Such observations suggest at least two conclusions: first, that the atoms that constitute the GB volume in nanophase materials have sufficient mobility during cluster consolidation to accommodate themselves into relatively low energy GB configurations; and second, that the local driving forces for grain growth are relatively small, despite the large amount of energy stored in the many GBs

in these materials. These conclusions also have an impact upon the grain size stability of nanophase materials, as described in section 16.3.3.

Discussion

The combined data reviewed here, particularly those from HREM, do not appear to support nanophase GB disorder of the type and extent suggested by earlier X-ray [47, 49, 50] studies, but rather suggest that GBs in nanophase materials (with grain sizes of the order of 5–10 nm) have structures

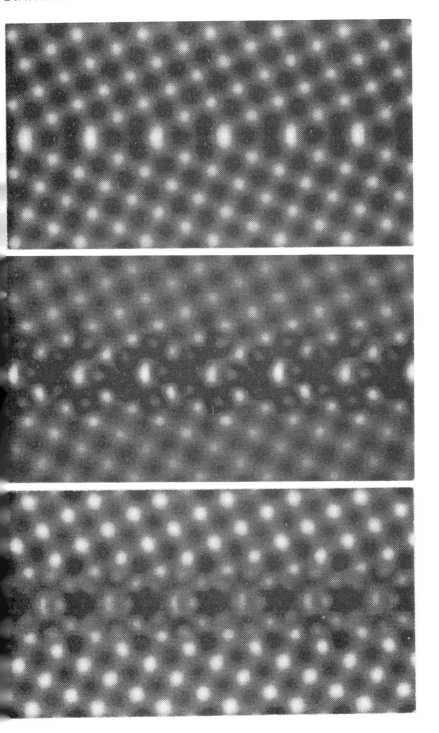

Fig. 16.20 Image simulations for a $\Sigma 5$ symmetric $\langle 001 \rangle$ tilt boundary in 7.6 nm thick Pd using microscope parameters and imaging conditions consistent with those used in HREM experimental observations [19, 57]. (a) 'perfect' structure with no atomic displacements; (b) randomly disordered structure of the GB with a maximum atomic displacement of 0.5 nnd and average of 0.25 nnd; (c) randomly disordered structure of the GB with a maximum atomic displacement of 0.25 nnd and average of 0.125 nnd.

fundamentally similar to those in conventional coarser-grained polycrystalline materials. However, their study by a number of other experimental methods is complicated by the presence in as-consolidated nanophase samples of large numbers of voids (porosity) with their associated free surfaces and GB junctions as well.

It is well known that significant atomic relaxations occur at surfaces (including internal void surfaces) and GBs; they presumably are at least as prevalent in GB junctions, although little is known of their particular structures. An example of the theoretically expected [61] atomic relaxations at a GB is shown in Fig. 16.21, which demonstrates that even for an infinite bicrystal the local boundary structure represents a wide distribution of inter-atomic distances relative to the perfect crystal. An example of the considerable atomic relaxation (reconstruction) that can occur at a (110) surface of TiO$_2$ has been recently elucidated by scanning tunneling microscopy [71]. The various atomic relaxations from normal lattice sites, which in the case of as-consolidated nanophase materials involve the atoms in GBs, at void surfaces, and in GB junctions (hence a significant fraction of the atoms in the material), can be expected to give rise to observations of 'non-lattice' contributions to several types of experimental observations that are sensitive to such relaxations, such as EXAFS [49, 50, 53] and Mössbauer spectroscopy [48, 52]. It is clear that quantitative theoretical predictions of the effects to be expected from such relaxations as those shown in Fig. 16.21 and those known for free surfaces need to be compared with experimental observations in order to ascertain whether any unexpected structural behavior remains to be attributed to nanophase grain boundaries or their junctions [72].

Beyond such comparisons between experiment and theory of static structural behavior, the dynamic behavior of atoms in interfaces and on surfaces in nanophase materials will also have to be taken into account. While HREM and X-ray diffraction yield time averaged information relative to the time scale of atomic vibrations and diffusion, EXAFS and Mössbauer measurements probe experimental times commensurate with such dynamic events. For

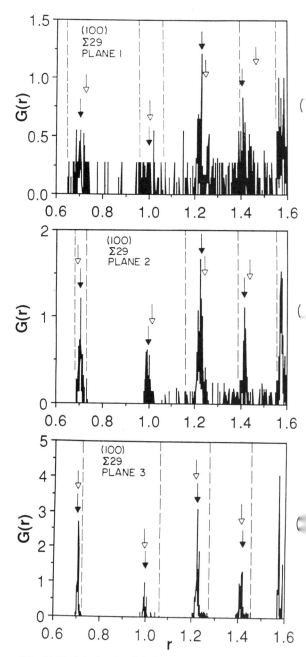

Fig. 16.21 Calculated radial distribution functions G(r) for the three lattice planes nearest to (and on either side of) a Σ29 GB in Cu. Full arrows indicate the corresponding perfect-crystal peak positions; open arrows show the average value of r in a given shell, denoted by the dashed lines. Whereas the atoms in the plane nearest the interface (a) are very strongly affected by the presence of the interface, the atoms in the third-nearest plane (c) are found in an almost perfect-crystal environment. From [61].

example, Stern and coworkers [73, 74] have recently observed local liquid-like dynamic behavior around Hg [73] and Sn [74] impurities in Pb using, respectively, EXAFS (and also XANES, X-ray absorption near-edge structure) measurements and Mössbauer spectroscopy. Whether similar dynamic high-entropy effects at the many GB, surface, and boundary-junction sites in nanophase materials could have made significant contributions to the previous EXAFS and Mössbauer measurements on them needs to be considered. It may also be that such dynamic behavior is responsible for the enhanced specific heat and entropy measured in nanophase materials [75, 76] that have been attributed to their GBs alone. The limited available information on Debye–Waller factors from X-ray measurements [49–51, 53] indicates that the integrated average mean-square atomic displacements are larger in nanophase materials than in conventional polycrystals.

In addition to their unique properties, nanophase materials should also be a valuable resource for studying the average properties of GBs and interfaces in general. The high number density of such defects in these materials enhance their influence on macroscopic properties, allowing these effects to be studied by a variety of experimental techniques. Indeed, a number of the effects observed in nanophase materials to date that have been deemed unusual may simply result from so many GBs being available for study in a sample for the first time. For the careful study of GB properties to be successful in the future, however, specimen porosity will need to be removed, via consolidation at elevated temperature and/or pressure, so that its property contributions can be eliminated. By varying their grain size, the effects from interfaces and junctions in nanophase materials could be effectively separated in such future studies.

16.3.3 Stability

With so many GBs present in nanophase materials, and their concomitantly large stored interface energy, the question of stability against grain growth is important to address from both techno-

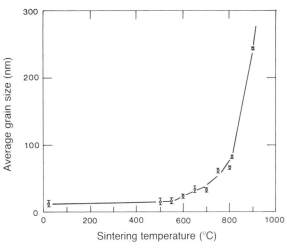

Fig. 16.22 Variation of average grain size with sintering temperature (0.5 h at each) for a nanophase TiO_2 (rutile) sample compacted to 1.4 GPa at room temperature, as determined by dark-field TEM. After [21].

logical and scientific viewpoints. Interestingly, nanophase materials assembled from atomic clusters appear to possess an inherent stability against grain growth. Their grain sizes, as measured by TEM, remain rather deeply metastable to elevated temperatures. For example, as shown in Fig. 16.22 for nanophase TiO_2 [21], the 12 nm initial average grain diameter for the distribution shown in Fig. 16.6 changes little with annealing to elevated temperatures until about 40–50% of the absolute melting temperature (T_m) of TiO_2 is reached. This behavior appears to be rather typical for the nanophase oxides already investigated [16] and for nanophase metals as well [77], as shown in the Arrhenius plot in Fig. 16.23. In the case of the TiO_2, rapid grain growth only develops above the temperature at which the mean bulk diffusion distance $(D_{Ti}t)^{\frac{1}{2}}$ of Ti becomes comparable to the mean grain size, at which temperature any local barriers to grain growth would cease to be significant.

Given the observations of the grain size distributions, grain morphologies, and GB structures in nanophase materials, it seems likely that the resistance to grain growth observed for nanophase materials results primarily from frustration [78]. It

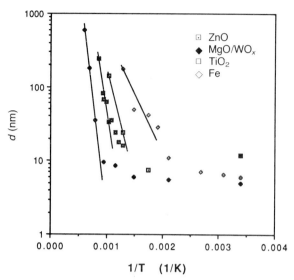

Fig. 16.23 Arrhenius plot of the variation of average grain size, measured by dark-field TEM, with sintering temperature for nanophase Fe [77], TiO_2 [21], MgO/WO_x [16], and ZnO [16]. The oxide samples were annealed for 0.5 h in air at each temperature; the Fe for 10 h in vacuum. From [78].

appears that the narrow grain size distributions normally observed in these cluster-assembled materials coupled with their relatively flat GB configurations (and also enhanced by their multiplicity of GB junctions) place these nanophase structures in a local minimum in energy from which they are not easily extricated. They are thus analogous to a variety of closed-cell foam structures, which are stable (really, deeply metastable) despite their large stored surface energy. Under such conditions, only at temperatures above which bulk diffusion distances are comparable to or greater than the grain size, as in the case of nanophase TiO_2 cited above, will this metastability give way to global energy minimization via rapid grain growth.

Such diffusion controlled grain-growth behavior is apparent in Fig. 16.23. The effective activation energy of this high temperature limiting behavior, however, is only about $9kT_m$, approximately one half that for self-diffusion. Exceptions to this frustrated grain growth behavior could be expected if considerably broader grain size distributions were present in a sample, which would allow a few

larger grains to grow at the expense of smaller ones, or if significant GB contamination were present, allowing enhanced stabilization of the small grain sizes to further elevated temperatures. One could, of course, intentionally stabilize against grain growth by appropriate doping or composite formation. It has also been recently suggested [79] that the porosity present in nanophase ceramics may assist in their stability against grain growth.

16.4 PROPERTIES

16.4.1 Sinterability

Nanophase materials have a variety of properties that are different and often considerably improved in comparison with those of conventional coarse-grained structures. For example, nanophase TiO_2 (rutile) exhibits significant improvements in both sinterability and resulting mechanical properties relative to conventionally synthesized coarser-grained rutile [21, 45, 80, 81]. Nanophase TiO_2 with a 12 nm initial mean grain diameter has been shown [21] to sinter under ambient pressures at 400–600°C lower temperatures than conventional coarse-grained rutile, and without the need for any compacting or sintering aid, such as polyvinyl alcohol, which is usually required. This behavior is shown in Fig. 16.24. Furthermore, it has been recently demonstrated [45] that sintering the same nanophase material under pressure (1 GPa), or with appropriate dopants such as Y, can further reduce the sintering temperatures, while suppressing grain growth as well. The resulting fracture characteristics [79, 80, 82] developed for sintered nanophase TiO_2 are as good or in some aspects improved relative to those for conventional rutile. It may not be so surprising that nanophase ceramics, with their ultrafine grain sizes in the nanometer regime, clean cluster surfaces, and high GB purity, will sinter at much lower temperatures than conventional coarser-grained ceramics. However, it is unique that they can also retain their ultrafine grain sizes after sintering to full density and can continue to exhibit superior mechanical properties as well.

Just as PAS can be a useful tool in the study

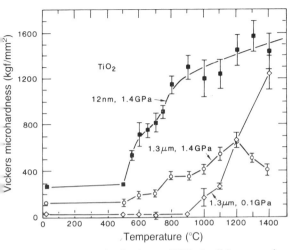

Fig. 16.24 Vickers microhardness of TiO_2 (rutile) measured at room temperature as a function of 0.5 h sintering at successively increased temperatures. Results for a nanophase sample (filled squares) with an initial average grain size of 12 nm consolidated at 1.4 GPa are compared with those for coarser-grained samples with 1.3 µm initial average grain size sintered with (diamonds) or without (circles) the aid of polyvinyl alcohol from commercial powder consolidated at 0.1 GPa and 1.4 GPa, respectively. After [21].

is also observed for the coarser-grained samples investigated, but as expected, the densification proceeds more slowly in these latter samples and the average pore sizes are larger according to the larger values of τ_2. The redistribution of void or pore sizes can also be monitored by means of BET measurements as demonstrated by the work of Hahn *et al.* [45] shown in Fig. 16.8.

The structure of nanophase TiO_2 in its as-consolidated state and as a function of sintering in air has also been followed by SANS [59]. As described in section 16.3.2, SANS contrast can yield information regarding the nature of the inter-grain nanophase boundaries, particularly their average density vis-à-vis the grain density [59, 60, 66]. However, SANS can also yield information about the presence of voids (or porosity) and their removal during the sintering process. Scattering data from nanophase TiO_2 as a function of sintering time at 550°C and the results of a maximum entropy analysis of these SANS data were shown in Figs. 16.17 and 16.18, respectively. The distributions of scattering centers for various times during sintering, taken along with the PAS results [21] shown in Figs. 16.9 and 16.10, indicate that at least the first peak in these scattering-center distributions results from voids, whose number diminishes with sintering. However, the question of further void contributions to these and similar SANS investigations is still open, as discussed in section 16.3.2.

16.4.2 Mechanical properties

Ceramics

Nanophase ceramics are easily formed, as has been clearly evident from the sample compaction process [14, 21] and from demonstrations via deformation [83]. However, the degree to which nanophase ceramics are truly ductile is only beginning to be understood. Nanoindenter measurements on nanophase TiO_2 [81] and ZnO [84] have recently demonstrated that a dramatic increase of strain rate sensitivity occurs with decreasing grain size, as shown in Fig. 16.25. Since this strong grain-size dependence is found for sets of samples in which

of the ultrafine-scale porosity inherent in as-consolidated nanophase compacts [21, 42, 43] (see section 16.3.1.), it can as well probe such porosity as a function of sintering temperature, to observe densification via the removal of voids. An example of PAS lifetime results [21] used to follow the sintering behavior of nanophase TiO_2 was already shown in Figs. 16.9 and 16.10. The positron lifetime spectra in Fig. 16.9 exhibit the long tails representative of void-trapped positrons until annealing above 500°C significantly reduces the porosity in the nanophase TiO_2. In the deconvoluted data shown in Fig. 16.10, the intensity I_2 of the lifetime (τ_2) signal corresponding to positron annihilation from void-trapped states in the nanophase sample is seen to decrease rapidly during sintering above 500°C as a result of the densification of this ultrafine-grained ceramic, even though rapid grain growth does not set in until above 800°C. Furthermore, the variation of τ_2 with sintering indicates that there is a redistribution of void sizes accompanying this densification. Similar behavior

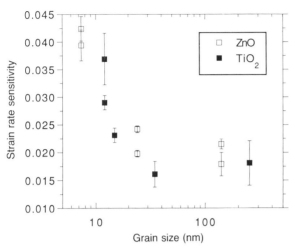

Fig. 16.25 Strain rate sensitivity of nanophase TiO₂ [81] and ZnO [84] as a function of grain size. The strain rate sensitivity was measured by a nanoindentation method [81] and the grain sizes were determined by dark-field TEM. After [8].

the porosity is changing very little, it appears to be an intrinsic property of these ultrafine-grained ceramics.

The strain rate sensitivity (m) values at the smallest grain sizes yet investigated (12 nm in nanophase TiO₂ and 7 nm in ZnO) indicate ductile behavior of these nanophase ceramics, as well as a significant potential for increased ductility at even smaller grain sizes and elevated temperatures. The maximum strain rate sensitivities measured in these studies, about 0.04, are approximately one-quarter that for Pb at room temperature, for example. However, no superplasticity has yet been observed in nanophase materials at room temperature, which would yield m values about an order of magnitude higher than the maximum observed. Nevertheless, it already seems clear that in the future, at smaller grain sizes and/or at elevated temperatures, superplasticity of these materials will indeed be observed.

The possibilities for plastic-forming nanophase ceramics to near net shape appear to be well on their way to realization. Karch and Birringer [85] have recently demonstrated that nanophase TiO₂ could be readily formed to a desired shape with excellent detail below 900°C, and the fracture toughness was found to increase by a factor of two

as well. The ability to extensively deform nanophase TiO₂ at elevated temperatures (*ca.* 800°C) without cracking or fracture has been demonstrated by Hahn and coworkers [86, 87]. While these demonstrations have been accompanied by significant grain growth in the samples at the elevated temperatures employed, it can be expected that lower temperature studies (below 0.4–$0.5\,T_m$) in the future will also allow for near net shape forming of nanophase ceramics with both their ultrafine grain sizes and their attendant properties retained.

Metals

The predominant mechanical property change resulting from reducing the grain sizes of nanophase metals is the significant increase in their strength. While the microhardness of as-consolidated nanophase oxides is reduced relative to their fully-dense counterparts (Fig. 16.24), owing to significant porosity in addition to their ultrafine grain sizes, the case for nanophase metals is quite different. Figs. 16.26, 16.27, and 16.28 show recent microhardness and stress–strain results for nanophase Pd and Cu compared with similar results for their coarser-grained counterparts [44, 88, 89]. In their as-consolidated state, nanophase Pd samples with 5–10 nm grain sizes exhibit up to about a 500% increase in hardness over coarser-grained (*ca.* 100 μm) samples (Fig. 16.26), with concomitant increases in yield stress σ_y (Fig. 16.28). Similar results have been observed in nanophase Cu as well (Fig. 16.27). As shown in Fig. 16.26, the hardness of nanophase Pd measured at room temperature falls only slowly with annealing up to about 50% of its absolute melting temperature (T_m) commensurate with the rather deep observed grain size metastability in these materials cited in section 16.3.2.

Nanophase metals and alloys produced via mechanical attrition also exhibit significantly enhanced strength. For example, Koch and coworkers [90, 91] have found hardness increases of factors of 4 to 5 in nanophase Fe and a factor of about 1.2 in nanophase Nb₃Sn when the grain size drops from 100 nm to 6 nm. On the other hand, Chokshi and

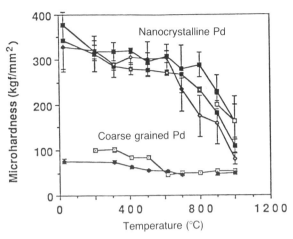

Fig. 16.26 Microhardness of three nanocrystalline (5–10 nm) Pd samples and two coarse-grained (100 μm) Pd samples as a function of annealing temperature. All samples were annealed for 100 min in 0.16 Pa vacuum and then measured at room temperature. From [88].

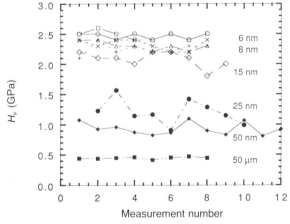

Fig. 16.27 Vickers microhardness measurements at a number of positions across several nanophase Cu samples ranging in grain size from 6 to 50 nm, compared with similar measurements from an annealed conventional 50 μm grain size Cu sample. After [44].

coworkers [92] have reported an apparent softening with decreasing grain size in the nanometer regime for cluster-assembled Cu and Pd samples, which was rationalized in terms of their expectation [83, 93] of room temperature diffusional creep in these ultrafine-grained metals.

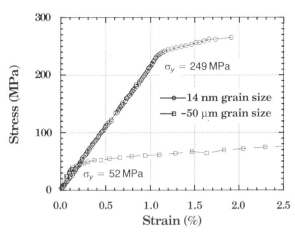

Fig. 16.28 Stress–strain curve for a nanophase (14 nm grain size) Pd sample compared with that for a coarse-grained (50 μm) Pd sample. The strain rate $\dot{\varepsilon} \approx 2 \times 10^{-5}\,\mathrm{s}^{-1}$. After [89].

The rapid atomic diffusion observed in nanophase materials (section 16.4.3), along with their nanometer grain sizes, has suggested that a large creep enhancement might result in these materials, even at room temperature [83, 93]. However, recently completed constant-stress creep measurements on nanophase Pd and Cu [44, 89] show that the observed creep rates at room temperature are at least three orders of magnitude smaller than predicted on the basis of a Coble creep model, in which the creep rate varies as D_b/d^3, where D_b is the grain boundary diffusivity and d is the mean grain size. Such creep resistance will need to be explored further at elevated temperatures in these and other nanophase materials.

Comparisons between ceramics and metals

The increased strength observed in ultrafine-grained nanophase metals, although apparently analogous to conventional Hall–Petch strengthening observed with decreasing grain size in coarser-grained metals, must result from fundamentally different mechanisms. The grain sizes here, after all, are smaller than the necessary critical bowing lengths for Frank–Read dislocation sources to operate at the stresses involved and smaller also than the normal spacings between dislocations in a pile-up. An adequate description of the mechanisms responsible for the increased strength observed in nanophase metals will clearly need to accommodate to the ultrafine grain-size scale in these materials. It appears that as this scale is reduced, and dislocation generation and migration become increasingly difficult, the energetic hierarchy of microscopic deformation mechanisms or paths is successively accessed. Thus, easier paths (such as dislocation generation from Frank–Read sources) become frozen out at sufficiently small grain sizes and more costly paths become necessary to effect deformation.

The enhanced strain rate sensitivity at room temperature found in the nanophase ceramics TiO_2 and ZnO [81, 84] appears to result from increased GB sliding in this material, aided by the presence of porosity, ultrafine grain size, and probably rapid short-range diffusion as well. The increased strength of nanophase metals, on the other hand, indicates that dislocation generation, as well as dislocation mobility, may become significantly difficult in ultrafine-grained metals. It may thus be that the increased strength of nanophase metals and the increased ductility of nanophase ceramics indicate a convergence of the mechanical response of these two classes of materials as grain sizes enter the nanometer size range. In such a case, GB sliding mechanisms, accompanied by short-range diffusion assisted healing events, would be expected to increasingly dominate the deformation of nanophase materials, and enhanced forming and even superplasticity in a wide range of nanophase materials including metals and alloys, intermetallic compounds, ceramics, and semiconductors could result. As such, increased opportunities for high deformation or superplastic near net-shape forming of a very wide range of even conventionally rather brittle and difficult to form materials could become available.

16.4.3 Diffusion and doping

Atomic diffusion in nanophase materials, which can have a significant bearing on their mechanical properties, such as creep and superplasticity, and electrical properties as well, has been found to be very rapid. Measurements of self-diffusion and

impurity diffusion [80, 94–98] in as-consolidated nanophase metals (Cu, Pd) and ceramics (TiO_2) indicate that atomic transport can be orders of magnitude faster in these materials than in coarser-grained polycrystalline samples, exhibiting 'surface-diffusion-like' behavior. However, the very rapid diffusion in as-consolidated nanophase materials appears to be intrinsically coupled with the porous nature of the interfaces in these materials.

It has been recently shown in at least one case (Hf in TiO_2) that the rapid 'surface-like' diffusivities can be suppressed back to conventional values by sintering samples to full density [80], as demonstrated by a comparison between the Hf diffusion observations shown in Figs. 16.29(a) and (b) before and after pressure-assisted sintering of nanophase TiO_2, respectively. Nonetheless, there exist considerable possibilities for efficiently doping nanophase materials at relatively low temperatures via the rapid diffusion available along their ubiquitous GB networks and interconnected porosity, with only short diffusion paths remaining into their grain interiors, to synthesize materials with tailored optical, electrical, or mechanical properties.

16.4.4 Electrical properties

Little work has been carried out so far on the electrical properties of nanophase materials. However, there appear to be interesting prospects, if the results shown in Fig. 16.30 are any indication. Nanophase TiO_2 was doped at about the 1% level with Pt diffused in from the surface, as shown by Rutherford backscattering measurements [16]. After annealing in air for 4 h at about 500°C, the ac conductivity of the sample was measured as a function of temperature. The strongly nonlinear, and reversible, electrical response shown in Fig. 16.30, caused presumably by the Pt doping into the band gap of this wide band gap (3.2 eV) semiconductor, suggests that the rather easy doping of nanophase electroceramics may lead to a wide range of interesting device applications in the future. However, much work remains to be done in this area.

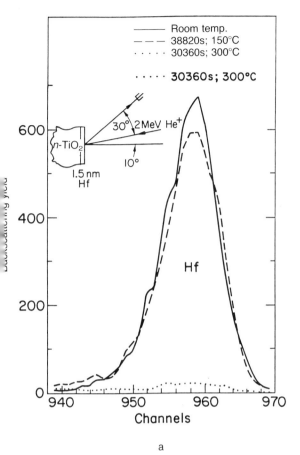

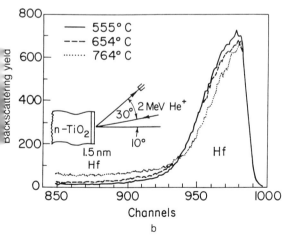

Fig. 16.29 Diffusion profiles of Hf in nanophase TiO₂ measured by Rutherford backscattering (a) after sintering in air at atmospheric pressure at 100°C [96] or (b) after pressure-assisted sintering in air at 1 GPa at 550°C [80] with subsequent Hf deposition on the sample surface.

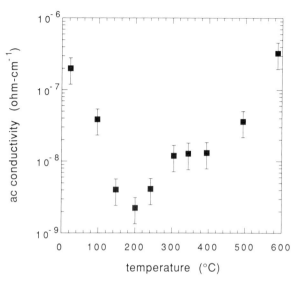

Fig. 16.30 The ac conductivity of Pt-doped nanophase TiO₂ as a function of temperature. The sample was pre-annealed in air at about 500°C for 4 h prior to the conductivity measurements; the electrical response is reversible with temperature. From A. Narayanasamy, J. A. Eastman, and R. W. Siegel, unpublished results.

16.5 CONCLUSIONS

The rapidly developing opportunities for synthesizing nanophase materials via the assembly of atom clusters are just beginning to make an impact on materials science and technology. However, based upon the limited but growing knowledge that has already been accumulated, the future appears to hold great promise for these new materials. The cluster sizes accessed to date (down to about 5 nm) indicate that the high reactivities and short diffusion distances available in these cluster-assembled materials can have profound effects upon their processing characteristics. Furthermore, the property changes that are found to occur when grain sizes are scaled down into the nanometer regime and interfaces take up significant volume fractions of the material can make significant differences in the range of use of a variety of materials. Examples are the increased ductility of nanophase ceramics and the enhanced strength of nanophase metals. Such characteristics as these should be further enhanced as even smaller and more uniformly sized clusters become available in sufficiently large numbers to effect their assembly into usable materials.

The enhanced diffusivities available along the GB networks and interconnected porosity of nanophase materials, with only few atomic jumps separating grain interiors from GBs, should enable efficient impurity doping of these materials. Nanophase insulators and semiconductors, for example, could be easily doped with impurities at relatively low temperatures, thus allowing efficient introduction of impurity levels into their band gaps and control over their electrical and optical properties. Moreover, the ability to produce via cluster assembly fully dense ultrafine-grained nanophase ceramics that are formable and exhibit ductility can have a significant technological impact in a wide variety of applications. Near net-shape forming of nanophase ceramic parts with complex and ultrafine detail would seem to be possible. Subsequent controlled grain growth can then be used to alter the grain-size dependent properties of these ceramics.

Research on cluster-assembled nanophase materials is currently being carried out in only a few laboratories, although this appears now to be changing rapidly, and considerable work still remains to be done. An understanding of the structure of these new interfacial materials is now being developed and a number of their interesting properties are being disclosed. It can be expected that numerous relationships between their atomic and electronic structures and properties will soon become clear, and that these relationships will lead to new materials science. For this to happen, the nature of the local structure of the GBs and interfaces in these materials, which comprise such a large fraction of their volumes and affect their properties so dramatically, must become better understood. While progress has been made in this area recently, as described in this chapter, further work is still needed to fully understand which interface characteristics and properties are specific to those in nanophase materials and which are simply representative of interfaces in general and only observable in nanophase materials owing to the uniquely large volume fraction of material that

they command. The contributions from atomic scale porosity must be clearly elucidated in this regard. With porosity removed, nanophase materials should offer a unique environment for the study of the properties of internal interfaces which has heretofore been unavailable to the materials research community.

Finally, further research on the synthesis of a broader range of nanophase materials, encompassing metals, alloys, ceramics, semiconductors, and composites, is needed. Also, a variety of measurements of the electrical, optical, magnetic, and mechanical properties of these new materials will certainly help in developing an understanding of how great an impact cluster-assembled nanophase materials will eventually have on materials technology.

ACKNOWLEDGMENTS

This work was supported by the US Department of Energy, BES-Materials Sciences, under Contract W-31-109-Eng-38. The author wishes to thank his many collaborators at Argonne National Laboratory and elsewhere, without whose effort and contributions this work would not have been possible.

REFERENCES

1. B. H. Kear, L. E. Cross, J. E. Keem, R. W. Siegel, F. Spaepen, K. C. Taylor, E. L. Thomas, and K.-N. Tu, *Research Opportunities for Materials with Ultrafine Microstructures*, National Academy, Washington, DC (1989), Vol. NMAB-454.
2. R. P. Andres, R. S. Averback, W. L. Brown, L. E. Brus, W. A. Goddard, III, A. Kaldor, S. G. Louie, M. Moskovits, P. S. Peercy, S. J. Riley, R. W. Siegel, F. Spaepen, and Y. Wang, *J. Mater. Res.* **4** (1989), 704.
3. M. Blander and J. L. Katz, *Geochim. Cosmochim. Acta*, **31** (1967), 1025.
4. M. Blander and M. Abdel-Gawad, *Geochim. Cosmochim. Acta*, **33** (1969), 701.
5. L. Grossman, *Geochim. Cosmochim. Acta*, **36** (1972), 597.
6. H. Gleiter, in *Deformation of Polycrystals: Mechanisms and Microstructures* (eds N. Hansen *et al.*), Risø National

Laboratory, Roskilde (1981), p. 15.

7. R. Birringer, U. Herr, and H. Gleiter, *Suppl. Trans. Jpn. Inst. Met.*, **27** (1986), 43.

8. R. W. Siegel, *Ann. Rev. Mater. Sci.*, **21**, (1991), 559.

9. K. Kimoto, Y. Kamiya, M. Nonoyama, and R. Uyeda, *Jpn. J. Appl. Phys.*, **2** (1963), 702.

10. C. G. Granqvist and R. A. Buhrman, *J. Appl. Phys.*, **47** (1976), 2200.

11. A. R. Thölen, *Acta Metall.*, **27** (1979), 1765.

12. R. Birringer, H. Gleiter, H.-P. Klein, and P. Marquardt, *Phys. Lett. A*, **102** (1984), 365.

13. R. W. Siegel and H. Hahn, in *Current Trends in the Physics of Materials* (ed M. Yussouff), World Scientific Publ. Co., Singapore, (1987), p. 403.

14. H. Hahn, J. A. Eastman, and R. W. Siegel, in Ceramic Transactions, *Ceramic Powder Science*, (eds G. L. Messing *et al.*), Vol. 1, Part B, American Ceramic Society, Westerville (1988), p. 1115.

15. R. Birringer and H. Gleiter, in *Encyclopedia of Materials Science and Engineering*, Suppl. Vol. 1 (ed R. W. Cahn), Pergamon Press, Oxford (1988), p. 339.

16. J. A. Eastman, Y. X. Liao, A. Narayanasamy, and R. W. Siegel, *Mater. Res. Soc. Symp. Proc.*, **155** (1989), 255.

17. H. Gleiter, *Prog. in Mater. Sci.*, **33** (1989), 223.

18. R. W. Siegel, in *Materials Science and Technology – A Comprehensive Treatment*, (ed R. W. Cahn), Vol. 15: *Processing of Metals and Alloys*, VCH, Weinheim (1991), 583.

19. G. J. Thomas, R. W. Siegel, and J. A. Eastman, *Scripta Metall. et Mater.*, **24** (1990), 201; *Mater. Res. Soc. Symp. Proc.*, **153** (1989), 13.

20. R. W. Siegel and J. A. Eastman, *Mater. Res. Soc. Symp. Proc.*, **132** (1989), 3.

21. R. W. Siegel, S. Ramasamy, H. Hahn, Z. Li, T. Lu, and R. Gronsky, *J. Mater. Res.*, **3** (1988), 1367.

22. R. S. Bowles, J. J. Kolstad, J. M. Calo, and R. P. Andres, *Surf. Sci.*, **106** (1981), 117.

23. C. Hayashi, *J. Vac. Sci. Technol. A*, **5** (1987), 1375.

24. M. Uda, *Nanostructured Mater.*, **1** (1992), 101.

25. H. Oya, T. Ichihashi, and N. Wada, *Jpn. J. Appl. Phys.*, **21** (1982), 554.

26. S. Yatsuya, K. Yamauchi, T. Kamakura, A. Yanagada, H. Wakaiyama, and K. Mihama, *Surf. Sci.*, **156** (1985), 1011.

27. S. Yatsuya, T. Kamakura, K. Yamauchi, and K. Mihama, *Jpn. J. Appl. Phys. Part 2*, **25** (1986), L42.

28. H. Hahn and R. S. Averback, *J. Appl. Phys.*, **67** (1990), 1113.

29. G. M. Chow, R. L. Holtz, A. Pattnaik, A. S. Edelstein, T. E. Schlesinger, and R. C. Cammerata, *Appl. Phys. Lett.*, **56** (1990), 1853.

30. S. Iwama, E. Shichi, and T. Sahashi, *Jpn. J. Appl. Phys.*, **12** (1973), 1531.

31. S. Iwama, K. Hayakawa, and T. Arizumi, *J. Cryst. Growth*, **56** (1982), 265; ibid., **66** (1984), 189.

32. S. Iwama and K. Hayakawa, *Surf. Sci.*, **156** (1985), 85.

33. B. Günther and A. Kumpmann, *Nanostructured Mater.*, **1** (1992), 27.

34. A. Matsunawa and S. Katayama, in *Laser Welding, Machining and Materials Processing, Proc. ICALEO '85* (ed C. Albright), IFS (1985), 205.

35. K. Baba, N. Shohata, and M. Yonezawa, *Appl. Phys. Lett.*, **54** (1989), 2309.

36. A. E. Berkowitz and J. L. Walter, *J. Mater. Res.*, **2** (1987), 277.

37. E. Hellstern, H. J. Fecht, Z. Fu, and W. L. Johnson, *J. Appl. Phys.*, **65** (1989), 305.

38. I. A. Aksay, G. L. McVay, and D. R. Ulrich (eds), *Processing Science of Advanced Ceramics, Mater. Res. Soc. Symp. Proc.*, **155** (1989).

39. P. C. Rieke, P. D. Calvert, and M. Alper (eds), *Materials Synthesis Utilizing Biological Processes, Mater. Res. Soc. Symp. Proc.*, **174** (1990).

40. R. S. Averback, D. L. Nelson, and J. Bernholc (eds), *Clusters and Cluster-Assembled Materials, Mater. Res. Soc. Symp. Proc.*, **206** (1991).

41. B. H. Kear and R. W. Siegel (eds), *Proc. Acta Metallurgica Conf. on Materials with Ultrafine Microstructures, Nanostructured Mater.*, **1** (1992), Numbers 1–3.

42. H. E. Schaefer, R. Würschum, M. Scheytt, R. Birringer, and H. Gleiter, *Mater. Sci. Forum*, **15–18** (1987), 955.

43. H. E. Schaefer, R. Würschum, R. Birringer, and H. Gleiter, *Phys. Rev. B.*, **38** (1988), 9545.

44. G. W. Nieman, J. R. Weertman, and R. W. Siegel, *J. Mater. Res.*, **6** (1991), in press.

45. H. Hahn, J. Logas, and R. S. Averback, *J. Mater. Res.*, **5** (1990a), 609.

46. S. J. Gregg and K. S. W. Sing, *Adsorption, Surface Area and Porosity*, Academic Press, New York (1982).

47. X. Zhu, R. Birringer, U. Herr, and H. Gleiter, *Phys. Rev. B.*, **35** (1987), 9085.

48. U. Herr, J. Jing, R. Birringer, U. Gonser, and H. Gleiter, *Appl. Phys. Lett.*, **50** (1987), 472.

49. T. Haubold, R. Birringer, B. Lengeler, and H. Gleiter, *J. Less-Common Metals*, **145** (1988), 557.

50. T. Haubold, R. Birringer, B. Lengeler, and H. Gleiter, *Phys. Lett. A*, **135** (1989), 461.

51. M. R. Fitzsimmons, J. A. Eastman, M. Müller-Stach, G. Wallner, *Phys. Rev. B*, in press (1991).

52. J. Jiang, S. Ramasamy, R. Birringer, U. Gonser, and H. Gleiter, *Solid State Comm.*, **80** (1991), 525.

53. J. A. Eastman, M. R. Fitzsimmons, M. Müller-Stach,

G. Wallner, and W. T. Elam, *Nanostructured Mater.*, **1** (1992), 47; W. T. Elam, J. A. Eastman, K. H. Kim, E. F. Skelton, J. Lupo, and M. J. Sabochick, to be published.

54. C. A. Melendres, A. Narayanasamy, V. A. Maroni, and R. W. Siegel, *J. Mater. Res.*, **4** (1989), 1246.

55. J. C. Parker and R. W. Siegel, *J. Mater. Res.*, **5** (1990), 1246.

56. J. C. Parker and R. W. Siegel, *Appl. Phys. Lett.*, **57** (1990), 943.

57. G. J. Thomas, R. W. Siegel, and J. A. Eastman, *Mater. Res. Soc. Symp. Proc.*, **153** (1989), 13.

58. G. J. Thomas and R. W. Siegel, to be published.

59. J. E. Epperson, R. W. Siegel, J. W. White, T. E. Klippert, A. Narayanasamy, J. A. Eastman, and F. Trouw, *Mater. Res. Soc. Symp. Proc.*, **132** (1989), 15.

60. J. E. Epperson, R. W. Siegel, J. W. White, J. A. Eastman, Y. X. Liao, and A. Narayanasamy, *Mater. Res. Soc. Symp. Proc.*, **166** (1990), 87.

61. D. Wolf and J. F. Lutsko, *Phys. Rev. Lett.*, **60** (1988), 1170.

62. D. Wolf and J. F. Lutsko, *J. Mater. Res.*, **4** (1989), 1467.

63. S. R. Phillpot, D. Wolf, and S. Yip, *MRS Bulletin*, **XV** (10) (1990), 38.

64. G. J. Exarhos, in *Characterization of Semiconductor Materials*, Vol. 1 (ed G. E. McGuire), Noyes Publ., Park Ridge (1989), p. 242.

65. G. J. Exarhos and M. Aloi, *Thin Solid Films*, **193–194** (1990), 42.

66. E. Jorra, H. Franz, J. Peisl, G. Wallner, W. Petry, R. Birringer, H. Gleiter, and T. Haubold, *Phil. Mag. B*, **60**, (1989), 159.

67. S. K. Ganapathi and D. A. Rigney, *Scripta Metall. et Mater.*, **24** (1990), 1675.

68. M. L. Trudeau, A. Van Neste, and R. Schultz, *Mater. Res. Soc. Symp. Proc.*, **206** (1991), 487.

69. W. Wunderlich, Y. Ishida, and R. Maurer, *Scripta Metall. et Mater.*, **24** (1990), 403.

70. M. J. Mills and M. S. Daw, *Mater. Res. Soc. Symp. Proc.*, **183** (1990), 15.

71. G. S. Rohrer, V. E. Henrich, and D. A. Bonnell, *Science*, **250** (1990), 1239.

72. R. W. Siegel and G. J. Thomas, *Mater. Res. Soc. Symp. Proc.*, **209** (1991), 15.

73. E. A. Stern and Z. Ke, *Phys. Rev. Lett.*, **60** (1988), 1872.

74. H. Shechter, E. A. Stern, Y. Yacoby, R. Brener, and Z. Zhe, *Phys. Rev. Lett.*, **63** (1989), 1400.

75. J. Rupp and R. Birringer, *Phys. Rev. B*, **36** (1987), 7888.

76. D. Korn, A. Morsch, R. Birringer, W. Arnold, and H. Gleiter, *J. de Phys. C5*, **49** (1988), 769.

77. E. Hort, Diploma Thesis, Universität des Saarlandes, Saarbrücken (1986).

78. R. W. Siegel, *Mater. Res. Soc. Symp. Proc.*, **196** (1990), 59.

79. H. J. Höfler and R. S. Averback, *Scripta Metall. et Mater.*, **24** (1990), 2401.

80. R. S. Averback, H. Hahn, H. J. Höfler, J. L. Logas, and T. C. Chen, *Mater. Res. Soc. Symp. Proc.*, **153** (1989), 3.

81. M. J. Mayo, R. W. Siegel, A. Narayanasamy, and W. D. Nix, *J. Mater. Res.*, **5** (1990), 1073.

82. Z. Li, S. Ramasamy, H. Hahn, and R. W. Siegel, *Mater. Lett.*, **6** (1988), 195.

83. J. Karch, R. Birringer, and H. Gleiter, *Nature*, **330** (1987), 556.

84. M. J. Mayo, R. W. Siegel, Y. X. Liao, and W. D. Nix, *J. Mater. Res.*, **7** (1992), 973.

85. J. Karch and R. Birringer, *Ceramics Int.*, **16** (1990), 291.

86. H. Hahn, J. Logas, H. J. Höfler, P. Kurath, and R. S. Averback, *Mater. Res. Soc. Symp. Proc.*, **196** (1990), 71.

87. M. Guermazi, H. J. Höfler, H. Hahn, and R. S. Averback, *J. Amer. Cer. Soc.*, **74** (1991), 2672.

88. G. W. Nieman, J. R. Weertman, and R. W. Siegel, *Scripta Metall.*, **23** (1989), 2013.

89. G. W. Nieman, J. R. Weertman, and R. W. Siegel, *Scripta Metall. et Mater.*, **24** (1990), 145.

90. J. S. C. Jang and C. C. Koch, *Scripta Metall. et Mater.*, **24** (1990), 1599.

91. C. C. Koch and Y. S. Cho, *Nanostructured Mat.*, **1** (1992), in press.

92. A. H. Chokshi, A. Rosen, J. Karch, and H. Gleiter, *Scripta Metall.*, **23** (1989), 1679.

93. R. Birringer, H. Hahn, H. Höfler, J. Karch, and H. Gleiter, *Defect and Diffusion Forum*, **59** (1988), 17.

94. J. Horváth, R. Birringer, and H. Gleiter, *Solid State Commun.*, **62** (1987), 319.

95. J. Horváth, *Defect and Diffusion Forum*, **66–69** (1989), 207.

96. H. Hahn, H. Höfler, and R. S. Averback, *Defect and Diffusion Forum*, **66–69** (1989), 549.

97. S. Schumacher, R. Birringer, R. Straub, and H. Gleiter, *Acta Metall.*, **37** (1989), 2485.

98. T. Mütschele and R. Kirchheim, *Scripta Metall.*, **21** (1987), 135; ibid. (1987), 1101.

PART III:
Role of Interface Chemistry

CHAPTER SEVENTEEN

Interfacial segregation, bonding, and reactions

C. L. Briant

Introduction · The segregation process · Grain boundary segregation · Grain boundary reactions · Applications · Conclusions

17.1 INTRODUCTION

Grain boundary segregation in metals refers to the process by which a solute element in an alloy becomes enriched at the grain boundary (GB). It occurs when the solute element diffuses through the alloy, encounters a GB, and becomes trapped there. It is now recognized that this segregation can affect many metallurgical processes including fracture, corrosion, recrystallization, grain growth, creep, environmental embrittlement, and GB diffusion.

The last two decades have seen a great amount of research in this field. This activity has been a result of two factors. The first was the development during this time of a number of experimental techniques that could examine GB segregation in metals much more completely than was possible in the past. Of these, Auger electron spectroscopy [1] has been of special importance and usefulness. The second factor was the continuing number of engineering problems in which GB segregation and its consequences played a major role. These ranged from the ones that received wide-scale, public attention, and consequent research funding, such as the fracture of components in large steam turbines [2, 3] and intergranular stress corrosion cracking of stainless steel recirculation pipes in

nuclear reactors [4] to the more common, unpublicized occurrences such as failure of thermocouple leads [5] and fractures during processing of metal castings [6].

As a result of this research it is now possible to write down certain general observations about this process. These include the following.

1. The solute elements that become most enriched at the GBs and that seem to have the greatest effect on metallurgical properties are impurity elements. These are elements that usually enter the material through the scrap. They are present in the scrap because it is not economically feasible to refine them out. In the alloy specifications there may be a maximum concentration for these impurities that cannot be exceeded, but except for special applications there is no additional control on them. Thus, these elements usually have bulk concentration of 10–200 wppm, but at the GBs their local concentrations may exceed 5 at.%. Common among these elements are sulfur, tin, antimony, and especially phosphorus. Alloying elements can also be enriched at the GBs, but their increase over the bulk concentration is rarely as great as that of impurities.
2. The segregated layer is usually confined to one

or two atom layers on either side of the GB. There are no wide concentration profiles extending many nanometers into the bulk.

3. Segregation is maximized over a fairly narrow temperature range. For example, in iron and iron-based body-centered-cubic alloys segregation builds up most rapidly at temperature between 450 and 600°C [7]. In nickel and its alloys and face-centered-cubic iron-base alloys segregation is greatest at 600 to 750°C [8, 9]. The reason for this behavior is that below these temperature regimes diffusion occurs so slowly that segregation can only be observed after very long times. Above these temperature regimes entropic effects limit segregation by allowing desegregation to occur more easily. In practice, segregation most often occurs when a part is either slowly cooled through the temperature regimes or when it is put in service at a temperature where segregation can occur.

4. Although segregation is maximized in these temperature ranges, the total amount of segregation that has occurred at any given time in a material does depend on the complete thermal history that the material has received. Segregation usually occurs first when a sample is cooled after solidification. Then as the sample is worked and annealed segregation and desegregation can occur. The extent of these two processes will depend on the times and temperatures of the processing and service operations.

Now that we have made these general comments we would like to consider the process of segregation and its effects on metallurgical problems in much more detail. We will begin by considering a simple model that can be used to think about the segregation process. Then we will consider a number of specific examples of GB segregation. Here we will begin with binary alloys composed of the segregant and the host metal and proceed to compositionally more complex systems. We will then examine the effect of this segregation on various metallurgical processes. Finally, we will consider several engineering problems where GB chemistry has been of overriding importance.

17.2 THE SEGREGATION PROCESS

The process by which a segregant arrives at the GB and becomes entrapped there can be represented by two chemical reactions. The first of these simply represents the diffusion of the atom to the GB and can be written as

$$I_b \rightleftarrows I_{GB} \tag{17.1}$$

where I stands for the diffusing element (impurity), b stands for bulk, and GB stands for grain boundary. The second reaction represents the structural accommodation that occurs in the GB once the atom arrives there. It can be written as

$$(M_x)_{GB} + I_{GB} \rightleftarrows (M_xI)_{GB} \tag{17.2}$$

where $(M_x)_{GB}$ stands for a group of atoms of the host metal in the GB that does not yet contain a segregated atom I, and $(M_xI)_{GB}$ stands for the same group of atoms after the segregant enters it.

Let us now consider these two reactions in more detail. Equation (17.1) is analogous to an equation describing the heterogeneous equilibrium that is set up when a solute is distributed between two immiscible phases. A simple example of this equilibrium is the distribution of an organic acid between water and benzene [10]. Water and benzene are completely immiscible but the acid, which will be denoted as HA, can be dissolved in either of them. To describe the distribution of the acid between H_2O and C_6H_6 as it is added to this system we would write the equation

$$HA_{H_2O} \rightleftarrows HA_{C_6H_6} \tag{17.3}$$

For this equation we can write an expression for the equilibrium constant

$$K = \frac{[HA]_{C_6H_6}}{[HA]_{H_2O}} \tag{17.4}$$

where the square brackets denote concentrations. A more accurate description would be given by

$$K = \frac{a_{C_6H_6}^{HA}}{a_{H_2O}^{HA}} \tag{17.5}$$

where a denotes the activity of the acid in the phase.

If we continue the analogy between this type of heterogeneous equilibrium and GB segregation we can write down an equilibrium constant for segregation as

$$K = \frac{a_{GB}^i}{a_b^i} \cong \frac{[I]_{GB}}{[I]_b} \qquad (17.6)$$

The activity of an element in the GB is difficult to measure, but the concentration can be measured by Auger electron spectroscopy. When this is done, one can then measure K; it is the same as the enrichment ratio used by Hondros and Seah [11] to describe segregation.

Based on our knowledge of chemical systems we would expect K to change as the temperature at which segregation occurs is changed. However, for a given temperature we would expect this ratio of activities to be independent of the bulk activity. Figures 17.1(a) and (b) show these two effects. To make these plots we have collected data for phosphorus segregation in iron. Figure 17.1(a) plots the effect of temperature on K. The results show that as the temperature increases the value of K decreases, although the magnitude of this effect appears to depend on the bulk concentration of the segregant. This result is not surprising because as the temperature is raised the increase in entropy will cause desegregation. The results in Fig. 17.1(b) show that for a given temperature K decreases with increasing bulk concentration of the segregant. This decrease probably results from the following effect. We have calculated K as a ratio of concentrations. A more accurate treatment would be a ratio of activities. As the impurity concentration on the GB builds up, it is possible that an increase in concentration begins to cause a larger change in activity because of impurity–impurity interactions or the fact that the segregant must occupy sites that are of increasingly higher energy if the concentration at the boundary is to increase. The true equilibrium constant, defined as a ratio of activities, would not change but the ratio of concentrations would decrease.

We now wish to turn our attention to the question of what physical factors determine the value of K for a given system. Most of the discussions in the past concerning causes of GB

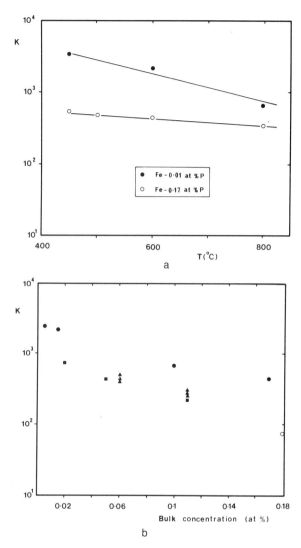

Fig. 17.1 (a) The ratio, K, of the GB concentration of phosphorus to the bulk concentration of phosphorus plotted as a function of ageing temperature. All data were taken from Ref. [20]. (b) Values of K plotted as a function of the bulk concentration of phosphorus. The closed circles represent data taken from reference [20], the closed squares represent data taken from Ref. [75], the closed triangles represent data taken from Ref. [22], and the open circle represents data taken from Ref. [45].

segregation have simply suggested that segregation lowers the energy of the system [12–14]. This idea is undoubtedly correct, but we would like to know the important contributions to this reduction of energy. Several authors have suggested that misfit of the atoms in the matrix would be a primary driving force for segregation [13, 14]. Another factor that must be considered is the different chemical bond geometries available to the segregant at the GB that are not present in the matrix. The formation of these more favorable bonds will certainly help trap the segregant in the GB. One can see that these factors are very similar to those discussed by Hume-Rothery [15] in his consideration of the limits of solid solubility. He proposed that solubility would decrease as the misfit of the atom in the matrix increases, as the valency difference between the solute and solvent increases, and as the electronegativity difference between the solute and solvent increases. The first of these is identical to the size effect mentioned above, and the latter two are similar to the chemical bond effect that we described. Hondros and Seah [11] have pointed out that there is generally an inverse relationship between solid solubility and K, and, based on the above discussion, this idea seems quite reasonable. Figure 17.2 plots the value of K for various systems as a function of solid solubility. One can observe that, although there is a lot of scatter, there is a general increase in K with decreasing solid solubility.

The relationship between solubility and segregation can also be understood in another way. Burton and Machlin [16] found that a solute which forms a eutectic with the solvent also segregates to the surface of the solid solvent in a high vacuum. They suggested that the same forces that would tend to expel the solute from the solid phase into the liquid phase during solidification (thus creating a eutectic diagram) are the same ones that would tend to expel the solute from the solid phase to its free surface. They argued, and one could make similar statements for the GBs, that certain similarities exist between a surface and a liquid as compared with the crystalline matrix. These similarities include lower coordination, lower symmetry, and lower strain. Indeed all of the documented GB segregants in iron and nickel alloys form eutectics with the matrix element, and thus would be rejected from the solid when solidification occurs.

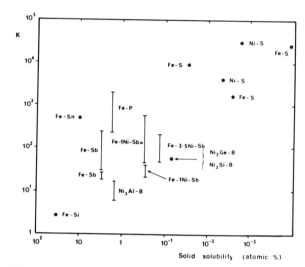

Fig. 17.2 The ratio, K, of the GB concentration of an element to the bulk concentration of that element for various impurity elements in iron and nickel base alloys plotted as a function of the solid solubility of the impurity at the temperature where segregation occurred. The data were taken from Refs. [11] and [76].

We stated above that one reason a segregant will remain in the GB once it has arrived there is because it can form bonds with the host metal in the GB that would not be allowable in the matrix. This bond formation is the step represented by eq. (17.2) above. The segregant arrives in the GB and diffuses within the boundary if necessary until it finds a suitable site. This site may already exist within the GB or it may form by movement of atoms in response to the presence of the segregant. Within this site we would expect the chemical bonds that have formed to be more favorable for the segregant than those allowed in the matrix.

To enlarge on this discussion, we must first describe the basic structure of a GB. The most useful atomic picture of a GB that has been developed is the structural unit model [17, 18]. This model developed as a result of careful examination of the structures of various clusters of atoms that compose the boundary. The results showed that there were several atomic units that appeared again and again in boundaries. These include the tetrahedron, the pentagonal bi-pyramid, the capped trigonal prism, and the Archimedian anti-

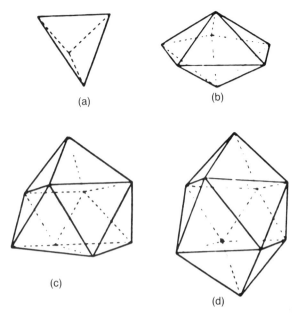

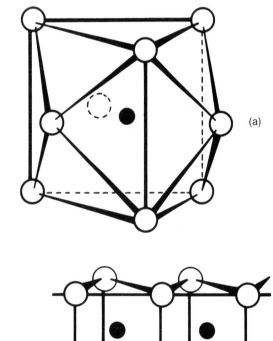

Fig. 17.3 Structural units that are commonly found in GBs. (a) The tetrahedron, (b) the pentagonal bi-pyramid, (c) the capped trigonal prism, and (d) the capped Archimedean anti-prism.

Table 17.1 Structures of metal-rich metal impurity compounds

Impurity	Compound	Structure	Reference
B	Ni_3B	Fe_3C	[70]
S	Ni_3S_2	Complex triangular bi-pyramid with a triangle of Ni atoms and a S at the apex	[71]
S	FeS	NiAs	[15]
P	Fe_3P Ni_3P	Fe_3P	[70]
Sb	NiSb FeSb	NiAs	[15]
Sn	FeSn	NiAs	[15]
Te	FeTe	NiAs	[15]

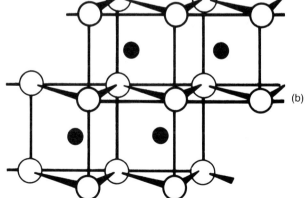

Fig. 17.4 (a) The Fe_3C structure, (b) the NiAs structure.

prism. They are shown in Fig. 17.3. If one allows some distortion in these structures, plus some additional atoms to connect them up, one can construct a wide variety of GB structures with these atomic clusters.

Let us now consider how a segregant might interact with these structures. One of the most useful approaches for predicting the structures that adsorbates will form on free surfaces has been to examine the structure of simple molecules that form between the adsorbate and the host metal. One often finds that the surface structure mimics this molecular structure. The structures of the metal-rich compounds that many segregating impurities form with typical host metals are listed in Table 17.1 and depicted in Fig. 17.4. These structures tend to have a unit in common, namely a trigonal prism, sometimes capped, of metal atoms surrounding the non-metal (impurity). These structures are very similar to the various distorted versions of capped trigonal prisms that are such common structural units of GBs [17, 18]. The fact that at a GB the segregant may find an atomic

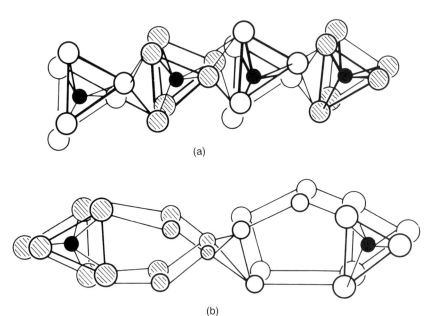

(a)

(b)

Fig. 17.5 Two schematics of GBs. The first is composed of trigonal prisms. The shaded atoms are used to denote that the two sets of prisms are in different planes. The center dark atoms denote segregated atoms. The second GB contains two trigonal prisms plus connective regions. The segregant is only in the trigonal prisms.

geometry much more compatible with the bond geometry that it apparently wishes to form with the host metal should provide a strong contribution to keeping a segregant in the GB. Furthermore, because of the more open structure of the boundary it seems possible that the impurity might also attempt to rearrange the atomic configuration of the boundary to a more favorable structure.

Therefore, the picture of segregation that we have developed is the following. For a variety of reasons, such as size or various chemical differences between the impurity and the matrix, an impurity may prefer to reside at a GB instead of in the matrix. As the segregation takes place, bonds will break and form as the atom moves through the solid. When the atom reaches the GB it can form a set of bonds that are more favorable than those it could form in the bulk. The impurity thus establishes a molecular unit which in the ideal case is probably very similar in both structure and bonding to the compound that an impurity would try to form with host. The segregated boundary can then be thought of as a string of these molecular units with connecting regions between them as shown schematically in Fig. 17.5. In Fig. 17.5(a) the boundary is made up of a series of capped

trigonal prisms which are staggered up and down in the plane of the page. Each prism is drawn with a central impurity atom. In Fig. 17.5(b) the two trigonal prisms that contain impurities are separated by a large amount of connective material. With this basic model for GB segregation we can now review the specific studies of segregation that have been made.

17.3 GRAIN BOUNDARY SEGREGATION

In this section we will review experimental studies that have measured GB segregation. We are concerned with the process described by eq. (17.1). As changes are made in the alloy that change the activity of the segregant, we will find that the amount of segregation is also affected. We begin our discussion of GB segregation by considering studies of binary alloys composed of the host metal and the segregant. Note that all of the data reported below were obtained with Auger electron spectroscopy. A review of this technique and its application to GB segregation is given in Ref. [19].

A number of researchers have studied the segregation of phosphorus in iron [20–23]. The

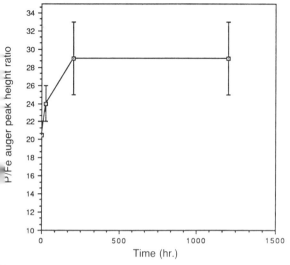

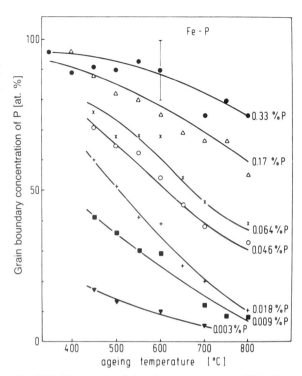

Fig. 17.6 The amount of phosphorus segregation on GBs in iron plotted as a function of ageing time. The ageing temperature was 480°C. The samples were held at that temperature for a given time, quenched, and then analyzed by Auger electron spectroscopy. The peak height ratio is directly correlated with the amount of phosphorus on the GBs. Data taken from Ref. [22].

Fig. 17.7 The amount of phosphorus segregation to GBs in iron plotted as a function of ageing temperature. The different curves represent data for samples with different bulk concentrations of phosphorus. The data are taken from Ref. [20].

kinetics of segregation in this system are relatively rapid. Figure 17.6 shows that at 480°C equilibrium is achieved within 200 hours. At higher temperatures equilibrium can be achieved even more rapidly, but at temperatures below 400°C the times required to reach equilibrium go well beyond 2000 hours. Figure 17.7 shows the effect of the bulk phosphorus concentration and ageing temperature on grain boundary concentration. (The ageing temperature refers to the temperature to which the sample was heated so that segregation could occur. After ageing the samples are usually quenched to room temperature and then analyzed.) The plot shows that as the amount of phosphorus in the bulk increases and the ageing temperature decreases the amount of phosphorus on the GB increases.

The research on Fe–Sn alloys has not been as extensive as that on Fe–P alloys, but the results do show that as the bulk concentration increases the amount of segregation also increases [24, 25]. However, as can be seen in Fig. 17.8 tin segregation does not show as strong a temperature dependence as does phosphorus segregation.

One should note that for these results and all others that will be reported below there are large error bars associated with each data point. These arise from the fact that the amount of segregation measured from different GBs in a polycrystalline material may vary as much as ±30% [9, 26–29]. This variability is thought to arise from differences in GB structures. As we showed schematically in Fig. 17.5, a GB that has a large density of sites favorable to the segregant will have a high concentration of the segregant. A GB with a low density of favorable sites will not.

The two systems described above had concentrations of the segregant that were well below the solubility of the segregant. We now wish to consider what happens when the segregating element can also enter into a precipitate with the host metal. A good example of this behavior can be seen in the study of sulfur segregation in

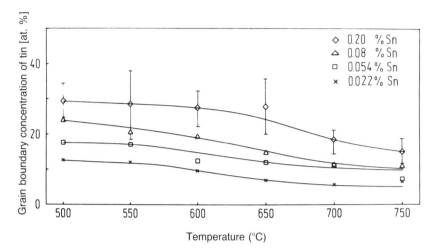

Fig. 17.8 The amount of tin segregation to GBs in iron plotted as a function of ageing temperature. The different curves are for alloys with different tin concentrations. The data are taken from Ref. [24].

iron [30]. Sulfur has a very low solubility in iron; for example, at 600°C it is approximately 0.001 at.% [31]. Once the concentration exceeds this solubility limit iron sulfide forms.

Figure 17.9 shows the segregation behavior of sulfur in iron for an alloy which contained 0.0035 at.% sulfur. This plot shows the equilibrium amount of segregation for different temperatures. The interesting point to note is that at ageing temperatures up to approximately 675°C the amount of segregation increases with increasing temperature, in contrast to the behavior noted for tin and phosphorus. Above this temperature the amount of sulfur segregation decreases with increasing temperature.

These results can be interpreted in the following way. The temperature at which all sulfur should be in solution in the 0.0035 at.% sulfur alloy is approximately 675°C. Below this temperature part of the sulfur will be precipitated as iron sulfide. Therefore, as the temperature is raised there will be two competing effects until this solubility limit is crossed. There will be the usual tendency to desegregate with increasing temperature because of entropic effects, but at the same time increasing the temperature will increase the amount of sulfur that is free in solution and available to segregate because of the dissolution of precipitates. This increase in the concentration, or activity, of the dissolved sulfur will tend to increase segregation,

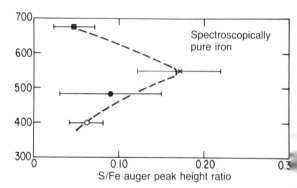

Fig. 17.9 The amount of sulfur segregation in iron plotted a a function of ageing temperature. The data are taken from Ref. [30].

just as an increase in bulk concentration did fo phosphorus and tin. The results show that below the solubility limit this effect is dominant and w observe an increase in segregation with increasing temperature. Above the temperature at which a sulfur is in solution, only the entropic effect o segregation will be observed. Consequently, th slope of the curve in Fig. 17.9 changes sign at th solubility limit and the amount of segregatio decreases with increasing temperature.

Sulfur also has a very low solubility in nickel and its segregation in nickel has been studied b Mulford [9]. He found that, in contrast with iron as the temperature decreased, the amount c segregation continued to increase even though sulfides were precipitating. Therefore, in the Ni–S system the temperature effect dominates segregation even though precipitation is lowering the bulk activity of sulfur.

We now wish to consider results for ternary alloys. These will show the more complicated behavior that can occur when an additional element is added to the system and will demonstrate why it can be difficult to interpret segregation data obtained on multi-component alloys. The possible effects of a third element include the following.

(a) it can decrease the amount of segregation through formation of a precipitate with the segregating element in the matrix;

(b) it can decrease the amount of segregation below that found in the binary alloy by segregating and competing with the other elements for sites at the GB;

(c) it can increase the amount of segregation above that observed in the binary alloy;

(d) it can produce no change in the amount of segregation from that observed in the binary alloy.

Let us first consider the effect of a third element that reduces segregation by precipitation of the segregant. An excellent example of this can be observed when titanium is added to an Fe–P alloy [32]. Figure 17.10 shows the effect of titanium additions on phosphorus segregation in an alloy containing 0.04% phosphorus that had been aged at 550°C for 24 hours. The addition of titanium causes a decrease in the amount of phosphorus segregation. Titanium phosphide is a stable compound and precipitates of this compound were found in this alloy after ageing. The precipitation of these phosphides lowers the activity of phosphorus in solution and thus decreases segregation.

Another good example of this effect can be observed for Fe–S alloys. Additions of chromium and manganese both greatly reduce the segregation of sulfur through the formation of chromium and manganese sulfides [33, 34]. This precipitation lowers the activity of sulfur in the matrix and decreases segregation. Because most commercial

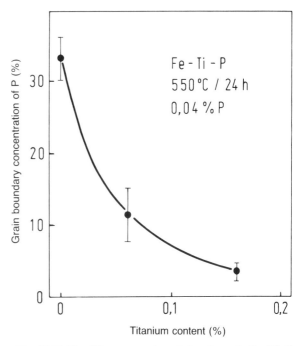

Fig. **17.10** The GB concentration of phosphorus in Fe–Ti–P alloys plotted as a function of titanium concentration. The ageing time and temperature are given in the legend. Data taken from Ref. [32].

alloys contain one or both of these elements, sulfur segregation is rarely observed in them.

The next effect that we wish to consider is that of site competition. This effect has received a great deal of attention in recent years because it has been found that small amounts of carbon can effectively compete with many elements for sites at the GB [20–23, 35–38]. This result has two important implications. The first is that if accurate measurements of GB segregation are to be made for elements such as phosphorus, antimony, and sulfur, essentially all carbon must be removed from the alloy. Secondly, since all steels contain carbon, we must understand this effect if we are to understand segregation in steels.

Table 17.2 lists all well documented examples of site competition and Figs. 17.11 and 17.12 show two examples of the type of behavior that has been observed in these systems. Figure 17.11 shows the evidence for the competition between phosphorus

Table 17.2 Examples of competitive segregation

Competing Elements	Reference
P–S	[33, 73, 74]
P–C	[20–23, 35, 36]
S–C	[36, 37]
N–S	[37]
P–N	[72]
C–Sb	[38]
C–Ni	[38]

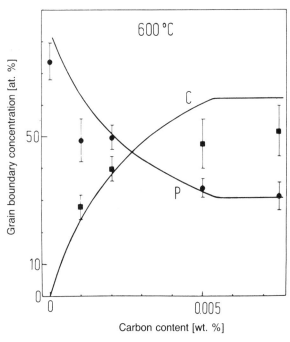

Fig. 17.11 Site competition between phosphorus and carbon at GBs in iron. The data are taken from Ref. [20].

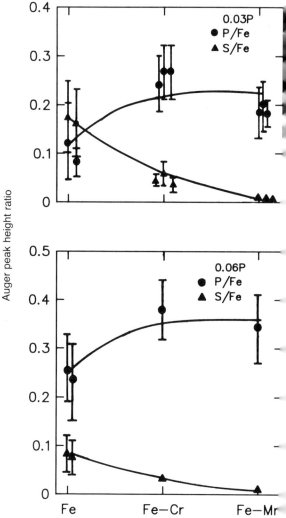

Fig. 17.12 Site competition between phosphorus and sulfur a GBs in iron. The upper curve is for alloys containing 0.03 wt.% phosphorus and the lower curves are for alloys containin 0.06 wt.% phosphorus. The circles represent phosphorus seg regation and the triangles represent sulfur segregation. Th activity of sulfur in the matrix was changed by additions c chromium or manganese. Data are taken from Ref. [38].

and carbon [20]. As the carbon concentration in the bulk and on the GBs increases, the amount of phosphorus segregation decreases. The bulk concentration of carbon in these experiments was established by carburization at well defined carbon activities in CH_4–H_2 mixtures. At about 50 at.ppm the maximum solid solubility of carbon in equilibrium with cementite is reached, and a further increase in the bulk content of carbon does not change the amount in solid solution. Thus, the

equilibrium segregation of carbon and phosphoru are expected to be constant for bulk concentration of carbon greater than this value, and the data sho that a plateau is reached.

Figure 17.12 shows the competition betwee phosphorus and sulfur [30]. For these experimen

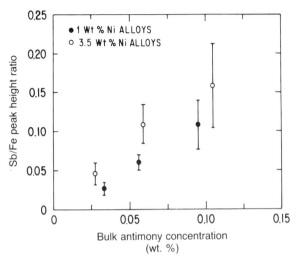

Fig. 17.13 The amount of antimony segregation to GBs of Fe–Ni–Sb alloys plotted as a function of the bulk antimony concentration. Data taken from Ref. [38].

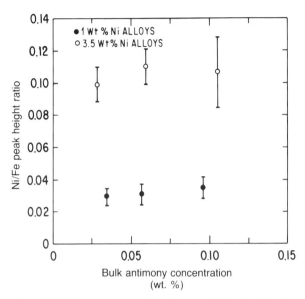

Fig. 17.14 The segregation of nickel to GBs in Fe–Ni–Sb alloys as a function of the bulk antimony concentration. Data taken from Ref. [38].

six alloys were used. Two of them were Fe–P–S alloys, two were Fe–P–S alloys with 0.1 wt.% Cr, and two were Fe–P–S alloys that contained 0.1% Mn. Within each of these pairs, one alloy contained 0.03 wt.% phosphorus and the other contained 0.06 wt.% phosphorus. The elements chromium and manganese form stable sulfides. The addition of chromium will greatly reduce the amount of sulfur free in solution that is available to segregate through the precipitation of the sulfide. The addition of manganese will reduce the amount of sulfur in solution even more because manganese sulfide has a higher free energy of formation than chromium sulfide. The results in Fig. 17.12 show that as less sulfur is available to segregate, as a result of formation of the sulfides, there is more phosphorus segregation.

The third category that we wish to consider is where addition of the third element causes an increase in segregation. The only well documented example of this effect is in Fe–Ni–Sb alloys [38–42]. Antimony segregates in iron, but it has been demonstrated in a number of studies that nickel additions cause an increase in antimony segregation. Figure 17.13 presents results which show that this effect occurs. In this figure the

antimony concentration on the GBs is plotted as a function of the bulk antimony concentration. One set of data is for alloys that contained 1 wt.% nickel and the other is for alloys that contain 3.5 wt.% nickel. The amount of segregation in the alloys that contained 3.5 wt.% nickel is greater than in the alloys that contained 1 wt.% nickel even though the bulk concentrations of antimony in the two sets of alloys are very similar. However, it is important to note that changes in the antimony concentration at these low levels have no effect on the amount of nickel segregation, as shown in Fig. 17.14. For the three alloys that contain 1 wt.% nickel, the results are all identical. The amount of segregation is greater in the 3.5 wt.% nickel alloys than in the 1 wt.% nickel alloys, but within this group the amount of nickel segregation is identical. We can interpret these results in the following way. Nickel additions at the levels of 1 to 3.5 wt.% increase the activity of antimony in solution and thus increase the amount of segregation [42–44]. However, additions of 100 to 1000 wppm antimony have no affect on the nickel activity and do not change the segregation of nickel.

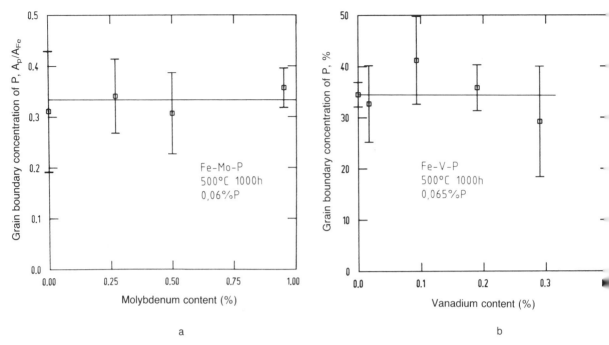

Fig. 17.15 The effect of (a) Mo and (b) V on phosphorus segregation in iron. Data taken from Ref. [32].

The final situation that we wish to consider is that where the third element has no effect on the amount of segregation. All of the studies that have presented this type of result have been concerned with Fe–P–X alloys where X stands for the third element. Nickel, molybdenum, vanadium, and chromium have been found to leave the amount of phosphorus segregation unchanged in these alloys [32, 45]. Examples are shown in Fig. 17.15.

In quaternary and higher order alloys many interactions among the elements are possible. These can be understood and predicted to a great extent based on our knowledge obtained on the binary and ternary alloys. The main problem is to consider all the possibilities that might occur and weigh each of them carefully. To illustrate these points we will simply give two examples. The first of these involves phosphorus segregation in Fe–C–Cr–P alloys.

We showed in an example above that carbon and phosphorus compete for sites at GBs in Fe–C–P alloys. Also work has shown that Cr has no effect on phosphorus segregation in Fe–Cr–P alloys,

as demonstrated by the data in Fig. 17.16(b). However, as shown in Fig. 17.16(a), when chromium is added to Fe–C–P alloys an increase is observed in phosphorus segregation. Investigations [20] determined that this increase occurred because chromium precipitated more of the carbon as Cr-carbides. This precipitation lowered the activity of carbon in the matrix, which decreased the amount that would segregate. There was less competition between carbon and phosphorus for sites at the GBs, and more phosphorus segregation was observed. These effects are displayed in Fig. 17.16.

Another example is the segregation of antimony in Fe–Ni–Cr–C–Sb alloys. In the alloy containing all of these elements, antimony segregates to the GB, as shown in Fig. 17.17. If nickel or chromium is removed from the alloy, the antimony segregation decreases. This effect is also shown in Fig. 17.17. The removal of nickel causes a decrease in antimony segregation because, as discussed above, nickel additions increase the activity of antimony in solution. When chromium is removed

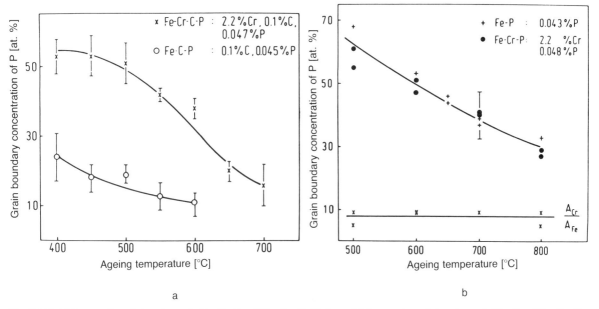

Fig. 17.16 (a) Segregation of phosphorus in Fe–C–P and Fe–Cr–C–P alloys. (b) Segregation of phosphorus in Fe–P and Fe–Cr–P alloys and Cr segregation in Fe–Cr–P alloys.

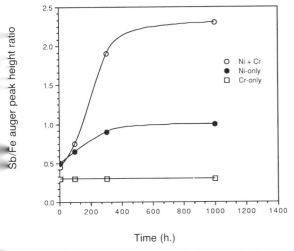

Fig. 17.17 Antimony segregation in Fe–Ni–Cr–C–Sb, Fe–Ni–C–Sb (Ni- only on log) and Fe–Cr–C–Sb (Cr-only on log) plotted as a function of ageing time. The data were taken from Ref. [39].

there is a decrease in segregation below that observed in the Fe–Ni–Cr–C–Sb alloy, but the effect is not as great as when Ni is removed. The decrease caused by chromium removal comes from

two factors. The first is that chromium additions cause a slight increase in the antimony activity and this effect is now missing. But chromium removal also raises the activity of carbon and sulfur in solution because the chromium carbides and sulfides will no longer be present. These elements will compete with antimony for sites at the GB and lower the amount of segregation [46].

17.4 GRAIN BOUNDARY REACTIONS

In the preceding two sections we have described GB segregation. The picture that we have presented is one where the segregant diffuses to the GB and becomes part of a structural unit that is at the GB. We have suggested that the driving forces for segregation, which can be formally represented by the activity of the segregant in solution, are size differences between the atom and the matrix and chemical incompatibility between the segregant and the matrix. We also proposed that once in the boundary the segregant has formed chemical bonds within the structural unit and that these bonds may be more favorably arrayed or stronger than those in the matrix. Therefore, as we begin to consider reactions that can occur at a segregated GB, we must think in terms of breaking these chemical bonds.

There are a number of reactions that can occur at the GB that are of interest to metallurgists. These include fracture, corrosion, grain growth, recrystallization, GB diffusion, and creep. We will begin our discussion by considering fracture, since this process has been the most thoroughly examined from the chemical bond point of view. We will then suggest how this same approach might be valid for considering some of the other processes listed above.

To consider GB fracture, let us return to reactions (17.1) and (17.2) above. The first of these reactions simply described segregation of the segregant from the matrix to the GB. The second reaction represented the formation of a molecular unit in the GB. Therefore, intergranular fracture would simply involve the rupture of these bonds at the GB. This can be represented by the following chemical reaction

$$(M_xI)_{GB} \overset{\sigma}{\rightleftarrows} (M_yI)_{fs} + (M_{x-y})_{fs} \qquad (17.7)$$

where $(M_yI)_{fs}$ and $(M_{x-y})_{fs}$ are the two fragments of $(M_xI)_{GB}$ that now exist on the fracture surface and σ is the applied stress. Therefore, the application of this applied stress disrupts the molecular unit. This type of reaction is exactly analogous to chemical reactions in which molecules are broken up by an applied external energy source. An example would be the photodissociation of oxygen as given by the reaction

$$O_2 \overset{h\nu}{\rightleftarrows} O + O \qquad (17.8)$$

In this reaction the external energy is light instead of stress, but again this additional energy is enough to disrupt the bond between the two oxygens. If instead of having a diatomic molecule, we had a polyatomic molecule, then we would expect that the light would first break the weakest bonds in the molecule. Similarly, in the case of intergranular fracture, we would expect the applied stress to

cause a rupture of the weakest bonds at the GBs. We now wish to consider what types of bonds these might be.

The first piece of evidence that we need to note is that the GBs are very strong in most high purity metals. It is only when segregation occurs that they become weak and become the path for brittle fracture. Therefore, the presence of the impurity must produce some weak bonds that can be easily broken. The question that we must address is how these weak bonds are created and which bonds they are.

Some insight has been gained into this question by the application of full-scale quantum mechanical calculations to this problem [44, 47–54]. Although the details of the calculations used by different researchers have differed, the same general picture appears to result in each case. What has been found is that the element that segregates and weakens the GB is always electronegative with respect to the host metal. The chemical bonds the impurity forms with the host metal in the GB tend to be heteropolar and electronic charge is drawn from the surrounding atoms onto the impurity. The result of this electronic charge transfer is that there is less electronic charge in the metal–metal bonds surrounding the embrittling element. Consequently, these bonds are weakened, and it is the weakening of these bonds that causes embrittlement.

With this model we have a mechanism of how reaction (17.7) could occur, and we propose that this embrittlement is a direct consequence of the change in the chemical bonds at the GB. However, it should be noted that there is also evidence that two other factors could contribute to the bond weakening. One of these is simply size. If the atom places a strain on the GB, it will effectively shift the bonds along their potential curve, and less energy will be required to cause the fracture [55]. Secondly, there is evidence that the segregated atom may try to rearrange the structural units at the GB to make the bonds between the impurity and the matrix more favorable [51]. This movement could also put strain on the metal–metal bonds and thus reduce the stress required to break them.

Intergranular fracture is the only GB reaction that has been explained in terms of the changes in the chemical bonds that form at a GB as a result of segregation. But it is certainly easy to suggest within this framework how changes in chemical bonds could affect other reactions. It is well known that segregation to GBs can make them much more susceptible to corrosion [19]. If we consider the basic corrosion reaction

$$M \rightleftharpoons M^+ + e^- \qquad (17.9)$$

we see that a metal atom M must be transferred from the solid metal into solution, and that as it makes the transfer it must be ionized by giving up an electron e^-. This reaction clearly involves breaking of bonds as well as charge transfer and it is certainly reasonable to expect that the ease with which this reaction can take place will depend on the strength and type of bonds that are present in the metal. Also, at the segregated GB the impurity itself could be very susceptible to undergoing this reaction.

Processes of recrystallization, grain growth, diffusion, and creep all involve movement of atoms in the GBs. Again, it would seem reasonable that as the chemical bonds change in the GB as a result of segregation, the ease with which these processes could occur would be affected. For example, the reason why an impurity could pin a moving GB could simply be that the bonds that it forms in the GB are very stable, and it takes much more energy for the boundary to break them and move on to a new position than if the impurity were not there.

Thus we would propose that all of these metallurgical effects of GB segregation can be viewed as chemical reactions and that the final correct interpretation of these effects will be in terms of changes in chemical bonds that the segregant induces at the grain boundary.

17.5 APPLICATIONS

The importance of all of the studies and ideas discussed above rests solely on the fact that they have application to engineering problems. In this section we wish to discuss several of these problems

and show how the problem prompted studies of GB segregation and how the information obtained from these studies was then used in the solution to the problem.

The first problem that we wish to consider is intergranular fracture in steel. It has been known for many years that a steel can undergo brittle intergranular fracture in a notched bar mechanical test if it is slowly cooled from the tempering temperature or if, after tempering, it is reheated to a temperature in the range 400–550°C. This fracture mode is always accompanied by an increase in the brittle-to-ductile transition temperature. In a steel that has been tempered and rapidly quenched this transition temperature is usually near −100°C. Thus in most standard applications the steel would be ductile. However, if GB embrittlement has occurred this transition temperature can be well above room temperature. Measurements above 200°C have been recorded [39, 46].

The problem of intergranular brittleness was known to blacksmiths [56, 57] and was a problem for ordnance and armor manufacturers during World War I [57]. During the 1950s this problem received additional attention because of its occurrence in rotor forgings of large steam turbines [2, 3, 58]. This brittleness in the turbines could occur in two ways. First, as the energy demands grew, the turbines were designed to be larger and larger. The cores of the huge castings that were required could not be rapidly quenched. The slow cool that they underwent made them brittle and if the stress on the piece became high enough, a crack would nucleate there that could lead to failure. Also, the temperature of parts of the turbine were high enough so that embrittlement could occur during use. A piece of steel that was ductile at room temperature when it was put into service was brittle at room temperature after several years of service at an elevated temperature. These pieces would often fail when the rotor was shut off, allowed to cool to room temperature, and then started up again. The high stresses on the cold pieces during start up would cause part failures.

Early research demonstrated that this brittleness could be related to the presence of impurity elements in the steel such as phosphorus, sulfur, and tin [59]. The development of Auger electron spectroscopy allowed researchers to confirm that the intergranular fracture occurred when these elements had segregated to the GBs [60]. Thus either the slow cool after tempering or the use of the part at an elevated temperature allowed this segregation to take place and gave rise to the observed embrittlement. As a result of these problems many studies were carried out on GB segregation in steels and its effects on GB embrittlement. As an understanding of the embrittlement process was obtained, new specifications were set for impurity concentrations in rotor steels and there was a renewed interest in the manufacturing of clean steels [61–63].

Another related problem in the turbine industry was the occurrence of intergranular stress corrosion cracking in the steam turbines. This problem received considerable attention after the Hinkley Point Power Plant Failure in England [64]. In this failure the crack initiated in a keyway in the rotor, and then an intergranular crack spread rapidly from that point. Analysis of the failure showed that the initial crack was caused by stress corrosion cracking in the keyway, and it was suggested that this occurred because of crevice conditions that existed. The rest of the fracture was simply of a brittle, intergranular nature. Auger analysis of this material showed that phosphorus had segregated to the GBs to produce this fracture [14]. Since the material had been ductile at room temperature when the piece was installed, it was concluded that this segregation occurred during use.

The question arose as to whether or not phosphorus or some other segregated impurity could enhance the intergranular corrosion that nucleated this catastrophic crack. The need to answer this question became more urgent as stress corrosion cracks were also found in other turbines throughout the world. Research on this problem showed that in a caustic environment phosphorus could certainly enhance the corrosion process [65]. If the environment had a neutral pH, phosphorus appeared to have little effect [66]. Therefore, to be certain that stress corrosion cracking would not occur, two approaches were taken. One was to use

steels with lower phosphorus concentrations. The other was to redesign the rotors to lower stresses and to remove crevices.

Another example of GB chemistry playing a crucial role in an engineering problem concerns grain growth in Fe-3%Si transformer steels. This material is processed to develop a very strong (110)[001] sheet texture. This texture optimizes magnetic permeability in the rolling direction of the sheet, and it is produced by secondary recrystallization. For this recrystallization to occur, normal grain growth must be suppressed.

It had been recognized that the use of second phase particles such as MnS to inhibit normal grain growth were insufficient to produce high permeability products [67, 68]. Processes were developed based on the assumption that solute segregation to the GBs would inhibit the grain growth, and although the products worked well, it was difficult to determine the mechanism of the inhibition. The development of Auger electron spectroscopy allowed this problem to be investigated. The results of these studies showed that nitrogen segregation to the GBs inhibited the normal grain growth and thus its presence was a requirement to produce the high permeability products required for transformers [69].

These problems are simply given as examples of ways in which the fundamental studies of GBs have had a large impact on engineering problems. Many more examples could be cited. The main point to be made is that as each fundamental study is done it is important to consider its impact on these applied problems.

17.6 CONCLUSIONS

The conclusions to a chapter such as this one are different from that those that end a standard scientific paper. What we have tried to present is not only an overview of the experimental work that has been done in this field but also a way in which to think about these problems. Therefore, the conclusions that follow are more concerned with the general picture that we have developed. They are the following.

1. Grain boundary segregation can be thought of as a two-step process. In the first step the solute diffuses from the bulk to the GB. In the second step the solute diffuses within the GB until it finds or creates a suitable location. The chemical bonds within this location are expected to be more favorable than those that can be formed in the bulk.

2. An increase in the activity of the segregant in the matrix will give rise to an increase in GB segregation. One can use this simple idea to interpret many experiments that have been performed.

3. Many factors can affect the activity of the solute in the matrix. One of these is simply the bulk concentration. Others include the size of the segregant and its chemical incompatibility with the matrix. These are the same factors that govern solid solubility. The activity may also be affected by all other elements present in the alloy.

4. The various reactions that take place at a segregated GB such as corrosion or fracture should be interpreted in terms of the chemical bonds that are formed at the GB. If an element makes these processes occur more easily or with more difficulty, then one should recognize that this is a result of different chemical bonds.

5. The final importance of all of these studies arises from their usefulness in engineering problems.

REFERENCES

1. L. A. Harris, *J. Appl. Physics*, **39** (1968), 1419.
2. D. L. Newhouse, *Metal Progress*, **93** (1967).
3. H. D. Greenberg, *Metal Progress*, **87** (1967).
4. J. A. Board, *J. Inst. Metals*, **101** (1973), 241.
5. K. Fritz, General Electric Company, private communication, 1978.
6. W. Yamaguchi, H. Kobayashi, T. Matsumiya, and S. Hayami, *Metal Tech.* (1979), 171.
7. C. L. Briant and S. K. Banerji, *Int. Metal. Rev.* **23** (1978), 164.
8. C. L. Briant, *Metall. Trans. A*, **18A** (1987), 691.
9. R. A. Mulford, *Metall. Trans. A*, **14A**, (1983), 865.
10. E. A. Molwyn-Hughes, *Physical Chemistry*, Pergamon Press, Oxford (1961), p. 1051.
11. E. D. Hondros and M. P. Seah, *Scripta Metall.*, **6** (1972), 1007.
12. C. L. White and W. A. Coughlan, *Metall. Trans. A*, **8A** (1977), 1403.
13. D. McLean, *Grain Boundaries in Metals*, Clarendon Press, Oxford (1957).
14. M. P. Seah, *Surf. Sci.*, **53** (1975), 168.
15. W. Hume-Rothery and G. V. Raynor, *The Structure of Metals and Alloys*, Institute of Metals, London (1962).
16. J. J. Burton and E. S. Machlin, *Phys. Rev. Lett*, **37** (1976), 1433.
17. M. F. Ashby, F. Spaepen, and S. Williams, *Acta Metall.* **26** (1978), 1647.
18. A. P. Sutton and V. Vitek, *Phil. Trans. Roy. Soc. Lond. A*, **309** (1983), 1.
19. C. L. Briant, in *Auger Electron Spectroscopy* (eds C. L. Briant and R. P. Messmer), Academic Press, Orlando, (1988), p. 111.
20. H. Erhart and H-J. Grabke, *Metal Science*, **15** (1981), 401.
21. S. Suzuki, M. Obata, K. Abiko, and H. Kimura, *Scripta Metall.*, **17** (1985), 1325.
22. Ya-Qing Weng and C. J. McMahon, Jr., *Materials Sci. and Tech.*, **3** (1986), 207.
23. K. Abiko, S. Suzuki, and H. Kimura, *Trans. Japan Inst. Metals*, **23** (1982), 43.
24. J. Jager, H. J. Grabke, and Yu Jin, *Proc. 2nd Internatl. Conf. on Creep and Fracture of Engineering Materials and Structures* (eds B. Wiltshire and D. R. J. Owen), Pineridge Press, Swansea (1984).
25. W. Jager, H. J. Grabke, and R. Moller, *Proc. Internatl. Conf. on Residuals and Trace Elements in Iron and Steel*, Portoroz, Okt. (1985).
26. C. L. Briant, *Acta Metall.*, **31** (1983), 257.
27. Jun Kameda and C. J. McMahon, Jr., *Metall, Trans. A*, **12A** (1981), 31.
28. T. Ogura, C. J. McMahon, Jr., H. C. Feng, and V. Vitek, *Acta Metall.*, **26** (1978), 1317.
29. R. A. Mulford, C. L. Briant and R. G. Rowe, *Scanning Electron Microscopy/1980/I*, SEM Inc. (1980), p. 487.
30. C. L. Briant, *Acta Metall.*, **33** (1985), 1241.
31. W. H. Herrnstein, III, F. H. Beck, and M. G. Fontana, *Trans. TMS-AIME*, **242** (1968), 1559.
32. H. J. Grabke, R. Moller, H. Erhart, and S. S. Brenner, *Surface and Interfaces Analysis*, **10** (1987), 202.
33. C. L. Briant, *Acta Metall*, **36** (1988), 1805.
34. B. J. Schultz and C. J. McMahon, Jr., *Metall. Trans.*, **4** (1973), 2485.
35. S. Suzuki, K. Abiko, and H. Kimura, *Scripta Metall*, **15** (1981), 1139.
36. S. Suzuki, S. Tanii, K. Abiko, and H. Kimura, *Metall. Trans. A*, **18A**, (1987), 1109.

37. G. Tauber and H. J. Grabke, *Ber. Bunsenges Phys. Chem.*, **82** (1978), 298.

38. C. L. Briant, *Acta Metall.*, **35** (1987), 149.

39. R. A. Mulford, C. J. McMahon, Jr., D. P. Pope and H. C. Feng, *Metall. Trans. A*, **7A** (1976), 1269.

40. H. Ohtani, H. C. Feng, C. J. McMahon, Jr., and R. A. Mulford, *Metall. Trans. A*, **7A** (1976), 87.

41. P. Gas, H. Bernardini, and M. Guttmann, *Acta Metall.*, **30** (1982), 1309.

42. C. L. Briant and A. M. Ritter, *Acta Metall.*, **32** (1984), 2031.

43. M. Nagesawararao, C. J. McMahon, Jr. and H. Herman, *Metall. Trans.*, **5** (1974), 1061.

44. C. L. Briant and R. P. Messmer, *Acta Metall.*, **32** (1984), 2043.

45. D. Y. Lee, E. V. Barrerra, J. P. Stark, and H. L. Marcus, *Metall. Trans. A*, **15A** (1984), 1415.

46. C. L. Briant, *Materials Science and Technology*, **4** (1988), 956.

47. C. L. Briant and R. P. Messmer, *Phil. Mag. B*, **42** (1980), 569.

48. R. P. Messmer and C. L. Briant, *Acta Metall.*, **30** (1982), 457.

49. M. E. Eberhart, R. M. Latanision, and K. H. Johnson, *Acta Metall.*, **32** (1984), 955.

50. W. Losch, *Acta Metall.*, **27** (1979), 1885.

51. H. Hashimoto, Y. Ishida, R. Yamamoto, and M. Doyama, *Acta Metall.* **32** (1984), 1.

52. G. S. Painter and F. W. Averill, *Phys. Rev. Lett*, **58** (1987), 234.

53. M. Mori and Y. Ishida, in *Grain Boundary Structure and Related Phenomena, Proc. JIMIS-4*, Suppl. to the *Trans. Japan Inst. Metals*, **27** (1986), p. 361.

54. D. D. Vvedensky and M. E. Eberhart, *Phil Mag. Lett*, **55** (1987), 157.

55. R. P. Messmer and C. L. Briant, in *Hydrogen Degradation of Ferrous Alloys* (eds R. A. Oriani, J. P. Hirth, and M. Smialowski), Noyes Publications, Park Ridge, New Jersey (1985) p. 140.

56. H. M. Howe, *Proc. Instn Mech. Engrs* (Jan–May, 1919), p. 405.

57. John H. Hollomon, *Trans ASM*, **36** (1946), 473.

58. R. L. Bodnar and R. F. Cappelini, Bethlehem Steel Corporation, Research Dept. Publication (1986).

59. W. Steven and K. Balajiva, *J. Iron Steel Inst.*, **193** (1959), 141.

60. D. F. Stein, A. Joshi, and R. P. LaForce, *Trans ASM*, **62** (1969), 776.

61. D. L. Newhouse, *Amer. Foundrymans Soc. Trans.*, **85** (1977), 389.

62. S. Miyana, *Metal Progress*, **130** (1986), 27.

63. R. I. Jaffe, *Metall. Trans. A*, **17A** (1985), 755.

64. J. M. Hodge and I. L. Mogford, *Proc. Instn Mech. Engrs*, **193** (1979), 93.

65. N. Bandyopadhyay and C. L. Briant, *Metall. Trans. A*, **14A** (1983), 2005.

66. N. Bandyopadhyay and C. L. Briant, EPRI Report 85-SRD-005, 1985.

67. H. C. Fiedler, *Metall. Trans. A*, **8A** (1977), 1307.

68. H. E. Grenoble, *IEEE Trans. Mag.*, **13** (1977), 1424.

69. R. G. Rowe, *Metall. Trans. A*, **10A** (1979), 997.

70. B. Aronsson, T. Lundstrom, and R. Rundquist, *Borides, Silicides and Phosphides*, Methuen and Co., London (1985).

71. John B. Parise, *Acta Crystall.*, **36** (1980), 1179.

72. H. Erhart and H. J. Grabke, *Scripta Metall.*, **15** (1981), 531.

73. Ph. Doumoulin, M. Guttmann, M. Foucault, M. Palmier, M. Wayman, and M. Biscondi, *Metal Science*, **14** (1980), 1.

74. H. Sato and K. Sato, *J. Japan Inst. Metals*, **41** (1977), 458.

75. C. L. Briant, *Scripta Metall.*, **15** (1981), 1013.

76. C. L. Briant, *Metall. Trans. A*, (1990), 2339.

CHAPTER EIGHTEEN

Physics and chemistry of segregation at internal interfaces

R. Kirchheim

Introduction · Distribution of segregation energies · High concentrations with solute–solute interaction · On the correlation between segregation and solubility · Interstitial diffusion in GBs · Experimental results on H-segregation at GBs · Phase separation in nanocrystalline Pd-H · H-diffusion in nanocrystalline Pd · H-segregation at metal/oxide interfaces · Conclusion

18.1 INTRODUCTION

As internal interfaces we will treat explicitly grain boundaries (GBs) and phase boundaries, although the laws of statistical thermodynamics applied to the segregation of foreign atoms could be used for other interfaces, as well as surfaces or stacking faults. It will be shown that the knowledge of the chemical potential of the foreign atoms and its dependence on concentration is sufficent to describe the partitioning of the foreign atoms between lattice sites in the grains (or adjacent phases) and sites within the interface. In order to calculate the chemical potential the energy distribution of sites or site exchanges for interstitially or substitutionally dissolved foreign atoms, respectively, is introduced. This concept is applicable to defected solids in a general way including deformed solids containing dislocations, disordered alloys, and metallic and oxidic glasses [1].

It is generally accepted and proven by computer simulations (i.e. [2]) that internal interfaces have a variety of interstitial and substitutional sites and, therefore, a variety of segregation energies, too.

Nevertheless, the Langmuir–McLean equation [3] based on a single segregation energy is generally used, in order to describe the relationship between the coverage of an interface by foreign atoms and their concentration within the grains. A more general treatment including a discrete spectrum of segregation energies for GBs was given by White and Coghlan [4]. Their results are a special case of the treatment presented in this study.

In comparison with GBs we would expect that phase boundaries (for phases of different chemical composition) are defects where chemical inhomogeneities are present besides structural ones. The variety of the latter may be minimized by a well defined orientational relationship between the two phases as in coincidence GBs. Thus impurity segregation to phase boundaries could be more complicated and/or more pronounced due to the presence of several chemical elements similar to cosegregation of different impurities to GBs [5].

There are only a few studies, mostly qualitative with respect to interfacial composition, on impurity segregation to phase boundaries [6] and the numerous studies on GB segregation were conducted at high impurity coverages of usually more than 10%, because the common technique of exposing the GBs by interfacial fracture requires large coverages. However, at these coverages the energetics of solute–solute interaction at the interface may affect or even dominate the energetics of segregation. It will be shown in this study that this interaction can explain the correlation [7] between the impurity enrichment at GBs and the terminal solubility of the impurity atoms within the grains. The treatment of solute–solute interaction is one of the most difficult problems in statistical thermodynamics and often the most simple approximation is used where a term proportional to the impurity concentration is added to the chemical potential. Well known examples of this approach in connection with interfaces are the treatment of surface adsorption by Fowler [8] and of GB segregation by Guttmann [9].

Experimental results of hydrogen segregation to GBs in Pd [10] and to Pd–metal oxide phase boundaries [11, 12] will be described in some detail, because they allow to distinguish between segregation at low coverages without H–H interaction where the Langmuir–McLean equation is not valid due to a distribution of segregation energies and segregation at high coverages with H–H interaction. Hydrogen segregation to the oxide phase boundary is strongly dependent on the composition of the interface.

18.2 DISTRIBUTION OF SEGREGATION ENERGIES

Disturbances of the lattice periodicity within an interface lead to a variety of different interstitial and substitutional sites which can be occupied by impurity atoms giving rise to different free energies of segregation.

The distribution of impurity atoms among the different sites is governed by the laws of statistical thermodynamics [13, 14], where all the sites are divided into different ensembles. Each ensemble contains sites of the same energy and a site is considered to be a subsystem. The partition function of the grand canonical ensemble, Ξ, of the whole system will be equal to the product

$$\Xi = \prod_i \xi_i^{M_i} \tag{18.1}$$

of the partition functions, $\xi_i^{M_i}$, of the single ensembles where M_i is the number of subsystems within the ensemble labeled i. ξ_i can be written as a sum of canonical partition functions κ [13]

$$\xi_i = \kappa_{iA}\lambda_A + \kappa_{iB}\lambda_B \tag{18.2}$$

where λ_A and λ_B are the activities of the matrix atoms A and the impurity atoms B. Under equilibrium the average number of B-atoms, N_B, in the whole system is given by a standard formula of statistical thermodynamics [13]

$$N_B = \lambda_B \left[\frac{\partial \ln \Xi}{\partial \lambda_B} \right]_{M_i, T} = \sum \frac{M_i}{1 + \kappa_{iA}\lambda_A/\kappa_{iB}\lambda_B} \tag{18.3}$$

and the average number of B-atoms within an ensemble, i.e. the occupancy of sites labeled i, is

$$\theta_{Bi} = \frac{N_{Bi}}{M_i} = \frac{1}{1 + \kappa_{iA}\lambda_A/\kappa_{iB}\lambda_B} \tag{18.4}$$

where N_{Bi} is the number of B-atoms in i-sites. The total number of sites may be N, leading to the constraint

$$N = \sum M_i \tag{18.5}$$

For sites in the crystalline lattice we choose $i = 0$ and for sites within the interface $i = 1, 12, \ldots$. The θ_{B0} corresponds to the molar fraction x_{B0} of the impurity within the crystalline lattice (for phase boundaries adjacent phases are different and the ratio of the different molar fractions of B within these two phases has to be calculated by thermodynamics). For small concentrations of B, Raoult's law, $\lambda_A \simeq 1 - x_B$, can be used and eq. (18.4) yields for $i = 0$:

$$\lambda_B = \frac{\kappa_{A0}}{\kappa_{B0}} x_{B0} \tag{18.6}$$

Thus the ratio of the partition functions, κ, equivalent with the activity coefficient of B crystalline A, γ_{B0}, or is related to the free energy dissolution of B into crystalline A, G_{B0}, by the equation:

$$\gamma_{B0} = \frac{\kappa_{A0}}{\kappa_{B0}} = \exp\left(\frac{G_{B0}}{RT}\right) \tag{18.7}$$

where R is the gas constant and T is temperature. In a similar way the ratio of the partition functions for sites of type i can be related to a free energy of solution of B-atoms into i-sites:

$$\frac{\kappa_{Ai}}{\kappa_{Bi}} = \exp\left(\frac{G_{Bi}}{RT}\right) \tag{18.8}$$

The eqs. (18.6) and (18.7) are slightly different for interstitially dissolved impurities (as $\lambda_A \simeq 1$, the special case of interstitials is treated in Ref. [1]). For an equilibrium distribution of the impurity atoms among the different sites the activity λ of A- and B-atoms has to be the same at the interface and within the crystalline phase. Thus eqs. (18.4) and (18.6) to (18.8) yield

$$\frac{\theta_{Bi}}{1 - \theta_{Bi}} = \frac{x_{B0}}{1 - x_{B0}} \exp\left(-\frac{\Delta G_{Bi}}{RT}\right),$$
$$i = 0, 1, 2, \ldots \tag{18.9}$$

where $\Delta G_{Bi} = G_{Bi} - G_{B0}$ is the free energy difference for B-atoms between an interface site of type i and a site in the crystal ($i = 0$). Equation (18.9) is equivalent with eq.(3) in Ref. [4] despite the fact that ΔG_{Bi} represents a difference of free energies rather than energies. If only one type of interfacial site is present, eq. (18.9) represents the Langmuir–McLean equation [3].

Equation (18.9) is useful if the impurity concentrations (x_{B0} and θ_{Bi}) for the various sites are known. However, in the usual experiment where the interface is exposed by intercrystalline fracture only a weighted average, θ_{Ba}, over the occupied sites in the interface is determined which is defined in terms of the nomenclature of this study by

$$\theta_{Ba} = \frac{1}{N - M_0} \sum M_i \theta_{Bi} \quad \text{with } i = 1, 2, \ldots \tag{18.10}$$

which depends on the chemical potentials, $\mu_A = \mu_A^0 + RT \ln \lambda_A$ and $\mu_B = \mu_B^0 + RT \ln \lambda_B$, as determined by eqs. (18.4) and (18.8):

$$\theta_{Ba} = \frac{1}{N - M_0} \sum_{i=1} \frac{M_i}{1 + \exp\left(\dfrac{G_{Bi} - \mu_B + \mu_B^0 + \mu_A - \mu_A^0}{RT}\right)} \tag{18.11}$$

where μ_A^0 and μ_B^0 are the chemical potentials for the standard states of B- and A-atoms. In our case λ_B and λ_A are both unity in their standard states, i.e. the pure elements are chosen as standard states. The ratio $1/[1 + \exp((G_{Bi} - \mu_B + \mu_B^0 + \mu_A - \mu_A^0))]$ describes the thermal occupancy and it is similar to the Fermi–Dirac function, i.e. it can be approximated by a step function becoming 0 for $G_{Bi} - \mu_B + \mu_B^0 + \mu_A - \mu_A^0 < 0$ and 1 for $G_{Bi} - \mu_B + \mu_B^0 + \mu_A - \mu_A^0 > 0$. If we take into account that $\mu_A - \mu_A^0$ is approximately zero for dilute solution of B in A, eq. (18.11) reduces to:

$$\theta_{Ba} = \frac{1}{N - M_0} \sum_{i=1} \frac{M_i}{1 + \exp\left(\dfrac{G_{Bi} - \mu_B + \mu_B^0}{RT}\right)} \tag{18.12}$$

Choosing the ideal solution of B in crystalline A as a standard state for B the chemical potential becomes $\mu_B = G_{B0} + \mu_B^0 + RT \ln x_{B0} = \mu_B^{is} + RT \ln x_{B0}$ and the last equation changes to

$$\theta_{Ba} = \frac{1}{N - M_0} \sum_{i=1} \frac{M_i}{1 + \exp\left(\dfrac{\Delta G_{Bi} - \mu_B + \mu_B^{is}}{RT}\right)} \tag{18.13}$$

which states that within the limits of the step function, approximation sites at the interface are occupied if their segregation free energy is smaller than $\mu_B - \mu_B^{is}$ and empty vice versa.

The total content of B, x_B, is the weighted sum of B-atoms at the interface and the ones within the grains forming the interfaces:

$$x_B = \frac{M_0}{N} x_{B0} + \frac{N - M_0}{N} \theta_{Ba} \tag{18.14}$$

or by using eqs. (18.13) and $\mu_B = \mu_B^{is} + RT \ln x_{B0}$

$$x_B = \frac{M_0}{N} \frac{1}{\exp\left(\dfrac{\mu_0^{is} - \mu_B}{RT}\right)}$$

$$+ \frac{1}{N} \sum_{i=1} \frac{M_i}{1 + \exp\left(\dfrac{\Delta G_{Bi} - \mu_B + \mu_B^{is}}{RT}\right)}$$

$$(18.15)$$

If the discrete distribution of ensembles is replaced by a continuous one, the last equation can be written as follows:

$$x_B = \frac{1 - c_t}{\exp\left(\dfrac{\mu_B^{is} - \mu_B}{RT}\right)}$$

$$+ c_t \int_{-\infty}^{\infty} \frac{n(\Delta G_{Bi})\, d\Delta G_{Bi}}{1 + \exp\left(\dfrac{\Delta G_{Bi} - \mu_B + \mu_B^{is}}{RT}\right)}$$

$$(18.16)$$

with

$$\int n(\Delta G_{Bi})\, d\Delta G_{Bi} = 1 \quad \text{and}$$

$$\theta_{Ba} = \int_{-\infty}^{\infty} \frac{n(\Delta G_{Bi})\, d\Delta G_{Bi}}{1 + \exp\left(\dfrac{\Delta G_{Bi} - \mu_B + \mu_B^{is}}{RT}\right)}$$

where the number of subsystems M_i having a free energy of segregation ΔG_{Bi} is substituted by $n(\Delta G_{Bi})\, d\Delta G_{Bi}$, and $n(\Delta G_{Bi})$ stands for the distribution function of segregation energies which may be also called density of sites function, and c_t is the fraction of sites belonging to the interface. If the B-atoms are dissolved interstitially eq. (18.16) is slightly different:

$$c_B = \frac{1 - c_t}{1 + \exp\left(\dfrac{\mu_B^{is} - \mu_B}{RT}\right)}$$

$$+ c_t \int_{-\infty}^{\infty} \frac{n(\Delta G_{Bi})\, d\Delta G_{Bi}}{1 + \exp\left(\dfrac{\Delta G_{Bi} - \mu_B + \mu_B^{is}}{RT}\right)}$$

$$(18.17)$$

because c_B is no longer a molar fraction but the fraction of interstices occupied by B-atoms (cf. Ref.

[1]). Usually the second term on the right-hand side of eqs. (18.16) and (18.17) representing the fraction of B-atoms at the interface is small when compared with the first term. However, this is no longer true if the grain diameter is small as in the case of hydrogen in nanocrystalline palladium discussed in this study.

18.3 HIGH CONCENTRATIONS WITH SOLUTE–SOLUTE INTERACTION

The onset of solute–solute interaction and following phase transitions at higher solute concentrations are among the most difficult problems of statistical mechanics [13, 14]. By a distribution of site energies as introduced before, the problem becomes even more difficult. Therefore, only a simple mean field or quasichemical approach will be used in the following to account for solute-solute interaction, where a term is added to the chemical potential which is proportional to the solute concentration [13, 14], i.e. for interaction within the crystalline grains and for interstitial solutions:

$$\mu_B = \mu_{id} + W c_{B0} = \mu_B^{is} + RT \ln\frac{c_{B0}}{1 - c_{B0}} + W c_{B0}$$

$$(18.18)$$

where μ_{id} is the chemical potential without solute-solute interaction and W corresponds to an interaction energy. If the total concentration is small but locally increased by segregation (i.e. as in grain or phase boundaries), x_{B0} in eq. (18.18) has to be replaced by the local concentration θ_{Ba}. For interstitial solutions which will be discussed later eq. (18.17) then becomes [1, 15]

$$c_B = \frac{1 - c_t}{1 + \exp\left(\dfrac{\mu_B^{is} - \mu_B}{RT}\right)}$$

$$+ c_t \int_{-\infty}^{\infty} \frac{n(\Delta G_{Bi})\, d\Delta G_{Bi}}{1 + \exp\left(\dfrac{\Delta G_{Bi} - \mu_B + \mu_B^{is} + W\theta_{Ba}}{RT}\right)}$$

$$(18.19)$$

For $W < 0$ (attractive interaction) the coverage of the interface (second term) increases. This is equivalent with the statement that the segregation energies ΔG_{Bi} decrease by the term $W\theta_{Bi}$, yielding more negative energies. In the mean field approximation the solute–solute interaction occurs among solutes in different types of sites. One should also expect that the site energies of normal sites belonging to the grains ($i = 0$) but, being adjacent to the GBs, are affected by larger average coverages, θ_{Ba}, at the interface as well. This effect can be heuristicly included in eq. (18.19) by reducing the fraction of normal sites $1 - c_t$ by the sites next to the GBs and including these sites into the distribution function $n(\Delta G_{Bi})$ with an energy $W\theta_{Ba}$.

Using eq. (18.19) and a Gaussian distribution of site energies has led to some interesting results on phase changes in amorphous metal–hydrogen systems [15], where a miscibility gap does not occur, when the width of the energy distribution is larger than the interaction energy, W. Under these circumstances the B-atoms gain more energy by occupying the low energy sites of the distribution instead of forming a high concentration phase where they gain interaction energy but have to occupy less favorable sites of the distribution. If we transfer this result to GBs we would expect that a phase transformation in a large angle GB has a lower probability to occur due to a broad distribution of site energies when compared with a high coincidence GB.

18.4 ON THE CORRELATION BETWEEN SEGREGATION AND SOLUBILITY

According to Seah and Hondros [7], the segregation of a solute is correlated with its terminal solubility by the relation

$$\beta = \frac{K}{x_{Bs}} \tag{18.20}$$

where K is a constant between 1 and 10, x_{Bs} is the terminal solubility where a second phase of either pure B or a compound AB_y is formed and β is the

enrichment factor of solute atoms at the GB defined by

$$\beta = \frac{\theta_{Ba}}{x_{B0}} \tag{18.21}$$

If an average, apparent segregation energy ΔG_{Ba} is defined in analogy to eq. (18.9) by

$$\Delta G_{Ba} = -RT \ln \frac{\theta_{Ba}}{x_{B0}} \tag{18.22}$$

and the free energy change ΔG_{Bs} for the precipitation of B by the reaction

B (dissolved in A) $\rightarrow$ B (in pure B or in AB_y

formed in the grains)

(18.23)

is introduced, we obtain

$$\Delta G_{Bs} = RT \ln x_{Bs} \tag{18.24}$$

and by comparison with eqs. (18.20–18.22)

$$\Delta G_{Ba} = \Delta G_{Bs} - RT \ln K \tag{18.25}$$

For $T = 800\,K$ the second term on the right-hand side of eq. (18.25) varies between 0 and 15 KJ-mol B and, therefore, we can conclude that in most cases

$$\Delta G_{Ba} \approx \Delta G_{Bs} \tag{18.26}$$

Furthermore the local concentration at the GBs or the average coverage, respectively, was rather high for those measurements leading to the empirical equation (18.20), because intercrystalline fracture occurs for $\theta_{Ba} > 0.1$ only. Thus we conclude from eq. (18.26) that the formation of a precipitate AB_y (including pure B with $y = \infty$) within the grains having a local concentration of $\theta = y/(1 + y)$ yields about the same free energy as the transition from a pure GB ($\theta_{Ba} = 0$) to a covered GB ($\theta_{Ba} = 0.1$ to 1). This implies that the structural and compositional differences between the precipitated compound and the highly covered GB do not play a major role for their free energies of formation from the solid solution of B in A. This is in accordance with the observation that the transition from an amorphous alloy to its crystalline counterpart is accompanied by small energy changes only and that the free energy of formation of compounds AB_y with different values of y is often very similar if it is referred to one mole of B.

Let us consider a very dilute solution of B where the average coverage of the GBs is still very small (i.e. $\theta_{Ba} < 0.05$). By increasing the concentration, x_{B0}, close to the terminal solubility, x_{Bs}, B–B interaction energies become important when compared to the changes in configurational entropy and, therefore, the coverage of the GBs is increased remarkably. This is also obvious from an inspection of eq. (18.19), where the second term on the right-hand side describes the fraction of B in the grain boundaries and W the B–B interaction energy. It has been tacitly assumed that the parameters W are about the same for the grains and the GBs. Evidence for this assumption is provided by the measurements of hydrogen segregation at GBs of Pd (see the corresponding section in this work). On the other hand, eqs. (18.20) and (18.21) show that for high coverages $\theta_{Ba} \approx 1$ the concentration of B is indeed close to the terminal solubility.

Thus the correlation between segregation and solubility is not surprising at all, as the experimental data were obtained with highly covered GBs and concentrations within the grains being close to the terminal solubility where segregation may be considered to be a precursor of precipitation.

Other attempts to explain the correlation are either similar to the concept described here [7] or misleading [2, 16]. The use of the BET adsorption isotherms in Ref. [7] is a different form of including B–B interaction which, however, is rather unrealistic because usually there is no formation of several layers of B by segregation at GBs. In Refs. [2] and [16] the change of the GB energy $\Delta\gamma_b^{N_b}$ was calculated and the condition

$$\Delta\gamma_b^{N_b} < \Delta H_m^B - \Delta E_c^{AB} \qquad (18.27)$$

was derived for GB segregation, where ΔH_m^B is the enthalpy of mixing of A and B and ΔE_c^{AB} is the difference of cohesive energies between A and B. Due to the appearance of the enthalpy of mixing in eq. (18.27) a correlation between segregation and solubility was predicted. However, it appears to be

more reasonable using the solid solution of B in A as a reference state rather than pure, crystalline A as it was done in eq. 4 of Ref. [16]. Then in eq. 5 of Ref. [16] the difference of the enthalpy between the two solid solutions, one with and one without the additional B-atom, has to be considered. This additional term is in the notation of Ref. [16] equal to $-\Delta H_m^B + \Delta E_c^{AB}$ and the change of the newly defined GB energy $\Delta\gamma$ is

$$\Delta\gamma = \Delta\gamma_b^{N_b} - \Delta H_m^B + \Delta E_c^{AB} \qquad (18.28)$$

The following eqs. 6 to 8 of Ref. [16] remain unchanged and the familiar equality $\Delta\gamma = E_s^{N_B}$ is obtained, where $E_s^{N_B}$ is the segregation energy and the condition for segregation becomes $\Delta\gamma < 0$ being independent of the enthalpy of mixing.

18.5 INTERSTITIAL DIFFUSION IN GBs

In the following, interstitial diffusion in GBs is described by a model which was developed for amorphous alloys [17] and successfully applied to various cases of hydrogen in disordered lattices [1]. An extension of this model to substitutional diffusion will be possible if the distribution of vacancies (or excess volume) among the different sites of the GB is taken into account. A comparison of the final result of the model with predictions of irreversible thermodynamics seems to indicate a rather general validity of the model.

In order to demonstrate the applicability of the model and its limitations the major assumptions and equations will be repeated (for a detailed analysis see Ref. [17]). The jump rate, I_{ik}, which is defined as the number of jumps out of site i into site k per unit of time, shall depend on temperature according to an Arrhenius Law:

$$I_{ik} = I_0(c_k^0 - c_k) \exp\left(-\frac{Q_{ik}}{RT}\right) \qquad (18.29)$$

where I_0 is a constant prefactor and Q_{ik} an activation energy for changing the site from i to k. The first term in brackets with the partial concentration c_k describes the probability for site k being empty. Thus blocking of sites is taken into account.

A flux of interstitial atoms, $\mathcal{J}$, occurs, if a concentration gradient is present. This flux is used to define the intrinsic diffusion coefficient, D, via Fick's First Law:

$$\mathcal{J} = -D\frac{\partial c}{\partial x} \qquad (18.30)$$

This law can be derived in an atomistic model by considering jump processes back and forth through an imaginary plane at $x + l/2$ which is perpendicular to the x-axes. In this way one starts with a kind of master equation like:

$$\mathcal{J} = f\sum_i \sum_k (c_k I_{ki})_x - (c_k I_{ki})_{x+l} \qquad (18.31)$$

where l is the jump distance and f is a geometry factor depending on the jump geometry and the dimensionality of the lattice.

Equation (18.31) must not be used for potential traces having a variable saddle point energy without further consideration, because correlations between jumps are not taken into account. This will become more obvious in the following example. Assume an atom is next to a saddle point of low energy. Here it will jump back and forth several times without contributing to long-range diffusion. In order to avoid this case of correlation, the severe assumption of constant saddle-point energies is often made [17–20]. Then the activation energy, Q_{ik}, in eq. (18.29) will become:

$$Q_{ik} = Q^0 - E_i \qquad (18.32)$$

where E_i is the site energy and Q^0 is the constant saddle point energy. Assuming further on that the various sites of the disordered matrix are occupied with atoms according to Fermi–Dirac Statistics (eq. (18.11)), the following relations are obtained for the self-diffusion or tracer–diffusion coefficient D^*

$$D^* = D^0\gamma_B(1 - c_B)^2 \qquad (18.33)$$

where $\gamma_B = \dfrac{\lambda_B}{c_B} = \dfrac{1}{c_B}\exp[(\mu_B - \mu_B^0)/RT]$ is the activity coefficient. For the intrinsic or chemical diffusion coefficient, D, the following equation can be derived for low coverages [17]:

$$D = D^0\frac{\partial \lambda_B}{\partial c_B} = D^0\frac{\partial}{\partial c_B}\exp[(\mu_B - \mu_B^0)/RT] \qquad (18.34)$$

with $D^0 = fl^2 \exp(-\Delta Q^0/RT)$ and ΔQ^0 being an average activation energy.

Using eq. (18.34) and eqs. (18.17) or (18.19), we have implicit equations for the concentration and temperature dependence of D for the cases with and without B–B interaction. Inserting eq. (18.34) in eq. (18.30) yields

$$\mathcal{J} = -D^0\frac{\partial \lambda_B}{\partial x} \qquad (18.35)$$

which states that the activity gradient rather than the concentration gradient is the driving force for the atomic flux in agreement with irreversible thermodynamics.

18.6 EXPERIMENTAL RESULTS ON H-SEGREGATION AT GBs

For a distribution of segregation energies the proportionality between GB and bulk concentration (strictly speaking between $\theta_{Ba}/(1 - \theta_{Ba})$ and $x_{B0}/(1 - x_{B0})$) as predicted by the Langmuir–McLean equation (eq. (18.9) with $i = 1$) no longer holds, because at low concentration, sites with low segregation energies are occupied first leading to a large enrichment factor β. Thus the GB coverage increases strongly as a function of total concentration for low values of x_B and less pronounced for higher values of x_B, as then low energy sites in the boundaries are saturated and sites of higher energy have to be filled. This is shown for a Gaussian distribution of segregation energies

$$n(\Delta G_{Bi}) = \frac{1}{\pi\sqrt{\sigma}}\exp\left[-\left(\frac{\Delta G_{Bi} - \Delta G_B^0}{\sigma}\right)^2\right] \qquad (18.36)$$

where numerical integration in both eqs. (18.16) and (18.17) yields the same result for $c_t \ll 1$ which is shown in Fig. 18.1. It can be seen that the plot of $\theta_{Ba}/(1 - \theta_{Ba})$ versus $x_{B0}/(1 - x_{B0})$ does not give a straight line as predicted by the Langmuir–McLean equation but a curved line as observed for

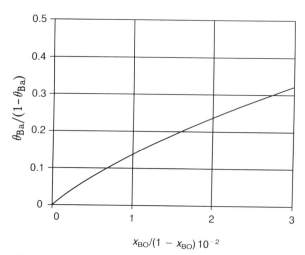

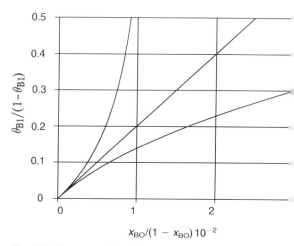

Fig. 18.1 Average coverage of the GBs, θ_{Ba}, as defined by eq. 18.16 for a Gaussian distribution of segregation energies (cf. eq. 18.36) versus concentration within the grains, x_{BO}. The curve was calculated with $\Delta G_B^0/RT = -2$ and $\sigma/RT = 2$.

Fig. 18.2 Coverage of the GBs, θ_{B1}, as defined by eq. 18.9 with one type of site, $i = 1$ (Langmuir–McLean equation) and a segregation energy which changes with coverage θ_{B1} according to the equation $\Delta G_{B1} = \Delta G_{B1}(\theta_{B1} = 0) + W\theta_{B1}$ versus concentration within the grains, x_{BO}. The curves were calculated with $\Delta G_{B1}(\theta_{B1} = 0)/RT = -3$ and $W/RT = -3$ (upper) 0 (middle) and 3 (lower curve).

instance for phosphorus segregation at iron GBs (cf. Fig. 5b in Ref. [21]).

However, for the concentration range of Fig. 18.1 the same curvature can be obtained by using the Langmuir–McLean equation and including a positive (repulsive) interaction term $W\theta_{Ba}$ as shown in Fig. 18.2. Thus measurements at coverages above 0.05 cannot distinguish whether the curvature in plots like the ones in Figs. 18.1 and 18.2 arises from a distribution of segregation energies or a repulsive B–B interaction at the GBs. As mentioned before, lower coverages are difficult to access experimentally, because intercrystalline fracture does no longer occur and some of the surface analytical techniques reach their detection limit. A possible approach for measuring at lower coverages may be an embrittlement of the GBs by adding hydrogen [22] and the use of secondary ion mass spectrometry.

Different to the conventional technique of a direct determination of the GB concentration, the segregation of hydrogen to Pd GBs was studied by an indirect way of measuring the total concentration as a function of the chemical potential [10] and eqs. (18.17) and (18.19) were used to evaluate the experimental results. In comparison, hydrogen

solubility was measured in a single crystal of Pd where $c_t = 0$ and μ_H^{is} is obtained from eq. (18.17). In polycrystalline samples the fraction of sites in the grain boundaries depends on the grain diameter, d, and the thickness of the grain boundary, δ. For a spherical or cubic shape of grains the volume fraction of GB regions is

$$c_t = \frac{3\delta}{d} \qquad (18.37)$$

which is assumed to be equal to the site density fraction. For a grain size of $d = 1\,\mu m$ and $\delta \approx 0.7\,nm$ the fraction of sites within the GBs will be $c_t = 2.1 \times 10^{-3}$. Strong segregation of hydrogen at or trapping by GBs could have been observed by the applied, sensitive electrochemical techniques. However, the increase of H-concentration by the second term in eq. (18.17) was rather small. By using a polycrystalline Pd with an average grain size of about 8 nm (called nanocrystalline Pd in the following), the amount of hydrogen sitting in GBs was much larger than the one within the grains as can be seen in Fig. 18.3, where the chemical potential of hydrogen is presented as a function of

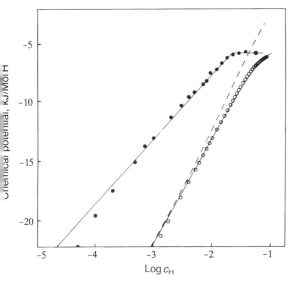

Fig. 18.3 Chemical potential of hydrogen at 293 K in a nano-crystalline Pd-sample (open circles) and in a single crystal of Pd (closed circles) as a function of H-concentration c_H. The curves are calculated from eq. (18.17) (dashed line: without H–H interaction) and eq. (18.19) (solid line: with H–H interaction) by using a Gaussian distribution of segregation energies (eq. 18.36). The straight line through the solid data points corresponds to an ideal dilute solution of hydrogen in α-Pd and the $\alpha + \beta$ two phase region.

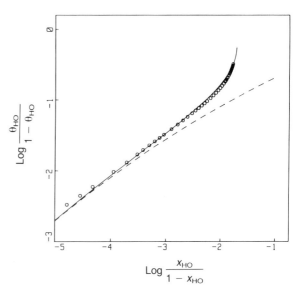

Fig. 18.4 Average H-coverage in the GBs θ_{Ha} and in the grains x_{H0} evaluated from the results in Fig. 18.3. The curves are calculated for a Gaussian distribution of segregation energies using eqs. (18.17) and (18.19) (dashed line: without- and solid line: with H–H interaction). The procedure for calculating the curve is described in the text.

the total concentration for both single crystalline and nanocrystalline Pd. Nanocrystalline Pd samples were prepared by the group of H. Gleiter, University of Saarbrücken, by pressing nm sized Pd crystallites which were formed by condensation of Pd-vapor in a helium atmosphere. TEM micrographs yielded an average value of 8 nm for the grain diameter.

If we assume that the sites in the grains of the nanocrystalline Pd are the same as in the single crystal (= octahedral interstices), the H-concentration in the grains, c_{H0}, is equal to the one in the single crystal, c_H^{sc} at the same chemical potential. In the following hydrogen will be the impurity which segregates and, therefore, the subscript B in eqs. (18.2) to (18.26) is replaced by H. Weighing c_{H0} with $(1 - c_t)$ gives the first term on the right-hand side of eq. (18.19) and if this is subtracted from the total measured H-concentration c_H of the nanocrystalline Pd, the second term is obtained,

which describes the fraction of hydrogen being segregated at the GBs, $c_t \theta_{Ha}$. Thus the average coverage of the GBs, θ_{Ha}, can be calculated by assuming a value of c_t (0.27 has been used in agreement with eq. (18.37)). For the case of a degeneracy of site energies in the GBs eq. (18.9) is in agreement with the Langmuir–McLean equation and becomes

$$\frac{\theta_{Ha}}{1 - \theta_{Ha}} = \frac{c_{H0}}{1 - c_{H0}} \exp\left(-\frac{\Delta G_{Ha}}{RT}\right) \qquad (18.38)$$

Thus the values obtained for θ_{Ha} and c_{H0} by the procedure described before should give a straight line with slope 1 in the plot shown in Fig. 18.4. However, the slope of the experimental results is about 0.5 for $c_{H0} < 10^{-3}$ and independent of other choices for c_t, because a different value of c_t will move the data points parallel to each other at low concentration. The deviation from a Langmuir–McLean behavior occurs at coverages as low as 10^{-3} and, therefore, they cannot arise from H–H interaction, because this would lead to a very high and positive value for the interaction parameter W and a stronger curvature in Fig. 18.4. In the following a distribution of segregation energies and eqs. (18.17) and (18.19) will be used to describe the data and explain the failure of the Langmuir–McLean equation.

For the single crystal c_t is zero and eq. (18.19) yields

$$\mu_H = \mu_H^{is} + RT \ln \frac{c_H^{sc}}{1 - c_H^{sc}} \qquad (18.39)$$

which was used to determine μ_H^{is}. Equation (18.39) predicts a slope of $2.3 \times RT$ for the single crystalline data in Fig. 18.3 in excellent agreement with experiment for $c_H^{sc} > 10^{-4}$. In order to calculate the relationship between μ_H and c_H an assumption has to be made about the distribution function of segregation energies. Regarding the many different GBs in a nanocrystalline material which was produced by compacting the small crystallites a Gaussian distribution as given by eq. (18.36) was used. Three parameters, c_t, σ, and ΔG_H^0, could be varied in order to obtain good agreement with the experimental data. To minimize the ambiguity of the fitting, $c_t = 0.27$ was kept constant (this corresponds to a GB width of 0.7 nm according to eq. (18.37) and it was determined from the narrowing of the miscibility gap [23]; see also following section). During the numerical evaluation of eq. (18.17) a set of arbitrary values for $\mu_H - \mu_H^{is}$ was chosen in order to calculate $(1 - c_t)c_{H0}$ (first term on the right-hand side of eq. 18.17), $c_t\theta_{Ha}$ (second term on the right-hand side of eq. 18.17) and $c_H(\mu_H)$ (sum of both terms) and compare it with the experimental data in Figs. 18.3 and 18.4. By this procedure it was impossible to obtain good agreement between the numerical analysis and the experiment over the whole range of concentrations. This could be due to either the distribution function of segregation energies being different from a Gaussian one or the onset of H–H interaction at high coverages. Only the latter case was taken into consideration, because the range of low coverages could be fitted very well and deviations became remarkable at coverages of a few percent, where one would expect

them to occur, if H–H interaction plays a role, because then H–H distances become rather short.

Figure 18.4 shows that the experimentally determined coverages are larger than the ones calculated for an ideal dilute solution of H-atoms in the GBs and, therefore, an attractive H–H interaction or strictly speaking a negative value for W in eq. (18.18) will be responsible for the deviations. In the quantitative treatment of H–H interaction eq. (18.19) was applied and with a value of $W = -30$ kJ/mol H good agreement was obtained with the experimental results (cf. Figs. 18.3 and 18.4). It is interesting to note that this value is in between two reported values for W [24, 25] in polycrystalline Pd of large grains, where only a negligible amount of hydrogen is in the GBs and H–H interaction takes place in the grains. This similarity of the W-values may be considered as evidence for the arguments used before on the correlation between solubility and segregation. During the numerical evaluation of eq. (18.19) a set of arbitrary values for $\mu_H - \mu_H^{is} + W\theta_{Ha}$ was chosen to calculate the second term on the right-hand side of eq. (18.19) ($= c_t\theta_{Ha}$) with the distribution of eq. (18.36) by numerical integration. Then $\mu_H - \mu_H^{is}$ could be determined from the original set of $\mu_H - \mu_H^{is} + W\theta_{Ha}$ which allowed to calculate the first term on the right-hand side of eq. (18.19) ($= (1 - c_t)c_{H0}$).

The total distribution of site energies including the sites within the grains is shown in Fig. 18.5. The sites within the grains have all the same free energy, G_0, as in the single crystal which can be calculated from the e.m.f. values as 3.9 kJ/mol H with respect to gaseous hydrogen at 1 atm and 25°C. The values for the other parameters are $\sigma = 15$ kJ/mol H and $\Delta G_B^0 = 5.3$ kJ/mol H.

The width, σ, is about the same as in amorphous Pd–Si alloys [1, 17], indicating that the distribution of interstices from all the different GBs in nanocrystalline Pd is similar to an amorphous structure. The positive value for the average segregation energy is surprising, because from computer simulations [26] the types of polyhedra (octahedra, trigonal prisms, cubes etc.) obtained for tilt boundaries in a fcc metal have a volume which is equal or larger than that of a normal octahedral site in the grain yielding a smaller density of GBs. Simple models [27, 28] relating the size of an interstice to its site energy for an H-atom would predict that the segregation energies should be all negative which contradicts the distribution shown in Fig. 18.5 containing energies above G_0. The contradiction may be caused by the simplifications made in the models or by the possibility that GBs in nanocrystalline palladium may be different when compared with simulated GBs based on geometrical reasoning [26]. On the other hand, the Gaussian distribution shown in Fig. 18.5 is filled in this study only up to 30%, i.e. only for energies $G < G_0$, and the sites above this energy were experimentally not accessible. Due to the difficulties discussed, a quantitative relationship between the measured distribution of segregation energies and the structure of GBs could not be established.

18.7 PHASE SEPARATION IN NANOCRYSTALLINE Pd–H

If a phase transformation occurs in metal hydrogen systems, the chemical potential or the hydrogen pressure remains constant within the two-phase region, although the total hydrogen concentration increases. This behavior can be observed in crystalline metals but not in amorphous metals [1, 15] which can be understood theoretically if the width of the site distribution is larger than the H–H interaction energy [15].

However, the experimental results presented in Fig. 18.6 show that the β-phase is formed in nanocrystalline Pd as a plateau occurs at the same hydrogen activity where the β-phase is formed in polycrystalline Pd (grain diameter >20 µm). But less hydrogen is absorbed before the activity rises again, which is attributed to the exclusion of a considerable volume fraction of the sample from the phase transformation. This volume fraction is assumed to belong to the distorted regions around the GBs. The β-phase forms within the grains of nanocrystalline Pd because they offer the same octahedral sites for hydrogen occupation as in a single crystal of Pd. Then the electronic and/or

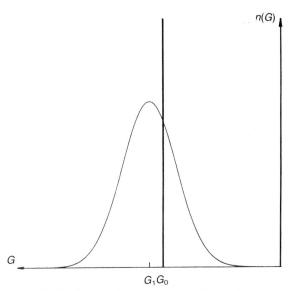

Fig. 18.5 Distribution of site energies for H-atoms in a nanocrystalline metal with a Gaussian distribution for the GBs and a delta function $\delta(G - G_0)$ for the interior of the grains. The average segregation energy is $\Delta G_H^0 = G_1 - G_0$.

elastic H–H interaction causes the β-phase to form at the same hydrogen activity, whereas the distorted regions at the GBs do not perform a phase transformation analogously to the behavior in amorphous metals. The local concentration within the GBs remains constant during the phase transformation in the grains because the chemical potential does not change. Accepting these explanations, the volume fraction of GBs can be calculated straightforwardly from the different width of the two-phase region (miscibility gap) for nano- and polycrystalline Pd [23]. For the example shown in Fig. 18.6 the volume fraction of the GB regions is 0.27, which corresponds to an average width of the GBs of 0.7 nm.

Arguments based on the analogy between GBs and amorphous metals are rather weak as the evidence is increasing that GBs may be composed of a few numbers of structural units which would lead to a correspondingly small number of interstices and the distribution function $n(\Delta G)$ is the sum of a few delta functions. Then in a special GB some of these interstices may form a two-dimensional lattice with a distance especially

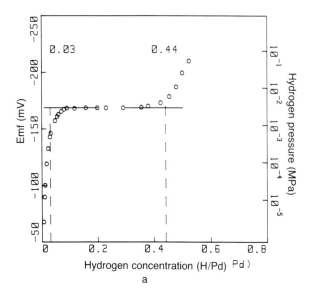

a

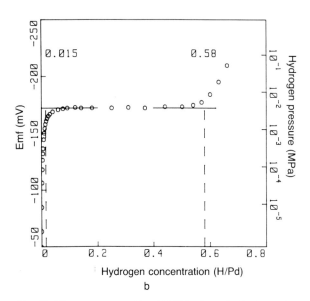

b

Fig. 18.6 Pressure composition 333 K-isotherm for hydrogen in (a) nanocrystalline Pd and (b) polycrystalline Pd (grain diameter ≈20 μm). The narrower two-phase region (= length of the pressure plateau) in the nanocrystalline sample is explained by an exclusion of the GBs from the α to β phase transformation.

favorable for H–H interaction. For this boundary a two-dimensional ordered phase would be formed at lower H-activities compared to the α/β transition in the bulk. However, in nanocrystalline Pd the

analogy may be appropriate because there are many different GBs, they are considerably curved and they extend over short distances only.

It is interesting to note that the miscibility gap for hydrogen in thin Pd films is also considerably decreased [29] which can be explained in the same way by an increase of the terminal solubility due to H-segregation at the GBs narrowing the gap from the left-hand side and a shift at the right-hand side due to an exclusion of GBs from the phase transformation. The narrowing of the miscibility gap seen in Fig. 18.6 is in between the values for a 6 and 18 nm thick, slightly annealed film [29] as one would expect if the grain size of these films is about the same as their thickness.

18.8 H-DIFFUSION IN NANOCRYSTALLINE Pd

H-diffusion in a polycrystalline metal may be inter- or intracrystalline in general. However, in nanocrystalline Pd a large fraction of the hydrogen atoms are in the low energy sites of the GBs for concentrations within the α-phase region. The grains offer only sites of higher energy and, therefore, are like inclusions which have to be circumvented by the hydrogen atoms. Thus the values obtained from the electrochemical time-lag measurements [1, 10] and shown in Fig. 18.7 are in a first order approximation GB diffusion coefficients of hydrogen neclecting the effect of the embedded grains and the two-dimensional character of the transport (both are counteracting effects!).

The GB diffusion coefficient of H in Pd is smaller than the value for the single crystal at low hydrogen concentrations. For high H-concentrations the values go through a flat maximum and are larger than the value of the single crystal.

The distribution of site energies as obtained for nanocrystalline palladium from measurements of the chemical potential (cf. Fig. 18.5) allows a qualitative interpretation of the concentration dependence of the diffusion coefficient. At small hydrogen concentrations the H-atoms are trapped within the energetically deep sites of the Gaussian distribution and the diffusion coefficient becomes

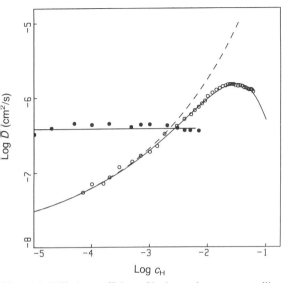

Fig. 18.7 Diffusion coefficient of hydrogen in a nanocrystalline Pd-sample (open circles) and in a single crystal of Pd (closed circles) as a function of H-concentration c. The curves were calculated using eq. (18.40) and the distribution of site energies shown in Fig. 18.5 (dashed line: without- and solid line: with H–H interaction).

even lower than in single-crystalline palladium. However, with increasing hydrogen concentration sites of higher energy have to be occupied and, therefore, the diffusion coefficient increases due to a decrease of the average activation energy. Thus D-values can become larger than in single-crystalline palladium. For high coverages of the GBs ($\theta_{Ha} > 0.1$) the negative H–H interaction energy counteracts the increase of the diffusivity and causes finally a decrease of the diffusion coefficient.

During the quantitative treatment of this problem a more general form than eq. (18.34) was derived [17] which includes the effect of blocking of sites:

$$D = D^0 \frac{\partial}{\partial \theta_{Ha}} \left[(1 - \theta_{Ha})^2 \exp\left(\frac{\mu - G_1}{RT}\right) \right]$$
(18.40)

where D^0 is the diffusion coefficient of a reference state which contains only sites of the same free partial energy $G_1 = \mu_H^{is} + \Delta G_H^0$ (mean value of the

Gaussian distribution in Fig. 18.5) and a constant saddle point energy Q^0. For the numerical calculation of D from eqs. (18.19) and (18.40) the same distribution function of site energies and the same H–H interaction energy were used as for the calculation of the chemical potential. The unknown parameter D^0 causes only a parallel movement of the calculated log D/log c curves in Fig. 18.7 and, therefore, does not affect the concentration dependence.

The good agreement between measured and calculated values for the thermodynamic quantity μ and the kinetic quantity D can be considered as a proof of the validity of the model and of the reliability of the parameter values for σ and W (G_1 and D^0 do not change the dependence on concentration).

It is generally accepted that impurity diffusion in GBs is always faster than in the grains due to a lower atomic density leading to a lower energy for 'vacancy' formation (excess volume) in the GBs. However, the results for hydrogen in Pd demonstrate that this is not the case for interstitial diffusion at low concentrations because vacancies are always present in the lattice of interstitial sites and the distribution of site energies provides low energy sites which act as traps for hydrogen diffusion. After saturation of the low energy sites H-atoms can take advantage of the more open structure of the GBs and migrate faster through the boundaries. Similar consideration may apply to substitutional impurities at low coverages when they are sitting mostly within the low energy sites and, therefore, are handicapped during competition with neighboring matrix atoms for an adjacent vacancy.

18.9 H-SEGREGATION AT METAL/OXIDE INTERFACES

By internal oxidation of Pd–X (X = Mg, Al, Zn, and Zr) alloys at 1000°C, oxides of the alloying addition are formed which have a well defined orientational relationship with the Pd-matrix [11, 12]. Structure, shape, and orientational relationship were especially simple for MgO, which was formed as octahedrally shaped particles in its well known NaCl structure. The planes of the regular octahedra are parallel to (111) planes in both MgO and fcc–Pd [30]. Because of the small size of the oxide precipitates the interfacial area was large ($\approx 100\,\mathrm{m^2/mol}$ Pd [12]) and, therefore, the interaction between hydrogen being dissolved in the Pd-matrix and the interfaces was rather pronounced.

There are some fundamental differences between the GBs and the Pd/MgO phase boundary used as examples for internal boundaries and their interaction with hydrogen. The GBs developed during the compaction of randomly oriented nanometer sized crystallites, whereas a well defined orientational relationship exists between Pd-matrix and the precipitated magnesium oxide. Therefore, a broad spectrum of different interstices was expected for the GBs. However, at the Pd/MgO phase boundary a discrete spectrum of new sites is created (new in comparison with the octahedral site of Pd). These sites are formed because of the mismatch between the two lattices and/or because of the different thermal expansion coefficients (the interface was formed at 1000°C but studied at room temperature). In addition, the different chemical species Mg or O are present to form bonds with hydrogen.

The behavior of hydrogen in internally oxidized Pd-samples was quite different when compared with nanocrystalline Pd. Deep trap sites had to be saturated first, before in a narrow concentration range hydrogen activity and mobility were raised by orders of magnitude and became high enough to be detected by the electrochemical technique. Whereas measurements could be reproduced with the same sample of nanocrystalline Pd after removal of the hydrogen by anodic oxidation, a part of the hydrogen was trapped irreversibly during the first run of measurements in internally oxidized samples. The concentration of irreversible trap sites corresponded to $1.0(\pm0.5) \times 10^{15}\,\mathrm{H/cm^2}$, i.e. a monolayer of H-atoms at the interface.

During irreversible trapping of hydrogen at the Pd–MgO interface the partial molar volume was $3.5\,\mathrm{cm^3/mol}$ H which is about twice as large as in Pd ($1.7\,\mathrm{cm^3/mol}$ H). This is considered to be strong

evidence that the segregation of hydrogen at th interface is not elastic in nature. On the other han a large trapping (segregation) energy of abo $-90\,\mathrm{kJ/mol}$ H was calculated from the desorptic kinetics at 300°C, which is another indication of chemical interaction.

In a recent high resolution electron microsco study [31] some experimental evidence was giv that oxygen atoms form the terminating layer at CdO/Ag interface which was produced by intern oxidation of Ag–Cd alloys in air. Taking th high affinity between oxygen and hydrogen in account, it was concluded [11] that the formatic of O–H bonds at the interface causes the stro segregation of hydrogen. From the observation th the irreversible trapping of hydrogen does n appear when the samples were annealed at 800° in vacuum or in an environment of low oxyge activity it was further concluded that the oxygen the interface (or part of it) does not belong to th oxide but is segregated itself. Irreversible trappi of hydrogen at the interfaces could be 'switched and off' by annealing in high or low oxygen e vironments. Thus hydrogen acts as a pro detecting the coverage of the oxides with exce (segregated) oxygen.

Therefore, irreversible trapping of hydrogen the Pd–oxide interfaces is an effect of cosegr gation rather than segregation and similar to th well known H-embrittlement of GBs of stee which are covered with non-metals like phosphor [32]. The reversible part of hydrogen trapping observed after saturation of the irreversible traps after removal of the excess oxygen and its amount about $5 \times 10^{14}\,\mathrm{H/cm^2}$, i.e. somewhat less than monolayer. The reversible segregation of hydrog can be described by a single segregation free ener [11] of about $-29\,\mathrm{kJ/mol}$ H for the Pd/Al$_2$C interface.

In order to maintain stoichiometry for the stab oxides and to account for excess oxygen, we assum that the terminating oxygen layer of an oxid particle MeO$_x$ partly belongs to the oxide an partly forms a palladium oxide. The latter one called excess oxygen and it will trap hydrog irreversibly according to the following rea tion:

$$MeO_x + xPdO + xH_2 \rightarrow Me(OH)_{2x} + Pd$$

Then the reaction enthalpy divided by $2x$ corresponds to the binding or trap energy of the irreversible traps. Thus trap energies of -130 (Me = Mg), -110 (Me = Al), and -106 kJ/mol H (Me = Zn) were calculated from the thermodynamic data [12]. All estimated energies are nearly equal to the formation energy of water $(= -119 \text{ kJ/mol H})$ and in good agreement with the irreversible trap energy of -90 kJ/mol H estimated from the desorption at 300°C.

We assume that the oxide particles increase their volume by the formation of hydroxide at the interface, where the amount of hydroxide is given by the coverage of the interface during irreversible trapping. Then this volume change ΔV_{part} of the particles is

$$\Delta V_{part} = \begin{cases} A\delta \\ \dfrac{A\theta}{L}\left[\dfrac{V_{hydroxide}}{2x} - \dfrac{V_{oxide}}{x}\right] \end{cases} \quad (18.41)$$

where A is the total interfacial area, δ is the increase of spacing between the oxygen layer of the oxide and the adjacent Pd layer during H incorporation, θ is the coverage of the boundary by hydrogen, L is Avogadro's number, and $V_{hydroxide}$ and V_{oxide} are the molar volumes of the hydroxide $Me(OH)_{2x}$ and the oxide MeO_x, respectively. From the last equations the increase of phase boundary 'thickness' δ was calculated and included in Table 18.1.

In order to compare these values with the measured partial molar volumes of hydrogen V_H the volume change of the oxide particles ΔV_{part} has to be related to the volume change of the sample $\Delta V = A\theta V_H/L$ due to the irreversible trapping of $A\theta/L$ moles of H. According to Eshelby's theory this is done by the following equation [33]:

$$\Delta V_{part} = \frac{1 + v}{3(1 - v)}\frac{A\theta}{L}V_H \quad (18.42)$$

where v is Poisson's ratio ($= 0.39$ for Pd). Using the first line in eq. (18.41), we obtain the increase of the phase boundary thickness δ from eq. (18.42):

$$\delta = \frac{1 + v}{3(1 - v)}\frac{\theta}{L}V_H \quad (18.43)$$

Values calculated from the last equation are also included in Table 18.1. They agree very well with the values obtained from eq. (18.41).

18.10 CONCLUSION

By the indirect measurements of impurity segregation described in this work the coverage of the GBs could be varied over a wide range beginning with much smaller coverages than in other studies. Therefore, the transition from an ideal dilute behavior in the GBs to the behavior of a regular solution with H–H interaction could be demonstrated for the first time. This transition occurs for hydrogen in palladium at coverages of a few percent. All measurements of segregation energies are usually conducted at higher coverages and, therefore, most probably contain a large contribution from the solute–solute interaction energy, which may be responsible for the observed correlation between solubility and the enrichment at GBs. The experimental results for the ideal dilute regime demonstrate also for the first time that a spectrum of segregation energies is more appropriate in describing the phenomenon of solute segregation at GBs.

The interaction of hydrogen with metal–oxide interfaces is strongly affected by the composition of the interface which changes with oxygen activity.

Table 18.1 Interfacial density (coverage) of irreversible and reversible traps (θ_{irr} and θ_{rev}) and increase of the spacing between the adjacent layers at the interface evaluated from experimental results by eq. (18.43) (δ_{exp}) and calculated from the second line in eq. (18.41) (δ_{theor})

	T (°C)	t (h)	Coverage (10^{15} cm^{-2})		Spacing (Å)	
			θ_{rev}	θ_{irr}	δ_{exp}	δ_{theor}
Al$_2$O$_3$	1000	24	0.9	1.5	0.5	0.5
MgO	1000	44	0.7	0.7	0.4	0.3
ZnO	1000	60	0.3	3.0	0.9	0.9

ACKNOWLEDGMENT

The author is grateful for financial support provided by the Deutsche Forschungsgemeinschaft (SFB 270).

REFERENCES

1. R. Kirchheim, *Prog. Mat. Sci.*, **32** (1988), 261.
2. A. P. Sutton and V. Vitek, *Acta Metall.*, **30** (1982), 2011.
3. D. McLean, *Grain Boundaries in Metals*, Clarendon Press, Oxford (1957).
4. C. L. White and W. A. Coghlan, *Met. Trans. A*, **8** (1977), 1403.
5. M. Guttmann and D. McLean, in *Interfacial Segregation* (ed W. C. Johnson and J. M. Blakely), ASM, Metals Park, OH (1979).
6. W. C. Johnson, in *Interfacial Segregation* (ed W. C. Johnson and J. M. Blakely), ASM, Metals Park, OH (1979).
7. M. P. Seah and E. D. Hondros, *Scripta Metall.*, **7** (1973), 735; see also E. D. Hondros and M. P. Seah, *Int. Metals Rev.* (1977), 262.
8. R. H. Fowler, *Statistical Mechanics*, Cambridge University Press (1966), p. 825 ff.
9. M. Guttmann, *Metal Sci.*, **10** (1976), 337.
10. T. Mütschele and R. Kirchheim, *Scripta Metall.*, **21** (1987), 135.
11. X. Y. Huang, W. Mader, J. A. Eastman and R. Kirchheim, *Scripta Metall.*, **22** (1988), 1109.
12. X. Y. Huang, W. Mader and R. Kirchheim, *Acta Metall.*, **39** (1991), 893.
13. R. H. Fowler and E. A. Guggenheim, *Statistical Thermodynamics*, Cambridge University Press (1960), p. 242 ff.
14. T. L. Hill, *Introduction to Statistical Thermodynamics*, Addison-Wesley, London (1962), p. 431
15. R. Griessen, *Phys. Rev. B*, **27** (1983), 7575.
16. V. Vitek and G. J. Wang, *J. de Physique*, **43** (1982), C6–147.
17. R. Kirchheim and U. Stolz, *J. Non-crystall. Solids*, **70** (1985), 323.
18. R. Kirchheim, *Acta Metall.*, **30** (1982), 1069.
19. J. W. Haus, K. W. Kehr and J. W. Lyklema, *Phys. Rev. B*, **25** (1982), 2905.
20. J. B. Leblond and D. Dubois, *Acta Metall.*, **31** (1983), 1459.
21. C. L. Briant, *Scripta Metall.*, **15** (1981), 1013.
22. R. M. Latanision and H. Oppenhauer Jr., *Met. Trans. A*, **5** (1974), 223.
23. T. Mütschele and R. Kirchheim, *Scripta Metall.*, **21** (1987), 1101.
24. E. Wicke and J. Blaurock, *Ber. Bunsenges. Phys. Chem.*, **85** (1981), 1091.
25. T. Kuji, W. A. Oates, B. S. Bowerman and T. B. Flanagan *J. Phys. F*, **13** (1983), 1785.
26. H. J. Frost, M. F. Ashby and F. Spaepen, *Technical Report* Harvard University, Cambridge, MA (1982).
27. R. Kirchheim, F. Sommer and G. Schluckebier, *Acta Metall.*, **30** (1982), 1059.
28. P. M. Richards, *Phys. Rev. B*, **27** (1983), 2095.
29. M. W. Lee and R. Glosser, in *Hydrogen in Disordered and Amorphous Solids* (eds G. Bambakidis and R. C. Bowman, Jr.), NATO ASI series, Plenum Press, New York (1986), p. 351.
30. W. Mader and B. Maier, *J. de Physique*, **51** (1990), C1–867
31. G. Necker and W. Mader, *Phil. Mag. Lett.*, **58** (1988), 205.
32. J. Kameda and C. J. McMahon, Jr., *Met. Trans. A*, **14** (1983), 903.
33. J. D. Eshelby, in *Solid State Physics* (ed F. Seitz and D. Turnbull), Academic Press Inc. New York 1956.

CHAPTER NINETEEN

Atomic resolution study of solute-atom segregation at grain boundaries: experiments and Monte Carlo simulations

S. M. Foiles and D. N. Seidman

Introduction · Methodology · Monte Carlo computer simulations · Atom-probe observations of solute-atom segregation · Conclusion

9.1 INTRODUCTION

In an alloy, the equilibrium composition near an inhomogeneity, such as a grain boundary (GB), will generally be different from the composition in the bulk. This was first recognized by Gibbs [1]. This change in local composition may affect the structure and mechanical properties of the boundary. As an example, experiments by Sass and coworkers [2, 3] have shown that the Burger's vector of the dislocations present in Fe–Au twist boundaries depends on the Au concentration near the boundary. This demonstrates that the structure does, in fact, depend on the local composition of the alloy at the interface.

In this chapter both theoretical and experimental efforts to understand this segregation at GBs **on an atomic level** will be described. The theoretical studies are based on atomistic computer simulations and the experimental studies use atom-probe field-ion microscopy (APFIM). These studies address two questions. First, what is the composition of the boundary region as a function of the bulk composition and temperature? Second, what is the spatial distribution of the two species in the boundary? It should be noted that the comparison of experiment and calculations, where possible, is crucial. Such comparisons can provide guidance in the experimental analysis and the comparison provides a stringent test of the reliability of the simulation methods.

The paper is organized as follows. Following the introduction, the simulation and atom-probe field-ion-microscopy techniques are described. In the next section the results of computer simulation studies will be presented. The segregation in Ni–Cu alloys at a sessile dislocation and twist GBs is determined. Also, the effect of bulk composition on the local compositional ordering at the GBs in the alloy Ni_3Al is addressed and the variation of the segregation to Pt-1 at.% Au twist boundaries as a function of twist angle is presented. The final section contains experimental studies of the Pt–Ni system, along with corresponding simulation results, and of W–Re alloys.

19.2 METHODOLOGY

19.2.1 Computer simulation

The atomistic calculation of the equilibrium structure and composition of interfaces is a two-part problem. First, a model for the energetics of the interface as a function of the location and chemical identity of the constituent atoms is required. This model must have sufficient computational speed so that the energetics of numerous configurations can be evaluated while retaining a physically correct description of the energetics. The approach that we will pursue here is the embedded atom method (EAM) developed by Daw and Baskes [4]. The second part of the problem is to solve the statistical mechanics of the compositional equilibrium between the bulk and the interfacial region. This second aspect is addressed using Monte Carlo computer simulation techniques. The combination of using the EAM in conjunction with Monte Carlo computer simulations has been very successful in computing segregation of alloys at surfaces where a great deal of experimental information is available for comparison. This work has been reviewed by Foiles [5].

Embedded atom method

The EAM is a semi-empirical technique for computing the total energy of an arbitrary arrangement of atoms [4, 6]. It has been used successfully for a wide range of problems involving transition metals with filled or nearly filled *d*-bands. These applications include surface structural and dynamic properties (see reviews by Daw [7] and by Foiles [5]), point defect properties [6], GB structure [8, 9] and properties [10], and thermodynamic properties [11]. A thorough comparison of this approach to other similar approximations can be found in a review by Carlsson [12].

The EAM models the energetics of the metal by the sum of two terms. The first is the embedding energy, which is the energy to place an atom into the electron gas provided by the surrounding atoms. This background electron density is approximated by the superposition of atomic densities. The second term is a pair interaction which incorporates electrostatic interactions between the atoms. The total energy is then written

$$E = \sum_i F_i \left(\sum_{j \neq i} \rho_j^a(R_{ij}) \right) + \frac{1}{2} \sum_{ij, i \neq j} U_{ij}(R_{ij}) \quad (19.1)$$

Here F_i is the energy to place atom i into the electron density given by the sum of the contributions from the atomic electron densities, ρ_j^a, and U_{ij} is the pair interaction. In practice the functions F and U are determined empirically by fitting to bulk properties of the metals and alloys. This has been done for the elements in the Ni and Cu columns of the periodic table by Foiles *et al.* [6].

The main advantage of the embedded atom method over pair potential treatments is that it incorporates the trend that bond strengths increase and bond lengths decrease as the coordination of atoms decreases. However, the embedded atom method does not describe covalent bonding effects. The embedded atom form of the energetics has been derived from first principles by Jacobsen *et al.* [13] and also by Daw [14]. The derivation by Daw does not consider the details of the band structure and the derivation by Jacobsen *et al.* only leads to the above expression if certain terms which are significant for partially filled *d*-bands are neglected. Therefore, it is expected that this approach will be most reliable for materials with either filled or empty *d*-bands.

Monte Carlo computer simulations

The thermal equilibrium associated with the EAM energetics are determined using atomistic Monte Carlo simulations [15]. These simulations keep track of both the position and the chemical identity of each atom. Two types of variations are included in these simulations. First, each atom is allowed to displace from its current position. Second, the chemical species of each atom can be changed. (In the simulations, the chemical potential difference between the two elements is held fixed; the bulk composition is selected by the choice of the chemical potential difference.) This type of simulation has several advantages. First, thermal equilibrium is determined correctly within the limits of the statistical sampling. In addition, the atomic structure of the boundary is allowed to relax so as to lower the energy. Since atomic relaxation is included in the simulations, the strain energy contributions to the energetics are included automatically. Finally, the vibrational contributions to the energetics are included. The inclusion of atomic transmutation is an important computational convenience. It allows the local composition to rearrange without having to follow the slow physical diffusion processes. The price of sidestepping the diffusion problem is that the simulations do not yield any information about the kinetics of the equilibration process. Thus one needs to be careful when comparing these simulation results to experiment, to be certain that the experimental results are for thermodynamic equilibrium. A more complete description of the simulation method can be found in a review by Foiles [5] of the application of this technique to the calculation of surface segregation.

19.2.2 Atom-probe field-ion-microscopy determination of chemical compositions

The determination of the chemical composition of an internal interface on an atomic scale presents a major experimental challenge because of the simple fact that it is buried within a solid. The atom-probe field-ion microscope (APFIM) is, at present, the only technique that allows one to study the chemical composition of an internal interface with atomic spatial resolution. In the past extensive use has been made of scanning Auger electron spectroscopy to determine the chemical composition of GBs, albeit on a scale significantly coarser than on an atomic scale [16, 17, 18]. However, the use of an APFIM to study the chemical composition of an internal interface offers many advantages over electron optical techniques, i.e. Auger, electron-energy loss, or energy dispersive X-ray spectroscopies. First, the FIM forms atomic resolution images in direct lattice space of a surface of a sharply-pointed tip: ≈ 10–60 nm radius. The lateral resolution for

chemical composition – i.e. in a surface – is ≈ 0.3–0.5 nm; and the depth resolution is equal to the interplanar spacing of the region being analyzed and this can be <0.1 nm. The chemical identity of an individual atom is determined by measuring the mass-to-charge state ratio (m/q), employing a special time-of-flight mass spectrometer technique. With this time-of-flight technique the chemical identity of all the elements in the periodic table are readily identified with the same sensitivity for each element. The mass resolution of the atom-probe technique can be made to approach 10 000 using a so-called reflectron lens [19, 20, 21]. Thus it is possible to determine a chemical composition profile associated with an internal interface that has atomic spatial resolution for the particular segregant being studied. The only possible correction to the experimental data obtained is a purely geometric one, as the volume analyzed may include atoms from the matrix; it is, however, possible to choose a geometry such that a matrix correction is unnecessary.

Basic physics of the atom probe technique

An FIM allows for the routine imaging of individual atoms in direct lattice space. The FIM is a point projection microscope and therefore *no* lenses are required for the formation of an image. A basic FIM consists of a vacuum system (high to ultra-high vacuums are employed), a liquid-helium cryostat for cooling the specimen to temperatures in the range 4.2–300 K, a well-stabilized high-voltage power supply (0–30 kV), an internal image intensification system (a 7.5 cm diameter channel electron multiplier array with a 3–6 mm diameter probe hole and a phosphor screen), and a gas train to supply an imaging gas. The FIM specimen is mounted pointing towards a channel electron multiplier array [22]. The background pressure in an FIM is typically less than 10^{-8} torr before it is backfilled with an imaging gas – helium or neon or a mixture of these gases – to a pressure of 10^{-5}–10^{-4} torr; to obtain excellent mass spectra in an atom probe a pressure of $\approx 3 \times 10^{-10}$ torr is necessary.

An FIM specimen is fabricated by electroetching

or electropolishing a sharply-pointed tip on an 8–10 mm long wire with a diameter of 125–200 μm. The atoms on the surface of this sharply-pointed tip are observed via an imaging gas, as the messenger [23, 24, 25]. The tip is placed at a positive potential (typically between 5 and 15 kV d.c.) with respect to the channel electron multiplier array which amplifies the ion current. The helium or neon gas atoms are ionized by a tunneling mechanism; the outermost electron of a gas atom tunnels through the deformed electron potential energy barrier into an unfilled electron energy level at or near the Fermi level of the specimen. The positively charged helium or neon ions created by this process are accelerated along an electric field line to the channel electron multiplier array, where the energy of each ion is ultimately converted into a visible light signal. A simple physical picture of the surface of a metal in the presence of a high electric field – required for FIM – is the so-called ionic model [26, 27, 28, 29, 30]. In this model, the free-electron gas is pushed back slightly by the field into the bulk of the metal to create positively-charged ions on the surface of the specimen – i.e. polarized atoms. Thus each individual metal ion on the surface of the specimen is a source of a small pencil of gas ions; each pencil of gas ions ultimately creates a field-ion image of an individual atom. The overall image is produced by all the pencils of gas ions associated with the surface atoms. To first order the image produced is a stereographic projection of the surface atoms. The geometry of a field-ion image – but not all the detailed contrast variations – can be understood employing the so-called Moore model [31, 32]. The high symmetry inherent in a field-ion image of a perfect crystal is broken by the presence of an internal interface, and it is this broken symmetry that is used to detect the presence of an interface [33, 34, 35, 36].

The physical process of field evaporation is fundamental to the use of the FIM. Field evaporation is the the controlled sublimation of ions from a specimen under the influence of an electric field [26, 27, 28, 29, 30]. The basic mechanism involves the thermally-activated evaporation of an ion over a small Schottky hump that is created in the ionic potential curve of an ion residing on the surface, as a result of the applied electric field: ≈2–5.5 V/Å, depending on the specific material. The field-evaporation process can be controlled with great precision by applying short high-voltage pulses – 1–10 ms in width – on top of the steady-state d.c. voltage that is used to image the atoms. This technique, called pulsed field-evaporation, enables dissection of an atomic plane at a rate of one to three atoms per pulse or even less [26, 27, 28, 29, 30]. And it is this pulsed dissection technique that enables the chemical composition of an interface to be determined on an atomic plane-by-atomic plane basis.

Chemical analysis of an internal interface

There are two basic well-defined geometries for determining the composition of an internal interface by the APFIM technique. In the first geometry the interface is aligned such that its plane is parallel to the plane of the image intensification system (Fig. 19.1(a)). And for this geometry one determines a concentration profile that is normal to the interface plane; for this arrangement there is *no* correction to the data. Also for this case the analysis is performed with the probe hole covering as much of the interface as is possible to maximize the signal. The second method involves aligning the plane of the interface parallel, as much as possible, to the axis of the APFIM. The trace of the interface plane is thereby centered in the projection of the probe hole on the surface of an FIM tip (Fig. 19.1(b)). For this second geometric arrangement atoms from the interface, as well as the matrix, are collected and analyzed; thus the measured integral profiles obtained must be corrected for the contribution of the matrix; it should be noted that this correction is purely a geometric one. The equation relating the measured concentration to the maximum concentration for an assumed linearly decaying solute-atom segregation profile is [37]:

$$\langle C_s^{gb} \rangle^* = \langle C_s^{gb} \rangle_u \left(\frac{\pi D_a}{8\eta} \right) + \langle C_s \rangle \left(1 - \frac{\pi D_a}{8\eta} \right)$$

$$(19.2)$$

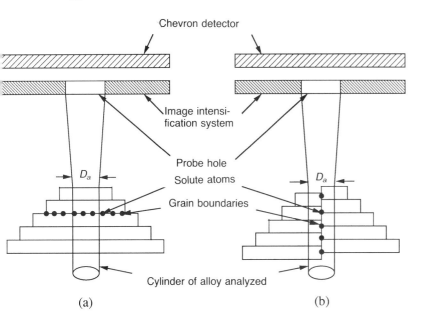

Chevron detector

Image intensi-
fication system

Probe hole

Solute atoms

D_a

Grain boundaries

D_a

Cylinder of alloy analyzed

(a)   (b)

Fig. 19.1 Schematic diagram illustrating the two basic geometric configurations for analyzing a GB by the APFIM technique. The distance between the Chevron detector and the FIM tip is ≈ 2200 mm, while the distance from the tip to the image intensification system is variable and is typically in the range 40–100 mm. The diameter of the probed region, D_a, is a variable quantity that depends on the local radius of the tip, and the distance from the analyzed region to the internal image intensification system. The latter consists of a channel electron multiplier array and a phosphor screen.

where $\langle C_s^{\mathrm{gb}} \rangle^*$ is the maximum concentration of solute at the interface; $\langle C_s^{\mathrm{gb}} \rangle_u$ is the uncorrected concentration, i.e. the measured value; $\langle C_s \rangle$ is the mean solute concentration of the matrix; D_a is the measured projected diameter of the probe hole; and η is the half-width of the linear distribution, measured at the point where the linear concentration profile intersects the constant $\langle C_s \rangle$. The value of D_a is measured from FIM images that show the projection of the probe hole on the tip of the specimen [11]. The value of η can be determined directly from the first geometry (Fig. 19.1(a)), or obtained indirectly by measuring $\langle C_s^{\mathrm{gb}} \rangle_u$ at two different values of D_a and then solving eq. (19.2) for η and $\langle C_s^{\mathrm{gb}} \rangle^*$. If the value of η can not be measured by this latter technique, and is therefore unknown, then we simply use the magnitude of the Burger's vector of the primary grain boundary dislocations (PGBDs) that comprise the interface; this is a reasonable first order assumption, as the Monte Carlo simulations show that the solute atoms segregate mainly to the cores of the PGBDs in the case of pure twist or tilt boundaries in the Pt(Au) system [38, 39].

Grain boundary crystallography via transmission electron microscopy

The five macroscopic degrees of freedom (DOFs) of an interface [40] are considered to be geometric thermodynamic state variables in the local phase rule [41] for an interface in a bicrystal. A total of 6 + C macroscopic state variables is required to specify the thermodynamic state of a bicrystal – C is the number of chemical components in the single phase alloy; this number of state variables is for a locally relaxed interface [41]. Therefore for a bicrystal binary single-phase alloy an eight-dimensional hyperspace is involved. Each point in this hyperspace represents a thermodynamic state of an interface. The eight state variables are the five macroscopic DOF, temperature, pressure, and bulk composition. Experimentally it is possible to fix or measure all of these eight variables for an interface.

The five DOFs of an interface are determined via transmission electron microscopy (TEM) of FIM specimens. These five DOFs are specified by the unit vector (c) about which one grain is rotated with respect to a second, the rotation angle (θ) about c, and the outward normal n to the plane of the interface. The approach utilized typically involves electropolishing an FIM specimen – that has been annealed to induce solute-atom segregation at interfaces – so that it is electron transparent, and then searching for an interface that is almost perpendicular to the long axis of a specimen. Once such an interface is observed it is analyzed using Kikuchi patterns of adjacent grains at the same tilt angle; the procedure is repeated for several different tilt angles. The Kikuchi patterns are then matched with computer-simulated Kikuchi patterns to determine the directions of the electron beam. The value of θ of an interface is determined from the known beam directions. Four Kikuchi patterns, two for each grain, are required to determine c and θ; in practice, however, three sets of Kikuchi patterns are used to check for self consistency. The beam direction and tilt axis are used as invariant directions to calculate c and θ. The total angular spread in c or θ is less than 5°. The vector n is determined by rotating and tilting the specimen until a minimum in the projected width of an interface is achieved; bright-field images and Kikuchi patterns are then recorded

and analyzed to calculate n. The quantities n_1 and n_2 are the outward normals to an interface plane is grains 1 and 2; they are related to one another via the rotation matrix $\mathbf{R}$, i.e. they are related by the equation:

$$n_1 = \mathbf{R}\, n_2 \tag{19.3}$$

In principle, once n_1 and $\mathbf{R}$ are determined experimentally n_2 is calculated employing eq. (19.3). In practice n_2 is also determined experimentally and it is compared with the calculated vector as a safeguard. The difference between the two values of n_2 is within 2°, and in most cases <1°. The disorientation of an interface is given by a c/θ pair. For cubic symmetry there are 1152 equivalent c/θ pair descriptions. The disorientation is for the smallest θ with the c vector pointing in the standard stereographic triangle [42].

19.3 MONTE CARLO COMPUTER SIMULATIONS

19.3.1 Segregation phenomena in Ni–Cu alloys

Sessile edge dislocation

The first application of this approach to problems directly related to GB segregation is the calculation of segregation at a sessile edge dislocation in the Ni–Cu alloy system [43]. This is relevant here since a low angle tilt boundary is an array of edge dislocations. The particular dislocation studied has a Burger's vector of $(a/2)$ [110] and the dislocation line is along the [1$\bar{1}$2] direction. In an fcc lattice this dislocation separates into two partials, with Burger's vectors $(a/6)\langle 112 \rangle$. For these calculations periodic boundary conditions are applied both along the dislocation line, [1$\bar{1}$2], and in the direction between the partials, [110], with periodic lengths of 13.3 Å and 62.5 Å, respectively. Free surfaces are used in the remaining two directions. The chemical potential difference was chosen to correspond to Ni-10 at.% Cu and the temperature used in the simulations was 800 K.

The average composition as a function of position within the plane perpendicular to the dislocation line is presented in Fig. 19.2. The most dramati

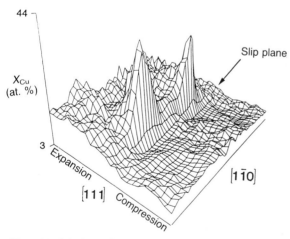

Fig. 19.2 Calculated Cu concentration of an (a/2) [110] edge dislocation in NiCu(10 at.%) at 800 K as a function of position perpendicular to the dislocation core. The length of the [111] axis is 50 Å and the length of the [110] axis is 62.5 Å. Note that the Cu enrichment occurs predominantly at the partial dislocation cores on the expansive side of the slip plane.

and important feature of these results is the two peaks in the composition. The peaks are located at the centers of the two partial dislocations and have compositions of around 40 at.% Cu. This is a significant enhancement over the bulk Cu content of 10 at.%. The cliff on the front side of the peaks (as shown in the figure) is located at the slip plane of the dislocation. On the other side of the slip plane there is a small depletion of Cu. The Cu enhancement occurs on the expanded side of the slip plane and the depletion on the compressed side. Note that the details of the composition profile away from the cores are not reliable in these calculations since the boundary conditions do not reflect the correct long-range strain field. It is important to note that there is significant segregation to the core of the dislocation. This is the region where treatment based on elasticity theory, such as the concept of a Cottrell atmosphere [44], cannot be applied. The ability to study the core regions is the main power of this approach.

(001) Twist boundaries

The segregation to three different (001) twist boundaries in Ni–Cu has also been computed [45]. In particular, the $\Sigma = 5$ (36.9°), $\Sigma = 13$ (22.6°), and $\Sigma = 61$ (10.4°) twist boundaries are simulated for bulk Cu concentrations of 10, 50, and 90 at.%. The simulations were performed at 800 K. The results for the overall composition of each of the three planes adjacent to the boundary are presented in Table 19.1. In all cases the GB region is enriched in Cu relative to the bulk Cu concentration and the change in composition is confined to the region within three to four atomic planes of the boundary. In addition, Table 19.1 lists the net expansion normal to the boundary. This is defined here as the difference in the distance between two planes on opposite sides of the boundary for the system with the GB and the distance for the same number of interlayer spacings at the bulk lattice constant. In all cases, there is an expansion of the boundary region and the amount of expansion is greater than can be accounted for by the increased concentration of Cu at the boundary. (Cu has a somewhat larger (3%) lattice constant than Ni.) This

Table 19.1 Composition of the first three planes adjacent to the GB computed using by Monte Carlo simulations using the embedded atom method. The net expansion of the GB normal to the interface is also listed. The statistical uncertainty in the compositions is $\pm 1\%$.

	First plane	Second plane	Third plane	Net expansion
NiCu(10%) $\Sigma 5$	74% Cu	22% Cu	11% Cu	0.60 Å
NiCu(10%) $\Sigma 13$	62% Cu	24% Cu	11% Cu	0.35 Å
NiCu(10%) $\Sigma 61$	41% Cu	27% Cu	15% Cu	0.44 Å
NiCu(50%) $\Sigma 5$	78% Cu	48% Cu	42% Cu	0.42 Å
NiCu(50%) $\Sigma 13$	74% Cu	52% Cu	44% Cu	0.29 Å
NiCu(50%) $\Sigma 61$	66% Cu	57% Cu	51% Cu	0.30 Å
NiCu(90%) $\Sigma 5$	95% Cu	90% Cu	90% Cu	0.32 Å
NiCu(90%) $\Sigma 13$	95% Cu	90% Cu	89% Cu	0.27 Å
NiCu(90%) $\Sigma 61$	92% Cu	91% Cu	90% Cu	0.20 Å

expansion is a general feature of GBs as discussed in Chapter 3 by Wolf and Merkle.

There are two trends which are apparent from the results in Table 19.1. First, the segregation is strongest for the higher angle boundaries. A similar trend is found for the Pt–Au twist boundaries discussed later in this chapter. Second, the segregation is strongest for the Ni-rich alloy and weakest for the Cu-rich alloys. This latter observation is consistent with calculations of the dilute segregation energies performed for the $\Sigma 5$ boundaries. The dilute segregation energy is computed by comparing the energy of a single substitutional impurity located at a position at the GB compared to its energy in the bulk material. (The atomic positions of all atoms are allowed to relax when computing these energies.) For the case of a Cu impurity in Ni, the Cu is bound by 0.22 eV to the coincident sites of the boundary which comprise 1/5 of the boundary sites and is bound by 0.13 eV to the four equivalent non-coincident sites which comprise the remaining 4/5 of the boundary sites. For the case of a Ni impurity in Cu, the Ni is repelled from the boundary plane by 0.07 eV for both the coincident and non-coincident sites. These energies indicate that the segregation of Cu to the boundary is stronger for the case of pure Ni than for the case of pure Cu consistent with the trend observed for the concentrated alloys.

The segregation at the boundary in the concentrated alloys, though, cannot be determined simply by these dilute segregation energies. If one ignores interactions between the sites at the GB the above energies predict an average concentration of 48% Cu in the boundary plane for the case of the Σ5 NiCu (10%) boundary at 800 K using the segregation expression derived for this case by McLane [46]. This is substantially smaller than the value of 74% Cu obtained in the simulation. The sense of this difference is consistent with the fact that the Ni–Cu alloy system is a clustering alloy. Thus the enhancement of the Cu concentration due to the presence of the boundary is complemented by the tendency of the Cu atoms to cluster together.

In addition to the average composition of each plane, the simulations determine the composition variations within each plane. This has been studied in detail for the case of the NiCu (10%) Σ61 boundary. This low angle boundary can be viewed as a square array of screw dislocations in the plane of the boundary [44]. (The dislocation picture of twist boundaries is discussed further below.) In Fig. 19.3, the projected atomic positions of the two planes on either side of the boundary are shown for a randomly chosen configuration from the simulation. The filled circles are Ni and the open circles are Cu atoms. The boundary clearly breaks up into regions of good match separated by a square array of poor match. These areas of poor match are the screw dislocations.

In Fig. 19.4, the average composition of the first three planes on one side of the boundary are shown as a function of position. The plots in Fig. 19.4 correspond to the central unit cell of Fig. 19.3. In particular, the screw dislocations are located along the diagonal lines that connect the midpoints of adjacent sides of this cell. (In computing these results, an average was taken over the four unit cells and the average was required to have the symmetry of the unit cell. Raw data have much more statistical noise but are qualitatively similar.) For the planes two and three layers away from the boundary, the Cu concentration is highest in the center and at the corners. These regions correspond to the areas furthest away from the screw dislocations, i.e. the areas of good match

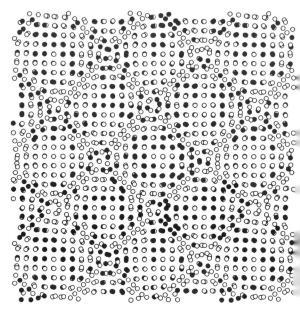

Fig. 19.3 Atomic positions projected on the boundary plane for the Σ61 boundary in the NiCu(10%) alloy. The first two atomic layers on either side of the boundary are shown. The open circles are Ni and the filled circles are Cu. The screw dislocation network is located in the regions of poor match between the two crystals.

between the crystals. This can be understood qualitatively as follows. There is a net expansion of the lattice near the boundary as discussed above. However, in the areas of good match, one would expect that the bulk lattice spacing would be preferred. Thus, these areas are in effect under tensile strain and Cu is enhanced in regions of tensile strain. The composition variation in the plane adjacent to the boundary is different and more complicated. There the Cu concentration has minimums at the center, corners, and along the outside of the dislocations. The Cu concentration is largest at regions which are offset towards the center of the cell from the intersections of the screw dislocations.

19.3.2 Segregation to GBs in Ni₃Al

The alloy Ni₃Al has received a great deal of attention in the last few years due to its high temperature mechanical properties. The GBs in this

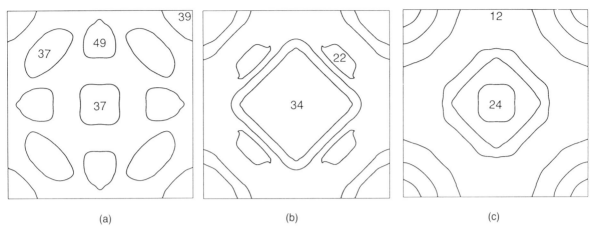

Fig. 19.4 Contour plots of the Cu concentration as a function of position in the plane for the first (a), second (b), and third (c) atomic layers from the boundary. These plots correspond to the central unit cell of Fig. 19.1. The numbers indicate the composition at extrema in atomic present Cu. The contour spacing is 4.4%.

alloy are of particular interest since one of the main drawbacks to this alloy is the tendency towards brittle intergranular fracture. It has been discovered that this problem can be reduced by both adding B and using alloys rich in Ni relative to the ideal 3 : 1 ratio. It has been speculated that these changes in the composition produce a compositional disordering of the boundary which would allow for easier transmission and adsorption of dislocations at the boundary and so increase ductility [47]. The simulation techniques discussed here have been applied to GBs in Ni–Al alloys [48, 49]. The intent of the simulations described here is to determine the effect of variations in the Ni to Al ratio on the GB structure and compositional ordering. The effect of B on the boundary has been investigated in simulations by Chen *et al.* [50].

Three different boundary geometries were considered in this study, a $\Sigma = 5$ and a $\Sigma = 13$ (001) twist boundary and a $\Sigma = 5$ (210) symmetric tilt boundary. The simulations were performed both at 00 K and at 1000 K. Simulations were performed for three sets of chemical potential differences corresponding to a Ni-rich bulk, an Al-rich bulk and an ideal stoichiometric bulk.

The compositional structure determined by the simulations for the twist boundaries at ideal stoichiometry is very simple. The (100) planes of the Ni$_3$Al structure alternate between pure Ni and an equal mix of Ni and Al. At the two [100] twist GBs, this alternating pattern of the (100) planes parallel to the boundary continues uninterrupted through the boundary. Thus the compositional ordering is the same that would be obtained by taking an ideal Ni$_3$Al crystal and rotating the two halves to form the boundary. This result is not surprising since the interactions favor the presence of Ni–Al nearest neighbor pairs and this structure accomplishes that. The structure of the tilt boundary is more complicated. The atoms originally in the (210) plane on each side of the boundary combine to form one dense plane with little compositional order. The atoms in the next (210) plane on each side of the boundary have positions close to those in the bulk crystal.

The overall composition of the boundary region was studied as a function of the bulk composition. At the ideal 3 : 1 bulk composition, the average composition of the GBs was also 3 : 1. For Ni-rich samples, the GB region is found to have a further enhanced Ni concentration relative to the bulk, and for the Al-rich samples the boundary is enhanced in Al. This effect can be quantified in terms of the interfacial excess of one of the two components. The interfacial excess of Ni is defined as the difference per unit area of the boundary between the total number of Ni atoms in the sample (including the interface) and the number that would be present if the bulk composition is assumed throughout. The Ni excesses computed at 1000 K were similar for the different boundaries studied and are approximately $-0.015/\text{Å}$ for a bulk composition of 74 at.% Ni, 0 for 75 at.% Ni and $0.012/\text{Å}^2$ for 76 at.% Ni. This behavior can be qualitatively understood by considering the energy to create anti-site defects near the boundary. (An anti-site defect corresponds to placing a Ni atom in a lattice site normally occupied by an Al atom or vice versa.) The anti-site formation energies for sites near the boundary are reduced by 0.2 to 0.4 eV for both Ni atoms in Al sites and Al atoms in Ni sites. Thus it is energetically favorable to create either kind of anti-site defect at the boundary rather than in the bulk material. Therefore, the change in composition of the boundary region can be thought of as being due to the binding of anti-site defects to the boundary region.

In addition to the overall composition of the boundary region, the compositional order of the boundary is also affected by the bulk composition. This is most easily seen for the twist boundaries. For boundaries that are Al-rich, the excess Al could either be placed in the Ni plane or in the Ni sites of the mixed composition plane. The simulations show that in fact most of the Al goes to the Ni plane. For the Ni-rich case, one would simply expect to replace the Al atoms in the mixed composition plane with Ni atoms. However, the simulations indicate a reduction in the ordering within the mixed composition plane. For the $\Sigma = 13$ boundary at a bulk composition of 24.5 at.%, there are three times more anti-site defects in the mixed composition plane than are required by the reduced Al concentration in that plane. In addition, the Ni plane in this boundary is found to contain 4 at.% Al atoms even though the system is deficient in Al. Thus the compositional ordering is reduced near the boundary in the case of Ni-rich alloys. It is important to note that this disordering effect is very localized. Only the planes immediately adjacent to the boundary are affected.

There is significant interest in whether the GBs in this alloy are compositionally ordered or dis-

ordered. King and Yoo [47] have pointed out that a compositionally disordered boundary can more easily absorb or transmit dislocations, and it has been suggested that the reason that boron addition combined with excess Ni content ductilize the alloy is due to a compositional disordering of the boundary. While there have been no experimental studies of boron-free alloys as studied above, high resolution electron microscopy has been used to search for such a disordered region in boron-doped alloys. The results are inconclusive. Mackenzie and Sass [51] reported a large (~ 40 Å thick) region of compositional disorder at the GBs. Mills [52] does not find a large region of disorder. His observations indicate that if there is compositional disorder at the boundary, it is confined to one or two planes on either side of the boundary. This conclusion is consistent with the very localized disordering seen in the above simulations of the boron-free material.

19.3.3 (001) twist boundaries in Pt-1 at.% Au alloys

A pure twist boundary is the simplest GB [53, 54]. For this type of boundary the two crystals are rotated around an axis normal to the interface. The (001) twist boundary in fcc metals has been studied experimentally employing TEM and X-ray diffraction [55, 56, 57, 58, 59], and by computer simulation [60] including detailed comparisons of the computed and experimental atomic structure [8, 57]. In terms of a dislocation model a twist boundary consists of pairs of orthogonal screw dislocations whose Burger's vectors (b) are of the type $b = (a/2) \langle 110 \rangle$ in a face-centered cubic lattice. The spacing (d) between these screw dislocations is given by the classical Read–Shockley equation [53, 54]: $d = |b|/[2 \sin (\theta/2)]$. The value of d decreases monotonically as the twist angle θ increases. Furthermore, as θ increases the dislocation cores overlap and a continuous layer of 'bad material' is formed between the two halves of the bicrystal. 'Bad material' implies that linear elasticity theory can not be used to describe the displacements of atoms. This core structure can formally be treated in terms of a dislocation model

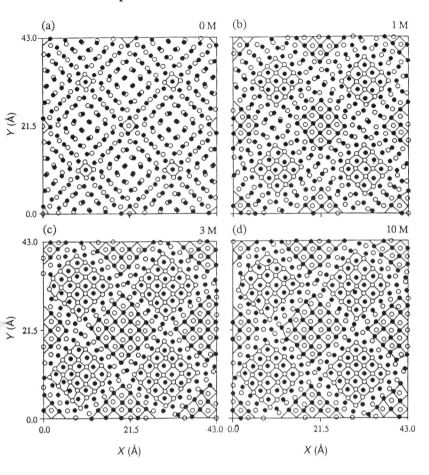

(a) 0 M

(b) 1 M

(c) 3 M

(d) 10 M

Fig. 19.5 (a) An **unrelaxed** $\theta = 10.4°$ ($\Sigma = 61$) [001] twist boundary in a 4880 atom face-centered cubic bicrystal of Pt-1 at.% Au. The atoms in the top (002) plane are denoted by filled solid circles and the atoms in the bottom (002) plane are indicated by open circles. This is a 0 K GB with no atomic relaxations present. (b) A snapshot of the same boundary after 10^6 Monte Carlo steps at 850 K. (c) A snapshot after a total of 3×10^6 Monte Carlo steps at 850 K. (d) A snapshot after a total of 10^7 Monte Carlo steps at 850 K. Note the atomic relaxations associated with the atoms at the interface and the presence of an orthogonal grid of two pairs of screw dislocations with Burger's vector (**b**) of the type **b** = $(a/2)\langle 110 \rangle$.

over the entire angular range [61, 62]. Transmission electron microscopy and X-ray diffraction studies on [001] twist boundaries in Au bicrystals, with θ varying between approximately 1° and 45°, have demonstrated the existence of a primary dislocation array of orthogonal screw dislocations with d values given by the Read–Shockley equation. Twist boundaries can also be described in terms of the structural unit/grain boundary dislocation model [63, 64, 65]. This description is useful when the cores of the dislocations overlap, and the identification of individual dislocations at the interface becomes difficult.

To study the effects of GB structure on segregation behavior, as a function of temperature, we constructed a series of twist boundaries with θ in the range 0° to 45° [66]. The particular values of θ

chosen are 5.0°, 10.4°, 16.3°, 22.6°, 28.1°, 33.9°, 36.9°, 41.1°, and 43.6°. These angles correspond to the $\Sigma = 265$, 61, 25, 13, 17, 289, 5, 73, and 29 coincident-site-lattice (CSL) orientations. The alloy studied is a single phase Pt-1 at.% Au alloy in the temperature range 850–1900 K. Great care was taken to stay within the primary single-phase field.

Figure 19.5(a) is a projection of two (002) planes of an unrelaxed (001) twist boundary – $\theta = 10.4°$ ($\Sigma = 61$) – at 0 K. These planes are separated by one-half the lattice parameter. The interface plane is parallel to them and lies half-way between them; there are no atoms at the interface plane. The atoms in the top (002) plane are indicated by solid filled circles, while the atoms in the bottom (002) plane are denoted by open circles. The model bicrystal used for the Monte Carlo simulations

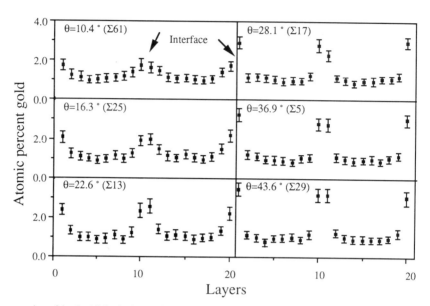

Fig. 19.6 The concentration of Au (at.%) in the (002) planes versus the distance normal to the interface for $\theta = 10.4°$ ($\Sigma = 61$), $\theta = 16.3°$ ($\Sigma = 25$), $\theta = 22.6°$ ($\Sigma = 13$), $\theta = 28.1°$ ($\Sigma = 17$), $\theta = 36.9°$ ($\Sigma = 5$), and $\theta = 43.6°$ ($\Sigma = 29$) [001] twist boundaries. The interfaces are indicated by arrows. There is more than one interface present because of the periodic boundary conditions employed. The concentrations are averaged over the (3 to 5) $\times 10^6$ Monte Carlo steps and the error bars have a total length equal to four standard deviations.

consists of ≈ 5000 atoms and it has 20 (002) planes parallel to the interface. Figure 19.5(b), (c), and (d) are snap-shots after a total of 10^6, 3×10^6 and 10^7 Monte Carlo steps at 850 K. The atomic configurations achieved are the result of both structural relaxations and thermal vibrations. This figure shows that the configurations achieved after 3×10^6 and 10^7 Monte Carlo steps are essentially the same. In Fig. 19.5(d) note that the region of poor lattice matching corresponds to the cores of pairs of orthogonal screw dislocations.

Figure 19.6 exhibits the Au concentration for each (002) plane for Monte Carlo simulations performed at 850 K. The positions of the interface are indicated by arrows – multiple interfaces are a result of the periodic boundary conditions. This figure demonstrates that Au segregation occurs at all the twist boundaries studied; mainly the two planes that adjoin each interface are enriched in Au atoms. The segregating Au atoms sit at substitutional sites in these planes.

An average segregation enhancement factor (S_{ave}) is defined to be the ratio of the solute con-

centration in the two planes that adjoin the interface, divided by the solute concentration in the bulk. Plots of S_{ave} versus $\sin(\theta/2)$ at 850, 900, 1000, 1300, 1500, and 1900 K – $\sin(\theta/2)$ is proportional to the dislocation density – were calculated. The results demonstrate that $S_{ave} = S_{ave}(T, \theta)$. A fixed temperature the value of S_{ave} increases monotonically, as θ increases up to $\sin(\theta/2) \approx 0.3$, while at fixed θ the value of S_{ave} decreases exponentially as temperature increases.

An Arrhenius plot of S_{ave} yields straight lines for all the twist boundaries. A classical thermodynamic analysis of the Arrhenius plots yields a binding enthalpy (Δh^b_{s-gb}) and a binding entropy (Δs^b_{s-gb}) of a solute atom at a twist boundary. The range of Δh^b_{s-gb} is from 0.011 ± 0.009 to 0.121 ± 0.009 eV/atom, as θ is increased from 5.0 to 36.9°. The largest value of Δh^b_{s-gb} is for the $\Sigma = 5$ twist boundary. The range of Δs^b_{s-gb} is from 0.07 ± 0.10 K for $\theta = 16.3°$ to 0.46 ± 0.09 K for $\theta = 36.9°$ ($\Sigma = 5$). The values of Δs^b_{s-gb} for $\theta = 5.0°$ to 16.3° are essentially the same -0.1 K – when the uncertainty (± 0.1 K) is taken into account. The fact that all values of Δs_{s-gb}^b are positive – with the exception of the 5° boundary – implies that within the context of an Einstein solid [67] the vibrational frequencies associated with a Au atom at a GB are higher than those of a Au atom in the bulk. This result is qualitatively consistent with the idea that Au is an oversized substitutional atom in the cores of the dislocations.

An analysis of the two-dimensional spatial distribution of Au atoms reveals that they sit mainly in the cores of the primary GB dislocations. (No evidence was found for the formation of atmospheres, due to the elastic inhomogeneity interaction, around individual screw dislocations as envisaged in the linear elasticity theory model of solute-atom segregation at dislocations [44].) These observations are the basis of a model for the Monte Carlo simulation results. We divide the GB interface into two regions. The first region contains the cores of the dislocations and the second one is a region of good atomic fit. The structure of the core region is taken to be the same for all θ's studied since the b of the primary GB dislocations is identical for all values of θ. The average Au concentration at an interface is the sum of the Au concentrations in the region of good atomic fit plus the concentration in the cores of the dislocations. This model leads to a linear equation for S_{ave}, as a function of θ. The linear dependence holds up to $\theta \approx 35°$ since beyond this value S_{ave} is approximately independent of θ, i.e. all material at the interface is at core sites, and it is not possible to increase the fraction of sites in the cores. For this model the segregation factor for the core (S_{core}) has a single value for all twist boundaries. An Arrhenius plot of S_{core} has a single value of the binding enthalpy of a Au atom to the core, 0.095 ± 0.01 eV/atom, and a single value of the binding entropy, $0.49 \pm 0.10 k$; these values are for a core radius equal to $0.8|b|$.

The physical picture that emerges from the Monte Carlo simulations is that Au segregation occurs primarily at the cores of the primary GB dislocations, and therefore the Au concentration at the interface depends on the fraction of atoms in a bicrystal that are located in the cores. This explains why the value of S_{ave} is a function of θ, and why it

saturates when the cores of the primary dislocations overlap. When the cores overlap all of the atoms at the interface are in the cores of the dislocations, and increasing θ does not significantly change the fraction of sites in the bicrystal at the cores. It is noteworthy that Wolf has shown that the energy of GBs, including high angle boundaries, can also be described in terms of a dislocation model [68].

19.4 ATOM-PROBE OBSERVATIONS OF SOLUTE–ATOM SEGREGATION

The atom probe technique, in conjunction with TEM, has been employed to systematically study solute–atom segregation at individual GBs in both fcc and bcc cubic alloys. In this section we discuss recent examples of this experimental work.

19.4.1 Oscillatory Ni profile at a $\Sigma5/(202)$ interface in a Pt(Ni) alloy

Oscillatory solute-atom segregation profiles have been observed: (a) at the solid-vacuum (111) and (110) surfaces of Pt–Ni alloys via low-energy electron diffraction and calculations [69, 70, 71, 72, 73]; (b) at the solid-vacuum surfaces of Ni–Cu, Pt–Rh, Pt–Ru, Pt–Rh–S, and Pt–Ru–S alloys by means of atom-probe field-ion microscopy [74, 75, 76]; and (c) at the Si–Ge (100) 2×1 solid–vacuum surface via Monte Carlo simulations [74, 75, 76]. We have recently observed for the first time an oscillatory segregation profile at an internal interface in a Pt-3 at.% Ni alloy, and thereby observed the analog for solid–solid interfaces of what has been observed at solid–vacuum interfaces [77]. Thus a segregation profile at an internal interface need not decay monotonically to the bulk value.

Both Monte Carlo simulations and atom-probe field-ion-microscopy studies were performed on a bicrystal of a Pt-3 at.% Ni alloy. The experimental specimen was annealed at 850 K for 24 h to induce the segregation of Ni. The orientation of the two crystals in the experimental sample was determined via TEM and found to be within 3.7° of an asymmetric $\Sigma5/(202)$ coincident site lattice boundary.

This experimentally characterized boundary was next studied by the atom-probe technique.

The boundary is analyzed chemically in three different regions, as shown in Fig. 19.7 for the interface just discussed. In all three cases the geometry of Fig. 19.1(b) with D_a = 1.4 nm was employed. In Fig. 19.7 each plot is an integral profile, i.e. the cumulative number of Ni atoms versus the cumulative number of Pt plus Ni atoms for all the events detected from the cylinder of alloy analyzed. The smallest vertical or horizontal line represents a **single** atom. Figure 19.7(a) is a control run of the matrix concentration far from the boundary. The slope of the straight line is equal to $\langle C_{Ni} \rangle$. The local fluctuations about $\langle C_{Ni} \rangle$ are random solid-state fluctuations, due to the size of the sample; for these data $\langle C_{Ni} \rangle$ is 2.9 ± 0.2 at.%. Figure 19.7(b) was recorded with the probe hole centered symmetrically with respect to the boundary plane; the mean slope is $\langle C_{Ni}^{gb} \rangle_u$ = 8.9 ± 1.4 at.% Ni; this value is ≈3.07 times greater than $\langle C_{Ni} \rangle$. This is a minimum value for S_{ave} because of the matrix contribution. Figure 19.7(c) was recorded with the boundary placed just at the periphery of the probe hole; the measured concentration is 4.0 ± 0.8 at.% Ni. The latter value is ≈(33%) greater than $\langle C_{Ni} \rangle$, and this implies that the concentration profile associated with this boundary is broadened, i.e. it is not a delta function.

To understand this broadened profile we utilized the Monte Carlo methodology to simulate Ni segregation, at 850 K, for an asymmetric Σ = 5 coincidence site lattice orientation with (202) and (10 −6 8) planes on either side of the boundary. The computer-generated bicrystal consists of 20 (202) planes with each plane containing 100 atoms (G1), and 100 (10 −6 8)planes with each plane containing 20 atoms (G2); this bicrystal consists of 4000 atoms.

Figure 19.8 shows the calculated Ni concentration profile of the region in the immediate vicinity of the interface. This profile shows that the Ni concentration at the (202) plane immediately to the right of GB (G1) has a concentration of 25.1 at.% Ni, and the Ni concentration falls monotonically to the matrix value in the second (202) plane. The

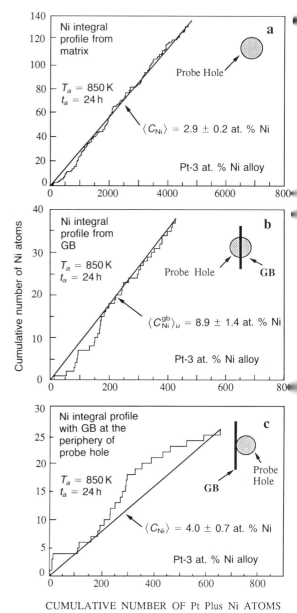

Fig. 19.7 Three Ni integral profiles obtained employing a pulse fraction of 15%, a specimen temperature of 45 K, and a vacuum of 6.7×10^{-8} Pa in the APFIM; the probe hole diameter 1.4 nm.

(10 −6 8) planes immediately to the left of the GB plane (G2) have a concentration of 57.4 at.% Ni. Instead of the Ni concentration falling monotonically to the matrix value it oscillates for at least

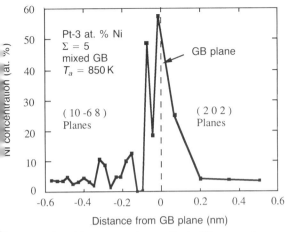

Fig. 19.8 The Ni concentration profile (at.%) versus distance (nm) from the GB plane for the bicrystal in the immediate vicinity of the GB plane.

15 (10 − 6 8) planes (≈0.416 nm) before reaching ≈3 at.% Ni. This oscillatory Ni concentration profile is the first observation of such segregation behavior at a GB. The other interesting feature is that the Ni profile is asymmetric with respect to the boundary plane; this asymmetry demonstrates that the structure of a boundary affects its segregation behavior.

Finally the Monte Carlo results allow us to interpret the measured concentration values. The computed compositions of each plane allow the average compositions corresponding to Fig. 19.7(b) and 19.7(c) to be computed. For the probe hole centered on the boundary the calculated Ni concentration is 9.75 at.%, as compared to an experimental value of 8.9 ± 1.4 at.% Ni. For the probe hole adjacent to the boundary, the calculated Ni concentration is 4.1 at.%, as compared to the experimental value of 4.0 ± 0.7 at.% Ni. Thus the calculated Ni concentrations are in good agreement with the experimental values. While the APFIM results do not directly show the oscillatory profile, the agreement between theory and experiment lends credence to the existence of an oscillatory Ni concentration profile, as observed in the Monte Carlo simulations.

19.4.2 Solute atom segregation at a GB in a W(Re) alloy via APFIM

We now consider the direct determination of boundary segregation in a bcc W-25 at.% Re alloy [78] which was studied experimentally by both TEM and the atom probe. Direct evidence is presented for two-dimensional segregation at a GB, and the relationship between the chemical compositions of GBs and their structure is discussed. In addition, systematic evidence is presented for a direct relationship between GB structure and its chemical composition, as measured by the atom probe technique, for a series of almost pure twist GBs in this alloy.

A bulk specimen of this single-phase solid-solution alloy was heat treated at 1913 K for 5 h to induce Re segregation. A boundary was located in this specimen by TEM and five macroscopic degrees of the interface were determined. The misorientation of grain 1 with respect to grain 2 is 88.7° about the rotation vector [0.70 0.72 0.01], and the disorientation is [221]/61.93°; the normal to the plane of the grain boundary is close to ⟨156⟩. The boundary is almost a pure twist boundary with a small tilt component. The nearest coincident-site-lattice is Σ = 17 ([110]/86.63°); the deviation from exact coincidence is small enough such that this boundary can be termed a 'special' boundary using Brandon's criterion [79].

Next the specimen was electrolytically thinned to place this GB in the tip of a sharply pointed APFIM specimen. The geometry exhibited in Fig. 19.1(a) was employed to chemically analyze this boundary. The depth resolution is the interplanar spacing of the (156) planes, ≈0.04 nm. Figure 19.9 is a plot of the Re concentration in each atomic plane, as a function of the number atomic planes analyzed. The Re concentration is observed to increase from ⟨C_{Re}⟩ = 29.2 ± 5.64 at.% Re to 73.8 ± 5.45 at.% Re in *one* interplanar spacing and then to decrease to ⟨C_{Re}⟩ = 23.1 ± 5.23 at.% Re in *one* interplanar spacing; note carefully the planes numbered 9, 10, and 11. The mean concentration of Re in the matrix is 25.2 ± 0.98 at.% Re averaged over 29 planes, excluding the interface plane. This result demonstrates that for this boundary the Re

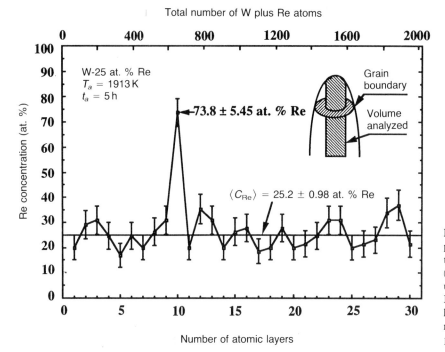

Fig. 19.9 The Re concentration profile perpendicular to the plane of the interface of a $\Sigma \approx 17$ ([70 72 1]/88.7°) GB. The geometry used for the analysis is that of Fig. 19.1(a) and the inset on the right hand side of this figure. The total number of W plus Re atoms per plane is 65.

segregation is localized at one atomic plane, and the Re concentration in the planes that adjoin this interface are at the matrix concentration. The data imply an $S_{ave} = 2.9$. It is emphasized that this result represents direct evidence for two-dimensional segregation, and that *no* model-dependent deconvolution of the experimental data is required to obtain this information.

Figure 19.10 is a plot of the segregation enhancement factor (S_{ave}) versus $\sin(\theta/2)$ for a number of GBs in specimens which had been annealed at 1913 K for 5 h to induce Re segregation [78]. The angle θ is the twist angle about the vector c which is close to $\approx[011]$ for all the boundaries analyzed; thus the GBs are all twist boundaries with a small tilt component; the full symmetry of the (011) twist boundary is exhibited in the range 0° to 90°. The general trend is that the amount of Re segregation increases with increasing $\sin(\theta/2)$ – with the exception of the boundary for which $\Sigma \approx 3$ – or alternatively with increasing dislocation density at the GB. This result is qualitatively consistent with the results for the Pt–Au alloy. There the Au

concentration at the [001] twist boundaries in the Pt-1 at.% Au alloy increases with increasing twist angle between 0° and 45°. The simplest physical explanation of this result is that the fraction of sites at the cores of the interface dislocations increases as the twist angle increases and that the dominant sites for Re segregation are substitutional sites in the cores of the dislocations.

In the case of this bcc alloy it is not yet possible to perform Monte Carlo simulations, as suitable potentials are unavailable.

19.5 CONCLUSIONS

In this chapter we have summarized some recent research using computer simulation methods and atom-probe field-ion-microscopy to study equilibrium segregation at GBs in binary alloys. The segregation at an isolated edge dislocation was studied and it was found that there is substantial segregation to the core region of the dislocation which is not amenable to linear elasticity theory.

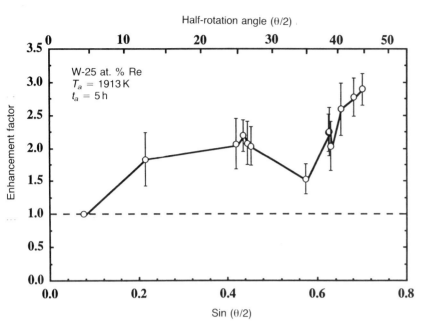

Fig. 19.10 The segregation enhancement factor (S_{ave}) versus sin ($\theta/2$) for a number of GBs in a W-25 at.% Re alloy which had been annealed at 1913 K for 5 h to induce Re segregation. The angle θ is the twist angle for these GBs; they are all almost pure twist with a small tilt component and the vector (c) is close to [011] for all the boundaries.

This demonstrates the need for the current computer simulation approach. Next, the systematics of the segregation to a series of twist boundaries in the solid solution alloys Ni–Cu and Pt-1 at.% Au was determined. The results showed that the trends of the segregation behavior can be understood in terms of the density of the dislocations that comprise the boundary. Next, the segregation at an asymmetric mixed tilt–twist boundary in fcc Pt-3 at.% Ni was studied both by computer simulation and experimentally using atom-probe field-ion-microscopy. There is good quantitative agreement found between the compositions measured with the APFIM and those obtained in the computer simulations. In addition, the simulations indicate the possibility of oscillatory composition profiles in the vicinity of the boundary. Then APFIM evidence for two dimensional Re segregation at an almost pure twist boundary in a bcc W-25 at.% Re alloy was presented, as well as evidence for the effect of dislocation structure on the amount of Re segregation at a series of almost pure twist boundaries between 0° and 90°.

Certain general features have emerged from the limited number of results available to date. First, the segregation is a rather localized effect; the composition returns to the bulk value within a few ångstroms of the boundary. Also, as the dislocation density of the boundary increases, the degree of the segregation is also seen to increase. There are still significant differences between the systems studied. First, the distribution of segregant in the plane of the boundary is different for the Pt–Au and Ni–Cu alloys studied here. This difference is not currently understood. Second, both monotonic and oscillatory variations of the composition near the boundary have been obtained in the simulations. Thus there is still substantial work to be done before the systematics of GB segregation are completely understood. Finally, the importance of combined experimental and theoretical studies should be emphasized. The ability to compare results from both calculations and experiments will ultimately lead both to improved understanding of the experimental results and to more reliable theoretical methods.

ACKNOWLEDGMENTS

This research is supported by the US Department of Energy, Office of Basic Energy Sciences, Division of Materials Science (SMF), and the National Science Foundation through grant No. DMR-8819074 (DNS). The Monte Carlo simulations were performed at Sandia National Laboratory at Livermore, CA (SMF), and the National Science Foundation funded National Center for Supercomputing Applications at the University of Illinois at Urbana-Champaign, IL (DNS). DNS wishes to thank his co-workers Dr Y. Oh, Mr J. G. Hu, and Mr A. Seki for permission to quote from unpublished research.

REFERENCES

1. J. W. Gibbs, *The Collected Works of J. Willard Gibbs* (Yale University Press, New Haven (1948).
2. K. E. Sickafus and S. L. Sass, *Acta Metall.*, **35** (1987), 69.
3. J. R. Michael, C.-H. Lin and S. L. Sass, *Scripta Metall.*, **22** (1988), 1121.
4. M. S. Daw and M. I. Baskes, *Phys. Rev. B*, **29** (1984), 6443.
5. S. M. Foiles, in *Surface Segregation and Related Phenomena*, Vol. 1 (ed P. A. Dowben and A. Miller), CRC (1990).
6. S. M. Foiles, M. I. Baskes and M. S. Daw, *Phys. Rev. B*, **33** (1986), 7983.
7. M. S. Daw, in *Reconstruction of Solid Surfaces* (eds K. Christman and K. Heinz), Springer-Verlag, Berlin (in press).
8. S. M. Foiles, *Acta Metall.*, **37** (1989), 2815.
9. I. Majid, P. D. Bristowe and R. W. Balluffi, *Phys. Rev.*, **40** (1989), 2779.
10. D. Wolf, J. Lutsko and M. Kluge, in *Atomistic Simulation of Materials: Beyond Pair Potentials* (eds V. Vitek and D. J. Srolovitz), Plenum Press, New York (1989), p. 245.
11. S. M. Foiles and J. B. Adams, *Phys. Rev. B*, **40** (1989), 5909.
12. A. E. Carlsson, in *Solid State Physics*, Vol. 43 (eds H. Ehrenreich and D. Turnbull), Academic Press, New York (1990), p. 1.
13. K. W. Jacobsen, J. K. Nørskov and M. J. Puska, *Phys. Rev. B*, **35** (1987), 7423.
14. M. S. Daw, *Phys. Rev. B*, **39** (1989), 7441.
15. S. M. Foiles, *Phys. Rev. B*, **32** (1985), 7685.
16. A. Joshi, in *Interfacial Segregation* (eds W. C. Johnson and J. M. Blakely), American Society for Metals, Metals Park, OH (1979), p. 39.
17. D. F. Stein and L. A. Heldt, in *Interfacial Segregation* (eds W. C. Johnson and J. M. Blakely), American Society for

Metals, Metals Park, OH (1979), p. 239.
18. E. D. Hondros and M. P. Seah, in *Physical Metallurgy* (eds R. W. Cahn and P. Haasen), North-Holland, Amsterdam (1983), p. 856.
19. W. Drachsel, L. V. Alvensleben and A. J. Melmed, *Colloque de Physique*, C8 (1989), 541.
20. H. J. Neusser, U. Boesl, R. Weinkauf and E. W. Schlag, *Int. J. Mass Spectrom. Ion Phys.*, **60** (1984), 147.
21. B. A. Mamyrin, V. I. Karataev, D. V. Shmikk and V. A. Zagulin, *Sov. Phys. JETP*, **3** (1973), 45.
22. W. B. Colson, J. McPherson and F. T. King, *Rev. Sci. Instrum.*, **44** (1973), 1694.
23. E. W. Müller, *Z. Physik*, **131** (1951), 136.
24. M. G. Inghram and R. Gomer, *J. Chem. Phys.*, **22** (1954), 1279.
25. E. W. Müller and K. Bahadur, *Phys. Rev.*, **102** (1956), 624.
26. E. W. Müller, *Phys. Rev.*, **102** (1956), 618.
27. J. Gomer, *J. Chem. Phys.*, **31** (1959), 341.
28. R. Gomer and L. W. Swanson, *J. Chem. Phys.*, **38** (1963), 1613.
29. R. Gomer and L. W. Swanson, *J. Chem. Phys.*, **39** (1963), 2813.
30. D. G. Brandon, *Brit. J. Appl. Phys.*, **14** (1963), 474.
31. A. J. W. Moore, *Phys. Chem.*, **23** (1962), 907.
32. A. J. W. Moore, in *Field-Ion Microscopy* (eds J. J. Hren and S. Ranganathan), Plenum, New York (1968), p. 69.
33. D. G. Brandon, B. Ralph, S. Ranganathan and M. Wald, *Acta Metall.*, **12** (1964), 813.
34. S. Ranganathan, in *Field-Ion Microscopy* (eds J. J. Hren and S. Ranganathan), Plenum, New York (1968), Ch. 8.
35. K. M. Bowkett and D. A. Smith, *Field-Ion Microscopy*, North-Holland, Amsterdam (1970).
36. R. Wagner, *Field-Ion Microscopy*, Springer-Verlag, Berlin (1982).
37. R. Herschitz and D. N. Seidman, *Acta. Metall.*, **33** (1985), 1547.
38. A. Seki, D. N. Seidman, Y. Oh and S. M. Foiles, *Aeta Metall.*, **39** (1991), 3179.
39. Y. Oh, M. Ueno and D. N. Seidman, submitted for publication (1990).
40. W. T. Read Jr. and W. Shockley, *Phys. Rev.*, **78** (1950), 275.
41. J. W. Cahn, *J. Phys. (Paris), Colloq.*, **43** (1982), C6.
42. W. Bollmann, *Crystal Lattices, Interfaces, Matrices* Imprimerie des Bergues, Geneva (1982).
43. S. M. Foiles, in *Computer-Based Microscopic Description of the Structure and Properties of Materials* (eds J. Broughton, W. Krakow and S. T. Pantelides), Materials Research Society, Pittsburgh (1986), p. 61.
44. J. P. Hirth and J. Lothe, *Theory of Dislocations*, 2nd edn, John Wiley and Sons, New York (1982).

45. S. M. Foiles, *Phys. Rev. B*, **40** (1989), 11502.

46. D. McLane, *Grain Boundaries in Metals* Clarendon, Oxford (1957).

47. A. H. King and M. H. Yoo, *Scripta Metall.*, **21** (1987), 1115.

48. S. M. Foiles, in *High-Temperature Ordered Intermetallic Alloys II*, *Vol. 81* (eds N. S. Stoloff, C. C. Koch, C. T. Liu and O. Izumi), Materials Research Society, Pittsburgh (1987), p. 51.

49. S. M. Foiles and M. S. Daw, *J. Mater. Res.*, **2** (1987), 5.

50. S. P. Chen, D. J. Srolovitz and A. F. Voter, *J. Mater. Res.*, **4** (1989), 62.

51. R. A. D. Mackenzie and S. L. Sass, *Scripta Metall.*, **22** (1988), 1807.

52. M. J. Mills, *Scripta Metall.*, **23** (1989), 2061.

53. W. T. J. Read and W. Shockley, *Phys. Rev.*, **78** (1950), 275.

54. W. R. Read Jr., *Dislocations in Crystals* McGraw-Hill, New York (1953).

55. T. R. Schober and R. W. Balluffi, *Phil. Mag.*, **21** (1970), 109.

56. R. W. Balluffi, T. R. Schober and Y. Komem, *Surf. Sci.*, **31** (1972), 68.

57. I. Majid, P. D. Bristowe and R. W. Balluffi, *Phys. Rev.*, **40** (1989), 2779.

58. W. Gaudig and S. L. Sass, *Phil. Mag. A*, **39** (1979), 725.

59. M. R. Fitzsimmons, K. Burkel and S. L. Sass, *Phys. Rev. Lett.*, **61** (1988), 2237.

60. D. Wolf, *Acta Metall.*, **37** (1989), 1983.

61. S. L. Sass, T. Y. Tan and R. W. Balluffi, *Phil. Mag.*, **31** (1975), 559.

62. W. Gaudig, D. Y. Guan and S. L. Sass, *Phil. Mag.*, **34** (1976), 923.

63. A. P. Sutton and V. Vitek, *Phil. Trans. Roy. Soc. Lond.*, **A309** (1983), 1.

64. P. D. Bristowe and R. W. Balluffi, *J. Phys. (Paris)*, **46** (1985), C4.

65. A. P. Sutton, *J. Phys. (Paris)*, **51** (1990), C1.

66. A. Seki, D. N. Seidman, Y. Oh and S. M. Foiles, *Acta Metall.*, **39** (1991), 3167.

67. H. B. Huntington, G. A. Shirn and E. S. Wadja, *Phys. Rev.*, **99** (1955), 1085.

68. D. Wolf, *Scripta Metall.*, **23** (1989), 1713.

69. Y. Gauthier, R. Joly, R. Baudoing and J. Rundgren, *Phys. Rev. B*, **31** (1985), 6216.

70. R. Baudoing, Y. Gauthier, M. Lundberg and J. Rundgren, *J. Phys. C: Solid State Phys.*, **19** (1986), 2825.

71. G. Treglia and B. Legrand, *Phys. Rev. B*, **35** (1987), 7876.

72. Y. Gauthier, R. Baudoing, M. Lundberg and J. Rundgren, *Phys. Rev. B*, **35** (1987), 7876.

73. M. Lundberg, *Phys. Rev. B*, **36** (1987), 4692.

74. Y. S. Ng, T. T. Tsong and S. B. Mclane Jr., *Phys. Rev. Lett.*, **42** (1979), 588.

75. M. Ahmad and T. T. Tsong, *J. Chem. Phys.*, **83** (1985), 388.

76. T. T. Tsong, D. M. Ren and M. Ahmad, *Phys. Rev. B*, **38** (1988), 7428.

77. S.-M. Kuo, A. Seki, Y. Oh and D. N. Seidman, *Phys. Rev. Lett.*, **65** (1990), 199.

78. J. G. Hu and D. N. Seidman, *Phys. Rev. Lett.*, **65** (1990), 1615.

79. D. G. Brandon, *Acta Metall.*, **14** (1966), 1479.

CHAPTER TWENTY

Amorphization by interfacial reactions

W. L. Johnson

Introduction · Review of experimental results · Phase formation and growth in diffusion couples – theoretical considerations · Relationship to other melting phenomena

20.1 INTRODUCTION

The subject of amorphous phase formation by interfacial reactions can be viewed as a special case of the more general problem of the evolution of diffusion couples. On the other hand, it is also a special case of the general problem of crystal-to-amorphous or crystal-to-glass transformations. In the present chapter, we will discuss the subject in both contexts. The formation and growth of an amorphous interlayer in an initially crystalline diffusion couple was first observed in 1983 [1]. Thin film diffusion couples consisting of alternating polycrystalline layers of the metals gold and lanthanum were observed to react at relatively low temperatures ($\sim$100°C) to form amorphous alloy interlayers along each interface between the metals. When the individual layer thicknesses were chosen sufficiently small, growth of the amorphous interlayers was found to consume the polycrystalline metal layers entirely, yielding a single phase amorphous product. At about the same time and independently, it was found that certain polycrystalline metals in contact with amorphous silicon could be reacted to yield an amorphous metal silicide interlayer [2]. That these observations aroused the interest of the thin film physics community was not surprising. It was commonly

accepted that amorphous phases are metastable For an amorphous material of given composition there always exists a crystalline phase or mixture of several crystalline phases (of differing composition) which is thermodynamically more stable than the amorphous phase. Furthermore, when an amorphous phase is heated to temperature which permit sufficient atomic mobility of its constituent atoms, it tends to crystallize into these more stable crystalline phases. As such, it was surprising to find that two crystalline phases could spontaneously react along a common interface to form an amorphous material. Such a diffusion reaction requires significant atomic mobility, and one would expect that formation of thermodynamically more stable crystalline product phase would be favored.

Within a few years of the initial discovery of solid state amorphizing reactions (SSARs), it became apparent that the phenomena were not restricted to a few pathological cases, but was rather common to a broad class of binary diffusion couples. By 1986 at least 15 examples of this type of reaction had been observed, as described in two review article [3]. At present, the list of binary diffusion couple exhibiting this type of reaction numbers in the dozens. Furthermore, it has been found that the reaction can be driven not only by heating, but also by mechanical deformation of two metals in contact, by irradiation of an interface with charged energetic particles, and other methods. In section 20.2, a review of experimental studies of SSAR will be presented. This is followed in section 20.3 by a discussion of the principles which govern the formation and growth of new phases during interfacial chemical reactions. In particular, the microscopic principles which result in the formation of metastable amorphous phases are discussed. Section 20.4 discusses SSAR as a special case of the more general phenomena of melting. Here, it is argued that the study of SSAR and other types of crystal to glass transformations can yield new fundamental insights into the nature of the melting transition.

20.2 REVIEW OF EXPERIMENTAL RESULTS

20.2.1 Amorphization in metallic bilayers and multilayers – a summary of experimental observations

The earliest studies of SSAR were carried out on thin film diffusion couples consisting of either a single bilayer of two metals, or multilayers consisting of alternating layers of two metals. These bilayers and multilayers are prepared by sequential deposition of the elements from the vapor phase. Thermal evaporation, electron beam evaporation, and sputtering have all been used to prepare specimens. Individual layer thicknesses typically range from 5 to 300 nm. The individual metal layers are polycrystalline with grain sizes ranging from 5 to 100 nm. Generally speaking, the crystallite grain size of the pure metals is of the order of or less than the thickness of the individual layers. As such, the individual layers of the diffusion couples generally contain a variety of grain boundaries (GBs). Upon heating, the amorphous phase forms along the interface(s) separating the individual metal layers. It is frequently found that even during the earliest stages of the reaction the amorphous phase appears as a uniform planar interlayer. This morphology suggests initial nucleation of the amorphous phase is quickly followed by rapid spreading along the

Table 20.1 Binary metal–metal systems which have been observed to exhibit solid state amorphization by interdiffusion or other type of driven mixing. The table lists the type of experiment, the typical reaction temperature, T_R (for the case of interdiffusion reaction), and references. B = thin film bilayer diffusion couples, M = thin film multilayer diffusion couple, S = interdiffusion of polycrystalline layer of one component with single crystal of other component, MA = mechanical alloying of the metals, MAT = thermal reaction of a mechanically deformed composite, IM = ion mixing of layers.

System	Type of experiment	T_R (C)	References
Late transition metal–rare earth systems			
Au–La	B, M	80	[1, 3]
Cu–Er	MA, MAT	150	[3, 48]
Ni–Er	MA, MAT	150	[48]
Au–Y	B, M	120	[3, 12]
Late transition metal–early transition metal systems			
Au–Zr	B, M, IM	250	[13]
Co–Zr	B, M	300	[3, 22, 82]
Cu–Ti	MA	–	[14]
Cu–Y	M, B	200	[31]
Cu–Zr	MA, MAT	–	[16]
Fe–Zr	B, M, MA	300	[17]
Ni–Hf	B, M, IM	350	[3, 23, 68]
Ni–Ti	B, M, MA	350	[3, 20, 34]
Ni–Zr	B, M, MA, MAT, IM, S	320	[3, 21, 24, 25, 28–30] [33, 47, 51–53]
Simple metal–transition metal systems			
Al–Mo	B	250	15
Al–Mn	M	220	10
Al–Pt	M, IM	200	11
Co–Sn	M	50	19
Ni–Mg	M	50	18

interfaces. Growth of the amorphous phase along the interface seems much more rapid than in the direction normal to the interface.

Table 20.1 gives a summary of binary metal–metal systems in which SSAR has been reported in thin film diffusion couples. The entries in the table are divided into several groups. These include binary systems which contain a rare earth (RE) and a late transition metal (LTM). This, the RE–LTM group, includes the La–Au system as its prototype along with other similar systems such as Er–Ni. A second group consists of an early transition metal (ETM) together with an LTM. The ETM–LTM group includes binary pairs such as Zr–Ni, Zr–

Table 20.2 Binary semiconductor/metal systems which have been observed to exhibit solid state amorphization by interdiffusion reactions. Type of experiment, typical reaction temperature (T_R), maximum thickness of the amorphous layer (X_c), and references are listed. S = interdiffusion of polycrystalline metal with semiconductor single crystal; A = interdiffusion of amorphous semiconductor layers with polycrystalline metal.

System	Type of experiment	$T_R (C)$	$X_c (nm)$	Reference
Si–Nb	S	450	5.5	[39]
Si–Ni	A	280	120	[35, 37, 38]
Si–Rh	A	250	<10	[2]
Si–Ti	A, S	350	5–15	[32, 38, 39]
Si–V	S	380	2–3	[39]
Si–Zr	S	350	17	[39]
Te–Ag[a]	A	100	>100	[36]
GaAs–Ni	S	180	–	[40]
GaAs–Co	S	280	20–30	[41]
GaAs–Pd	S	–	5	[43]
GaAs–Pt	S	60	5	[40]
InP–Pd	S	200	<10	[42]
InP–Ni	S	200	<10	[42]

[a] Diffusion of polycrystalline Ag layer in an amorphous–Te film

Co, Y–Cu, Hf–Ni, etc. A third group consists of simple metals (e.g. Sn, Pb, etc.) reacted with a late transition metal (e.g. Co, Au, etc.) and is referred to later as the SM–LTM group. All of the above types of binary systems have certain features in common. For example, the phenomenon of 'anomalous fast diffusion' (AFD) has been found to occur in experiments where the second member of the pair diffuses into single crystals of the first member. For instance, Au is an anomalous fast diffuser in crystalline La, and Ni and Co are anomalous fast diffusers in crystalline Zr and Hf. As will be seen below, the fast diffusion phenomenon is also common to other non-metal–metal systems which exhibit SSAR (Table 20.2). For instance, Au, Ni, Co, etc. are anomalous fast diffusers in crystalline Si. ADF is defined as solute diffusion with a diffusion constant exceeding the self diffusion constant of the host solvent by at least several orders of magnitude. The phenomena have been studied extensively by numerous investigators and was well known prior to the discovery of SSAR. Comprehensive reviews have been published

by Le Claire [4] and Warburton [5]. For anomalous fast diffusion systems, it has been found that the activation energy for diffusion of solute in the matrix is generally significantly smaller than the activation energy for self diffusion of matrix atoms. This has lead to considerable debate as to the microscopic mechanism of ADF. For most ADF systems, the solute atom has an atomic radius which is at least 10% smaller than the atomic radius of the host atom. It is currently accepted that the fast diffusing solute tends to occupy interstitial sites in the host matrix. The small activation energy is then associated with the existence of easily accessible excited configurations of the interstitial atom. In one model, a substitutional solute atom pairs with an interstitial solute to form a 'dumbbell'. This dumbbell can undergo an easily activated rotational motion which permits diffusion. For the present purposes, it is sufficient to note that ADF systems are characterized by a fast moving solute species and a slow moving host matrix. As will be seen shortly, this phenomenon has an analog in the SSAR process.

A second feature common to systems which exhibit SSAR is a negative free energy of mixing. SSAR systems tend to form alloys and/or intermetallic compounds in equilibrium. It is this feature of binary systems which provides the thermodynamic driving force for chemical reaction. For many of the SSAR systems, the negative free energy of mixing is quite large and consists of an enthalpic and an entropic contribution. In those systems with large negative free energies of mixing (e.g. 10–150 kJ/mole), the enthalpic contribution dominates. Systems which exhibit SSAR are thus typically characterized by large negative enthalpies of mixing (note that SSAR has been observed in systems with near zero enthalpy of mixing such as Co–Sn). Since the enthalpy of mixing depends somewhat on the structure of the product phase, it is useful to be more specific. In particular, the liquid (or amorphous phase) is often used as reference state. This chapter makes frequent reference to the negative enthalpy of mixing of the liquid–amorphous phase. This is a natural reference quantity since SSAR involves an amorphous product phase.

SSAR is observed when the binary diffusion couples listed in Table 20.1 are heated to an appropriate reaction temperature, T_R. Examples of typical values of T_R are given in Table 20.1. It is well known that amorphous metallic alloys tend to crystallize in laboratory time scales upon heating to temperatures close to their glass transition temperature, T_g [6]. For a typical practical time scale (e.g. minutes) one can define a crystallization temperature [6] as the temperature at which a significant fraction of an amorphous sample undergoes crystallization in the specified time. The glass transition temperature is the kinetic freezing temperature of an undercooled melt. Upon cooling to near T_g, undercooled melts exhibit a dramatic rise in viscosity, η, over a narrow range of temperatures. Typically as temperature is lowered to near T_g, η rises from values of order 10^{-2} poise (characteristic of a viscous liquid) to values of 10^{16} poise (conventionally taken to indicate a solid) over a temperature range of a few tens of degrees K. Inversely, as the temperature of an amorphous material (glass) is raised to near or above T_g, the viscosity drops rapidly. This drop in viscosity is accompanied by an inversely proportional increase in atomic mobility and atomic diffusion constants. Typically D_{oi}, the intrinsic diffusion constant of the ith component of the glass rises from values of 10^{-16} to $10^{-18}\,\mathrm{cm}^2/\mathrm{s}$ at temperatures of the order of 100°C below T_g to values of 10^{-5} to $10^{-7}\,\mathrm{cm}^2/\mathrm{s}$ above T_g. Near but above T_g, the atomic diffusion constant follows the Vogel–Fulcher law $D_{oi} = A_o \exp[-B/(T - T_{go})]$ where A and B are positive constants [7]. In the low temperature regime (well below T_g) atomic diffusion in amorphous alloys is thermally activated (as is typical of solids) and roughly follows Arrhenius behavior with $D_{oi} = D_{oi}^o \exp(-Q/K_B T)$ [8–9]. The nucleation and growth of more stable crystalline phases over laboratory time scales (seconds to days) requires atomic mobility over distances of the order of the size of a critical crystalline nucleus. This requirement is generally fulfilled somewhere in the vicinity of T_g as T_g is approached from below. As such, it is apparent that formation and growth of amorphous interlayers in diffusion couples requires $T_R \lesssim T_g$. Near or above T_g, a product amorphous phase would crystallize on the time scale of a reaction experiment and the crystalline product would be observed instead of the amorphous phase.

The above factors lead to the concept of a 'kinetic window'. The Arrhenius law for atomic diffusion is applicable for temperatures below T_g. It implies that when the temperature is too low, no diffusion will occur. The system is kinetically frozen. No reaction occurs. Near or above T_g, crystallization of an amorphous product to thermodynamically preferred crystalline phases occurs in relatively short times. Reactions at these temperatures produce crystalline products. SSAR is observed when an intermediate temperature range exists. In this temperature range, there is sufficient atomic mobility to permit atomic interdiffusion to form and grow an amorphous interlayer, but not so much mobility as to permit nucleation and growth of more stable crystalline phases within that layer. This temperature interval thus presents a 'kinetic window' of opportunity with respect to amorphous interlayer growth. This kinetic window will be further characterized in section 20.3.

Figure 20.1(a) shows a cross-sectional electron micrograph of an amorphous interlayer grown in a bilayer diffusion couple of Zr and Ni [21]. The interlayer was grown at a temperature of 300°C for a time period of 6 h. In the micrograph, the amorphous interlayer appears as a featureless gray zone between the polycrystalline layers of Ni and Zr. Note that the amorphous phase forms a rather uniform planar interlayer. At a later time (12 h at 300°C), further evolution of the diffusion couple has occurred as shown in Fig. 20.1(b). A second crystalline interlayer has appeared separating the amorphous phase from the polycrystalline Zr. The second phase has the composition ZrNi and is the equiatomic intermetallic compound found in the equilibrium phase diagram of Zr–Ni. Figure 20.2 (taken from Schroeder *et al.* [22]) shows a similar micrograph illustrating the formation of many amorphous interlayers in a multilayer diffusion couple of Zr–Co. Here the initial layers of Zr and Co are rather thin (15 nm and 7.5 nm respectively). One sees that amorphous phase formation and growth occurs at each interface in the multilayer. This latter micrograph was the earliest direct evi-

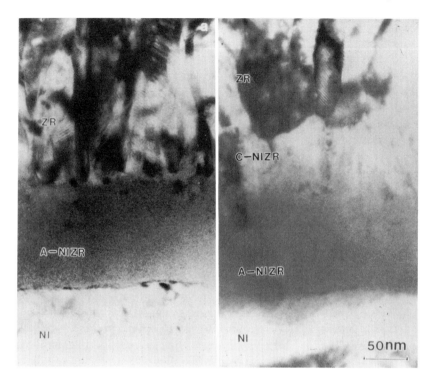

Fig. 20.1 Cross-sectional electron micrograph of a Ni–Zr diffusion couple reacted at (a) 300°C for 6 hours, showing ~80 nm thick amorphous interlayer which has grown by interdiffusion. (b) The same diffusion couple reacted at 300°C for 18 hours, showing both an amorphous interlayer and a layer of the crystalline intermetallic compound Ni–Zr (after [21]).

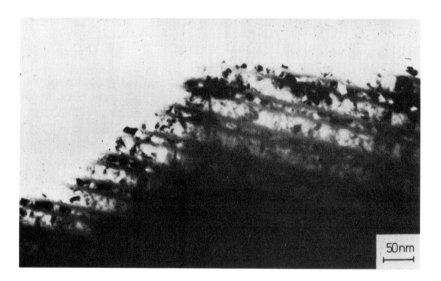

Fig. 20.2 Cross-sectional electron micrograph of Co–Zr multilayer diffusion couple consisting of many alternating layers of Co and Zr. The thin layers showing bright and dark contrast are Ni while the thicker layers are Zr. The period of the multilayer structure is about 50 nm. The diffusion couple has been reacted for 2 hours at 210°C. The gray layers separating the Co the Zr are amorphous (after [22]).

dence of the uniform planar layer morphology typical of SSAR. In early TEM studies, selected area electron diffraction was employed to prove the amorphous structure of these featureless inter-

layers. Bright field/dark field pairs of electron micrographs were also used to demonstrate the lack of crystalline diffraction in these amorphous inter layers. These studies allayed early criticism that the reaction products of SSAR might be micro-crystalline. High resolution electron micrographs reveal other interesting features of SSAR. For example, Fig. 20.3 shows a high resolution image of one interface in a Zr–Ni diffusion couple [21]. One notes that the interface separating the poly-crystalline metal layers and the amorphous phase is atomically sharp. This is characteristic of the interface between a crystal and its corresponding melt or the interface between crystals which nucleate and grow from an amorphous matrix during crystallization [6]. This provides additional evidence that the amorphous product of SSAR is indeed a new phase and not some highly disordered crystalline material. Sharp interfaces are character-istic of phases separated by a first order phase transition such as melting.

Figure 20.1 illustrates another feature of SSAR. There exists, at any given reaction temperature, a maximum critical thickness X_m, to which the amorphous interlayer will grow. For the case of Zr–Ni, this maximum thickness is determined by the onset of growth of the more stable crystalline intermetallic compound Zr–Ni. Generally speak-ing, growth of the amorphous interlayer is con-trolled by atomic diffusion across the interlayer. This, as will be seen in section 20.3, leads in the simplest case to a growth law of the form $X = (\alpha D(T)t)^{\frac{1}{2}}$ where α is a constant of order unity, X is the thickness at time t, and $D(T)$ is a temperature dependent diffusion constant which increases with T in a thermally activated manner. Careful studies using the Rutherford backscattering technique for composition profiling have been used to study the growth law [23, 24, 25]. Figure 20.4 (taken from [24]) shows the thickness of an amorphous inter-layer as a function of time at several temperatures for a Zr–Ni diffusion couple. X, as measured by Rutherford backscattering, has been plotted as a function of $t^{\frac{1}{2}}$ yielding a straight line over most of the time scale of growth. One notes that at later times (typically as X approaches X_m) one observes a characteristic 'slow down' of the reaction. A downward deviation from the '$t^{\frac{1}{2}}$-law' is seen. This has been attributed to stress relaxation in the dif-fusion couple and/or to stress relaxation effects

within the growing amorphous interlayer [24, 26]. This will be further discussed in section 20.3.

The Rutherford backscattering technique has been further used to study atomic diffusion in SSAR. In particular, the method can be used to carry out 'Kirkendall Marker' experiments from which one can identify the moving atomic species during interdiffusion. Early experiments of this type were carried out by Cheng *et al.* [27] and by Barbour *et al.* [28]. Later, more detailed experi-ments were reported by Hahn *et al.* [29]. These marker experiments clearly established that one of the two atomic species comprising the diffusion couple acted as the dominant moving species during SSAR. In the case of Zr–Ni [27–29], it was shown that the atomic mobility of Ni is at least one and probably several orders of magnitude greater than the atomic mobility of Zr during SSAR. Figure 20.5 (taken from [27]) shows the result of a Kirkendall marker experiment in which the position of the amorphous interface is measured with respect to an inert marker. The two dashed lines indicate the theoretical positions of the marker under the assumptions that Zr has no mobility (only Ni diffuses) or Ni has no mobility (only Zr diffuses). To within experimental error, the actual marker shifts agree with the curve predicted by the assumption that only Ni moves. Since these original experiments, several studies of this type have demonstrated that SSAR generally involves a dominant moving species. Further, the dominant diffusing species of a binary pair of metals is in-variably that which exhibits ADF behavior as dis-cussed above. The non-moving species is that which serves as the host metal in ADF. Naturally, ADF involves diffusion in the crystalline state whereas the growth of an amorphous interlayer during SSAR involves diffusion within the growing amorphous phase. Apparently, there is a general relationship between the ADF phenomenon and the existence of a fast moving species in SSAR. This relationship will be taken up again in section 20.3.

There is one additional phenomenon associated with the existence of a dominant moving species in SSAR. This is the formation of 'Kirkendall' voids along the interface of the diffusion couple. In poly-

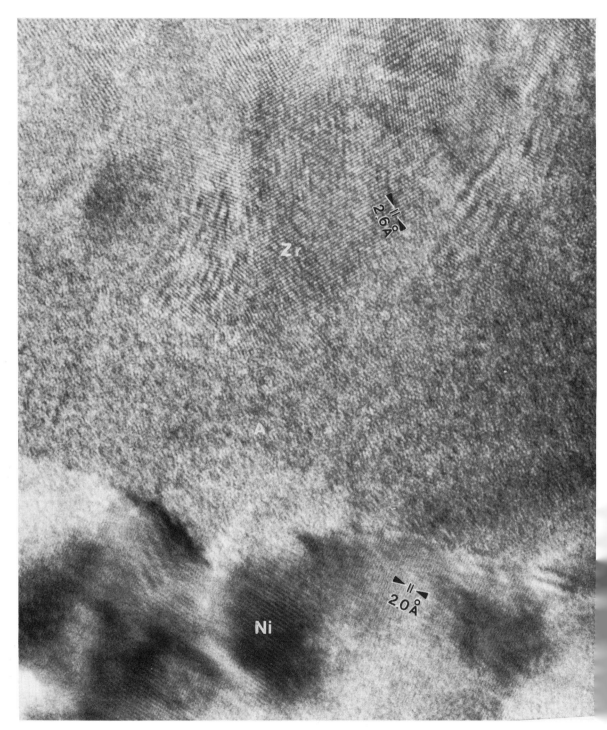

Fig. 20.3 High resolution image of an amorphous interlayer during the early stages of growth from a Ni–Zr diffusion couple. Notice the morphology of the amorphous layer at this stage of growth. Also note that the amorphous–crystalline interfaces are atomically sharp

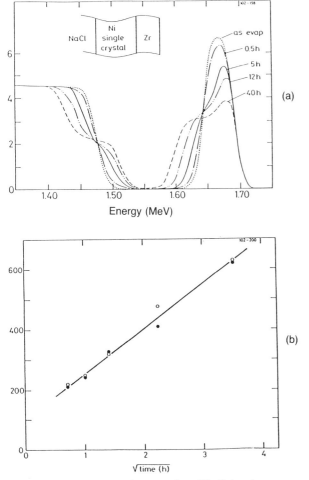

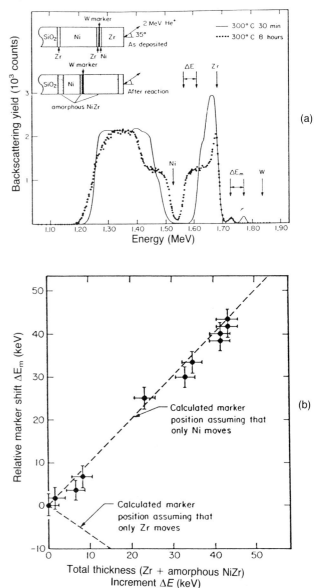

Fig. 20.4 The thickness of an amorphous Ni–Zr interlayer as a function of time as oberved by Rutherford backscattering spectroscopy. The layer growth follows a '$t^{\frac{1}{2}}$' time dependence. Upper part shows RBS spectrum. Lower plot shows thickness of amorphous layer versus '$t^{\frac{1}{2}}$' (after [24]).

Fig. 20.5 Result of a Kirkendall marker experiment carried out by Rutherford backscattering spectroscopy. The upper part shows the marker shifts (ΔE_m and ΔE) for the tungsten marker and the Zr edge from the backscattering spectrum. The lower part compares the experimental data with the expected behavior of the marker shift when only one species moves (dashed lines). Within experimental error, it is found that only Ni moves (see [27] for details).

crystalline bilayer and multilayer diffusion couples, these voids appear along the interface separating the amorphous interlayer from the fast diffusing metal (e.g. Ni, Co, etc.). The voids form during the early stages of the reaction and grow as the reaction proceeds. Their appearance in the diffusion couple can be directly related to the existence of a fast moving species (e.g. Ni in Zr–Ni couples) [21, 30]. It is argued that the diffusion of Ni atoms in the growing amorphous layer is accompanied by a counter flux of 'vacancy-like' defects. The vacancy-like defects are transported to the amorphous–Ni interface. Here, they can condense to form Kirkendall voids or are transported as vacancies into the Ni layer where they ultimately condense at a free surface or other nucleation site for voids. The formation of voids at the Ni–amorphous interface is suppressed by elimination of GBs along the Ni side of the diffusion couple. No voids are formed when a single crystal of Ni is reacted with polycrystalline Zr. Voids form by a process of nucleation. Grain boundaries apparently serve as heterogeneous nucleation sites for voids. In the absence of Ni GBs along the Ni–amorphous interface, voids do not nucleate at the interface. Instead, vacancies are transported to the free surface of the Ni layer [21].

To this point, the discussion has focused on the growth of a relatively uniform amorphous interlayer separating two initially polycrystalline layers. In a few instances, other types of morphologies have been observed during SSAR. For instance, when Y–Cu multilayers with elemental layer thicknesses in the range of 4–30 nm were reacted at temperatures in the range of 60–100°C, the formation of a thin amorphous layer was found to occur along grain boundaries of the polycrystalline Y as well as along the interface between Y and Cu grains [31]. This leads to a morphology in which individual crystallites of Y are coated with a Cu-rich amorphous layer. Presumably, the rapid amorphous phase growth along Y GBs is related to rapid GB diffusion of Cu. The fact that the amorphous interlayer wets the GB suggests that the interfacial free energy of two Y–amorphous interfaces is less than half that of the original GB which they replace when an amorphous interlayer wets the boundary. This in turn suggests an Y-amorphous interface of relatively low interfacial free energy. Similar GB amorphization has been observed in other multilayer systems such as Ti–Si and Ti–Ni [32, 17]. Here the initial silicon layers are amorphous while the Ti layers are polycrystalline and rather thin (~5–15 nm). An amorphous layer is observed to rapidly wet the GBs of Ti during the early stages of the reaction leading to a morphology in which Ti grains are surrounded

by amorphous material. As the reaction proceeds, the Ti crystallites are gradually consumed as the Ti–amorphous interface moves into the grain interiors.

The role of GBs within the two reacting layers in the initial formation of an amorphous phase has been studied in some detail. In the well studied case of Zr–Ni diffusion couples, experiments were carried out with single crystals of both Zr and Ni. In the first case [21, 33], large single crystals of Zr were prepared by recrystallization of a Zr foil. The free surface of these crystals had a typical (002) orientation (with close packed planes parallel to the surface). The recrystallized foil consists of a mosaic of such crystals (the crystals are columnar and span the thickness of the foil) with typical sizes (in the plane of the foil) ranging from 20 μm to 100 μm depending on the details of the recrystallization process. The individual crystals are typically separated by low angle (low energy) GBs. The surface of the recrystallized foil was sputter cleaned and thermally treated under UHV conditions to produce a relatively contaminant-free surface. Polycrystalline Ni was subsequently deposited onto the clean surface. When this 'single crystal Zr' diffusion couple was heated to temperatures between 300°C and 375°C for time periods of hours, no observable reaction occurred along the interface. Similar thermal treatments of polycrystalline bilayers result in amorphous interlayer growth to a thickness of 100 nm. Figure 20.6(a) shows a cross-sectional high resolution electron micrograph of the interface separating a large (~50 μm) single crystal of Zr from a smaller polycrystal of Ni following a heat treatment at 300°C. The interface remains atomically sharp with no observable indication of any reaction. Upon heating to higher temperatures (>400°C), sudden formation and growth of crystalline Zr–Ni is observed. Above 400°C, the crystalline Zr–Ni grows to micron scale thicknesses in time scales of minutes. The absence of GBs on the Zr side of the diffusion couple suppresses the formation of an amorphous interlayer. The low angle tilt boundaries which remain are apparently also ineffective in initiating amorphous phase formation. Apparently, the high angle Zr GBs present in a sputtered or vapor-deposited

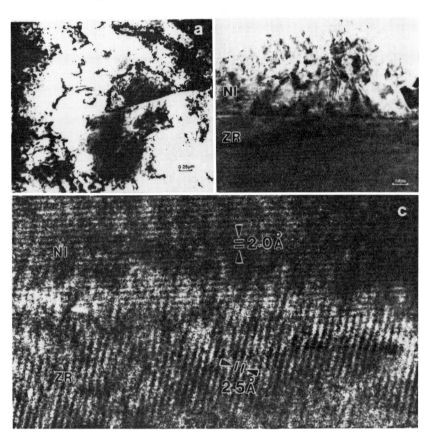

Fig. 20.6 Electron micrographs of a Ni–Zr diffusion couple consisting of a mosaic of large (25–75 μm) single crystals of Zr (a recrystallized foil) and polycrystalline Ni (sputter deposited). The diffusion couple was reacted at 300°C for 6 hours. (a) plane view showing typical microstucture of the mosaic of Zr crystals, (b) bright field cross-sectional micrograph showing no reaction after 6 hours at 300°C, (c) high resolution image of reacted interface showing Ni (111) and Zr (101) lattice fringes. The interface remains atomically sharp (after [21]).

polycrystalline Zr layer provide a catalytic or heterogeneous nucleation site for initiation of amorphous phase growth. Without these high energy GBs, SSAR is not observed. One can also form diffusion couples with Ni single crystals [21]. A single crystal of Ni can be epitaxially grown on (200) crystals of NaCl. When polycrystalline Zr is deposited onto such single crystals of Ni and the diffusion couple reacted at 300°C, an amorphous interlayer is again readily observed to grow. The absence of GBs on the Ni side of the diffusion couple does not suppress SSAR. There are subtle differences between the Ni single crystal experiment and the polycrystalline Ni–polycrystal Zr experiments. In the latter case, Kirkendall voids are observed to nucleate and grow along the original Zr–Ni interface (as mentioned above). In the case of reaction of polycrystalline Zr with single crystal Ni, no such Kirkendall voids are

observed. As already mentioned, Ni GBs serve as nucleation sites for the voids. In the absence of such GBs, the excess volume associated with a counter flux of vacancy-like defects (counter to the diffusion of Ni atoms) is transported as Ni vacancies to a free surface of the Ni crystal or other void nucleation site.

Finally, a related set of experiments are of interest. Clemens [34] has studied multilayers of Ni–Ti. When these layers are grown with sufficiently large spatial periods at ambient temperature, it is observed that the Ni layers grow with an epitaxial registration on the Ti layers. As such, partially coherent multilayers can be grown. On the other hand, when the individual layer thickness is too small (typically of order 1.5 nm), this epitaxial relationship gives way to an incoherent interface. Clemens used glancing angle X-ray scattering techniques to study the coherency of the interfaces. For multilayer periods of less than about 5 nm, where coherency is lost, the entire multilayer structure takes on a very disordered structure as observed by X-ray diffraction. Clemens explained these results in terms of the elastic strain energy associated with forming coherent interfaces. He argues that for sufficiently small layer spacings the elastic strain energy becomes so large that it drives a transformation to an incoherent interface with amorphous structure. These experiments suggest that coherency strains may play a key role in the formation of an amorphous nucleus.

20.2.2 Amorphization at metal–non-metal interfaces

The SSAR phenomenon has also been reported to occur at the interface separating a crystalline metallic layer from a non-metallic layer. The majority of these studies have concerned the interface formed between various polycrystalline metals and amorphous or crystalline silicon [2, 35–40]. Recently, examples of SSAR have also been reported at the interfaces of metals with a III–IV semiconductor (e.g. GaAs) [41–43]. The formation of metal silicides upon thermal reaction of metal layers with silicon crystals is a problem of long standing interest. This problem is of critical importance to the semiconductor device industry owing to the widespread need to form low impedance metallic contacts on devices. As such much attention has been given to the problem of understanding the sequence of phases which form and grow during thermal reaction of a metal with a silicon surface. It is interesting to note that, despite widespread early studies of silicide formation and growth, it was not until 1983 [2] that clear evidence of amorphous silicide formation was reported. This is probably due to the early absence of suitable techniques for studying interlayer growth at nanometer distances. The development of cross-sectional, high-resolution TEM methods for studying transverse sections of diffusion couples did not occur until the early 1980s. Prior to that time, techniques in use (e.g. X-ray diffraction, depth profiling by backscattering, and plane view transmission microscopy) lacked sufficient depth re-

solution to detect ultra thin (thickness of order of nanometers) interlayers in diffusion couples. It is noteworthy that the concept of an amorphous interlayer in metal–single crystal silicon diffusion couples was introduced years before its experimental observation. In their efforts to develop a model to predict which silicide would form first during interdiffusion, Walser and Bene [44] suggested that the interface between a silicon crystal and a metal could be viewed as an 'amorphous membrane'. They then suggested that the nucleation of a crystalline silicide occurred within this 'amorphous membrane'. In particular, they argued that the first phase to form would be the congruent melting silicide lying near the deepest eutectic in the binary phase diagram and having the highest melting point (when two congruent melting silicides surround the same eutectic). It is ironic not only that an amorphous interlayer phase actually does form in certain metal–silicon diffusion couples, but that it actually grows to significant thicknesses prior to the formation of a crystalline silicide. In such systems, the Walser–Bene membrane is more than a theoretical construct.

When silicon is deposited from the vapor phase at ambient temperature, amorphous silicon is formed. Vapor-deposited bilayers and multilayers of silicon with metals thus consist of polycrystalline metal and amorphous silicon. The earliest observations of amorphous silicide formation by SSAR were made on such diffusion couples [2, 32]. Similar results were also obtained earlier by Hauser when Au was diffused into amorphous Te [36]. Most silicide formation studies have involved either amorphous silicon layers (prepared by vapor deposition) or silicon single crystals onto which a metal layer is deposited. As already seen, SSAR tends to be suppressed when the non-diffusing component of the diffusion couple is initially in the form of a single crystal (recall the single crystal Zr experiment described in the previous section). For SSAR silicon–metal systems, the metal is frequently the dominant diffusing species. The absence of silicon GBs could tend to suppress the formation of an amorphous phase. This would be attributed (as in the case of Ni–Zr diffusion couples) to the absence of a suitable nucleation site for an amorphous silicide.

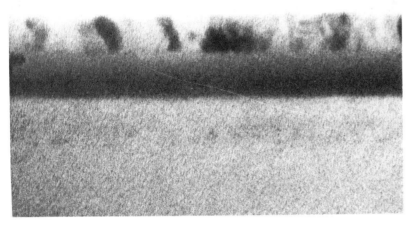

Fig. 20.7 Cross-sectional electron micrographs of a polycrystalline Rh film reacted with a (100) single crystal of Si. The top layer is polycrystalline Rh, the gray center layer an amorphous Rh-silicide, and the lower layer the single crystal of Si. In (A) the layer was reacted at 200°C. In (B), the layer was further reacted at 260°C for an equal time. The amorphous interlayer is seen to thicken as the reaction proceeds. See ref. [2b] for details.

Figure 20.7 shows an example of an amorphous silicide formed by reaction of a (100) oriented Si crystal with a polycrystalline Rh film (taken from [2b]). As in the case of typical metal–metal systems, the amorphous layer is planar and uniform. It is also interesting that the interface between silicon and the amorphous silicide appears to be atomically sharp. This suggests that silicon (a covalently bonded non-metallic phase with four-fold co-ordinated silicon atoms) is distinctly different from an amorphous silicide (a metallically bonded system with higher atomic coordination number). These two phases are apparently connected by a discontinuous phase transformation. As such, when in contact, they form a sharp interface as opposed to a continuous diffuse interface. The amorphous silicide and the crystalline silicon are distinct phases as are the crystalline and amorphous phases in metal–metal diffusion couples.

An example of amorphous silicide growth in a diffusion couple consisting of a Ni film deposited on a layer of amorphous Si (initially deposited on a Si substrate) is shown in Fig. 20.8. The results are taken from ref. [35]. The diffusion couple is reacted by rapid thermal annealing at 350°C for 2–10 sec. The electron micrograph reveals that an amorphous interlayer grows concurrently with a crystalline Ni_2Si layer. The two phases thicken simultaneously as the diffusion couple evolves. In this case, the formation of an amorphous interlayer is immediately followed by formation of a crystalline intermetallic layer. Both phases are always observed at each stage of growth.

Following the analogy with metallic bilayers, one would expect that SSAR is possible with silicon single crystals in the case where silicon becomes the moving species (recall the Ni single crystal–Zr experiments described in the last section). In fact, solid state amorphization has been observed in diffusion couples consisting of silicon single

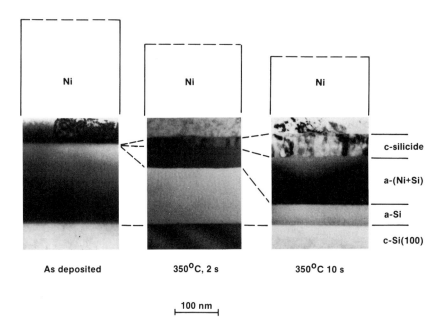

As deposited   350°C, 2 s   350°C 10 s

├─ 100 nm ─┤

Fig. 20.8 Concurrent growth of amorphous interlayer and crystalline Ni$_2$Si silicide layer (the layers are adjacent) in an amorphous–Si–polycrystalline–Ni diffusion couple (after [35]).

crystals reacted with several polycrystalline metals. In particular, when an early transition metal (e.g. Nb, Zr, Ta) is chosen where Si becomes the dominant moving species, one observes solid state amorphization with single crystals of Si [39]. Note, however, that the maximum thickness of the amorphous layer in these cases appears to be very small (2–5 nm). Table 20.2 summarizes these results together with other results on silicon metal systems. The table is divided into four sections. The first refers to amorphous silicon or single crystal Si reacted with a metal film. The second involves an early experiment on amorphous Te. The third and fourth sections deal with experiments on compound semiconductors (III–V systems) reacted with metals. In this latter section, one notes, for example, that late transition metals such as Co, Pt, and Pd form amorphous phases when reacted with certain III–V semiconductors such as GaAs and InP [41–43]. These reactions are of particular interest in that the reaction appears to be initiated (ref. [41]) at near or slightly above room temperature. When a clean surface of a single crystal of GaAs is covered by deposition of a poly-crystalline platinum film at room temperature, an amorphous interlayer is spontaneously formed. Much higher reaction temperatures are required to

form crystalline interlayer phases. This low reaction temperature suggests that the interface separating the III–V semiconductor and the metal is rather unstable. Such experiments suggest that certain heterointerfaces may be intrinsically unstable against collapse into a glass. Again, this is reminiscent of the Walser–Bene amorphous membrane concept. Further, it is related to the observation by Clemens [34] that coherent interfaces become unstable with respect to disordered interfaces when elastic energies of multilayers become sufficiently large.

20.2.3 Externally driven reactions – mechanical alloying, irradiation, etc.

The interdiffusion reaction which leads to formation of an amorphous interlayer in diffusion couples can be induced by methods other than simple heating. It can be argued to varying degrees that such processes are athermal. That is, the role of increased temperature alone can be argued to be less important than other processes which are induced by external driving forces in leading to formation of the amorphous interlayer. Here, we give examples of two types of such 'externally driven' diffusion couples. These include mech anically driven systems and irradiation driven systems.

Mechanical code formation of a physical mixture of metal powders has been observed to lead to the formation of an amorphous alloy. Experiments of this type fall into two categories depending upon the intensity of the mechanical deformation process which induces the reaction of the constituents. We refer to these experiments as either slow deformation experiments (e.g. cold rolling) or fast deformation experiments (e.g. high energy ball milling). The experiments can be at least qualitatively distinguished according to whether the process of deformation is isothermal or adiabatic. It is well known that deformation at relatively low strain rates ($\dot{\varepsilon} \sim 10^{-6}\,\text{s}^{-1}-10^{2}\,\text{s}^{-1}$) is roughly isothermal (local temperature changes within the deforming metal are typically less than 100 K) while deformation at higher strain rates ($\dot{\varepsilon} \sim 10^{3}\,\text{s}^{-1}-10^{6}\,\text{s}^{-1}$) leads to formation of localized adiabatic shear bands in which temperatures may rise substantially ($\Delta T \sim 100-3000\,\text{K}$) [45]. For example, shock waves can produce local melting in rather refractory metals [46]. We begin with a discussion of experiments at relatively low strain rates.

Atzmon et al. [47, 48] carried out relatively slow deformation experiments in which thin sheets of two metals (e.g. Ni–Zr and Cu–Er) were codeformed in a conventional rolling mill to form a laminated composite. Successive folding and rolling of the composite leads to an ultrafine laminated microstructure (layer thickness in the range of 10–50 nm) consisting of alternating layers of the metals. As the codeformation process proceeds an amorphous interlayer is observed to form and grow along the interfaces separating the metals despite the fact that the rolling process is carried out at ambient temperature. An analysis carried out by Atzmon et al. suggests that the local temperatures achieved in the vicinity of highly strained regions during this relatively 'slow' deformation of the composite are far lower than T_R (Table 20.1), the temperature at which thermally treated diffusion couples display a reaction. However, this point is somewhat controversial. In more recent studies of cold rolled Zr–Ni composites, Bordeaux et al. have argued that rather larger temperature

increases (of order 150 K) may occur [49]. Atzmon et al. argued that the high density of dislocations produced during plastic deformation may lead to enhanced interdiffusion of the metals at ambient temperature. Dislocations are known to act as short circuit paths for diffusion. Furthermore, deformation to large plastic strains is known to result in a comminution of the grain size of metals. This is a consequence of the collapse of dislocations moving through single crystals to form dislocation walls. These walls are in fact low angle GBs. With further deformation, additional dislocations are attracted to these boundaries. Ultimately, high angle GBs are formed, and these may act as short circuit pathways for diffusion. The activation energy for GB diffusion in metals is known to be typically half the activation energy from bulk diffusion [50]. This may enhance the interdiffusion reaction. In any case, the microstructure observed and the uniformity of the amorphous interlayer formed in the above experiments are qualitatively similar to those observed in thin film multilayers. Wong et al. [51, 52], Schultz et al. [53, 54], and Bordeaux et al. [49] have all shown that such composites can be codeformed to obtain layer thickness in the range of 10–50 nm and then subsequently thermally reacted at temperatures near 300°C. The subsequent reaction leads to growth of the already formed amorphous interlayer following a growth law nearly identical to that found in thin film bilayers and multilayers upon heating [54]. In Cu–Er laminated composites, Atzmon et al. were able to demonstrate complete amorphization of the bulk composite following extensive and repeated deformation of the sample at ambient temperature [48]. To explain these results apparently requires some form of defect-enhanced diffusion at low temperatures. It is thus some combination of defect-enhanced diffusion together with deformation-induced local temperature increases that leads to amorphous interlayer growth.

The process of amorphous interlayer formation along a single Zr–Ni interface during codeformation of a bulk Zr and Ni in contact was recently studied in greater detail by Mazonni et al. [55]. They used an impact-loading machine to deform a nickel plate sandwiched between two Zr plates. The plates have thicknesses in the mm range.

By varying the load, they were able to produce deformation at varying strain rates. Furthermore, by using the junction of the two plates as a thermocouple, they were able to measure the average temperature along the interface during deformation of the plates. The loads are applied during a time scale of order 50 ms and produce deformation of both the Zr and Ni plates at a typical strain rate of order $10-100 \, s^{-1}$. Along the interface, a thin and rather uniform amorphous layer (thickness ~5–10 nm) was observed to form. The average interface temperature was observed to rise 60–80 K above room temperature during the deformation. The authors argue, based on the measured temperature increase, that thermal interdiffusion alone cannot explain their results. Further in a series of experiments at differing loads and strain rates, they show that growth rate of the amorphous interlayer is correlated to the strain rate. This, they argue, shows that interdiffusion is enhanced during plastic deformation by an amount which increases with the deformation rate. Their experiments provide direct and convincing evidence of the role of defect-enhanced diffusion in deformation induced amorphous interlayer growth.

High energy ball milling of a physical mixture of metal powders also leads to amorphous phase formation as first reported by Yermakov *et al.* [56] and Koch *et al.* [57]. In these experiments, metal powder mixtures alloyed during collisions of centimeter size hardened steel or tungsten carbide balls. The balls move at typical velocities of the order of 1–10 m/s. Powder grains are deformed, cold welded together, fractured, etc. during the process. This produced particles consisting of laminated domains of the two metals which become progressively refined as the milling proceeds. Deformation of individual grains at strain rates of $10^3 \, s^{-1}$ or higher leads to rather localized shear bands (typical widths of order 1 μm) in which temperatures may rise appreciably during a deformation event. The magnitude of the temperature increase within the shear bands is controversial. Estimates range from tens of degrees K to hundreds of degrees [58, 59]. If the process is stopped at intermediate stages, amorphous interlayers are found to form at the interfaces of the two metals. Protracted ball milling leads to comminution of the metal laminae, growth of the amorphous interlayers, and ultimately to a completely amorphous product powder. The relative roles played by thermal diffusion and defect-enhanced diffusion in the growth of the amorphous layers is the subject of ongoing debate. The results may depend on the mechanical properties of the two metals. More experiments of the type performed by Martelli *et al.* [55] are required to clarify these points.

Finally, we mention radiation driven systems. Ion beam irradiation of thin film diffusion couples is known to produce atomic mixing along the interface of two metals [60]. The atomic mixing process is induced by collision cascades in which atoms are displaced by interatomic collisions. In earlier models of this process, analysis of the mixing process was based on ballistic models. These models assume that the majority of atomic displacement occur as a result of collisions of relatively high energy particles [61]. When an atom is displaced this leaves a vacant site. The displaced atom either comes to rest at an interstitial site or ultimately recombines with a vacancy. A vacancy–interstitial pair is referred to as a Frenkel defect or Frenkel pair. More recent models are based on the evolution of the collision cascade into a thermal spike [62, 63] as originally proposed by Vineyard [64]. The latter models predict that the majority of atomic displacement occurs during the late stages of cascade development when small (of the order of 3–10 nm regions become thermalized at temperatures of the order of $10^4 \, K$. These hot regions are subsequently quenched to ambient temperature by heat conduction to the surrounding medium at cooling rates ranging from 10^{10} to $10^{12} \, K/s$. In this picture atoms in local regions along the interface are suddenly heated to very high temperature, atom configurations evolve at this temperature for very short times ($\sim 10^{-11} \, s$) and are then quenched ambient temperature at very high rates. The entire process can be described by an effective diffusion constant which reflects the number of atomic displacements induced per incident ion. The process can be viewed as 'athermal diffusion' since the ambient temperature has little to do with the amount of actual diffusive mixing that occurs. For the present purposes, we can view the process as externally driven athermal mixing of atoms along the interface between two materials. In general, one distinguishes two regimes of ambient temperature in ion mixing. When defects produced by the collision cascade (the Frenkel pairs) are mobile at the ambient temperature of the experiment, one observes thermally activated migration of the defects. This is referred to as radiation-enhanced diffusion. When the ambient temperature is sufficiently low, the Frenkel defects produced in the cascade are configurationally frozen and the chemical configuration is fixed following the cool-down phase of the thermal spike. Under these conditions, diffusional mixing does not occur at the ambient temperature and the process is referred to as 'temperature independent mixing'.

Ion mixing of metal layers along an interface frequently leads to formation of an amorphous interlayer [65–69]. As a general rule, metals which do not form extended terminal solid solutions will tend to form an amorphous interlayer when mixing is carried out at low temperatures [60]. As will be seen in the next section, this can be explained in terms of a polymorphic melting diagram [70] in which the 'melting point' of a solid solution with fixed composition is determined as a function of the composition. Ion mixing produces metastable solid solutions with compositions lying beyond the equilibrium solubility limits. For sufficiently extended solutions, the 'polymorphic melting point' falls below the ambient temperature and below the T_g of the liquid phase. Under these circumstances, the solution can 'melt' to a liquid below its T_g. In other words, the solution undergoes a crystal-to-glass transformation.

20.3 PHASE FORMATION AND GROWTH IN DIFFUSION COUPLES – THEORETICAL CONSIDERATIONS

The formation and growth of a new phase in a binary diffusion couple involves both thermodynamic and kinetic principles. One can describe the thermodynamics of the problem in terms of free energy diagrams. Such diagrams give the free energy of all relevant phases as functions of temperature, composition, and pressure. An analysis of these diagrams permits one to determine the change in bulk Gibbs free energy associated with

changes of phase as functions of composition and temperature. All phase changes and processes which lead to a reduction in the free energy of the diffusion couple are thermodynamically allowed. This is discussed in section 20.3.1. To predict what phases actually form and how they grow, one must also analyze the kinetic aspects of the problem. This is discussed in sections 20.3.2 and 20.3.3.

20.3.1 Thermodynamics, free energy diagrams, and polymorphic diagrams

The construction of free energy diagrams for solid and liquid phases is discussed in many texts [71, 72]. For the present purposes, we will use the binary Ni–Zr system as an example for the purposes of illustration. The equilibrium phase diagram at ambient pressure for this system is shown in Fig. 20.9. Free energy diagrams for the various phases have been computed by Saunders and Miodownik [73] using the Calphad approach. Figure 20.10 shows the free energies of mixing of the various solid solutions (hcp Zr-base, bcc Zr-base, and fcc Ni-base) as functions of composition at a temperature of 550 K. The free energy of mixing is measured with respect to the free energy of a physical mixture of Ni and Zr. This temperature is typical of the conditions under which an amorphous interlayer grows. The intermetallic compounds $NiZr_2$, $NiZr$, etc., are depicted as line compounds in the free energy diagram. This amounts to ignoring the homogeneity range of the compounds. The free energy of the liquid phase is also shown as a function of composition at $T = 550$ K. In fact, the liquid is a glass at this temperature since the T_g of liquid Ni–Zr typically lies above 550 K. One readily sees that the free energy of an amorphous phase near the equiatomic composition is lower than that of a physical mixture of Ni and Zr with the same overall composition by an amount $\Delta G = -40$ kJ/mole. An hcp solution would have $\Delta G \sim -20$ kJ/mole at the same composition. A thermodynamic driving force exists for forming both phases from a physical mixture of Ni and Zr. Both processes are thermodynamically allowed. Formation of the intermetallic compound NiZr is accompanied by a free energy drop, $\Delta G \sim -50$ kJ/mole. This is the lowest free energy state for an

Ni-Zr Phase Diagram

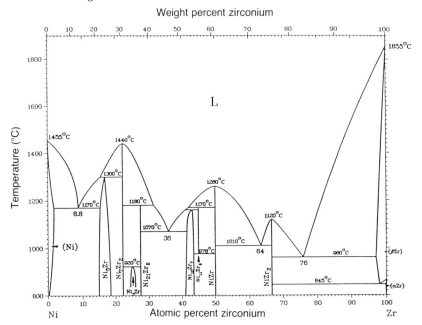

Fig. 20.9 Equilbrium phase diagram of the Ni–Zr system.

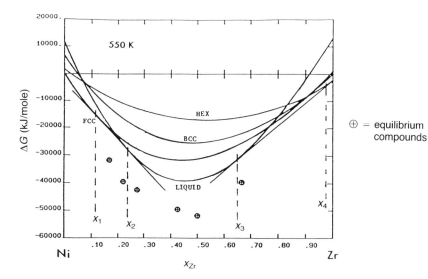

⊕ = equilibrium compounds

Fig. 20.10 Free energy diagram for the Ni–Zr system at 300°C showing the free energy of mixing as a function of composition for the hcp, bcc, fcc, and liquid–amorphous phases. Intermetallic compounds are shown by dots and are assumed to be line compounds. The common tangents for the hcp–liquid, and liquid–fcc metastable equilbrium are shown. These define compositions x_1, x_2, x_3, (after [73]).

equiatomic mixture of Ni and Zr and thus represents thermodynamic equilibrium.

Now suppose that intermetallic compounds are suppressed in Fig. 20.10. Consider the metastable equilibrium between an amorphous phase and the terminal solutions of α-Zr (bcc Zr) and Ni. Under these conditions, the common tangent construction yields a metastable equilibrium state which consists of a single phase α-solution for $x_{Zr} > x_4$, a two phase region of liquid (glass)/α-solution for x_3

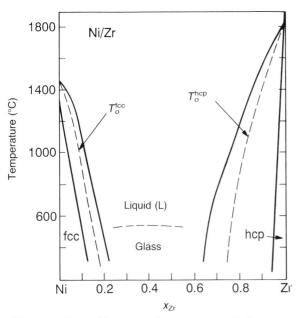

Fig. 20.11 Metastable eutectic diagram for the Ni–Zr system assuming that the intermetallic compounds are suppressed. The metastable liquidus and solidus curves are defined by x_1, x_2, etc. of Fig. 20.10. Also shown are the $T_o(x)$ curves for the hcp and fcc phases. These curves define the polymorphic phase diagram.

cleation barrier for the compound is related to the interfacial free energy of the between the inter-metallic compounds and the phase with which it is in contact. Under suitable conditions, the formation of intermetallic compounds is suppressed by this nucleation barrier. When this is true and when chemical interdiffusion is still possible, then the metastable simple eutectic diagram may be appropriate.

It is often useful to define another type of non-equilibrium phase diagram. For sufficiently low temperatures, chemical diffusion is suppressed and a system can fall out of chemical equilibrium. Starting with a chemically homogeneous phase, one can consider phase transformations in the absence of chemical diffusion. Such transformations are referred to as polymorphic phase transformations and must occur without change of composition. Under this type of metastable equilibrium, the homogeneous phase of lowest free energy becomes the metastable equilibrium phase. If one again ignores intermetallic compounds, one finds (Fig. 20.10) that the single phase of lowest free energy is determined by the crossing of the free energy curves of the terminal solution phases and that of the liquid (glass) phase. When these crossing points are taken as function of temperature, one obtains $T_o(x)$ curves for the terminal solution phases. These are also illustrated in Fig. 20.11 for the Ni–Zr system. The $T_o(x)$ curve for a solid solution defines the melting point of the solution in the absence of diffusion. Above $T_o(x)$, a solid solution of composition x will 'melt' without composition change to a liquid (glass). In the case that the liquid is a glass, we have a crystal-to-glass transformation. Polymorphic melting curves depend not only on composition but also on pressure. The general form of these curves is discussed by Fecht *et al.* [70].

20.3.2 Diffusion controlled growth of an amorphous layer

Given the free energy diagrams and metastable diagrams of Fig. 20.10 and the metastable phase diagram of Fig. 20.11, one is in a position to describe the thermodynamic aspects of amorphous interlayer growth. When an amorphous interlayer

$x_{Zr} < x_4$, a single phase liquid (glass) for $x_2 < x_{Zr} < x_3$, a two-phase Ni–liquid (glass) region for $x_1 < x_{Zr} < x_2$, and a single phase fcc Ni-solution for $x_{Zr} < x_1$. Thus, an α-Zr solid solution of composition x_4 can be in chemical equilibrium with an amorphous phase of composition x_3 etc. A metastable phase diagram based on the assumption that intermetallic compounds do not form is shown in Fig. 20.11. The diagram is of the simple eutectic type. It differs from ordinary equilibrium simple eutectic diagrams in that the eutectic temperature T_E is negative. Greer has discussed such meta-stable eutectic diagrams in detail [74]. This negative eutectic feature is characteristic of binary systems in which the free energy of mixing of the amorph-ous phase is more negative than that of both solid solution phases over some range of composition. The diagram assumes that intermetallic compound phases *cannot form*! In fact, when reaction occurs in a diffusion couple, an intermetallic compound must form by a process of nucleation. The nu-

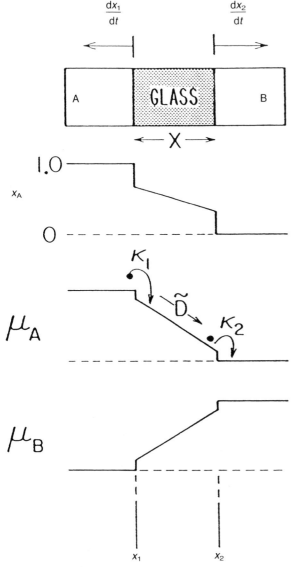

Fig. 20.12 Schematic illustration of the various quantities used to describe the growth of an amorphous interlayer. The composition profiles, chemical potential profile, etc. are shown. The quantity x_A is the concentration of element A in the diffusion couple (in the text $x_A = 1 - x$) while X_1 and X_2 are the positions of the amorphous–crystal interfaces (after [77]).

forms in a diffusion couple in the absence of intermetallic compound formation, one uses the metastable phase diagram of Fig. 20.11 to describe the thermodynamics. The situation is depicted in

Fig. 20.12. Here A and B refer to the pure metals (e.g. Ni and Zr). Several features of the problem are illustrated in the figure. The composition profile shows discontinuities at the interface separating the glass (amorphous) layer from the crystalline metals. If the interfaces are in chemical equilibrium, then the compositions of the solutions and amorphous phases along the interface are given by x_1, x_2, x_3, and x_4 of Fig. 20.10 (also Fig. 20.11). These values depend on the temperature at which growth takes place (Fig. 20.11). The growth of the interlayer can be limited by the mobility of the two interfaces or by diffusive transport over the amorphous interlayer [3, 75–78]. The interface mobility can be described by mobility parameters κ_1 and κ_2. These can be defined in terms of the deviation of the interface from chemical equilibrium (this is the driving force for interface motion). In particular, one can define

$$dX_1/dt = \text{velocity of interface No. 1}$$
$$= \kappa_1(x - x_2)$$

where x is the composition of the amorphous layer at the interface with A, and x_1 is the composition which the amorphous layer has when in equilibrium with the A solution (Fig. 20.10). One can similarly define κ_2. The chemical potential drop of A and B atoms ($\Delta\mu_A$ and $\Delta\mu_B$) along the interface (Fig. 20.12) is proportional to $(x - x_2)$ and κ is a response coefficient to this chemical potential drop. The chemical potential gradient over the amorphous interlayer can be determined from Fig 20.10. For interfaces near chemical equilibrium the drop in chemical potential of an A atom over the amorphous layer

$$\mu_a^A(X_1) - \mu_a^A(X_2) = \Delta\mu_a^A$$

is given in Fig. 20.10. Here $\mu_a^A(X_1)$ is the chemical potential of A atoms in the amorphous layer along interface No. 1 and similarly for $\mu_a^A(X_2)$. The chemical potential variations provide the driving force for atomic diffusion. In the simple case where $\kappa_1/x_1 = \kappa_2/x_4$ (x_1 and x_4 given as in Figs. 20.10 and 20.11), interdiffusion over the growing interlayer is described by the diffusion equation

$$\partial x/\partial t = \tilde{D}\partial^2 x/\partial X^2 \qquad (20.1$$

where $\widetilde{D}$ is the chemical interdiffusion constant, while interface motion is governed by the continuity equation

$$\widetilde{D}\partial x/\partial X = x(X_2)\partial X_2/\partial t$$

and the equation [77]

$$\partial X_2/\partial t = (x(X_2) - x_2)\kappa_2 \qquad (20.2)$$

The solutions of these equations have the following asymptotic forms [75–78]:

$$X_2 = -\widetilde{D}/\kappa_2 + (2a\widetilde{D}t)^{\frac{1}{2}}, \quad \text{as } t \to \infty \qquad (20.3)$$

and

$$X_2 = A\kappa_2 t, \quad \text{as } t \to 0 \qquad (20.4)$$

where a and A are numerical constants having values near unity. The interlayer grows linearly at short times with a rate determined by the interface response parameter κ. For long times, a $t^{\frac{1}{2}}$ growth law is observed. These regimes are referred to as interface-limited and diffusion-limited growth. Since κ has the dimensions of velocity, while $\widetilde{D}$ has the dimensions (length^2/s), the ratio $\widetilde{D}/\kappa_2$ has the dimensions of length. The latter is the characteristic thickness at which the interlayer changes form interface-limited to diffusion-limited growth. Experiments on amorphous interlayer growth have shown that this characteristic length scale is rather small (on the order of several nm). Thus, diffusion-limited growth is generally found over observable ranges of thickness. An example has already been shown in Fig. 20.4 for the NiZr system. Here the thickness of the amorphous layer is plotted as a function of $t^{\frac{1}{2}}$, yielding a straight line over a broad range of growth times. The thickness was determined by Rutherford Backscattering Spectroscopy (RBS). According to eq (20.4), the slope of the line can be related to $\widetilde{D}$, the interdiffusion constant. For the Ni–Zr system, Kirkendall marker experiments have shown that Ni is the dominant moving species (Fig. 20.5). As such, $\widetilde{D}$ can be related to the intrinsic diffusion constant of Ni in the growing amorphous layer.

Assuming $a = 1$, one can determine the diffusion constant of Ni in the amorphous layer. Hahn and

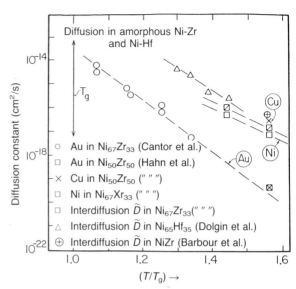

Fig. 20.13 Summary of chemical interdiffusion constants $\widetilde{D}_{\text{Ni}-\text{Zr}}$, and $\widetilde{D}_{\text{Ni}-\text{Hf}}$, and intrinsic Ni diffusion constant D_{Ni} obtained from RBS studies of amorphous interlayer growth. Also shown are intrinsic diffusion constants (after Hahn and Averback and Cantor) for Ni, Cu, Au, etc. in amorphous layers of average composition equal to that of the amorphous interlayer formed by SSAT.

Averback [29], Cantor *et al.* [79], Dolgin *et al.* [77], Barbour *et al.* [25], and Van Rossum *et al.* [23] have used this method to determine interdiffusion constants in growing interlayers of amorphous Ni–Zr and Ni–Hf. Hahn and Averback have also measured the intrinsic diffusion constant of Ni, Au, and Cu, in as deposited amorphous layers of Ni–Zr of similar composition. These data are summarized in Fig. 20.13. The figure is an Arrhenius plot of the diffusion constants versus T_g/T. Here, T_g for amorphous Ni–Zr near the equiatomic composition is roughly 550°C. T_g is about 600°C for amorphous NiHf near the equiatomic composition. One notes that the interdiffusion constants $\widetilde{D}$ obtained for amorphous interlayer growth in Ni–Zr are in very close agreement with the intrinsic diffusion constant of Ni in an amorphous layer of the same average composition as the growing interlayer. One obtains an activation energy for interdiffusion (e.g. also for intrinsic diffusion of Ni) of about 1.2 eV/atom for Ni in amorphous $\text{Ni}_{\sim 50}\text{Zr}_{\sim 50}$.

Such studies as those described above using RBS have convincingly shown that amorphous interlayers grow by diffusion-limited growth. Further, such experiments have been useful in determining the magnitude of the chemical interdiffusion constant $\widetilde{D}$, as well as the intrinsic diffusion constant of the fast moving species (e.g. D_{Ni}^o) in amorphous alloys.

Growth of amorphous interlayers has also been studied by differential scanning calorimetry (DSC) [51–52, 80–81]. In fact, these studies represent the first use of the DSC technique to study thin film reactions. In the DSC technique, a diffusion couple is heated at a constant heating rate. One measures the rate at which heat (enthalpy) is evolved by the diffusion couple during the interdiffusion reaction, $\dot{H}$. When an amorphous interlayer is growing, the heat released corresponds to the enthalpy of formation of the growing amorphous layer from the elemental components (e.g. Ni and Zr). In other words, the experiment measures the enthalpy of mixing of the components in the amorphous phase. In practice, a multilayer stack consisting of many diffusion couples is used in order to obtain an adequate signal from the calorimeter. Cotts *et al.* [80], and later Highmore *et al.* [81] have studied Ni–Zr diffusion couples in this manner. In such an experiment, the total integrated enthalpy released by the diffusion couple is given by

$$H = \int_{t=0}^{t} \dot{H}(t')\,dt' \qquad (20.5)$$

and is proportional to the total thickness of the amorphous layer which has grown.

On the other hand $\dot{H}$ is proportional to the rate of growth of the amorphous layer. According to eq. (20.3),

$$H \sim (2a\widetilde{D})^{\frac{1}{2}} t^{\frac{1}{2}} \sim X \qquad (20.6)$$

while

$$\dot{H} \sim \tfrac{1}{2}(2a\widetilde{D})^{\frac{1}{2}} t^{-\frac{1}{2}} \sim X^{-1} \qquad (20.7)$$

so that the product $H\dot{H}$ has the form

$$H\dot{H} \sim a\widetilde{D} = a\widetilde{D}_o e^{-Q/k_B T} \qquad (20.8)$$

Thus, if one plots $\ln(H\dot{H})$ versus T^{-1} from DSC data, one should obtain an Arrhenius plot giving

the activation energy for interdiffusion. This has indeed been found to be the case. Figure 20.14, taken from [80], illustrates this behavior. A linear fit to the data yields an activation energy of 1.05 eV/atom, in good agreement with that obtained by RBS studies of the interdiffusion reaction. This shows that reaction calorimetry can be used to directly monitor amorphous interlayer growth. The straight line obtained in Fig. 20.14 further verifies that growth is diffusion limited.

In addition to allowing measurements of $\widetilde{D}$ calorimetry studies also allow direct determination of the total enthalpy evolved when Ni and Zr layers are completely converted to amorphous phase. This is the enthalpy of mixing of the amorphous phase (referred to the starting pure metals). Cotts *et al.* [51–52, 80] showed, for example, that the enthalpy of mixing of an amorphous alloy of composition $Ni_{68}Zr_{32}$ is 35 ± 5 kJ/mole. Since the entropy of mixing of this alloy at a temperature of 300°C (the typical growth temperature) is much smaller, one can approximate

$$\Delta G_{mix} = \Delta H_{mix} + T\Delta S_{mix} \approx \Delta H mix \qquad (20.9)$$

Direct comparison with the value of ΔG_{mix} obtained by the Calphad calculation of Saunders and Miodownik (Fig. 20.10) gives excellent agreement. It is thus directly established that the large negative heat of chemical mixing provides the primary thermodynamic driving force for amorphous interlayer growth.

Other types of measurements have been used to monitor amorphous interlayer growth. For instance, one can monitor the in-plane resistance of a thin film diffusion couple. It can be shown that the change in resistance due to the growth of an amorphous interlayer is proportional to the interlayer thickness. Samwer *et al.* [82], Dolgin [77] and Schwarz *et al.* [83] have used this method to monitor amorphous interlayer growth. They were able to again verify that growth is diffusion limited.

In conclusion, it is now well established that the growth of planar amorphous interlayers is essentially a diffusion-limited process. Here we have focused on the Ni–Zr system. Such experiments have in fact been performed for a variety of systems and show similar results. Monitoring of diffusion

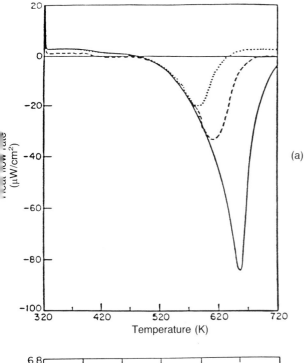

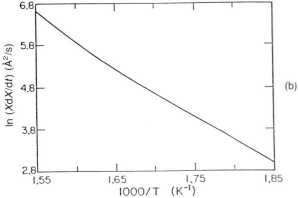

Fig. 20.14 (a) Differential scanning calorimetry scans of various Ni–Zr multilayer diffusion couples heated at a constant rate of 20°C/s. The heat flow rate, $\dot{H}$, has been normalized by the total Ni–Zr interfacial area in the diffusion couple. Dotted line corresponds to an individual Ni layer thickness of 30 nm and an individual Zr layer thickness of 45 nm, the dashed line to 50 nm–80 nm, and the solid line to 100 nm–100 nm, respectively. (b) A plot of ln (XX) (note that X ~ H and X are obtained from experiment) versus $(1/T)$ for the third sample in the upper figure. See text for further explanation. The slope of the curve gives the activation energy for interdiffusion of Ni and Zr in the amorphous layer (after [80]).

limited interlayer growth provides a direct measurement of the interdiffusion constant $\tilde{D}$. One can, for instance, obtain the activation energy for interdiffusion from such experiments. On the other hand, the description of the growth law for an amorphous layer leaves unanswered a more fundamental question. Why does an amorphous layer grow instead of a more stable intermetallic compound? To answer this question requires that we examine the nucleation behavior of competing phases in the diffusion couple. This is explored in the next section.

20.3.3 Kinetic constraints and the phase sequence problem

In section 20.2.1, it was shown that many possible reactions can occur in a diffusion couple. Any new phase for which $\Delta G < 0$ is thermodynamically allowed. In fact, the thermodynamic driving force for intermetallic compounds (e.g. crystalline NiZr) is greater than that for formation of an amorphous phase

$$\Delta G_{c-\mathrm{NiZr}} < \Delta G_{a-\mathrm{NiZr}} \qquad (20.10)$$

where ΔG is the negative free energy change associated with forming crystalline NiZr ($c - \mathrm{NiZr}$) and amorphous NiZr ($a - \mathrm{NiZr}$). Why then does the amorphous phase form? Condition (20.10) is equivalent to saying that the amorphous phase is metastable. It can lower its free energy by crystallizing to the intermetallic compound. Why should a metastable phase form during the evolution of a diffusion couple? To answer this question, one must first examine the nucleation behavior of the competing phases. Other factors such as interfacial mobilities may also play an important role in the phase sequence problem. We shall examine these effects in turn.

A new phase in a diffusion couple can nucleate either homogeneously or heterogeneously. The existence of an initial hetero-interface separating the reactants offers a natural site for heterogeneous nucleation. This interface is typically incoherent since the crystal structures of the reactants are generally not well lattice matched. A typical situation for both homogeneous and heterogeneous

Homogeneous Nucleation

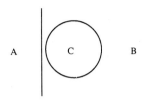

Heterogeneous nucleation – interface

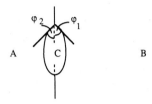

Heterogeneous nucleation – grain boundary

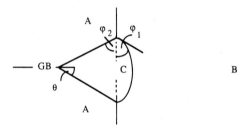

Fig. 20.15 Schematic illustration of the geometries for homogeneous and heterogeneous nucleation of a new phase, C, in a diffusion couple of A and B. See text for further explanation.

nucleation is depicted schematically in Fig. 20.15. Heterogeneous nucleation along an interface is illustrated as well as a situation where a GB exists in one of the reactant crystals along the interface. The classical theory of nucleation for such situations is well studied [84]. Here, only the salient features of heterogeneous nucleation are emphasized as they relate to the phase selection problem. The critical contact angles for heterogeneous nucleation along an interface shown in the center diagram of Fig. 20.15 are φ_1 and φ_2. These are determined by the interfacial free energies σ_{AB} = interfacial energy between initial phases A and B, σ_{AC}, and σ_{BC}, the latter being the corresponding interfacial energies

between the new phase C and the existing phases A and B. For simplicity and illustration, we can assume $\sigma_{AC} = \sigma_{BC}$ and $\sigma_{AC} < \sigma_{AB}$. Then $\varphi_1 = \varphi_2 = \varphi$, and

$$\cos\varphi = \sigma_{AB}/2\sigma_{AC} \qquad (20.11)$$

If $\sigma_{AC} = \sigma_{AB}/2$, then $\varphi = 0$ and the new phase C wets the interface. If $\sigma_{AC} < \sigma_{AB}/2$, the formation of C will be a driven process. Under either circumstance, the nucleation barrier for the new phase C will vanish. On the other hand, if $\sigma_{AB} \ll \sigma_{AC}$, then $\varphi \rightarrow \pi/2$, and we recover the case of homogeneous nucleation. This would be the case if the original interface were nearly coherent, i.e. A–B is an epitaxial structure. Heterogeneous nucleation would be effectively suppressed in this case. In general, the nucleation barrier for C, ΔG^C_{Nuc}, depends on φ. It achieves a maximum for $\varphi = \pi$ and vanishing when $\varphi \rightarrow 0$. These general features are applicable irrespective of whether C is a metastable amorphous phase or a crystalline intermetallic phase. The lower diagram of Fig. 20.15 illustrates a more complex case. The existence of a GB in A, perpendicular to the interface, permits a more complex shaped heteronucleus. With the assumptions above for σ_{AB}, σ_{AC}, and σ_{BC}, and the introduction of a GB energy σ_{GB}, the problem involves several angles, φ_1, φ_2, and θ. Generally speaking the nucleation barrier becomes a function of two of these angles. The nucleation barrier for C will vanish when either

$$\sigma_{AC} \leqslant \sigma_{AB}/2 \qquad (20.12a)$$

or

$$\sigma_{AC} \leqslant \sigma_{GB}/2 \qquad (20.12b)$$

In general, for given values of the interfacial energies, the existence of the GB leads to a lower nucleation barrier than for a simple interface. The greater the GB free energy σ_{GB}, the more favored heterogeneous nucleation will be. High energy GBs are more potent catalytic sites for heterogeneous nucleation than low energy GBs. Finally, inter sections of three GBs (each perpendicular to the original interface), provide still more favorable heterogeneous nucleation sites. As a final note some comments regarding magnitude of inter facial tensions are in order. The interfacial tension between a liquid metal and crystal of the same metal σ_{xL} (the crystal melt interfacial tension) is comparatively small for pure metals. For example, for the metal Cu, $\sigma_{xL} \approx 150 \, \text{mJ/m}^2$ [85]. A high energy GB in Cu at temperatures near the melting point has a typical energy $\sigma_{GB} \approx 600 \, \text{mJ/m}^2$. Such numbers suggest that a high energy GB may be a very potent site for nucleation of a liquid/amorphous phase. In section 20.2, it was noted that GB amorphization occurs spontaneously in certain binary diffusion couples containing high energy GBs in the reactant layers (e.g. recall the case of Cu–Y and Ti–Si referred to in sections 20.2.1 and 20.2.2). Apparently, this occurs because the free energy of a high angle GB energy in Y (for example) is more than double the interfacial energy of an amorphous Cu–Y phase in contact with an yttrium crystal. There is no nucleation barrier for amorphous phase formation in this case.

Based on the above discussion, one can understand how the formation of an amorphous interlayer is influenced by the nature of the interface separating the two reactants. The rate at which a new phase nucleates is given in general by

$$\dot{N} = \text{nucleation rate (nuclei/s} - \text{cm}^3)$$
$$= KDe^{-\Delta G_N/k_BT} \qquad (20.13)$$

where ΔG_N is the nucleation barrier and D is an atomic diffusion constant which determines the rate at which atoms rearrange diffusively. The constant, K, is determined by the geometry of the problem and other details which we shall neglect here. In the case of an amorphous phase, it is natural to assume that D is related to the interdiffusion constant, $\tilde{D}$, of Ni and Zr since formation of a nucleus of the amorphous phase requires a local composition fluctuation but may assume a variety of topological forms. In the case of nucleation of a crystalline intermetallic, this identification may be less clear. To form a nucleus of an crystalline intermetallic compound requires atomic rearrangement of both species (e.g. Ni and Zr) both topologically and chemically. This will likely require atomic mobility of both species whereas formation of an amorphous nucleus is

topologically far less restrictive. Both D and ΔG_N are required to determine the nucleation rate.

Westendorp *et al.* [33] observed no amorphous interlayer growth in Ni–Zr diffusion couples consisting of a single crystal of Zr in contact with a polycrystalline Ni layer. Meng *et al.* [21] similarly observed that polycrystalline Ni reacted with a mosaic of large ($\sim 50 \, \mu$m) Zr crystals separated by low angle tilt boundaries results in no amorphous interlayer growth. By contrast, amorphous interlayer growth is observed when polycrystalline Zr is reacted with either polycrystalline Ni or single crystal Ni. These experiments suggest that Zr GBs characterized by significant misorientation are required to nucleate the amorphous phase. Kamenetsky *et al.* [86] and Koster *et al.* [87] have reported evidence that triple junctions of Zr GBs along the interface with Ni are the likely nucleation sites for the amorphous phase. They further report that once nucleated at these GB intersections, the amorphous phase spreads rapidly along the Ni–Zr interface. The growth of an amorphous nucleus parallel to the interface apparently proceeds at much higher velocity than growth normal to the interface. This ultimately leads to rather uniform planar interlayer growth. Based on these studies, one concludes that heterogeneous nucleation of the type illustrated in Fig. 20.15 is responsible for the initial formation of the amorphous phase. The fact that high energy GBs of Zr are required (the low angle tilt boundaries in the experiments of Meng *et al.* [21] were not effective nucleation sites) suggests that a large value of σ_{GB} (Fig. 20.15(b)) is required to sufficiently lower the nucleation barrier for the amorphous phase to allow nucleation in the observed temperature ranges ($\sim 300°$C). In the case of Cu–Y [16, 31], the observation of an amorphous phase along the GBs of Y in Cu–Y multilayers sputtered at ambient temperature but otherwise unreacted suggests either a very low or near zero nucleation barrier along GBs. This in turn suggests that condition (20.12b) is close to being satisfied for high-energy Y GBs.

Assuming that nucleation of an amorphous phase is possible, there remains a second important problem. The thermodynamic driving force for competing intermetallic compounds is greater than that for amorphous phase formation (Fig. 20.10). Classical nucleation theory predicts that nucleation barriers have the general form

$$\Delta G_N \sim A(\sigma^3/\Delta g^2) \qquad (20.14)$$

where A is a geometrical constant, σ an interfacial energy, and Δg the drop in Gibbs free energy per unit volume accompanying the phase change. Since Δg is greater for intermetallic compounds, it follows that the nucleation barrier should tend to be smaller if σ were comparable. This suggests that the interfacial free energy of intermetallic phases with the reactants (e.g. Ni and Zr) tends to be greater than the interfacial free energy of an amorphous phase with the reactants. In turn, it can be argued that the free energy of hetero-interface between two crystals depends greatly on lattice matching. If mutual orientations of two crystalline phases exist along which coherent, or near coherent, interfaces are formed, then one expects a low interfacial energy. Incoherent interfaces are expected to have much higher energy. For nucleation kinetics to favor formation of an amorphous phase, there must be an absence of low energy interfaces between the intermetallic compounds and the parent crystalline layers.

A second explanation for the suppression of the nucleation of intermetallic compounds would involve the value of D in eq. (20.13). The marker experiments discussed earlier show that Ni is the dominant moving species in Ni–Zr diffusion couples and suggest that the intrinsic diffusion constant of Ni is at least an order of magnitude or more larger than that of Zr at the temperatures where interdiffusion is typically observed. The formation of the nucleus of an intermetallic compound may require atomic rearrangement of both species (Ni and Zr) over significant distances, whereas formation of an amorphous nucleus may occur through only atomic rearrangement of Ni. As such, intermetallic nucleation could be suppressed by the absence of mobility of Zr atoms. In effect, the D in eq. (20.13) would represent different atomic rearrangement processes for compound nucleation than for amorphous phase nucleation. The asymmetry in atomic mobilities discussed earlier would then enter the nucleation problem

in a manner which favors amorphous phase nucleation. We see that initial nucleation of an amorphous phase (as opposed to the crystalline intermetallic) can either be a consequence of a lower ΔG_n, or of a different rate limiting D, or some combination of both.

Meng et al. [21, 80] and Highmore et al. [81] have considered the problem of nucleation of a second intermediate phase along a moving interface separating the first nucleated phase from the parent reactants. In particular, they have examined the formation of the intermetallic compound NiZr along the interface separating a growing amorphous layer from crystalline Zr. Meng et al. proposed that heterogeneous nucleation of crystalline NiZr along this interface must involve a time scale which is related to the characteristic dimension of a critical nucleus R_c and to the velocity of the interface $v = \mathrm{d}X/\mathrm{d}t$ by

$$\tau = R_c/v \qquad (20.15)$$

where τ is the time required for the moving interface to advance over the length scale of a critical nucleus. This leads, at a given temperature, to a critical velocity for the moving interface. Below this velocity, nucleation of the intermetallic compound will occur. Diffusion limited growth at a fixed temperature leads to an interface velocity which falls like $(1/L)$ where L is the total thickness of the amorphous interlayer. Since L increases with time like $t^{\frac{1}{2}}$, one expects the velocity of the Zr-amorphous interface to decrease like $t^{-\frac{1}{2}}$. At a critical thickness L_c, nucleation of the compound becomes possible. Highmore et al. [81] have developed a similar model called the 'transient nucleation model' which makes essentially the same prediction. Both models predict a critical thickness for the amorphous layer growth. Both predictions are based on a minimum growth velocity below which nucleation and growth of the intermetallic compound replaces further growth of the amorphous layer.

Desre et al. [88] have proposed a mechanism for the suppression of nucleation of intermetallics in the case that an amorphous layer has already formed. In this model, nucleation of the intermetallic is impeded by the composition gradient in the growing amorphous interlayer. According to Figs. 20.10 and 20.12, this composition gradient is given by

$$G = \text{composition gradient} = (x_3 - x_2)/(X_2 - X_1)$$
$$= (x_3 - x_2)/L \qquad (20.16)$$

where L is the total thickness of the amorphous interlayer which has grown. This gradient decreases like $L^{-1} \sim t^{-\frac{1}{2}}$ in the case of diffusion limited growth. Desre *et al.* [88] show that this gradient increases the nucleation barrier for intermetallic compounds within the amorphous layer. Their argument applies to both homogeneous nucleation of the intermetallic phase within the amorphous layer as well as heterogeneous nucleation of the intermetallic along the interfaces separating the amorphous layer from the reactant phases (interfaces at X_1 and X_2) in Fig. 20.12. Their analysis is involved but essentially requires the evaluation of the influence of the composition gradient on the overall free energy change required to form a critical nucleus of the compound. They show that the composition gradient leads to an enhancement of the nucleation barrier of the intermetallic which increases with increasing G. In turn, the nucleation barrier for intermetallic compounds is thus expected to decrease as the amorphous interlayer thickens. When the nucleation barrier is sufficiently lowered, one expects the intermetallic to form. As in the case of the 'transient nucleation model', the model of Desre *et al.* [88] predicts that nucleation of the crystalline intermetallic NiZr will occur when the amorphous layer reaches a critical thickness.

It is difficult experimentally to distinguish between these various models for the critical thickness of an amorphous interlayer. In the case of Ni–Zr diffusion couples, the amorphous interlayer is observed to reach thicknesses of $L \sim 80-120\,\text{nm}$ prior to formation of the intermetallic NiZr. This critical thickness, defined as the thickness of the amorphous interlayer where nucleation of NiZr occurs, was measuring directly from cross-sectional TEM micrographs and found by Meng [21] to depend slightly on the growth temperature, as

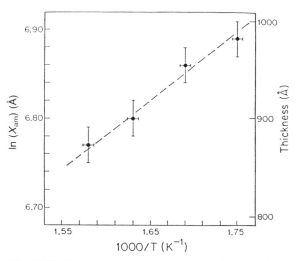

Fig. 20.16 The maximum thickness of an amorphous interlayer obtained by SSAT, a various reaction temperatures in a Ni–Zr diffusion bilayer diffusion couple (after [21]).

shown in Fig. 20.16. The figure shows the critical thickness (and its logarithm) plotted versus T^{-1}. Meng argued that this dependence was consistent with the transient nucleation model. Highmore *et al.* [81] have argued similarly. Desre *et al.* [88] have claimed that the 'composition gradient' model is also consistent with the experimental data.

It is noteworthy that models not based on a nucleation barrier have been proposed for understanding both the phase sequence and critical thicknesses of intermediate phases in diffusion couples. For example, a model by Gosele and Tu [78] is based on 'kinetic barriers'. In this model, the lack of interfacial mobility (as measured by the parameter κ in Fig. 20.12) could also be a rate-limiting step in the growth of an intermediate phase. In this picture, the initial formation of the amorphous phase is assumed. It is then argued that the interfacial mobility of the amorphous–Zr interface is greater than that of the NiZr–Zr interface. Crystalline NiZr appears only when the amorphous interlayer growth velocity, $\dot{L}$, slows sufficiently that it falls below the interface-limited velocity of NiZr. In the Gosele and Tu model, it is assumed that nucleation of intermediate phases is not the limiting step in their formation, but rather their interfacial mobilities. Experimental results discussed earlier have shown that formation of an amorphous interlayer certainly does involve a nucleation barrier. Furthermore, studies of the time dependence of amorphous interlayer growth demonstrate clearly the validity of the growth law given by eq. (20.3) with $\widetilde{D}/\kappa \sim 1\text{--}3\,\text{nm}$. As such we must conclude that interface-limited growth is confined to very small interlayer thicknesses. If this is true, then the Gosele and Tu model does not seem to be applicable. Nucleation barriers, and not 'kinetic barriers', seem to more likely control the phase sequence in Ni–Zr diffusion couples.

Experimentally, the transient nucleation of the crystalline intermetallic NiZr occurs heterogeneously along the interface separating Zr from the growing amorphous layer. The intermetallic has an orthorhombic structure ('CrB-type') and tends to nucleate with a preferred orientation along the Zr interface. Both sputtered and evaporated Zr layers tend to have a strong (002) c-axis texture. The preferred orientation of the intermetallic compound is apparently related to the existence of a low energy hetero-interface with the (002) Zr surface.

20.4 RELATIONSHIP TO OTHER MELTING PHENOMENA

20.4.1 Metastable eutectic melting

When the intermetallic compounds are eliminated from the phase diagram of a binary system, the phase diagram obtained is either of the continuous solution type or of the eutectic type. For solids with restricted solubility in the solid state, the metastable eutectic diagram of Fig. 20.11 is obtained. Section 20.3.2 shows that, in practice, the formation of intermetallic compounds can be suppressed during interdiffusion of solids by the existence of kinetic constraints which favor amorphous phase formation. Under these constraints, interdiffusion reactions lead to a product which can be described by Fig. 20.11. When the combined layer thicknesses of the reactants (e.g. Ni and Zr) are sufficiently small (less than the critical thickness of

the amorphous layer, L_c), an amorphous interlayer will consume one or both of the reactants prior to the formation of a crystalline intermetallic. If one of the reactants is entirely consumed while an excess of the other remains, a metastable equilibrium will ultimately be established between the amorphous phase and the remaining reactant. Amorphous interlayer growth will cease, the composition gradient within the amorphous layer and any solute composition gradient in the remaining reactant will disappear with time. A true metastable equilibrium will be established in accordance with Fig. 20.11. The metastable eutectic system will be single phase or two phase depending on whether the initial overall composition of the diffusion couple is less than x_1 (a solution of reactant B in reactant A), between x_1 and x_2 (reactant A in metastable equilibrium with an amorphous phase) between x_2 and x_3 (a single amorphous phase) between x_3 and x_4 (amorphous phase in metastable equilibrium with reactant B), and greater than x_5 (a solution of reactant A in reactant B). Final products of these types were observed in reaction of thin Au–La multilayers in the earliest observation of solid state amorphization reactions [1]. Similar results have been found in all other systems for which the reactant layer thicknesses are sufficiently small.

Metastable eutectic diagrams have been classified according to whether the metastable eutectic temperature T_E is greater than, equal to, or less than the glass transition temperature $T_g(x_E)$ of the liquid at the metastable eutectic composition [74]. As mentioned earlier, T_E may in fact be negative! this represents an extreme case of $T_E \ll T_g(x_E)$. Solid state amorphizing reactions are thermodynamically allowed in those systems for which $T_E < T_g(x_E)$ whenever the reaction temperature is above T_E but below $T_g(x_E)$. The temperature range $T_g(x_E) < T < T_E$ is a kinetic 'window of opportunity' for solid state amorphization. Within this temperature range, two reactant crystals will tend to react to form a glass. When $T_E > T_g(x_E)$ the metastable eutectic represents equilibrium with a metastable liquid (as opposed to a glass). Below T_E, there is no driving force for 'melting', Above T_E, melting may occur but the product phase is liquid. Since atomic mobilities in a liquid are very high, one would expect nucleation of intermetallic compounds to occur easily under these conditions. This is particularly true since the liquid is in contact with two crystalline reactants. It is well known that nucleation and growth of intermetallics from an undercooled liquid are generally quite rapid (rapid solidification techniques are required to suppress nucleation in sufficiently undercooled liquid metallic alloys). Heterogeneous nucleation along the interfaces with the reactants enhances the nucleation rate with respect to homogeneous nucleation. Observation of metastable eutectic melting would thus tend to be very difficult when $T_E > T_g(x_E)$. Metastable eutectic melting is most easily observable in those systems where the product phase is a glass, i.e. in those systems with $T_E < T_g(x_E)$.

20.4.2 Polymorphic melting and catastrophic amorphization

The metastable eutectic diagram of Fig. 20.11 is appropriate to problems in which the nucleation and growth of crystalline intermediate phases is suppressed. The absence of nucleation of crystalline intermetallic phases is a kinetic constraint which prevents achievement of thermodynamic equilibrium in a diffusion couple. To understand amorphization and its relationship to melting, it is useful to consider other types of kinetic constraints. One such constraint is the absence of thermal diffusion. In the case of the diffusion couple, the absence of thermal interdiffusion implies that the diffusion couple does not evolve at all. In practice, this constraint could be maintained by holding the ambient temperature so low that thermally activated interdiffusion is suppressed. At first sight, this seems like an uninteresting constraint since essentially nothing happens. This is true in the absence of any other atomic mixing mechanism. On the other hand, in section 20.2.3 we considered externally driven mechanisms of atomic mixing. For example, when a diffusion couple is irradiated with high energy ions or neutrons, atomic mixing occurs in collisional cascades. This 'athermal' mixing causes the composition profile of the dif-

fusion couple to evolve during the collisional cascade, but leaves the profile 'frozen' at the ambient temperature. This leads one to naturally consider melting and amorphization in the absence of thermal diffusion. Under these conditions, a local region of the sample may not change composition. Under what conditions will such a sample amorphize or melt? Thermodynamically, amorphization or melting are allowed when the transformation is accompanied by a drop in Gibbs free energy. This occurs when the composition and temperature of the solid fall outside of its $T_o(x)$-curve. The metastable eutectic diagram is replaced by the polymorphic phase diagram. The $T_o(x)$-curve defines melting at constant composition or 'polymorphic melting'.

Whether polymorphic melting actually occurs will be determined by the nucleation barrier of the liquid–amorphous phase. Since it is assumed that the sample is at very low temperature (sufficient to suppress atomic diffusion), it follows that the nucleation barrier for amorphization must be very small if it is to be overcome. Further, one can assume that the temperature lies below the glass transition temperature of the liquid. Were this not the case, there would be sufficient atomic mobility to allow atomic diffusion! The nucleation barrier for amorphization could be overcome for one of two reasons. Heterogeneous nucleation sites could reduce the barrier, or homogeneous nucleation could occur if the interfacial free energy of the glass–crystal interface became very small. Both cases may be relevant to real systems as discussed below.

The disappearance of the nucleation barrier for amorphization at heterogeneous nucleation sites has already been mentioned in section 20.2.2. There, it was pointed out that when metal films (e.g. Pt and Ni) are deposited onto single crystal surfaces of GaAs at ambient temperature, a thin amorphous layer is spontaneously formed. There it was mentioned that the metal–GaAs interface may be intrinsically unstable. This corresponds to the disappearance of the nucleation barrier for amorphous phase formation. A heterointerface is a natural site for easy nucleation of an amorphous phase. Grain boundaries, stacking faults, and other two-dimensional defects may also be favored sites at which the nucleation barrier for glass formation disappears. Solute segregation to GBs in a solid solution may tend to reduce the barrier to amorphization since the local composition of the GB may be much higher than that of the bulk solution. The enhanced composition of the GB can favor amorphization.

To reduce the homogeneous nucleation barrier for amorphous phase formation within a crystal requires a reduction of the amorphous–crystal interfacial free energy, σ_{ax}. This would occur, for instance, if amorphization (melting) became a continuous phase transition. In that case, σ_{ax} would vanish at the melting point (e.g. at $T_o(x)$). The crystalline phase would then become unstable with respect to amorphization at $T_o(x)$. Melting becomes catastrophic. The disappearance of the nucleation barrier for amorphization can generally be associated with the development of an instability in the crystalline phase.

Motivated by the above considerations, several authors have considered the conditions under which a crystalline phase becomes unstable with respect to amorphization. In fact, early theories of melting were based on the concept of crystal instability. Lindemann's melting criterion [89], for example, is based on the notion that a crystal becomes unstable when thermally induced atomic displacements reach a critical fraction of the typical distance between neighboring atoms in the crystal. Following Lindemann, Born proposed that melting might be triggered by a shear instability in the crystalline phase [90]. Born suggested that one of the independent shear moduli of the crystal vanishes at the melting point. In real crystals, this turns out to not be the case. At the melting point, the mechanical moduli remain finite. A modified form of the Born hypothesis was proposed by Tallon and his collaborators [91]. They noticed that when the shear moduli of cubic crystals is plotted as a function of the crystal volume, one of the independent shear moduli, $(C_{11} - C_{12})/2 = \mu_2$, extrapolates to zero at a volume equal to that of the liquid phase at the thermodynamic melting point. In fact, the crystal itself would become shear unstable at a temperature well above the actual

melting point (recall that crystals typically expand on melting). This is illustrated in Fig. 20.18 (taken from Tallon *et al.* [91]). This work led Tallon and his co-workers to postulate that melting is related to an underlying shear instability. Recently, Okamoto *et al.* [92] have investigated the relationship between amorphization of crystals and shear instability. They show that disordering of a crystal by irradiation results in a swelling of the crystal volume followed by amorphization. The volume increase of the crystal is accompanied by a drop in the average shear modulus of the crystal as measuring by Brillouin scattering. A plot of the shear modulus of the crystal as a function of its volume shows that the average shear modulus extrapolates to near zero at a volume equal to that of the amorphous phase which is formed by irradiation to high doses. This suggests that shear softening of a crystal prior to amorphization follows the same behavior as shear softening in ordinary melting.

When one of the elastic shear moduli of a crystal vanishes, the crystal will exhibit spontaneous shear fluctuations which grow without bound. For a small but finite shear modulus, shear fluctuations will be thermally driven. For sufficiently high T, these thermally driven fluctuations may still render the crystal unstable. This type of shear instability is of the type originally discussed by Gibbs. Instability of a crystal with respect to amorphization may take on other forms as well.

Fecht and Johnson have discussed other criteria for crystal stability [93]. For pure metals, the heat capacity at constant pressure of the crystalline solid increases at high temperatures (near T_m) due to anharmonicity, point defect formation, etc. On the other hand, the heat capacity of the liquid metal tends to fall with temperature in the vicinity of the melting point. Empirically, it is found that the two heat capacity curves tend to cross in the vicinity of the melting point. As such, the heat capacity of the crystal exceeds that of the liquid above T_m. Therefore, the entropy of the crystal increases more rapidly than that of the liquid for temperatures above T_m. Based on an analysis of several pure metals, Fecht and Johnson were able to show that the entropy of a superheated crystalline metal exceeds that of the liquid above an upper isentropic

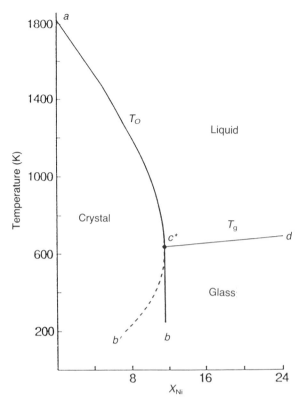

Fig. 20.17 A polymorphic phase diagram for the hcp Zr-base solid solution of the Ni–Zr binary system. The diagram shows the critical concentration c^* ($c^* = X_{crit}$) (at a temperature T^*) where the isentropic temperature (versus composition) crosses the T_o-line for the hcp solid solution. The glass transition temperature (T_g versus composition) must also pass through this point (see text for further discussion) (after [70]).

temperature, T_S. They then argued that this isentropic temperature is a natural limit for superheating a crystal. Above this temperature, melting would become exothermic (as opposed to ordinary endothermic melting). Once begun, melting would lead to a temperature increase. This, in turn, would accelerate the melting transformation leading to a thermal runaway effect. The isentropic temperature is thus a natural limit to crystal stability. The argument is similar to Kauzmann's argument regarding the glass transition [94]. Kauzmann showed that the entropy of an undercooled liquid would fall to a value equal to that of the corresponding solid if the liquid could be sufficiently

undercooled. He defined a lower isentropic temperature, T'_S, for liquid undercooling. He further argued that some form of transformation must intervene as the liquid is undercooled in order to avert this paradoxical situation. He identified this transformation as the glass transition and postulated that it was a natural limit to undercooling. When these arguments are extended to the case of a binary alloy such as Ni–Zr, one finds that the upper and lower isentropic temperatures become composition dependent. For a given crystalline phase (e.g. the α-Zr solid solution (Fig. 20.10)), one can estimate the entropy of the solid solution at a given composition and compare this with that of a liquid at the same composition. The crossing of the two curves defines $T_S(x)$ and $T'_S(x)$. For the α-Zr solid solution, a calculation leads to the diagram shown in Fig. 20.17. At a critical composition x_{crit}, one finds that $T_S(x_{crit}) = T'_S(X_{crit}) = T_o(x_{crit}) = T_{crit}$. The existence of such a point (x_{crit}, T_{crit}) follows if one assumes that the entropy of a stable liquid must always be greater than the entropy of a stable solid. Violation of this condition is taken as the criterion for instability. For the case of Ni–Zr, Fecht and Johnson [93] estimated that $x_{crit} \approx 0.12$. This suggests that the α-Zr solid solution is unstable at $x > 0.12$. Recently, Tallon [95] has attempted to combine his earlier arguments based on shear instability with the Fecht–Johnson argument to show that the isentropic and shear stability arguments are related. He argues that shear instability is likely to be a more restrictive stability requirement than the entropy condition of Fecht and Johnson.

All of the above arguments suggest that stability of a crystalline material against amorphization requires that the crystal composition and temperature fall within limits determined by a curve in the (x, T) plane (pressure is assumed to be fixed). More generally, if pressure is allowed to vary, stability requires that the crystal lie within a surface in (x, T, P) space. Desre *et al.* have extended the Clapeyron equation to the case of polymorphic melting [70] and have discussed the calculation of the $T_o(x, P)$ surface. The above discussion suggests that there is a corresponding surface which describes the limits of stability of a crystal

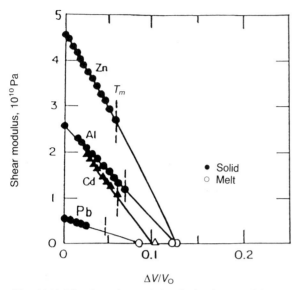

Fig. 20.18 The dependence of the elastic shear modulus on volume distension for several elements. The shear modulus plotted is the lesser of the independent shear moduli. The open circle is the volume of the melt at the thermodynamic melting point. All volumes are referenced to the volume of crystal at $T = 0\,\mathrm{K}$ (after [91]).

with respect to amorphization. In particular, one could define $T_i(x, P)$ as the temperature above which a crystalline solution of composition x at pressure P would become unstable with respect to amorphization–melting. When $T_i(x, P)$ is equal to or less than $T_g(x, P)$, the unstable crystal would spontaneously collapse into a glass. This could be referred to as spontaneous amorphization. Referring to our earlier discussion of the nucleation barrier for amorphization, one can identify $T_i(x, P)$ as the temperature at which a solid solution of composition x and at pressure P transforms to an amorphous phase with no nucleation barrier. We see that the condition that the nucleation barrier disappears can be associated with a crystal which is driven to a state that lies outside of the $T_i(x, P)$ surface. At a given pressure, one then expects (Fig. 20.17) that the crystal will have limits of composition (x_{crit}) beyond which it is strictly unstable. Beyond this limit, amorphization is spontaneous and does not require thermal activation.

For externally driven processes like ion mixing of a diffusion couple, the composition of a solid solution formed can be driven athermally to exceed the limit, x_{crit}. When such mixing is carried out at very low temperatures (where no thermal diffusion is possible), the solid solution will be driven beyond its stability limit. We see that amorphous interlayer growth during ion mixing may be a spontaneous process.

The above arguments lead to some final observations regarding amorphization by thermal interdiffusion reactions. Consider thermal interdiffusion of two crystalline solids under conditions whereby no intermediate phase may nucleate (including the amorphous phase). Experimentally, this situation is achieved below $T = 400\,°\mathrm{C}$ when polycrystalline Ni is interdiffused with a single crystal of Zr as discussed in section 20.2. Under these circumstances, the nucleation barrier for amorphization (at a single crystal Zr interface) cannot be overcome. Nevertheless, at $T = 350\,°\mathrm{C}$, the intrinsic diffusion constant of Ni in α-Zr is large [4], $D \sim 10^{-10}\,\mathrm{cm^2/s}$, and should permit extensive thermal interdiffusion. This will lead to two terminal solutions of composition determined by the common tangent rule. In the case of terminal solutions having a negative free energy of mixing (as is the case for the Saunders–Miodownik calculation for the hcp–Zr and fcc–Ni solid solutions of Fig. 20.10), one would predict extensive solubility of the reactants. From Fig. 20.10, combined with the fact that Ni is the dominant moving species, one would expect extensive dissolution of Ni into Zr. In fact, the solubility of Ni in single crystal Zr is rather very limited. This indicates that the free energy curve for the hcp–Zr solid solution in Fig. 20.10 is probably incorrect. Were this not the case, one would expect Ni to dissolve into hcp–Zr to high concentrations. Above, we estimated that an hcp–Zr solid solution would become unstable if the Ni concentration exceeded $x \sim 0.12$. The dissolution of Ni into the Zr single crystal would result in spontaneous amorphization! No nucleation barrier would exist for formation of the amorphous phase! The amorphous phase does not form (recall that Zr GBs were found to be necessary for heterogeneous nucleation of the amorphous phase). A such, we must conclude that the free energy curve of Saunders and Miodownik for the solid α-Zr solution is wrong. The heat of mixing for the solid solution should in fact be positive since only then will the solid solubility of Ni in α-Zr be restricted as observed. Further, only then can the α-Zr solid solution remain stable with respect to Ni concentration.

REFERENCES

1. R. B. Schwarz and W. L. Johnson, *Phys. Rev. Lett.*, **51** (1983), 415.
2. (a) S. R. Herd and K. N. Tu, *Appl. Phys. Lett.*, **42** (1983), 597; (b) K. N. Tu, S. R. Herd, and K. Gosele, *Phys. Rev. B*, **43**, 1198 (1991).
3. For reviews see W. L. Johnson, *Prog. in Mat. Sci.*, **30** (1986), 81: also K. Samwer, *Physics Reports*, **161** (1988), 1.
4. A. D. Le Claire, *J. Nucl. Mater.* **69/70** (1978), 70.
5. W. K. Warburton and D. Turnbull, in *Diffusion in Solids – Recent Developments*, Academic Press, New York (1975), Chapter 4.
6. U. Koster, in *Glassy Metals I* (ed H. Guntherodt and H. Beck), Springer-Verlag, Heidelberg (1981), Chapter 10.
7. F. Spaepen and D. Turnbull, in *Metallic Glasses* (eds J. J. Gilman and H. J. Leamy), American Society for Metals, Metals Park, OH (1978), Chapter 5.
8. B. Cantor and R. W. Cahn, in *Amorphous Metallic Alloys* (ed F. Luborsky), Butterworth Press, London (1983), Chapter 25.
9. H. Mehrer, Review of Diffusion in amorphous alloys
10. J. M. Frigerio and J. Rivory, *J. de Physique C*, **4**, Suppl. to No. 14, **51** (1990), 163.
11. B. Blainpain, J. M. Legresy, and J. W. Mayer, *J. de Phys., Coll. C*, **4**, Suppl. to No. 14, **51** (1990), 131: also L. Hung, M. Nastasi, and J. W. Mayer, *Appl. Phys. Lett.*, **42** (1983), 672.
12. R. B. Schwarz, K. L. Wong, and W. L. Johnson, *J. Non Cryst. Sol.*, **61/62** (1984), 129.
13. F. R. Ding, P. R. Okamoto, and L. E. Rehn, in *Beam Solid Interactions and Transient Thermal Processing, MRS Symp. Proc.*, **100** (1987), 69.
14. C. Politis and W. L. Johnson, *J. Appl. Phys.*, **60** (1986): also G. Cocco *et al.*, *J. de Physique C*, **4** (1990), 181.
15. E. Ma, C. W. Nieh, M.-A. Nicolet, and W. L. Johnson, *J. Mater. Res.*, **4** (1989), 1299.
16. M. Atzmon, J. R. Veerhoven, E. R. Gibson, and W. L. Johnson, *Appl. Phys. Lett.*, **45** (1984), 1052: also L. Schultz, *Rapidly Quenched Metals V* (eds S. Steeb and H. Warlimont), North-Holland, Amsterdam (1985), p. 1585.
17. H. U. Krebs, D. J. Webb, and A. Marshall, *Phys. Rev. B.*, **35** (1987), 5392: C. Michaelsen, M. Piepenbring, and H. U. Krebs, *J. de Physique C*, **4** (1990), 151: H. U. Krebs, D. J. Webb, and A. F. Marshall, *J. Less Comm. Met.*, **140** (1988), 17: also E. Hellstern and L. Schultz, *J. Appl. Phys.*, **63** (1988), 1408.
18. T. Ben Ameur and A. R. Yavari, *J. de Physique C*, **4** (1990), 219.
19. P. Guilmin, P. Guyot, and G. Marchal, *Phys. Lett.*, **109A**, (1985), 174.
20. G. J. Van der Kolk, A. R. Miedema, and A. K. Niessen, *J. Less Comm. Met.*, **145** (1988), 1: also P. I. Loeff, A. W. Weeber, and A. R. Miedema, *J. Less Comm. Met.*, **140** (1988), 299.
21. W. J. Meng, C. W. Nieh, and W. L. Johnson, *Appl. Phys. Lett.*, **51** (1987), 1693: also W. J. Meng, E. J. Cotts, and W. L. Johnson, *Mater. Res. Soc. Symp. Proc.*, **77** (1987), 223: W. J. Meng, C. W. Nieh, E. Ma, B. Fultz, and W. L. Johnson, *Mat. Sci. & Eng.*, **97** (1988), 87: also W. J. Meng, Ph.D thesis, California Institute of Technology Dept. of Applied Physics, 1987.
22. H. Schroeder, K. Samwer, and U. Koster, *Phys. Rev. Lett.*, **54** (1985), 197: also K. Samwer, in *Amorphous Metals and Non-Equilibrium Processing* (ed M. Von Allmen), Editions de Physique, Couteabeouf, France (1984), p. 123; also H. U. Krebs and K. Samwer, *Europhys. Lett.*, **2** (1986), 141.
23. M. Van Rossum, M.-A. Nicolet, and W. L. Johnson, *Phys. Rev. B.*, **29** (1984), 5498.
24. K. Pampus, K. Samwer, and J. Bottiger, *Europhys. Lett.*, **3** (1987), 581.
25. J. C. Barbour, F. W. Saris, M. Nastasi, and J. W. Mayer, *Phys. Rev. B.*, **32** (1985), 1363.
26. W. L. Johnson, *Rapidly Quenched Metals VI* (eds J. Strom Olsen and R. W. Cochrane), *Mat. Sci. & Eng.*, **97** (1988), 1.
27. Y. T. Cheng, M.-A. Nicolet, and W. L. Johnson, *Appl. Phys. Lett.*, **47** (1985), 800.
28. J. C. Barbour, *Phys. Rev. Lett.*, **55** (1985), 2872: Also see J. C. Barbour, Ph.D thesis, Cornell Univ., Dept. of Mat. Sci. (1986).
29. H. Hahn, R. S. Averback, and S. J. Rothman, *Phys. Rev. B.*, **33** (1986), 8825.
30. S. B. Newcomb and K. N. Tu, *Appl. Phys. Lett.*, **48** (1986), 1436.
31. R. W. Johnson, C. C. Ahn, and E. R. Ratner, *Phys. Rev. B.*, **40** (1989), 8139: also R. W. Johnson, C. C. Ahn, and E. R. Ratner, *Appl. Phys. Lett.*, **54** (1989), 795.
32. K. Halloway and Robert Sinclair, *J. of the Less Comm. Met.*, **140** (1988), 139: also K. Halloway, P. Moine,

J. Delage, R. Bormann, L. Capuano, and R. Sinclair, *Mat. Res. Soc. Symp. Proc.*, **187** (1990), 71.

33. J. F. M. Westendorp, Ph.D thesis, Univ. of Utrecht, The Netherlands (1986); also A. M. Vredenberg, J. F. M. Westerdorp, F. W. Saris, and N. M. van der Pers, *J. Mater. Res.*, **1** (1986), 774.

34. B. M. Clemens, *Phys. Rev. B.*, **33** (1986), 7615.

35. E. Ma, W. J. Meng, W. L. Johnson, and M.-A. Nicolet, *Appl. Phys. Lett.*, **53** (1988), 2033.

36. J. J. Hauser, *J. Phys. (Paris), Colloq.*, **42** C4–943 (1981).

37. D. M. Van der Walker, *Appl. Phys. Lett.*, **48** (1986), 707.

38. C. V. Thompson, L. A. Clevenger, R. DeAvillez, E. Ma, and H. Miura, *Mat. Res. Soc. Symp. Proc.*, **187** (1990), 61.

39. J. Y. Cheng, M. H. Wang, and L. J. Chen, *Mat. Res. Soc. Symp. Proc.*, **187** (1990), 77.

40. V. A. Ushkow, A. B. Fedotov, E. A. Eroteeva, A. I. Rodionov, and D. T. Dzhumakulov, *Izv. Akad. Nauk SSSR, Neorgan. Mater.* **23** (1987), 186: also D. H. Ko and R. Sinclair, *Appl. Phys. Lett.*, in press (1990).

41. F. Y. Shiau and Y. A. Chang, *Appl. Phys. Lett.*, **55** (1989), 1510: also F. Y. Shiau and Y. A. Chang, *Mat. Res. Soc. Symp. Proc.*, **187** (1990), 89.

42. T. Sands, C. C. Chang, A. S. Kaplan, V. G. Keramidas, K. M. Krishman, and J. Washburn, *Appl. Phys. Lett.*, **50** (1987), 1346: also R. Caron-Popowich, J. Washburn, T. Sands, and A. S. Kaplan, *J. Appl. Phys.*, **64** (1988), 4909.

43. F. Y. Shiau and Y. A. Chang, *Appl. Phys. Lett.*, in press (1990).

44. R. M. Walser and R. W. Bene, *Appl. Phys. Lett.*, **28** (1976), 624.

45. K. A. Hartley, J. Duffy, and R. H. Hawley, *J. Mech. and Phys. Solids*, **35** (1987), 283.

46. T. Vreeland Jr., P. Kasiraj, A. H. Mutz, and N. N. Thadhani, in *Proc. Int. Conf. on Metallurgical Applications of Shock Wave and High Strain Rate Phenomena* (ed L. E. Murr, K. P. Staudhammer, and M. A. Meyers), Marcel Dekker, New York (1986), p. 231.

47. M. Atzmon, J. R. Veerhoven, E. R. Gibson, and W. L. Johnson, *Appl. Phys. Lett.*, **45** (1984), 1052.

48. M. Atzmon, K. Unruh, and W. L. Johnson, *J. Appl. Phys.*, **58** (1985), 3865.

49. F. Bordeaux, E. Gaffet, and A. R. Yavari, *Europhys. Lett.*, **12** (1990), 63.

50. H. Mehrer, private communication.

51. G. C. Wong, W. L. Johnson, and E. J. Cotts, *J. Mater. Res.*, **5** (1990), 488.

52. E. J. Cotts, G. C. Wong, and W. L. Johnson, *Phys. Rev. B.*, **37** (1988), 9049.

53. L. Schultz, *Rapidly Quenched Metals V* (eds S. Steeb and H. Warlimont), North-Holland, Amsterdam (1985), p.

1585.

54. L. Schultz, *Mat. Res. Soc. Symp. Proc.*, **80** (1987), 97.

55. S. Martelli, G. Mazzone, A. Montone, and M. V. Antisari, *J. de Physique, Colloque No. C*, **4** (1990), 241.

56. A. Y. Yermakov, Y. Y. Yurchikov, and V. A. Barinov, *Phys. Met. Metall.*, **52** (1981), 50.

57. C. C. Koch, O. B. Cavin, C. G. McKamey, and J. O. Scarbrough, *Appl. Phys. Lett.*, **43** (1983), 1017.

58. R. B. Schwarz and C. C. Koch, *Appl. Phys. Lett.*, **49** (1986), 146.

59. G. Martin and E. Gaffet, *J. de Physique*, Colloque C4, **51** C471 (1990).

60. W. L. Johnson, Y. T. Cheng, M. Van Rossum, M.-A. Nicolet, *Nucl. Inst. & Meth. B.*, **7/8** (1985), 657.

61. P. Sigmund and A. Gras-Marti, *Nucl. Inst. & Meth.* **182/183** (1981), 25.

62. T. W. Workman, Y. T. Cheng, W. L. Johnson, and M.-A. Nicolet, *Appl. Phys. Lett.*, **50** (1987), 1485.

63. T. Diaz de la Rubia, R. S. Averback, R. Benedek, and W. E. King, *Phys. Rev. Lett.*, **59** (1987), 1930.

64. G. H. Vineyard, *Radiat. Effects*, **29** (1976), 245.

65. B. Y. Tsaur, S. S. Lau, S. Hung, and J. W. Mayer, *Nucl. Inst. Meth.* **182/183** (1981), 67: also B. Y. Tsaur, Ph.D. thesis, Calif. Inst. of Tech. (1980).

66. B. X. Liu, W. L. Johnson, and M.-A. Nicolet, *Appl. Phys. Lett.*, **42** (1983), 45.

67. L. S. Hung, M. Nastasi, J. Gyulai, and J. W. Mayer, *Appl. Phys. Lett.*, **42** (1983), 672: also M. Nastasi, Ph.D thesis, Cornell University (1985).

68. M. Van Rossum, U. Shreter, W. L. Johnson, and M.-A. Nicolet, *Mat. Res. Soc. Symp. Proc.*, **27** (1984).

69. Y. T. Cheng, W. L. Johnson, and M.-A. Nicolet, *Mat. Res. Soc. Symp. Proc.*, **37** (1985), 565.

70. H. J. Fecht, P. Desre, and W. L. Johnson, *Phil. Mag.*, **59** (1989), 577.

71. R. A. Swalin, *Thermodynamics of Solids*, John Wiley & Sons, New York (1962), Chapter 11.

72. L. Kaufman and H. Bernstein, *Computer Calculations of Phase Diagrams*, Academic Press, New York (1970).

73. N. Saunders and A. P. Miodownik, *J. Mater. Res.* **1** (1986), 38.

74. A. L. Greer and R. J. Highmore, *Nature*, **339** (1989), 363.

75. B. E. Deal and A. S. Grove, *J. Appl. Phys.*, **36** (1985), 3770.

76. Y. Y. Geguzin, Y. S. Kagonouskly, L. M. Paritskoya, and V. I. Solunskiy, *Phy. Met. and Metall.*, **47** (1980), 127.

77. B. Dolgin, Ph.D thesis, Calif. Inst. of Tech., Dept. of Applied Physics, (1985); see also Ref.3 for a description of interdiffusion reaction kinetics.

78. U. Gosele and K. N. Tu, *J. Appl. Phys.*, **53** (1982), 3252.

79. R. W. Cahn, *J. de Physique, Colloque C4*, **51** (1990), 3: also

B. Cantor, in *Amorphous Metals and Semiconductors* (eds P. Haasen and R. I. Jaffee), (1986), p. 108.

80. E. J. Cotts, W. J. Meng, and W. L. Johnson, *Phys. Rev. Lett.*, **57** (1986), 2295.

81. R. J. Highmore, J. E. Evetts, A. L. Greer, and R. E. Somekh, *Appl. Phys. Lett.*, **50** (1987), 566.

82. H. Schroder and K. Samwer, *J. Mater. Res.*, **3** (1988), 461.

83. R. B. Schwarz and W. L. Johnson, *J. Less Comm. Met.*, **140** (1988), 1.

84. J. W. Christian, *The Theory of Transformations in Metals and Alloys*, Pergamon Press, London (1965), Chapter X.

85. E. S. Machlin, *Thermodynamics and Kinetics in Materials Science*, Giro Press, Croton on Hudson, NY (1991), Chapter IV.

86. E. Kamenetsky, W. J. Meng, L. Tanner, and W. L. Johnson, in *Analytical Electron Microscopy* (ed D. C. Joy), San Francisco Press, San Francisco (1987), p. 83; see also W. J. Meng, Ph.D thesis, Calif. Inst. of Tech., 1987.

87. U. Koster, R. Pries, G. Bewernick, B. Schuhmacher, and M. Blank-Bewersdorff, *J. de Physique*, Colloque C4, (1990), 121.

88. P. J. Desre and A. R. Yavari, *Phys. Rev. Lett.*, **64** (1990), 1533.

89. A. R. Ubbelohde, *Melting and Crystal Structure*, Oxford University Press (1965).

90. M. Born and K. Huang. *Dynamical Theory of Crystal Lattices*, Oxford University Press (1962).

91. J. F. Tallon, *J. Phys. Chem. Sol.*, **41** (1984), 837.

92. D. Wolfe, P. R. Okamoto, S. Yip, J. F. Lutsko, and M. Kluge, *J. Mater. Res.*, **5** (1990), 286.

93. H. J. Fecht and W. L. Johnson, *Nature* **334** (1988), 50.

94. W. Kauzmann, *Chem. Rev.*, **43** (1948), 219.

95. J. Tallon, *Nature*, in press (1990).

CHAPTER TWENTY-ONE

Relationship between structural and electronic properties of metal–semiconductor interfaces

R. T. Tung

Introduction · Interface states · Electron transport of Schottky barriers · Evidence for SBH inhomogeneity: non-epitaxial MS interfaces · Epitaxial MS interfaces · Conclusions

21.1 INTRODUCTION

Metal–semiconductor (MS) structures are an essential part of virtually all electronic and opto-electronic devices. Much progress has been made in our understanding of the chemistry and metallurgy at these interfaces. However, as far as electronic properties of MS interfaces are concerned, little is firmly understood or experimentally proven. The formation mechanism of the Schottky barrier (SB) at a MS interface, despite much investigation and debate, still remains at the stage of speculation [1, 2, 3, 4, 5]. A major obstacle which has thus far prevented a more in-depth understanding of the SB problem has to do with the complexity of the structure of available MS interfaces. Polycrystallinity and other crystalline imperfections make the atomic structures at an ordinary MS interface too complicated and literally impossible to study by experimental means. Without information on the structure of real MS interfaces, no first principles calculation of the interface electronic properties may be carried out. Since the SB problem seems intractable through direct experimental

and theoretical investigations based on real structures, existing SB models have to resort to mechanisms not directly linked to the structure of the MS interface. Presently, the absence of a strong dependence of the experimentally observed Schottky barrier height (SBH) on the metal work function is thought to be due to the presence of a high density of interface states which pin the Fermi-level (FL) position [6, 7]. Within the FL pinning model, the SBH of a MS system does not have a first-order dependence on the atomic structure of the interface.

Recently, several **epitaxial** MS structures have been fabricated. In particular, epitaxial silicide–Si interfaces have shown very high structural perfection [8, 9]. Since the best silicide structures are single-crystalline, there is but one atomic structure for the entire MS interface. State of the art microscopic and spectroscopic techniques have been used to determine the atomic structure of silicide–Si interfaces [9, 10], which, in turn, have been used in direct calculations of the interface electronic structure [11, 12]. Epitaxial MS interfaces provide an unprecedented opportunity to apply first principles in solving the SB mystery. A dependence of the SBH on epitaxial orientation has been observed for NiSi$_2$ [13], and has tentatively been accounted for by theoretical calculations of the intrinsic MS electronic properties [11, 12]. Results from epitaxial MS interfaces show the importance of the role played by interface structure in the formation of the SBH.

The fact that entirely different conclusions were drawn in various studies concerning the role played by interface structure is but a small part of the many mysterious phenomena and their, often conflicting, explanations found in the SBH literature. Whether a connection exists between the interface structure and the SBH is one of the most pressing questions in the study of the formation mechanism of the SB. This chapter will address this specific question from two angles. We begin with a brief discussion of the traditional view of interface states. Then, for reasons which will become obvious, the electron transport at inhomogeneous SBs is discussed. Existing experimental data will then be discussed and shown to agree well with the presence of SBH inhomogeneity, and to be inconsistent with existing theories based on interface states. The presence of SBH inhomogeneity in the majority of polycrystalline MS interfaces is inconsistent with the FL pinning concept. Next, we examine the SB problem from a different angle. Recent experimental data from high quality epitaxial silicide–Si and other MS interfaces will be examined to show that, under very well-defined experimental conditions, there is a clear dependence of the SBH on the specifics of the interface. Finally, the implications of the demonstrated relationship between the structure and the SBH of MS interfaces will be discussed.

21.2 INTERFACE STATES

21.2.1 Interface state models and FL pinning

According to Schottky's original description [14], the SBH between a metal with a work function of ϕ_m and a n-type semiconductor with an electron affinity of χ should be

$$\Phi_{Bn,o} = \phi_m - \chi \tag{21.1}$$

as schematically shown in Fig. 21.1(a). However, experimentally observed SBHs are less dependent on the metal work function than predicted by eq. (21.1) [15]. This led to proposals of the presence of interface states in the band gap which pin the FL. Present models of interface states can be approximately divided into two groups, namely, those which involve an interfacial insulating layer and those which do not. Bardeen's proposal called for pinning by states on the semiconductor surface [6] prior to the semiconductor being brought into contact with the metal. When the contact is made, there is simply a lineup of the two FLs, leading to a SBH which is independent of the metal, as schematically shown in Fig. 21.1(b). Bardeen's model involves no interaction at the interface, and hence is, strictly speaking, only applicable when a thin dielectric layer separates the metal and the semiconductor. By artificially inserting a thin dielectric layer with a fixed thickness at the MS interface, the band diagram for a MS interface can then be deduced as a function of the metal work function [16, 17, 18]. Because the interface states are assumed to be independent of the metal, the dependences of the SBH on the metal work function and on the applied bias may be obtained within one analysis. Under the assumption of a uniform distribution of interface states, the expected SBH as a function of the metal work function may be expressed as [16]

$$\Phi_{Bn,o} = \gamma_{is}(\phi_m - \chi) + (1 - \gamma_{is})(E_g - \phi_o) \tag{21.2}$$

where the parameter γ_{is}, defined by

$$\gamma_{is} = \left(1 + \frac{D_{ss}d}{\varepsilon_i}\right)^{-1} \tag{21.3}$$

is related to the famous S-parameter [19, 20] frequently discussed, and ϕ_o is the charge neutrality level of the interface states. One sees that a critical parameter in such an analysis is the product of the interface state density, D_{ss}, and the thickness of the dielectric layer, d.

Subsequent to Bardeen's proposal, Heine [7] pointed out that for intimate MS contacts, elec-

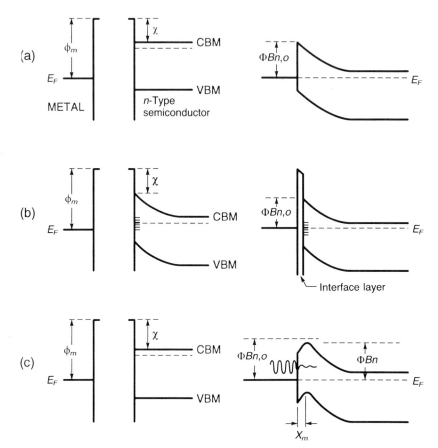

Fig. 21.1 Band diagrams before (left panel) and after (right panel) the metal and the *n*-type semiconductor are brought into contact. (a) corresponds to the non-interacting Schottky model, where eq. (21.1) holds. (b) represents the situation when the semiconductor FL is pinned by surface states before metallization. A dielectric interface layer is assumed to be present at the MS interface to make up the difference between the two FLs. (c) is an intimate MS interface, with short-range band bending due to penetration of MIGS into the semiconductor. In (c), the effective electrical interface (position of maximum electric potential) is located in the semiconductor, at some distance x_m from the metallurgical MS interface. (c) is sometimes referred to as the negative charge model. Image force lowering is not included in these drawings (the upward band bending at the interface in (c) is not due to image-force lowering).

tronic interaction between the metal and the semiconductor is unavoidable, even in the absence of any metallurgical reaction. Metal wave functions penetrate into the semiconductor and lead to MIGSs and the formation of an interface dipole [7, 21, 22, 23]. For an analysis of the interface electronic properties, an exponential decay of the (negative) charge into the semiconductor, with a single decay length λ_s of ~5–20 Å, is usually assumed. It is understood that the negative charge extending into the semiconductor is offset by an equal positive charge residing in the metal. However, since the charge is assumed to penetrate a negligible distance into the metal compared to its spatial extension into the semiconductor, potential variation inside the metal is usually ignored. Therefore, once λ_s and D_{ss} are known, the band-bending at an intimate MS interface may be uniquely determined, as in Fig. 21.1(c) [24, 25]. An expression almost identical to eq. (21.2) may be deduced [26], under these usual assumptions of the MIGS models, for the SBH at intimate MS interfaces.

The mechanism proposed by Spicer *et al.* [27, 28, 29], i.e. interface FL position pinning by defect states, is independent of whether a dielectric layer is present at the interface. However, in the absence of a dielectric layer, a fixed separation of the defect states from the metal, of the order of 5–10 Å, is usually assumed [30, 31] for an estimation of the defect density.

21.2.2 Lack of evidence for interface states models

Because of various experimental difficulties, it has not been possible to unequivocally deduce the distribution of interface states by experimental means. Techniques such as photoemission [32], junction capacitance [33, 34, 35], photoelectric spectroscopy [36], and optical absorption [37] measurements, thus far employed for this purpose, all involve arbitrary assumptions, the validity of which is difficult to evaluate. While there is hardly any question that electronic states particular to the MS interface are present, the question is whether they lead to the formation of the SBH in the fashion proposed by existing models. Even though experimental evidence for a causal relationship between interface gap states and the SBH is generally lacking, FL pinning by interface states seems to be a concept which has already been widely accepted. This likely has to do with a (mis)conception that interface states are able to account for both a lack of dependence of the SBH on metal work function and, at the same time, a host of non-ideal behaviors routinely observed from SBH experiments. Actually, the fact that interface state models have implications on both the SBH's dependence on metal work function and its dependence on the applied bias, has been shown to lead to a dilemma [38]. Furthermore, the presence of states at intimate MS interfaces cannot satisfactorily account for experimental observations on both *n*- and *p*-type semiconductors, as will be discussed. Interface states are usually assumed to have a distribution with respect to the semiconductor bands, and that their population is governed by the FL position at the interface. However, because of the inherent ambiguity in how quasi-FLs should be drawn at a MS interface when a bias is applied, it is somewhat uncertain how the interface states should be populated under bias.

Present models of the interface states involve many assumptions. The most obvious, and the most questionable, simplification of existing models is the omission of the structural dependence of the electronic states. As a result, interface states are assumed to have a uniform lateral distribution. According to eq. (21.3), both the density and the effective depth of the interface states have to remain constant for the local FL position to remain laterally uniform. At real MS interfaces, it seems unlikely that these conditions are met. For instance, the native oxide layer suspected of being present at some MS interfaces is most certainly non-uniform in its thickness and/or its composition. Structural defects are certainly not, as present models invariably assume, uniformly distributed at a constant distance from the MS interface. As will be discussed, many consequences of an inhomogeneous SBH cannot be derived from transport equations involving homogeneous SBs. A fixed separation between interface states and the metal is usually assumed at intimate MS interfaces. This assumption, which is necessary for the interface charge and the interface dipole to vary with the semiconductor quasi-FL, is unfounded. It appears that the very assumptions which have made the interface state models easy to analyze may have made them too simplified to describe real experimental situations.

What has been generally overlooked in interface states models so far is that the distribution of states with energies lying within the semiconductor band gap represents only a small portion of the total redistribution of charge at the MS interface. Recent theoretical calculations [11, 12] based on the atomic structures at real MS interfaces showed that the interaction of the conduction band of the metal and the valence band of the semiconductor plays a more important role in establishing the dipole at the MS interface than do states in the band gap. The ranges/depths of electrons involved in such an interaction, which have energies below the valence band maximum (VBM) of the semiconductor, are similar to that usually argued to be the decay length of MIGS, $\sim 5\,\text{Å}$. However, the rapid variation of the electric potential, due to the rearrangement of charge at the interface, occurs in a much more complicated manner than the monotonic behavior assumed in existing phenomenological models of the interface gap states. The distribution of the MIGS was found to depend on the atomic structure of the MS interface, contrary to the popular view of MIGS [12]. It has not been possible to identify the role played by MIGS in the formation of the SBH [12]. It thus appears that, in order to understand the formation mechanism of the SBH, one should not concentrate exclusively on MIGS and ignore the role played by bonding electrons which depend on the local structure of the MS interface.

21.3 ELECTRON TRANSPORT OF SCHOTTKY BARRIERS

21.3.1 The issue of SBH inhomogeneity

If the SBH depends on atomic structure, then at an ordinary, polycrystalline, MS interface, the FL is expected to be inhomogeneous. If, on the other hand, the FL were pinned, SBH is not expected to vary locally within one SB diode. Therefore, the question of whether the FL position varies at ordinary, polycrystalline, MS interfaces may have a direct bearing on the formation mechanism of the SBH. A homogeneity of the SBH has so far been implicitly assumed in the analyses of electrical data obtained from SBH measurements. As a result, there has not been much experimental evidence for a local variation of the SBH and, therefore, the question concerning SBH inhomogeneity is often not raised or not addressed. However, as was recently pointed out, the lack of report on inhomogeneous SBs is not due to an absence of SBH inhomogeneity, but is due to a failure to recognize it from common experimental data. Actually, the assumption of SBH homogeneity is invalid for the majority of SBs studied in routine experiments. A correct interpretation of transport properties of MS interfaces requires a knowledge of the expected behavior of inhomogeneous SBs. For this reason, our discussion of the junction

current is generalized to cover the situation at inhomogeneous SBs. Transport at homogeneous SBs is just a special case in the many possible forms of SBH distributions.

21.3.2 Non-interacting model of inhomogeneous SBs

The current–voltage $(I–V)$ relationship of a homogeneous SB junction has been described by the thermionic emission theory as [1, 2]

$$I(V_a) = I_s \left[\exp\left(\frac{qV_a}{nk_BT}\right) - 1 \right] \qquad (21.4)$$

where q is the electronic charge, k_B the Boltzmann constant, T the absolute temperature, V_a the applied bias, and I_s is the saturation current, defined by

$$I_s = A^*AT^2 \exp\left(-\frac{q\Phi_B}{k_BT}\right) \qquad (21.5)$$

where A^* is the Richardson constant and A is the area of the diode. When the SB is homogeneous Φ_B is the SBH of the junction. The ideality factor n, in eq. (21.4) is a fit to the slope of an experimentally obtained (semi-logarithmic) $I–V$ curve. An ideality factor of 1 is predicted by the thermionic emission theory. Since a formula like eq. (21.5) is used to analyze all $I–V$ data, it is not surprising that a single parameter Φ_B is always obtained which is then regarded as the SBH. The verification of the homogeneity of the SBH, which is necessary for eq. (21.5) to be valid, is usually absent.

The junction current of an inhomogeneous SB has been proposed to be described by a parallel conduction model [39, 40]. In such a model, the junction current is a linear sum of the contributions from every individual area, namely,

$$I(V_a) = A^*T^2 [\exp(\beta V_a) - 1] \sum_i \exp(-\beta\Phi_i)A_i \qquad (21.6)$$

where $\beta = q/k_BT$, and A_i and Φ_i are, respectively, the area and the SBH of the i-th 'patch'. Such theory has been applied to describe mixed-phase diodes [41, 42, 43, 44, 45, 46] and also been shown

[47, 48] to lead to an increase of the apparent SBH with temperature. As previously pointed out [49, 50], the parallel conduction model [40] is in significant error when the SBH varies spatially on a scale less than, or comparable to, the width of the space charge region. The error arises because eq. (21.6) fails to take into account the interactions between neighboring patches with different SBHs [44]. For example, the conduction path in front of a small patch with a low SBH should be pinched-off if surrounded by high-SBH patches. 'Pinch-off' is a terminology often used to describe the operation of a field effect transistor. In its present use, an area is said to be pinched-off if majority carriers originating from outside the space charge region need to go over a potential barrier, higher than the band-edge position at the MS interface, in order to reach the MS interface. Such an interaction (potential smoothing) of neighboring patches with different SBH necessitates significant modifications to eq. (21.6).

21.3.3 Potential at inhomogeneous MS interfaces

When the SBH varies locally at a MS interface, the potential also varies from region to region. The solution to such a problem is usually obtained by solving Poisson's equation, with the SBH contours supplied as the boundary condition. Obviously, the presence or absence of SBH inhomogeneity is closely related to the formation mechanism of the SBH, which will be addressed at a later section. Numerical solutions to this boundary value problem in two convenient geometries all showed the interesting phenomenon of potential pinch-off [49, 51, 52, 53]. The two geometries studied are the most interesting forms of possible SBH inhomogeneities, namely, small areas of the MS interface with a low SBH, embedded in an interface with an otherwise uniform higher SBH. One is in the form of a circular patch [49, 50, 53], the 'patch' geometry, with a small radius, R_o, and the other consists of a strip [51, 52, 53], the 'strip' geometry, with a width of L_o, placed at the origin. The SBH is $\Phi_{Bn,o}^{mean}$ everywhere except for the patch/strip which has a lower but constant SBH of $\Phi_{Bn,o}^{mean} - \Delta$.

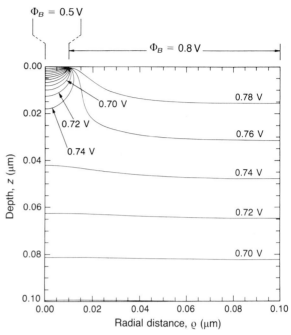

Fig. 21.2 Numerically determined potential contours at zero applied bias along a slice of constant azimuthal angle for a diode which contains a low-SBH circular patch at the origin. The semiconductor is assumed to be n-type Si with doping $1 \times 10^{15}\,\mathrm{cm}^{-3}$, and the simulation temperature is 300 K. The SBH of the diode is 0.8 V except for the low-SBH patch, which, with a radius $R_o = 0.01\,\mu\mathrm{m}$, has an SBH of 0.5 V. The saddle point is at $\rho = 0$ and $z \approx 0.03\,\mu\mathrm{m}$ (after [53]).

The phenomenon of potential pinch-off in the vicinity of a low-SBH patch is shown in Fig. 21.2 with a two-dimensional contour plot generated by numerical calculation [53]. A 'saddle-point' in front of the low-SBH patch, whose potential is a local maximum along the z-axis, but is a local minimum along the lateral axis, may be identified on Fig. 21.2. Obviously, the height and the distribution of the potential at the saddle-points are of vital importance to understanding the physics at inhomogeneous SBs, because they control the electron transport to and from the low-SBH patches/strips. However, the complicated dependences of the potential distribution near the saddle-points on the patch/strip characteristics, the applied bias, the doping level, the temperature, etc. make a conceptual grasp of the pinch-off phenomenon difficult through numerical simulations [49, 52]. In addition, it is impractical to solve the boundary value problems for a large number of parameters in order to explain the electrical data from inhomogeneous SBs. Fortunately, an analytic solution to the potential (and the electron transport) at inhomogeneous SBs was recently obtained [54]. Recent numerical calculations have provided excellent proof for the validity and the accuracy of this analytic theory [53]. For instance, the conduction band minimum (CBM) potential for the patch geometry is [54]

$$V_{\text{patch}}(0, 0, z) = V_d\left(1 - \frac{z}{W}\right)^2 + V_n + V_a$$
$$- \Delta\left[1 - \frac{z}{(z^2 + R_o^2)^{\frac{1}{2}}}\right] \quad (21.7)$$

and, for the strip geometry, is

$$V_{\text{strip}}(x, y, z) = V_d\left(1 - \frac{z}{W}\right)^2 + V_n + V_a$$
$$- \frac{\Delta}{\pi}\tan^{-1}\frac{|x| + L_o/2}{z}$$
$$+ \frac{\Delta}{\pi}\tan^{-1}\frac{|x| - L_o/2}{z} \quad (21.8)$$

where W is the depletion width and V_n is the Fermi energy (the difference between FL and CBM in neutral semiconductor). Equations (21.7) and (21.8), and subsequent discussions, also apply to p-type semiconductors, with a change of appropriate subscripts. The near perfect agreement between the analytic expressions and numerical solutions generated by computer simulations can be seen in Fig. 21.3, under different applied biases.

21.3.4 Electron transport of inhomogeneous SBs

Numerical simulations of electron transport at inhomogeneous SBs, taking into account the pinch-off effect, have existed for some time [50, 52]. However, it was not until recently that the importance of the saddle-point potential has been recognized and that the concept of assigning an effective SBH to each low-SBH area has been developed [54]. Numerical simulations [53] gave excellent

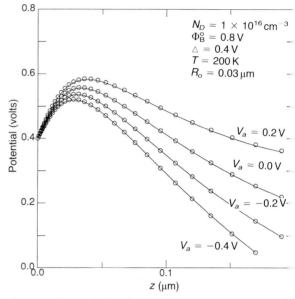

Fig. 21.3 Numerically simulated potential distribution [53] (shown as circles) along $\rho = 0$ of the low-SBH patch at different biases across the MS contact. The potential based on the analytic theory [54], Equation (21.7), is shown as the solid lines.

support to expressing the total current as a modified sum of currents flowing in each patch [54],

$$I(V_a) = A^*T^2\left[\exp(\beta V_a) - 1\right]$$
$$\sum_i A_{i,\text{eff}} \exp(-\beta\Phi_{i,\text{eff}}) \quad (21.9)$$

where the effective SBH, $\Phi_{i,\text{eff}}$, is simply the height of the saddle-point in front of the i-th low-SB patch. The effective area of a patch, $A_{i,\text{eff}}$, is related to the rate of lateral ascent of the potential near the saddle-point. Since the analytic theory describes the potential near the saddle-point extremely well, it is not surprising that the numerical simulated currents [53] showed quantitative agreement with the predictions of the analytic theory using only saddle-point potentials, i.e. eq. (21.9). This discovery obviates the need for complicated computer calculations and makes physical interpretations of the experimental data transparent.

It is probably already obvious from the discussion of the potential that the A_{eff}s and the Φ_{eff}s depend on bias, temperature, doping level, geometry, etc. As will be discussed, almost all the abnormal behavior observed from SBH experiments may be simply explained by these dependences. For a low-SBH circular patch, the effective SBH and the effective area are,

$$\Phi_{\text{eff}} = \Phi_{Bn,o}^{\text{mean}} - 3\Gamma V_d = \Phi_{Bn,o}^{\text{mean}} - \frac{\gamma V_d^{\frac{1}{3}}}{\eta^{\frac{1}{3}}} \quad (21.10)$$

and

$$A_{\text{eff}} = \tfrac{4}{3}\pi\lambda_D^2\Gamma = \frac{4\pi\gamma\eta^{\frac{2}{3}}}{9\beta V_d^{\frac{2}{3}}} \quad (21.11)$$

respectively, according to the analytic theory [54]. The constant η is ε_s/qN_d, and the parameters Γ and γ, which measure the strength of a low-SBH patch, are defined as

$$\gamma \equiv 3\left(\frac{\Delta R_o^2}{4}\right)^{\frac{1}{3}} = 3\Gamma\eta^{\frac{1}{3}}V_d^{\frac{2}{3}} \quad (21.12)$$

It is clear that since the saddle-point potential increases with forward bias (eq. (21.10) and Fig. 21.3), the component of current flowing to a low-SBH area has an ideality factor greater than 1. Specifically,

$$n \approx 1 + \Gamma = 1 + \frac{\gamma}{3\eta^{\frac{1}{3}}V_d^{\frac{2}{3}}} \quad (21.13)$$

for the current flowing to a low-SBH patch. The validity of eqs. (21.10–21.13) has recently been verified by computer simulation [53], as illustrated in Fig. 21.4.

At a real MS interface, SBH variations may occur at a variety of shapes and spatial frequencies. The observed I–V behavior from an inhomogeneous diode depends very much on the characteristics of SBH variation. A random variation of the SBH at a MS interface may be simulated by a statistical distribution of the patch–strip characteristics. For instance, one may assume a Gaussian distribution of the patches

$$N(\gamma) = \frac{c_1}{\sigma}\exp\left[-\frac{\gamma^2}{2\sigma^2}\right] \quad \gamma > 0 \quad (21.14)$$

where $N(\gamma)\,d\gamma$ is the density of circular patches with a characteristic γ between γ and $\gamma + d\gamma$. In eq. (21.14), σ is the standard deviation of the normal

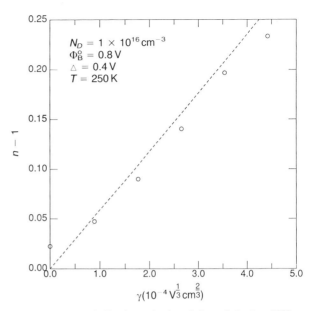

Fig. 21.4 Numerically determined variation of the low-SBH patch ideality factor with γ. The circles represent the points obtained from numerical simulations [53], and the dashed line is the expected result based on an analytic theory [54].

distribution. It has been shown [54] that the current flowing in such a composite inhomogeneous SB diode is made up of two components. One component is the current over the entire diode, which is governed by a uniform SBH of $\Phi_{Bn,o}^{\text{mean}}$, and the other is an additional current due to the presence of the low-SBH patches. The combined effect of all the low-SBH patches is as if there were a big low-SBH region in the diode with an effective SBH of

$$\Phi_{\text{eff}} = \Phi_{Bn,o}^{mean} - \frac{\beta\sigma^2 V_d^{\frac{2}{3}}}{2\eta^{\frac{2}{3}}} \quad (21.15)$$

and an 'overall ideality factor' of

$$n_{\text{eff}} \approx 1 + \frac{\beta\sigma^2}{3\eta^{\frac{2}{3}}V_d^{\frac{1}{3}}} \quad (21.16)$$

Eqs. (21.15) and (21.16) are fascinating. They show that, even though the effective SBH and the ideality factor of each individual patch are roughly temperature-independent, put together in the same diode, they may be represented by a temperature-dependent effective SBH and a temperature-dependent ideality factor! Analytic expressions qualitatively similar to those derived in eqs. (21.10)–(21.16) have also been obtained for the strip geometry [54]. The main difference between the two geometries is in the dependence of the effective SBHs on bias: $\Phi_{Bn,o}^{mean} - \Phi_{eff}$ is proportional to $V_d^{\frac{1}{3}}$ for the patch geometry and proportional to $V_d^{\frac{1}{4}}$ for the strip geometry. Hence, the currents for both patch and strip geometries may be expressed in a generalized function form:

$$I_{total} = AA^* T^2 f(\beta, V_d) [\exp(\beta V_a) - 1]$$
$$\times \exp(- \beta \Phi_{Bn,o}^{mean} + \beta \kappa_1 V_d^{\xi/2} + \beta^2 \kappa_2 V_d^{\xi}) \quad (21.17)$$

where f is a slowly varying function in comparison to the exponential function, and the exponent ξ is $\frac{2}{3}$ and $\frac{1}{2}$, respectively, for the patch and strip geometry. The constants κ_1 and κ_2 depend on the doping level and the distribution of SBHs, cf. eq. (21.15).

21.4 EVIDENCE FOR SBH INHOMOGENEITY: NON-EPITAXIAL MS INTERFACES

21.4.1 Leakages and edge-related currents

Experimentally observed $I–V$ curves have frequently been analyzed and shown to be comprised of two or more components of current [55, 56], as illustrated in Fig. 21.5(a). At small biases, the forward current is sometimes dominated by a 'soft', or 'leaky', component which leads to both a curvature in the $I–V$ curve and tangential slopes corresponding to ideality factors much in excess of 1. As bias increases, the $I–V$ relationship becomes semi-logarithmic with an ideality factor not far from unity, before at very large biases, when the current turns over due to series resistance. The presence of a leaky component is particularly easy to detect at lower temperatures. It is customary to attribute the linear portion of the $I–V$ curve to the main conduction mechanism, e.g. thermionic emission over the SB, and the leakage current at

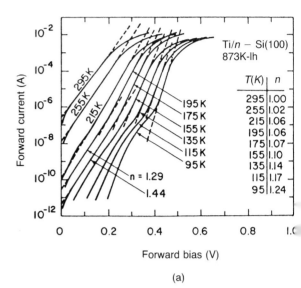

(a)

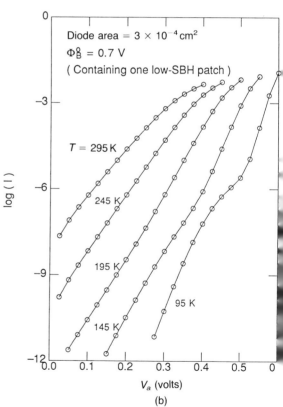

(b)

Fig. 21.5 (a) Forward current-voltage characteristics of Ti on *n*-type Si(100), showing the presence of more than one current component (after [56]), and (b) simulated forward current voltage characteristics from a SB diode which contains a low SBH region.

low biases to a different mechanism. Generation and recombination in the space charge region [57] and edge-related conduction [42, 58, 59, 60] are the mechanisms most frequently thought to lead to the leakage currents. Because edge-related currents scale with the peripheral length of the diode, and not with the area of the diode as would be expected from common junction conduction mechanisms, they may be unambiguously identified by using diodes with different sizes [60, 61]. Until recently, the additional current component associated with edges was thought to be due to an alleged larger electric field at the diode edges which leads to increased tunneling and/or increased generation-recombination [57]. Because the oxide–silicon interface is usually unpinned, the increase in the electric field near the edges is not significant enough, at least for Si, to explain the magnitude of the observed edge current by tunneling. Sometimes, detailed analyses showed that edge-related currents have ideality factors similar to [60], or smaller than [61, 62], that associated with the center portions of the diodes. All these observations indicate that edge-related currents cannot be indiscriminately attributed to the generation-recombination process. It has been so customary to attribute edge-related leakage current to generation-recombination that sometimes this is done even when the activation energy of such current component was shown to be far from one half of the semiconductor band gap [56, 63]. When the leakage current is not clearly related to edges, generation-recombination is still the explanation most commonly invoked. However, the fact that leakage currents were observed to be clearly dominating in some diodes and completely missing in others, among diodes on the same sample [61], seems inconsistent with a uniform distribution of generation-recombination centers.

Experimentally observed leakage currents and edge-related currents are both consistent with SBH inhomogeneity. The presence of a few large low-SBH regions (with their large γs and, hence, large ideality factors) in the SB diode can certainly lead to the observation of a leaky component in the junction current, as shown in Fig. 21.5(b). The small slope of this current (high ideality factor)

and effects due to series resistance limit the predominance of this leakage component to small forward biases. Because of a lower effective SBH, the presence of even a single low-SBH region can lead to the observation of leakage current. Experimentally observed diode-to-diode variations of the leakage current are in much better agreement with isolated 'leakage spots' due to low local SBH rather than with a distribution of recombination centers in the space-charge region. The existence of a current component which is proportional to the perimeter of the diode is also in good agreement with SBH inhomogeneity. A low-SBH patch is less effectively pinched-off when it is situated in close proximity (on the order of the depletion width) to the edge than when it is at the central portion of a diode, as has been clearly demonstrated by computer simulations [53]. The reason for this effect is an increased electric field at the edges. When an SB diode contains a uniform distribution of low-SBH patches, the number of patches which are found near the edges is proportional to the perimeter. Therefore, edge-related currents are consistent with the presence of SBH inhomogeneity. The fact that the edge-related currents sometimes have low ideality factors is also consistent with SBH inhomogeneity, because low-SBH patches near the edges are less pinched-off. Finally, the structure of the MS interface near the edges may be different from that at the central portion of the diode, resulting in a locally different SBH and an edge-related current.

21.4.2 Greater-than-unity ideality factors

Even though the forward $I-V$ plots experimentally observed from SBs are almost always semi-logarithmic, in agreement with the thermionic emission theory, the slopes of such traces often differ from theoretical prediction. This necessitates the inclusion of an empirical parameter, the ideality factor n [64], in the description of the junction current, eq. (21.4). An ideality factor greater than 1 has no direct explanation in the thermionic emission theory, and is generally attributed to a SBH which is bias-dependent. Image force lowering [65], generation-recombination, interface states (negative charge) [18, 66], and thermionic field emission (TFE) [67, 68] have all been discussed as possible mechanisms which could lead to a greater-than-unity ideality factor. Since the image-force lowering and TFE may be calculated and the generation-recombination contribution can be distinguished experimentally, the maximum ideality factor due to these mechanisms may be accurately estimated. Observed ideality factors often far exceed these estimates, prompting the proposal that interface states are a main origin of greater-than-unity ideality factors. The increase of the ideality factor with the doping level, which usually occurs at a faster rate than that predicted by image-force and TFE, has thus far also been attributed to the existence of interface states.

Two entirely different mechanisms have been proposed to explain the ideality factor on interface states: the interface layer (the tunnel MIS diode) approach and the intimate MIGS (negative charge) approach. In the presence of an interfacial dielectric layer (Fig. 21.1(b)), the charge at the dielectric–semiconductor interface becomes more negative (or less positive) with applied forward bias, leading to an increase of the SBH with bias and, hence, a large ideality factor. However, the interpretation based on an interface layer is not consistent with the work function dependence of the SBHs, as already discussed [38]. In addition, annealed MS interfaces, which show no evidence for an interface dielectric layer, often display large ideality factors. At intimate SB interfaces, interface states can also lead to large ideality factors. The upward bending of the semiconductor bands near the MS interface, due to the spatial extension of the (negative) charge, results in different turning points of the potential (different effective SBHs) for different electric fields. This dependence of the SBH on the electric field, shown in Fig. 21.1(c), is the mechanism with which ideality factors are explained by interface states at an intimate MS junction. However, there is a major consequence of such an explanation which has already been violated. The short-range band bending at the MS

interface is independent of the semiconductor doping type, and, therefore, may be used to explain only effects on one type of semiconductor. Specifically, for a particular MS system, MIGS can lead to large ideality factors only on n-type, or p-type, semiconductor, but not on both types. Experimental results show large ideality factors on both n- and p-type semiconductors, in disagreement with the interface state mechanism. There are other experimental observations which are not consistent with interface states. For instance, the ideality factors are often found to vary significantly with processing, or from diode to diode on the same sample, while the SBHs are essentially the same. These results are hard to explain with interface states, because they are usually assumed to decide both the magnitude of the SBH and the ideality factor of a SB diode. Occasionally, the ideality factor seems to correlate mysteriously with the magnitude of the observed SBH: among identically prepared diodes, higher ideality factors were often found to accompany lower observed SBHs [41, 69].

The bias dependence of the effective SBHs (saddle-point potential) of an inhomogeneous SB can explain all the observed behavior of the ideality factors. Since the ideality factor depends on the characteristic of the low-SBH patches, the slope of the I–V curves can vary when the saturation current remains relatively unchanged. When the doping level increases (a smaller η), the Φ_{eff} of a low-SBH patch with a fixed γ decreases and its ideality factor increases, in good agreement with experimental observations. The correlation of SBH and ideality factor [41, 69] is expected when the local SBH varies about the same mean SBH, but with different amplitudes and/or periods. The diode-to-diode variation of the ideality factor and the dependence on processing are also consistent with variations in the distribution of local SBH in the diodes under study. The observation of large ideality factors when the diode is in a state of maximum confusion [70, 71] is also in good agreement with the interpretation of ideality factor based on SBH inhomogeneity. Recently, greater-than-unity ideality factors were correctly speculated to be related to SBH inhomogeneities [72].

even though the underlying reason, namely potential pinch-off, was not discussed.

21.4.3 T_o anomaly and other dependences of ideality factor

Many different temperature dependences of the ideality factor have been experimentally observed. Most frequently, the ideality factor of a diode increases when the sample temperature is lowered. At many MS interfaces, the deduced SBH and ideality factor are found to vary with the measurement temperature in a fashion generally known as the 'T_o anomaly' [73, 74, 75, 76]. Such a phenomenon has been observed from all types of SBs, on elemental semiconductors [73] and compound semiconductors [74, 77] alike. A diode is said to display the T_o effect if its junction current may be expressed as

$$I = A^{**}AT^2 \exp\left(-\frac{q\Phi}{k_{\mathrm{B}}(T + T_o)}\right)$$

$$\left[\exp\left(\frac{qV_a}{k_{\mathrm{B}}(T + T_o)}\right) - 1\right] \qquad (21.18)$$

where T_o is a constant, typically $10-60$ Kelvin. Demonstration of the T_o effect is usually accomplished by plotting $nk_{\mathrm{B}}T$ (the inverse slope of an $I-V$ curve) against $k_{\mathrm{B}}T$ and observing a straight line, with a slope of unity, which does not extrapolate through the origin [73], as illustrated by line 3 of Fig. 21.6. In addition, by changing the abscissa of the Richardson plot from $1/T$ to $1/nT$, a straight line should be observed in cases displaying the T_o anomaly. Levine proposed that the T_o anomaly was due to an exponential distribution of the density of interface states [78]. However, such an analysis [78] depends on the existence of an interface layer [79] and, therefore, cannot explain the T_o effect, which is frequently observed at intimate SBs. Furthermore, there is no experimental evidence for an exponential distribution of states at MS interfaces. The fact that the measured T_o varies significantly among similarly fabricated diodes [80] and the proposal that T_o varies locally in a large diode [81] are suggestive that the T_o anomaly is not directly related to the formation

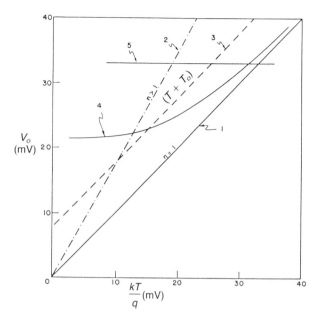

Fig. 21.6 Plot of inverse slope V_o ($=nk_BT$) versus k_BT/q, showing the five basic categories of the temperature dependence of an ideality factor. Line 1 is an ideal SB which follows the prediction of the thermionic emission theory. Line 2 shows a temperature-independent, greater-than-unity, ideality factor. Line 3 displays the T_o effect. Lines 4 and 5 represent the behaviors when the conduction is dominated by, respectively, TFE and FE (after [73]).

mechanism of the SBH. Crowell [79] pointed out that the T_o anomaly was consistent with band bending such as that arising from opposite-type doping at the interface [82]. However, deliberate doping modifications brought out changes [83] in the apparent T_o exactly opposite to that expected from Crowell's proposal. The suggestion that the T_o anomaly is a result of the temperature dependence of the work function [73] seems numerically off by at least an order of magnitude. It thus appears that the T_o anomaly has thus far not been explained adequately, [1, 4, 5] except by the presence of SBH inhomogeneity to be discussed below.

It is usually assumed that a study of the dependence of ideality factor on temperature can reveal the conduction mechanism of a particular SB diode. The T_o phenomenon is only one of five distinctive temperature dependences according to the original categorization by Saxena [73], as schematically shown in Fig. 21.6. A temperature-independent, large ideality factor, line 2 in Fig. 21.6, has often been observed experimentally. So has a dependence similar to line 4 in Fig. 21.6, which is usually attributed to a domination of the conduction mechanism by TFE [67, 68]. Good quantitative agreements of experimental data with the TFE theory have been observed [67]. However, the occasional observations of data similar to line 4 in Fig. 21.6, under experimental conditions where tunneling should be negligible [47], suggest that the interpretation of the conduction mechanism based on the temperature dependence of the ideality factor may not be unique. Frequently, the ideality factor of a SB diode is shown to follow different behaviors at different temperature ranges. For instance, a diode may exhibit a high (low) ideality at high temperature and a low (high) ideality at low temperatures. Furthermore, just like the magnitude of the ideality factor, the temperature dependence of the ideality factor of a diode also varies with processing. Both the diode-to-diode variation and the variation with processing are suggestive that the many distinctive dependences of the ideality factor on temperature are not unrelated.

A temperature-independent, greater-than-unity ideality factor, often observed experimentally [76, 84], is consistent with a SB diode whose current transport is dominated by low-SBH regions with narrowly distributed γs (since Γs, and hence ns, do not have a strong dependence on temperature). An increase of ideality factor with decreasing temperature, as is the case for the majority of SB diodes, is consistent with the presence of many low-SBH regions with a distribution in their low-SBH characteristics (γs). The current of a random, inhomogeneous, SB diode is described by eq. (21.17), which, as shown explicitly [54], may be expressed phenomenologically in a form identical to eq. (21.18). Therefore, SBH inhomogeneity offers the only valid explanation of the T_o anomaly: as the temperature is lowered, the junction current is dominated by fewer low-SBH regions with lower effective SBHs and larger ideality factors. As an example, the inverse slope, in an I–V simulation,

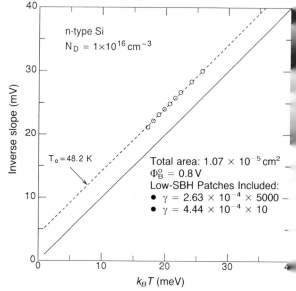

Fig. 21.7 The inverse slope of the simulated I–V plot as a function of $k_B T$ for an inhomogeneous SB diode composed of low-SBH patches as indicated on the figure (γ is in units of $V^{\frac{1}{3}} cm^{\frac{2}{3}}$). The T_o anomaly is reproduced by the presence of just two types of low-SBH patches (after [53]).

of a SB diode with low-SBH patches of just two distinct γs is plotted in Fig. 21.7 as a function of temperature, to illustrate the origin of the T_o mystery. The empirical constant T_o, which depends on how the ideality factor is evaluated experimentally, is related to σ and the doping level [54]. Since the fluctuation of SBH likely varies for different diodes, the inconsistency of the measured T_os [80, 81] and the doping dependence [83] are all naturally explained.

A decrease of the ideality factor with cooling occasionally observed [85], is consistent with the presence of general SBH inhomogeneity about some mean SBH and, in addition, a small number of low-SBH regions which are large enough that they are not pinched-off. This type of temperature dependence [85], which is a contradiction to all existing models of the ideality factor based on interface states or tunneling, has been reproduced in recent simulations of inhomogeneous SBH [53]. The apparent switch-overs, between different ideality factor categories (switching between curve shown in Fig. 21.6), observed at different temperature ranges [85, 86] have also been demonstrated by computer simulations of inhomogeneous SB diodes [53]. Since ideality factors are simply a manifestation of the SBH uniformity, it is not surprising that it may be improved by improving the uniformity of the layer, which presumably leads to more uniform interface structures [87, 88]. Nor should one find it odd that the ideality factor is the largest when the layer is the most non-uniform [71]. The perfect explanation of almost the entire spectrum of the observed temperature-dependence of the ideality factor by SBH inhomogeneity rules out interface states as being of any real significance to the ideality factor.

21.4.4 'Soft' reverse characteristic

It is a universal observation that the current from any SB diode never truly saturates at large reverse bias. These 'soft' reverse characteristics are observed even when the utmost care is taken to eliminate possible effects due to edges of the diode [58]. Image-force lowering is capable of explaining the reverse currents at some SB diodes [65, 89]. However, the majority of SB diodes show reverse currents which far exceed that predicted by the image-force mechanism alone [57, 90]. Andrews and Lepselter [91], who did an extensive investigation of the reverse characteristics of SBs, proposed that, in addition to SBH lowering due to image-force $\delta\Phi_{img}$, there is a SBH lowering, $\delta\Phi_{is}$, which is proportional to the electric field, i.e.

$$\delta\Phi = \delta\Phi_{img} + \delta\Phi_{is} = \frac{1}{2}\left(\frac{qE_{max}}{\pi\varepsilon_s}\right)^{\frac{1}{2}} + \alpha E_{max} \quad (21.19)$$

This additional SBH lowering is thought to arise from the upward bending of the semiconductor band due to MIGS [7, 24], identical to the mechanism proposed (and presently ruled out) to explain greater-than-unity ideality factors. In eq. (21.19), the constant α is thought to be related to the density and depth of the interface states [24]. Experimental results in agreement with the prediction of this model have been observed in many studies [91, 92, 93]. For example, the good fit of

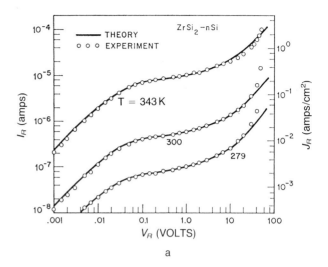

a

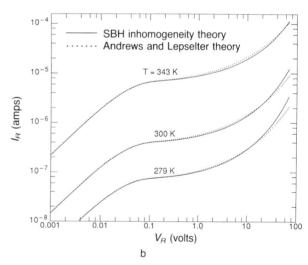

b

Fig. 21.8 (a) Reverse characteristics of a $ZrSi_2$ Schottky diode formed in *n*-type silicon. Andrews and Lepselter's theory [91] was used (eq. (21.19) with $\alpha = 10\text{Å}$) to generate the solid lines (after [91]). (b) Comparison of reverse currents due to SBH inhomogeneity with that according to Andrews and Lepselter's theory.

experimental data, over many decades, with the prediction of eq. (21.19) is shown in Fig. 21.8. However, even though the functional form of the experimental reverse current may be explained by this model, other consequences of this model, due to interface states, have not been borne out by experiments. For instance, it is not clear why very different αs are found for similar MS interfaces. Also, the proposed mechanism of SBH lowering is completely absent in some diodes [65, 89]. But, as before, the most severe problem with the interface state model is that the proposed SBH lowering should be proportional to the electric field perpendicular to the interface, and *not* to its absolute value. In other words, for a particular MS system, the SBH lowering mechanism due to interface states, $\delta\Phi_{is}$, may be operative on either the n-type or the p-type semiconductor, but not on both. Experimental results [92, 94], including those shown by Andrews and Lepselter in their original paper [91], indicated that the soft SB characteristics for many metals/silicides occurred with similar magnitudes on both types of Si substrate. This contradiction suggests that interface states are not the main reason for the observed bias dependence of the reverse currents.

Due to the limited range of available reverse bias, a reliable determination of the functional form of the current of any particular SB diode is difficult enough. In addition, the reverse currents of different diodes often show different behaviors. Therefore, even though eq. (21.19) may account for the observed reverse currents of some diodes, it is by no means the only lowering mechanism which is capable of explaining the experimental data, nor is it able to explain all the experimental observations. Any SBH lowering mechanism which varies with the electric field more rapidly than $E_{max}^{\frac{1}{2}}$, could lead to a very satisfactory fit with the experimentally observed reverse currents. As discussed, the reverse characteristic of an inhomogeneous SB depends critically on the actual variation of the SBH. A variety of SBH lowerings on the reverse bias, such as proportional to the $\frac{1}{2}$ and $\frac{2}{3}$ powers of V_d, are possible in inhomogeneous SB diodes with a random distribution of low-SBH patches. SBH inhomogeneities are perfectly capable of explaining the bias dependence of experimentally observed reverse characteristics. As shown in Fig. 21.8(a), Andrews and Lepselter's theory [91] tends to slightly overestimate the current at very-large reverse biases at high temperatures and underestimates it at low temperatures. The experimental results are actually in better agreement with the behavior expected of inhomogeneous SBs, shown in Fig. 21.8(b). The wide range of behaviors of reverse currents experimentally observed from various SBs suggests that each specific diode may have its own individuality. Such a scenario is in complete accord with SBH inhomogeneity and is not easily reconciled with the interface state model [91].

21.4.5 Dependence of SBH on measurement technique

The SBH measured by the I–V technique, Φ_{I-V}, often decreases with increasing doping level, while the SBH measured by the C–V method, Φ_{C-V}, remains constant. Frequently, the SBHs depend on the technique of measurement, namely, Φ_{C-V} sometimes significantly exceeds Φ_{I-V} and the SBH derived from PR techniques, Φ_{PR} [95, 96, 97]. Identical to the proposed explanations of the ideality factor, lowering of the SBH by image-force, interface states, and TFE have been frequently invoked to explain the doping-level-dependence of Φ_{I-V}s [98, 99, 100]. While SBH investigations, especially those where the doping-dependence of the SBH has been studied, have generally concentrated on n-type semiconductors it is known that SBH lowerings on p-type semi conductors also routinely exceed that predicted by image-force alone. The argument invoked in the last few sections concerning the sign of SBH lowering due to MIGS, namely, it is not applicable to both n- and p-type semiconductors, also applies to the dependence of Φ_{I-V} on doping level. Thus it is clear that the observed doping-dependence of moderately-doped, intimate, SB contacts may not be attributed to interface states as previously thought. As already pointed out [40, 48, 49, 101] a dependence of SBH on the measurement tech nique is in agreement with the presence of SBH inhomogeneities. Current transport at inhomo geneous SBs is dominated by low-SBH patches leading to the deduction by I–V and PR tech niques of apparent SBHs which are lower than the arithmetic average of the entire diode. Since under usual circumstances, the C–V technique yields an average SBH for the whole diode [54], the experimentally observed dependence of SBH on the technique of measurement is likely due to SBH inhomogeneity.

A common occurrence in $C-V$ experiments is the deduction of an apparent SBH which is even higher than the true arithmetic average. This phenomenon is often associated with experimentation problems, namely, how to determine the true space charge capacitance from raw $C-V$ data. There are many well-known phenomena which may lead to the measurement of apparent capacitances which are not the true junction capacitances, as previously discussed [102]. Besides possible explanations from deep levels [103, 104], doping variations [105], edges [106], interface states [107], etc., one notes that a large series resistance may lead to such an observation [102, 108]. The effect due to series resistance is particularly noticeable at high modulation frequencies and when the magnitude of C is large [102]. When the SBH is uniform, the true junction capacitance may be deconvoluted, using a correct equivalent circuit diagram, and the measured value of the series resistance. But in the case of an inhomogeneous SB, the series resistance issue is difficult to handle because of the lateral inhomogeneity of the in-phase current. Since the series resistance is not a fixed parameter for an inhomogeneous diode, an appropriate equivalent circuit diagram cannot be drawn. It is difficult to utilize the $C-V$ technique for SBH deduction under such conditions.

Other $C-V$ anomalies exist, e.g. the occasional observation of a downward curvature in $C^{\frac{1}{2}}$ plots has been explained in terms of an 'excess capacitance (C_o)' and attributed to the presence of interface states [109, 110, 111]. Excess capacitances observed at an SB in the forward bias have also been attributed to interface states [33, 34, 35]. The explanation of excess capacitance in terms of interface states depends on either the existence of an interfacial layer or of a fixed separation between the charge on the semiconductor side and that on the metal side of the MS interface [112]. For intimate SBs, these assumptions have no physical basis, and one expects a negative capacitance (inductance) contribution from interface states [112]. It is likely that the observed excess capaci-

tance may be dominated by minority carriers [113]. In the presence of SBH inhomogeneity, the minority carrier injection ratio far exceeds that predicted based on homogeneous SB, for the same average SBH, and effects due to minority carriers are greatly enhanced [113]. One further notes that, in the presence of an inhomogeneous SBH, excessive capacitances may be observed simply because charges stored in the potential 'pockets', in front of low-SBH patches, modulate with the applied bias! In many respects, such a fluctuation of charge is indistinguishable from capture and emission processes due to interface states. It is interesting to note that much higher excess capacitances are observed from interfaces which display higher ideality factors in $I-V$ measurements [35], in good agreement with the behavior expected of inhomogeneous SBHs.

21.4.6 Inconsistencies in SBH data

Different preparations of the semiconductor surface lead to a difference in the amounts of chemical impurities, oxides, and structural defects and, hence, different electronic properties of the eventual MS interface. One of the earliest discoveries concerning the influence of surface treatment on SBH was the observation of 'ageing' (variation of SBH characteristics with time) on chemically etched surfaces [114], and not on surfaces cleaved in vacuum. Apparently not all etched surfaces behaved the same way, as ageing was found to be totally absent in many studies [115, 116]. The SBHs of deposited metals sometimes show dependences on other properties of the semiconductor, such as the orientation [117] and the doping level [98, 99]. Another inconsistency which often exists is a difference in SBHs between diodes prepared under similar conditions by different laboratories. For instance, as deposited Ni shows SBHs ranging from ~0.7 eV [114, 118, 119] to ~0.5 eV on n-type Si [120]. Significant variations in the SBHs have also been observed on cleaved surfaces [95, 121, 122]. It has been demonstrated that the quality of cleaved surfaces depends much on the particular cleave, and may even vary considerably from region to region on one cleaved surface. This fact led to large fluctuations and inconsistencies in the observed SBHs [122, 123, 124]. Variations and inconsistencies in the observed SBHs are usually larger for as-deposited metals than for well annealed junctions. For instance, it is generally concluded that various silicides of one metal have similar SBHs with Si [69]. However, some minor variations in the magnitude of the SBH and in other junction characteristics, such as the ideality factor, the difference between $I–V$ and $C–V$ measurements, the reverse currents, etc. are almost always present. Occasionally, the SBH is found to depend on the orientation of the semiconductor substrate [41], the thickness of the silicide layer [62, 71, 125], the degree of epitaxy [126], and the technique of measurements [127].

The various dependences just described are not consistent with the simple FL pinning picture, but are in agreement with a dependence of the SBH on the structure of the interface. Different preparations can certainly lead to different structures at the MS interface. Since annealed MS interfaces usually show less fluctuations than as-deposited interfaces, it has been proposed that this phenomenon is due to defects on the original surfaces which are subsequently annealed out. Since defects, impurities, etc. are nothing more than part of the structure of a surface, perhaps a more general way to view an as-deposited MS interface is to say that it has a less stable structure, which is reminiscent of the original surface and certainly may fluctuate from region to region, than the structure of an annealed interface, which is more a result of thermodynamics. Direct evidence for the inhomogeneity in the SBHs of various silicides has occasionally been recognized [128, 129, 130], further supporting the dependence of the SBH on the structure. In particular, evidence for SBH inhomogeneity has been reported by ballistic electron emission microscopy (BEEM) [131]. Although the interpretation of BEEM image is not trivial [54, 132], it does offer exciting prospects for a simultaneous study of the structure and the SBH of a MS interface, on a small lateral scale.

21.4.7 SBH trends and summary

When the SBHs of a large number of metals are summarized onto a table or a plot, there are usually interesting correlations and systematics which one may point to. Guided by the Schottky–Mott relationship [133], (eq. (21.1)), early correlations have mostly been made to the electronegativity [19, 134] and the work function [135] of the metal. Although the SBHs are usually higher for metals with larger work functions or electronegativities, the fitted slope of such a plot is far less than that predicted by eq. (21.1) [19, 15]. This led to the suggestion of surface/interface states, as discussed earlier. Andrews and Phillips [136] noted that, with few exceptions, the silicide SBHs exhibit a linear relationship with the heat of formation of the silicides. It has also been pointed out that these silicide SBHs may be correlated with the eutectic temperature of the metal–Si binary system [137], which led to the proposal that an interfacial (amorphous) layer determines the SBH. The use of effective work functions, characteristics of an interfacial MSi_4 stoichiometry, have also produced a correlation with the observed silicide SBHs [138]. Recent experiments have shown the abruptness of most silicide interfaces, with no evidence for amorphous or additional interface layers. Tersoff [23] proposed that the SBHs are determined by the charge neutrality level of the MIGS, which implies that the SBH does not have a first order dependence on the metal. A refined treatment [139] considered the screening of the interface dipole by MIGS and derived a relationship of the SBH with the metal electronegativity. Since the charge neutrality of the entire MS interfacial region, which includes about 10 Å of metal and 20 Å of semiconductor, is already approximately satisfied, it is likely that the insistence of MIGS charge neutrality [23] is an unnecessary over constraint. Schmid plotted the silicide SBH against the Miedema electronegativities of the metals and produced a relationship which is reminiscent of the theoretical predicted FL pinning behavior [30] by a finite density of interface states. This result is further expanded to stress the importance of considering effects due to both defects and MIGS [26].

It has been a general practice for SBH investigators to attribute changes in the electronic properties of the MS interface to changes in parameters such as interface states density and interfacial layer thickness. The obvious reason is that one may apply these quantities directly in some models one assumes for the interface states, such as those shown in Fig. 21.1. Possible roles played by the almost certain variations in the chemical nature and the physical structure of the interfaces during these SBH changes have been largely ignored. It should be kept in mind that the common attribution of SBH variations to interface states originated from convenience, rather than through scientific deduction. Generally speaking, the simplicity and the rigidity of the FL pinning concept are hard to reconcile with the diversity of phenomena observed at MS interfaces. In the light of massive evidence for SBH inhomogeneities in polycrystalline SB diodes discussed earlier, the very approach of concentrating only on the magnitudes of the experimental SBHs, in order to assess the validity of a SB theory, seems inappropriate. If a MS interface has an inhomogeneous SBH, then the SBH experimentally obtained from this interface is just an averaged value of some weighted distribution of different FL positions. Such an average does not necessarily have physical significance and certainly should not be the only data used for a deduction of the SB mechanism.

The most direct explanation of inhomogeneous SBHs is that the SB mechanism depends on some local parameters of the MS interface. It is the variation of these local specifics which causes the local FL to change. The true formation mechanism of the SB is, of course, the one which determines the local FL position based on these local specifics. However, the possible relationship between interface structure and the SBH is not easily studied in MS systems discussed so far. The reason is that the structure at an ordinary, polycrystalline, MS interface is too complicated to allow a meaningful correlation with the observed electronic structure. Fortunately, high quality epitaxial MS interfaces

have become available which have made the study of the correlation of structure with FL position considerably easier, as will be discussed in detail.

21.5 EPITAXIAL MS INTERFACES

21.5.1 Advantages of epitaxy

Epitaxially fabricated MS systems offer the best opportunity to understand the dependence of the electronic properties on the structure of a MS interface. Not only can the atomic structure of a single crystal MS interface be obtained by state-of-the-art experimental techniques, but the electronic structures may also be calculated based on the observed atomic structures. A comparison of the experimental and theoretical SBH results represents the best hope of understanding the formation mechanism of the SB. Therefore, even though the formation mechanism of the SBH at ordinary MS interfaces is of the most scientific and technological interest, its deduction almost certainly has to come from the simpler, epitaxial, 'model' MS systems.

21.5.2 NiSi$_2$ on Si(111)

Fabrication and structures

On Si(111), two epitaxial orientations are possible for NiSi$_2$ (and CoSi$_2$). The type A silicide has the same orientation as the silicon substrate, and the type B silicide shares the surface normal $\langle 111 \rangle$ axis with Si, but is rotated 180° about this axis with respect to the Si [140]. High quality single crystals of NiSi$_2$ may be grown on Si(111), with either type A or type B orientation, by a proper choice of template growth condition [8, 141, 142, 143]. The epitaxial orientation of thin NiSi$_2$ layers depends on the amount of deposited nickel. When ~16–20 Å nickel is deposited at room temperature, subsequent annealing leads to the growth of type A NiSi$_2$ [8, 144, 145]. Even though there are no dislocations in thin layers, steps, due to the accidental misorientation of the Si(111) wafer, are

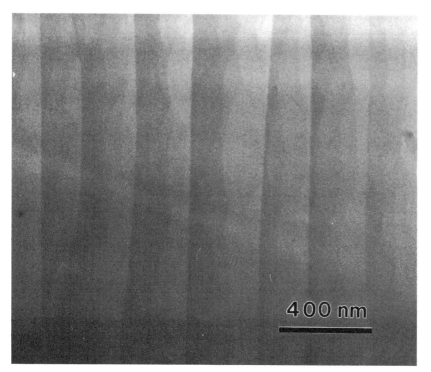

Fig. 21.9 Plan view, (200) dark-field, TEM image of a 63 Å thick typ A NiSi$_2$ layers on Si(111). This layer contains no observable dislocations and pinholes. However, the (unintentional) misorientation of the Si(111) substrate is ~0.4–0.5°, which leads to regular arrays of steps at the interface and on the surface of the layer. This slight modulation of the film thickness is revealed by TEM.

invariably present at type A NiSi$_2$ interfaces, leading to the terrace-like transmission electron microscopy (TEM) images, as shown in Fig. 21.9. Deposition of Si or co-deposition of NiSi$_2$ are common techniques employed for growth of purely type B oriented NiSi$_2$ [145, 146, 147]. Recently, it was discovered that single crystal type B NiSi$_2$ layers can be grown at room temperature by (pre-)deposition of a suitable amount, ~2 Å, of Ni and co-deposition of NiSi$_2$ [146]. A typical room-temperature-grown NiSi$_2$ layer is shown in Fig. 21.10(a). Although single crystal growth is demonstrated at room temperature, the density of defects contained in the grown layers is high, evidenced by a fuzziness in their TEM images, as shown in Fig. 21.10(a). Upon annealing at above 500°C the defects in these NiSi$_2$ layers are reduced, evidenced by an improvement of the channeling characteristics and the observation of much more robust TEM images, as shown in Fig. 21.10(b).

The atomic structure of both type A and type B NiSi$_2$–Si(111) interfaces have been studied by high resolution electron microscopy (HREM) and foun to have the 7-fold structure [148, 149, 150]. A hig resolution image [150] of the type A NiSi$_2$ interfac is shown in Fig. 21.11. The terminology for th structure of a silicide interface, e.g. 7-fold, is base on the number of nearest Si neighbor atoms to metal atom at the interface. In a bulk disilicid lattice, each metal atom has a coordination numbe of 8. Structural models of the two 7-fold NiSi$_2$ Si(111) interfaces are schematically shown i Fig. 21.12. These early HREM results were late confirmed by X-ray standing wave (XSW) [15 152] medium energy ion scattering (MEIS) [153 and X-ray interference [154] investigations.

Electronic properties

An intriguing dependence of the SBH on th epitaxial orientation has been observed at th epitaxial NiSi$_2$–Si(111) interfaces: Type A ar type B NiSi$_2$ have distinctively different SBF [101, 155]. Typical *I–V* characteristics fro

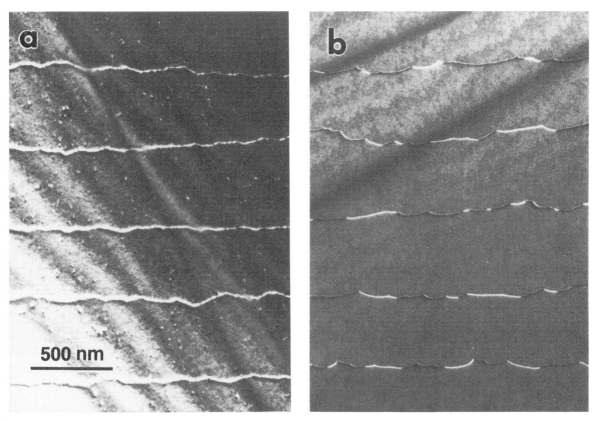

Fig. 21.10 Plan view, (220) weak beam TEM micrographs of ~80 Å thick, type B, NiSi$_2$ layers on Si(111). (a) is a layer grown at room temperature by deposition of ~2 Å of Ni and the codeposition of NiSi$_2$. (b) is a similarly grown layer, after a 500°C anneal.

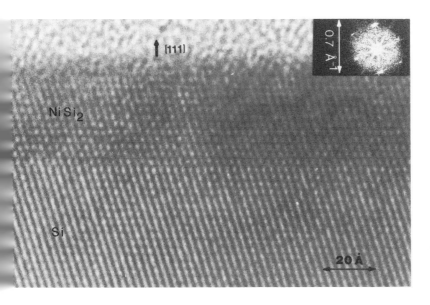

Fig. 21.11 Cross-sectional HREM image of the lattice at a type A NiSi$_2$/Si(111) interface, viewed in the [1$\bar{1}$0] direction (after [150]).

(a) Type A NiSi$_2$–Si (III)

(b) Type B NiSi$_2$–Si (III)

(c) Type B CoSi$_2$–Si (III)

(d) NiSi$_2$–Si (100)

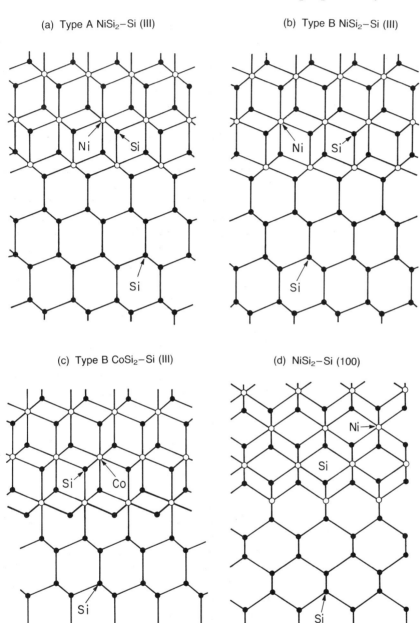

Fig. 21.12 Ball and stick models of epitaxial silicide–Si interfaces, viewed in the [1$\bar{1}$0] direction. (a) the 7-fold type A NiSi$_2$–Si(111) interface, (b) the 7-fold type B NiSi$_2$–Si(111) interface, (c) the 8-fold type B CoSi$_2$–Si(111) interface and (d) the 6-fold NiSi$_2$–Si(100) interface.

NiSi$_2$–Si(111) interfaces are shown in Fig. 21.13 to illustrate the significant difference between the two orientations. This dependence of SBH on the epitaxial orientation was briefly challenged. However, very extensive studies by various groups [156, 157, 158] have since fully confirmed the original findings [155] that type B NiSi$_2$ has a SBH about 0.14 eV higher than type A NiSi$_2$ on *n*-type Si(111). The origin of the initial disagreement is now understood to be a surface boron co-

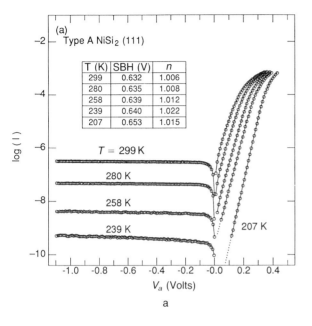

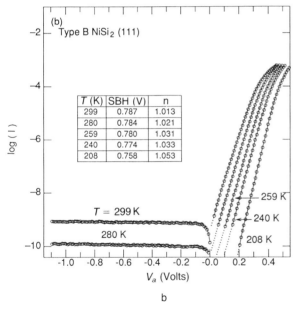

Fig. 21.13 Current-voltage characteristics from epitaxial NiSi$_2$ diodes fabricated on *n*-type Si(111). (a) type A orientation, (b) type B orientation.

tamination problem associated with the experimental conditions employed in the study of Liehr *et al.* Under careful procedures which avoid this boron problem, a low SBH, ~0.65 eV, for type

A NiSi$_2$ on *n*-type Si(111) is very consistently measured by every laboratory. *I–V* studies of type A NiSi$_2$ diodes on *n*-type Si(111), such as that shown in Fig. 21.13(a), indicate that the ideality factor for high quality type A diodes is very close to 1.00, independent of the measurement temperature. From the earlier discussion of inhomogeneous SBHs, this observation is consistent with a homogeneous SBH.

For the type B NiSi$_2$ interface, some variation of the SBH data has appeared in the literature, which suggests that the SBH may depend on growth, diode processing, and the method of SBH measurements [156, 157, 158, 159]. The value of 0.79 eV, originally reported [155] for this interface, has since been quantitatively reproduced by one group [157]. One report claimed that the SBH of type B NiSi$_2$ depended on film thickness [160] and attributed this dependence to a variation of the dislocation density [160, 161]. However, the dislocation density at a type B NiSi$_2$ interface is not a simple function of the film thickness, but is strongly dependent on the growth procedure. One notes that the samples studied by Kikuchi *et al.* [160] were grown by co-deposition of NiSi$_2$ at ~400°C, a technique which likely leads to very rough layer morphology. Inclined facets of a type B NiSi$_2$ layer are incoherent twin boundaries with structures different from the planar, 7-fold structure. Therefore, non-planar type B NiSi$_2$ interfaces may well have inhomogeneous SBHs. Of course, the dislocations present at type B interfaces may have a locally different SBH which also affects the electron transport. Type B NiSi$_2$ SB diodes usually show a slightly larger ideality factor than type A diodes grown on the same substrate, e.g. Fig. 21.13(b). This is suggestive of some minor inhomogeneity at the type B interface.

As already mentioned, the most important advantage of an epitaxial MS interface, in terms of SBH investigations, is the possibility to conduct theoretical calculations based on real atomic structures. Because of their high structural perfection and the intriguing variation of SBH with orientation, type A and type B NiSi$_2$ interfaces represent the best testing ground for a theoretical investigation of the formation mechanism of the SB.

Not surprisingly, there have already been many calculations of the electronic structure and SBH properties at these interfaces. Recent calculations, involving very large supercells [11, 12, 162], have all yielded SBH results which are in excellent agreement with experimental results [155]. An *n*-type SBH for the type B $NiSi_2$ interface which is 0.14–0.15 eV higher than the type A interface was obtained in two independent calculations [11, 12] (to be compared with the 0.14 eV difference experimentally observed [155]). The importance of using a large enough supercell to obtain convergence was clearly noted [12, 163]. Even though the validity of these calculations is not beyond suspicion, it is, nevertheless, informative to examine the calculated electronic structures for clues to the SBH mechanism. Although the densities of MIGS are high at both interfaces [12], there is no easily identifiable feature in the energetic distribution of these states to suggest a 'pinning' of the FL. Actually, the role played by MIGS in determining the FL position was not clearly revealed by these calculations [11, 12]. One notes that there are significant differences in the distribution of MIGS at these two interfaces, suggesting a dependence of MIGS on the interface structure. This dependence indicates that the formalisms previously constructed for MIGS [7, 24, 23] are too simplified for real MS interfaces. Furthermore, since MIGSs depend on interface structure, one may not apply a charge neutrality criterion [23] based on the semiconductor band structure alone.

An examination of the potential due to all charges at the two interfaces [12], as shown in Fig. 21.14, reveals several other surprising results. First of all, the charge due to MIGS is only a small part of the total redistributed charge at the interface. Since the difference in the calculated SBHs originates from the difference in the interface dipoles for the two orientations, that this difference in dipoles is found to be due largely to states below the VBM is contradictory to the concept of FL pinning. Local structures and bonding geometries seem more important than MIGS in contributing to the difference in charge distributions at the $NiSi_2$–Si interface. Furthermore, the local charge

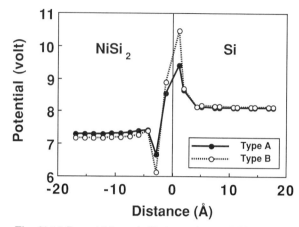

Fig. 21.14 Potential in each Si site at the two $NiSi_2$–Si(111) interfaces. Potential was calculated by Madelung constants and charges in other atomic spheres. The center line is the interface (after [12]).

oscillates rapidly at the MS interface and extends roughly an equal distance, ~6–9 Å, into both the Si and the $NiSi_2$ sides of the interface. The potentials at these interfaces [12], shown in Fig. 21.14, bear no resemblance to the monotonic and smooth band diagram usually assumed for MIGS [7, 24], shown in Fig. 21.1(c). Also absent from the calculated potential of the $NiSi_2$–Si interfaces are the exponential decay of the interface charge and the corresponding band-bending, both of which are central in models of interface states [7, 24]. Without the exponentially decaying band-bending, explanations based on interface states to observations such as the ideality factor, the soft reverse current, the dependence of SBH on the measurement techniques, etc. also lose their basis. The formation mechanism of the SBH at epitaxial $NiSi_2$ interfaces seems, in spirit, close to the charge transfer and chemical bonding mechanism of Andrews and Phillips [136] and others. However, it is the specific charge redistribution due to the *interfacial* bonding, and not silicide bulk bonding, that seems to be more relevant. Perhaps the formation mechanism of the SBH should best be described as one which depends, in a complicated way, on the local atomic structure and bonding configuration. It is simply amazing that the minute difference in the interface structures

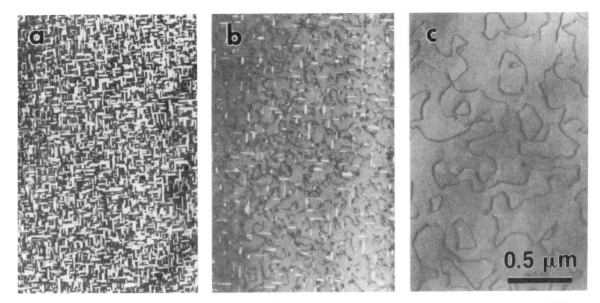

Fig. 21.15 Plan view, (002) dark field, TEM images of ~80 Å thick NiSi$_2$ layers grown on Si(100). A facet bar shows up as a bright streak under this imaging condition. Dark lines are defects with characters related to $\frac{1}{4}$(111), which decorate steps with an odd number of atomic planes at the interface. (a) a layer which is nearly completely faceted, (b) a layer with mixed-morphology, (c) a uniform layer.

between A and B interfaces is enough to produce a significant difference in the SBHs.

21.5.3 NiSi$_2$ on Si(100)

Fabrication and structures

Layer uniformity is the main issue facing the growth of NiSi$_2$ on Si(100). The nucleation of NiSi$_2$ on Si(100) requires ~350°C, but its homo-epitaxial growth may take place at room temperature [164]. Epitaxial NiSi$_2$ layers grown at temperatures below 700°C often contain a high density, 10^8–10^{10} cm^{-2}, of 'facet bars' and therefore are very non-uniform in thickness [165]. A facet bar is a slender and straight NiSi$_2$ protrusion bound by two inclined $\langle 111 \rangle$ facets, usually ~100 Å in its depth and height and a few thousand angstroms in length [165]. The density and dimensions of facet bars are conveniently studied by plan view TEM, where under (020)-type dark field these facet bars appear as short, bright streaks, as shown in Fig. 21.15(b). In addition

to facet bars, NiSi$_2$ layers contain a high density of dislocations. For films less than 200 Å thick, most dislocations have $\frac{1}{4}\langle 111 \rangle$-related characters. They are not driven by misfit stress but are a result of steps at the interface. Symmetry requires that across such a dislocation the height of the interface must change by an odd number of atomic planes [165]. It was recently shown that high temperature (>700°C) anneals significantly reduce the density of dislocations and may completely eliminate facet bars in NiSi$_2$ layers [61], as shown in Fig. 21.15(c). It was also discovered that the deposition of a layer (~10 Å) of Ni on a thin NiSi$_2$ layer and the subsequent annealing at ~400–600°C reproducibly lead to the fabrication of almost completely faceted interfaces [61], as shown in Fig. 21.15(a). An examination of the three micrographs shown in Fig. 21.15 demonstrates that the morphology of the epitaxial NiSi$_2$–Si(100) interface may be controlled.

The atomic structure of the planar NiSi$_2$–Si(100) interface has been studied experimentally by HREM [150, 166]. Rigid shifts measurements

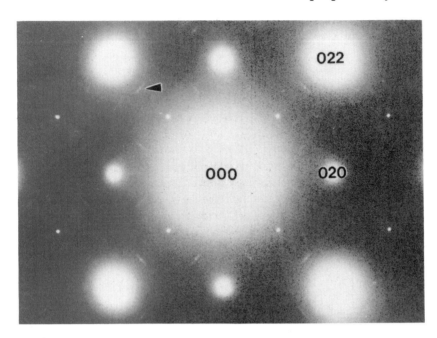

Fig. 21.16 Transmission electron diffraction pattern from a ~50 Å thick NiSi$_2$ layer which was annealed at 700°C. Arrow points to a $(\frac{1}{2}\frac{1}{2}0)$-related streak.

were used to conclude that the NiSi$_2$–Si(100) interface was 6-fold coordinated, as depicted in Fig. 21.12(d), although these experimental results were also consistent with an 8-fold coordination model. Actually, recent theoretical calculations are suggestive that the coordination number at this interface is higher than 6 [167, 168]. In addition to the uncertainty in the 6-fold model, there is now new evidence for the presence of additional structures at this interface [61]. Shown in Fig. 21.16 is a transmission electron diffraction pattern of a thin NiSi$_2$ layer grown on Si(100). In addition to diffraction spots expected of the fcc NiSi$_2$, strong $\langle 011 \rangle$ spots and weak streaks at $\langle 0\frac{1}{2}\frac{1}{2} \rangle$-related locations are seen. The $\langle 011 \rangle$ diffracted beams originate from an abrupt termination of the crystal at the silicide interface and are an indication of a very flat interface. It has been verified that the streaky $\langle 0\frac{1}{2}\frac{1}{2} \rangle$ beams are associated with the planar NiSi$_2$ interfaces and do not originate from facet bars [61]. The intensity and diffuseness of these $\langle 0\frac{1}{2}\frac{1}{2} \rangle$ spots depend on the preparation of the NiSi$_2$ layers. The presence of streaked intensities at these superperiodicity positions and the absence of any $\langle 010 \rangle$-related spots suggest that

(part of) the NiSi$_2$–Si(100) interface may have a structure which is similar to the 2 × 1 reconstruction at the (8-fold) CoSi$_2$–Si(100) interface [169]. It seems reasonable to tentatively attribute the presence of $\langle 0\frac{1}{2}\frac{1}{2} \rangle$ streaks to some derivatives of a 1 × 2 reconstruction at the NiSi$_2$–Si(100) interface. The observation of this reconstruction suggests that the structure of even the most uniform, single crystal NiSi$_2$–Si(100) interfaces may be inhomogeneous.

Electronic properties

It has long been shown that the NiSi$_2$–Si(100) interface could have a much lower SBH on *n*-type Si than the SBHs at NiSi$_2$–Si(111) interfaces [170, 171]. However, with the everpresence of facet bars in the early studies, no conclusion on the SBH could be drawn [171]. Kikuchi *et al.* [160] suggested that the SBH of NiSi$_2$–Si(100) might not be different from that observed at the type A NiSi$_2$–Si(111) interface. These early issues were addressed in a recent study [132], where layers with a variety of morphologies were employed (Fig. 21.15). It was demonstrated that the SBH

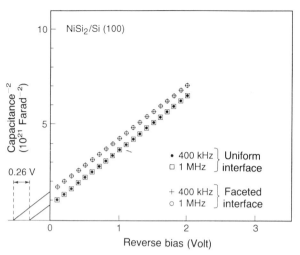

Fig. 21.17 Capacitance-voltage plots of NiSi$_2$ layers on 1.6×10^{15} cm^{-3} phosphorus-doped (n-type) Si(100). Measurements are performed at $T = 198$ K at frequencies of 400 kHz and 1 MHz. Diode area is 1.6×10^{-3} cm^2.

measured at a NiSi$_2$–Si(100) interface depended very much on the observed morphology of the particular layer. Typically C–V result from a faceted and a uniform NiSi$_2$–Si(100) diode are shown in Fig. 21.17 to illustrate the dependence of SBH on interface morphology. Interfaces which are almost completely $\langle 111 \rangle$ faceted were shown to have SBHs similar to that found at a type A NiSi$_2$–Si(111) interface [132]. This result is in good agreement with Kikuchi *et al.* [160], since the growth procedures used by them invariably lead to nearly completely faceted interfaces [61]. Inclined facets at the NiSi$_2$–Si(100) interface are simply sections of a type A, 7-fold, NiSi$_2$–Si(111) interface. It is not surprising that the SBH found for a NiSi$_2$–Si(100) interface made up entirely of facets is identical to that found at a planar type A NiSi$_2$–Si(111) interface [155].

Uniform NiSi$_2$–Si(100) interfaces show an SBH which is much lower on n-type Si than either of the two NiSi$_2$–Si(111) interfaces. Since the NiSi$_2$–Si(100) interface has an entirely different atomic structure from either of the two NiSi$_2$–Si(111) interfaces, it is perhaps not surprising that the SBH is also different. One notes that the low SBH of 0.4 eV measured from uniform NiSi$_2$ layers is also

very different from the value of 0.6–0.7 eV usually observed for all phases of polycrystalline nickel silicides on Si [120, 69]. On n-type substrates, ideality factors for I–V measurements were found to be good ($n < 1.03$) and very consistent results were obtained from either I–V, C–V, or activation energy studies [132]. There was no observed dependence of measured SBH on the doping level [132]. However, on p-type Si(100), planar NiSi$_2$ diodes showed evidence for a slight inhomogeneity of the SBH, such as large ideality factors, $n \geqslant 1.08$, and discrepancies between SBHs measured by I–V and C–V techniques [132]. Therefore, even very uniform NiSi–Si(100) interfaces may in fact be electrically inhomogeneous. The origin of this SBH inhomogeneity is likely related to the 1×2 reconstruction which has been shown to be present at this interface [61].

The presence of a few 'facet bars' at NiSi$_2$–Si(100) interfaces, which are otherwise flat, was found to have little effect on n-type SBH, but had a strong influence on the measured SBH on p-type Si. The I–V deduced p-type SBH decreased rapidly as the density of facet bars increased, while a slower, but noticeable, decrease of the C–V SBH was concurrently observed. As a result, the C–V measured SBH for any specific diode significantly exceeded that deduced from I–V. Mixed-morphology p-type diodes were 'leaky', having poor ideality factors $n > 1.1$ in forward bias and displaying reverse currents which did not saturate. There was also a clear dependence of the electron transport on the substrate doping level. NiSi$_2$ layers with similar densities of facet bars, but grown on p-type Si with different doping levels, showed almost identical SBHs as determined by C–V, but a sharp decrease of the SBHs determined by I–V with doping level. As discussed previously, these experimental results are a clear indication of an inhomogeneous SB. Applying the analytic theory for inhomogeneous SBH, a semi-quantitative explanation of these electrical data was demonstrated [132]. In detail, a local n-type Φ_B^o of 0.65 eV (0.47 eV on p-type Si) was assigned to areas occupied by facet bars and 0.40 eV (0.72 eV on p-type Si) to the flat areas. Since flat areas usually occupied the majority of the interface in a mixed

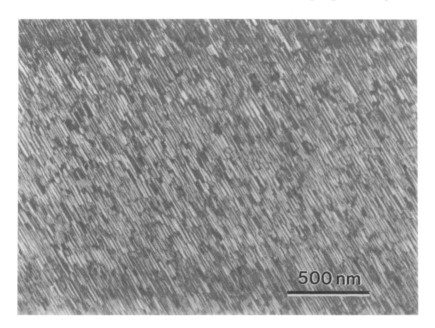

Fig. 21.18 Plan view, (002) dark-field, TEM image of a ~70 Å thick NiSi$_2$ layer on Si(110). Long streaks running along the [1$\bar{1}$0] direction indicate a near complete ⟨111⟩ faceting.

morphology film, an almost constant SBH was observed by $I–V$ on n-type Si. On a p-type substrate, almost all the current originated from the small, isolated facet bars which were partially pinched-off. The effective SBH for a low-SBH 'strip' sandwiched between high-SBH regions was recently derived [54],

$$\Phi_{\text{eff,strip}} = \Phi_{B,o}^{\text{mean}} - 2\left(\frac{L_o \Delta E_{\max}}{\pi}\right)^{\frac{1}{2}} \quad (21.20)$$

where L_o is the width of a low-SBH strip, and Δ is the difference between the SBHs. Based on the observed average width of larger facet bars, $L_o \approx 150$ Å, and $\Delta = 0.25$ eV, the saddle-point potential for a typical facet bar was evaluated as a function of the doping level. The apparent SBH of the NiSi$_2$ diodes was calculated using the experimentally observed typical areal density of 10%, and shown to explain the observed SBHs. Such a (crude) estimate of the saddle-point potential semi-quantitatively reproduced the experimentally observed dependence on doping level. Results from the NiSi$_2$–Si(100) interface are in good agreement with the predictions of a recent theory on SBH inhomogeneity [54]. Because of the uncertainty concerning the atomic structure at planar NiSi$_2$–

Si(100) interfaces, the electronic properties of this interface have not been conclusively studied with theoretical means. Very recent calculations by Fujitani and Asano [167] showed that the SBH at an 8-fold planar NiSi$_2$–Si(100) interface resembled that of the type A NiSi$_2$–Si(111) interface. However, very different redistributions of charge were found for these two interfaces [167].

21.5.4 NiSi$_2$ on Si(110)

The growth of single crystal NiSi$_2$ on Si(110) is very easy; however, interface faceting is an unavoidable problem [172]. Due to the poor quality of epitaxy and the lack of obvious applications, the NiSi$_2$–Si(110) epitaxy has not been carefully studied. Essentially uniform NiSi$_2$ layer may be grown by deposition of Ni and Si, or by co deposition of NiSi$_2$, at room temperature and annealing at ~500°C. However, upon annealing to high temperature (>550°C), the interface breaks up into inclined ⟨111⟩ facets. Indeed, interfacial faceting of this system is so complete that, in most well-annealed samples, the entire NiSi$_2$–Si interface is made up of inclined ⟨111⟩ facets, as shown in Fig. 21.18. At the NiSi$_2$–Si(110) interface a $\frac{1}{4}\langle 111 \rangle$ type dislocation is required at the boundaries between any two $\langle 111 \rangle$ facets [172]. The fact that not even a small portion of flat [110]–Si interface has been observed suggests that this interface may well be unstable. Since the $NiSi_2$–Si(110) interface is actually made up of $\langle 111 \rangle$ facets, it is not surprising that a SBH similar to that of type A $NiSi_2$–Si(111) interface, ~0.65 eV on *n*-type Si, is usually found [170, 173].

21.5.5 CoSi$_2$ on Si(111)

Type B is the dominant epitaxial orientation observed for $CoSi_2$ layers grown on Si(111) [174, 175, 176]. Growth techniques which involve the deposition of Si routinely produce $CoSi_2$ films which are purely type B oriented [177, 178]. Pinholes in $CoSi_2$ layers, somewhat infamous because they led to a well-publicized, yet erroneous, claim of ballistic electron transistors [179, 180], epitomize the layer non-uniformity problem of this epitaxial system. An important driving force for pinhole formation in epitaxial $CoSi_2$ films was identified as a change in the energetics associated with different surface structures of $CoSi_2$(111). As a result of this discovery, pinholes may be totally suppressed in a fashion completely independent of the deposition schedule of the $CoSi_2$ film. The most reliable way to grow a high quality, uniform $CoSi_2$ layer is by predeposition of Co and co-deposition at room temperature [181], followed by annealing at >400°C. Dislocations found at these silicide interfaces are only those required by symmetry, due to the presence of steps on the original Si(111) surface.

The high stress in a uniform, thin $CoSi_2$ layer with a low density of dislocations occasionally induces a novel interfacial transformation. When this transformation occurs, it involves only about three triple layers of $CoSi_2$ at the interface which go through a seemingly shear-induced crystallographic change at below ~100°C. This transformation is reversible as the fluorite lattice is recovered at over 100°C [182]. The most prominent demonstration of this phase transformation is by plan view TEM observation of patches of different contrasts under weak beam conditions [183, 184],

as shown in Fig. 21.19(a). The three equivalent azimuthal directions (the three $[\bar{2}11]$s) for the shear to occur lead to different phase shifts with respect to the operating diffracted beam, which leads to the observation of the domain-like structure in TEM [182]. Other evidences for this interfacial transformation have come from Rutherford backscattering (RBS) and channeling, which shows an interface dechanneling peak, and from X-ray scattering, which show extra diffraction spots [184, 182]. A HREM [182] image of the interfacial layer is shown in Fig. 21.20. However, the structure of this transformed phase is still not fully understood. By a slight change in the procedures of silicide growth, it is possible to prepare $CoSi_2$ layers which either undergo, or do not undergo such a phase transition upon cooling to room temperature. Other than the fact that it is only observed in thin, ~20–40 Å, $CoSi_2$ layers with a low density of dislocations, circumstances for this interface transformation are not fully established. The transformation seems to correlate with a Si-rich stoichiometry at the interface and it may be removed by ion beam bombardment [182, 185], as shown in Fig. 21.19(b).

The atomic structure of the type B $CoSi_2$–Si(111) interface, despite years of investigation, is not fully resolved. Early studies by HREM [186] XSW [187, 188], and MEIS [189] all concluded the interface to have the 5-fold model. One notes that the rigid shifts associated with the 5-fold model are indistinguishable from that of the 8-fold models. Two theoretical papers [190, 191] pointed to the high interfacial free energy for the 5-fold model in comparison to the 8-fold model. Detailed investigations by HREM showed that the 8-fold model is more likely the structure experimentally observed [192, 193]. However, it is also known that the interfacial structure of type B $CoSi_2$–Si(111) may vary according to preparation: evidence for 7-fold coordinated structure has been obtained from layers which have only been annealed to low temperatures, <500°C [185]. Besides interfacial structural transformation, annealed type B $CoSi_2$ layers often contain other mysterious defects. For instance, dark-field images of some $CoSi_2$ layers, taken with the inclined $\frac{1}{3}\langle 511 \rangle$ reflections, which originate from inclined $\langle 11\bar{1} \rangle$

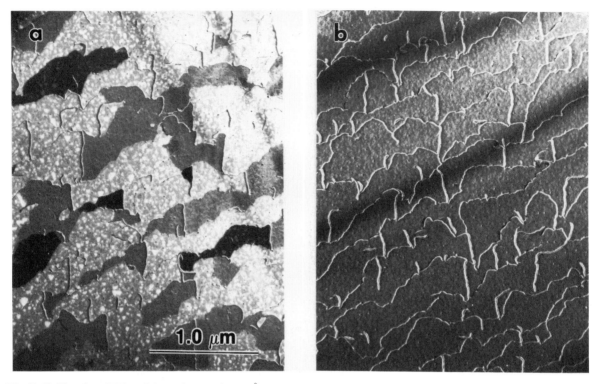

Fig. 21.19 Plan view, (220) weak beam, images of a ~30 Å thick type B $CoSi_2$ layer, grown at 600°C on Si(111). (a) shows the layer as grown. Patch-like contrasts are due to the presence of an interfacial structural transformation. (b) shows the same layer, after the sample had been irradiated with 2 MeV helium ions to a dose of 5×10^{14} cm^{-2}.

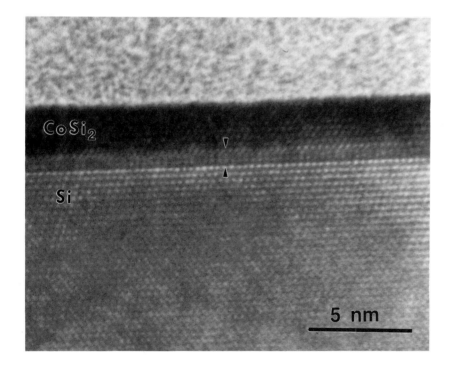

Fig. 21.20 HREM image of the interface of a thin B-type $CoSi_2$–Si(111) interface, showing the structural transformation which has taken place (after [182]).

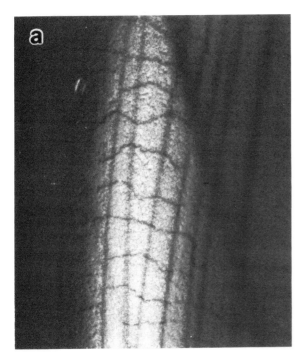
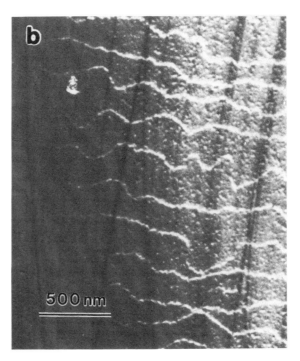

Fig. 21.21 Dark-field plan view images of a CoSi$_2$ layer on Si(111). The ~14 Å thick CoSi$_2$ template, used to grow this layer, had been grown by deposition of ~4 Å Co and ~4 Å Si at room temperature followed by annealing to ~480°C. (b) is a $\langle 2\bar{2}0 \rangle$ weak beam image and (a) was taken with a $\frac{1}{3}[\bar{5}11]$ near the [255] pole. This latter diffraction is a type B related $\langle 1\bar{1}1 \rangle$.

planes of the type B crystals, often reveal defects in the form of very fine lines, as shown in Fig. 21.21. These defects, tentatively identified as of $\frac{1}{4}[11\bar{1}]$ character, are probably related to boundaries between 7-fold structured regions and 8-fold structured regions. Type A CoSi$_2$–Si(111) is thought to be 7-fold coordinated [192, 193], although 8-fold coordinated sections have also been observed occasionally [193].

The structure of type B CoSi$_2$–Si(111) is complicated because of the presence of a possible phase transformation, a variation of atomic structure, and a high density of line defects. The SBH of type B CoSi$_2$ layers grown at ~600°C by SPE is usually in the range 0.65–0.70 eV on n-type Si [101, 194, 195]. It seems reasonable to attribute this SBH to an interface which has the 8-fold structure, assuming the effect due to dislocations is secondary. On the other hand, CoSi$_2$ layers grown at lower temperatures, which have a low density of dis-

locations, show considerable variation in their SBH [185]. SBHs as low as ~0.50 eV have been observed from these layers on n-type Si(111). However, these results have not been consistent enough to allow a conclusion to be drawn. Since a variation of atomic structure and the existence of a phase transformation at this MS interface have already been suggested by experiments, it is perhaps not surprising that the electronic properties of the type B CoSi$_2$ interface show such variations. This is a system where theoretical calculations may shed some light on the explanation of the SBHs. An early tight-binding calculation showed that the SBH of type B CoSi$_2$–Si(111) depends on whether a 5-fold or a 8-fold structure is assumed for the interface atomic structure [196]. Calculations involving large supercells [197] showed that, on n-type Si, the 7-fold coordinated interface should have a SBH which is higher than the 8-fold interface.

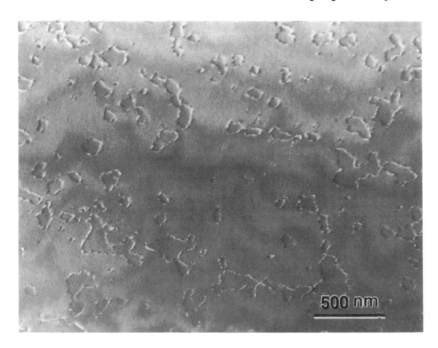

Fig. 21.22 Plan view, (022) dark-field, TEM image of a thin CoSi$_2$ layer on Si(100).

21.5.6 CoSi$_2$ on Si(100)

There are two major competing orientations for CoSi$_2$–Si(100) epitaxy: CoSi$_2$(100) ∥ Si(100) and CoSi$_2$(110) ∥ Si(100) [198, 199, 200]. Probably due to nucleation of type B CoSi$_2$ on inclined $\langle 111 \rangle$ facets, sometimes the CoSi$_2$(22$\bar{1}$) ∥ Si(100) relationship is also observed for some grains [201]. There are two variants of the [110] epitaxy, related by a 90° rotation. Most of the dislocations seen in (100)-oriented CoSi$_2$ areas are related to $\frac{1}{4}\langle 111 \rangle a_o$ and are associated with steps of odd atomic height. Epitaxial CoSi$_2$ films on Si(100) may be grown with very large anti-phase domains which seem to mimic the terrace structure of the original Si surface. It is possible to completely eliminate [110]-oriented grains by deposition of cobalt and co-deposition of CoSi$_2$, as shown in Fig. 21.22. The careful preparation of an atomically clean Si(100) surface is crucial to the growth of single crystal CoSi$_2$ layers. However, for reasons which are still unclear, the growth of pure [100] oriented CoSi$_2$ by this method is not easily reproducible. Recently, it is reported that co-deposition of a cobalt-rich material CoSi$_x$, $x < 2$, directly on Si(100) at

~500°C reproducibly leads to the growth of purely (100) oriented CoSi$_2$ layers [202]. However, such layers invariably contain a high density of dislocations. Moreover, it is not likely that thin (<50 Å) and continuous layers can be grown by co-deposition at elevated temperature.

A 2 × 1 reconstruction is often observed at the annealed interfaces of CoSi$_2$–Si(100). The occurrence of this reconstruction is dependent on the preparation of the silicide layer, being most prominently seen in samples which have been annealed to higher temperatures. A transmission electron diffraction pattern from a thin CoSi$_2$ layer is shown in Fig. 21.23. The $\langle 0\frac{1}{2}\frac{1}{2} \rangle$-related spots are from this superstructure at the interface. The atomic structure responsible for this reconstruction has been proposed as due to a dimerization of excess Si at CoSi$_2$–Si(100) interface [169], as shown in Fig. 21.24. One notes that, with this proposed model, full 8-fold coordination is fulfilled at the CoSi$_2$–Si(100) interface. The proposal of the dimerization model [169] was largely based on the intensities of the diffracted beams from the superstructure. There are two domains, separated by a 90° rotation (or an odd number of interfac

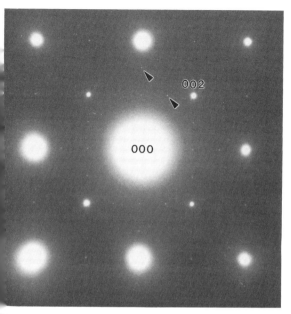

Fig. 21.23 Transmission electron diffraction pattern (120 keV) from a CoSi$_2$ layer grown on Si(100). Arrows indicate $\langle \frac{1}{2}\frac{1}{2}0 \rangle$-related spots.

planes) for dimerization to occur. Recently, it was argued that a 'missing row' model is more likely the structure for the observed 2×1 reconstruction, because it better fits the experimentally observed HREM images [203]. With such a model, of which there are also two possible domains, the coordination number of interface Co is 7, instead of 8. One notes that there is significant difference in the

preparation of CoSi$_2$ layers in these two TEM studies [169, 203] which may account for the discrepancy between these two investigations. Further studies are needed to resolve this issue. The electronic properties of single crystal CoSi$_2$ interfaces on Si(100) has only been briefly studied [61, 201], where a SBH of ~0.7 eV was indicated on *n*-type Si.

21.5.7 CoSi$_2$ on Si(110)

CoSi$_2$ grows with the regular epitaxial orientation on Si(110) [204]. Uniform layers of CoSi$_2$ may be grown by the deposition of cobalt and the co-deposition of CoSi$_2$ at room temperature, followed by annealing at above 300°C. TEM images of a thin CoSi$_2$ layer grown on Si(110) are shown in Fig. 21.25. Most line defects observed at the CoSi$_2$–Si(110) interfaces are boundaries of a unique domain structure at this interface [205]. The CoSi$_2$ crystal is apparently shifted laterally with respect to the Si lattice. Because regions of the interface may have two possible lateral shifts, differing by $\frac{1}{4}[1\bar{1}0]$ [205], domain structures and different contrasts are observed under various TEM imaging conditions. The line defects serve as the boundaries of domains, as shown in Fig. 21.25. One notes that there is no phase difference across a step at an otherwise planar CoSi$_2$–Si(110) interface, hence these domain structures are neither related to steps at the interface nor required by symmetry. Furthermore, they are not generated to

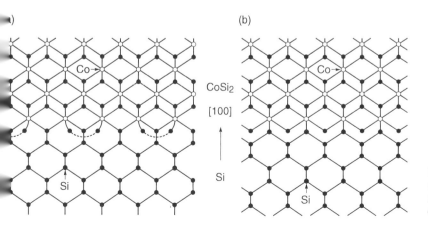

Fig. 21.24 Two views (from [110] and [1$\bar{1}$0] directions) of the 'dimer' model of the 1×2 reconstructed CoSi$_2$–Si(100) interface (after [169]).

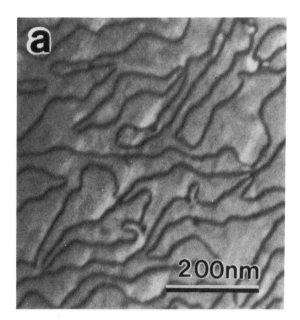 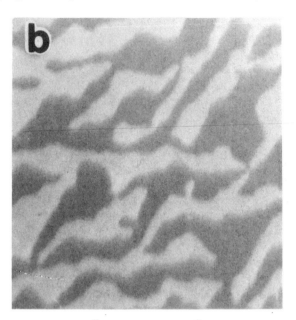

Fig. 21.25 Weak beam TEM images of a thin (~20 Å) CoSi₂ layer grown on Si(110). (a) (2$\bar{2}$0) dark-field and (b) (3$\bar{1}$1) dark-field images show the correlation of the location of the two phases and the boundaries of these anti-phase domains.

relieve misfit strain. RBS and channeling spectra from thin CoSi₂ layers are shown in Fig. 21.26. A clear dechanneling peak, which only consists of a signal from Si atoms at the silicide interface, is observed for the CoSi₂–Si(110) sample shown in Fig. 21.26(b). This interface peak is consistent with a lateral relaxation between the two crystals. Such a peak is usually absent at other silicide interfaces, such as that, shown in Fig. 21.26(a), from a CoSi₂–Si(100) sample.

The lateral shift at the CoSi₂–Si(110) interface, indicated by plan view TEM and ion channeling experiments [205], is obviously part of the structure at this interface. Presently, it seems reasonable to view such lateral shifts as being driven by the energetics of the interface atomic structure. Since the phase difference between the two domains seems to be $\frac{1}{4}[1\bar{1}0]$, a pair of shifts $v + \frac{1}{8}[1\bar{1}0]$ and $v - \frac{1}{8}[1\bar{1}0]$, where v is any vector, could lead to the observed TEM contrasts [205]. However, a v such as $\frac{1}{8}[001]$ brings the Co atoms at the interface directly over the 'bridge site' of the interface Si, as shown in Fig. 21.27. This allows full 8-fold coordination for the interface Co, which, in view

of the experimental and theoretical results from the CoSi₂–Si(111) system, may be the energetically favored configuration of the CoSi₂–Si(110) interface. One should bear in mind that the model shown in Fig. 21.27 has not been confirmed by experiments and should be viewed as a speculation. The junction characteristics of epitaxial CoSi₂–Si(110) have been studied, where a SBH of ~0.7 eV on n-type Si and a SBH of ~0.4 eV on p-type Si have been measured [61]. There have not been any theoretical investigations of the energetics and the SBH of this epitaxial interface.

21.5.8 Other epitaxial metals and silicides on Si

Recently, a correlation of the observed SBH with the starting surface structure has been observed from the Pb–Si(111) system [206]. A SBH of ~0.7 eV was measured when the surface had a starting (7 × 7)-Pb structure, and 0.93 eV was measured when the ($\sqrt{3} \times \sqrt{3}$)R30°-Pb structure was present [206]. This dependence has been attributed to the difference in the Pb–Si interface structures, originating from the starting surfaces

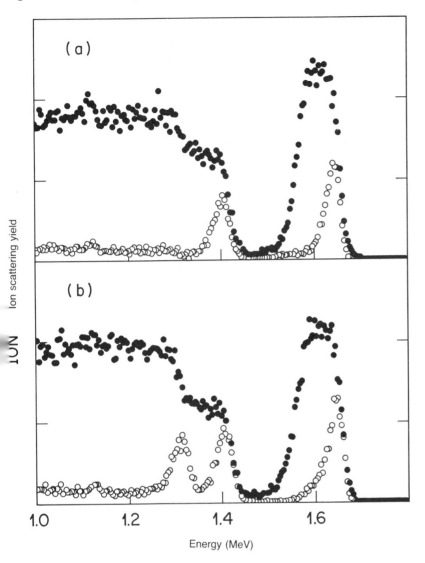

Fig. 21.26 Channeling (open circles) and random (closed circles) ion scattering spectra, obtained with a glancing exit angle, of thin $CoSi_2$ layers. (a) a 100 Å thick layer on Si(100) and (b) a 110 Å thick layer on Si(110).

This FL difference has also been observed by photoemission [207]. However, the preservation of these surface reconstructions at the eventual Pb–Si interface was recently questioned [208]. In any event, the structures of the two Pb–Si interfaces, albeit possibly changed from those on the original surfaces, are still expected to be different. Hence, the observed difference in SBH may still be due to a structural difference. One notices that the high ideality factors and the difference between I–V and C–V measurements may be suggestive of some SBH inhomogeneity at one or both of these interfaces [206].

21.5.9 Epitaxial elemental metals on compound semiconductors

Epitaxial metals have grown on III–V compound semiconductors. Most notable of these are Al, Ag, and Fe. A large nominal lattice mismatch exists between GaAs and either Al or Ag. With a 45° azimuthal rotation, good lattice

(a)   (b)

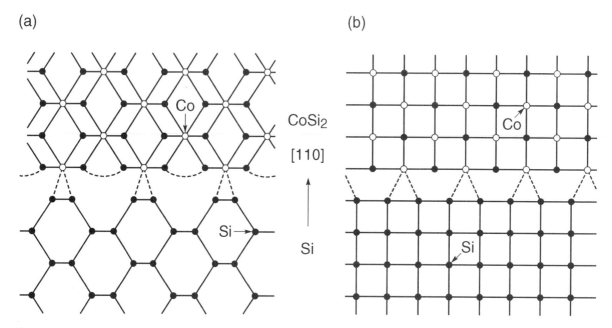

Fig. 21.27 Ball and stick model of a likely structure of the $CoSi_2$–$Si(110)$ interface, viewed from (a) $[1\bar{1}0]$ and (b) $[010]$.

matching conditions may be established for these two metals on GaAs(100). Such is indeed the most common epitaxial orientation for Al(100) grown on GaAs(100) [209, 210]. However, (110) oriented Al has also been observed [211]. It has been recently pointed out that the reconstruction of the initial GaAs surface structure may influence the epitaxial orientation of the Al films [212]. Silver grows with the (100) orientation on GaAs(100) at elevated temperatures, and with the pure (110) orientation at room temperature [213, 214]. Surprisingly, there is no azimuthal rotation for epitaxial Ag(100) on GaAs(100). The azimuthal orientation of the (110)-oriented Ag and Al appears to be related to the prevalent dangling bond direction on the GaAs(100) surface. Depending on the growth conditions, an Al–Ga exchange reaction takes place at the Al–GaAs interface [215]. There is also a noted reaction at the interface of Fe–GaAs. The interfaces between epitaxial Al and GaAs(100) have been studied by a number of techniques [216, 217]. MBE prepared GaAs(100) surfaces may have a variety of reconstructions which are associated with different surface stoichiometry. The periodicities

of some of these superstructures are found to be preserved at the Al–GaAs interface [218]. It has been discovered that the Schottky barrier height between epitaxial Al or Ag layers and GaAs is a function of the original GaAs surface reconstruction/stoichiometry [209, 213, 219], which is suggestive of a possible correlation of the interface structure and SBH. However, this correlation has been questioned in other studies [220, 221, 222]. Recently, a direct correlation of the SBH with the experimentally observed superstructures, 1×6 and 1×4, at the Sb–GaAs(100) interface has been demonstrated [223], where the difference in the atomic structure has been offered as a likely explanation of the difference, $\sim 0.1\,eV$, between the observed SBHs.

21.5.10 Epitaxial metallic compounds on III–V semiconductors

Except for refractory metals, most elemental metal/III–V compound semiconductor structures are thermodynamically unstable. Annealing, even at moderate temperatures, often leads to interdiffusion and formation of other compound phases. Two groups of intermetallic compounds are found to have stable interfaces with III–V compound semiconductors with good lattice matching conditions [224]. Hence, these are good candidates for the formation of stable, single crystal, MS interfaces with GaAs. NiAl [224, 225], CoGa [226], NiGa [227], CoAl, and related compounds form one group. They have the cubic CsCl crystal structure and lattice parameters ~1–2% larger than one half the lattice parameter of GaAs [224]. Careful control of the initial reaction on AlAs(100) has led to the successful growth of single crystal (100)CoAl and NiAl [225]. Rare-earth (RE) monopnictides, mostly with the NaCl structure, form the other group of intermetallic compounds suitable for epitaxial growth on III–V semiconductors. Here, an abundance of systems exist which are closely lattice-matched to GaAs [228]. ErAs [229], YbAs [230], LuAs [231], as well as alloyed (ternary) compounds, $ErP_{1-x}As_x$ [232], $Sc_{1-x}Er_xAs$ [228], have been grown on GaAs(100). The usual (100) epitaxial orientation is found for these epitaxial systems. RE metals are highly reactive which leads to uniform growth of their arsenides.

The interfaces between GaAs(AlAs) and a number of epitaxial intermetallic compounds have been examined by HREM [225, 231, 226, 233, 234]. However, the atomic structure of these interfaces has not been properly modeled. Evidence for a reconstruction has been observed at the interface between NiAl and overgrown GaAs. The SBH of various structures involving epitaxial metallic compounds has been studied in detail. It was found that buried NiAl has a different SBH with respect to the overgrown (AlAs)GaAs from that on an (AlAs)GaAs substrate [235]. The observed SBH also seems to depend on the thickness of the NiAl and CoAl layers [236]. Furthermore, the ideality factors of the I–V measurements are noticeably higher than unity, and the measured SBHs depend on the measurement technique [236]. From the previous discussion of such behaviors, a non-uniformity of the various SBHs is suggested. The most spectacular result from these epitaxial metallic compound interfaces is the

Table 21.1 Schottky barrier heights of epitaxial silicides

Silicide	Orientation	Substrate	Interface structure	SBH (eV)	
				n-type	*p*-type
$NiSi_2$	Type A	Si(111)	7-fold	0.65	0.47
$NiSi_2$	Type B	Si(111)	7-fold	0.79	0.33
$NiSi_2$	(100)	Si(100)	6-fold[a]	0.40	0.73
$NiSi_2$	(100)	Si(100)	7-fold {111}A	0.65	0.45
$NiSi_2$	(110)	Si(110)	7-fold {111}A	0.65	0.45
$CoSi_2$	Type B	Si(111)	8-fold[c]	0.67	0.44
$CoSi_2$	Type B	Si(111)	7-fold[c]	0.4	0.7
$CoSi_2$	(100)	Si(100)	8-fold[b]	0.71	0.41
$CoSi_2$	(110)	Si(110)	8-fold[c]	0.70	0.42

[a] With partial 1 × 2 reconstruction.
[b] 1 × 2 reconstructed.
[c] Tentative.

observation of a dependence of the SBH on epitaxial orientation, observed at the lattice-matched $Sc_{1-x}Er_xAs$–GaAs systems [237]. It was shown that annealing may lead to significant changes in the observed SBH. The SBH of the ternary compound $Sc_{0.32}Er_{0.68}As$ varies by as much as 0.4 eV when the *n*-type GaAs orientation is changed from (100) to $(\overline{111})$ [237]. It is possible that this significant variation of the SBH is related to the different atomic structures at these MS interfaces.

21.5.11 Summary of SBHs at epitaxial interfaces

The SBHs of epitaxial silicides are summarized in Table 21.1. The three planar $NiSi_2$–Si interfaces have been studied extensively by various experimental techniques, and shown to be of very high structural perfection. There is good experimental and theoretical evidence to suggest that the three distinctive SBHs observed at these $NiSi_2$ interfaces, spanning over a range over one third of the Si band gap, are related to the atomic structures seen at these interfaces. These $NiSi_2$ results are not consistent with FL theories. Interface structure also seems to influence the SBH of $CoSi_2$–Si(111). However, the structure of $CoSi_2$–Si interfaces are less clearly understood and, therefore, much more work is needed to clarify the role played by the interface structure on the observed SBH. A dependence of the SBH on interface specifics is also observed from other epitaxial MS systems.

High quality, epitaxial MS interfaces display nearly ideal diode electrical behavior. Deliberately fabricated inhomogeneous $NiSi_2$ SB diodes display non-ideal behaviors similar to that commonly observed from polycrystalline MS interfaces. This further suggests that the electronic structure of polycrystalline MS interfaces is non-uniform and not suited for a reliable deduction of the SBH formation mechanism. Experimental and theoretical results from epitaxial MS interfaces are not consistent with FL pinning, but strongly support a causal relationship between the structural properties and the electronic properties of the MS interface.

21.6 CONCLUSIONS

Recent theoretical and experimental works have shed considerable light on the relationship between the structure of the MS interface and the SBH. In this chapter, two different angles have been used to illustrate the key role played by interface structure on the SBH. First, it was shown that the vast majority of electrical data obtained from polycrystalline MS interfaces contained evidence for inhomogeneity. Second, results from epitaxial MS interfaces were discussed in detail to illustrate the numerous correlations and dependences of the SBH on specifics of the MS structure. In both cases, it was concluded that a relationship between structure and SBH indeed exists. This view is actually in conflict with that adopted by the majority of researchers in the SBH–interface physics community. However, a careful examination of existing data suggested that the concept of FL pinning at MS interface, which may be of value to some specific systems, was too simple to cover the wide range of behaviors exhibited by metallurgically different MS systems. It was also pointed out that existing interface state models were not, as widely thought, capable of explaining most of the non-ideal SBH behaviors. Ideas derived from the FL pinning concept, such as an automatic

assumption of the homogeneity of any SB, should no longer be taken for granted and should be examined closely. It was suggested that the rearrangement of all charges at the interface, not just those in the semiconductor band gap, is the immediate cause of SBH formation. Direct evidence for a causal relationship between the atomic structure and the interface dipole has been repeatedly demonstrated from high-quality epitaxial MS interfaces. It seems obvious, at this stage, that a careful study of this relationship may be the most beneficial approach to solving the long standing SBH mystery.

ACKNOWLEDGMENTS

I am indebted to J. P. Sullivan for discussions and suggestions. I also thank my colleagues F. Schrey, J. M. Gibson, A. F. J. Levi, D. J. Eaglesham, C. J. Palmstrom, M. O. Aboelfotoh, A. N. Saxena and J. M. Andrews.

REFERENCES

1. E. H. Rhoderick and R. H. Williams, *Metal-Semiconductor Contacts*, Clarendon Press, Oxford (1988).
2. S. M. Sze, *Physics of Semiconductor Devices*, Wiley (1981).
3. L. J. Brillson, *Surface Sci. Rep.* **2** (1982), 123.
4. M. S. Tyagi, in *Metal-Semiconductor Schottky Barrier Junctions and Their Applications* (ed B. L. Sharma), Plenum, New York (1984).
5. G. Y. Robinson, in *Physics and Chemistry of III–V Compound Semiconductor Interfaces* (ed C. W. Wilmsen), Plenum, New York (1985).
6. J. Bardeen, *Phys. Rev.*, **771** (1947), 717.
7. V. Heine, *Phys. Rev. A*, **138** (1965), 1689.
8. R. T. Tung, J. M. Gibson and J. M. Poate, *Phys. Rev. Lett.*, **50** (1983), 429.
9. J. M. Gibson, R. T. Tung, and J. M. Poate, *MRS Symp. Proc.*, **14** (1983), 395.
10. D. Cherns, G. R. Anstis, J. L. Hutchison and J. C. H. Spence, *Phil. Mag. A*, **46** (1982), 849.
11. G. P. Das, P. Blöchl, O. K. Andersen, N. E. Christensen, and O. Gunnarsson, *Phys. Rev. Lett.*, **63** (1989), 1168.
12. H. Fujitani and S. Asano, *Phys. Rev. B*, **42** (1990), 1696.
13. R. T. Tung, *Phys. Rev. Lett.*, **52** (1984), 461.
14. W. Schottky, *Z. Phys.* **113** (1939), 367.

15. C. A. Mead and W. G. Spitzer, *Phys. Rev.*, **134** (1964), A713.

16. A. M. Cowley and S. M. Sze, *J. Appl. Phys.*, **36** (1965), 3212.

17. R. J. Archer and M. M. Atalla, *Ann. NY Acad. Sci.*, **101** (1963), 697.

18. C. R. Crowell and G. I. Roberts, *J. Appl. Phys.*, **40** (1969), 3726.

19. S. Kurtin, T. C. McGill, and C. A. Mead, *Phys. Rev. Lett.*, **22** (1970), 1433.

20. L. J. Brillson, *Phys. Rev. Lett.*, **40** (1978), 260.

21. C. Tejedor, F. Flores, and E. Louis, *J. Phys. C*, **10** (1977), 2163.

22. S. G. Louie, J. R. Chelikowsky, and M. L. Cohen, *Phys. Rev. B*, **15** (1977), 2154.

23. J. Tersoff, *Phys. Rev. Lett.*, **52** (1984), 465.

24. G. H. Parker, T. C. McGill, C. A. Mead, and D. Hoffman, *Solid-St. Electron.*, **11** (1968), 201.

25. C. R. Crowell, *J. Vac. Sci. Technol.*, **11** (1974), 951.

26. W. Mönch, *Phys. Rev. Lett.*, **58** (1987), 1260.

27. W. E. Spicer, I. Lindau, P. Skeath, C. Y. Su, and P. W. Chye, *Phys. Rev. Lett.*, **44** (1980), 420.

28. M. S. Daw and D. L. Smith, *Solid-St. Commun.*, **37** (1981), 205.

29. O. F. Sankey, R. E. Allen, and J. D. Dow, *Solid St. Commun.*, **49** (1984), 1.

30. A. Zur, T. C. McGill, and D. L. Smith, *Phys. Rev. B*, **28** (1983), 2060.

31. C. B. Duke and C. Mailhiot, *J. Vac. Sci. Technol. B*, **3** (1985), 1170.

32. G. W. Rubloff and P. S. Ho, *Thin Solid Films*, **93** (1982), 19.

33. C. Barret and P. Muret, *Appl. Phys. Lett.*, **42** (1983), 890.

34. P. S. Ho, E. S. Yang, H. L. Evans, and X. Wu, *Phys. Rev. Lett.*, **56** (1986), 177.

35. P. Muret and D. Elguennouni, M. Missous, and E. H. Rhoderick, *Appl. Phys. Lett.*, **58** (1991) 155.

36. A. Deneuville and B. K. Chakraverty, *Phys. Rev. Lett.*, **28** (1972), 1258.

37. T. Flohr and M. Schulz, *Appl. Phys. Lett.*, **48** (1986), 1534.

38. J. L. Freeouf, *Appl. Phys. Lett.*, **41** (1982), 285.

39. C. Canali, F. Catellani, S. Mantovani, and M. Prudenziati, *J. Phys. D*, **10** (1977), 2481.

40. I. Ohdomari and K. N. Tu, *J. Appl. Phys.*, **51** (1980), 3735.

41. I. Ohdomari, T. S. Kuan, and K. N. Tu, *J. Appl. Phys.*, **50** (1979) 7020.

42. D. Dascalu, Gh. Brezeanu, P. A. Dan and C. Dima, *Solid-St. Electron.*, **24** (1981), 897.

43. K. N. Tu, R. D. Thompson, and B. Y. Tsaur, *Appl. Phys. Lett.*, **38** (1981), 626.

44. R. D. Thompson and K. N. Tu, *J. Appl. Phys.*, **53** (1982), 4285.

45. M. V. Schneider, A. Y. Cho, E. Kollberg, and H. Zirath, *Appl. Phys. Lett.*, **43** (1983), 558.

46. T. Q. Tuy and I. Mojzes, *Appl. Phys. Lett.*, **56** (1990), 1652.

47. V. B. Bikbaev, S. C. Karpinskas, and J. J. Vaitkus, *Phys. Stat. Sol. (a)* **75** (1983), 583.

48. H. H. Güttler and J. H. Werner, *Appl. Phys. Lett.*, **56** (1990), 1113.

49. J. L. Freeouf, T. N. Jackson, S. E. Laux, and J. M. Woodall, *Appl. Phys. Lett.*, **40** (1982), 634.

50. J. L. Freeouf, T. N. Jackson, S. E. Laux, and J. M. Woodall, *J. Vac. Sci. Technol.*, **21** (1982), 570.

51. I. Ohdomari and H. Aochi, *Phys. Rev. B*, **35** (1987), 682. (However, a faulty boundary condition in this paper led to very significant errors in its estimation of the saddle-point potential.)

52. A. I. Bastys, V. B. Bikbaev, J. J. Vaitkus, and S. C. Karpinskas, *Litovskii Fizicheskii Sbornik*, **28** (1988), 191.

53. J. P. Sullivan, R. T. Tung, M. Pinto, and W. R. Graham, submitted *J. Appl. Phys.*, **70** (1991), 7403.

54. R. T. Tung, *Appl. Phys. Lett.*, **58** (1991), 2821.

55. J. H. Werner, *Appl. Phys. A*, **47** (1988), 291.

56. M. O. Aboelfotoh, *J. Appl. Phys.*, **64** (1988) 4046.

57. A. Y. C. Yu and E. H. Snow, *J. Appl. Phys.*, **39** (1968), 3008.

58. M. P. Lepselter and S. M. Sze, *Bell Syst. Tech. J.*, **47** (1968), 195.

59. Z. Liliental-Weber, R. Gronsky, J. Washburn, N. Newman, W. E. Spicer, and E. R. Weber, *J. Vac. Sci. Technol. B*, **4** (1986), 912.

60. D. Dascalu, Gh. Brezeanu, P. A. Dan and C. Dima, *Solid-St. Electron.*, **24** (1981), 897.

61. R. T. Tung, J. P. Sullivan, F. Schrey and W. R. Graham, *J. Vac. Sci. Technol.* (1992) (in press).

62. F. La Via, P. Lanza, O. Viscuso, G. Feria, and E. Rimini, *Thin Solid Films*, **161** (1988), 13.

63. M. Wittmer, *Phys. Rev. B*, **42** (1990), 5249.

64. M. M. Atalla and R. W. Soshea, *Scientific Report No. 1* (1962), Contract No. AF 19 (628)-1637. Hewlett-Packard Associates.

65. S. M. Sze, C. R. Crowell, and D. Kahng, *J. Appl. Phys.*, **35** (1964), 2534.

66. H. C. Card and E. H. Rhoderick, *J. Phys. D*, **4** (1971), 1589.

67. F. A. Padovani and R. Stratton, *Solid-St. Electron.*, **9** (1966), 695.

68. V. L. Rideout and C. R. Crowell, *Solid-St. Electron.*, **13** (1970), 993.

69. G. Ottaviani, K. N. Tu, and J. W. Mayer, *Phys. Rev. B*, **24** (1981), 3354.

70. H. H. Hosack, *Appl. Phys. Lett.*, **2** (1972), 256.

71. N. Toyama, T. Takahashi, H. Murakami, and H. Horiyama, *Appl. Phys. Lett.*, **46** (1985), 557.

72. J. H. Werner and H. H. Güttler, *J. Appl. Phys.*, **69** (1991), 1522.

73. A. N. Saxena, *Surface Sci.*, **13** (1969), 151.

74. F. A. Padovani and G. G. Sumner, *J. Appl. Phys.*, **36** (1965), 3744.

75. S. Ashok, J. M. Borrego, and R. J. Guttmann, *Solid-St. Electron.*, **22** (1979), 621.

76. M. O. Aboelfotoh, A. Cros, B. G. Svensson, and K. N. Tu, *Phys. Rev. B*, **41** (1990), 9819.

77. B. Tuck, G. Eftekhari, and D. M. de Cogan, *J. Phys. D*, **15** (1982), 457.

78. J. D. Levine, *J. Appl. Phys.*, **42** (1971), 3991.

79. C. R. Crowell, *Solid-St. Electron.*, **20** (1977), 171.

80. F. A. Padovani, in *Semiconductors and Semimetals* (eds R. K. Willardson and A. C. Beer), Vol. 7A, Academic Press, New York (1971).

81. G. S. Visweswaran and R. Sharan, *Proc. IEEE*, **67** (1979), 436.

82. J. Basterfield, J. M. Shannon, and A. Gill, *Solid-St. Electron.*, **18** (1975), 290.

83. B. Studer, *Solid-St. Electron.*, **23** (1980), 1181.

84. M. O. Aboelfotoh, *J. Appl. Phys.*, **66** (1989), 262.

85. M. O. Aboelfotoh and K. N. Tu, *Phys. Rev. B*, **34** (1986), 2311, Fig. 7.

86. See, for example, Fig. 13(a) of Ref. 84 and Fig. 3 of Ref. 76.

87. C.-A. Chang, *J. Appl. Phys.*, **59** (1986), 3116.

88. B.-Y. Tsaur, D. J. Silversmith, R. W. Mountain, L. S. Hung, S. S. Lau, and T. T. Sheng, *J. Appl. Phys.*, **51** (1981), 5243.

89. A. Y. C. Yu and C. A. Mead, *Solid-St. Electron.*, **13** (1970), 97.

90. V. W. L. Chin, J. W. V. Storey, and M. A. Green, *J. Appl. Phys.*, **68** (1990), 4127.

91. J. M. Andrews and M. P. Lepselter, *Solid-St. Electron.*, **13** (1970), 1011.

92. J. M. Andrews and F. B. Koch, *Solid-St. Electron.*, **14** (1971), 901.

93. C. J. Kircher, *Solid-St. Electron.*, **14** (1971), 507.

94. A. M. Cowley, *Solid-St. Electron.*, **12** (1970), 403.

95. A. Thanailakis and A. Rasul, *J. Phys. C*, **9** (1976), 337.

96. N. Newman, M. van Schilfgaarde, T. Kendelwicz, M. D. Williams, and W. E. Spicer, *Phys. Rev. B*, **33** (1986), 1146.

97. A. B. McLean and R. H. Williams, *J. Phys. C*, **21** (1988), 783.

98. R. J. Archer and T. O. Yep, *J. Appl. Phys.*, **41** (1970), 303.

99. D. Kahng, *Solid-St. Electron.*, **6** (1963), 281.

100. R. F. Broom, *Solid-St. Electron.*, **14** (1971), 1087.

101. R. T. Tung, *J. Vac. Sci. Technol. B*, **2** (1984), 465.

102. A. M. Goodman, *J. Appl. Phys.*, **34** (1963), 329.

103. G. L. Miller, D. V. Lang, and L. C. Kimerling, *Ann. Rev. Mat. Sci.*, **7** (1977), 377.

104. F. J. Bryant, J. M. Majid, C. G. Scott and D. Shaw, *Solid State Commun.*, **63** (1987), 9.

105. H. Grinolds and G. Y. Robinson, *J. Vac. Sci. Technol.*, **14** (1977), 75.

106. J. A. Copeland, *IEEE Trans. Elec. Dev.*, **ED-17** (1970), 404.

107. A. M. Cowley, *J. Appl. Phys.*, **37** (1966), 3024.

108. J. D. Wiley and G. L. Miller, *IEEE Trans. Elec. Dev.*, **ED-22** (1975), 265.

109. P. K. Vasudev, B. L. Mattes, E. Petras and R. H. Bube, *Solid-St. Electron.*, **19** (1976), 557.

110. J. M. Borrego, R. J. Gutmann and S. Ashok, *Solid-St. Electron.*, **20** (1977), 125.

111. B. Pellegrini and G. Salardi, *Solid-St. Electron.*, **21** (1978), 465.

112. J. H. Werner, *Metallization and Metal-Semiconductor Interfaces* (ed I. P. Batra), Plenum, New York (1989), p. 35.

113. J. H. Werner, A. F. J. Levi, R. T. Tung, M. Anzlowar and M. Pinto, *Phys. Rev. Lett.*, **60** (1988), 53.

114. M. J. Turner and E. H. Rhoderick, *Solid-St. Electron.*, **11** (1968), 291.

115. T. Arizumi and M. Hirose, *Jpn. J. Appl. Phys.*, **8** (1969), 749.

116. M. Hirose, N. Altaf, and T. Arizumi, *Japan. J. Appl. Phys.*, **9** (1970), 260.

117. P. Gutknecht and M. J. O. Strutt, *Appl. Phys. Lett.*, **21** (1972), 405.

118. K. E. Sundström, S. Petersson, and P. A. Tove, *Phys. Stat. Sol. (a)* **20** (1973), 653.

119. D. J. Coe and E. H. Rhoderick, *J. Phys. D*, **9** (1976), 965.

120. P. E. Schmid, P. S. Ho, H. Föll, and T. Y. Tan, *Phys. Rev. B*, **28** (1983), 4593.

121. C. R. Crowell, W. G. Spitzer, L. E. Howarth, and E. E. LaBate, *Phys. Rev.*, **127** (1962), 2006.

122. J. D. van Otterloo, *Surface Sci.*, **104** (1981), L205.

123. J. D. van Otterloo and J. G. De Groot, *Surface Sci.*, **57** (1976), 93.

124. R. W. Soshea and R. C. Lucas, *Phys. Rev. A*, **138** (1965), 1182.

125. J. Silverman, P. Pellegrini, J. Comer, A. Golvbovic, M. Weeks, J. Mooney, J. Fitzgerald, *MRS Symp. Proc.*, **54** (1986), 515.

126. R. C. McKee, *IEEE Trans. Elec. Dev.*, **31** (1984), 968.
127. E. Calleia, J. Garrido, J. Piqueras, and A. Martinez, *Solid-St. Electron.*, **23** (1980), 591.
128. T. Okumura and K. N. Tu, *J. Appl. Phys.*, **54** (1983), 922.
129. O. Engstrom, H. Pettersson, and B. Sernelius, *Phys. Stat. Sol. (a)* **95** (1986), 691.
130. A. Tanabe, K. Konuma, N. Teranishi, S. Tohyama, and K. Masubuchi, *J. Appl. Phys.*, **69** (1991), 850.
131. W. J. Kaiser and L. D. Bell, *Phys. Rev. Lett.*, **60** (1988), 1406.
132. R. T. Tung, A. F. J. Levi, J. P. Sullivan, and F. Schrey, *Phys. Rev. Lett.*, **66** (1991), 72.
133. N. F. Mott, *Proc. Cambridge Phil. Soc.*, **34** (1938), 568.
134. M. Schlüter, *Phys. Rev. B*, **17** (1978), 5044.
135. W. Mönch, *Surface Sci.*, **21** (1970), 443.
136. J. M. Andrews and J. C. Phillips, *Phys. Rev. Lett.*, **35** (1975), 56.
137. G. Ottaviani, K. N. Tu, and J. W. Mayer, *Phys. Rev. Lett.*, **44** (1980), 284.
138. J. L. Freeouf, *Solid State Commun.*, **33** (1980), 1059.
139. J. Tersoff, *Phys. Rev. B*, **32** (1985), 6968.
140. R. T. Tung, J. M. Poate, J. C. Bean, J. M. Gibson and D. C. Jacobson, *Thin Solid Films*, **93** (1982), 77.
141. E. J. van Loenen, A. E. M. Fischer, J. F. van der Veen, and F. LeGoues, *Surface Sci.* **154** (1985), 52.
142. B. D. Hunt, L. J. Schowalter, N. Lewis, E. L. Hall, R. J. Hauenstein, T. E. Schlesinger, T. C. McGill, M. Okamoto, and S. Hashimoto, *Mat. Res. Soc. Symp. Proc.*, **54** (1986), 479.
143. H. von Känel, T. Graf, J. Henz, M. Ospelt, and P. Wachter, *J. Cryst. Growth*, **81** (1987), 470.
144. R. T. Tung, J. M. Gibson and J. M. Poate, *Appl. Phys. Lett.*, **42** (1983), 888.
145. R. T. Tung, *J. Vac. Sci. Technol. A*, **5** (1987), 1840.
146. R. T. Tung and F. Schrey, *Appl. Phys. Lett.*, **55** (1989), 256.
147. A. Ishizaka and Y. Shiraki, *Surface Sci.*, **174** (1986), 671.
148. H. Föll, *Phys. Stat. Sol. (a)* **69** (1982), 779.
149. D. Cherns, G. R. Anstis, J. L. Hutchison, and J. C. H. Spence, *Phil. Mag. A*, **46** (1982), 849.
150. J. M. Gibson, R. T. Tung, and J. M. Poate, *Mater. Res. Soc. Symp. Proc.*, **14** (1983), 395.
151. E. Vlieg, A. E. M. J. Fischer, J. F. van der Veen, B. N. Dev and G. Materlik, *Surface Sci.*, **178** (1986), 36.
152. J. Zegenhagen, M. A. Kayed, K.-G. Huang, W. M. Gibson, J. C. Phillips, L. J. Schowalter and B. D. Hunt, *Appl. Phys. A*, **44** (1987), 365.
153. E. J. van Loenen, J. W. M. Frenken, J. F. van der Veen and S. Valeri, *Phys. Rev. Lett.*, **54** (1985), 827.
154. I. K. Robinson, R. T. Tung, and R. Feidenhans'l, *Phys.*

Rev. B, **38** (1988), 3632.
155. R. T. Tung, *Phys. Rev. Lett.*, **52** (1984), 461.
156. R. J. Hauenstein, T. E. Schlesinger, T. C. McGill, B. D. Hunt and L. J. Schowalter, *Appl. Phys. Lett.*, **47** (1985), 853.
157. M. Ospelt, J. Henz, L. Flepp, and H. von Känel, *Appl. Phys. Lett.*, **52** (1988), 227.
158. J. Vrijmoeth, J. F. van der Veen, D. R. Heslinga, and T. M. Klapwijk, *Phys. Rev. B*, **42** (1990), 9598.
159. Y. Shiraki, T. Ohshima, A. Ishizaka, and K. Nakagawa, *J. Cryst. Growth*, **81** (1987), 476.
160. A. Kikuchi, T. Ohshima, and Y. Shiraki, *J. Appl. Phys.*, **64** (1988), 4614.
161. A. Kikuchi, *Phys. Rev. B*, **40** (1989), 8024.
162. D. R. Hamann, in *Metallization and Metal-Semiconductor Interfaces* (ed I. P. Batra), Plenum, New York (1988).
163. H. Fujitani and S. Asano, *Appl. Surface Sci.*, **41/42** (1989), 164.
164. R. T. Tung, F. Schrey, and S. M. Yalisove, *Appl. Phys. Lett.*, **55** (1989), 2005.
165. J. L. Batstone, J. M. Gibson, R. T. Tung and A. F. J. Levi, *Appl. Phys. Lett.*, **52** (1988), 828.
166. D. Cherns, C. J. D. Hetherington, and C. J. Humphreys, *Phil. Mag. A*, **49** (1984), 165.
167. H. Fujitani and S. Asano, *Appl. Surface Sci.*, **56–58** (1992), 408.
168. C. C. Matthai, N. V. Rees, and T. H. Shen, ibidi.
169. D. Loretto, J. M. Gibson, and S. M. Yalisove, *Phys. Rev. Lett.*, **63** (1989), 298.
170. R. T. Tung, *Mat. Res. Soc. Symp. Proc.*, **37** (1985), 345.
171. J. L. Batstone, J. M. Gibson, R. T. Tung, A. F. J. Levi, and C. A. Outten, *Mat. Res. Soc. Symp. Proc.*, **82** (1987), 335.
172. R. T. Tung, S. Nakahara, and T. Boone, *Appl. Phys. Lett.*, **46** (1985), 895.
173. R. T. Tung and J. M. Gibson, *J. Vac. Sci. Technol. A*, **3** (1985), 987.
174. R. T. Tung, J. M. Gibson, J. C. Bean, J. M. Poate and D. C. Jacobson, *Appl. Phys. Lett.*, **40** (1982), 684.
175. Y. C. Kao, M. Tejwani, Y. H. Xie, T. L. Lin and K. L. Wang, *J. Vac. Sci. Technol. B*, **3** (1985), 596.
176. F. Arnaud D'Avitaya, S. Delage, E. Rosencher and J. Derrien, *J. Vac. Sci. Technol, B*, **3** (1985), 770.
177. R. T. Tung, A. F. J. Levi, and J. M. Gibson, *Appl. Phys. Lett.*, **48** (1986), 635.
178. B. D. Hunt, N. Lewis, L. J. Schowalter, E. L. Hall, and L. G. Turner, *Mat. Res. Soc. Symp. Proc.*, **77** (1987), 351.
179. E. Rosencher, S. Delage, Y. Campidelli, and F. Arnaud D'Avitaya, *Electron. Lett.*, **20** (1984), 762.
180. E. Rosencher, S. Delage, F. Arnaud D'Avitaya, C.

D'Anterroches, K. Belhaddad, and J. C. Pfister, *Physica B*, **134** (1985), 106.

180. F. Hellman and R. T. Tung, *Phys. Rev. B*, **37** (1988), 10786.

181. R. T. Tung and F. Schrey, *Appl. Phys. Lett.*, **54** (1989), 852.

182. D. J. Eaglesham, R. T. Tung, R. L. Headrick, I. K. Robinson, and F. Schrey, *Mat. Res. Soc. Symp. Proc.*, **159** (1990), 141.

183. R. T. Tung, *J. Vac. Sci. Technol. A*, **7** (1989), 599.

184. R. T. Tung, and F. Schrey, *Mat. Res. Soc. Symp. Proc.*, **122** (1988), 559.

185. R. T. Tung, A. F. J. Levi, F. Schrey, and M. Anzlowar, *NATO ASI Series B: Physics*, **203** (1989), 167.

186. J. M. Gibson, J. C. Bean, J. M. Poate and R. T. Tung, *Appl. Phys. Lett.*, **41** (1982), 818.

187. A. E. M. J. Fischer, E. Vlieg, J. F. van der Veen, M. Clausnitzer and G. Materlik, *Phys. Rev. B*, **36** (1987), 4769.

188. J. Zegenhagen, K.-G. Huang, B. D. Hunt and L. J. Schowalter, *Appl. Phys. Lett.*, **51** (1987), 1176.

189. A. E. M. J. Fischer, T. Gustafsson, and J. F. van der Veen, *Phys. Rev. B*, **37** (1988), 6305.

190. D. R. Hamann, *Phys. Rev. Lett.*, **60** (1988), 313.

191. P. J. van den Hoek, W. Ravenek, and E. J. Baerends, *Phys. Rev. Lett.*, **60** (1988), 1743.

192. C. W. T. Bulle-Lieuwma, A. F. de Jong, A. H. van Ommen, J. F. van der Veen, and J. Vrijmoeth, *Appl. Phys. Lett.*, **55** (1989), 648.

193. R. Hull, Y. F. Hsieh, K. T. Short, A. E. White, and D. C. Cherns, *Mat. Res. Soc. Symp. Proc.*, **183** (1990), 91.

194. E. Rosencher, S. Delage and F. Arnaud D'Avitaya, *J. Vac. Sci. Technol. B*, **3** (1985), 762.

195. Y. C. Kao, Y. Y. Wu, and K. L. Wang, *Proc. 1st Intl Symp. Si MBE* (ed J. C. Bean) The Electrochem. Society (1985) p. 261.

196. N. V. Rees and C. C. Matthai, *J. Phys. C*, **21** (1988), L981.

197. H. Fujitani and S. Asano, *Mat. Res. Soc. Symp. Proc.*, **193** (1990), 77.

198. R. T. Tung, J. L. Batstone, and S. M. Yalisove, *Mat. Res. Soc. Symp. Proc.*, **102** (1988), 265.

199. A. H. van Ommen, C. W. T. Bulle-Lieuwma, and D. Langereis, *J. Appl. Phys.*, **64** (1988), 2706.

200. S. M. Yalisove, R. T. Tung, and D. Loretto, *J. Vac. Sci. Technol. A*, **7** (1989), 599.

201. J. R. Jimenez, L. M. Hsiung, R. D. Thompson, S. Hashimoto, K. V. Ramanathan, R. Arndt, K. Rajan, S. S. Iyer, and L. J. Schowalter, *MRS Symp. Proc.*, **160** (1990), 237.

202. J. R. Jimenez, L. M. Hsiung, K. Rajan, L. J. Schowalter,

S. Hashimoto, R. D. Thompson and S. S. Iyer, *Appl. Phys. Lett.*, **57** (1990), 2811.

203. C. W. T. Bulle-Lieuwma, A. H. Van Ommen, D. E. Vandenhoudt, J. J. Ottenheim and A. F. de Jong, *J. Appl. Phys.*, **70** (1991), 3093.

204. S. M. Yalisove, D. J. Eaglesham, and R. T. Tung, *Appl. Phys. Lett.*, **55** (1989), 2075.

205. D. J. Eaglesham, R. T. Tung, and S. M. Yalisove, to be published.

206. D. R. Heslinga, H. H. Weitering, D. P. van der Werf, T. M. Klapwijk, and T. Hibma, *Phys. Rev. Lett.*, **64** (1990), 1585.

207. G. Le Lay, K. Hricovini, and J. E. Bonnet, *Appl. Surf. Sci.*, **41/42** (1989), 25.

208. G. Le Lay and K. Hricovini, *Phys. Rev. Lett.*, **65** (1990), 807; also see the reply by Weitering *et al.* ibid.

209. A. Y. Cho and P. D. Dernier, *J. Appl. Phys.*, **49** (1978), 3328.

210. R. Ludeke, G. Landgren and L. L. Chang, *Vide. Couches Minces, Suppl.*, **201** (1980), 579.

211. R. Ludeke, L. L. Chang, and L. Esaki, *Appl. Phys. Lett.*, **23** (1973), 201.

212. S. K. Donner, K. P. Caffey, and N. Winograd, *J. Vac. Sci. Technol. B*, **7** (1989), 742.

213. R. Ludeke, T.-C. Chiang, and D. E. Eastman, *J. Vac. Sci. Technol.*, **21** (1982), 599.

214. J. Massies, P. Delescluse, P. Etienne, and N. T. Linh, *Thin Solid Films*, **90** (1982), 113.

215. G. Landgren and R. Ludeke, *Solid St. Comm.*, **37** (1981), 127.

216. W. C. Marra, P. Eisenberger, and A. Y. Cho, *J. Appl. Phys.*, **50** (1979), 6927.

217. C. J. Kiely, D. Cherns, and D. J. Eaglesham, *Phil. Mag.* (1988).

218. J. Mizuki, K. Akimoto, I. Hirosawa, K. Hirose, T. Mizutani, and J. Matsui, *J. Vac. Sci. Technol. B*, **6** (1988), 31.

219. W. I. Wang, *J. Vac. Sci. Technol. B*, **1** (1983), 574.

220. C. Barret and J. Massies, *J. Vac. Sci. Technol. B*, **1** (1983), 819.

221. S. P. Svensson, G. Landgren, and T. G. Andersson, *J. Appl. Phys.*, **54** (1983), 4474.

222. M. Missous, E. H. Rhoderick, and K. E. Singer, *J. Appl. Phys.*, **54** (1983), 4474.

223. K. Hirose, K. Akimoto, I. Hirosawa, J. Mizuki, T. Mizutani and J. Matsui, *Phys. Rev. B*, **43** (1991), 4538.

224. T. Sands, *Appl. Phys. Lett.*, **52** (1988), 197.

225. J. P. Harbison, T. Sands, N. Tabatabaie, W. K. Chan, L. T. Florez, and V. G. Keramidas, *Appl. Phys. Lett.*, **53** (1988), 1717.

226. C. J. Palmstrom, B.-O. Fimland, T. Sands, K. C.

Garrison, and R. A. Bartynski, *J. Appl. Phys.*, **65** (1989), 4753.

227. A. Guivarc'h, R. Guerin, and M. Secoue, *Electron. Lett.*, **23** (1987), 1004.

228. C. J. Palmstrom, S. Mounier, T. G. Finstad, and P. F. Miceli, *Appl. Phys. Lett.*, **56** (1990), 382.

229. C. J. Palmstrom, N. Tabatabaie, and S. J. Allen, Jr., *Appl. Phys. Lett.*, **53** (1988), 2608.

230. H. J. Richter, R. S. Smith, N. Herres, M. Seelmann-Eggebert, and P. Wennekers, *Appl. Phys. Lett.*, **53** (1988), 99.

231. C. J. Palmstrom, K. C. Garrison, S. Mounier, T. Sands, C. L. Schwartz, N. Tabatabaie, S. J. Allen, Jr., H. L. Gilchrist, and P. F. Miceli, *J. Vac. Sci. Technol. B*, **7** (1989), 747.

232. A. Le Corre, J. Caulet, and A. Guivarc'h, *Appl. Phys. Lett.*, **55** (1989), 2298.

233. N. Tabatabaie, T. Sands, J. P. Harbison, H. L. Gilchrist, and V. G. Keramidas, *Appl. Phys. Lett.*, **53** (1988), 2528.

234. J. G. Zhu, C. B. Carter, C. J. Palmstrom, and K. C. Garrison, *Appl. Phys. Lett.*, **55** (1989), 39.

235. T. L. Cheeks, T. Sands, R. E. Nahory, J. P. Harbison, N. Tabatabaie, H. L. Gilchrist, B. J. Wilkens and V. G. Keramidas, *Appl. Phys. Lett.*, **56** (1990), 1043.

236. T. Sands, C. J. Palmstrom, J. P. Harbison, V. G. Keramidas, N. Tabatabaie, T. L. Cheeks, R. Ramesh and Y. Silberberg, *Mat. Sci. Rep.*, **5** (1990), 99.

237. C. J. Palmstrom, T.-L. Cheeks, H. L. Gilchrist, J. G. Zhu, C. B. Carter, and R. E. Nahory, *Mat. Res. Soc. EA*, **21** (1990), 63.

CHAPTER TWENTY-TWO

Electronic properties of semiconductor–semiconductor interfaces and their control using interface chemistry

D. W. Niles and G. Margaritondo

Surface techniques in heterojunction physics · Microscopic control of interface parameters · Futur
directions

22.1 SURFACE TECHNIQUES IN HETEROJUNCTION PHYSICS

The interface between two semiconducting materials is a very interesting system because of its present and potential applications in microelectronics. The same system poses also a challenging and interesting problem in fundamental materials science. The problem arises from the presence of two band structures that must be somehow connected at the interface. Since the two semiconductors have different forbidden gap widths, the connection produces discontinuities in the conduction and valence band edges across the junction. Such discontinuities are, arguably, the most important factor in the performance of heterojunction devices [1].

The problem of how the two band structures are connected and the band discontinuities established appears deceivingly simple, but it has in fact occupied several generations of solid-state theorists without producing a universal consensus [1]. Such a problem is, in fact, quite difficult both

conceptually and computationally. It is furthe
complicated for 'real' heterojunction interfaces b
a number of factors: microscopic diffusion, loca
chemical reactions, the general morphology of th
interface, etc.

Progress towards the solution of this funda
mental problem requires, of course, a good bas
of experimental data. A semiconductor–semi
conductor interface is a localized system, extendin
a few atomic planes in the direction perpendicula
to the interface – two planes for 'sharp' interfaces
Thus, its study is greatly helped by the use c
experimental techniques with high spatial reso
ution [2]. Not surprisingly, since the late 197C
surface-sensitive techniques have played an impor
ant role in heterojunction research, with phote
emission spectroscopy in the leading position. Fc
example, such techniques provided the first hir
that a widely believed assumption about the ban
lineups for the prototypical GaAs–AlAs an
GaAs–$Ga_{1-x}Al_xAs$ interfaces was, in fact, wror
[1, 3, 4].

The widespread use of surface-sensitive tecl

niques has produced three important results in heterojunction physics. First, it has provided an alternative way of measuring band discontinuities, in addition to transport measurements and other techniques. This has eliminated, in particular, some of the mistakes [1, 3, 4] produced by other approaches – and enabled the experimentalists to study interfaces not suitable for other techniques such as those with large lattice mismatch [1]. Second, it has produced a large data base, making it possible to identify general trends in the discontinuities of different interfaces, and a correlation between heterojunction discontinuities and Schottky barriers [5].

Third, and perhaps most important, surface-ensitive techniques have played a fundamental role in the microscopic control of heterojunction band discontinuities, that was successfully achieved in the past three years [6, 7]. The key to the control is interface chemistry, specifically the controlled contamination over a distance small with respect to the typical carriers' diffusion length. This control is the main topic of the present review.

Understanding the ways to microscopically control heterojunction band discontinuities requires a reasonable background understanding of the problem of band lineups – both theoretically and experimentally. Such an understanding is beyond the scope of the present review, primarily because a number of reviews have recently appeared in the literature on this subject [1]. The reader is therefore referred to such reviews for a complete treatment; we will limit our discussion to the elements necessary to understand the specific problem of band discontinuity control.

2.1.1 Surface-sensitive experimental approaches

The most important background knowledge necessary to understand this review is that concerning the photoemission techniques to measure heterojunction band discontinuities [1]. The elementary aspects of such techniques can be understood with the help of Fig. 22.1. A typical experiment begins with an ultraclean surface (under ultrahigh vacuum) of one of the two semi-conductors that are involved in the interface to be investigated. From the leading edge of the valence-band photoemission spectrum, and after subtracting the photon energy $h\nu$, one obtains the position in energy of the valence band edge for the clean surface of the material. Note that this position does not coincide with the bulk valence-band edge, because of band bending near the surface; photoemission spectroscopy is extremely surface-sensitive, and it only probes the edge at the surface.

The second step of the experiments is the creation of an interface by depositing a thin overlayer of the second material on top of the first one. 'Thin' here means that the thickness is comparable to the surface sensitivity of the photoemission probe, and also that the band bending is negligible over such a thickness. The leading edge of the new valence-band spectra reveals the position in energy of the valence-band edge of the overlayer. The valence band discontinuity ΔE_v, however, is **not** given by the difference between the overlayer's valence-band edge position and the substrate's one previously determined. The reason is that, in general, the substrate band bending of the substrate is modified by the deposited overlayer.

This band-bending modification must, therefore, be determined. This can be done by monitoring the shifts in energy of substrate core-level peaks in the photoemission spectrum; the procedure is not straightforward, since such positions are affected not only by the band bending changes, but also by overlayer-induced changes in the chemical shifts of the core-level peaks. Therefore, the determination of the band-bending changes requires a careful comparative analysis of several core-level peaks [1]. Recently, a very elegant solution of this crucial problem was proposed by Xiaohua Yu and co-workers, based on experiments at different temperatures [8].

The problems associated to the changes in band bending are eliminated, in a few cases, by the direct observation of the two edges after the overlayer has been deposited. This observation is made impossible, in most cases, by the fact that the two edges are too close in energy to each other to be resolved; it becomes possible to resolve them,

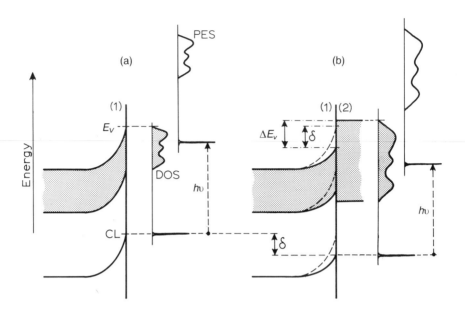

Fig. 22.1 Measurements of heterojunction valence band discontinuities, ΔE_v, using photoemission spectroscopy. (a) Experiments conducted on the ultraclean surface of the first semiconductor (1) that forms the heterojunction. The electronic density of states (DOS) is reproduced, in first approximation, by the photoelectron spectrum (PES); the PES is shifted upwards in energy by $h\nu$, the photon energy. Note the leading edge of the PES and of the DOS that correspond to the valence-band edge E_v. Also note the DOS and PES peaks that correspond to one of the core levels of the substrate (CL). Due to the high surface sensitivity of the photoemission experiment, and to the presence of band bending, from the PES one derives the **surface** positions in energy of E_v and CL derived from the PES, that are different from the bulk positions. (b) After depositing a thin overlayer of the second semiconductor (2), the valence-band edge derived from the PES is that of the overlayer. The difference between this valence-band edge position and the valence-band edge position derived in step (A) does **not** directly give ΔE_v, because of the overlayer-induced change in band bending, δ. The value of δ can be derived from the core-level peak shift; the picture presented here is oversimplified, however, since it assumes no other factors i the core-level shift. A more complete discussion can be found in the text and in [1].

however, when ΔE_v is unusually large. Such is the case of many interfaces involving at least one large-gap semiconductor, for example ZnSe–Ge [9].

Extensive experiments conducted in the past decade have demonstrated that photoemission techniques of this kind can measure ΔE_v for all types of heterojunction interfaces [1]. Variations of these techniques have been able to directly measure the conduction band discontinuity [10], and also to study 'buried' interfaces [8, 11]. It should also be possible to measure band discontinuities with another surface-sensitive technique: scanning tunnel microscopy.

The use of the techniques previously outlined

has led to the creation of a large data base fc valence band discontinuities [1]. This data ba⸱ has been extensively used to test theoretic⸱ models of the band lineups – and also, as we ha▾ seen, to search for general trends [1]. Furthe⸱ more, it has been used to derive an empiric⸱ approach for the estimate of valence band d⸱ continuities [12]. This consists of taking the d⸱ ference of the valence-band edge positions of t⸱ corresponding materials; in turn, these positio⸱ were empirically determined by Katnani a⸱ Margaritondo [12] from the photoemission d⸱ base on ΔE_v. The fundamental implications the success of this approach have been discuss⸱ elsewhere [1, 12].

22.2 MICROSCOPIC CONTROL OF INTERFACE PARAMETERS

We are now ready to turn our attention to the main topic of this review: novel approaches to control the band lineups at semiconductor–semiconductor interfaces [6, 7]. Roughly speaking, the approach consists of inserting an ultrathin intralayer at the interface; 'ultrathin' means shorter than the carrier's diffusion length, so that the entire interface including the intralayer still acts as if it had a sharp discontinuity in transport phenomena. In practice, significant modification of ΔE_v has been achieved even with submonolayer intralayer thicknesses [6, 7].

The experiments discussed in this review use intralayers with only one element. It should be mentioned, however, that heterogeneous intralayers offer tremendous technological possibilities in this area. For example, McKinley *et al.* [13] have obtained preliminary evidence that ΔE_v can be controlled by double-monolayer intralayers, exploiting the corresponding additional interface dipole as suggested by Baroni and Resta [14].

In the case of one-element intralayers, we repeatedly and consistently found that such a 'contamination' at the interface between two semiconductors can strongly influence the band lineup [15, 16]. The 'contamination' can be a thin layer ($\sim 1\,\text{Å}$) of metallic atoms such as Al or Cs, or saturation coverages of H or O. In some cases, the change in the valence band discontinuity can be as large as 0.5 eV, and it can be an increase or a decrease depending on the material of the intralayer. These results demonstrate that the nature of the interface is important to the band lineup. Band alignment theories which rely only on intrinsic bulk-derived quantities are clearly at odds with these results, since they imply that ΔE_v is determined by the bulks of the two semiconductors alone.

After many years of theoretical work, two main classes of theories which predict band lineups have emerged. Theories based on the charge neutrality level explain band lineups in terms of the states created by tunneling across the junction [1, 16–18]. The unified defect model attributes the properties of the interface to Fermi-level pinning by interface defects [19, 20]. However, both classes of theories are limited due to approximations. In particular, the simplest versions of both kinds of theories imply that the Schottky barrier height and the heterojunction valence band offset are **intrinsic** quantities which cannot be altered by changes in the properties of the interface. These implications contradict experimental results.

Recently, Flores [17] developed a version of the charge-neutrality-level approach which depends on the **extrinsic** interface properties, suggesting a way to eliminate the discrepancy. Also, Mönch demonstrated that both defect states and gap states must be taken into account while explaining interface between Si and different metals, and that the Schottky barrier height is influenced by a correction term dependent on the electronegativity of the metal atoms [21]. He suggested that to a first order approximation the charge neutrality level of Si aligned to the Fermi level of the metal, but then a small amount of excess charge transfer due to the different electronegativities of the Si and the metal slightly changed the alignment. Similarly, Tersoff used a Schottky-like correction term to explain the discrepancy between the midgap energy predictions and the observed correlations between heterojunction band lineups and Au Schottky barriers [22].

We review here experimental results of the effects of intralayers at the interfaces of semiconductor–semiconductor contacts. The intralayers significantly alter the chemical status of the atoms at the interface. Changes in the band lineup correlate with chemical changes in the interface. The changes in the band lineup result from intralayer-induced dipole at the interface, which can be understood qualitatively by two different mathematical procedures. First, Schottky-like correction terms to band lineup theories, such as the midgap energy theory or the unified defect model, predict intralayer induced changes in the band lineup. Second, atomic dipoles due to the transfer of charge among the intralayer and the component semiconductors can be added to band lineup theories. Both procedures predict with reasonable accuracy the induced changes in the band lineup.

22.2.1 Experimental results on 'sharp' interfaces

Our group has performed photoemission experiments on three different systems: Ge–ZnSe(110), Ge–CdS($10\bar{1}0$), and Si–GaP(110). In all cases, the compound semiconductor was the substrate and the group IV material was the overlayer. Al was used as an intralayer for all three systems.

The photoemission experiments were performed at the Aladdin storage ring of the University of Wisconsin-Madison, on the Mark II grazing incidence 'Grasshopper' beamline and the University of Wisconsin–General Motors 'Extended Range Grasshopper' beamline. The photon energies for all the spectra are in the range $h_v = 40$–300 eV. The photon flux from the monochromator are measured by a 90% transmission copper mesh mounted in the sample chamber. The photoemitted electrons were collected with a double pass cylindrical mirror analyzer. The overall experimental resolution (analyzer and monochromator) was 0.2–0.5 eV. Data acquisition was controlled by a PDP-11 computer and a CAMAC crate.

The base pressure of the sample chamber was 5×10^{-11} torr. The II–VI and III–V substrates were cleaved *in situ*, and the Al, Si, and Ge were deposited on the freshly cleaved surfaces. The Al and Ge was evaporated from a tungsten basket and the Si from an electron bombardment source where electrons of 6 keV energy were directed against a Si single crystal. The overlayers were deposited on room temperature substrates and the thicknesses were measured with a quartz oscillator. Typical evaporation rates were 0.5 to 5 Å per minute. During the evaporations, the pressure in the chamber rose to at most 4×10^{-10} torr.

For the three systems investigated, Si–Al–GaP, Ge–Al–CdS, and Ge–Al–ZnSe, we measured the valence band discontinuity as a function of the thickness of the Al intralayer. Therefore, for each measurement of ΔE_v we had to first cleave the substrate, then deposit the Al intralayer, and then build up layers of the group IV semiconductor.

The complete measurement of each value of ΔE_v for a given intralayer thickness required ~12 hours of synchrotron radiation. For each individual experiment, we monitored the following core levels (where applicable) in addition to the valence band: Si2p, Ge3d, Al2p, Ga3d, Cd4d, Zn4d, S2p, Se2p, and P2p.

From the core levels, we deduce information about the chemical species formed at the interface as well as band bending changes due to the overlayer. We selected photon energies for all of the core levels so that the photoemitted electrons would fall near the minimum of the universal escape depth curve. The binding energy (E_B) scale is referred for all of the spectra **to the Fermi level**, as determined by the emission cut-off for a fresh thick Al film. The position in energy of the valence band maximum is determined by a linear extrapolation of the leading edge of the valence band photoemission spectrum. References [1] and [23] give more details about the accuracy of this method for determining valence band discontinuities.

Note that we are most concerned here not in the absolute value of the valence band discontinuity, but in the relative change as a function of the thickness of the Al intralayer. Therefore, it is possible to use core level binding energies for an estimation of the induced change, provided that the core levels do not show appreciable chemical shifts resulting from changes in bonding. Figure 22.2 outlines the method for the Ge–CdS heterojunction with Al intralayers. From the figure, we see that a change in ΔE_v appears as a change in the binding energy difference E_B(Ge3d) $- E_B$(Cd4d), as long as neither the Cd4d nor the Ge3d show chemical shifts. In all cases, we first measured ΔE_v directly with the leading edge technique discussed in section 22.1, and then we derived the changes in ΔE_v with the technique of Fig. 22.2.

Results for Ge–Al–CdS

The compositional changes in the interfacial region brought about by the intralayer can belong to one of two classes. Either the intralayer diffuses into the heterojunction components, or it remains at the interface and separates them. A possible expla-

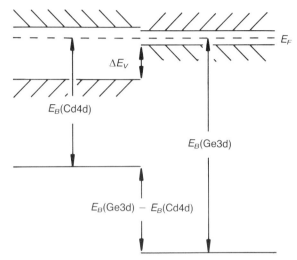

Fig. 22.2 Core level method for extracting a change in the valence band discontinuity induced by the Al intralayer.

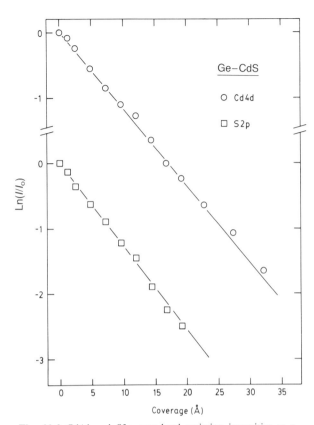

Fig. 22.3 Cd4d and S2p core level emission intensities as a function of Ge overlayer for the Ge–CdS heterojunction. The vertical axis is the natural logarithm of the intensity normalized to the value from the cleaved CdS surface.

nation for the induced changes brought about by the intralayer is that the intralayer diffuses into the component semiconductors and, though doping, changes the Fermi level positions on the two sides of the interface. If this were true, then the appropriate description of the induced changes in the band lineup would have to deal with the doping effects of the intralayer. Thus, a very important concept concerning the intralayers is whether or not they diffuse into the component semiconductors. In particular, we want to know if the intralayer changes the width of the heterojunction interface. To do this, we studied the core level intensities of the Ge–CdS heterojunction with and without an Al intralayer. We find that the Al reacts strongly with the S and forms a layer of Al_2S_3, but that the interface is atomically abrupt in both cases. The main effect of the intralayer is to change the chemical status of the interface region, so that the dipole leading to the change in the band lineup exists in a region very near the interface. The Ge–CdS heterojunction has been studied previously, and the magnitude of the valence band offset is $\Delta E_v = 1.75\,\text{eV}$ [15]. We find that the insertion of a thin Al intralayer increases the valence band discontinuity by $\sim 0.2\,\text{eV}$, and that this change correlates with the consumption of the surface S atoms in Al_2S_3.

Figure 22.3 shows the attenuations of the Cd4d and the S2p core level emissions as a function of overlayer thickness for the Ge–CdS heterojunction without an Al intralayer. The x-axis is the Ge thickness and the y-axis is the natural logarithm of the intensity normalized to that of the cleaved surface. For a layer-by-layer growth of the overlayer, the substrate core level emissions should attenuate as $\ln(I/I_o) = -d/\lambda$, where I is the intensity, I_o is the intensity for the cleaved surface, d is in first approximation, the overlayer thickness, and λ is the escape depth for photoelectrons in the overlayer. As seen from the Fig. 22.3, both core levels attenuate exponentially with escape depths of $\lambda \sim 8.6\,\text{Å}$ and $\lambda \sim 8.0\,\text{Å}$ for Cd4d and S2p. The kinetic energies of the core emitted electrons are

$E_K(\text{Cd4d}) = 29\,\text{eV}$ and $E_K(\text{S2p}) = 34\,\text{eV}$. The minimum of the escape depth curve occurs for kinetic energies near 50–60 eV, and escape depths rise sharply as kinetic energy decreases. These values and the trend with kinetic energy agree with the universal escape depth curve.

Neither the Cd4d nor the S2p core level emission lineshapes show any evidence for a strong interaction with the Ge overlayers. However, the Ge3d core emission for very thin coverages ($\sim 1\,\text{Å}$) consists of two peaks separated in energy by $\sim 1.2\,\text{eV}$. The Cd4d core emission lineshape does not give evidence for a metallic-like component, so that the Ge–S bonds must not be able to disrupt the stronger Cd–S bonds. The important point from Fig. 22.3 and the lineshapes is that the Ge–CdS without an intralayer is atomically abrupt.

The Al intralayer does not change the abruptness of the interface. Figure 22.4 shows the attenuations of the substrate and the intralayer core levels as a function of coverage for the case of 2.4 Å Al. For the S2p and the Cd4d core level emissions, the total overlayer coverage is the thickness of the Ge layer plus the thickness of the Al intralayer, whereas for the Al2p it is just the thickness of the Ge overlayer. With the first deposition of 2.4 Å Al, the Cd4d core level takes a sharp decrease while the S2p decreases only slightly. With increasing coverages of Ge, both the S2p and the Cd4d follow exponential attenuation with an escape depth of $\lambda \sim 7.2\,\text{Å}$. The kinetic energies for the Cd4d and the S2p core emitted electrons are again $E_K(\text{Cd4d}) = 29\,\text{eV}$ and $E_K(\text{Ge3d}) = 34\,\text{eV}$.

The sharp decrease in the Cd4d core level emission intensity and the lack of decrease in the S2p intensity is consistent with the chemical picture of the interface. Core level photoemission data show that the Al reacts with the S to form Al_2S_3 and frees the Cd. A possible process is that the Al pulls the S from the substrate to form the compound and leaves the Cd below. For this to occur the Cd would effectively decrease more quickly than expected since it is buried by both the S and the Al, and the S would not decrease since it has been pulled out into the Al layer.

The most important observation from Fig. 22.4 is that the Al2p core level intensity also attenuates

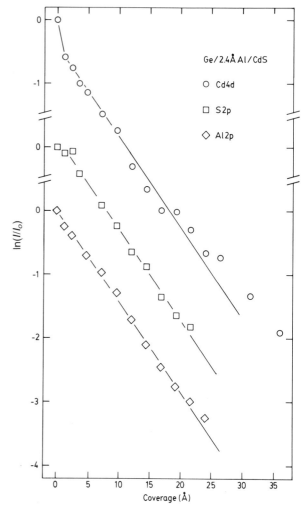

Fig. 22.4 Cd4d, S2p, and Al2p emission intensities as a function of overlayer thickness for the Ge–CdS heterojunction with a 2.4 Å Al intralayer.

as expected for homogeneous overlayer formation without any diffusion. We measure an escape depth of $\lambda \sim 7.2\,\text{Å}$ from the slope of the attenuation, in accord with a kinetic energy of $E_K \sim 30\,\text{eV}$ and the escape depths of the Cd4d and S2p core-emitted electrons. This behavior shows that the Al does not diffuse into the Ge overlayer, so that the interface is atomically abrupt. We emphasize the importance of this result, since the appropriate model for the interface induced changes in the band lineup must

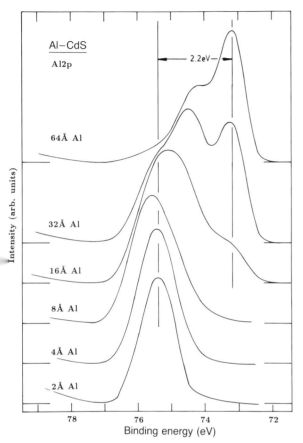

Fig. 22.5 Al2p core level emission lineshape as a function of Al thickness for the Al–CdS interface.

account for local changes in chemical species at the interface rather than diffusion of the intralayer.

The insertion of the Al intralayer does not change the abruptness of the Ge–CdS interface, but it does make substantial changes in the morphology local to the interface. In particular, we argued that it forms a reacted layer of Al_2S_3. To prove this we show in Fig. 22.5 the Al2p core emission for increasing thicknesses of Al on CdS. This figure shows the strength of the Al–S reaction. The 64 Å Al film shows emission from both metallic Al and reacted Al. The binding energy of the metallic component of the core level is $E_B = 73.2$ eV. The spin-orbit splitting is 0.4 eV, and we do not resolve it here. At coverages less than 8 Å Al, the Al2p core level emission has only

one component at ∼2.2 eV below the metallic level. Brillson *et al.* [24] argued that this component results from the formation of Al_2S_3. At a coverage of 16 Å Al, the emission begins to broaden strongly and show some metallic Al. We see three distinct components at a coverage of 32 Å Al. At a coverage of 64 Å Al, the metallic component appears to dominate.

The complicated lineshapes of the Al2p for coverages near 32 Å result from three distinct components, one from the metallic Al, one from the Al_2S_3, and one from a metastable sulfide, possibly AlS. Figure 22.5 shows an attempt to decompose the lineshape of the 32 Å Al emission into three peaks. We performed the decomposition in the following manner. First, we scaled the spectrum for the 2 Å Al film shown in Fig. 22.6 so that its intensity at energy $E_B \sim 76.5$ eV was equal to the intensity of the 32 Å Al film at the same energy, producing peak 'A'. We then substracted it so that the remaining curve was the sum of peaks 'B' and 'C' in the 32 Å Al spectrum. We then used the same procedure for the 64 Å Al spectrum. Then by matching peak 'C' of the 32 Å Al spectrum to the corresponding peak 'C' for the 64 Å Al spectrum, we could extract the relative heights and lineshapes of peaks 'B' and 'C'. Thus, peaks 'A', 'B', and 'C' in Fig. 22.6 are not computer-generated lineshapes that we tried to fit to the data, but rather are experimentally determined lineshapes deduced from subtraction routines. The sum of peaks 'A', 'B', and 'C' are exactly equal to the spectrum taken from the 32 Å Al film.

The lineshapes of the three peaks, shown in Fig. 22.6, strongly suggest three components to the core level emission. Peaks 'A' and 'C' represent Al_2S_3 and metallic Al. Peak 'B' is intermediate, and most likely results from the formation of a metastable sulfide such as AlS. This component may be formed because of lack of sufficient S to form Al_2S_3. From a chemical viewpoint, the thick Al overlayers change the stoichiometry of the CdS by pulling S out into the Al film. However, for thin Al intralayers there is not enough Al to deplete the surface region of CdS of S.

Figure 22.7 shows the intensities of the Cd4d and the S2p core level emissions as a function of

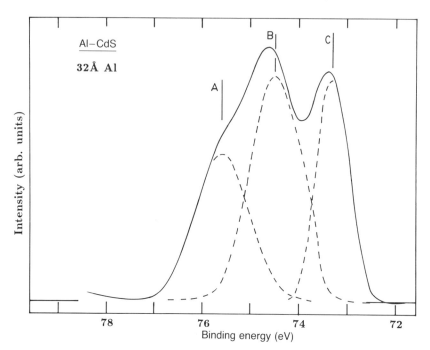

Fig. 22.6 Lineshape analysis of the 32 Å Al spectrum from Fig. 22.4. The decomposition procedure is described in the text.

the thickness of the Al overlayer for the Al–CdS interface. For a nondiffusive interface, both intensities should attenuate exponentially as previously discussed. From Fig. 22.7 we see that the Cd4d core emission does decrease exponentially with a reasonable escape depth, but the S2p core level emission does not. The slow attenuation indicates that the S atoms diffuse into the Al film, in support of the arguments made from the lineshape of the Al2p core level emission. In fact, we can make a rough estimate of the amount of S in the Al film from the attenuation of the S2p emission intensity. At a coverage of 32 Å Al, the intensity of the S2p core emission has decreased from that of the cleaved surface by a factor of $\sim$0.4. Since 50% of the CdS is S, approximately $0.4 \times 50\%$ or 20% of the overlayer film is S. This is in reasonable agreement with the intensity of peak 'C' in Fig. 22.6.

The thin Al intralayer forms a reacted region of Al_2S_3. The Ge overlayers then form an abrupt interface with the reacted region of Al_2S_3. The excess Cd liberated by the process remains buried beneath the reacted Al intralayer. The chemical status of the first layer of Ge atoms is drastically

changed by the presence of the Al. Figure 22.8 shows the Ge3d core level emission as a function of Ge overlayer thickness for the Ge–CdS interface. For a 32.2 Å Ge overlayer, the Ge3d core level emission is sharp although the spin-orbit splitting is not resolved. The centroid of the core level is at a binding energy of $E_B = 29.39$ eV and the width is $E_{fwhm} = 1.18$ eV. For a 1.2 Å Ge overlayer, the Ge3d emission is significantly broader and raised to higher binding energy. In addition, the lineshape implies two chemically different components to the emission.

The insertion of a 2.4 Å Al intralayer completely changes the chemical status of the first layers of Ge. Figure 22.9 shows the Ge3d core emission as a function of the thickness of the Ge overlayer for the case of a 2.4 Å Al intralayer. For a thickness of 33.6 Å Ge, the Ge3d is identical in line shape and binding energy to the 32.2 Å Ge signal without the intralayer. This suggests that the Al intralayer does not have an effect on the chemical status of Ge away from the interface. The Al only changes the status of the Ge in the first layer near the interface. The lineshape of the Ge3d core

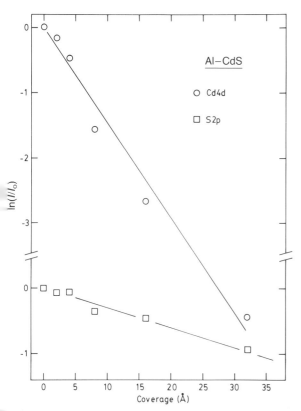

Fig. 22.7 Cd4d and S2p core level emission intensities as a function of Al coverage for the Al–CdS interface.

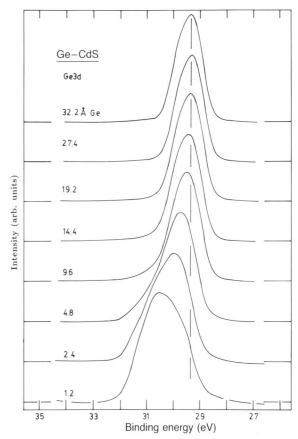

Fig. 22.8 Ge3d core emission lineshape as a function of Ge coverage for the Ge–CdS heterojunction.

emission from the 1.2 Å Ge film is entirely different with the Al intralayer. It is not nearly as broad and binding energy of the centroid is not raised to as high a binding energy.

Figure 22.10 compares the lineshapes of the Ge3d core emission for a 1 Å Ge film with and without a 2 Å Al intralayer. The lineshape without the Al intralayer appears to consist of two components, one from bulk-like Ge and a second reacted one. We have decomposed the lineshape into two components by the subtraction method described for the Al2p core level emission. The dashed curve at lower binding energy represents the signal from thick film of Ge. The intensity has been scaled so that it matches the intensity of the solid line at a binding energy of $E_B = 29.5$ eV. After subtracting the bulk-like component from the total signal, we obtain the dashed curve at

higher binding energy. The difference in binding energy between the two components of the Ge3d core level emission is 1.2 eV. Hollinger *et al.* found for bulk samples that the energy of the Ge3d core level in GeS$_2$ was 1.2 eV below that in bulk Ge [25]. Therefore, our results suggest that the first layer of Ge forms GeS$_2$.

Of the total signal for the 1 Å Ge overlayer on cleaved CdS, the reacted and the bulk-like components represent 59% and 41%. Therefore, 0.59 Å Ge are consumed in the formation of GeS$_2$. If we define the monolayer as 'one Ge atom for each surface atom on CdS($10\bar{1}0$)', then the one monolayer equivalent is 1.6 Å. For only the first layer of S atoms to participate in the formation of GeS$_2$, the Ge–S reaction should saturate at a

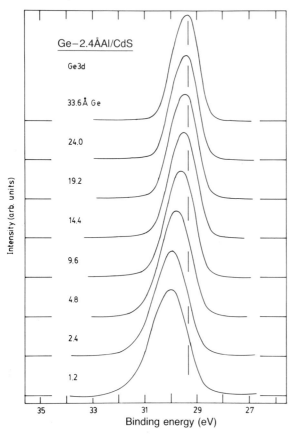

Fig. 22.9 Ge3d core emission lineshape as a function of Ge coverage for the Ge–CdS heterojunction with a 2.4 Å Al intralayer.

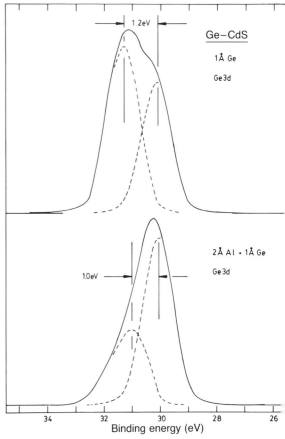

Fig. 22.10 Lineshape analysis of the Ge3d core level for 1 Å C on CdS without and with a 2.4 Å Al intralayer.

coverage of 1/4 ML, or 0.4 Å. Therefore, the data imply that the first layer and half of the second layer of the CdS surface react with the Ge. This is consistent with a picture of a somewhat disordered but atomically abrupt interface.

The bottom spectrum of Fig. 22.10 shows that the insertion of a 2 Å Al intralayer eliminates much of the Ge–S interaction. We have decomposed the lineshape in the same manner as for the top spectrum. The lineshape still shows evidence for bonds between the Ge and the S, but the amount has been substantially reduced. The reacted and the bulk-like components represent 30% and 70% of the total signal. Therefore, the Al layer has decreased the number of Ge–S bonds

by a factor of 2. The fact that there is any inter action at all confirms that either the Al_2S_3 does no uniformly cover the substrate or that the surfac region is disordered.

The chemical status of the Al atoms is n influenced by the Ge overlayer. Figure 22.1 shows the lineshape of the Al2p core level as function of the thickness of the Ge overlayer. Th intensity of the Al2p core level emission was di cussed previously. The vertical arrow marks th centroid of the spin-orbit split components of th Al2p core level emission from a metallic Al filr The energy shift increase of ~2.5 eV shows th the Al is reacted with the S. The sharpness of t emission suggests that the Al exists in only o:

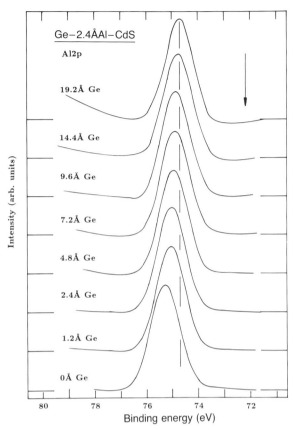

Fig. 22.11 Al2p core level emission as a function of Ge overlayer thickness for the Ge–CdS heterojunction with a 2.4 Å Al intralayer.

band discontinuity increases rapidly with the Al intralayer thickness. The arrow shows the value obtained for a thick (64 Å) intralayer, corresponding to an increase of approximately 0.25 eV. This value was **measured** by taking the difference between the two Schottky barrier heights for the two interfaces, both obtained from the photoemission data. This thick intralayer gives a 'back-to-back Schottky barrier' configuration. We see, however, that the saturation value is reached well before having a truly metallic intralayer. In fact, the data suggest saturation at a submonolayer coverage.

One can argue that the saturation value of ΔE_v occurs when the Al layer is sufficiently thick to consume all of the surface S atoms that participated in the bonding to Ge in a compound. The Ge3d lineshape of Fig. 22.10 suggests that the Al must be thick enough to involve all of the first layer and half of the second layer of S atoms in Al_2S_3. This would require minimum thickness of $\frac{1}{3}$ monolayer Al to consume the S atoms in the first layer, and $\frac{1}{6}$ monolayer to consume half of the S atoms in the second layer, for a total of $\frac{1}{2}$ monolayer. The solid line in Fig. 22.12 is not a fit to the data but rather the expected behavior from this simple picture of the intralayer. It was drawn assuming saturation at $\frac{1}{2}$ monolayer and a linear relation between the band lineup change and the number of S atoms bound to Al. The fit of the data to this expected behavior is remarkably good, and indicates that the interpretation of the induced change in ΔE_v as a result of changes in the chemical status of the interface is correct.

Results for Ge–Al–ZnSe and Si–Al–GaP

The Ge–ZnSe(110) heterojunction with Al intralayers is chemically very similar to the Ge–Al–CdS(10$\bar{1}$0) system. In fact, we can make the same general statements as for the Ge–Al–CdS system by simply replacing Zn for Cd and Se for S. The motivation for studying this system is that it is one of the prototypical lattice-matched heterojunctions. Previous experiments showed that the insertion of a thin Al intralayer increased the valence band discontinuity by 0.2–0.3 eV. Durán *et al.* modeled this system and in a tight binding framework

chemical environment, and that the Al–S bonds are much stronger than the Ge–S bonds. The slow shift in energy with increasing Ge coverage is due to changes in the position of the Fermi level. The Cd4d core level emission follows that same gradual decrease.

The Al intralayer substantially changes the local chemistry of the interfacial region, and these changes correlate to the changes in the valence band discontinuity. Figure 22.12 shows the height of the valence band discontinuity for the Ge–Al–CdS system as a function of the thickness of the Al intralayer. Without the intralayer, we measure $\Delta E_v = 1.75 \pm 0.1$ eV, in good agreement with the results of previous experiments [15]. The valence

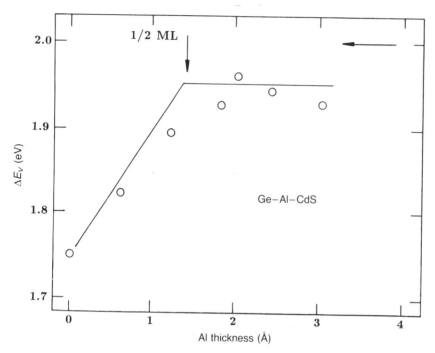

Fig. 22.12 Ge–CdS($10\bar{1}0$) valence band discontinuity as a function of Al intralayer thickness. The horizontal arrow marks the value for 'back-to-back Schottky barriers'. Saturation occurs for submonolayer intralayer thicknesses, as emphasized by the solid line that was drawn assuming saturation at $\frac{1}{2}$ monolayer.

showed that one monolayer of Al increased the valence band discontinuity by 0.35 eV [26]. The reason for the increase was a change in the charge neutrality level [19].

From a chemical point of view, the major difference between the Ge–Al–ZnSe and Ge–Al–CdS systems is that less Al_2Se_3 is formed than Al_2S_3. Otherwise, everything is very similar. The Ge–ZnSe is atomically abrupt, and the insertion of an Al intralayer does not increase the width of the interface by promoting diffusion. The Ge3d core level emission for thin Ge layers gives evidence for chemical processes identical to that for the Ge–CdS system. The first layer of Ge bonds strongly to the Se, and the Al intralayer consumes the surface Se and prohibits the bonding of Ge to Se.

A significant difference between this system and the previous one is that the Al is not as effective at extracting Se as it is S. The Al–ZnSe contact is not abrupt, but not as extended as the Al–CdS contact. Figure 22.13 shows the Al2p core level emission for increasing layers of Al on ZnSe. For a 64 Å Al film, the Al2p core level shows the 0.4 eV spin-orbit splitting. For thin films, the

Al2p emission shows one broad peak at ~2.3 eV below the metallic level. In analogy with the formation of Al_2S_3 and Al_2O_3, which increase the binding energy by 2.4 eV and 2.7 eV, we assign this peak to the formation of Al_2Se_3. An interesting difference here is that even for thin (0.3 Å) Al layers, the Al2p core level emission indicates the presence of some metallic or only partially reacted Al. At a coverage of only 2.4 Å, the signal already shows evidence for numerous chemically different environments for the Al, in contrast with the single sharp peak for the same amount of Al on CdS. The lower binding energy shoulder on the 2.4 Å spectrum of Fig. 22.13 does not exactly align with the metallic level. Most likely, the Al is unable to extract sufficient Se from the substrate to form Al_2Se_3, so that instead a metastable selenide such as AlSe forms.

For a 2.4 Å Al overlayer on ZnSe, the Al is not entirely reacted in the form of Al_2Se_3, but the addition of small amounts of Ge on top of the Al layer seem to promote the Al–Se reaction. Figure 22.14 shows the results of a 2.4 Å Al intralayer experiment. The bottom spectrum is the Al2p

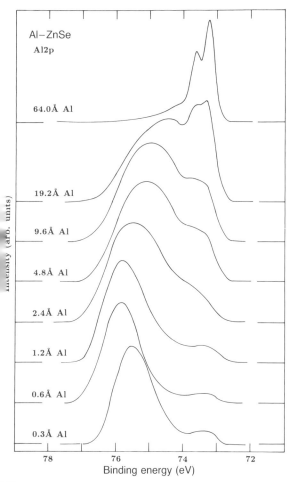

Fig. 22.13 Al2p core level emission lineshape as a function of Al thickness for the Al–ZnSe interface.

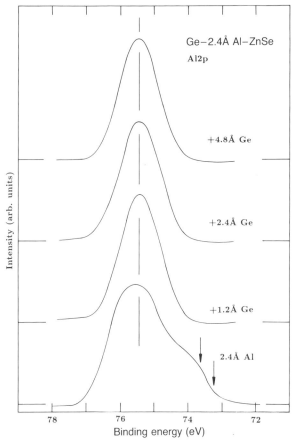

Fig. 22.14 Al2p core level emission for the Ge–Al–ZnSe interface. The vertical arrows mark the position spin-orbit components of the Al2p emission from a thick Al film.

core level emission from a 2.4 Å Al layer on the ZnSe, and the top three spectra are for increasing thickness of Ge on the Al layer. The horizontal arrow marks the positions of the spin-orbit split components of the Al2p core emission for the 64 Å thick layer of Al shown in Fig. 22.13. Without Ge, the core level emission is broad with a shoulder to higher binding energy, as also seen in Fig. 22.13. The shoulder emission does not align with the spin-orbit split components of the metallic level as marked by the two vertical arrows, and most likely results from AlSe. The deposition of 1.2 Å Ge onto the Al film removes the shoulder emission completely. The Ge may either assist in the formation

of Al$_2$Se$_3$ or bind to the Al of the metastable Al–Se compounds. Further increase in Ge coverage does not produce an additional change in the chemical status of the Al.

Figure 22.15 shows the magnitude of the valence band discontinuity for the Ge–Al–ZnSe system. Without the intralayer, we measure $\Delta E_v = 1.45 \pm 0.1$ eV, in good agreement with the results of previous experiments [2]. Once again, ΔE_v changes rapidly with the intralayer thickness, and the value at less than one monolayer is indistinguishable, within experimental uncertainty, from that given by a 'back-to-back Schottky barriers' configuration, 1.75 eV (horizontal arrow;

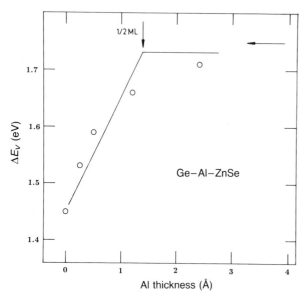

Fig. 22.15 Ge–ZnSe(110) valence band discontinuity as a function of Al intralayer thickness. The horizontal arrow marks the value for 'back-to-back Schottky barriers'. Saturation occurs for submonolayer intralayer thicknesses, as emphasized by the solid line that was drawn assuming saturation at $\frac{1}{2}$ monolayer.

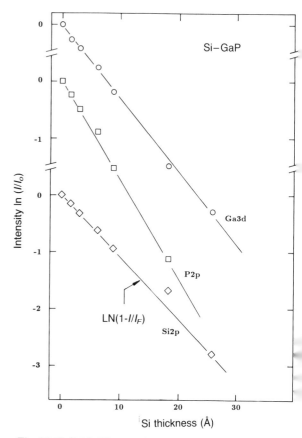

Fig. 22.16 Ga3d, P2p, and Si2p core level emission intensitie for the Si–GaP(110) heterojunction as a function of the S overlayer thickness.

this was again derived by taking the difference of the measured Schottky barrier heights for the two interfaces). The solid line in the figure is not a fit to the data, but rather the expected behavior from the simple intralayer model described for the Ge–Al–CdS system. The correlation between the data points and the suggested behavior is remarkable. Figure 22.15 suggests that the band lineup changes saturate at a coverage near $\frac{1}{2}$ monolayer, consistent with the engagement of all the surface Se atoms in chemical bonds as observed in Figs. 22.13 and 22.14.

The Si–GaP(110) is another prototypical lattice-matched heterojunction. Al is extremely reactive on GaP, since it is energetically favorable to dislodge the P from the substrate to form free Ga and AlP. Despite the strong Al–Ga exchange reaction, ΔE_v is insensitive to the presence of Al intralayers. A major difference between this system and the other two is that the interface is wider, since Ga tends to diffuse into the Si overlayer.

Figure 22.16 shows the intensities of the Ga3d,

P2p, and Si2p core level emissions as a functio of Si overlayer thickness for the Si–GaP hetero junction without an Al intralayer. For the Ga3 and the P2p emissions, we have plotted $\ln(I/I_o)$ whereas for the Si2p emission we have plotte the increase as $\ln(1 - I/I_F)$, where I_F is the fin al intensity of the Si2p core level emission measure for thick Si films. The evidence indicates tha a small amount of Ga diffuses into the Si film The P2p core level emission intensity follows th expected behavior for homogeneous overlaye growth, with an escape depth of $\lambda \sim 5.6\,\text{Å}$. Th Ga3d emission intensity also follows exponentia attenuation, but the escape depth is significantl longer, $\lambda \sim 8.0\,\text{Å}$. Note that the kinetic energies c

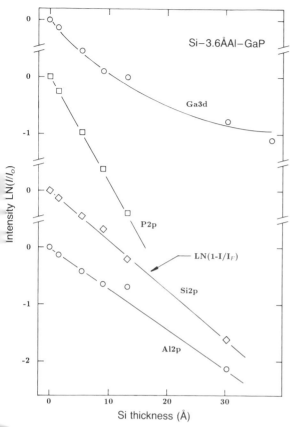

Fig. 22.17 Ga3d, P2p, Al2p, and Si2p core level emission intensities for the Si–GaP(110) heterojunction with a 3.6 Å Al intralayer as a function of the thickness of the Si overlayer.

he core emitted electrons for both the P2p and the Ga3d are $E_K = 30 \pm 1$ eV, so that the dependence of the escape depth on kinetic energy cannot explain the discrepancy. The most likely explanation is that a small amount of Ga diffuses into the Si overlayer.

The Si2p intensity increase also does not have the expected escape depth. We estimate an escape depth of $\lambda \sim 9.1$ Å. The kinetic energy of the core emitted electron is again $E_K = 30 \pm 1$ eV, so the difference can not be explained by the universal escape depth curve. The slow increase in the intensity of the Si2p core level emission may result from the outdiffusion of the Ga. As some Ga mixes with the Si, some of the Si is effectively buried by the Ga.

The insertion of a 3.6 Å Al intralayer leads to the formation of AlP, and an increase in the diffusion of Ga into the Si overlayer. Despite the formation of AlP, the P2p core level does not show any broadening due to multiple components. This suggests that the binding energy of the P2p core level in GaP is nearly equal to that in AlP. Figure 22.17 shows the attenuations of the intensities of the Ga3d, P2p, and Al2p core level emissions as well as the increase in the Si2p core emission. The kinetic energies of all the core emitted electrons are $E_K = 30 \pm 1$ eV. The P2p follows the expected exponential attenuation with an escape depth of $\lambda \sim 5.7$ Å, consistent with the escape depth observed without the Al intralayer. However, the Ga3d core emission no longer shows exponential decrease, and actually begins to level off for thick Si coverages. This leveling-off is strong evidence that the Ga diffuses into the Si film. The Al2p core level seems to decrease exponentially, but with a long escape depth of $\lambda \sim 13$ Å. Therefore, it must diffuse into the Si overlayer. The Si2p core level intensity also does not increase as quickly as expected, most likely due to the outdiffusion of Ga and Al.

Figure 22.18 shows the lineshape of the Ga3d core level as a function of the thickness of the Al intralayer and the Si overlayer. For the cleaved surface, the emission is composed of one peak centered at $E_B = 20.13$ eV. The deposition of 3.6 Å Al strongly perturbs the chemical environment of the Ga atoms, as seen by the presence of two resolved peaks in the lineshape. The peak at higher binding energy is from Ga in GaP, and the one at lower binding comes from Ga in a metallic state. With increasing Si coverages, both of these components seem to disappear and a third intermediate component from the Ga which has diffused into the Si overlayer dominates.

The Al2p core level emission shows again that the Al reacts strongly with the substrate, and does not exhibit metallic behavior. The vertical arrow in Fig. 22.19 marks the position of the metallic Al2p core level measured from a thick Al film. Without a Si overlayer, the core level emission shows two unresolved components separated by 1 eV, with the shallower one centered ~0.9 eV

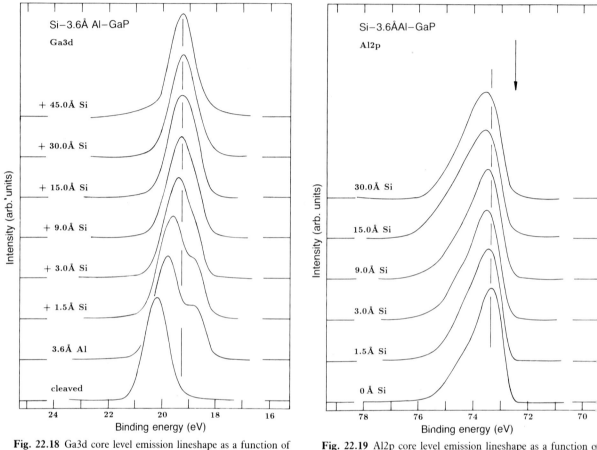

Fig. 22.18 Ga3d core level emission lineshape as a function of overlayer coverage for the Si–Al–GaP(110) system.

Fig. 22.19 Al2p core level emission lineshape as a function of overlayer coverage for the Si–Al–GaP(110) system.

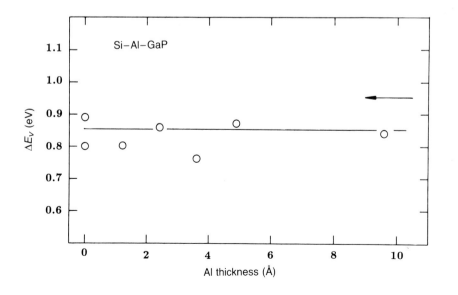

Fig. 22.20 Si–GaP(110) valence band discontinuity as a function of Al intralayer thickness. The horizontal arrow marks the value for 'back-to-back Schottky barriers'. The horizontal line is the average of all the data points.

below the metallic level. With increasing Si coverage, the lineshape broadens, showing that there is an interaction between the Si and Al atoms. As we have seen from the intensity, the Al atoms do not remain at the interface but instead seem to diffuse into the Si overlayer. The broadening of the lineshape is most likely the result of Al atoms bound to Si atoms in the overlayer.

The Al intralayers do not make a measurable change in the band lineup despite reacting strongly with the anion of the substrate. Figure 22.20 shows the magnitude of the valence band discontinuity for this system. Even when the 'back-to-back Schottky barriers' configuration is reached, $\Delta E_v = 0.95$ eV (horizontal arrow) is, within experimental uncertainty, equal to the value measured without an intralayer, $\Delta E_v = 0.85$ eV. We reiterate that this system is different from the previous two in that it does show a significant amount of interdiffusion. This may ultimately be the reason the Al intralayers do not affect the valence band discontinuity.

22.2.2 Theoretical models

We have seen that the Al intralayers induce changes in the band lineups of the Ge–CdS and Ge–ZnSe interfaces, but not in the Si–GaP interface. This, as we have seen, cannot be explained by theories which predict that the band lineup is an intrinsic property which is not affected by the details of the interfacial region. An explanation can be sought within the framework of either the unified defect model [16, 17] or the charge-neutrality level, [18–20] by considering extrinsic interface properties, and their effect on the local dipoles. Mönch, for example, suggested that a difference in electronegativities between the component semiconductors leads to an addition dipole which offsets the band lineup from the value expected by the alignment of the 'intrinsic' charge-neutrality levels [21]. Tersoff, on the other hand, argued that the same dipole can be calculated from a Schottky-like correction term [22].

We analyze the Schottky-like correction term in the framework of the charge-neutrality-level or midgap-energy rule. However, the analysis also applies to the unified defect model, replacing the midgap energy with the defect level. The midgap energy rule was first proposed by Tersoff [6] to explain Schottky barrier and heterojunction formation. Early versions of the midgap energy rule stated that the valence band discontinuity at a heterojunction interface is

$$\Delta E_v = E_M^1 - E_M^2 \qquad (22.1)$$

where E_M^i is the energy of the midpoint of the dielectric energy gap measured from the valence band maximum. The analog of eq. (22.1) for a metal–semiconductor interface is

$$\phi_B = E_M \qquad (22.2)$$

where ϕ_B is the Schottky barrier (assuming a p-type material) and E_M is the midgap energy. One of the strongest points of the midgap energy approach is that it treats heterojunctions and Schottky barriers on similar footing. The insertion of a thick intralayer of Al is conceptually equivalent to the formation of 'back-to-back Schottky barriers'. Such a system has an effective valence band discontinuity $\Delta E_v(\infty)$ equal to the difference in the two Schottky barriers, $\phi_B^1 - \phi_B^2$. Thus, without any correction term, the midgap energy rule (expressed by eqs. (22.1) and (22.2)) does not predict a change in ΔE_v for thick intralayers, contrary to our experimental results. A similar conclusion is valid for the defect model.

Recent versions of the midgap energy rule [19] incorporate a correction term. The corrected expression for the valence band discontinuity is

$$\Delta E_v = E_M^1 - E_M^2 + \bar{S}((\chi^1 + E_G^1 - E_M^1) - (\chi^2 + E_G^2 - E_M^2)) \qquad (22.3)$$

where i labels the semiconductor, χ^i is the electron affinity, and E_G^i is the optical energy gap. The 'pinning strength parameter', $\bar{S}$, is proportional to $1/\varepsilon_\infty$, the inverse of the dielectric constant applicable to the interface. In first approximation, ε_∞ can be estimated from the average of the dielectric constants of the two component semiconductors. The corresponding equation for Schottky barriers is

$$\phi_B^i = E_M^i - S^i(\chi^i + E_G^i - E_M^i - \psi) \qquad (22.4)$$

where ϕ_B^i is the Schottky barrier height (p-type material) and ψ is the work function of the metal. As before, S^i is proportional to the reciprocal of the dielectric constant of the semiconductor.

Since the midgap energy rule handles Schottky barriers and heterojunction band offsets on the same footing, there is a correlation between the two. To a first order approximation, it predicts a linear relationship between the difference in Schottky barrier heights for two semiconductors and a given metal and the valence band discontinuity between the two semiconductors. Since we have found that the thickness increases of the Al intralayer are irrelevant to the change in band lineup, provided the thickness is above $\sim\frac{1}{2}$ mono-layer, we expect to be able to use the correlation between heterojunctions and Schottky barriers to predict a change in the valence band discontinuity for thin intralayers.

In order to proceed, one must make some assumptions on the screening constants. Since we have seen that S^i should scale linearly with the inverse of the dielectric constant $1/\varepsilon_\infty^i$, it should also scale linearly with the energy of the dielectric midpoint. Therefore, we assume that in eq. (22.3) $S^i = aE_M^i + b$ and that $\bar{S} = (S^1 + S^2)/2$. By substracting two Schottky barriers of the form of eq. (22.4), one can show that

$$\Delta E_v(\infty) = \phi_B^1 - \phi_B^2 = \Delta E_v(0) \times (1 + K) + K^* \tag{22.5}$$

where

$$K = \frac{a}{2}((\chi^1 + E_G^1 - E_M^1 - \psi + (\chi^2 + E_G^2 - E_M^2 - \psi)) \tag{22.6}$$

and

$$K^* = -\frac{a}{2}\bar{S}((\chi^1 + E_G^1 - E_M^1 - \psi)^2 - (\chi^2 + E_G^2 - E_M^2 - \psi)^2) \tag{22.7}$$

A typical magnitude for K^* is 0.001 eV, so we will neglect it. Equation (22.5) predicts a linear relationship between Schottky barrier heights and valence band discontinuities. We should be able to use eq. (22.5) to estimate an effective valence band

Table 22.1

Material	χ (eV)	E_G (eV)	E_M (eV)
Si	4.0	1.1	0.2
Ge	4.1	0.7	0.0
GaAs	4.1	1.4	0.6
GaP	3.6	2.3	0.7
InP	4.4	1.4	0.9
ZnSe	4.1	2.7	1.4
CdTe	4.2	1.5	0.8
CdS	4.8	2.4	1.7[a]

[a] The midgap energy of wurzite CdS is not available from the literature. We estimated such a value from the measured band lineup of CdS–Ge, assuming that, in first approximation, the midgap energy correctly predicts the band lineup. $\psi^{Al} = 4.2$ eV; $\psi^{Au} = 5.2$ eV; $\psi^{Cs} = 4.2$ eV.

discontinuity for thick intralayers. As we have seen from the data, this discontinuity coincides with that given by a thin intralayer.

To make use of eq. (22.5), we need to know the value of K from eq. (22.6). Of course, it will depend on the properties of the semiconductors involved and the work function of the metal. However, it will also depend on a. We have used the parameters of Ref. [20] to make estimates of $a = 0.068$ and $b = 0.061$. Table 22.1 lists the parameters needed to perform the test. The work function of Al is $\psi^{Al} = 4.2$ eV. The parameters for Table 22.1 were taken from Refs. [20], [27], and [28]. The use of these parameters gives values of $K = +0.035$, $K = +0.063$, and $K = +0.06$ for the Ge–Al–CdS, the Ge–Al–ZnSe, and the Si–Al–GaP systems. Then eq. (22.5) predicts that:

1. Al increases the Ge–CdS ΔE_v by $+0.06$ eV
2. Al increases the Ge–ZnSe ΔE_v by $+0.09$ eV
3. Al increases the Si–GaP ΔE_v by $+0.05$ eV

The most important result is that the Schottky-like correction term to the midgap energy rule predicts increases in all the valence band discontinuities with Al intralayers. For the Si–Ga heterojunction, however, the predicted increase is small, and could not be detected beyond the experimental uncertainty.

For the systems with the II–VI substrates, w

do see increases, although the experimental increases are ~ 3 times larger than the predicted increases. Thus, our approach based on the midgap energy rules predicts the correct trend for the Al intralayers, but misses the magnitude. The most likely source of the error is the assumption that $S = 1/\varepsilon_\infty$. Note that the interface is a highly localized system in one direction. Therefore, the use of a dielectric constant to describe screening is questionable, and this could explain the above discrepancies.

A similar approach, also based on the Schottky-like correction term to the midgap energy rule, solves another problem related to the correlation between Schottky barriers and heterojunctions. In Ref. [6], we used the simplest version of the midgap energy rule to predict a correlation between Au Schottky barriers and heterojunction valence band discontinuities of the form of eq. (22.5) with $K = 0$. However, the data suggested that instead $K = -0.2$. Use of the Au work function ($\psi^{Au} = 5.2$) and the parameters of Table 22.1 shows that indeed K is less than zero, but a typical value is only $K = -0.07$. Therefore, the Schottky-like correction term does predict the correct trend for the correlation between Schottky barrier heights and valence band discontinuities, but again the magnitude is in error by approximately a factor of 3. The fact that K is positive for Al and negative for Au reflects the corresponding work functions. For metals with a large work function such as Au, K is negative; but it is positive for metals with a small work function like Al.

Mönch has shown that the midgap energy rule must be explicitly corrected for an interface dipole, but he used a different formalism than the Schottky-like correction term [21]. His argument is that when two materials in contact have different electronegativities, charge will transfer from the material of the lower electronegativity to the material of the higher one. This transfer of charge is done through the tails of the virtual gap states (VGS).

The condition for no charge transfer at a heterojunction interface is that the charge neutrality levels of semiconductors must align. In order to have a transfer of charge from, say, semiconductor 1 to semiconductor 2, the charge neutrality level of

1 must be higher in electron energy than the charge neutrality level of 2. We have attempted to model the induced changes brought about by the intralayer using electronegativity based dipole calculation first described by Sanderson for the formation of molecules [29].

The tenet of the Sanderson model is that to form a molecule, charge transfers from atoms of lesser electronegativity to atoms of higher electronegativity until equal electronegativity is achieved. The electronegativity of a molecule, S_m, is given by

$$S_m = \left(\prod_{i=1}^{N} S_i \right)^{1/N} \tag{22.8}$$

where S_i is the electronegativity of the ith among the N atoms forming the molecule. The charge transfer (in electron charges) on the ith atom is given by

$$\rho = \frac{S_m - S_i}{2.08 S_i^{\frac{1}{2}}} \tag{22.9}$$

The values of S_i are given in Ref. [15]. To apply this model to a calculation, we must first consider the dipole between the two semiconductors before the insertion of the intralayer, and then consider it with the intralayer. The basic mechanism is that since the two semiconductors have different electronegativities, the semiconductor of higher electronegativity will take more charge from the intralayer than the semiconductor of lower electronegativity. This imbalance of charge transfer leads to a net dipole due to the intralayer. To calculate the dipole from the amount of charge transferred, we assume a very simple 'parallel plate capacitor' model.

The straight application of this model gives remarkably good results for the Ge–ZnSe and Ge–CdS heterojunctions, but not for Si–GaP. The changes in the valence band discontinuity for the three are:

1. Al increases ΔE_v by 0.23 eV for Ge–CdS.
2. Al increases ΔE_v by 0.09 eV for Ge–ZnSe.
3. Al decreases ΔE_v by 0.53 eV for Si–GaP.

The poor agreement of the Si–GaP system is not surprising in light of the diffusion of the Ga and

Table 22.2

System	Intra (ML)	V_D	SCT	$\delta(\Delta E_v)$
Ge–CdS	0.4 Al	+0.10	+0.06	+0.20
Si–CdS	0.4 Al	−0.20	+0.12	+0.15
Si–CdS	sat H	−0.62	–	−0.50
Ge–ZnSe	0.8 Al	+0.07	+0.09	+0.25
Ge–GaAs	0.4 Al	+0.02	+0.03	+0.00
Si–SiO$_2$	0.2 Cs	−0.13	–	+0.25
Si–SiO$_2$	sat H	−1.27	–	−0.50
Si–GaP	sat H	−0.65	–	−0.40
Si–GaP	0.2 Cs	−0.07	+0.17	+0.10
Si–GaP	0.4 Al	−0.22	+0.05	+0.00

Al into the Si overlayer. The straightforward dipole calculation assumes an abrupt interface and no intralayer diffusion.

Table 22.2 lists the experimental changes in the valence band discontinuity for all of the systems studied to date. The columns are:

(a) system – the convention is overlayer/substrate;
(b) intralayer material and thickness in monolayers;
(c) estimated change from the dipole calculation;
(d) estimated change from the Schottky-like correction term;
(e) experimental change in ΔE_v.

We were unable to calculate Schottky-like correction terms for all cases since we do not know the midgap energy point of SiO$_2$ and H does not have a work function. The most striking result of the table is the agreement between the Schottky-like correction terms and the experimental values. Note that in all cases the Schottky-like correction term gets the correct sign for the change in the band lineup, and in some cases magnitude is remarkably good.

The accuracy of the dipole calculation is less good, and in some cases the sign of the estimated change disagrees with experiment. We were able to bring the results of the calculation into better agreement with experiment by limiting charge transfer between some atoms. For instance, if the predominant charge transfer for the Si–SiO$_2$ heterojunction with the Cs intralayer is between the Si of the overlayer and the Cs, then the dipole calculation predicts an increase in ΔE_v by 0.2 eV, in accord with experiment.

In summary, we have shown that intralayers can produce changes in valence band lineups significant in magnitude to affect the performances of heterojunction devices. For nondiffusive intralayers, a Schottky-like corrective term to the midgap energy rule for band lineups appears to explain qualitatively the induced changes. To improve the quantitative results of the theory, a better understanding of the interface pinning strength parameter, S, is needed. Unfortunately, a general theory of the dielectric properties of interfaces, and hence the nature of S, is an extremely complex problem. A simplified treatment of the interface in terms of dipoles resulting from charge transfer also gives remarkably accurate predictions.

22.3 FUTURE DIRECTIONS

The initial successes in controlling heterojunction band lineups open the way to a new chapter in microelectronics research. In the long run, there is the possibility of tailoring the device parameters to specific functions, using band lineup control and other methods of bandgap engineering. The gateway to these applications, however, is a complete understanding of the phenomena that govern the band lineup control. As we have seen, substantial progress has been made, but we are still far from such a complete knowledge of these complex phenomena.

One important point is that the 'sharp', i.e. reactive interfaces explored until now are relatively simple systems, not representative of heterojunction interfaces in general. We foresee, therefore, additional research on similar phenomena in other kinds of interfaces, both reactive and unreactive. In the latter, microscopic diffusion processes could become very important, for example by establishing extended interface dipoles due to charged impurities. Very little is known about these phenomena, and they must be explored extensively and in detail.

The other obvious avenue in the future of this field is the move from one-element intralayers to many-element intralayers: a true step into sophisticated bandgap engineering. The first direction will be, of course, the study of bilayers. Theoretical models for bilayer effects have already been provided [14] and, as we have seen, some preliminary experimental results have been obtained [13]. By and large, however, this is a virgin field, and one with tremendous potentiality for applications.

22.3.1 From spectroscopy to spectromicroscopy

A third avenue of development of this research area will be stimulated by advances in the instrumentation. In a broad sense, the advances already achieved are in large part a product of using photoemission techniques for interface studies [2]. In turn, this has been made possible by the advances in photoemission instrumentations of the 1970s, primarily in electron analysis and photon sources (synchrotron radiation) – as well as by previous advances in ultrahigh vacuum technology. We are now at the threshold of yet another jump ahead in instrumentation: the commissioning of a new generation of ultrabright synchrotron radiation sources, such as ALS at Berkeley and ELETTRA at Trieste.

In our opinion, the greatest impact of ultrahigh brightness on photoemission spectroscopy will be the capability of controlling and exploring parameters that have not been exploited in the past. A similar evolution occurred in the mid-1970s, when instrumentation advances made it possible to control new parameters in the photoelectric process, primarily the direction of emission of the photoelectrons (with the development of angle-resolved photoemission) [2]. Several parameters are still largely unexploited: the most important are the coordinates on the emitting surface's plane, spin, and time. The main reason for not exploring them is lack of signal, and that is where the ultrabright sources can have the greatest impact.

Substantial progress has been already made, for example, in exploiting the surface coordinates, such as in performing photoemission experiments with high lateral resolution [30]. Electron-imaging photoelectron microscopes have been implemented at Stanford, Minnesota, Argonne and Wisconsin-

Milwaukee. The first two scanning instruments have been implemented at Brookhaven and Wisconsin-Madison. Figure 22.21, for example, shows a total-yield photoelectron micrograph of a metal-covered cleaved GaAs substrate, taken with the Wisconsin scanning instrument MAXIMUM, that was developed in collaboration with the Berkeley Center for X-ray Optics, with the participation of Minnesota, Xerox, and Frascati [31]. Note the pattern of the metal overlayer, created by depositing the overlayer through a mesh; also note the cleavage steps, whose imaging demonstrates the topographic capabilities of this novel technique.

The advent of lateral resolution transforms a spectroscopy like photoemission into a spectromicroscopy: while conserving all the spectroscopic capabilities such as those of detecting elements and their valence state, photoemission spectromicroscopy also achieves lateral resolution. This novel capability is potentially very relevant to the kind of research discussed in our review. Consider specifically the band lineups (or, similarly, the Schottky barriers): in the past, they have been studied as 'average' properties, and very little is known about their variations over the interface. Yet, such variations could have a great impact on the performances of the corresponding devices.

Can photoemission spectromicroscopy really contribute to heterojunction interface research? The main question, of course, is that of lateral resolution. At present, the best lateral resolution is of the order of one-half micron, and rapidly improving. No fundamental limitations are present before a level of a few hundred ångstrom. The relevant reference for studying the lateral dependence of band lineups is the Debye length, that depends on the local doping. A resolution of hundreds of ångstroms, although not sufficient for the cases of high doping, is still capable of studying lightly-doped and in some cases medium-doped systems.

Photoemission spectromicroscopy, therefore, can contribute to heterojunction interface research. With its use, we can expect a better understanding of the band lineups and of their correlation to the local chemistry and to the local interface morphology. With the parallel contribution of

MAXIMUM

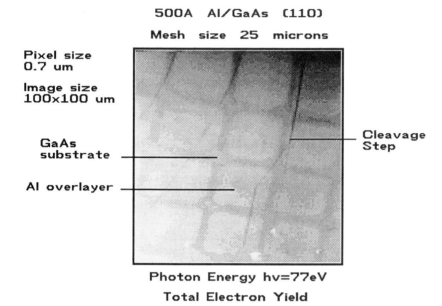

500A   Al/GaAs   (110)

Mesh   size   25   microns

Pixel size
0.7 um

Image size
100x100 um

GaAs
substrate

Al overlayer

Cleavage
Step

Photon Energy hν=77eV

Total Electron Yield

Fig. 22.21 Photoelectron micrograph of a cleaved GaAs(110) substrate covered by an Al overlayer patterned by a mesh. The micrograph was taken with the undulator photoelectron spectromicroscope MAXIMUM at the Wisconsin Synchrotron Radiation Center ([31]).

other powerful instruments such as the scanning tunnel microscope, we expect not only progress in the fundamental understanding of heterojunctions, but also progress in our capabilities of controlling their parameters, along the lines already initiated and discussed in this chapter.

ACKNOWLEDGMENTS

Our own research in this field, described by this review, was performed in collaboration with Ming Tang, Jim McKinley, Roberto Zanoni, Claudio Quaresima, and Paolo Perfetti, supported by the National Science Foundation and by the Wisconsin Alumni Research Foundation, and using the University of Wisconsin Synchrotron Radiation Center (also supported by the National Science Foundation). We are grateful to Jerry Tersoff for several stimulating interactions.

REFERENCES

1. For a review of this field and of its history, see: F. Capasso and G. Margaritondo, *Heterojunctions: Band Discontinuities and Device Applications*, North-Holland, Amsterdam (1987), and G. Margaritondo, *Electronic Structure of Semiconductor Heterojunctions* (Jaca, Milan, and Kluwer, Dordrecht, 1988); F. Flores and C. Tejedor, *J. Phys. C*, **20** (1987), 145 and the references therein.

2. See, for example, G. Margaritondo, *Introduction to Synchrotron Radiation* Oxford University Press, New York (1988).

3. R. Dingle, W. Weigmann and C. H. Henry, *Phys. Rev. Lett.*, **33** (1974), 827; R. People, K. W. Wecht, K. Alavi and A. Y. Cho, *Appl. Phys. Lettt.*, **43** (1983), 118; A. C. Gossard, W. Brown, C. L. Allyn and W. Wiegmann, *J. Vac. Sci. Technol.*, **20** (1982), 694.

4. J. R. Waldrop, S. P. Kowalczyk, R. W. Grant, E. A. Kraut and D. L. Miller, *J. Vac. Sci. Technol.*, **19** (1981), 573; A. D. Katnani and R. S. Bauer, *Phys. Rev. B*, **33** (1986), 1106.

5. D. W. Niles, M. Tang, J. McKinley, R. Zanoni and G. Margaritondo, *J. Vac. Sci. Technol. A*, **7** (1989), 2464.

6. D. W. Niles, G. Margaritondo, C. Quaresima, and M. Capozi, *Appl. Phys. Lett.*, **47** (1985), 1092.

7. P. Perfetti, C. Quaresima, C. Coluzza, C. Fortunato and G. Margaritondo, *Phys. Rev. Lett.*, **57** (1986), 2065.

8. Xiaohua Yu, A. Raisanen, G. Haugstad, G. Ceccone, N. Troullier and A. Franciosi, *Phys. Rev. B*, **42** (1990), 1872.

9. G. Margaritondo, C. Quaresima, F. Patella, F. Sette, C. Capasso, A. Savoia and P. Perfetti, *J. Vac. Sci. Technol. A*, **2** (1984), 508.

10. P. P. Perfetti, F. Patella, F. Sette, C. Quaresima, C. Capasso, A. Savoia and G. Margaritondo, *Phys. Rev. B*, **29** (1984), 5941.

11. D. W. Niles, B. Lai, J. T. McKinley, G. Margaritondo, G. Wells, F. Cerrina, G. J. Gualtieri and G. P. Schwartz, *J. Vac. Sci. Technol. B*, **5** (1987), 1286.

12. A. D. Katnani and G. Margaritondo, *J. Appl. Phys.*, **54** (1983), 2522.

13. J. McKinley and G. Margaritondo, unpublished.

14. S. Baroni and R. Resta, private communication.

15. D. W. Niles, G. Margaritondo, E. Colavita, P. Perfetti, C. Quaresima, and M. Capozi, *J. Vac. Sci. Technol. A*, **4** (1986), 962.

16. J. Tersoff, *Phys. Rev. B*, **30** (1984), 4874; W. A. Harrison and J. Tersoff, *J. Vac. Sci. Technol. B*, **4** (1986), 1068.

17. R. Perez, A. Munoz and F. Flores, *Surface Sci.*, **226** (1990), 371 and private communication.

18. M. Cardona and N. E. Christensen, *Phys. Rev. B*, **35** (1987), 6182.

19. W. E. Spicer, P. W. Chye, P. R. Skeath, C. Y. Su, and I. Lindau, *J. Vac. Sci. Technol.*, **16** (1979), 1422.

20. W. E. Spicer, I. Lindau, P. Skeath, and C. Y. Su, *J. Vac. Sci. Technol.*, **17** (1980), 1019.

21. W. Mönch, *Phys. Rev. Lett.*, **58** (1987), 1260.

22. J. Tersoff, private communication.

23. A. D. Katnani and G. Margaritondo, *Phys. Rev. B*, **28** (1983), 1944.

24. L. Brillson, R. Bauer, R. Bachrach, and J. McMenamin, *J. Vac. Sci. Technol.*, **17** (1980), 476.

25. G. Hollinger, R. Kumurdjian, J. M. MacKrowski, P. Pertosa, L. Porte, and Tran Minh Duc, *J. Electron Spectros. and Rel. Phenom.*, **5** (1974), 237.

26. J. Durán, A. Mūnoz, and F. Flores, *Phys. Rev. B*, **35** (1987), 7721.

27. *CRC Handbook of Chemistry and Physics*, edition 61.

28. J. L. Freeouf and J. M. Woodall, *Appl. Phys. Lett.*, **39** (1981), 727.

29. R. T. Sanderson, *Chemical Bonds and Bond Energy*, Academic Press, New York (1971).

30. G. Margaritondo and F. Cerrina, *Nucl. Instr. Meth. A*, **291** (1990), 26 and the references therein.

31. F. Cerrina, B. Lai, C. Gong, A. Ray-Chaudhuri, G. Margaritondo, M. A. Green, H. Höchst, R. Cole, D. Crossley, S. Collier, J. Underwood, L. J. Brillson and A. Franciosi, *Rev. Sci. Instrum.*, **60** (1989), 2249; F. Cerrina, S. Crossley, D. Crossley, C. Gong, J. Guo, R. Hansen, W. Ng, A. Ray-Chaudhuri, G. Margaritondo, J. H. Underwood, R. Perera and J. Kortright, *J. Vac. Sci. Technol. A*, **8** (1990), 2563; W. Ng, A. K. Ray-Chadhuri, R. K. Cole, S. Crossley, D. Crossley, C. Gong, M. Green, J. Guo, R. W. C. Hansen, F. Cerrina, G. Margaritondo, J. H. Underwood, J. Korthright and R. C. C. Perera, *Phys. Scripta*, **41** (1990), 758; Gelsomina De Stasio, W. Ng, A. K. Ray-Chaudhuri, R. K. Cole, Z. Y. Guo, J. Wallace, G. Margaritondo, F. Cerrina, J. Underwood, R. Perera, J. Kortright, Delio Mercanti and M. Teresa Ciotti, *Nucl. Instrum. Methods in Physics Section*, **294** (1990), 351.

CHAPTER TWENTY-THREE

Microscopic nature of metal–polymer interfaces

P. S. Ho, B. D. Silverman and
Shih-Liang Chiu

Introduction · Molecular structure and morphology of polyimides · Surface and interface chemistry · Diffusion and interface formation · Thermal stress and interfacial fracture · Summary

23.1 INTRODUCTION

Polymers, as a class of materials, have been used extensively in microelectronics. Applications include photoresists, metal–polymer laminate boards, and protective coatings. In the past several years, polymers have found an important application as dielectric layers in metal–polymer multi-layered structures for chip interconnect and packaging modules [1, 2]. Polyimides have been prime candidates for such applications in micro-electronics due to a combination of attractive properties [3]. First, in their precursor form, the polyamic acids, they are easy to work with; to shape, to spin, and to coat. They planarize readily. After curing they form a tough, mechanically strong material over a wide range of temperatures and provide specific applications with desirable electrical properties, e.g. low dielectric constant and good electrical insulation. The wide range of different aromatic units that may be chosen for the polymeric backbone provides an additional degree of freedom (DOF) with respect to the specific tailoring of these properties for particular applications.

Consider the commonly used polyimide; PMDA–ODA (pyromellitic dianhydrideoxydianiline) has a monomeric repeat unit with length of approximately 18 Å and in common applications is part of a polymeric chain having a molecular weight of 20 000–30 000. This corresponds to a coil diameter of about 200 Å. These dimensions, the repeat unit and coil dimension, are one and two orders of magnitude, respectively, larger than the lattice parameter of a crystalline solid. When the polymer is fused with a metal, such as Cu, to form an interface such as schematically illustrated in Fig. 23.1, the interfacial structure contrasts sharply with the interface formed when metals are deposited on crystalline solids. In particular, the lattice match/mismatch of a crystalline interface which controls the nature of defects at the interface, e.g. vacancies and dislocations, and consequently many of its thermal and mechanical properties, can no longer be meaningfully defined. Instead, one has to consider two structural aspects: first, the local bonding of the Cu atoms with the chemical entities of the polyimide, e.g. the aromatic rings and the carbonyl ligands; second, the chain structure of the polymer at and near the interface. The latter extends several coil diameters from the interface providing the bulk and surface polymeric molecular structure and morphology with an important role in determining the interfacial properties. The

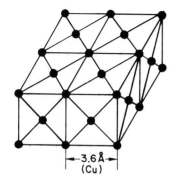

Metal crystalline structure

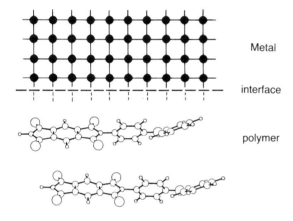

Fig. 23.1 Schematic presentation of a Cu–PMDA-ODA interface. The monomeric unit of PMDA–ODA is shown approximately in scale to that of Cu.

of the interface, which gives rise to a time- and temperature-dependent behavior unique to this class of materials. This feature has been extensively investigated for many polymers, although the information available for polyimides is rather limited, particularly for metal–polyimide layered structures.

These particular interfacial properties present challenging basic questions regarding the nature of the metal–polymer interface. Although the current state of understanding has not reached a molecular level, we will attempt to focus on what is presently known about the microscopic nature of the interface, and in particular upon aspects relating to the thermal and mechanical properties. The present discussion will emphasize two polyimides, PMDA–ODA and BPDA–PDA (bisphenyldianhydride-phenyldiamine). Although closely related in chemistry, the BPDA–PDA polyimide has a molecular structure that is different from PMDA–ODA, and with a better thermal match to form relatively stable interfaces involving various conducting metals. A comparison between the two materials will provide an interesting example to demonstrate the effect of structure on interfacial properties. In view of the importance of molecular morphology, we will first summarize results of recent studies of the molecular structure of polyimides and the influence of the molecular structure on thermal and mechanical properties. The chemistry and bonding characteristics of the interface will then be addressed. This aspect has been extensively investigated for PMDA–ODA and some model compounds related to BPDA–PDA using X-ray (XPS) and UV (UPS) photoemission spectroscopies. In addition to the photoemission studies, we will include recent results obtained from vibrational and energy loss spectroscopies. This will be followed by a discussion of the results of radio-tracer diffusion studies of metal atoms deposited onto the surface of the polyimide and the consequent morphology developed after annealing. Results of computer simulation will be discussed in connection with the radio-tracer diffusion studies. Finally, the recent results of studies of stress relaxation and fracture behavior of the interface will be reported.

wealth of information regarding the atomic structure and preparation of many crystalline surfaces, e.g. the (111) 7 × 7 Si surface, stands in stark contrast with the limited information we have about polymers. For the polyimides, the surface structure as well as the bulk molecular morphology near the interface are not well understood. It is known, however, that the polymeric backbone structure together with certain processes, e.g. the curing steps, are important in determining the chain morphology, in the bulk as well as near the surface. In addition to this structural aspect, the viscoelasticity of the polymer is important in controlling the thermal and mechanical properties

23.2 MOLECULAR STRUCTURE AND MORPHOLOGY OF POLYIMIDES

The aromatic polyimides are characterized by a backbone structure consisting of cyclic imide groups interconnected to aromatic dianhydrides and diamines. Various chemical elements and groups can be incorporated into the diamine and/or the dianhydride moieties to provide a torsional or flexural DOF to the rigid aromatic ring structures [4]. Depending on the combination of the particular diamine and dianhydride groups, the polyimide structure can have a considerable range of flexibility. The semiflexible nature of the polymeric chain gives rise to a molecular order in orientation and conformation for the noncrystalline state of the polyimide. Since the high-temperature properties, such as thermal stability and heat resistance, arise from the imide and aromatic ring structures, it is possible to synthesize aromatic polyimides having a wide range of thermal and mechanical properties.

The polyimide precursor, polyamic acid (PAA), is synthesized by condensation and polymerization of aromatic diamine and tetracarboxylic dianhydride in N-methyl-2-pyrrolidone (NMP) and xylene. The polyimide film is usually prepared by spin-coating the polyamic acid (PAA) on a substrate, following by drying and curing steps in a nitrogen atmosphere. Drying is carried out at a temperature range from 75 to 100°C. This yields a glassy film which can be removed from the substrate and put in a free-standing state or mounted in a frame with uniaxial or biaxial confinement for the curing process. Curing usually consists of two steps, one at about 200°C to imidize the polyamic acid, and the other at 350–400°C to drive out the water and solvent produced during imidization. The annealing temperature and the heating rate used in the curing process have been found to be important in controlling the molecular order and morphology.

The curing process has been investigated for a number of polyimides. Detailed studies have been carried out for PMDA–ODA using wide-angle and small-angle X-ray diffraction [5–8] and dynamic mechanical thermal analysis [9, 10]. Upon drying, NMP–PAA complexes with a ratio of about 2 to 1

are formed in the condensed state [9]. X-ray diffraction has shown that the condensed state of the NMP–PAA complex has an ordered glassy structure with a nearly fully extended chain conformation along the film plane [5]. The observed repeat distance of 13–14 Å is close to the projected length of the monomeric unit along the chain direction. The chains are packed in a manner such that the monomeric units of a given chain are in registry with those of neighboring chains. Following drying, subsequent heating leads to a sequence of processes consisting of decomplexation–plasticization of PAA, and imidization of the polyamic acid upon increasing temperature. Dynamic mechanical thermal analysis has shown that the first process of curing consists of decomplexation and plasticization starting at about 110°C with subsequent imidization at somewhat below 200°C [9]. This can be seen from the dynamic modulus and loss angle observed as a function of curing temperature and rate, as shown in Figs. 23.2(a) and 23.2(b). The variation of these parameters indicates that the amount of change is rate-dependent and that the molecular order is usually reduced by plasticization. The change is restored to a large extent by the imidization process. X-ray diffraction has confirmed the reduction in molecular order for a free-standing film, although for an oriented film confined to a substrate, the chain orientation becomes more pronounced after imidization [5]. During the second curing step at 300–400°C, no additional imidization was observed; instead changes in the molecular packing occurred in the polyimide. Separate experiments performed under the conditions of isothermal annealing indicated some improvement in molecular order. This was found to occur via a thermally activated process but at a rate significantly slower than that of the imidization process [9]. X-ray diffraction confirmed this finding. In addition, the extent of the effect was found to depend upon the molecular order at the completion of imidization. With an oriented film on a substrate, the change was not appreciable, although for unoriented films, the improvement was more significant.

Molecular structure and chain flexibility are expected to be important in determining some

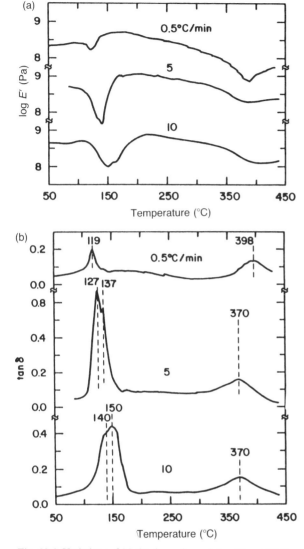

Fig. 23.2 Variations of (a) the dynamic modulus E' and (b) the loss angle δ observed during thermal curing of a PMDA–ODA polyamic acid film as a function of temperature for three heating rates of 0.5, 5, and 20°C/min (from [9]).

has been clearly demonstrated in a study of the dependence of the thermal expansion coefficient and Young's modulus on the molecular packing coefficient [11, 12]. The value of the packing coefficient of many polyimides is in the same range as the value for standard polymers. There are, however, two exceptions: the rod-like polyimides and those polyimides with side groups. The former has a high packing coefficient while the latter has a low coefficient due to the steric restriction of the side groups. The results of this study are summarized in Figs. 23.3(a) and 23.3(b) where the thermal expansion and elastic modulus are seen to correlate with the packing coefficient for aromatic polyimides. This is also true for rod-like polyimides, but not for polyimides with side groups. This has been further confirmed by a separate measurement [14] in which polyimides with rigid and flexible chains were uniaxially stretched. The thermal expansion and elastic modulus were found to change only for the flexible-chain polyimides. This trend is consistent with increasing chain linearity.

In this regard, it is interesting to compare the properties of PMDA–ODA and BPDA–PDA. Their molecular structures are shown in Fig. 23.4. The PMDA-ODA structure is characterized by a relatively rigid PMDA structure and a flexible ODA fragment with two benzene rings twisted relative to the PMDA unit and linked by the ether oxygen atom at an angle of about 120°. The BPDA–PDA structure can be thought of as being derived from the PMDA–ODA repeat unit by removing the ether oxygen in ODA while inserting the separated benzene ring into the PMDA fragment. This produces a monomer with less flexibility in the backbone structure, since the flexibility of the monomeric unit arises mainly from the relative twist of the two halves of the BPDA fragment. As summarized in Table 23.1, the effect of the molecular structure is clearly reflected in the physical properties of these two polyimides. In particular, the elastic modulus, the in-plane thermal expansion and the moisture uptake of the BPDA–PDA polyimide are about 300–500% different from the PMDA–ODA polyimide although their densities are about the same.

of the thermal and mechanical properties of the polyimide. This aspect has been investigated using a systematic combination of diamine and dianhydride chemical groups [11–14]. Results of these studies revealed that except for properties such as solvent and moisture uptake, polyimide properties are closely correlated with chain morphology instead of with the packing coefficient or free volume. This

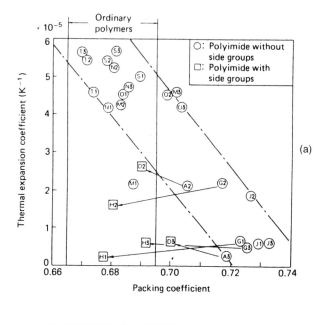

(a)

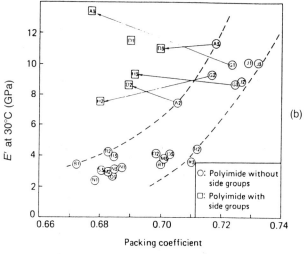

(b)

Fig. 23.3 Correlations between packing coefficient and (a) thermal expansion coefficient and (b) Young's modulus for various polyimides (From [11]).

23.3 SURFACE AND INTERFACE CHEMISTRY

23.3.1 Photoemission studies

X-ray photoemission spectroscopy has been one of the important experimental techniques in providing evidence concerning details of the surface and

PMDA-ODA   (a)

BPDA-PDA   (b)

Fig. 23.4 Molecular structures of the monomeric units of (a) PMDA–ODA and (b) BPDA–PDA.

Table 23.1 Physical properties of PMDA–ODA and BPDA–PDA

	PMDA–ODA	*BPDA–PDA*
Density (g/cm³)	1.42–1.44	1.44–1.47
Young's modulus (GPa)	2.5–3.0	9.5–13
Tensile strength (MPa)	180–200	480–600
Strain to break (%)	60–80	30–50
Thermal expansion coefficient (ppm/°C)	25–40	5–7
Tg (°C)	400	>400
NMP uptake (wt. % at 85°C)	~30–40	~1–3
Water uptake (wt. % at 22°C)	3–4	1–1.5

interface chemistry of the polyimides [15, 16]. There have been early studies of the interface prior to metal deposition as well as studies of the effects of metal deposition and in particular examination of shifts in the XPS line positions after or during the deposition of metals [17, 18]. The interpretation of the undeposited material requires some care since the polymeric chain consists of strong electron withdrawing groups on an aromatic backbone [19, 20]. A simple picture emerges concerning the XPS profile one observes or expects to observe for different polyimides. Usually there are several carbon 1s XPS lines at different positions depending upon the degree of electron charge transferred from the carbon atom. An example of the carbon 1s spectrum of the PMDA–ODA is shown in Fig. 23.5(a). The highest binding energy contribution to the C1s spectrum is associated with the carbonyl carbon atoms, at −288.5 eV for the PMDA–ODA polyimide. Fluorine groups will usually yield C1s lines at even higher binding energies [21, 22] but such groups are generally not present in the most commonly investigated polyimides. Carbon atoms singly bonded to either electronegative oxygen or nitrogen atoms provide a set of lines several electron volts lower in binding energy than the C1s lines of the carbonyl carbon atoms. This corresponds to the peak at −284.7 eV in Fig. 23.5(a). Near such lines arising from singly bonding the carbon atoms to either oxygen or nitrogen are the aromatic carbon atoms that are nearest neighbors to the carbonyl groups. The relative position of these aromatic carbon atom C1s lines depends upon the number of carbonyl groups adjacent to the aromatic benzene ring. For the PMDA moiety there are four carbonyl groups adjacent to this ring and the C1s spectra from this ring lie approximately 1 eV below the aliphatic carbon atom positions (the peak at −285.7 eV). Actually, there is a simple rule-of-thumb that can be used to determine the shift to higher binding energy of the benzene ring C1s positions as a function of the number of carbonyl groups adjacent to the benzene ring. Each carbonyl group can be assumed to shift the C1s lines originating from the benzene ring carbon atoms by about one-quarter of an electron volt. In comparison, the BPDA moiety of the

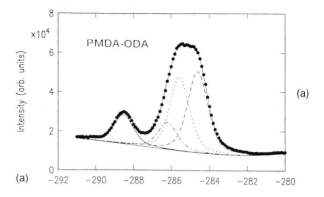

(a)

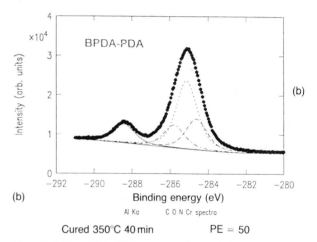

(b)

Cured 350°C 40 min   PE = 50

Fig. 23.5 The C(1s) XPS spectra observed for the clean (a) PMDA–ODA and (b) BPDA–PDA polyimide surfaces. The deconvoluted contributions reflect the different chemical environment of the carbon atoms (from [23] and [24]).

BPDA–PDA polymer exhibits a narrower major C1s peak (Fig. 23.5(b)) since the aromatic carbon atoms on the ring are adjacent to only two carbonyl groups and are therefore not shifted out of the instrumentally broadened major peak. The instrumental broadening is approximately 0.5 eV which is about the intrinsic broadening of the major C1s peak in BPDA–PDA. The XPS shifts calculated and observed for the BPDA–PDA polymer [23], shown in Fig. 23.5(b), are similar to those previously reported [20] for the N-phenylphthalidomide model compound.

It is of interest to note that for most of the polyimides the areas under the C1s peaks are not in close accord with what one might expect from a stoichiometric polymer and counting the various contributions from the different functional units as integral contributions. If one fits intensities to the major C1s peak then one observes a so-called, 'carbonyl deficiency'. Before attempting to understand the origin of such deficiency it is necessary to correct the data in at least two different ways. First, any significant shake-up features involving the carbonyl carbon atoms must have this intensity subtracted from the main carbonyl C1s peak [16, 24]. Calculations of relative XPS intensities should also take into account final state effects [25, 26]. With all such corrections, however, there still appears to be a carbonyl deficiency as evidenced by the reduced peak area under the carbonyl C1s peak. It is tempting to speculate that this reduced intensity is related to binding of the carbonyl groups to other adjacent groups on other polymer strands. If so, this would then represent a 'covalent contribution' to the interaction energy and consequently to the binding between different polymer chains. This together with Van der Waals interactions and the entanglement between strands would contribute to the overall cohesiveness of the polymer.

The chemistry of the metal–polyimide interface has been extensively studied. Results have been reviewed for interfaces formed with metals deposited on cured polyimide surfaces [15, 16] and for interfaces formed by depositing the polyimide on metal surfaces [27, 28]. The chemistry of these two types of interfaces is significantly different, particularly for the interface formed by curing the polyamic acid on a metal surface. Chemical reaction can occur between the polyamic acid and the metal to form an interface with compound particulates (mostly oxides) dispersed into the polyimide. Even for the interface formed with vapor-deposition of polyimide fragments, the chemistry of the initial deposition is distinct since the interface is formed by attaching the polyimide fragments to a metallic surface [27, 28]. Here we limit our discussions to the interface formed by metal deposition on a cured polyimide surface.

Upon metallization of PMDA–ODA polyimide with the transition metal, chromium, one infers from the changes in the C1s spectrum that the metal interacts solely with the PMDA unit of the polymer. This is unambiguously inferred upon initial deposition since the C1s contribution from carbon atoms on the ODA group that are not bound to either the ether oxygen atom or to nitrogen atoms are initially unaffected [29, 30]. Furthermore, the nitrogen 1s XPS peak is unshifted upon initial deposition. These features can be seen from the series of C1s spectra for the initial Cr coverage of PMDA–ODA shown in Fig. 23.6. This is consistent with a picture of binding between the metal atom and polyimide that results from charge transfer from the metal to aromatic ligand involving the lowest unoccupied orbital (LUMO) of the ligand [31]. The orbital amplitude at the nitrogen atoms for this LUMO is essentially zero. This contrasts with results obtained for BPDA–ODA. For this polymer it has been observed [23] that the nitrogen N1s XPS peaks shift in energy at relatively low depositions. One surmises that this might be a consequence of the reduced symmetry of the BPDA unit and resultant orbital amplitude on the nitrogen of the LUMO of this moiety.

Studies on other 3d-metals reveal a consistent chemical trend in which the bonding strength and the chemical reactivity depend upon the number of the d-electrons in the metal [15]. At one end of the 3d metals, Ti is similar to Cr as it interacts strongly with PMDA–ODA, while at the other end, Ni is similar to Cu although its reactivity is somewhat higher. It is of interest that at low deposition, the type of XPS changes observed upon copper deposition are similar to those observed upon chromium deposition even though the changes are reduced in magnitude. One might be tempted to ascribe such shifts resulting from copper deposition as due to a similar type of binding site or sites expected for chromium. The filled copper 3d shell as well as the results of preliminary *ab-initio* studies suggest, however, that copper atoms will bind to a polyimide substrate in a very different manner than the chromium atoms. One feature does, however, seem to apply equally well to these systems, i.e. that the atoms of both metals initially (at low

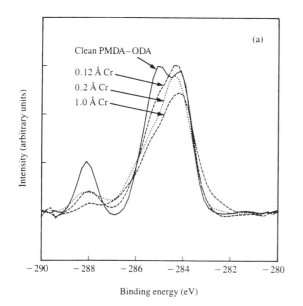

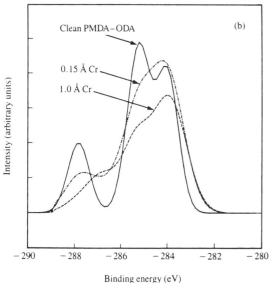

Fig. 23.6 (a) A series of C(1s) spectra observed with increasing Cr coverage on PMDA–ODA surface. Note that 0.3 Å corresponds to approximately one chromium atom per monomeric unit; (b) Calculated C(1s) spectra with increasing Cr coverage. For 0.15 and 1.0 Å coverages, Cr atoms were distributed over the five- and six-member rings of PMDA (from [24]).

is more complex due to the difficulty to delineate the spectral contributions from various metal–polymer complexes. Furthermore, diffusional reaction can occur at a thicker coverage which alters the chemical nature of the interface and causes additional complications to spectroscopic studies. There is a considerable debate concerning the nature of the Cr–PMDA–ODA complex, particularly regarding the possibility of formation of metallic compounds, e.g. Cr oxide [29, 32, 33]. Diffusional reactions have been observed on PMDA–ODA surfaces upon deposition at 300°C for Al [34] and upon annealing for Cu [35]. In spite of the interest, an interpretation of the results of these studies is beyond the scope of this chapter and will not be discussed.

A correspondence between strongly bonding metals such as chromium can be made with the magnitude of XPS shifts observed of the polymeric constituents upon metal deposition. For instance, C1s shifts observed upon deposition of copper onto a PMDA–ODA surface are much smaller than observed for chromium deposited upon the same type of substrate. For PMDA–ODA polyimide, this trend has been further demonstrated between the strength of XPS shifts and measured adhesion by peel tests that yield a qualitative measure of the binding of the metal to polymer. Changes in the XPS spectrum upon deposition of chromium metal are the largest and this metal is found to bind most strongly to the polymer [36]. The results of the adhesion studies will be discussed in greater detail later.

Ultraviolet photoemission studies have also been performed on polyimides to assist in understanding the surface and interface chemistry. Measurements on the clean surface have been interpreted with the results of theoretical electronic structure calculations [37, 38]. The various electronic levels have been assigned to the different regions of the UPS spectrum. As expected, one finds the orbitals associated with excitations from low lying levels to be localized to the various atomic constituents whereas the higher lying levels are delocalized with the highest lying orbitals having either pi character or being lone-pair orbitals. If it is assumed that each orbital contributes equally to the spectrum the

deposition) bind first to the PMDA functionality of the PMDA–ODA polyimide.

The interpretation for the bonding characteristics beyond the initial coverage of submonolayer agreement between the experimental result and the calculated spectrum is quite close.

It is generally stated that changes in the UPS spectrum upon deposition of a metal or interaction of a diffusant such as CO_2 or H_2O should in principle yield more detailed information concerning the bonding of such constituents to the polyimide substrate than changes observed in the XPS spectrum. One expects this since the orbitals actively involved in the UPS are the higher lying delocalized orbitals which in principle should contain a greater amount of information concerning the details of interaction between the polymer and interacting atom or molecule than the lower lying tightly bound XPS inner shell levels. This has not been, however, generally true in practice. Whereas one can observe changes in the UPS that arise from the interaction of the polyimide with another molecular or atomic species such changes are difficult to interpret. Delocalization of the orbitals over the different elemental constituents of the polymer has made it almost impossible to identify the local region of interaction.

This has been an advantage of XPS measurements, i.e. contributions from the different elemental species are so well separated in energy. As an example of this, even though copper binds weakly to the PMDA–ODA polyimide, one can infer from the XPS measurements that upon initial deposition, i.e. at very low levels of copper concentration it does initially bond only to the PMDA functionality [39]. Effects of copper deposition on the UPS spectrum are observed [35]; however, the most we have learned is that the copper contribution to the spectrum is at a position that we would expect it to be and that there is a further washing-out of the PMDA–ODA polyimide spectrum as the deposition of copper is increased. The measurements have not enabled us to infer at what position on the polyimide structure such interaction occurs. Similar information has been obtained for water uptake in polyimide [35]. The water bands are observed superimposed upon the PMDA–ODA UPS spectrum and the process of water uptake in the polyimide has been shown to be reversible.

Part of the difficulty of UPS studies can be circumvented, as demonstrated in recent experiments using synchrotron UV sources [16, 37, 40]. By tuning the photon energy, it is possible to modulate the cross-section of the interfacial atomic species to reduce the intensity overlap of their spectral lines. This allows one to delineate more clearly the various spectral intensities of the interfacial elements. An example is shown in Fig. 23.7 for two valence-band spectra taken at photon energies of 150 and 500 eV on an interface formed by depositing PMDA–ODA polyamic acid on a Cu surface [16]. The spectrum at 150 eV comes mainly from the polyimide constituents while the Cu 3d states near the Fermi edge are largely suppressed. The opposite is shown in the spectrum taken at 500 eV photon energy where the Cu valence states emerge clearly near the Fermi edge. Nevertheless, a proper interpretation of these spectra, to deduce the nature of the interfacial chemistry, still requires detailed theoretical analysis.

23.3.2 Lattice vibrations of the polyimides

High resolution electron energy loss spectrum of the PMDA–ODA polyimide has been utilized to characterize the vibrations of the polymer at and near the surface of the polymeric film [41]. In the dipolar scattering regime, bulk infrared vibrational data [42] can be used to assist with the interpretation of the mode assignments. Assignment of these vibrational frequencies can be important in identifying those chemical functionalities that are actively involved in binding a metallic species during and after film deposition. Initial studies of the deposition of chromium on the PMDA–ODA polymer [43] have been interpreted as suggesting that chromium binds preferentially to the carbonyl functionality of the PMDA moiety. To make assignments of the different vibrational modes, molecular orbital studies were carried out for the two backbone polymeric functional units, the PMDA and ODA units [44]. All vibrations observed in the FTIR studies [42] that had been assigned with medium to strong intensities and with frequencies greater than 1000 wave numbers were assigned and the eigen vectors of the modes explicitly illustrated [44]. The simulated FTIR spectrum is shown in

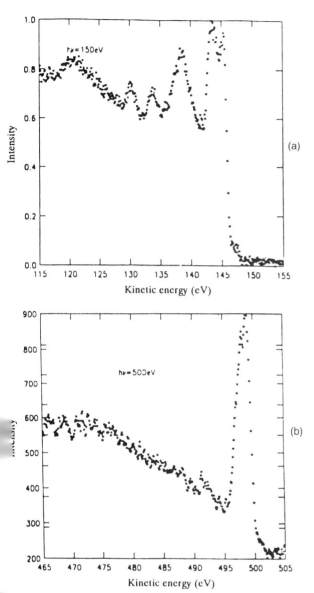

chromium to the PMDA moiety yielded a set of vibrational modes of reduced frequency, as a result of charge transfer from the chromium atom to the aromatic functionality. Second, it was shown that an infrared mode of very low intensity prior to complexing with chromium could exhibit a significant increase in dipolar intensity as a result of charge transfer to just those particular atoms with motions that contribute most significantly to the dipolar matrix element of the mode. Figure 23.9 shows the simulated HREELS spectrum prior to complexing with chromium and after such complexing. One sees the shift to lower frequencies as previously discussed. One furthermore sees a general increase in intensity that is associated with increasing dipolar moment on the aromatic ligands as a consequence of charge transfer from the 3d-transition metal. The calculations have also indicated significant frequency shifts for modes that were infrared inactive. The eigen vector of one such mode is shown in Fig. 23.10. Such shifts upon complexing with a metal might be observed by Raman or other spectral techniques. What is to be expected generally, for the modes associated with the polymer, is that upon complexing with a metal and a consequent charge transfer to the aromatic backbone, one expects an increase of all infrared active intensities. This has been shown in Fig. 23.8. Further FTIR measurements [45] have shown mode softening upon deposition of chromium onto a PMDA–ODA substrate. Unfortunately, observed intensities are such that coverages of a monolayer or greater are required before one achieves significant observable changes. This precludes the possibility of monitoring interesting changes that occur during the initial stages of film coverage prior to the formation of continuous or semi-continuous metallic films.

23.4 DIFFUSION AND INTERFACE FORMATION

The metal–polyimide interface can be formed in one of two ways. Polyamic acid can be deposited onto the metal substrate or metal can be deposited upon the imidized polyimide surface. In the former

Fig. 23.8. In order to understand effects that might be observed during interface formation with a metal, the effect of the binding of a single chromium metal atom upon the calculated spectra was examined. Several interesting effects were found. First, as generally expected, binding of

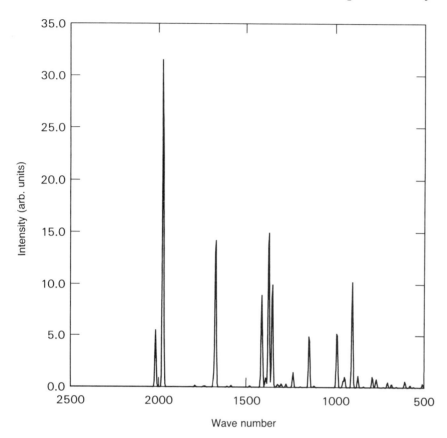

Fig. 23.8 Simulated FTIR spectrum for PMDA–ODA polyimide (from [44]).

case complex chemistry occurs at the interface [16, 36], yielding an adhesive bond that is greater than achieved with the latter type of interfacial synthesis. Thermally evaporating metal onto the polyimide surface provides a very different scenario for interface formation. For this type of interface, the chemical reactivity of the metal has been found to be important in determining the mode of formation and morphology of the metal layer on the polyimide surface. The interfacial morphology has been investigated for a series of metals by transmission electron microscopy (TEM) [46, 47] and medium-energy ion scattering [48]. Figures. 23.11(a)–11(d) show the cross-sectional micrographs of interfaces formed by depositing Ti, Al, Ni, and Cu respectively, with a slow rate of about three monolayers per minute on a PMDA–ODA surface at 300°C. The correlation between chemical reactivity and interfacial morphology is readily observed as the metal layers change from uniform coverage for Ti to cluster formation with Cu.

The clustering of Cu at the polyimide interface results from the weak binding of copper atoms to the polymeric substrate. Copper atoms cluster at the surface as well as migrate into the interior of the material. This has been confirmed by medium-energy ion scattering [48], which revealed that annealing enhances such migration and copper globules form at distances into the film that are comparable with the film thickness. Metals that bind more strongly to the polymer such as chromium do not exhibit such 'fuzzy' interface, but show a well-defined separation between metal and polymer on a nanometer scale. Aluminum binding to the PMDA–ODA polyimide is intermediate to that of copper and chromium [34]. XPS analysis indicates

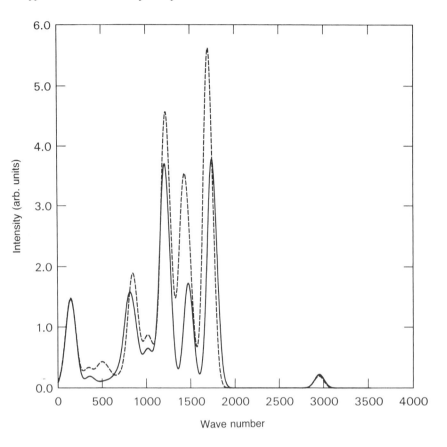

Fig. 23.9 Simulated HREELS spectra for PMDA–ODA before complexing with Cr (solid line) and after complexing with Cr (dashed line) (from [44]).

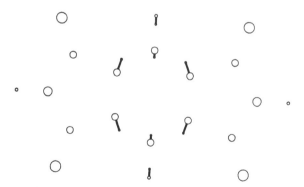

Fig. 23.10 Eigen vector of infrared inactive mode with significantly enhanced Raman intensity upon complexation with Cr (from [44]).

that upon deposition at 300°C an intermixed Al/polyimide interface layer is formed up to a coverage of about 3.5 monolayers of Al. With further deposition at the elevated temperature, island for-

mation occurs, which is consistent with the TEM observation shown in Fig. 23.11(c).

The effect of chemical binding on morphology has also been investigated by radio-tracer diffusion studies [49, 50]. One-quarter of a monolayer of radioactive copper was flash-evaporated onto a polyimide substrate. The films so generated were annealed for extended times at elevated temperatures. The interfacial layers were then sputtered with low energy ions for serial sectioning. Depth resolution of this technique yielded copper penetration profiles good to within approximately 50 Å. All annealed films showed copper as the predominant species at depths of approximately 700 Å into the film. The copper concentration near the film surface suggested that it had clustered upon annealing. With increasing anneal, greater amounts of copper were found at interior regions of the film. Such significant penetration of copper

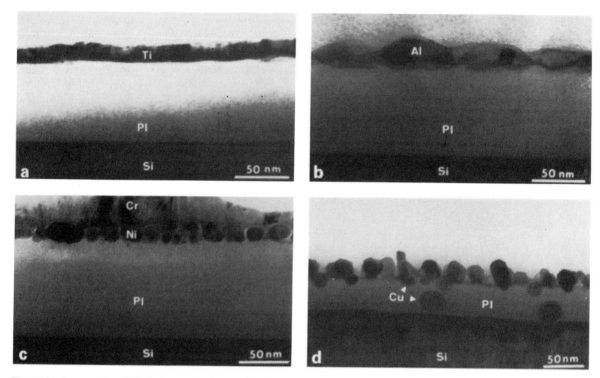

Fig. 23.11 Cross-sectional TEM micrographs for interfaces formed on PMDA–ODA polyimide at 300°C with (a) Ti, (b) Al, (c) Ni and (d) Cu (from [47]).

into the polyimide substrate is consistent with the results obtained in ion scattering experiments as well as in the TEM studies. One additional feature provided by the radio-tracer study is a Fickian diffusion behavior revealed by a straight-line diffusion profile as a function of the squared distance from the film surface. Such Fickian behavior was observed beyond the cluster zone near the surface, for all of the films that were subject to different annealing temperatures. An Arrhenius plot of the temperature dependence of the diffusion constant for Cu in PMDA–ODA shows an apparent curvature in the temperature range studied (Fig. 23.12). The convex curvature has been attributed to the free volume effect on diffusion in polyimides [49]. This concept follows the model originally proposed for diffusion of simple gases in elastomers [51] which has been extended for diffusion of small molecules in glassy polymers below the glass transition [52]. Accordingly, the diffusion behavior

of a small molecule would depend on the local distribution of the free volume in the polymer. For large diffusion distance over an extended temperature range, this model predicts a convex curvature in the Arrhenius plot. Applying this concept to metal diffusion would necessitate a weak interaction between the metal and the polyimide. Subsequent to the studies of copper diffusion in the PMDA–ODA polyimide, similar radio-tracer diffusion studies were performed with silver [53, 54]. Silver is known to interact weakly with PMDA–ODA and qualitatively similar results to that of copper were found. However, the diffusion coefficient of silver was found to be less than that of copper and its Arrhenius plot exhibited greater convexity. This was attributed to the relative atomic sizes of these two elements, suggesting that Ag with a larger atomic size, diffuses more slowly in the same polyimide. The morphology of the Ag/PMDA–ODA interface has been examined by

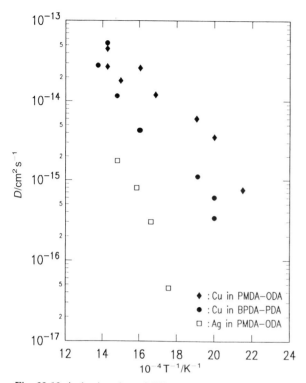

Fig. 23.12 Arrhenius plots of diffusional coefficients measured for Cu and Ag radiotracers in PMDA–ODA and BPDA–PDA polyimides (from [49], [50], and [55]).

cross-sectional TEM [54]. The morphology observed was similar to that of Cu, with the formation of Ag agglomerates of about 35 Å in size.

Recently, the diffusion of Cu has been measured in BPDA–PDA by the radio-tracer technique [55]. The result is plotted in Fig. 23.12. Compared with PMDA–ODA, the diffusion constant of Cu in BPDA–PDA is about an order of magnitude less and its Arrhenius plot does not show as clear a convex curvature as in PMDA–ODA. According to the free-volume concept, one may expect some correlation between metallic diffusion and the packing coefficient of the molecular chains in the polyimides. Interestingly, the packing coefficient, as estimated from the density, is about the same for these two polyimides. Therefore, it seems that density is not a sensitive measure of molecular packing and other details concerning the free

volume distribution are required in order to understand the diffusion of metals in the polyimides.

Metal diffusion into the polyimide is also affected by the chemical interaction between metal and polymer. Presently, quantitative measurement of the diffusion constant of metals having a strong interaction with the polymer has not been reported, although, as expected, a preliminary measurement of Cr and Co diffusion [56] in PMDA–ODA indicated a rate significantly less than Cu.

Monte Carlo studies have been performed to assist in interpreting the radio-tracer diffusion studies as well as the TEM [49, 57]. Parameters used in the studies were chosen to be consistent with experimental data. The simulations show that upon initial deposition and annealing, metal atom migration at the interface results in metallic agglomeration in the vicinity of the interface with a small amount of metal atom migration into the substrate. The morphology of the interface is significantly modified during the anneal and consequent metal atom migration; however, the amount of material diffusing into the substrate is a small fraction of the amount of material deposited. The ratio of the number of atoms escaping into the interior of the substrate to the number bound at the surface, obtained from the simulation, is in order of magnitude agreement with the ratio determined from the radio-tracer diffusion studies [49, 50].

Another feature of interest clearly delineated is the globular nature of the metal agglomerates that are formed during interfacial annealing. The subject of diffusion limited aggregation has received a lot of attention recently. Many of the structures generated and investigated have exhibited a ramified fractal like structure. In contrast, the metallic structures revealed by the TEM studies show a nonfractal globular like structure. One expects this to be due predominantly to thermodynamic relaxation of the clusters during the agglomeration process. The range of the metal atom interaction potential has also been shown to hold important consequences for the shape of the metallic particles [57].

Calculations were performed for metal atoms interacting in two different ways, i.e. with a short nearest-neighbor interaction as well as with an

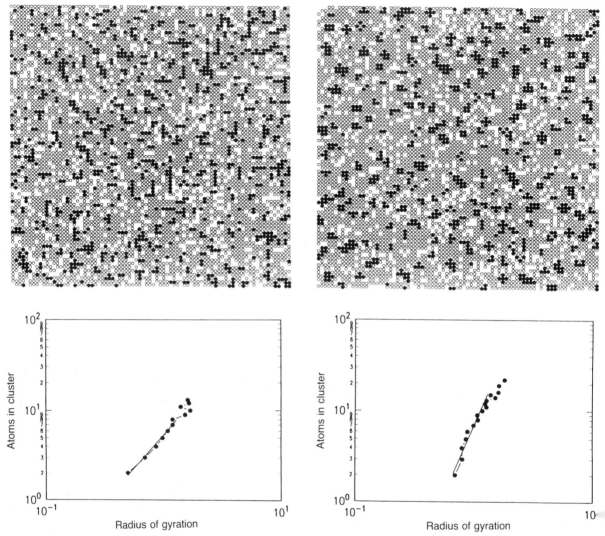

Fig. 23.13 Top view of a polymer surface after annealing of $\frac{1}{4}$ monolayer of metal deposition. Nearest-neighbor interaction only. The lower figure shows cluster size versus radius of gyration; fractal dimensionality $D = 1.5$ (from [57]).

Fig. 23.14 Top view of a polymer surface after annealing of $\frac{1}{4}$ monolayer of metal deposition. Nearest-neighbor and next nearest neighbor interactions. The lower figure shows cluster size versus radius of gyration; fractal dimensionality $D = 2.98$ (from [57]).

interaction between next-nearest neighbors. Simulation of annealing after flash evaporation has shown that the two different interaction potentials yield very different distributions of small cluster shapes for the one-quarter of a monolayer deposition. Figure 23.13 is a top view of the surface after annealing with an interaction energy between

metal atoms that treats only nearest neighbors. This should be compared with Fig. 23.14 which is a top view of the surface obtained when next as well as nearest neighbor metal atom interactions are included. One sees that even at the low depositions considered, a distinction between the cluster shapes is apparent. The inset in the figures is a plot of cluster size as a function of the radius of gyration yielding the fractal dimensionality. One sees that the more ramified fractal shape of the clusters arises from the shorter range interaction. The irregular morphology of this type of interface has recently been addressed in more formal studies of percolation in a concentration gradient [58].

23.5 THERMAL STRESS AND INTERFACIAL FRACTURE

The chemical bonding and interfacial morphology discussed so far are expected to be important in determining the thermal–mechanical properties of the interface. Consider the delamination of a metal–polymer interface. When interfacial separation occurs, the metal and the polymer layers sustain a large deformation where the strain field around the crack front will extend into the metal and the polymer layers. The delamination behavior and the fracture toughness of the interface are controlled by the deformation characteristics of the materials within the strain field. As a softer material, the polymer will deform more extensively near the interface than the metal. A deformation in the range of one half to one micron has been observed at the Cr–PMDA–ODA interface [59]. Within such a range, the energy dissipation process for crack propagation is expected to be controlled by the molecular morphology of the polymer, particularly by the entanglement of the polymer chains. The viscoelasticity of the polymer, which depends upon molecular structure, exhibits additional complexity through the time and temperature dependence of the delamination process. This aspect of this polymeric property should also be considered in studying fracture and stress relaxation in metal-polymer structures.

The stress generated in a metal–polymer structure during thermal cycling provides concern for the reliability of such materials in microelectronic applications. Consider a structure of metal–polyimide bilayer on a ceramic substrate. The magnitude of the stress is determined by the mismatch in thermal expansion and elastic modulus of the materials. For a thick substrate, as encountered in many applications, the deformation

of the metal and the polymer is constrained by the substrate. This constraint gives rise to a non-uniform distribution of stress and strain normal to the plane of the film in the structure, particularly in the region near the interface. A number of techniques have been developed for stress measurements in polymer multilayered structures and the results have been reviewed recently [60, 61]. A comparative study has been carried out recently for PMDA–ODA and BPDA–PDA utilizing a bending beam technique [62]. With Cu as the metal layer to minimize the effect of interfacial chemistry, this study focused on the effect of the polymer properties. Results of a Cu(0.53 µm)–PMDA–ODA(2.9 µm)–quartz structure are shown in Figs. 23.15(a) and (b), where the stress and strain for the Cu and the PMDA–ODA layers are deduced from the observed beam bending during thermal cycling [62]. In the trilayered structure, the overall deformation is constrained by the thermal expansion of the quartz substrate, so the strains of the metal and the polymer layers have to accommodate to that of the quartz substrate. Under this constraint condition, the thermal stress generated in Cu is higher than in PMDA–ODA by more than an order of magnitude while the reverse holds for the thermal strain. It is worth noting that the stress levels in the Cu–PMDA–ODA–quartz trilayer cannot be obtained simply by a superposition of the elastic stresses from the equivalent bilayer structures. Instead, the observed stress in Cu is considerably less, indicating that the polyimide undergoes some plastic deformation in order to reduce the overall stress of the structure. The stress reduction occurs, however, at the expense of a relatively large deformation of the polyimide. This stress buffering effect was investigated as a function of PMDA–ODA thickness and was found to saturate with about 3 µm of the polymer for 0.5 µm of Cu. The saturation thickness provides an indirect measure of the range of the non-uniform deformation in the polyimide near the interface. Results obtained for a Cu–BPDA–PDA–quartz structure under similar thermal cycling are shown in Figs. 23.16(a)–(c). The value of the thermal expansion coefficient of BPDA–PDA is between that of Cu and quartz, or

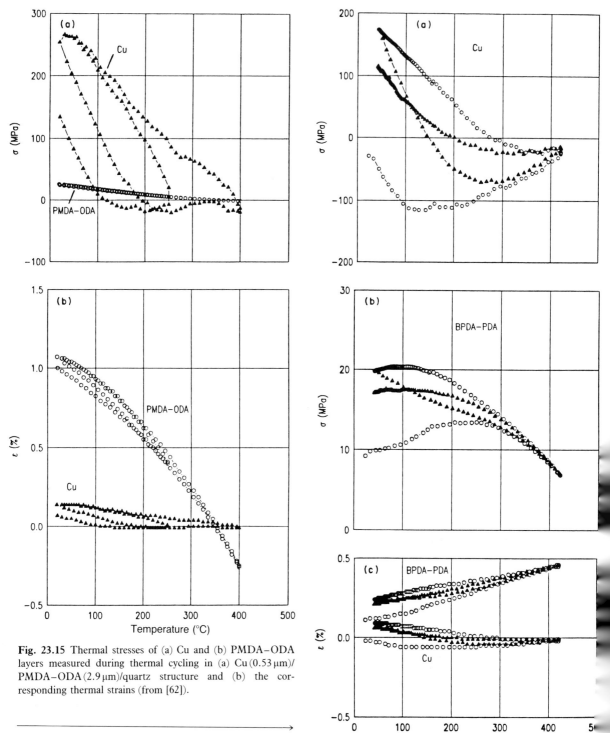

Fig. 23.15 Thermal stresses of (a) Cu and (b) PMDA–ODA layers measured during thermal cycling in (a) Cu(0.53 μm)/PMDA–ODA(2.9 μm)/quartz structure and (b) the corresponding thermal strains (from [62]).

Fig. 23.16 Thermal stresses of (a) Cu and (b) BPDA–PDA layers measured during thermal cycling in a Cu(0.5 μm)/BPDA–PDA (1.57 μm)/quartz structure and (c) the corresponding thermal strains (from [62]).

about 5 × less than PMDA–ODA. When applied as an interlayer between Cu and quartz, BPDA–PDA provides a thermal matching superior to that of PMDA–ODA. This factor, in combination with a higher Young's modulus, enables BPDA–PDA to reduce the thermal stress in Cu by about half, with comparable reduction in the thermal strain. Overall, the BDPA–PDA is more effective than PMDA–ODA for buffering the stress in the metal-polymer structure, and the saturation thickness was found to be less, about 1 μm.

Results from this comparative study demonstrate the effect of thermal stress behavior as a result of different polyimide properties. Although the deformation mechanism is not clearly understood at this time, the observed effect can be attributed to the different molecular structures of these two polyimides. Recently, the inelastic deformation behavior of a Cu–PMDA–ODA bilayer has been studied in tension up to fracture [63]. The relaxation of the stress in the polyimide indicated different activated rate processes below or above a strain of about 20%. In the lower strain range, the rate process can be characterized by an activation volume corresponding to a shear/slip deformation in the polyimide. This deformation mode was confirmed by the appearance of slip bands in the polyimide near the microcracks in the Cu film, where the load transferred from Cu to the polyimide is concentrated. This process occurs in a range of strain relevant to that during thermal cycling and may well be the controlling mechanism for thermal deformation of the layered structures investigated. Similar studies on other polyimides with different molecular structures would be valuable for understanding the role of molecular structure in the deformation of the metal-polymer structures.

The stress generated due to thermal mismatch, if sufficiently large, can cause interfacial fracture. For layered structures formed with blanket thin films, the magnitude of the stress, as shown for the Cu-polyimide structures in Figs. 23.15 and 23.16, is about 1%, probably insufficient to cause interfacial delamination. In microelectronic structures, however, the local stress near the small wiring geometries can be substantially higher. In a case study, Lacombe [64] computed the stress distribution using a finite-element analysis for an isolated Cu via contact (a circular contact to be filled by metal for wiring interconnection) formed in a PMDA–ODA layer on a silicon substrate. For a 2 μm via, the normal and shear stresses can increase 250% at the via corner during thermal cycling between 25°C and 400°C. This raises the stress level to about 4% of the Young's modulus, a magnitude probably sufficient to cause local fracture of the polyimide. In this analysis, the deformation of the polyimide was taken to be linear and elastic; the viscoelasticity of the polyimide which should be important and is expected to reduce the magnitude of the thermal stress was not included. This result shows that interfacial fracture is of concern, particularly for structures with small geometries where local stress concentration can be significant.

The toughness of a metal–polymer interface has often been evaluated by its adhesion strength. The testing methods for various material combinations have been reviewed recently [65–67]. For compliant and ductile polymeric films, it was concluded that many conventional methods are not well-suited for measuring the interfacial fracture toughness. For example, the peel test would require a correction of more than 50% in the peel force in order to account for the large deformation sustained by the metal and the polymer in the delamination process [68]. For this purpose, a stretch deformation method has been developed to measure the fracture energy of the metal–polymer interface [69]. In contrast to the peel test where the fracture is mode I controlled, this method applies a mode II fracture, which makes it possible to measure the shear strength of the interface in addition to the fracture energy. Examples of measurements performed on Cu–PMDA–ODA and Cu–Cr–PMDA–ODA interfaces are shown in Figs. 23.17(a) and (b). It is clear that the deformation behavior is significantly altered by a 800 Å Cr interlayer. Not only is the initial slope in the stress–strain curve increased by about a factor of two, the onset of delamination and the deformation behavior are substantially affected. The fracture energy, obtained by integrating the area between the two deformation curves for the delamination process, yields a value of 33 J/m^2 for the Cu–Cr

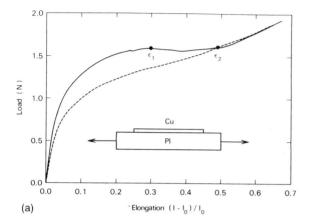

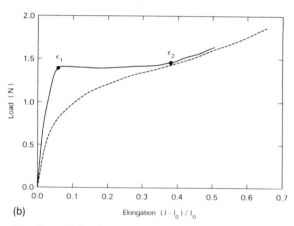

Fig. 23.17 (a) Load versus elongation curves for Cu–PMDA–ODA (solid line) and PMDA–ODA (dashed line) structures. Inserts show the sample geometry for stretch deformation test. (b) Load versus elongation curves for Cu–Cr–PMDA–ODA (solid line) and PMDA–ODA (dashed line) structures. Layer thicknesses: Cu 4300 A, Cr 800 A, and PMDA–ODA 5.3 µm. e_1 and e_2 indicate the starting and end points of delamination (from [69]).

interface, as compared with 9.6 J/m^2 for the Cu interface.

The different fracture energies observed show that chemical bonding is important in controlling the fracture behavior of the interface. To answer the question regarding to what extent bonding contributes to adhesion, the cohesion energy of a Cr–PMDA–ODA interface was estimated [24, 70]. Assuming a sharp interface with an energy of 3–4 eV per Cr bond to the six-membered or the five-membered ring in the PMDA or the ODA fragment, the interfacial bonding energy is about 3 J/m^2. This corresponds to 10% of the measured fracture energy. The estimated value can be increased by using different interfacial configuration or to include some Cr penetration below the first Cr layer. Nevertheless, even with the increase, this can account for only a fraction of the cohesive energy. This indicates that a major portion of the energy for delamination is dissipated by plastic deformation of the polyimide and probably transmitted through the Cr bonds to the molecular structure. Recently, stretch deformation tests have been performed on several metal–BPDA–PDA interfaces [71]. The preliminary data show that the fracture energy is about 2–4 × higher than the equivalent PMDA–ODA interface. This result can not be accounted for by the difference in chemical bonding, instead, a major part has to be attributed to the deformation behavior due to the different molecular structures.

Recently, studies have been carried out on Au–Cr–PMDA–ODA [72, 73] and Au–Cr–BPDA–PDA [74] layered structures to investigate the effects of metal geometry and polyimide structure on interfacial fracture. The metal line width varied from 2 to 16 µm while the thickness of the Au layer changed from 0.25 to 1 µm and of the Cr layer from 50 to 500 Å. The delamination behavior shows a clear trend in geometrical dependence. With a thickness increase or a width decrease, the value of the strain at failure increases while the progression of fracture becomes more abrupt. This can be seen in Fig. 23.18 where the deformation characteristics for an Au–Cr–BPDA–PDA layered structure are shown as a function of the line width. The fracture morphology and energy dissipation rate also show an apparent dependence on the metal dimensions. With increasing metal thickness, the crack morphology changed from circular to flat and an increase was observed in the energy dissipation rate (Fig. 23.19). The values of the fracture energy determined from the energy dissipation rate are in reasonable agreement with those calculated from an energy balance approach. Within the error limits of the experiment (*ca* 50%), the fracture energy seems to be independent

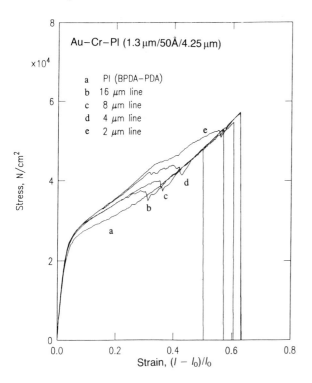

Fig. 23.18 Deformation curves for Au–Cr–BPDA–PDA line structures as a function of the line width (from [74]).

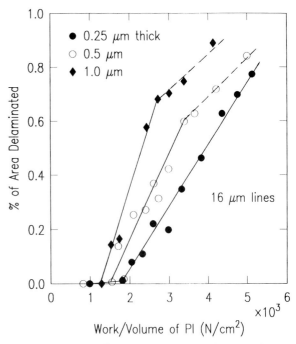

Fig. 23.19 Results of delaminated areas as a function of work done per unit polyimide volume for different metal line thicknesses. The slopes of these curves give the energy dissipation rates for crack propagation (from [73]).

of the metal dimension, suggesting that the effect of the metal geometry is primarily on the crack formation mode and morphology of the fracture process. These results can be explained by considering the change in the stress gradients near the interface due to the confinement of metal on polymer. This has been confirmed by a finite element analysis, where the plastic deformation of the polyimide was found to be important and has to be taken into account. On the other hand, the fracture energy clearly depends on the molecular structure of the polyimide, with a factor of about 3–4× higher for the BPDA–PDA. This result shows that the molecular structure of the polymer plays an important role in controlling the amount of energy dissipation in the fracture process and

can be optimized to control the adhesion strength of the interface.

23.6 SUMMARY

In this chapter we have attempted to assess the current understanding of the metal–polymer interface from a microscopic viewpoint. We have emphasized from the beginning the distinct features of molecular structure and morphology of the polymer. This distinguishes the metal–polymer interface from other crystalline solid interfaces, concerning characteristics of chemical bonding, interfacial structure, and metallic diffusion. The chain structure of the polymer necessitates the consideration of an amorphous solid with a viscoelastic response, two aspects important for studying the thermal and mechanical properties of the interface. These material characteristics have contributed to an understanding on a molecular level of the metal–polymer interface. Furthermore, the wide range of molecular structures provides an opportunity for molecular engineering of polymeric properties to suit specific application requirements.

This chapter began with a discussion on the curing process and its effect on the molecular order and morphology. This evolved into discussions of the chemical bonding, interfacial morphology, metal diffusion, stress and fracture behavior of metal–polymer structures. Most of the results are related to the polyimides, particularly to PMDA–ODA and BPDA–PDA. These two polyimides are closely related in chemistry and structure, thus serving as prototype materials to illustrate the extent of the control achievable on the interfacial properties by structural modification of the polymer. The picture of the metal–polyimide interface that emerges is that the formation mode and the morphology depend on the chemical reactivity which, in turn, is controlled by the chemical bond of metal to polymer as well as by the molecular structure of the polyimide. For a weakly interacting metal, such as Cu, the interface is characterized by an irregular morphology with the presence of metal agglomerates, while for a strongly interacting metal, such as Cr, the interface is abrupt with a well-defined morphology. For the polyimides, chemical bonding has been investigated in atomistic detail using spectroscopic techniques and quantum chemistry calculations. On the other hand, stress and fracture behavior have been studied by a continuum mechanics approach. In between these two ranges, a quantitative molecular model has yet to be developed to better understand the nature of the metal–polymer interface. With the technological importance of this class of polymers, there is increasing interest in studying diverse aspects of the interfacial phenomena at present unexplored. Significant progress is to be expected for the future understanding of this metal–polymer interface.

ACKNOWLEDGMENT

The authors would like to thank many coworkers and colleagues for their contributions and valuable discussions in our study of the metal–polymer interfaces. In particular, we acknowledge M. Moske, C. Feger, F. Faupel, D. Shinozaki, S. T. Chen, D. Gupta, S. Anderson, J. Leu, A. Rossi, R. Haight, R. Saraf, and T. W. Poon. The support and discussions with R. Acosta and A. Wagner of Yorktown Research Center and M. Haley, K. Srikrishnan, B. Agarwala, N. Koopman and H. M. Tong of the IBM General Technology Division are also greatly appreciated.

REFERENCES

1. *Microelectronics Packaging Handbook* (eds R. R. Tummala and E. J. Rymaszewski), Van Nostrand Reinhold, New York (1989).
2. *Principles of Electronic Packaging* (eds D. P. Seraphim, R. Lasky and C. Y. Li), McGraw Hill, New York (1989).
3. C. Feger, M. M. Khojasteh and J. E. McGrath, *Polyimides: Materials, Chemistry and Characterization*, Elsevier, Amsterdam (1989).
4. M. I. Bessonov, M. M. Koton, C. C. Kudryavtsev and L. A. Laius, *Polyimides: Thermally Stable Polymers*, Consultants Bureau, New York (1987).
5. N. Takahashi, D. Y. Yoon and W. Parrish, *Macromolecules*, **17** (1984), 2583.
6. T. P. Russel, *J. Poly. Science: Poly. Phys.*, **22** (1984), 1105.
7. T. P. Russel and H. R. Brown, *J. Poly. Science: Poly. Phys.*, **25** (1987), 1129.
8. R. F. Boehme and G. S. Cargill, in *Polyimides, Vol. 1* (ed K. L. Mittal), Plenum Publishing (1984).
9. C. Feger, *Poly. Eng. Science*, **29** (1989), 347.
10. M. J. Brekner and C. Feger, *J. Poly. Science: Poly. Chem.*, **25** (1987), 2005; ibid. **25** (1987), 2479.
11. S. Numata, K. Fujisaki and N. Kinjo, *Polymer*, **28** (1987), 2282.
12. S. Numata, S. Oohara, K. Fujisaki, J. Imaizumi and N. Kinjo, *J. Appl. Poly. Science*, **31** (1986), 101.
13. S. Numata, N. Kinjo and D. Makino, *Poly. Eng. Science*, **28** (1988), 906.
14. S. Numata and T. Miwa, *Polymer*, **30** (1989), 1170.
15. P. S. Ho, R. Haight, R. C. White, B. D. Silverman, and F. Faupel, in *Fundamentals of Adhesion* (ed L. Lee), Plenum Publishing (1991).
16. S. Kowalczyk, in *Metallization of Polymers* (eds E. Sacher,

J. Pireaux, and S. Kowalczk, ACS Symposium Series 440, American Chemical Society, Washington, DC (1990), p. 10.

17. H. J. Leary and D. S. Campbell, *Surf. Interface Anal.*, **1** (1979), 75.

18. J. M. Burkstrand, *J. Vac. Sci. Technol.* **16** (1979), 363; *J. Appl. Phys.*, **52** (1981), 4795.

19. P. L. Buchwalter and A. I. Baise, in *Polyimides: Synthesis, Characterization, and Applications* (ed K. Mittal), Plenum, New York, Vol. 1 (1987), p. 537.

20. B. D. Silverman, P. N. Sanda, P. S. Ho, and A. R. Rossi, *J. Polym. Sci.: Pol. Chem.*, **23** (1985), 2857.

21. L. P. Buchwalter, B. D. Silverman, L. Witt, and A. R. Rossi, *J. Vac. Sci. Technol. A*, **5** (1987), 226.

22. B. D. Silverman, P. N. Sanda, J. G. Clabes, P. S. Ho, D. C. Hofer, and A. R. Rossi, *J. Poly. Sci.: Polymer Chem.*, **26** (1988), 1199.

23. S. Anderson, J. Leu, B. D. Silverman and P. S. Ho, *MRS Symp. on Mat. Sci. of High Temp. Poly. for Microel.*, Anaheim, CA (1991).

24. R. Haight, R. C. White, B. D. Silverman and P. S. Ho, *J. Vac. Sci. Technol. A*, **6** (1988), 2188.

25. P. S. Bagus, D. Coolbaugh, S. P. Kowalczk, G. Pacchioni, and F. Parmigiani, *J. Elecron Spectr. and Related Phen.*, **51** (1990), 69.

26. A. R. Rossi and B. D. Silverman, in *Polymeric Materials for Electronic Packaging and Interconnection* (eds J. H. Lupinski and R. S. Moore), *ACS Symposium Series* Vol. 407, American Chemical Society, Washington, DC (1989).

27. M. Grunze and R. N. Lamb, *J. Vac. Sci. Technol. A*, **5** (1987), 1685; M. Grunze, W. N. Unertl, S. Gnanarajan and J. French, *Mater. Res. Soc. Sym. Proc.*, **108** (1988), 189.

28. R. N. Lamb, J. Baxter, M. Grunze, C. W. Kong and W. N. Unertl, *Langmuir*, **4** (1988), 249.

29. C. A. Kovac, J. L. Jordan-Sweet, M. J. Goldberg, J. G. Clabes, A. Viehbeck, and R. A. Pollack, *IBM Journal of Research and Development*, **32** (1988), 603.

30. P. S. Ho, B. D. Silverman, R. A. Haight, R. C. White, P. N. Sanda, and A. R. Rossi, *IBM Journal of Research and Development*, **32** (1988), 658.

31. A. R. Rossi, P. N. Sanda, B. D. Silverman and P. S. Ho, *Organometallics*, **6** (1987), 580.

32. J. G. Clabes, M. J. Goldberg, A. Viehbeck and C. A. Kovac, *J. Vac. Sci. Technol. A*, **6** (1988), 985.

33. J. G. Clabes, *J. Vac. Sci. Technol. A*, **6** (1988), 2887.

34. J. W. Bartha, P. O. Hahn, F. LeGoues, and P. S. Ho, *J. Vac. Sci. Technol. A*, **3** (1985), 1390.

35. P. O. Hahn, G. W. Rubloff, and P. S. Ho, *J. Vac. Sci. Technol. A*, **2** (1984), 756.

36. J. Kim, S. P. Kowalczyk, Y. H. Kim, N. J. Chou, and T. S. Oh, *Mat. Res. Soc. Symp. Proc.*, **167** (1990), 137.

37. S. P. Kowalczyk, S. Stafstrom, J. L. Bredas, W. R.

Salaneck, and J. L. Jordan-Sweet, *Phys. Rev. B*, **41** (1990), 1645.

38. P. S. Ho, P. O. Hahn, J. W. Bartha, G. W. Rubloff, F. K. LeGoues, and B. D. Silverman, *J. Vac. Sci. Technol. A*, **3** (1985), 739.

39. P. O. Hahn, G. W. Rubloff, J. W. Bartha, F. LeGoues and P. S. Ho, *Mat. Res. Soc. Symp. Proc.*, **40** (1985), 251.

40. S. P. Kowalczyk and J. L. Jordan-Sweet, *Chemistry of Materials*, **1** (1989), 592.

41. J. J. Pireaux, M. Vermeersch, C. Gregoire, P. A. Thiry, and R. Caudano, *J. Chem. Phys.*, **88** (1988), 3353.

42. H. Ishida, S. T. Wellinghoff, E. Baer, and J. L. Koenig, *Macromolecules*, **13** (1980), 826.

43. N. J. DiNardo, J. E. Demuth, and T. C. Clarke, *J. Chem. Phys.* **85** (1986), 6739.

44. B. D. Silverman, *Macromolecules*, **22** (1989), 3768.

45. D. S. Dunn and J. L. Grant, *J. Vac. Sci. Technol. A*, **7** (1989), 253.

46. Y.-H. Kim, J. Kim, G. F. Walker, C. Feger, and S. P. Kowalczyk, *J. Adhes. Sci. Technol.*, **2** (1988), 95.

47. F. K. LeGoues, B. D. Silverman and P. S. Ho, *J. Vac. Sci. Technol. A*, **6** (1988), 2200.

48. R. Tromp, F. K. LeGoues and P. S. Ho, *J. Vac. Sci. Technol. A*, **3** (1985), 782.

49. F. Faupel, D. Gupta, B. D. Silverman, and P. S. Ho, *Appl. Phys. Lett*, **55** (1989) 357.

50. F. Faupel, D. Gupta, B. D. Silverman, and P. S. Ho, *Proc. Europe Conf. Adv. Mater. Processes, Aachen, 1989*; Deutsche Gesellschaft f. Materialkunde, Oberursel, (1990).

51. H. L. Frisch and S. A. Stern, *CRC Critical Review* in *Solid State and Mater. Sci.* **II**, (1983), 123.

52. J. S. Vrentas and J. L. Duda, *J. Appl. Poly. Sci.*, **22** (1978), 2325.

53. F. Faupel, *Advanced Materials*, **2** (1990), 266.

54. A. Foitzik and F. Faupel, *Mat. Res. Soc. Symp. Proc.*, **203** (1991).

55. K. Vieregge and D. Gupta, to be published.

56. D. Gupta, private communication.

57. B. D. Silverman, *Macromolecules*, **22**, 3768 (1989).

58. M. Kolb, T. Gobron, J. F. Gouyet, and B. Sapoval, *Europhys. Lett*, **11** (1990), 601.

59. S. T. Chen, C. H. Yang, F. Faupel and P. S. Ho, *J. Appl. Phys.*, **64** (1988), 6690.

60. H. M. Tong and K. L. Saenger, in *New Characterization Techniques for Thin Polymer Films* (eds H. M. Tong and L. T. Nguyen), Wiley-Interscience, New York (1990), Ch. 2.

61. C. L. Bauer and R. J. Farris, in *Polyimides: Materials, Chemistry and Characterization*, (eds C. Feger, M. M. Khojasteh and J. E. McGrath), Elsevier Amsterdam (1989), p. 549.

62. M. Moske, J. E. Lewis and P. S. Ho, in *Proc. Soc. Plastic Eng.*, Montreal, Canada (1991).

63. D. M. Shinozaki, A. Klauzner and P. C. Cheng, to appear in *Mater. Sci. and Eng.* (1991).

64. R. H. Lacombe in *Surface and Colloid Science in Computer Technology* (ed K. L. Mittal), Plenum Press, New York (1987), p. 178.

65. P. Buchwalter, *J. Adhes. Sci. Technol.*, **4** (1990), 697.

66. K. S. Kim, *Mat. Res. Soc. Symp.*, **119** (1988), 31.

67. K. S. Kim, *Mat. Res. Soc. Symp.*, **203** (1991).

68. K. S. Kim and J. Kim, *Trans. ASME*, **110** (1988), 266.

69. F. Faupel, C. H. Yang, S. T. Chen and P. S. Ho, *J. Appl. Phys.*, **65** (1989), 1911; P. S. Ho and F. Faupel, *Appl. Phys. Lett.*, **53** (1988), 1602.

70. P. S. Ho, *Appl. Surf. Science*, **41/42** (1989), 559.

71. J. Leu, S. L. Chiu and P. S. Ho, *Mat. Res. Soc. Symp.*, April 27–May 1, San Francisco, CA (1992).

72. S. L. Chiu, Y. H. Jeng, R. E. Acosta and P. S. Ho, *Mat. Res. Soc. Symp.*, **167** (1990), 129.

73. S. L. Chiu, and P. S. Ho, *Mat. Res. Soc. Symp. Proc.*, **203** (1991).

74. J. Leu, S. L. Chiu and P. S. Ho, *Mat. Res. Soc. Symp.*, April 28–May 2, Anaheim, CA (1991).

PART IV:
Fracture Behavior

CHAPTER TWENTY-FOUR

Tensile strength of interfaces

A. S. Argon and V. Gupta

Introduction · Cracks at interfaces · Measurement of strength of interfaces · Measurement of toughness of interfaces · Measurement of strength and toughness of carbon fibers · Conclusions

24.1 INTRODUCTION

In many aligned fiber-reinforced composites, the structural service requirements are almost entirely met by the volume fraction of stiff and strong but brittle fibers. The matrix then acts merely to position the fibers in space and to impart to the composite a minimum level of transverse tensile and longitudinal shear properties. It is now well recognized, however, that in such composites the evolution of subcritical damage under stress by correlated fiber fractures is governed by the mechanical coupling between the fibers through the matrix. When the interface transmits all tractions fully and the coupling between fibers is too good, isolated fractures in fibers with small variability in strength tend to spread more readily to surrounding fibers, and hasten the development of a supercritical damage cluster [1], [2]. In such instances, the strength of the composite is often less than the average strength of an unbonded bundle of similar fibers of equal length [2]. These composites can be made more damage tolerant by decoupling fractured fibers from their neighbors through controlled delamination of their interfaces.

The delamination or fracture toughness of interfaces play a key role in all composites, but particularly so in metal matrix composites, where unwanted reactions at the interface between the matrix and the fiber can produce reaction products with positive material misfit that can severely damage the fiber. This was first observed in Metcalfe's elegant experiments [3] in the boron fiber–aluminum systems. Since, during the production of composites, processing histories that give proper wetting of the fibers by the molten matrix and good adhesion are often in conflict with the requirements to limit reaction damage, it has been proposed by us earlier that it is desirable to separate functions of proper wetting from careful control of interface mechanical properties. This can be accomplished by tailoring the desired mechanical properties of the interface between the protective coating and the reinforcing fiber, which can be controlled during fiber preparation, while accepting some wetting-related reaction damage between the matrix and the outer surface of the protective coating of the fiber. When properly controlled, this permits the pedigreed key interface between the coating and the fiber to act as a mechanical fuse to decouple the fiber from its surrounding by initiating delaminations along it.

The nature of the interface cracking problem that is of interest is depicted in Fig. 24.1, which shows on an enlarged basis a crack in a hard coating such as SiC, terminating on the interface between the coating and the fiber. The crack deflection process along the interface requires the following

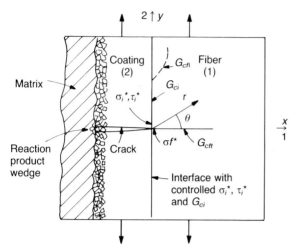

Fig. 24.1 Crack terminating perpendicular to the interface between a stiff coating and its substrate.

conditions to be satisfied near the tip of the crack [4]:

1. The ratio of the fiber strength to the interface normal strength should be greater than the ratio of the crack tip stresses responsible for fiber fracture ($\sigma_{\theta\theta}$ at $\theta = 0$) and interface separation ($\sigma_{\theta\theta}$ at $\theta = \pi/2$). This will lead to the delamination of the interface in tension when $\sigma_{\theta\theta}$ there equals σ_i^*, the interface tensile strength. This condition leads to an acceptable upper bound for the interface normal strength.

2. The ratio of the fiber strength to the interface shear strength should be greater than the ratio of $\sigma_{\theta\theta}$ at $\theta = 0$ to $\sigma_{r\theta}$ at the interface. This will ensure the separation of the interface in shear when $\sigma_{r\theta}$ there equals τ_i^*, the interface shear strength. This bounds the interface shear strength. When the ratio of the interface normal strength to the interface shear strength is less than the ratio of $\sigma_{\theta\theta}/\sigma_{r\theta}$ at the interface, tensile separation will be preferred.

Alternatively, for two possible directions of growth of the crack, i.e. across the fiber versus along the interface, if the ratio of the energy release rate for growth across the fiber, to growth along the interface is less than the ratio of the work of fracture G_{cft} of transversely across the fiber to work of separation of the interface G_{ci}, then the fracture

will follow the interface. Hence, it is evident from the above discussion that in order to make effective use of the primary interfaces in composites, it is necessary that the following be done:

(a) determine the stress field for cracks terminating at the interface incorporating the elastic anisotropy of the fiber;
(b) experimentally determine the strength and toughness of the tailored interfaces;
(c) experimentally determine the strength and fracture work in fibers both across the fiber axis as well as along the fiber axis to find to what extent the above goals have been achieved.

In this chapter we give the main results of our recent findings on this subject and briefly indicate the test procedures developed to achieve the above goals. More detailed descriptions of these points can be found elsewhere [5–8]. While the results that we will present are general and applicable to most pairs of hard materials, we will discuss more specifically the results on the case of amorphous SiC coatings on anisotropic Pitch-55 type carbon ribbons.

24.2 CRACKS AT INTERFACES

24.2.1 The crack tip stress field

The stress field for a crack terminating perpendicular to an interface between two aligned orthotropic media was determined by representing the crack as a line of distributed edge dislocations satisfying the traction-free conditions of the crack. The resulting singular integral equation was solved by using the known solution of an edge dislocation in the vicinity of an interface. The details of the solution can be found elsewhere [5]; here we give the final results and their dependence on the various elastic constants of the bi-material pair. The stress and deformation fields derived for such cases were shown to depend on the individual material constants λ and ρ for the two adjoining media, and the two bi-material constants α and β. The above

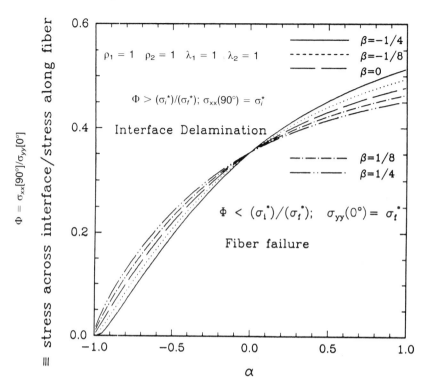

Fig. 24.2 Interface delamination diagram.

parameters depend only on the elastic compliances s_{ij} of the two media as

$$\lambda = \frac{s_{22}}{s_{11}} \tag{24.1a}$$

$$\rho = \frac{(2s_{12} + s_{66})}{2\gamma} \tag{24.1b}$$

$$\alpha = \frac{(\gamma_2 - \gamma_1)}{(\gamma_2 + \gamma_1)} \tag{24.1c}$$

$$\beta = \frac{(\gamma + s_{12})_2 - (\gamma + s_{12})_1}{\sqrt{H_{11}H_{22}}} \tag{24.1d}$$

where

$$\gamma = \sqrt{s_{11}s_{22}} \tag{24.1e}$$

$$H_{11} = (2n\lambda^{\frac{1}{4}}\gamma)_1 + (2n\lambda^{\frac{1}{4}}\gamma)_2 \tag{24.1f}$$

$$H_{22} = (2n\lambda^{-\frac{1}{4}}\gamma)_1 + (2n\lambda^{-\frac{1}{4}}\gamma)_2 \tag{24.1g}$$

$$n = \sqrt{\frac{(1 + \rho)}{2}} \tag{24.1h}$$

where the subscripts 1 and 2 refer to the fiber (or ribbon) and 2 to the cracked coating. The parameters λ and ρ take the value of unity as the material becomes isotropic. It was noted that the crack deflection criterion in (a) above derived from the current solutions was relatively insensitive to the variation of λ and ρ of the two media over practical ranges of these material parameters. Therefore, it becomes possible to present meaningful results based on the bi-material coefficients α and β, which characterize the elastic dissimilarity of the two media. As shown in Fig. 24.2, it is then possible to determine the level of interface strength required in fiber composites in order to enhance their overall toughness by interface delamination. Furthermore, this diagram can be used for the fiber–coating or the matrix–coating interface, depending on which interface is of critical interest for crack deflection. For the appropriate material pair, the values of α and β can be calculated by using the expressions given in eqs. (24.1a–h) by incorporating the elastic anisotropy of either medium. For the specific pair of α and β, the maximum permissible ratio of the interface strength to the fiber strength can be calculated from the delamination diagram. Finally, the desirable value of the interface normal strength can be calculated from the above ratio for a fiber of given axial strength. As we will discuss in more detail below, for the SiC coating–Pitch-55 fiber system, the interface strength should be less than 200 MPa, based on a consideration for an average axial tensile strength of 1.3 GPa for the Pitch-55 fiber measured by Li [9].

24.2.2 Optimizing toughness and transverse strength in composites

In addition to high toughness, it is required that the composite sustain adequate transverse loads in service. Unfortunately, maximizing these two properties demands satisfying conflicting requirements on the interface strength. High interface strength is required in order to transfer transverse loads on the composite over to the fibers via a small interface area. On the other hand, the interface strength needs to be limited at a definite level to enhance the toughness of the composite by crack arrest at such interfaces.

The interface delamination diagram of Fig. 24.2 shows the attainable possibilities for composite axial tensile strength and transverse strength for a tough behavior, promoting crack arrest at coating–fiber interfaces. Consider first what is achievable with a high modulus amorphous SiC coating on a Pitch-55 fiber with an interface tensile strength of 220 MPa, as measured with the laser spallation method discussed in section 24.3 below. If the interface stress concentration on a transverse stress σ_T is taken as 2.4 as determined by Zywicz and Parks [10] for a quasi-hexagonal packing of parallel fibers with a volume fraction in the range of 0.4, then the composite transverse strength would be 91.7 MPa. With this transverse strength limitation, for tough behavior due to effective crack arrest at the coating–fiber interfaces, it is necessary that the fiber strength exceed a certain lower limit. For Pitch-55 fiber with an axial modulus of 380 GPa and a transverse modulus of only 13 GPa, and for

the isotropic SiC coating of an average modulus of 400 GPa the material constants will be $\gamma_1 = 1.42 \times 10^{-2}$ and $\gamma_2 = 2.5 \times 10^{-3} \, \text{GPa}^{-1}$ respectively, giving the so-called Dundurs parameter $\alpha = -0.7$ for which the required ratio of interface strength to fiber strength must be less than 0.14; or the fiber tensile strength must exceed 1.57 GPa. Since, however, the average tensile strength of the Pitch-55 fiber is only 1.3 GPa it is necessary to reduce the tensile strength of the interface to less than 182 MPa by controlled interface embrittlement. This would, of course, lower the allowable transverse composite strength to 75.8 MPa.

Consider now a second approach that starts with the same basic fiber tensile strength of 1.3 GPa, but takes another route by modifying the modulus of the amorphous SiC coating, which is controllable over a very wide range by changing the coating conditions of the plasma-assisted chemical vapor deposition process explored by Landis [11] (see also [12]). Through such changes in coating conditions it is possible to achieve amorphous SiC coatings of modulus as low as 20 GPa giving $\gamma_2 = 5 \times 10^{-2} \, \text{GPa}^{-1}$ (albeit with increased porosity but no important loss in interface toughness [12]). For these conditions the so-called Dundurs parameter $\alpha = 0.56$ and from the delamination diagram of Fig. 24.2 the required corresponding ratio of interface strength to fiber strength must be less than 0.43. This permits an interface strength not to exceed 560 MPa, making a transverse composite strength up to $\sigma_T = 233$ MPa possible. Naturally this requires substantial improvement in the interface tensile strength from 220 MPa which should be technically achievable through control of interface imperfections since interface strengths of up to 7.0 GPa have been measured between SiC and pyrolytic graphite [7].

The above excursion into composite property design through control of interface strength and coating stiffness serves to demonstrate the large ranges of degrees of freedom, provided: (a) the properties of coating and interface are indeed controllable by conditions of fiber processing; and (b) that these changed properties are measurable.

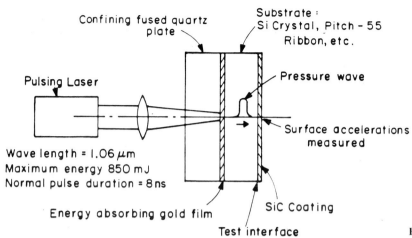

Confining fused quartz plate

Substrate: Si Crystal, Pitch-55 Ribbon, etc.

Pulsing Laser

Pressure wave

Surface accelerations measured

Wave length = 1.06 μm
Maximum energy 850 mJ
Normal pulse duration = 8 ns

Energy absorbing gold film

SiC Coating

Test interface

Fig. 24.3 Laser spallation experiment setup.

24.3 MEASUREMENT OF STRENGTH OF INTERFACES

24.3.1 The laser spallation experiment

The tensile strength of the interface plays a principal role in governing the path of the crack at the interface. The measurement of such strengths in macroscopic experiments has been attempted by many investigators but is not free of problems. In most of the reported cases [13, 14], part of the interface is subjected to the high stresses of an inhomogeneous deformation field. The magnitudes of these stresses need to be determined through the solutions of complex boundary value problems of elastic and plastic deformation. While these attempts have often given operationally useful answers, the factor of uncertainty in them has been considerable because of the inadequacy of the solutions of the local deformation problems. The measurement of strength between a substrate and a thin coating presents particular difficulties in the application of the stress by an inhomogeneous local deformation field where premature failure elsewhere is likely, before the desired interface can be probed. Here we present the results of a new laser pulse induced spallation experiment that has been developed specifically for the measurement of tensile strength of interfaces between thin coatings and hard substrates [6, 7].

The laser pulse induced spallation experiment has been specifically developed to measure the strength of interfaces by circumventing the above difficulties, provided that the interface to be probed can be obtained in planar form.

24.3.2 Experimental arrangement

Since the actual interfaces of interest are between a cylindrical fiber and its coating, which are not readily accessible to measurement by the spallation technique, the experimental approach to be presented here utilizes a planar arrangement of a model substrate and coating combination which is shown in Fig. 24.3. The collimated laser pulse from a Nd-YAG laser (1.06 μm wavelength) is made to impinge on a thin energy-absorbing gold film sandwiched between the back surface of the substrate of interest and a fused quartz confining plate, transparent to the laser wavelength. On absorption of the laser energy, the gold film undergoes a sudden thermal expansion leading to the generation of a compressive shock wave directed towards the test coating–substrate interface. It is the reflection of the compressive stress pulse from the free surface of the test coating that gives rise to a tensile pulse and leads to the removal of the coating, if the amplitude of the pulse is high enough. Based on a figure of merit analysis that captures the essence of the thermoelastic stress generation for various alternative energy-absorbing materials, it was concluded that gold is the most effective material that maximizes the generated stress pulse amplitudes.

A key element of the experiment is to determine the amplitude of the tension wave that is formed by reflection of the pressure wave from the free surface of the coating attached to the rear surface of the substrate. The preferred method of accomplishing this is to measure the time rate of change of displacement of the free surface of the coating as the main compression pulse is reflected. This technique, which is relatively uninfluenced by internal dissipation in the coating, is usually done by laser Doppler interferometry used widely in plate impact research [15]. In the experiments to be reported here, however, only pulsing lasers of limited power were available, requiring to focus the beam over relatively small areas of roughly 1 mm^2. This limits the planar portion of the pressure pulse to be reflected from the free surface of the test coating to a similarly small area, and makes the accurate measurements of the accelerations and decelerations of these small planar portions of the free surface rather difficult by the laser Doppler interferometry. In view of this, the measurements of the interface strength have made use of a three part strategy.

24.3.3 The three-part strategy of interface strength measurement

The first part of the strategy of evaluation of experimental results has been the development of a finite element computer simulation of the conversion of the laser light pulse into a pressure pulse based on the simultaneous solution of the transient heat transfer and elastic pressure wave generation problems. Next, the traverse of the pressure pulse through any desired substrate of interest is modeled and finally the resulting history of the tensile stress acting across the desired interface is determined as the stress wave is reflected from the free surface of the coating of given properties and thickness.

In the second part of the strategy, the pressure pulses were measured in a piezo-electric micro-electronics device in which the conditions of the computer simulation study were experimentally

achieved. In this device the substrate and its test coating were replaced by an X-cut piezo-electric (PE) crystal equipped on its back face with an energy-absorbing gold film and on the front surface with a very thin gold electrode for signal pickup. The PE crystal with its attached fused quartz confining plate then became a pressure transducer operating in the short circuit mode, capable of pressure wave determination with a time resolution of 0.7 ns. The measured pressure signal profiles were then compared with those obtained with the computer simulation in which the substrate is appropriately given the elastic and thermal properties of the X-cut PE quartz crystal. This permits verifying and fine tuning of the computer simulation.

In the third part of the strategy, actual spallation experiments were carried out. The laser fluence necessary for the removal of the probed portion of the coating at the interface was recorded, and the tensile stress across the interface that accomplishes this was determined from the computer program. The computer simulation along with its fine tuning via the micro-electronics device is discussed in detail elsewhere [6, 7] and will not be presented here. However, to indicate the accuracy of the computer program, we compare in Fig. 24.4 the stress pulse predicted in the piezoelectric crystal by the computer program with that measured experimentally by the micro-electronics device. Because of this good agreement, the resulting history of the tensile stress generated at the interface due to the reflection of the compressive stress pulse was determined solely from the computer program. One such computed stress history for a SiC coating on single crystal Si is shown in Fig. 24.5. The amplitude of such plots at the threshold laser fluence gives the interface strength.

Note added in proof: The initial computer simulation reported in Ref. [7] was based on a one-dimensional pressure pulse. Later examination of conditions of spallation has disclosed the need for 3-D pressure 'diffusion'. Corrections to the measurements reported in [7] are now in the press [20].

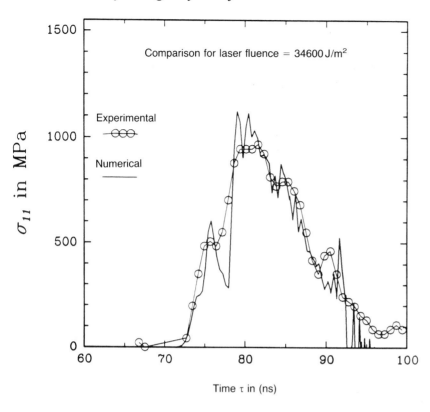

Fig. 24.4 Comparison of the predicted and measured profiles of the generated stress pulse in the PE crystal probe.

24.3.4 Results of spallation experiments

The spallation technique was applied to three substrates consisting of Si single crystal wafers with (100) plane surfaces; pyrolytic graphite (PG) platelets with the principal axis of layer normals lying in the plane of the platelet, and Pitch-55 (P-55) type carbon ribbons of 600 μm width and 35 μm thickness and having a meso-phase morphology very similar to the Pitch-55 carbon fibers. Detailed description of the morphology of the Pitch-55 ribbon and PG can be found elsewhere [16]. In all these cases, the coatings to be removed were amorphous SiC coatings deposited by the plasma-assisted chemical vapor deposition technique developed by Landis [11]. The coatings had a thickness of about 1–3 μm. They were deposited in a nearly stress-free manner by maintaining the substrates at a certain temperature that prevented the entrapment of significant concentrations of hydrogen, that otherwise would have resulted in high residual compressive stresses and premature delamination in a manner described in detail by us earlier [12].

The coating could be successfully spalled off in all cases with appropriate levels of laser pulse energies. The threshold laser energy at which spallation occurs was also recorded using a light power meter. Since the amplitude of the stress pulse should depend upon the absorbed laser fluence, the diameter of the laser beam was also recorded on a photographic film for the tabulation of the threshold laser fluence.

Si–SiC interface

Figure 24.6 shows a micrograph of a typical spot from where the SiC coating has spalled off. The laser fluence necessary to achieve this spallation was 7.39 J/m². The interface strength corre-

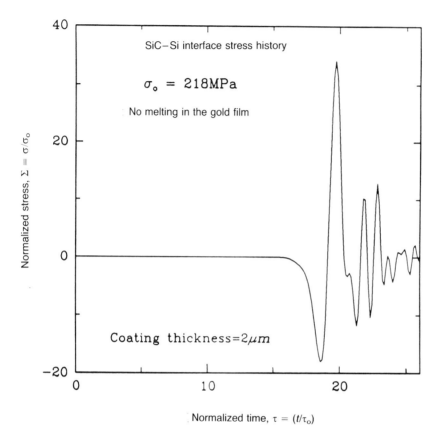

Fig. 24.5 Stress history at the interface between a SiC coating and Si substrate.

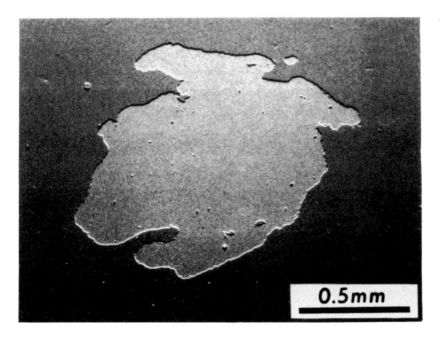

Fig. 24.6 Spalled spot of SiC coating from a Si single crystal surface.

sponding to this laser fluence as calculated from the computer simulation was 5.29 GPa (see previous note added in proof). The thickness of the amorphous SiC coating of low modulus, deposited at low ion beam energy, was 1.5 μm. The islands of SiC within the spalled spots are due to the statistical variability of the interface strength over the delaminated spot. The measured interface strength was found to vary from 10.19 to 10.88 GPa; this spread is surprisingly low (only 7%) for a brittle interface.

The clean flat structure of the interface shows that the SiC coating has come off uniformly from the substrate at the interface. In order to verify this, Auger electron spectroscopy (AES) was performed on the spalled spot to determine the location of the failure. It was possible to successfully delaminate coatings of thickness ranging from 0.3 to 3 μm. Finally, the actual interface strength values for the various SiC–Si systems produced by using different deposition parameters (of the SiC coating) was found to vary from 3.7 to 10.88 GPa. While such a variation may appear disappointingly large, this degree of freedom, if controllable, is exactly what is required to carry out the delamination scheme in composites as outlined in section 24.2.

PG–SiC interface

In another case SiC coatings of 2.1 μm thickness were spalled-off from PG surfaces. The substrate disk was 1.69 mm thick. The spalled areas were elongated along the edges of the graphitic planes that terminate perpendicular to the surface. This is the stiffest direction in that plane. For this case the average interface strength was determined to be 3.68 GPa. These calculations have taken account of the anisotropic character of the PG substrate. Since the surface analytical techniques available were not able to distinguish the carbon of the SiC from that of the PG substrate, the depth of the crater as determined by a mechanical profilometer (to an accuracy of 2.5 to 10 nm) compared remarkably well with the thickness of the deposited SiC coatings, thereby confirming that the failure was at the interface. Interface strength values ranging from 3.68 to 7.48 GPa were obtained for different SiC coatings.

Pitch-55 ribbon–SiC interface

Due to the presence of inhomogeneities and the weak (transverse) strength of these ribbons across

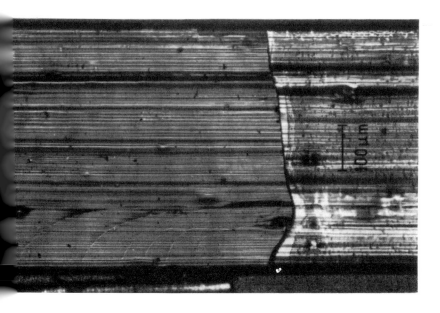

Fig. 24.7 Spalled spot of SiC coating from the Pitch-55 ribbon.

lamellae, the failure was predominantly observed to be within the ribbon. However, in some cases, failure at the interface was also observed. Figure 24.7 shows the edge of a spalled coating. The coating is intact on the right of the micrograph and is removed from the left side. For this sample, an average interface strength value of 0.23 GPa was obtained. These calculations again took full account of the anisotropic character of the ribbon material. A part of the same ribbon was also tested without the coating in order to determine the transverse strength of the ribbon. A value of 0.26 GPa was obtained. As expected, this value is higher than the interface strength observed on the same ribbon–coating system. Interface strength values ranging from 0.22 to 0.24 GPa were obtained for the Pitch-55 ribbon–SiC system. In all the above cases, the location of the plane of failure was confirmed by matching the coating thickness to the depth of the spalled spot determined from a high precision profilometer. Clearly, the interface strength determined from this test is significantly lower than that obtained from the PG–SiC system. This is attributed to the structural inhomogeneities of the ribbon surface [8], which act as sizable interface flaws and lead to significantly lower strength values. In view of the poor structural integrity of the ribbon material, the PG–SiC system is probably a better candidate for determining the properties of the Pitch-55 fiber–SiC interface. Nevertheless, the values obtained from both the systems are greater than the desired level of interface strength of 182 MPa for interface delamination discussed in section 24.2 for the case of the stiff coating. Thus, it is necessary to impair the interface strength by implanting embrittling agents at the interface during the coating deposition process. This could be achieved by planting a small atom concentration of, for example, Sb or As, that are known to be very effective sources of degradation of cohesion in steel. Recently, Rice and Wang [17] have theoretically explored the effect of such foreign agents on the interface toughness. Similar effects need to be experimentally probed in the present context to bring down the interface strength levels. This is presently under active consideration [18].

Since the interface is loaded by a stress pulse that is external to the material system, the laser spallation experiment is capable of determining the interface strength for any thin film interface provided the plastic dissipation in the substrate is negligible. To permit wide applicability of the laser spallation experiment, the results of the computer code have been generalized into interface stress charts. This should make such strength measurements possible in most hard substrate–coating pairs of interest. For a discussion of such charts the reader is referred to Gupta *et al.* [7].

24.4 MEASUREMENT OF TOUGHNESS OF INTERFACES

The toughness of interfaces is an important additional dimension in the strategy of controlled delamination. Such toughness measurements are important in designing the structure of the interface that will satisfy the various mechanical criteria outlined in section 24.2 above. We have demonstrated previously that the spontaneous delamination phenomenon of thin misfitting coatings that are under residual tensile or compressive stresses can be used very effectively to determine the toughness of the interface between the coating and the substrate [12]. When the elastic strain energy per unit area due to the misfit in the coating begins to exceed the intrinsic interface toughness, i.e. the specific work of separation of the coating from the substrate along the interface, the coating delaminates starting from any interface flaw. Since no instability is involved, the delamination, once started, can propagate under quasistatic conditions. Figure 24.8 shows a SEM micrograph of a fiber outgassed with SiC coating of 0.33 μm thickness, initially deposited under conditions of low ion beam energy. The coating showed some small through-the-thickness cracks, but no delamination during several weeks of storage in a desiccator after the regular outgassing treatment of 30 min at 600°C to remove the entrapped hydrogen. When it was examined again after several months of further storage in laboratory air with the usual level of relative humidity, it

Fig. 24.8 Delaminated flakes of SiC coating on a Pitch-55 fiber.

the vicinity of 60%, copious delamination of the SiC coating was found, as shown in Fig. 24.8. The initial bi-axial tensile misfit strain ε_m between the coating and the fiber could be determined from the ratio of the average gap size between flakes to the dimensions of the flakes. With knowledge of this, and the assumption that the Young's modulus of the coating was the same as that for similar coatings on Si wafers deposited under the same ion beam conditions, the total elastic strain energy release per unit area of interface could be determined from the sample. An elementary misfit analysis presented in more detail in [12] gives the intrinsic toughness G_{co} of the interface to be

$$G_{co} = \frac{E\varepsilon_m^2 t_c}{2}\left(\frac{F}{H^2}\right) \qquad (24.2)$$

where ε_m is the bi-axial misfit strain between coating and fiber, E the Young's modulus of the SiC coating, t_c is the thickness of the coating, and F and H are complicated functions of the ratios of Young's moduli, and the coating thickness to fiber radius that are given in [12].

As a specific example, we evaluate the critical energy release rate of the coating related to the case of Fig. 24.8 of a coating applied to a fiber under low ion beam conditions in the plasma-assisted CVD apparatus. This should result in a coating with Young's modulus of $E = 16$ GPa. The initial misfit strain, as measured from the gaps between the fragments of the coatings, was found to be $\varepsilon_m = 2.7 \times 10^{-2}$ in the axial direction. Since the axial and radial Young's moduli E_z and E_r of the fiber were 385 GPa and 14 GPa respectively, $E/E_r = 1.14$ and $E/E_z = 4.16 \times 10^{-2}$. Furthermore, the thickness of the SiC coating was 0.33 µm and the fiber radius 5 µm, giving $t/R = 6.6 \times 10^{-2}$. Thus, for assumed Poisson's ratios of 0.3, the values of the functions F and H can be calculated to be 1.53 and 0.734 respectively. This gave for the intrinsic interface toughness $G_{co} = 5.5$ J/m², which is very close to the corresponding value measured for the delamination of the interface between SiC and a Si substrate. For the Si–SiC system a value of 5.95 J/m² was obtained [12].

24.5 MEASUREMENT OF STRENGTH AND TOUGHNESS OF CARBON FIBERS

The strategy of deflecting cracks along interfaces in a controlled manner in composites also requires experimental determination of the strength and toughness of the reinforcing fibers. Since such measurements are difficult to perform in fibers that are only 10 μm in diameter, specially prepared Pitch-55 ribbons of 600 μm width and 35 μm thickness were obtained for doing the tests. As with carbon fibers, these ribbons show strong anisotropy in their elastic properties. Hence, the toughness along and across the graphitic planes were determined. Average longitudinal tensile strengths of these ribbons were determined to be 0.6 GPa. Due to the size effect and inhomogeneities on the ribbon surface, this value is considerably lower than that for the fiber (1.3 GPa). Average toughness of the ribbons across the graphitic planes was determined to be 3.5 J/m² by conventional single edge notched fracture mechanics tests.

Special *in situ* experiments were designed in the scanning electron microscope to measure the toughness along the graphitic lamellae for cracks propagating parallel to the axis. The set-up involved inserting a knife edge to prop open the flanks of a pre-existing crack and recording the total crack length at the critical crack-tip opening displacement (CTOD) on a micrograph. The observed CTOD and the crack length is related to the fracture toughness by employing the principles of linear elastic fracture mechanics. Unusually high toughness values of 166 J/m² were obtained for such cracks propagating along the graphitic lamellae. These high values result from the very low interlamellar plastic shear resistance between the graphitic lamellae lying parallel to the crack plane. An asymptotic elastic–plastic analysis for single crystals, following the basic developments given by Rice [19], was used to estimate the extent of inelastic energy dissipated. These estimates were quite close to the experimentally observed values. Such high values suggest that in composite systems involving carbon fibers the plastic deformation in the fibers should be considered in all the micromechanical modelling exercises involving interface failure.

24.6 CONCLUSIONS

We have developed the required tools to tailor the mechanical properties of interfaces between fibers and coatings in order to use them as mechanical fuses for damage containment. A brief outline of these tools was presented above. It still remains to be demonstrated that when the above stated principles are systematically applied that tough composites will result.

ACKNOWLEDGMENTS

This research leading to the above results has been supported initially by IST/SDIO through the ONR under Contract No. N00014-85-K-0645 and more recently by ONR under Grant No. N00014-89-J-1609. For this support and his continued keen interest in this research we are grateful to Dr S. Fishman of that agency. We are also grateful to the MIT George Harrison Spectroscopy Laboratory for the use of their facilities for the laser experiments and finally to our colleagues Dr J. A. Cornie for his general collaboration and Mr B. Chambers for providing the SiC coatings.

REFERENCES

1. A. S. Argon, in *Treatise on Material Science and Technology*, Vol. 1 (ed H. Herman), Academic Press, NY, (1972), p. 79.
2. A. S. Argon, in *Composite Materials: Fracture and Fatigue*, Vol. 5 (ed L. J. Broutman), Academic Press, NY, (1974), p. 153.
3. A. G. Metcalfe, in *Composite Materials: Interfaces in Metal Matrix Composites*, Vol. 1 (ed A. G. Metcalfe), Academic Press, NY, (1974), p. 1.
4. V. Gupta, A. S. Argon and J. A. Cornie, *J. Mater. Sci.*, **24** (1989), 2031.
5. V. Gupta, A. S. Argon and Z. Suo, *J. Appl. Mech.*, in press.
6. V. Gupta, A. S. Argon, J. A. Cornie and D. M. Parks, *Mater. Sci. Eng. A*, **126** (1990), 105.
7. V. Gupta, A. S. Argon, D. M. Parks and J. A. Cornie, *J. Mech. Phys. Solids*, **40** (1992), 141.
8. V. Gupta and A. S. Argon, *J. Mater. Sci.*, **27** (1992), 777.
9. Q. Li, S. M. Thesis, Department of Materials Science and Engineering, MIT, Cambridge, MA (1987).
10. E. Zywicz and D. M. Parks, *Composites Sci. Technol.*, **33** (1988), 295.
11. H. S. Landis, Ph.D Thesis, Department of Materials Science and Engineering, MIT, Cambridge, MA (1987).

12. A. S. Argon, V. Gupta, H. S. Landis and J. A. Cornie, *J. Mater. Sci.*, **24** (1989), 1207.
13. J. Ann, K. L. Mittal, and R. H. MacQueen, in *ASTM STP 640*, **134** (1978).
14. S. D. Chiang, D. B. Marshall and A. G. Evans, in *Materials Science Research*, Vol. 14 (eds J. Pask and A. G. Evans), Plenum Press, NY, (1981), p. 603.
15. R. J. Clifton, *J. Appl. Phys.*, **41** (1978), 5335.
16. V. Gupta, Ph.D Thesis, Department of Mechanical Engineering, MIT, Cambridge, MA (1989).
17. J. R. Rice and J. S. Wang, *Mater. Sci. Eng. A*, **107** (1989), 227.
18. J. A. Cornie, private communication 1990.
19. J. R. Rice, *Mech. Mater.*, **6** (1987), 317.
20. A. S. Argon, D. M. Parks, V. Gupta, and J. A. Cornie, submitted to *J. Mech. Phys. Solids*.

CHAPTER TWENTY-FIVE

Microstructure and fracture resistance of metal–ceramic interfaces

A. G. Evans and M. Rühle

Introduction · Atomistic structure · Bonding models · Defects at interfaces · The work of adhesion · Fracture resistance

25.1 INTRODUCTION

Metal–ceramic interfaces play an important, sometimes controlling, role in composites, multi-layer substrates, capacitors, electron tubes and automotive power sources. Bonding and adhesion between the ceramic and metal are often critical to the performance of components. The interface geometry and chemistry play a dominant role in determining the mechanical and electrical integrity of composites. Furthermore, unique properties may be developed from multilayer ceramic–metal structures. Systematic studies of metal–ceramic interfaces started in the early 1960s. Such studies were directed towards identification of general rules that govern bonding and interface behavior both theoretically and experimentally, including the thermodynamics of interfacial reactions and crystallographic relationships and for the evaluation of atomistic structure at the interface. This chapter summarizes results concerning the inter-relation between atomistic structure and the macroscopic fracture resistance of metal–ceramic interfaces. For more details the reader is referred to a recent conference proceedings [1].

25.2 ATOMISTIC STRUCTURE

The determination of atomistic structures of metal–ceramics interfaces is, in general, complicated, since the two materials that have to be matched exhibit different atoms (ions) and possess different crystal symmetries, crystal structures, and lattice parameters. The adjacent lattices are not commensurate, the two different structures can be described being just quasiperiodic [2] However, there exist examples wherein the lattice mismatch is small and both components possess the same lattice symmetry. Ag–MgO and Nb–Al$_2$O interfaces are examples which serve as model systems for experimental studies, as well as theoretical calculations. The interfaces can be formed either by diffusion-bonding [3], internal oxidation [4], or epitaxial film growth [5].

The atomistic structures can, in principle, be determined by high resolution electron microscope (HREM) [6]. A new generation of electron microscopes allows direct imaging of atomic columns by aligning the electron beam along the plane of an interface possessing only a tilt component. Each column of atoms is imaged in one spot giving one

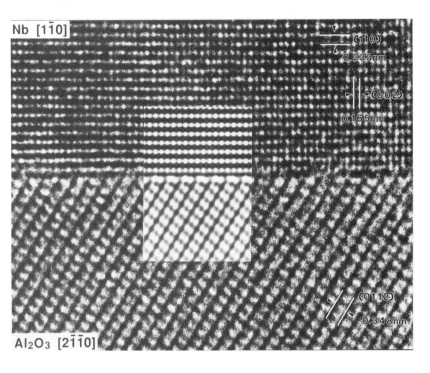

Nb [1$\bar{1}$0]

(110)
0.233 nm

(002)
0.165 nm

(01$\bar{1}$2)
0.348 nm

Al$_2$O$_3$ [2$\bar{1}\bar{1}$0]

Fig. 25.1 High resolution electron micrograph of a Nb–Al$_2$O$_3$ interface. Close-packed planes of both systems are parallel $(0001)_S \parallel (110)_{Nb}$ and $[0110]_S \parallel [001]_{Nb}$. The simulated micrograph which shows best agreement to the experimentally obtained micrographs is included in the inlet (D. Knauß and W. Mader, unpublished research).

projection of atomic arrangements in the interface. However, to derive reliable positions of atomic columns, a series of observed images taken at different focus settings must be matched quantitatively with computed images based on an assumed set of atomic coordinates. For matching purposes, image computation and position adjustment is repeated until the best possible fit is reached for the entire focus series. The interpretation is most difficult close to the interface where deviations from the perfect lattice are most extreme. Quantitative studies for one projected structures have been performed on interfaces between CdO and Ag as well as Al$_2$O$_3$ and Nb. However, HREM allows only the determination of a **projected** structure of the interface. The determination of the three-dimensional interface structure requires a reconstruction from at least two projected structures taken under different orientations. Recently, results have been obtained for several metal–ceramic systems; most convincingly for the system Nb–Al$_2$O$_3$ [3, 4, 7].

Figure 25.1 shows a high resolution image of regions close to a diffusion-bonded interface between Nb and Al$_2$O$_3$. The corresponding computed image is included as an inset in Fig. 25.1. A comparison between simulated and observed images allows the determination of the translation state of the Nb–Al$_2$O$_3$ interface as well as atomistic relaxations.

25.3 BONDING MODELS

Elucidation of the essential issues in bonding requires examination at all levels. The eventual objective would be the judicious coupling of information obtained from the most rigorous, but computer-bound, quantum mechanical supercell approaches with the results of cluster calculations and of simple continuum thermodynamic formulisms. Whereas the continuum models only allow for the prediction of trends, atomistic models provide a quantitative measure for the bonding issue. *Ab initio* calculations thus seem to be essential for a full understanding. Recently, Blöchl *et al.*

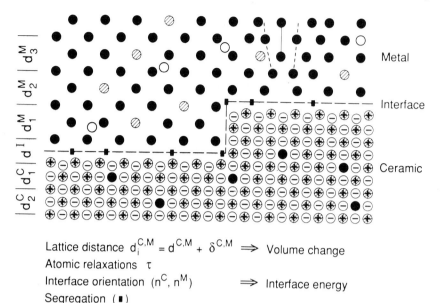

Lattice distance $d_i^{C,M} = d^{C,M} + \delta^{C,M}$ $\Rightarrow$ Volume change
Atomic relaxations τ
Interface orientation (n^C, n^M) $\Rightarrow$ Interface energy
Segregation (∎)

Fig. 25.2 Schematic drawing of a real metal–ceramic interface which includes possible structural and chemical defects.

[8] performed calculations for MgO–Ag interfaces. The calculations resulted in the determination of the energetically most favorable translation state, including the extraction of the volume increase. Also a separation of bonding into different contributions (ionic, covalent, and polarization) is possible.

These rather promising calculations should not give the impression that bonding at an arbitrary metal–ceramic interface could be evaluated by performing *ab initio* calculations. The calculations are restricted to systems which contain at the most 40 atoms and serve as a benchmark, whereby bonding tendencies in different systems can be developed. More phenomenological studies may be required for assessing bonding in complicated, low periodicity systems.

25.4 DEFECTS AT INTERFACES

HREM images of interfaces reveal many different types of defects [9] (Fig. 25.2). The defects can be classified as either chemical or structural (Table 25.1). **Chemical defects** occur if any deviation from the exact composition of the two phases exists

Table 25.1 Defects at or close to metal–ceramic interfaces

Chemical defects	Segregation
	Chemical reactions
	Chemical gradients
	Interphases (reaction products)
Structural defects	Facets/steps
	Misfit dislocations
	Lattice dislocations
	Impurity pores precipitates

at or near the interface. The defects may be caused either by segregation, by reactivity at the interface, or as a reaction product between the metallic and ceramic components. The type of defect that develops depends on the specific system investigated. **Segregation** of rare-earth ions is often observed at the interface between an oxide scale formed by oxidation of a metallic matrix [10] **Reactions** may occur at the interface by dissolving one phase in the other as exemplified by the Nb–Al_2O_3 interface, wherein Al_2O_3 is dissolved in Nb until the equilibrium concentration is reached for Al and O in Nb [11]. Concentration profiles have been determined on cross sections of rapidly cooled

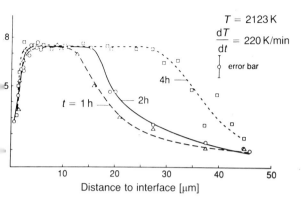

$T = 2123\,\mathrm{K}$

$\dfrac{dT}{dt} = 220\,\mathrm{K/min}$

error bar

4h

$t = 1\,\mathrm{h}$ — 2h

Distance to interface [µm]

Fig. 25.3 Measured concentration profiles of Al in Nb as a function of the distance to the interface for various bonding times. Diffusion bonding conditions: 2073 K, dynamic vacuum (10^{-4} torr), cooling rate 215 K min^{-1}.

specimens (Fig. 25.3). The studies revealed that, close to the interface, the concentration of Al is below the limit of detectability. However, with increasing distance from the interface the concentration of Al in Nb increases to a saturation value and decreases again with increasing distance from the interface. The total amount of dissolved Al depends on the bonding time and temperature. Diffusion bonding for long times results in extended Al diffusion into the Nb. The corresponding oxygen content is below the limit of detectability of the electron energy loss spectrometer (~ 1 at.%). However, very small concentrations of oxygen should be present (<60 ppm), due to the high oxygen mobility at the bonding temperature.

For systems that form interphases it is important to be able to predict the product phases. However, even if all thermodynamic data are known so that the different phase fields and the connecting tielines can be calculated, the preferred product phase still cannot be unambiguously determined [12], because kinetic considerations are involved. Specifically, the diffusion paths in phase space are controlled by different diffusion coefficients and consequently interphase composition depends also on the diffusivity ratios. Sometimes, small changes in the initial conditions can influence the reaction path dramatically, as exemplified by the Ni–Al$_2$O$_3$ system [13]. It was shown that a threshold concentration of ~ 120 at.ppm of O (dissolved in Ni) is required for the formation of a spinel interphase. Higher concentrations of dissolved oxygen results in a spinel (NiAl$_2$O$_4$) layer, with the thickness of this layer dependent on the oxygen concentration. Similar results are obtained for bonding of other fcc materials (Cu, Co, Ni, and Al) to Al$_2$O$_3$. Bonding of Ti to Al$_2$O$_3$ results in formation of complex intermetallic phases (TiAl or Ti$_3$Al).

Structural defects (Table 25.1) can also be identified. Steps or facets are observed if the interface plane does not coincide with a low indexed crystallographic plane of the metal or ceramic. The step height varies between the atomistic level (Fig. 25.4) and macroscopic values introduced by a surface roughness prior to bonding. The roughness and/or facet height increases the fracture resistance of the interface [14]. Dislocations are frequently observed at or close to an interface. Misfit dislocations may be generated during interface formation or during relaxations of thermal stresses. Residual pores or non-wetting impurity precipitates could be identified at the interface [11].

Both chemical and structural defects have an important influence on the fracture resistance.

25.5 THE WORK OF ADHESION

Thermodynamic considerations define the work of adhesion as the work required to reversibly separate the interface between the metal and the ceramic,

$$W_{\mathrm{ad}} = \gamma_{\mathrm{c}} + \gamma_{\mathrm{m}} - \gamma_{\mathrm{mc}}$$

where γ_{c} and γ_{m} are the free energies of the relaxed surfaces of ceramic and metal, respectively, and γ_{mc} represents the energy of a relaxed interface between the metal and the ceramic. The quantity W_{ad} is thus the reversible work required per unit area of interface to separate the bond into two free surfaces (Fig. 25.6). It has not been possible to measure the work of adhesion directly [15]. Therefore, in practice, W_{ad} is deduced by measuring the equilibrium contact angle Θ established by a solid metal in contact with the ceramic

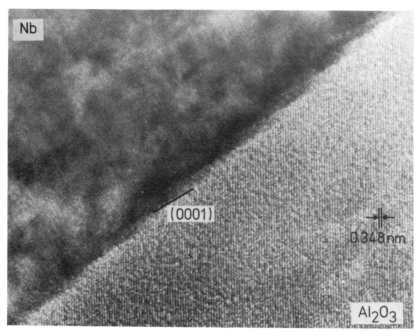

Fig. 25.4 Facets at a Nb–Al$_2$O$_3$ interface. An angular deviation between the interface orientation and the basal plane exists.

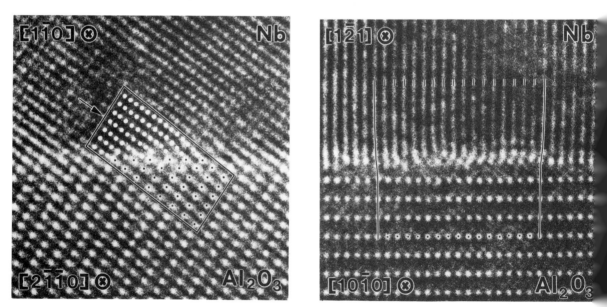

Fig. 25.5 Misfit dislocation near a Nb–Al$_2$O$_3$ interface (MBE grown film on sapphire substrate). The projection of the core of the misfit dislocation line lies parallel to the electron beam. Different orientation. The core of the dislocation line is inclined with respect to the electron beam. The region of no matching of corresponding lattice planes is marked. This region corresponds to the projection width of the additional lattice planes.

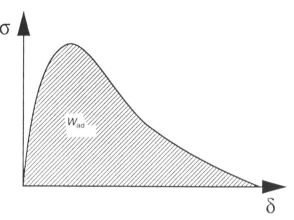

Fig. 25.6 Definition of the work of adhesion W_{ad}, as the reversible work done to separate the interfaces at a boundary.

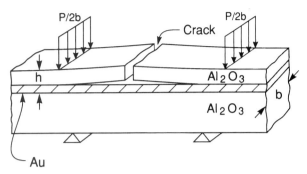

Fig. 25.7 Schematic of the test specimen. The metal (Au) is bonded between two Al_2O_3 single crystals.

$$W_{ad} = \gamma_m (1 + \cos \Theta)$$

Adequate measurements of Θ and γ_m constitute a non-trivial experimental task. Often γ_m is anisotropic and, hence, the crystallography of a surface has to be determined. Furthermore, true equilibrium has to be established by allowing sufficient mass transport and the associated morphological evolution in both the metal and ceramic. The work of adhesion is, of course, strongly influenced by both chemical and structural defects at interfaces [16].

25.6 FRACTURE RESISTANCE

The fracture energy of interfaces between dissimilar materials Γ_i can be determined by various mechanical testing experiments [17] (Fig. 25.7). A small crack is introduced at the interface and Γ_i determined from the load at which the crack extends along the interface. A large discrepancy exists between the work of adhesion, W_{ad} ($\sim 1\,J\,m^{-2}$) and the fracture resistance which is typically $> 50\,W_{ad}$ (Table 25.2). Often the interface fracture resistance is higher than the fracture resistance of the adjoining ceramic. This differential involves dissipation processes which accompany crack propagation. One process derives from the nonplanarity of a interface [14]. Roughness-

Table 25.2 Work of adhesion and fracture resistance of metal–ceramic interfaces

System	W_{ad} ($J\,m^{-2}$)	Fracture resistance Γ_i ($J\,m^{-2}$)	
		Without interphase	*With interphase*
Nb–Al_2O_3	0.8	10–300	–
Au–Al_2O_3	0.5	50	–
Ni–Al_2O_3	0.2–0.8	80	0–60
Cu–Al_2O_3	0.4–0.7	120	0–45

related shielding can be characterized by a parameter dependent on the amplitude and the wavelength of the roughness, as well as the shear modulus [14]. Plasticity occurring in one of the bond layers can cause dissipation as governed by the yield strength and the thickness of the ductile layer [18]. The plastic dissipation depends multiplicatively on the work of adhesion provided that the ductile material has viscoplastic properties that allow a strong singularity to be retained at the crack tip. Otherwise the dissipation term has no dependence on W_{ad}. Contributions to Γ_i must be evaluated and understood in order to control the mechanical behavior of interfaces in composites, coatings, bonds etc.

A few experimental studies exist on systems in which chemical reactions are absent and no interface layer is formed. The system Al_2O_3–Au is of particular relevance [19]. Furthermore, the transparency of sapphire allows *in-situ* observation

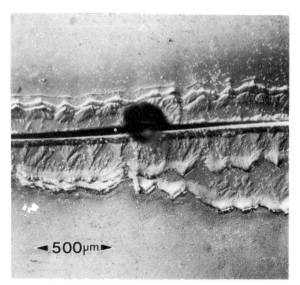

Fig. 25.8 Crack front profiles that illustrate the nature of crack extension.

of crack propagation along the interface, as needed to elucidate the dominant fracture mechanism. The fracture energy has been measured using a mixed mode four-point bending specimen [17] (Fig. 25.8). The specimen has the dual advantage that precracking can be conducted with good control and that, when the interface crack is between the inner loading points, the energy release is essentially crack-length independent. In addition the crack orientation in this specimen facilitates *in-situ* observations. Studies on this system revealed that subcritical crack extension occurs upon testing in air. However, crack extension is intermittent and erratic (Fig. 25.8). Nevertheless, the crack extends at an essentially uniform mean velocity when a constant load is applied and furthermore the mean velocity increases when the load is increased. A quantitative evaluation [17] revealed that the fracture energy is about $50 \, J \, m^{-2}$. The magnitude of this energy has been related to plastic dissipation in Au.

Further evaluation requires knowledge of dislocation densities. A direct determination of this density has not yet been possible by direct imaging techniques, e.g. TEM, since the dislocation density is strongly modified during specimen preparation.

Other techniques capable of characterizing dislocations generated by crack tip propagation are being tested, such as the line width of either Kossel (Kikuchi) [20] or X-ray lines [21]. Thus, line widths of electron channeling patterns have been used for such determinations. The line broadening is calibrated on uniaxially deformed crystals. The highest dislocation densities ($\sim 10^{13} \, cm^{-2}$) occurred at the fracture interface and decreased with increasing distance from the interface. These densities are connected with measured values of plastic dissipation. However, further research is needed to relate trends in Γ_i to the plastic properties of the metal.

REFERENCES

1. M. Rühle, A. G. Evans, M. F. Ashby and J. P. Hirth (eds) *Metal/Ceramic Interfaces*, Pergamon Press, Oxford (1990).
2. A. P. Sutton, *Phase Transitions*, **16/17** (1989), 563.
3. M. Florjancic, W. Mader, M. Rühle and M. Turwitt, *J. de Physique*, **46** (1985), C4–129
4. W. Mader, *Z. Metallkunde*, **80** (1989), 139.
5. C. P. Flynn in Ref. 1, p. 168.
6. J. C. H. Spence, *Experimental High Resolution Electron Microscopy*, 2nd edn, Oxford University Press (1988).
7. J. Mayer, C. P. Flynn and M. Rühle, *Ultramicroscopy*, **33** (1990), in press.
8. P. Blöchl, G. P. Das, H. F. Fischmeister and U. Schönberger, in Ref. 1, p. 9.
9. W. Mader and M. Rühle, *Acta Metall. Mater.*, **37** (1988), 253.
10. W. E. King (ed), *The Reactive Element Effect on High Temperature Oxidation After Fifty Years*, Trans Tech Publications, Aedermannsdorf (1989).
11. K. Burger and M. Rühle, *Ceram. Eng. Sci. Proc.*, **10** (1989), 1549.
12. M. Backhaus-Ricoult, *Ber. Bunsenges. Phys. Chem.*, **89** (1989), 1323.
13. K. P. Trumble and M. Rühle, in Ref. 1, p. 144.
14. A. G. Evans and J. W. Hutchinson, *Acta Metall. Mater.*, **3** (1989), 909.
15. B. J. Derjaguim, *Recent Advances in Adhesion*, Gordon and Breach, New York, (1971), 513.
16. M. Nicholas, *J. Mater. Sci.*, **3** (1968), 571.
17. P. G. Charalambides, J. Lund, R. M. McMeeking and A. G. Evans, *J. Appl. Mech.*, **111** (1989), 77.
18. A. G. Evans, B. J. Dalgleish, P. G. Charalambides and M. Rühle, *Met. Trans.*, **A126** (1990), 53.

19. I. E. Reimanis, B. J. Dalgleish, M. Brahy, M. Rühle and A. G. Evans, *Acta Metall. Mater.*, **38** (1990), 2645.

20. D. J. Dingley, in *Scanning Electron Microscopy*, SEM Inc., A MF O'Hare (1984), p. 569.

21. T. Ungar, H. Mughrabi, D. Rönnpagel and M. Wilkens, *Acta Metall. Mater.*, **32** (1984), 334.

CHAPTER TWENTY-SIX

Role of interface dislocations and surface steps in the work of adhesion

D. Wolf and J. A. Jaszczak

Introduction · Interfacial decohesion: from interface dislocations to surface steps · Computer simulations · Core and elastic strain-field effects in free surfaces and GBs · A broken-bond model for interfacial decohesion · Summary and conclusions

26.1 INTRODUCTION

Although crack extension is usually accompanied by plasticity and other highly complex, non-equilibrium kinetic phenomena, the ideal brittle-fracture energy nevertheless represents a useful lower thermodynamic limit for the work of adhesion, W^{ad}, particularly since it is often assumed that both the plastic work and the energy due to kinetic processes increase in proportion with W^{ad} [1]. Brittle interfacial decohesion in the sense of Griffith [2], i.e. the reversible transformation of an internal interface into two free surfaces, hence requires as a minimum an understanding of the equilibrium energies of the initial and final states, from which the ideal cleavage-fracture energy,

$$W^{ad} = \gamma_1 + \gamma_2 - \varepsilon \qquad (26.1)$$

can be determined. Here the energy of the internal interface is denoted by ε, while the two free-surface energies are denoted by γ_1 and γ_2, respectively. To elucidate ideal-cleavage decohesion of internal interfaces hence necessitates a better understanding, from a common viewpoint, of the energies of internal interfaces and external surfaces. In the following, by investigating the zero-temperature work of adhesion, we hope to develop such a unified framework.

Much of the high-resolution-microscopy work of recent years has shown that the structure of internal interfaces generally consists of areas of localized misfit, so-called misfit **dislocations**, which are separated by elastically strained regions of relatively good match across the interface [1, 3–5]. Similarly, the structure of free surfaces is usually characterized in terms of localized **steps** or **ledges** which are separated by regions of 'flat' surface material [6]. The ideal-cleavage decohesion of an internal interface into two free surfaces may hence be viewed as the reversible transformation of misfit dislocations into surface steps. Since both interface dislocations and surface steps consist of highly distorted core regions surrounded by elastic strain fields, interfacial decohesion therefore involves the transformation of dislocation cores into the cores of surface steps, and of the long-range elastic strain fields near dislocations into the much shorter-ranged strain fields surrounding the surface steps. It is the purpose of this chapter to investigate the underlying core and elastic phenomena accompanying this transformation by means of atomistic computer simulations.

Without loss of generality, throughout this chapter we will consider the decohesion of grain boundaries (GBs) as simple model systems for elucidating the transformation of dislocations into steps. As discussed, for example, in Chapter 1 of this volume [7], GBs generally have **five** macroscopic degrees of freedom (DOFs), i.e. three more than free surfaces (with only the two DOFs associated with the surface normal, $\hat{n}$). However, to emphasize the geometrical similarity between GBs and free surfaces, these DOFs will be chosen such that **four** are associated with the GB-plane normal, characterized by the unit vectors $\hat{n}_1$ and $\hat{n}_2$ in the two halves of the bicrystal (corresponding to the two fracture surfaces obtained upon decohesion), while the remaining one is represented by the twist angle, θ. A non-vanishing value of θ adds a twist component (i.e. screw dislocations) to the GB, in addition to the tilt component (i.e. edge dislocations) already present once the GB plane has been fixed. Then, to focus on the similarity between surface steps and **edge** dislocations (to be elaborated upon below), but yet still capture the essential core and elastic phenomena involved, we limit ourselves to the following two types of GBs whose structure contains edge dislocations only.

1. By investigating pure **tilt** boundaries, from the outset the twist component is eliminated altogether. Like free surfaces, the GBs are fully characterized by only the DOFs associated with the interface plane (four in the case of asymmetrical tilt GBs, and two in the symmetrical case). Their dislocation structure should therefore be most intimately connected with the structure of the steps in the corresponding fracture surfaces.

2. For large twist angles, θ, in the Read–Shockley sense [8], the cores of the **screw** dislocations overlap completely, and a simple model may be formulated in which all interactions across the interface are assumed to be entirely random. This model of a 'random grain boundary' (RGB) [9–11] thus represents an idealized theoretical model suitable for **high-angle twist** boundaries.

Because in the RGB model the twist angle is eliminated as a DOF, both pure tilt boundaries and

RGBs are characterized fully by only the DOFs associated with the GB plane. However, in the RGB model screw dislocations – albeit with completely overlapping cores – are, in principle, present. A comparison of the work of adhesion of RGBs with that of pure tilt boundaries should therefore elucidate the role of screw dislocations in the transformation of interfacial edge dislocations into surface steps.

Given that (i) the appearance of steps in surfaces is directly connected with the crystallographic orientation of the surface **plane** and (ii) the strictly edge-type dislocations in both of the above types of GBs are fully characterized solely by the DOFs associated with the GB **plane**, a comparison of the energies and structures of these GBs with those of free surfaces should also provide insight into the importance of the role of the interface plane in the work of adhesion. Moreover, since the bulk **ideal-crystal** brittle-fracture energy, 2γ (eq. (26.1)) is also a very sensitive function of the fracture plane, such a comparison of free surfaces and GBs also provides information on the crystallographic anisotropy of the perfect-crystal cleavage energy.

Two conceptually different types of interatomic potentials will be used in our computer simulations. To provide some insight into the role of many-body effects, results obtained via a semiempirical many-body embedded-atom-method (EAM) potential [12] will be compared with simulations involving a conceptually simpler Lennard–Jones (LJ) pair potential. EAM potentials have the advantage over pair potentials that they incorporate, at least conceptually, the many-body nature of metallic bonding, while being relatively efficient computationally. In these potentials the strength of the interaction between atoms depends on the local volume or, in another interpretation, on the local electron density 'sensed' by every atom. Also, while at zero temperature any equilibrium pair potential automatically satisfies the Cauchy relation for the elastic constants, $C_{12} = C_{44}$, these many-body potentials permit all three elastic constants of a cubic metal to be determined (or used in the fitting). An iterative energy-minimization algorithm ('lattice statics') is used to compute the fully relaxed zero-temperature atomic structure and energy of free surfaces and GBs, including the volume expansion at the internal interfaces [13, 14].

It should be noted that the absolute values of GB and free-surface energies presented throughout this chapter are probably not very reliable. Even for the many-body potentials there exist discrepancies (in some cases up to a factor of two [12]) between the computed free-surface energies, on the one hand, and the values obtained from experiments and by means of electronic-structure methods on the other. We nevertheless believe that a comparison of the **relative** energies of GBs and free surfaces is meaningful, particularly when the same generic behavior is obtained by means of conceptually different interatomic potentials.

The chapter is organized as follows. In section 26.2 we offer a formal description of the work of adhesion for symmetrical tilt boundaries (STGBs) in terms of the line energies of the underlying edge dislocations and surface steps, including their elastic interactions. In section 26.3, our computer-simulation results for the energies of STGBs, the RGB model and free surfaces will be compared, including the work of adhesion of the internal interfaces. An analysis of the crystallographic anisotropy of the work of adhesion is also presented in section 26.3. This systematic investigation of the role of the interface plane in ideal cleavage decohesion leads naturally to the distinction between 'special', 'vicinal', and 'high-angle' interfaces, with qualitatively rather different behaviors of the work of adhesion. Then, in section 26.4 the core and elastic-interaction energies of steps and dislocations are determined directly by considering in detail the energies of STGBs and surfaces in the vicinity of two energy cusps. Finally, in section 26.5 it is shown that only the cores of the steps and dislocations, but not their surrounding strain fields, give rise to broken bonds; this finding leads naturally to an elucidation of the role of broken bonds in interfacial deco-hesion and, hence, the intrinsic limitations of broken-bond and structural-unit models.

26.2 INTERFACIAL DECOHESION: FROM INTERFACE DISLOCATIONS TO SURFACE STEPS

26.2.1 Surface steps

With only the two macroscopic DOFs associated with the surface normal, free surfaces conceptually represent the simplest of all planar defects (Chapter 1 in this volume [7]). Given that these two DOFs are usually chosen to represent the orientation of the surface normal, the appearance of steps in surfaces is clearly connected with the crystallographic orientation of the surface **plane**. Figure 26.1 illustrates the orientation of a 'vicinal' surface (i.e. one containing steps [6, 7]), with unit normal $\hat{n}_v$, relative to a 'special' surface (i.e. one giving rise to an energy cusp), with unit normal $\hat{n}_{cusp}$. In both Figs. 26.1(a) and (b), the vicinal surface is assumed to lie in the x–y-plane, with the surface normal defining the z-direction. While Fig. 26.1(a) shows a view down the y-axis (i.e. parallel to the steps), Fig. 26.1(b) represents a view onto the x–y-plane, with the steps, separated by the di

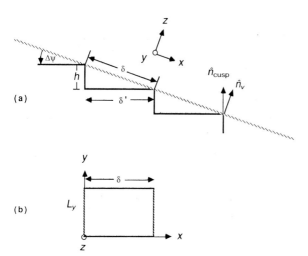

Fig. 26.1 Orientation of a 'vicinal' surface (i.e. one containing steps of height h [6, 7]), with unit normal $\hat{n}_v$ which is rotated by the angle $\Delta\psi$ relative to a 'special' surface (i.e. one giving rise to an energy cusp), with unit normal $\hat{n}_{cusp}$. In both (a) and (b), the vicinal surface is assumed to lie in the x–y plane, with the surface normal defining the z direction. While (a) shows a view down the y axis (i.e. parallel to the steps), (b) represents a view onto the x–y plane, with the steps, separated by the distance δ, forming the left and right edges of the planar unit cell.

tance δ, forming the left and right edges of the planar unit cell.

According to the figure, the detailed geometry of the steps in the vicinal surface may be described in terms of a rotation of the unit vector $\hat{n}_{cusp}$ about an axis perpendicular to both $\hat{n}_{cusp}$ and $\hat{n}_v$ by the angle $\Delta\psi$; the latter is obviously determined by the dot product

$$\cos \Delta\psi = (\hat{n}_v \cdot \hat{n}_{cusp}) \qquad (26.2)$$

Thus, the spacing, δ, between the steps of height h is given by

$$\delta = h/\sin \Delta\psi \qquad (26.3)$$

If one assumes the increased energy of the vicinal surface to be entirely due to the introduction of steps into the cusped surface, its energy per unit area, $\gamma_v \equiv \gamma(\Delta\psi)$, may be written as follows [15]:

$$\gamma(\Delta\psi) = [\gamma_{cusp} A_{cusp} + \Gamma(\delta)L_y]/A(\Delta\psi) \qquad (26.4)$$

The second contribution on the right-hand side represents the total line energy of the steps per unit surface area, with $\Gamma(\delta)$ denoting the energy per unit length of the steps (i.e. their line energy) and L_y representing the total step length in the planar unit cell (Fig. 26.1(b)). Because of the interaction between neighboring steps, the line energy Γ is a function of their distance δ. The first contribution to $\gamma(\Delta\psi)$ in eq. (26.4) arises from the energy of the cusped surface, $\gamma_{cusp} \equiv \gamma(\Delta\psi = 0)$, reduced however by its projection onto the larger planar unit-cell area, $A(\Delta\psi)$, of the vicinal surface relative to the unit-cell area, $A_{cusp} = A(\Delta\psi = 0)$, of the cusped orientation. As seen from Fig. 26.1(b), $A(\Delta\psi)$ is given by

$$A(\Delta\psi) = L_y\delta = L_yh/\sin \Delta\psi \qquad (26.5)$$

from which it follows that

$$A_{cusp}/A(\Delta\psi) = \delta'/\delta = \cos \Delta\psi \qquad (26.6)$$

with δ' defined in Fig. 26.1(a).

Inserting eqs. (26.5) and (26.6) into eq. (26.4), and using eq. (26.3) to replace $\Delta\psi$ by δ, we obtain:

$$\gamma(\delta) - \gamma_{cusp}(1 - h^2/\delta^2)^{\frac{1}{2}} = \Gamma(\delta)/\delta \qquad (26.7a)$$

Sometimes it is more convenient to express γ directly in terms of $\Delta\psi$ and write instead:

$$\gamma(\Delta\psi) - \gamma_{cusp} \cos \Delta\psi = [\Gamma(\delta)/h] \sin \Delta\psi \qquad (26.7b)$$

As discussed further in section 26.4.1, the mutual repulsion between steps arising from their overlapping elastic strain fields may formally be incorporated in eqs. (26.7(a)) and (b) by decomposing the line energy Γ as follows:

$$\Gamma(\delta) = \Gamma_{core}^{\infty} + \Gamma_{el}(\delta) \qquad (26.8)$$

where Γ_{core}^{∞} and Γ_{el} represent, respectively, the 'core' and strain-field (i.e. elastic) energies of the step. In writing eq. (26.8), it was assumed that only the elastic contribution to the line energy varies as a function of δ while the core energy of the steps is essentially independent of δ. The validity of this assumption will be tested in section 26.4.1.

The elastic energy in eq. (26.8) may be broken down further into the line energies of isolated (i.e. non-interacting) steps, $\Gamma_{el}^{\infty} = \Gamma_{el}(\delta \to \infty)$, and the step–step interaction energy, $\Gamma_{el}^{s-s}(\delta)$, according to

$$\Gamma_{el}(\delta) = \Gamma_{el}^{\infty} + \Gamma_{el}^{s-s}(\delta) \qquad (26.9)$$

As illustrated schematically in Fig. 26.2, the elastic step–step interaction energy is determined by the (shaded) area of overlap of the elastic displacement fields surrounding each individual step, and therefore varies as a function of δ. According to Marchenko and Parshin [16], the consequent elastic **repulsion** between two identical steps decreases inversely with the square of their distance, according to

$$\Gamma_{el}^{s-s}(\delta) = G_{el}^{s-s}/\delta^2 \qquad (26.10)$$

where G_{el}^{s-s} is a constant characterizing the elastic strength of the step–step interaction. Inserting eqs. (26.8)–(26.10) into eq. (26.7(a)), we thus obtain:

$$\gamma(\delta) - \gamma_{cusp}(1 - h^2/\delta^2)^{\frac{1}{2}} = \Gamma^{\infty}/\delta + G_{el}^{s-s}/\delta^3 \qquad (26.11a)$$

where Γ^{∞} contains the δ-independent contributions, according to

$$\Gamma^{\infty} = \Gamma_{core}^{\infty} + \Gamma_{el}^{\infty} \qquad (26.12)$$

By definition the line energy of an isolated step, Γ^{∞}, thus includes contributions from both the fully

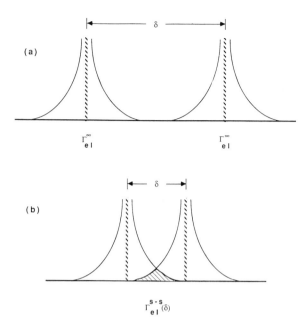

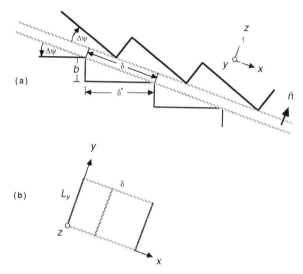

Fig. 26.2 Elastic strain fields between steps (schematic). (a) Non-interacting limit (for large separations δ) in which the strain fields do not overlap. (b) The interaction between steps is given by the area of overlap between the strain fields of neighboring steps.

Fig. 26.3 Conceptually a symmetrical tilt grain boundary (STGB) may be formed by bringing two identical free surfaces into contact, thus transforming the two parallel sets of surface steps, each of height h, into two parallel sets of edge dislocations, each with Burger's vector b (compare Fig. 26.1). In forming such a tilt bicrystal, the relative positions of its two halves, characterized by the rigid-body-translation vector $T = (T_x, T_y, T_z)$, are variables.

relaxed cores and elastic displacement fields of the steps, while $\Gamma_{el}^{s-s}(\delta)$ arises solely from the displacement fields (Fig. 26.2).

Instead of expressing the elastic energy directly as a function of δ, it may alternatively be expressed as a function of $\Delta\psi$, i.e. $\Gamma(\delta) \rightarrow \Gamma(\Delta\psi)$; eq. (26.3). Equations (26.8)–(26.10) may then, instead, be written as follows

$$\Gamma(\Delta\psi) \quad = \Gamma_{core} + \Gamma_{el}(\Delta\psi) \qquad (26.13)$$

$$\Gamma_{el}(\Delta\psi) \ = \Gamma_{el}^{\infty} + \Gamma_{el}^{s-s}(\Delta\psi) \qquad (26.14)$$

$$\Gamma_{el}^{s-s}(\Delta\psi) = (G_{el}^{s-s}/h^2) \sin^2 \Delta\psi \qquad (26.15)$$

and eq. (26.11a) becomes (eq. (26.7b)):

$$\gamma(\Delta\psi) - \gamma_{cusp} \cos \Delta\psi$$
$$= (\Gamma^{\infty}/h) \sin \Delta\psi + (G_{el}^{s-s}/h^3) \sin^3 \Delta\psi \quad (26.11b)$$

The validity of eqs. (26.11(a)) and (b), and the assumptions made in their derivation, will be tested in some detail in section 26.4.1.

26.2.2 Interface dislocations

It is interesting to observe the geometrical similarity of this picture of a vicinal-free surface with the geometry of an STGB. As illustrated in Fig. 26.3, an STGB may conceptually be formed by bringing two identical free surfaces into contact, thus transforming the two parallel sets of surface steps, each of height h, into two parallel sets of edge dislocations, each with Burger's vector b. In forming such a tilt bicrystal, the relative positions of its two halves are obviously variables. The three components of the related rigid-body translation vector, $T = (T_x, T_y, T_z)$, are known as the three microscopic or translational DOFs of the GB [17]; in practice they are governed by the condition that the bicrystal be in a state of minimum free energy or, at least, in a metastable translational state for a given choice of the macroscopic DOFs.

Assuming some arbitrary, but well-defined rigid-body translation to be present at the GB, in analogy to eq. (26.7a) the energy per unit area, $\varepsilon(\Delta\psi)$, of a 'vicinal' GB (i.e. one in the vicinity of a 'special' or 'cusped' GB which, by this definition, is entirely dislocation free) may be written as follows (Figs. 26.1 and 26.3):

$$\varepsilon(\delta) - \varepsilon_{\text{cusp}}(1 - h^2/\delta^2)^{\frac{1}{2}} = 2\Lambda(\delta)/\delta \qquad (26.16a)$$

or, in analogy to eq. (26.7b),

$$\varepsilon(\Delta\psi) - \varepsilon_{\text{cusp}} \cos \Delta\psi = 2[\Lambda(\delta)/b] \sin \Delta\psi \qquad (26.16b)$$

where $\varepsilon_{\text{cusp}}$ is the energy per unit area of the cusped GB while $\Lambda(\delta)$ denotes the total energy per unit length of the dislocations, i.e. their total line energy. Notice that in writing eq. (26.16b), the step height, h, in eq. (26.7b) was replaced by the magnitude of the Burger's vector, b (Fig. 26.3(a)).

If the tilt boundary in Fig. 26.3 were asymmetrical, the two sets of edge dislocations would still be parallel to the tilt axis; however, their spacings, δ_1 and δ_2 (which are related to angles $\Delta\psi_1$ and $\Delta\psi_2$), and Burgers vectors, b_1 and b_2, would now differ, and eqs. (26.16a,b), would have to be modified accordingly.

As in the case of the steps, the mutual repulsion between dislocations arising from their overlapping elastic strain fields may formally be incorporated in eq. (26.16) by decomposing Λ as follows (eq. (26.8)):

$$\Lambda(\delta) = \Lambda_{\text{core}}^{\infty} + \Lambda_{\text{el}}(\delta) \qquad (26.17)$$

where $\Lambda_{\text{core}}^{\infty}$ and Λ_{el} represent, respectively, the core and elastic strain-field energies of the two sets of dislocations. As for the steps, in writing eq. (26.17) it was assumed that only the elastic contribution to the line energy depends on δ while the dislocation-core energy is essentially independent of their spacing, an assumption to be tested in section 26.4.2.

The elastic energy in eq. (26.17) may be broken down further into the line energy of an isolated (i.e. non-interacting) dislocation, $\Lambda_{\text{el}}^{\infty} = \Lambda_{\text{el}}(\delta \rightarrow \infty)$, and the energy of interaction between different dislocations, $\Lambda_{\text{el}}^{\text{d-d}}(\delta)$, according to (Fig. 26.2)

$$\Lambda_{\text{el}}(\delta) = \Lambda_{\text{el}}^{\infty} + \Lambda_{\text{el}}^{\text{d-d}}(\delta) \qquad (26.18)$$

As in the case of the steps, the latter is determined by the area of overlap of the elastic displacement fields surrounding each individual dislocation, and therefore varies as a function of δ. According to Read and Shockley [8], the dislocation–interaction energy depends logarithmically on their distance, according to

$$\Lambda_{\text{el}}^{\text{d-d}}(\delta) = -L_{\text{el}}^{\text{d-d}} \ln(b/\delta) \qquad (26.19)$$

where the constant $L_{\text{el}}^{\text{d-d}}$ characterizes the elastic strength of this repulsive interaction. Inserting eqs. (26.17)–(26.19) into eq. (26.16a), analogous to eq. (26.11a) we obtain:

$$\varepsilon(\delta) - \varepsilon_{\text{cusp}}(1 - h^2/\delta^2)^{\frac{1}{2}} = 2[\Lambda^{\infty}/\delta - (L_{\text{el}}^{\text{d-d}}/\delta)\ln(b/\delta)] \qquad (26.20a)$$

where Λ^{∞} contains the δ-independent contributions, according to (compare eq. (26.12))

$$\Lambda^{\infty} = \Lambda_{\text{core}}^{\infty} + \Lambda_{\text{el}}^{\infty} \qquad (26.21)$$

As is well known, the elastic contribution to the line energy of an isolated lattice dislocation in an infinite crystal, $\Lambda_{\text{el}}^{\infty}$, diverges whereas the core contribution, $\Lambda_{\text{core}}^{\infty}$, remains finite. However, for the long-range-ordered arrays of dislocations in GBs (with mutually opposite Burger's vectors; Fig. 26.3), these divergences in the far-away strain fields of the individual dislocations cancel one another, with a consequently finite and small residual value of $\Lambda_{\text{el}}^{\infty}$ determined by the geometry of the array. The line energy in eq. (26.21) is therefore probably dominated by that of the dislocation **cores**. Interestingly, the total line energy of the steps in eq. (26.12) is also dominated by the core energy; however, for an entirely different reason, namely the weakness of the step–step interaction (section 26.4.1 below). A comparison of the line energies of steps and dislocations therefore provides mostly information on the related core energies while the information on their elastic displacement fields is largely contained in the interaction-energy term, i.e. the magnitudes of $G_{\text{el}}^{\text{s-s}}$ and $L_{\text{el}}^{\text{d-d}}$.

As for the steps, eq. (26.20a) may alternatively be

written in terms of $\Delta\psi$ as follows (eqs. (26.16b) and (26.11b)):

$$\varepsilon(\Delta\psi) - \varepsilon_{\text{cusp}} \cos \Delta\psi$$
$$= 2[(\Lambda^{\infty}/b) \sin \Delta\psi$$
$$- (L_{\text{el}}^{\text{d-d}}/b) \sin \Delta\psi \, \ln(\sin \Delta\psi)] \qquad (26.20b)$$

The validity of eqs. (26.20a,b), and the assumptions made in their derivation, will be tested in some detail in section 26.4.2.

26.2.3 Work of adhesion

If we assume that the free surfaces in Fig. 26.1 and the STGBs in Fig. 26.3 show energy cusps for the same special crystallographic planes (an assumption to be tested below), eqs. (26.11) and (26.20) may be inserted into eq. (26.1) to obtain the ideal cleavage-fracture energy,

$$W^{\text{ad}}(\delta) = (2\gamma_{\text{cusp}} - \varepsilon_{\text{cusp}})(1 - h^2/\delta^2)^{\frac{1}{2}}$$
$$+ 2(\Gamma^{\infty} - \Lambda^{\infty})/\delta$$
$$+ (2/\delta)[G_{\text{el}}^{\text{s-s}}/\delta^2 + L_{\text{el}}^{\text{d-d}}\ln(b/\delta)] \qquad (26.22a)$$

or, in terms of $\Delta\psi$,

$$W^{\text{ad}}(\Delta\psi) = (2\gamma_{\text{cusp}} - \varepsilon_{\text{cusp}}) \cos \Delta\psi$$
$$+ 2(\Gamma^{\infty}/h - \Lambda^{\infty}/b) \sin \Delta\psi$$
$$+ 2 \sin \Delta\psi[(G_{\text{el}}^{\text{s-s}}/h^3) \sin^2 \Delta\psi$$
$$+ (L_{\text{el}}^{\text{d-d}}/b) \ln(\sin \Delta\psi)] \qquad (26.22b)$$

According to these expressions, a high cleavage energy therefore requires (i) a large difference in the cusped energies, $2\gamma_{\text{cusp}} - \varepsilon_{\text{cusp}}$, and/or (ii) a large line energy of the steps compared to that of the dislocations, and/or (iii) a high elastic energy of interaction between steps by comparison with dislocations. For a better understanding of the work of adhesion it is therefore necessary to investigate the core and elastic properties of steps and dislocations.

Finally, exactly at the cusp (i.e. for $\Delta\psi = 0$, or $\delta \to \infty$), eqs. (26.22a,b) reproduce the trivial result that for a symmetrical GB,

$$W^{\text{ad}}(\Delta\psi = 0) = (2\gamma_{\text{cusp}} - \varepsilon_{\text{cusp}}) \qquad (26.23)$$

Because the cusped surfaces and interfaces are free of steps and dislocations, respectively, the related work of adhesion therefore provides no information

on either the line energies of steps and dislocations or on their interaction.

26.2.4 Distinction between 'special', 'vicinal', and 'high-angle' interfaces

The above discussion illustrates the distinct effects due to the cores and elastic strain fields of interface dislocations and surface steps in the work of adhesion. These effects lead naturally to a distinction between three types of interfaces, with a rather different physical behavior as far as the work of adhesion is concerned. In analogy to the distinction between low- and high-angle GBs, we distinguish 'vicinal' interfaces, whose work of adhesion is governed by both the elastic and the core effects, from 'high-angle' interfaces, in which the core effects dominate completely. Because they are entirely free of dislocations or steps, the 'special' interfaces represent a class all by themselves; similar to the 'high-angle' interfaces, however, no elastic effects therefore contribute to their work of adhesion.

Since the distinction between special and vicinal interfaces was discussed earlier in this section, here we only consider further what we refer to as the 'high-angle' interfaces. Because in these interfaces the cores of the dislocations and of the corresponding surface steps overlap, their work of adhesion can be expected to be dominated by the core energies. Given that (i) the elastic interaction between line defects is mediated by the strained perfect-crystal-like regions between them, and (ii) virtually no such regions remain when the cores overlap, the interaction energies, $\Gamma_{\text{el}}^{\text{s-s}}(\delta)$ and $\Lambda_{\text{el}}^{\text{d-d}}(\delta)$, should vanish as $\delta \to h$ or $\delta \to b$, respectively (i.e. for $\Delta\psi \to 90°$; eq. (26.2) and Figs. 26.1 and 26.3). While the Read–Shockley equation (26.19) automatically satisfies this requirement, the Marchenko–Parshin expression (26.10) approaches a finite value of $G_{\text{el}}^{\text{s-s}}/h^2$. In our further analysis of high-angle interfaces, this unphysical prediction of eq. (26.10) will be corrected by eliminating this remaining, finite-energy elastic energy for $\delta \to h$ from the relevant expressions.

Ignoring thus all effects arising from the elastic

strain fields near the steps and dislocations, the expressions derived in sections 26.2.1–26.2.3 simplify considerably. For example, the free-surface energy in eqs. (26.11a,b) becomes

$$\gamma_{\text{core}} \equiv \gamma(\Delta\psi = 90°) = \Gamma_{\text{core}}^{\infty}/h \qquad (26.24)$$

while the GB energy in eqs. (26.20a,b) simplifies as follows:

$$\varepsilon_{\text{core}} \equiv \varepsilon(\Delta\psi = 90°) = 2\Lambda_{\text{core}}^{\infty}/b \qquad (26.25)$$

The ratio,

$$f_{\varepsilon} \equiv \varepsilon_{\text{core}}/2\gamma_{\text{core}} = (\Lambda_{\text{core}}^{\infty}/b)/(\Gamma_{\text{core}}^{\infty}/h) \qquad (26.26)$$

therefore provides a direct measure for the relative magnitudes of the **core** energies of isolated steps and dislocations. As illustrated in section 26.4, the simulation of free surfaces and STGBs far away from any of the energy cusps permits determination of f_{ε}. The above relations yield the following expression for the work of adhesion:

$$W_{\text{core}}^{\text{ad}} \equiv W^{\text{ad}}(\Delta\psi = 90°) = 2(\Gamma_{\text{core}}^{\infty}/h - \Lambda_{\text{core}}^{\infty}/b) \qquad (26.27)$$

Equations (26.24)–(26.27) have two noteworthy properties. First, a positive work of adhesion implies automatically that the core energy of the steps exceeds that of the dislocations. As discussed in the next section, this requirement finds a natural explanation within the framework of a broken-bond model. Second, the energies of the cusped surfaces and GBs are remarkably absent in all three, i.e. the energy and the ideal cleavage-fracture energy of a high-angle interface do not depend on the energies of any of the cusped orientations, and neither do the energies of the related free surfaces. Instead, these energies are governed solely by the core energies of the underlying line defects and by the magnitude of the geometrical discontinuity at the line defects (i.e. the step height and Burger's vector). The ratio,

$$\begin{aligned} f_{\text{ad}} &\equiv W_{\text{core}}^{\text{ad}}/2\gamma_{\text{core}} = 1 - (\Lambda_{\text{core}}^{\infty}/b)/(\Gamma_{\text{core}}^{\infty}/h) \\ &= 1 - f_{\varepsilon} \end{aligned} \qquad (26.28)$$

therefore provides another direct measure for the relative magnitudes of the core energies of steps and dislocations.

26.2.5 Role of broken bonds

Broken-bond models have been used for the calculation of surface energies with a great a deal of success for at least 50 years [6]. Such models naturally give rise to cusps in energy-orientation plots and lead to faceted crystal shapes [6]. Since broken-bond models do not take into account elastic-interaction effects, it appears that such effects are small for surfaces – an assumption that will be quantified in section 26.4.1. Broken-bond models can also be useful in predicting structure–property correlations in GBs [18]. While, based on the Read–Shockley model [8], elastic-interaction effects are expected to be important in the GB case, broken-bond models are nevertheless useful in predicting the energy contributions of the cores, since it is only the dislocation cores which contribute to miscoordination but not their elastic strain fields (for details see section 26.6).

The coordination coefficient, $C(\alpha, \hat{n})$, for the α-th nearest neighbors (*nn*), defined by [19]

$$C(\alpha, \hat{n}) = \sum_{n} |K_{\text{id}}(\alpha) - K_{\text{n}}(\alpha, \hat{n})|/A(\hat{n})$$
$$(\alpha = 1, 2, \dots) \qquad (26.29)$$

represents a convenient measure per unit area of how well, on average, an interface of orientation $\hat{n}$ is coordinated (or, perhaps better, miscoordinated). Here $A(\hat{n})$ is the planar unit-cell area (assuming, for convenience, that the interface is commensurate [7]); the dimensions of $C(\alpha, \hat{n})$ are therefore $(\text{length})^{-2}$, and its values are usually given in units of a^{-2}, where a is the lattice parameter. We define as α-th neighbors all those atoms n within a minimum radius $(R_{\alpha} + R_{\alpha-1})/2$ and a maximum radius of $(R_{\alpha+1} + R_{\alpha})/2$ of a given atom, i.e. all atoms within a radius half way between corresponding neighboring shells. R_{α} here denotes the radius of the α-th *nn* shell in a perfect crystal. For every atom we thus determine the deviation, $\Delta K_{\text{n}}(\alpha, \hat{n}) = K_{\text{id}}(\alpha) - K_{\text{n}}(\alpha, \hat{n})$, of its number of α-th nearest neighbors from that of a perfect fcc crystal, $K_{\text{id}}(\alpha)$, and subsequently summing over the absolute values. For $\alpha = 1$, all atoms between $R_0 = 0$ and the half-way point between nearest and 2nd-nearest neighbors are included [19].

To illustrate eq. (26.29) by an example, we consider unrelaxed free surfaces in monatomic structures for which the miscoordination in eq. (26.29) may be given analytically as a function of the orientation $\hat{n}$ as follows [6]:

$$C^{surf}(\alpha, \hat{n}) = (\rho/2) \sum_i |\hat{n} \cdot \boldsymbol{B}(\alpha)_i| \qquad (26.30)$$

where the $\boldsymbol{B}(\alpha)_i$ are the 'bond vectors' from an atom to its α-th neighbors, and ρ is the number of atoms per volume; the summation over i involves all the neighbors of type α. For example, for surfaces vicinal to (100) and normal to the [001] pole axis in the fcc structure, the first three mis-coordination coefficients are given by (in units of a^{-2})

$$C^{surf}(1, \Delta\psi) = 8 \cos \Delta\psi + 4 \sin \Delta\psi \qquad (26.\,31a)$$

$$C^{surf}(2, \Delta\psi) = 4 \cos \Delta\psi + 4 \sin \Delta\psi \qquad (26.\,31b)$$

$$C^{surf}(3, \Delta\psi) = 32 \cos \Delta\psi \qquad (26.\,31c)$$

assuming $\Delta\psi \leqslant 26.57°$ for which the (210) plane is reached.

The similarity of these expressions to eq. (26.7b) is rather striking. In fact, in complete analogy with eq. (26.7b), for vicinal surfaces eq. (26.30) may generally be rewritten in terms of the step model as follows:

$$C^{surf}(\alpha, \Delta\psi) = C_{cusp}^{surf}(\alpha) \cos \Delta\psi \\ + [C_{step}(\alpha)/h] \sin \Delta\psi \qquad (26.32)$$

where $C_{cusp}^{surf}(\alpha)$ is the α-th neighbor miscoordina-tion per unit area of the cusped surface (at $\Delta\psi = 0$) while $C_{step}(\alpha)$ is the α-th neighbor miscoordination per unit length of the steps on the surface.

To quantify the similarity between eqs. (26.32) and (26.7b), we assume that the elastic strain fields surrounding the steps indeed do not cause any broken bonds, an assumption to be validated in section 26.5.2. Both the cusped energy and the line-energy contribution due to the isolated steps in eq. (26.11b) may then be assumed to be prop-ortional to the corresponding miscoordination per unit area. The broken-bond contribution, γ_{b-b}, to

the total surface energy may then be written as follows:

$$\gamma_{b-b}(\Delta\psi) \equiv \sum_\alpha \beta(\alpha)C^{surf}(\alpha, \Delta\psi) \\ = \gamma_{cusp} \cos \Delta\psi + (\Gamma^\infty/h) \sin \Delta\psi \\ (26.33)$$

with (eq. (26.32))

$$\gamma_{cusp} \equiv \sum_\alpha \beta(\alpha)C_{cusp}^{surf}(\alpha) \qquad (26.34)$$

and

$$\Gamma^\infty/h = \sum_\alpha \beta(\alpha) \, C_{step}(\alpha)/h \qquad (26.35)$$

The proportionality constants, $\beta(\alpha)$, are obviously determined by the strengths of the bonds that were broken. In an unrelaxed surface interacting via a pair potential, $\beta(\alpha)$ is identical to the corre-sponding α-th *nn* perfect-crystal bond energy; by contrast, in a fully relaxed surface $\beta(\alpha)$ obviously represents an average over the slightly varying bond lengths in a given neighbor shell. In contrast to a perfect crystal, in a crystal containing lattice defects the magnitude of $\beta(\alpha)$ may therefore be expected to depend on the local environment of the defects.

Inserting eqs. (26.33) into eq. (26.11b), we obtain

$$\gamma(\Delta\psi) = \gamma_{b-b}(\Delta\psi) + (G_{el}^{s-s}/h^3) \sin^3 \Delta\psi \quad (26.36)$$

with (eqs. (26.33)–(26.35))

$$\gamma_{b-b}(\Delta\psi) = \cos \Delta\psi \sum_\alpha \beta(\alpha)C_{cusp}^{surf}(\alpha) \\ + \sin \Delta\psi \sum_\alpha \beta(\alpha)C_{step}(\alpha)/h \quad (26.37)$$

Similarly, for internal interfaces, if we assume that (i) the elastic strain fields surrounding the dislocations do not create any broken bonds (an assumption also to be validated in section 26.5.2) and (ii) that the line energy in eq. (26.21) is, indeed, dominated by the core term, in analogy to eq. (26.32) we may write:

$$C^{int}(\alpha, \Delta\psi) = C_{cusp}^{int}(\alpha) \cos \Delta\psi \\ + 2[C_{disl}(\alpha)/b] \sin \Delta\psi \qquad (26.38)$$

where, in analogy to the steps, $C_{\text{disl}}(\alpha)/b$ denotes the α-th *nn* miscoordination per unit length of the interface dislocations while $C_{\text{cusp}}^{\text{int}}(\alpha)$ is the coordination coefficient of the cusped interface. The broken-bond contribution, $\varepsilon_{\text{b-b}}$, to the total interface energy may then be defined as follows:

$$\varepsilon_{\text{b-b}}(\Delta\psi) \equiv \sum_{\alpha} \beta(\alpha) C^{\text{int}}(\alpha, \Delta\psi)$$
$$= \varepsilon_{\text{cusp}} \cos \Delta\psi + 2(\Lambda^{\infty}/b) \sin \Delta\psi \tag{26.39}$$

with (eq. (26.20b))

$$\varepsilon_{\text{cusp}} \equiv \sum_{\alpha} \beta(\alpha) C_{\text{cusp}}^{\text{int}}(\alpha) \tag{26.40}$$

$$(\Lambda^{\infty}/b) = \sum_{\alpha} \beta(\alpha) C_{\text{disl}}(\alpha)/b \tag{26.41}$$

Inserting these expressions into eq. (26.20b), in analogy to eq. (26.36) we obtain

$$\varepsilon(\Delta\psi) = \varepsilon_{\text{b-b}}(\Delta\psi) - 2(L_{\text{el}}^{\text{d-d}}/b) \sin \Delta\psi \ln(\sin \Delta\psi) \tag{26.42}$$

with the broken-bond contribution, $\varepsilon_{\text{b-b}}(\Delta\psi)$, given by (compare eq. (26.37))

$$\varepsilon_{\text{b-b}}(\Delta\psi) = \cos \Delta\psi \sum_{\alpha} \beta(\alpha) C_{\text{cusp}}^{\text{int}}(\alpha)$$
$$+ 2 \sin \Delta\psi \sum_{\alpha} \beta(\alpha) C_{\text{disl}}(\alpha)/b \tag{26.43}$$

Equations (26.36) and (26.42) may be inserted into eq. (26.1) to express W^{ad} in terms of its broken-bond (i.e. core) and elastic strain-field contributions, according to (eq. (26.22b))

$$W^{\text{ad}}(\Delta\psi) = W_{\text{b-b}}^{\text{ad}}(\Delta\psi) + 2 \sin \Delta\psi \, [(G_{\text{el}}^{\text{s-s}}/h^3)$$
$$\sin^2 \Delta\psi + (L_{\text{el}}^{\text{d-d}}/b) \ln(\sin \Delta\psi)] \tag{26.44}$$

Here the broken-bond work of adhesion is given by

$$W_{\text{b-b}}^{\text{ad}}(\Delta\psi) = 2\gamma_{\text{b-b}}(\Delta\psi) - \varepsilon_{\text{b-b}}(\Delta\psi)$$
$$= \sum_{\alpha} \beta(\alpha)[2C^{\text{surf}}(\alpha, \Delta\psi) - C^{\text{int}}(\alpha, \Delta\psi)] \tag{26.45a}$$

or, using eqs. (26.40) and (26.41),

$$W_{\text{b-b}}^{\text{ad}}(\Delta\psi) = \sum_{\alpha} \beta(\alpha)\{[2C_{\text{cusp}}^{\text{surf}}(\alpha) - C_{\text{cusp}}^{\text{int}}(\alpha)]$$
$$+ 2[C_{\text{step}}(\alpha)/h - C_{\text{disl}}(\alpha)/b]\} \tag{26.45b}$$

As noted in the preceding section, the energies of the 'special' and the 'high-angle' interfaces are, by definition, independent of elastic interaction effects. Their work of adhesion is therefore governed completely by the broken-bond contribution in eq. (26.44). For the 'special' interfaces we thus obtain (eq. (26.23))

$$W^{\text{ad}}(\Delta\psi = 0) \equiv W_{\text{b-b}}^{\text{ad}}(\Delta\psi = 0) = (2\gamma_{\text{cusp}} - \varepsilon_{\text{cusp}})$$
$$= \sum_{\alpha} \beta(\alpha)[2C_{\text{cusp}}^{\text{surf}}(\alpha) - C_{\text{cusp}}^{\text{int}}(\alpha)] \tag{26.46}$$

while for the 'high-angle' interfaces eq. (26.45b) yields (eq. (26.27))

$$W^{\text{ad}}(\Delta\psi = 90°) = 2 \sum_{\alpha} \beta(\alpha)[C_{\text{step}}(\alpha)/h$$
$$- C_{\text{disl}}(\alpha)/b] \tag{26.47}$$

These expressions imply that a positive work of adhesion necessitates the dislocation cores in the internal interface to be better coordinated than the cores of the steps in the corresponding two fracture surfaces, i.e. $C_{\text{step}}(\alpha)/h > C_{\text{disl}}(\alpha)/b$. That this is, indeed, the case will be shown in section 26.5 in which the miscoordination per unit length of steps and dislocations will be determined by means of computer simulations.

26.3 COMPUTER SIMULATIONS

As already mentioned, two conceptually different types of interatomic potentials will be used throughout in order to provide some insight into the role of many-body effects. For the fcc metals, results obtained via a semi-empirical embedded-atom-method (EAM) potential fitted to represent Au [12] will be compared with simulations involving the well-known Lennard–Jones (LJ) potential, with only two adjustable parameters, σ and ε, defining the length and energy scales, respectively. Although the LJ potential was fitted

to the lattice parameter and melting point of Cu
(with $\varepsilon = 0.167$ eV and $\sigma = 2.315$ Å), the relative
energies of different interfaces are the same for
any LJ system if all energies and distances are
expressed in units of ε and σ, respectively.

To enable a most direct comparison of free
surfaces with GBs, only energies of **symmetrical**
tilt boundaries (STGBs) and random grain boundaries (RGBs) will be discussed here. The two
fracture surfaces are then identical and, as for the
free surfaces, both of these types of GBs are then
characterized by only the two DOFs associated
with the GB plane, thus permitting a direct investigation of how one particular type of GB edge
dislocation is transformed into surface steps.

26.3.1 Energies of free surfaces and GBs

The energies of free surfaces, STGBs and RGBs
with a common $\langle 110 \rangle$ and $\langle 100 \rangle$ pole (or tilt)
axis are plotted, respectively, in Figs. 26.4 and 26.5
against the pole (or tilt) angle, $\psi = 2\Delta\psi$ (Fig. 26.1).
While these figures show only the EAM results
obtained for Au, the LJ potential yields qualitatively identical results.

According to these figures, the energy of the
RGB configuration on a given plane is always
lower than that of the related two free surfaces.
Considering that some of the bonds broken during
bulk cleavage fracture are recovered when the
two free surfaces are brought back into contact,
irrespective of their relative orientation, this result
is not surprising. Also, in all instances investigated
the STGB configuration on a given plane has a
much smaller energy than the RGB on the same
plane, with a consequently smaller volume expansion. This difference arises from the local
interlocking of the lattice planes in the STGBs,
which is possible because of (i) their additional two
translational DOFs, (T_x, T_y), parallel to the GB,
and (ii) their very small planar unit cells (with only
one atom per (hkl) plane, as in the perfect crystal
[20]). We note that, because the STGB configuration on the (100) and (110) plane is identical to
the perfect crystal [20], the corresponding STGB
energies vanish. Also, although finite, the energy
of the STGB on the (111) plane (the so-called (111)

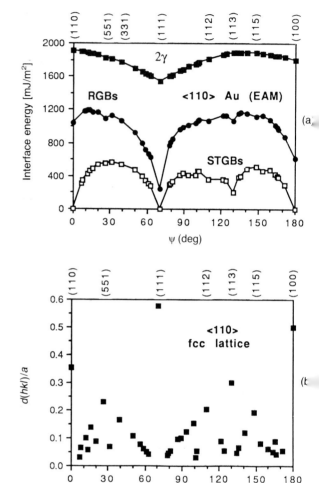

Fig. 26.4 (a) Energies (in units of mJ/m²) of fully relaxed free
surfaces (2γ), STGBs and RGBs on planes perpendicular to the
$\langle 110 \rangle$ pole axes for the Au(EAM) potential. $\psi = 2\Delta\psi$ is the tilt
rotation angle about $\langle 110 \rangle$ (see also Fig. 26.1). The lines
connecting the data points are merely a guide to the eye. (b)
Related interplanar lattice spacings, $d(hkl)$, in units of the lattice
parameter a.

twin boundary), is very small for both potentials
(Table 26.1). A realistic comparison of the relative
energies of STGBs, RGBs and free surfaces should
therefore involve the less-dense planes.

We should mention that the Au(EAM) energies
are substantially lower than the measured **average**
surface energy of about 1500 mJ/m² of gold [21].
Similarly large (~60–80%) discrepancies between

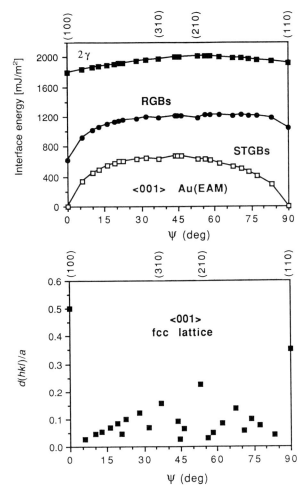

Fig. 26.5 Same as Fig. 26.4, but for interface planes perpendicular to ⟨001⟩.

computed and measured surface energies are also found for other fcc metals [12]. As already mentioned, we nevertheless hope that a comparison of relative energies of surfaces and GBs which uses these potentials is meaningful.

A comparison of the energies in Figs. 26.4(a) and 26.5(a) with the interplanar spacings, $d(hkl)$, shown in Figs. 26.4(b) and 26.5(b) (see also Table 26.2) demonstrates that the appearance of cusps in the energies of free surfaces, RGBs and STGBs is closely connected with relatively large values of the interplanar spacings $d(hkl)$. In contrast with the free surfaces, however, which show only the three cusps associated with the three densest planes, the GBs show minor cusps all the way up to about the 10th- or 11th-densest fcc plane. (Notice, however, that in the case of the STGBs, the (100) and (110) interplanar spacings are irrelevant because the STGBs on these planes are identical to the perfect crystal ([20]).) As a consequence, the energies of the free surfaces vary much more smoothly as a function of the pole angle ψ. This difference between GBs and free surfaces is thought to be due to the absence of the three translational DOFs in the latter. In STGBs, by contrast, the possible rigid-body translations in **T** enable a much more effective minimization of the GB energy, particularly for GBs with the smallest planar unit cells, i.e. those on planes with the largest $d(hkl)$ values, which consequently give rise to energy cusps. With only one translational DOF (that associated with volume expansion at the GB), RGBs fall somewhere

Table 26.1 Comparison of the surface energies, γ, and nearest-neighbor coordination coefficients, $C(1)/a^2$ (see eq. (26.29)) obtained for the two potentials for surface normals in the four principal cubic directions [19]. For comparison, the corresponding energies and coordination coefficients of STGBs [23] and RGBs [9] are also listed. All energies are in units of mJ/m² (=erg/cm²)

	(hkl)	γ	$C(1)/a^2$	ε^{RGB}	$C(1)/a^2$	ε^{STGB}	$C(1)/a^2$
Cu(LJ)	(111)	839	6.93	432	3.17	2	0
Au(EAM)	(111)	767	6.93	249	1.80	2	0
Cu(LJ)	(100)	892	8.00	810	7.90	0	0
Au(EAM)	(100)	897	8.00	616	6.30	0	0
Cu(LJ)	(110)	957	8.48	1329	12.76	0	0
Au(EAM)	(110)	957	8.48	1041	11.43	0	0
Cu(LJ)	(113)	961	8.44	1431	13.73	293	2.41
Au(EAM)	(113)	944	8.44	1069	9.46	211	2.41

Table 26.2 Interplanar spacing, $d(hkl)$ (in units of the lattice parameter a), for the 11 most widely spaced planes in the fcc lattice. These planes also correspond to the ones with the highest planar density of atoms, i.e. the smallest planar repeat unit cells

No.	(hkl)	$d(hkl)/a$
1	(111)	0.5774
2	(100)	0.5000
3	(110)	0.3535
4	(113)	0.3015
5	(331)	0.2294
6	(210)	0.2236
7	(112)	0.2041
8	(115)	0.1925
9	(513)	0.1690
10	(221)	0.1667
11	(310)	0.1581

Table 26.3 Core energies of steps and edge dislocations in STGBs and RGBs in fcc metals extracted from the simulation results in the 'high-angle' limit via eqs. (26.26) or (26.27); see also Figs. 26.6(a) and (b) and Fig. 26.8

From	STGBs $(\Lambda^{\infty}_{core}/b)/(\Gamma^{\infty}_{core}/h)$		RGBs $(\Lambda^{\infty}_{core}/b)/(\Gamma^{\infty}_{core}/h)$	
	$\varepsilon_{core}/2\gamma_{core}$	$W^{rad}_{core}/2\gamma_{core}$	$\varepsilon_{core}/2\gamma_{core}$	$W^{rad}_{core}/2\gamma_{core}$
Cu(LJ)	0.48	0.48	0.85	0.86
Au(EAM)	0.33	0.33	0.60	0.62

between STGBs and free surfaces as far as their sensitivity towards rigid-body translations is concerned, with a consequently more smoothly varying energy plot than that of the STGBs. (For a quantitative analysis of the role of the interplanar spacing in the above energies, see section 26.3.3.)

That the energies in Figs. 26.4(a) and 26.5(a) are intimately connected with the number of nearest-neighbor bonds per unit area broken upon creation of the free surface or GB, $C(1)$ (see eq. (26.29)), is demonstrated in Table 26.1. It is interesting to note that in the case of the free surfaces, the coordination coefficients extracted from the simulations were the same for both the unrelaxed and fully relaxed structures and for both potentials. As discussed further in section 26.5, this suggests that the elastic strain fields surrounding the surface steps do not cause any broken bonds in the sense defined above; instead, similar to the case of dislocations, all broken bonds arise from the highly disordered **cores** of the steps. We note, however, that by choosing the primitive planar unit cell of the free surface, reconstruction was systematically discouraged in our simulations. If reconstruction were to take place, one would in principle expect differences in the average atom coordination between the unrelaxed and relaxed structures.

As discussed in section 26.2.4, for the 'high-angle' free surfaces and GBs, i.e. those well outside

of any cusps in which the cores overlap completely, one would expect a linear relationship between the free-surface and GB energies [see eqs. (26.24)–(24.26)]. To investigate this relationship, in Figs. 26.6(a) and (b) the energies of STGBs and RGBs are plotted against 2γ. As expected, the energies associated with vicinal orientations scatter widely but, as indicated by the dashed lines, are systematically related to the energies of the densest (cusped) planes. As shown in the inserts, however, for all the remaining ('high-angle') interfaces a reasonably good correlation exists between the energies of STGBs and RGBs, on the one hand, and the energy of free surfaces, on the other. According to eq. (26.26), the slopes of the solid lines through the origin give the ratio, $f_\varepsilon \equiv \varepsilon_{core}/2\gamma_{core}$, of the core energies of steps and dislocations. According to the values listed in Table 26.3, this ratio is less than one in all cases, consistent with the positive work of adhesion of the internal interfaces (see below). As discussed further in section 26.4, the origin of the qualitatively different behavior of the 'special' and 'vicinal' interfaces in Fig. 26.6 compared to the 'high-angle' interfaces lies in the fundamental difference between the short-ranged elastic strain fields near the steps, which are contrasted by the long-ranged strain fields near the dislocations.

26.3.2 Cleavage-fracture energies

The above simulation data for free surfaces and GBs may be combined to determine the work of adhesion for the internal interfaces. The plots thus

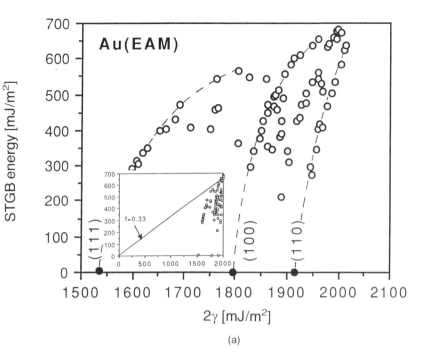

(a)

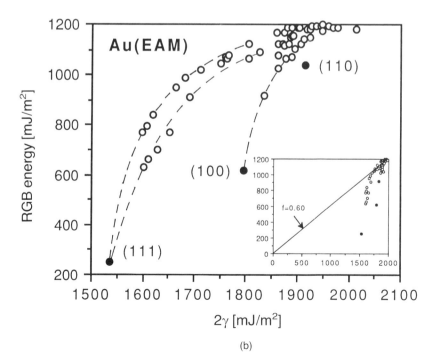

(b)

Fig. 26.6 Energies of (a) STGBs and (b) RGBs plotted against the bulk-cleavage energy, 2γ (in mJ/m^2) in the fcc lattice. Dashed lines correlate 'vicinal' orientations with their corresponding 'cusped' orientations (solid symbols). The two sets of (111) vicinals correspond to tilting toward (100) and (110) surfaces respectively. The solid lines in the inserts show a fit to the energies of the 'high-angle' interfaces (i.e. those far from any of the cusps) through the origin (eqs. (26.24)–(26.26)). The slopes, f_c, obtained for the two potentials are listed in Table 26.3.

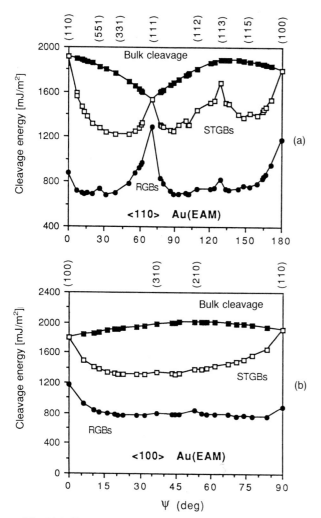

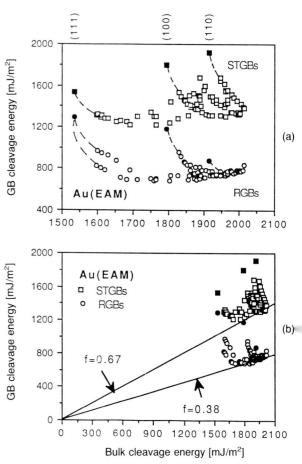

Fig. 26.7 Cleavage energies for the STGBs in Figs. 26.4(a) and 26.5(a). For comparison, the bulk ideal-crystal cleavage energies ($= 2\gamma$) are also shown.

Fig. 26.8 Plots analogous to Figs. 26.6(a) and (b) showing (a) the cleavage energies of the STGBs and the RGBs plotted against the bulk-cleavage energy, 2γ (in mJ/m^2). Solid symbols mark the cusped orientations and the dashed lines connect corresponding vicinal orientations. The two sets of (111) vicinals correspond to tilting toward (100) and (110) surfaces respectively (Fig. 26.7(a)). The solid lines in the blow-up in (b) show a fit to the cleavage energies of the 'high-angle' interfaces through the origin. According to eq. (26.28), their work of adhesion is proportional to the energy of the related two fracture surfaces. For both potentials, values of $f_c = 1 - f_{ad}$ as obtained from the slopes, f_{ad}, are also listed in Table 26.3.

obtained from Figs. 26.4(a) and 26.5(a) are shown in Figs. 26.7(a) and (b), respectively. A comparison of the STGB and RGB data shows qualitatively the same behavior for both types of GBs. This similarity is particularly remarkable in the vicinity of the (100) and (110) planes for which, in both lattices, the STGB energy vanishes identically while the RGB energy remains finite. From these figures, the correlation between a large value of $d(hkl)$ and a **large** work of adhesion is rather apparent, as is the correlation between large $d(hkl)$ values and **small** values of 2γ.

This correlation between the **cusps** in the bulk ideal-crystal cleavage energy, $E^{cl} = 2\gamma$, and the **peaks** in the GB work of adhesion, W^{ad}, is particularly interesting. Intuitively one would expect the opposite behavior: if the corresponding free-surface energy is particularly small one would think that it should be easier to separate a particular bicrystal. However, a comparison of the corresponding interface energies in Figs. 26.4(a) and 26.5(a) shows that while 2γ may be small for a given cusp orientation, the corresponding GB energy is significantly smaller still. Since the energetics of the interfaces at cusped orientations are dominated by miscoordination (section 26.2.5), the origin of this behavior is largely crystallographic (section 26.5.2). Furthermore, the elastic-interaction energy and the high miscoordination of the dislocations in GBs results in a more rapid increase in the GB energies, as compared to free-surface energies. This leads to relatively higher GB energies, and correspondingly smaller works of adhesion, for the high-angle orientations.

To enable a more quantitative analysis of the relationship between the bulk ideal-crystal cleavage energy, E^{cl}, and the GB work of adhesions, W^{ad}, Fig. 26.8 shows the two plotted against one another. These figures suggest the distinction of two types of GBs as far as their work of adhesion is concerned. First, for the 'special' GBs and their vicinals, W^{ad} **decreases** with increasing E^{cl}. This behavior is due to the elastic energy of the vicinals. Second, for all remaining GBs (i.e. those that are neither 'special' nor 'vicinal'), W^{ad} **increases** with increasing E^{cl}, as seen from Figs. 26.8(a) and (b).

26.3.3 Role of the interplanar lattice spacing: vicinal versus special interfaces

The simulation results presented in the preceding suggest a special role of the densest planes in the energies of free surfaces, STGBs and RGBs, and therefore in the work of adhesion. Since both the STGBs and the symmetrical RGBs are fully characterized macroscopically by only the two DOFs associated with the GB normal, a comparison of their energies with those of free surfaces enables a systematic investigation of the role of the interface **plane** in both their energy and work of adhesion.

To analyze the role of the interface plane quantitatively, in Figs. 26.9(a) and (b) the energies of the free surfaces and GBs are plotted against

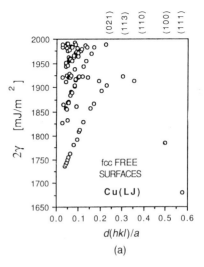

(a)

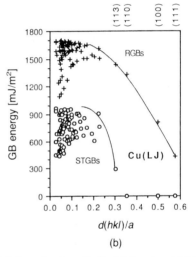

(b)

Fig. 26.9 Energies (in mJ/m²) of fully relaxed (a) free surfaces (2γ) and (b) STGBs and RGBs plotted against $d(hkl)/a$. The densest surface planes (Table 26.2) are indicated on the top.

$d(hkl)$. These plots demonstrate that the criterion that a large interplanar spacing gives rise to a particularly low GB energy is valid only for the few densest lattice planes, i.e. for the 'special' interfaces giving rise to the cusps. For the smaller values of $d(hkl)$ the simulation data scatter widely, indicating that the GB energy is not governed by the interplanar spacing exclusively.

That the energies of the **vicinal** interfaces *cannot* be governed by $d(hkl)$ becomes obvious if we consider, for example, the free surfaces and GBs in the vicinity of the cusps in Figs. 26.4(a) and 26.5(a). Their energies are obviously governed by the energy and interplanar spacing of the free surface or GB at the bottom of the related cusp. As a cusp is approached, the corresponding value of $d(hkl)$ approaches zero (Figs 26.4(b) and 26.5(b)). Then, at the cusp, the interplanar spacing suddenly jumps to a finite and large value associated with the cusp. For example, the (41, 41, 42) plane is very close to the (111) plane, however with a practically vanishing value of $d(hkl)$ (because the latter is proportional to $(h^2 + k^2 + l^2)^{-\frac{1}{2}}$). The energies of the free surface, the STGB and the RGB on the (41, 41, 42) plane of the fcc lattice are therefore practically the same as those on the (111) plane. Consequently, while the energy decreases smoothly towards that of the cusped orientation, $d(hkl)$ does not vary steadily as a function of ψ, which is the reason for the absence of a direct relation between the two for the vicinal but not the cusped interface-plane orientations.

A detailed analysis of the distribution of all the data points in Figs. 26.9(a) and (b) confirms that for every plane with a relatively high $d(hkl)$ value, there is a point at $d(hkl) = 0$ (associated with some infinitesimally close vicinal interface) with identical energy. All other surfaces associated with the same cusp then fall on a smooth line emanating from this point as $d(hkl)$ increases slowly from zero.

As is well known, the distinction between vicinal and special interfaces is very useful for describing the underlying atomic structure of surfaces and GBs. For the case of the STGBs, Weins *et al.* [24] demonstrated 20 years ago that the structure of those STGBs in the vicinity of a major cusp may be decomposed into the polyhedral units of, and the angular deviation from, the 'special' STGB at the bottom of that cusp. (These structural units correspond to the dislocation cores in the Read–Shockley model considered in section 26.2.2.) By comparison, for the case of free surfaces it has been known for well over half a century that the structure and energy of 'vicinal' surface orientations is governed by the number of and distance between the steps introduced into the nearby 'flat'

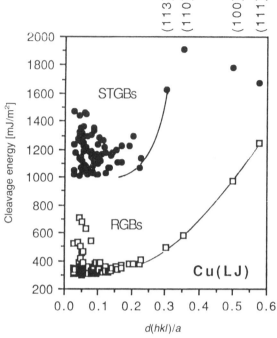

Fig. 26.10 Work of adhesion (in mJ/m²) of fully relaxed STGBs and RGBs in fcc metals against $d(hkl)/a$ (see also Table 26.4).

(i.e. step-free) principal surface (section 26.2.1) [6].

Finally, by subtracting corresponding data points in Figs. 26.9(a) and (b), the role of $d(hkl)$ in the work of adhesion is readily analyzed (Fig. 26.10). As expected, the distinction between **vicinal** and **special** GB-plane orientations is the same in the cleavage energies as that discussed above for the underlying GB and free-surface energies. It is interesting to note, however, that the particularly **low** energy of both the free surfaces and GBs on the densest planes is accompanied by a particularly **high** work of adhesion. This counter-intuitive behavior of the special interface planes is due to the very low degree of miscoordination at the special GB interfaces (section 26.5). While therefore the interplanar spacing of the lattice planes parallel to the interface should be of importance only for the cusped orientations, it is obviously irrelevant for vicinal ones.

In summary, it appears that in assessing the rol

Table 26.4 Work of adhesion (see eq. (26.1)) for STGBs and RGBs for the four densest planes of the fcc lattice obtained from Table 26.1 (in units of mJ/m^2). For comparison, the bulk ideal-crystal cleavage-fracture energy ($=2\gamma$) is also listed

(hkl)	Cu(LJ)			Au(EAM)		
	STGBs	RGBs	bulk (2γ)	STGBs	RGBs	bulk (2γ)
(111)	1676	1246	1678	1532	1287	1534
(100)	1784	974	1784	1794	1178	1794
(110)	1914	585	1914	1914	873	1914
(113)	1629	491	1922	1677	819	1888

of the GB plane in the energies of free surfaces, STGBs and RGBs, and therefore in the work of adhesion, one should distinguish between 'special' and 'vicinal' interface planes. While the energies of the **special** interfaces are governed by the interplanar spacing of the lattice planes parallel to the interface plane, the value of $d(hkl)$ is irrelevant for the **vicinal** ones. Plots such as Figs. 26.9 and 26.10 provide a quick and simple method for separating the two types of interfaces for a given crystal structure and interatomic potential used in the simulations.

26.4 CORE AND ELASTIC STRAIN-FIELD EFFECTS IN FREE SURFACES AND GBs

We are now ready to test some of the assumptions made in sections 26.2.1 and 26.2.2 concerning the core energies of and elastic interactions between steps, on the one hand, and dislocations, on the other. For that purpose, the energies of free surfaces and STGBs perpendicular to $\langle 100 \rangle$ determined for the Au(EAM) potential (Fig. 26.5(a)) will be analyzed in some detail.

26.4.1 Surface steps

We first consider the line energies of surface steps. According to Fig. 26.5(a), two cusped orientations, associated with the (100) and (110) surfaces, are encountered when rotating about a $\langle 001 \rangle$ pole axis. For a detailed investigation of the line energies of the corresponding steps, a number of surfaces in the close vicinity of these cusps were simulated in addition to those shown in Fig. 26.5(a). For reasons to become evident below, these most vicinal surfaces are needed for a reliable determination of the line energies of isolated steps, i.e. in the limit for $\delta \rightarrow \infty$. According to Table 26.5, the largest separation considered for the (100) steps is $\delta = 21.54a$, by comparison with $\delta = 15.95a$ for the (110) step. By investigating step separations down to about $1.12a$ (Table 26.5), we hope to gain insight into the nature of the elastic interaction between the steps as a function of their distance.

The relaxed and unrelaxed energies of all surfaces perpendicular to $\langle 001 \rangle$ considered here are shown in Fig. 26.11. We note that both sets of energies are plotted against $\Delta\psi_1$, the angular

Table 26.5 Geometrical parameters associated with the steps considered in detail in this section. n is the step height, while a denotes the lattice parameter.

n	Pole axis	Plane	h/a	$\Delta\psi_n$ [deg]	Vicinal plane	δ/a	Vicinal plane	δ/a
1	$\langle 100 \rangle$	(100)	0.5000	0.00–26.56	(0 4 3 1)	21.54	(0 2 1)	1.12
2	$\langle 100 \rangle$	(100)	0.3535	0.00–18.44	(0 23 22)	15.95	(0 2 1)	1.12

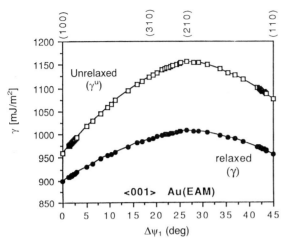

Fig. 26.11 Unrelaxed (γ^u) and relaxed (γ) energies of free surfaces of Au perpendicular to $\langle 001 \rangle$ against $\Delta\psi_1$, with $\Delta\psi_2 = 45° - \Delta\psi_1$. Notice that by comparison to Fig. 26.5(a), considerably more vicinal surfaces close to the (100) and (110) planes are considered here (Table 26.5). Notice that $\Delta\psi_1 = \psi/2$.

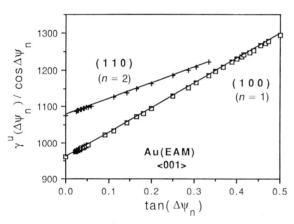

Fig. 26.12 Unrelaxed energies of Fig. 26.11, $\gamma^u/\cos \Delta\psi$ (in mJ/m²), plotted against $\tan \Delta\psi$ (eq. (26.48)) for the (100) and (110) vicinals perpendicular to $\langle 001 \rangle$. The line energies of the unrelaxed steps extracted from the slopes of the solid lines are listed in Table 26.6.

deviation from the (100) cusp. The results associated with the (110) cusp should actually be plotted against, and will be interpreted in terms of the angle $\Delta\psi_2 = 45° - \Delta\psi_1$.

To determine the line energy of the isolated steps, it is useful to first consider the unrelaxed surface energy, γ^u. Without relaxation, elastic

strain fields cannot develop near the steps; γ^u is therefore governed completely by core effects (i.e. broken bonds), according to (eqs. (26.11b) and (26.12)):

$$\gamma^u(\Delta\psi) = \gamma^u_{\text{cusp}} \cos \Delta\psi + (\Gamma^u_{\text{core}}/h) \sin \Delta\psi$$
(26.48)

where γ^u_{cusp} denotes the unrelaxed energy of the cusped surface, while Γ^u_{core} is the unrelaxed core energy per unit length of the steps. To test the validity of eq. (26.48), in Fig. 26.12 the values of $\gamma^u(\Delta\psi)/\cos \Delta\psi$ obtained from Fig. 26.11 are plotted against $\tan \Delta\psi$. The extremely good linearity exhibited in Fig. 26.12 is proof that, down to the smallest values of δ, the unrelaxed energy is determined by a δ-independent core energy of the steps and, hence, by the number of broken bonds per unit surface area (eqs. (26.32) and (26.37)). The unrelaxed core energies obtained from the related slopes are listed in Table 26.6.

Next, in order to focus entirely on the elastic contribution to the line energy, it is useful to define the (positive) relaxation energy (eqs. (26.11b) and (26.48)),

$$\Delta\gamma(\Delta\psi) \equiv \gamma^u(\Delta\psi) - \gamma(\Delta\psi)$$
$$= (\gamma^u_{\text{cusp}} - \gamma_{\text{cusp}}) \cos \Delta\psi$$
$$+ [(\Gamma^u_{\text{core}} - \Gamma^\infty)/h] \sin \Delta\psi$$
$$- (G^{s-s}_{\text{el}}/h^3) \sin^3 \Delta\psi$$
(26.49)

where, according to eq. (26.12), the relaxed line energy of the isolated steps, Γ^∞, contains contributions from both the relaxed cores and elastic strain fields; therefore

$$\Gamma^u_{\text{core}} - \Gamma^\infty = \Gamma^u_{\text{core}} - \Gamma^\infty_{\text{core}} - \Gamma^\infty_{\text{el}}$$
(26.50)

Using the data in Fig. 26.11, the relaxation energies shown in Fig. 26.13 are readily obtained. In this representation of the simulation data, the discontinuity in the slope at the (210) plane delimiting the two cusps is particularly noticeable (Table 26.5).

To extract the step-step interaction energy from the data in Fig. 26.13 by means of eq. (26.49), we take advantage of the fact that for a large separation between the steps (i.e. for the smallest values of $\Delta\psi$), their interaction energy in the third term on the right-hand side of eq. (26.49) should be

Table 26.6 Step parameters determined for the Au (EAM) potential (in units of mJ/m^2)

Pole axis	Cusped plane	Height (h/a)	γ^u_{cusp}	γ_{cusp}	Γ^u_{core}/h	$(\Gamma^u_{\text{core}} - \Gamma^\infty)/h$	Γ^∞/h	$(G^{s-s}_{\text{el}}/h^3)$	(G^{s-s}_{n-1}/h^3)
$\langle 100 \rangle$	(100)	0.5	960.6	898.6	669.5 ± 0.5	265.4 ± 0.5	404.1 ± 1.0	570 ± 60	280 ± 10
$\langle 100 \rangle$	(100)	0.3535	1077.0	958.5	431.4 ± 0.5	144.7 ± 0.5	286.7 ± 1.0	540 ± 50	280 ± 10

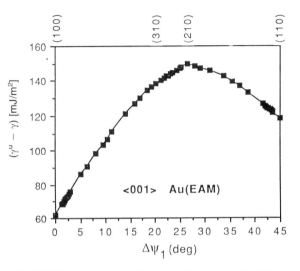

Fig. 26.13 Relaxation energies, $\Delta\gamma = \gamma^u - \gamma$ (eq. (26.49)), for the free surfaces of Fig. 26.11. The discontinuity in the slope at the (210) orientation, delimiting the two cusps at $\Delta\psi_1 = 0°$ and 45°, respectively, is clearly visible.

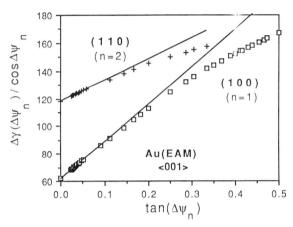

Fig. 26.14 Relaxation energies of Fig. 26.13 plotted against $\tan \Delta\psi$ (eq. (26.49)). From the slopes of the straight-line fits to the **low-angle** data (solid lines), the line energies of the isolated, non-interacting steps can be extracted; these are listed in Table 26.6.

negligibly small compared to the line-energy contribution in the second term (eqs. (26.10) and (26.15)). Similar to Fig. 26.12 we therefore plot $\Delta\gamma(\Delta\psi)/\cos \Delta\psi$ against $\tan \Delta\psi$ (Fig. 26.14); the slope of a straight-line fit to the low-angle data in Fig. 26.14 (indicated in the figure), should therefore yield the line energies of the isolated, non-interacting steps. According to Fig. 26.14, for values of $\tan \Delta\psi$ less than about 0.05, these plots are, indeed, very well represented by straight lines. Their slopes ($\Gamma^u_{\text{core}} - \Gamma^\infty)/h$, obtained from a linear fit through the origin for only the few most vicinal surfaces are listed in Table 26.6. The convergence of these fits was tested by including variable numbers of the closest few vicinals in the fit, with no discernible change in the slope, confirming that the largest separations between

steps considered here (Table 26.5) are, indeed, large enough to enable a reliable determination of the line energies of non-interacting steps.

Finally, if we attribute any deviations from the straight lines in Fig. 26.14 to the elastic interaction between the steps, i.e. if we plot the difference (eq. (26.49))

$$
\begin{aligned}
\gamma^{s-s}_{\text{el}}(\Delta\psi)/\cos\Delta\psi \\
\equiv -\Delta\gamma(\Delta\psi)/\cos \Delta\psi + (\gamma^u_{\text{cusp}} - \gamma_{\text{cusp}}) \\
+ [(\Gamma^u_{\text{core}} - \Gamma^\infty)/h] \tan \Delta\psi \\
\overset{?}{=} (G^{s-s}_{\text{el}}/h^3) \sin^2 \Delta\psi \tan \Delta\psi
\end{aligned}
\tag{26.51}
$$

against $\sin^2 \Delta\psi \tan \Delta\psi$, we expect a straight line with a slope of G^{s-s}_{el}/h^3 if the interaction is, indeed, proportional to $1/\delta^2$ (eqs. (26.10) and (26.15)) [16]. According to Figs. 26.15(a) and 26.16(a), a reasonably good linear behavior is, indeed, obtained for the surfaces with the largest separations between the steps (typically $\delta \leqslant 10a$ for the (100) vicinals, and $\delta \leqslant 7a$ for the (110) vicinals). The

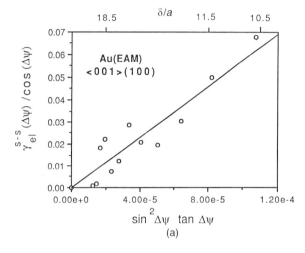

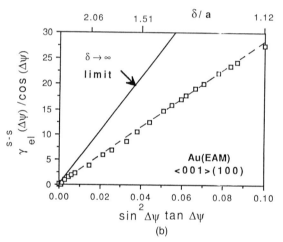

Fig. 26.15 Step–step-interaction contribution to the surface energy, $\gamma_{el}^{s-s}(\Delta\psi)/\cos\Delta\psi$ (eq. (26.51)) for (a) the largest and (b) the smallest values of δ (indicated on the top) for the (100) vicinals plotted against $\sin^2\Delta\psi\,\tan\Delta\psi$ (in mJ/m^2). The slope of the solid line in (a), G_{el}^{s-s}/h^3, obtained from a least-squares fit through the origin (eq. (26.51)), is listed in Table 26.6 together with the slope of the dashed line, G_{n-l}^{s-s}/h^3, which represents a least-squares fit to the small-δ data alone.

slopes of the solid lines, obtained from a least-squares fit through the origin, are listed in Table 26.6. This $1/\delta^2$ variation of the step–step inter-action energy in Au is in good agreement with recent simulations for Si [25] in which the same behavior was found for both steps on a flat surface and vicinal surfaces.

Not too surprisingly, for smaller step separations significant deviations from this linear behavior are apparent: As illustrated in Figs. 26.15(b) and 26.16(b), the step–step repulsion does not increase as rapidly with decreasing δ (indicated in the tops of these figures) as predicted by the Marchenko–Parshin formula (eqs. (26.10), (26.15), and (26.51)). That the step–step repulsion *cannot* increase indefinitely is intuitively obvious because the perfect-crystal-like regions mediating this elastic interaction are virtually eliminated when the step cores start to overlap. Another reason for the failure of eq. (26.49) for small values of δ can be expected to arise from a δ-dependence of the relaxed core energy, Γ_{core}, as the cores start to interact.

Because no theoretical predictions are available for the required modification, at small separations, of either the elastic step–step interaction or the core energy, at present the two effects causing deviations from eq. (26.11(b)) cannot be separated. It appears, however, that Figs. 26.15(b) and 26.16(b) provide a clue. As is evident from these figures, even for the smallest separations between the steps their interaction energy varies linearly with $1/\delta^2$ (see the dashed lines in the figures). By comparison with the large-δ Marchenko–Parshin limit, however (derived from isotropic linear continuum-elasticity theory; see the solid lines in these figures), the slopes of the dashed lines labeled (G_{n-l}^{s-s}/h^3), are smaller by about a factor of two (Table 26.6).

Two interpretations of the observed small-δ behavior appear possible. First, linear continuum elasticity theory used to derive the Marchenko–Parshin expression in eq. (26.10) obviously break down when the steps get too close. Although a non-linear-elastic extension of eq. (26.8) is not available, the above results would suggest that the non-linear effects give rise to a reduced prefactor G_{n-l}^{s-s}, without affecting the basic functional form of the Marchenko–Parshin formula.

Second, as the steps get rather close, one could envision a modification of the core energies of the steps, i.e. a non-elastic core–core interaction. Such an interaction – which, according to the above results, would be attractive – could be incorporated

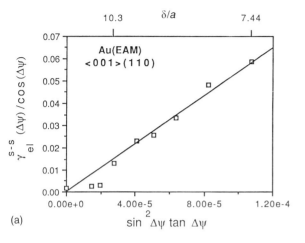

(a)

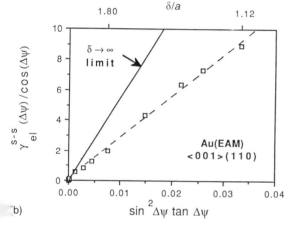

(b)

Fig. 26.16 Same as Fig. 26.15 for the (110) vicinals. The least-squares-fit parameters of the solid and dashed lines are listed in Table 26.6.

into the theory of section 26.2.1 by replacing the line energy in eq. (26.8) by

$$\Gamma(\delta) = \Gamma_{core}(\delta) + \Gamma_{el}(\delta) \qquad (26.52)$$

analogous to eq. (26.9), one can then define the core–core interaction energy, $\Gamma_{core}^{s-s}(\delta)$, as follows:

$$\Gamma_{core}(\delta) = \Gamma_{core}^{\infty} + \Gamma_{core}^{s-s}(\delta) \qquad (26.53)$$

and eq. (26.11b) becomes instead:

$$\gamma(\Delta\psi) - \gamma_{cusp}\cos\Delta\psi = (\Gamma^{\infty}/h)\sin\Delta\psi$$
$$+ (G_{el}^{s-s}/h^3)\sin^3\Delta\psi = [\Gamma_{core}^{s-s}(\delta)/h]\sin\Delta\psi$$
$$(26.54)$$

If one were to assume the Marchenko–Parshin contribution to remain unchanged even at the smaller separations, $\Gamma_{core}^{s-s}(\delta)$ could readily be extracted from the simulation data in Figs. 26.15(b) and 26.16(b) by subtracting the data points from the corresponding solid lines. The attractive core–core interaction energies thus obtained would obviously still follow a $1/\delta^2$ dependence. However, to understand – or at least rationalize – this result seems difficult.

To summarize and, in particular, to elucidate the overall magnitude of the step–step interaction relative to the other contributions to the surface energy, we return to the original expression for γ in eq. (26.11b), which may more generally be rewritten as follows:

$$\gamma(\Delta\psi) = \gamma_{cusp}\cos\Delta\psi + \gamma^{\infty}(\Delta\psi) + \gamma^{s-s}(\Delta\psi)$$
$$(26.55)$$

Here the contribution due to the total line energy of non-interacting steps is given by (eq. (26.11b))

$$\gamma^{\infty}(\Delta\psi) = (\Gamma^{\infty}/h)\sin\Delta\psi \qquad (26.56)$$

while, for the present purpose, no particular functional form needs to be assumed for the (elastic or non-elastic) contribution, $\gamma^{s-s}(\Delta\psi)$, due to the step–step interaction. Using the values of γ_{cusp} and (Γ^{∞}/h) listed in Table 26.6 together with the simulation data in Figs. 26.15(b) and 26.16(b), the three contributions to γ in eq. (26.55) may be plotted separately against $\Delta\psi_1$ for both types of steps (Fig. 26.17).

Figure 26.17 demonstrates that the overall surface energy is dominated by the energies, $\gamma_{cusp}\cos\Delta\psi$, of the cusped orientations projected onto the vicinal planes. For both types of steps, the contributions due to the line energies of the isolated steps are much smaller by comparison. Finally, on the scale of the first two contributions, the effects due to the step–step interactions appear entirely negligible. As illustrated in the next section, the relative proportions of these three contributions are rather different for the case of interface dislocations.

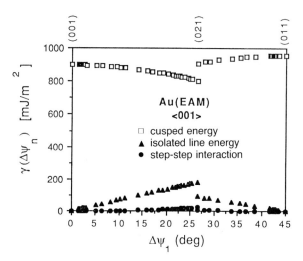

Fig. 26.17 Comparison of the relative magnitudes of the three contributions in eq. (26.55) to the free-surface energy, plotted separately against $\Delta\psi_1$ for both types of steps.

26.4.2 Grain boundary dislocations

To compare the core and strain-field energies of steps with those of dislocations, we now analyze the GB results in a similar manner. For simplicity, as in the case of the steps we again limit ourselves to interfaces in Au perpendicular to an $\langle 001 \rangle$ tilt (or pole) axis.

According to Fig. 26.5(a), two major cusped orientations, associated with the (100) and (110) planes, are encountered when rotating about a $\langle 001 \rangle$ tilt axis. As discussed elsewhere [7, 20], the STGBs on these two planes are identical to the related perfect-crystal configurations, with consequently vanishing cusped energies, by contrast with free surfaces and RGBs (Fig. 26.5(a)). By contrast to the free surfaces, however, for both the STGBs and RGBs several minor cusps appear between these major cusps; particularly noticeable are the ones associated with the (310) and (210) planes. These additional cusps, positioned in the 'high-angle' region of the two major cusps, make it exceedingly difficult to reliably extract core energies for the underlying main dislocations, particularly since we have no information on how these minor cusps affect the background energy associated with the main cusps. Moreover, as

in the case of the steps one might contemplate consideration of additional vicinal STGBs, i.e. nearer to the bottoms of the two main cusps and hence far from these minor cusps. Unfortunately, however, because of the total domination for small values of $\Delta\psi$ of the strain-field over the dislocation-core contribution (as evidenced by the well-known logarithmic shapes of GB cusps; see for example, eq. (26.20b) and Fig. 26.5(a)), consideration of more-vicinal STGB planes provides no more information on the magnitude of the core energy.

To nevertheless gain some insight into the relative magnitudes of the core and strain-field contributions to the GB energy, we have performed a least-squares fit of the Read–Shockley equation (26.20b) to the simulation data for the STGBs in Fig. 26.5(a). Values thus obtained for the line energy, Λ^∞, and for the strength of the dislocation–dislocation interaction, $L_{el}^{d\text{-}d}$, are listed in Table 26.7. How well these parameters represent the actual simulation data for the two types of dislocations is illustrated in Fig. 26.18, which also shows the breakdown of the total energy in eq. (2.20b) (solid lines, with the simulation data of Fig. 26.11(a) superimposed) into the dominating contribution due to the dislocation–dislocation interaction and the one associated with their line energy (dotted lines).

Because of the particular pairwise geometrical grouping of the edge dislocations in STGBs (with mutually opposite Burgers vectors; Fig. 26.3), the line energies listed in Table 26.7 are thought to be governed by the dislocation **cores**. A comparison with the corresponding line energies of the steps in Table 26.6, Γ^∞/d, shows that the latter are typically three to four times larger than the core energies Λ^∞/b, of the dislocations. Because of the weakness of the step–step interaction, it appears reasonable

Table 26.7 Dislocation parameters determined for the Au (EAM) potential (in units of mJ/m²)

Tilt axis	Cusped plane	Burger's vector, b/a	Λ^∞/b	$L_{el}^{d\text{-}d}/b$
$\langle 100 \rangle$	(100)	0.5	80 ± 25	$960 \pm$
$\langle 100 \rangle$	(110)	0.3535	110 ± 25	$800 \pm$

to assume that the total line energy of the steps, $\Gamma^{\infty} = \Gamma^{\infty}_{\text{core}} + \Gamma^{\infty}_{\text{el}}$ (eq. (26.12)), is also dominated by the core energy. The magnitude of the ratio, $(\Lambda^{\infty}/b)/(\Gamma^{\infty}/d) \approx 0.25-0.35$ obtained from Tables 26.10 and 26.11 therefore indicates a significantly larger core energy of the steps than of the dislocations.

If the above interpretation is correct, this value for $(\Lambda^{\infty}/b)/(\Gamma^{\infty}/d)$ should be comparable in magnitude to the related core–energy ratios, $(\Lambda^{\infty}_{\text{core}}/b)/(\Gamma^{\infty}_{\text{core}}/h)$, determined in Figs. 26.7(a) and 26.9 from the direct simulation of the 'high-angle' interfaces, in which all elastic effects vanish, and which are listed in Table 26.3. The value of 0.33 obtained for the latter (for the Au(EAM) potential; Table 26.3) is in remarkable agreement with the above results, suggesting that (i) the line energies are, indeed, governed by the core contributions and (ii) the core energies of GB dislocations are substantially lower than those of surface steps.

A comparison between Figs. 26.17 and 26.18 shows another fundamental difference between GBs and free surfaces. While in the STGBs (Fig. 26.18) the elastic interaction energy vastly outweighs the core energy [8], the opposite is true for the free surfaces, in which the elastic contribution to the line energy is negligible compared to the core contribution (Fig. 26.17). As discussed further in the next section, this difference is responsible for the fact that a broken-bond model is so successful for free surfaces while for internal interfaces its usefulness is limited to the 'high-angle' limit in which elastic effects are irrelevant.

We conclude by pointing out that the vastly different amounts of elastic energy associated with the strain fields of GB dislocations, on the one hand, and surface steps, on the other, by comparison with the corresponding core energies have important consequences for the fracture behavior of internal interfaces in general. When fracturing an interface situated in the vicinity of an energy cusp, elastic work has to be done to convert the long-range strain fields near the dislocations into the short-range strain fields near the steps. When fracturing a **high-angle** boundary, by contrast, no sizeable elastic work has to be performed because the strain-field energies in both the GB and free surface are negligible due to the virtually complete core overlap. Among the three types of interfaces defined in section 26.2.4, the 'high-angle' interfaces therefore represent the group with the smallest work of adhesion, followed by the 'vicinal' and finally the 'cusped' or 'special' interfaces.

26.5 A BROKEN-BOND MODEL FOR INTERFACIAL DECOHESION

In the discussion of the role of broken bonds in section 26.2.5 it was argued that the energies of surfaces and interfaces, and therefore the work of adhesion, may be decomposed into a broken-bond contribution – associated with the cores of steps and dislocations in addition to those of the cusps – and the elastic energy of interaction between these line defects. To study the role of broken bonds, in all our simulations the *nn* and 2nd-*nn* miscoordination coefficients defined in eq. (26.29) were determined. Their analysis in this section will enable us to test the underlying basic assumptions, and to investigate the limitations of a broken-bond model for interfacial decohesion.

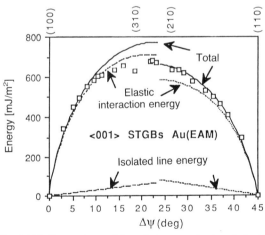

Fig. 26.18 Comparison of the relative magnitudes of the line energy and elastic–interaction energy contributions to the total GB energy in the Read–Shockley eq. (26.20b) (solid lines). The values for the line energy, Λ^{∞}, and for the strength of the dislocation–dislocation interaction, $L^{\text{d–d}}_{\text{el}}$, obtained from a least-squares fit of eq. (26.20b) to the simulation data in Fig. 26.5(a) (squares), are listed in Table 26.7.

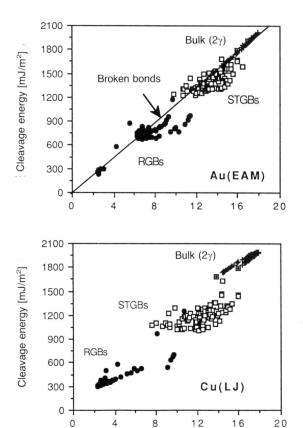

Fig. 26.19 Simulation data for STGBs, RGBs, and bulk perfect-crystal cleavage (2γ) obtained by means of the two fcc potentials against miscoordination difference, $(2C^{\mathrm{surf}}(1, \Delta\psi) - C^{\mathrm{int}}(1, \Delta\psi))$, between the two fracture surfaces and the internal interface (eqs. (26.44) and (26.45a)).

26.5.1 Work of adhesion

According to eq. (26.17a), the contribution to the work of adhesion due to bond breaking should be governed by the difference in the number of bonds broken in the two free surfaces and in the internal interface. To investigate the validity of this prediction, in Figs. 26.19(a) and (b) all the simulation data for STGBs, RGBs, and bulk cleavage (2γ) obtained by means of the two potentials are plotted against $[2C^{\mathrm{surf}}(1, \Delta\psi) - C^{\mathrm{int}}(1, \Delta\psi)]$. As seen from the figure, both the GB and free-surface energies are, indeed, reasonably well correlated with the

difference in nearest-neighbor miscoordination, although the GB data scatter systematically towards smaller cleavage energies. By contrast, the scatter of the free-surface energies alone is much smaller; the scatter in this case was shown to disappear completely when the breaking of 2nd-*nn* bonds is also considered in the analysis [19].

As discussed in section 26.2.5, the total work of adhesion contains, in addition to broken bonds, an elastic contribution due to the conversion of the long-ranged dislocation strain fields into the short-ranged strain fields surrounding the steps (eq. (26.44)). However, following our discussion in sections 26.4.1 and 26.4.2, for all the vicinal interfaces the (negative) contribution in eq. (26.44) due to the dislocations dominates over the positive contribution due to the steps. Consequently, for the same number of broken bonds a 'vicinal' interface has a lower work of adhesion than either a 'special' or a 'high-angle' interface. The vicinal interfaces therefore appear to be the reason for the systematic scatter of the GB data in Figs. 26.19(a) and (b) towards lower energies. A broken-bond description of interfacial decohesion (see, for example, the solid line in Fig. 26.19(a)) therefore represents an upper limit for W^{ad}, with the elastic effects in vicinal interfaces causing a lowering of the work of adhesion due to bond breaking alone.

26.5.2 Broken bonds in steps and dislocations

In the discussion of the role of broken bonds in section 26.2.5 it was argued that the relatively small elastic displacements of the atoms situated near steps and dislocations should not appear in the miscoordination per unit interface area, $C(\alpha, \hat{n})$, defined in eq. (26.29). In writing eqs. (26.36) and (26.42) it was therefore assumed that only the cores of the steps and dislocations cause additional broken bonds over the cusped interfaces and, hence, contribute to $C(\alpha, \hat{n})$. In the following, this assumption will finally be tested.

Starting with the free surface, and recalling that for the Au(EAM) potential the 2nd-nearest-neighbor contribution to the surface energy is negligibly small [19], in Fig. 26.20 we have plotted the coordination difference (eq. (26.32)),

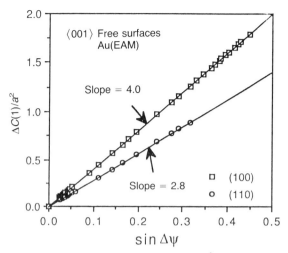

Fig. 26.20 Coordination difference, $\Delta C(1)/a^2$, defined in eq. (26.57) against $\sin \Delta \psi$ for the two steps considered in detail in section 26.4.1. The slopes of the solid lines, $C_{\text{step}}(1)/h$, obtained from a least-squares fit through the origin, are listed in Table 26.8.

Table 26.8 Nearest-neighbor broken-bond parameters defined in section 26.2.5 (eqs. (26.32) to (26.35)) for the two steps considered in section 26.4.1 for the Au (EAM) potential (Fig. 26.20)

Pole axis	Cusped plane	h/a	$C_{\text{cusp}}^{\text{surf}}(1)/a^2$	$C_{\text{step}}(1)/(ha^2)$
$\langle 100 \rangle$	(100)	0.5	8.00	4.0
$\langle 100 \rangle$	(110)	0.3535	8.48	2.8

$$\Delta C(1) \equiv C^{\text{surf}}(1, \Delta \psi) - C_{\text{cusp}}^{\text{surf}}(1) \cos \Delta \psi$$
$$\stackrel{?}{=} [C_{\text{step}}(1)/h] \sin \Delta \psi \qquad (26.57)$$

against $\sin \Delta \psi$ for the two steps considered in detail in section 26.4.1. The excellent linearity of the plots, with slopes $C_{\text{step}}(1)/h$ listed in Table 26.8, confirms that the additional number of broken bonds per unit area introduced into the cusped surface is governed by the total length of the steps in the vicinal surface.

To demonstrate that only the cores are responsible for the broken bonds, in Fig. 26.21 the related total-energy increase from the cusp, $\gamma(\Delta \psi) - \gamma_{\text{cusp}} \cos \Delta \psi$ (eq. (26.11b)), is plotted against $\Delta C(1)$ from Fig. 26.20. If both the step

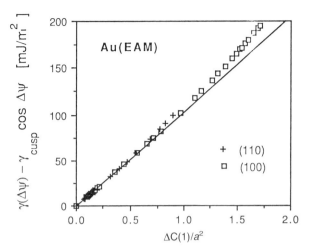

Fig. 26.21 Total energy increase from the cusp, $\gamma(\Delta \psi) - \gamma_{\text{cusp}} \cos \Delta \psi$ (eq. (26.11b)) against $\Delta C(1)$ from Fig. 26.20. The slight upwards curvature of the simulation data arises from the fact that only the cores of the steps give rise to broken bonds.

cores *and* their elastic strain fields would cause broken bonds, a perfectly linear plot would be expected in Fig. 26.20. However, the slight upwards curvature clearly evident in the figure is proof that not all of the energy increase can be accounted for by bond-breaking alone. Given that both the sign and magnitude of the deviation from linear behavior (the latter obtained from a detailed analysis of the simulation data) are in quantitative agreement with the repulsive step–step interactions in Figs. 26.15 and 26.16, we conclude that, indeed, only the cores of the steps give rise to broken bonds.

It is interesting to note that the different slopes obtained in Fig. 26.20 for the (100) and (110) steps scale quantitatively with the corresponding line energies, Γ^∞/h, listed in Table 26.7, with the larger energy per unit height of the (100) steps giving rise to a proportionally larger number of broken bonds. This leads to the virtual overlap of the (100) and (110) data in Fig. 26.21, which represents further proof for the validity of a broken-bond description of the core energies, independent of the origin of the broken bonds.

To perform a similar analysis for the GB dislocations, in Fig. 26.22 we have plotted the coordination coefficients determined for the $\langle 001 \rangle$

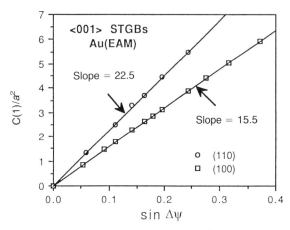

Fig. 26.22 Coordination coefficients, $C(1)/a^2$, for the $\langle 001 \rangle$ STGBs in Fig. 26.19, against $\sin \Delta \psi$ for the two types of dislocations considered in detail in section 26.4.2. The slopes of the solid lines, $C_{\mathrm{disl}}(1)/h$, obtained from a least-squares fit through the origin, are listed in Table 26.9.

STGBs in Fig. 26.18 against $\sin \Delta \psi$ for the two types of GB dislocations considered in section 26.4.2. Since the cusped energies vanish in this particular case (Fig. 26.18), according to eq. (26.57), $\Delta C(1) \equiv C(1)$. By contrast with Fig. 26.20, in Fig. 26.22 we have therefore plotted the total value of $C(1)$ against $\sin \Delta \psi$. In accordance with eq. (26.32), Fig. 26.22 exhibits excellent linearity between the number of broken *nn* bonds per unit area and the total length of the dislocation cores in the different GBs. If the strain fields were to contribute broken bonds also, their contribution would be expected to be logarithmic rather than sinusoidal. We therefore interpret Fig. 26.22 as proof that only the dislocation **cores** cause broken bonds.

Table 26.9 Nearest-neighbor broken-bond parameters defined in section 26.2.5 (eqs. (26.36) to (26.38)) for the dislocations considered in section 26.4.2 for the Au (EAM) potential (Fig. 26.22)

Pole axis	Step plane	b/a	$C_{\mathrm{cusp}}^{\mathrm{int}}(1)/a^2$	$C_{\mathrm{disl}}(1)/(ba^2)$
$\langle 100 \rangle$	(100)	0.5	0.00	15.5
$\langle 100 \rangle$	(110)	0.3535	0.00	22.5

As for the steps, the different slopes in Fig. 26.21 associated with the (100) and (110) steps scale quantitatively with the corresponding line energies, Λ^∞/b, listed in Table 26.8, with the larger energy of the dislocations introduced into the (100) cusp giving rise to a proportionally larger number of broken bonds. While the cusped orientations have a very low degree of miscoordination in the GBs as compared to the free surfaces, it is interesting to note the much larger miscoordination per unit length of the dislocations compared to that of the steps (cf. Figs. 26.22 and 26.20). This result is particularly surprising if we recall the approximately three times larger line energy of the steps (cf. Tables 26.6 and 26.7). The origin of this apparent discrepancy is not understood yet; it appears, however, that the assumption of a single interaction-strength parameter, $\beta(\alpha)$, for the relaxed cores of both steps (eq. (26.33)) and dislocations (eq. (26.39)) may be an oversimplification. The existence of broken bonds due to large elastic strains, as well as the relative contribution of higher-order-neighbor bonds, may also be of importance in the latter.

We finally mention that, in a sense, the broken-bond description discussed here may be viewed as little more than a quantified polyhedral-unit model [24]; while the structural units of a free surface are represented by the steps, the corresponding building blocks of the structure of an internal interface are the dislocation cores. Such structural-unit models have been criticized for their inability to (i) provide a quantifiable description of the interface structure and (ii) incorporate the systematic distortion of the structural units as a function, for example, of a steadily varying misorientation angle. Because the small elastic strains near steps and dislocations are usually not transmitted into broken bonds, the broken-bond model suffers from the same shortcoming in (ii) as do structural-unit models. However, by contrast with structural-unit models, a broken-bond description provides a quantitative characterization, in terms of the miscoordination per unit area of the related polyhedral building blocks, of the core contributions to various physical properties of the interface.

26.6 SUMMARY AND CONCLUSIONS

In this chapter we have taken the viewpoint that interfacial decohesion involves the transformation of the very long-ranged elastic strain fields near dislocations into the much shorter-ranged strain fields surrounding the surface steps. Our main goals have been to (i) formulate a comprehensive theoretical framework in which interfacial ideal-cleavage decohesion is viewed as the reversible transformation of misfit dislocations into surface steps, (ii) test the validity of the basic assumptions made in deriving the relevant expressions by means of atomistic computer simulations at zero temperature, and (iii) elucidate the distinct roles of broken bonds and of the elastic interactions between interface dislocations and surface steps during interfacial decohesion.

Within this framework, only the highly distorted core regions of the steps and dislocations were shown to give rise to broken bonds, while the overlapping elastic strain fields surrounding the two types of line defects are the source of their elastic interaction. Our simulations thus not only provide insight into the very different magnitudes of the elastic strain fields surrounding steps and dislocations, respectively, but also permit extraction of all relevant core parameters.

The main conclusions that have been drawn from our investigation may be summarized as follows.

1. The elastic strain-field energies per unit length of interface dislocations and surface steps differ dramatically. While the long-ranged elastic-interaction energy between dislocations shows the well-known logarithmic (Read–Shockley [8]) variation as a function of δ^{-1} (where δ is the separation between the line defects), the interaction between steps is of much shorter range, falling off proportionally to δ^{-2}. By comparison, the related core energies differ by much less, with the steps showing a typically two to three times larger core energy per unit length than the dislocations.

2. Depending on the relative magnitudes of the elastic strain-field energies by comparison with the related core energies, three types of internal interfaces and free surfaces may be distinguished, namely 'special', 'high-angle', and 'vicinal' interfaces. While the energies of both special and high-angle interfaces, and their work of adhesion, are dominated by the broken bonds, the behavior of vicinal interfaces, i.e. those in the vicinity of an energy cusp, is governed by both elastic and core effects. A broken-bond description of interfacial decohesion is therefore limited to special and high-angle interfaces.

3. A systematic investigation of the role of the interface plane demonstrates that the energies of surfaces and GBs on the **special**, i.e. most widely spaced, planes, and the ideal cleavage-fracture energy of the GBs, are governed by the interplanar lattice spacing, $d(hkl)$, parallel to the interface plane. By contrast, on the **vicinal** planes the value of $d(hkl)$ is irrelevant in both the energy and work of adhesion.

4. The comparison between interfacial and ideal-crystal brittle decohesion shows that **cusps** in the bulk ideal-crystal cleavage-fracture energy coincide with **peaks** in the GB work of adhesion. Responsible for this interesting behavior is the significantly smaller miscoordination of the **special** GBs as compared to the corresponding **flat** free surfaces.

5. Although the magnitudes of the works of adhesion of tilt and twist boundaries differ dramatically, the comparison of symmetrical tilt boundaries, which contain only edge dislocations, with high-angle twist boundaries, represented by the RGB model in which all screw-dislocation cores are assumed to overlap, shows a qualitatively similar behavior as far as core and elastic strain-field effects are concerned. This similarity suggests that our basic conclusions remain the same even when a twist component is added to a tilt component already present in the interface. The net effect is a lowering of the work of adhesion when screw dislocations are added to the edge dislocations.

In concluding we mention that, although the absolute values of GB and free-surface energies presented throughout this paper are probably not very reliable when compared with experimental results, we hope that the comparison of the **relative** energies of GBs and free surfaces presented here is meaningful, particularly since throughout our investigation the same generic behavior was observed for conceptually rather different (pair versus many-body) interatomic potentials.

ACKNOWLEDGMENTS

We have benefited from discussions with Simon Phillpot and Sidney Yip. JAJ gratefully acknowledges support from the Office of Naval Research, under Contract No. N00014-88-F-0019. This work was supported by the US Department of Energy, BES-Materials Sciences under Contract W-31-109-Eng-38.

REFERENCES

1. See, for example, Chapter 25 in this volume.
2. A. A. Griffith, *Phil. Trans. R. Soc. Lond. A*, **221** (1920), 163.
3. See, for example, Chapter 2 in this volume.
4. See, for example, Chapter 3 in this volume.
5. See, for example, Chapter 4 in this volume.
6. See, for example, C. Herring, in *Structure and Properties of Solid Surfaces* (eds R. Gomer and C. S. Smith), University of Chicago Press (1953), p. 4, and references therein.
7. Chapter 1 in this volume.
8. W. T. Read and Shockley, *Phys. Rev.*, **78** (1950), 275.
9. D. Wolf, *J. Mater. Res.*, **5** (1990), 1708.
10. D. Wolf, *Phil. Mag. A*, **63** (1991), 1117.
11. A. P. Sutton, *Phil. Mag. A*, **63** (1991), p. 000.
12. M. S. Daw and M. I. Baskes, *Phys. Rev. B*, **29** (1986), 6443.
13. D. Wolf, *Physica B*, **131** (1985), 53.
14. D. Wolf, *Acta Metall.*, **37** (1989), 1983.
15. See, for example, P. G. Shewman and W. M. Robertson, in *Structure and Properties of Solid Surfaces* (eds R. Gomer and C. S. Smith), University of Chicago Press (1953), p. 67.
16. V. I. Marchenko and A. Y. Parshin, *Sov. Phys.-JETP*, **52** (1980), 129.
17. See, for example, C. Goux, *Can. Metall. Quarterly*, **13** (1974), 9.
18. D. Wolf, *J. Appl. Phys.*, **69** (1991), 185.
19. D. Wolf, *Surf. Sci.*, **226** (1990), 389.
20. D. Wolf, *J. de Physique*, **46** (1985), 197; D. Wolf and J. F. Lutsko, *Z. Kristallographie*, **189** (1989), 239.
21. S. A. Lindgren, L. Wallden, J. Rundgren and P. Westrin, *Phys. Rev. B*, **29** (1984), 576.
22. S. M. Foiles, M. I. Baskes, and M. S. Daw, *Phys. Rev. B*, **33** (1986), 7983.
23. D. Wolf, *Acta Metall. Mater.*, **38** (1990), 781.
24. M. Weins, H. Gleiter, and B. Chalmers, *Scripta Metall.*, **4** (1970), 235, and *J. Appl. Phys.*, **42** (1971), 2639.
25. T. W. Poon, S. Yip, P. S. Ho and F. F. Abraham, *Phys. Rev. Lett.*, **65** (1991), 2161.

CHAPTER TWENTY-SEVEN

Microstructural and segregation effects in the fracture of polycrystals

D. J. Srolovitz, W. H. Yang,
R. Najafabadi, H. Y. Wang, and
R. LeSar

Introduction · Microstructural effects · Segregation effects on GB fracture · Final remarks

27.1 INTRODUCTION

The fracture properties of many materials are known to depend sensitively on microstructure and the properties of the defects that compose the microstructure. For example, stiff inclusions are known to repel cracks and thereby make the crack path tortuous. As increasing the aspect ratio of such inclusions increases the magnitude of the crack deflection, the addition of high aspect ratio whiskers may increase the fracture toughness of a material. Microstructure can also have deleterious effects on fracture toughness. For example, weak grain boundaries (GBs) in a material provide an easy, 'short-circuit' path for crack propagation.

In classical fracture mechanics, changes in microstructure are usually accounted for by simply modifying the values of a fracture toughness parameter. However, from a materials science point of view, what is required is a more direct, mechanistic understanding of how microstructure affects fracture properties. In many cases, this need has been addressed by performing careful analyses of the interaction between individual cracks and individual defects. While this approach has proven to be quite useful and successful, it ignores the fact that microstructure consists not of single defects but of large, non-random ensembles or networks of defects.

One of the aims of this chapter is to rectify this shortcoming by demonstrating a new method by which the fracture properties of materials with realistic microstructures can be investigated. In order to account for realistic microstructures, however, this method makes a number of approximations to simplify the mechanics and crack propagation criteria. In section 27.2, below, we present the microstructural mechanics model and apply it to understanding the effect of microstructure on the fracture of polycrystalline, elastic materials. In particular, we examine the effects of grain size and GB cohesion on the transition between intergranular and transgranular failure.

The microstructural mechanics results, presented below, demonstrate that GB cohesion plays a major role in determining fracture morphology. However, since GB cohesion enters the microstructural mechanics model as a parameter, it cannot provide a means for investigating the nature of boundary cohesion or methods for modifying it. Such information can only be obtained from electronic calculations or atomistic simulation.

One of the best known methods for modifying GB cohesion is to alloy the material with a species that segregates to GBs. Rice [1] has demonstrated that the role of segregation on fracture may be understood in terms of the difference in free energy between the GB and the surfaces produced when it is fractured. Unfortunately, determining the appropriate free energies is not a simple matter. The first difficulty is obtaining the free energies of the appropriate defects. The second difficulty is to determine the equilibrium segregation profile around such defects. We have recently introduced a new atomistic simulation method, known as the free energy simulation method, that addresses both of these difficulties [2].

In the free energy simulation method, the atomic positions and the composition profile are determined by minimizing an approximate free energy functional with respect to these variables. The free energy functional is based upon a local harmonic model for the individual atomic vibrations and includes a point approximation to the configurational entropy. These free energy minimizations are typically performed within the framework of the grand canonical ensemble. This method yields the equilibrium atomic structure of GBs and surfaces self-consistently with the interfacial segregation profile and the interfacial thermodynamic

properties. Below, we apply the free energy simulation method to the problem of determining the interfacial free energies appropriate for slow fracture, where the solute atoms move quickly relative to the crack, and to fast fracture, where the GB is compositionally equilibrated but the surfaces produced have the same composition profile as the GB.

Our goal in presenting both microstructural and atomistic level results on the fracture of polycrystalline materials is to clarify the role of each and identify ways in which simulations can be used to span the many length scales associated with complex phenomena such as fracture.

27.2 MICROSTRUCTURAL EFFECTS

27.2.1 Microstructural mechanics model

The mechanics model employed in the simulations described below is based upon the elastic properties of a network of springs. The model consists of a triangular array of lattice points which are connected by bonds, as indicated in Fig. 27.1. The energy of this array of bonds consists of a bond stretching and a bond bending term:

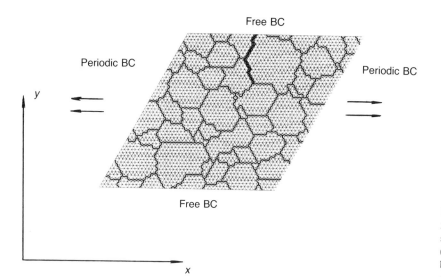

Fig. 27.1 The geometry of the microstructural mechanics model showing the triangular grid of spring (thin light lines), the locations of grain boundaries (thin dark lines) and a crack (wide light line).

$$E = \frac{1}{2} \sum_{i=1}^{N} \sum_{j}^{n.n.} \Phi_{ij}(R_{ij}) + \frac{1}{6} \sum_{i=1}^{N} \sum_{j} \sum_{k} \Psi_{ijk}(R_{ij}, R_{ik})$$

$$(27.1)$$

The sums on i are over all N sites in the system, the sums on j are over all sites which are nearest neighbors of site i, the sum on k is over all sites that are nearest neighbors of sites i and j, and R_{ij} is the distance between sites i and j. The bond stretching Φ and bending functions Ψ may be chosen to represent linear elastic springs and watch springs, respectively. Hence, the bond stretching and bending terms may be written as:

$$\Phi_{ij} = \frac{1}{2} k (R_{ij} - a_o)^2 \qquad (27.2a)$$

$$\Psi_{ijk} = \frac{1}{2} c \left(\frac{R_{ij} \cdot R_{ik}}{|R_{ij}| |R_{ik}|} - b_o \right)^2 \qquad (27.2b)$$

where k and c are constants that scale the stiffness of the bond stretching and bond bending interactions, θ_{ijk} is the angle between the ij and ik bonds, and a_o and b_o are related to the equilibrium bond lengths and bond angles, respectively.

For a uniform solid all of the k_{ij} and c_{ijk} are identical ($k_{ij} = k$ and $c_{ijk} = c$). In this case, the elastic constants are given by

$$C_{11} = C_{22} = \frac{9}{8} k a_o^2 + \frac{21}{16} c \qquad (27.3a)$$

$$C_{33} = \frac{3}{8} k a_o^2 + \frac{21}{16} c \qquad (27.3b)$$

$$C_{12} = \frac{3}{8} k a_o^2 - \frac{21}{16} c \qquad (27.3c)$$

$$\text{All other } C_{ij} = 0 \qquad (27.3d)$$

The symmetry of the elastic constant tensor is that of an isotropic continuum. As in an isotropic solid, there are only two independent elastic constants and the symmetry is that of the triangular lattice (e.g. [3]). The Lamé constant λ, shear modulus μ, Poisson's ratio ν, Young's modulus E, and bulk modulus B may also be written in terms of the C_{ij} or c and k as:

$$\lambda = C_{12} = \frac{3}{8} k a_o^2 - \frac{21}{16} c \qquad (27.4a)$$

$$\mu = C_{33} = \frac{1}{2}(C_{11} - C_{12}) = \frac{3}{8} k a_o^2 + \frac{21}{16} c \qquad (27.4b)$$

$$\nu = C_{12}/C_{11} = \frac{2 k a_o^2 - 7c}{6 k a_o^2 + 7c} \qquad (27.4c)$$

$$E = 2\mu(1 + \nu) = 3 k a_o^2 \frac{2 k a_o^2 + 7c}{6 k a_o^2 + 7c} \qquad (27.4d)$$

$$B = \frac{E}{3(1 - 2\nu)} = k a_o^2 \frac{2 k a_o^2 + 7c}{2 k a_o^2 + 21c} \qquad (27.4e)$$

The $\alpha\beta$ component of the stress at site i in the model may be written as

$$\sigma_i^{\alpha\beta} = \frac{1}{2\Omega} \sum_j \left\{ \frac{\partial \Phi_{ij}}{\partial R_{ij}} \frac{\partial R_{ij}}{\partial R_{ij}^{\beta}} R_{ij}^{\alpha} \right.$$

$$\left. + \sum_k \left[\frac{\partial \Psi_{ij}}{\partial R_{ij}} \frac{\partial R_{ij}}{\partial R_{ij}^{\beta}} R_{ij}^{\alpha} + \frac{\partial \Psi_{ij}}{\partial R_{ij}} \frac{\partial R_{ij}}{\partial R_{ij}^{\alpha}} R_{ij}^{\beta} \right] \right\}$$

$$(27.5)$$

where Ω is the area associated with site i, α and β are Cartesian directions that take on the values x and y (or 1 and 2), R_{ij}^{α} is the α component of the distance between sites i and j. The macroscopic stress in the sample is obtained by averaging $\sigma_i^{\alpha\beta}$ over all sites.

Microstructural features may be incorporated into the model by associating different properties with each bond. For example, a polycrystalline body may be simulated by identifying a bond as a bulk bond if it separates two adjacent lattice sites within the same grain and as a GB bond if it separates two adjacent lattice sites belonging to different grains. The fracture properties of the model may also be incorporated via a bond type dependent failure criterion. In the present study, a bond breaks irreversibly when its strain energy

$$E(ij) = \Phi_{ij}(R_{ij}) + \frac{1}{2} \sum_k \Psi_{ijk}(R_{ij}, R_{ik}) \qquad (27.6)$$

exceeds a critical strain energy $E_c(ij)$, where the magnitude of E_c is different for bulk and GB bonds. A multiphase material may be modeled by choosing the elastic parameters k_{ij} and c_{ijk} to depend on the phases of sites i, j, and k. Misfit or eigenstrain strain may be included by choosing the constants a_o and b_o to also be site dependent (i.e. a_{ij} and b_{ijk}).

The simulations described in the present chapter correspond to strain controlled tensile tests. The test sample (Fig. 27.1) is periodic in the direction parallel to the tensile axis and has free surfaces in the directions normal to the applied load. Hence, the simulation is performed on a sample with effectively infinite gauge length. The strain is applied by defining the distance over which the sample is periodic L as $L = L_o(1 + n\Delta\varepsilon)$, where L_o is the length of the unstrained sample, $\Delta\varepsilon$ is the strain step, and n is the number of strain steps applied. The model is equilibrated in the following manner: (i) apply a small finite strain increment; (ii) relax the total energy of the system with respect to the site coordinates using a double precision conjugate gradient algorithm; (iii) identify the bond with the maximum energy; (iv) if the bond energy is greater than the local $E_c(ij)$, break the bond with the highest energy $E(ij)/E_c(ij)$ and return to step (ii); (v) if the model is not completely fractured, return to step (i). In the simulations presented here, a lattice of $10^4 (= 100 \times 100)$ sites was employed and the following parameters were employed: $k = 1$, $c = (\frac{1}{7})$, $a_o = 1$, $b_o = \frac{1}{2}$ and $E_c = 10^{-4}$ for the bulk.

27.2.2 Polycrystalline microstructure simulation

In order to simulate fracture in a polycrystalline material, a realistic polycrystalline microstructure must first be mapped onto the microstructural mechanics model. Such a polycrystalline microstructure may be produced using the Monte Carlo simulation procedure introduced by Srolovitz and co-workers [4, 5]. This procedure has been shown to produce microstructures with grain size and grain topology distributions in excellent agreement with experiment.

In short, a continuum microstructure is mapped onto a two-dimensional triangular lattice containing 10 000 sites. Each lattice site is assigned a number, S_i, which corresponds to the orientation of the grain in which it is embedded. The number of distinct grain orientations is Q. Lattice sites which are adjacent to neighboring sites having dif-

ferent grain orientations are regarded as being adjacent to a GB, while a site surrounded by sites with the same grain orientation is in the bulk or grain interior. The GB energy is specified by associating a positive energy with GB bonds and zero energy for bonds in the grain interior, according to

$$E_i = -J \sum_{j}^{nn(i)} (\delta_{S_i S_j} - 1) \qquad (27.7)$$

where δ_{ij} is the Kronecker delta, the sum is taken over nearest neighbor (nn) sites of site i, and J is a positive constant that sets the energy scale of the simulation. The kinetics of boundary motion are simulated via a Monte Carlo technique in which a lattice site is selected at random and its orientation is randomly changed to one of the other grain orientations. The change in energy associated with the change in orientation is evaluated. If the change in energy is less than or equal to zero, the reorientation is accepted. However, if the energy is raised, the reorientation is rejected.

The microstructures shown below were produced by initially assigning a random value of the grain orientation to each site ($1 \leq S_i \leq Q$) and then running the Monte Carlo simulation procedure until the desired grain size was produced. The resultant two-dimensional polycrystalline microstructures are in good agreement with those found from taking cross-sections through three-dimensional polycrystalline materials and three-dimensional simulations [5]. In addition to producing an accurate representation of observed microstructures, this simulation procedure has the advantage of producing microstructures on exactly the same lattice as that employed for the microstructural mechanics simulations.

27.2.3 Microstructural fracture results and discussion

The propagation of a crack through a polycrystalline microstructure of linear grain size $r = 15.3$ (where the equilibrium spring length is unity) and the ratio of the GB to bulk critical energy $R = E_c^{GB}/E_c^B = 0.28$ (i.e. weak GBs) is shown in Fig. 27.2. In this case a small pre-crack was nucleated at the top of the sample microstructure by breaking

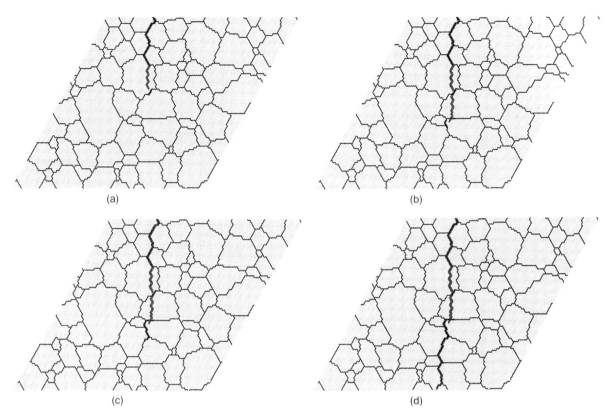

Fig. 27.2 The progression of a crack through a microstructure with $R = 0.28$.

one bond (parallel to the load) prior to straining the sample. This was necessary since, prior to the development of the crack, the sample was elastically homogeneous. The crack begins by propagating along the low cohesion GB network until it finds itself in a situation in which the GB turns parallel to the load. At this point, the driving force for crack propagation along the GB network is greatly reduced, and the crack begins to propagate through the interior of a grain (Fig. 27.2(a)). Since the grain boundaries are weak compared with the interior of the grain, the stress field of the crack causes a new crack to nucleate at the GB on the opposite side of the grain. This new crack then propagates back to the original crack. The crack continues to propagate in a mixed inter-/trans-granular mode until the entire sample is fractured.

The sample shown in Fig. 27.2 fails in a brittle manner with a stress–strain curve which is linear up to the fracture stress and then drops vertically to zero. However, the fact that each spring is linearly elastic/perfectly brittle is not sufficient to imply a stress–strain curve of this form. For example, a random distribution of voids or a distribution of internal stresses (non-uniform values of a_o and b_o in eq. (27.2)) can lead to nonlinear stress–strain curves. In these cases, the nonlinearity is associated with damage accumulation. The value of the fracture stress in such studies is statistically distributed since it depends sensitively on the details of the microstructure in the vicinity of the pre-crack.

The effect of GB cohesion (relative to the bulk) $R = E_c^{GB}/E_c^B$ may be seen in Fig. 27.3, where we show the crack path at four different values of R in the same microstructure and with the same pre-crack location. At $R = 0.2$ (Fig. 27.3(a)), the fracture is completely intergranular. At $R = 0.25$

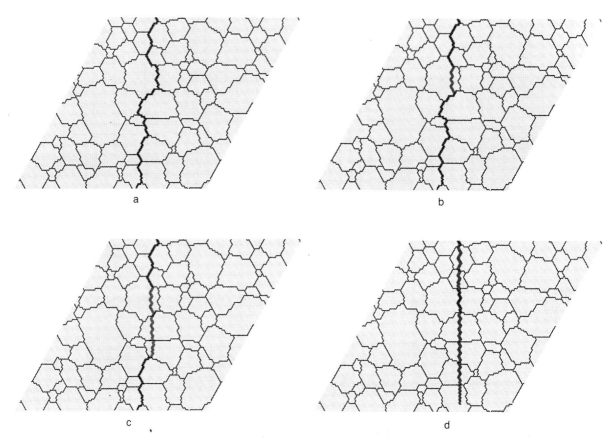

Fig. 27.3 Crack paths in the same microstructure at different values of the grain boundary cohesion parameter R: (a) $R = 0.20$, (b) $R = 0.25$, (c) $= 0.30$, and (d) $R = 0.70$.

(Fig. 27.3(b)), the crack path is predominantly along GBs but with a relatively small amount of bulk fracture (i.e. one grain). When $R = 0.3$ (Fig. 27.3(c)), the fracture mode is approximately evenly mixed between intergranular and transgranular. At $R = 0.7$, however, the fracture mode is almost completely transgranular. In this case, the crack path is perfectly straight, with the exception of a slight oscillation which is due to the discreteness of the lattice and the orientation of the springs relative to the loading direction.

The results of the simulations on the effects of GB cohesion on fracture morphology are summarized in Fig. 27.4, where we plot the ratio of the number of bulk bonds broken to the total number of bonds broken L_B/L_C during the fracture simu-lation versus $E_c^{GB}/E_c^B = R$. L_B/L_C is effectively the percent transgranular fracture. The three curves in this figure correspond to three different locations of the pre-crack. L_B/L_C is approximately zero for $E_c^{GB}/E_c^B \leqslant 0.2$, indicating almost pure intergranular fracture at low GB cohesion. L_B/L_C then rises to a plateau value for R in the range $0.2 \leqslant R \leqslant 0.5$ corresponding to mixed mode failure. Above $R = 0.5$, the fracture mode is transgranular and further increases in R have little effect on the fracture mode. In the transgranular failure regime L_B/L_C is not one since even a crack going straight through the microstructure must still cross and break some GB bonds. The transgranular value of L_B/L_C reflects the discretization of the lattice.

A self-similar microstructure, such as those

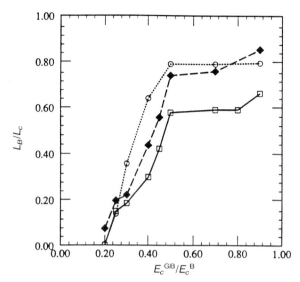

Fig. 27.4 Fraction intergranular fracture versus GB cohesion. L_C and L_B are the total number of bonds broken and the number of bulk bonds broken during fracture. The three curves correspond to three fracture simulation in the same microstructure with different pre-crack locations.

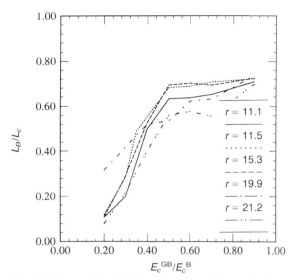

Fig. 27.5 Same as Fig. 27.4, but for different linear grain sizes r. Each curve represents an average over three simulations.

found in grain growth, are completely described by the microstructure observed at any time and the overall scale or mean grain size. The effect of grain microstructure, *per se*, on fracture was described above. In order to complete the analysis, we must account for the effect of grain size on fracture morphology. We have performed a series of simulations in which both the grain size and GB cohesion were varied. Figure 27.5 shows L_B/L_C versus E_c^{GB}/E_c^B for linear grain sizes in the range of $11 < r < 22$, corresponding to a factor of four variation in mean grain area. In this plot, each data point shown is averaged over three simulations on the same microstructure with three different pre-crack locations. To within the statistical error in the data, we find that grain size plays no role in determining the fracture mode.

The relatively small range of E_c^{GB}/E_c^B values over which the fracture mode changes from intergranular to transgranular suggests that small changes in the GB cohesion will, in most cases, have little effect on fracture morphology. This is true, of course, unless the GB cohesion happens to be very near the transition range of E_c^{GB}/E_c^B.

The transition from intergranular to transgranular fracture in a brittle polycrystalline material begins at $E_c^{GB}/E_c^B \approx 0.2$ and has an average value ($\approx 50\%$ transgranular) around $E_c^{GB}/E_c^B \approx 0.35$. Recently, He and Hutchinson [6, 7] analyzed the problem of a crack in an elastic medium interacting with an interface with a different toughness than the bulk. They found that for a crack impinging on a GB at right angles, the crack would be transmitted if $E_c^{GB}/E_c^B > 0.25$ and deflected into the interface if $E_c^{GB}/E_c^B < 0.25$. If the crack is incident at an angle of less than 90°, the critical energy ratio is greater than 0.25. While He and Hutchinson's results do not apply directly to the polycrystalline case, they are consistent with the simulation results where transgranular fracture began at $E_c^{GB}/E_c^B \approx 0.2$ and has an average value near 0.35. The fact that the average value of E_c^{GB}/E_c^B for the fracture mode transition in the polycrystalline material is higher than 0.25 reflects the range of angles for which a crack is incident on GBs in the polycrystal.

The present results suggest that GB properties play a crucial role in determining fracture morphology. Therefore, anything that modifies the GB cohesion will likely play an important role in fracture. Atomistic computer simulations show that grain misorientation, segregation and temperature all play an important role in determining the energy of GBs. Of these three, segregation is known to play a dominant role in its ability to modify interfacial thermodynamics and, hence, the fracture properties of brittle materials.

27.3 SEGREGATION EFFECTS ON GB FRACTURE

27.3.1 Thermodynamic basis

In order to determine the effects of GB segregation on fracture, we must first establish the thermodynamic basis for obtaining fracture data in systems that exhibit impurity segregation. If plastic flow is negligible, then the strain energy release rate, G, associated with crack growth along a GB is given in the Griffith model as

$$G = \frac{K^2}{E} = f_{s1} + f_{s2} - f_{gb} = 2\gamma_{int} \qquad (27.8)$$

where K is the stress intensity factor, E is Young's modulus, f_{gb} is the excess free energy of the GB per unit area, and f_{s1} and f_{s2} are the excess free energies of the two free surfaces produced when the boundary is fractured. Following Rice and Wang [8], we can analyze the effects of impurity segregation by examining the effects of segregation on $2\gamma_{int}$.

There are two limiting cases which may be distinguished. The first case is when fracture is fast and the solute does not have sufficient time to redistribute on the free surfaces produced by the high velocity crack. This is a non-equilibrium situation (i.e. the chemical potential is non-uniform) in which the solute concentration on each surface is simply one half of its equilibrium concentration on the GB. In this case [8, 9], $2\gamma_{int}$ is given by

$$(2\gamma_{int})_{\Gamma=const} = 2f_s(\Gamma/2) - f_{gb}(\Gamma) \qquad (27.9)$$

where Γ is the constant segregant concentration and we have assumed that the two surfaces are identical. $(2\gamma_{int})_\Gamma$ was shown [1] to be an upper bound on the work of separation.

The second limiting case applies to the situation where the segregant has a high mobility such that the solute concentration at the GB and surface is in equilibrium. According to Rice and Wang [8], this situation probably applies to H in steel at low temperatures and to sulphur in steel at high temperatures. In this situation, the chemical potential is uniform and there is no simple relation between the segregant concentration on the GB and the surfaces. Since the concentration is not constant, in this case, the free energies must be evaluated within the grand canonical ensemble [8, 9]

$$(2\gamma_{int})_{\mu=const} = 2\gamma_s(\mu) - \gamma_{gb}(\mu) \qquad (27.10)$$

where $\gamma = f - \mu\Gamma$. $(2\gamma_{int})_\mu$ is a lower bound on the work of separation.

The remainder of this section focuses on new atomistic simulation procedures for obtaining the GB and surface free energy data needed to evaluate $2\gamma_{int}$ from eqs. (27.9) and (27.10). This new simulation method allows for the self-consistent determination of interface segregation profiles and free energies at finite temperatures.

27.3.2 Free energy simulation method

The free energy of alloy systems consists of three parts: the atomic bonding, the atomic vibrations, and the configurational entropy. We describe the atomic bonding using embedded atom method (EAM) potentials [10]. The effect of atomic vibrations is included within the framework of the local harmonic (LH) model [11], which is given by

$$A_v = k_B T \sum_{i=1}^{N} \sum_{\beta=1}^{3} \ln\left(\frac{h\omega_{i\beta}}{2\pi k_B T}\right) \qquad (27.11)$$

where $k_B T$ is the thermal energy, h is Planck's constant, N is the total number of atoms in the system, and ω_{i1}, ω_{i2}, and ω_{i3} are the three vibrational eigenfrequencies of atom i. These frequencies may be determined in terms of the local dynamical matrix of each atom $D_{i\alpha\beta} = (\partial^2 E/\partial x_{i\alpha}\,\partial x_{i\beta})$, where the $x_i\beta$ correspond to atomic displacements of atom i in some coordinate system [12].

Within the point approximation, the configurational entropy [2] may be written as

$$S_c = -k_B \sum_{i=1}^{N} \{c_a(i) \ln[c_a(i)] + c_b(i) \ln[c_b(i)]\}$$

(27.12)

where $c_a(i)$ is the concentration of a-atoms and $c_b(i)$ is the concentration of b-atoms on site i. Since we are interested in equilibrium properties, these concentrations may be viewed as the time averaged composition of each atomic site in a system where the atoms are free to diffuse. In this sense, the atoms are 'effective' or 'mean-field' atoms. Since we replace real atoms by effective atoms, the internal energy E which is defined in terms of the interatomic potential must also be suitably averaged over the composition of each atom and its interacting neighbors. A method for performing these averages for the EAM potentials is described in Ref. [2].

In the simulations described below, we employ a reduced (baby) grand canonical ensemble, where the total number of atoms remains fixed but the relative amounts of each atomic species varies. The appropriate thermodynamic potential for this type of ensemble is the grand potential and is given by

$$\Omega = A + \Delta\mu \sum_{i=1}^{N} c_a(i)$$
$$= E + A_v - TS_c + \Delta\mu \sum_{i=1}^{N} c_a(i) \quad (27.13)$$

where $\Delta\mu$ is the difference in chemical potential between the a- and b-atoms.

The calculation of the equilibrium segregation around an interface is performed in steps. First, the properties of the perfect, uniform composition crystal are determined. This is done by choosing a composition and then minimizing the free energy, at the temperature and pressure of interest, with respect to the lattice parameter. Differentiating this equilibrium free energy with respect to composition yields the chemical potential difference $\Delta\mu$. For the Cu–Ni alloy system, we have verified that the equilibrium structure is a solid solution at the temperature, pressure, and composition of interest by minimizing the grand potential (eq. 27.13)) with respect to lattice parameter and local concentration and verifying that the resultant concentration profile was uniform. Since, at equilib-

rium, the chemical potential of a component is everywhere constant, we fix the chemical potentials at their bulk values, introduce the appropriate interface, and minimize the grand potential with respect to the concentration and position of each site. The geometry of the cells employed in the GB and surface simulations were described in Ref. [12]. The segregation simulation results presented below were all obtained on the $\Sigma 13$ [001] (22.62°) twist boundary in $Cu_x Ni_{1-x}$ at 800 K, where $x = 0.1$ or 0.5.

27.3.3 Segregation results and discussion

If fracture occurs sufficiently slowly, compared with the rate at which the segregation profile equilibrates, then it is appropriate to determine the critical strain energy release rate or $2\gamma_{int}$ in terms of the free energy of the equilibrated GB and surfaces, i.e. $(2\gamma_{int})_\mu$. The equilibrated segregant concentration profiles for the GBs and free surfaces are shown in Fig. 27.6(a) and (b), respectively, for the $Cu_{10}Ni_{90}$ and the $Cu_{50}Ni_{50}$ alloys. In all cases, Cu is observed to segregate very strongly to the interfaces. The degree of segregation to the first atomic layer closest to the interface is nearly independent of the Cu concentration in the bulk (Fig. 27.6 and Table 27.1). On the other hand, the degree of segregation is higher for the free surface than for the GB.

While segregation to the atomic layer adjacent to the interfaces is strong, the region of strong segregation is effectively limited to a single atomic layer with buried atomic layers generally showing a deficit of the segregating element. Another measure of the degree of segregation is obtained by summing the solute concentration excess (relative to the bulk solute concentration) over all atomic planes (for the symmetric GB case, only the atomic planes on one side of the boundary are summed over). Since the excess must go to zero for layers deep within the bulk, this summation yields a finite excess (quoted in Table 27.1 in terms of monolayers of Cu). (Note the total excess solute need not be zero since we are working in the grand canonical ensemble.) As may be seen in Table 27.1 and Fig. 27.6, when the bulk solute concentration is low

Table 27.1 Interface segregation data

	10% Cu		50% Cu	
	Segregated	Unsegregated	Segregated	Unsegregated
γ_{gb} (mJ/m^2)	690	901	619	699
γ_s (mJ/m^2)	1068	1560	1029	1284
$(2\gamma_{int})_\mu$ (mJ/m^2)	1446	2219	1440	1869
$\gamma_s^\star$ (mJ/m^2)	1137	–	1120	–
$(2\gamma_{int})_\Gamma$ (mJ/m^2)	1584	–	1621	–
1st Layer GB Cu %	79%	–	81%	–
1st Layer Surf. Cu %	98%	–	98%	–
Excess GB Cu	0.83	–	0.19	–
Excess Surf. Cu	0.84	–	−0.33	–

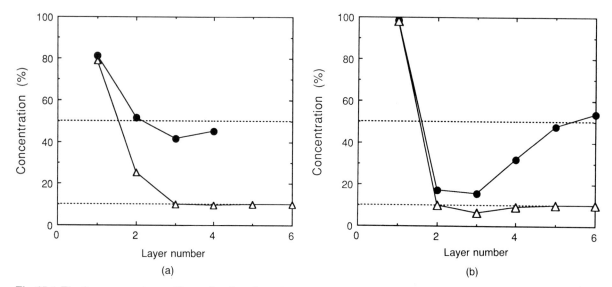

Fig. 27.6 The Cu concentration profile as a function of atomic layer number (a) on one side of the equilibrated symmetric twist GB and (b) adjacent to the (001) free surface. Layer number one is adjacent to the interface. The solid circles are for Cu$_{50}$Ni$_{50}$ and the open triangles are for Cu$_{10}$Ni$_{90}$.

there is a total or net segregation of Cu to the interfaces. However, in the Cu$_{50}$Ni$_{50}$ alloy the total Cu segregation is positive for the GB and negative for the free surface. Therefore, depending on the definition chosen for describing segregation, Cu either does or does not segregate to the (001) free surface in Cu$_{50}$Ni$_{50}$.

The excess free energies (i.e. grand potentials) of these GBs γ_{gb} and surfaces γ_s are shown in Table 27.1. In the Cu$_{10}$Ni$_{90}$ alloy, the GB free energy drops by approximately 25% on segregation, while the drop is only of order 10% in the Cu$_{50}$Ni$_{50}$ alloy. The surface free energy in the Cu$_{10}$Ni$_{90}$ drops by approximately 30% on segregation and, correspondingly, the drop in the Cu$_{50}$Ni$_{50}$ case is $\approx$20%. These changes in interfacial free energy upon segregation are consistent with the view that the largest changes in interfacial free energy occur with the largest changes in interfacial composition. It is interesting to note that while the free energies of the unsegregated interfaces at 10% and 50% Cu differ widely, the 10% and 50% bulk Cu alloys show nearly equal interfacial free energies upon segregation. This is undoubtedly a result of the fact that both the GB and surface first layers have nearly the same composition.

These interfacial free energy results may be combined, as per eq. (27.10), to yield the constant μ, or equilibrated segregation, value of $2\gamma_{int}$. The results for $2\gamma_{int}$ both with and without interfacial segregation are shown in Table 27.1. For the unsegregated interfaces, $2\gamma_{int}$ is larger for the 10% Cu alloy than for the 50% Cu alloy. This is simply indicative of $2\gamma_{int}(Ni) > 2\gamma_{int}(Cu)$. Upon segregation, we find that $(2\gamma_{int})\mu$ for the two alloys are essentially identical. This too may be attributed to the fact that both the GB and the surface first layers have nearly the same composition. The $Cu_{10}Ni_{90}$ alloy exhibits a 35% decrease in $2\gamma_{int}$ upon segregation, while the $Cu_{50}Ni_{50}$ alloy shows only a 23% drop. Since the magnitude of the change in Cu concentration at the interfaces is much greater in the Ni rich alloy, it is not surprising that this alloy shows the greatest change in $2\gamma_{int}$ upon segregation.

Additional simulations were performed in which the concentration profile of the segregated GBs was held fixed while the GB was pulled apart, creating two free surfaces. The free energy of these new surfaces γ_s^* was determined and then used to obtain the constant composition value of $2\gamma_{int}$, namely $(2\gamma_{int})_\Gamma$. The values of γ_s^* and $(2\gamma_{int})_\Gamma$ are also shown in Table 27.1. The $(2\gamma_{int})_\Gamma$ values for the two alloys only differ from each other by approximately 2%. In both cases, we find that $(2\gamma_{int})_\Gamma$ is larger than $(2\gamma_{int})_\mu$ by approximately 10%.

These results demonstrate that Cu tends to embrittle GBs in Ni. As $(2\gamma_{int})_\Gamma$ provides an upper bound on $2\gamma_{int}$ and $(2\gamma_{int})_\mu$ provides a lower bound on this quantity, we now have a reasonably good estimate for $2\gamma_{int}$ for interfacial fracture in this system. If the Cu–Ni alloys exhibited no plasticity, then $2\gamma_{int}$ would be equal to the critical strain energy release rate G_c. However, since these alloys do show significant plasticity, these results may

be viewed as input to a theory which may account for the plastic portion of the strain energy release rate. Another interesting conclusion that may be drawn from the present results is that the solute concentration on the atomic layers directly adjacent to the GB dominates the surface and GB free energies and hence the value of $2\gamma_{int}$. These results also show that even in a system with very similar atoms (Ni–Cu), the effects of segregation on fracture properties are strong. This leads us to suspect that in cases where the solute and solvent atoms are very different (e.g. atom size or bonding type) segregation effects could be immense. This is consistent with long established experimental observations.

27.4 FINAL REMARKS

The inherent difficulty in simulating the fracture of real materials is the large range of length scales inherent in the process. These range from the macroscopic, including such effects as specimen geometry and loading – the realm of traditional fracture mechanics, to the microstructural, to the atomistic and electronic. While it is possible to perform simulations of fracture appropriate for any of these length scales in isolation, this is clearly insufficient. The present report outlined new approaches for attacking this problem at the microstructural and atomic scale. The challenge is now to put these together.

On the most simple level, we may view the output of the atomistic simulations as providing input to the microstructural mechanics model. The atomistic simulation method, since it self-consistently predicts the degree of segregation and the interfacial thermodynamics, may be used to predict the effect of temperature and bulk composition on the interfacial cohesion $2\gamma_{int}$. It is this parameter which enters directly into the microstructural mechanics model in terms of E_c^{GB} in $R = E_c^{GB}/E_c^B$. While we do not believe that the atomistic predictions of E_c^{GB} are quantitative in the absolute sense, it is reasonable to expect them to be semi-quantitative and provide reliable predictions on the effects of changes in external conditions (e.g. temperature, bulk composition).

A true simulation of microstructural effects in fracture should include the variability which is inherent within the microstructure itself. For example, the degree of segregation and, consequently, the values of E_c^{GB}/E_c^B and $2\gamma_{int}$, will undoubtedly vary with GB misorientation in polycrystalline materials. To perform atomistic simulations on all possible GBs would represent a formidable task. It is probably more appropriate to use the simulations simply to predict the range of variability of these parameters as input to the microstructural mechanics model. Taken *in toto*, the results presented above should be viewed as simply first steps in learning how to span the range of length scales inherent in complex materials phenomena.

ACKNOWLEDGMENTS

We gratefully acknowledge the support of the Air Force Office of Scientific Research's URI program (contract AFOSR-90-0112) for the research on microstructural effects in fracture (DJS and WHY) and the support of the Division of Materials Science of the Office of Basic Energy Sciences of the United States Department of Energy (Grant No. FG02-88ER45367) for the atomistic simulation research (DJS, RN and HYW). The work of RL was performed under the auspices of the US Department of Energy.

REFERENCES

1. J. R. Rice, in *Chemistry and Physics of Fracture* (eds R. M. Latanision and R. H. Jones), Martinus Nijhoff, Dordrecht (1987), p. 27.
2. R. Najafabadi, H. Y. Wang, D. J. Srolovitz, and R. LeSar, *Acta Metall. Mater.*, **39** (1992), 3071.
3. J. F. Nye, *Physical Properties of Crystals* Clarendon Press (Oxford), 1985.
4. D. J. Srolovitz, M. P. Anderson, G. S. Grest and P. S. Sahni, *Scripta Metall.* **17** (1983), 241.
5. G. S. Grest, M. P. Anderson and D. J. Srolovitz, *Philo. Mag. B*, **59** (1988), 293.
6. M. Y. He and J. W. Hutchinson, *Int. J. Solids and Structures*, **25** (1989), 1053.
7. M. Y. He and J. W. Hutchinson, *J. App. Mec.* **56** (1989), 270.
8. J. R. Rice and J. S. Wang, *Materials Sci. Eng A*, **107** (1989), 23.
9. P. M. Anderson, J. S. Wang and J. R. Rice, in *Innovation in Ultrahigh-Strength Steel Technology* (eds G. B. Olson, M. Azrin and E. S. Wright), Sagamore Army Materials Research Conference Proceedings, Vol. 34. (1990), p. 619.
10. S. M. Foiles, M. I. Bakes, M. S. Daw, *Phys. Rev. B*, **33** (1986), 7983.
11. R. LeSar, R. Najafabadi, and D. J. Srolovitz, *Phys. Rev. Lett.*, **63** (1989), 624.
12. R. Najafabadi, D. J. Srolovitz and R. LeSar, *J. Mater. Res.*, **5** (1990), 2663.

Materials index

Bicrystals, superlattices and plane interfaces are indicated by a solidus (/). Constituents of alloys are separated by a dash (−). The components of a system are listed alphabetically and not according to the usual conventions, although the order is preserved for multilayer structures. The components of alloys are in alphabetical order. Segregants follow the segregating system, separated by a colon, e.g. Al_2O_3/Pd : Pd.

Abbreviations: BPDA-PDA, bisphenyldianhydride-phenyldiamine; PMDA-ODA, pyromellitic dianhydride-oxydianiline; Gr, graphite. Elements are represented by their atomic symbols, which are listed alphabetically.

Ag, Ag/Ag bicrystals 198
Ag-containing systems
  Ag/Au 229
  Ag/CdO:H segregation 494
  Ag/Co superlattice 356, 361
  Ag−Cu 263
  Ag/Fe 277−8, 279−81
  Ag/GaAs epitaxial interface 583−4
  Ag/MgO 282−3, 654, 656
  Ag/Pd superlattice 356
  Ag/PMDA-ODA 628−9
  Ag/Si 309−10
  Ag/Te 518
  Ag/V 278
  Ag/Xe 347
  Al:Ag segregation 202
Al 202, 546
  Al/Al bicrystals 197, 198−9, 200, 203−4, 240
  embrittled with Ga−Hg 270
  free surface 240
  GB migration 217, 221
  structure−energy correlation 194, 196−200
  twin boundaries 68, 69, 96
  volume expansion 135, 245
Al compounds
  AlAs
    AlAs/GaAs 294, 585, 592

AlAs/GaAs superlattice 356
AlGaAs, AlGaAs−GaAs 331
AlN
  AlN/NbN superlattice 356
  AlN/Y_2O_3 composites 152
  AlN/ZrN superlattice 356, 361
  ceramic substrates 168−70, 173, 174, 175, 176
  YAG interphase 169, 172
Al_2O_3
  Al_2O_3/Au 659−60
  Al_2O_3/Cu 659
  Al_2O_3/Nb 655, 656−7, 658, 659
  Al_2O_3/Ni 657, 659
  Al_2O_3/Pd:H segregation 494−5
  Al_2O_3/SiC composites 185−6
  Al_2O_3/spinel 176, 179
  Al_2O_3/TiO_2 9
  Al_2O_3/ZrO_2 175, 184, 185
  dislocations 156
  GBs 159−64, 165, 166, 167
  penetration and wetting 267−8, 269
  sapphire *see* Sapphire-containing systems
AlRu, nanophase 439
Al_2S_3 599−603
Al_2Se_3 604, 605
FaAl amorphization 252

NiAl amorphization 250
Ni_3Al:B and C segregation 92−3
Zr_3Al amorphization 248−9, 250
Al-containing systems
  Al:Ag segregation 202
  Al−Cu
    Al−Cu:Al segregation 198
    triple junctions 206
  Al/Cu bicrystals 204
  Al/GaAs 318−19, 613, 614
    epitaxial interface 72, 583−4
  Al−Mg:Al segregation 198
  Al/Mn 517
  Al/Mo 517
  Al−Ni segregation 504−6
  Al/PMDA-ODA 623, 626−7, 628
  Al/Pt 517
  Al/Si 299−306
  Al:Zn segretation 197−8
  B-reinforced Al 641
  CdS/Al/Ge 596−603, 610, 611−12
  Cu/Al bicrystals 201
  Cu/Al superlattice 356
  Ga penetration 267
  GaAs/Al/Ge 612
  GaP/Al/Si 596, 606−9, 610, 611−12
  Ge/Al/ZnSe 596, 603−6, 610, 611−12

Alkali metals
  intercalation 415, 416
  intercalation with Gr 418–19, 420,
    422, 424–5
Aniline, intercalation with TaS₂ 426,
  427
Ar/Gr monolayer 338, 343–4, 347
As-containing systems
  As/Si 307–8
  Fe/As 290
Au
  Au/Au bicrystals 121–8, 129, 130,
    136–8, 195, 507
  Au/Au reconstruction 338–9,
    349–50, 351, 352
  computer simulations 367–9,
    373–94, 671–89
  GB dislocations 213
  GB migration 218–19, 221
  heterojunctions 595, 611
  structure–energy correlation for
    GBs 95–146
  twinning 72–3, 214
Au-containing systems
  Au/Au 229
  Al₂O₃/Au 659–60
  Au–Fe:Au segregation 93
  Au/Cr superlattice 356, 358, 360
  Au/Cr/polyimide 634–5
  Au–Cu 220–1
  Au/Cu superlattice 356
  Au/Fe 227–8, 279–81, 497
  Au/Fe/Cr 278
  Au/GaAs 313–14
  Au/La 516, 517, 518, 542
  Au/MoS₂ 68–70
  Au/NaCl 102, 318
  Au/Ni superlattice 356
  Au–Pb 202
  Au–Pt:Au segregation 192,
    506–9, 513
  Au/Si 310–11
  Au/Si₃N₄ 223, 224
  Au–Ta:Au segregation 198
  Au/Te 526
  Au/V 278
  Au/Y 517
  Au/ZnSe 293
  Au/Zr 517
  Ni/Zr, Au diffusion 535

B-containing systems
  B-reinforced Al 641
  B/C/SiC 173, 177
  B/Si 324–5

  segregation of B in Ni-based alloys
    466–7
Ba compounds
  BaF₂
    BaF₂/Ge 318
    BaF₂/InP 318
  BaTiO₃ 164–5, 167, 168, 169
Bi compound, Bi₂O₃–ZnO 265, 266
Bi-containing systems
  Bi–Cu 195
  Bi–Cu:Bi segregation 195–6, 197,
    201–2
  Bi/GaAs 313
BPDA-PDA 618–22
  Au/Cr/BPDA-PDA 634–5
  Cu/BPDA-PDA/quartz 631–3
Br/Gr intercalation 337, 341–3, 344,
  411, 416, 419–20, 421, 424–5
β-Brass–Ga polycrystals 202

C fiber 651–2
  SiC-coated 644, 646–7, 649–51
C-containing systems
  B/C/SiC 173, 177
  diamond semiconductors 70
  segregation in Fe-based alloys
    471–2, 474–5
Ca, Ca/Gr intercalation 416
Ca compounds
  Ca aluminosilicate
    penetration of Al₂O₃ 267–8
    SiC fiber composite 186
  Ca²⁺ segregation from ferrites 203
  CaF₂/Si thin films 331
  calcite/NaNO₃ 316
Cd 245, 546
Cd compounds
  Ag/CdO:H segregation 494
  CdS
    CdS/Al/Ge 596–603, 610,
      611–12
    CdS/Ge 597–8, 600, 601, 602,
      603
    CdS/H/Si 612
    interface parameters 610
  CdTe, interface parameters 610
Clay, intercalation 407, 408–9, 410
Co, magnetic properties 277–9, 281
Co compounds
  CoAl/GaAs epitaxial interface 585
  CoCl₂
    CoCl₂/FeCl₃/Gr intercalation
      410
    CoCl₂/Gr intercalation 418, 420,
      425, 426

CoSi₂
  CoSi₂/GaAs epitaxial interface
    585–6
  CoSi₂/Si 318, 320–2, 323, 325,
    331, 332
  epitaxial interface 577–82, 583,
    584
  Si/CoSi₂/Si superlattice 320–4
Co-containing systems
  Ag/Co superlattice 356, 361
  Co/Cu 278–9, 280, 281
  Co/Cu superlattice 356
  Co/GaAs 518, 528
  Co/Sn 517
  Co/ZnSe 293
  Co/Zr 517–18, 519–20
Collidine, intercalation with TaS₂
  426
Cr, magnetic properties 277–9
Cr-containing systems
  Au/Cr 278
  Au/Cr superlattice 356, 358, 360
  Au/Cr polyimide 634–5
  Cr/Fe/Cr 278
  Cr–Ni stainless steel 204
  Cr/PMDA-ODA 622, 623,
    624–5, 631, 634
  Cu/Cr/PMDA-ODA 633–4
  Fe–C–Cr–Ni–Sb:Sb segregation
    474–5
  Fe–C–Cr–P:P segregation 474,
    475
  Fe–C–Cr–Sb:Sb segregation
    474–5
  Nb:Cr segregation 197
  W:Cr segregation 192
Cs compound, CsCl/NaCl 331
Cs-containing systems
  Cs/GaAs 314
  Cs/Gr intercalation 419
  GaP/Cs/Si 612
  Si/Cs/SiO₂ 612
Cu
  computer simulations 367–9,
    373–9, 671–8, 685–6
  Cu/Cu bicrystals 199, 230–1,
    233–4, 237–9, 240
  GB dislocations 212
  GB migration 217
  melting 230–1, 233–4, 237–9,
    240, 241–3, 245–6, 539
  multiple junctions 207
  nanophase 443, 444, 446, 450,
    454–5, 456
  polycrystals 204–5

structure–energy correlation 24–32, 96–120, 125–6, 130–2, 139–44, 194, 196–200, 199
triple junctions 206
twinning 73–4, 96
twist boundaries 75, 104–9, 131–2, 139–44
Cu compound, $CuCl_2$/Gr intercalation 414, 415
Cu-containing systems
  Ag–Cu 263
  Al–Cu 198, 206
  Al/Cu bicrystals 204
  Al–Cu:Al segregation 198
  Al_2O_3/Cu 659
  Au–Cu, GB migration 220–1
  Bi–Cu 195, 201–2
  Co/Cu 278–9, 280, 281
  Co/Cu superlattice 356
  Cu/Al bicrystals 204
  Cu/Al superlattice 356
  Cu/Au superlattice 356
  Cu:Bi segregation 195–6, 197
  Cu/BPDA-PDA/quartz 631–3
  Cu/Cr/PMDA-ODA 633–4
  Cu/Er 517, 529
  Cu/Nb superlattice 356, 360, 361
  Cu/Ni superlattice 356, 359
  Cu–Ni:Cu segregation 502–4, 505, 513, 699–701
  Cu–Ni/Ni 700–1
  Cu/Pd superlattice 356, 395
  Cu/PMDA-ODA 617, 622–3, 624, 625, 626, 627–9, 631, 633–4
  Cu/PMDA-ODA/quartz 631–3
  Cu/PMDA-ODA/Si 633
  Cu/Si 194, 310
  Cu–Si, triple junctions 206
  Cu/Ti 517
  Cu/Y 517–18, 524, 539
  Cu–Zn 200
  Cu/ZnSe 293
  Cu/Zr 517
  Fe/Cu 277–8, 279
  Fe/Cu superlattice 356
  Ni/Zr, Cu diffusion 535
  Pb:Cu segregation 97

Er compounds
  ErAs/GaAs epitaxial interface 585
  Fe_2Er amorphization 249
Er-containing systems
  Cu/Er 517, 529

Ni/Er 517
Eu, Eu/Gr intercalation 417, 425, 426

F, F/Gr intercalation 414, 424–5
Fe
  Fe/Fe bicrystals 195
  magnetic properties 277–8
  nanophase 440–3, 446, 452, 454
  structure–energy correlation for GBs 96, 103–20, 130–3, 141
Fe compounds
  Fe_2Er, FeAl and FeTi amorphization 249
  Fe_3C 467
  $FeCl_3$, $FeCl_3$/Gr intercalation 408
  FeF_2 442–3
  Fe_2O_3
    hematite/sapphire 178, 183
    twin boundaries 161
  Fe_3P:P segregation 467
  FeSb:Sb segregation 467
  FeSn:Sn segregation 467
  FeS:S segregation 467
  FeTe:Te segregation 467
  MnZn ferrites: Ca^{2+} segregation 203
Fe-containing systems
  Ag/Cr/Fe 278
  As/Fe 290
  Au/Fe 497
  Au/Fe/Cr 278
  $CoCl_2$/$FeCl_3$/Gr intercalation 410
  Cr–Ni stainless steel 204
  Cu/Fe superlattice 356
  Fe:Au segregation 3
  Fe/GaAs 289–90, 314
  Fe/metal overlayers 277–81
  Fe/MgO 283–7
  Fe/Pd superlattice 356
  Fe–P:P segregation 465, 466, 468–9, 488
  Fe–P–S:P and S segregation 472–3
  Fe:S segregation 193
  Fe/Se 290
  Fe/Si 192, 221
  Fe–Si 268
  Fe:Si segregation 97
  Fe–Sn:Sn segregation 202
  Fe/ZnSe 289–94
  Fe/Zr 517
  segregation of Fe-based alloys 464, 466, 469–75, 478
  stainless steel 202, 205, 207

Ga compounds
  GaAs 214–15, 610
    Ag/GaAs 583–4
    Al/GaAs 72, 318–19, 583–4, 613, 614
    AlAs/GaAs 294, 585, 592
    AlAs/GaAs superlattice 356
    AlGaAs–GaAs 331
    Au/GaAs 313–14
    Bi/GaAs 313
    Co/GaAs 518, 528
    CoAl/GaAs 585
    $CoSi_2$/GaAs 585–6
    Cs/GaAs 314
    ErAs/GaAs 585
    Fe/GaAs 289–90, 314
    GaAs/Al/Ge 612
    GaAs/$Ga_{1-x}Al_xAs$ 592
    GaAs/$GaAs_{1-x}P_x$ superlattice 326–7, 330
    GaAs–GaAsSb 331
    GaAs/Ge 331
    GaAs/InGaAs 326, 328, 330
    GaAs/Mo 319
    GaAs/Nb 319
    GaAs/Ni 518, 543
    GaAs/NiAl 585
    GaAs/$NiSi_2$ 585–6
    GaAs/Pd 518, 528
    GaAs/Pt 518, 528, 543
    GaAs/Sb 313, 584
    GaAs/$Sc_{1-x}Er_xAs$ 585
    GaAs/$Sc_{1-x}Er_xAs$/GaAs superlattice 322
    GaAs/Si 320, 330, 331
    GaAs/YbAs 585
    GaAs/$GaAs_{1-x}P_x$ superlattice 326–7, 330
    $GaCl_3$/Gr intercalation 422
  GaP
    GaP/Al/Si 596, 606–9, 610, 611–12, 612
    GaP/Cs/Si 612
    GaP/H/Si 612
    interface parameters 610
Ga-containing systems
  Ga–Hg embrittling Al 270
  Ga/Si 299, 306–7
  penetration of Al 267
Garnet 156, 169, 172
Ge
  dislocations 201
  Ge/Ge bicrystals 325
  interface parameters 610
  STGB 68

Ge *continued*
  twin boundaries 70, 71
  volume expansion 135
Ge-containing systems
  BaF_2/Ge 318
  CdS/Al/Ge 596–603, 610, 611–12
  CdS/Ge 597–8, 600, 601, 602, 603
  GaAs/Al/Ge 612
  GaAs/Ge 331
  Ge/Al/ZnSe 596, 603–6, 610, 611–12
  Ge/GeAs 328
  Ge/Nb multilayers 426
  Ge/Pb monolayer 336–7, 340–1
  Ge/Si 319–20, 325–6, 328, 330
  Ge/ZnSe 594
GeSi, GeS/Si 328
Glass, borosilicate 270–1
Gr intercalation compounds 407–27
  Ar/Gr monolayer 338, 343–4, 347
  Br/Gr 337, 341–3, 344, 411, 416, 419–21, 424–5
  Ca/Gr 416
  $CoCl_2$/$FeCl_3$/Gr 410
  $CoCl_2$/Gr 418, 420, 425, 426
  Cs/Gr 419
  $CuCl_2$/Gr 414, 415
  Eu/Gr 417, 425, 426
  F/Gr 414, 424–5
  $FeCl_3$/Gr 408
  Gr/HNO_3 422
  Gr/K 409, 416, 417, 420–1, 427
  Gr/KH 417, 424, 575
  Gr/KHg 411, 412, 417, 421, 424, 427
  Gr/K_xRb_{1-x} 411
  Gr/Kr 337, 337–8, 343–7, 348
  Gr/Li 410, 417
  Gr/N monolayer 343–4
  Gr/Na 416
  Gr/Rb 416, 420–1
  Gr/$SbCl_5$ 417, 421
  Gr/SiC 646–7, 649
Graphite *see* Gr

H-containing systems
  Ag/CdO:H segregation 494
  Al_2O_3/Pd:H segregation 494–5, 495
  CdS/H/Si 612
  GaP/H/Si 612
  H segregation in steel 698
  MgO/Pd:H segregation 493–4, 495
  Pd:H segregation 486, 488–93, 495

Pd/ZnO:H segregation 495
SiO_2/H/SiO_2 612
He, wetting 263
Hematite *see* Fe_2O_3
Hf-containing systems
  diffusion in TiO_2 456, 457
  Hf/Ni 517–18, 535
Hg-containing systems
  Ga–Hg embrittling Al 270
  Hg impurity in Pb 451
HNO_3/Gr intercalation 415, 422
H_2O
  H_2O/polyimide 624
  ice
    amorphization 252
    ice/ice bicrystals 200
  sapphire/water 257, 258
  wetting of quartz 270, 271

Ice *see* H_2O
In compounds
  InAs/InP 294–6
  In_2AsP 295
  InGaAs, InGaAs/GaAs thin films 326, 328, 330
  InP 610
    BaF_2/InP 318
    InAs/InP 294–6
    InP/Ni 518, 528
    InP/Pt 518, 528
  InSb/α–Sn 331
In-containing systems
  In/Si 299, 301, 306
  Ni:In segregation 192
Ir, Ir:O_2 segregation 192

K compounds
  KCl wetting of mica 268–9, 270
  KHg/Gr intercalation 411, 412, 417, 421, 424, 427
  K_xRb_{1-x}/Gr intercalation 411
K-containing systems
  Gr/K intercalation 416, 417, 420–1, 427
  Gr/K superlattice 409
  K/Si 308
Kr/Gr 337, 343–5, 348

La compound, $LaAlO_3$ 164, 165, 170
La-containing system, Au/La 516, 517, 518, 542
Li-containing systems
  Li/Gr intercalation 410, 417
  Li/Si 308
Lu compound, LuAs/GaAs 585

Mg, nanocrystalline 366
Mg compound
  MgO
    Ag/MgO 282–3, 654, 656
    Fe/MgO 283–7
    free surface 282
    GB dislocations 75, 213
    MgO/TiO_x 452
    Pd/MgO:H segregation 493–4, 495
    twin boundaries 161
    $YBa_2Cu_3O_{7-x}$/MgO 152, 155, 171, 179–81, 184
Mg-containing systems
  Al–Mg:Al segregation 198
  Ni/Mg 517
Mica, wetting by KCl 268–9, 270
Mn compound, MnZn ferrite:Ca^{2+} segregation 203
Mn-containing system, Al/Mn 517
Mo, structure–energy correlation 95–7, 103–4, 108, 109–16, 124, 130–3, 141
Mo compound, Au/MoS_2 68–70
Mo-containing systems
  Al/Mo 517
  Fe–Mo–P:P segregation 474
  GaAs/Mo 319
  Mo/Ni superlattice 356, 357, 359, 360
  Mo:Re segregation 192
  Mo/Ta superlattice 356

N
  N/Gr monolayer 343–4
  segregation 92, 472, 478
Na compounds
  NaCl
    Au/NaCl 102, 318
    CsCl/NaCl 331
  $NaNO_3$/calcite 316
Na-containing systems
  Na/Gr intercalation 416
  Na/Ru overlayer 348
Nb compounds
  AlN/NbN superlattice 356
  nanophase Nb_3Sn 454
  Nb_3Ir amorphization 250
  Nb/Nb bicrystals 202, 203, 204
Nb-containing systems
  Al_2O_3/Nb 655, 656–7, 658, 659
  Cr:Nb segregation 197
  Cu/Nb superlattice 356, 360, 361
  GaAs/Nb 319
  Ge/Nb multilayers 426
  Nb/Si 356, 360–1, 518, 527

Ni
magnetic properties 277
melting simulation 240
multiple junctions 207
Ni/Ni bicrystals 195, 201
polycrystals 93, 202, 205–6
triple junctions 206, 207
twist boundaries 75
Ni compounds
amorphization 249–50, 531–42
Ni silicides/Si 331–2
NiAl$_2$O$_4$ 153
Ni$_3$Al:B and C segregation 92–3, 466
NiAl/GaAs 585
NiAs 467
Ni$_3$B:B segregation 467
NiFe$_2$O$_4$ 153
Ni$_3$Ge:B segregation 466
NiO
ceramics 159–60
NiO/spinel 153, 154, 156, 158
NiO/ZrO$_2$ interface 72, 73
structure 55, 129
structure–energy correlation 194
Ni$_3$P:P segregation 467
NiS$_2$/Si 331
NiSb:Sb segregation 467
Ni$_2$Si 526, 527
NiSi$_2$
GaAs/NiSi$_2$ 585–6
NiSi$_2$/Si 325, 551, 567–77
Ni$_3$Si:B segregation 466
Ni$_3$S$_2$:S segregation 467
Ni-containing systems
Ag/Ni 278
Al–Ni segregation 504–6
Al$_2$O$_3$/Ni 657, 659
Au/Ni superlattice 356
Cr–Ni stainless steel 204
Cu/Ni superlattice 356, 359
Cu–NiCu segregation 502–4, 505, 513, 699–701
Cu–Ni/Ni 700–1
Er/Ni 517
GaAs/Ni 518, 543
Hf/Ni 517–18, 535
InP/Ni 518, 528
Mg/Ni 517
Mo/Ni superlattice 356, 357, 359, 360
Ni$_3$Al:B and C segregation 92–3
Ni-based alloys 464
Ni:In segregation 192
Ni/PMDA-ODA 622, 628

Ni/Pt segregation 509–11, 513
Ni/Pt superlattice 356
Ni-S 196
Ni:S segregation 93, 195, 205–6, 466, 470–1
Ni/Si 310, 331, 518, 526, 527, 565
Ni/Ti 517, 524, 525–6
Ni–Ti amorphization 248
Ni/V superlattice 356, 358
Ni/Zr 517–18, 519, 520, 521–5, 529–30, 535
phase diagrams 531–42, 545–7
segregation of Ni-based alloys 466, 472–5

O, Ir:O segregation 192
Octadecylamine, intercalation with TaS$_2$ 426
Olivine/spinel 10, 11

P
segregation
Fe-based alloys 465–9, 471–5, 488
P compound, Ni$_3$P:P 467
Pb 546
GB migration 217
melting 240
nanophase 451
Pb/Pb bicrystals 198
structure–energy correlation 96–7, 199–200
twin boundaries 96
volume expansion 245
Pb-containing systems
Au–Pb 202
Ge/Pb 336–7, 340–1
Pb:Cu segregation 97
Pb/Si 582–3
Pb–Sn 202, 217
Pd, nanophase 366, 433, 434, 441–9, 454–5, 456, 486, 488–93
Pd-containing systems
Al$_2$O$_3$/Pd:H segregation 494–5
Cu/Pd superlattice 356, 395
Fe/Pd 279, 280–1
Fe/Pd superlattice 356
GaAs/Pd 518, 528
Pd:H segregation 486, 488–93, 495
Pd/MgO:H segregation 493–4, 495
Pd/Si 311–13, 325

Pd/ZnO:H segregation 495
PMDA-ODA 618–22, 624–31
Ag/PMDA-ODA 628–9
Al/PMDA-ODA 623
Au/Cr/PMDA-ODA 634–5
Cr/PMDA-ODA 622, 623, 624–5, 631, 634
Cu/Cr/PMDA-ODA 633–24
Cu/PMDA-ODA 617, 622–3, 624, 625, 626, 627–9, 631, 633–4
Cu/PMDA-ODA/quartz 631–3
Cu/PMDA-ODA/Si 633
Ni/PMDA-ODA 622, 628
PMDA-ODA/Ti 622, 626, 628
Polyimides
metal interfaces 616–36
see also BPDA-PDA, PMDA-ODA
Pt, Pt/Pt reconstruction 338–9, 349–52
Pt-containing systems
Al/Pt 517
diffusion in TiO$_2$ 456, 457
GaAs/Pt 518, 528, 543
InP/Pt 518, 528
Ni–Pt segregation 509–11, 513
Ni/Pt superlattice 356
Pt:Au segregation 192, 506–9, 513
Pt/ZnSe 293
Pyridine, intercalation with TaS$_2$ 426

Quartz 261, 270, 271
Cu/polyimide/quartz 631–3

Rare-earth oxides 161
Rare gases, intercalation with Gr 421
Rb/Gr intercalation 416, 420–1
Re-containing systems
Mo:Re segregation 192
Re–W 63–6, 511–12, 513
Rh compound, Zr$_3$Ph amorphization 249
Rh-containing system, Rh/Si 518, 527
Ru, nanophase 439
Ru-containing systems
Fe/Ru 279
Na/Ru 348

S
segregation 93, 195, 205–6
Fe–C–S:C and S 472
Fe–N–S:N and S 472
Fe–P–S:P and S 472–3

S *continued*
    Fe–S:S 193, 466, 467, 469–70,
      471
    Ni–S 196
    Ni_3S_2:S 467
    Ni–S:S 466, 470–1
    segregation in steel 698
Sapphire-containing systems
    hematite/sapphire 178, 183
    sapphire/water 257, 258
Sb-containing systems
    GaAs/Sb 313, 584
    segregation
      Fe–C–Cr–Ni–Sb:Sb 474–5
      Fe–C–Cr–Sb:Sb 474–5
      Fe–C–Ni–Sb:Sb 474–5
      Fe–C–Sb:C and Sb 472
      Fe–Ni–Sb:Ni 473
      Fe–Ni–Sb:Sb 466
      Fe–Sb:Sb 466
      FeSb:Sb 467
      NiSb:Sb 467
    Si/Sb 307–8
$SbCl_5$, Gr/$SbCl_5$ intercalation 417,
    421
$Sc_{1-x}Er_xAs$ 322, 585
Se-containing system, Fe/Se 290
Si
    amorphization 248, 252
    GB migration 217
    interface parameters 610
    melting simulation 231, 233–7,
      240, 243
    multiple junctions 207
    n-type 555–6
    Si/Si bicrystals 203, 231, 233–7
    Si/Si epitaxy 324–5
    Si/Si reconstruction 338
    twin boundaries 70, 71
    volume expansion 135
Si compounds
    metal silicides 516, 526, 526–8,
      550–1, 566–7
    Ni silicides/Si 331–2
    $NiSi_2$/Si 325, 331
    SiC
      Al_2O_3/SiC composites 185–6
      B/C/SiC 173, 177
      fiber composite in glass 186
      Gr/SiC 646–7, 649
      SiC-coated C fiber 644, 646–7,
        649–51
      Si/SiC 646–9
    $Si_{1-x}Ge_x$, $Si_{1-x}Ge_x$–Si 326, 329,
      331

SiGe–Si 331
Si_3N_4 173
    Au/Si_3N_4 223, 224
    see also Sialons
SiO_2
    quartz *see* Quartz
    Si/Cs/SiO_2 612
    Si/H/SiO_2 612
Si/$ZrSi_2$ 563
Si-containing systems
    Ag/Si 309–10
    Al/Si 299–306
    As/Si 307–8
    Au/Si 310–11
    B/Si 324–5
    CaF_2/Si thin films 331
    CdS/H/Si 612
    $CoSi_2$/Si 318, 320, 321, 322, 325,
      331, 332, 577–82, 583, 584
    Cu/PMDA-ODA/Si 633
    Cu/Si 310
    Cu–Si 206
    Fe/Si 221
    Fe–Si 268
    Fe/Si bicrystals:Si, P and N
      segregation 192
    Fe–Si:N segregation 478
    Fe–Si:Si segregation 97, 466
    GaAs/Si 320, 330, 331
    GaP/Al/Si 596, 606–9, 610,
      611–12
    GaP/Cs/Si 612
    GaP/H/Si 612
    Ga/Si 299, 306–7
    Ge/Si 319–20, 325–6, 328, 329,
      330
    GeSi/Si 328
    In/Si 299, 301, 306
    K/Si 308
    Li/Si 308
    metal interfaces 595
    metal silicides/Si 550–1, 566–7
    Nb/Si 518, 527
    Nb/Si superlattice 356, 360–1
    Ni/Si 310, 331, 518, 526, 527,
      565
    Pb/Si 582–3
    Pd/Si 311–13, 325
    Rh/Si 307–8
    segregation 92
    Si/$CoSi_2$/Si superlattice 320–4
    Si/Cs/SiO_2 612
    $Si_{1-x}Ge_x$–Si 331
    $Si_{1-x}Ge_x$/Si 326, 329, 330
    Si/H/SiO_2 612

Si/SiC 646–9
Si–SiGe 331
Si/Ti 518, 524, 558
Si/V 518
Si/Zr 518, 527
Sialons 176–7, 180, 263–4
    see also Si_3N_4
Sn, Sn/Sn bicrystals 199
Sn-containing systems
    Co/Sn 517
    Fe–Sn:Sn segregation 202, 466,
      467, 469, 470
    impurity in Pb 451
    InSb/Sn 331
    Pb–Sn 202, 217
Spinels
    Al_2O_3/spinel 176, 179
    Al_2O_3/spinel/Ni 657
    ceramics 159–62, 163
    GBs 156–7
    NiO/spinel 153, 154, 156, 158
    olivine/spinel 10, 11
Sr compounds
    $SrTiO_3$ 178, 181
    strontium barium aluminum
      silicate 268, 269
Stainless steel
    Cr–Ni 204
    multiple junctions 207
    type 304 202, 205
Steel 477–8, 494, 698

Ta-containing system, Mo/Ta
    superlattice 356
TaS_2, intercalation with organic
    compounds 426, 427
Te-containing systems
    Ag/Te 518
    Au/Te 526
    Fe–Te:Te segregation 467
Ti compounds
    FeTi amorphization 250
    TiO_x, MgO/TiO_x 452
    TiO_2, nanophase 436–8, 439,
      440, 441, 444–6, 447, 450,
      451–4, 456–7
    Ti_2O_3, TiO_2/Al_2O_3 interface
      9
Ti-containing systems
    Cu/Ti 517
    Fe–Ti–P:P segregation 471
    Ni/Ti 517, 524, 525–6
    Ni–Ti amorphization 248
    Si/Ti 518, 524
    Ti/Ni superlattice 356

Ti/PMDA-ODA 622, 626, 628
Ti/Si 558
Transition-metal compounds
  intercalation dichalcogenides 407,
    408–9, 410, 426
  halides 410, 417–18, 421–2,
    425–6

V-containing systems
  Ag/V 278
  Au/V 278
  Fe–V–P:P segregation 474
  Ni/V superlattice 356, 358
  Si/V 518

W-containing systems
  W:Cr segregation 192
  W–Re, segregation 511–12, 513
  W/Re bicrystals 63–6
Water *see* H_2O

Xe-containing systems
  Xe/Ag 347
  Xe/Gr 337–8, 343–4, 345–7

Y compounds
  $YBa_2Cu_3O_7$, structure–energy
    correlation 195

$YBa_2Cu_3O_{7-x}$ 164, 165–6, 170–2,
    174, 177
$YBa_2Cu_3O_{7-x}$/MgO 152, 155,
    171, 179–81, 184
$YBa_2Cu_3O_{7-x}$/SrTiO$_3$ 178, 181
Y_2O_3
    AlN/Y_2O_3 composites 152
    YAG interphase 169, 172
Y-containing systems
  Au/Y 517
  Cu/Y 517–18, 524, 539
Yb compound, YbAs/GaAs 585

Zn 546
  volume expansion 245
  Zn/Zn bicrystals 201
Zn compounds
  ZnO 255
    Bi_2O_3–ZnO 265, 266
    nanophase 452, 453–4, 456
    Pd/ZnO:H segregation 495
  ZnSe
    Fe/ZnSe 289–94
    Ge/Al/ZnSe 596, 610, 611–12
    Ge/ZnSe 594
    interface parameters 610
Zn-containing systems

Al:Zn segregation 197–8
Cu–Zn 200
MnZn ferrites:Ca^{2+} segregation
  203
Zr compounds
  NiZr$_2$ amorphization 249–50,
    531–42
  Zr$_3$Al amorphization 248–9, 250
  Zr$_3$Rh amorphization 249
  ZrN, AlN/ZrN superlattice 356,
    361
  ZrO_2 152
    Al_2O_3/ZrO_2 184, 185
    NiO/ZrO_2 interface 72, 73
    phase transformation 152, 153
    toughened ceramics 175,
      181–4, 185
  ZrSi$_2$/Si 563
Zr-containing systems
  Au/Zr 517
  Co/Zr 517–18, 519–20
  Cu/Zr 517
  Fe/Zr 517
  Ni/Zr 517–18, 519, 520, 521–5,
    529–30, 535
  phase diagrams 531–42, 545–7
  Si/Zr 518, 527

Subject index

The following abbreviations are used: CSL, coincident site lattice; STM, scanning tunneling microscopy; STGB, symmetrical tilt grain boundary; TEM, transmission electron microscopy.

Adhesion
  metal–ceramic interfaces 657–9
  metal–polyimide interfaces 633–4
  work of adhesion 633–4, 657–9,
      662–9, 673–8, 685–6,
      691–701
Adsorption 270–1
Alloys
  amorphization *see* Amorphization
  segregation 463, 468–75, 477–9,
      502–3
  see also Individual alloys in
      Materials index
Alpha-fringe (displacement) contrast
      microscopy 67–8, 98
Amorphization 229, 246–51,
      516–47
  by ion-beam irradiation 247,
      530–1
  hydrogen-induced 247, 248–9
  mechanical alloying 528–30
  relationship with other melting
      phenomena 542–7
Annealing of metal–polyimide
      systems 629–31
Atom probe methods 76–9, 97, 340
  see also Field-ion microscopy;
      Rutherford back-scattering
Atomic-level geometry 1–55
  ceramic substrates 168–70, 172–3
  definition 1–2
  determination for internal
      interfaces 61–6

free surfaces 34, 35
grain structure of thin films 221–3
intercalation compounds 407–10,
      412–23
metal–polyimide interfaces
      616–36
monolayers 338–9, 347
morphology of GB migration
      218–21
morphology of nanophase
      materials 439–52
planar stacking 29–35
superlattices 354–62
symmetry of epitaxial interfaces
      330–1, 347–8
topology of GBs in metals 126–7,
      128
wetting of GBs 262–3, 265–7
Auger electron spectroscopy of
      segregating systems 463, 468,
      478

Ball milling 530
Bardeen's model 288, 551
Bicrystals xv, 3–4
  DOFs 17–21, 35–52, 58–76
  internal interface, experimental
      characterization 58–82
  manufacture 100–1
  structure–corrosion correlation
      203–4
  tilt 120–7
Bonding

at polyimide interfaces 620–5
in ceramics 151–2
cohesion of metal–polyimide
      interfaces 634–5
models for metal–ceramic
      interfaces 656–7
segregation at GBs 476–7
see also Broken bond model
Born instablity 241–2, 243–4,
      245–6
Bravais lattice 13–16, 29–35
  lattice-parameter changes 380–6,
      389–94, 396–400, 401–2, 403
Broken bond model 53–5, 93–4
  computer simulation 139–44
  interfacial decohesion 669–71,
      685–8
Bulk interfaces xv–xvi, 3–4
  computer simulation techniques
      94–7
  interfacial stress 7

Cavitation of GBs 202
Ceramics 151–87
  atomic structure 151–9
  GB dislocations 75, 156
  heterophase interfaces 72, 174–86
  metal interfaces
    magnetic behavior 281–7
    structure–fracture resistance
        correlation 654–60
  nanophase 452, 453–4, 455–6
  wetting 255–6, 263–4

see also Individual ceramics in Materials index
Charge transfer across interfaces 421, 423–4, 476
Chemical analysis of internal interfaces 80–1, 498–501, 502, 509–12
Cleavage fracture 27–8, 35, 662–89
Clusters
    at metal–polyimide interfaces 626–31
    in nanophase materials 431–3, 434–8, 458
Coherency 8–11, 136–9
    effect on elastic behavior 394–6, 398–400
    epitaxial interfaces 318–19, 326
    role in amorphization 525–6
Coincident site lattice (CSL) 35–8, 46, 88–90, 158–9
    characterization of GBs 39–52, 125, 129–30, 133, 190–2
    CSL misorientation scheme 36–8, 43–7, 60–1, 89, 102–7, 213–15
        special properties of sigma GBs 190–208
    determination of misorientation for internal interfaces 61–6
Cold rolling 529
Commensurability 10–11, 13–16
    commensurate–incommensurate transition in monolayers 337, 341–3, 344–5, 348
    cubic lattice planes 32, 33
    epitaxial interfaces 318
    intercalation compounds 416–20
    STGBs 145–6
Computer simulation
    annealing of metal–polyimide interface 629–31
    border conditions 229–30, 231–2, 240, 370
    broken bond model 139–44, 685–8
    elastic behavior
        model for fracture 692–8
        superlattices 360–2, 389–403
        thin films 365–404
    fracture of polycrystals 691–702
    GB migration and dislocation 221, 223
    HREM images
        metals 100, 101
        nanophase materials 446, 449
    interatomic potentials 230–1

see also Finnis–Sinclair potentials; Lennard–Jones potentials
    interface geometry 231
    metal diffusion in polyimides 625
    metal–ceramic interfaces 656–7
    Monte Carlo 509–11, 629, 694
    planar disorder and melting 232–3, 242–3
    premelting 239–41
    segregation in alloys 502–9, 698–701
    solid–state amorphization 249–50
    spallation of interfaces 646–7
    structure–energy correlation
        in GBs 671–8, 692–8
        in metals 94–7, 101–27, 146
    techniques *see* Computer simulation techniques
    volume expansion at GBs 130–5
Computer simulation techniques xvii–xviii
    elastic behavior 370–2, 692–4
    GB superlattice models 389–93
    GBs 47
    HREM observations of metals 99–100, 101
    molecular dynamics 229–33, 251–2
    Monte Carlo 498–9
    structure–property correlation 275–7
    two-state model for thin films 386–9
    see also Embedded atom method
Convergent beam electron diffraction 70–3
Corrosion
    at GBs 203–6, 207
    and segregation 477, 478
Cracking
    ceramic interfaces, crack front profile 172, 181–4, 660
    crack tip stress field 642–4
    stress corrosion 478
    see also Fracture
Critical thickness 10
    amorphization 521, 540
    epitaxial interfaces 318, 326–8
Crystallography *see* Atomic-level geometry

Daumas–Herold domain model 411, 413, 416
Decohesion *see* Fracture
Degrees of freedom 16–29

macroscopic 17–21, 35–52, 88–92, 501–2
    bicrystal 58–66
    correlation with energy 101–27
    for decohesion of grain boundaries 663
microscopic 21–2, 369–70
    bicrystal 66–76
    correlation with energy 126–35
    in misorientation phase space 25–9
Delamination *see* Fracture toughness
Dielectric interlayer 257, 258
    in metal–semiconductor systems 551–3, 560, 595–612
Differential scanning calorimetry 536, 537
Diffusion
    at GBs 197–8
    at metal–polyimide interfaces 625–31
    at semiconductor–semiconductor interfaces 606–9
    diffusion couples in amorphization 516–47
    growth control in amorphization 533–7
    in nanophase materials 456, 492–3
    role in segregation 464–5, 486–7
Disinclinations 206
Dislocations
    at epitaxial interfaces 326–31
    at metal–ceramic interfaces 657, 658, 660
    in ceramics 156, 178–84
        climb-associated 160, 161
    comparison with steps 667, 683–5
    computer simulation of segregation phenomena 502–3, 504–9
    determination of translation vector 74–6
    misfit 326–31, 658
        role in interfacial decohesion 662–4, 683–5, 686–7, 688
    Read–Shockley *see* Read–Shockley dislocation model
    role in GB migration 198, 201, 214–21
Disorder
    computer simulation of planar disorder and melting 232–3
    structural *see* Structural disorder
Displacement-shift-complete (DSC) lattice 158, 190–1, 213–5
Doping 424, 456, 458

Elastic behavior
  crack tip stress field 642–4
  experimental determination 355,
    645–51
  lattice-parameter changes 380–6,
    389–94, 396–400, 401–2,
    403
  microstructural mechanics model
    for fracture 692–8
  nanophase ceramics 453–4
  shear instability in amorphization
    250–1
  superlattices 354–62, 389–403
  tensile strength of interfaces
    641–52
  thermal stress at metal–polyimide
    interfaces 631–3
  thin films 376–89
Electrical properties
  at GBs 202–3
  intercalation compounds 423–4
  nanophase materials 456, 457
Electronic properties
  correlation with elastic behavior
    365–6
  intercalation compounds 423–5
  metal–ceramic interfaces 284–7
  metals 278–9, 280, 282–4
  semiconductor interfaces 294–6,
    550–586, 593–614
  semiconductor superlattices
    290–4, 365–6
Electrostatic double layer 257, 258
Embedded atom method 498
  and interfacial fracture 663–4,
    671–8, 698–701
  potentials 95–6, 230, 368–9
Energy–structure correlation *see*
  Structure–energy correlation
Epitaxy 8–11, 316–17
  epitaxial growth 154–5, 178–81,
    317–25, 326–31
  epitaxial interfaces xv, xvi, 4–5, 9
    electronic properties 568–73,
      574–6, 582, 585–6
    interfacial stress 7
    metal–semiconductor 299–314,
      316–33, 567–86
  migration 215–6
  orientation 330–1, 340–51
  reconstruction 322–5, 574,
    580–1
  stability 325–6, 331–2

Faceting
  at GBs 195–6

  at metal–ceramic interfaces 657,
    658
  in ceramics 156–7, 163, 165, 166
  effect on GB migration 200
  epitaxial metal–semiconductor
    interfaces 573, 575, 576–7
  in nanophase materials 447–8
  of tilt bicrystals 120–7
  and wetting of GBs 268–70
Fermi-level pinning 287–8, 551–4,
    586, 595
Fiber-reinforced composites 184–6,
    641–52
Field-ion microscopy 61–6, 76–81,
    499–502, 509–12
Finnis–Sinclair potentials 95–7,
    230, 368–9
Fracture
  brittleness of steel 463, 477–8,
    494
  chemical reactions and segregation
    at GBs 465, 474–5
  GBs 201–2
  Griffith 662–89, 698–701
  metal–ceramic interfaces, fracture
    resistance 659–60
  metal–polyimide interfaces 633–5
  polycrystals 691–702
  role of dislocations and steps
    662–89
  tensile strength of interfaces
    641–52
  see also Cracking
Fracture toughness (delamination)
  631–5, 641–4, 650–2
Free energy of mixing 531–2, 533–4
Free surfaces
  atomic-level geometry 34, 35
  DOFs 22
  effect of structure on epitaxial
    growth 319–25
  elastic behavior 372, 374, 375,
    376
  fracture 662–6, 669–73, 678–83
  as homophase interface systems
    22–3
  magnetic behavior of metal
    surfaces 277–9, 281
  melting at 240–1
  STM of metal–semiconductor
    systems 299–314
  stress effects 379–80
  wetting 258–62
Frenkel defects 530–1
Fresnel fringe contrast microscopy
  73–4

Glass transition temperature 252,
  519, 542–3
Grain boundaries (GBs) 157–9
  antiphase GBs in ceramics 169–70
  asymmetrical 89–90, 116–20
  asymmetrical tilt 18–19, 91
    atomic-level geometry 38,
      39–44
    structure–energy correlation
      116–20
  in ceramics 156–74
  CSL misorientation scheme
    description 35–8, 39–52,
      125, 129–30, 133, 190–2
  general 18–9, 20, 91–2
    atomic-level geometry 38,
      39–41
    structure–energy correlation
      116–20
  growth *see* Grain growth
  interface place scheme description
    17–21, 38–52
  in nanophase materials 431, 433,
    438–51
  random (RGBs) 663, 672–8, 679
    model 109–13
  role in amorphization 524–5
  special 23–5, 511, 664, 668,
    674–8
    structure–energy correlation
      109–13
  stability in nanophase materials
    451–2
  structure–energy correlation,
    metals 87–147
  superlattice models 389–93
  symmetrical 19–20, 89–90, 91–2
    atomic-level geometry 44–52
    structure–energy correlation
      26–9, 102–16
  symmetrical tilt (STGB) *see*
    Symmetrical tilt grain
    boundaries
  symmetrical twist 20, 44–5
    structure–energy correlation
      104–9, 194–5
  tilt 18
    coherency 136–9
    structure–energy correlation
      120–7, 195
    structure–transport property
      correlation 170–2
  twin *see* Twinning
  twist 18
    computer simulation 231, 663,
      671–8, 679

in segregating systems 503–4, 505, 506–9, 513
  structure–energy correlation 195
vicinal 23–5, 664–6, 668, 674–9
  structure–energy correlation 109–13
see also Interfaces
Griffith fracture 662–89, 698–701
Growth
  epitaxial 154–5, 178–81, 319, 326
  of GBs 451–2, 478
  of interlayers 521, 533–7
  recrystallization at phase interface 219–21
  suppression 478, 540–1
see also Nucleation

Hard sphere model 53, 144
Hardening of GBs 202
Hexatic phases 337, 345, 347, 422
High-resolution electron microscopy 88, 146
  atomic structure of GBs 70–1, 97–107
  coherency of interfaces 136–9
  identification of planar facets 120–7
  metal–ceramic interfaces 655
  structural periodicity of GBs 145–6
  volume expansion 131–5

Impurity segregation 463
see also Segregation
Intercalation compounds 407–27
  commensurability 416–20
  host materials 408–11
  magnetic interfaces 425–6
  superconducting interfaces 426–7
  ternary 411, 421
Intercalation process 410–11
Interface plane scheme 16, 17–21, 39, 46
Interfaces xiv–xvi, 3–7
  bulk xv–xvi, 3–4
  dielectric interlayer 257, 258, 551–3, 560, 595–612
  epitaxial *see* Epitaxy
  experimental characterization 58–82
  heterophase 72
    ceramic 174–86
    formation 152–5
    semiconductor *see* Semiconductors

structure–property correlation 275–96, 550–86
high-angle 44, 213, 668–9
  in ceramics 156, 159–61
  computer simulation 231, 663, 671–8, 679
  structure–energy correlation 107–8, 109–13
  TEM 219
homophase 22–3
low-angle 44, 45, 213
  in ceramics 156, 159–61
  structure–energy correlation *see* Read–Shockley dislocation model
  TEM 218–9
magnetic 425–6
models for Schottky barrier formation 551–4
semi-bulk *see* Semi-bulk interfaces
superconducting 426–7
thin film *see* Thin films
see also Free surfaces
Interplanar spacing
  correlation with GB energy 110–11
  planar stacking in Bravais lattice 29–35
  role in interfacial decohesion 674–9
  STGBs 50–2, 103–4, 674–8
Inverse volume density *see* Sigma
Island epitaxial growth 319–25

Kikuchi patterns 62–6, 98, 502

Langmuir–McLean equation 481, 483, 487, 490–1
Laser spallation 645–50
Lennard–Jones potentials 96–7, 230, 368–9, 663–4, 671–8
Lifshitz theory of dipole interaction 256–7

Magnetic behavior
  metal surfaces and interfaces 277–81
  semiconductor superlattices 290–4
Melting 228–53
  curves 244–7
  intercalation compounds 419–20, 421–2
  mechanical 229, 241–4, 251–2
  in monolayers 345–7, 349–51
  polymorphic 543–7
  premelting 239–41

theories 544
thermodynamic (interface-induced) 228–9, 237–9, 243–4, 251–2
  computer simulation 233–41
see also Amorphization
Metals
  amorphization 247–51, 517–47
  complete wetting 263
  grain structure of thin films 221–3
  hydrogen segregation in internally oxidized metals 493–5
  interfaces
    electronic properties 278–9, 280, 282–7
    metal–ceramic 654–60
    metal–matrix 641–52
    metal–polyimide 616–36
  magnetic behavior 277–87
  nanophase 454–6
  phase behavior of monolayers 340–1, 347–52
  segregation 463, 468–75, 477–9, 502–13, 698–701
  STM of metals on semiconductors 299–314
  structure–elastic behavior correlation 372–403, 454–6
  structure–energy correlation 87–147, 193–7
  structure–fracture resistance correlation 654–60
  structure–property correlation 197–206
  tensile strength 641–52
see also Individual metals in Materials index
Microscopy techniques
  alpha–fringe contrast 67–8, 98
  for ceramic interfaces 155–6, 655
  Fresnel fringe contrast 73–4
  for heterojunction interfaces 613–14
  Moiré fringe contrast 68–70, 179–81, 184
  visible-light 155
see also Transmission electron microscopy
Midgap energy rule 609–12
Migration of GBs 198–200, 212–25, 237–8
Miller indices 19, 31
  expression of STGB geometry 48–52
Misorientation
  CSL 61–6
see also Coincident site lattice

Misorientation *continued*
  misorientation phase space 25–9,
    102–27
Moiré fringe contrast microscopy
  68–70, 179–81, 184
Monolayers *see* Thin films
Morphology *see* Atomic-level
  geometry

Nanophase materials 7–8
  cluster formation 431–3, 434–8,
    458
  diffusion and segregation of
    hydrogen 487–93
  elastic behavior 365–6
  grain size distribution 431, 433–4,
    438, 439, 447, 452
  pore structure 439, 440, 441, 445,
    452–3
  stability 451–2
  structure–property correlation
    431–58
  synthesis 431–9
Nucleation
  in amorphization 537–42, 543–4
  epitaxial interfaces 319–25, 328–9

O-lattice 159

Periodicity of structural units 31,
  136–8, 145–6, 299–314
Phase contrast microscopy 70
Phase transitions
  in ceramics 153, 164–7, 168, 169
  epitaxial metal–semiconductor
    interfaces 579
  epitaxial phase stability 331–2
  hydrogen segregation at
    boundaries 481–95
  intercalation compounds 419–20,
    421–2
  metal–ceramic interphases 657
  monolayers 336–53
  phase diagrams 244–7, 531–4
  see also Amorphization; Melting
Photoemission spectroscopy 593–4,
  596, 612–14
  polyimide interfaces 620–4
Planar stacking 29–35
Planar structure factor 232–3
Poisson effect 372, 374, 379, 381,
  384–5, 386–9
Polycrystals 7–8
  fracture of polycrystalline materials
    691–702

GB migration 221–3
  structure–property correlation
    204–7
  see also Nanophase materials
Polyhedral unit model 53, 93, 144
  segregant interaction 466–8
Polyimides 616–36
Premelting 239–41

Radial distribution function 53–5,
  139–40, 366, 367, 376, 379
Random grain boundary model
  109–13
Read–Shockley dislocation model
  44–5, 93, 107–9, 191, 213
  applied to STGBs 111–13, 684
  and broken bond model 94, 141–3
  wetting of GBs 268, 269
Reconstruction
  in epitaxial growth 322–5
  metal monolayers 338–9, 349–51
Recrystallization of phase interface
  219–21
Rutherford back-scattering 521, 523

Scanning electron microscopy, for
  ceramic interfaces 155
Scanning tunneling microscopy 97,
  299–314
  intercalation compounds 415, 422
  nanophase materials 450
Schottky barrier 287–8, 595
  capacitance–voltage behavior
    564–5
  current–voltage behavior 556–64
  formation 550–4, 609
  ideality barrier 556–63
  inhomogeneity 554–67
  semiconductor superlattices 292–3
  theory 609–12
Segregation
  at GBs 192–3, 194–5, 202,
    463–79, 481–95
  at metal–ceramic interfaces 656–7
  computer simulation 502–9
  and corrosion 477
  diffusion model 486–7
  effect on GB migration 223–5
  free energy at interface 482–6,
    487–91, 494–5
  of hydrogen 482, 485, 487–91
  investigation by atom probe
    techniques 79–82
  in metal alloys 463, 468–75,
    477–9

process 463–8
  role in interfacial fracture 476–7,
    692, 698–701
  solute–solute interaction 482,
    484–5
Semi-bulk interfaces xvi, 3–5
  coherency and epitaxy 8–9
Semiconductors
  interfaces and heterojunctions
    287–96
  metal–semiconductor systems
    amorphization 518
    epitaxy 299–314, 316–33
    STM 299–314
    structure–electronic property
      correlation 550–86
  modulation doping 424
  semiconductor–semiconductor
    interfaces 294–6
  electronic properties 592–614
  superlattices 289–96
Sigma 60, 89, 159, 190
  STGB configuration 50–2
Sintering 153
  effect on nanophase materials 439,
    440, 441, 445, 447, 452–3
Sliding of GBs 210
Solubility, effect on segregation
  484–6
Solvation forces 257–8
Spallation 645–50
Stacking
  faults
    at bicrystal boundary 66–7
    atomic-level geometry 2, 3,
      34–5
    DOFs 21
    planar 29–35
Staging in intercalation compounds
  409–16
Steps
  at metal–semiconductor interfaces
    567–8
  comparison with dislocations 667,
    683–5
  formation in epitaxial interfaces
    215–6
  GB migration mechanism 200
  in nanophase materials 447–8
  role in fracture 662–6, 668–71,
    678–83, 686–7, 688
Stillinger–Weber potentials 231
Strain effects
  epitaxial interfaces 318, 326–30
  GB migration 199

semiconductor interfaces 294–6
Stress effects 5–7, 365
  on atomic structure 369–70
  lattice-parameter changes 380–6,
    389–94, 396–400, 401–2
  on structure and elastic behavior
    free surfaces 379–80
    thin films 379–86
Structural disorder
  atomic-level 53, 126–7, 128, 139
  and elastic behavior 366–8, 380,
    384, 390–403
  nanophase materials 448–50
Structure
  ceramics 151–9, 168–70
  interfaces 1
    atom probe microscopy 76–82
    correlation with properties *see*
      Structure–energy correlation;
      Structure–property
      correlation
    high-resolution electron
      microscopy 97–101, 136–9
    metal-ceramic 654–9
    models 52–3, 93–4
    see also Bonding; Broken bond
      model; Hard sphere model;
      Polyhedral unit model;
      Read–Shockley dislocation
      model and segregation
      phenomena 464, 465–8,
      502–13
    thin metal films 373–6
    triple junctions 206–7
  metals on semiconductors
    299–314, 568, 573–4, 576–86
  periodicity 31, 136–8, 145–6,
    299–314
  polyimides 618–19, 620
  see also Atomic-level geometry
Structure–energy correlation
  at sigma GBs 193–7
  fcc lattices 24–5
  GBs in metals 87–147
  interfacial decohesion 664–85
  metal–ceramic interfaces 282–3
  in misorientation phase space
    25–9
Structure–property correlation
    52–4, 173–4
  at heterophase interfaces 275–96
  at sigma GBs 190–208
  corrosion 203–6
  elastic behavior
    nanophase ceramics 453–4

superlattices 354–62, 364–5,
    389–404
  thin films 364–5, 376–89
electronic properties 550–86
  in intercalation compounds 417,
    423–7
  migration properties 225
  in nanophase materials 431–58
  in polyimides 618–19, 620
  transport properties 170–2
Superconducting interfaces 426–7
Superheating 241–3
Superlattices xvi, 5, 6, 354
  coherency 11–2
  commensurability 13–6
  elastic behavior 389–403
  in intercalation compounds
    409–10, 421
  semiconductor 289–96, 330
  structure–elastic behavior
    correlation 354–62
Supermodulus effect 364–5, 403,
    404
Symmetrical tilt grain boundaries
    20–1
  atomic-level geometry 34–5,
    41–52, 127–9
  commensurability 145–6
  dislocations and work of adhesion
    663, 666–71, 672–83
  DOFs 21–2, 41–7, 91–2
  in segregating system 505
  structure–energy correlation
    102–4, 105, 113–4, 194–5
  unit cell volume 105–7

T_0 anomaly 561–3
Tensile strength of fiber-reinforced
    composites 641–52
Thermal effects
  amorphization 519, 546
  curing of polyimides 618–19
  disorder 53, 232–3
  on segregation 464, 465, 469–71,
    507–8
  stress at metal–polyimide
    interfaces 631–3
  see also Melting
Thin films
  elastic behavior, computer
    simulation 372–94
  epitaxial 299–314, 316–33
  grain structure 221–3, 224
  interfaces xv, xvi, 4, 5, 6
    coherency 10–12

formation in ceramics 152–5
  interfacial stress 6–7
  magnetic behavior 279–81
  phase behavior of monolayers
    336–53
  solvation force 258
  stability 325–6, 331–2
  STM of metals on semiconductors
    299–314
  wetting by adsorption 260
  see also Monolayers
Tilt inclination scheme 41–4, 47,
    90–2
Topotaxy 8, 9, 10
Toughening of ceramics 172–3,
    181–4, 185
Toughness, fracture (delamination)
    631–5, 641–4, 650–2
Transmission electron microscopy
  for ceramic interfaces 155, 156
  GB dislocations 75
  GB migration and structure
    218–21, 223, 224
  internal interfaces 67–76, 98–100,
    120–7, 501–2
Twinning 47–8, 49, 127, 128
  at sigma GBs 195–6
  during recrystallization 219–21
  effect on grain orientation 212
  migration of twin boundary 214–5
  twin GBs in ceramics 159, 161–7,
    170, 171

Valence band discontinuity 595–609
  theory 609–12
van der Waals forces 256–7, 258
Visible-light microscopy 155
Volume expansion
  at GBs 92–3, 130–5, 267–8
  and elastic behavior 392, 393
  hydrogen trapping at interface 495
  for melting 244, 245–6
  segregation in alloys 503
  for solid-state amorphization 246,
    247–8, 250
  and structural disorder at
    interfaces 365, 366

Walser–Bene membrane 526
Wetting 255–71
  at free surfaces 258–62, 271
  at GBs 255–6, 262–71
  effect on epitaxial growth 319–20
  long-range forces 256–8
  partial 260–2, 263, 265

and penetration 267–8
and surface chemistry 270–1
Work of adhesion *see* Adhesion

X-ray scattering
    characterization of monolayers
    337, 339–40

determination of superlattice
    structures 360

LIVERPOOL
UNIVERSITY
LIBRARY

FIAT LVX